Spon's Mechanical and Electrical Services Price Book

Edited by

A=COM

2019

Fiftieth edition

First edition 1968
Fiftieth edition published 2019
by CRC Press
2 Park Square, Milton Park, Abingdon, Oxon, OX14 4RN

and by CRC Press
Taylor & Francis, 6000 Broken Sound Parkway, NW, Suite 300, Boca Raton, FL 33487

CRC Press is an imprint of the Taylor & Francis Group, an informa business

British Library Cataloguing in Publication Data
A catalogue record for this book is available from the British Library

ISBN: 978-1-138-61205-1
Ebook: 978-0-429-46446-1
ISSN: 0305-4543

Typeset in Arial by Taylor & Francis Books

MIX
Paper from
responsible sources
FSC® C013056
www.fsc.org

Printed and bound in Great Britain by
TJ International Ltd, Padstow, Cornwall

Contents

PART 4: MATERIAL COSTS/MEASURED WORK PRICES – ELECTRICAL INSTALLATIONS

PART 5: RATES OF WAGES

PART 6: DAYWORK

Preface to the Fiftieth Edition

The inaugural 1968 edition of *Spon's Mechanical and Electrical Services Price Book* was published in response to the trend for tender enquiries to include mechanical and electrical Bills of Quantities and as a result of the growth of cost planning and cost control in the design stage of building works. It was recognized that without data obtained from detailed pricing, it was not possible to 'pre-plan expenditure for these services with the same degree of confidence as for the remainder of the building work' when building services 'account for a large and increasing proportion of building costs – 25 to 40% for many building types.'

The intention of the 1968 edition was for the information provided to allow a Bill of Quantities to be priced and provide a 'reasonably accurate estimate of the likely cost of a project and to give supplementary information which will enable the reader to make adjustments to suit his own requirements'. Whilst not unrecognizable from the current day editions, it differed significantly in that it contained separate sections for material prices and measured works whilst the approximate estimating and cost modelling sections were noticeably brief as a result of the newly emergent role that building services cost planning was playing in pre-contract cost management.

Whilst MEP Bills of Quantities are no longer commonly used, building services cost planning has become an integral part of pre-contract cost management and detailed supply and installation costs for plant and materials a necessity for the pricing of increasingly complex MEP installations.

Subsequent editions of the *Spon's Mechanical and Electrical Services Price Book* have built on the structure and information within the first edition and this 50th edition continues to cover the widest range and depth of engineering services, reflecting the many alternative systems and products that are commonly used today alongside current industry trends.

During 2018, MEP output continued to outperform the general construction industry, but there was a noticeable tempering of forecast growth levels, certainly within the London market but also around the UK in general. However, the inflation drivers experienced in previous years have continued with tender price inflation of 2–4%.

Key influences

- Continued skills shortage
- UK Sterling exchange rate against the Euro and US Dollar
- Commodity price volatility

2019 Wage increases have been agreed for Electrical trades, however Mechanical trades have yet to finalize their final uplift at the time of publishing. The editors have however been given an indication by the relevant organization and this has been used to calculate this year's wage increases.

Before referring to prices or other information in the book, readers are advised to study the `Directions' which precede each section of the Materials Costs/Measured Work Prices. As before, no allowance has been made in any of the sections for Value Added Tax.

The order of the book reflects the order of the estimating process, from broad outline costs through to detailed unit rate items.

The approximate estimating section has been revised to provide up to date key data in terms of square metre rates, all-in rates for key elements, and selected specialist activities and elemental analyses on a comprehensive range of building types.

The prime purpose of the Materials Costs/Measured Work Prices part is to provide industry average prices for mechanical and electrical services, giving a reasonably accurate indication of their likely cost. Supplementary information is included which will enable readers to make adjustments to suit their own requirements. It cannot be emphasized too strongly that it is not intended that these prices should be used in the preparation of an actual tender without adjustment for the circumstances of the particular project in terms of productivity, locality, project size and current market conditions. Adjustments should be made to standard rates for time, location, local conditions, site constraints and any other factor likely to affect the costs of a specific scheme. Readers are referred to the build up of the gang rates, where allowances are included for supervision, labour related insurances, and where the percentage allowances for overhead, profit and preliminaries are defined.

Readers are reminded of the service available on the Spon's website detailing significant changes to the published information: www.pricebooks.co.uk/updates

As with previous editions the Editors invite the views of readers, critical or otherwise, which might usefully be considered when preparing future editions of this work.

Whilst every effort is made to ensure the accuracy of the information given in this publication, neither the Editors nor the Publisher in any way accept liability for loss of any kind resulting from the use made by any person of such information.

In conclusion, the Editors record their appreciation of the indispensable assistance received from the many individuals and organizations who helped in compiling this book.

AECOM Ltd
Aldgate Tower
2 Leman Street
London E1 8FA

Special Acknowledgements

The Editors wish to record their appreciation of the special assistance given by the following organizations in the compilation of this edition.

HOTCHKISS AIR SUPPLY

Hampden Park Industrial Estate
Eastbourne
East Sussex
BN22 9 AX
Tel: 01323 501234
Email: info@Hotchkiss.co.uk
www.Hotchkiss.co.uk

23–24 Riverside House
Lower Southend Road
Wickford
Essex
SS11 8BB
Tel: 01268 572116
Fax: 01268 572117
Email: general@abbeythermal.com

T.Clarke
BUILDING SERVICES GROUP

45 Moorfields
London
EC2Y 9AE
Tel: 020 7997 7400
Email: info@tclarke.co.uk
www.tclarke.co.uk

DORNAN

Dornan Engineering Ltd
114a Cromwell Road
Kensington
London
SW7 4ES
Tel: 020 7340 1030
Email: info@dornangroup.com
www.dornan.ie

SPON'S 2019 PRICEBOOKS from AECOM

Spon's Architects' and Builders' Price Book 2019

Editor: AECOM

Now with Semi and Automatic pedestrian doors -- Revolving, Sliding, and Swing doors; an expanded range of industrial shutter doors; industrial docks and shelters; and an expanded range of aluminium gutters.

Hbk & VitalSource® ebook 840pp approx.
978-1-138-61201-3 £160
VitalSource® ebook
978-0-429-46449-2 £160
(inc. sales tax where appropriate)

Spon's Civil Engineering and Highway Works Price Book 2019

Editor:AECOM

This year gives more emphasis to sampling from major projects of £20-£100m in value run under NEC contracts. Prices are based more strongly on global structural steel grades and rates and extrapolated accordingly. And prices are now given for pipe lining for maintenance drainage works.

Hbk & VitalSource® ebook 712pp approx.
978-1-138-61202-0 £180
VitalSource® ebook
978-0-429-46448-5 £180
(inc. sales tax where appropriate)

Spon's External Works and Landscape Price Book 2019

Editor:AECOM in association with LandPro Ltd
Landscape Surveyors

Now including Corten edgings, glass balustrades; decking pedestals for roof gardens; stone copings; and tier system cladding.

Hbk & VitalSource® ebook 672pp
approx. 978-1-138-61203-7 £150
VitalSource® ebook
978-0-429-46447 £150
(inc. sales tax where appropriate)

Spon's Mechanical and Electrical Services Price Book 2019

Editor:AECOM

This NRM edition includes an updated engineering features section and a section on smart building technology – along with new and significantly developed items: tuneable white luminaires; wireless lighting control; PV cells; and battery storage systems.

Hbk & VitalSource® ebook 872pp approx.
978-1-138-61205-1 £160
VitalSource® ebook
978-0-429-46446-1 £160
(inc. sales tax where appropriate)

Receive our VitalSource® ebook free when you order any hard copy Spon 2019 Price Book

Visit www.pricebooks.co.uk

To order:
Tel: 01235 400524
Post: Taylor & Francis Customer Services, Bookpoint Ltd, 200 Milton Park, Abingdon, Oxon,
OX14 4SB, UK
Email: book.orders@tandf.co.uk
A complete listing of all our books is on www.crcpress.com

CRC Press
Taylor & Francis Group

Acknowledgements

The editors wish to record their appreciation of the assistance given by many individuals and organizations in the compilation of this edition.

Manufacturers, Distributors and Subcontractors who have contributed this year include:

Aquality Trading & Consulting Ltd
6 Wadsworth Road
Perivale, Greenford
London UB6 7JJ
Tel: 020 8991 3725
Email: info@aqua-lity.co.uk
www.aqua-lity.co.uk
Water Management

Aquatech-Pressmain
A G M House
London Road
Copford
Colchester CO6 1GT
Tel: 01206 215121
www.aquatechpressmain.co.uk
LTHW Pressurization Units

Aquilar Ltd
Weights and Measures House
20 Barttelot Road
Horsham
West Sussex RH12 1DQ
Tel: 01403 216100
Email: info@aquilar.co.uk
www.aquilar.co.uk
Leak Detection

Babcock Wanson
7 Elstree Way
Borehamwood
Hertfordshire WD6 1SA
Tel: 020 8953 7111
www.babcock-wanson.co.uk
Packaged Steam Generators

Balmoral Tanks
Rathbone Square Office 5B
24 Tanfield Road
Croydon
Surrey CR0 1AL
Tel: 01208 665410
www.balmoral-group.com
Sprinkler Tanks

Biddle Air Systems Ltd
St Mary's Road, Nuneaton
Warwickshire CV11 5AU
Tel: 02476 384233
Email: sales@biddle-air.co.uk
www.biddle-air.co.uk
Air Curtains

Bodet Ltd
4 Sovereign Park
Cleveland Way
Industrial Estate
Hemel Hempstead HP2 7DA
Tel: 01442 418800
www.bodet.co.uk
Clocks

Design Intent
Unit 4 Eaton Court
Colmworth Business Park
Eaton Socon
St Neotts
Cambridgeshire, PE19 8ER
General Lighting

Dewey Waters Ltd
The Heritage Works
Winterstoke Rd
Weston-Super-Mare BS24 9 AN
Tel: 01934 421477
www.deweywaters.co.uk
Cold Water Storage Tanks

Diffusion
47 Central Avenue
West Molesey
Surrey KT8 2QZ
Tel: 020 8783 0033
Email: diffusion@etenv.co.uk
www.diffusion-group.com
Fan Coil Units

DMS Flow Measurement and Controls Ltd
X-Cel House
Chrysalis Way
Langley Bridge
Eastwood
Nottinghamshire
NG16 3RY
Tel: 01773 534555
Email: sales@dmsltd.com
www.dmsltd.com
Chilled Water Plant and Energy Meters

Dormakaba Ltd
Lower Moor Way
Tiverton
Devon EX16 6SS
Tel: 01884 256464
Email: info.gb@dormakaba.com
www.dormakaba.com
Access Control Barriers, Revolving Doors

Dunham-Bush Ltd
8 Downley Road
Havant
Hampshire PO9 2JD
Tel: 02392 477700
www.dunham-bush.com
Convectors and Heaters

EA-RS Fire Engineering Ltd
4 Swanbridge Industrial Park
Black Croft Road
Witham
Essex, CM8 3YN
Tel: 01376 503680
Email: onesolution@ea-rsgroup.com
www.ea-rsgroup.com
Fire Detection and Alarms

ETAP Lighting
7 Progress Business Centre
Whittle Park Way
Slough
SL1 6DQ
www.etaplighting.com
Emergency Lighting

Eton Associates
9 Quebec Wharf
Thomas Road
London E14 7 AF
Tel: 020 7068 7900
Email: enquiries@etonassociates.com
www.etonassociates.com
BMS

Frontline Security Solutions Ltd
Reflex House
The Vale
Chalfont St Peter
Bucks
SL9 9RZ
Tel: 01753 482 248
www.fsslimited.com/
Access Control and Security Detection Alarm

Harlequin Manufacturing Ltd
21 Clarehill Road
Moira, County Armagh
Northern Ireland
BT67 0PB
Tel: 028 9261 1077
Email: info@harlequin-mfg.com
www.harlequinplastics.co.uk/
Fuel Oil Storage Tanks

Hawker Siddeley Switchgear (HSS) Ltd
Unit 3, Blackwood Business Park
Newport Road, Blackwood
South Wales NP12 2XH
Tel: 01495 223001
www.hss-ltd.com
HV Circuit Breakers

Hoval Ltd
Northgate
Newark
Notts NG24 1JN
Tel: 01636 672711
www.hoval.co.uk
Storage Cylinders/Calorifiers/Commercial Oil Boilers

HRS Hevac Ltd
10–12 Caxton Way
Watford Business Park
Watford
Herts WD18 8JY
Tel: 01923 232335
Email: mail@hrshevac.co.uk
www.hrshevac.co.uk
Heat Exchangers

Hudevad
Record Hall Business Centre,
Rm 215 Hatton Garden
London, EC1N 7RJ
Tel: 02476 881200
hudevad.com/
Heat Emitters & Radiators

Hydrotec (UK) Ltd
Hydrotec House
5 Mannor Courtyard
Hughenden Avenue
High Wycombe HP13 5RE
Tel: 01494 796040
www.hydrotec.com
Cleaning and Chemical Treatment

Industrial Engineering Plastics
Passfield Mill Business Park
Passfield
Liphook
Hampshire GU30 7QU
Tel: 0121 771 2828
Email: sales@ieplastics.co.uk
www.iep-ltd.co.uk
Plastic

Jaga
Jaga House
Orchard Business Park
Bromyard Road
Ledbury
HR8 1LG
Tel: 01531 631533
Email: jaga@jaga.co.uk
www.jaga.co.uk
Heat Emitters, Fan Convectors & Trench Heating

JCB Power Products Broadcrown Ltd
Air Field Industrial Estate
Hixon
Stafford
Staffordshire ST18 0PF
Tel: 01889 272200
Email: generator.sales@jcb.com
www.broadcrown.com
Standby Generators

Nick Dunford Associates
Building Three
Watchmoor Park
Camberley, Surrey
GU15 3YL
Tel: 01189 078 622
Email: nickdunford@ndaconsulting.co.uk
www.ndaconsulting.co.uk
BMS

Ormandy Group
Atlas Works, Gibbet Street
Halifax, HX1 4DB
Tel: 01422 350111
www.ormandygroup.com
Calorifiers

Purified Air
Lyon House
Lyon Road
Romford
RM1 2BG
Tel: 0800 018 4000
Email: info@purifiedair.com
www.purifiedair.com
Air Filtration

Rock Clean Energy
Unit 5, Buckholt Business Centre
Buckholt Drive
Worcester WR4 9ND
Tel: 0330 223 4566
Email: enquire@rockcleanenergy.co.uk
www.rockcleanenergy.co.uk
Battery Storage Systems

Rung Heating
Unit 4a, Henstridge Trading Estate
Henstridge
Somerset BA8 0TG
Tel: 01963 364600
Email: sales@rung-heating.co.uk
www.rungheating.co.uk
Flue Systems

Safelincs Ltd
Farlesthorpe Road Industrial Estate
Alford
Lincolnshire
LN13 9PS
Tel: 01507 463288
Email: support@safelincs.co.uk
www.safelincs.co.uk
Fire Protection Equipment

Schneider Electric
2nd Floor
80 Victoria Street
London
SW1E 5JL
Tel: 0870 608 8608
Email: gb-customerservices@schneider-electric.com
www.schneider-electric.co.uk/en/
LV Switch & Distribution Boards/Breakers & Fusers

Simmtronic Ltd
Waterside
Charlton Mead Lane
Hoddesdon
Hertfordshire EN11 0QR
Tel: 01992 456869
www.simmtronic.com
Lighting Controls

Socomec Ltd
Knowl Piece
Wilbury Way
Hitchin
Hertfordshire SG4 0TY
Tel: 01462 440033
www.socomec.com
Automatic Transfer Switches

Swegon Air Management
Stourbridge Road
Stropshire
Bridgnorth WV15 5BB
Tel: 01746 761921
Email: sales@actionair.co.uk
www.swegonair.co.uk
Dampers and Access Hatches

Utile Engineering Company Ltd
Irthlingborough
Northants NN9 5UG
Tel: 01933 650216
www.utileengineering.com
Gas Boosters

Vent Axia
Fleming Way
Crawley
West Sussex
RH10 9YX
Tel: 0344 856 0591
Email: sales@vent-axia.com
www.vent-axia.com
Fans

Waterloo Air Products PLC
Mills Road
Aylesford
Kent ME20 7NB
Tel: 01622 711511
Email: sales@waterloo.co.uk
www.waterloo.co.uk
Grilles, Diffusers, Louvres, VAV Boxes

Whitecroft Lighting
Burlington Street
Ashton-under-Lyne
Lancashire
OL7 0AX
Tel: 0161 330 6811
Email: email@whitecroftlight.com
www.whitecroftlighting.com
Emergency Lighting

It is one thing to imagine a better world.

It's another to deliver it.

Creating iconic structures. Planning entire communities. Connecting people with transport systems. Rescuing historic buildings. Our team of circa 900 cost and project managers collaborate to turn imagination into reality, providing innovative solutions to complex challenges across the globe.

Imagine it. Delivered.

New Aspects of Quantity Surveying Practice, 4th edition

Duncan Cartlidge

In this fourth edition of New Aspects of Quantity Surveying Practice, renowned quantity surveying author Duncan Cartlidge reviews the history of the quantity surveyor, examines and reflects on the state of current practice with a concentration on new and innovative practice, and attempts to predict the future direction of quantity surveying practice in the UK and worldwide.

The book champions the adaptability and flexibility of the quantity surveyor, whilst covering the hot topics which have emerged since the previous edition's publication, including:

*   the RICS 'Futures' publication;
*   Building Information Modelling (BIM);
*   mergers and acquisitions;
*   a more informed and critical evaluation of the NRM;
*   greater discussion of ethics to reflect on the renewed industry interest;
*   and a new chapter on Dispute Resolution.

As these issues create waves throughout the industry whilst it continues its global growth in emerging markets, such reflections on QS practice are now more important than ever. The book is essential reading for all Quantity Surveying students, teachers and professionals. It is particularly suited to undergraduate professional skills courses and non-cognate postgraduate students looking for an up to date understanding of the industry and the role.

December 2017: 234 x 156 mm: 282 pp
Pb: 978-1-138-67376-2 : £45.99

To Order: Tel: +44 (0) 1235 400524 Fax: +44 (0) 1235 400525
or Post: Taylor and Francis Customer Services,
Bookpoint Ltd, Unit T1, 200 Milton Park, Abingdon, Oxon, OX14 4TA UK
Email: book.orders@tandf.co.uk

For a complete listing of all our titles visit:
www.tandf.co.uk

PART 1

Engineering Features

This section on Engineering Features deals with current issues and/or technical advancements within the industry. These shall be complimented by cost models and/or itemized prices for items that form part of such.

The intention is that the book shall develop to provide more than just a schedule of prices to assist the user in the preparation and evaluation of costs.

Contractual Procedures in the Construction Industry, 7th edition

Allan Ashworth and Srinath Perera

Contractual Procedures in the Construction Industry 7th edition aims to provide students with a comprehensive understanding of the subject, and reinforces the changes that are taking place within the construction industry. The book looks at contract law within the context of construction contracts, it examines the different procurement routes that have evolved over time and the particular aspects relating to design and construction, lean methods of construction and the advantages and disadvantages of PFI/PPP and its variants. It covers the development of partnering, supply chain management, design and build and the way that the clients and professions have adapted to change in the procurement of buildings and engineering projects.

Key features of the new edition include:

- A revised chapter covering the concept of value for money in line with the greater emphasis on added value throughout the industry today.
- A new chapter covering developments in information technology applications (building information modelling, blockchains, data analytics, smart contracts and others) and construction procurement.
- Deeper coverage of the strategies that need to be considered in respect of contract selection.
- Improved discussion of sustainability and the increasing importance of resilience in the built environment.
- Concise descriptions of some the more important construction case laws.

March 2018: 246 x 174 mm: 458 pp
Pb: 978-1-138-69393-7 : £45.99

To Order: Tel: +44 (0) 1235 400524 Fax: +44 (0) 1235 400525
or Post: Taylor and Francis Customer Services,
Bookpoint Ltd, Unit T1, 200 Milton Park, Abingdon, Oxon, OX14 4TA UK
Email: book.orders@tandf.co.uk

For a complete listing of all our titles visit:
www.tandf.co.uk

Building Information Modelling (BIM)

The publication of the Government Construction Strategy in 2011 has helped bring about a digital construction revolution in the UK and the rest of the world. At the centre is a package of measures with Building Information Modelling, focused on digital collaboration achieving best value across the project timeline. Following the UK Government's BIM mandate in 2016 and a Scottish BIM policy the following year, BIM has seen a steady rise in its use within the construction industry.

BIM has become an agent for change. A central proposition of BIM is the concept of working collaboratively in order to improve productivity and reduce waste. It also plays an important role in achieving the ambitious targets outlined in the Government Construction Strategy 2025 that aims to sustainably create more assets that are built faster and are more cost efficient by moving towards offsite manufacturing. Working in conjunction with a number of other initiatives such as new procurement models, benchmarking and early contractor engagement, the UK Government has demonstrated considerable rewards using BIM, such as a reduction in construction costs by 20% on public sector projects. This is achieved by unlocking savings in capital expenditure and also the operational stages of the asset lifecycle. However it is not just the public sector that has much to gain, but the private sector has also demonstrated considerable benefits using BIM processes.

So what is BIM?

BIM is a collaborative process, enabled by technology. BIM exploits the potential in computer-based modelling technologies to provide a new way of designing buildings and managing design and construction processes. It requires new ways of working that need cultural, behavioural and technological changes in order to deliver better value and better project outcomes. Information models and structured data are created, shared and exchanged throughout the project lifecycle in a managed way.

It requires clients to be clear about their information requirements. This means communicating to the supply chain stipulating information they need, when they need it and who is responsible for it, in order to make key decisions at important points in the project. Essentially the increased access to information allows the client to confirm that the asset meets its performance expectation. BIM also requires teams to come together earlier on in the process in order to bridge the gap between design, construction and operation. This ensures that the operational performance is optimized and the asset performs as intended, an area which the industry needed to improve upon.

To embed and prepare the construction industry for BIM adoption, a phased approach was undertaken. A chosen staging post, known as BIM Level 2 was selected for the 2016 mandate. This is a series of domain and collaborative federated models. (Federation means that models and information are linked together to create a single model of the asset). Models, also known as information models, are a combination of geometrical and non-graphical data, prepared by different parties during the project lifecycle. Thus, improving the ability to interrogate, repurpose and verify data via graphical models and databases. Dynamic modelling software can be used to develop and manipulate these digital models to refine the design and to test and validate its potential performance across a range of criteria, including buildability, energy performance-in-use, whole life costing, etc.

Collaborative working

A key theme of BIM is the sharing of information and data within a central online space, known as a Common Data Environment or CDE. Significant benefits are to be gained from this approach, as it ensures everyone has access to the most up-to-date data, decreasing risk and increasing accuracy. It also improves design coordination, reduces design costs and improves communications throughout the design and construction process. By working together to develop sophisticated, coordinated, information models, the different design and construction disciplines can 'prototype' projects before they are built. Designs can be developed and tested 'virtually' so that the performance and cost are optimized. They can be coordinated so that many potential problems are either designed-out or avoided altogether. Ultimately, the benefits not only lie in more efficient and effective design and construction processes, but better and more certain project outcomes – better buildings that are more fit-for-purpose and meet their brief requirements and design intent.

How does BIM work?

The potential for all key project information to be stored and manipulated on a computer is what sets BIM apart from more conventional approaches, and BIM-based design solutions differ from their traditional counterparts in that they:

- Are created and developed on digital databases which enable collaboration and effective data exchange between different disciplines;
- Allow change to be managed through these databases, so that changes in one part of the database are reflected in (and coordinated through) changes in other parts; and
- Capture and preserve information for reuse by all members of the design and construction team, including facilities management (FM), and user operation and management.

Conventionally, a good deal of design and construction work is document based. Information is communicated and stored via a variety of drawings and reports that, despite being stored and distributed in digital form, are essentially 'unstructured' and thus of limited use. Not only is this information unstructured, it is also held in a variety of forms and locations that are not formally coordinated (information on individual building components, for example, are contained in drawings, specifications, bills of quantity descriptions, etc.). Such an approach has considerable potential for data conflicts and redundancy as well as risks to data integrity and security.

BIM opens up a wide range of possibilities for improvement that includes better ways of generating, exchanging, storing and reusing project information that greatly improve communications between different design and construction disciplines through the life of the asset. As well as procuring a physical asset, the client is also procuring a 'digital asset', one which can be used for forward planning, investment decisions and operational and maintenance activities.

Of course, digitally structuring information can take many forms, from simple temporary models created for a specific purpose (i.e. a room schedule with key room dimensions in spreadsheet form) through to shared 3D whole-building models containing architectural, structural, servicing and other data all in the same place. The greater the degree of information-sharing and collaboration in the development of the models, the more accurate and complete the models will be and therefore the benefits of using BIM are altogether clearer. As shared models are developed, problems – such as clashes between structure and services – tend to get ironed out, and designs become more consistent and coordinated.

BIM benefits

Generally, BIM is recognized as providing a wide range of valuable benefits including:

- **Design:** improved coordination of design and deliverables between disciplines: improved project understanding through visualization; improved design management and control, including change control; and improved understanding of design changes and implications through parametric modelling.
- **Compliance:** ability to perform simulation and analysis for regulatory compliance; and ability to simulate and optimize energy and wider sustainability performance.
- **Costing/economics:** ability to perform cost analysis as the design develops, and to check for adherence to budget/cost targets; ability to understand cost impacts of design changes; and improved accuracy of cost estimates.
- **Construction:** reduction of construction risks through identification of constructability issues early in the design process; early detection and avoidance of clashes; ability to model impact of design changes on schedule and programme; and ability to integrate contractor/subcontractor design input directly to the model.

- **Operation and management:** creation of an FM database directly from the project (as built) model; ability to perform FM costing and procurement from the model; and ability to update the model with real-time information on actual performance through the life of the building.

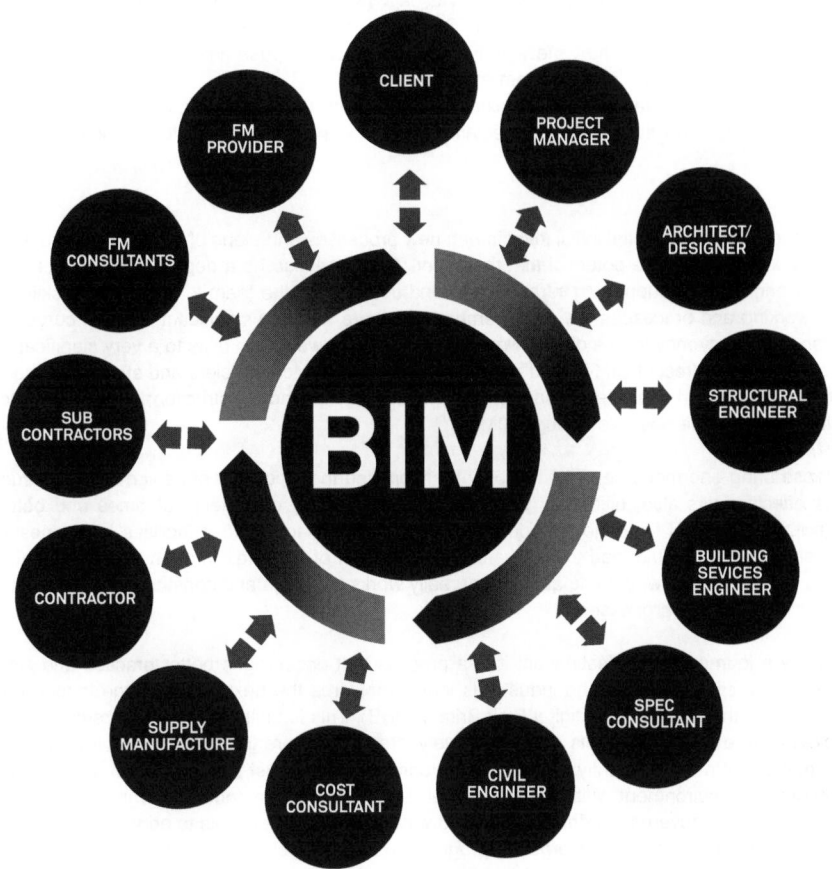

Ongoing BIM issues and risks

There are a number of uncertainties and potential risks in its adoption including:

- **Data Security:** The increased use of digital data means new threats have arisen around cyber security. The management of sensitive projects, data and information requires permissions and polices in place concerning who can have access and who can allow permissions. This is a key area that was recognized during the development of BIM Level 2, and a standard PAS 1192:5 has been published which addresses these issues.
- **Legal, contractual and insurance issues:** BIM Levels 1 and 2 require little change to the fundamental building blocks of copyright law, contracts or insurance. Standard forms of building contract – including design and consultant appointments – often do not make specific provision for BIM, however they may cover areas such as collaborative working. A BIM Protocol can be used as a supplementary legal agreement, appended to existing building contracts. The protocol covers the ethos of collaboration by placing obligations on parties to provide and share information at defined stages of a project. As the industry moves towards the concept of a single central model, which all parties contribute directly to, a new contractual landscape will be required.

- **IT and software:** BIM Level 2 allows industry to use the 'best of breed' software, however, it does require particular information about the facilities management and operational maintenance of an asset to be supplied in a common file format, so that it can be accessed by everyone, and also in the future. The move towards greater interoperability of systems and data (where all data and information can be read by all software) will require greater use of Open Standards such as the implementation and use of Industry Foundation Classes (IFC). The variety of systems in use can impose heavy training burdens on firms who need to operate with some or all of them.
- **Awareness and expectations:** Paradoxically, BIM suffers from relatively low levels of client and sector awareness combined with very high expectations typical of new information technology developments.

A balanced view

BIM inevitably requires a substantial initial investment; new processes and plans of work typically require an upfront investment in order to realize the potential that they offer. There will also be a degree of investment in technology and software, along with the supporting infrastructure and upskilling of the team to acquire new skills. While these new ways of working and processes are slowly embedded, there will be a significant learning curve, during which productivity and efficiency may take a dip. However the benefits outweigh the risks to a very significant degree and the sheer transformative effect of BIM should not be underestimated. More efficient and effective design processes; greatly improved information quality and coordination between design and construction; and better prototyping prior to construction are all made possible through BIM.

Potentially these bring enormous benefits, not only by improving the efficiency of design and construction, but by improving its effectiveness also, ultimately providing greater certainty of project outcomes and better buildings. More than that, the potential for project models to support the management of facilities in the post-construction phase is considerable and could lead to more effective operation of buildings through the whole lifecycle. One of these benefits is to provide new learning about what really works in design and construction which can be fed back to inform and improve these processes.

BIM is ultimately a journey, not a destination; it is a process that encourages better practice and aims to deliver better results for the client. Already, the industry is looking towards the next step and the longer-term trajectory towards Level 3 with the formation of Digital Built Britain (DBB). This is built around a programme that will help us plan new infrastructure and built assets more effectively, create (or more likely assemble) it at a lower cost, and operate and maintain it more efficiently. The Level 3 programme is not just looking at BIM, but at the wider digitization of the building environment. With the BIM Level 2 standards and mandate now in place, part of the DBB programme is supporting government departments and wider public-sector clients to adopt Level 2 as 'Business As Usual'. Following this, the Level 2 Convergence programme looks at how we can start to connect construction to the smart city world of service, consumers and the citizen, using the rich data sources such as IoT. This will require the use of advanced manufacturing such as offsite manufacturing, standardization and optimization and changes to the current delivery model.

In conclusion

The most important decision clients will face is the selection and appointment of the design and construction team. Finding the right team of people with a positive approach to using BIM, to sharing information and, above all, to learning from the experience is key. In this regard, clients should expect all their consultants, contractors and specialists to be familiar with BIM and its requirements; to be positively engaged in its adoption; and to be actively developing ways in which processes can be made more value-adding and effective. Many projects have seen the adoption of BIM for clash detection and coordination purposes with a BIM consultant being appointed to manage the process. Greater understanding from the design teams and demand from clients will see more traditionally based tasks being achieved through BIM in subsequent years to come.

Smart Building Technology

Whichever Smart building technology is deployed, to the users, the technology must deliver tangible benefits over and above those delivered by a traditional solution. Users shall include:

- Employees working within the building
- Visitors and guests
- Building owners, facility managers and maintenance teams.

These benefits must be able to be quantified and measured to ensure that any Smart building enhancements provide improvements in Comfort, Productivity and or Operational Efficiency. These benefits need to be quantified against any enhancement cost variations, both Capex and Opex.

The Smart building enhancement benefits will vary across each user group, building type and industry sector.

The following discusses some of the Smart building technologies presently available to building owners and operators. Depending upon the aspirations of the client, the Smart building system will often consist of a number of technologies integrated into a single Smart building package linked to cause and effect of the building operation.

Employee and visitor apps

App and portal based technologies are being deployed within the workplace to enhance the workplace experience. Often used to support both employees and visitors to the site, a workplace App can be used to:

- Check on travel options to the building
- View weather predictions
- Pre-book visitor car parking
- Pre-book with security notifying them of their planned visit
- 1st level security check in, which can be verified against by secondary level security verification code
- In-building wayfinding
- Desk booking
- Meeting room booking
- HVAC controls interface
- Light controls
- Building amenity and facilities such as a canteen portal detailing menu options, nutrition values and allergy considerations. Location sensor linked to the App could display the canteen queue time
- Security travel updates
- General travel updates
- Facility managers can be provided with real time building performance information
- Maintenance staff can receive works orders and once completed can close down the task and await details of the next work order.

These are just some of the uses for work-based Apps, delivered to a mobile device acting as the user's single pane of glass.

Automatic number plate recognition and Smart parking systems

If traveling by car, at the car park entrance, Automatic Number Plate Recognition (ANPR) cameras match the registration with the vehicle listed in the visitor's profile and allow access. The same could be applied to visitors who have registered their visit via the business App or portal.

An intelligent parking system then directs visitors to their reserved spot.

Information from the above can be shared with Security making them aware of arrivals.

Enhanced security check-in

As visitors enter reception they are identified by facial recognition cameras linked to their profiles. Depending upon the security procedures and present security level, before they are allowed further a second level of verification may be required. This could take the form of either:

- Security matrix barcode previously sent to their mobile devices
- Smart ID card
- Finger print or retina recognition.

Vertical transportation with port destination

Knowing their meeting room or hot desk destination, vertical transport port destination can select and call the lift car. The open door to the lift awaits, poised for the journey skyward.

The vertical transport system can also detect the number of persons entering a lift car, and check numbers against the required occupancy. The system can be integrated to the CCTV monitoring system and could activate a cause and effect process and commence recording of CCTV images, alerting security that tailgating has taken place.

Location sensing and wayfinding

There are many systems available today that provide location sensing. Systems can be linked to mobile phones, WiFi pendants, employee and visitor badges. In addition to providing instant location sensing they can also be used to monitor people flow through buildings and wayfinding.

Linked to the workplace App, In-building wayfinding allows users to:

- Navigate through the building, finding quickest routes to meeting rooms or allocated hot desks
- Find the nearest print station, toilet or breakout area
- Identify workplace locations for individuals or workplace neighbourhoods.

Integrated Smart meeting rooms

Room booking systems not only allow the booking and scheduling of the meeting space but integrated solutions can also provision the AV connections matched to users' preferences, set up external video conference links and via the integrated intelligent Building Management System ensure that HVAC and lighting levels are set to the optimum level for the meeting type.

Meeting room monitoring

IoT sensors within the room can monitor the room environment, such as:

- Lighting levels
- Ambient noise
- CO_2
- VOC
- Power usage
- Occupancy.

Linked to the iBMS the data were used to set and adjust controls to ensure users' wellbeing and comfort. Many of the systems on the market utilize WiFi enabling quick and easy deployment with limited cabled infrastructure.

The above information can also be sent back to the facilities manager where real time environment occupancy data can be stored and analysed.

Automated and semi-automated meeting room controls

The abovementioned occupancy sensors can be used to detect when a meeting room is no longer occupied and, linked to the iBMS, place the room into standby mode adjusting the heating, cooling and light levels until the next scheduled booking.

There is also the option of incentivizing users to release rooms early. This could take the form of cashless vending credits.

Power over ethernet

Within the IT sector Power over Ethernet (PoE) has been adopted for many years to power DC devices such as CCTV cameras, and WiFi Access Points.

The technology is now being adopted for office lighting. A structured cabling grid system is installed at a high level to facilitate connection of PoE enabled devices including lighting. Lighting arrays also connect to other integrated sensors, including environmental, occupancy and footfall enabling clients to identify where tangible business and colleague wellbeing benefits can be delivered.

Systems can use Bluetooth low energy (BLE) technology to track people's movements by facilitating wireless communication between luminaires within a lighting control system and moveable objects, e.g. security passes.

Data can then be presented in a number of ways allowing decisions about how building controls are adapted and integrated to, thus leading to the more efficient use of office space and allowing more informed decisions on how space is used in the future.

Desk booking and utilization systems

With more and more companies moving to Agile Activity Based Working, where users are no longer allocated permanent desks and users are encouraged to select their own work place location or work group neighbourhood; desk booking systems are now becoming a common offering in the modern office.

Users can easily book a desk in either their preferred location or work group neighbourhood. In addition to improving productivity it also removes user frustration and time spent 'walking the floor(s)' looking for an available desk.

Desk monitoring and occupancy systems provide facility managers and building owners with real time floor and desk utilization data. Trend data can then be analysed and built into predictive analysis models, thus providing potential answers to the following questions:

- What is our workspace utilization?
- Could we reduce our floor plate whilst still maintaining the same level of operation?
- Does the trend information show patterns when workspace utilization falls?
- Could we shut a floor on Fridays?
- Are certain areas regularly underutilized? If so, why?
- Could occupancy levels be improved if the layout was changed?
- New workspace areas and layouts could be trialled with real time occupancy data provided to the facilities managers.

In-building cellular and Wi-Fi

Some would question if In-Building Cellular and Wi-Fi is a Smart technology, but without properly designed in-building mobile communications the adoption of many Smart technologies would not be possible.

We know the deployment of Wi-Fi will continue to proliferate and become evermore integral within buildings. Traditional Wi-Fi, based around IEEE 802.11 standards, is poised to take a significant step forward with the introduction of 802.11ax, the next standard for wireless, for a connected world where upload and download traffic will be equally important.

There will be various speeds of up to 4.8 Gbps on the 5 GHz radio frequency and 1.15 Gbps on 2.4 GHz.

Whilst ratification of 802.11ax isn't expected until mid-2019, things are far enough along now that the specifications are unlikely to change and we would expect roll-out around the end of the year.

New and emerging technologies for In-Building Cellular (2G, 3G, 4G and the future potential for 5G) provide users with cellular coverage in the following scenarios:

- Within buildings designed with high performing solar glass that protect users and optimize building performance but where the glass provides a shield to the transmission of cellular services
- High rise buildings where signal levels are poor
- Basements where there is unlikely to be any coverage.

New and emerging alternative technologies including low power devices and LiFi

Devices working over lower power Bluetooth and RFID are already being deployed. Within hospitals portable equipment can be tagged allowing staff to quickly and efficiently identify equipment locations. This can include RFID tagged documents. The same technology can be applied to the office.

LiFi, a high speed bi-directional mobile networked and communications platform delivered using light to transmit data will make an impact in the Smart building arena. This is a technology seen to deliver high speed secure communications, a concern that many users have when deploying traditional Wi-Fi

Intelligent Building Management Systems (iBMS)

An iBMS can integrate many of these technologies into a single system with a single operator interface, often referred to as a single pane of glass.

In addition to providing users with a common window into the system the iBMS systems allow services & devices to be monitored, controlled and optimized, running analytics improving building performance whilst driving down energy loads and costs.

Data on device performance can be collected and analysed and compared against trend data. Variants can then be reviewed and actioned upon.

As an example:

A motor fitted with IoT sensors is showing a higher than average level of vibration and noise indicating a potential bearing failure. Maintenance repairs or replacement can then be scheduled in advance often providing both labour cost savings and reductions in system downtime.

As the use of Artificial Intelligence increases many of these repetitive big data number crunching processes can be undertaken using AI, with optimized alerts sent to operators and maintenance staff.

Are Smart buildings secure?

Any system that connects over the internet or uses an Internet Protocol (IP) enabled interface connected to LAN switch is potentially at risk of network penetration, which everyone within the industry accepts, therefore it is important that the correct Cyber Security systems and processes are put in place in advance to protect the Smart building.

Many traditional systems also fall into this category, for example:

- The majority of traditional in-building security systems solutions rely on IP enabled CCTV cameras and IP connected access controllers connected over LAN switches.
- Traditional BMS often provide remote monitoring facilities connected over the internet.
- Many meeting room booking systems can be booked remotely over the internet.

These examples, and many more, show that both traditional and Smart building solutions are both potentially at risk of network penetration/hack and therefore it is essential that a full analysis review is undertaken to minimize risk.

Risks can be minimized by:

- Undertaking a risk review process for each connected system and discipline. This needs to review the level of risk and the type of data that is associated with each discipline. For example:
  - Security systems will contain staff personnel data and video footage which could be used to affect business operation.
  - Cashless vending systems will contain staff personnel bank details.
  - BMS or iBMS control critical operational plant which in the case of an illegal penetration/hack could be targeted.
- Review of device security and passwords. Devices are often sent out with default passwords which need to be changed.
- Physical network separation between in-building systems, technology disciplines and the corporate network.
- Physical security for building ingress and exit paths.
- Physical security on main and secondary equipment rooms and building entrance facilities housing the external telecoms links.
- Review of firewall policy.
- Review of policy for soft and firmware updates.
- Penetration testing.
- Policy and action plan for dealing with an illegal penetration/hack. There are many documented examples where companies and organizations are aware of an illegal penetration/hack but have failed to respond or advise affected third parties.
- Ongoing review of the above.

These are just some of the activities and processes that need be undertaken to minimize penetration risk.

Summary

These are just a sample of the smart technologies that could be deployed within the business workplace today.

The building design, construction, fit-out and handover is often undertaken over several years. A three to five year period is not uncommon. Yet at the same time new and emerging IoT devices and Smart building technologies are evolving at a rapid rate. IoT is being labelled as the 'Next Industrial Revolution' but it is largely already here.

With the growth in the IoT and Smart building arena it is very difficult to accurately predict what a Smart building will look like in 5-10 years. We know the technologies that are available today, we can evaluate new technologies as they come to market and we can analyse technology trends and make predictions.

We do know, and history has shown, that the infrastructure, be that wired, wireless or light, is the enabler for the deployment of new and emerging technologies. Without investment in the correct infrastructure, the latest technologies cannot be deployed and taken advantage of.

Quantity Surveyor's Pocket Book, 3rd Edition

D. Cartlidge

The third edition of the Quantity Surveyor's Pocket Book has been updated in line with NRM1, NRM2 and NRM3, and remains a must-have guide for students and qualified practitioners. Its focused coverage of the data, techniques and skills essential to the quantity surveying role makes it an invaluable companion for everything from initial cost advice to the final account stage.

Key features and updates included in this new edition:
- An up-to-date analysis of NRM1, 2 and 3;
- Measurement and estimating examples in NRM2 format;
- Changes in procurement practice;
- Changes in professional development, guidance notes and schemes of work;
- The increased use of NEC3 form of contract;
- The impact of BIM.

This text includes recommended formats for cost plans, developer's budgets, financial reports, financial statements and final accounts. This is the ideal concise reference for quantity surveyors, project and commercial managers, and students of any of the above.

March 2017; 186 × 123 mm, 466 pp
Pbk: 978-1-138-69836-9; £20.99

To Order: Tel: +44 (0) 1235 400524 Fax: +44 (0) 1235 400525
or Post: Taylor and Francis Customer Services,
Bookpoint Ltd, Unit T1, 200 Milton Park, Abingdon, Oxon, OX14 4TA UK
Email: book.orders@tandf.co.uk

For a complete listing of all our titles visit:
www.tandf.co.uk

Renewable Energy Options

This article focuses on building-integrated options rather than large-scale utility solutions such as wind farms, which are addressed separately, and provides an analysis of where they may be best installed.

The legislative background, imperatives and incentives

In recognition of the causes and effects of global climate change, the Kyoto protocol was signed by the UK and other nations in 1992, with a commitment to reduce the emission of greenhouse gases relative to 1990 as the base year.

The first phase of European Union Emission Trading Scheme (EU ETS) covered the power sector and high-energy users such as oil refineries, metal processing, mineral and paper pulp industries. From 1 January 2005, all such companies in the EU had to limit their CO_2 emissions to allocated levels in line with Kyoto. The EU ETS is now in its third phase, running from 2013 to 2020 and covers the 27 EU Member States, as well as Iceland, Liechtenstein, Norway and Croatia. The main change from previous phases being that additional sectors and gases are now included. Key principles of the EU ETS are that participating organizations can:

- Meet the targets by reducing their own emissions (e.g. by implementing energy efficiency measures, using renewable energy sources), or
- Exceed the targets and sell or bank their excess emission allowances, or
- Fail to meet the targets and buy emission allowances from other participants.

The EU ETS is designated as a 'cap and trade' system, where participation is mandatory for the sectors covered and it accepts credits from emission-saving projects carried out under the Kyoto Protocol's Clean Development Mechanism (CDM) and Joint Implementation instrument (JI).

In the UK, the Utilities Act (2000) requires power suppliers to provide some electricity from renewables, starting at 3% in 2003 and rising to 15% by 2015. In a similar way to the EU ETS, generating companies receive and can trade Renewables Obligation Certificates (ROCs) for the qualifying electricity they generate.

The focus is on Greenhouse Gases (GHG) and Carbon Dioxide (CO_2) in particular as the main direct contributor to the greenhouse effect. The goals set by the UK government are:

- 34% GHG emissions reduction by 2020 (below 1990 baseline).
- Around 30% of electricity from renewables by 2020.
- 80% GHG emissions reduction by 2050 (below 1990 baseline).

Four years after the introduction of ROCs, it was estimated that less than 3% of UK electricity was being generated from renewable sources. A 'step change' in policy was required, and the Office of the Deputy Prime Minister (ODPM) published 'Planning Policy Statement 22 (PPS 22): Renewable Energy' in order to promote renewable energy through the UK's regional and local planning authorities. PPS 22 was replaced by the National Planning Policy Framework, published in March 2012 by the Department for Communities and Local Government (DCLG), which encourages the use of renewable resources.

Local planning & building regulations

More than 100 local authorities embraced PPS 22 by adopting pro-renewables planning policies typically requiring a percentage (e.g. 10%) of a development's electricity or thermal energy needs to be derived from renewable sources. Government is developing further guidance to support local authorities addressing the sustainability of planned developments and to ensure a level playing field across the country.

Government is also committed to successive improvements in national new-build standards through changes to the Building Regulations, Part L, Conservation of Fuel and Power. In October 2010, new regulations introduced a 25% improvement on 2006 standards.

Energy performance certificates (EPCs)

A Recast of the 2002 EU Directive on the Energy Performance of Buildings (EPBD) was published in 2010. It requires that EPCs be produced for buildings constructed, sold, or rented out to new tenants. For buildings occupied by public authorities and frequently visited by the public, the EPC must be displayed in a prominent place clearly visible to the public. These provisions enable prospective buyers and tenants to be informed of a building's energy performance, ensure the public sector leads by example, and raise public awareness. The Recast was transposed in the UK by national Governments, for example in England & Wales through the Energy Performance of Buildings (England & Wales) Regulations 2012 and other regulations. Other requirements of the Recast include: adopting a method to calculate the energy performance of buildings, setting minimum energy performance requirements and for new buildings to be Nearly Zero Energy Buildings by 2020.

Assessing the regulated carbon emissions associated with new buildings is now an important part of the design and building permitting process with the regulatory approach set out in Part L of the Building Regulations (Conservation of Fuel & Power), the associated National Calculation Methodology (NCM) Modelling Guide and the Standard Assessment Procedure (SAP). On-site renewable energy sources are taken into account and there are limits on design flexibility to discourage inappropriate trade-offs such as buildings with poor insulation standards offset by renewable energy systems.

Technology options and applications

- **Wind generators** – In a suitable location, wind energy can be an effective source of renewable power. Without grant, an installed cost range of £3500 to £5500 per kW of generator capacity may be achieved for small building-mounted turbines. A common arrangement was for a turbine with three blades on a horizontal axis, all mounted on a tower or, for small generators in inner city areas, on a building. Such arrangements typically compare poorly against other renewable options, as they are highly dependent on wind speed at the turbine, obstructions (e.g. nearby buildings and trees), turbulences, the elevation of the turbine above ground, and mitigating other impacts such as aesthetics consideration for planning permission, noise, and vibrations. With suitable conditions, average site wind speeds of 4 m/s can produce useful amounts of energy from a small generator up to about 3 kW, but larger generators require at least 7 m/s. A small increase in average site wind speed will typically result a large increase in output. There will be a need for inverters, synchronizing equipment, and metering for a grid connection.

  Larger, stand-alone turbines typically compare more favourably than smaller building-mounted turbines. Third party provision through an Energy Service Company (ESCo) can be successful for larger (stand-alone) installations located within or close to the host building's site, especially in industrial settings where there may be less aesthetic or noise issues than inner city locations. The ESCo provides funding, installs and operates the plant and the client signs up for the renewable electrical energy at an agreed price for a period of time.

- **Building integrated photovoltaics (BIPV)** – Photovoltaic materials, commonly known as solar cells, generate direct current electrical power when exposed to light. Solar cells are constructed from semiconducting materials that absorb solar radiation; electrons are displaced within the material, thus starting a flow of current through an external connected circuit. PVs are available in a number of forms including monocrystalline, polycrystalline, amorphous silicon (thin film) or hybrid panels that are mounted on or integrated into the roofs or facades of buildings. Conversion efficiency of solar energy to electrical power is improving with advances in technology and ranges from 10% to 20%. In practice, allowing for UK weather conditions, an installation of 7 m² of monocrystalline modules (south facing at 30° from horizontal) typically produces 1000 watts peak (1 kWp), yielding about 800 kWh in a year. Installed costs range from £1300 to £2750/kWp.

- **Ground source heat pumps (heating & cooling)** – At a particular depth (about 10 m in the UK), the ground temperature remains substantially constant throughout the year. Heat (and coolth) may be extracted through either an 'open' system – discharging groundwater to river or sewer after passing it through a heat exchanger, or a 'closed' system – circulating a fluid (often water) through a heat exchanger and (typically) vertical pipes extending below the groundwater table. An electrically driven heat pump is then used to raise the fluid temperature via the refrigeration cycle, and low temperature hot water is delivered to the building.

  Most inner city ground heating and cooling systems consist of a cluster of pipes inserted into vertical holes typically 50 to 100 metres deep depending on space and ground conditions. Horizontal systems can be used where site circumstances allow. Costs for the drilling operation vary according to location, site accessibility, and ground conditions. Geological investigations are recommended to confirm ground conditions, reduce risks, and improve design and cost certainty.

  Such systems may achieve a Coefficient of Performance (COP = heat output/ electrical energy input) of between 3 and 4, achieving good savings of energy compared with conventional fossil fuels based systems. Installed costs are in the range £650 to £1800/kW depending on system type (vertical or horizontal), its size and complexity.

  Note that there is some debate on the status of ground source heat pumps as a renewable source of energy as it requires an external source of power which may not be renewable, typically electricity from the grid.

- **Solar water heating** – The basic principle is to collect heat from the sun via a fluid which is circulated in a roof solar panel or 'collector'. The heated fluid is then used to preheat hot water for space heating or domestic hot water, either in a separate tank or a twin coil hot water cylinder. Purpose-designed 'evacuated tube collectors' were developed to increase performance against the typical 'flat plate collectors'. A typical residential 'evacuated tube collectors' system has a cost ranging from £750 to £1100/m^2 depending on pipe runs and complexity. Such system may produce approximately 500 to 800 kWh/m^2 per year. Commercial systems are larger and more complex, and may achieve similar performance, providing there is sufficient hot water demand. Low-density residential, retail and leisure developments with washrooms and showers may also be suitable applications providing adequate demand for hot water.

- **Biomass boilers** – Wood chips or pellets derived from waste or farmed coppices or forests are available commercially and are considered carbon neutral, having absorbed carbon dioxide during growth. With a suitable fuel storage hopper and automatic screw drive and controls, biomass boilers can replace conventional boilers with little technical or aesthetic impact. However, they do depend on a viable source of fuel, and there are requirements for fuel deliveries access (in particular for inner city or remote locations), fuel storage, ash removal/disposal, as well as periodic de-coking. In individual dwellings, space may be a problem because a biomass boiler does not integrate readily into a typical modern kitchen. However, communal systems (serving multiple dwellings/flats) may be a viable domestic application. Biomass boilers are available in a wide range of domestic and commercial sizes. For a large installation, biomass boilers are more likely to form part of a modular system rather than to displace conventional boilers entirely. There is a cost premium for the biomass storage and feed system, and the cost of the fuel is currently comparable with other solid fuels. Installed costs of a biomass boiler range from £220 to £400/kW.

- **Biomass combined heat & power (CHP)** – Conventional CHP installations consist of either an internal combustion engine or a gas turbine driving an alternator, with maximum recovery of heat, particularly from the exhaust system. For best efficiency, there needs to be a convenient and constant requirement for the heat energy output and the generated electricity should also be utilized locally, with any excess exported to the grid.

  Considering the cost implications for biomass storage and handling as described for biomass boilers above, biomass CHP would only be viable in specific circumstances, with installed system costs in the order of £2600 to £3700/kW (electrical). Note that, at the time of writing, the authors are not aware of any small scale biomass CHP system successfully operated in the UK over any significant period.

Estimating and Tendering for Construction Work,
5th edition

Martin Brook

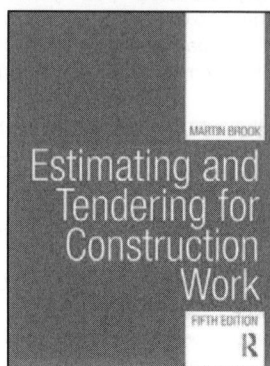

Estimators need to understand the consequences of entering into a contract, often defined by complex conditions and documents, as well as to appreciate the technical requirements of the project. Estimating and Tendering for Construction Work, 5th edition, explains the job of the estimator through every stage, from early cost studies to the creation of budgets for successful tenders.

This new edition reflects recent developments in the field and covers:
- new tendering and procurement methods
- the move from basic estimating to cost-planning and the greater emphasis placed on partnering and collaborative working
- the New Rules of Measurement (NRM1 and 2), and examines ways in which practicing estimators are implementing the guidance
- emerging technologies such as BIM (Building Information Modelling) and estimating systems which can interact with 3D design models

With the majority of projects procured using design-and-build contracts, this edition explains the contractor's role in setting costs, and design statements, to inform and control the development of a project's design.

Clearly-written and illustrated with examples, notes and technical documentation, this book is ideal for students on construction-related courses at HNC/HND and Degree levels. It is also an important source for associated professions and estimators at the outset of their careers.

July 2016; 246 × 189 mm, 334 pp
Pbk: 978-1-138-83806-2; £34.99

To Order: Tel: +44 (0) 1235 400524 Fax: +44 (0) 1235 400525
or Post: Taylor and Francis Customer Services,
Bookpoint Ltd, Unit T1, 200 Milton Park, Abingdon, Oxon, OX14 4TA UK
Email: book.orders@tandf.co.uk

For a complete listing of all our titles visit:
www.tandf.co.uk

Taylor & Francis
Taylor & Francis Group

Greywater Recycling and Rainwater Harvesting

The potential for greywater recycling and rainwater harvesting for both domestic residential and for various types of commercial building, considering the circumstances in which the systems offer benefits, both as stand-alone installations and combined.

Water usage trends

Water usage in the UK has increased dramatically over last century or so and it is still accelerating. The current average per capita usage is estimated to be at least 150 litres per day, and the population is predicted to rise from 60 million now to 65 million by 2017 and to 75 million by 2031, with an attendant increase in loading on water supply and drainage infrastructures.

Even at the current levels of consumption, it is clear from recent experience that long, dry summers can expose the drier regions of the UK to water shortages and restrictions. The predicted effects of climate change include reduced summer rainfall, more extreme weather patterns, and an increase in the frequency of exceptionally warm dry summers. This is likely to result in a corresponding increase in demand to satisfy more irrigation of gardens, parks, additional usage of sports facilities and other open spaces, together with additional needs for agriculture. The net effect, therefore, at least in the drier regions of the UK, is for increased demand coincident with a reduction in water resource, thereby increasing the risk of shortages.

Water applications and reuse opportunities

Average domestic water utilization can be summarized as follows, as a percentage of total usage (Source: Three Valleys Water)

- Wash hand basin 8%
- Toilet 35%
- Dishwasher 4%
- Washing machine 12%
- Shower 5%
- Kitchen sink 15%
- Bath 15%
- External use 6%

Water for drinking and cooking makes up less than 20% of the total, and more than a third of the total is used for toilet flushing. The demand for garden watering, although still relatively small, is increasing year by year and coincides with summer shortages, thereby exacerbating the problem.

In many types of building it is feasible to collect rainwater from the roof area and other surface areas and to store it, after suitable filtration, in order to meet the demand, for example, toilet flushing, cleaning, washing machines and outdoor use – thereby saving in many cases 50% of the water demand. The other often complementary recycling approach is to collect and disinfect 'greywater' – the waste water from baths, showers and washbasins. Hotels, leisure centres, care homes and apartment blocks generate large volumes of waste water and therefore present a greater opportunity for recycling. With intelligent design, even offices can make worthwhile water savings by recycling greywater, not necessarily to flush all of the toilets in the building but perhaps just those in one or two primary cores, with the greywater plant and distribution pipework dimensioned accordingly.

Intuitively, rainwater harvesting and greywater recycling seem like 'the right things to do' and rainwater harvesting is already common practice in many countries in Northern continental Europe. In Germany, for example, some 60,000 to 80,000 systems are being installed every year – compared with perhaps 2,000 systems in the UK. Greywater recycling systems, which are less widespread than rainwater harvesting, have been developed over the last 20 years. The 'state of the art' system is to use biological and UV (ultra-violet) disinfection rather than chemicals, and to reduce the associated energy use through advanced technologies such as membrane filtration.

In assessing the environmental credentials of new developments, the sustainable benefits are recognized by the Building Research Establishment Environmental Assessment Method (BREEAM) whereby additional points can be gained for efficient systems – those designed to achieve enough water savings to satisfy at least 50% of the relevant demand. Furthermore, rainwater harvesting systems from several manufacturers are included in the 'Energy Technology Product List' and thereby qualify for Enhanced Capital Allowances (ECAs). Claims are allowed not only for the equipment, but also to directly associated project costs including:

- Transportation – the cost of getting equipment to the site.
- Installation – cranage (to lift heavy equipment into place), project management costs and labour, plus any necessary modifications to the site or existing equipment.
- Professional fees – if they are directly related to the acquisition and installation of the equipment.

Rainwater harvesting

A typical rainwater harvesting system can be installed at reasonable cost if properly designed and installed at the same time as building the development/building. The collection tank can either be buried or installed in a basement area. Rainwater enters the drainage system through sealed gullies and passes through a pre-filter to remove leaves and other debris before passing into the collection tank. A submersible pump, under the control of the monitoring and sensing panel, delivers recycled rainwater on demand. The non-potable distribution pipework to the washing machine, cleaner's tap, outside tap and toilets etc. could be either a boosted system or configured for a header tank in the loft, with mains supply back-up, monitors and sensors located there instead of at the control panel.

Calculating the collection tank size brings into play the concept of system efficiency – relating the water volume saved to the annual demand. In favourable conditions – ample rainfall and large roof collection area – it would be possible in theory to achieve almost 100%. In practice, systems commonly achieve 50 to 70% efficiency, with enough storage to meet demand for typically 18 days or 2–3 weeks (BS 8525:10), though this is subject to several variables. As well as reducing the demand for drinking quality mains supply water, rainwater harvesting tanks act as an effective storm water attenuator, thereby reducing the drainage burden and the risk of local flooding which is a benefit to the wider community. Many urban buildings are located where conditions are unfavourable for rainwater harvesting – low rainfall and small roof collection area. In these circumstances it may still be worth considering water savings through greywater recycling or communal systems, as well as considering the use of different collection surfaces.

Greywater recycling

In a greywater recycling system, waste water from baths, showers and washbasins is collected via a separate conventional fittings and pipework, to enter a pre-filtration which removes the larger dirt particles. This is followed by the aerobic treatment tank in which cleaning bacteria ensure that all biodegradable substances are broken down. The water then passes onto a third tank, where an ultra-filtration membrane removes all particles larger than 0.05 microns (this includes viruses and bacteria) effectively disinfecting the recycled greywater. The clean water is then stored in the fourth tank from where it is pumped on demand under the control of monitors and sensors in the control panel. Recycled greywater may then be used for toilet and urinal flushing, for laundry and general cleaning, and for outdoor use such as vehicle washing and garden irrigation – a substantial water saving for premises such as hotels. If the tank becomes depleted, the distribution is switched automatically to the mains water back-up supply. If there is insufficient plant room space, then the tanks may be buried but with adequate arrangements for maintenance access.

Combined rainwater/greywater systems

Rainwater can be integrated into a greywater scheme with very little added complication other than increased tank size with the exception of green roofs, where additional polishing (i.e. colour treatment) may be required. In situations where adequate rainwater can be readily collected and diverted to pretreatment, then heavy demands such as garden irrigation can be met more easily than with greywater alone. An additional benefit is that they reduce the risk of flooding by keeping collected storm water on site instead of passing it immediately into the drains.

Indicative system cost and payback considerations

Rainwater Harvesting Scenario: Office building in Leeds having a roof area of 2000 m² and accommodating 435 people over 3 floors. Local annual rainfall is 875 mm and the application is for toilet and urinal flushing. An underground collection tank of 25,000 litres has been specified to give normally 18 days storage.

Rainwater harvesting cost breakdown	Cost £
System tanks and filters and controls	21,000
Mains water back-up and distribution pump arrangement	5,000
Non-potable distribution pipework	1,000
Connections to drainage	1,000
Civil works and tank installation (assumption normal ground conditions)	8,000
System installation & commissioning	2,000
TOTAL COST	38,000

Rainwater harvesting payback considerations

Annual water saving: 1100 m³ @ average cost £2.30/m³ = £2,530

Annual maintenance and system energy cost = £800

Indicative payback period = 38,000/1,730 = 22 years

Greywater recycling scenario: Urban leisure hotel building with 200 bedrooms offers little opportunity for rainwater harvesting but has a greywater demand of up to 12000 litres per day for toilet and urinal flushing, plus a laundry. The greywater recycling plant is located in a basement plant room.

Greywater recycling cost breakdown	Cost £
System tanks and controls	37,000
Mains water back-up and distribution pump arrangement	4,000
Non-potable distribution pipework	4,000
Greywater waste collection pipework	5,000
Connections to drainage	1,000
System installation & commissioning	6,000
TOTAL COST	57,000

Greywater recycling payback considerations

Annual water saving: 4260 m³ @ average cost £2.30/m³ = £9,798

Annual maintenance and system energy cost = £2,200

Indicative payback period = 57,000/7,598 = 7.5 years

Exclusions

- Site organization and management costs other than specialist contractor's allowances
- Contingency/design reserve
- Main contractor's overhead and profit or management fee
- Professional fees
- Tax allowances
- Value added tax

Conclusions

There is a justified and growing interest in saving and recycling water by way of both greywater recycling and rainwater harvesting. The financial incentive at today's water cost is not great for small or inefficient systems but water costs are predicted to rise and demand to increase – not least as a result of population growth.

The payback periods for the above scenarios are not intended to compare to potential payback periods of rainwater harvesting against greywater recycling but rather to illustrate the importance of choosing 'horses for courses'. The office building with its relatively small roof area and limited demand for toilet flushing results in a fairly inefficient system. Burying the collection tank also adds a cost so that payback exceeds 20 years. Payback periods of less than 10 years are feasible for buildings with large roofs and a large demand for toilet flushing or for other uses. Therefore sports stadia, exhibition halls, supermarkets, schools and similar structures are likely to be suitable.

The hotel scenario is good application for greywater recycling. Many hotels and residential developments will generate more than enough greywater to meet the demand for toilet flushing etc. and in these circumstances large quantities of water can be saved and recycled with attractive payback periods.

Groundwater Cooling

The use of groundwater cooling systems considering the technical and cost implications of this renewable energy technology.

The application of groundwater cooling systems is quickly becoming an established technology in the UK with numerous installations having been completed for a wide range of building types, both new build and existing (refurbished).

Buildings in the UK are significant users of energy, accounting for 60% of UK carbon emissions in relation to their construction and occupation. The drivers for considering renewable technologies such as groundwater cooling are well documented and can briefly be summarized as follows:

- Government set targets – The Energy White Paper, published in 2003, setting a target of producing 10% of UK electricity from renewable sources by 2010 and the aspiration of doubling this by 2020.
- The proposed revision to the Building Regulations Part L 2006, in raising the overall energy efficiency of non-domestic buildings, through the reduction in carbon emissions, by 27%.
- Local Government policy for sustainable development. In the case of London, major new developments (i.e. City of London schemes over 30,000 m^2) are required to demonstrate how they will generate a proportion of the site's delivered energy requirements from on-site renewable sources where feasible. The GLA's expectation is that, overall, large developments will contribute 10% of their energy requirement using renewables, although the actual requirement will vary from site to site. Local authorities are also likely to set lower targets for buildings which fall below the GLA's renewables threshold.
- Company policies of building developers and end users to minimize detrimental impact to the environment.

The ground as a heat source/sink

The thermal capacity of the ground can provide an efficient means of tempering the internal climate of buildings. Whereas the annual swing in mean air temperature in the UK is around 20 K, the temperature of the ground is far more stable. At the modest depth of 2 m, the swing in temperature reduces to 8 K, while at a depth of 50 m the temperature of the ground is stable at 11–13°C. This stability and ambient temperature therefore makes groundwater a useful source of renewable energy for heating and cooling systems in buildings.

Furthermore, former industrial cities like Nottingham, Birmingham, Liverpool and London have a particular problem with rising groundwater levels due to a reduction in groundwater abstraction for use in manufacturing, particularly since the 1970s. The use of groundwater for cooling is therefore encouraged by the Environment Agency in areas with rising groundwater levels as a means of combating this problem.

System types

Ground and groundwater cooling systems may be defined as either open or closed loop systems.

Open loop systems

Open loop systems generally involve the direct abstraction and use of groundwater, typically from aquifers (porous water bearing rock), although there also are systems which utilize water in former underground mineral workings. Water is abstracted via one or more boreholes and passed through a heat exchanger and then is returned to the aquifer via a separate borehole or boreholes, discharged to foul water drainage or released into a suitable available source such as a river. Typical groundwater supply temperatures are in the range 8–12°C and typical re-injection temperatures for a cooling mode operation is 12–18°C.

The properties of the aquifer are important in ensuring the necessary flow rates to achieve the energy requirements. Whilst reasonable flow rates normally are obtained from the major aquifers, such as the Chalk and the Sherwood Sandstones, there is a small risk that boreholes in these strata will not provide the required yield. In addition, reinjection of the abstracted water tends to be more difficult than abstraction and it often is necessary to install more than one recharge borehole to accept the discharge from a single abstraction borehole. This has obvious cost implications. This issue is often site-specific and an assessment of the hydrogeological conditions, in particular the depth to groundwater, should be obtained before progressing with the scheme.

All open loop abstractions from groundwater require an abstraction licence. If the water is being returned to the aquifer, it is designated as non-consumptive use. The system also requires an environmental permit to discharge the water back into the aquifer. Where it is a simple abstraction, it is regarded as a consumptive use and the abstraction charges are higher. In order to protect groundwater from the discharges from open loop GSHC systems, the Environment Agency restrict the change in temperature between the abstracted and discharged waters to no greater than +/- 8°C and the outflow water temperature must not exceed 25°C.

The efficiency of an open loop system can deteriorate if a large proportion of the reinjected water is contained within the water pumped from the abstraction borehole. This can occur if the abstraction and recharge boreholes are positioned too close to each other, allowing recirculation of the recharged water. This is not an uncommon event and in many circumstances cannot be avoided due to the size of the development. In order to minimize this effect, the abstraction and recharge boreholes should be separated by as large a distance as possible.

Open loop systems fed by groundwater at 8°C, can typically cool water to 12°C on the secondary side of the heat exchanger to serve conventional cooling systems.

Open loop systems are thermally efficient but over time can suffer from blockages caused by silt, and corrosion due to dissolved salts. As a result, additional cost may be incurred in having to provide filtration or water treatment, before the water can be used in the building.

There is also a risk that minerals can precipitate out of the groundwater, especially where the water is iron and/or manganese rich and there is either and/or a change in pressure or exposure of the water to air. This could require cleaning of the pipework and the borehole lining or some form of filtration as indicated above. Several water supply boreholes suffer from this problem, which can cause a reduction in borehole yield.

An abstraction licence and environmental permit/discharge consent need to be obtained for each installation, and this together with the maintenance and durability issues can significantly affect whole life operating costs, making this system less attractive.

A final point to consider is that access by plant will be required to the boreholes for pump replacement, maintenance and hence the boreholes should be located where access for plant is easily achievable.

Closed loop systems

Closed loop systems do not rely on the direct abstraction of water, but instead comprise a continuous pipework loop buried in the ground. Water circulates in the pipework and provides the means of heat transfer with the ground. Since groundwater is not being directly used, closed loop systems therefore suffer fewer of the operational problems of open loop systems, being designed to be virtually maintenance free, but do not contribute to the control of groundwater levels.

There are two types of closed loop system:

Vertical Boreholes – Vertical loops are inserted as U tubes into pre-drilled boreholes, typically less than 150 mm in diameter. These are backfilled with a high conductivity grout to seal the bore, prevent any cross contamination and to ensure good thermal conductivity between the pipe wall and surrounding ground. Vertical boreholes have the highest performance and means of heat rejection, but also have the highest cost due to associated drilling and excavation requirements.

Alterative systems to the U tubes are available on the market with larger diameter propriety pipework systems with flow and return pipes. The larger pipework can have better heat transfer properties, reduced resistance and therefore achieve better efficiencies or less pipework and therefore less ground area requirement than a traditional U tube system which may outweigh the additional cost per m length of the pipework.

As an alternative to having a separate borehole housing the pipe loop, it can also be integrated with the piling, where the loop is encased within the structural piles. This obviously saves on the costs of drilling and excavation since these would be carried out as part of the piling installation. The feasibility of this option would depend on marrying up the piling layout with the load requirement, and hence the number of loops, for the building.

Horizontal Loops – These are single (or pairs) of pipes laid in 2 m deep trenches, which are backfilled with fine aggregate. These obviously require a greater physical area than vertical loops but are cheaper to install. As they are located closer to the surface where ground temperatures are less stable, efficiency is lower compared to open systems. Alternatively, coiled pipework can also be used where excavation is more straightforward and a large amount of land is available. Although performance may be reduced with this system as the pipe overlaps itself, it does represent a cost effective way of maximizing the length of pipe installed and hence overall system capacity.

The case for heat pumps

Instead of using the groundwater source directly in the building, referred to as passive cooling, when coupled to a reverse cycle heat pump, substantially increased cooling loads can be achieved.

Heat is extracted from the building and transferred by the heat pump into the water circulating through the loop. As it circulates, it gives up heat to the cooler earth, with the cooler water returning to the heat pump to pick up more heat. In heating mode the cycle is reversed, with the heat being extracted from the earth and being delivered to the HVAC system.

The use of heat pumps provides greater flexibility for heating and cooling applications within the building than passive systems. Ground source heat pumps are inherently more efficient than air source heat pumps, their energy requirement is therefore lower and their associated CO_2 emissions are also reduced, so they are well suited for connection to a groundwater source.

Closed loop systems can typically achieve outputs of 50 W/m (of bore length), although this will vary with geology and borehole construction. When coupled to a reverse cycle heat pump, 1 m of vertical borehole will typically deliver 140 kWh of useful heating and 110 kWh of cooling per annum, although this will depend on hours run and length of heating and cooling seasons.

Key factors affecting cost

• The cost is obviously dependent on the type of system used. Deciding on what system is best suited to a particular project is dependent on the geological conditions, the peak cooling and heating loads of the building and its likely load profile. This in turn determines the performance required from the ground loop, in terms of area of coverage in the case of the horizontal looped system, and in the case of vertical boreholes, the depth and number or bores. The cost of the system is therefore a function of the building load.

- In the case of vertical boreholes for a closed loop system, drilling costs are significant factor, as ground conditions can be variable, and there are potential problems in drilling through sand layers, pebble beds, gravels and clay, which may mean additional costs through having to drill additional holes or the provision of sleeving. The costs of drilling make the vertical borehole solution significantly more expensive than the equivalent horizontal loop.
- The thermal efficiency of the building is also a factor. The higher load associated with a thermally inefficient building obviously results in the requirement for a greater number of boreholes or greater area of horizontal loop coverage, however in the case of boreholes the associated cost differential between a thermally inefficient building and a thermally efficient one is substantially greater than the equivalent increase in the cost of conventional plant. Reducing the energy consumption of the building is cheaper than producing the energy from renewables and the use of renewable energy only becomes cost effective, and indeed should only be considered, when a building is energy efficient.
- With open loop systems, the principal risk in terms of operation is that the user is not in control of the quantity or quality of the water being taken out of the ground, this being dependent on the local ground and groundwater conditions. Reduced performance due to blockage (silting etc.) may lead to the system not delivering the design duties whilst bacteriological contamination may lead to the expensive water treatment or the system being taken temporarily out of operation. In order to mitigate the above risk, it may be decided to provide additional means of heat rejection and heating by mechanical means as a back up to the borehole system, in the event of operational problems. This obviously carries a significant cost. If this additional plant were not provided, then there are space savings to be had over conventional systems due to the absence of heating, heat rejection and possibly refrigeration plant.
- With the design of open loop systems close analysis can be undertaken of existing boreholes in the area and also detailed desktop studies can be undertaken by a Hydrogeologist to determine the feasibility of the scheme and predict the yield for the proposed borehole. As noted above the yield of the borehole is subject to the local hydrogeological conditions in the location where the borehole is located and as such the delivered yield will only be validated after the borehole is completed and tested. There is a risk that you can get less (you could get more) than was predicted requiring additional boreholes (and cost) to be delivered to obtain the required yield.
- Open loop systems may lend themselves particularly well to certain applications increasing their cost effectiveness, i.e. in the case of a leisure centre, the removal of heat from the air-conditioned parts of the centre and the supply of fresh water to the swimming pool.
- In terms of the requirements for abstraction and disposal of the water for open loop systems, there are risks associated with the future availability and cost of the necessary licences; particularly in areas of high forecast energy consumption, such as the South East of England, which needs to be borne in mind when selecting a suitable system. Provided that the system is designed to be non-consumptive, a licence restriction is unlikely.
- Closed loop systems and non-consumptive open loop systems generally require balanced heating and cooling loads (If there is moving groundwater this is not as critical) in order to not cause the ground to heat up or cool down over time. If the heating and cooling loads (kWh) over the year are not balanced this can cause the ground to heat up or cool down over time causing a reduction in the efficiency of the system over time which will have a detrimental effect to the seasonal COP's and operational running costs of the system over time.
- Heat pump systems use electricity to power them and winter triad electrical costs should be considered in life cycle analysis for commercial and industrial users and appropriate controls storage systems put in place to mitigate costs.

Whilst open loop systems would suit certain applications or end user clients, for commercial buildings the risks associated with this system tend to mean that closed loop applications are the system of choice. When coupled to a reversible heat pump, the borehole acts simply as a heat sink or heat source so the problems associated with open loop systems do not arise.

Typical costs

Table 1 gives details of the typical borehole cost to an existing site in Central London, using one 140 m deep borehole working on the open loop principle, providing heat rejection for the 600 kW of cooling provided to the building. The borehole passes through rubble, river gravel terraces, clay and finally the Chalk, and is lined above the Chalk to prevent the hole collapsing. The breakdown includes all costs associated with the provision of a working borehole up to the well head, including the manhole chamber and manhole. The costs of any plant or equipment to deliver the water from the borehole are not included.

Heat is drawn out of the cooling circuit and the water is discharged into the Thames at an elevated temperature. In this instance, although the boreholes are more expensive than the dry air cooler alternative, the operating cost is significantly reduced as the system can operate at around three times the efficiency of conventional dry air coolers, so the payback period is a reasonable one. Additionally, the borehole system does not generate any noise, does not require rooftop space and does not require as much maintenance.

This is representative of a typical cost of providing a borehole for an open loop scheme within the London basin. There are obviously economies of scale to be had in drilling more than one well at the same time, with two wells saving approximately 10% of the comparative cost of two separate wells and four wells typically saving 15%.

Table 2 provides a summary of the typical range of costs that could expected for the different types of system based on current prices.

Table 1: Breakdown of the Cost of a Typical Open Loop Borehole System

Description	Cost £
General Items	
• Mobilization, insurances, demobilization on completion	21,000
• Fencing around working area for the duration of drilling and testing	2,000
• Modifications to existing LV panel and installation of new power supplies for borehole installation	15,000
Trial Hole	
• Allowance for breakout access to nearest walkway (Existing borehole on site used for trial purposes, hence no drilling costs included)	3,000
Construct Borehole	
• Drilling, using temporary casing where required, permanent casing and grouting	32,000
Borehole Cap and Chamber	
• Cap borehole with PN16 flange, construct manhole chamber in roadway, rising main, header pipework, valves, flow meter	12,000
• Permanent pump	14,000
Samples	
• Water samples	1,000
Acidization	
• Mobilization, set up and removal of equipment for acidization of borehole, carry out acidization	12,000
Development and Test Pumping	
• Mobilize pumping equipment and materials and remove on completion of testing	4,000
• Calibration test, pretest monitoring, step testing	4,000
• Constant rate testing and monitoring	20,000
• Waste removal and disposal	3,000
Reinstatement	
• Reinstatement and making good	2,000
Total	**145,000**

Table 2: Summary of the Range of Costs for Different Systems

	Range			
System	**Small – 4 kWth**	**Medium – 50 kWth**	**Large – 400 kWth**	**Notes**
Heat pump (per unit)	£ 3,000–5,000	£ 30,000–42,000	£ 145,000–175,000	
Slinky pipe (per installation) including excavation	£ 3,000 –£ 4,000	£ 42,000–52,000	£ 360,000–390,000 [1]	[1] Based on 90 nr 50 m lengths
Vertical, closed (per installation) using structural piles	N/A	£ 42,000–63,000	Not available	Based on 50 nr piles. Includes borehole cap and header pipework but excludes connection to pump room and heat pumps
Vertical, closed (per installation) including excavation	£ 2,000–3,000	£ 63,000–85,000	£ 370,000–400,000	Includes borehole cap and header pipework but excludes connection to pump room and heat pumps
Vertical, open (per installation) including excavation	£ 2,000–3,000	£ 45,000–65,000	£ 335,000–370,000	Excludes connection to pump room and heat exchangers

Building Services Handbook, 9th Edition

Fred Hall and Roger Greeno

The ninth edition of Hall and Greeno's leading textbook has been reviewed and updated in relation to the latest building and water regulations, new technology, and new legislation. For this edition, new updates includes: the reappraisal of CO_2 emissions targets, updates to sections on ventilation, fuel, A/C, refrigeration, water supply, electricity and power supply, sprinkler systems, and much more.

Building Services Handbook summarises the application of all common elements of building services practice, technique and procedure, to provide an essential information resource for students as well as practitioners working in building services, building management and the facilities administration and maintenance sectors of the construction industry. Information is presented in the highly illustrated and accessible style of the best-selling companion title *Building Construction Handbook*.

THE comprehensive reference for all construction and building services students, Building Services Handbook is ideal for a wide range of courses including NVQ and BTEC National through Higher National Certificate and Diploma to Foundation and three-year Degree level. The clear illustrations and complementary references to industry Standards combine essential guidance with a resource base for further reading and development of specific topics.

May 2017; 234 × 156 mm, 786 pp
Pbk: 978-1-138-24435-1; £32.99

To Order: Tel: +44 (0) 1235 400524 Fax: +44 (0) 1235 400525
or Post: Taylor and Francis Customer Services,
Bookpoint Ltd, Unit T1, 200 Milton Park, Abingdon, Oxon, OX14 4TA UK
Email: book.orders@tandf.co.uk

For a complete listing of all our titles visit:
www.tandf.co.uk

Water Network Charges

This article gives guidance on water suppliers, new connections, infrastructure charges and network charges.

Existing appointees' and new appointments & variations (NAVs)

Most customers in England and Wales currently receive their water and sewerage services from one of 22 appointed 'water' and 'sewerage and water only' suppliers that were in existence when the sectors were privatized. These suppliers are known as 'existing appointees' or statutory providers.

New Appointments and Variations or NAVs refer to small competitor water companies that are not subject to full price controls like the statutory providers. The regulator 'Ofwat' recognizes that NAVs can provide services, such as their own water infrastructure and that these services can create competition in the market place when customers are purchasing new connections.

New appointments provide challenges to existing appointees. This drives efficiencies, stimulates innovation and reveals information. They have the potential to benefit all customers through:

- lower prices;
- improved service;
- greater choice of supplier for developers and large user customers.

Providing new connections

The term 'new connection' is used to describe where a customer requires either or both:

- Access to the existing public water supply or sewerage system by means of a service pipe or lateral drain.
- A new water main or public sewer.

A customer may choose their own contractor to do the work, which is then known as 'self-lay'. The statutory provider will take over responsibility for (adopt) all self-laid infrastructure that meets the terms of its agreement with the owner, developer or self-lay organization that carries out the work.

The Water Industry Act 1991 (WIA91) places a number of duties on statutory 'water only' and 'water and sewerage' companies in providing or enabling new connections for an individual property or development site.

If a property requires a new water main, sewer, service pipe or lateral drain for domestic purposes (cooking, cleaning, central heating or sanitary facilities), the owner or developer may ask the local statutory provider to install the infrastructure. For water mains and public sewers this is often referred to as 'requisitioning' the infrastructure.

Infrastructure charges

If a supplier is providing new water supplies for domestic purposes (such as to new housing developments) it is legally entitled to levy infrastructure charges. These can be raised where premises are connected to the water company's water supply (or to the sewerage company's sewers) for the first time. These charges are intended to provide a contribution towards the costs of developing or enhancing local networks to serve new customers.

A customer will have to pay infrastructure charges when a property is connected to the water and/or wastewater networks for the first time. This is in addition to the charges for making the actual physical connection to the water main and/or public sewer. When a customer makes a connection to a network the operator is entitled to charge the customer in accordance with its published Charging Arrangements for the connection works in addition to raising infrastructure charges. Infrastructure charges apply for premises where the supply of water or provision of sewerage services is intended for domestic purposes. Water for domestic purposes refers to usage for drinking, washing, cooking, central heating and sanitary purposes for which water is supplied to premises. All other purposes (including supplies for a laundry business or for a food or drink take away business), are regarded as non-domestic purposes. **The definition is about the usage of the water and not the type of property being supplied.** Domestic sewerage purpose refers to the removal of the contents of lavatories, water which has been used for cooking or washing or surface water from the premises and associated land, with the exception of laundries and take away restaurants. **As with a potable water connection, the definition is not about the type of property.** Please note that a single development may include a combination of supplies for domestic and non-domestic purposes. For example, a development may include a number of flats (expected to be for domestic purposes), retail units (expected to be for domestic purposes), a fire-fighting supply (non-domestic purposes) and landlord supplies (non-domestic purposes). Similarly a separate supply requested for a swimming pool or a garden tap is considered to be for non-domestic purposes even if within a residential property.

The relevant multiplier

A Relevant Multiplier is used to calculate the amount of an infrastructure charge for the connection of premises other than dwelling houses or flats, if water is provided by a supply larger than the standard size used for new connections of houses – 25 mm or 32 mm external dia. pipe. For other properties, such as student housing, offices or care homes the regulator has introduced a pricing calculator known as a relevant multiplier. This tool is utilized to measure a cost that reflects the increased impact on a provider's network. The wastewater infrastructure charge is calculated on the same basis as the water infrastructure charge unless the customer is able to show that waste and surface water flows are not being discharged to the public sewer.

The Relevant Multiplier (RM) is a way of working out infrastructure charges for the following types of property:

- Residential properties with a single, shared supply pipe and which are subject to a 'common billing agreement'; this includes sheltered housing, student accommodation and high-rise flats.
- Non-residential properties where the supply pipe is larger than the standard size, such as office blocks.

How the relevant multiplier is calculated

Each water fitting (wash basin, bath, shower, etc.) is given a 'loading unit' based on the amount of water it uses. The average number of units per property is taken as 24, equal to an RM of 1.00. The operator can use this as the basis for calculating the RM for each property on a development where the RM applies. The operator will do this by adding up the loading units for all the water fittings on a development. It divides this by the number of properties to give the average loading units per property. It divides this again by 24 (the average loading units) to give the RM for each property. Details of the number of loading units assigned to each water fitting are shown below. For properties subject to a common billing agreement, the RM can be more or less than 1.00. For other properties the minimum is 1.00.

Using the relevant multiplier to calculate infrastructure charge

Operator will utilize the RM multiplied by the standard charge to provide the infrastructure charge for that property.

Example of a residential development

- The development consists of 20 flats with a common billing agreement and the total loading units are 460.
- The total loading units (460) are divided by the number of properties (20) and again by the average (24). This gives an RM for each flat of 0.96 (460 ÷ 20 ÷ 24 = 0.96).
- The infrastructure charge for each flat is the RM of 0.96 multiplied by the standard charges.
- The infrastructure charge for the whole development is the RM multiplied by the standard charge multiplied by the number of properties.

The water infrastructure charge for the development is therefore

RM 0.96 × 20 properties × £140 standard charge = £2,688.00

The wastewater infrastructure charge for the development is therefore

RM 0.96 × 20 properties × £210 standard charge = £4,032.00

Network charges may also be payable in respect of supplies for non-domestic purposes within the development such as the irrigation supply and bin store.

Example of a commercial development

- The development consists of one office and the total loading units are 340.
- The total loading units (340) are divided by the number of properties (1) and again by the average (24). This gives an RM for the office of 14.17 (340 ÷ 1 ÷ 24 = 14.17).
- The infrastructure charge for the office is the RM of 14.17 multiplied by the standard charges.

The water infrastructure charge for the development is therefore

RM 14.17 × £140 standard charge = £1,983.80

The wastewater infrastructure charge for the development is therefore

RM 14.17 × £210 standard charge = £2,975.70

Network charges may also be payable in respect of supplies for non-domestic purposes within the development such as the irrigation supply and bin store.

Infrastructure charges and bulk supply agreements

Generally, infrastructure charges are dealt with in bulk supply agreements, with new appointees agreeing to levy such charges on their customers and pass them through to the existing appointee. Ofwat will support this approach in that it is generally the existing appointee that owns the relevant network that may need to be enhanced.

The Water Industry Act 1991 (WIA91) does not explicitly set out the timing of infrastructure charge payments. The regulators view is that it considers infrastructure charges are payable to a new appointee, when supply is made available – that is, when the first time connection is made.

For bulk supplies, timing of the payment of the amount equivalent to infrastructure charges by the customer to the existing appointee should be covered within a bulk supply agreement.

Supplies for non-domestic purposes are subject to the payment of network charges instead of infrastructure charges to reflect the different demands on the statutory provider's network.

Infrastructure credits

The operator may reduce the total infrastructure charge for a redeveloped site, if there were any properties connected to the water main or wastewater system during the five years before development began. For example, if a block of 15 flats is replaced by a block of 20 flats, the developer only pays infrastructure charges for the extra five flats.

If the previous connection was not a house or flat, the operator will calculate a credit based on the average annual consumption of the previously connected property compared to the average annual consumption for a dwelling. This gives the operator a benchmark of infrastructure charges (subject to a minimum of one) to credit against the infrastructure charges payable for the new development.

Network charges

Network charges help pay for developing an operator's network to meet the increased demands of connections for non-domestic purposes. The customer must pay network charges on top of the cost of connecting a property to the water main and/or wastewater system. The developer pays the charges for any new properties. Due to the requirement of increasing the size of the supply pipe, existing customers will have to pay the charges for an increase in demand at their property. The charge is payable before the property is connected. The charge will often be based on the size of the new meter.

Income offset or asset value payments

Where new water mains are provided for developments, the operator will take into account the future revenue from connected properties associated with the new mains. It will discount the amount payable by developers for the site specific mains that they ask to be provided ('requisitioned mains') by way of an income offset. The operator may also make an asset payment to developers/Self-Lay Providers (SLP) for self-laid mains that are adopted by them. The gross payment for self-laid mains is the equivalent of the income offset that the operator would apply had the mains been requisitioned. This is subject to the maximum amount being equivalent to the overall cost of the water mains had they been requisitioned. The cost of providing the site specific works is avoided by the operator when a NAV becomes the supplier of water services. The operator will pay a NAV income offsets for each occupied property that becomes connected to its water infrastructure through the NAV's network during the charging year. The income offsets will only be paid in respect of NAV bulk supply agreements made after 31 March 2018 and they will be offset against the wholesale charges made as part of the bulk supply agreements. Operators will change the way they calculate the income offset amount and this will affect:

- The asset payment made to SLPs when they provide the new water mains for developments
- The payment made to NAVs when they connect new properties to our network via their networks
- The 'discount' that is applied to the cost of new water mains if an operator is asked to provide them

Regulatory information within this document has been referenced directly from the regulator Ofwat. This information relates to its current guidance on infrastructure and network charging and is correct as of April 2018.

Fuel Cells

The application of fuel cell technology within buildings

Fuel cells for buildings are a form of combined heat and power (CHP) based around electrochemical energy in fuel and electrical and heat energy directly, without combustion, with high electrical efficiency and low pollutant emissions. They represent a new type of power generation technology that offers modularity, efficient operation across a wide range of load conditions, and opportunities for integration into co-generation systems. With the publication of the energy white paper, the Government confirmed its commitment to the development of fuel cells as a key technology in the UK's future energy system, as the move is made away from a carbon based economy.

There are currently very few fuel cells available commercially, and those that are available are not financially viable. Demand has therefore been limited to niche applications, where the end user is willing to pay the premium for what they consider to be the associated key benefits. Indeed, the UK currently has only 2–3 fuel cells in regular commercial operation. Whilst research and development of fuel cell technology continues (especially in the automotive industry), stationary building based fuel cells remain high compared to similar technologies.

Fuel cell technology

A fuel cell is composed of an anode (a negative electrode that repels electrons), an electrolyte membrane in the centre, and a cathode (a positive electrode that attracts electrons). As hydrogen flows into the cell on the anode side, a platinum coating on the anode facilitates the separation of the hydrogen gas into electrons and protons. The electrolyte membrane only allows the protons to pass through to the cathode side of the fuel cell. The electrons cannot pass through this membrane and flow through an external circuit to form an electric current.

As oxygen flows into the fuel cell cathode, another platinum coating helps the oxygen, protons, and electrons combine to produce pure water and heat.

The voltage from a single cell is about 0.7 volts, just enough for a light bulb. However by stacking the cells, higher outputs are achieved, with the number of cells in the stack determining the total voltage, and the surface area of each cell determining the total current. Multiplying the two together yields the total electrical power generated.

In a fuel cell the conversion process from chemical energy to electricity is direct. In contrast, conventional energy conversion processes first transform chemical energy to heat through combustion and then convert heat to electricity through some form of power cycle (e.g. gas turbine or internal combustion engine) together with a generator. On this basis, hydrogen powered fuel cells have the capability to operate much more efficiently than other forms of CHP, however if powered by fossil fuels, the auxiliary systems used to create the hydrogen for the fuel cell to reduce efficiency such that they are similarly efficient to turbine or reciprocating engine CHPs.

Fuel cell systems

In addition to the fuel cell itself, the system comprises the following subsystems:

- A fuel processor – This allows the cell to operate with available hydrocarbon fuels, by cleaning the fuel and converting (or reforming) it as required.
- A power conditioner – This regulates the DC electricity output of the cell to meet the application, and to power the fuel cell auxiliary systems.
- An air management system – This delivers air at the required temperature, pressure and humidity to the fuel stack and fuel processor.
- A thermal management system – This heats or cools the various process streams entering and leaving the fuel cell and fuel processor, as required.
- A water management system – Pure water is required for fuel processing in all fuel cell systems, and for dehumidification in the PEMFC.

The overall electrical conversion efficiency of a fuel cell system (defined as the electrical power out divided by the chemical energy into the system, taking into account the individual efficiencies of the subsystems) ranges from 35–55%. Taking into account the thermal energy available from the system, the overall or cogeneration efficiency is 75–90%.

Also, unlike most conventional generating systems (which operate most efficiently near full load, and then suffer declining efficiency as load decreases), fuel cell systems can maintain high efficiency at loads as low as 20% of full load.

Fuel cell systems also offer the following potential benefits:

- At operating temperature, they respond quickly to load changes, the limiting factor usually being the response time of the auxiliary systems.
- They are modular and can be built in a wide range of outputs. This also allows them to be located close to the point of electricity use, facilitating cogeneration systems.
- Fuel cells operate near silently, however the associated pumps fans and auxiliary systems generate noise similar to other commercial ventilation and pumping installations. Consideration should be given to the acoustic design of these buildings.
- Commercially available systems are designed to operate unattended and manufactured as packaged units.
- Consideration need to be given to allow for the replacement of the fuel stack every 3–5 days. Due to the packaged nature of plan installations, this may require the replacement of the auxiliary systems as well as the fuel cell stack.
- Fuel cell stacks fuelled by hydrogen produce only water, therefore the fuel processor is the primary source of emissions, and these are significantly lower than emissions from conventional combustion systems.
- Since fuel cell technology generates 50% more electricity than the conventional equivalent without directly burning any fuel, CO_2 emissions are significantly reduced in the production of the source fuel.
- Potentially zero carbon emissions when using hydrogen produced from renewable energy sources.
- The facilitation of embedded generation, where electricity is generated close to the point of use, minimizing transmission losses.
- The fast response times of fuel cells offer potential for use in UPS systems, replacing batteries and standby generators.

Types of fuel cell

There are four main types of fuel cell technology that are applicable for building systems, classed in terms of the electrolyte they use. The chemical reactions involved in each cell are very different.

Phosphoric Acid Fuel Cells (PAFCs) are the dominant current technology for large stationary applications and have been available commercially for some time. There is less potential for PAFC unit cost reduction than for some other fuel cell systems, and this technology may be superseded in time by the other technologies.

The Solid Oxide Fuel Cell (SOFC) offers significant flexibility due to its large power range and wide fuel compatibility. SOFCs represent one of the most promising technologies for stationary applications. There are difficulties when operating at high temperatures with the stability of the materials, however, significant further development and cost reduction is anticipated with this type.

The relative complexity of Molten Carbonate Fuel Cells (MCFCs) has tended to limit developments to large scale stationary applications, although the technology is still very much in the development stages.

The quick start-up times and size range make Polymer Electrolyte Membrane Fuel Cells (PEMFCs) suitable for small to medium sized stationary applications. They have a high power density and can vary output quickly, making them well suited for transport applications as well as UPS systems. The development efforts in the transport sector suggest there will continue to be substantial cost reductions over both the short and long term.

All four technologies remain the subject of extensive research and development programmes to reduce initial costs and improve reliability through improvements in materials, optimization of operating conditions and advances in manufacturing.

The market for fuel cells

The stationary applications market for fuel cells can be split as follows:

- Distributed generation/CHP – For large scale applications, there are no drivers specifically advantageous to fuel cells, with economics (and specifically initial cost) therefore being the main consideration. So, until cost competitive and thoroughly proven and reliable fuel cells are available, their use is likely to be limited to niche applications such as environmentally sensitive areas.
- Domestic and small scale CHP – The drivers for the use of fuel cells in this emerging market are better value for customers than separate gas and electricity purchase, reduction in domestic CO_2 emissions, and potential reduction in electricity transmissioncosts. However, the barriers of resistance to distributed generation, high capital costs and competition from Stirling engines needs to be overcome.
- Small generator sets and remote power – The drivers for the use of fuel cells are high reliability, low noise and low refuelling frequencies, which cannot be met by existing technologies. Since cost is often not the primary consideration, fuel cells will find early markets in this sector. Existing PEMFC systems are close to meeting the requirements in terms of cost, size and performance. Small SOFC's have potential in this market, but require further development.

Costs

Fuel cell technologies are still significantly more expensive than the existing technologies typically and, dependent on the fuel cell type, between £3,000 – £5,000/kW. To extend fuel cell application beyond niche markets, their cost needs to reduce significantly. The successful and wide-spread commercial application of fuel cells is dependent on the projected cost reductions indicated, with electricity generated from fuel cells being competitive with current centralized and distributed power generation.

Conclusions

Despite significant growth in recent years, fuel cells are still at a relatively early stage of commercial development, with prohibitively high capital costs preventing them from competing with the incumbent technology in the market place. The fuel cell industry have been predicting falling costs for the past 10 years or more, however there is little evidence this is going to happen without wider market uptake in order to move from a niche application into a mass produced product.

In order for these projected cost reductions to be achieved, customers need to be convinced that the end product is not only cost competitive but also thoroughly proven, and Government support represents a key part in achieving this.

The Governments of Canada, USA, Japan and Germany have all been active in supporting development of the fuel cell sector through integrated strategies, however the UK has been slow in this respect, and support has to date been small in comparison. It is clear that without Government intervention, fuel cell applications may struggle to reach the cost and performance requirements of the emerging fuel cell market.

BIM and Quantity Surveying
S Pittard *et al.*

The sudden arrival of Building Information Modelling (BIM) as a key part of the building industry is redefining the roles and working practices of its stakeholders. Many clients, designers, contractors, quantity surveyors, and building managers are still finding their feet in an industry where BIM compliance can bring great rewards.

This guide is designed to help quantity surveying practitioners and students understand what BIM means for them, and how they should prepare to work successfully on BIM compliant projects. The case studies show how firms at the forefront of this technology have integrated core quantity surveying responsibilities like cost estimating, tendering, and development appraisal into high profile BIM projects. In addition to this, the implications for project management, facilities management, contract administration and dispute resolution are also explored through case studies, making this a highly valuable guide for those in a range of construction project management roles.

Featuring a chapter describing how the role of the quantity surveyor is likely to permanently shift as a result of this development, as well as descriptions of tools used, this covers both the organisational and practical aspects of a crucial topic.

December 2015: 234 x 156 mm: 258 pp
Pbk: 978-0-415-87043-6; £28.99

To Order: Tel: +44 (0) 1235 400524 Fax: +44 (0) 1235 400525
or Post: Taylor and Francis Customer Services,
Bookpoint Ltd, Unit T1, 200 Milton Park, Abingdon, Oxon, OX14 4TA UK
Email: book.orders@tandf.co.uk

For a complete listing of all our titles visit:
www.tandf.co.uk

Biomass Energy

The potential for biomass energy systems, with regards to the adequacy of the fuel supply and the viability of various system types at different scales.

Biomass heating and combined heat and power (CHP) systems have become a major component of the low-carbon strategy for many projects, as they can provide a large renewable energy component at a relatively low initial cost. Work by the Carbon Trust has demonstrated that both large and small biomass systems were viable even before recent increases in gas and fuel oil prices, so it is no surprise that recent research by South Bank University into the renewables strategies to large London projects has found that 25% feature biomass or biofuel systems.

These proposals are not without risk, however. Although the technology is well established, few schemes are in operation in the UK and the long-term success depends more on the effectiveness of the local supply chain than the quality of the design and installation.

How the biomass market works

Biomass is defined as living or recently dead biological material that can be used as an energy source. Biomass is generally used to provide heat, generate electricity or drive CHP engines. The biomass family includes biofuels, which are being specified in city centre schemes, but which provide lower energy outputs and could transfer farmland away from food production.

In the UK, much of the focus in biomass development is on the better utilization of waste materials such as timber and the use of set-aside land for low-intensity energy crops such as willow, rather than expansion of the biofuels sector. There are a variety of drivers behind the development of a biomass strategy. In addition to carbon neutrality, another policy goal is the promotion of the UK's energy security through the development of independent energy sources. A third objective is to address energy poverty, particularly for off-grid energy users, who are most vulnerable to the effects of high long-term costs of fuel oil and bottled gas.

Biomass' position in the zero-carbon hierarchy is a little ambiguous in that its production, transport and combustion all produce carbon emissions, albeit most is offset during a plant's growth cycle. The key to neutrality is that the growing and combustion cycles need to occur over a short period, so that combustion emissions are genuinely offset. Biomass strategy is also concerned about minimizing waste and use of landfill, and the ash produced by combustion can be used as a fertilizer.

Dramatic increases in fossil fuel prices have swung considerations decisively in favour of technologies such as biomass. Research by the Carbon Trust has demonstrated that, with oil at $35 (£25) a barrel, rates of return of more than 10% could be achieved with both small and large heating installations. CHP and electricity-only schemes have more complex viability issues linked to renewable incentives, but with oil currently trading at over ($75) £50 a barrel and a plentiful supply of source material, it is argued that biomass input prices will not rise and so the sector should become increasingly competitive.

The main sources of biomass in the UK include:

- Forestry crops, including the waste products of tree surgery industry
- Industrial waste, particularly timber, paper and card: timber pallets account for 30% of this waste stream by weight
- Woody energy crops, particularly those grown through 'short rotation' methods such as willow coppicing
- Wastes and residues taken from food, agriculture and manufacturing

Biomass is an emerging UK energy sector. Most suppliers are small and there remains a high level of commercial risk associated with finding appropriate, reliable sources of biomass. This is particularly the case for larger-scale schemes such as those proposed for Greater London, which will have sourced biomass either from multiple UK suppliers or from overseas. Many have adopted biofuels as an alternative.

The UK's only large-scale biomass CHP in Slough has a throughput of 180,000 tonnes of biomass per year requiring the total production of more than 20 individual suppliers – not a recipe for easy management or product consistency. However, Carbon Trust research has identified significant potential capacity in waste wood (5 million to 6 million tonnes) and short-rotation coppicing, which could create the conditions for wider adoption of small and large-scale biomass.

Biomass technologies

A wide range of technologies have been developed for processing various forms of biomass, including anaerobic digesters and gasifiers. However, the main biomass technology is solid fuel combustion, as a heat source, CHP unit or energy source for electricity generation.

Solid fuel units use either wood chippings or wood pellets. Wood chippings are largely unprocessed and need few material inputs, other than seasoning, chipping and transport. Wood pellets are formed from compressed sawdust. As a result they have a lower moisture content than wood chippings and consistent dimensions, so are easier to handle but are about twice as expensive.

Solid fuel burners operate in the same way as other fossil fuel-based heat sources, with the following key differences:

- Biomass heat output can be controlled but not instantaneously, so systems cannot respond to rapid load changes. Solutions to provide more flexibility include provision of peak capacity from gas-fired systems, or the use of thermal stores that capture excess heat energy during off-peak periods, enabling extended operation of the biomass system itself
- Heat output cannot be throttled back by as much as gas-fired systems, so for heat-only installations it may be necessary to have an alternative summer system for water heating, such as a solar collector
- Biomass feedstock is bulky and needs a mechanized feed system as well as extensive storage
- Biomass systems are large, and the combustion unit, feed hopper and fuel store take up substantial floor area. A large unit with an output of 500 kW has a footprint of 7.5 m × 2 m
- Biomass systems need maintenance related to fuel deliveries, combustion efficiency, ash removal, adding to the lifetime cost
- Fuel stores need to be physically isolated from the boiler and the rest of the building in order to minimize fire risk. The fuel store needs to be sized to provide for at least 100 hours of operation, which is approximately 100 m² for a 500 kW boiler. The space taken up by storage and delivery access may compromise other aspects of site planning
- Fire-protection measures include anti-blowback arrangements on conveyors and fire dampers, together with the specification of elements such as flues for higher operating temperatures
- Collocation of the fuel source and burner at ground level require larger, free-standing flues

As a result of these issues, which drive up initial costs, affect development efficiency and add to management overheads, take-up of biomass has initially been mostly at the small-scale, heat-only end of the market, based on locally sourced feedstock. In such systems the initial cost premium of the biomass boiler can be offset against long-term savings in fuel costs.

Sourcing biomass

Compared with solar or wind power installations, the initial costs of biomass systems are low, the technology is well established and energy output is dependable. As a result, the real challenge for successful operation of a biomass system is associated with the reliable sourcing of feedstock.

Heat-only systems themselves cost between £150 and £750 per kW (excluding costs of storage), depending on scale and technology adopted. This compares with a typical cost of £50 to £300 per kW for a gas-fired boiler – which does not require further investment in fuel or thermal storage bunkers.

As a high proportion of lifetime cost is associated with the operation of a system, availability of good-quality, locally sourced feedstock is essential for long-term viability – particularly in areas where incentivization through policies like the Merton rule is driving up demand.

Research funded by BioRegional in connection with medium-scale biomass systems in the South-east shows that considerable feedstock is already in the system but far more is required to respond to emerging requirements.

Fortunately, the scale of the UK's untapped resource is considerable, with 5 million to 6 million tonnes of waste wood going to landfill annually, and 680,000 ha of set-aside land that could be used for energy crops without affecting agricultural output.

Based on these figures, it is estimated that 15% of the UK's building-related energy load could be supported without recourse to imported material. However, the supply chain is fragmented in terms of producers, processors and distributors – presenting potential biomass users with a range of complexities that gas users simply do not need to worry about. These include:

- Ensuring quality. Guaranteeing biomass quality is important for the assurance of performance and reliability. Variation in moisture content affects combustion, while inconsistent woodchip size or differences in sawdust content can result in malfunction. The presence of contaminants in waste wood causes problems too. High-profile schemes including the 180,000 tonne generator in Slough have had to shutdown because of variations in fuel quality. Use of pellets reduces to risk, but they are more expensive and require more energy for processing and transport
- Functioning markets. The scale of trade in biomass compares unfavourably with gas or oil, in that there are no standard contracts, fixed-price deals or opportunities for hedging which enable major users to manage their energy cost risk
- Security of supply. The potential for competing uses could lead to price inflation. Biofuels carry the greatest such risk, but many biomass streams have alternative uses. Lack of capacity in the marketplace is another security issue, with no mechanism to encourage strategic stockpiling for improved response to crop failure or fluctuations in demand
- Installation and maintenance infrastructure. The different technologies used in biomass systems creates maintenance requirements not yet met by a readily available pool of skilled system engineers

Optimum uses of biomass technology

Biomass is a high-grade, locally available source of energy that can be used at a range of scales to support domestic and commercial use. Following increases in fossil fuel prices, one of the main barriers to adoption is, now, the capability of the supply chain.

The Carbon Trust's biomass sector review, completed before the large energy price rises in recent years, drew the following key conclusions about the most effective application of the technologies:

- Returns on CHP and electricity-generating systems depend heavily on government incentives such as renewable obligations certificates. Under the present arrangements, large CHP systems provide the best returns
- Heat-only systems are very responsive to changes in fuel prices
- Small-scale heat-only plants produce the best returns, because the cost of the displaced fuel (typically fuel oil) is more expensive
- Small-scale electricity and large-scale heat-only installations produce very poor returns
- There is little difference in the impact of fuel type in the returns generated by projects

The study also concluded that heat installations at all scales had the greater potential for carbon saving, based on a finite supply of biomass. This is because heat-generating processes have the greatest efficiency and, in the case of small-scale systems in isolated, off-grid dwellings, displace fuels such as oil that have the greatest carbon intensity. Ninety percent of the UK's existing biomass resource of 5.6 m tonnes per annum could be used in displacing carbon-intensive off-grid heating, saving 2.5 m tonnes of carbon emissions.

Small-scale systems are well established in Europe and the existing local supply chain suits the demand pattern. In addition, since the target market is in rural, off-grid locations, affected dwellings are less likely to suffer space constraints related to storage. As fuel costs continue to rise, the benefits of avoiding fuel poverty, combined with the effective reduction of carbon emissions from existing buildings, mean smaller systems are likely to offer the best mid-term use of the existing biomass supply base, with large-scale systems being developed as the supply chain matures and expands.

Large-scale systems also offer the opportunity to generate significant returns, but the barriers that developers or operators face are significant, particularly if there is an electricity supply component, which requires a supply agreement. However, while developers are required to delivery renewable energy on site, biomass in the form of biofuels, has the great attraction of being able to provide a scale of renewable energy generation that other systems such as ground source heating or photovoltaics simply cannot compete with.

Whether biomass plant should be used on commercial schemes in urban locations is potentially a policy issue. Sizing of both CHP and heat-only systems should be determined by the heat load, which for city-centre schemes may not be that large – affecting the potential for the CHP component. The costs of a district heating element on these schemes may also be prohibitively high, and considerations of biomass transport and storage also make it harder to get city-centre schemes to stack up.

It may be a more appropriate policy to encourage industrial users or large scale regenerators to take first call on the expanding biomass resource, rather than commercial schemes. The launch of a 45 MW biomass power station in Scotland illustrates this trend. Data show that 50% of the market potential for industrial applications of CHP could utilize 100% of the UK's available biomass resource. The issues that city-centre biomass schemes face in connection with storage, transport, emissions and supply chain management might be better addressed by industrial users or their energy suppliers in low-cost locations rather than by developers in prime city-centre sites.

Indicative costs

System	Indicative load kW	Capital cost £/kWh
Gas-fired boiler	50	90
	400	50
Biomass-fired boiler	50	530
	500	265
Biomass-fired CHP	1,000	480

Allowance for stand-alone boiler house and fuel store £30,000–60,000 for 50 kWh system indicative costs exclude flues and plant room installation

Fuel costs (typical bulk prices at small commercial scale – February 2012)

Wood chip: 2.9 p/kWh

Wood pellet: 4.2 p/kWh

Fuel oil: 6.0 p/kWh

Natural gas: 4.8 p/kWh

Bottled LPG: 7.6 p/kWh

Capital Allowances

Introduction

Capital allowances provide tax relief by prescribing a statutory rate of depreciation for tax purposes in place of that used for accounting purposes. They are utilized by government to provide an incentive to invest in capital equipment, including assets within commercial property, by allowing the majority of taxpayers a deduction from taxable profits for certain types of capital expenditure, thereby reducing or deferring tax liabilities.

The capital allowances most commonly applicable to real estate are those given for capital expenditure on existing commercial buildings in disadvantaged areas, and plant and machinery in all buildings other than residential dwellings. Relief for certain expenditure on industrial buildings and hotels was withdrawn from April 2011, although the ability to claim plant and machinery remains.

Enterprise Zone Allowances are also available for capital expenditure within designated areas only where there is a focus on high value manufacturing. Enhanced rates of allowances are available on certain types of energy and water saving plant and machinery assets, whilst reduced rates apply to 'integral features' and items with an expected economic life of more than 25 years.

The Act

The primary legislation is contained in the Capital Allowances Act 2001. Major changes to the system were announced by the Government in 2007 and there have been further changes in subsequent Finance Acts.

The Act is arranged in 12 parts (plus two addenda) and was published with an accompanying set of Explanatory Notes.

Plant and machinery

The Finance Act 1994 introduced major changes to the availability of Capital Allowances on real estate. A definition was introduced which precludes expenditure on the provision of a building from qualifying for plant and machinery, with prescribed exceptions.

List A in Section 21 of the 2001 Act sets out those assets treated as parts of buildings:

- *Walls, floors, ceilings, doors, gates, shutters, windows and stairs.*
- *Mains services, and systems, for water, electricity and gas.*
- *Waste disposal systems.*
- *Sewerage and drainage systems.*
- *Shafts or other structures in which lifts, hoists, escalators and moving walkways are installed.*
- *Fire safety systems.*

Similarly, List B in Section 22 identifies excluded structures and other assets.

Both sections are, however, subject to Section 23. This section sets out expenditure, which although being part of a building, may still be expenditure on the provision of Plant and Machinery.

List C in Section 23 is reproduced below:

Sections 21 and 22 do not affect the question whether expenditure on any item in List C is expenditure on the provision of Plant or Machinery.

1. Machinery (including devices for providing motive power) not within any other item in this list.
2. Gas and sewerage systems provided mainly –
    a.   to meet the particular requirements of the qualifying activity, or
    b.   to serve particular plant or machinery used for the purposes of the qualifying activity.
3. Omitted.
4. Manufacturing or processing equipment; storage equipment (including cold rooms); display equipment; and counters, checkouts and similar equipment.
5. Cookers, washing machines, dishwashers, refrigerators and similar equipment; washbasins, sinks, baths, showers, sanitary ware and similar equipment; and furniture and furnishings.
6. Hoists.
7. Sound insulation provided mainly to meet the particular requirements of the qualifying activity.
8. Computer, telecommunication and surveillance systems (including their wiring or other links).
9. Refrigeration or cooling equipment.
10. Fire alarm systems; sprinkler and other equipment for extinguishing or containing fires.
11. Burglar alarm systems.
12. Strong rooms in bank or building society premises; safes.
13. Partition walls, where moveable and intended to be moved in the course of the qualifying activity.
14. Decorative assets provided for the enjoyment of the public in hotel, restaurant or similar trades.
15. Advertising hoardings; signs, displays and similar assets.
16. Swimming pools (including diving boards, slides & structures on which such boards or slides are mounted).
17. Any glasshouse constructed so that the required environment (namely, air, heat, light, irrigation and temperature) for the growing of plants is provided automatically by means of devices forming an integral part of its structure.
18. Cold stores.
19. Caravans provided mainly for holiday lettings.
20. Buildings provided for testing aircraft engines run within the buildings.
21. Moveable buildings intended to be moved in the course of the qualifying activity.
22. The alteration of land for the purpose only of installing Plant or Machinery.
23. The provision of dry docks.
24. The provision of any jetty or similar structure provided mainly to carry Plant or Machinery.
25. The provision of pipelines or underground ducts or tunnels with a primary purpose of carrying utility conduits.
26. The provision of towers to support floodlights.
27. The provision of –
    a.   any reservoir incorporated into a water treatment works, or
    b.   any service reservoir of treated water for supply within any housing estate or other particular locality.
28. The provision of –
    a.   silos provided for temporary storage, or
    b.   storage tanks.
29. The provision of slurry pits or silage clamps.
30. The provision of fish tanks or fish ponds.
31. The provision of rails, sleepers and ballast for a railway or tramway.
32. The provision of structures and other assets for providing the setting for any ride at an amusement park or exhibition.
33. The provision of fixed zoo cages.

Capital allowances on plant and machinery are given in the form of writing down allowances at the rate of 18% per annum on a reducing balance basis. For every £100 of qualifying expenditure £18 is claimable in year 1, £14.76 in year 2 and so on until either all the allowances have been claimed or the asset is sold.

Integral features

The category of qualifying expenditure on 'integral features' was introduced with effect from April 2008. The following items are integral features:

- An electrical system (including a lighting system)
- A cold water system
- A space or water heating system, a powered system of ventilation, air cooling or air purification, and any floor or ceiling comprised in such a system
- A lift, an escalator or a moving walkway
- External solar shading

A reduced writing down allowance of 8% per annum is available on integral features.

Thermal insulation

For many years the addition of thermal insulation to an existing industrial building has been treated as qualifying for plant and machinery allowances. From April 2008 this has been extended to include all commercial buildings but not residential buildings.

A reduced writing down allowance of 8% per annum is available on thermal insulation.

Long-life assets

A reduced writing down allowance of 8% per annum is available on long-life assets. Allowances were given at the rate of 6% before April 2008.

A long-life asset is defined as plant and machinery that can reasonably be expected to have a useful economic life of at least 25 years. The useful economic life is taken as the period from first use until it is likely to cease to be used as a fixed asset of any business. It is important to note that this likely to be a shorter period than an item's physical life.

Plant and machinery provided for use in a building used wholly or mainly as dwelling house, showroom, hotel, office or retail shop or similar premises, or for purposes ancillary to such use, cannot be long-life assets.

In contrast plant and machinery assets in buildings such as factories, cinemas, hospitals and so on are all potentially long-life assets.

Case law

The fact that an item appears in List C does not automatically mean that it will qualify for capital allowances. It only means that it may potentially qualify.

Guidance about the meaning of plant has to be found in case law. The cases go back a long way, beginning in 1887. The current state of the law on the meaning of plant derives from the decision in the case of *Wimpy International Ltd and Associated Restaurants Ltd v Warland* in the late 1980s.

The Judge in that case said that there were three tests to be applied when considering whether or not an item is plant.

1. Is the item stock in trade? If the answer yes, then the item is not plant.
2. Is the item used for carrying on the business? In order to pass the business use test the item must be employed in carrying on the business; it is not enough for the asset to be simply used in the business. For example, product display lighting in a retail store may be plant but general lighting in a warehouse would fail the test.

3.   Is the item the business premises or part of the business premises? An item cannot be plant if it fails the premises test, i.e. if the business use is as the premises (or part of the premises) or place on which the business is conducted. The meaning of part of the premises in this context should not be confused with the law of real property. The Inland Revenue's internal manuals suggest there are four general factors to be considered, each of which is a question of fact and degree:

   •   Does the item appear visually to retain a separate identity
   •   With what degree of permanence has it been attached to the building
   •   To what extent is the structure complete without it
   •   To what extent is it intended to be permanent or alternatively is it likely to be replaced within a short period.

There is obviously a core list of items that will usually qualify in the majority of cases. However, many other still need to be looked at on a case-by-case basis. For example, decorative assets in a hotel restaurant may be plant but similar assets in an office reception area would almost certainly not be.

One of the benefits of the integral features rules, apart from simplification, is that items that did not qualify by applying these rules, such as general lighting in a warehouse or an office building, will now qualify albeit at a reduced rate.

Refurbishment schemes

Building refurbishment projects will typically be a mixture of capital costs and revenue expenses, unless the works are so extensive that they are more appropriately classified a redevelopment. A straightforward repair or a 'like for like' replacement of part of an asset would be a revenue expense, meaning that the entire amount can be deducted from taxable profits in the same year.

Where capital expenditure is incurred that is incidental to the installation of plant or machinery then Section 25 of the 2001 Act allows it to be treated as part of the expenditure on the qualifying item. Incidental expenditure will often include parts of the building that would be otherwise disallowed, as shown in the Lists reproduced above. For example, the cost of forming a lift shaft inside an existing building would be deemed to be part of the expenditure on the provision of the lift.

The extent of the application of Section 25 was reviewed for the first time by the Special Commissioners in December 2007 and by the First Tier Tribunal (Tax Chamber) in December 2009, in the case of JD Wetherspoon. The key areas of expenditure considered were overheads and preliminaries where it was held that such costs could be allocated on a pro-rata basis; decorative timber panelling which was found to be part of the premises and so ineligible for allowances; toilet lighting which was considered to provide an attractive ambience and qualified for allowances; and incidental building alterations of which enclosing walls to toilets and kitchens and floor finishes did not qualify but tiled splash backs, toilet cubicles and drainage did qualify along with the related sanitary fittings and kitchen equipment.

Annual investment allowance

The annual investment allowance is available to all businesses of any size and allows a deduction for the whole of the first £500,000 from 19 March 2014 (£250,000 before 19 March 2014) of qualifying expenditure on plant and machinery, including integral features and long-life assets. The annual investment allowance will return to £25,000.00 from January 2016.

The enhanced capital allowances scheme

The scheme is one of a series of measures introduced to ensure that the UK meets its target for reducing greenhouse gases under the Kyoto Protocol. 100% first year allowances are available on products included on the Energy Technology List published on the website at www.eca.gov.uk and other technologies supported by the scheme. All businesses will be able to claim the enhanced allowances, but only investments in new and unused Machinery and Plant can qualify.

There are currently 15 technologies with multiple sub-technologies currently covered by the scheme:

- Air-to-air energy recovery
- Automatic monitoring and targeting (AMT)
- Boiler equipment
- Combined heat and power (CHP)
- Compressed air equipment
- Heat pumps
- Heating ventilation and air conditioning (HVAC) equipment
- High speed hand air dryers
- Lighting
- Motors and drives
- Pipework insulation
- Radiant and warm air heaters
- Refrigeration equipment
- Solar thermal systems
- Uninterruptible power supplies

The Finance Act 2003 introduced a new category of environmentally beneficial plant and machinery qualifying for 100% first-year allowances. The Water Technology List includes 14 technologies:

- Cleaning in place equipment
- Efficient showers
- Efficient taps
- Efficient toilets
- Efficient washing machines
- Flow controllers
- Greywater recovery and reuse equipment
- Leakage detection equipment
- Meters and monitoring equipment
- Rainwater harvesting equipment.
- Small scale slurry and sludge dewatering equipment
- Vehicle wash water reclaim units
- Water efficient industrial cleaning equipment
- Water management equipment for mechanical seals

Buildings and structures and long-life assets as defined above cannot qualify under the scheme. However, following the introduction of the integral features rules, lighting in any non-residential building may potentially qualify for enhanced capital allowances if it meets the relevant criteria.

A limited payable ECA tax credit equal to 19% of the loss surrendered was also introduced for UK companies in April 2008.

From April 2012 expenditure on plant and machinery for which tariff payments are received under the renewable energy schemes introduced by the Department of Energy and Climate Change (Feed-in Tariffs or Renewable Heat Incentives) will not be entitled to enhanced capital allowances.

Enterprise zones

The creation of 11 new enterprise zones was announced in the 2011 Budget. Additional zones have since been added bringing the number to 24 in total. Originally introduced in the early 1980s as a stimulus to commercial development and investment, they had virtually faded from the real estate psyche.

The original zones benefited from a 100% first year allowance on capital expenditure incurred on the construction (or the purchase within two years of first use) of any commercial building within a designated enterprise zone, within 10 years of the site being so designated. Like other allowances given under the industrial buildings code the building has a life of 25 years for tax purposes.

The majority of these enterprise zones had reached the end of their 10-year life by 1993. However, in certain very limited circumstances it may still be possible to claim these allowances up to 20 years after the site was first designated.

Enterprise zones benefit from a number of reliefs, including a 100% first year allowance for new and unused non-leased plant and machinery assets, where there is a focus on high-value manufacturing.

Flat conversion allowances

Tax relief is available on capital expenditure incurred on or after 11 May 2001 on the renovation or conversion of vacant or underused space above shops and other commercial premises to provide flats for rent.

In order to qualify the property must have been built before 1980 and the expenditure incurred on, or in connection with:

- Converting part of a qualifying building into a qualifying flat.
- Renovating an existing flat in a qualifying building if the flat is, or will be a qualifying flat.
- Repairs incidental to conversion or renovation of a qualifying flat.
- The cost of providing access to the flat(s).

The property must not have more than four storeys above the ground floor and it must appear that, when the property was constructed, the floors above the ground floor were primarily for residential use. The ground floor must be authorized for business use at the time of the conversion work and for the period during which the flat is held for letting. Each new flat must be a self-contained dwelling, with external access separate from the ground-floor premises. It must have no more than 4 rooms, excluding kitchen and bathroom. None of the flats can be 'high value' flats, as defined in the legislation. The new flats must be available for letting as a dwelling for a period of not more than 5 years.

An initial allowance of 100% is available or, alternatively, a lower amount may be claimed, in which case the balance may be claimed at a rate of 25% per annum in subsequent a years. The allowances may be recovered if the flat is sold or ceases to be let within 7 years.

At Budget 2011 the Government announced that the relief would be abolished from April 2013.

Business premises renovation allowance

The business premises renovation allowance (BPRA) was first announced in December 2003. The idea behind the scheme is to bring long-term vacant properties back into productive use by providing 100% capital allowances for the cost of renovating and converting unused premises in disadvantaged areas. The legislation was included in Finance Act 2005 and was finally implemented on 11 April 2007 following EU state aid approval.

The legislation is identical in many respects to that for flat conversion allowances. The scheme will apply to properties within the areas specified in the Assisted Areas Order 2007 and Northern Ireland.

BPRA is available to both individuals and companies who own or lease business property that has been unused for 12 months or more. Allowances will be available to a person who incurs qualifying capital expenditure on the renovation of business premises.

An announcement to extend the scheme by a further five years to 2017 was made within the 2011 Budget, along with a further 11 new designated Enterprise Zones.

Legislation was introduced in Finance Bill 2014 to clarify the scope of expenditure qualifying for relief to actual costs of construction and building work and for certain specified activities such as architectural and surveying services. The changes will have effect for qualifying expenditure incurred on or after 1 April 2014 for businesses within the charge to corporation tax, and 6 April 2014 for businesses within the charge to income tax.

Other capital allowances

Other types of allowances include those available for capital expenditure on Mineral Extraction, Research and Development, Know-How, Patents, Dredging and Assured Tenancy.

Enhanced Capital Allowances (ECAs), The Energy Technology List (ETL) and The Water Technology List (WTL)

Background

This bulletin is intended to raise awareness of the existence, location and relevance of the ETL and the WTL which are often referred to for simplicity as 'The Technology Lists'. The lists comprise the technologies and products which qualify for the UK Government's Enhanced Capital Allowances (ECA) scheme. Also included in the lists are the energy-saving and water-saving performance criteria for each product or technology.

What are ECAs?

Enhanced Capital Allowances are a form of tax relief enabling a business to claim 100% first-year capital allowances on their spending on qualifying plant and machinery instead of the normal reducing balance writing down allowances of 8% for integral features and 18% for main pool plant. There are two building services-related schemes for ECAs:

- Energy-saving plant and machinery
- Water conservation plant and machinery

Businesses can write off the whole of the capital cost of their investment in these technologies against their taxable profits of the period during which they make the investment. This can deliver a helpful cash flow boost and a shortened payback period via both the energy, or water, saved and the allowances claimed. Alternatively a loss making business can surrender the ECA value of the loss to obtain a 19% tax credit.

ECAs are only available to investors of qualifying plant and machinery and not developers. Developers are 'traders' and they incur revenue expenditure and not capital expenditure and so cannot claim capital allowances. However the purchaser of a newly constructed development for investment can still make use of ECAs provided they are sold unused and the purchaser is a taxpayer. This can be a valuable selling point for developers if highlighted in the marketing literature.

Interest in ECAs has been highlighted by the changes to the existing capital allowances regime which have reduced the value of allowances for many qualifying items. The introduction of Energy Performance Certificates and the Carbon Reduction Commitment Energy Saving Scheme will also promote the types of technologies that could qualify for ECAs.

Who manages the technology lists and where are they?

The lists are part of the ECA scheme which was developed by the Treasury, HM Revenue & and the Department of Energy & Climate Change (DECC). The ETL is managed and maintained by the Carbon Trust and can be found at http://etl.decc.gov.uk/. The WTL is managed by Business Link and the Department for Environment Food & Rural Affairs (DEFRA) and can be found at http://wtl.defra.gov.uk.

What technologies and products are included in the ETL? (as at 2013)

- Air-to-air energy recovery
- Automatic monitoring and targeting
- Boiler equipment
- Combined heat and power (CHP)
- Compact heat exchangers
- Compressed air equipment
- Heat pumps
- Heating ventilation and air conditioning (HVAC) equipment
- High speed hand air dryers
- Lighting (high efficiency lighting units, lighting controls and LEDs)
- Motors and drives
- Pipework insulation
- Radiant and warm air heaters
- Refrigeration equipment
- Solar thermal systems
- Thermal screens
- Uninterruptible power supplies
- Waste heat to electricity conversion equipment

And the WTL?

- Cleaning in place equipment
- Efficient showers
- Efficient taps
- Efficient toilets
- Efficient washing machines
- Flow controllers
- Leakage detection equipment
- Meters and monitoring equipment
- Rainwater harvesting equipment
- Small scale slurry and sludge dewatering equipment
- Vehicle wash water reclaim units
- Water efficient industrial cleaning equipment
- Water management equipment for mechanical seals
- Water reuse

Eligibility criteria

It is not possible to summarize the eligibility criteria here because each technology has different definitions and considerations. Most of the above technologies have specific products listed that are eligible, but others are defined by detailed performance criteria. The key technologies worthy of note which operate on a Performance criteria are Lighting, CHP, Pipework Insulation, Automatic Monitoring & Targeting equipment and Water Reuse. Qualification is generally extremely strict. To give a brief example; Lighting does not merely concentrate on the lamp type used but also has very strict requirements for the light fitting itself, having to meet high Light Output Ratios, colour rendering requirements and use high frequency control equipment. CHP and AMT equipment is also heavily guided.

Scope of ECA claims

For products on the technology lists, claims will be considered for the cost of the equipment itself, and other costs directly involved in installing it. These include:

- Transportation – the cost of getting equipment to the site.
- Installation – cranage (to lift heavy equipment into place), project management costs and labour, plus any necessary modifications to the site or existing equipment.
- Preliminary costs and oncosts – if they are directly related to the acquisition and installation of the equipment.
- Professional fees – if they are directly related to the acquisition and installation of the equipment.

Making a claim for ECAs

If investing in eligible technology claimants should submit their ECA claims as part of their normal Income or Corporation Tax return. The best advice we can give clients on relevant projects is to highlight the fact that their investment may be eligible, and introduce AECOM Banking Tax & Finance to the client as early as possible in the design process so that they can offer a view on the potential for worthwhile tax relief before, during and after the investment programme.

Building Regulations Pocket Book

Ray Tricker and Samantha Alford

This handy guide provides you with all the information you need to comply with the UK Building Regulations and Approved Documents. On site, in the van, in the office, wherever you are, this is the book you'll refer to time and time again to double check the regulations on your current job.

The *Building Regulations Pocket Book* is the must have reliable and portable guide to compliance with the Building Regulations.

- Part 1 provides an overview of the Building Act
- Part 2 offers a handy guide to the dos and don'ts of gaining the Local Council's approval for Planning Permission and Building Regulations Approval
- Part 3 presents an overview of the requirements of the Approved Documents associated with the Building Regulations
- Part 4 is an easy to read explanation of the essential requirements of the Building Regulations that any architect, builder or DIYer needs to know to keep their work safe and compliant on both domestic or non-domestic jobs

This book is essential reading for all building contractors and sub-contractors, site engineers, building engineers, building control officers, building surveyors, architects, construction site managers and DIYers. Homeowners will also find it useful to understand what they are responsible for when they have work done on their home (ignorance of the regulations is no defence when it comes to compliance!).

February 2018: 186 x 123 mm: 456 pp

Pb: 978-0-8153-6838-0 : £19.99

To Order: Tel: +44 (0) 1235 400524 Fax: +44 (0) 1235 400525
or Post: Taylor and Francis Customer Services,
Bookpoint Ltd, Unit T1, 200 Milton Park, Abingdon, Oxon, OX14 4TA UK
Email: book.orders@tandf.co.uk

For a complete listing of all our titles visit:
www.tandf.co.uk

LED Lighting

Background

LED lighting technology is now sufficiently developed to be widely accepted as the light source of choice for most lighting applications. The technology offers benefits including energy efficacy, high quality light appearance, light colour adjustability, device connectivity and controllability. This makes LED luminaires suitable for applications ranging from street lighting to office and residential lighting.

Currently available types of LED luminaries

Light emitting diodes (LEDs) are small high efficacy light sources that can be packaged into a range of products suitable to both new build and retrofit applications. In new build applications LEDs are often built-in to the luminaire and form an integral part of it, which can offer some technical benefits. Whereas in retrofit applications LEDs are packaged into lamp formats that can easily be inserted into existing light fixtures to replace traditional light sources. There are many types of retrofit LED lamps, from opalescent and filament LED bulbs that are an alternative to halogen and compact fluorescent bulbs to linear LED tubes that are an alternative to linear fluorescent lamps.

LED efficacy

LEDs are commercially available with an efficacy of over 120 lumens per watt. This makes them far more efficient than halogen lamps that have a typical efficacy range of 5–15 lumens per watt and are now banned in the UK, and more efficient than both fluorescent and discharge technology lamps that have an efficacy range of 40–100 lumens per watt.

LED light colour

LEDs are available in a wide range of light colour appearances. Very warm white with a colour appearance of 1800 K to very cool white with a colour appearance of over 10,000 K is available. In addition, many saturated colour options are available that can be blended together to create millions of colour options.

Typical LED colour rendering characteristics vary from CRI 65–95 according to a product's priority for efficacy or colour appearance fidelity. In addition specialized LED products are available with light spectral properties tuned to specific applications including human centric circadian lighting and retail lighting requirements.

LED control

In both new build and retrofit applications a wide range of LED control options are available, both wired and wireless. This makes it possible to dim LEDs and create programmable colour change effects in both new build and retrofit applications.

Cost implications

Manufacturers and suppliers are keen to highlight potential energy saving from LED lighting technology. Although the principle of realizable savings is completely valid, detailed study of potential energy savings is essential to ensure the most beneficial and appropriate LED technologies and design approaches can be implemented.

Payback

LED lamps/luminaires are typically more expensive than traditional lighting alternatives across most types. However, the cost variance and energy saving potential for different LED lamps/luminaires types is significantly varied. Therefore, the return on investment from different LED lighting typologies and implementation approaches is not constant and requires detailed study to forecast in detail.

The following provides a return on investment overview for a sample LED Luminaire typology and excludes the cost of installation labour that is assumed to be consistent:

Power cost (£)/kWh	0.12
Days in use/year	260
Hours in use/day	12

	Typical LED Downlight	Compact Fluorescent Downlight	LV Halogen (IRC) Downlight
Input Power (watts)	6	12	35
Lifetime (hours)	50,000	12,000	4,000
Replacement Lamp Cost including labour (£)	–	15.00	15.00
Annual Energy Cost per lamp (£)	2.26	4.53	13.21
Total Energy Cost for 25 luminaires per year (£)	56.50	113.25	330.28
Luminaire Unit Cost (£)	110.00	60.00	80.00

Supply only cost of 25 luminaires (£)	2,750.00	1,500.00	2,000.00
Allowance for 4nr emergency battery packs (£)	320.00	320.00	320.00
Lamps (£)	Included	375.00	375.00
Total cost of luminaries/emergencies/ lamps (£)	3,070.00	2,195.00	2,695.00

Extra over cost of LED luminaires and emergency luminaires compared to traditional luminaires is (£)		875.00	375.00
Yearly cost of energy and yearly re-lamping allowance (compact fluorescent @ every 4 years; halogen @ every 1.5 year) (£)	56.50	207.00	580.28
The calculated yearly energy and re-lamping saving of using LED luminaires when compared to traditional luminaires is (£)		150.50	523.78
Therefore time taken in years to 'pay' for the additional cost of the LED luminaires based on the energy and re-lamping costs of 'traditional' luminaires in **years** is		**5.81**	**0.72**

Power cost (£)/kWh	0.12
Days in use/year	260
Hours in use/day	12

	Recessed LED Office Luminaire	Recessed Fluorescent Office Luminaire
Input Power (watts)	37.5	42
Typical Lifetime (hours)	50,000	12,000
Replacement Lamp Cost including labour (£)	–	20.00
Annual Energy Cost per lamp (£)	14.15	15.85
Total Energy Cost for 25 luminaires per year (£)	353.93	396.28
Luminaire Unit Cost including lamps (£)	185.00	165.00

Supply only cost of 25 luminaires including lamps (£)	4,625.00	4,125.00
Allowance for 4nr emergency battery packs (£)	480.00	400.00
Total cost of luminaries/emergencies/lamps (£)	5,105.00	4,525.00

Extra over cost of LED luminaires and emergency luminaires compared to fluorescent luminaires is (£)		580.00
Yearly cost of energy and yearly re-lamping allowance (fluorescent @ every 4 years) (£)	353.93	521.28
The calculated yearly energy and re-lamping saving of using LED luminaires when compared to fluorescent luminaires is (£)		167.35
Therefore time taken in years to 'pay' for the additional cost of the LED luminaires based on the energy and re-lamping costs of fluorescent luminaires in **years** is		**3.47**

Notes and conclusions

1. LED luminaires do save energy.
2. LED luminaires are generally twice the price of a 'traditional' downlight but LED luminaire prices are becoming more competitive as the technology develops.
3. Some LED luminaires may not be converted to emergency so 'additional' luminaires may be required.
4. LED replacement lamps may have completely different light emitting characteristics. And, the luminaire may require modification to accept the retrofit LED lamp.

Getting the Connection

Details the process involved in the provision of a new electrical supply connection offer to a site.

The provision of an electricity connection has both a physical and contractual element. Physically, it involves the design, planning and construction of electrical infrastructure (cables, switchgear, civil works), whilst contractually, it requires legal agreements to be drawn up and agreed (construction, connection, adoption).

Planning

The Planning Stage, typically RIBA Stage 2, is where the site's developer (the Developer) should be formulating their plans for the scheme and, in doing so, should consult the local distribution network operator's (the Host)[1] long-term development statement, which will identify potential connection point opportunities. This information is normally readily available from the Host and should provide an early indication of whether the Host's network may need to be reinforced before the development can be connected.

As the Planning Stage progresses, the Developer should discuss its proposals with the Host. Relatively simple connections for single building supplies are straight forward, however schemes of a more complex nature may require some form of feasibility study to be carried out to assess connection options and provide indicative costs for the contestable[2] and non-contestable[3] work elements.

Design

The Design Stage, typically RIBA Stage 3, is the point at which the Developer submits its formal connection request to the Host. It is important that this is completed in accordance with the specific procedures of the Host, since if it does not include all supporting information required by the Host's application process, there is likely to be a delay in the processing of a firm connection offer. The Host is expected to provide a documented application process to assist the applicant make a complete application.

The convention is to request a Section 16/16 A connection[4] where the terms are standard and non-negotiable. The alternative is a Section 22 connection[4] offer, where the terms are fully negotiable. If the Section 22 route is chosen then caution is required, as disputes over Section 22 agreements cannot be referred to the regulator for determination after the connection agreement has been signed; a post contract dispute will have to be pursued as a civil action. It is always advised that a Section 16/16 A offer is requested before considering the Section 22 option.

On receipt of the Host's firm offer[5], the Developer generally has 90 days to accept its terms and to undertake its own review of the offer to ensure it meets its requirements. If for any reason the Developer and Host are unable to reach agreement of the terms, it is recommended that the Developer seeks specialist advice. It should be noted that in extreme circumstances, it may be necessary to refer an issue to regulator for determination but, be warned, this can be a lengthy process, taking up to 16 weeks to conclude.

1. *A distribution network operator (DNO) is a company that is responsible for the design, construction operation and maintenance of a public electricity distribution network. The host DNO is the electricity distribution network to which the development site will directly connect.*
2. *Contestable is work in providing the connection that can be carried out by an accredited independent party.*
3. *Non-contestable is work that can only be carried by the Host.*
4. *The Electricity Act 1989.*
5. *Unless the offer becomes interactive, i.e. another connection scheme is vying for network capacity. Where the connection request becomes interactive the Developer has 30 days to accept the offer. The Host will inform the Developer if the connection request is interactive.*

Competition

One of the key decisions affecting the way in which the connection process proceeds is whether the Developer wishes to introduce competition into the procurement process by appointing a third party to design and construct the contestable connection works. In these circumstances, the Developer can requisition a non-contestable quotation from the Host or ask the third party to do so on its behalf. The Host is obliged to provide information within standardized time frames against which its performance is monitored by the industry regulator (Ofgem).

Contestable works are those that may be carried out either by the Host or by an approved contractor, on the Developer's behalf. This contrasts with non-contestable works, which can only be carried out by the Host. Broadly speaking, the Host will make the connection to its network for the new supply and undertake any upstream network reinforcement works. All works downstream from the point of connection to the Host network into the site and to each building are contestable works. Therefore, the extent of the contestable works can vary significantly depending on where the point of connection is designated by the Host. Alternatively, the Developer can request an independent licensed network operator (IDNO) to adopt assets constructed by the third party.

If the Developer decides to contract with a third party to construct the contestable works, it is the Developer's responsibility to ensure that the construction works meet the Host's network adoption requirements. This can be achieved by appointing a NERS (National Electrical Registration Scheme – via Lloyds Register) accredited Independent Connection Provider (ICP) to carry out the works whilst ensuring that responsibility for achieving adoption of the assets by the DNO (or IDNO) remains with the ICP.

Network reinforcement

Reinforcement works may be required to increase the capacity of the network to enable the connection to meet a site's projected demand. In terms of the capacity made available by reinforcement, the following possible scenarios arise:

- Where reinforcement works are necessary for sole use by the Developer, the Developer is charged the full cost of the works. In some circumstances the most economic method of reinforcement may introduce spare network capacity in excess of the Developer's requirements. In such circumstances the Developer can receive a rebate where this spare capacity is absorbed by subsequent developments. However, the possibility of a rebate is time constrained and the original development will only qualify for a rebate for up to 5 years after the connection is completed.
- Where reinforcement works are necessary but also cater for the Host's future network requirements, the Developer is charged for a proportion of the cost of the works in the form of a capital contribution. This is calculated using cost apportionment factors for security (the ratio of capacity requested to that which is made available) and fault level (the ratio of fault level contribution of connection to that which is made available).

The charges levied for reinforcement works are attributable to those reinforcement works undertaken at one voltage level above the connection voltage only. That is, if connection voltage is 400 V then costs are chargeable for reinforcement works undertaken at the next highest voltage, which is usually 11 kV, and not for works conducted at the next highest voltage, which is usually 33 kV[6]. These deeper network reinforcement costs are recovered as part of the system charges built into the electricity supply tariffs.

Costs

The cost of the Developer's connection depends on the nature and extent of the works to be undertaken. The distance between the site and the Host's network, the size of the customer demand in relation to available capacity (and hence the potential need for reinforcement), customer-specific timescales and to an extent market value of raw material and labour, are all significant factors that will affect the cost.

6.   *Some DNOs use voltages of 6.6 kV and 22 kV.*

As part of the firm offer received from the Host, the Developer is provided with a charging statement, which includes the charges to be levied for the following items:

- Assessment and Design – to identify and design the most appropriate point on the existing network for the connection.
- Design Approval – to ensure design of a connection meets the safety and operation requirements of the Host.
- Non-contestable Works/Reinforcement – to include circuits and plant forming part of the connection that can be undertaken by the Host only, including land rights issues and consents.
- Contestable Connection Works – to include circuits and plant forming part of the connection that can be undertaken by approved contractors or the Host.
- Inspection of Works – to ensure that works are being constructed in accordance with the design requirements of the Host.
- Commissioning of Works – to include circuit outages and testing that will ensure that connection is safe to be energized.

In some circumstances, in addition to the cost of the physical connection works there may be chargeable costs associated with operation, maintenance repair and replacement of the new or modified connection. These are known as O&M costs and are chargeable as capitalized up-front costs where the Developer requests a solution that is in excess of the minimum necessary to provide the connection. For example, where extra resilience is requested for a connection, over and above that which the Host is obligated to provide, then O&M costs can be levied but only for the extra resilience element of the connection, not the total cost of the connection.

The Streetwise Subbie, 4th Edition

Barry J Ashmore

Very few books explore the problems which are particular to the relationship between Specialist Contractors and the Main Contractor, or Clients with whom they are in contract. Fewer still provide solutions in such a down to earth no-nonsense way as The Streetwise Subbie does.

The Streetwise Subbie is a highly regarded practical guide to contractual matters. Its original author Jack Russell was well known for his 'contractual terrier' column in Electrical Times. Now, Barry Ashmore has updated and revised this work by drawing on his 46 years of construction experience and professional expertise gained at the sharp end resolving disputes and solving contractual problems for Specialist Contractors.

Thousands of subbies have already benefited from the insights and the streetwise approach to avoiding or resolving contractual problems, and the clarity of thought and advice the book provides. The fourth edition features all the old favourites such as payment, delay and disruption, extension of time and the all-important checklists and site records. But it has now been brought bang up to date to reflect the importance of the 2011 revisions to the Construction Act and the emergence of adjudication as the pre-eminent means of dispute resolution.

It is an easy to read, practical, and essential guide, aimed at Specialist Contractors of all sizes and specialisations, be they sole traders, company directors, or any member of the subbie's team that has to handle the commercial and contractual aspects of the projects they undertake.

It's the kind of book that you keep handy, because it has so many answers that you can refer to it over and over again

May 2018: 234 x 156 mm: 218 pp
Pb: 978-1-138-30016-3 : £16.99

To Order: Tel: +44 (0) 1235 400524 Fax: +44 (0) 1235 400525
or Post: Taylor and Francis Customer Services,
Bookpoint Ltd, Unit T1, 200 Milton Park, Abingdon, Oxon, OX14 4TA UK
Email: book.orders@tandf.co.uk

For a complete listing of all our titles visit:
www.tandf.co.uk

Taylor & Francis
Taylor & Francis Group

Feed-In Tariffs (FITs)

The Feed-In Tariff (FIT) is a UK government incentive to encourage the development of small-scale renewable energy technology; the tariff has a maximum capacity of 5 MW.

Original FIT roll-out

FITs were initially introduced in April 2010 by the Department of Energy and Climate Change (DECC), and are effectively payments which are made for every kilowatt-hour (kWh) of energy generated from a renewable energy source. As FITs are paid based on generation, owners of installations also have the opportunity to benefit financially by consuming the energy generated by the renewable energy technology, displacing energy which would have been supplied by the grid.

The Feed-in Tariffs (Amendment) (No. 2) Order 2010 came into effect on 1 August 2012. The schemes policy and tariff rates are set by the UK Government with the scheme being administered by energy suppliers and Ofgem.

Feed-in Tariffs are paid by electricity suppliers. Generators between 50 kW and 5 MW should apply to Ofgem for accreditation. FIT payments are made based on meter readings taken from compliant metering and submitted to the elected FIT Licensee. Microgenerators can obtain accreditation via the Microgeneration Certification Scheme (MCS), before then applying to a supplier for a FIT agreement. All generation will be metered and FIT payments will be made in accordance with the Electricity Act 1989. The consumer (owner or beneficiary) will be responsible for the capital of the plant; the necessary access/connections to the electricity and distribution systems and organization of the payment receipts.

FITs review

The first review of the Feed-in Tariff scheme was scheduled to be carried out in 2012 however it was brought forward by a year due to the surge in uptake which placed a huge strain on the FITs budget. The surge was due to the sharp fall in the cost of solar PV installations and also the creation of large-scale solar farms, which went against the ethos of the scheme which was to promote small-scale installations.

February 2012 saw the DECC announce wide ranging reforms to the FIT scheme which aimed to provide a better framework for consumers and communities. There were five key areas that DECC has identified as needing refinements;

- The tariff for micro-CHP installations will be increased to recognize the benefits this technology could bring and to encourage its development.
- A tariff of 21p/kWh took effect from 1 April 2012 for domestic-size solar panels with an eligibility date on or after 3 March 2012.
- Installations of solar PV panels on or after 1 April 2012 will be required to produce an Energy Performance Certificate rating of 'D' or above to qualify for a full FIT.
- 'Multi-installation' tariff rates set at 80% of the standard tariffs will be introduced for solar PV installations where a single individual or organization is already receiving FITs for other solar PV installations.
- Individuals or organizations with 25 or fewer installations will still be eligible for the individual rate. The department for Business, Energy and Industrial Strategy (formerly DECC) is consulting on a proposal that social housing, community projects and distributed energy schemes be exempt from multi-installation tariff rates.

Feed-in tariffs (FITs)

Following the FITs review the Department of Energy and Climate Change (DECC) announced the first scheduled reduction in tariff rates for all PV installations with an eligibility date on or after 1 November 2012. The new tariff rates were published by Ofgem by 31 August 2015. Tariff rates will be continuously reviewed as set out in the degression mechanism as follows:

- Degression will take place on a quarterly basis, with tariffs changing on the first day of the first month of the quarter for new installations with an eligibility date from that date.
- Tariffs will be published at least two months before the start of the new quarter, and will be based on deployment of new PV generating capacity in the previous quarter.
- The baseline degression rate will be 3.5% per quarter.
- Degression will be skipped if deployment is below a floor threshold set by BEIS.
- Deployment statistics will be published on a monthly basis by BEIS.

Eligibility

Small-scale (up to a maximum capacity of 5 MW) low-carbon electricity generating technologies eligible for FITs are:

- Wind
- Solar photovoltaics (PV)
- Hydro
- Anaerobic digestion
- Domestic scale micro-combined heat & power (CHP), eligible up to 2 kWe.

From 1 April 2010, microgenerators (<50 kW) in Anaerobic Digestion, Hydro, Solar PV and Wind are ineligible for support under the Renewable Obligation (RO) due to FIT inclusion. Refurbished/renovated installations will be ineligible. Small generators that applied for accreditation on or after 15 July 2009 and before 1 April 2010 are eligible for transfer to FITs.

Any existing systems installed before 15 July 2009 will only qualify for the FIT scheme if they are under 50 kW and registered with the Renewable Obligation Certificates.

As of 1 December 2012, provisions were introduced for schools and some community organizations.

Tariffs

The Feed-in Tariffs (FITs) provide two main financial benefits to the consumer:

- A 'generation' tariff (a rate per kWh dependent on technology and size banding) based on the total generation (these bandings are set out in the table below).
- An 'export' tariff (5.03p/kWh) for any surplus energy produced exported into the grid.

The most up to date tariffs are available at the web address below:
https://www.OFGEM.gov.uk/environmental-programmes/fit

Feed-in Tariff Banding

Generation Technology	Capacity Band
Anaerobic digestion	≤250 kW
Anaerobic digestion	>250 kW–500 kW
Anaerobic digestion	>500 kW–5 MW
Hydro generating station	≤15 kW
Hydro generating station	>15 kW–100 kW
Hydro generating station	>100 kW–500 kW
Hydro generating station	>500 kW–2 MW
Hydro generating station	>2 MW–5 MW
Combined heat & power (CHP)	≤2 kW
Solar photovoltaic	≤4 kW
Solar photovoltaic	>4 kW–10 kW
Solar photovoltaic	>50 kW–100 kW
Solar photovoltaic	>100 kW–150 kW
Solar photovoltaic	>150 kW–250 kW
Solar photovoltaic	>250 kW–5 MW
Wind	≤1.5 kW
Wind	>1.5 kW–15 kW
Wind	>15 kW–100 kW
Wind	>100 kW–500 kW
Wind	>500 kW–1.5 MW
Wind	>1.5 MW–5 MW

An Introduction to Electrical Science, 2nd Edition

Adrian Waygood

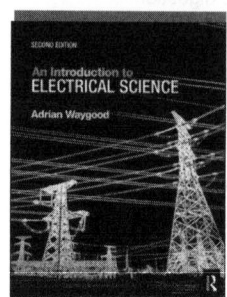

Heavily updated and expanded, this second edition of Adrian Waygood's textbook provides an indispensable introduction to the science behind electrical engineering.
While fully matched to the electrical science requirements of the 2330 levels 2 and 3 Certificates in Electrotechnical Technology from the City and Guilds (Electrical Installation), the main purpose of this book is to develop an easy understanding of the how and why within each topic. It is aimed for those starting careers in electronics, as well as any hobbyists, with an array of new material to reflect changes in the industry.

New chapters include:
* Electrical Drawings
* Practical Resistors
* Measuring Instruments
* Basic Motor Action
* Practical Inductors
* Basic Transformer Theory
* The Electricity Supply Industry
…and more

The author details the historical context of each main principle and offers a wealth of examples, images and diagrams, all whilst maintaining his signature conversational and accessible style. And there is also a companion site with interactive multiple choice quizzes for each chapter and more, at www.routledge.com/cw/waygood

August 2018: 276 × 219 mm: 384 pp
Pb: 978-0-8153-9181-4: £39.99

To Order: Tel: +44 (0) 1235 400524 Fax: +44 (0) 1235 400525
or Post: Taylor and Francis Customer Services,
Bookpoint Ltd, Unit T1, 200 Milton Park, Abingdon, Oxon, OX14 4TA UK
Email: book.orders@tandf.co.uk

For a complete listing of all our titles visit:
www.tandf.co.uk

Taylor & Francis
Taylor & Francis Group

Renewable Obligation Certificates (ROCs)

The Renewable Obligation (RO) is the main support scheme for renewable electricity projects in the UK, placing an obligation on licensed UK suppliers of electricity to increase their proportion of electricity production from renewable sources, or result in a penalty.

Since its introduction in 2002, it has succeeded in tripling the level of renewable electricity in the UK from 1.8% to 5.4% and is currently worth around £1 billion/year in support to the renewable electricity industry. The RO applies to all powered plant with a power capacity greater than 5 MW. The target started at 3% and is presently at 11.4% rising incrementally to 15.4% by 2015. It is likely to be extended to 20% by 2020.

In April 2009, the introduction of banding under the Renewables Obligation Order 2009 meant different technologies receive different levels of support, providing a greater incentive to those that are further from the market.

The RO was extended from its current end date of 2027 to 2037, in April 2010, for new projects with a view to providing greater long-term certainty for investors and an increase in support for offshore wind projects.

The RO is administered by the Office of the Gas and Electricity Markets (Ofgem) and suppliers of electricity have to prove they have met this obligation, producing Renewable Obligation Certificates (ROCs) to renewable electricity generators at the end of each year.

A Renewable Obligation Certificate (ROC) is a green certificate that is issued by Ofgem to an accredited generator for eligible renewable electricity generated within the UK and supplied to customers in the UK by a licensed supplier. A ROC is issued for each megawatt hour (MWh) of eligible renewable output generated.

Failing to meet the obligation results in 'buy-out' fines being paid to Ofgem on the shortfall of every MWh sold that was not renewable. Ofgem then distributes the funds to all electricity supply companies possessing ROCs, the amount received being in proportion to the number of ROCs held (at the end of the year). If a supplier meets part or all of its RO, but other companies do not, the supplier who has ROCs will be rewarded with a share of the fines.

Previously, 1 ROC was issued for each megawatt hour (MWh) of eligible generation, regardless of technology. Since April 2009, the reforms introduced means that new generators joining the RO now receive different numbers of ROCs, depending on their costs and potential for large-scale deployment. For example, onshore wind continues to receive 1 ROC/MWh, whereas offshore wind and energy crops currently receive 2 ROCs/MWh.

Obligation periods are valid for a year, beginning on 1 April to 31 March. Supply companies have until 31 September following the period to submit sufficient ROCs to cover their obligation, or submit sufficient payment to Ofgem to cover their shortfall.

Buy-out price

Suppliers can meet all, or part of their obligations by making a buy-out payment. The buy-out price set by Ofgem for the compliance period of 2017–2018 is £45.58 per Renewables Obligation Certificate (ROC). The buy-out price sets the rate which suppliers must pay if they fail to meet their obligations under the scheme and is adjusted annually in accordance with the Retail Prices Index (RPI).

Buy-out fund redistribution

At the end of the year, the funds made to Ofgem are distributed to all the electricity suppliers possessing ROCs, with the amount received in proportion to the number of ROCs held. If a supplier meets all or part of its RO, it will be rewarded with a share of the buy-out fines.

Pricing

Due to ROCs having the potential to save the supplier from having to commit to a buy-out payment, it increases the price of the electricity. When the renewable generator sells the electricity to a supplier it is not uncommon for the ROC to be sold in addition ultimately forcing the cost of electricity upwards. Also, due to the fact that ROCs entitles suppliers to a share of the 'buy-out' fund at the end of year, increases its value. Electrical suppliers can benefit financially by participating in the RO system due to the renewable targets set by the Government likely to be under-fulfilled and the fact that the RO is not over-subscribed will result in the ROCs and their recycled values being worth more than the £42.02 per MWh.

e-ROC

The most efficient method of buying and selling Renewable Obligation Certificates (ROCs) is through the e-ROC on-line auctions. They offer renewable generators access to the whole supplier market in the UK, delivering high ROC prices for low fees. The average price of ROCs sold through the auctions 25 May 2017 was £47.91 and with the fees set at only 50p per ROC (subject to a minimum fee of £50), indicates a profitable return for those in par-ticipation. Auctions are operated by NFPAS, a subsidiary of the Non-Fossil Purchasing Agency Limited (NFPA) and are usually held four times a year. NFPAS runs regular e-ROC on-line auctions for the sale of Renewable Obliga-tion Certificates (ROCs).

Eligibility

The reforms stated in Renewable Obligations Order 2009 introduced the concept of 'banding' for the Renewable Obligation Certificates (ROCs). The aim of ROC banding is to establish the number of ROCs per MWh that can be obtained according to the type of technology that is used to generate the renewable electricity.

There are 28 renewable technologies covered by ROCs Banding, resulting in an increasingly complex regulatory environment for technology providers, project developers and finance providers to navigate.

Band	Renewable Technology	Level of Banding (ROCs/MWh)
Established 1	Landfill Gas	0.25
Established 2	Sewage Gas Co-Firing of Non-Energy Crops (Regular) Biomass	0.5
Reference	Onshore Wind Hydro-Electric Co-Firing of Energy Crops Co-Firing of Biomass with CHP Energy from Waste with CHP Geo Pressure Pre-Banded Gasification Pre-Banded Pyrolysis Standard Gasification Standard Pyrolysis	1
Post-Demonstration	Offshore Wind Dedicated Regular Biomass Co-Firing of Energy Crops with CHP	1.5

Band	Renewable Technology	Level of Banding (ROCs/MWh)
Emerging	Wave Tidal Steam Advanced Gasification Advanced Pyrolysis Anaerobic Digestion Dedicated Energy Crops Dedicated Energy Crops with CHP Dedicated Regular Biomass with HP Solar Photovoltaic Geothermal Tidal Lagoons Tidal Barrages	2

Electrical Circuit Theory and Technology, 6th edition

John Bird

A fully comprehensive text for courses in electrical principles, circuit theory and electrical technology, providing 800 worked examples and over 1,350 further problems for students to work through at their own pace. This book is ideal for students studying engineering for the first time as part of BTEC National and other pre-degree vocational courses, as well as Higher Nationals, Foundation Degrees and first-year undergraduate modules.

March 2017: 276 x 219 mm: 858 pp
Pb: 978-1-138-67349-6 : £38.99

To Order: Tel: +44 (0) 1235 400524 Fax: +44 (0) 1235 400525
or Post: Taylor and Francis Customer Services,
Bookpoint Ltd, Unit T1, 200 Milton Park, Abingdon, Oxon, OX14 4TA UK
Email: book.orders@tandf.co.uk

For a complete listing of all our titles visit:
www.tandf.co.uk

RICS SKA Rating

The RICS SKA Rating is an assessment method that focuses on promoting environmental and sustainable good practice in the fit-out and refurbishment of existing buildings. The RICS SKA Rating is now seven years old and has grown from an idea into a credible, widely adopted environmental assessment method with 5,761 users and 554 certified projects on the system.

Existing non-domestic building stock represents 18% of UK carbon emissions and up to 40 fit-outs can take place over a building's life. Fit-out and refurbishment activity represents roughly 10% of UK construction spend and until SKA was launched in 2009, it was something of a sustainability blind spot.

SKA provides a set of good practice measures that can be implemented on a project and has been widely adopted by property occupiers in both the office and retail sectors including Bank of China, Yell, BBC, GE, Westpac Banking, Derwent London, Nationwide and Lush.

From a standing start, SKA has become a widely recognized assessment method that is mentioned in the same breath as BREEAM and LEED and is referenced in documents such as the UK Government's Low Carbon Construction Action Plan and the British Council for Offices 'Guide to Fit-out'. Figure 1 shows the growth over the last 4 years.

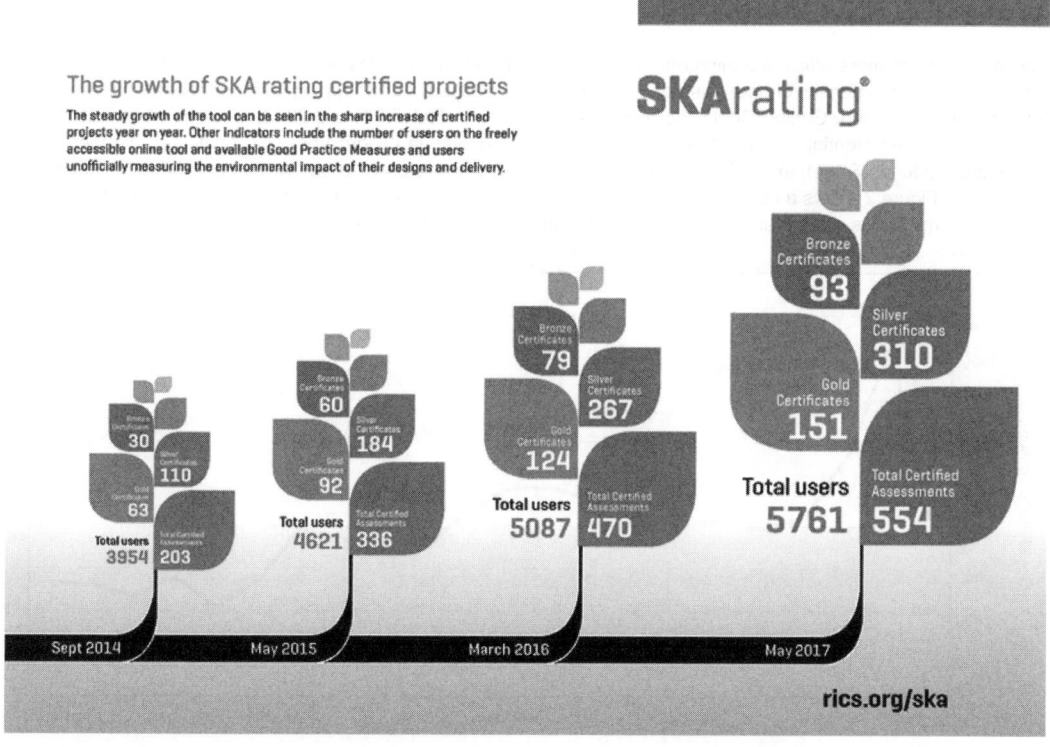

Figure 1: Diagram showing the growth of SKA rating certified projects over last 4 years.

The key to SKA's growth is that it has been designed to be much cheaper and simpler to implement than other environmental assessment methods. SKA is entirely on-line with the guidance, assessment methodology and even the certification generated through the tool. The aim is to promote good practice in the fit-out marketplace, so the on-line tool and the associated good practice guidance can be accessed by anyone at sska-tool.rics.org. Also, anyone can train as an assessor and RICS provides a one day Foundation course either undertaken via face-to-face training or via the RICS Online Academy. Then there are separate online modules for the Offices, Retail and the new Higher Education Scheme.

There are currently over 200 accredited SKA assessors and the numbers are steadily growing. Qualified SKA assessors can assess projects and can generate certificates once they have demonstrated that the project meets the SKA criteria.

SKA covers the following topics and issues:

- Reducing energy and water use by selecting efficient equipment and promoting metering etc.
- Selecting materials with lower environmental impact.
- Reducing construction site impacts, including waste.
- Reducing pollutants such as refrigerant leakage.
- Promoting health and wellbeing in the working environment, including improving internal air quality, daylight, etc.
- Rewarding more sustainable project delivery, such as registration to the Considerate Constructors Scheme and seasonal commissioning.

These issues are addressed by providing a long list of good practice measures that are rewarded if the project demonstrates implementation. For example, SKA aims to reduce energy use in the completed fit-out by promoting the use of energy efficient equipment.

The installation of more efficient equipment means that the fit-out can use less energy, providing it is well managed and the use (or the intensity of use) has not changed. A British Council of Shopping Centres' report on Low Carbon Fit-out (Cutting Carbon, Cutting Costs: Achieving Performance in Retail Fit-outs 2013) includes a case study that demonstrates the potential savings. Nationwide moved its premises in Oxford from one location to another on the same street (Queen Street) in Oxford. The fit-out was specifically designed to be low carbon and achieved a SKA Gold rating. Figure 2 shows a comparison between the energy use of the old and new store. The case study shows that the energy use is more than 40% lower after the fit-out of the new store.

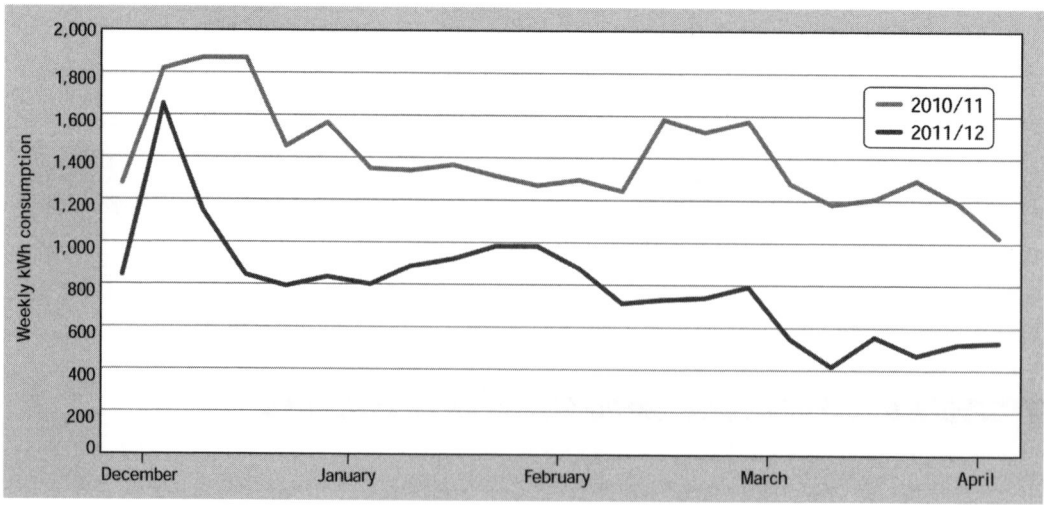

Figure 2: Comparison of energy use before and after fit-out of a Nationwide branch.

Fit-out projects are typically of a short duration and vary widely in scope from a quick refresh to a minor refurbishment or a comprehensive fit-out programme. Other environmental assessment tools tend to employ a 'one size fits all' approach and they assess the whole building or rely upon the basebuild to complete the assessment. SKA has a flexible scope as it assesses the project rather than the whole building. This means that the project team can get rewarded for implementing a good practice fit-out, even if the building has poor transport links or if it has inefficient central air conditioning plant.

SKA broke new ground when the offices scheme was launched in 2009 and again in 2011 when the retail version was launched. The aim is to increase the adoption of SKA, particularly in retail and to continue to push out sustainable good practice into other sectors.

Higher education

RICS worked with the Association of Directors of Estates (AUDE) to develop SKA Higher Education. The scheme is based on SKA rating methodology and has been evolved to reflect the specific requirements of a Higher Education estate including:

- lecture theatres
- laboratories
- and other specialist educational uses

Over 30 universities have been using the SKA on-line tool for fit-outs of their office areas. So the launch of the scheme will allow other parts of the estate to be formally assessed and certified. The scheme has been designed by the industry, for the industry to allow the sector to benchmark their fit-outs against each other and promote good practice.

Product labelling

Product selection is a key part of SKA. This includes everything from procuring energy efficient technologies through to using furniture that have low environmental impacts during manufacture and high levels of recycled material. Grigoriou Interiors, a SKA development partner, has a label and directory for products that are proven to be SKA-compliant. This will allow designers to find and specify products much more quickly and it will make the assessment process even simpler than it is now. For more information see: http://www.rics.org/uk/knowledge/ska-rating-/ska-rating-product-compliance-label/.

Conclusion

SKA has been a great success story to date and has helped to promote sustainable good practice in fit-out and refurbishment projects. It can provide tangible and reputational benefits to occupiers and other property stakeholders and the ultimate aim of SKA is to get it widely adopted to the point that it is embedded in all projects as part of the established process.

For more information see: http://www.rics.org/uk/knowledge/ska-rating-/.

Electricians' On-Site Companion

Christopher Kitcher

This book contains everything electricians need to know about working on site, covering not only the health and safety aspects of site work, but also the techniques and testing knowledge required from the modern-day electrician. Regulations issues are included alongside step-by-step instructions for each task, after which testing information, checklists and example forms are given so that site workers can ensure they have done everything required of them.

October 2017: 234 x 156 mm: 174 pp
Pb: 978-1-138-68332-7 : £24.99

To Order: Tel: +44 (0) 1235 400524 Fax: +44 (0) 1235 400525
or Post: Taylor and Francis Customer Services,
Bookpoint Ltd, Unit T1, 200 Milton Park, Abingdon, Oxon, OX14 4TA UK
Email: book.orders@tandf.co.uk

For a complete listing of all our titles visit:
www.tandf.co.uk

The Aggregates Levy

The Aggregates Levy came into operation on 1 April 2002 in the UK, except for Northern Ireland where it has been phased in over five years from 2003.

It was introduced to ensure that the external costs associated with the exploitation of aggregates are reflected in the price of aggregate, and to encourage the use of recycled aggregate. There continues to be strong evidence that the levy is achieving its environmental objectives, with sales of primary aggregate down and production of recycled aggregate up. The Government expects that the rates of the levy will at least keep pace with inflation over time, although it accepts that the levy is still bedding in.

The rate of the levy will continue to be £2.00 per tonne from 1 April 2017 and is levied on anyone considered to be responsible for commercially exploiting 'virgin' aggregates in the UK and should naturally be passed by price increase to the ultimate user.

All materials falling within the definition of 'Aggregates' are subject to the levy unless specifically exempted.

It does not apply to clay, soil, vegetable or other organic matter.

The intention is that it will:

- Encourage the use of alternative materials that would otherwise be disposed of to landfill sites
- Promote development of new recycling processes, such as using waste tyres and glass
- Promote greater efficiency in the use of virgin aggregates
- Reduce noise and vibration, dust and other emissions to air, visual intrusion, loss of amenity and damage to wildlife habitats

Definitions

'Aggregates' means any rock, gravel or sand which is extracted or dredged in the UK for aggregates use. It includes whatever substances are for the time being incorporated in it or naturally occur mixed with it.

'Exploitation' is defined as involving any one or a combination of any of the following:

- Being removed from its original site, a connected site which is registered under the same name as the originating site or a site where it had been intended to apply an exempt process to it, but this process was not applied
- Becoming subject to a contract or other agreement to supply to any person
- Being used for construction purposes
- Being mixed with any material or substance other than water, except in permitted circumstances

The definition of 'aggregate being used for construction purposes' is when it is:

- Used as material or support in the construction or improvement of any structure
- Mixed with anything as part of a process of producing mortar, concrete, tarmacadam, coated roadstone or any similar construction material

Incidence

It is a tax on primary aggregates production – i.e. 'virgin' aggregates won from a source and used in a location within the UK territorial boundaries (land or sea). The tax is not levied on aggregates which are exported or on aggregates imported from outside the UK territorial boundaries.

It is levied at the point of sale.

Exemption from tax

An 'aggregate' is exempt from the levy if it is:

- Material which has previously been used for construction purposes
- Aggregate that has already been subject to a charge to the Aggregates Levy
- Aggregate which was previously removed from its originating site before the start date of the levy
- Aggregate which is moved between sites under the same Aggregates Levy Registration
- Aggregate which is removed to a registered site to have an exempt process applied to it
- Aggregate which is removed to any premises where china clay or ball clay will be extracted from the aggregate
- Aggregate which is being returned to the land from which it was won provided that it is not mixed with any material other than water
- Aggregate won from a farm land or forest where used on that farm or forest
- Rock which has not been subjected to an industrial crushing process
- Aggregate won by being removed from the ground on the site of any building or proposed building in the course of excavations carried out in connection with the modification or erection of the building and exclusively for the purpose of laying foundations or of laying any pipe or cable
- Aggregate won by being removed from the bed of any river, canal or watercourse or channel in or approach to any port or harbour (natural or artificial), in the course of carrying out any dredging exclusively for the purpose of creating, restoring, improving or maintaining that body of water
- Aggregate won by being removed from the ground along the line of any highway or proposed highway in the course of excavations for improving, maintaining or constructing the highway otherwise than purely to extract the aggregate
- Drill cuttings from petroleum operations on land and on the seabed
- Aggregate resulting from works carried out in exercise of powers under the New Road and Street Works Act 1991, the Roads (Northern Ireland) Order 1993 or the Street Works (Northern Ireland) Order 1995
- Aggregate removed for the purpose of cutting of rock to produce dimension stone, or the production of lime or cement from limestone
- Aggregate arising as a waste material during the processing of the following industrial minerals:
    - anhydrite
    - ball clay
    - barytes
    - calcite
    - china clay
    - clay, coal, lignite and slate
    - feldspar
    - flint
    - fluorspar
    - fuller's earth
    - gems and semi-precious stones
    - gypsum
    - any metal or the ore of any metal
    - muscovite
    - perlite
    - potash
    - pumice
    - rock phosphates

- sodium chloride
- talc
- vermiculite
- spoil from the separation of the above industrial minerals from other rock after extraction
- material that is mainly but not wholly the spoil, waste or other by-product of any industrial combustion process or the smelting or refining of metal

Anything that consists 'wholly or mainly' of the following is exempt from the levy (note that 'wholly' is defined as 100% but 'mainly' as more than 50%, thus exempting any contained aggregates amounting to less than 50% of the original volumes:

- clay, soil, vegetable or other organic matter
- drill cuttings from oil exploration in UK waters
- material arising from utility works, if carried out under the New Roads and Street Works Act 1991

However, when ground that is more than half clay is mixed with any substance (for example, cement or lime) for the purpose of creating a firm base for construction, the clay becomes liable to Aggregates Levy because it has been mixed with another substance for the purpose of construction.

Anything that consists completely of the following substances is exempt from the levy:

- Spoil, waste or other by-products from any industrial combustion process or the smelting or refining of metal – for example, industrial slag, pulverized fuel ash and used foundry sand. If the material consists completely of these substances at the time it is produced it is exempt from the levy, regardless of any subsequent mixing
- Aggregate necessarily arising from the footprint of any building for the purpose of laying its foundations, pipes or cables. It must be lawfully extracted within the terms of any planning consent
- Aggregate necessarily arising from navigation dredging
- Aggregate necessarily arising from the ground in the course of excavations to improve, maintain or construct a highway or a proposed highway
- Aggregate necessarily arising from the ground in the course of excavations to improve, maintain or construct a railway, monorail or tramway

Relief from the levy either in the form of credit or repayment is obtainable where:

- it is subsequently exported from the UK in the form of aggregate
- it is used in an exempt process
- where it is used in a prescribed industrial or agricultural process
- it is waste aggregate disposed of by dumping or otherwise, e.g. sent to landfill or returned to the originating site

The Aggregates Levy Credit Scheme (ALCS) for Northern Ireland was suspended with effect from 1 December 2010 following a ruling by the European General Court.

An exemption for aggregate obtained as a by-product of railway, tramway and monorail improvement, maintenance and construction was introduced in 2007.

Exemptions to the levy were suspended on 1 April 2014, following an investigation by the European Commission into whether they were lawful under State aid rules. The Commission announced its decision on 27 March 2015 that all but part of one exemption (for shale) were lawful and all the exemptions have, therefore, been reinstated apart from the exemption for shale. The effective date of reinstatement is 1 April 2014, which means that businesses which paid Aggregates Levy on materials for which the exemption has been confirmed as lawful may claim back the tax they paid while the exemption was suspended.

However, under EU law the UK government is required to recover unlawful State aid with interest from businesses that benefited from it. HM Revenue & Customs therefore initiated a process in 2015 of clawing back the levy, plus compound interest, for the deliberate extraction of shale aggregate for commercial exploitation from businesses it believes may have benefited between 1 April 2002 and 31 March 2014.

Discounts

Water which is added to the aggregate after the aggregate has been won (washing, dust dampening etc.) may be discounted from the tax calculations. There are two accepted options by which the added water content can be calculated.

The first is to use HMRC's standard added water percentage discounts listed below:

- washed sand 7%
- washed gravel 3.5%
- washed rock/aggregate 4%

Alternatively a more exact percentage can be agreed for dust dampening of aggregates.

Whichever option is adopted, it must be agreed in writing in advance with HMRC.

Impact

The British Aggregates Association suggested that the additional cost imposed by quarries is more likely to be in the order of £3.40 per tonne on mainstream products, applying an above average rate on these in order that by-products and low grade waste products can be held at competitive rates, as well as making some allowance for administration and increased finance charges.

With many gravel aggregates costing in the region of £20.00 per tonne, there is a significant impact on construction costs.

Avoidance

An alternative to using new aggregates in filling operations is to crush and screen rubble which may become available during the process of demolition and site clearance as well as removal of obstacles during the excavation processes.

Example: Assuming that the material would be suitable for fill material under buildings or roads, a simple cost comparison would be as follows (note that for the purpose of the exercise, the material is taken to be 1.80 tonne per m³ and the total quantity involved less than 1,000 m³):

Importing fill material:	£/m³	£/tonne
Cost of 'new' aggregates delivered to site	37.10	20.16
Addition for Aggregates Tax	3.60	2.00
Total cost of importing fill materials	40.70	22.61

Disposing of site material:	£/m³	£/tonne
Cost of removing materials from site	26.63	14.79

Crushing site materials:	£/m³	£/tonne
Transportation of material from excavations or demolition to stockpiles	0.88	0.49
Transportation of material from temporary stockpiles to the crushing plant	2.36	1.31
Establishing plant and equipment on site; removing on completion	2.36	1.31
Maintain and operate plant	10.62	5.90
Crushing hard materials on site	15.34	8.52
Screening material on site	2.36	1.31
Total cost of crushing site materials	33.92	18.84

From the above it can be seen that potentially there is a great benefit in crushing site materials for filling rather than importing fill materials.

Setting the cost of crushing against the import price would produce a saving of £6.78 per m³. If the site materials were otherwise intended to be removed from the site, then the cost benefit increases by the saved disposal cost to £33.41 per m³.

Even if there is no call for any or all of the crushed material on site, it ought to be regarded as a useful asset and either sold on in crushed form or else sold with the prospects of crushing elsewhere.

Specimen unit rates	Unit	£
Establishing plant and equipment on site; removing on completion		
crushing plant	trip	1,400.00
screening plant	trip	700.00
Maintain and operate plant		
crushing plant	week	8,500.00
screening plant	week	2,100.00
Transportation of material from excavations or demolition places to temporary stockpiles	m³	3.50
Transportation of material from temporary stockpiles to the crushing plant	m³	2.80
Breaking up material on site using impact breakers		
mass concrete	m³	16.50
reinforced concrete	m³	19.00
brickwork	m³	7.00

Specimen unit rates	Unit	£
Crushing material on site		
mass concrete not exceeding 1000 m³	m³	15.00
mass concrete 1000–5000 m³	m³	14.00
mass concrete over 5000 m³	m³	13.00
reinforced concrete not exceeding 1000 m³	m³	18.00
reinforced concrete 1000–5000 m³	m³	16.00
reinforced concrete over 5000 m³	m³	15.00
brickwork not exceeding 1000 m³	m³	14.00
brickwork 1000–5000 m³	m³	13.00
brickwork over 5000 m³	m³	12.00
Screening material on site	m³	2.50

More detailed information can be found on the HMRC website (www.hmrc.gov.uk) in Notice AGL 1 Aggregates Levy published 1 April 2014 (updated 9 February 2017).

Value Added Tax

Introduction

Value Added Tax (VAT) is a tax on the consumption of goods and services. The UK introduced a domestic VAT regime when it joined the European Community in 1973. The principal source of European law in relation to VAT is Council Directive 2006/112/EC, a recast of Directive 77/388/EEC, which is currently restated and consolidated in the UK through the VAT Act 1994 and various Statutory Instruments, as amended by subsequent Finance Acts.

VAT Notice 708: Buildings and construction (August 2016) provides HMRC's interpretation of the VAT law in connection with construction works, however, the UK VAT legislation should always be referred to in conjunction with the publication. Recent VAT tribunals and court decisions since the date of this publication will affect the application of the VAT law in certain instances. The Notice is available on HM Revenue & Customs website at www.hmrc.gov.uk.

The scope of VAT

VAT is payable on:

- Supplies of goods and services made in the UK;
- By a taxable person;
- In the course or furtherance of business; and
- Which are not specifically exempted or zero-rated.

Rates of VAT

There are three rates of VAT:

- A standard rate, currently 20% since January 2011;
- A reduced rate, currently 5%; and
- A zero rate of 0%.

Additionally some supplies are exempt from VAT and others are considered outside the scope of VAT.

Recovery of VAT

When a taxpayer makes taxable supplies he must account for VAT, known as output VAT at the appropriate rate of 20%, 5% or 0%. Any VAT due then has to be declared and submitted on a VAT submission to HM Revenue & Customs and will normally be charged to the taxpayer's customers.

As a VAT registered person, the taxpayer is entitled to reclaim from HM Revenue & Customs, commonly referred to as input VAT the VAT incurred on their purchases and expenses directly related to its business activities in respect of a standard-rated, reduced-rated and zero-rated supplies. A taxable person cannot however reclaim VAT that relates to any non-business activities (but see below) or depending on the amount of exempt supplies they made input VAT may be restricted or not recoverable.

At predetermined intervals the taxpayer will pay to HM Revenue & Customs the excess of VAT collected over the VAT they can reclaim. However if the VAT reclaimed is more than the VAT collected, the taxpayer who will be a net repayment position can reclaim the difference from HM Revenue & Customs.

Example

X Ltd constructs a block of flats. It sells long leases to buyers for a premium. X Ltd has constructed a new building designed as a dwelling and will have granted a long lease. This first sale of a long lease is VAT zero-rated supply. This means any VAT incurred in connection with the development which X Ltd will have paid (e.g. payments for consultants and certain preliminary services) will be recoverable. For reasons detailed below the contractor employed by X Ltd will not have charged VAT on his construction services as these should be zero-rated.

Use for business and non-business activities

Where a supply relates partly to business use and partly to non-business use then the basic rule is that it must be apportioned on a fair and reasonable basis so that only the business element is potentially recoverable. In some cases VAT on land, buildings and certain construction services purchased for both business and non-business use could be recovered in full by applying what is known as 'Lennartz' accounting to reclaim VAT relating to the non-business use and account for VAT on the non-business use over a maximum period of 10 years. Following an ECJ case restricting the scope of this approach, its application to immovable property was removed completely in January 2011 by HMRC (business brief 53/10) when UK VAT law was amended to comply with EU Directive 2009/162/EU.

Taxable persons

A taxable person is an individual, firm, company etc. who is required to be registered for VAT. A person who makes taxable supplies above certain turnover limits is compulsorily required to be VAT registered. From 1 April 2017, the current registration limit known as the VAT threshold is £85,000. If the threshold is exceeded in any 12 month rolling period, or there is an expectation that the value of the taxable supplies in a single 30 day period, or you receive goods into the UK from the EU worth more than the £85,000, then you must register for UK VAT.

A person who makes taxable supplies below the limit is still entitled to be registered on a voluntary basis if they wish, for example, in order to recover input VAT incurred in relation to those taxable supplies, however output VAT will then become due on the sales and must be accounted for.

In addition, a person who is not registered for VAT in the UK but acquires goods from another EC member state, or make distance sales in the UK, above certain value limits may be required to register for VAT in the UK.

VAT exempt supplies

Where a supply is exempt from VAT this means that no output VAT is payable – but equally the person making the exempt supply cannot normally recover any of the input VAT on their own costs relating to that exempt supply.

Generally commercial property transactions such as leasing of land and buildings are exempt unless a landlord chooses to standard-rate its interest in the property by a applying for an option to tax. This means that VAT is added to rental income and also that VAT incurred, on, say, an expensive refurbishment, is recoverable.

Supplies outside the scope of VAT

Supplies are outside the scope of VAT if they are:

- Made by someone who is not a taxable person;
- Made outside the UK; or
- Not made in the course or furtherance of business.

In course or furtherance of business

VAT must be accounted for on all taxable supplies made in the course or furtherance of business with the corresponding recovery of VAT on expenditure incurred.

If a taxpayer also carries out non-business activities then VAT incurred in relation to such supplies is generally not recoverable.

In VAT terms, business means any activity continuously performed which is mainly concerned with making supplies for a consideration. This includes:

- Any one carrying on a trade, vocation or profession;
- The provision of membership benefits by clubs, associations and similar bodies in return for a subscription or other consideration; and
- Admission to premises for a charge.

It may also include the activities of other bodies including charities and non-profit making organizations.

Examples of non-business activities are:

- Providing free services or information;
- Maintaining some museums or particular historic sites;
- Publishing religious or political views.

Construction services

In general the provision of construction services by a contractor will be VAT standard rated at 20%, however, there are a number of exceptions for construction services provided in relation to certain relevant residential properties and charitable buildings.

The supply of building materials is VAT standard rated at 20%, however, where these materials are supplied and installed as part of the construction services the VAT liability of those materials follows that of the construction services supplied.

Zero-rated construction services

The following construction services are VAT zero-rated including the supply of related building materials.

The construction of new dwellings

The supply of services in the course of the construction of a new building designed for use as a dwelling or number of dwellings is zero-rated other than the services of an architect, surveyor or any other person acting as a consultant or in a supervisory capacity.

The following basic conditions must ALL be satisfied in order for the works to qualify for zero-rating:

1. A qualifying building has been, is being or will be constructed;
2. Services are made 'in the course of the construction' of that building;
3. Where necessary, you hold a valid certificate;
4. Your services are not specifically excluded from zero-rating.

The construction of a new building for 'relevant residential or charitable' use

The supply of services in the course of the construction of a building designed for use as a Relevant Residential Purpose (RRP) or Relevant Charitable Purpose (RCP), is zero-rated other than the services of an architect, surveyor or any other person acting as a consultant or in a supervisory capacity.

A 'relevant residential' use building means:

1.   A home or other institution providing residential accommodation for children;
2.   A home or other institution providing residential accommodation with personal care for persons in need of personal care by reason of old age, disablement, past or present dependence on alcohol or drugs or past or present mental disorder;
3.   A hospice;
4.   Residential accommodation for students or school pupils;
5.   Residential accommodation for members of any of the armed forces;
6.   A monastery, nunnery, or similar establishment; or
7.   An institution which is the sole or main residence of at least 90% of its residents.

A 'relevant residential' purpose building does not include use as a hospital, a prison or similar institution or as a hotel, inn or similar establishment.

A 'relevant charitable' purpose means use by a charity in either or both of the following ways:

1.   Otherwise than in the course or furtherance of a business; or
2.   As a village hall or similarly in providing social or recreational facilities for a local community.

Non-qualifying use which is not expected to exceed 10% of the time the building is normally available for use can be ignored. The calculation of business use can be based on time, floor area or head count subject to approval being acquired from HM Revenue & Customs.

The construction services can only be zero-rated if a certificate is given by the end user to the contractor carrying out the works confirming that the building is to be used for a qualifying purpose i.e. for a 'relevant residential or charitable' purpose. It follows that such services can only be zero-rated when supplied to the end user and, unlike supplies relating to dwellings, supplies by subcontractors cannot be zero-rated.

The construction of an annex used for a 'relevant charitable' purpose

Construction services provided in the course of construction of an annexe for use entirely or partly for a 'relevant charitable' purpose can be zero-rated.

In order to qualify the annexe must:

1.   Be capable of functioning independently from the existing building;
2.   Have its own main entrance; and
3.   Be covered by a qualifying use certificate.

The conversion of a non-residential building into dwellings or the conversion of a building from non-residential use to 'relevant residential' use where the supply is to a 'relevant' housing association

The supply to a 'relevant' housing association in the course of conversion of a non-residential building or non-residential part of a building into:

1.   A new eligible dwelling designed as a dwelling or number of dwellings; or
2.   A building or part of a building for use solely for a relevant residential purpose, of any services related to the conversion other than the services of an architect, surveyor or any person acting as a consultant or in a supervisory capacity are zero-rated.

A 'relevant' housing association is defined as:

1. A private registered provider of social housing;
2. A registered social landlord within the meaning of Part I of the Housing Act 1996 (Welsh registered social landlords);
3. A registered social landlord within the meaning of the Housing (Scotland) Act 2001 (Scottish registered social landlords); or
4. A registered housing association within the meaning of Part II of the Housing (Northern Ireland) Order 1992 (Northern Irish registered housing associations).

If the building is to be used for a 'relevant residential' purpose the housing association should issue a qualifying use certificate to the contractor completing the works. Subcontractors services that are not made directly to a relevant housing association are standard-rated.

The development of a residential caravan parks

The supply in the course of the construction of any civil engineering work 'necessary for' the development of a permanent park for residential caravans of any services related to the construction are zero-rated when a new permanent park is being developed, the civil engineering works are necessary for the development of the park and the services are not specifically excluded from zero-rating. This includes access roads, paths, drainage, sewerage and the installation of mains water, power and gas supplies.

Certain building alterations for disabled persons

Certain goods and services supplied to a disabled person, or a charity making these items and services available to disabled persons can be zero-rated. The recipient of these goods or services needs to give the supplier an appropriate written declaration that they are entitled to benefit from zero rating.

The following services (amongst others) are zero-rated:

1. The installation of specialist lifts and hoists and their repair and maintenance;
2. The construction of ramps, widening doorways or passageways including any preparatory work and making good work;
3. The provision, extension and adaptation of a bathroom, washroom or lavatory; and
4. Emergency alarm call systems.

Approved alterations to protected buildings

The zero rate for approved alterations to protected buildings was withdrawn from 1 October 2012, other than for projects where a contract was entered into or where listed building consent (or equivalent approval for listed places of worship) had been applied for before 21 March 2012.

Provided the application was in place before 21 March 2012, zero rating will continue under the transitional rules until 30 September 2015.

All other projects will be subject to the standard rate of VAT on or after 1 October 2012.

Sale of reconstructed buildings

Since 1 October 2012 a protected building shall not be regarded as substantially reconstructed unless, when the reconstruction is completed, the reconstructed building incorporates no more of the original building than the external walls, together with other external features of architectural or historical interest. Transitional arrangements protect contracts entered into before 21 March 2012 for the first grant of a major interest in the protected building made on or before 20 March 2013.

DIY builders and converters

Private individuals who decide to construct their own home are able to reclaim VAT they pay on goods they use to construct their home by use of a special refund mechanism made by way of an application to HM Revenue & Customs. This also applies to services provided in the conversion of an existing non-residential building to form a new dwelling.

The scheme is meant to ensure that private individuals do not suffer the burden of VAT if they decide to construct their own home.

Charities may also qualify for a refund on the purchase of materials incorporated into a building used for non-business purposes where they provide their own free labour for the construction of a 'relevant charitable' use building.

Reduced-rated construction services

The following construction services are subject to the reduced rate of VAT of 5%, including the supply of related building materials.

Conversion – changing the number of dwellings

In order to qualify for the 5% rate there must be a different number of 'single household dwellings' within a building than there were before commencement of the conversion works. A 'single household dwelling' is defined as a dwelling that is designed for occupation by a single household.

These conversions can be from 'relevant residential' purpose buildings, non-residential buildings and houses in multiple occupation.

A house in multiple occupation conversion

This relates to construction services provided in the course of converting a 'single household dwelling', a number of 'single household dwellings', a non-residential building or a 'relevant residential' purpose building into a house for multiple occupation such as a bed sit accommodation.

A special residential conversion

A special residential conversion involves the conversion of a 'single household dwelling', a house in multiple occupation or a non-residential building into a 'relevant residential' purpose building such as student accommodation or a care home.

Renovation of derelict dwellings

The provision of renovation services in connection with a dwelling or 'relevant residential' purpose building that has been empty for two or more years prior to the date of commencement of construction works can be carried out at a reduced rate of VAT of 5%.

Installation of energy saving materials

A reduced rate of VAT of 5% is paid on the supply and installation of certain energy saving materials including insulation, draught stripping, central heating, hot water controls and solar panels in a residential building or a building used for a relevant charitable purpose.

Buildings that are used by charities for non-business purposes, and/or as village halls, were removed from the scope of the reduced rate for the supply of energy saving materials under legislation introduced in Finance Bill 2013.

Grant-funded installation of heating equipment or connection of a gas supply

The grant-funded supply and installation of heating appliances, connection of a mains gas supply, supply, installation, maintenance and repair of central heating systems, and supply and installation of renewable source heating systems, to qualifying persons. A qualifying person is someone aged 60 or over or is in receipt of various specified benefits.

Grant-funded installation of security goods

The grant-funded supply and installation of security goods to a qualifying person.

Housing alterations for the elderly

Certain home adaptations that support the needs of elderly people were reduced rated with effect from 1 July 2007.

Building contracts

Design and build contracts

If a contractor provides a design and build service relating to works to which the reduced or zero rate of VAT is applicable then any design costs incurred by the contractor will follow the VAT liability of the principal supply of construction services.

Management contracts

A management contractor acts as a main contractor for VAT purposes and the VAT liability of his services will follow that of the construction services provided. If the management contractor only provides advice without engaging trade contractors his services will be VAT standard rated.

Construction management and project management

The project manager or construction manager is appointed by the client to plan, manage and coordinate a construction project. This will involve establishing competitive bids for all the elements of the work and the appointment of trade contractors. The trade contractors are engaged directly by the client for their services.

The VAT liability of the trade contractors will be determined by the nature of the construction services they provide and the building being constructed.

The fees of the construction manager or project manager will be VAT standard rated. If the construction manager also provides some construction services these works may be zero or reduced rated if the works qualify.

Liquidated and ascertained damages

Liquidated damages are outside of the scope of VAT as compensation. The employer should not reduce the VAT amount due on a payment under a building contract on account of a deduction of damages. In contrast an agreed reduction in the contract price will reduce the VAT amount.

Similarly, in certain circumstances HM Revenue & Customs may agree that a claim by a contractor under a JCT or other form of contract is also compensation payment and outside the scope of VAT.

Spon's Asia Pacific Construction Costs Handbook, Fifth Edition

LANGDON & SEAH

In the last few years, the global economic outlook has continued to be shrouded in uncertainty and volatility following the financial crisis in the Euro zone. While the US and Europe are going through a difficult period, investors are focusing more keenly on Asia. This fifth edition provides overarching construction cost data for 16 countries: Brunei, Cambodia, China, Hong Kong, India, Indonesia, Japan, Malaysia, Myanmar, Philippines, Singapore, South Korea, Sri Lanka, Taiwan, Thailand and Vietnam.

April 2015: 234X156 mm: 452 pp
Hb: 978-1-4822-4358-1: £195.00

Typical Engineering Details

In addition to the Engineering Features, Typical Engineering Details are included. These are indicative schematics to assist in the compilation of costing exercises. The user should note that these are only examples and cannot be construed to reflect the design for each and every situation. They are merely provided to assist the user with gaining an understanding of the Engineering concepts and elements making up such.

ELECTRICAL

- Urban Network Mainly Underground
- Urban Network Mainly Underground with Reinforcement
- Urban Network Mainly Underground with Substation Reinforcement
- Typical Simple 11 kV Network Connection For LV Intakes Up To 1000 kVA
- Typical 11 kV Network Connections For HV Intakes 1000 kVA To 6000 kVA
- Static UPS System – Simplified Single Line Schematic For a Single Module
- Typical Data Transmission (Structured Cabling)
- Typical Networked Lighting Control System
- Typical Standby Power System, Single Line Schematic
- Typical Fire Detection and Alarm Schematic
- Typical Block Diagram – Access Control System (ACS)
- Typical Block Diagram – Intruder Detection System (IDS)
- Typical Block Diagram – Digital CCTV

MECHANICAL

- Fan Coil Unit System
- Displacement System
- Chilled Ceiling System (Passive System)
- Chilled Beam System (Passive System)
- Variable Air Volume (VAV)
- Variable Refrigerant Volume System (VRV)
- Reverse cycle heat pump

Urban network mainly underground

Details: Connection to small housing development, namely 10 houses, 60 m of LV cable from local 11 kV/400 V substation route in footpath and verge, 10 m of service cable to each plot in verge

Supply Capacity: 200 kVA

Connection Voltage: LV

3 phase supply

Breakdown of detailed cost information

	Labour	Plant	Materials	Overheads	Total
Mains Cable	£300	£100	£ 1,200	£320	£1,920
Service Cable	£300	£100	£500	£180	£1,080
Jointing	£1,200	£400	£1,500	£300	£3400
Termination	Incl.	Incl.	Incl.	Incl.	Incl.
Trench/Reinstate	£1,800	£600	£1,500	£780	£4,680
Substation	–	–	–	–	–
HV Trench & Joint	–	–	–	–	–
HV Cable Install	–	–	–	–	–
Total Calculated Price	£3,600	£1,200	£4,700	£1,580	£11,080

Total Non-contestable Elements and Associated Charges	£3,000
Grand Total Calculated Price excl. VAT	£14,080

Urban network mainly underground with reinforcement

Details: Connection to small housing development, namely 10 houses, 60 m of LV cable from local 11 kV/400 V substation route in footpath and verge, 10 m of service cable to each plot in unmade ground. Scheme includes reinforcement of LV distribution board at substation

Supply Capacity: 200 kVA

Connection Voltage: LV

Single phase supply

Breakdown of detailed cost information

	Labour	Plant	Materials	Overheads	Total
Mains Cable	£300	£100	£ 1,200	£320	£1,920
Service Cable	£300	£100	£500	£180	£1,080
Jointing	£1,200	£400	£1,500	£300	£3400
Termination	Incl.	Incl.	Incl.	Incl.	Incl.
Trench/Reinstate	£1,800	£600	£1,500	£780	£4,680
Substation	–	–	–	–	–
HV Trench & Joint	–	–	–	–	–
HV Cable Install	–	–	–	–	–
Total Calculated Price	£3,600	£1,200	£4,700	£1,580	£11,080
Total Non-contestable Elements and Associated Charges					£13,000
Grand Total Calculated Price excl. VAT					£24,080

Urban network mainly underground with substation reinforcement

Details: Connection to small housing development, namely 10 houses, 60 m of LV cable from local 11 kV/400 V substation route in footpath and verge, 10 m of service cable to each plot in verge. Scheme includes reinforcement of LV distribution board and new substation and 20 m of HV cable

Supply Capacity: 200 kVA

Connection Voltage: LV

Single phase

Breakdown of detailed cost information

	Labour	Plant	Materials	Overheads	Total
Mains Cable	£300	£100	£ 1,500	£380	£2,280
Service Cable	£300	£100	£500	£180	£1,080
Jointing	£1,200	£400	£1,500	£300	£3400
Termination	Incl.	Incl.	Incl.	Incl.	Incl.
Trench/Reinstate	£1,800	£600	£1,500	£780	£4,680
Substation	£10,000	£3,000	£80,000	£18,600	£111,600
HV Trench & Joint	£1,800	£600	£1,000	£680	£4,080
HV Cable Install	£300	£100	£1,125	£305	£1,830
Total Calculated Price	£15,700	£4,900	£87,125	£21,225	£128,950
Total Non-contestable Elements and Associated Charges					£15,000
Grand Total Calculated Price excl. VAT					£143,950

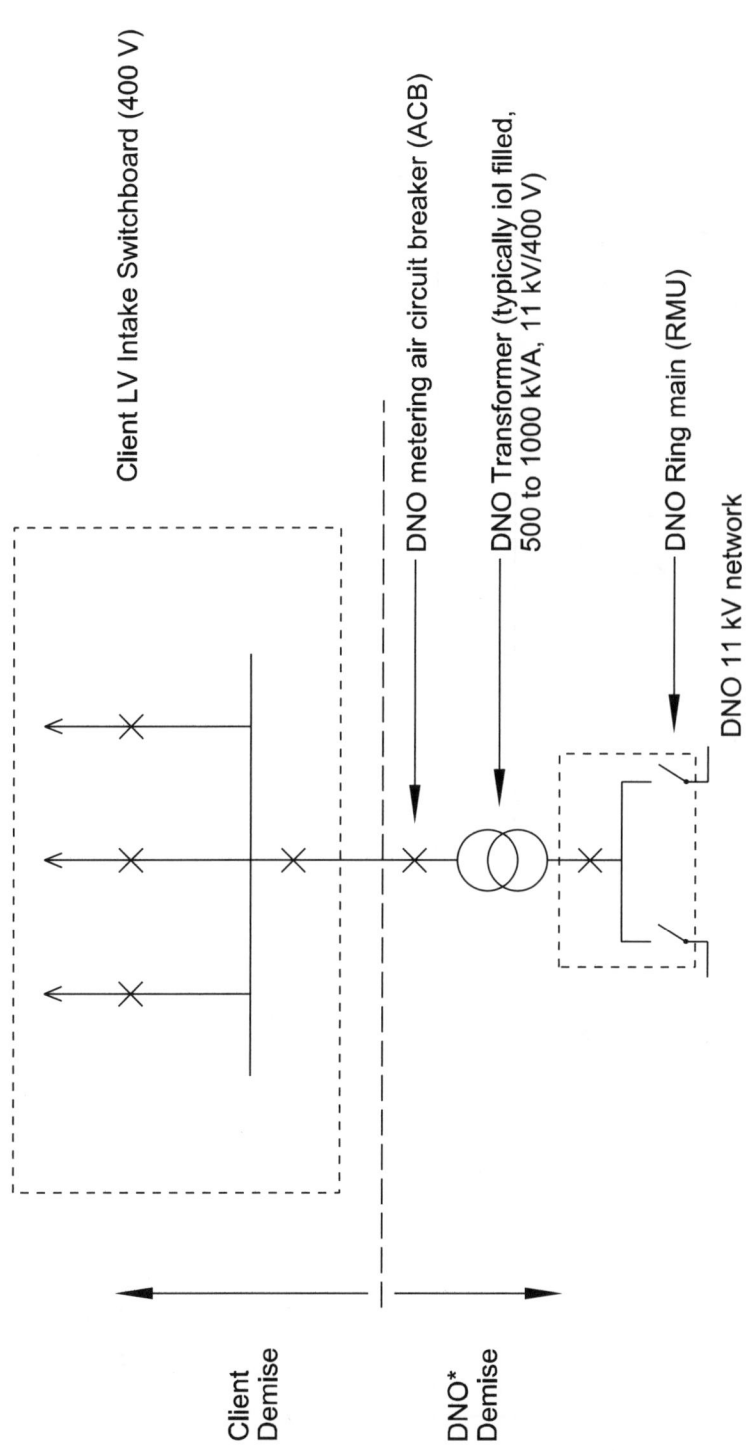

Client LV Intake Switchboard (400 V)

DNO metering air circuit breaker (ACB)

DNO Transformer (typically ioI filled, 500 to 1000 kVA, 11 kV/400 V)

DNO Ring main (RMU)

DNO 11 kV network

Client Demise

DNO* Demise

Note: *DNO – Distribution Network Operator

Typical Simple 11 kV Network Connection for LV Intakes up to 1000 kVA

93

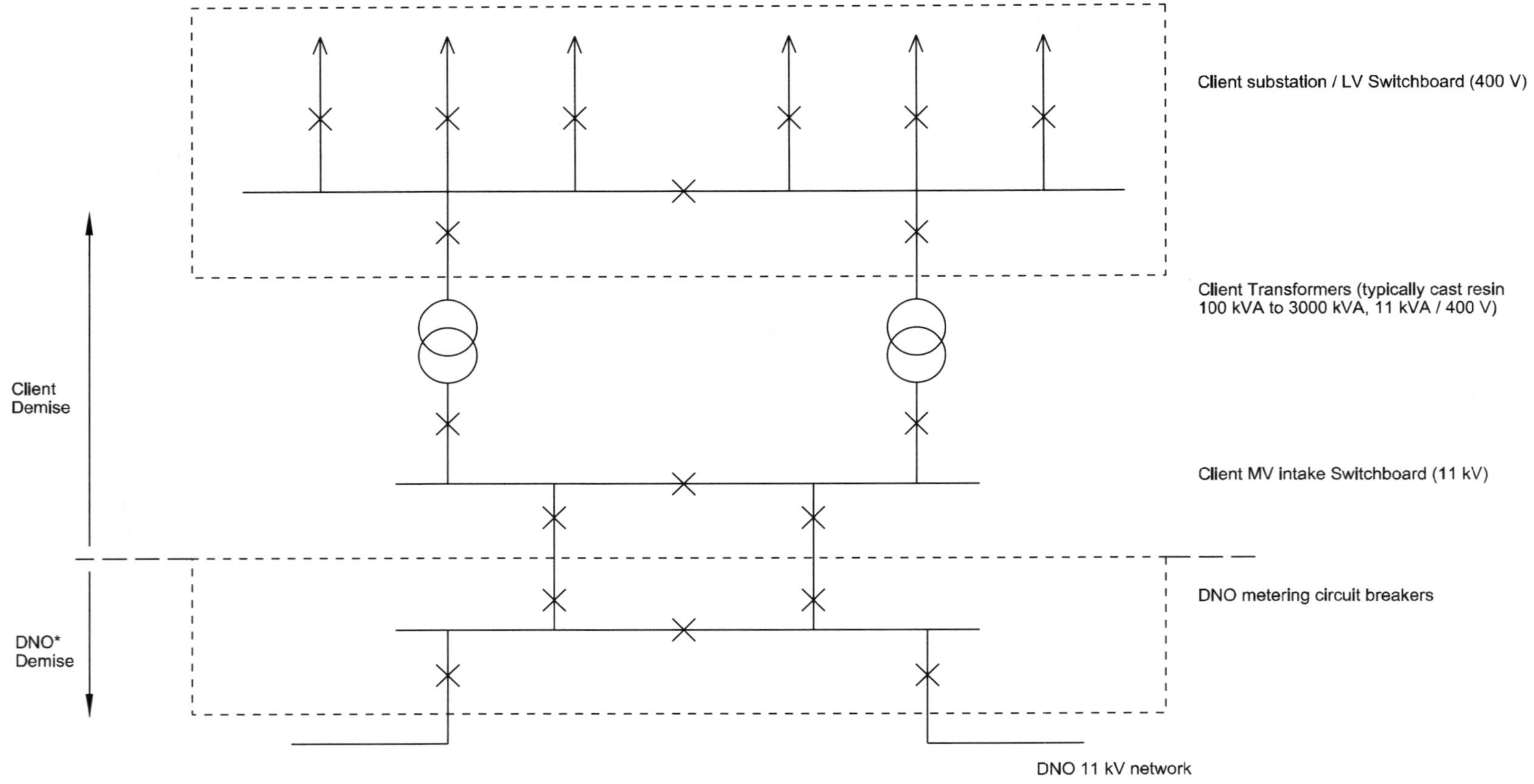

Typical 11 kV Network Connection for MV Intakes 1000 kVA up to 6000 kVA

Client substation / LV Switchboard (400 V)

Client Transformers (typically cast resin 100 kVA to 3000 kVA, 11 kVA / 400 V)

Client MV Intake Switchboard (11 kV)

DNO metering circuit breakers

DNO 11 kV network

Client Demise

DNO* Demise

Note: *DNO - Distribution Network Operator

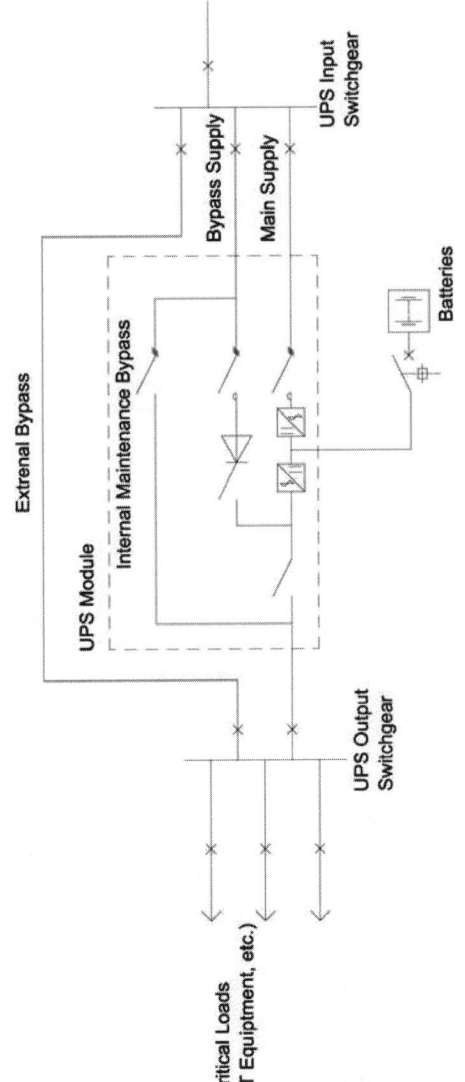

Static UPS System – Simplified Schematic For Single Module

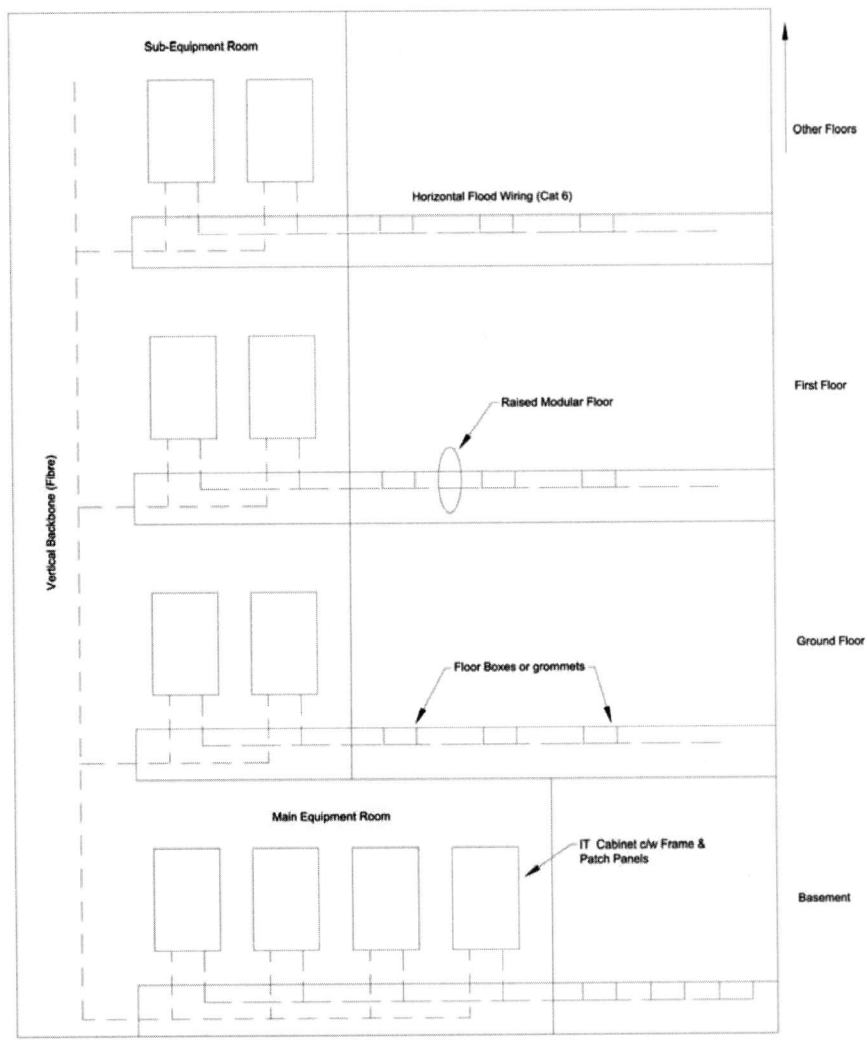

Typical Data Transmission (Structured Cabling)

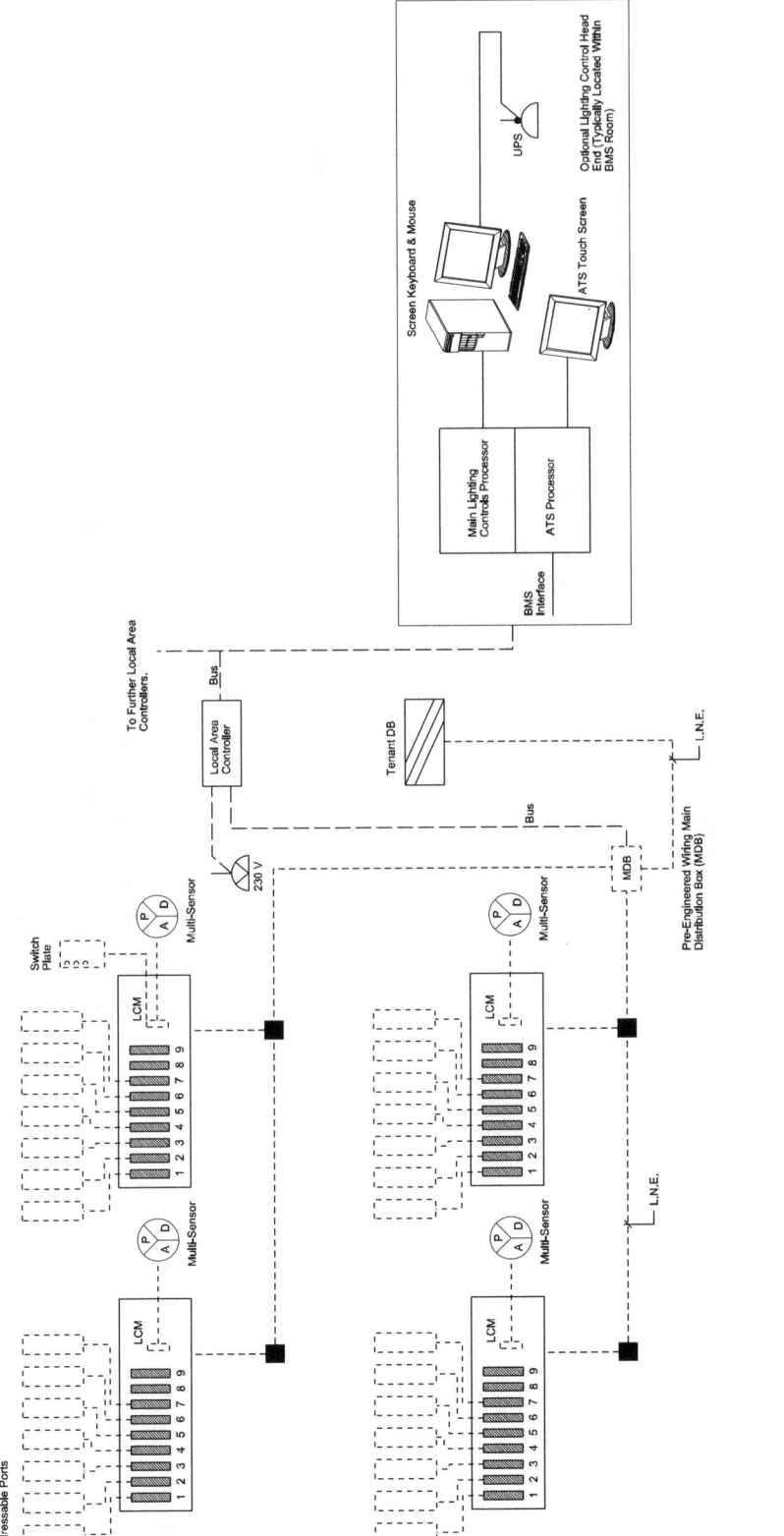

Typical Networked Lighting Control System

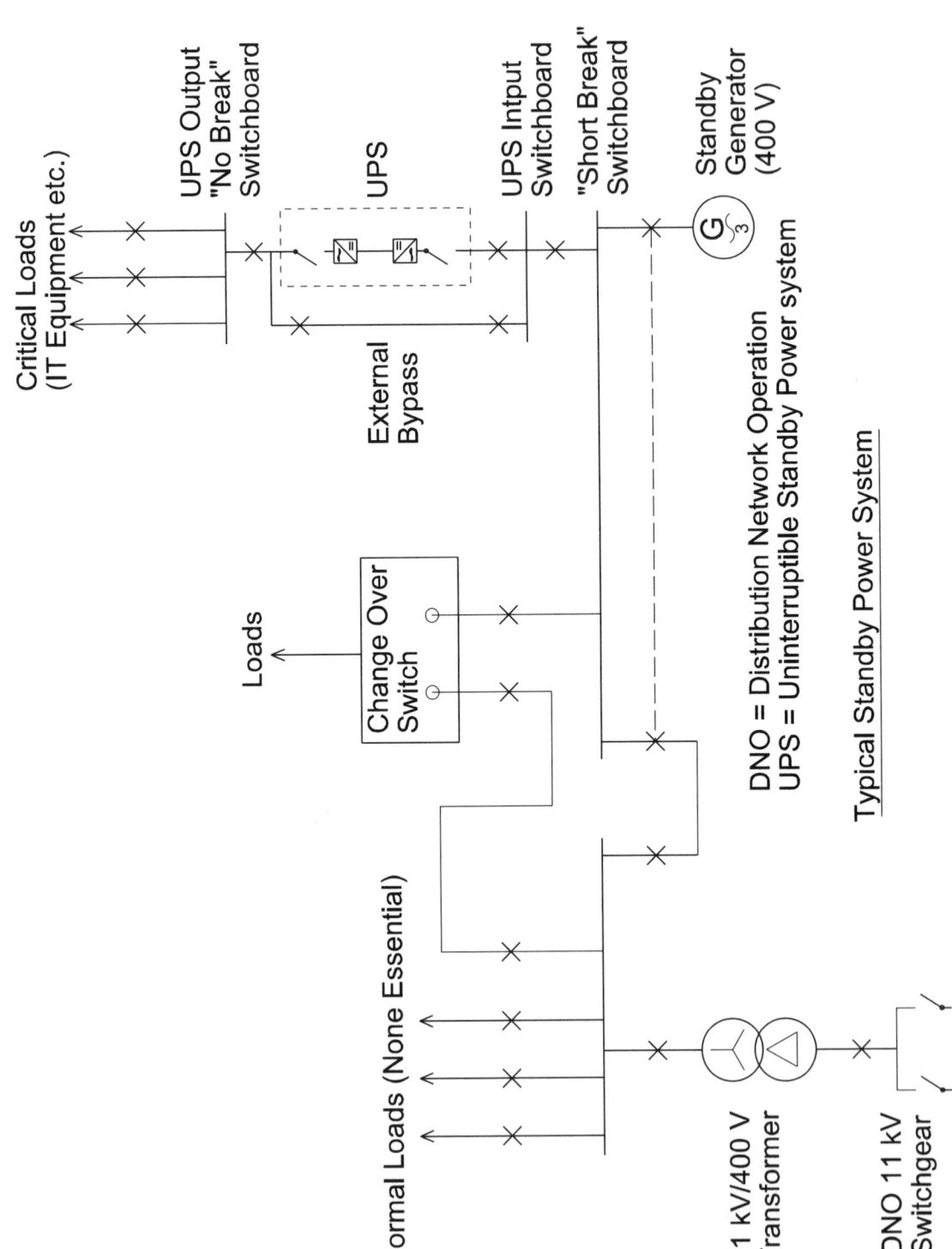

Critical Loads
(IT Equipment etc.)

UPS Output "No Break" Switchboard

UPS

External Bypass

UPS Intput Switchboard

"Short Break" Switchboard

Standby Generator (400 V)

Loads

Change Over Switch

Normal Loads (None Essential)

11 kV/400 V Transformer

DNO 11 kV Switchgear

DNO = Distribution Network Operation
UPS = Uninterruptible Standby Power system

Typical Standby Power System

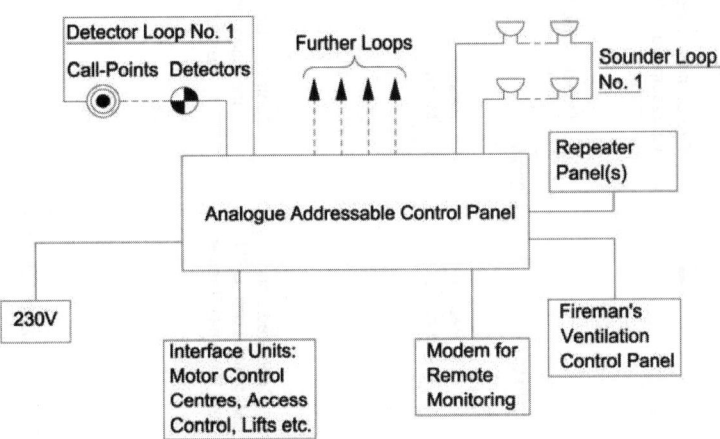

Typical Fire Detection and Alarm Schematic

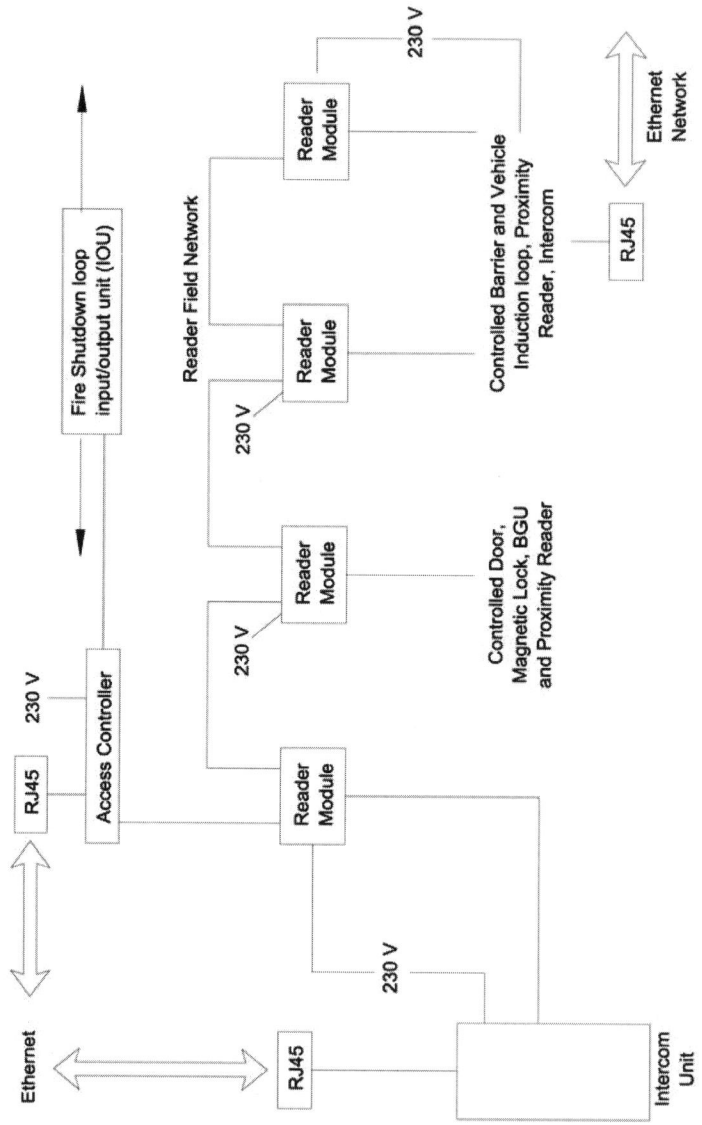

Typical Block Diagram – Access Control System

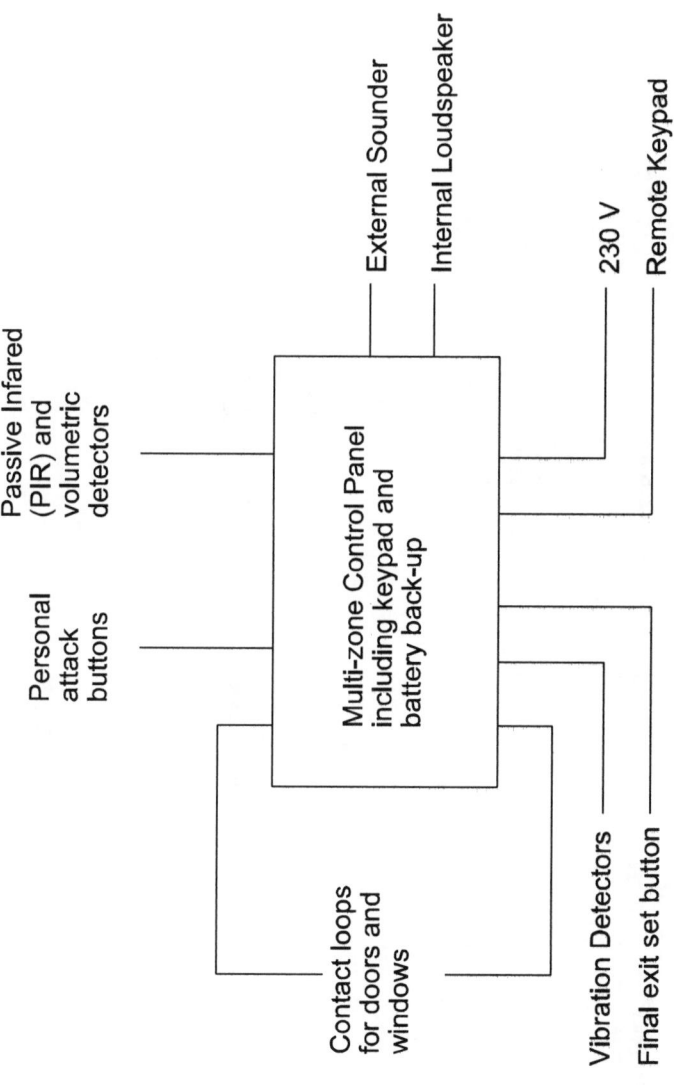

Passive Infared (PIR) and volumetric detectors

Personal attack buttons

Contact loops for doors and windows

Multi-zone Control Panel including keypad and battery back-up

External Sounder

Internal Loudspeaker

230 V

Remote Keypad

Vibration Detectors

Final exit set button

Typical Block Diagram – Intruder Detection System (IDS)

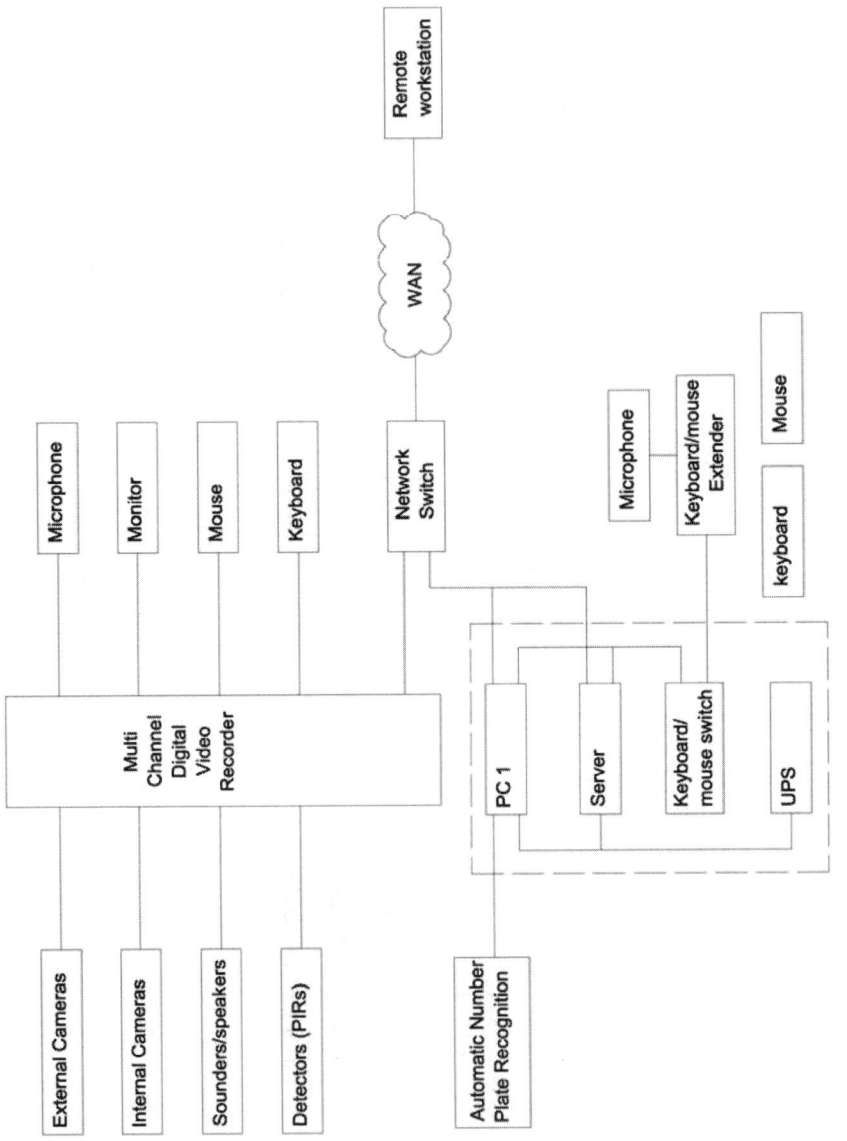

Typical Block Diagram – Digital CCTV

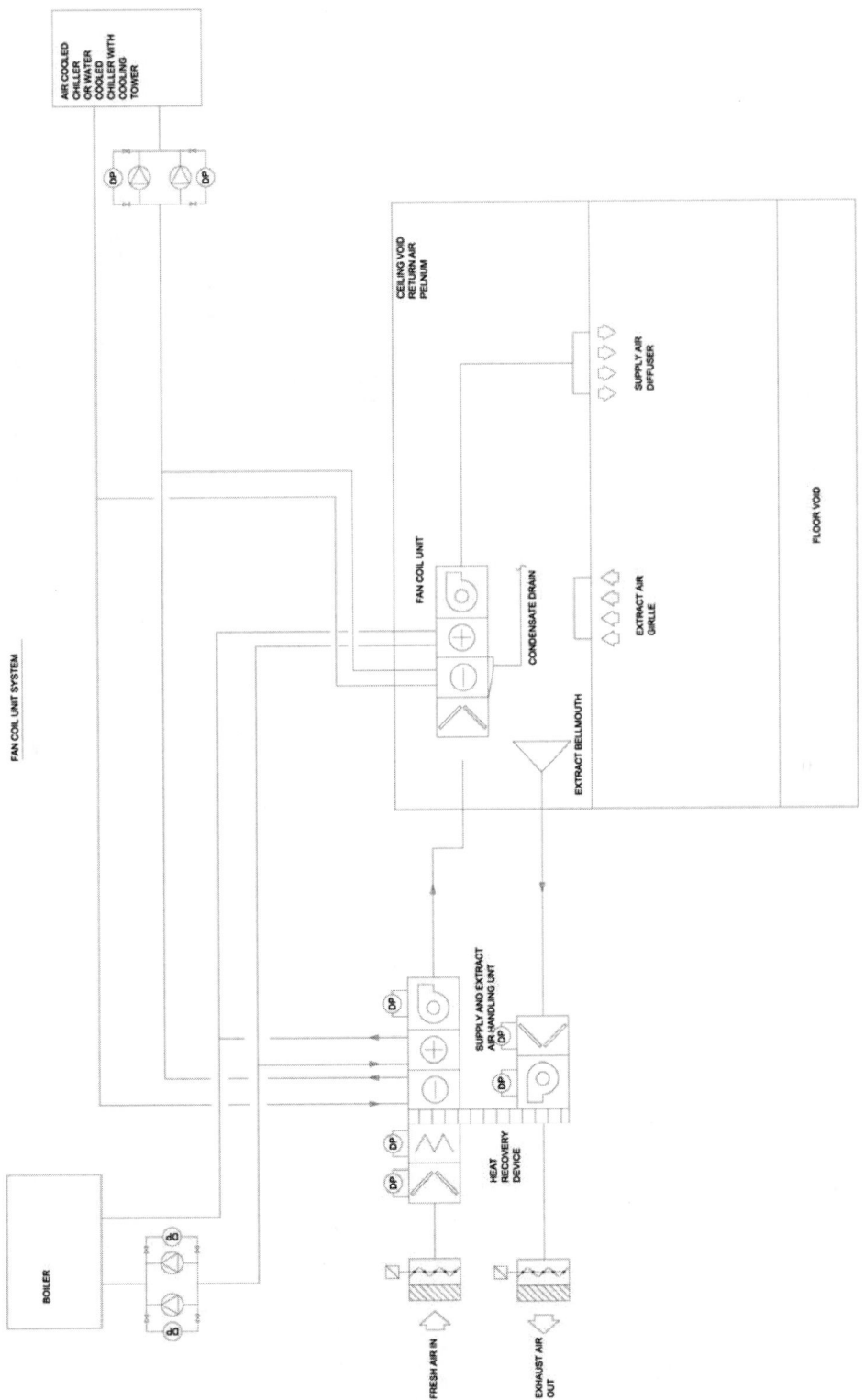

FAN COIL UNIT SYSTEM

AIR COOLED CHILLER OR WATER COOLED CHILLER WITH COOLING TOWER

BOILER

CEILING VOID RETURN AIR PELNUM

FAN COIL UNIT

SUPPLY AIR DIFFUSER

CONDENSATE DRAIN

EXTRACT AIR GIRLLE

FLOOR VOID

EXTRACT BELLMOUTH

SUPPLY AND EXTRACT AIR HANDLING UNIT

HEAT RECOVERY DEVICE

FRESH AIR IN

EXHAUST AIR OUT

103

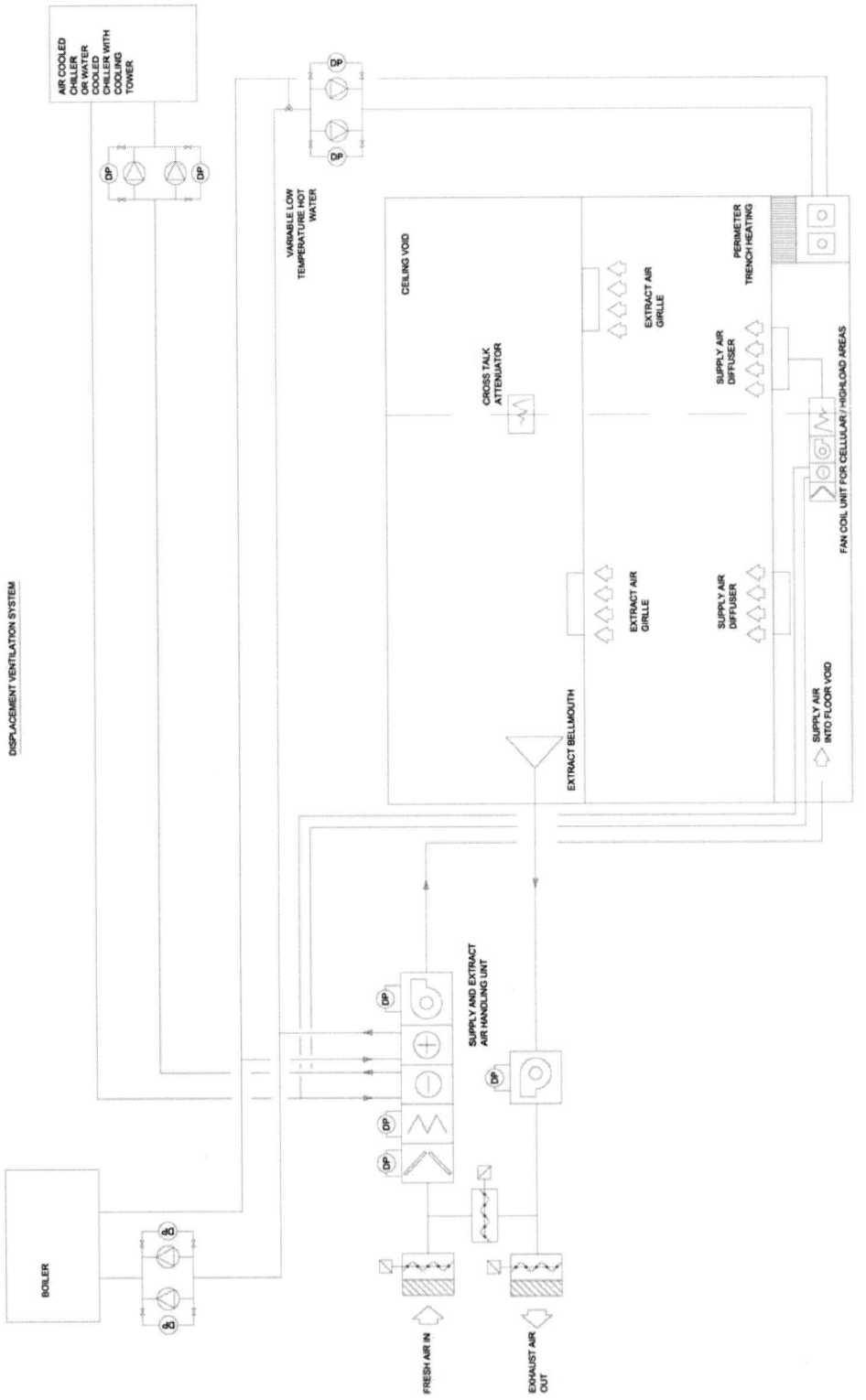

DISPLACEMENT VENTILATION SYSTEM

AIR COOLED CHILLER OR WATER COOLED CHILLER WITH COOLING TOWER

BOILER

VARIABLE LOW TEMPERATURE HOT WATER

CEILING VOID

CROSS TALK ATTENUATOR

EXTRACT AIR GIRLLE

EXTRACT AIR GIRLLE

SUPPLY AIR DIFFUSER

SUPPLY AIR DIFFUSER

PERIMETER TRENCH HEATING

FAN COIL UNIT FOR CELLULAR / HIGH LOAD AREAS

EXTRACT BELLMOUTH

SUPPLY AIR INTO FLOOR VOID

SUPPLY AND EXTRACT AIR HANDLING UNIT

FRESH AIR IN

EXHAUST AIR OUT

DP

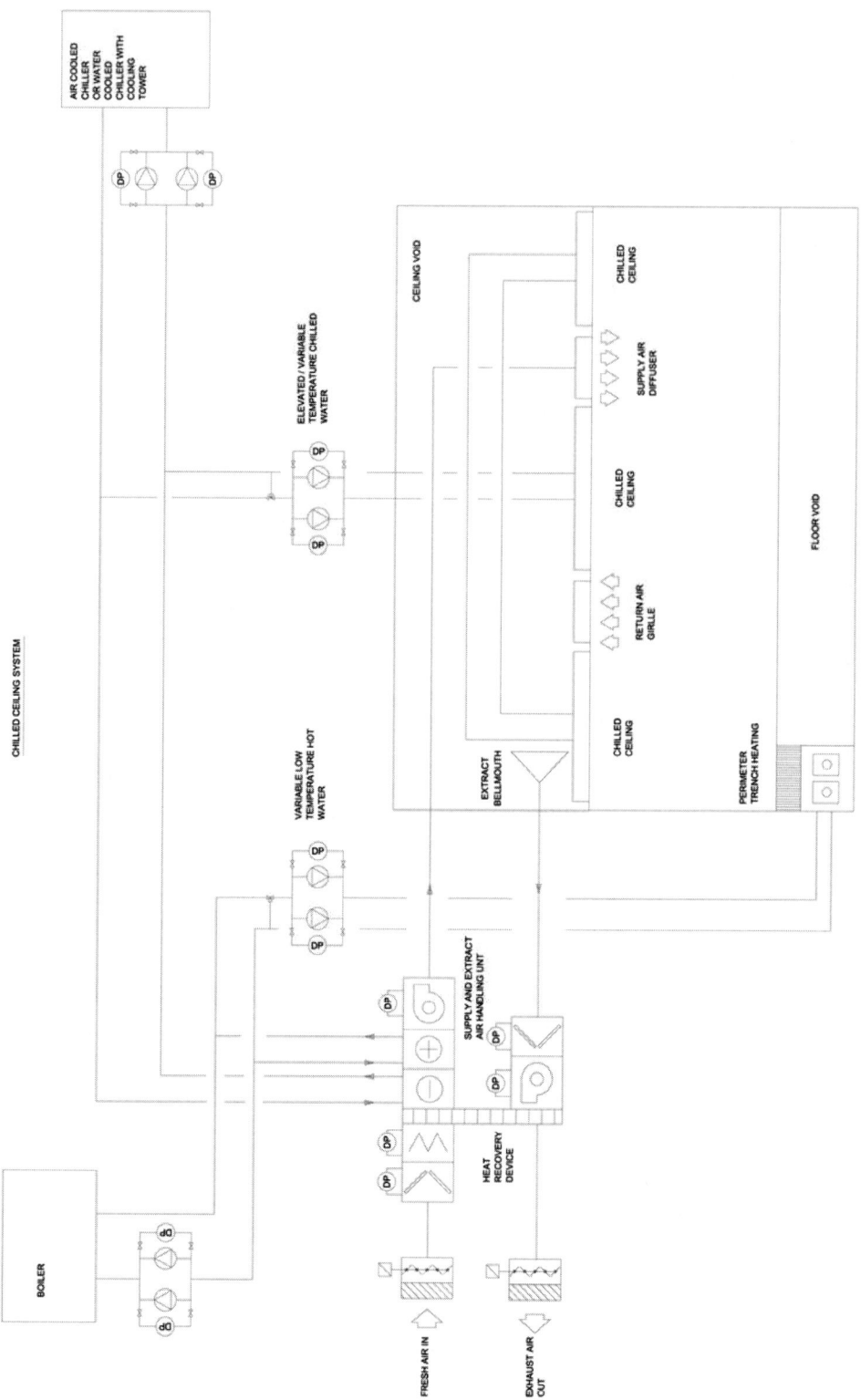

CHILLED CEILING SYSTEM

AIR COOLED CHILLER OR WATER COOLED CHILLER WITH COOLING TOWER

BOILER

CEILING VOID

ELEVATED / VARIABLE TEMPERATURE CHILLED WATER

VARIABLE LOW TEMPERATURE HOT WATER

CHILLED CEILING

SUPPLY AIR DIFFUSER

CHILLED CEILING

RETURN AIR GIRLLE

CHILLED CEILING

FLOOR VOID

PERIMETER TRENCH HEATING

EXTRACT BELLMOUTH

SUPPLY AND EXTRACT AIR HANDLING UNT

HEAT RECOVERY DEVICE

FRESH AIR IN

EXHAUST AIR OUT

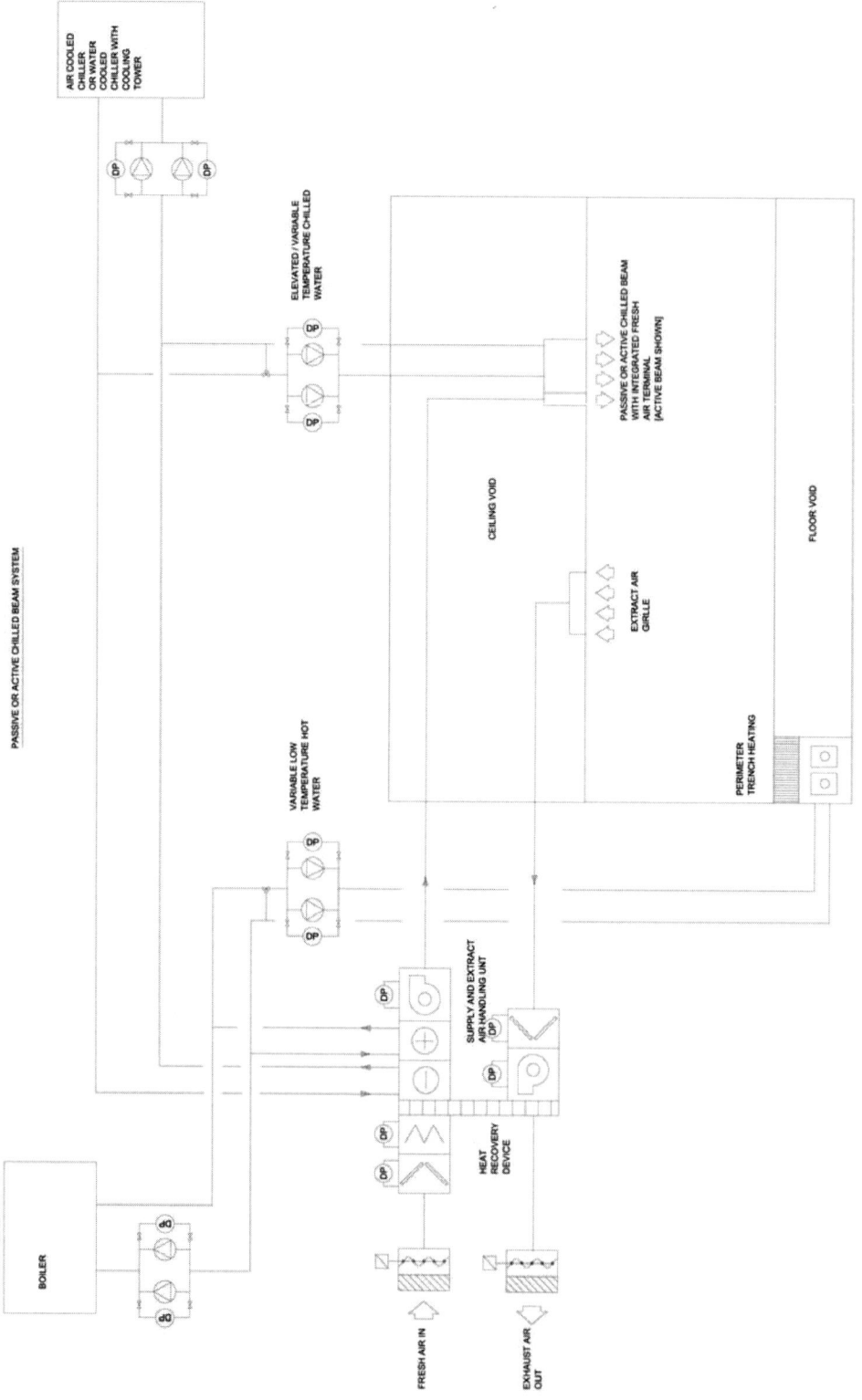

PASSIVE OR ACTIVE CHILLED BEAM SYSTEM

AIR COOLED CHILLER OR WATER COOLED CHILLER WITH COOLING TOWER

ELEVATED / VARIABLE TEMPERATURE CHILLED WATER

PASSIVE OR ACTIVE CHILLED BEAM WITH INTEGRATED FRESH AIR TERMINAL [ACTIVE BEAM SHOWN]

CEILING VOID

EXTRACT AIR GRILLE

FLOOR VOID

VARIABLE LOW TEMPERATURE HOT WATER

PERIMETER TRENCH HEATING

SUPPLY AND EXTRACT AIR HANDLING UNIT

HEAT RECOVERY DEVICE

BOILER

FRESH AIR IN

EXHAUST AIR OUT

DP

VARIABLE AIR VOLUME SYSTEM

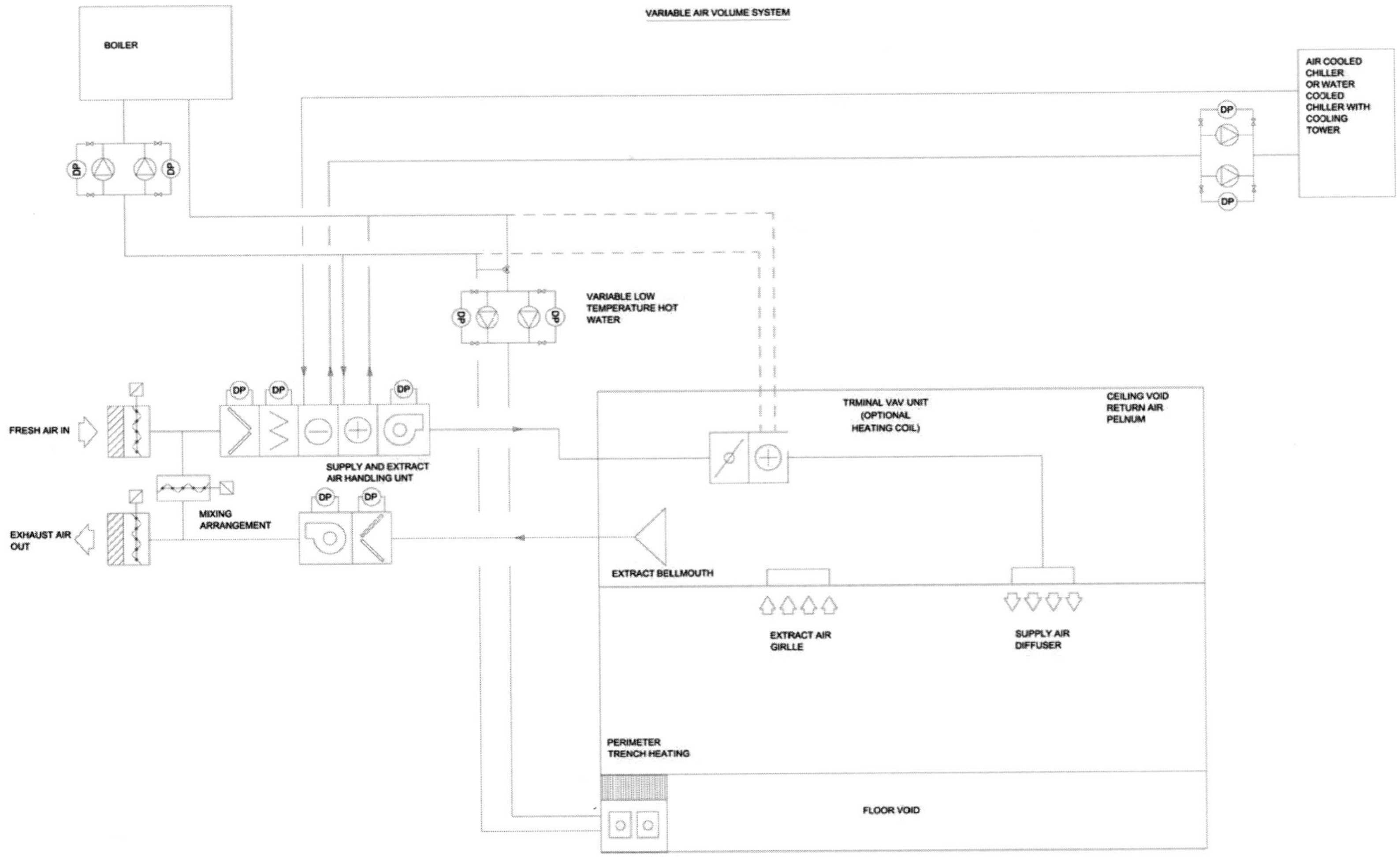

AIR COOLED
CHILLER
OR WATER
COOLED
CHILLER WITH
COOLING
TOWER

CEILING VOID
RETURN AIR
PELNUM

SUPPLY AIR
DIFFUSER

TRMINAL VAV UNIT
(OPTIONAL
HEATING COIL)

FLOOR VOID

VARIABLE LOW
TEMPERATURE HOT
WATER

EXTRACT AIR
GIRLLE

EXTRACT BELLMOUTH

PERIMETER
TRENCH HEATING

DP   DP

SUPPLY AND EXTRACT
AIR HANDLING UNIT

DP

DP

DP

DP   DP

BOILER

DP

DP

MIXING
ARRANGEMENT

FRESH AIR IN

EXHAUST AIR
OUT

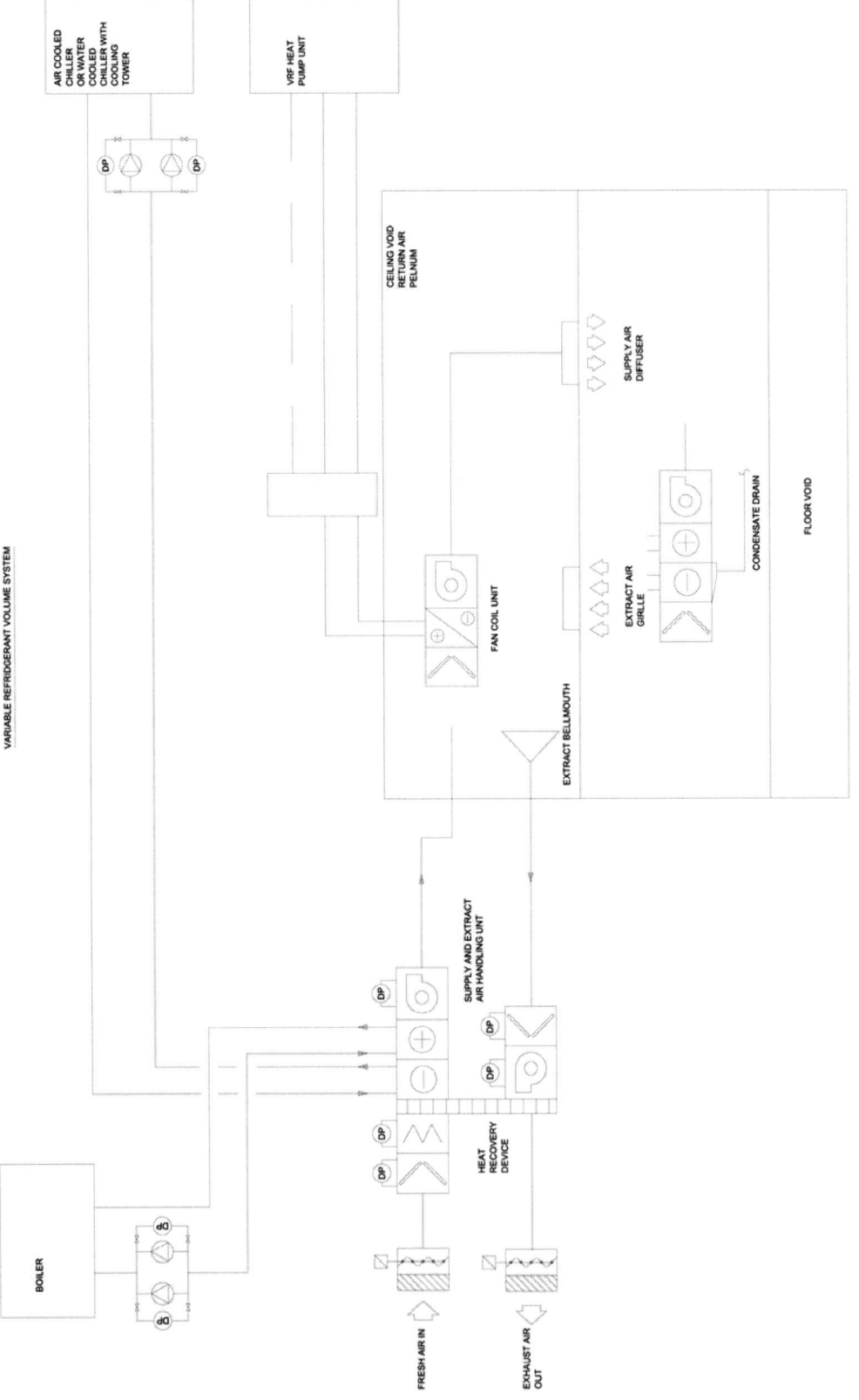

VARIABLE REFRIGERANT VOLUME SYSTEM

AIR COOLED CHILLER OR WATER COOLED CHILLER WITH COOLING TOWER

VRF HEAT PUMP UNIT

DP

CEILING VOID RETURN AIR PELNUM

SUPPLY AIR DIFFUSER

CONDENSATE DRAIN

FLOOR VOID

FAN COIL UNIT

EXTRACT AIR GRILLE

EXTRACT BELLMOUTH

SUPPLY AND EXTRACT AIR HANDLING UNIT

DP

DP

DP

DP

DP

HEAT RECOVERY DEVICE

BOILER

DP

DP

FRESH AIR IN

EXHAUST AIR OUT

REVERSE CYCLE HEAT PUMP

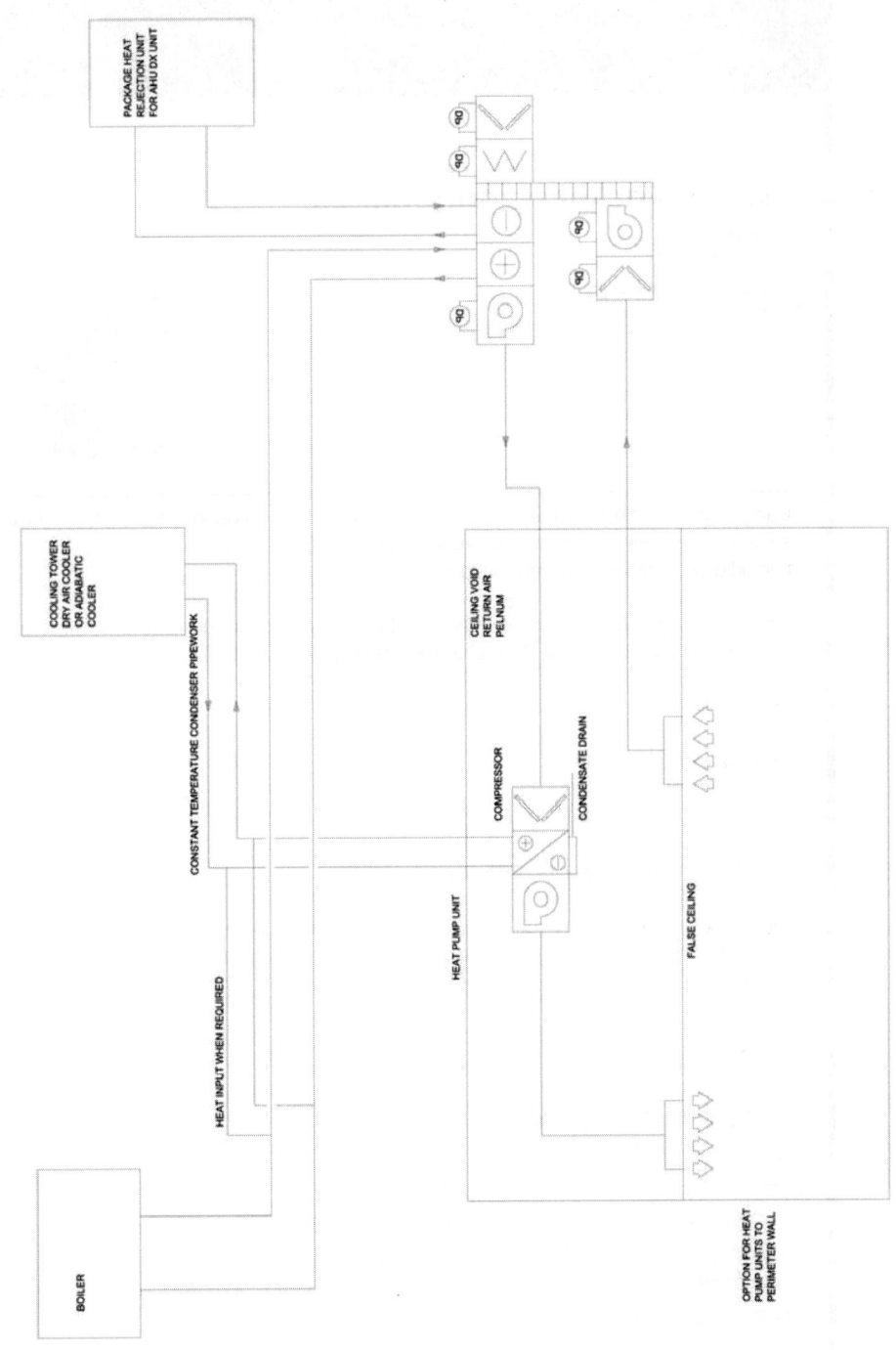

PACKAGE HEAT
REJECTION UNIT
FOR AHU DX UNIT

COOLING TOWER
DRY AIR COOLER
OR ADIABATIC
COOLER

CONSTANT TEMPERATURE CONDENSER PIPEWORK

CEILING VOID
RETURN AIR
PELNUM

COMPRESSOR

CONDENSATE DRAIN

HEAT INPUT WHEN REQUIRED

HEAT PUMP UNIT

FALSE CEILING

BOILER

OPTION FOR HEAT
PUMP UNITS TO
PERIMETER WALL

Watts Pocket Handbook, 29th Edition

Trevor Rushton

Back in print for the first time in years, the Watts Pocket Handbook renews its commitment to share industry knowledge by providing technical and legal information across a comprehensive spread of property and construction topics.

Compiled by the Watts Technical Director, the Handbook provides specialist information and guidance on a vast selection of construction related subjects including:

- Contracts and procurement

- Insurance

- Materials and defects

- Environmental and sustainability issues

Watts Pocket Handbook remains the must-have reference book for professionals and students engaged in construction, building surveying, service engineering, property development and much more.

April 2016; 186 x 123 mm; 324 pp
Pb: 978-1-138-66595-8; £22.99

To Order: Tel: +44 (0) 1235 400524 Fax: +44 (0) 1235 400525
or Post: Taylor and Francis Customer Services,
Bookpoint Ltd, Unit T1, 200 Milton Park, Abingdon, Oxon, OX14 4TA UK
Email: book.orders@tandf.co.uk

For a complete listing of all our titles visit:
www.tandf.co.uk

PART 2

Approximate Estimating

Faber & Kell's Heating & Air-conditioning of Buildings
11th Edition
D. Oughton *et al.*

For over 70 years, Faber & Kell's has been the definitive reference text in its field. It provides an understanding of the principles of heating and air-conditioning of buildings in a concise manner, illustrating practical information with simple, easy-to-use diagrams, now in full-colour.

This new-look 11th edition has been re-organised for ease of use and includes fully updated chapters on sustainability and renewable energy sources, as well as information on the new Building Regulations Parts F and L. As well as extensive updates to regulations and codes, it now includes an introduction that explains the role of the building services engineer in the construction process. Its coverage of design calculations, advice on using the latest technologies, building management systems, operation and maintenance makes this an essential reference for all building services professionals.

January 2015: 246 x 189: 968pp
Pbk: 978-0-415-52265-6: £115.00

To Order: Tel: +44 (0) 1235 400524 Fax: +44 (0) 1235 400525
or Post: Taylor and Francis Customer Services,
Bookpoint Ltd, Unit T1, 200 Milton Park, Abingdon, Oxon, OX14 4TA UK
Email: book.orders@tandf.co.uk

For a complete listing of all our titles visit:
www.tandf.co.uk

DIRECTIONS

The prices shown in this section of the book are average prices on a fixed price basis for typical buildings tendered during the second quarter of 2018. Unless otherwise noted, they exclude external services and professional fees.

The information in this section has been arranged to follow more closely the order in which estimates may be developed, in accordance with RIBA work stages.

a) Cost Indices and Regional Variations – These provide information regarding the adjustments to be made to estimates taking into account current pricing levels for different locations in the UK.

b) Feasibility Costs – These provide a range of data (based on a rate per square metre) for all-in engineering costs, excluding lifts, associated with a wide variety of building types. These would typically be used at work stage 0/1 (feasibility) of a project.

c) Elemental Rates – The outline costs for offices have been developed further to provide rates for the alternative solutions for each of the services elements. These would typically be used at work stage 2, outline proposal.

Where applicable, costs have been identified as Shell and Core and Fit Out to reflect projects where the choice of procurement has dictated that the project is divided into two distinctive contractual parts.

Such detail would typically be required at work stage 3, detailed proposals.

d) All-in Rates – These are provided for a number of items and complete parts of a system, i.e. boiler plant, ductwork, pipework, electrical switchgear and small power distribution, together with lifts and escalators. Refer to the relevant section for further guidance notes.

e) Elemental Costs – These are provided for a diverse range of building types; offices, laboratory, shopping mall, airport terminal building, supermarket, performing arts centre, sports hall, luxury hotel, hospital and secondary school. Also included is a separate analysis of a building management system for an office block. In each case, a full analysis of engineering services costs is given to show the division between all elements and their relative costs to the total building area. A regional variation factor has been applied to bring these analyses to a common London base.

Prices should be applied to the total floor area of all storeys of the building under consideration. The area should be measured between the external walls without deduction for internal walls and staircases/lift shafts, i.e. GIA (Gross Internal Area).

Although prices are reviewed in the light of recent tenders it has only been possible to provide a range of prices for each building type. This should serve to emphasize that these can only be average prices for typical requirements and that such prices can vary widely depending on variations in size, location, phasing, specification, site conditions, procurement route, programme, market conditions and net to gross area efficiencies. Rates per square metre should not therefore be used indiscriminately and each case needs to be assessed on its own merits.

The prices do not include for incidental builder's work nor for profit and attendance by a Main Contractor where the work is executed as a subcontract: they do however include for preliminaries, profit and overheads for the services contractor. Capital contributions to statutory authorities and public undertakings and the cost of work carried out by them have been excluded.

Where services works are procured indirectly, i.e. ductwork via a mechanical Subcontractor, the reader should make due allowance for the addition of a further level of profit, etc.

COST INDICES

The following tables reflect the major changes in cost to contractors but do not necessarily reflect changes in tender levels. In addition to changes in labour and materials costs, tenders are affected by other factors such as the degree of competition in the particular industry, the area where the work is to be carried out, the availability of labour and the prevailing economic conditions. This has meant in recent years that when there has been an abundance of work, tender levels have tended to increase at a greater rate than can be accounted for solely by increases in basic labour and material costs and, conversely, when there is a shortage of work this has tended to result in keener tenders. Allowances for these factors are impossible to assess on a general basis and can only be based on experience and knowledge of the particular circumstances.

In compiling the tables the cost of labour has been calculated on the basis of a notional gang as set out elsewhere in the book. The proportion of labour to materials has been assumed as follows:

Mechanical Services – 30:70, Electrical Services – 50:50, (1976 = 100)

Mechanical Services					
Year	*First Quarter*	*Second Quarter*	*Third Quarter*	*Fourth Quarter*	*Annual Average* *Annual Change*
2015	99.8	99.7	99.3	101.0	100.0	
2016	101.0	102.0	102.6	104.3	102.5	2.5%
2017	105.6	105.6	106.6	108.6	106.6	4.0%
2018	109.3 (P)	109.6 (F)	109.9	112.1	110.2	3.4%
2019	112.4	112.9	113.2	115.5	113.5	3.0%
2020	115.9	116.4	116.6	118.8	116.9	3.0%
Electrical Services					
Year	*First Quarter*	*Second Quarter*	*Third Quarter*	*Fourth Quarter*	*Annual Average* *Annual Change*
2015	100.3	100.2	99.9	99.7	100.0	
2016	102.4	102.5	102.5	102.9	102.6	2.6%
2017	104.8	105.1	105.0	105.3	105.1	2.4%
2018	107.7 (P)	108.1 (F)	108.2	108.3	108.1	2.9%
2019	111.4	111.5	111.6	111.9	111.6	3.3%
2020	115.2	115.3	115.5	115.7	115.4	3.4%

(P = Provisional)

(F = Forecast)

COST INDICES

Regional variations

Prices throughout this Book apply to work in the London area (see Directions at the beginning of the Mechanical Installations and Electrical Installations sections). However, prices for mechanical and electrical services installations will of course vary from region to region, largely as a result of differing labour costs but also depending on the degree of accessibility, urbanization and local market conditions.

The following table of regional factors is intended to provide readers with indicative adjustments that may be made to the prices in the Book for locations outside of London. The figures are of necessity averages for regions and further adjustments should be considered for city centre or very isolated locations, or other known local factors.

Inner London	1.05	North East	0.84
Outer London	1.00	North West	0.89
South East	0.97 (Excl GL)	Yorkshire and Humberside	0.87
South West	0.90	Scotland	0.88
East of England	0.92	Wales	0.86
East Midlands	0.87	Northern Ireland	0.76
West Midlands	0.89		

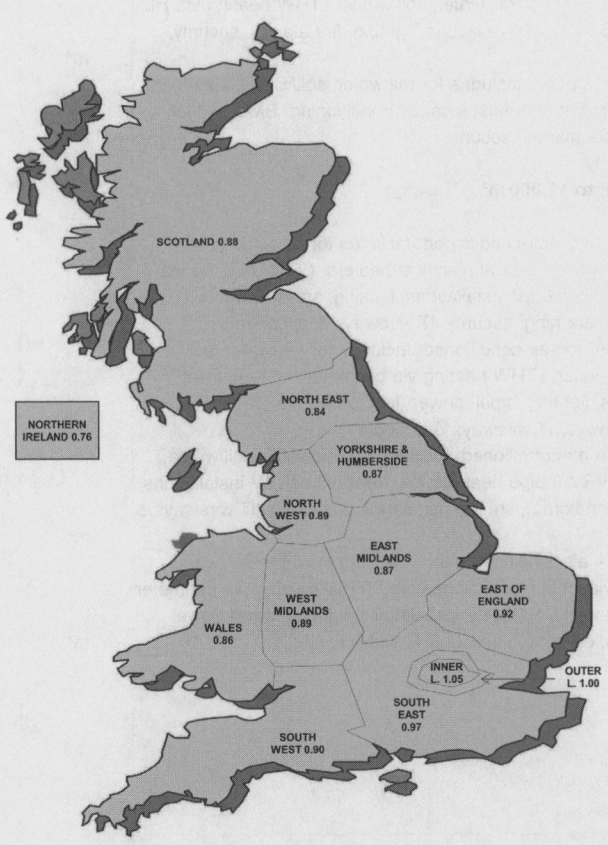

Approximate Estimating

RIBA STAGE 0/1 PREPARATION

Item	Unit	Range £
Typical Square Metre Rates for Engineering Services The following examples indicate the range of rates within each building type for engineering services, excluding lifts etc., utilities services and professional fees. Based on Gross Internal Area (GIA).		
Industrial Buildings Factories		
Owner occupation: Includes for rainwater, soil/waste, LTHW heating via HL radiant heaters, BMS, LV installations, lighting, fire alarms, security, earthing	m²	120.00 to 150.00
Owner occupation: Includes for rainwater, soil/waste, sprinklers, LTHW heating via HL gas fired heaters, local air conditioning, BMS, HV/LV installations, lighting, fire alarms, security, earthing	m²	175.00 to 220.00
Warehouses		
High bay for owner occupation: Includes for rainwater, soil/waste, LTHW heating via HL gas fired heaters, BMS, HV/LV installations, lighting, fire alarms, security, earthing	m²	100.00 to 120.00
High bay for owner occupation: Includes for rainwater, soil/waste, sprinklers, LTHW heating via HL radiant heaters, local air conditioning, BMS, HV/LV installations, lighting, fire alarms, security, earthing	m²	200.00 to 240.00
Distribution Centres		
High bay for letting: Includes for rainwater, soil/waste, LTHW heating via HL gas fired heaters, BMS, HV/LV installations, lighting, fire alarms, security, earthing	m²	125.00 to 160.00
High bay for owner occupation: Includes for rainwater, soil/waste, sprinklers, LTHW heating via HL radiant heaters, local air conditioning, BMS, HV/LV installations, lighting, fire alarms, security	m²	235.00 to 255.00
Office Buildings 5,000 m² to 15,000 m² Offices for Letting		
Shell & Core and Cat A non-air-conditioned: Includes for rainwater, soil/waste, cold water, hot water via local electrical heaters, LTHW heating via radiator heaters, toilet extract, LV installations, lighting, small power (landlords), fire alarms, earthing, security, IT wireways & ripcontrols	m²	310.00 to 360.00
Shell & Core and Cat A non-air-conditioned: Includes for rainwater, soil/waste, cold water, hot water, LTHW heating via perimeter heaters, toilet extract, LV installations, lighting, small power (landlords), fire alarms, earthing, security wireways, IT wireways & controls	m²	320.00 to 375.00
Shell & Core and Cat A air-conditioned: Includes for rainwater, soil/waste, cold water, hot water, VRV 3 pipe heat pumps, toilet extract, LV installations, lighting, small power (landlords), fire alarms, earthing, security, IT wireways & controls	m²	470.00 to 600.00
Shell & Core and Cat A air-conditioned: Includes for rainwater, soil/waste, cold water, hot water via local electrical heaters, LTHW heating via perimeter heaters, 2 pipe, toilet extract, BMS, LV installations, lighting, small power (landlords), fire alarms, earthing, security & IT wireways	m²	600.00 to 680.00

RIBA STAGE 0/1 PREPARATION

Item	Unit	Range £
Offices for Owner Occupation		
Non-air-conditioned: Includes for rainwater, soil/waste, cold water, hot water via local electrical heaters, LTHW heating via radiator heaters, toilet extract, LV installations, lighting, small power (landlords), dry risers, fire alarms, earthing, security, IT wireways & controls	m²	320.00 to 370.00
Non-air-conditioned: Includes for rainwater, soil/waste, cold water, hot water, dry risers, LTHW heating via perimeter heaters, toilet extract, LV installations, lighting, small power (landlords), dry risers, fire alarms, earthing, security, IT wireways & controls	m²	340.00 to 390.00
Air-conditioned: Includes for rainwater, soil/waste, cold water, hot water via centralized system, dry risers, 4 pipe air conditioning, toilet extract, BMS, LV installations, life safety standby generators, lighting, small power, sprinkler protection, fire alarms, earthing, lightning protection, security, IT wireways, fire alarms L1/P1 & controls	m²	580.00 to 680.00
Health and Welfare Facilities		
District General Hospitals		
Natural ventilation: Includes for rainwater, soil/waste, cold water, hot water, dry risers, sprinklers, medical gases, LTHW heating, WC/kitchen extract, BMS, LV installations, standby generation, lighting, small power, fire alarms, earthing/lightning protection, nurse call systems, security, IT wireways	m²	1080.00 to 1160.00
Natural ventilation: Includes for rainwater, soil/waste, cold water, hot water, dry risers, sprinklers, medical gases, LTHW heating, localized VAV air conditioning and mechanical ventilation, WC/kitchen extract, BMS, LV installations, standby generation, lighting, small power, fire alarms, earthing/ lightning protection, nurse call systems, security, IT wireways	m²	1160.00 to 1260.00
Private Hospitals		
Natural ventilation: Includes for rainwater, soil/waste, cold water, hot water, dry risers, sprinklers, medical gases, LTHW heating, localized VAV air conditioning, WC/kitchen extract, BMS, LV installations, standby generation, lighting, small power, fire alarms, earthing/lightning protection, nurse call systems, security, IT wireways	m²	1185.00 to 1290.00
Air-conditioned: Includes for rainwater, soil/waste, cold water, hot water, dry risers, sprinklers, medical gases, LTHW heating, supply and extract ventilation, 4 pipe air conditioning, WC/kitchen extract, BMS, LV installations, standby generation, lighting, small power, fire alarms, earthing/lightning protection, nurse call systems, security, IT wireways	m²	1350.00 to 1460.00
GP Surgery		
Natural ventilation: Includes for rainwater, soil/waste, cold water, hot water, LTHW heating, WC/kitchen extract, BMS, LV installations, lighting, small power, fire alarms, earthing/lightning protection, nurse call systems, security, IT wireways	m²	745.00 to 850.00
Air-conditioned: Includes for rainwater, soil/waste, cold water, hot water, LTHW heating, supply and extract, BMS, LV installations, lighting, small power, fire alarms, earthing/lightning protection, nurse call systems, security, IT wireways	m²	825.00 to 1050.00

RIBA STAGE 0/1 PREPARATION

Item	Unit	Range £	
Typical Square Metre Rates for Engineering Services – cont			
Entertainment and Recreation Buildings			
Non-Performing			
Natural ventilation: Includes for rainwater, soil/waste, cold water, central hot water, dry risers, LTHW heating, toilet extract, kitchen extract, controls, LV installations, lighting, small power, fire alarms, earthing, security, IT wireways	m²	340.00 to	420.00
Comfort cooled: Includes for rainwater, soil/waste, cold water, hot water via electrical heaters, sprinklers/dry risers, LTHW heating, DX air conditioning, kitchen/toilet extract, BMS, LV installations, lighting, small power, fire alarms, earthing, security, IT wireways	m²	625.00 to	750.00
Performing Arts (with Theatre)			
Natural ventilation: Includes for rainwater, soil/waste, cold water, central hot water, sprinklers/dry risers, LTHW heating, toilet extract, kitchen extract, controls, LV installations, lighting, small power, fire alarms, earthing, security, IT wireways	m²	570.00 to	690.00
Comfort cooled: Includes for rainwater, soil/waste, cold water, central hot water, sprinklers/dry risers, LTHW heating, air conditioning, kitchen/toilet extract, BMS, LV installations, lighting including enhanced dimming/scene setting, small power, fire alarms, earthing, security, IT wireways including production, audio and video recording	m²	780.00 to	940.00
Sports Halls			
Natural ventilation: Includes for rainwater, soil/waste, cold water, hot water gas fired heaters, LTHW heating, toilet extract, BMS, LV installations, lighting, small power, fire alarms, earthing, security	m²	260.00 to	310.00
Comfort cooled: Includes for rainwater, soil/waste, cold water, hot water via LTHW heat exchangers, LTHW heating, air conditioning via AHUs to limited areas, toilet extract, BMS, LV installations, lighting, small power, fire alarms, earthing, security	m²	360.00 to	440.00
Multi-Purpose Leisure Centre			
Natural ventilation: Includes for rainwater, soil/waste, cold water, hot water gas fired heaters, LTHW heating, toilet extract, BMS, LV installations, lighting, small power, fire alarms, earthing, security, IT wireways	m²	360.00 to	440.00
Comfort cooled: Includes for rainwater, soil/waste, cold water, hot water LTHW heat exchangers, LTHW heating, air conditioning via AHU to limited areas, pool hall supply/extract, kitchen/toilet extract, BMS, LV installations, lighting, small power, fire alarms, earthing, security, IT wireways	m²	575.00 to	730.00
Retail Buildings			
Open Arcade			
Natural ventilation: Includes for rainwater, soil/waste, cold water, hot water, sprinklers/dry risers, LTHW heating, toilet extract, smoke extract, BMS, LV installations, life safety standby generators, lighting, small power, fire alarms, public address, earthing/lightning protection, security, IT wireways	m²	340.00 to	420.00
Enclosed Shopping Mall			
Air-conditioned: Includes for rainwater, soil/waste, cold water, hot water, sprinklers/dry risers, LTHW heating, air conditioning via AHUs, toilet extract, smoke extract, BMS, LV installations, life safety standby generators, lighting, small power, fire alarms, public address, earthing/lightning protection, CCTV/ security, IT wireways people counting systems	m²	520.00 to	625.00

RIBA STAGE 0/1 PREPARATION

Item	Unit	Range £	
Department Stores			
Air-conditioned: Includes for sanitaryware, soil/waste, cold water, hot water, sprinklers/dry risers, LTHW heating, air conditioning via AHUs, toilet extract, smoke extract, BMS, LV installations, life safety standby generators, lighting, small power, fire alarms, public address, earthing, lightning protection, CCTV/ security, IT installation wireways	m²	385.00 to	470.00
Supermarkets			
Air-conditioned: Includes for rainwater, soil/waste, cold water, hot water, sprinklers, LTHW heating, air conditioning via AHUs, toilet extract, BMS, LV installations, lighting, small power, fire alarms, earthing, lightning protection, security, IT wireways, refrigeration	m²	560.00 to	690.00
Educational Buildings			
Secondary Schools (Academy)			
Natural Ventilation: Includes for rainwater, soil/waste, cold water, central hot water, LTHW heating, toilet extract, BMS, LV installations, lighting, small power, fire alarms, earthing, security, IT wireways	m²	405.00 to	470.00
Natural vent with comfort cooling to selected areas (BB93 compliant): Includes for rainwater, soil/waste, cold water, central hot water, LTHW heating, DX air conditioning, general supply/extract, toilet extract, BMS, LV installations, lighting, small power, fire alarms, earthing, security, IT wireways	m²	470.00 to	535.00
Scientific Buildings			
Educational Research			
Comfort cooled: Includes for rainwater, soil/waste, cold water, central hot water, dry risers, compressed air, medical gasses, LTHW heating, 4 pipe air conditioning, toilet extract, fume, BMS, LV installations, lighting, small power, fire alarms, earthing, lightning protection, security, IT wireways	m²	720.00 to	890.00
Air-conditioned: Includes for rainwater, soil/waste, laboratory waste, cold water, central hot water, specialist water, dry risers, compressed air, medical gasses, steam, LTHW heating, VAV air conditioning, Comm's room cooling, toilet extract, fume extract, BMS, LV installations, UPS, standby generators, lighting, small power, fire alarms, earthing, lightning protection, security, IT wireways	m²	1275.00 to	1560.00
Commercial Research			
Air-conditioned; Includes for rainwater, soil/waste, laboratory waste, cold water, central hot water, specialist water, dry risers, compressed air, medical gasses, steam, LTHW heating, VAV air conditioning, Comm room cooling, toilet extract, fume extract, BMS, LV installations, UPS, standby generators, lighting, small power, fire alarms, earthing, lightning protection, security, IT wireways	m²	1400.00 to	1700.00
Residential Facilities			
Hotels			
1 to 3 Star: Includes for rainwater, soil/waste, cold water, hot water, dry risers, LTHW heating via radiators, refrigerate cooling to bedrooms, toilet/bathroom extract, kitchen extract, BMS, LV installations, lighting, small power, fire alarms, earthing, lightning protection, security, IT wireways	m²	470.00 to	600.00
4 to 5 Star: Includes for rainwater, soil/waste, cold water, hot water, sprinklers, dry risers, 4 pipe air conditioning, kitchen extract, toilet/bathroom extract, BMS, LV installations, life safety standby generators, lighting, small power, fire alarms, earthing, security, IT wireways	m²	660.00 to	825.00

RIBA STAGE 2/3 DESIGN

Item	Unit	Range £	
Elemental Rates for Alternative Engineering Services Solutions – Offices			
The following examples of building types indicate the range of rates for alternative design solutions for each of the engineering services elements based on Gross Internal Area for the Shell and Core and Net Internal Area for the Fit Out. Fit Out is assumed to be to Cat A Standard. Consideration should be made for the size of the building, which may affect the economies of scale for rates, i.e. the larger the building the lower the rates.			
5 Services			
Shell & Core			
5.1 Sanitary Installations			
Building up to 5,000 m²	m² GIA	10.00 to	14.00
Building over 5,000 m² up to 15,000 m² (low rise)	m² GIA	8.00 to	12.00
5.3 Disposal Installations			
Building up to 5,000 m²	m² GIA	18.00 to	22.00
Building over 5,000 m² up to 15,000 m²	m² GIA	17.00 to	20.00
5.4 Water Installations			
Building up to 5,000 m²	m² GIA	23.00 to	27.00
Building over 5,000 m² up to 15,000 m²	m² GIA	21.00 to	25.00
5.6 Space Heating and Air Conditioning			
LPHW heating installation; including gas installations			
Building up to 5,000 m²	m² GIA	40.00 to	45.00
Building over 5,000 m² up to 15,000 m²	m² GIA	35.00 to	40.00
Comfort cooling			
2 pipe fan coil for building up to 5,000 m²	m² GIA	85.00 to	95.00
2 pipe fan coil for building over 5,000 m² up to 15,000 m²	m² GIA	75.00 to	85.00
2 pipe variable refrigerant volume (VRV) for building up to 5,000 m²	m² GIA	60.00 to	70.00
Full air conditioning			
4 pipe fan coil for building up to 5,000 m²	m² GIA	110.00 to	140.00
4 pipe fan coil for building over 5,000 m² up to 15,000 m²	m² GIA	100.00 to	130.00
4 pipe variable refrigerant volume for building up to 5,000 m²	m² GIA	95.00 to	110.00
Ventilated (active) chilled beams for building over 5,000 m² up to 15,000 m²	m² GIA	120.00 to	140.00
Chilled beam exposed services for building over 5,000 m² to 15,000 m²	m² GIA	130.00 to	150.00
Concealed passive chilled beams for building over 5,000 m² up to 15,000 m²	m² GIA	120.00 to	140.00
Chilled ceiling for building over 5,000 m² to 15,000 m²	m² GIA	120.00 to	140.00
Chilled ceiling/perimeter beams for building over 5,000 m² up to 15,000 m²	m² GIA	130.00 to	150.00
Displacement for building over 5,000 m² up to 15,000 m²	m² GIA	140.00 to	160.00
5.7 Ventilation Systems (excluding smoke extract)			
Building up to 5,000 m²	m² GIA	30.00 to	45.00
Building over 5,000 m² up to 15,000 m²	m² GIA	25.00 to	35.00
5.8 Electrical Installations			
LV installations			
Standby generators (life safety only)			
Buildings over 5,000 m² up to 15,000 m²	m² GIA	12.00 to	15.00

RIBA STAGE 2/3 DESIGN

Item	Unit	Range £	
LV distribution			
Buildings up to 5,000 m²	m² GIA	28.00 to	35.00
Buildings over 5,000 m² up to 15,000 m²	m² GIA	35.00 to	45.00
Lighting installations (including lighting controls and luminaires)			
Buildings up to 5,000 m²	m² GIA	18.00 to	23.00
Buildings over 5,000 m² up to 15,000 m²	m² GIA	16.00 to	21.00
Small power			
Buildings up to 5,000 m²	m² GIA	6.00 to	7.00
Buildings over 5,000 m² to up to 15,000 m²	m² GIA	5.00 to	6.00
Electrical installations for mechanical plant			
Buildings up to 5,000 m²	m² GIA	7.00 to	9.00
Buildings over 5,000 m² to up to 15,000 m²	m² GIA	8.00 to	10.00
5.11 Fire and Lightning Protection			
Fire protection over 5,000 m² to 15,000 m²			
Dry risers	m² GIA	3.00 to	5.00
Sprinkler installation	m² GIA	20.00 to	25.00
Protective installations			
Earthing			
Buildings up to 5,000 m²	m² GIA	2.00 to	3.00
Buildings over 5,000 m² to 15,000 m²	m² GIA	2.00 to	3.00
Lightning protection			
Buildings up to 5,000 m²	m² GIA	2.00 to	3.00
Buildings over 5,000 m² to 15,000 m²	m² GIA	2.00 to	3.00
5.12 Communication Security and Control Systens			
Fire alarms (single stage)			
Buildings up to 5,000 m²	m² GIA	8.00 to	12.00
Buildings over 5,000 m² to 15,000 m²	m² GIA	12.00 to	15.00
Fire alarms (phased evacuation)			
Buildings over 5,000 m² to 15,000 m²	m² GIA	14.00 to	17.00
IT (wireways only)			
Buildings up to 5,000 m²	m² GIA	3.00 to	4.00
Buildings over 5,000 m² to 15,000 m²	m² GIA	3.00 to	4.00
Security			
Buildings up to 5,000 m²	m² GIA	6.00 to	8.00
Buildings over 5,000 m² to 15,000 m²	m² GIA	8.00 to	10.00
BMS controls; including MCC panels and control cabling			
Full air conditioning			
Buildings up to 5,000 m²	m² GIA	20.00 to	25.00
Buildings up to 15,000 m²	m² GIA	22.00 to	27.00
Fit Out			
5.6 Space Heating and Air Conditioning			
LPHW heating installation			
Building up to 5,000 m²	m² NIA	50.00 to	60.00
Building over 5,000 m² to 15,000 m²	m² NIA	45.00 to	55.00
Comfort cooling			
2 pipe fan coil for building up to 5,000 m²	m² NIA	105.00 to	125.00
2 pipe fan coil for building over 5,000 m² to 15,000 m²	m² NIA	90.00 to	105.00
2 pipe variable refrigerant volume (VRV) for building up to 5,000 m²	m² NIA	80.00 to	90.00

RIBA STAGE 2/3 DESIGN

Item	Unit	Range £	
Elemental Rates for Alternative Engineering Services Solutions – Offices – cont			
Fit Out – cont			
Full air conditioning			
4 pipe fan coil for building up to 5,000 m²	m² NIA	170.00 to	200.00
4 pipe fan coil for building over 5,000 m² to 15,000 m²	m² NIA	170.00 to	200.00
4 pipe variable refrigerant volume for building up to 5,000 m²	m² NIA	150.00 to	170.00
Ventilated (active) chilled beams for building over 5,000 m² to 15,000 m²	m² NIA	180.00 to	205.00
Chilled beam exposed services for building over 5,000 m² to 15,000 m²	m² NIA	200.00 to	230.00
Concealed passive chilled beams for building over 5,000 m² to 15,000 m²	m² NIA	150.00 to	170.00
Chilled ceiling for building over 5,000 m² to 15,000 m²	m² NIA	220.00 to	265.00
Chilled ceiling/perimeter beams for building over 5,000 m² to 15,000 m²	m² NIA	235.00 to	290.00
Displacement for building over 5,000 m² to 15,000 m²	m² NIA	90.00 to	120.00
5.8 Electrical Installations			
Lighting installations (including lighting controls and luminaires)			
Buildings up to 5,000 m²	m² NIA	65.00 to	75.00
Buildings over 5,000 m² to 15,000 m²	m² NIA	75.00 to	90.00
Electrical installations for Mechanical Plant			
Buildings up to 5,000 m²	m² NIA	7.00 to	10.00
Buildings over 5,000 m² to 15,000 m²	m² NIA	7.00 to	10.00
5.11 Fire and Lightning Protection			
Fire protection over 5,000 m² to 15,000 m²			
Sprinkler installation	m² NIA	22.00 to	30.00
Earthing			
Buildings up to 5,000 m²	m² NIA	2.00 to	3.00
Buildings over 5,000 m² to 15,000 m²	m² NIA	2.00 to	3.00
5.12 Communication, Security and Control Systems			
Fire alarms (single stage)			
Buildings up to 5,000 m²	m² NIA	12.00 to	15.00
Buildings over 5,000 m² to 15,000 m²	m² NIA	12.00 to	15.00
Fire alarms (phased evacuation)			
Buildings over 5,000 m² to 15,000 m²	m² NIA	15.00 to	20.00
BMS controls; including MCC panels and control cabling			
Full air conditioning			
Buildings up to 5,000 m²	m² NIA	22.00 to	26.00
Buildings up to 15,000 m²	m² NIA	20.00 to	24.00

RIBA STAGE 2/3 DESIGN

Item	Unit	Range £	
Elemental Rates for Alternative Engineering Services Solutions – Hotels			
The following examples of building types indicate the range of rates for alternative design solutions for each of the engineering services elements based on Gross Internal Area for the Shell and Core and Net Internal Area for the Fit Out. Fit Out is assumed to be to Cat A Standard.			
Consideration should be made for the size of the building, which may affect the economies of scale for rates, i.e. the larger the building the lower the rates.			
5 Services			
Shell & Core			
5.1 Sanitary Installations			
2 to 3 star	m² GIA	27.00 to	33.00
4 to 5 star	m² GIA	42.00 to	50.00
5.3 Disposal Installations			
2 to 3 star	m² GIA	29.00 to	35.00
4 to 5 star	m² GIA	30.00 to	36.00
5.4 Water Installations			
2 to 3 star	m² GIA	39.00 to	48.00
4 to 5 star	m² GIA	53.00 to	68.00
5.6 Space Heating and Air Conditioning			
LPHW heating installation; including gas installations			
2 to 3 star	m² GIA	42.00 to	51.00
4 to 5 star	m² GIA	42.00 to	53.00
Air conditioning; including ventilation			
2 to 3 star – 4 pipe fan coil	m² GIA	240.00 to	300.00
4 to 5 star – 4 pipe fan coil	m² GIA	250.00 to	305.00
2 to 3 star – 3 pipe variable refrigerant volume	m² GIA	150.00 to	190.00
4 to 5 star – 3 pipe variable refrigerant volume	m² GIA	150.00 to	190.00
5.8 Electrical Installations			
LV installations			
Standby generators (life safety only)			
2 to 3 star	m² GIA	16.50 to	20.00
4 to 5 star	m² GIA	17.00 to	21.00
LV distribution			
2 to 3 star	m² GIA	35.00 to	43.00
4 to 5 star	m² GIA	47.00 to	57.00
Lighting installations			
2 to 3 star	m² GIA	25.00 to	30.00
4 to 5 star	m² GIA	37.00 to	46.00
Small power			
2 to 3 star	m² GIA	9.00 to	10.50
4 to 5 star	m² GIA	14.00 to	18.00
Electrical installations for mechanical plant			
2 to 3 star	m² GIA	7.00 to	9.00
4 to 5 star	m² GIA	7.00 to	9.00

RIBA STAGE 2/3 DESIGN

Item	Unit	Range £	
Elemental Rates for Alternative Engineering Services Solutions – Hotels – cont			
Shell & Core – cont			
5.11 Fire and Lightning Protection			
Fire protection			
2 to 3 star – Dry risers	m² GIA	11.00 to	14.00
4 to 5 star – Dry risers	m² GIA	11.00 to	14.00
2 to 3 star – Sprinkler installation	m² GIA	25.00 to	30.00
4 to 5 star – Sprinkler installation	m² GIA	28.00 to	34.00
Earthing			
2 to 3 star	m² GIA	2.00 to	2.50
4 to 5 star	m² GIA	2.00 to	2.50
Lightning protection			
2 to 3 star	m² GIA	2.00 to	3.00
4 to 5 star	m² GIA	2.00 to	3.00
5.12 Communication, Security and Control Systems			
Fire alarms			
2 to 3 star	m² GIA	16.50 to	21.00
4 to 5 star	m² GIA	16.50 to	23.00
IT			
2 to 3 star	m² GIA	14.00 to	17.00
4 to 5 star	m² GIA	14.00 to	17.00
Security			
2 to 3 star	m² GIA	20.00 to	27.00
4 to 5 star	m² GIA	21.00 to	27.00
BMS controls; including MCC panels and control cabling			
2 to 3 star	m² GIA	13.00 to	17.00
4 to 5 star	m² GIA	32.00 to	38.00

RIBA STAGE 2/3 DESIGN

Item	Unit	Range £	
Elemental Rates for Alternative Engineering Services Solutions – Apartments			
The following examples of building types indicate the range of rates for alternative design solutions for each of the engineering services elements based on Gross Internal Area for the Shell and Core and Net Internal Area for the Fit Out. Fit Out is assumed to be to Cat A Standard. Consideration should be made for the size of the building height, sales point (£/ft²) and location, which may affect the economies of scale for rates, i.e. the larger the building the lower the rates.			
5 Services			
Shell & Core			
5.1 Sanitary Installations			
Affordable	m² GIA	0.50 to	0.75
Private	m² GIA	2.50 to	3.00
5.3 Disposal Installations			
Affordable	m² GIA	19.00 to	24.00
Private	m² GIA	23.00 to	29.00
5.4 Water Installations			
Affordable	m² GIA	19.00 to	23.00
Private	m² GIA	32.00 to	41.00
5.5 Heat Source			
Affordable	m² GIA	9.00 to	11.00
Private	m² GIA	11.00 to	14.00
5.6 Space Heating and Air Conditioning			
Affordable	m² GIA	23.00 to	28.00
Private	m² GIA	74.00 to	95.00
5.7 Ventilation Systems			
Affordable	m² GIA	17.00 to	20.00
Private	m² GIA	22.00 to	26.00
5.8 Electrical Installations			
Affordable	m² GIA	41.00 to	51.00
Private	m² GIA	63.00 to	80.00
5.9 Fuel Installations (Gas)			
Affordable	m² GIA	2.50 to	3.00
Private	m² GIA	4.00 to	5.00
5.11 Fire and Lightning Protection			
Affordable	m² GIA	21.00 to	25.00
Private	m² GIA	30.00 to	37.00
5.12 Communications, Security Control Systems			
Affordable	m² GIA	32.00 to	39.00
Private	m² GIA	45.00 to	55.00
5.13 Specialist Installations			
Affordable	m² GIA	9.00 to	11.00
Private	m² GIA	38.00 to	50.00

RIBA STAGE 2/3 DESIGN

Item	Unit	Range £	
Elemental Rates for Alternative Engineering Services Solutions – Apartments – cont			
Fit Out			
5.1 Sanitary Installations			
Affordable	m² NIA	32.00 to	39.00
Private	m² NIA	95.00 to	115.00
5.3 Disposal Installations			
Affordable	m² NIA	10.00 to	12.00
Private	m² NIA	24.00 to	30.00
5.4 Water Installations			
Affordable	m² NIA	27.00 to	32.00
Private	m² NIA	50.00 to	62.00
5.6 Space Heating and Air Conditioning			
Affordable	m² NIA	73.00 to	90.00
Private	m² NIA	200.00 to	245.00
5.7 Ventilation Systems			
Affordable (whole house vent)	m² NIA	35.00 to	42.00
Private (whole house vent)	m² NIA	50.00 to	60.00
5.8 Electrical Installations			
Affordable	m² NIA	57.00 to	70.00
Private	m² NIA	120.00 to	150.00
5.11 Fire and Lightning Protection			
Affordable	m² NIA	21.00 to	26.00
Private	m² NIA	27.00 to	34.00
5.12 Communication, Security and Control Systems			
Affordable	m² NIA	18.00 to	23.00
Private	m² NIA	98.00 to	120.00
5.13 Specialist Installations			
Affordable	m² NIA	11.00 to	13.00
Private	m² NIA	30.00 to	37.00

APPROXIMATE ESTIMATING RATES – 5 SERVICES

Item	Unit	Range £	
5.3 Disposal Installations			
ABOVE GROUND DRAINAGE			
Soil and waste	point	380.00 to	475.00
5.4 Water Installations			
Cold Water	point	380.00 to	475.00
Hot Water	point	450.00 to	550.00
Pipework			
Hot and cold water			
Excludes insulation, valves and ancillaries, etc.			
Light gauge copper tube to EN1057 R250 (TX) formerly BS 2871 Part 1 Table X			
with joints as described including allowance for waste, fittings and supports			
assuming average runs with capillary joints up to 54 mm and bronze welded			
thereafter			
Horizontal High Level Distribution			
15 mm	m	35.00 to	42.00
22 mm	m	36.00 to	43.00
28 mm	m	45.00 to	55.00
35 mm	m	51.00 to	61.00
42 mm	m	59.00 to	72.00
54 mm	m	75.00 to	92.00
67 mm	m	90.00 to	115.00
76 mm	m	107.00 to	132.00
108 mm	m	163.00 to	208.00
Risers			
15 mm	m	20.00 to	24.00
22 mm	m	23.00 to	28.00
28 mm	m	35.00 to	41.00
35 mm	m	38.00 to	46.00
42 mm	m	47.00 to	57.00
54 mm	m	54.00 to	66.00
67 mm	m	95.00 to	112.00
76 mm	m	105.00 to	130.00
108 mm	m	135.00 to	160.00
Toilet Areas, etc., at Low Level			
15 mm	m	59.00 to	71.00
22 mm	m	72.00 to	87.00
28 mm	m	96.00 to	117.00
LTHW and chilled water			
Excludes insulation, valves and ancillaries, etc.			
Black heavy weight mild steel tube to BS 1387 with joints in the running length,			
allowance for waste, fittings and supports assuming average runs			
Horizontal Distribution – Basements, etc.			
15 mm	m	42.00 to	50.00
20 mm	m	45.00 to	55.00
25 mm	m	51.00 to	62.00

APPROXIMATE ESTIMATING RATES – 5 SERVICES

Item	Unit	Range £	
5.4 Water Installations – cont			
LTHW and chilled water – cont			
32 mm	m	59.00 to	72.00
40 mm	m	67.00 to	82.00
50 mm	m	83.00 to	100.00
65 mm	m	88.00 to	107.00
80 mm	m	116.00 to	140.00
100 mm	m	155.00 to	185.00
125 mm	m	215.00 to	265.00
150 mm	m	295.00 to	350.00
200 mm	m	390.00 to	470.00
250 mm	m	480.00 to	590.00
300 mm	m	570.00 to	690.00
Risers			
15 mm	m	26.00 to	32.00
20 mm	m	31.00 to	37.00
25 mm	m	34.00 to	41.00
32 mm	m	38.00 to	46.00
40 mm	m	43.00 to	51.00
50 mm	m	53.00 to	64.00
65 mm	m	65.00 to	80.00
80 mm	m	89.00 to	106.00
100 mm	m	110.00 to	135.00
125 mm	m	140.00 to	165.00
150 mm	m	180.00 to	220.00
200 mm	m	280.00 to	350.00
250 mm	m	365.00 to	445.00
300 mm	m	410.00 to	510.00
On Floor Distribution			
15 mm	m	41.00 to	49.00
20 mm	m	46.00 to	56.00
25 mm	m	50.00 to	62.00
32 mm	m	57.00 to	70.00
40 mm	m	65.00 to	80.00
50 mm	m	77.00 to	94.00
Plantroom Areas, etc.			
15 mm	m	43.00 to	51.00
20 mm	m	47.00 to	57.00
25 mm	m	54.00 to	66.00
32 mm	m	62.00 to	75.00
40 mm	m	70.00 to	85.00
50 mm	m	83.00 to	102.00
65 mm	m	92.00 to	107.00
80 mm	m	135.00 to	16500
100 mm	m	155.00 to	190.00
125 mm	m	230.00 to	280.00
150 mm	m	310.00 to	370.00
200 mm	m	400.00 to	490.00
250 mm	m	510.00 to	615.00
300 mm	m	600.00 to	725.00

APPROXIMATE ESTIMATING RATES – 5 SERVICES

Item	Unit	Range £	
5.5 Heat Source			
Gas fired boilers including gas train and controls	kW	36.00 to	43.00
Gas fired boilers including gas train, controls, flue, plantroom pipework, valves and insulation, pumps and pressurization unit	kW	99.00 to	124.00
5.6 Space Heating and Air Conditioning			
Chilled Water			
Air cooled R134a refrigerant chiller including control panel, anti-vibration mountings	kW	150.00 to	200.00
Air cooled R134a refrigerant chiller including control panel, anti-vibration mountings, plantroom pipework, valves, insulation, pumps and pressurization units	kW	240.00 to	295.00
Water cooled R134a refrigerant chiller including control panel, anti-vibration mountings	kW	85.00 to	130.00
Water cooled R134a refrigerant chiller including control panel, anti-vibration mountings, plantroom pipework, valves, insulation, pumps and pressurization units	kW	180.00 to	210.00
Absorption steam medium chiller including control panel, anti-vibration mountings, plantroom pipework, valves, insulation, pumps and pressurization units	kW	315.00 to	380.00
Heat Rejection			
Open circuit, forced draft cooling tower	kW	75.00 to	92.00
Closed circuit, forced draft cooling tower	kW	67.00 to	82.00
Dry Air	kW	37.00 to	46.00
Pumps			
Pumps including flexible connections, anti-vibration mountings	kPa	38.00 to	47.00
Pumps including flexible connections, anti-vibration mountings, plantroom pipework, valves, insulation and accessories	kPa	96.00 to	116.00
Ductwork			
The rates below allow for ductwork and for all other labour and material in fabrication, fittings, supports and jointing to equipment, stop and capped ends, elbows, bends, diminishing and transition pieces, regular and reducing couplings, volume control dampers, branch diffuser and 'snap on' grille connections, ties, 'Ys', crossover spigots, etc., turning vanes, regulating dampers, access doors and openings, hand-holes, test holes and covers, blanking plates, flanges, stiffeners, tie rods and all supports and brackets fixed to structure.			
Rectangular galvanized mild steel ductwork as HVCA DW 144 up to 1000 mm longest side	m²	62.00 to	75.00
Rectangular galvanized mild steel ductwork as HVCA DW 144 up to 2500 mm longest side	m²	67.00 to	82.00
Rectangular galvanized mild steel ductwork as HVCA DW 144 up to 3000 mm longest side and above	m²	92.00 to	110.00
Circular galvanized mild steel ductwork as HVCA DW 144	m²	67.00 to	82.00
Flat oval galvanized mild steel ductwork as HVCA DW 144 up to 545 mm wide	m²	67.00 to	82.00
Flat oval galvanized mild steel ductwork as HVCA DW 144 up to 880 mm wide	m²	75.00 to	92.00
Flat oval galvanized mild steel ductwork as HVCA DW 144 up to 1785 mm wide	m²	85.00 to	105.00

APPROXIMATE ESTIMATING RATES – 5 SERVICES

Item	Unit	Range £
5.6 Space Heating and Air Conditioning – cont		
Packaged Air Handling Units		
Air handling unit including LPHW preheater coil, pre-filter panel, LPHW heater coils, chilled water coil, filter panels, inverter drive, motorized volume control dampers, sound attenuation, flexible connections to ductwork and all anti-vibration mountings	m³/s	7500.00 to 10000.00
5.7 Ventilation Systems		
Extract Fans		
Extract fan including inverter drive, sound attenuation, flexible connections to ductwork and all anti-vibration mountings	m³/s	2800.00 to 3300.00
5.8 Electrical Installations		
HV/LV Installations		
The cost of HV/LV equipment will vary according to the electricity supplier's requirements, the duty required and the actual location of the site. For estimating purposes, the items indicated below are typical of the equipment required in a HV substation incorporated into a building.		
Ring Main Unit		
Ring Main Unit, 11 kV including electrical terminations	point	16500.00 to 20000.00
Transfomers		
Oil filled transformers, 11 kV to 415 kV including electrical terminations	kVA	18.00 to 22.00
Cast resin transformers, 11 kV to 415 kV including electrical terminations	kVA	20.00 to 23.00
Midal filled transformers, 11 kV to 415 kV including electrical terminations	kVA	21.00 to 25.00
HV Switchgear		
Cubicle section HV switchpanel, Form 4 type 6 including air circuit breakers, meters and electrical terminations	section	17000.00 to 25000.00
LV Switchgear		
LV switchpanel, Form 3 including all isolators, fuses, meters and electrical terminations	isolator	2500.00 to 3150.00
LV switchpanel, Form 4 type 5 including all isolators, fuses, meters and electrical terminations	isolator	3925.00 to 4900.00
External Packaged Substation		
Extra over cost for prefabricated packaged substation housing, excludes base and protective security fencing	each	26500.00 to 32500.00
Standby Generating Sets		
Diesel powered including control panel, flue, oil day tank and attenuation		
Approximate installed cost, LV	kVA	310.00 to 370.00
Approximate installed cost, HV	kVA	350.00 to 420.00
Uninterruptible Power Supply		
Rotary UPS including control panel and choke transformer (excludes distribution)		
Approximate installed cost (range 1000 kVA to 2500 kVA)	kVA	485.00 to 585.00
Static UPS including control panel, automatic bypass, DC isolator and batteries for 10 minutes standby (excludes distribution)		
Approximate installed cost (range 500 kVA to 1000 kVA)	kVA	200.00 to 270.00

APPROXIMATE ESTIMATING RATES – 5 SERVICES

Item	Unit	Range £	
Small Power Approximate prices for wiring of power points of length not exceeding 20 m, including accessories, wireways but excluding distribution boards			
13 amp Accessories Wired in PVC insulated twin and earth cable in ring main circuit			
Domestic properties	point	61.00 to	74.00
Commercial properties	point	83.00 to	100.00
Industrial properties	point	83.00 to	100.00
Wired in PVC insulated twin and earth cable in radial circuit			
Domestic properties	point	83.00 to	100.00
Commercial properties	point	99.00 to	124.00
Industrial properties	point	99.00 to	124.00
Wired in LSF insulated single cable in ring main circuit			
Commercial properties	point	99.00 to	125.00
Industrial properties	point	99.00 to	125.00
Wired in LSF insulated single cable in radial circuit			
Commercial properties	point	120.00 to	145.00
Industrial properties	point	120.00 to	145.00
45 amp Wired in PVC insulated twin and earth cable			
Domestic properties	point	110.00 to	140.00
Low Voltage Power Circuits Three phase four wire radial circuit feeding an individual load, wired in LSF insulated single cable including all wireways, isolator, not exceeding 10 metres; in commercial properties Cable size			
1.5 mm²	point	200.00 to	250.00
2.5 mm²	point	220.00 to	265.00
4 mm²	point	230.00 to	275.00
6 mm²	point	250.00 to	320.00
10 mm²	point	300.00 to	370.00
16 mm²	point	330.00 to	400.00
Three phase four core radial circuit feeding an individual load item, wired in LSF/ SWA/XLPE insulated cable including terminations, isolator, clipped to surface, not exceeding 10 metres; in commercial properties Cable size			
1.5 mm²	point	150.00 to	185.00
2.5 mm²	point	170.00 to	200.00
4 mm²	point	175.00 to	220.00
6 mm²	point	200.00 to	245.00
10 mm²	point	295.00 to	366.00
16 mm²	point	375.00 to	460.00

APPROXIMATE ESTIMATING RATES – 5 SERVICES

Item	Unit	Range £	
5.8 Electrical Installations – cont			
Lighting			
Approximate prices for wiring of lighting points including rose, wireways but excluding distribution boards, luminaires and switches			
Final Circuits			
Wired in PVC insulated twin and earth cable			
Domestic properties	point	44.00 to	54.00
Commercial properties	point	55.00 to	67.00
Industrial properties	point	55.00 to	67.00
Wired in LSF insulated single cable			
Commercial properties	point	72.00 to	88.00
Industrial properties	point	72.00 to	88.00
Electrical Works in Connection with Mechanical Services			
The cost of electrical connections to mechanical services equipment will vary depending on the type of building and complexity of the equipment.			
Typical rate for power wiring, isolators and associated wireways	m²	7.00 to	10.00
Fire Alarms			
Cost per point for two core FP200 wired system including all terminations, supports and wireways			
Call point	point	270.00 to	325.00
Smoke detector	point	230.00 to	280.00
Smoke/heat detector	point	270.00 to	325.00
Heat detector	point	250.00 to	310.00
Heat detector and sounder	point	230.00 to	270.00
Input/output/relay units	point	325.00 to	390.00
Alarm sounder	point	250.00 to	310.00
Alarm sounder/beacon	point	300.00 to	370.00
Speakers/voice sounders	point	300.00 to	370.00
Speakers/voice sounders (weatherproof)	point	325.00 to	390.00
Beacon/strobe	point	230.00 to	280.00
Beacon/strobe (weatherproof)	point	340.00 to	420.00
Door release units	point	340.00 to	420.00
Beam detector	point	970.00 to	1175.00
For costs for zone control panel, battery chargers and batteries, see 'Prices for Measured Work' section.			
External Lighting			
Estate Road Lighting			
Post type road lighting lantern 70 watt CDM-T 3000k complete with 5 m high column with hinged lockable door, control gear and cut-out including 2.5 mm two core butyl cable internal wiring, interconnections and earthing fed by 16 mm² four core XLPE/SWA/LSF cable and terminations. Approximate installed price per metre road length (based on 300 metres run) including time switch but excluding builder's work in connection			
Columns erected at 30 m intervals along road (cost per m of road)	m	76.00 to	95.00

APPROXIMATE ESTIMATING RATES – 5 SERVICES

Item	Unit	Range £
Bollard Lighting		
Bollard lighting fitting 26 watt TC-D 3500k including control gear, all internal wiring, interconnections, earthing and 25 metres of 2.5 mm² three core XLPE/SWA/LSF cable		
Approximate installed price excluding builder's work in connection	each	1100.00 to 1350.00
Outdoor Flood Lighting		
Wall mounted outdoor flood light fitting complete with tungsten halogen lamp, mounting bracket, wire guard and all internal wiring and containment, fixed to brickwork or concrete and connected		
Installed price 500 watt	point	220.00 to 275.00
Installed price 1000 watt	point	265.00 to 325.00
Pedestal mounted outdoor flood light fitting complete with 1000 watt MBF/U lamp, control gear, contained in weatherproof steel box, all internal wiring and containment, interconnections and earthing, fixed to brickwork or concrete and connected		
Approximate installed price excluding builder's work in connection	each	1100.00 to 1350.00
Building Management Installations		
Category A Fit-Out		
Option 1 – 185 Nr four pipe fan coil – 740 points		
1.0 Field Equipment		
Network devices; Valves/actuators; Sensing devices	point	74.00 to 76.00
2.0 Cabling		
Power – from local isolator to DDC controller; Control – from DDC controller to field equipment	point	43.00 to 45.00
3.0 Programming		
Software – central facility; Software – network devices; Graphics	point	23.00 to 25.00
4.0 On site testing and commissioning		
Equipment; Programming/graphics; Power and control cabling	point	24.00 to 26.00
Total Option 1 – Four pipefan coil	point	160.00 to 170.00
Cost/FCU	each	670.00 to 690.00
Option 2 – 185 Nr two pipe fan coil system with electric heating – 740 points		
1.0 Field Equipment		
Network devices; Valves/actuators/thyristors; Sensing devices	point	97.00 to 100.00
2.0 Cabling		
Power – from local isolator to DDC controller; Control – from DDC controller to field equipment	point	45.00 to 47.00
3.0 Programming		
Software – central facility; Software – network devices; Graphics	point	24.00 to 26.00
4.0 On site testing and commissioning		
Equipment; Programming/graphics; Power and control cabling	point	25.00 to 27.00
Total Option 2 – Two pipe fan coil with electric heating	point	190.00 to 200.00
Cost/FCU	each	760.00 to 780.00

APPROXIMATE ESTIMATING RATES – 5 SERVICES

Item	Unit	Range £	
5.8 Electrical Installations – cont			
Option 3 – 180 Nr chilled beams with perimeter heating – 567 points			
1.0 Field Equipment			
Network devices; Valves/actuators; Sensing devices	point	93.00 to	95.00
2.0 Cabling			
Power – from local isolator to DDC controller; Control – from DDC controller			
to field equipment	point	63.00 to	65.00
3.0 Programming			
Cost/Point	point	31.00 to	33.00
4.0 On site testing and commissioning			
Equipment; Programming/Graphics; Power and control cabling	point	33.00 to	35.00
Total Option 3 – Chilled beams with perimeter heating	point	220.00 to	225.00
Cost/Chilled Beam	each	700.00 to	720.00
Shell & Core Only			
Main Plant – Cost/BMS Points	each	730.00 to	740.00
Landlord FCUs – Cost/Terminal Units	each	730.00 to	740.00
Trade Contract Preliminaries	%	20.00 to	30.00
Notes: The following are included in points rates			
– DDC Controllers/Control Enclosures/Control Panels			
– Motor Control Centre (MCC)			
– Field Devices			
– Control and Power Cabling from DDC Controllers/MCC			
– Programming			
– On Site Testing and Commissioning			

APPROXIMATE ESTIMATING RATES – 5 SERVICES

Item	Unit	Range £
5.10 Lift and Conveyor Systems		
Lift Installations		
The cost of lift installations will vary depending upon a variety of circumstances. The following prices assume a car height of 2.2 m, manufacturer's standard car finish, brushed stainless steel 2 panel centre opening doors to BSEN81–20 and Lift Regulations 2016.		
Passenger Lifts, machine room above		
Electrically operated AC drive serving 2 levels with directional collective controls and a speed of 1.0 m/s		
8 Person	item	65000.00 to 79000.00
10 Person	item	70000.00 to 85000.00
13 Person	item	74000.00 to 90000.00
17 Person	item	84000.00 to 101000.00
21 Person	item	93000.00 to 114000.00
26 Person	item	104000.00 to 130000.00
Electrically operated AC drive serving 4 levels and a speed of 1.0 m/s		
8 Person	item	76000.00 to 93000.00
10 Person	item	80000.00 to 98000.00
13 Person	item	85000.00 to 104000.00
17 Person	item	94000.00 to 114000.00
21 Person	item	105000.00 to 130000.00
26 Person	item	125000.00 to 152000.00
Electrically operated AC drive serving 6 levels and a speed of 1.0 m/s		
8 Person	item	86000.00 to 105000.00
10 Person	item	90000.00 to 108000.00
13 Person	item	94000.00 to 114000.00
17 Person	item	105000.00 to 130000.00
21 Person	item	125000.00 to 152000.00
26 Person	item	135000.00 to 162000.00
Electrically operated AC drive serving 8 levels and a speed of 1.0 m/s		
8 Person	item	96000.00 to 119000.00
10 Person	item	101000.00 to 119000.00
13 Person	item	105000.00 to 130000.00
17 Person	item	125000.00 to 152000.00
21 Person	item	130000.00 to 157000.00
26 Person	item	152000.00 to 184000.00
Electrically operated AC drive serving 10 levels and a speed of 1.0 m/s		
8 Person	item	105000.00 to 130000.00
10 Person	item	108000.00 to 135000.00
13 Person	item	108000.00 to 135000.00
17 Person	item	130000.00 to 157000.00
21 Person	item	146000.00 to 173000.00
26 Person	item	168000.00 to 206000.00
Electrically operated AC drive serving 12 levels and a speed of 1.0 m/s		
8 Person	item	108000.00 to 135000.00
10 Person	item	125000.00 to 152000.00
13 Person	item	125000.00 to 152000.00
17 Person	item	141000.00 to 173000.00
21 Person	item	157000.00 to 195000.00
26 Person	item	179000.00 to 216000.00

APPROXIMATE ESTIMATING RATES – 5 SERVICES

Item	Unit	Range £
5.10 Lift and Conveyor Systems – cont		
Passenger Lifts, machine room above – cont		
Electrically operated AC drive serving 14 levels and a speed of 1.0 m/s		
8 Person	item	130000.00 to 157000.00
10 Person	item	130000.00 to 157000.00
13 Person	item	141000.00 to 173000.00
17 Person	item	157000.00 to 195000.00
21 Person	item	168000.00 to 206000.00
26 Person	item	195000.00 to 238000.00
Add to above for:		
Increase speed from 1.0 m/s to 1.6 m/s		
8 Person	item	4400.00 to 5400.00
10 Person	item	4600.00 to 5700.00
13 Person	item	4700.00 to 5800.00
17 Person	item	4700.00 to 5800.00
21 Person	item	4700.00 to 5800.00
26 Person	item	4700.00 to 5800.00
Increase speed from 1.6 m/s to 2.0 m/s		
8 Person	item	6200.00 to 7500.00
10 Person	item	6600.00 to 8100.00
13 Person	item	7300.00 to 8800.00
17 Person	item	3300.00 to 3900.00
21 Person	item	3300.00 to 4100.00
26 Person	item	3300.00 to 4100.00
Increase speed from 2.0 m/s to 2.5 m/s		
8 Person	item	2300.00 to 2700.00
10 Person	item	2300.00 to 2700.00
13 Person	item	2700.00 to 3300.00
17 Person	item	2700.00 to 3300.00
21 Person	item	3300.00 to 3900.00
26 Person	item	3300.00 to 3900.00
Enhanced finish to car – Centre mirror, flat ceiling, carpet		
8 Person	item	3200.00 to 3800.00
10 Person	item	3400.00 to 4200.00
13 Person	item	3400.00 to 4200.00
17 Person	item	3800.00 to 4700.00
21 Person	item	4500.00 to 5400.00
26 Person	item	5100.00 to 6300.00
Bottom motor room		
8 Person	item	7700.00 to 9300.00
10 Person	item	7700.00 to 9300.00
13 Person	item	7700.00 to 9300.00
17 Person	item	9400.00 to 11400.00
21 Person	item	9400.00 to 11400.00
26 Person	item	9600.00 to 11900.00

APPROXIMATE ESTIMATING RATES – 5 SERVICES

Item	Unit	Range £	
Firefighting control			
8 Person	item	6200.00 to	7500.00
10 Person	item	6200.00 to	7500.00
13 Person	item	6200.00 to	7500.00
17 Person	item	6200.00 to	7500.00
21 Person	item	6200.00 to	7500.00
26 Person	item	6200.00 to	7500.00
Glass back			
8 Person	item	2800.00 to	3400.00
10 Person	item	3200.00 to	3800.00
13 Person	item	3700.00 to	4600.00
17 Person	item	4500.00 to	5400.00
21 Person	item	4500.00 to	5400.00
26 Person	item	4500.00 to	5400.00
Glass doors			
8 Person	item	21600.00 to	26000.00
10 Person	item	21600.00 to	26000.00
13 Person	item	23300.00 to	28100.00
17 Person	item	24900.00 to	30800.00
21 Person	item	23300.00 to	28100.00
26 Person	item	23300.00 to	28100.00
Painting to entire pit			
8 Person	item	2300.00 to	2700.00
10 Person	item	2300.00 to	2700.00
13 Person	item	2300.00 to	2700.00
17 Person	item	2300.00 to	2700.00
21 Person	item	2300.00 to	2700.00
26 Person	item	2300.00 to	2700.00
Dual seal shaft			
8 Person	item	4500.00 to	5400.00
10 Person	item	4500.00 to	5400.00
13 Person	item	4500.00 to	5400.00
17 Person	item	5400.00 to	6600.00
21 Person	item	5400.00 to	6600.00
26 Person	item	5400.00 to	6600.00
Dust sealing machine room			
8 Person	item	880.00 to	1070.00
10 Person	item	880.00 to	1070.00
13 Person	item	1500.00 to	1800.00
17 Person	item	1500.00 to	1800.00
21 Person	item	1500.00 to	1800.00
26 Person	item	1500.00 to	1800.00
Intercom to reception desk and security room			
8 Person	item	440.00 to	540.00
10 Person	item	440.00 to	540.00
13 Person	item	440.00 to	540.00
17 Person	item	440.00 to	540.00
21 Person	item	440.00 to	540.00
26 Person	item	440.00 to	540.00

APPROXIMATE ESTIMATING RATES – 5 SERVICES

Item	Unit	Range £	
5.10 Lift and Conveyor Systems – cont			
Add to above for – cont			
Heating, cooling and ventilation to machine room			
8 Person	item	1110.00 to	1380.00
10 Person	item	1110.00 to	1380.00
13 Person	item	1110.00 to	1380.00
17 Person	item	1110.00 to	1380.00
21 Person	item	1110.00 to	1380.00
26 Person	item	1110.00 to	1380.00
Shaft lighting/small power			
8 Person	item	3840.00 to	4700.00
10 Person	item	3840.00 to	4700.00
13 Person	item	3840.00 to	4700.00
17 Person	item	3840.00 to	4700.00
21 Person	item	3840.00 to	4700.00
26 Person	item	3840.00 to	4700.00
Motor room lighting/small power			
8 Person	item	1410.00 to	1710.00
10 Person	item	1410.00 to	1710.00
13 Person	item	1760.00 to	2140.00
17 Person	item	1760.00 to	2140.00
21 Person	item	1760.00 to	2140.00
26 Person	item	1760.00 to	2140.00
Lifting beams			
8 Person	item	1680.00 to	2030.00
10 Person	item	1680.00 to	2030.00
13 Person	item	1680.00 to	2030.00
17 Person	item	1870.00 to	2300.00
21 Person	item	1870.00 to	2300.00
26 Person	item	1870.00 to	2300.00
10 mm equipotential bonding of all entrance metalwork			
8 Person	item	880.00 to	1070.00
10 Person	item	880.00 to	1070.00
13 Person	item	880.00 to	1070.00
17 Person	item	880.00 to	1070.00
21 Person	item	880.00 to	1070.00
26 Person	item	880.00 to	1070.00
Shaft secondary steelwork			
8 Person	item	6400.00 to	7900.00
10 Person	item	6600.00 to	8000.00
13 Person	item	6900.00 to	8400.00
17 Person	item	7000.00 to	8600.00
21 Person	item	7000.00 to	8600.00
26 Person	item	7000.00 to	8600.00
Independent insurance inspection			
8 Person	item	2110.00 to	2570.00
10 Person	item	2110.00 to	2570.00
13 Person	item	2110.00 to	2570.00
17 Person	item	2110.00 to	2570.00
21 Person	item	2110.00 to	2570.00
26 Person	item	2110.00 to	2570.00

APPROXIMATE ESTIMATING RATES – 5 SERVICES

Item	Unit	Range £
12 month warranty service		
8 Person	item	1140.00 to 1380.00
10 Person	item	1140.00 to 1380.00
13 Person	item	1140.00 to 1380.00
17 Person	item	1140.00 to 1380.00
21 Person	item	1140.00 to 1380.00
26 Person	item	1140.00 to 1380.00
Passenger Lifts, machine room less		
Electrically operated AC drive serving 2 levels with directional collective controls and a speed of 1.0 m/s		
8 Person	item	59000.00 to 71000.00
10 Person	item	65000.00 to 80000.00
13 Person	item	69000.00 to 84000.00
17 Person	item	83000.00 to 100000.00
21 Person	item	90000.00 to 108000.00
26 Person	item	100000.00 to 119000.00
Electrically operated AC drive serving 4 levels and a speed of 1.0 m/s		
8 Person	item	67000.00 to 83000.00
10 Person	item	74000.00 to 90000.00
13 Person	item	78000.00 to 96000.00
17 Person	item	93000.00 to 114000.00
21 Person	item	101000.00 to 119000.00
26 Person	item	108000.00 to 135000.00
Electrically operated AC drive serving 6 levels and a speed of 1.0 m/s		
8 Person	item	78000.00 to 96000.00
10 Person	item	84000.00 to 101000.00
13 Person	item	88000.00 to 107000.00
17 Person	item	101000.00 to 119000.00
21 Person	item	108000.00 to 135000.00
26 Person	item	119000.00 to 141000.00
Electrically operated AC drive serving 8 levels and a speed of 1.0 m/s		
8 Person	item	87000.00 to 106000.00
10 Person	item	92000.00 to 108000.00
13 Person	item	98000.00 to 119000.00
17 Person	item	108000.00 to 135000.00
21 Person	item	119000.00 to 141000.00
26 Person	item	130000.00 to 162000.00
Electrically operated AC drive serving 10 levels and a speed of 1.0 m/s		
8 Person	item	97000.00 to 119000.00
10 Person	item	100000.00 to 119000.00
13 Person	item	108000.00 to 130000.00
17 Person	item	119000.00 to 141000.00
21 Person	item	130000.00 to 162000.00
26 Person	item	152000.00 to 184000.00
Electrically operated AC drive serving 12 levels and a speed of 1.0 m/s		
8 Person	item	108000.00 to 130000.00
10 Person	item	108000.00 to 130000.00
13 Person	item	119000.00 to 141000.00
17 Person	item	130000.00 to 162000.00
21 Person	item	141000.00 to 173000.00
26 Person	item	157000.00 to 195000.00

APPROXIMATE ESTIMATING RATES – 5 SERVICES

Item	Unit	Range £
5.10 Lift and Conveyor Systems – cont		
Passenger Lifts, machine room less – cont		
Electrically operated AC drive serving 14 levels and a speed of 1.0 m/s		
8 Person	item	119000.00 to 141000.00
10 Person	item	119000.00 to 141000.00
13 Person	item	125000.00 to 152000.00
17 Person	item	152000.00 to 184000.00
21 Person	item	157000.00 to 195000.00
26 Person	item	179000.00 to 216000.00
Add to above for:		
Increase speed from 1.0 m/s to 1.6 m/s		
8 Person	item	2870.00 to 3510.00
10 Person	item	2870.00 to 3510.00
13 Person	item	3030.00 to 3680.00
17 Person	item	4650.00 to 5620.00
21 Person	item	5350.00 to 6590.00
26 Person	item	6810.00 to 8320.00
Enhanced finish to car – Centre mirror, flat ceiling, carpet		
8 Person	item	3140.00 to 3780.00
10 Person	item	3300.00 to 4050.00
13 Person	item	3300.00 to 4050.00
17 Person	item	3780.00 to 4650.00
21 Person	item	4380.00 to 5350.00
26 Person	item	5190.00 to 6270.00
Firefighting control		
8 Person	item	6050.00 to 7350.00
10 Person	item	6050.00 to 7350.00
13 Person	item	6050.00 to 7350.00
17 Person	item	6050.00 to 7350.00
21 Person	item	6050.00 to 7350.00
26 Person	item	6050.00 to 7350.00
Painting to entire pit		
8 Person	item	2190.00 to 2680.00
10 Person	item	2190.00 to 2680.00
13 Person	item	2190.00 to 2680.00
17 Person	item	2190.00 to 2680.00
21 Person	item	2190.00 to 2680.00
26 Person	item	2190.00 to 2680.00
Dual seal shaft		
8 Person	item	4430.00 to 5400.00
10 Person	item	4430.00 to 5400.00
13 Person	item	4430.00 to 5400.00
17 Person	item	5350.00 to 6590.00
21 Person	item	5350.00 to 6590.00
26 Person	item	5350.00 to 6590.00

APPROXIMATE ESTIMATING RATES – 5 SERVICES

Item	Unit	Range £	
Shaft lighting/small power			
8 Person	item	3840.00 to	4700.00
10 Person	item	3840.00 to	4700.00
13 Person	item	3840.00 to	4700.00
17 Person	item	3840.00 to	4700.00
21 Person	item	3840.00 to	4700.00
26 Person	item	3840.00 to	4700.00
Intercom to reception desk and security room			
8 Person	item	440.00 to	540.00
10 Person	item	440.00 to	540.00
13 Person	item	440.00 to	540.00
17 Person	item	440.00 to	540.00
21 Person	item	440.00 to	540.00
26 Person	item	440.00 to	540.00
Lifting beams			
8 Person	item	1650.00 to	2030.00
10 Person	item	1650.00 to	2030.00
13 Person	item	1650.00 to	2030.00
17 Person	item	1870.00 to	2300.00
21 Person	item	1870.00 to	2300.00
26 Person	item	1870.00 to	2300.00
10 mm equipotential bonding of all entrance metalwork			
8 Person	item	880.00 to	1070.00
10 Person	item	880.00 to	1070.00
13 Person	item	880.00 to	1070.00
17 Person	item	880.00 to	1070.00
21 Person	item	880.00 to	1070.00
26 Person	item	880.00 to	1070.00
Shaft secondary steelwork			
8 Person	item	6270.00 to	7560.00
10 Person	item	6380.00 to	7890.00
13 Person	item	6590.00 to	8100.00
17 Person	item	6810.00 to	8320.00
21 Person	item	6810.00 to	8320.00
26 Person	item	6810.00 to	8320.00
Independent insurance inspection			
8 Person	item	2030.00 to	2460.00
10 Person	item	2030.00 to	2460.00
13 Person	item	2030.00 to	2460.00
17 Person	item	2030.00 to	2460.00
21 Person	item	2030.00 to	2460.00
26 Person	item	2030.00 to	2460.00
12 month warranty service			
8 Person	item	1070.00 to	1300.00
10 Person	item	1070.00 to	1300.00
13 Person	item	1070.00 to	1300.00
17 Person	item	1070.00 to	1300.00
21 Person	item	1070.00 to	1300.00
26 Person	item	1070.00 to	1300.00

APPROXIMATE ESTIMATING RATES – 5 SERVICES

Item	Unit	Range £
5.10 Lift and Conveyor Systems – cont		
Goods Lifts, machine room above		
Electrically operated two speed serving 2 levels to take 1000 kg load, prime coated internal finish and a speed of 1.0 m/s		
2000 kg	item	108000.00 to 135000.00
2250 kg	item	119000.00 to 141000.00
2500 kg	item	119000.00 to 141000.00
3000 kg	item	130000.00 to 162000.00
Electrically operated two speed serving 4 levels and a speed of 1.0 m/s		
2000 kg	item	119000.00 to 141000.00
2250 kg	item	130000.00 to 162000.00
2500 kg	item	135000.00 to 162000.00
3000 kg	item	152000.00 to 184000.00
Electrically operated two speed serving 6 levels and a speed of 1.0 m/s		
2000 kg	item	135000.00 to 162000.00
2250 kg	item	152000.00 to 184000.00
2500 kg	item	152000.00 to 184000.00
3000 kg	item	168000.00 to 206000.00
Electrically operated two speed serving 8 levels and a speed of 1.0 m/s		
2000 kg	item	152000.00 to 184000.00
2250 kg	item	157000.00 to 195000.00
2500 kg	item	168000.00 to 206000.00
3000 kg	item	189000.00 to 233000.00
Electrically operated two speed serving 10 levels and a speed of 1.0 m/s		
2000 kg	item	168000.00 to 206000.00
2250 kg	item	179000.00 to 216000.00
2500 kg	item	179000.00 to 216000.00
3000 kg	item	206000.00 to 249000.00
Electrically operated two speed serving 12 levels and a speed of 1.0 m/s		
2000 kg	item	179000.00 to 216000.00
2250 kg	item	189000.00 to 233000.00
2500 kg	item	189000.00 to 233000.00
3000 kg	item	227000.00 to 270000.00
Electrically operated two speed serving 14 levels and a speed of 1.0 m/s		
2000 kg	item	195000.00 to 238000.00
2250 kg	item	211000.00 to 260000.00
2500 kg	item	211000.00 to 260000.00
3000 kg	item	238000.00 to 292000.00
Add to above for:		
Increase speed of travel from 1.0 m/s to 1.6 m/s		
2000 kg	item	1410.00 to 1710.00
Enhanced finish to car – Centre mirror, flat ceiling, carpet		
2000 kg	item	3840.00 to 4700.00
2250 kg	item	3840.00 to 4700.00
2500 kg	item	3840.00 to 4700.00

APPROXIMATE ESTIMATING RATES – 5 SERVICES

Item	Unit	Range £
Bottom motor room		
2000 kg	item	9400.00 to 11340.00
Painting to entire pit		
2000 kg	item	2190.00 to 2680.00
2250 kg	item	2190.00 to 2680.00
2500 kg	item	2190.00 to 2680.00
3000 kg	item	2190.00 to 2680.00
Dual seal shaft		
2000 kg	item	4430.00 to 5400.00
2250 kg	item	4430.00 to 5400.00
2500 kg	item	4430.00 to 5400.00
3000 kg	item	4430.00 to 5400.00
Intercom to reception desk and security room		
2000 kg	item	450.00 to 540.00
2250 kg	item	450.00 to 540.00
2500 kg	item	450.00 to 540.00
3000 kg	item	450.00 to 540.00
Heating, cooling and ventilation to machine room		
2000 kg	item	1220.00 to 1490.00
2250 kg	item	1220.00 to 1490.00
2500 kg	item	1220.00 to 1490.00
3000 kg	item	1220.00 to 1490.00
Lifting beams		
2000 kg	item	1620.00 to 2000.00
2250 kg	item	1620.00 to 2000.00
2500 kg	item	1620.00 to 2000.00
3000 kg	item	1620.00 to 2000.00
10 mm equipotential bonding of all entrance metalwork		
2000 kg	item	880.00 to 1070.00
2250 kg	item	880.00 to 1070.00
2500 kg	item	880.00 to 1070.00
3000 kg	item	880.00 to 1070.00
Independent insurance inspection		
2000 kg	item	2080.00 to 2570.00
2250 kg	item	2080.00 to 2570.00
2500 kg	item	2080.00 to 2570.00
3000 kg	item	2080.00 to 2570.00
12 month warranty service		
2000 kg	item	1140.00 to 1380.00
2250 kg	item	1330.00 to 1600.00
2500 kg	item	1330.00 to 1600.00
3000 kg	item	1330.00 to 1600.00
Goods Lift, machine room less		
Electrically operated two speed serving 2 levels to take 1000 kg load, prime coated internal finish and a speed of 1.0 m/s		
2000 kg	item	97000.00 to 119000.00
2250 kg	item	101000.00 to 119000.00
2500 kg	item	108000.00 to 135000.00

APPROXIMATE ESTIMATING RATES – 5 SERVICES

Item	Unit	Range £
5.10 Lift and Conveyor Systems – cont		
Goods Lift, machine room less – cont		
Electrically operated two speed serving 4 levels and a speed of 1.0 m/s		
2000 kg	item	108000.00 to 135000.00
2250 kg	item	114000.00 to 141000.00
2500 kg	item	125000.00 to 152000.00
Electrically operated two speed serving 6 levels and a speed of 1.0 m/s		
2000 kg	item	125000.00 to 152000.00
2250 kg	item	125000.00 to 152000.00
2500 kg	item	141000.00 to 173000.00
Electrically operated two speed serving 8 levels and a speed of 1.0 m/s		
2000 kg	item	130000.00 to 162000.00
2250 kg	item	141000.00 to 173000.00
2500 kg	item	152000.00 to 184000.00
Electrically operated two speed serving 10 levels and a speed of 1.0 m/s		
2000 kg	item	146000.00 to 179000.00
2250 kg	item	152000.00 to 184000.00
2500 kg	item	157000.00 to 195000.00
Electrically operated two speed serving 12 levels and a speed of 1.0 m/s		
2000 kg	item	130000.00 to 162000.00
2250 kg	item	168000.00 to 206000.00
2500 kg	item	179000.00 to 216000.00
Electrically operated two speed serving 14 levels and a speed of 1.0 m/s		
2000 kg	item	168000.00 to 206000.00
2250 kg	item	179000.00 to 216000.00
2500 kg	item	189000.00 to 233000.00
Add to above for:		
Increase speed of travel from 1.0 m/s to 1.6 m/s		
2000 kg	item	5350.00 to 6600.00
2250 kg	item	8540.00 to 10500.00
Enhanced finish to car – Centre mirror, flat ceiling, carpet		
2000 kg	item	4540.00 to 5600.00
2250 kg	item	5940.00 to 7300.00
2500 kg	item	8750.00 to 1070.00
Painting to entire pit		
2000 kg	item	2220.00 to 2700.00
2250 kg	item	2220.00 to 2700.00
2500 kg	item	2220.00 to 2700.00
Dual seal shaft		
2000 kg	item	4430.00 to 5400.00
2250 kg	item	4430.00 to 5400.00
2500 kg	item	4430.00 to 5400.00
Intercom to reception desk and security room		
2000 kg	item	450.00 to 540.00
2250 kg	item	450.00 to 540.00
2500 kg	item	450.00 to 540.00

APPROXIMATE ESTIMATING RATES – 5 SERVICES

Item	Unit	Range £		
Lifting beams				
2000 kg	item	1680.00	to	2030.00
2250 kg	item	1680.00	to	2030.00
2500 kg	item	1680.00	to	2030.00
10 mm equipotential bonding of all entrance metalwork				
2000 kg	item	880.00	to	1070.00
2250 kg	item	880.00	to	1070.00
2500 kg	item	880.00	to	1070.00
Independent insurance inspection				
2000 kg	item	2110.00	to	2570.00
2250 kg	item	2110.00	to	2570.00
2500 kg	item	2110.00	to	2570.00
12 month warranty service				
2000 kg	item	1140.00	to	1380.00
2250 kg	item	1140.00	to	1380.00
2500 kg	item	1140.00	to	1380.00
Escalator Installations				
30Ø Pitch escalator with a rise of 3 to 6 metres with standard balustrades				
1000 mm step width	item	88000.00	to	107000.00
Add to above for:				
Balustrade lighting	item	2700.00	to	3400.00
Skirting lighting	item	10800.00	to	13500.00
Emergency stop button pedestals	item	4900.00	to	6000.00
Truss cladding – stainless steel	item	27600.00	to	33500.00
Truss cladding – spray painted steel	item	24300.00	to	29700.00

APPROXIMATE ESTIMATING RATES – 5 SERVICES

Item	Unit	Range £
5.11 Fire and Lightning Protection		
Sprinkler Installations		
Recommended maximum area coverage per sprinkler head:		
Extra light hazard, 21 m² of floor area		
Ordinary hazard, 12 m² of floor area		
Extra high hazard, 9 m² of floor area		
Equipment		
Sprinkler equipment installation, pipework, valve sets, booster pumps and water storage	item	74000.00 to 90000.00
Price per sprinkler head; including pipework, valves and supports	point	190.00 to 230.00
Hose Reels and Dry Risers		
Wall mounted concealed hose reel with 36 m hose including approximately 15 m of pipework and isolating valve:		
Price per hose reel	point	1925.00 to 2375.00
100 mm dry riser main including 2 way breeching valve and box, 65 mm landing valve, complete with padlock and leather strap and automatic air vent and drain valve:		
Price per landing	point	2300.00 to 2750.00
5.12 Communication, Security and Control Systems		
Access Control Systems		
Door mounted access control unit inclusive of door furniture, lock plus software; including up to 50 m of cable and termination; including documentation testing and commissioning		
Internal single leaf door	point	1150.00 to 1400.00
Internal double door	point	1250.00 to 1550.00
External single leaf door	point	1350.00 to 1675.00
External double leaf door	point	1575.00 to 1950.00
Management control PC with printer software and commissioning up to 1000 users	point	16000.00 to 19000.00
CCTV Installations		
CCTV equipment inclusive of 50 m of cable including testing and commissioning		
Internal camera with bracket	point	1025.00 to 1225.00
Internal camera with housing	point	1225.00 to 1500.00
Internal PTZ camera with bracket	point	1900.00 to 2350.00
External fixed camera with housing	point	1350.00 to 1650.00
External PTZ camera dome	point	2700.00 to 3250.00
External PTZ camera dome with power	point	3300.00 to 4000.00

APPROXIMATE ESTIMATING RATES – 5 SERVICES

Item	Unit	Range £
Turnstiles		
Physical Access Control Barrier system – standard security level comprising unit caseworks in stainless steel finish, standard lane width – 650 mm, restricting panels standard 900 mm high, standard level detection sensor system, including provision to integrate Access Control Card Readers (issued by others), including LED Pictogram – Green Arrow/Red Cross, closed base for ease of installation and cable management		
Including delivery, installation and commissioning to a site in London; budget cost land configurations		
single lane	item	1000.00 to 11000.00
double lane	item	16500.00 to 17500.00
triple lane	item	23000.00 to 24000.00
Physical Access Control Barrier system – medium security level comprising unit caseworks in stainless steel finish, standard lane width – 650 mm, restricting panels standard 900 mm high, medium level detection sensor system, including provision to integrate Access Control Card Readers (issued by others), including LED Pictogram – Green Arrow/Red Cross, closed base for ease of installation and cable management		
Including delivery, installation and commissioning to a site in London; budget cost land configurations		
single lane	item	10500.00 to 12000.00
double lane	item	18000.00 to 19000.00
triple lane	item	25500.00 to 26520.00
Physical Access Control Barrier system – high security level comprising unit caseworks in stainless steel finish, standard lane width – 650 mm, restricting panels standard 1600 mm or 1800 mm high (same cost), higher level detection sensor system, including provision to integrate Access Control Card Readers (issued by others), including LED Pictogram – Green Arrow/Red Cross, closed base for ease of installation and cable management		
Including delivery, installation and commissioning to a site in London; budget cost land configurations		
single lane	item	12000.00 to 13000.00
double lane	item	20500.00 to 22500.00
triple lane	item	29500.00 to 31000.00
High security full height security revolving door – 2300 mm high, four wing T25 sections, no centre column, positioning drive with horizontal and vertical safety strips at door leaves, controlled via electronic Access Control Card Readers (supplied by others), sensor system in ceiling monitors door segments, secure simultaneous bi-directional use possible, finish – standard finish – PPC to RAL colour, card reader mounting boxes, 4 LED lights in ceiling, rubber matting Including delivery, installation and commissioning to a site in London		
Budget cost – 1800 mm dia., per door	item	27000.00 to 28500.00
budget cost – 2000 mm dia., per door	item	27000.00 to 28500.00
Additional cost – stainless steel finish	item	2000.00 to

APPROXIMATE ESTIMATING RATES – 5 SERVICES

Item	Unit	Range £	
5.12 Communication, Security and Control Systems – cont			
IT Installations			
Data Cabling			
Complete channel link including patch leads, cable, panels, testing and documentation (excludes cabinets and/or frames, patch cords, backbone/ harness connectivity as well as containment)			
Low Level			
Cat 5e (up to 5,000 outlets)	point	52.00 to	64.00
Cat 5e (5,000 to 15,000 outlets)	point	42.00 to	51.00
Cat 6 (up to 5,000 outlets)	point	62.00 to	77.00
Cat 6 (5,000 to 15,000 outlets)	point	58.00 to	71.00
Cat 6a (up to 5,000 outlets)	point	80.00 to	97.00
Cat 6a (5,000 to 15,000 outlets)	point	73.00 to	90.00
Cat 7 (up to 5,000 outlets)	point	94.00 to	115.00
Cat 7 (5,000 to 15,000 outlets)	point	90.00 to	105.00
Note: LSZH cable based on average of 50 metres false floor low level installation assuming 1 workstation in 2.5 m × 2.5 m (to 3.2 m × 3.2 m) density, with 4 data points per workstation. Not applicable for installations with less than 250 No. outlets.			
High Level			
Cat 5e (up to 500 outlets)	point	62.00 to	77.00
Cat 5e (over 500 outlets)	point	52.00 to	64.00
Cat 6 (up to 500 outlets)	point	73.00 to	90.00
Cat 6 (over 500 outlets)	point	69.00 to	84.00
Cat 6a (up to 500 outlets)	point	90.00 to	105.00
Cat 6a (over 500 outlets)	point	84.00 to	105.00
Cat 7 (up to 500 outlets)	point	105.00 to	125.00
Cat 7 (over 500 outlets)	point	100.00 to	120.00
Note: High level at 10 × 10 m grid.			
5.13 Specialist Installations			
Photovoltaic Panels			
Assuming roof mounted array, rate includes panels, associated framework, cabling, fixings, controls, delivery and commissioning			
High efficiency	m²	255.00 to	295.00
Standard efficiency	m²	210.00 to	250.00

BUILDING MODELS – ELEMENTAL COST SUMMARIES

Item	Unit	Range £	
AIRPORT TERMINAL BUILDING New build airport terminal building, premium quality, located in the South East of England, handling both domestic and international flights with a gross internal floor area (GIFA) of 25,000 m². These costs exclude baggage handling, check-in systems, pre-check in and boarding security systems, vertical transportation and travellators, pre-conditioned air systems to aircraft, services to stands and visual docking systems. Costs assume that the works are undertaken under landside access/logistics environment.			
5 Services			
5.1 Sanitary Installations	m²	6.30 to	9.50
5.3 Disposal Installations			
rainwater	m²	6.50 to	8.40
soil and waste	m²	8.50 to	12.60
condensate	m²	1.10 to	2.15
5.4 Water Installations			
domestic hot and cold water services	m²	17.85 to	21.00
5.5 Heat Source	m²	6.00 to	8.00
5.6 Space Heating and Air Conditioning			
LTHW heating system	m²	80.00 to	90.00
chilled water system	m²	88.00 to	100.00
supply and extract air conditioning system	m²	110.00 to	132.00
local cooling; DX systems to IT rooms	m²	6.00 to	8.00
5.7 Ventilation Systems			
mechanical ventilation to baggage handling and plantrooms	m²	27.00 to	32.00
toilet extract ventilation	m²	9.00 to	11.00
smoke extract installation	m²	10.00 to	13.25
kitchen extract system	m²	3.20 to	4.50
5.8 Electrical Installations			
main HV/MV Installations including switchgear, transformers (Cast Resin)	m²	65.00 to	80.00
Low Voltage (LV) – Incoming LV switchgear, distribution boards and distribution systems	m²	45.00 to	55.00
generators, life safety – Containerized stand-by diesel generators and 24 hour capacity belly tanks	m²	35.00 to	45.00
small power installations	m²	45.00 to	55.00
lighting installations; warehouse, office, ancillary and plant spaces	m²	120.00 to	145.00
emergency lighting installations	m²	12.00 to	18.00
power to mechanical services	m²	6.50 to	9.00
5.9 Fuel Installations			
gas mains to services	m²	2.20 to	3.30
5.11 Fire and Lightning Protection			
lightning protection	m²	0.70 to	0.75
earthing and bonding	m²	0.50 to	0.55
sprinkler installations	m²	35.00 to	44.00
dry riser and hose reel installations	m²	6.30 to	7.50
fire suppression to IT rooms	m²	4.00 to	5.50

BUILDING MODELS – ELEMENTAL COST SUMMARIES

Item	Unit	Range £	
AIRPORT TERMINAL BUILDING – cont			
5 Services – cont			
5.12 Communication, Security and Control Systems			
fire alarm systems; DDI	m²	35.00 to	44.00
voice/public address systems	m²	22.00 to	24.50
other alarm systems, e.g. disabled refuge	m²	2.00 to	3.00
security installations; CCTV, access control and intruder detection	m²	38.00 to	45.00
wireways for IT, comms and FA systems	m²	15.00 to	18.00
structured IT cabling; fibre optic backbone and copper Cat 6 to office and warehouse	m²	12.50 to	14.50
BMS systems	m²	58.00 to	65.00
5.13 Specialist Installations			
flight information display systems		28.00 to	35.00
Total Cost/m² (based on GIFA of 25,000 m²)	m²	967.00 to	1,170.00

Notes: The above includes; MEP preliminaries and OH&P allowances.
The above excludes; Main Contractor's preliminaries and OH&P, Utility
connections; electric, water, drainage and fibre connections, Contingency/risk
allowances, Foreign exchange fluctuations, VAT at prevailing rates

BUILDING MODELS – ELEMENTAL COST SUMMARIES

Item	Unit	Range £	
SHOPPING MALL (TENANT'S FIT OUT EXCLUDED) Natural ventilation shopping mall with approximately 33,000 m² two storey retail area and a 13,000 m² above ground, mechanically ventilated, covered car park, situated in a town centre in South East England.			
5 Services			
5.1 Sanitary Installations	m²	1.10 to	1.15
5.3 Disposal Installations			
rainwater	m²	6.70 to	7.10
soil, waste and vent	m²	6.70 to	7.10
5.4 Water Installations			
cold water installation	m²	6.70 to	7.10
hot water installation	m²	5.60 to	5.90
5.6 Space Heating and Air Conditioning			
condenser water system	m²	33.30 to	34.90
LTHW installation	m²	4.70 to	4.90
air conditioning system	m²	33.30 to	34.90
over door heaters at entrances	m²	1.10 to	1.15
5.7 Ventilation Systems			
public toilet ventilation	m²	1.10 to	1.15
plantroom ventilation	m²	4.70 to	4.90
supply and extract systems to shop units	m²	33.30 to	34.90
toilet extract systems to shop units	m²	2.25 to	2.35
smoke ventilation to Mall area	m²	11.45 to	12.10
service corridor ventilation	m²	2.25 to	2.35
other miscellaneous ventilation	m²	19.25 to	20.30
5.8 Electrical Installations			
LV distribution	m²	25.00 to	26.50
standby power	m²	4.70 to	5.00
general lighting	m²	68.70 to	71.80
external lighting	m²	5.80 to	6.00
emergency lighting	m²	13.85 to	14.60
small power	m²	11.25 to	11.90
mechanical services power supplies	m²	3.70 to	3.85
general earthing	m²	1.10 to	1.15
UPS for security and CCTV equipment	m²	1.10 to	1.15
5.9 Fuel Installations (gas)			
gas supply and boilers	m²	2.20 to	2.30
gas supplies to anchor (major) stores	m²	1.10 to	1.15
5.11 Fire and Lightning Protection			
lightning protection	m²	1.10 to	1.15
sprinkler installations	m²	15.60 to	16.50

BUILDING MODELS – ELEMENTAL COST SUMMARIES

Item	Unit	Range £	
SHOPPING MALL (TENANT'S FIT OUT EXCLUDED) – cont			
5.12 Communication, Security and Control Systems			
fire alarm installation	m²	9.00 to	9.60
public address/voice alarm	m²	5.80 to	6.00
security installation	m²	12.30 to	12.80
general containment	m²	13.50 to	14.25
5.13 Specialist Installations			
BMS/Controls	m²	19.25 to	20.10
Total Cost/m² (based on GIFA of 33,000 m²)	m²	370.00 to	390.00
CAR PARK – 13,000 m²			
5 Services			
5.3 Disposal Installations			
car park drainage	m²	5.50 to	5.70
5.7 Ventilation Systems			
car park ventilation (impulse fans)	m²	33.00 to	35.00
5.8 Electrical Installations			
LV distribution	m²	11.30 to	11.75
general lighting	m²	22.90 to	24.00
emergency lighting	m²	5.80 to	6.00
small power	m²	3.40 to	3.50
mechanical services power supplies	m²	5.50 to	5.90
general earthing	m²	1.10 to	1.15
ramp frost protection	m²	2.20 to	2.30
5.11 Fire and Lightning Protection			
sprinkler installation	m²	22.00 to	23.50
fire alarm installations	m²	22.00 to	23.50
5.12 Communication, Security and Control Systems			
security installation	m²	10.00 to	10.50
BMS/Controls	m²	6.70 to	7.00
5.13 Specialist Installations			
entry/exit barriers, pay stations	m²	6.50 to	6.90
Total Cost/m² (based on GIFA of 13,000 m²)	m²	160.00 to	170.00

BUILDING MODELS – ELEMENTAL COST SUMMARIES

Item	Unit	Range £	
SUPERMARKET Supermarket located in the South East with a total gross floor area of 4,000 m², including a sales area of 2,350 m². The building is on one level and incorporates a main sales, coffee shop, bakery, offices and amenities areas and warehouse.			
5 Services			
5.1 Sanitary Installations	m²	1.60 to	1.65
5.3 Disposal Installations	m²	3.90 to	4.10
5.4 Water Installations			
hot and cold water services	m²	28.10 to	29.10
5.6 Space Heating and Air Conditioning			
heating & ventilation with cooling via DX units	m²	11.20 to	11.75
5.7 Ventilation Systems			
supply and extract systems	m²	2.85 to	3.00
5.8 Electrical Installations			
panels/boards	m²	31.70 to	33.30
containment	m²	1.70 to	1.80
general lighting	m²	20.80 to	21.85
small power	m²	11.45 to	12.00
mechanical services wiring	m²	2.35 to	2.50
5.11 Fire and Lightning Protection			
lightning protection	m²	1.30 to	1.35
5.12 Communication, Security and Control Systems			
fire alarms, detection and public address	m²	2.85 to	3.00
CCTV	m²	2.85 to	3.00
intruder alarm, detection and store security	m²	2.85 to	3.00
telecom and structured cabling	m²	0.80 to	0.85
BMS	m²	7.80 to	8.20
data cabinet	m²	2.85 to	3.00
controls wiring	m²	4.20 to	4.40
5.13 Specialist Installations			
Refrigeration			
installation	m²	31.20 to	32.80
plant	m²	31.20 to	32.80
cold store	m²	11.45 to	12.10
cabinets	m²	84.25 to	88.40
Total Cost/m² (based on GIFA of 4,000 m²)	m²	300.00 to	315.00

BUILDING MODELS – ELEMENTAL COST SUMMARIES

Item	Unit	Range £	
OFFICE BUILDING			
Speculative 15 storey office in Central London for multiple tenant occupancy with a gross mounted water cooled chillers, located in basement.			
SHELL & CORE – 20,000 m²			
5 Services			
5.1 Sanitary Installations	m²	10.00 to	12.00
5.3 Disposal Installations			
rainwater/soil and waste	m²	16.00 to	20.00
condensate	m²	2.00 to	3.00
5.4 Water Installations			
hot and cold water services	m²	18.00 to	24.00
5.5 Heat Source	m²	8.00 to	12.00
5.6 Space Heating and Air Conditioning			
LTHW, plant and distribution	m²	14.00 to	18.00
chilled water, plant and distribution	m²	52.00 to	66.00
air, plant and distribution	m²	38.00 to	48.00
5.7 Ventilation Systems			
toilet extract ventilation	m²	6.00 to	8.00
basement extract	m²	15.00 to	20.00
miscellaneous ventilation systems	m²	15.00 to	20.00
5.8 Electrical Installations			
life safety generator, fuel and flue	m²	9.00 to	13.00
HV/LV supply/distribution	m²	45.00 to	60.00
general lighting (excluding external lighting) and lighting control	m²	30.00 to	35.00
general power	m²	6.00 to	9.00
electrical services for mechanical equipment	m²	3.00 to	6.00
voice and data (wireways)	m²	1.00 to	2.00
security (wireways)	m²	1.00 to	2.00
5.9 Fuel Installations (gas)	m²	1.00 to	2.00
5.11 Fire and Lightning Protection			
wet risers	m²	8.00 to	11.00
sprinklers	m²	20.00 to	24.00
earthing and bonding	m²	2.00 to	3.00
lightning protection	m²	2.00 to	3.00
5.12 Communication, Security and Control Systems			
fire and voice alarms	m²	12.00 to	16.00
disabled/refuge alarms	m²	2.00 to	3.00
CCTV/Access control/intruder detection	m²	3.00 to	5.00
BMS	m²	25.00 to	30.00
Total Cost/m² (based on GIFA of 20,000 m²)	m²	364.00 to	475.00

BUILDING MODELS – ELEMENTAL COST SUMMARIES

Item	Unit	Range £	
CATEGORY 'A' FIT OUT – 13,000 m² NIA			
5 Services			
5.6 Space Heating and Air Conditioning			
4 pipe fan coil units	m²	28.00 to	32.00
LTHW heating	m²	28.00 to	32.00
chilled water	m²	35.00 to	45.00
condensate	m²	10.00 to	15.00
ductwork distribution	m²	50.00 to	65.00
5.8 Electrical Installations			
tenant distribution boards	m²	3.00 to	5.00
lighting installation, lighting control and emergency	m²	70.00 to	90.00
supply only floor boxes	m²	3.00 to	4.00
earthing and bonding	m²	3.00 to	5.00
electrical services in connection with mechanical	m²	3.00 to	5.00
5.11 Fire and Lightning Protection			
sprinkler installation	m²	20.00 to	25.00
5.12 Communication, Security and Control Systems			
fire and voice alarms	m²	12.00 to	15.00
BMS	m²	22.00 to	26.00
Total Cost/m² (based on NIFA of 13,000 m²)	m²	287.00 to	364.00

BUILDING MODELS – ELEMENTAL COST SUMMARIES

Item	Unit	Range £	
BUSINESS PARK			
New build office in South East within M25 part of a speculative business park with a gross floor area of 10,000 m². A full air displacement system with roof mounted air cooled chillers, gas fired boilers and air handling plant. Total M&E services value approximately £2,500,000 (shell & core) and £1,000,000 (Cat A Fit Out).			
SHELL & CORE – 10,000 m² GIA			
5 Services			
5.1 Sanitary Installations	m²	8.00 to	10.00
5.3 Disposal Installations			
condensate	m²	2.00 to	3.00
rainwater, soil and waste	m²	11.00 to	13.00
5.4 Water Installations			
hot and cold water services	m²	12.00 to	16.00
5.5 Heat Source	m²	8.00 to	10.00
5.6 Space Heating and Air Conditioning			
LTHW heating; plantroom and risers	m²	12.00 to	16.00
chilled water, plantroom and risers	m²	30.00 to	35.00
ductwork; plantroom and risers	m²	60.00 to	70.00
5.7 Ventilation Systems			
toilet and miscellaneous ventilation	m²	12.00 to	16.00
5.8 Electrical Installations			
LV supply/distribution	m²	24.00 to	29.00
general lighting	m²	20.00 to	24.00
general power	m²	4.00 to	6.00
electrical services in connection with mechanical services	m²	2.00 to	4.00
security (wireways)	m²	1.00 to	3.00
voice and data (wireways)	m²	1.00 to	3.00
5.9 Fuel Installations (gas)	m²	1.00 to	3.00
5.11 Fire and Lightning Protection			
earthing and bonding	m²	2.00 to	4.00
lightning protection	m²	2.00 to	4.00
dry risers	m²	1.00 to	3.00
5.12 Communication, Security and Control Systems			
fire alarms	m²	9.00 to	11.00
BMS	m²	22.00 to	26.00
Total Cost/m² (based on GIFA of 10,000 m²)	m²	244.00 to	309.00

BUILDING MODELS – ELEMENTAL COST SUMMARIES

Item	Unit	Range £	
CATEGORY 'A' FIT OUT – 8,000 m² NIA			
5 Services			
5.6 Space Heating and Air Conditioning			
LTHW heating and perimeter heaters	m²	37.00 to	44.50
floor swirl diffusers and supply ductwork	m²	23.00 to	27.50
5.8 Electrical Installations			
distribution boards	m²	1.00 to	3.00
general lighting, recessed including lighting controls	m²	50.00 to	60.00
5.11 Fire and Lightning Protection			
earthing and bonding	m²	1.00 to	2.00
5.12 Communication, Security and Control Systems			
fire alarms	m²	7.00 to	9.10
BMS	m²	5.00 to	7.00
Total Cost/m² (based on NIFA of 8,000 m²)	each	124.00 to	153.00

BUILDING MODELS – ELEMENTAL COST SUMMARIES

Item	Unit	Range £	
PERFORMING ARTS CENTRE (MEDIUM SPECIFICATION)			
Performing Arts centre with a Gross Internal Area (GIA) of approximately 8,000 m², based on a medium specification with cooling to the Auditorium. The development comprises dance studios and a theatre auditorium in the outer London area. The theatre would require all the necessary stage lighting, machinery and equipment installed in a modern professional theatre (these are excluded from the model, as assumed to be FF&E, but the containment and power wiring is included).			
5 Services			
5.1 Sanitary Installations	m²	9.00 to	12.00
5.3 Disposal Installations			
soil, waste and rainwater	m²	15.00 to	19.00
5.4 Water Installations			
cold water installation	m²	14.00 to	19.00
hot water installation	m²	12.00 to	15.00
5.5 Heat Source	m²	11.00 to	14.00
5.6 Space Heating and Air Conditioning			
heating	m²	72.00 to	77.00
chilled water system	m²	62.00 to	82.00
supply and extract air systems	m²	113.00 to	134.00
5.7 Ventilation Systems			
ventilation and extract systems to toilets, kitchen and workshop	m²	21.00 to	26.00
5.8 Electrical Installations			
LV supply/distribution	m²	41.00 to	46.00
general lighting	m²	103.00 to	124.00
small power	m²	31.00 to	36.00
power to mechanical plant	m²	8.00 to	9.00
5.9 Fuel Installations (gas)	m²	3.00 to	5.00
5.11 Fire and Lightning Protection			
lightning protection	m²	3.00 to	4.00
5.12 Communication, Security and Control Systems			
fire alarms and detection	m²	26.00 to	31.00
voice and data complete installation (excluding active equipment)	m²	31.00 to	41.00
security; access; control; disabled alarms; staff paging	m²	31.00 to	41.00
BMS	m²	46.00 to	57.00
5.13 Specialist Installations			
theatre systems includes for containment and power wiring	m²	26.00 to	31.00
Total Cost/m² (based on GIFA of 8,000 m²)	m²	680.00 to	825.00

BUILDING MODELS – ELEMENTAL COST SUMMARIES

Item	Unit	Range £	
SPORTS HALL Single storey sports hall, located in the South East, with a gross internal area of 1,200 m² (40 m × 30 m).			
5 Services			
5.1 Sanitary Installations	m²	11.00 to	12.00
5.3 Disposal Installations			
rainwater	m²	3.00 to	4.00
soil and waste	m²	7.00 to	8.00
5.4 Water Installations			
hot and cold water services	m²	15.00 to	16.00
5.5 Heat Source			
boilers, flues, pumps and controls	m²	15.00 to	16.00
5.6 Space Heating and Air Conditioning			
warm air heating to sports hall area	m²	19.00 to	20.00
radiator heating to ancillary areas	m²	22.00 to	23.00
5.7 Ventilation Systems			
ventilation to changing, fitness and sports hall areas	m²	21.00 to	22.00
5.8 Electrical Installations			
main switchgear and sub-mains	m²	12.00 to	13.00
small power	m²	11.00 to	12.00
lighting and luminaires to sports areas	m²	26.00 to	27.00
lighting and luminaires to ancillary areas	m²	31.00 to	32.00
5.11 Fire and Lightning Protection			
lightning protection	m²	4.00 to	5.00
5.12 Communication, Security and Control Systems			
fire, smoke detection and alarm system, intruder detection	m²	12.00 to	13.00
CCTV installation	m²	14.00 to	15.00
public address and music systems	m²	7.00 to	8.00
wireways for voice and data	m²	3.00 to	4.00
Total Cost/m² (based on GIFA of 1,200 m²)	m²	233.00 to	250.00

BUILDING MODELS – ELEMENTAL COST SUMMARIES

Item	Unit	Range £	
STADIUM – NEW			
A three storey stadium, located in Greater London with a gross internal area of 85,000 m² and incorporating 60,000 spectator seats.			
5 Services			
5.1 Sanitary Installations	m²	11.20 to	11.90
5.3 Disposal Installations			
rainwater	m²	4.70 to	5.00
above ground drainage	m²	12.30 to	13.00
5.4 Water Installations			
hot and cold water services	m²	21.80 to	22.90
5.5 Heat Source	m²	10.10 to	10.70
5.6 Space Heating and Air Conditioning			
heating	m²	6.80 to	7.00
cooling	m²	15.60 to	16.40
5.7 Ventilation Systems	m²	62.40 to	65.50
5.8 Electrical Installations			
HV/LV supply	m²	11.20 to	11.90
LV distribution	m²	26.50 to	28.00
general lighting	m²	65.50 to	68.60
small power	m²	20.80 to	21.80
earthing and bonding	m²	1.10 to	1.15
power supply to mechanical equipment	m²	1.10 to	1.15
pitch lighting	m²	11.40 to	12.00
5.9 Fuel Installations (gas)	m²	5.20 to	5.50
5.11 Fire and Lightning Protection			
lightning protection	m²	1.10 to	1.15
hydrants	m²	2.25 to	2.35
5.12 Communication, Security and Control Systems			
wireways for data, TV, telecom and PA	m²	9.00 to	9.50
public address	m²	16.95 to	17.80
security	m²	15.80 to	16.60
data voice installations	m²	33.80 to	35.40
fire alarms	m²	9.10 to	9.60
disabled/refuse alarm/call systems	m²	3.40 to	3.50
BMS	m²	15.80 to	16.60
Total Cost/m² (based on GIFA of 85,000 m²)	m²	395.00 to	415.00
Total Cost/seat (based on 60,000 seats)	each	560.00 to	590.00

BUILDING MODELS – ELEMENTAL COST SUMMARIES

Item	Unit	Range £	
HOTELS 200 bedroom, four star hotel, situated in Central London, with a gross internal floor area of 16,500 m². The development comprises a ten storey building with large suites on each guest floor, together with banqueting, meeting rooms and leisure facilities.			
5 Services			
5.1 Sanitary Installations	m²	42.00 to	50.00
5.3 Disposal Installations			
rainwater	m²	4.00 to	5.00
soil and waste	m²	26.00 to	31.00
5.4 Water Installations			
hot and cold water services	m²	53.00 to	68.00
5.5 Heat Source			
condensing boiler, CHP and flues	m²	15.00 to	20.00
5.6 Space Heating and Air Conditioning			
air conditioning and space heating system; chillers, pumps, CHW and LTHW pipework, insulation, 4 pipe FCU, ductwork, grilles and diffusers, to guest rooms, public areas, meeting and banquet rooms	m²	220.00 to	260.00
5.7 Ventilation Systems			
general bathroom extract from guest suites, ventilation to kitchens and bathrooms, etc.	m²	45.00 to	60.00
smoke extract	m²	12.00 to	16.00
5.8 Electrical Installations			
HV/LV Installation, standby power, lighting, emergency lighting and small power to guest rooms and public areas, including earthing and bonding	m²	125.00 to	154.00
5.11 Fire and Lightning Protection			
dry risers and sprinkler installation	m²	39.00 to	48.00
lightning protection	m²	2.00 to	3.00
5.12 Communication, Security and Control Systems			
fire/smoke detection and fire alarm system	m²	17.00 to	23.00
security/access control	m²	20.00 to	27.00
integrated sound and AV system	m²	21.00 to	25.00
telephone and data and TV installation	m²	16.00 to	19.00
containment	m²	15.00 to	18.00
wire ways for voice and data (no hotel management and head end equipment)	m²	10.00 to	12.00
BMS	m²	32.00 to	38.00
Total Cost/m² (based on GIFA of 16,500 m²)	m²	714.00 to	877.00

BUILDING MODELS – ELEMENTAL COST SUMMARIES

Item	Unit	Range £	
PRIVATE HOSPITAL			
New build project building. The works consist of a new 80 bed hospital of approximately 15,000 m², eight storey with a plant room.			
All heat is provided from existing steam boiler plant, medical gases are also served from existing plant. The project includes the provision of additional standby electrical generation to serve the wider site requirements.			
This hospital has six operating theatres, ITU/HDU department, pathology facilities, diagnostic imaging, out-patient facilities and physiotherapy.			
5 Services			
5.1 Sanitary Installations	m²	34.00 to	36.00
5.3 Disposal Installations			
rainwater	m²	4.00 to	6.00
soil and waste	m²	37.00 to	39.00
5.4 Water Installations			
hot and cold water services	m²	85.00 to	88.00
5.5 Heat Source (included in 5.6)			
5.6 Space Heating and Air Conditioning			
LPHW heating	m²	70.00 to	72.00
chilled water	m²	54.00 to	57.00
steam and condensate	m²	34.00 to	37.00
ventilation, comfort cooling and air conditioning	m²	180.00 to	206.00
5.7 Ventilation Systems (included in 5.6)			
5.8 Electrical Installations			
HV distribution	m²	46.00 to	49.00
LV supply/distribution	m²	72.00 to	77.00
standby power	m²	43.00 to	47.00
UPS	m²	30.00 to	32.00
general lighting	m²	82.00 to	93.00
general power	m²	64.00 to	67.00
emergency lighting	m²	24.00 to	26.00
theatre lighting	m²	19.00 to	21.00
specialist lighting	m²	19.00 to	21.00
external lighting	m²	4.00 to	6.00
electrical supplies for mechanical services	m²	14.00 to	16.00
earthing and bonding	m²	5.00 to	7.00
5.9 Fuel Installations			
gas installations	m²	6.00 to	8.00
oil installations	m²	9.00 to	11.00
5.11 Fire and Lightning Protection			
dry risers	m²	4.00 to	6.00
sprinkler systems/gaseous fire suppression	m²	57.00 to	67.00
lightning protection	m²	2.00 to	4.00

BUILDING MODELS – ELEMENTAL COST SUMMARIES

Item	Unit	Range £	
5.12 Communication, Security and Control Systems			
fire alarms and detection	m²	34.00 to	36.00
voice and data	m²	52.00 to	54.00
data containment	m²	10.00 to	12.00
security and CCTV	m²	29.00 to	31.00
nurse call and cardiac alarm system	m²	45.00 to	47.00
TV systems	m²	19.00 to	21.00
disabled WC alarm	m²	5.00 to	7.00
BMS	m²	80.00 to	82.00
5.13 Specialist Installations			
pneumatic tube conveying system	m²	7.00 to	10.00
medical gases	m²	67.00 to	72.00
Total Cost/m² (based on GIFA of 15,000 m²)	m²	1346.00 to	1471.00

BUILDING MODELS – ELEMENTAL COST SUMMARIES

Item	Unit	Range £	
SCHOOL			
New build secondary school (Academy) located in Southern England, with a gross internal floor area of 10,000 m². Total M&E services value approximately £5,000,000. The building comprises a three storey teaching block, including provision for music, drama, catering, sports hall, science laboratories, food technology, workshops and reception/admin (BB93 compliant). Excludes IT cabling and sprinkler protection.			
5 Services			
5.1 Sanitary Installations			
toilet cores and changing facilities and lab sinks	m²	14.00 to	16.00
5.3 Disposal Installations			
rainwater installations	m²	5.00 to	6.00
soil and waste	m²	13.00 to	16.00
5.4 Water Installations			
potable hot and cold water services	m²	29.00 to	31.00
non-potable hot and cold water services to labs and art rooms	m²	10.00 to	11.00
5.5 Heat Source			
gas fired boiler installation	m²	12.00 to	15.00
5.6 Space Heating and Air Conditioning			
LTHW heating system (primary)	m²	36.00 to	39.00
LTHW heating system (secondary)	m²	10.00 to	12.00
DX cooling system to ICT server rooms	m²	5.00 to	7.00
mechanical supply and extract ventilation including DX type cooling to Music, Drama, Kitchen/Dining and Sports Hall	m²	62.00 to	67.00
5.7 Ventilation Systems			
toilet extract systems	m²	8.00 to	10.00
changing area extract systems	m²	5.00 to	7.00
extract ventilation from design/food technology and science labs	m²	10.00 to	12.00
5.8 Electrical Installations			
mains and sub-mains distribution	m²	36.00 to	40.00
lighting and luminaires including emergency fittings	m²	77.00 to	82.00
small power installation	m²	38.00 to	41.00
earthing and bonding	m²	2.00 to	3.00
5.9 Fuel Installations (gas)	m²	8.00 to	10.00
5.11 Fire and Lightning Protection			
lightning protection	m²	4.00 to	4.50
5.12 Communication, Security and Control Systems			
containment for telephone, IT data, AV and security systems	m²	5.00 to	7.00
fire, smoke detection and alarm system	m²	19.00 to	21.00
security installations including CCTV, access control and intruder alarm	m²	21.00 to	22.00
disabled toilet, refuge and induction loop systems	m²	3.00 to	5.00
BMS – to plant	m²	24.00 to	27.00
BMS – to opening vents/windows	m²	10.00 to	12.00
Total Cost/m² (based on GIFA of 10,000 m²)	m²	466.00 to	523.00

BUILDING MODELS – ELEMENTAL COST SUMMARIES

Item	Unit	Range £
AFFORDABLE RESIDENTIAL DEVELOPMENT An 8 storey, 117 affordable residential development with a gross internal area of 11,400 m² and a net internal area of 8,400 m², situated within the Central London area. The development has no car park or communal facilities and achieves a net to gross efficiency of 74%. Based upon radiator LTHW heating within each apartment, with plate heat exchanger, whole house ventilation, plastic cold and hot water services pipework, sprinkler protection to apartments. Excludes remote metering. Based upon 2 bed units with 1 bathroom.		
SHELL & CORE		
5 Services		
5.1 Sanitary Installations	m²	0.50 to 0.75
5.3 Disposal Installations	m²	19.00 to 24.00
5.4 Water Installations	m²	19.00 to 23.00
5.5 Heat Source	m²	9.00 to 11.00
5.6 Space Heating and Air Conditioning	m²	23.00 to 28.00
5.7 Ventilation Systems	m²	18.00 to 20.00
5.8 Electrical Installations	m²	41.00 to 51.00
5.9 Fuel Installations – Gas	m²	2.50 to 3.00
5.11 Fire and Lightning Protection	m²	21.00 to 25.00
5.12 Communication, Security and Control Systems	m²	32.00 to 39.00
5.13 Specialist Installations	m²	9.00 to 11.00
Total Cost/m² (based on GIFA of 11,400 m²)	m²	188.00 to 236.00
FITTING OUT		
5 Services		
5.1 Sanitary Installations	m²	32.00 to 39.00
5.3 Disposal Installations	m²	10.00 to 12.00
5.4 Water Installations	m²	27.00 to 32.00
5.5 Heat Source	m²	33.00 to 40.00
5.6 Space Heating and Air Conditioning	m²	40.00 to 50.00
5.7 Ventilation Systems	m²	35.00 to 42.00
5.8 Electrical Installations	m²	57.00 to 70.00
5.11 Fire and Lightning Protection	m²	21.00 to 26.00
5.12 Communication, Security and Control Systems	m²	18.00 to 23.00
5.13 Specialist Installations	m²	11.00 to 13.00
Total Cost/m² (based on NIA of 8,400 m²)	m²	284.00 to 345.00
Total cost per apartment – Shell & Core and Fit Out	each	39,000.00 to 50,000.00

BUILDING MODELS – ELEMENTAL COST SUMMARIES

Item	Unit	Range £	
PRIVATE RESIDENTIAL DEVELOPMENT			
A 24 storey, 203 apartment private residential development with a gross internal area of 50,000 m² and a net internal area of 33,000 m², situated within the London's prime residential area. The development has limited car park facilities, residents' communal areas such as gym, cinema room, self-stimulated, etc. Based upon LTHW underfloor heating, 4 pipe fan coils to principle rooms, with plate heat and cooling exchangers, whole house ventilation with summer house facility, copper hot and cold water services pipework, sprinkler protection to apartments, lighting control, TV/data outlets. Flood wiring for apartment, home automation and sound system only. Based upon a combination of 2/3 bedroom apartments with family bathroom and ensuite.			
SHELL & CORE			
5 Services			
5.1 Sanitary Installations	m²	2.50 to	3.00
5.3 Disposal Installations	m²	23.00 to	29.00
5.4 Water Installations	m²	32.00 to	41.00
5.5 Heat Source	m²	9.00 to	14.00
5.6 Space Heating and Air Conditioning	m²	74.00 to	95.00
5.7 Ventilation Systems	m²	22.00 to	26.00
5.8 Electrical Installations	m²	63.00 to	80.00
5.9 Fuel Installations – Gas	m²	4.00 to	5.00
5.11 Fire and Lightning Protection	m²	30.00 to	37.00
5.12 Communication, Security and Control Systems	m²	45.00 to	55.00
5.13 Specialist Installations	m²	38.00 to	50.00
Total Cost/m² (based on GIFA of 50,000 m²)	m²	340.00 to	440.00
FITTING OUT			
5 Services			
5.1 Sanitary Installations	m²	95.00 to	115.00
5.3 Disposal Installations	m²	24.00 to	30.00
5.4 Water Installations	m²	50.00 to	62.00
5.5 Heat Source	m²	50.00 to	60.00
5.6 Space Heating and Air Conditioning	m²	150.00 to	185.00
5.7 Ventilation Systems	m²	35.00 to	42.00
5.8 Electrical Installations	m²	120.00 to	150.00
5.11 Fire and Lightning Protection	m²	27.00 to	34.00
5.12 Communication, Security and Control Systems	m²	98.00 to	120.00
5.13 Specialist Installations	m²	30.00 to	37.00
Total Cost/m² (based on NIA of 33,000 m²)	m²	720.00 to	840.00
Total cost per apartment – Shell & Core and Fit Out	each	91,000.00 to	116,000.00

BUILDING MODELS – ELEMENTAL COST SUMMARIES

Item	Unit	Range £	
DISTRIBUTION CENTRE			
New build distribution centre located in the South East of England with a total gross floor area of 75,000 m², including a refrigerated cold box of 17,500 m². The building is on one level (no mezzanine decks) and incorporates a small office area at ground level, vehicle recovery unit, electric fork lift truck docking bay, gate house and associated plantrooms.			
5 Services			
5.3 Disposal Installations			
soil and waste	m²	6.80 to	7.10
rainwater	m²	2.25 to	2.36
5.4 Water Installations			
domestic hot and cold water services	m²	2.45 to	3.50
emergency drench showers & Cat 5 vehicle wash downs	m²	0.50 to	0.75
5.6 Space Heating and Air Conditioning			
heating with ventilation to offices, displacement system to main warehouse	m²	28.00 to	36.00
local cooling; DX systems to refrigerated cold box	m²	36.00 to	42.00
5.7 Ventilation Systems			
smoke extract system	m²	6.00 to	6.25
specialist extract to battery charging and maintenance areas	m²	1.50 to	2.00
5.8 Electrical Installations			
generator, life safety – containerized stand-by diesel generator and 24 hour capacity belly tanks	m²	4.70 to	5.85
main HV/MV installations including switchgear, transformers (Cast Resin)	m²	22.50 to	23.00
low voltage (LV) – incoming LV switchgear, distribution boards and distribution systems	m²	17.00 to	18.00
lighting installations; warehouse, office, ancillary and plant spaces	m²	14.80 to	17.85
small power installations inc. mech/HVAC plant power	m²	11.70 to	13.50
5.9 Fuel Installations			
gas mains to services	m²	0.75 to	1.00
5.11 Fire and Lightning Protection			
sprinklers (ESFR) including rack protection	m²	47.00 to	49.50
lightning protection	m²	0.75 to	0.75
earthing and bonding	m²	0.50 to	0.55
5.12 Communication, Security and Control Systems			
fire alarm systems; DDI	m²	12.00 to	13.00
security installations; CCTV, access control and intruder detection	m²	8.25 to	8.70
BMS systems	m²	6.00 to	6.25
other alarm systems, e.g. disabled refuge	m²	0.30 to	0.40
structured IT cabling; fibre optic backbone ad copper Cat 6 to office and warehouse	m²	12.50 to	14.50
5.13 Specialist Installations			
testing and commissioning	m²	2.45 to	2.70
Total Cost/m² (based on GIFA of 75,000 m²)	m²	247.00 to	275.00

Approximate Estimating

BUILDING MODELS – ELEMENTAL COST SUMMARIES

Item	Unit	Range £	
DATA CENTRE			
New build data centre in a single storey 'warehouse' type construction in the Greater London area/home counties proximity to the M25. Total Net Technical Area (NTA) of 2,000 m² (2 Nr data halls of 1,000 m² NTA each) with typically other areas of 300 m² for NOC/Security Room, office space, 350 m² ancillary space and 1,000 m² of internal plant areas. Total GIA of 3,650 m².			
Power and cooling to Technical Spaces designed to 1,500 W/m² of NTA = 3,000 kW IT load.			
Critical infrastructure built to UTI Tier III resilience standard (HVAC generally N +1) and Electrical on a dual 'A' and 'B' string supplies from MV/LV switchgear to IT cab (Critical UPSs N+1 on each string) with all heat rejection and stand-by generators (N+1) located in a secure, external plant farm.			
Total MEP services value of approximately £18,520,000 to £23,090,000.			
All costs expressed as £/m² against Nett Technical Area – NTA.			
5 Services			
5.3 Disposal Installations			
soil and waste	m²	15.00 to	17.00
rainwater	m²	11.00 to	13.00
condensate	m²	11.00 to	15.00
5.4 Water Installations			
domestic hot and cold water services	m²	23.00 to	30.50
5.5 Heat Source			
for office and ancillary spaces only	m²	17.00 to	25.00
5.6 Space Heating and Air Conditioning			
chilled water plant with redundancy of N+1 to technical space; packaged air-cooled Turbocor chillers with free cooling capability	m²	515.00 to	675.00
chilled water distribution systems; plant, header and primary distribution to technical space CRAC units, office and ancillary space cooling units; single distribution ring with redundancy N+1 on pumps, pressurization, dosing and buffer vessels	m²	1,040.00 to	1,190.00
chilled water final connections to technical space CRAC office/ancillary space cooling units	m²	136.00 to	168.00
computer room air conditioning (CRAC) units; 100 kW single coil units with 30% of units with humidification; N+2 redundancy	m²	250.00 to	315.00
CRAC units to critical ancillary space, e.g. UPS module and battery rooms	m²	44.00 to	63.00
cooling units to office, support and other spaces; single coil FCUs	m²	35.00 to	55.00
5.7 Ventilation Systems			
supply and extract ventilation systems to data halls, switchrooms and ancillary spaces including dedicated gas extract to gaseous fire suppression protected spaces	m²	95.00 to	140.00
computer room floor grilles; heavy duty 600 × 600 mm metal grilles with dampers	m²	135.00 to	160.00
General supply and extract to office, support and ancillary spaces	m²	53.00 to	85.00

BUILDING MODELS – ELEMENTAL COST SUMMARIES

Item	Unit	Range £	
5.8 Electrical Installations			
main HV/MV installations including switchgear, transformers (Cast Resin); dual MV supply from local authority (REC supply cost excluded)	m²	195.00 to	245.00
Low Voltage (LV) switchgear – incoming LV switchgear, UPS Input and Output LV boards, mechanical services and support/ancillary area panels and distribution boards	m²	1,010.00 to	1,110.00
LV Distribution including busbars, cabling and containment systems to provide full dual 'A' and 'B' supplies to data halls and supplies to mechanical services, office and ancillary areas	m²	765.00 to	890.00
generators – containerized stand-by diesel generators, DC rated with acoustic treatment and exhaust systems; on an N+1 redundancy for critical IT electrical and cooling loads including synchronization panel and 48 hour capacity belly tanks	m²	950.00 to	1,020.00
Uninterruptible Power Supplies (UPS) – Static UPS systems to provide 2 × (N+1) redundancy; critical IT loads – 600 kVA modules with 10 minute battery autonomy, 0.9 PFC, harmonic filtration and static by-pass; 500 kVA Static UPS with 10 minute battery autonomy for mechanical loads on N load basis	m²	795.00 to	925.00
Critical Power Distribution Units (PDUs) to data halls; dual 'A' and 'B' units to provide diverse supplies to each IT cabinet	m²	495.00 to	780.00
final critical IT power supplies. 2 Nr supplies to each cabinet from 'A' and 'B' PDUs	m²	420.00 to	570.00
lighting installations; data halls, office, support, ancillary and plant supplies	m²	136.00 to	170.00
5.9 Fuel Installations			
manual fill point and distribution pipework to generator belly tanks	m²	42.00 to	63.00
5.10 Lift and Conveyor Systems			
dock levellers and DDA access platform lifts only	m²	40.00 to	55.00
5.11 Fire and Lightning Protection			
gaseous fire suppression to technical, UPS and switchgear rooms	m²	310.00 to	350.00
lightning protection	m²	11.00 to	22.00
earthing and clean earth	m²	22.00 to	32.00
leak detection; to chilled water circuits within data hall service corridor	m²	55.00 to	75.00
5.12 Communication, Security and Control Systems			
fire alarm systems; DDI	m²	55.00 to	75.00
very Early Smoke Detection Alarm (VESDA) to technical areas	m²	125.00 to	160.00
other alarm systems, e.g. disabled refuge	m²	10.00 to	20.00
structured IT cabling; fibreoptic backbone and copper Cat 6 to office and support spaces only. IT cabling to data halls EXCLUDED (Client fit-out item)	m²	120.00 to	175.00
security installations; CCTV, Access Control and Intruder Detection	m²	525.00 to	730.00
BMS/EMS systems	m²	620.00 to	900.00
5.13 Specialist Installations			
factory acceptance testing, site acceptance testing, integrated systems testing and 'black building' testing @ 2% of MEP value	m²	180.00 to	225.00
Total Cost/m² (based on NTA of 2,000 m²)	m²	9261.00 to	11544.00
Total Cost/kW IT Load (based on 3,000 kW IT Load)	each	6174.00 to	7697.00

Note: The above includes; MEP preliminaries and OH&P allowances

The above excludes; Main contractor's preliminaries and OH&P; Utility connections; Electric, water, drainage and Fibre Connections; Contingency/Risk Allowances; Foreign Exchange fluctuations; VAT at prevailing rates

BUILDING MODELS – ELEMENTAL COST SUMMARIES

Item	Unit	Range £	
GYM			
Gym located in Central London on the ground floor of a mixed use development. This model is based on a total of 2,000 m², with costs based on connection to central plant within plant rooms. The model includes for an AHU dedicated to providing ventilation to the gym. Services provided also include for 4 Pipe Fan Coil Units providing Heating and Cooling, additional underfloor heating to changing rooms. Excluded from the gym fit-out costs are FFE, active IT equipment and gym equipment.			
5 Services			
5.1 Sanitary Installations			
sanitary appliances including WCs, wash hand basins, cleaner's sinks, water fountains, urinals, showers and the provision of disabled toilets and accessible showers	m²	35.00 to	40.00
5.3 Disposal Installations			
soil, waste and vent installation to all sanitary ware points, provision of stub stacks, condensate drainage for fan coil units including insulation, rainwater disposal excluded as part of base build installations	m²	15.00 to	20.00
5.4 Water Installations			
installation of mains cold water services including meter, storage tanks, pumps, electromagnetic water conditioner and connections to sanitary ware, hot water including connection to base built plate heat exchanger for hot water generation, bulk storage, distribution, pump sets and connections to sanitary ware, miscellaneous water points to drinking fountains and plant supplies	m²	25.00 to	30.00
5.5 Heat Source			
connection to base build plate heat exchanger, energy meter, buffer vessels, associated pump sets and primary distribution pipework and insulation	m²	20.00 to	25.00
5.6 Space Heating and Air Conditioning			
connection to base build plate heat exchanger, energy meter, buffer vessels, associated pump sets and CHW distribution pipework to AHU and FCUs secondary LTHW distribution to underfloor heating and FCUs Intake and exhaust air ductwork to AHU. Air handling unit and associated supply and extract ductwork distribution.	m²	230.00 to	275.00
5.7 Ventilation Systems			
Intake and exhaust air ductwork to fans, toilet and shower supply and extract system. Extract to stores and cleaner's cupboards. MVHR unit to office/staff rest room.	m²	30.00 to	35.00
5.8 Electrical Installations			
LV distribution system including switchgear, containment and cabling	m²	180.00 to	210.00
small power and lighting including emergency lighting and controls			
floor boxes to gym equipment			
enhanced lighting to lobbies, reception areas and toilet/changing areas			
earthing and bonding			
5.11 Fire and Lightning Protection			
connection to base build installations, zone valve, concealed sprinkler heads throughout	m²	25.00 to	30.00

BUILDING MODELS – ELEMENTAL COST SUMMARIES

Item	Unit	Range £	
5.12 Communication, Security and Control Systems fire alarm system, PA/VA installation, access control including card reader turnstiles at reception controlling entrance to the gym, security system comprising of security cameras, access control and intruder alarms to back of houses, TV and data installations to gym, security system comprising of security cameras, access control and intruder alarms to back of houses, TV and data installations to gym, Wi-fi installation, complete sound system including speakers, cabling and central music generation and smart device docks, installation of central building management system including central control panels and BMS to plant and equipment	m²	200.00 to	230.00
Total Cost/m² (based on GIFA of 2,000 m²)	m²	760.00 to	895.00

BUILDING MODELS – ELEMENTAL COST SUMMARIES

Item	Unit	Range £	
BAR Bar located in Central London on the 15th floor of a mixed use development. This model is based on a total GIA of 500 m², with costs based on for complete standalone plant. The model includes for a displacement ventilation installation, heating and cooling provided by a VRF installation via terminal units and an AHU. Excluded from the cost model is the bar and back bar unit, lifts with connections only provided for electrical, comms, water and drainage, FFE and active IT equipment.			
5 Services			
5.1 Sanitary Installations			
sanitary appliances including WCs, wash hand basins, urinals and the provision of disabled toilets	m²	55.00 to	65.00
5.3 Disposal Installations			
soil, waste and vent installation to all sanitary ware points, provision of stub stacks, capped connections to bar areas, condensate drainage for fan coil units including insulation, rainwater gullies to terrace areas	m²	25.00 to	30.00
5.4 Water Installations			
installation of mains cold water services including meter, storage tanks, pumps, electromagnetic water conditioner and connections to sanitary ware hot water generation, plant bulk storage, distribution, pump sets and connections to sanitary ware, capped connections to bar areas, miscellaneous water points and plant supplies	m²	130.00 to	150.00
5.6 Space Heating and Air Conditioning			
VRF installation serving the AHU and terminal units provide heating and cooling to bar areas, standalone cooling only VRF units to bar cellar and stores, specialist temperature control and humidity control units to wine cellar, Intake and exhaust air ductwork to AHU. Air handling unit and associated supply and extract ductwork distribution.	m²	255.00 to	290.00
5.7 Ventilation Systems			
Supply and extract installations to WC areas and back of house areas. Extract to cleaner's cupboard, stores and the like.	m²	55.00 to	65.00
5.8 Electrical Installations	m²	450.00 to	590.00
LV distribution system including switchgear, containment and cabling small power and lighting including emergency lighting and controls, provision of enhanced lighting to lobbies, bar areas and toilet areas, Allowance for statement lighting, full lighting control and scene setting, capped electrical supply to bar area, earthing and bonding			
5.10 Lift and Conveyor Systems			
excluded			
5.11 Fire and Lightning Protection			
connection to base build installations, zone valve, concealed sprinkler heads throughout	m²	30.00 to	35.00

BUILDING MODELS – ELEMENTAL COST SUMMARIES

Item	Unit	Range £
5.12 Communication, Security and Control Systems fire alarm system, PA/VA installation, security system comprising of security cameras, access control and intruder alarms to back of houses, TV and data installations to gym, Wi-fi installation, complete sound system including speakers, cabling and central music generation and smart device docks, installation of central building management system including central control panels and BMS to plant and equipment	m²	330.00 to 375.00
Total Cost/m² (based on GIFA of 500 m²)	m²	1,360.00 to 1,600.00

BUILDING MODELS – ELEMENTAL COST SUMMARIES

Item	Unit	Range £	
SPA			
High end luxury spa located in Central London within the basement level of a mixed use development. This model is based on a total GIA of 1,050 m², with costs based on connection to central plant within plant rooms. The model includes for an AHU which is only providing ventilation to the spa area and swimming pool hall. Services provided also include for 4 pipe fan coil units providing heating and cooling, displacement ventilation to the pool hall, additional underfloor heating to changing rooms. Included within the cost plan are costs for swimming pool plant, steam rooms and a sauna. Excluded from the spa fit-out costs are FFE and active IT equipment. Locating elements such as spas within the basements of developments will attract a cost premium due to the difficulty associated with bringing services into these areas. A holistic approach to costing must be considered for spa areas, as basement digs to accommodate swimming pools and the associated structural and water proofing required.			
5 Services			
5.1 Sanitary Installations			
sanitary appliances including WCs, wash hand basins, cleaner's sinks, water fountains, urinals, showers, ice fountain and the provision of disabled toilets and accessible showers	m²	90.00 to	110.00
5.3 Disposal Installations			
soil, waste and vent installation to all sanitary ware points, provision of stub stacks, condensate drainage for fan coil units including insulation, back wash drainage to pool plant, rainwater gullies to terrace areas	m²	25.00 to	30.00
5.4 Water Installations			
installation of mains cold water services including meter, storage tanks, pumps, electromagnetic water conditioner and connections to sanitary ware hot water including connection to base built plate heat exchanger for hot water generation, bulk storage, distribution, pump sets and connections to sanitary ware, softened water to experience shower, saunas and steam rooms, miscellaneous water points to drinking fountains and plant supplies	m²	55.00 to	65.00
5.5 Heat Source			
connection to base build plate heat exchanger, energy meter, buffer vessels, associated pump sets and primary distribution pipework and insulation	m²	30.00 to	35.00
5.6 Space Heating and Air Conditioning			
connection to base build plate heat exchanger for chilled water, energy meter, buffer vessels, associated pump sets and CHW distribution pipework to AHU and FCUs, VRF installation to comms/media rooms, secondary LTHW distribution to underfloor heating and FCUs, supply and extract installation to pool hall, Intake and exhaust air ductwork to AHU. Air handling unit and associated supply and extract ductwork distribution	m²	595.00 to	600.00
5.7 Ventilation Systems			
Intake and exhaust air ductwork to fans, toilet and shower supply and extract system. Extract to stores and cleaner's cupboards. MVHR unit to office/staff rest room. Extra over for smoke extract	m²	120.00 to	145.00

BUILDING MODELS – ELEMENTAL COST SUMMARIES

Item	Unit	Range £	
5.8 Electrical Installations			
LV distribution system including switchgear, containment and cabling small power and lighting including emergency lighting and controls, floor boxes to gym equipment, enhanced lighting to lobbies, reception areas and toilet/changing areas, allowance for statement lighting, full lighting control, fibre optic lighting with pool and scene setting, earthing and bonding	m²	330.00 to	385.00
5.11 Fire and Lightning Protection			
connection to base build installations, zone valve, concealed sprinkler heads throughout	m²	25.00 to	30.00
5.12 Communication, Security and Control Systems			
fire alarm system, PA/VA installation, access control including card reader turnstiles at reception controlling entrance to the gym, security system comprising of security cameras, access control and intruder alarms to back of houses, TV and data installations, Wi-Fi installation, complete sound system including speakers, cabling and central music generation and smart device docks, installation of central building management system including central control panels and BMS to plant and equipment	m²	330.00 to	375.00
5.13 Specialist Installations			
saunas and steam rooms, pool plant, including all pumps, water treatment, filters and controls	m²	505.00 to	550.00
Total Cost/m² (based on GIFA of 1,050 m²)	m²	2,105.00 to	2,385.00

Building Regulations in Brief, 9th edition

Ray Tricker and Samantha Alford

This ninth edition of the most popular and trusted guide reflects all the latest amendments to the Building Regulations, planning permission and the Approved Documents in England and Wales. This includes coverage of the new Approved Document Q on security, and a second part to Approved Document M which divides the regulations for 'dwellings' and 'buildings other than dwellings'. A new chapter has been added to incorporate these changes and to make the book more user friendly.

Giving practical information throughout on how to work with (and within) the Regulations, this book enables compliance in the simplest and most cost-effective manner possible. The no-nonsense approach of Building Regulations in Brief cuts through any confusion and explains the meaning of the Regulations. Consequently, it has become a favourite for anyone in the building industry or studying, as well as those planning to have work carried out on their home.

December 2017: 234 × 156 mm: 1302 pp
Pb: 978-1-138-28516-3 : £33.99

To Order: Tel: +44 (0) 1235 400524 Fax: +44 (0) 1235 400525
or Post: Taylor and Francis Customer Services,
Bookpoint Ltd, Unit T1, 200 Milton Park, Abingdon, Oxon, OX14 4TA UK
Email: book.orders@tandf.co.uk

For a complete listing of all our titles visit:
www.tandf.co.uk

PART 3

Material Costs/Measured Work Prices – Mechanical Installations

Get Qualified:
Inspection and Testing

Kevin Smith

The Get Qualified series provides clear and concise guidance for people looking to work within the electrical industry. This book outlines why the inspection and testing of electrical installations is important, and what qualifications are required in order to test, inspect and certify. All you need to know about the subject of inspection is covered in detail, making this book the ideal guide for those who are new to the subject and experienced professionals alike. There are also sections on exam preparation, revision exercises and sample questions.

July 2017: 198 x 129 mm: 168 pp
Pb: 978-1-138-18963-8 : £18.99

To Order: Tel: +44 (0) 1235 400524 Fax: +44 (0) 1235 400525
or Post: Taylor and Francis Customer Services,
Bookpoint Ltd, Unit T1, 200 Milton Park, Abingdon, Oxon, OX14 4TA UK
Email: book.orders@tandf.co.uk

For a complete listing of all our titles visit:
www.tandf.co.uk

Material Costs/Measured Work Prices

DIRECTIONS

The following explanations are given for each of the column headings and letter codes.

Unit — Prices for each unit are given as singular (i.e. 1 metre, 1 nr) unless stated otherwise.

Net price — Industry tender prices, plus nominal allowance for fixings (unless measured separately), waste and applicable trade discounts.

Material cost — Net price plus percentage allowance for overheads (7%), profit (5%) and preliminaries (12%).

Labour norms — In man-hours for each operation.

Labour cost — Labour constant multiplied by the appropriate all-in man-hour cost based on gang rate (See also relevant Rates of Wages Section) plus percentage allowance for overheads, profit and preliminaries.

Measured work — Material cost plus Labour cost.
Price (total rate)

MATERIAL COSTS

The Material Costs given are based at Second Quarter 2018 but exclude any charges in respect of VAT. The average rate of copper during this quarter is US$6,930/UK£5,131 per tonne. Users of the book are advised to register on the SPON's website www.pricebooks.co.uk/updates to receive the free quarterly updates – alerts will then be provided by email as changes arise.

MEASURED WORK PRICES

These prices are intended to apply to new work in the London area. The prices are for reasonable quantities of work and the user should make suitable adjustments if the quantities are especially small or especially large. Adjustments may also be required for locality (e.g. outside London – refer to cost indices in approximate estimating section for details of adjustment factors) and for the market conditions, e.g. volume of work secured or being tendered) at the time of use.

MECHANICAL INSTALLATIONS

The labour rate has been based on average gang rates per man hour effective from 1 October 2018. To this rate has been added 12% and 7% to cover preliminary items, site and head office overheads together with 5% for profit, resulting in an inclusive rate of £32.17 per man hour. The rate has been calculated on a working year of 2,016 hours; a detailed build-up of the rate is given at the end of these directions.

The rates and allowances will be subject to increases as covered by the National Agreements. Future changes will be published in the free Spon's quarterly update by registering on their website.

DIRECTIONS

DUCTWORK INSTALLATIONS

The labour rate basis is as per Mechanical above and to this rate has been added 12% plus 7% to cover site and head office overheads only (factory overhead is included in the material rate) and preliminary items together with 5% for profit, resulting in an inclusive rate of £30.72 per man hour. The rate has been calculated on a working year of 2,016 hours; a detailed build-up of the rate is given at the end of these directions.

In calculating the 'Measured Work Prices' the following assumptions have been made:

> (a)   That the work is carried out as a subcontract under the Standard Form of Building Contract.
> (b)   That, unless otherwise stated, the work is being carried out in open areas at a height which would not require more than simple scaffolding.
> (c)   That the building in which the work is being carried out is no more than six storey's high.

Where these assumptions are not valid, as for example where work is carried out in ducts and similar confined spaces or in multi-storey structures when additional time is needed to get to and from upper floors, then an appropriate adjustment must be made to the prices. Such adjustment will normally be to the labour element only.

Note: The rates do not include for any uplift applied if the ductwork package is procured via the Mechanical Sub-contractor

DIRECTIONS

LABOUR RATE – MECHANICAL & PUBLIC HEALTH

The annual cost of a notional twelve man gang

	FOREMAN	SENIOR CRAFTSMAN (+2 Welding skill)	SENIOR CRAFTSMAN	CRAFTSMAN	INSTALLER	MATE (Over 18)	SUB TOTALS
	1 NR	1 NR	2 NR	4 NR	2 NR	2 NR	
Hourly Rate from 1 October 2018	17.15	14.73	14.17	13.05	11.77	9.93	
Working hours per annum per man	1,680.00	1,680.00	1,680.00	1,680.00	1,680.00	1,680.00	
× Hourly rate × nr of men = £ per annum	28,812.00	24,746.40	47,611.20	87,696.00	39,547.20	33,364.80	261,777.60
Overtime Rate	24.16	20.75	19.98	18.38	16.61	13.99	
Overtime hours per annum per man	336.00	336.00	336.00	336.00	336.00	336.00	
× Hourly rate × nr of men = £ per annum	8,117.76	6,972.00	13,426.56	24,702.72	11,161.92	9,401.28	73,782.24
Total	36,929.76	31,718.40	61,037.76	112,398.72	50,709.12	42,766.08	335,559.84
Incentive schemes 0.00%	0.00	0.00	0.00	0.00	0.00	0.00	0.00
Daily Travel Time Allowance (15–20 miles each way)	10.18	10.18	10.18	10.18	10.18	10.18	
Days per annum per man	224.00	224.00	224.00	224.00	224.00	224.00	
× nr of men = £ per annum	2,280.32	2,280.32	4,560.64	9,121.28	4,560.64	4,560.64	27,363.84
Daily Travel Fare (15–20 miles each way)	15.60	15.60	15.60	15.60	15.60	15.60	
Days per annum per man	224.00	224.00	224.00	224.00	224.00	224.00	
× nr of men = £ per annum	3,494.40	3,494.40	6,988.80	13,977.60	6,988.80	6,988.80	41,932.80
Employer's Pension contributions at 1 October 2018:							
£ Contributions/annum	1,671.82	1,490.95	2,763.38	5,526.77	2,295.57	1,938.08	15,686.58
National Insurance Contributions:							
Gross pay – subject to NI	48,327.13	42,530.34	82,058.06	153,099.92	70,300.01	61,257.71	
% of NI Contributions	13.80	13.80	13.80	13.80	13.80	13.80	
£ Contributions/annum	5,222.79	4,422.83	8,431.30	15,342.36	6,808.69	5,560.85	45,788.83
Holiday Credit and Welfare contributions:							
Number of weeks	52	52	52	52	52	52	
Total weekly £ contribution each	89.46	83.50	75.64	70.43	64.49	56.00	
× nr of men = £ Contributions/annum	4,652.14	4,341.85	7,866.89	14,649.44	6,707.44	5,823.79	44,041.56
Holiday Top-up Funding including overtime	18.66	13.37	15.42	14.20	12.83	10.75	
Cost	970.51	695.36	1,603.97	2,952.88	1,334.01	1,118.40	8,675.14

SUBTOTAL		519,048.58
TRAINING (INCLUDING ANY TRADE REGISTRATIONS) – SAY	1.00%	5,190.49
SEVERANCE PAY AND SUNDRY COSTS – SAY	1.50%	7,863.59
EMPLOYER'S LIABILITY AND THIRD PARTY INSURANCE – SAY	2.00%	10,642.05
ANNUAL COST OF NOTIONAL GANG		542,744.70
THEREFORE ANNUAL COST PER PRODUCTIVE MAN		51,689.97
AVERAGE NR OF HOURS WORKED PER MAN = 2016		
THEREFORE ALL-IN MAN HOURS		25.64
PRELIMINARY ITEMS – SAY	12%	3.08
SITE AND HEAD OFFICE OVERHEADS AND PROFIT (7% & 5% RESPECTIVELY) – SAY	12%	3.45
THEREFORE INCLUSIVE MAN-HOUR RATE		32.17

MEN ACTUALLY WORKING = 10.5

DIRECTIONS

Notes:

1) The following assumptions have been made in the above calculations:
   a) Increase in Hourly Rate from 1 October 2018, as advised by BESA in advance of the latest Wage Agreement.
   b) The working week of 37.5 hours i.e. the normal working week as defined by the National Agreement.
   c) The actual hours worked are five days of 9 hours each.
   d) A working year of 2,016 hours.
   e) Five days in the year are lost through sickness or similar reason.
2) National insurance contributions are those effective from 5 April 2018.
3) Weekly Holiday Credit/Welfare Stamp values are those effective from 1 October 2018, as advised by BESA in advance of the latest Wage Agreement.
4) Rates are based from 1 October 2018.
5) Overtime rates are based on Premium Rate 1.
6) Fares with Oyster Card (New Malden to Waterloo + Zone 1 – Anytime fare) current at June 2018 (TfL – 0343 222 1234).

DIRECTIONS

LABOUR RATE – DUCTWORK

The annual cost of notional eight man gang

	FOREMAN	SENIOR CRAFTSMAN	CRAFTSMAN	INSTALLER	SUB TOTALS
	1 NR	1 NR	4 NR	2 NR	
Hourly Rate from 1 October 2018	**17.15**	**14.17**	**13.05**	**11.77**	
Working hours per annum per man	1,680.00	1,680.00	1,680.00	1,680.00	
x Hourly rate × nr of men = £ per annum	**28,812.00**	**23,805.60**	**87,696.00**	**39,547.20**	**179,860.80**
Overtime Rate	24.16	19.98	18.38	16.61	
Overtime hours per annum per man	336.00	336.00	336.00	336.00	
x hourly rate × nr of men = £ per annum	**8,117.76**	**6,713.28**	**24,702.72**	**11,161.92**	**50,695.68**
Total	**36,929.76**	**30,518.88**	**112,398.72**	**50,709.12**	**230,556.48**
Incentive schemes 0.00%	**0.00**	**0.00**	**0.00**	**0.00**	**0.00**
Daily Travel Time Allowance (15–20 miles each way)	10.18	10.18	10.18	10.18	
Days per annum per man	224.00	224.00	224.00	224.00	
x nr of men = £ per annum	**2,280.32**	**2,280.32**	**9,121.28**	**4,560.64**	**18,242.56**
Daily Travel Fare (15–20 miles each way)	15.60	15.60	15.60	15.60	
Days per annum per man	224.00	224.00	224.00	224.00	
x nr of men = £ per annum	**3,494.40**	**3,494.40**	**13,977.60**	**6,988.80**	**27,955.20**
Employer's Pension Contributions from 1 October 2018					
£ Contributions/annum	**1,671.82**	**1,381.69**	**5,526.77**	**2,295.57**	**10,875.85**
National Insurance Contributions:					
Gross pay – subject to NI	48,327.13	41,029.03	153,099.92	70,300.01	
% of NI Contributions	13.8	13.8	13.8	13.8	
£ Contributions/annum	**5,086.16**	**4,079.02**	**14,795.84**	**6,535.43**	**30,496.44**
Holiday Credit and Welfare contributions:					
Number of weeks	52	52	52	52	
Total weekly £ contribution each	89.46	75.64	70.43	64.49	
x nr of men = £ Contributions/annum	**4,652.14**	**3,933.45**	**14,649.44**	**6,575.92**	**29,942.46**
Holiday Top-up Funding including overtime	18.66	15.42	14.20	12.83	
Cost	**970.51**	**801.98**	**2,952.88**	**1,334.01**	**6,059.39**

SUBTOTAL		354,128.39
TRAINING (INCLUDING ANY TRADE REGISTRATIONS) – SAY	1.00%	3,541.28
SEVERANCE PAY AND SUNDRY COSTS – SAY	1.50%	5,365.05
EMPLOYER'S LIABILITY AND THIRD PARTY INSURANCE – SAY	2.00%	7,260.69
ANNUAL COST OF NOTIONAL GANG		370,295.41
MEN ACTUALLY WORKING = 7.5 THEREFORE ANNUAL COST PER PRODUCTIVE MAN		49,372.72
AVERAGE NR OF HOURS WORKED PER MAN = 2016		
THEREFORE ALL-IN MAN HOURS		24.49
PRELIMINARY ITEMS – SAY	12%	2.94
SITE AND HEAD OFFICE OVERHEADS AND PROFIT (7% & 5% RESPECTIVELY) – SAY	12%	3.29
THEREFORE INCLUSIVE MAN-HOUR RATE		30.72

DIRECTIONS

Notes:

1)   The following assumptions have been made in the above calculations:
   a)   Increase in Hourly Rate from 1 October 2018, as advised by BESA in advance of the latest Wage Agreement.
   b)   The working week of 37.5 hours i.e. the normal working week as defined by the National Agreement.
   c)   The actual hours worked are five days of 9 hours each.
   d)   A working year of 2,016 hours.
   e)   Five days in the year are lost through sickness or similar reason.
2)   National insurance contributions are those effective from 6 April 2018.
3)   Weekly Holiday Credit/Welfare Stamp values are those effective from 1 October 2018.
4)   Rates are based from 1 October 2018.
5)   Overtime rates are based on Premium Rate 1.
6)   Fares with Oyster Card (New Malden to Waterloo + Zone 1 – Anytime fare) current at June 2018 (TfL – 0343 222 1234).

33 DRAINAGE ABOVE GROUND

Item	Net Price £	Material £	Labour hours	Labour £	Unit	Total rate £
RAINWATER PIPEWORK/GUTTERS						
PVC-u gutters: push fit joints; fixed with brackets to backgrounds; BS 4576 BS EN 607						
Half round gutter, with brackets measured separately						
75 mm	3.92	4.92	0.69	22.19	m	**27.11**
100 mm	7.88	9.89	0.64	20.59	m	**30.48**
150 mm	9.10	11.41	0.82	26.36	m	**37.77**
Brackets: including fixing to backgrounds. For minimum fixing distances, refer to the Tables and Memoranda at the rear of the book						
75 mm; Fascia	1.54	1.93	0.15	4.83	nr	**6.76**
100 mm; Jointing	2.45	3.07	0.16	5.14	nr	**8.21**
100 mm; Support	0.99	1.24	0.16	5.14	nr	**6.38**
150 mm; Fascia	2.98	3.74	0.16	5.14	nr	**8.88**
Bracket supports: including fixing to backgrounds. For minimum fixing distances, refer to the Tables and Memoranda at the rear of the book						
Side rafter	3.90	4.89	0.16	5.14	nr	**10.03**
Top rafter	3.90	4.89	0.16	5.14	nr	**10.03**
Rise and fall	4.22	5.30	0.16	5.14	nr	**10.44**
Extra over fittings half round PVC-u gutter						
Union						
75 mm	2.28	2.86	0.19	6.10	nr	**8.96**
100 mm	2.45	3.07	0.24	7.72	nr	**10.79**
150 mm	9.17	11.50	0.28	9.00	nr	**20.50**
Rainwater pipe outlets						
Running: 75 × 53 mm dia.	5.27	6.61	0.12	3.86	nr	**10.47**
Running: 100 × 68 mm dia.	3.60	4.51	0.12	3.86	nr	**8.37**
Running: 150 × 110 mm dia.	16.82	21.10	0.12	3.86	nr	**24.96**
Stop end: 100 × 68 mm dia.	4.03	5.05	0.12	3.86	nr	**8.91**
Internal stop ends: short						
75 mm	2.28	4.17	0.09	2.90	nr	**7.07**
100 mm	1.16	2.74	0.09	2.90	nr	**5.64**
150 mm	6.07	7.62	0.09	2.90	nr	**10.52**
External stop ends: short						
75 mm	4.63	5.81	0.09	2.90	nr	**8.71**
100 mm	6.26	7.85	0.09	2.90	nr	**10.75**
150 mm	6.07	7.62	0.09	2.90	nr	**10.52**
Angles						
75 mm; 45°	6.22	7.81	0.20	6.44	nr	**14.25**
75 mm; 90°	6.22	7.81	0.20	6.44	nr	**14.25**
100 mm; 90°	4.29	5.38	0.20	6.44	nr	**11.82**
100 mm; 120°	4.74	5.95	0.20	6.44	nr	**12.39**
100 mm; 135°	4.74	5.95	0.20	6.44	nr	**12.39**
100 mm; Prefabricated to special angle	26.78	33.59	0.23	7.40	nr	**40.99**
100 mm; Prefabricated to raked angle	28.98	36.36	0.23	7.40	nr	**43.76**
150 mm; 90°	15.38	19.30	0.20	6.44	nr	**25.74**

33 DRAINAGE ABOVE GROUND

Item	Net Price £	Material £	Labour hours	Labour £	Unit	Total rate £
RAINWATER PIPEWORK/GUTTERS – cont						
Extra over fittings half round PVC-u gutter – cont						
Gutter adaptors						
100 mm; Stainless steel clip	2.20	2.76	0.16	5.14	nr	7.90
100 mm; Cast iron spigot	6.05	7.59	0.23	7.40	nr	14.99
100 mm; Cast iron socket	6.05	7.59	0.23	7.40	nr	14.99
100 mm; Cast iron Ogee spigot	6.14	7.71	0.23	7.40	nr	15.11
100 mm; Cast iron Ogee socket	6.14	7.71	0.23	7.40	nr	15.11
100 mm; Half round to Square PVC-u	10.98	13.78	0.23	7.40	nr	21.18
100 mm; Gutter overshoot guard	13.58	17.04	0.58	18.65	nr	35.69
Square gutter, with brackets measured separately						
120 mm square	3.71	4.66	0.82	26.36	m	31.02
Brackets: including fixing to backgrounds. For minimum fixing distances, refer to the Tables and Memoranda at the rear of the book						
Jointing	2.56	3.21	0.16	5.14	nr	8.35
Support	1.12	1.40	0.16	5.14	nr	6.54
Bracket support: including fixing to backgrounds. For minimum fixing distances, refer to the Tables and Memoranda at the rear of the book						
Side rafter	3.90	4.89	0.16	5.14	nr	10.03
Top rafter	3.90	4.89	0.16	5.14	nr	10.03
Rise and fall	4.22	5.30	0.16	5.14	nr	10.44
Extra over fittings square PVC-u gutter						
Rainwater pipe outlets						
Running: 62 mm square	3.61	4.52	0.12	3.86	nr	8.38
Stop end: 62 mm square	4.09	5.13	0.12	3.86	nr	8.99
Stop ends: short						
External	1.87	2.34	0.09	2.90	nr	5.24
Angles						
90°	4.64	5.82	0.20	6.44	nr	12.26
120°	11.62	14.57	0.20	6.44	nr	21.01
135°	5.43	6.81	0.20	6.44	nr	13.25
Prefabricated to special angle	27.47	34.46	0.23	7.40	nr	41.86
Prefabricated to raked angle	33.86	42.47	0.23	7.40	nr	49.87
Gutter adaptors						
Cast iron	17.62	22.10	0.23	7.40	nr	29.50
High capacity square gutter, with brackets measured separately						
137 mm	8.47	10.63	0.82	26.36	m	36.99

33 DRAINAGE ABOVE GROUND

Item	Net Price £	Material £	Labour hours	Labour £	Unit	Total rate £
Brackets: including fixing to backgrounds. For minimum fixing distances, refer to the Tables and Memoranda at the rear of the book						
Jointing	8.05	10.10	0.16	5.14	nr	15.24
Support	3.36	4.21	0.16	5.14	nr	9.35
Overslung	3.12	3.91	0.16	5.14	nr	9.05
Bracket supports: including fixing to backgrounds. For minimum fixing distances, refer to the Tables and Memoranda at the rear of the book						
Side rafter	4.79	6.00	0.16	5.14	nr	11.14
Top rafter	4.79	6.00	0.16	5.14	nr	11.14
Rise and fall	7.69	9.64	0.16	5.14	nr	14.78
Extra over fittings high capacity square UPV-C						
Rainwater pipe outlets						
Running: 75 mm square	13.19	16.54	0.12	3.86	nr	20.40
Running: 82 mm dia.	13.19	16.54	0.12	3.86	nr	20.40
Running: 110 mm dia.	11.54	14.47	0.12	3.86	nr	18.33
Screwed outlet adaptor						
75 mm square pipe	8.39	10.53	0.23	7.40	nr	17.93
Stop ends: short						
External	4.63	5.81	0.09	2.90	nr	8.71
Angles						
90°	11.54	14.47	0.20	6.44	nr	20.91
135°	13.58	17.04	0.20	6.44	nr	23.48
Prefabricated to special angle	39.22	49.20	0.23	7.40	nr	56.60
Prefabricated to raked internal angle	68.29	85.66	0.23	7.40	nr	93.06
Prefabricated to raked external angle	68.29	85.66	0.23	7.40	nr	93.06
Deep eliptical gutter, with brackets measured separately						
137 mm	4.59	5.76	0.82	26.36	m	32.12
Brackets: including fixing to backgrounds. For minimum fixing distances, refer to the Tables and Memoranda at the rear of the book						
Jointing	3.64	4.57	0.16	5.14	nr	9.71
Support	1.55	1.95	0.16	5.14	nr	7.09
Bracket support: including fixing to backgrounds. For minimum fixing distances, refer to the Tables and Memoranda at the rear of the book						
Side rafter	5.03	6.31	0.16	5.14	nr	11.45
Top rafter	5.03	6.31	0.16	5.14	nr	11.45
Rise and fall	8.08	10.14	0.16	5.14	nr	15.28

33 DRAINAGE ABOVE GROUND

Item	Net Price £	Material £	Labour hours	Labour £	Unit	Total rate £
RAINWATER PIPEWORK/GUTTERS – cont						
Extra over fittings deep eliptical PVC-u gutter						
Rainwater pipe outlets						
Running: 68 mm dia.	5.78	7.25	0.12	3.86	nr	11.11
Running: 82 mm dia.	5.48	6.88	0.12	3.86	nr	10.74
Stop end: 68 mm dia.	5.87	7.36	0.12	3.86	nr	11.22
Stop ends: short						
External	2.83	3.55	0.09	2.90	nr	6.45
Angles						
90°	5.83	7.31	0.20	6.44	nr	13.75
135°	6.92	8.68	0.20	6.44	nr	15.12
Prefabricated to special angle	20.26	25.41	0.23	7.40	nr	32.81
Gutter adaptors						
Stainless steel clip	4.97	6.24	0.16	5.14	nr	11.38
Marley deepflow	4.02	5.04	0.23	7.40	nr	12.44
Ogee profile PVC-u gutter, with brackets measured separately						
122 mm	4.83	6.06	0.82	26.36	m	32.42
Brackets: including fixing to backgrounds. For minimum fixing distances, refer to the Tables and Memoranda at the rear of the book						
Jointing	3.84	4.82	0.16	5.14	nr	9.96
Support	1.48	1.86	0.16	5.14	nr	7.00
Overslung	1.48	1.86	0.16	5.14	nr	7.00
Extra over fittings Ogee profile PVC-u gutter						
Rainwater pipe outlets						
Running: 68 mm dia.	5.48	6.88	0.12	3.86	nr	10.74
Stop ends: short						
Internal/External: left or right hand	2.78	3.48	0.09	2.90	nr	6.38
Angles						
90°: internal or external	5.55	6.97	0.20	6.44	nr	13.41
135°: internal or external	5.55	6.97	0.20	6.44	nr	13.41
PVC-u rainwater pipe: dry push fit joints; fixed with brackets to backgrounds; BS 4576/ BS EN 607						
Pipe: circular, with brackets measured separately						
53 mm	6.24	7.83	0.61	19.62	m	27.45
68 mm	6.50	8.15	0.61	19.62	m	27.77
Pipe clip: including fixing to backgrounds. For minimum fixing distances, refer to the Tables and Memoranda at the rear of the book						
68 mm	1.45	1.81	0.16	5.14	nr	6.95

33 DRAINAGE ABOVE GROUND

Item	Net Price £	Material £	Labour hours	Labour £	Unit	Total rate £
Pipe clip adjustable: including fixing to backgrounds. For minimum fixing distances, refer to the Tables and Memoranda at the rear of the book						
53 mm	2.22	2.79	0.16	5.14	nr	7.93
68 mm	3.11	3.90	0.16	5.14	nr	9.04
Pipe clip drive in: including fixing to backgrounds. For minimum fixing distances, refer to the Tables and Memoranda at the rear of the book						
68 mm	3.49	4.38	0.16	5.14	nr	9.52
Extra over fittings circular pipework PVC-u						
Pipe coupler: PVC-u to PVC-u						
68 mm	1.81	2.27	0.12	3.86	nr	6.13
Pipe coupler: PVC-u to Cast Iron						
68 mm: to 3" cast iron	7.11	8.92	0.17	5.47	nr	14.39
68 mm: to 3 3/4" cast iron	24.43	30.64	0.17	5.47	nr	36.11
Access pipe: single socket						
68 mm	11.35	14.24	0.15	4.83	nr	19.07
Bend: short radius						
53 mm: 67.5°	2.64	3.32	0.20	6.44	nr	9.76
68 mm: 92.5°	3.88	4.87	0.20	6.44	nr	11.31
68 mm: 112.5°	3.88	4.87	0.20	6.44	nr	11.31
Bend: long radius						
68 mm: 112°	3.68	4.61	0.20	6.44	nr	11.05
Branch						
68 mm: 92°	23.86	29.93	0.23	7.40	nr	37.33
68 mm: 112°	23.93	30.02	0.23	7.40	nr	37.42
Double branch						
68 mm: 112°	47.23	59.25	0.24	7.72	nr	66.97
Shoe						
53 mm	3.91	4.91	0.12	3.86	nr	8.77
68 mm	5.78	7.25	0.12	3.86	nr	11.11
Rainwater head: including fixing to backgrounds						
68 mm	11.35	14.24	0.29	9.33	nr	23.57
Pipe: square, with brackets measured separately						
62 mm	3.98	5.00	0.45	14.47	m	19.47
75 mm	7.56	9.49	0.45	14.47	m	23.96
Pipe clip: including fixing to backgrounds. For minimum fixing distances, refer to the Tables and Memoranda at the rear of the book						
62 mm	1.33	1.67	0.16	5.14	nr	6.81
75 mm	2.95	3.70	0.16	5.14	nr	8.84
Pipe clip adjustable: including fixing to backgrounds. For minimum fixing distances, refer to the Tables and Memoranda at the rear of the book						
62 mm	4.12	5.16	0.16	5.14	nr	10.30

33 DRAINAGE ABOVE GROUND

Item	Net Price £	Material £	Labour hours	Labour £	Unit	Total rate £
RAINWATER PIPEWORK/GUTTERS – cont						
Extra over fittings square pipework PVC-u						
Pipe coupler: PVC-u to PVC-u						
62 mm	1.95	2.44	0.20	6.44	nr	8.88
75 mm	3.03	3.80	0.20	6.44	nr	10.24
Square to circular adaptor: single socket						
62 mm to 68 mm	3.80	4.77	0.20	6.44	nr	11.21
Square to circular adaptor: single socket						
75 mm to 62 mm	4.91	6.16	0.20	6.44	nr	12.60
Access pipe						
62 mm	22.71	28.49	0.16	5.14	nr	33.63
75 mm	23.97	30.07	0.16	5.14	nr	35.21
Bends						
62 mm: 92.5°	3.51	4.40	0.20	6.44	nr	10.84
62 mm: 112.5°	2.56	3.21	0.20	6.44	nr	9.65
75 mm: 112.5°	6.40	8.03	0.20	6.44	nr	14.47
Bends: prefabricated special angle						
62 mm	22.83	28.64	0.23	7.40	nr	36.04
75 mm	31.71	39.78	0.23	7.40	nr	47.18
Offset						
62 mm	5.50	6.90	0.20	6.44	nr	13.34
75 mm	15.14	19.00	0.20	6.44	nr	25.44
Offset: prefabricated special angle						
62 mm	22.83	28.64	0.23	7.40	nr	36.04
Shoe						
62 mm	3.23	4.05	0.12	3.86	nr	7.91
75 mm	5.53	6.93	0.12	3.86	nr	10.79
Branch						
62 mm	7.98	10.01	0.23	7.40	nr	17.41
75 mm	38.91	48.81	0.23	7.40	nr	56.21
Double branch						
62 mm	42.81	53.70	0.24	7.72	nr	61.42
Rainwater head						
62 mm	35.90	45.04	0.29	9.33	nr	54.37
75 mm	37.77	47.38	0.35	110.90	nr	158.28
PVC-u rainwater pipe: solvent welded joints; fixed with brackets to backgrounds; BS 4576/ BS EN 607						
Pipe: circular, with brackets measured separately						
82 mm	12.25	15.37	0.35	11.26	m	26.63
Pipe clip: galvanized; including fixing to backgrounds. For minimum fixing distances, refer to the Tables and Memoranda at the rear of the book						
82 mm	4.98	6.25	0.58	18.65	nr	24.90

33 DRAINAGE ABOVE GROUND

Item	Net Price £	Material £	Labour hours	Labour £	Unit	Total rate £
Pipe clip: galvanized plastic coated; including fixing to backgrounds. For minimum fixing distances, refer to the Tables and Memoranda at the rear of the book						
82 mm	6.89	8.65	0.58	18.65	nr	27.30
Pipe clip: PVC-u including fixing to backgrounds. For minimum fixing distances, refer to the Tables and Memoranda at the rear of the book						
82 mm	3.70	4.64	0.58	18.65	nr	23.29
Pipe clip: PVC-u adjustable: including fixing to backgrounds. For minimum fixing distances, refer to the Tables and Memoranda at the rear of the book						
82 mm	6.78	8.50	0.58	18.65	nr	27.15
Extra over fittings circular pipework PVC-u						
Pipe coupler: PVC-u to PVC-u						
82 mm	6.21	7.80	0.21	6.75	nr	14.55
Access pipe						
82 mm	38.24	47.97	0.23	7.40	nr	55.37
Bend						
82 mm: 92, 112.5 and 135°	15.61	19.58	0.29	9.33	nr	28.91
Shoe						
82 mm	9.63	12.08	0.29	9.33	nr	21.41
110 mm	12.09	15.16	0.32	10.28	nr	25.44
Branch						
82 mm: 92, 112.5 and 135°	23.26	29.18	0.35	11.26	nr	40.44
Rainwater head						
82 mm	20.41	25.60	0.58	18.65	nr	44.25
110 mm	18.59	23.32	0.58	18.65	nr	41.97
Roof outlets: 178 dia.; Flat						
50 mm	20.40	25.59	1.15	36.99	nr	62.58
82 mm	20.40	25.59	1.15	36.99	nr	62.58
Roof outlets: 178 mm dia.; Domed						
50 mm	20.40	25.59	1.15	36.99	nr	62.58
82 mm	20.40	25.59	1.15	36.99	nr	62.58
Roof outlets: 406 mm dia.; Flat						
82 mm	39.86	50.00	1.15	36.99	nr	86.99
110 mm	39.86	50.00	1.15	36.99	nr	86.99
Roof outlets: 406 mm dia.; Domed						
82 mm	39.86	50.00	1.15	36.99	nr	86.99
110 mm	39.86	50.00	1.15	36.99	nr	86.99
Roof outlets: 406 mm dia.; Inverted						
82 mm	87.76	110.08	1.15	36.99	nr	147.07
110 mm	87.76	110.08	1.15	36.99	nr	147.07
Roof outlets: 406 mm dia.; Vent Pipe						
82 mm	57.64	72.31	1.15	36.99	nr	109.30
110 mm	57.64	72.31	1.15	36.99	nr	109.30

33 DRAINAGE ABOVE GROUND

Item	Net Price £	Material £	Labour hours	Labour £	Unit	Total rate £
RAINWATER PIPEWORK/GUTTERS – cont						
Extra over fittings circular pipework PVC-u – cont						
Balcony outlets: screed						
82 mm	34.12	42.80	1.15	36.99	nr	**79.79**
Balcony outlets: asphalt						
82 mm	33.93	42.56	1.15	36.99	nr	**79.55**
Adaptors						
82 mm × 62 mm square pipe	4.57	5.73	0.21	6.75	nr	**12.48**
82 mm × 68 mm circular pipe	4.57	5.73	0.21	6.75	nr	**12.48**
For 110 mm dia. pipework and fittings refer to 33 Drainage Above Ground						
Cast iron gutters: mastic and bolted joints; BS 460; fixed with brackets to backgrounds						
Half round gutter, with brackets measured separately						
100 mm	15.75	19.76	0.85	27.33	m	**47.09**
115 mm	16.42	20.60	0.97	31.19	m	**51.79**
125 mm	19.21	24.10	0.97	31.19	m	**55.29**
150 mm	32.85	41.20	1.12	36.03	m	**77.23**
Brackets; fixed to backgrounds. For minimum fixing distances, refer to the Tables and Memoranda at the rear of the book						
Fascia						
100 mm	3.69	4.63	0.16	5.14	nr	**9.77**
115 mm	3.69	4.63	0.16	5.14	nr	**9.77**
125 mm	3.69	4.63	0.16	5.14	nr	**9.77**
150 mm	4.67	5.86	0.16	5.14	nr	**11.00**
Rise and fall						
100 mm	7.34	9.21	0.39	12.54	nr	**21.75**
115 mm	7.34	9.21	0.39	12.54	nr	**21.75**
125 mm	7.54	9.45	0.39	12.54	nr	**21.99**
150 mm	7.67	9.62	0.39	12.54	nr	**22.16**
Top rafter						
100 mm	4.52	5.67	0.16	5.14	nr	**10.81**
115 mm	4.52	5.67	0.16	5.14	nr	**10.81**
125 mm	6.12	7.67	0.16	5.14	nr	**12.81**
150 mm	8.31	10.43	0.16	5.14	nr	**15.57**
Side rafter						
100 mm	4.52	5.67	0.16	5.14	nr	**10.81**
115 mm	4.52	5.67	0.16	5.14	nr	**10.81**
125 mm	6.12	7.67	0.16	5.14	nr	**12.81**
150 mm	8.31	10.43	0.16	5.14	nr	**15.57**

33 DRAINAGE ABOVE GROUND

Item	Net Price £	Material £	Labour hours	Labour £	Unit	Total rate £
Extra over fittings half round gutter cast iron BS 460						
Union						
100 mm	8.66	10.86	0.39	12.54	nr	23.40
115 mm	10.81	13.56	0.48	15.44	nr	29.00
125 mm	12.21	15.32	0.48	15.44	nr	30.76
150 mm	13.70	17.18	0.55	17.68	nr	34.86
Stop end; internal						
100 mm	4.42	5.54	0.12	3.86	nr	9.40
115 mm	5.72	7.18	0.15	4.83	nr	12.01
125 mm	5.72	7.18	0.15	4.83	nr	12.01
150 mm	7.94	9.96	0.20	6.44	nr	16.40
Stop end; external						
100 mm	4.42	5.54	0.12	3.86	nr	9.40
115 mm	5.60	7.02	0.15	4.83	nr	11.85
125 mm	5.72	7.18	0.15	4.83	nr	12.01
150 mm	7.94	9.96	0.20	6.44	nr	16.40
90° angle; single socket						
100 mm	13.13	16.48	0.39	12.54	nr	29.02
115 mm	13.51	16.95	0.43	13.83	nr	30.78
125 mm	15.93	19.98	0.43	13.83	nr	33.81
150 mm	29.12	36.52	0.50	16.08	nr	52.60
90° angle; double socket						
100 mm	15.92	19.97	0.39	12.54	nr	32.51
115 mm	16.89	21.19	0.43	13.83	nr	35.02
125 mm	21.86	27.42	0.43	13.83	nr	41.25
135° angle; single socket						
100 mm	13.41	16.82	0.39	12.54	nr	29.36
115 mm	13.54	16.98	0.43	13.83	nr	30.81
125 mm	20.01	25.10	0.43	13.83	nr	38.93
150 mm	29.68	37.23	0.50	16.08	nr	53.31
Running outlet						
65 mm outlet						
100 mm	12.82	16.08	0.39	12.54	nr	28.62
115 mm	13.94	17.48	0.43	13.83	nr	31.31
125 mm	15.93	19.98	0.43	13.83	nr	33.81
75 mm outlet						
100 mm	12.82	16.08	0.39	12.54	nr	28.62
115 mm	13.94	17.48	0.43	13.83	nr	31.31
125 mm	15.93	19.98	0.43	13.83	nr	33.81
150 mm	27.60	34.62	0.50	16.08	nr	50.70
100 mm outlet						
150 mm	27.60	34.62	0.50	16.08	nr	50.70
Stop end outlet; socket						
65 mm outlet						
100 mm	15.02	18.84	0.39	12.54	nr	31.38
115 mm	16.84	21.12	0.43	13.83	nr	34.95
75 mm outlet						
125 mm	15.02	18.84	0.43	13.83	nr	32.67
150 mm	31.58	74.23	0.50	16.08	nr	90.31

33 DRAINAGE ABOVE GROUND

Item	Net Price £	Material £	Labour hours	Labour £	Unit	Total rate £
RAINWATER PIPEWORK/GUTTERS – cont						
Extra over fittings half round gutter cast iron BS 460 – cont						
100 mm outlet						
150 mm	27.60	34.62	0.50	16.08	nr	**50.70**
Stop end outlet; spigot						
65 mm outlet						
100 mm	15.02	18.84	0.39	12.54	nr	**31.38**
115 mm	16.84	21.12	0.43	13.83	nr	**34.95**
75 mm outlet						
125 mm	15.02	18.84	0.43	13.83	nr	**32.67**
150 mm	31.58	39.61	0.50	16.08	nr	**55.69**
100 mm outlet						
150 mm	31.58	39.61	0.50	16.08	nr	**55.69**
Half round; 3 mm thick double beaded gutter, with brackets measured separately						
100 mm	16.04	20.12	0.85	27.33	m	**47.45**
115 mm	16.74	21.00	0.85	27.33	m	**48.33**
125 mm	19.20	24.08	0.97	31.19	m	**55.27**
Brackets; fixed to backgrounds. For minimum fixing distances, refer to the Tables and Memoranda at the rear of the book						
Fascia						
100 mm	3.69	4.63	0.16	5.14	nr	**9.77**
115 mm	3.69	4.63	0.16	5.14	nr	**9.77**
125 mm	3.69	4.63	0.16	5.14	nr	**9.77**
Extra over fittings half round 3 mm thick gutter BS 460						
Union						
100 mm	8.66	10.86	0.38	12.22	nr	**23.08**
115 mm	10.56	13.25	0.38	12.22	nr	**25.47**
125 mm	12.21	15.32	0.43	13.83	nr	**29.15**
Stop end; internal						
100 mm	4.42	5.54	0.12	3.86	nr	**9.40**
115 mm	5.72	7.18	0.12	3.86	nr	**11.04**
125 mm	5.74	7.20	0.15	4.83	nr	**12.03**
Stop end; external						
100 mm	4.42	5.54	0.12	3.86	nr	**9.40**
115 mm	5.72	7.18	0.12	3.86	nr	**11.04**
125 mm	5.74	7.20	0.15	4.83	nr	**12.03**
90° angle; single socket						
100 mm	13.65	17.12	0.38	12.22	nr	**29.34**
115 mm	13.83	17.35	0.38	12.22	nr	**29.57**
125 mm	15.93	19.98	0.43	13.83	nr	**33.81**
135° angle; single socket						
100 mm	13.41	16.82	0.38	12.22	nr	**29.04**
115 mm	13.51	16.95	0.38	12.22	nr	**29.17**
125 mm	16.85	21.13	0.43	13.83	nr	**34.96**

33 DRAINAGE ABOVE GROUND

Item	Net Price £	Material £	Labour hours	Labour £	Unit	Total rate £
Running outlet						
65 mm outlet						
100 mm	13.76	17.26	0.38	12.22	nr	29.48
115 mm	14.22	17.84	0.38	12.22	nr	30.06
125 mm	18.83	23.62	0.43	13.83	nr	37.45
75 mm outlet						
115 mm	14.22	17.84	0.38	12.22	nr	30.06
125 mm	16.50	20.70	0.43	13.83	nr	34.53
Stop end outlet; socket						
65 mm outlet						
100 mm	15.02	18.84	0.38	12.22	nr	31.06
115 mm	16.84	21.12	0.38	12.22	nr	33.34
125 mm	18.83	23.62	0.43	13.83	nr	37.45
75 mm outlet						
125 mm	19.17	24.05	0.43	13.83	nr	37.88
Stop end outlet; spigot						
65 mm outlet						
100 mm	15.02	18.84	0.38	12.22	nr	31.06
115 mm	16.84	21.12	0.38	12.22	nr	33.34
125 mm	18.83	23.62	0.43	13.83	nr	37.45
Deep half round gutter, with brackets measured separately						
100 × 75 mm	26.38	33.10	0.85	27.33	m	60.43
125 × 75 mm	34.09	42.76	0.97	31.19	m	73.95
Brackets; fixed to backgrounds. For minimum fixing distances, refer to the Tables and Memoranda at the rear of the book						
Fascia						
100 × 75 mm	12.46	15.64	0.16	5.14	nr	20.78
125 × 75 mm	15.35	19.25	0.16	5.14	nr	24.39
Extra over fittings deep half round gutter BS 460						
Union						
100 × 75 mm	14.49	18.18	0.38	12.22	nr	30.40
125 × 75 mm	15.35	19.25	0.43	13.83	nr	33.08
Stop end; internal						
100 × 75 mm	12.68	15.90	0.12	3.86	nr	19.76
125 × 75 mm	15.64	19.62	0.15	4.83	nr	24.45
Stop end; external						
100 × 75 mm	12.68	15.90	0.12	3.86	nr	19.76
125 × 75 mm	15.64	19.62	0.15	4.83	nr	24.45
90° angle; single socket						
100 × 75 mm	36.15	45.35	0.38	12.22	nr	57.57
125 × 75 mm	45.91	57.59	0.43	13.83	nr	71.42
135° angle; single socket						
100 × 75 mm	36.15	45.35	0.38	12.22	nr	57.57
125 × 75 mm	45.91	57.59	0.43	13.83	nr	71.42

33 DRAINAGE ABOVE GROUND

Item	Net Price £	Material £	Labour hours	Labour £	Unit	Total rate £
RAINWATER PIPEWORK/GUTTERS – cont						
Extra over fittings deep half round gutter BS 460 – cont						
Running outlet						
65 mm outlet						
100 × 75 mm	36.15	45.35	0.38	12.22	nr	**57.57**
125 × 75 mm	45.91	57.59	0.43	13.83	nr	**71.42**
75 mm outlet						
100 × 75 mm	36.15	45.35	0.38	12.22	nr	**57.57**
125 × 75 mm	33.32	41.80	0.43	13.83	nr	**55.63**
Stop end outlet; socket						
65 mm outlet						
100 × 75 mm	48.85	61.28	0.38	12.22	nr	**73.50**
75 mm outlet						
100 × 75 mm	48.85	61.28	0.38	12.22	nr	**73.50**
125 × 75 mm	48.85	61.28	0.43	13.83	nr	**75.11**
Stop end outlet; spigot						
65 mm outlet						
100 × 75 mm	48.85	61.28	0.38	12.22	nr	**73.50**
75 mm outlet						
100 × 75 mm	48.85	61.28	0.38	12.22	nr	**73.50**
125 × 75 mm	48.85	61.28	0.43	13.83	nr	**75.11**
Ogee gutter, with brackets measured separately						
100 mm	17.57	22.04	0.85	27.33	m	**49.37**
115 mm	19.32	24.24	0.97	31.19	m	**55.43**
125 mm	20.27	25.42	0.97	31.19	m	**56.61**
Brackets; fixed to backgrounds. For minimum fixing distances, refer to the Tables and Memoranda at the rear of the book						
Fascia						
100 mm	4.02	5.04	0.16	5.14	nr	**10.18**
115 mm	4.02	5.04	0.16	5.14	nr	**10.18**
125 mm	4.53	5.68	0.16	5.14	nr	**10.82**
Extra over fittings Ogee cast iron gutter BS 460						
Union						
100 mm	8.67	10.88	0.38	12.22	nr	**23.10**
115 mm	10.56	13.25	0.43	13.83	nr	**27.08**
125 mm	12.21	15.32	0.43	13.83	nr	**29.15**
Stop end; internal						
100 mm	4.52	5.67	0.12	3.86	nr	**9.53**
115 mm	5.84	7.32	0.15	4.83	nr	**12.15**
125 mm	5.84	7.32	0.15	4.83	nr	**12.15**
Stop end; external						
100 mm	4.52	5.67	0.12	3.86	nr	**9.53**
115 mm	5.84	7.32	0.15	4.83	nr	**12.15**
125 mm	5.84	7.32	0.15	4.83	nr	**12.15**

33 DRAINAGE ABOVE GROUND

Item	Net Price £	Material £	Labour hours	Labour £	Unit	Total rate £
90° angle; internal						
100 mm	13.70	17.18	0.38	12.22	nr	29.40
115 mm	14.86	18.64	0.43	13.83	nr	32.47
125 mm	16.21	20.34	0.43	13.83	nr	34.17
90° angle; external						
100 mm	13.70	17.18	0.38	12.22	nr	29.40
115 mm	14.86	18.64	0.43	13.83	nr	32.47
125 mm	16.21	20.34	0.43	13.83	nr	34.17
135° angle; internal						
100 mm	14.23	17.85	0.38	12.22	nr	30.07
115 mm	15.16	19.02	0.43	13.83	nr	32.85
125 mm	19.97	25.05	0.43	13.83	nr	38.88
135° angle; external						
100 mm	14.23	17.85	0.38	12.22	nr	30.07
115 mm	15.16	19.02	0.43	13.83	nr	32.85
125 mm	19.97	25.05	0.43	13.83	nr	38.88
Running outlet						
65 mm outlet						
100 mm	13.95	17.49	0.38	12.22	nr	29.71
115 mm	14.87	18.65	0.43	13.83	nr	32.48
125 mm	16.21	20.34	0.43	13.83	nr	34.17
75 mm outlet						
125 mm	16.21	20.34	0.43	13.83	nr	34.17
Stop end outlet; socket						
65 mm outlet						
100 mm	22.04	27.64	0.38	12.22	nr	39.86
115 mm	22.04	27.64	0.43	13.83	nr	41.47
125 mm	22.04	27.64	0.43	13.83	nr	41.47
75 mm outlet						
125 mm	22.04	27.64	0.43	13.83	nr	41.47
Stop end outlet; spigot						
65 mm outlet						
100 mm	22.04	27.64	0.38	12.22	nr	39.86
115 mm	22.04	27.64	0.43	13.83	nr	41.47
125 mm	22.04	27.64	0.43	13.83	nr	41.47
75 mm outlet						
125 mm	22.04	27.64	0.43	13.83	nr	41.47
Notts Ogee Gutter, with brackets measured separately						
115 mm	31.21	39.16	0.85	27.33	m	66.49
Brackets; fixed to backgrounds. For minimum fixing distances, refer to the Tables and Memoranda at the rear of the book						
Fascia						
115 mm	12.17	15.27	0.16	5.14	nr	20.41

33 DRAINAGE ABOVE GROUND

Item	Net Price £	Material £	Labour hours	Labour £	Unit	Total rate £
RAINWATER PIPEWORK/GUTTERS – cont						
Extra over fittings Notts Ogee Cast Iron Gutter BS 460						
Union						
115 mm	14.78	18.54	0.38	12.22	nr	**30.76**
Stop end; internal						
115 mm	12.17	15.27	0.16	5.14	nr	**20.41**
Stop end; external						
115 mm	12.17	15.27	0.16	5.14	nr	**20.41**
90° angle; internal						
115 mm	35.31	44.30	0.43	13.83	nr	**58.13**
90° angle; external						
115 mm	35.31	44.30	0.43	13.83	nr	**58.13**
135° angle; internal						
115 mm	35.98	45.14	0.43	13.83	nr	**58.97**
135° angle; external						
115 mm	35.98	45.14	0.43	13.83	nr	**58.97**
Running outlet						
65 mm outlet						
115 mm	42.33	53.10	0.43	13.83	nr	**66.93**
75 mm outlet						
115 mm	42.33	53.10	0.43	13.83	nr	**66.93**
Stop end outlet; socket						
65 mm outlet						
115 mm	54.49	68.35	0.43	13.83	nr	**82.18**
Stop end outlet; spigot						
65 mm outlet						
115 mm	54.49	68.35	0.43	13.83	nr	**82.18**
No 46 moulded Gutter, with brackets measured separately						
100 × 75 mm	30.18	37.86	0.85	27.33	m	**65.19**
125 × 100 mm	44.25	55.51	0.97	31.19	m	**86.70**
Brackets; fixed to backgrounds. For minimum fixing distances, refer to the Tables and Memoranda at the rear of the book						
Fascia						
100 × 75 mm	6.73	8.44	0.16	5.14	nr	**13.58**
125 × 100 mm	6.73	8.44	0.16	5.14	nr	**13.58**

33 DRAINAGE ABOVE GROUND

Item	Net Price £	Material £	Labour hours	Labour £	Unit	Total rate £
Extra over fittings						
Union						
100 × 75 mm	14.22	17.84	0.38	12.22	nr	**30.06**
125 × 100 mm	16.48	20.68	0.43	13.83	nr	**34.51**
Stop end; internal						
100 × 75 mm	12.73	15.97	0.12	3.86	nr	**19.83**
125 × 100 mm	16.48	20.68	0.15	4.83	nr	**25.51**
Stop end; external						
100 × 75 mm	12.73	15.97	0.12	3.86	nr	**19.83**
125 × 100 mm	16.48	20.68	0.15	4.83	nr	**25.51**
90° angle; internal						
100 × 75 mm	33.34	41.82	0.38	12.22	nr	**54.04**
125 × 100 mm	47.92	60.11	0.43	13.83	nr	**73.94**
90° angle; external						
100 × 75 mm	33.34	41.82	0.38	12.22	nr	**54.04**
125 × 100 mm	47.92	60.11	0.43	13.83	nr	**73.94**
135° angle; internal						
100 × 75 mm	33.98	42.63	0.38	12.22	nr	**54.85**
125 × 100 mm	47.92	60.11	0.43	13.83	nr	**73.94**
135° angle; external						
100 × 75 mm	33.98	42.63	0.38	12.22	nr	**54.85**
125 × 100 mm	47.92	60.11	0.43	13.83	nr	**73.94**
Running outlet						
65 mm outlet						
100 × 75 mm	33.98	42.63	0.38	12.22	nr	**54.85**
125 × 100 mm	47.92	60.11	0.43	13.83	nr	**73.94**
75 mm outlet						
100 × 75 mm	33.98	42.63	0.38	12.22	nr	**54.85**
125 × 100 mm	47.92	60.11	0.43	13.83	nr	**73.94**
100 mm outlet						
100 × 75 mm	33.98	42.63	0.38	12.22	nr	**54.85**
125 × 100 mm	47.92	60.11	0.43	13.83	nr	**73.94**
100 × 75 mm outlet						
125 × 100 mm	33.98	42.63	0.43	13.83	nr	**56.46**
Stop end outlet; socket						
65 mm outlet						
100 × 75 mm	64.40	80.79	0.38	12.22	nr	**93.01**
75 mm outlet						
125 × 100 mm	64.40	80.79	0.43	13.83	nr	**94.62**
Stop end outlet; spigot						
65 mm outlet						
100 × 75 mm	64.40	80.79	0.38	12.22	nr	**93.01**
75 mm outlet						
125 × 100 mm	64.40	80.79	0.43	13.83	nr	**94.62**

33 DRAINAGE ABOVE GROUND

Item	Net Price £	Material £	Labour hours	Labour £	Unit	Total rate £
RAINWATER PIPEWORK/GUTTERS – cont						
Box gutter, with brackets measured separately						
100 × 75 mm	48.94	61.39	0.85	27.33	m	88.72
Brackets; fixed to backgrounds. For minimum fixing distances, refer to the Tables and Memoranda at the rear of the book						
Fascia						
100 × 75 mm	7.76	9.73	0.16	5.14	nr	14.87
Extra over fittings Box Cast Iron Gutter BS 460						
Union						
100 × 75 mm	10.29	12.90	0.38	12.22	nr	25.12
Stop end; external						
100 × 75 mm	7.76	9.73	0.12	3.86	nr	13.59
90° angle						
100 × 75 mm	39.12	49.07	0.38	12.22	nr	61.29
135° angle						
100 × 75 mm	39.12	49.07	0.38	12.22	nr	61.29
Running outlet						
65 mm outlet						
100 × 75 mm	39.12	49.07	0.38	12.22	nr	61.29
75 mm outlet						
100 × 75 mm	39.12	49.07	0.38	12.22	nr	61.29
100 × 75 mm outlet						
100 × 75 mm	39.12	49.07	0.38	12.22	nr	61.29
Cast iron rainwater pipe; dry joints; BS 460; fixed to backgrounds: Circular						
Plain socket pipe, with brackets measured separately						
65 mm	28.33	35.54	0.69	22.19	m	57.73
75 mm	28.33	35.54	0.69	22.19	m	57.73
100 mm	38.65	48.48	0.69	22.19	m	70.67
Bracket; fixed to backgrounds. For minimum fixing distances, refer to the Tables and Memoranda at the rear of the book						
65 mm	10.35	12.98	0.29	9.33	nr	22.31
75 mm	10.39	13.04	0.29	9.33	nr	22.37
100 mm	10.52	13.19	0.29	9.33	nr	22.52
Eared socket pipe, with wall spacers measured separately						
65 mm	30.27	37.97	0.62	19.95	m	57.92
75 mm	30.27	37.97	0.62	19.95	m	57.92
100 mm	40.62	50.95	0.62	19.95	m	70.90
Wall spacer plate; eared pipework						
65 mm	6.68	8.38	0.16	5.14	nr	13.52
75 mm	6.80	8.53	0.16	5.14	nr	13.67
100 mm	10.52	13.19	0.16	5.14	nr	18.33

33 DRAINAGE ABOVE GROUND

Item	Net Price £	Material £	Labour hours	Labour £	Unit	Total rate £
Extra over fittings Circular Cast Iron Pipework BS 460						
Loose sockets						
Plain socket						
65 mm	8.18	10.26	0.23	7.40	nr	**17.66**
75 mm	8.18	10.26	0.23	7.40	nr	**17.66**
100 mm	12.65	15.87	0.23	7.40	nr	**23.27**
Eared socket						
65 mm	11.83	14.84	0.29	9.33	nr	**24.17**
75 mm	11.83	14.84	0.29	9.33	nr	**24.17**
100 mm	15.98	20.05	0.29	9.33	nr	**29.38**
Shoe; front projection						
Plain socket						
65 mm	25.46	31.94	0.23	7.40	nr	**39.34**
75 mm	25.46	31.94	0.23	7.40	nr	**39.34**
100 mm	34.33	43.06	0.23	7.40	nr	**50.46**
Eared socket						
65 mm	29.51	37.02	0.29	9.33	nr	**46.35**
75 mm	29.51	37.02	0.29	9.33	nr	**46.35**
100 mm	39.17	49.13	0.29	9.33	nr	**58.46**
Access pipe						
65 mm	45.97	57.67	0.23	7.40	nr	**65.07**
75 mm	48.25	60.52	0.23	7.40	nr	**67.92**
100 mm	85.60	107.37	0.23	7.40	nr	**114.77**
100 mm; eared	96.64	121.23	0.29	9.33	nr	**130.56**
Bends; any degree						
65 mm	18.39	23.07	0.23	7.40	nr	**30.47**
75 mm	21.91	27.48	0.23	7.40	nr	**34.88**
100 mm	30.96	38.84	0.23	7.40	nr	**46.24**
Branch						
92.5°						
65 mm	35.48	44.51	0.29	9.33	nr	**53.84**
75 mm	39.13	49.09	0.29	9.33	nr	**58.42**
100 mm	45.61	57.21	0.29	9.33	nr	**66.54**
112.5°						
65 mm	35.48	44.51	0.29	9.33	nr	**53.84**
75 mm	39.13	49.09	0.29	9.33	nr	**58.42**
135°						
65 mm	35.48	44.51	0.29	9.33	nr	**53.84**
75 mm	39.13	49.09	0.29	9.33	nr	**58.42**
Offsets						
75–150 mm projection						
65 mm	27.63	34.66	0.25	8.04	nr	**42.70**
75 mm	27.63	34.66	0.25	8.04	nr	**42.70**
100 mm	52.13	65.40	0.25	8.04	nr	**73.44**
225 mm projection						
65 mm	28.16	35.32	0.25	8.04	nr	**43.36**
75 mm	28.16	35.32	0.25	8.04	nr	**43.36**
100 mm	52.13	65.40	0.25	8.04	nr	**73.44**

33 DRAINAGE ABOVE GROUND

Item	Net Price £	Material £	Labour hours	Labour £	Unit	Total rate £
RAINWATER PIPEWORK/GUTTERS – cont						
Extra over fittings – cont						
305 mm projection						
65 mm	37.68	47.26	0.25	8.04	nr	**55.30**
75 mm	39.57	49.64	0.25	8.04	nr	**57.68**
100 mm	64.38	80.76	0.25	8.04	nr	**88.80**
380 mm projection						
65 mm	75.21	94.35	0.25	8.04	nr	**102.39**
75 mm	75.21	94.35	0.25	8.04	nr	**102.39**
100 mm	102.65	128.77	0.25	8.04	nr	**136.81**
455 mm projection						
65 mm	88.04	110.43	0.25	8.04	nr	**118.47**
75 mm	88.04	110.43	0.25	8.04	nr	**118.47**
100 mm	127.66	160.14	0.25	8.04	nr	**168.18**
Cast iron rainwater pipe; dry joints; BS 460; fixed to backgrounds: Rectangular						
Plain socket						
100 × 75 mm	80.02	100.37	1.04	33.45	m	**133.82**
Bracket; fixed to backgrounds. For minimum fixing distances, refer to the Tables and Memoranda at the rear of the book						
100 × 75 mm; build in holdabat	45.54	57.12	0.35	11.26	nr	**68.38**
100 × 75 mm; trefoil earband	36.29	45.52	0.29	9.33	nr	**54.85**
100 × 75 mm; plain earband	35.10	44.03	0.29	9.33	nr	**53.36**
Eared Socket, with wall spacers measured separately						
100 × 75 mm	81.35	102.04	1.16	37.31	m	**139.35**
Wall spacer plate; eared pipework						
100 × 75	10.52	13.19	0.16	5.14	nr	**18.33**
Extra over fittings Rectangular Cast Iron Pipework BS 460						
Loose socket						
100 × 75 mm; plain	33.94	42.57	0.23	7.40	nr	**49.97**
100 × 75 mm; eared	56.23	70.54	0.29	9.33	nr	**79.87**
Shoe; front						
100 × 75 mm; plain	90.03	112.93	0.23	7.40	nr	**120.33**
100 × 75 mm; eared	110.01	138.00	0.29	9.33	nr	**147.33**
Shoe; side						
100 × 75 mm; plain	109.23	137.02	0.23	7.40	nr	**144.42**
100 × 75 mm; eared	136.56	171.30	0.29	9.33	nr	**180.63**
Bends; side; any degree						
100 × 75 mm; plain	82.84	103.91	0.25	8.04	nr	**111.95**
100 × 75 mm; 135°; plain	84.61	106.13	0.25	8.04	nr	**114.17**
Bends; side; any degree						
100 × 75 mm; eared	104.74	131.39	0.25	8.04	nr	**139.43**

33 DRAINAGE ABOVE GROUND

Item	Net Price £	Material £	Labour hours	Labour £	Unit	Total rate £
Bends; front; any degree						
100 × 75 mm; plain	78.46	98.43	0.25	8.04	nr	106.47
100 × 75 mm; eared	88.90	111.52	0.25	8.04	nr	119.56
Offset; side						
Plain socket						
75 mm projection	109.14	136.91	0.25	8.04	nr	144.95
115 mm projection	113.52	142.40	0.25	8.04	nr	150.44
225 mm projection	141.42	177.40	0.25	8.04	nr	185.44
305 mm projection	163.00	204.47	0.25	8.04	nr	212.51
Offset; front						
Plain socket						
75 mm projection	83.04	104.16	0.25	8.04	nr	112.20
150 mm projection	91.77	115.11	0.25	8.04	nr	123.15
225 mm projection	113.61	142.51	0.25	8.04	nr	150.55
305 mm projection	135.16	169.55	0.25	8.04	nr	177.59
Eared socket						
75 mm projection	106.22	133.25	0.25	8.04	nr	141.29
150 mm projection	114.55	143.70	0.25	8.04	nr	151.74
225 mm projection	135.50	169.97	0.25	8.04	nr	178.01
305 mm projection	157.65	197.76	0.25	8.04	nr	205.80
Offset; plinth						
115 mm projection; plain	85.71	107.52	0.25	8.04	nr	115.56
115 mm projection; eared	110.38	138.47	0.25	8.04	nr	146.51
Rainwater heads						
Flat hopper						
210 × 160 × 185 mm; 65 mm outlet	65.21	81.80	0.40	12.87	nr	94.67
210 × 160 × 185 mm; 75 mm outlet	65.21	81.80	0.40	12.87	nr	94.67
250 × 215 × 215 mm; 100 mm outlet	77.29	96.95	0.40	12.87	nr	109.82
Flat rectangular						
225 × 125 × 125 mm; 65 mm outlet	90.66	113.72	0.40	12.87	nr	126.59
225 × 125 × 125 mm; 75 mm outlet	90.66	113.72	0.40	12.87	nr	126.59
280 × 150 × 130 mm; 100 mm outlet	125.17	157.01	0.40	12.87	nr	169.88
Rectangular						
250 × 180 × 175 mm; 75 mm outlet	84.53	106.03	0.40	12.87	nr	118.90
250 × 180 × 175 mm; 100 mm outlet	84.53	106.03	0.40	12.87	nr	118.90
300 × 250 × 200 mm; 65 mm outlet	117.74	147.69	0.40	12.87	nr	160.56
300 × 250 × 200 mm; 75 mm outlet	117.74	147.69	0.40	12.87	nr	160.56
300 × 250 × 200 mm; 100 mm outlet	117.74	147.69	0.40	12.87	nr	160.56
300 × 250 × 200 mm; 100 × 75 mm outlet	117.74	147.69	0.40	12.87	nr	160.56
Castellated rectangular						
250 × 180 × 175 mm; 65 mm outlet	84.53	106.03	0.40	12.87	nr	118.90

33 DRAINAGE ABOVE GROUND

Item	Net Price £	Material £	Labour hours	Labour £	Unit	Total rate £
DISPOSAL SYSTEMS						
Pricing note: degree angles are only indicated where material prices differ						
PVC-u overflow pipe; solvent welded joints; fixed with clips to backgrounds						
Pipe, with brackets measured separately						
19 mm	1.37	1.71	0.21	6.75	m	**8.46**
Fixings						
Pipe clip: including fixing to backgrounds. For minimum fixing distances, refer to the Tables and Memoranda at the rear of the book						
19 mm	0.53	0.66	0.18	5.79	nr	**6.45**
Extra over fittings overflow pipework PVC-u						
Straight coupler						
19 mm	1.45	1.81	0.17	5.47	nr	**7.28**
Bend						
19 mm: 91.25°	1.71	2.15	0.17	5.47	nr	**7.62**
19 mm: 135°	1.72	2.16	0.17	5.47	nr	**7.63**
Tee						
19 mm	1.86	2.33	0.18	5.79	nr	**8.12**
Reverse nut connector						
19 mm	0.68	0.85	0.15	4.83	nr	**5.68**
BSP adaptor: solvent welded socket to threaded socket						
19 mm × 3/4"	2.50	3.14	0.14	4.50	nr	**7.64**
Straight tank connector						
19 mm	2.27	2.84	0.21	6.75	nr	**9.59**
32 mm	1.11	1.39	0.28	9.00	nr	**10.39**
40 mm	1.11	1.39	0.30	9.64	nr	**11.03**
Bent tank connector						
19 mm	2.67	3.35	0.21	6.75	nr	**10.10**
Tundish						
19 mm	35.48	44.51	0.38	12.22	nr	**56.73**
MuPVC waste pipe; solvent welded joints; fixed with clips to backgrounds; BS 5255						
Pipe, with brackets measured separately						
32 mm	2.16	2.71	0.23	7.40	m	**10.11**
40 mm	2.67	3.35	0.23	7.40	m	**10.75**
50 mm	4.05	5.08	0.26	8.37	m	**13.45**
Fixings						
Pipe clip: including fixing to backgrounds. For minimum fixing distances, refer to the Tables and Memoranda at the rear of the book						
32 mm	0.45	0.56	0.13	4.18	nr	**4.74**
40 mm	0.50	0.63	0.13	4.18	nr	**4.81**
50 mm	0.68	0.85	0.13	4.18	nr	**5.03**

33 DRAINAGE ABOVE GROUND

Item	Net Price £	Material £	Labour hours	Labour £	Unit	Total rate £
Pipe clip: expansion: including fixing to backgrounds. For minimum fixing distances, refer to the Tables and Memoranda at the rear of the book						
32 mm	0.53	0.66	0.13	4.18	nr	**4.84**
40 mm	0.55	0.69	0.13	4.18	nr	**4.87**
50 mm	1.30	1.64	0.13	4.18	nr	**5.82**
Pipe clip: metal; including fixing to backgrounds. For minimum fixing distances, refer to the Tables and Memoranda at the rear of the book						
32 mm	1.94	2.43	0.13	4.18	nr	**6.61**
40 mm	2.31	2.90	0.13	4.18	nr	**7.08**
50 mm	2.94	3.68	0.13	4.18	nr	**7.86**
Extra over fittings waste pipework MuPVC						
Screwed access plug						
32 mm	1.26	1.58	0.18	5.79	nr	**7.37**
40 mm	1.26	1.58	0.18	5.79	nr	**7.37**
50 mm	1.81	2.27	0.25	8.04	nr	**10.31**
Straight coupling						
32 mm	1.35	1.69	0.27	8.68	nr	**10.37**
40 mm	1.35	1.69	0.27	8.68	nr	**10.37**
50 mm	2.47	3.10	0.27	8.68	nr	**11.78**
Expansion coupling						
32 mm	2.39	3.00	0.27	8.68	nr	**11.68**
40 mm	2.87	3.60	0.27	8.68	nr	**12.28**
50 mm	3.89	4.88	0.27	8.68	nr	**13.56**
MuPVC to copper coupling						
32 mm	2.39	3.00	0.27	8.68	nr	**11.68**
40 mm	2.87	3.60	0.27	8.68	nr	**12.28**
50 mm	3.89	4.88	0.27	8.68	nr	**13.56**
Spigot and socket coupling						
32 mm	2.39	3.00	0.27	8.68	nr	**11.68**
40 mm	2.87	3.60	0.27	8.68	nr	**12.28**
50 mm	3.89	4.88	0.27	8.68	nr	**13.56**
Union						
32 mm	5.66	7.10	0.28	9.00	nr	**16.10**
40 mm	7.42	9.31	0.28	9.00	nr	**18.31**
50 mm	8.45	10.60	0.28	9.00	nr	**19.60**
Reducer: socket						
32 × 19 mm	2.05	2.58	0.27	8.68	nr	**11.26**
40 × 32 mm	1.35	1.69	0.27	8.68	nr	**10.37**
50 × 32 mm	1.95	2.44	0.27	8.68	nr	**11.12**
50 × 40 mm	2.39	3.00	0.27	8.68	nr	**11.68**
Reducer: level invert						
40 × 32 mm	1.68	2.11	0.27	8.68	nr	**10.79**
50 × 32 mm	2.08	2.61	0.27	8.68	nr	**11.29**
50 × 40 mm	2.08	2.61	0.27	8.68	nr	**11.29**

33 DRAINAGE ABOVE GROUND

Item	Net Price £	Material £	Labour hours	Labour £	Unit	Total rate £
DISPOSAL SYSTEMS – cont						
Extra over fittings waste pipework – cont						
Swept bend						
32 mm	1.38	1.74	0.27	8.68	nr	**10.42**
32 mm: 165°	1.43	1.79	0.27	8.68	nr	**10.47**
40 mm	1.54	1.93	0.27	8.68	nr	**10.61**
40 mm: 165°	2.70	3.38	0.27	8.68	nr	**12.06**
50 mm	2.67	3.35	0.30	9.64	nr	**12.99**
50 mm: 165°	3.57	4.48	0.30	9.64	nr	**14.12**
Knuckle bend						
32 mm	1.26	1.58	0.27	8.68	nr	**10.26**
40 mm	1.39	1.75	0.27	8.68	nr	**10.43**
Spigot and socket bend						
32 mm	2.25	2.82	0.27	8.68	nr	**11.50**
32 mm: 150°	2.34	2.93	0.27	8.68	nr	**11.61**
40 mm	2.58	3.24	0.27	8.68	nr	**11.92**
50 mm	3.67	4.60	0.30	9.64	nr	**14.24**
Swept tee						
32 mm: 91.25°	1.84	2.31	0.31	9.97	nr	**12.28**
32 mm: 135°	2.21	2.78	0.31	9.97	nr	**12.75**
40 mm: 91.25°	2.35	2.95	0.31	9.97	nr	**12.92**
40 mm: 135°	2.92	3.66	0.31	9.97	nr	**13.63**
50 mm	4.57	5.73	0.31	9.97	nr	**15.70**
Swept cross						
40 mm: 91.25°	5.69	7.13	0.31	9.97	nr	**17.10**
50 mm: 91.25°	5.95	7.46	0.43	13.83	nr	**21.29**
50 mm: 135°	7.52	9.43	0.31	9.97	nr	**19.40**
Male iron adaptor						
32 mm	2.06	2.59	0.28	9.00	nr	**11.59**
40 mm	2.41	3.02	0.28	9.00	nr	**12.02**
Female iron adaptor						
32 mm	2.41	3.02	0.28	9.00	nr	**12.02**
40 mm	2.41	3.02	0.28	9.00	nr	**12.02**
50 mm	3.47	4.36	0.31	9.97	nr	**14.33**
Reverse nut adaptor						
32 mm	3.07	3.85	0.20	6.44	nr	**10.29**
40 mm	3.07	3.85	0.20	6.44	nr	**10.29**
Automatic air admittance valve						
32 mm	17.26	21.65	0.27	8.68	nr	**30.33**
40 mm	17.26	21.65	0.28	9.00	nr	**30.65**
50 mm	17.26	21.65	0.31	9.97	nr	**31.62**
MuPVC to metal adpator: including heat shrunk joint to metal						
50 mm	7.44	9.33	0.38	12.22	nr	**21.55**
Caulking bush: including joint to metal						
32 mm	3.24	4.07	0.31	9.97	nr	**14.04**
40 mm	3.24	4.07	0.31	9.97	nr	**14.04**
50 mm	3.24	4.07	0.32	10.28	nr	**14.35**

33 DRAINAGE ABOVE GROUND

Item	Net Price £	Material £	Labour hours	Labour £	Unit	Total rate £
Weathering apron						
50 mm	2.85	3.57	0.65	20.91	nr	**24.48**
Vent cowl						
50 mm	3.25	4.08	0.19	6.10	nr	**10.18**
ABS waste pipe; solvent welded joints;						
fixed with clips to backgrounds; BS 5255						
Pipe, with brackets measured separately						
32 mm	1.98	2.49	0.23	7.40	m	**9.89**
40 mm	2.47	3.10	0.23	7.40	m	**10.50**
50 mm	3.11	3.90	0.26	8.37	m	**12.27**
Fixings						
Pipe clip: including fixing to backgrounds. For						
minimum fixing distances, refer to the Tables						
and Memoranda at the rear of the book						
32 mm	0.34	0.43	0.17	5.47	nr	**5.90**
40 mm	0.42	0.53	0.17	5.47	nr	**6.00**
50 mm	1.26	1.58	0.17	5.47	nr	**7.05**
Pipe clip: expansion: including fixing to						
backgrounds. For minimum fixing distances,						
refer to the Tables and Memoranda at the rear						
of the book						
32 mm	0.34	0.43	0.17	5.47	nr	**5.90**
40 mm	0.42	0.53	0.17	5.47	nr	**6.00**
50 mm	1.26	1.58	0.17	5.47	nr	**7.05**
Pipe clip: metal; including fixing to						
backgrounds. For minimum fixing distances,						
refer to the Tables and Memoranda at the rear						
of the book						
32 mm	1.91	2.40	0.17	5.47	nr	**7.87**
40 mm	2.27	2.84	0.17	5.47	nr	**8.31**
50 mm	2.89	3.63	0.17	5.47	nr	**9.10**
Extra over fittings waste pipework ABS						
Screwed access plug						
32 mm	1.31	1.65	0.18	5.79	nr	**7.44**
40 mm	1.31	1.65	0.18	5.79	nr	**7.44**
50 mm	2.70	3.38	0.25	8.04	nr	**11.42**
Straight coupling						
32 mm	1.31	1.65	0.27	8.68	nr	**10.33**
40 mm	1.31	1.65	0.27	8.68	nr	**10.33**
50 mm	2.70	3.38	0.27	8.68	nr	**12.06**
Expansion coupling						
32 mm	2.70	3.38	0.27	8.68	nr	**12.06**
40 mm	2.70	3.38	0.27	8.68	nr	**12.06**
50 mm	4.13	5.19	0.27	8.68	nr	**13.87**
ABS to copper coupling						
32 mm	2.56	3.21	0.27	8.68	nr	**11.89**
40 mm	2.56	3.21	0.27	8.68	nr	**11.89**
50 mm	3.91	4.91	0.27	8.68	nr	**13.59**

33 DRAINAGE ABOVE GROUND

Item	Net Price £	Material £	Labour hours	Labour £	Unit	Total rate £
DISPOSAL SYSTEMS – cont						
Extra over fittings waste pipework – cont						
Reducer: socket						
40 × 32 mm	1.31	1.65	0.27	8.68	nr	**10.33**
50 × 32 mm	2.97	3.73	0.27	8.68	nr	**12.41**
50 × 40 mm	2.97	3.73	0.27	8.68	nr	**12.41**
Swept bend						
32 mm	1.31	1.65	0.27	8.68	nr	**10.33**
40 mm	1.31	1.65	0.27	8.68	nr	**10.33**
50 mm	2.70	3.38	0.30	9.64	nr	**13.02**
Knuckle bend						
32 mm	1.31	1.65	0.27	8.68	nr	**10.33**
40 mm	1.31	1.65	0.27	8.68	nr	**10.33**
Swept tee						
32 mm	1.87	2.34	0.31	9.97	nr	**12.31**
40 mm	1.87	2.34	0.31	9.97	nr	**12.31**
50 mm	4.94	6.19	0.31	9.97	nr	**16.16**
Swept cross						
40 mm	6.29	7.88	0.23	7.40	nr	**15.28**
50 mm	7.19	9.02	0.43	13.83	nr	**22.85**
Male iron adaptor						
32 mm	2.70	3.38	0.28	9.00	nr	**12.38**
40 mm	2.70	3.38	0.28	9.00	nr	**12.38**
Female iron adapator						
32 mm	2.70	3.38	0.28	9.00	nr	**12.38**
40 mm	2.70	3.38	0.28	9.00	nr	**12.38**
50 mm	4.04	5.06	0.31	9.97	nr	**15.03**
Tank connectors						
32 mm	1.86	2.33	0.29	9.33	nr	**11.66**
40 mm	2.06	2.59	0.29	9.33	nr	**11.92**
Caulking bush: including joint to pipework						
50 mm	3.19	4.00	0.50	16.08	nr	**20.08**
Polypropylene waste pipe; push fit joints; fixed with clips to backgrounds; BS 5254. Pipe, with brackets measured separately						
32 mm	1.15	1.44	0.21	6.75	m	**8.19**
40 mm	1.41	1.77	0.21	6.75	m	**8.52**
50 mm	2.36	2.96	0.38	12.22	m	**15.18**
Fixings						
Pipe clip: saddle; including fixing to backgrounds. For minimum fixing distances, refer to the Tables and Memoranda at the rear of the book						
32 mm	0.39	0.49	0.17	5.47	nr	**5.96**
40 mm	0.39	0.49	0.17	5.47	nr	**5.96**
Pipe clip: including fixing to backgrounds. For minimum fixing distances, refer to the Tables and Memoranda at the rear of the book						
50 mm	0.92	1.15	0.17	5.47	nr	**6.62**

33 DRAINAGE ABOVE GROUND

Item	Net Price £	Material £	Labour hours	Labour £	Unit	Total rate £
Extra over fittings waste pipework polypropylene						
Screwed access plug						
32 mm	1.10	1.38	0.16	5.14	nr	6.52
40 mm	1.10	1.38	0.16	5.14	nr	6.52
50 mm	1.87	2.34	0.20	6.44	nr	8.78
Straight coupling						
32 mm	1.10	1.38	0.19	6.10	nr	7.48
40 mm	1.10	1.38	0.19	6.10	nr	7.48
50 mm	1.87	2.34	0.20	6.44	nr	8.78
Universal waste pipe coupler						
32 mm dia.	1.85	2.32	0.20	6.44	nr	8.76
40 mm dia.	2.10	2.63	0.20	6.44	nr	9.07
Reducer						
40 × 32 mm	1.87	2.34	0.19	6.10	nr	8.44
50 × 32 mm	1.97	2.48	0.19	6.10	nr	8.58
50 × 40 mm	2.03	2.54	0.20	6.44	nr	8.98
Swept bend						
32 mm	1.10	1.38	0.19	6.10	nr	7.48
40 mm	1.10	1.38	0.19	6.10	nr	7.48
50 mm	1.87	2.34	0.20	6.44	nr	8.78
Knuckle bend						
32 mm	1.10	1.38	0.19	6.10	nr	7.48
40 mm	1.10	1.38	0.19	6.10	nr	7.48
50 mm	1.87	2.34	0.20	6.44	nr	8.78
Spigot and socket bend						
32 mm	1.10	1.38	0.19	6.10	nr	7.48
40 mm	1.10	1.38	0.19	6.10	nr	7.48
Swept tee						
32 mm	1.20	1.50	0.22	7.08	nr	8.58
40 mm	1.20	1.50	0.22	7.08	nr	8.58
50 mm	2.03	2.54	0.23	7.40	nr	9.94
Male iron adaptor						
32 mm	1.10	1.38	0.13	4.18	nr	5.56
40 mm	1.10	1.38	0.19	6.10	nr	7.48
50 mm	1.87	2.34	0.15	4.83	nr	7.17
Tank connector						
32 mm	1.10	1.38	0.24	7.72	nr	9.10
40 mm	1.10	1.38	0.24	7.72	nr	9.10
50 mm	1.87	2.34	0.35	11.26	nr	13.60
Polypropylene traps; including fixing to appliance and connection to pipework; BS 3943						
Tubular P trap; 75 mm seal						
32 mm dia.	4.45	5.58	0.20	6.44	nr	12.02
40 mm dia.	5.13	6.44	0.20	6.44	nr	12.88
Tubular S trap; 75 mm seal						
32 mm dia.	5.60	7.02	0.20	6.44	nr	13.46
40 mm dia.	6.61	8.29	0.20	6.44	nr	14.73

33 DRAINAGE ABOVE GROUND

Item	Net Price £	Material £	Labour hours	Labour £	Unit	Total rate £
DISPOSAL SYSTEMS – cont						
Polypropylene traps – cont						
Running tubular P trap; 75 mm seal						
32 mm dia.	6.82	8.56	0.20	6.44	nr	15.00
40 mm dia.	7.45	9.34	0.20	6.44	nr	15.78
Running tubular S trap; 75 mm seal						
32 mm dia.	8.21	10.30	0.20	6.44	nr	16.74
40 mm dia.	8.84	11.09	0.20	6.44	nr	17.53
Spigot and socket bend; converter from P to S Trap						
32 mm	1.75	2.20	0.20	6.44	nr	8.64
40 mm	1.86	2.33	0.21	6.75	nr	9.08
Bottle P trap; 75 mm seal						
32 mm dia.	4.96	6.23	0.20	6.44	nr	12.67
40 mm dia.	5.93	7.44	0.20	6.44	nr	13.88
Bottle S trap; 75 mm seal						
32 mm dia.	5.98	7.50	0.20	6.44	nr	13.94
40 mm dia.	7.26	9.11	0.25	8.04	nr	17.15
Bottle P trap; resealing; 75 mm seal						
32 mm dia.	6.17	7.74	0.20	6.44	nr	14.18
40 mm dia.	7.21	9.05	0.25	8.04	nr	17.09
Bottle S trap; resealing; 75 mm seal						
32 mm dia.	7.06	8.86	0.20	6.44	nr	15.30
40 mm dia.	8.19	10.27	0.25	8.04	nr	18.31
Bath trap, low level; 38 mm seal						
40 mm dia.	6.19	7.76	0.25	8.04	nr	15.80
Bath trap, low level; 38 mm seal complete with overflow hose						
40 mm dia.	9.55	11.98	0.25	8.04	nr	20.02
Bath trap; 75 mm seal complete with overlow hose						
40 mm dia.	9.52	11.94	0.25	8.04	nr	19.98
Bath trap; 75 mm seal complete with overflow hose and overflow outlet						
40 mm dia.	16.25	20.38	0.20	6.44	nr	26.82
Bath trap; 75 mm seal complete with overflow hose, overflow outlet and ABS chrome waste						
40 mm dia.	22.45	28.16	0.20	6.44	nr	34.60
Washing machine trap; 75 mm seal including stand pipe						
40 mm dia.	12.92	16.21	0.25	8.04	nr	24.25
Washing machine standpipe						
40 mm dia.	6.74	8.46	0.25	8.04	nr	16.50
Plastic unslotted chrome plated basin/sink waste including plug						
32 mm	8.12	10.18	0.34	10.94	nr	21.12
40 mm	10.86	13.62	0.34	10.94	nr	24.56

33 DRAINAGE ABOVE GROUND

Item	Net Price £	Material £	Labour hours	Labour £	Unit	Total rate £
Plastic slotted chrome plated basin/sink waste including plug						
32 mm	6.44	8.08	0.34	10.94	nr	19.02
40 mm	10.79	13.53	0.34	10.94	nr	24.47
Bath overflow outlet; plastic; white						
42 mm	5.66	7.10	0.37	11.91	nr	19.01
Bath overlow outlet; plastic; chrome plated						
42 mm	7.47	9.37	0.37	11.91	nr	21.28
Combined cistern and bath overflow outlet; plastic; white						
42 mm	11.61	14.56	0.39	12.54	nr	27.10
Combined cistern and bath overlow outlet; plastic; chrome plated						
42 mm	11.61	14.56	0.39	12.54	nr	27.10
Cistern overflow outlet; plastic; white						
42 mm	9.36	11.74	0.15	4.83	nr	16.57
Cistern overlow outlet; plastic; chrome plated						
42 mm	8.43	10.57	0.15	4.83	nr	15.40
PVC-u soil and waste pipe; solvent welded joints; fixed with clips to backgrounds; BS 4514/BS EN 607						
Pipe, with brackets measured separately						
82 mm	9.82	12.32	0.35	11.26	m	23.58
110 mm	10.02	12.57	0.41	13.18	m	25.75
160 mm	25.95	32.55	0.51	16.41	m	48.96
Fixings						
Galvanized steel pipe clip: including fixing to backgrounds. For minimum fixing distances, refer to the Tables and Memoranda at the rear of the book						
82 mm	4.08	5.12	0.18	5.79	nr	10.91
110 mm	4.23	5.31	0.18	5.79	nr	11.10
160 mm	10.19	12.78	0.18	5.79	nr	18.57
Plastic coated steel pipe clip: including fixing to backgrounds. For minimum fixing distances, refer to the Tables and Memoranda at the rear of the book						
82 mm	5.71	7.17	0.18	5.79	nr	12.96
110 mm	7.78	9.76	0.18	5.79	nr	15.55
160 mm	9.81	12.31	0.18	5.79	nr	18.10
Plastic pipe clip: including fixing to backgrounds. For minimum fixing distances, refer to the Tables and Memoranda at the rear of the book						
82 mm	3.02	3.79	0.18	5.79	nr	9.58
110 mm	5.81	7.29	0.18	5.79	nr	13.08

33 DRAINAGE ABOVE GROUND

Item	Net Price £	Material £	Labour hours	Labour £	Unit	Total rate £
DISPOSAL SYSTEMS – cont						
Fixings – cont						
Plastic coated steel pipe clip: adjustable; including fixing to backgrounds. For minimum fixing distances, refer to the Tables and Memoranda at the rear of the book						
82 mm	4.61	5.78	0.20	6.44	nr	12.22
110 mm	5.76	7.22	0.20	6.44	nr	13.66
Galvanized steel pipe clip: drive in; including fixing to backgrounds. For minimum fixing distances, refer to the Tables and Memoranda at the rear of the book						
110 mm	8.80	11.04	0.22	7.08	nr	18.12
Extra over fittings solvent welded pipework PVC-u						
Straight coupling						
82 mm	5.21	6.54	0.21	6.75	nr	13.29
110 mm	6.52	8.18	0.22	7.08	nr	15.26
160 mm	18.78	23.55	0.24	7.72	nr	31.27
Expansion coupling						
82 mm	7.80	9.79	0.21	6.75	nr	16.54
110 mm	7.98	10.01	0.22	7.08	nr	17.09
160 mm	23.99	30.09	0.24	7.72	nr	37.81
Slip coupling; double ring socket						
82 mm	16.20	20.32	0.21	6.75	nr	27.07
110 mm	20.29	25.45	0.22	7.08	nr	32.53
160 mm	32.21	40.41	0.24	7.72	nr	48.13
Puddle flanges						
110 mm	158.29	198.55	0.45	14.47	nr	213.02
160 mm	255.89	320.99	0.55	17.68	nr	338.67
Socket reducer						
82–50 mm	7.93	9.95	0.18	5.79	nr	15.74
110–50 mm	9.88	12.40	0.18	5.79	nr	18.19
110–82 mm	10.19	12.78	0.22	7.08	nr	19.86
160–110 mm	20.63	25.88	0.26	8.37	nr	34.25
Socket plugs						
82 mm	6.14	7.71	0.15	4.83	nr	12.54
110 mm	8.98	11.27	0.20	6.44	nr	17.71
160 mm	16.53	20.73	0.27	8.68	nr	29.41
Access door; including cutting into pipe						
82 mm	17.42	21.85	0.28	9.00	nr	30.85
110 mm	17.42	21.85	0.34	10.94	nr	32.79
160 mm	31.12	39.03	0.46	14.78	nr	53.81
Screwed access cap						
82 mm	12.34	15.48	0.15	4.83	nr	20.31
110 mm	14.53	18.22	0.20	6.44	nr	24.66
160 mm	27.36	34.32	0.27	8.68	nr	43.00

33 DRAINAGE ABOVE GROUND

Item	Net Price £	Material £	Labour hours	Labour £	Unit	Total rate £
Access pipe: spigot and socket						
110 mm	21.48	26.95	0.22	7.08	nr	34.03
Access pipe: double socket						
110 mm	21.48	26.95	0.22	7.08	nr	34.03
Swept bend						
82 mm	13.07	16.40	0.29	9.33	nr	25.73
110 mm	15.31	19.21	0.32	10.28	nr	29.49
160 mm	38.14	47.85	0.49	15.76	nr	63.61
Bend; special angle						
82 mm	25.19	31.60	0.29	9.33	nr	40.93
110 mm	30.04	37.68	0.32	10.28	nr	47.96
160 mm	50.73	63.64	0.49	15.76	nr	79.40
Spigot and socket bend						
82 mm	12.65	15.87	0.26	8.37	nr	24.24
110 mm	14.83	18.60	0.32	10.28	nr	28.88
110 mm: 135°	16.66	20.90	0.32	10.28	nr	31.18
160 mm: 135°	36.89	46.28	0.44	14.15	nr	60.43
Variable bend: single socket						
110 mm	28.71	36.02	0.33	10.62	nr	46.64
Variable bend: double socket						
110 mm	28.64	35.93	0.33	10.62	nr	46.55
Access bend						
110 mm	42.50	53.31	0.33	10.62	nr	63.93
Single branch: two bosses						
82 mm	18.28	22.93	0.35	11.26	nr	34.19
82 mm: 104°	19.49	24.45	0.35	11.26	nr	35.71
110 mm	20.26	25.41	0.42	13.51	nr	38.92
110 mm: 135°	21.13	26.51	0.42	13.51	nr	40.02
160 mm	43.02	53.96	0.50	16.08	nr	70.04
160 mm: 135°	43.99	55.18	0.50	16.08	nr	71.26
Single branch; four bosses						
110 mm	24.66	30.93	0.42	13.51	nr	44.44
Single access branch						
82 mm	59.27	74.35	0.35	11.26	nr	85.61
110 mm	34.68	43.50	0.42	13.51	nr	57.01
Unequal single branch						
160 × 160 × 110 mm	48.56	60.92	0.50	16.08	nr	77.00
160 × 160 × 110 mm: 135°	52.19	65.46	0.50	16.08	nr	81.54
Double branch						
110 mm	52.34	65.65	0.42	13.51	nr	79.16
110 mm: 135°	50.06	62.80	0.42	13.51	nr	76.31
Corner branch						
110 mm	88.43	110.92	0.42	13.51	nr	124.43
Unequal double branch						
160 × 160 × 110 mm	90.45	113.46	0.50	16.08	nr	129.54
Single boss pipe; single socket						
110 × 110 × 32 mm	5.47	6.87	0.24	7.72	nr	14.59
110 × 110 × 40 mm	5.47	6.87	0.24	7.72	nr	14.59
110 × 110 × 50 mm	5.78	7.25	0.24	7.72	nr	14.97

33 DRAINAGE ABOVE GROUND

Item	Net Price £	Material £	Labour hours	Labour £	Unit	Total rate £
DISPOSAL SYSTEMS – cont						
Extra over fittings – cont						
Single boss pipe; triple socket						
110 × 110 × 40 mm	8.88	11.14	0.24	7.72	nr	18.86
Waste boss; including cutting into pipe						
82–32 mm	7.16	8.98	0.29	9.33	nr	18.31
82–40 mm	7.16	8.98	0.29	9.33	nr	18.31
110–32 mm	7.16	8.98	0.29	9.33	nr	18.31
110–40 mm	7.16	8.98	0.29	9.33	nr	18.31
110–50 mm	7.42	9.31	0.29	9.33	nr	18.64
160–32 mm	10.12	12.69	0.30	9.64	nr	22.33
160–40 mm	10.12	12.69	0.35	11.26	nr	23.95
160–50 mm	10.12	12.69	0.40	12.87	nr	25.56
Self-locking waste boss; including cutting into pipe						
110–32 mm	9.56	12.00	0.30	9.64	nr	21.64
110–40 mm	9.99	12.53	0.30	9.64	nr	22.17
110–50 mm	12.91	16.20	0.30	9.64	nr	25.84
Adaptor saddle; including cutting to pipe						
82–32 mm	4.76	5.97	0.29	9.33	nr	15.30
110–40 mm	5.88	7.38	0.29	9.33	nr	16.71
160–50 mm	10.65	13.36	0.29	9.33	nr	22.69
Branch boss adaptor						
32 mm	2.85	3.57	0.26	8.37	nr	11.94
40 mm	2.85	3.57	0.26	8.37	nr	11.94
50 mm	4.07	5.11	0.26	8.37	nr	13.48
Branch boss adaptor bend						
32 mm	3.89	4.88	0.26	8.37	nr	13.25
40 mm	4.24	5.32	0.26	8.37	nr	13.69
50 mm	5.08	6.37	0.26	8.37	nr	14.74
Automatic air admittance valve						
82–110 mm	44.65	56.01	0.19	6.10	nr	62.11
PVC-u to metal adpator: including heat shrunk joint to metal						
110 mm	12.43	15.59	0.57	18.32	nr	33.91
Caulking bush: including joint to pipework						
82 mm	12.24	15.36	0.46	14.78	nr	30.14
110 mm	12.24	15.36	0.46	14.78	nr	30.14
Vent cowl						
82 mm	3.69	4.63	0.13	4.18	nr	8.81
110 mm	3.72	4.67	0.13	4.18	nr	8.85
160 mm	9.75	12.23	0.13	4.18	nr	16.41
Weathering apron; to lead slates						
82 mm	3.69	4.63	1.15	36.99	nr	41.62
110 mm	4.23	5.31	1.15	36.99	nr	42.30
160 mm	12.72	15.96	1.15	36.99	nr	52.95
Weathering apron; to asphalt						
82 mm	15.54	19.49	1.10	35.37	nr	54.86
110 mm	15.54	19.49	1.10	35.37	nr	54.86

33 DRAINAGE ABOVE GROUND

Item	Net Price £	Material £	Labour hours	Labour £	Unit	Total rate £
Weathering slate; flat; 406 × 406 mm						
82 mm	43.83	54.98	1.04	33.45	nr	**88.43**
110 mm	43.83	54.98	1.04	33.45	nr	**88.43**
Weathering slate; flat; 457 × 457 mm						
82 mm	44.97	56.41	1.04	33.45	nr	**89.86**
110 mm	44.97	56.41	1.04	33.45	nr	**89.86**
Weathering slate; angled; 610 × 610 mm						
82 mm	60.76	76.22	1.04	33.45	nr	**109.67**
110 mm	60.76	76.22	1.04	33.45	nr	**109.67**
PVC-u soil and waste pipe; ring seal joints; fixed with clips to backgrounds; BS 4514/ BS EN 607						
Pipe, with brackets measured separately						
82 mm dia.	8.66	10.86	0.35	11.26	m	**22.12**
110 mm dia.	8.73	10.95	0.41	13.18	m	**24.13**
160 mm dia.	31.30	39.27	0.51	16.41	m	**55.68**
Fixings						
Galvanized steel pipe clip: including fixing to backgrounds. For minimum fixing distances, refer to the Tables and Memoranda at the rear of the book						
82 mm	4.08	5.12	0.18	5.79	nr	**10.91**
110 mm	4.23	5.31	0.18	5.79	nr	**11.10**
160 mm	10.19	12.78	0.18	5.79	nr	**18.57**
Plastic coated steel pipe clip: including fixing to backgrounds. For minimum fixing distances, refer to the Tables and Memoranda at the rear of the book						
82 mm	5.71	7.17	0.18	5.79	nr	**12.96**
110 mm	7.78	9.76	0.18	5.79	nr	**15.55**
160 mm	9.81	12.31	0.18	5.79	nr	**18.10**
Plastic pipe clip: including fixing to backgrounds. For minimum fixing distances, refer to the Tables and Memoranda at the rear of the book						
82 mm	3.02	3.79	0.18	5.79	nr	**9.58**
110 mm	5.81	7.29	0.18	5.79	nr	**13.08**
Plastic coated steel pipe clip: adjustable; including fixing to backgrounds. For minimum fixing distances, refer to the Tables and Memoranda at the rear of the book						
82 mm	4.61	5.78	0.20	6.44	nr	**12.22**
110 mm	5.76	7.22	0.20	6.44	nr	**13.66**
Galvanized steel pipe clip: drive in; including fixing to backgrounds. For minimum fixing distances, refer to the Tables and Memoranda at the rear of the book						
110 mm	8.80	11.04	0.22	7.08	nr	**18.12**

33 DRAINAGE ABOVE GROUND

Item	Net Price £	Material £	Labour hours	Labour £	Unit	Total rate £
DISPOSAL SYSTEMS – cont						
Extra over fittings ring seal pipework PVC-u						
Straight coupling						
82 mm	5.67	7.11	0.21	6.75	nr	**13.86**
110 mm	6.36	7.97	0.22	7.08	nr	**15.05**
160 mm	21.48	26.95	0.24	7.72	nr	**34.67**
Straight coupling; double socket						
82 mm	10.02	12.57	0.21	6.75	nr	**19.32**
110 mm	9.09	11.40	0.22	7.08	nr	**18.48**
160 mm	21.48	26.95	0.24	7.72	nr	**34.67**
Reducer; socket						
82–50 mm	9.34	11.72	0.15	4.83	nr	**16.55**
110–50 mm	13.02	16.33	0.15	4.83	nr	**21.16**
110–82 mm	13.02	16.33	0.19	6.10	nr	**22.43**
160–110 mm	18.76	23.53	0.31	9.97	nr	**33.50**
Access cap						
82 mm	10.16	12.75	0.15	4.83	nr	**17.58**
110 mm	11.98	15.03	0.17	5.47	nr	**20.50**
Access cap; pressure plug						
160 mm	28.16	35.32	0.33	10.62	nr	**45.94**
Access pipe						
82 mm	24.95	31.29	0.22	7.08	nr	**38.37**
110 mm	24.90	31.24	0.22	7.08	nr	**38.32**
160 mm	52.29	65.59	0.24	7.72	nr	**73.31**
Bend						
82 mm	12.68	15.90	0.29	9.33	nr	**25.23**
82 mm; adjustable radius	28.62	35.90	0.29	9.33	nr	**45.23**
110 mm	13.65	17.12	0.32	10.28	nr	**27.40**
110 mm; adjustable radius	25.47	31.95	0.32	10.28	nr	**42.23**
160 mm	40.99	51.42	0.49	15.76	nr	**67.18**
160 mm; adjustable radius	53.77	67.45	0.49	15.76	nr	**83.21**
Bend; spigot and socket						
110 mm	13.78	17.28	0.32	10.28	nr	**27.56**
Bend; offset						
82 mm	11.17	14.01	0.21	6.75	nr	**20.76**
110 mm	17.92	22.48	0.32	10.28	nr	**32.76**
160 mm	39.54	49.59	0.32	10.28	nr	**59.87**
Bend; access						
110 mm	37.61	47.17	0.33	10.62	nr	**57.79**
Single branch						
82 mm	19.92	24.99	0.35	11.26	nr	**36.25**
110 mm	18.42	23.11	0.42	13.51	nr	**36.62**
110 mm; 45°	18.95	23.77	0.31	9.97	nr	**33.74**
160 mm	47.50	59.58	0.50	16.08	nr	**75.66**
Single branch; access						
82 mm	29.86	37.45	0.35	11.26	nr	**48.71**
110 mm	43.22	54.22	0.42	13.51	nr	**67.73**

33 DRAINAGE ABOVE GROUND

Item	Net Price £	Material £	Labour hours	Labour £	Unit	Total rate £
Unequal single branch						
160 × 160 × 110 mm	48.54	60.88	0.50	16.08	nr	**76.96**
160 × 160 × 110 mm; 45°	52.19	65.46	0.50	16.08	nr	**81.54**
Double branch; 4 bosses						
110 mm	48.22	60.49	0.49	15.76	nr	**76.25**
Corner branch; 2 bosses						
110 mm	88.41	110.90	0.49	15.76	nr	**126.66**
Multi-branch; 4 bosses						
110 mm	26.40	33.12	0.52	16.72	nr	**49.84**
Boss branch						
110 × 32 mm	5.46	6.85	0.34	10.94	nr	**17.79**
110 × 40 mm	5.46	6.85	0.34	10.94	nr	**17.79**
Strap on boss						
110 × 32 mm	6.09	7.64	0.30	9.64	nr	**17.28**
110 × 40 mm	6.09	7.64	0.30	9.64	nr	**17.28**
110 × 50 mm	6.13	7.69	0.30	9.64	nr	**17.33**
Patch boss						
82 × 32 mm	7.16	8.98	0.31	9.97	nr	**18.95**
82 × 40 mm	7.16	8.98	0.31	9.97	nr	**18.95**
82 × 50 mm	7.16	8.98	0.31	9.97	nr	**18.95**
Boss pipe; collar 4 boss						
110 mm	8.88	11.14	0.35	11.26	nr	**22.40**
Boss adaptor; rubber; push fit						
32 mm	3.33	4.18	0.26	8.37	nr	**12.55**
40 mm	3.33	4.18	0.26	8.37	nr	**12.55**
50 mm	3.76	4.72	0.26	8.37	nr	**13.09**
WC connector; cap and seal; solvent socket						
110 mm	9.26	11.61	0.23	7.40	nr	**19.01**
110 mm; 90°	14.88	18.67	0.27	8.68	nr	**27.35**
Vent terminal						
82 mm	3.69	4.63	0.13	4.18	nr	**8.81**
110 mm	3.72	4.67	0.13	4.18	nr	**8.85**
160 mm	9.77	12.25	0.13	4.18	nr	**16.43**
Weathering slate; inclined; 610 × 610 mm						
82 mm	60.76	76.22	1.04	33.45	nr	**109.67**
110 mm	60.76	76.22	1.04	33.45	nr	**109.67**
Weathering slate; inclined; 450 × 450 mm						
82 mm	44.97	56.41	1.04	33.45	nr	**89.86**
110 mm	44.97	56.41	1.04	33.45	nr	**89.86**
Weathering slate; flat; 400 × 400 mm						
82 mm	43.85	55.00	1.04	33.45	nr	**88.45**
110 mm	43.85	55.00	1.04	33.45	nr	**88.45**
Air admittance valve						
82 mm	44.65	56.01	0.19	6.10	nr	**62.11**
110 mm	44.65	56.01	0.19	6.10	nr	**62.11**

33 DRAINAGE ABOVE GROUND

Item	Net Price £	Material £	Labour hours	Labour £	Unit	Total rate £
DISPOSAL SYSTEMS – cont						
Cast iron pipe; nitrile rubber gasket joint with continuity clip BSEN 877; fixed vertically to backgrounds						
Pipe, with brackets and jointing couplings measured separately						
50 mm	22.11	27.73	0.31	9.97	m	37.70
75 mm	25.76	32.31	0.34	10.94	m	43.25
100 mm	28.62	35.90	0.37	11.91	m	47.81
150 mm	59.77	74.97	0.60	19.30	m	94.27
Fixings						
Brackets; fixed to backgrounds. For minimum fixing distances, refer to the Tables and Memoranda at the rear of the book						
50 mm	8.24	10.34	0.15	4.83	nr	15.17
75 mm	8.75	10.98	0.18	5.79	nr	16.77
100 mm	8.96	11.24	0.18	5.79	nr	17.03
150 mm	16.61	20.83	0.20	6.44	nr	27.27
Extra over fittings nitrile gasket cast iron pipework BS 416/6087, with jointing couplings measured separately						
Standard coupling						
50 mm	10.69	13.41	0.17	5.47	nr	18.88
75 mm	11.70	14.67	0.17	5.47	nr	20.14
100 mm	15.28	19.16	0.17	5.47	nr	24.63
150 mm	30.50	38.26	0.17	5.47	nr	43.73
Conversion coupling						
65 × 75 mm	12.40	15.56	0.60	19.31	nr	34.87
70 × 75 mm	12.40	15.56	0.60	19.31	nr	34.87
90 × 100 mm	15.97	20.04	0.67	21.55	nr	41.59
Access pipe; round door						
50 mm	31.25	39.20	0.41	13.18	nr	52.38
75 mm	44.93	56.36	0.46	14.78	nr	71.14
100 mm	47.23	59.25	0.67	21.59	nr	80.84
150 mm	78.56	98.55	0.83	26.72	nr	125.27
Access pipe; square door						
100 mm	93.07	116.75	0.67	21.59	nr	138.34
150 mm	142.46	178.71	0.83	26.72	nr	205.43
Taper reducer						
75 mm	25.68	32.21	0.60	19.31	nr	51.52
100 mm	32.60	40.89	0.67	21.55	nr	62.44
150 mm	63.48	79.63	0.83	26.72	nr	106.35
Blank cap						
50 mm	6.49	8.14	0.24	7.72	nr	15.86
75 mm	8.09	10.15	0.26	8.37	nr	18.52
100 mm	7.86	9.86	0.32	10.28	nr	20.14
150 mm	11.36	14.25	0.40	12.87	nr	27.12

33 DRAINAGE ABOVE GROUND

Item	Net Price £	Material £	Labour hours	Labour £	Unit	Total rate £
Blank cap; 50 mm screwed tapping						
75 mm	15.96	20.03	0.26	8.37	nr	28.40
100 mm	18.52	23.23	0.32	10.28	nr	33.51
150 mm	20.57	25.80	0.40	12.87	nr	38.67
Universal connector						
50 × 56/48/40 mm	9.64	12.10	0.33	10.62	nr	22.72
Change piece; BS416						
100 mm	20.57	25.80	0.47	15.12	nr	40.92
WC connector						
100 mm	33.09	41.51	0.49	15.76	nr	57.27
Boss pipe; 2" BSPT socket						
50 mm	26.42	33.14	0.58	18.65	nr	51.79
75 mm	38.66	48.50	0.65	20.91	nr	69.41
100 mm	46.19	57.94	0.79	25.41	nr	83.35
150 mm	71.96	90.27	0.86	27.66	nr	117.93
Boss pipe; 2" BSPT socket; 135°						
100 mm	55.65	69.81	0.79	25.41	nr	95.22
Boss pipe; 2 × 2" BSPT socket; opposed						
75 mm	51.23	64.27	0.65	20.91	nr	85.18
100 mm	56.73	71.16	0.79	25.41	nr	96.57
Boss pipe; 2 × 2" BSPT socket; in line						
100 mm	60.32	75.67	0.79	25.41	nr	101.08
Boss pipe; 2 × 2" BSPT socket; 90°						
100 mm	59.73	74.93	0.79	25.41	nr	100.34
Bend; short radius						
50 mm	18.69	23.44	0.50	16.08	nr	39.52
75 mm	21.14	26.52	0.60	19.31	nr	45.83
100 mm	25.85	32.42	0.67	21.55	nr	53.97
100 mm; 11°	22.28	27.94	0.67	21.55	nr	49.49
100 mm; 67°	25.85	32.42	0.67	21.55	nr	53.97
150 mm	46.19	57.94	0.83	26.72	nr	84.66
Access bend; short radius						
50 mm	46.03	57.74	0.50	16.08	nr	73.82
75 mm	49.93	62.63	0.60	19.31	nr	81.94
100 mm	54.70	68.61	0.67	21.55	nr	90.16
100 mm; 45°	54.70	68.61	0.67	21.55	nr	90.16
150 mm	77.66	97.42	0.83	26.72	nr	124.14
150 mm; 45°	77.66	97.42	0.83	26.72	nr	124.14
Long radius bend						
75 mm	35.27	44.24	0.60	19.31	nr	63.55
100 mm	41.88	52.54	0.67	21.55	nr	74.09
100 mm; 5°	25.85	32.42	0.67	21.55	nr	53.97
150 mm	91.28	114.50	0.83	26.72	nr	141.22
150 mm; 22.5°	95.69	120.03	0.83	26.72	nr	146.75
Access bend; long radius						
75 mm	62.59	78.51	0.60	19.31	nr	97.82
100 mm	70.74	88.74	0.67	21.55	nr	110.29
150 mm	124.58	156.27	0.83	26.72	nr	182.99

33 DRAINAGE ABOVE GROUND

Item	Net Price £	Material £	Labour hours	Labour £	Unit	Total rate £
DISPOSAL SYSTEMS – cont						
Extra over fittings – cont						
Long tail bend						
100 × 250 mm long	33.40	41.90	0.70	22.53	nr	**64.43**
100 × 815 mm long	106.18	133.19	0.70	22.53	nr	**155.72**
Offset						
75 mm projection						
75 mm	27.57	34.59	0.53	17.05	nr	**51.64**
100 mm	27.20	34.12	0.66	21.22	nr	**55.34**
115 mm projection						
75 mm	30.29	37.99	0.53	17.05	nr	**55.04**
100 mm	33.97	42.62	0.66	21.22	nr	**63.84**
150 mm projection						
75 mm	35.60	44.65	0.53	17.05	nr	**61.70**
100 mm	35.60	44.65	0.66	21.22	nr	**65.87**
225 mm projection						
100 mm	38.92	48.82	0.66	21.22	nr	**70.04**
300 mm projection						
100 mm	41.88	52.54	0.66	21.22	nr	**73.76**
Branch; equal and unequal						
50 mm	28.10	35.25	0.78	25.09	nr	**60.34**
75 mm	31.81	39.91	0.85	27.35	nr	**67.26**
100 mm	39.96	50.13	1.00	32.17	nr	**82.30**
150 mm	99.07	124.28	1.20	38.61	nr	**162.89**
150 × 100 mm; 87.5°	75.81	95.10	1.21	38.76	nr	**133.86**
150 × 100 mm; 45°	111.20	139.48	1.21	38.76	nr	**178.24**
Branch; 2" BSPT screwed socket						
100 mm	53.54	67.16	1.00	32.17	nr	**99.33**
Branch; long tail						
100 × 915 mm long	110.70	138.86	1.00	32.17	nr	**171.03**
Access branch; equal and unequal						
50 mm	61.21	76.79	0.78	25.09	nr	**101.88**
75 mm	61.21	76.79	0.85	27.35	nr	**104.14**
100 mm	68.82	86.33	1.02	32.82	nr	**119.15**
150 mm	146.48	183.75	1.20	38.61	nr	**222.36**
150 × 100 mm; 87.5°	123.11	154.43	1.20	38.61	nr	**193.04**
150 × 100 mm; 45°	133.45	167.40	1.20	38.61	nr	**206.01**
Parallel branch						
100 mm	41.88	52.54	1.00	32.17	nr	**84.71**
Double branch						
75 mm	47.23	59.25	0.95	30.55	nr	**89.80**
100 mm	49.43	62.00	1.30	41.82	nr	**103.82**
150 × 100 mm	139.20	174.61	1.56	50.18	nr	**224.79**
Double access branch						
100 mm	78.28	98.19	1.43	46.01	nr	**144.20**
Corner branch						
100 mm	70.28	88.16	1.30	41.82	nr	**129.98**

33 DRAINAGE ABOVE GROUND

Item	Net Price £	Material £	Labour hours	Labour £	Unit	Total rate £
Puddle flange; grey epoxy coated						
100 mm	46.99	58.95	1.00	32.17	nr	91.12
Roof vent connector; asphalt						
75 mm	57.19	71.74	0.90	28.95	nr	100.69
100 mm	44.66	56.02	0.97	31.19	nr	87.21
P trap						
100 mm	41.44	51.98	1.00	32.17	nr	84.15
P trap with access						
50 mm	63.42	79.55	0.77	24.76	nr	104.31
75 mm	63.42	79.55	0.90	28.95	nr	108.50
100 mm	70.30	88.19	1.16	37.32	nr	125.51
150 mm	122.74	153.97	1.77	56.93	nr	210.90
Bellmouth gully inlet						
100 mm	59.98	75.24	1.08	34.73	nr	109.97
Balcony gully inlet						
100 mm	181.10	227.17	1.08	34.73	nr	261.90
Roof outlet						
Flat grate						
75 mm	111.79	140.22	0.83	26.69	nr	166.91
100 mm	157.27	197.28	1.08	34.73	nr	232.01
Dome grate						
75 mm	111.79	140.22	0.83	26.69	nr	166.91
100 mm	176.41	221.29	1.08	34.73	nr	256.02
Top hat						
100 mm	238.83	299.59	1.08	34.73	nr	334.32
Cast iron pipe; EPDM rubber gasket joint with continuity clip; BS EN877; fixed to backgrounds						
Pipe, with brackets and jointing couplings measured separately						
50 mm	22.67	28.44	0.31	9.97	m	38.41
70 mm	26.50	33.24	0.34	10.94	m	44.18
100 mm	31.55	39.58	0.37	11.91	m	51.49
125 mm	50.62	63.49	0.65	20.91	m	84.40
150 mm	62.52	78.42	0.70	22.51	m	100.93
200 mm	104.46	131.04	1.14	36.67	m	167.71
250 mm	145.96	183.10	1.25	40.21	m	223.31
300 mm	181.76	228.00	1.53	49.21	m	277.21

33 DRAINAGE ABOVE GROUND

Item	Net Price £	Material £	Labour hours	Labour £	Unit	Total rate £
DISPOSAL SYSTEMS – cont						
Fixings						
Brackets; fixed to backgrounds. For minimum fixing distances, refer to the Tables and Memoranda at the rear of the book						
Ductile iron						
50 mm	9.63	12.08	0.10	3.21	nr	15.29
70 mm	9.63	12.08	0.10	3.21	nr	15.29
100 mm	11.13	13.97	0.15	4.83	nr	18.80
150 mm	20.62	25.86	0.20	6.44	nr	32.30
200 mm	76.47	95.93	0.25	8.04	nr	103.97
Mild steel; vertical						
125 mm	18.77	23.54	0.15	4.83	nr	28.37
Mild steel; stand off						
250 mm	40.87	51.26	0.25	8.04	nr	59.30
300 mm	45.02	56.47	0.25	8.04	nr	64.51
Stack support; rubber seal						
70 mm	38.22	47.95	0.55	17.68	nr	65.63
100 mm	42.52	53.33	0.65	20.91	nr	74.24
125 mm	47.16	59.16	0.74	23.80	nr	82.96
150 mm	67.30	84.43	0.86	27.66	nr	112.09
Wall spacer plate; cast iron (eared sockets)						
50 mm	8.20	10.28	0.10	3.21	nr	13.49
70 mm	8.20	10.28	0.10	3.21	nr	13.49
100 mm	8.20	10.28	0.10	3.21	nr	13.49
Extra over fittings EPDM rubber jointed cast iron pipework BS EN 877, with jointing couplings measured separately						
Coupling						
50 mm	10.30	12.92	0.10	3.21	nr	16.13
70 mm	11.34	14.22	0.10	3.21	nr	17.43
100 mm	14.76	18.51	0.10	3.21	nr	21.72
125 mm	18.34	23.00	0.15	4.83	nr	27.83
150 mm	29.58	37.11	0.30	9.64	nr	46.75
200 mm	66.16	82.99	0.35	11.26	nr	94.25
250 mm	94.77	118.88	0.40	12.87	nr	131.75
300 mm	109.66	137.56	0.50	16.08	nr	153.64
Plain socket						
50 mm	25.88	32.47	0.10	3.21	nr	35.68
70 mm	25.88	32.47	0.10	3.21	nr	35.68
100 mm	29.71	37.27	0.10	3.21	nr	40.48
150 mm	50.12	62.87	0.10	3.21	nr	66.08
Eared socket						
50 mm	26.65	33.43	0.25	8.04	nr	41.47
70 mm	26.65	33.43	0.25	8.04	nr	41.47
100 mm	32.29	40.50	0.25	8.04	nr	48.54
150 mm	52.95	66.42	0.25	8.04	nr	74.46

33 DRAINAGE ABOVE GROUND

Item	Net Price £	Material £	Labour hours	Labour £	Unit	Total rate £
Slip socket						
50 mm	33.52	42.04	0.25	8.04	nr	50.08
70 mm	33.52	42.04	0.25	8.04	nr	50.08
100 mm	39.15	49.11	0.25	8.04	nr	57.15
150 mm	58.92	73.91	0.25	8.04	nr	81.95
Stack support pipe						
70 mm	24.73	31.02	0.74	23.80	nr	54.82
100 mm	33.97	42.62	0.88	28.30	nr	70.92
125 mm	36.34	45.58	1.00	32.17	nr	77.75
150 mm	49.25	61.78	1.19	38.27	nr	100.05
Access pipe; round door						
50 mm	47.50	59.58	0.27	8.68	nr	68.26
70 mm	50.25	63.03	0.30	9.64	nr	72.67
100 mm	55.24	69.29	0.32	10.28	nr	79.57
150 mm	99.98	125.42	0.71	22.83	nr	148.25
Access pipe; square door						
100 mm	106.81	133.99	0.32	10.28	nr	144.27
125 mm	111.15	139.43	0.67	21.55	nr	160.98
150 mm	167.23	209.78	0.71	22.83	nr	232.61
200 mm	332.20	416.71	1.21	38.91	nr	455.62
250 mm	522.69	655.66	1.31	42.13	nr	697.79
300 mm	652.02	817.89	1.41	45.35	nr	863.24
Taper reducer						
70 mm	27.42	34.40	0.30	9.64	nr	44.04
100 mm	32.24	40.44	0.32	10.28	nr	50.72
125 mm	32.42	40.67	0.64	20.59	nr	61.26
150 mm	61.90	77.65	0.67	21.55	nr	99.20
200 mm	100.52	126.09	1.15	36.99	nr	163.08
250 mm	207.63	260.46	1.25	40.21	nr	300.67
300 mm	285.43	358.04	1.37	44.07	nr	402.11
Blank cap						
50 mm	7.18	9.00	0.24	7.72	nr	16.72
70 mm	7.58	9.51	0.26	8.37	nr	17.88
100 mm	8.81	11.05	0.32	10.28	nr	21.33
125 mm	12.46	15.64	0.35	11.26	nr	26.90
150 mm	12.74	15.98	0.40	12.87	nr	28.85
200 mm	56.24	70.55	0.60	19.30	nr	89.85
250 mm	114.95	144.19	0.65	20.91	nr	165.10
300 mm	163.83	205.51	0.72	23.16	nr	228.67
Blank cap; 50 mm screwed tapping						
70 mm	16.89	21.19	0.26	8.37	nr	29.56
100 mm	18.25	22.89	0.32	10.28	nr	33.17
150 mm	21.92	27.50	0.40	12.87	nr	40.37
Universal connector; EPDM rubber						
50 × 56/48/40 mm	14.65	18.38	0.30	9.64	nr	28.02
Blank end; push fit						
100 × 38/32 mm	16.36	20.52	0.32	10.28	nr	30.80
Boss pipe; 2" BSPT socket						
50 mm	40.16	50.38	0.27	8.68	nr	59.06
75 mm	40.16	50.38	0.30	9.64	nr	60.02
100 mm	49.08	61.57	0.32	10.28	nr	71.85
150 mm	80.02	100.37	0.71	22.83	nr	123.20

33 DRAINAGE ABOVE GROUND

Item	Net Price £	Material £	Labour hours	Labour £	Unit	Total rate £
DISPOSAL SYSTEMS – cont						
Extra over fittings – cont						
Boss pipe; 2 × 2" BSPT socket; opposed						
100 mm	63.41	79.54	0.32	10.28	nr	89.82
Boss pipe; 2 × 2" BSPT socket; 90°						
100 mm	63.41	79.54	0.32	10.28	nr	89.82
Manifold connector						
100 mm	97.90	122.81	0.64	20.59	nr	143.40
150 mm	136.41	171.11	1.00	32.17	nr	203.28
Bend; short radius						
50 mm	17.83	22.37	0.27	8.68	nr	31.05
70 mm	20.07	25.18	0.30	9.64	nr	34.82
100 mm	23.75	29.79	0.32	10.28	nr	40.07
125 mm	42.14	52.86	0.62	19.95	nr	72.81
150 mm	42.66	53.51	0.67	21.55	nr	75.06
200 mm; 45°	127.01	159.32	1.21	38.91	nr	198.23
250 mm; 45°	247.83	310.88	1.31	42.13	nr	353.01
300 mm; 45°	348.62	437.30	1.43	46.00	nr	483.30
Access bend; short radius						
70 mm	38.89	48.79	0.30	9.64	nr	58.43
100 mm	56.81	71.27	0.32	10.28	nr	81.55
150 mm	99.97	125.41	0.67	21.55	nr	146.96
Bend; long radius bend						
100 mm; 88°	60.37	75.72	0.32	10.28	nr	86.00
100 mm; 22°	44.52	55.84	0.32	10.28	nr	66.12
150 mm; 88°	173.06	217.09	0.67	21.55	nr	238.64
Access bend; long radius						
100 mm	73.49	92.19	0.32	10.28	nr	102.47
150 mm	178.53	223.94	0.32	10.28	nr	234.22
Bend; long tail						
100 mm	41.63	52.23	0.32	10.28	nr	62.51
Bend; long tail double						
70 mm	64.51	80.92	0.30	9.64	nr	90.56
100 mm	71.16	89.26	0.32	10.28	nr	99.54
Offset; 75 mm projection						
100 mm	36.51	45.80	0.32	10.28	nr	56.08
Offset; 130 mm projection						
50 mm	30.20	37.88	0.27	8.68	nr	46.56
70 mm	45.83	57.49	0.30	9.64	nr	67.13
100 mm	60.19	75.50	0.32	10.28	nr	85.78
125 mm	76.28	95.68	0.67	21.55	nr	117.23
Branch; equal and unequal						
50 mm	28.59	35.86	0.37	11.91	nr	47.77
70 mm	30.19	37.87	0.40	12.87	nr	50.74
100 mm	41.42	51.96	0.42	13.51	nr	65.47
125 mm	83.02	104.14	0.76	24.45	nr	128.59
150 mm	90.33	113.31	0.97	31.19	nr	144.50
200 mm	231.71	290.66	1.51	48.57	nr	339.23
250 mm	296.12	371.45	1.63	52.42	nr	423.87
300 mm	488.61	612.91	1.77	56.93	nr	669.84

33 DRAINAGE ABOVE GROUND

Item	Net Price £	Material £	Labour hours	Labour £	Unit	Total rate £
Branch; radius; equal and unequal						
70 mm	36.78	46.13	0.40	12.87	nr	59.00
100 mm	48.42	60.74	0.42	13.51	nr	74.25
150 mm	104.72	131.36	1.37	44.07	nr	175.43
200 mm	294.69	369.66	1.51	48.57	nr	418.23
Branch; long tail						
100 mm	135.08	169.44	0.52	16.72	nr	186.16
Access branch; radius; equal and unequal						
70 mm	53.94	67.66	0.40	12.87	nr	80.53
100 mm	73.12	91.72	0.42	13.51	nr	105.23
150 mm	174.56	218.97	0.97	31.19	nr	250.16
Double branch; equal and unequal						
100 mm	40.78	51.15	0.52	16.72	nr	67.87
100 mm; 70°	60.83	76.31	0.52	16.72	nr	93.03
150 mm	174.56	218.97	1.37	44.07	nr	263.04
200 mm	298.55	374.51	1.51	48.57	nr	423.08
Double branch; radius; equal and unequal						
100 mm	52.60	65.98	0.52	16.72	nr	82.70
150 mm	215.33	270.11	1.37	44.07	nr	314.18
Corner branch						
100 mm	106.39	133.46	0.52	16.72	nr	150.18
150 mm	117.63	147.56	0.52	16.72	nr	164.28
Corner branch; long tail						
100 mm	158.73	199.11	0.52	16.72	nr	215.83
Roof vent connector; asphalt						
100 mm	66.93	83.96	0.32	10.28	nr	94.24
Movement connector						
100 mm	85.02	106.65	0.32	10.28	nr	116.93
150 mm	157.38	197.42	0.67	21.55	nr	218.97
Expansion plugs						
70 mm	18.91	23.72	0.32	10.28	nr	34.00
100 mm	23.57	29.57	0.39	12.54	nr	42.11
150 mm	42.31	53.08	0.55	17.68	nr	70.76
P trap						
100 mm dia.	44.09	55.31	0.32	10.28	nr	65.59
P trap with access						
50 mm	67.34	84.47	0.27	8.68	nr	93.15
70 mm	67.34	84.47	0.30	9.64	nr	94.11
100 mm	72.93	91.48	0.32	10.28	nr	101.76
150 mm	130.40	163.58	0.67	21.55	nr	185.13
Branch trap						
100 mm	159.81	200.47	0.42	13.51	nr	213.98
Stench trap						
100 mm	310.78	389.84	0.42	13.51	nr	403.35
Balcony gully inlet						
100 mm	176.40	221.28	1.00	32.17	nr	253.45
Roof outlet						
Flat grate						
70 mm	108.89	136.60	1.00	32.17	nr	168.77
100 mm	153.18	192.15	1.00	32.17	nr	224.32

33 DRAINAGE ABOVE GROUND

Item	Net Price £	Material £	Labour hours	Labour £	Unit	Total rate £
DISPOSAL SYSTEMS – cont						
Extra over fittings – cont						
Dome grate						
70 mm	108.89	136.60	1.00	32.17	nr	**168.77**
100 mm	171.81	215.52	1.00	32.17	nr	**247.69**
Top Hat						
100 mm	232.62	291.79	1.00	32.17	nr	**323.96**
Floor drains; for cast iron pipework BS 416 and BS EN877						
Adjustable clamp plate body						
100 mm; 165 mm nickel bronze grate and frame	103.69	130.07	0.50	16.08	nr	**146.15**
100 mm; 165 mm nickel bronze rodding eye	120.67	151.37	0.50	16.08	nr	**167.45**
100 mm; 150 × 150 mm nickel bronze grate and frame	116.35	145.95	0.50	16.08	nr	**162.03**
100 mm; 150 × 150 mm nickel bronze rodding eye	120.67	151.37	0.50	16.08	nr	**167.45**
Deck plate body						
100 mm; 165 mm nickel bronze grate and frame	103.69	130.07	0.50	16.08	nr	**146.15**
100 mm; 165 mm nickel bronze rodding eye	120.67	151.37	0.50	16.08	nr	**167.45**
100 mm; 150 × 150 mm nickel bronze grate and frame	116.35	145.95	0.50	16.08	nr	**162.03**
100 mm; 150 × 150 mm nickel bronze rodding eye	120.67	151.37	0.50	16.08	nr	**167.45**
Extra for						
100 mm; Srewed extension piece	40.44	50.72	0.30	9.64	nr	**60.36**
100 mm; Grating extension piece; screwed or spigot	30.67	38.47	0.30	9.64	nr	**48.11**
100 mm; Brewary trap	1116.00	1399.91	2.00	64.32	nr	**1464.23**

38 PIPED SUPPLY SYSTEMS

Item	Net Price £	Material £	Labour hours	Labour £	Unit	Total rate £
COLD WATER PIPELINES: COPPER PIPEWORK						
Copper pipe; capillary or compression joints in the running length; EN1057 R250 (TX) formerly BS 2871 Table X						
Fixed vertically or at low level, with brackets measured separately						
12 mm dia.	1.67	2.09	0.39	12.54	m	14.63
15 mm dia.	1.87	2.34	0.40	12.87	m	15.21
22 mm dia.	3.77	4.73	0.47	15.12	m	19.85
28 mm dia.	4.78	5.99	0.51	16.41	m	22.40
35 mm dia.	11.33	14.21	0.58	18.65	m	32.86
42 mm dia.	13.80	17.32	0.66	21.22	m	38.54
54 mm dia.	17.74	22.25	0.72	23.16	m	45.41
67 mm dia.	23.20	29.10	0.75	24.12	m	53.22
76 mm dia.	32.81	41.16	0.76	24.45	m	65.61
108 mm dia.	47.20	59.20	0.78	25.09	m	84.29
133 mm dia.	61.24	76.82	1.05	33.77	m	110.59
159 mm dia.	97.07	121.77	1.15	36.99	m	158.76
Fixed horizontally at high level or suspended, with brackets measured separately						
12 mm dia.	1.67	2.09	0.45	14.47	m	16.56
15 mm dia.	1.87	2.34	0.46	14.78	m	17.12
22 mm dia.	3.77	4.73	0.54	17.37	m	22.10
28 mm dia.	4.78	11.98	0.59	18.98	m	30.96
35 mm dia.	11.33	14.21	0.67	21.55	m	35.76
42 mm dia.	13.80	17.32	0.76	24.45	m	41.77
54 mm dia.	17.74	22.25	0.83	26.69	m	48.94
67 mm dia.	23.20	29.10	0.86	27.66	m	56.76
76 mm dia.	32.81	41.16	0.87	27.99	m	69.15
108 mm dia.	47.20	59.20	0.90	28.95	m	88.15
133 mm dia.	61.24	76.82	1.21	38.91	m	115.73
159 mm dia.	97.07	121.77	1.32	42.45	m	164.22
Copper pipe; capillary or compression joints in the running length; EN1057 R250 (TY) formerly BS 2871 Table Y						
Fixed vertically or at low level with brackets measured separately (Refer to Copper Pipe Table X Section)						
12 mm dia.	2.16	2.71	0.41	13.18	m	15.89
15 mm dia.	3.08	3.86	0.43	13.83	m	17.69
22 mm dia.	5.43	6.81	0.50	16.08	m	22.89
28 mm dia.	7.01	8.79	0.54	17.37	m	26.16
35 mm dia.	10.23	12.84	0.62	19.95	m	32.79
42 mm dia.	12.41	15.57	0.71	22.83	m	38.40
54 mm dia.	21.16	26.54	0.78	25.09	m	51.63
67 mm dia.	28.10	35.25	0.82	26.36	m	61.61
76 mm dia.	41.06	51.51	0.60	19.30	m	70.81
108 mm dia.	57.01	71.51	0.88	28.30	m	99.81

38 PIPED SUPPLY SYSTEMS

Item	Net Price £	Material £	Labour hours	Labour £	Unit	Total rate £
COLD WATER PIPELINES: COPPER PIPEWORK – cont						
Copper pipe; capillary or compression joints in the running length; EN1057 R250 (TX) formerly BS 2871 Table X						
Plastic coated gas and cold water service pipe for corrosive environments, fixed vertically or at low level with brackets measured separtely						
15 mm dia. (white)	4.31	5.41	0.59	18.98	m	**24.39**
22 mm dia. (white)	7.64	9.59	0.68	21.87	m	**31.46**
28 mm dia. (white)	7.89	9.90	0.74	23.81	m	**33.71**
Fixings						
Saddle band						
6 mm dia.	0.06	0.08	0.05	1.65	nr	**1.73**
8 mm dia.	0.06	0.08	0.07	2.21	nr	**2.29**
10 mm dia.	0.09	0.11	0.09	2.74	nr	**2.85**
12 mm dia.	0.09	0.11	0.11	3.44	nr	**3.55**
15 mm dia.	0.09	0.11	0.13	4.18	nr	**4.29**
22 mm dia.	0.17	0.21	0.13	4.18	nr	**4.39**
28 mm dia.	0.09	0.11	0.16	5.14	nr	**5.25**
35 mm dia.	0.32	0.40	0.18	5.79	nr	**6.19**
42 mm dia.	0.67	0.84	0.21	6.75	nr	**7.59**
54 mm dia.	0.89	1.12	0.21	6.75	nr	**7.87**
Single spacing clip						
15 mm dia.	0.18	0.22	0.14	4.50	nr	**4.72**
22 mm dia.	0.18	0.22	0.15	4.83	nr	**5.05**
28 mm dia.	0.45	0.56	0.17	5.47	nr	**6.03**
Two piece spacing clip						
8 mm dia. Bottom	0.16	0.20	0.11	3.54	nr	**3.74**
8 mm dia. Top	0.16	0.20	0.11	3.54	nr	**3.74**
12 mm dia. Bottom	0.16	0.20	0.13	4.18	nr	**4.38**
12 mm dia. Top	0.16	0.20	0.13	4.18	nr	**4.38**
15 mm dia. Bottom	0.17	0.21	0.13	4.18	nr	**4.39**
15 mm dia. Top	0.17	0.21	0.13	4.18	nr	**4.39**
22 mm dia. Bottom	0.17	0.21	0.14	4.50	nr	**4.71**
22 mm dia. Top	0.18	0.22	0.14	4.50	nr	**4.72**
28 mm dia. Bottom	0.17	0.21	0.16	5.14	nr	**5.35**
28 mm dia. Top	0.31	0.39	0.16	5.14	nr	**5.53**
35 mm dia. Bottom	0.18	0.22	0.21	6.75	nr	**6.97**
35 mm dia. Top	0.44	0.55	0.21	6.75	nr	**7.30**
Single pipe bracket						
15 mm dia.	1.56	1.96	0.14	4.50	nr	**6.46**
22 mm dia.	1.58	1.98	0.14	4.50	nr	**6.48**
28 mm dia.	2.17	2.72	0.17	5.47	nr	**8.19**
Single pipe ring						
15 mm dia.	0.33	0.41	0.26	8.37	nr	**8.78**
22 mm dia.	0.40	0.50	0.26	8.37	nr	**8.87**
28 mm dia.	0.44	0.55	0.31	9.97	nr	**10.52**

38 PIPED SUPPLY SYSTEMS

Item	Net Price £	Material £	Labour hours	Labour £	Unit	Total rate £
35 mm dia.	0.58	0.73	0.32	10.28	nr	11.01
42 mm dia.	0.87	1.09	0.32	10.28	nr	11.37
54 mm dia.	1.35	1.69	0.34	10.94	nr	12.63
67 mm dia.	1.54	1.93	0.35	11.26	nr	13.19
76 mm dia.	1.81	2.27	0.42	13.51	nr	15.78
108 mm dia.	3.66	4.59	0.42	13.51	nr	18.10
Double pipe ring						
15 mm dia.	0.35	0.44	0.26	8.37	nr	8.81
22 mm dia.	0.63	0.80	0.26	8.37	nr	9.17
28 mm dia.	0.56	0.71	0.31	9.97	nr	10.68
35 mm dia.	0.60	0.75	0.32	10.28	nr	11.03
42 mm dia.	1.32	1.66	0.32	10.28	nr	11.94
54 mm dia.	1.42	1.78	0.34	10.94	nr	12.72
67 mm dia.	1.98	2.49	0.35	11.26	nr	13.75
76 mm dia.	2.60	3.26	0.42	13.51	nr	16.77
108 mm dia.	4.57	5.73	0.42	13.51	nr	19.24
Wall bracket						
15 mm dia.	3.09	3.88	0.05	1.60	nr	5.48
22 mm dia.	4.06	5.10	0.05	1.60	nr	6.70
28 mm dia.	4.84	6.07	0.05	1.60	nr	7.67
35 mm dia.	7.25	9.09	0.05	1.60	nr	10.69
42 mm dia.	9.60	12.04	0.05	1.60	nr	13.64
54 mm dia.	14.96	18.77	0.05	1.60	nr	20.37
Hospital bracket						
15 mm dia.	5.11	6.41	0.26	8.37	nr	14.78
22 mm dia.	6.05	7.59	0.26	8.37	nr	15.96
28 mm dia.	7.10	8.90	0.31	9.97	nr	18.87
35 mm dia.	7.51	9.42	0.32	10.28	nr	19.70
42 mm dia.	10.74	13.47	0.32	10.28	nr	23.75
54 mm dia.	14.37	18.02	0.34	10.94	nr	28.96
Screw on backplate, female						
All sizes 15 mm to 54 mm × 10 mm	1.63	2.05	0.10	3.21	nr	5.26
Screw on backplate, male						
All sizes 15 mm to 54 mm × 10 mm	2.18	2.73	0.10	3.21	nr	5.94
Pipe joist clips, single						
15 mm dia.	1.04	1.30	0.08	2.58	nr	3.88
22 mm dia.	1.04	1.30	0.08	2.58	nr	3.88
Pipe joist clips, double						
15 mm dia.	1.45	1.81	0.08	2.58	nr	4.39
22 mm dia.	0.79	0.99	0.08	2.58	nr	3.57
Extra over channel sections for fabricated hangers and brackets						
Galvanized steel; including inserts, bolts, nuts, washers; fixed to backgrounds						
41 × 21 mm	6.34	7.95	0.29	9.33	m	17.28
41 × 41 mm	7.60	9.53	0.29	9.33	m	18.86

38 PIPED SUPPLY SYSTEMS

Item	Net Price £	Material £	Labour hours	Labour £	Unit	Total rate £
COLD WATER PIPELINES: COPPER PIPEWORK – cont						
Extra over channel sections for fabricated hangers and brackets – cont						
Threaded rods; metric thread; including nuts, washers etc.						
10 mm dia. × 600 mm long for ring clips up to 54 mm	2.07	2.60	0.18	5.79	nr	8.39
12 mm dia. × 600 mm long for ring clips from 54 mm	3.20	4.01	0.18	5.79	nr	9.80
Extra over copper pipes; capillary fittings; BS 864						
Stop end						
15 mm dia.	1.51	1.89	0.13	4.18	nr	6.07
22 mm dia.	2.80	3.52	0.14	4.50	nr	8.02
28 mm dia.	5.02	6.29	0.17	5.47	nr	11.76
35 mm dia.	11.08	13.90	0.19	6.10	nr	20.00
42 mm dia.	19.09	23.95	0.22	7.08	nr	31.03
54 mm dia.	26.65	33.43	0.23	7.40	nr	40.83
Straight coupling; copper to copper						
6 mm dia.	1.58	1.98	0.23	7.40	nr	9.38
8 mm dia.	1.61	2.02	0.23	7.40	nr	9.42
10 mm dia.	0.83	1.04	0.23	7.40	nr	8.44
15 mm dia.	0.17	0.21	0.23	7.40	nr	7.61
22 mm dia.	0.48	0.60	0.26	8.37	nr	8.97
28 mm dia.	1.40	1.76	0.30	9.64	nr	11.40
35 mm dia.	4.53	5.68	0.34	10.94	nr	16.62
42 mm dia.	7.57	9.50	0.38	12.22	nr	21.72
54 mm dia.	13.97	17.53	0.42	13.51	nr	31.04
67 mm dia.	41.45	51.99	0.53	17.05	nr	69.04
Adaptor coupling; imperial to metric						
½" × 15 mm dia.	3.29	4.12	0.27	8.68	nr	12.80
¾" × 22 mm dia.	2.88	3.62	0.31	9.97	nr	13.59
1" × 28 mm dia.	5.65	7.09	0.36	11.58	nr	18.67
1 ¼" × 35 mm dia.	9.44	11.84	0.41	13.18	nr	25.02
1 ½" × 42 mm dia.	12.02	15.08	0.46	14.78	nr	29.86
Reducing coupling						
15 × 10 mm dia.	3.45	4.32	0.23	7.40	nr	11.72
22 × 10 mm dia.	5.06	6.35	0.26	8.37	nr	14.72
22 × 15 mm dia.	5.41	6.79	0.27	8.68	nr	15.47
28 × 15 mm dia.	6.22	7.81	0.28	9.00	nr	16.81
28 × 22 mm dia.	6.28	7.87	0.30	9.64	nr	17.51
35 × 28 mm dia.	8.99	11.28	0.34	10.94	nr	22.22
42 × 35 mm dia.	13.19	16.54	0.38	12.22	nr	28.76
54 × 35 mm dia.	23.13	29.02	0.42	13.51	nr	42.53
54 × 42 mm dia.	25.24	31.66	0.42	13.51	nr	45.17

38 PIPED SUPPLY SYSTEMS

Item	Net Price £	Material £	Labour hours	Labour £	Unit	Total rate £
Straight female connector						
15 mm × ½" dia.	3.43	4.30	0.27	8.68	nr	12.98
22 mm × ¾" dia.	4.96	6.23	0.31	9.97	nr	16.20
28 mm × 1" dia.	9.38	11.77	0.36	11.58	nr	23.35
35 mm × 1 ¼" dia.	16.22	20.35	0.41	13.18	nr	33.53
42 mm × 1 ½" dia.	21.04	26.39	0.46	14.78	nr	41.17
54 mm × 2" dia.	33.39	41.89	0.52	16.72	nr	58.61
Straight male connector						
15 mm × ½" dia.	2.92	3.66	0.27	8.68	nr	12.34
22 mm × ¾" dia.	5.22	6.55	0.31	9.97	nr	16.52
28 mm × 1" dia.	8.41	10.55	0.36	11.58	nr	22.13
35 mm × 1 ¼" dia.	14.78	18.54	0.41	13.18	nr	31.72
42 mm × 1 ½" dia.	19.04	23.88	0.46	14.78	nr	38.66
54 mm × 2" dia.	28.90	36.25	0.52	16.72	nr	52.97
67 mm × 2 ½" dia.	46.15	57.89	0.63	20.26	nr	78.15
Female reducing connector						
15 mm × ¾" dia.	8.51	10.67	0.27	8.68	nr	19.35
Male reducing connector						
15 mm × ¾" dia.	7.61	9.54	0.27	8.68	nr	18.22
22 mm × 1" dia.	11.58	14.53	0.31	9.97	nr	24.50
Flanged connector						
28 mm dia.	54.42	68.26	0.36	11.58	nr	79.84
35 mm dia.	68.89	86.42	0.41	13.18	nr	99.60
42 mm dia.	82.34	103.29	0.46	14.78	nr	118.07
54 mm dia.	124.48	156.15	0.52	16.72	nr	172.87
67 mm dia.	146.38	183.62	0.61	19.62	nr	203.24
Tank connector						
15 mm × ½" dia.	7.29	9.14	0.25	8.04	nr	17.18
22 mm × ¾" dia.	11.11	13.93	0.28	9.00	nr	22.93
28 mm × 1" dia.	14.60	18.31	0.32	10.28	nr	28.59
35 mm × 1 ¼" dia.	18.73	23.50	0.37	11.91	nr	35.41
42 mm × 1 ½" dia.	24.55	30.80	0.43	13.83	nr	44.63
54 mm × 2" dia.	37.53	47.07	0.46	14.78	nr	61.85
Tank connector with long thread						
15 mm × ½" dia.	9.44	11.84	0.30	9.64	nr	21.48
22 mm × ¾" dia.	13.45	16.87	0.33	10.62	nr	27.49
28 mm × 1" dia.	16.64	20.88	0.39	12.54	nr	33.42
Reducer						
15 × 10 mm dia.	1.16	1.46	0.23	7.40	nr	8.86
22 × 15 mm dia.	0.82	1.03	0.26	8.37	nr	9.40
28 × 15 mm dia.	3.07	3.85	0.28	9.00	nr	12.85
28 × 22 mm dia.	2.35	2.95	0.30	9.64	nr	12.59
35 × 22 mm dia.	8.72	10.94	0.34	10.94	nr	21.88
42 × 22 mm dia.	15.75	19.76	0.36	11.58	nr	31.34
42 × 35 mm dia.	12.18	15.28	0.38	12.22	nr	27.50
54 × 35 mm dia.	25.58	32.09	0.40	12.87	nr	44.96
54 × 42 mm dia.	22.06	27.68	0.42	13.51	nr	41.19
67 × 54 mm dia.	29.99	37.62	0.53	17.05	nr	54.67

38 PIPED SUPPLY SYSTEMS

Item	Net Price £	Material £	Labour hours	Labour £	Unit	Total rate £
COLD WATER PIPELINES: COPPER PIPEWORK – cont						
Extra over copper pipes – cont						
Adaptor; copper to female iron						
15 mm × ½" dia.	5.91	7.41	0.27	8.68	nr	16.09
22 mm × ¾" dia.	9.00	11.29	0.31	9.97	nr	21.26
28 mm × 1" dia.	12.68	15.90	0.36	11.58	nr	27.48
35 mm × 1 ¼" dia.	22.95	28.78	0.41	13.18	nr	41.96
42 mm × 1 ½" dia.	28.90	36.25	0.46	14.78	nr	51.03
54 mm × 2" dia.	34.77	43.61	0.52	16.72	nr	60.33
Adaptor; copper to male iron						
15 mm × ½" dia.	6.02	7.55	0.27	8.68	nr	16.23
22 mm × ¾" dia.	7.71	9.68	0.31	9.97	nr	19.65
28 mm × 1" dia.	12.88	16.16	0.36	11.58	nr	27.74
35 mm × 1 ¼" dia.	18.74	23.51	0.41	13.18	nr	36.69
42 mm × 1 ½" dia.	25.89	32.48	0.46	14.78	nr	47.26
54 mm × 2" dia.	34.77	43.61	0.52	16.72	nr	60.33
Union coupling						
15 mm dia.	8.15	10.23	0.41	13.18	nr	23.41
22 mm dia.	13.05	16.37	0.45	14.47	nr	30.84
28 mm dia.	19.04	23.88	0.51	16.41	nr	40.29
35 mm dia.	24.99	31.35	0.64	20.59	nr	51.94
42 mm dia.	36.50	45.79	0.68	21.87	nr	67.66
54 mm dia.	69.46	87.14	0.78	25.09	nr	112.23
67 mm dia.	117.62	147.54	0.96	30.87	nr	178.41
Elbow						
15 mm dia.	0.32	0.40	0.23	7.40	nr	7.80
22 mm dia.	0.84	1.05	0.26	8.37	nr	9.42
28 mm dia.	2.23	2.80	0.31	9.97	nr	12.77
35 mm dia.	9.71	12.19	0.35	11.26	nr	23.45
42 mm dia.	16.03	20.10	0.41	13.18	nr	33.28
54 mm dia.	33.13	41.56	0.44	14.15	nr	55.71
67 mm dia.	85.97	107.84	0.54	17.37	nr	125.21
Backplate elbow						
15 mm dia.	6.12	7.67	0.51	16.41	nr	24.08
22 mm dia.	13.16	16.51	0.54	17.37	nr	33.88
Overflow bend						
22 mm dia.	18.53	23.24	0.26	8.37	nr	31.61
Return bend						
15 mm dia.	9.17	11.50	0.23	7.40	nr	18.90
22 mm dia.	18.02	22.60	0.26	8.37	nr	30.97
28 mm dia.	23.03	28.88	0.31	9.97	nr	38.85
Obtuse elbow						
15 mm dia.	1.18	1.48	0.23	7.40	nr	8.88
22 mm dia.	2.45	3.07	0.26	8.37	nr	11.44
28 mm dia.	4.70	5.89	0.31	9.97	nr	15.86
35 mm dia.	14.66	18.39	0.36	11.58	nr	29.97
42 mm dia.	26.09	32.73	0.41	13.18	nr	45.91
54 mm dia.	47.16	59.16	0.44	14.15	nr	73.31
67 mm dia.	85.57	107.34	0.54	17.37	nr	124.71

38 PIPED SUPPLY SYSTEMS

Item	Net Price £	Material £	Labour hours	Labour £	Unit	Total rate £
Straight tap connector						
15 mm × ½" dia.	1.57	1.97	0.13	4.18	nr	6.15
22 mm × ¾" dia.	2.34	2.93	0.14	4.50	nr	7.43
Bent tap connector						
15 mm × ½" dia.	2.33	2.92	0.13	4.18	nr	7.10
22 mm × ¾" dia.	7.15	8.97	0.14	4.50	nr	13.47
Bent male union connector						
15 mm × ½" dia.	11.94	14.97	0.41	13.18	nr	28.15
22 mm × ¾" dia.	15.50	19.44	0.45	14.47	nr	33.91
28 mm × 1" dia.	22.19	27.83	0.51	16.41	nr	44.24
35 mm × 1 ¼" dia.	36.20	45.40	0.64	20.59	nr	65.99
42 mm × 1 ½" dia.	58.84	73.81	0.68	21.87	nr	95.68
54 mm × 2" dia.	92.95	116.59	0.78	25.09	nr	141.68
Bent female union connector						
15 mm dia.	11.94	14.97	0.41	13.18	nr	28.15
22 mm × ¾" dia.	15.50	19.44	0.45	14.47	nr	33.91
28 mm × 1" dia.	22.19	27.83	0.51	16.41	nr	44.24
35 mm × 1 ¼" dia.	36.20	45.40	0.64	20.59	nr	65.99
42 mm × 1 ½" dia.	58.84	73.81	0.68	21.87	nr	95.68
54 mm × 2" dia.	92.95	116.59	0.78	25.09	nr	141.68
Straight union adaptor						
15 mm × ¾" dia.	5.10	6.40	0.41	13.18	nr	19.58
22 mm × 1" dia.	7.24	9.08	0.45	14.47	nr	23.55
28 mm × 1 ¼" dia.	11.71	14.69	0.51	16.41	nr	31.10
35 mm × 1 ½" dia.	18.02	22.60	0.64	20.59	nr	43.19
42 mm × 2" dia.	22.76	28.55	0.68	21.87	nr	50.42
54 mm × 2 ½" dia.	35.13	44.07	0.78	25.09	nr	69.16
Straight male union connector						
15 mm × ½" dia.	10.17	12.76	0.41	13.18	nr	25.94
22 mm × ¾" dia.	13.19	16.54	0.45	14.47	nr	31.01
28 mm × 1" dia.	19.64	24.64	0.51	16.41	nr	41.05
35 mm × 1 ¼" dia.	28.30	35.50	0.64	20.59	nr	56.09
42 mm × 1 ½" dia.	44.47	55.79	0.68	21.87	nr	77.66
54 mm × 2" dia.	63.89	80.15	0.78	25.09	nr	105.24
Straight female union connector						
15 mm × ½" dia.	10.17	12.76	0.41	13.18	nr	25.94
22 mm × ¾" dia.	13.19	16.54	0.45	14.47	nr	31.01
28 mm × 1" dia.	19.64	24.64	0.51	16.41	nr	41.05
35 mm × 1 ¼" dia.	28.30	35.50	0.64	20.59	nr	56.09
42 mm × 1 ½" dia.	44.47	55.79	0.68	21.87	nr	77.66
54 mm × 2" dia.	63.89	80.15	0.78	25.09	nr	105.24
Male nipple						
¾ × ½" dia.	2.40	3.01	0.24	7.63	nr	10.64
1 × ¾" dia.	2.88	3.62	0.32	10.28	nr	13.90
1 ¼ × 1" dia.	3.16	3.96	0.37	11.91	nr	15.87
1 ½ × 1 ¼" dia.	11.75	14.74	0.42	13.51	nr	28.25
2 × 1 ½" dia.	24.04	30.15	0.46	14.78	nr	44.93
2 ½ × 2" dia.	31.31	39.28	0.56	18.01	nr	57.29

38 PIPED SUPPLY SYSTEMS

Item	Net Price £	Material £	Labour hours	Labour £	Unit	Total rate £
COLD WATER PIPELINES: COPPER PIPEWORK – cont						
Extra over copper pipes – cont						
Female nipple						
¾ × ½" dia.	3.72	4.67	0.19	6.10	nr	10.77
1 × ¾" dia.	5.81	7.29	0.32	10.28	nr	17.57
1 ¼ × 1" dia.	7.93	9.95	0.37	11.91	nr	21.86
1 ½ × 1 ¼" dia.	11.75	14.74	0.42	13.51	nr	28.25
2 × 1 ½" dia.	24.04	30.15	0.46	14.78	nr	44.93
2 ½ × 2" dia.	31.31	39.28	0.56	18.01	nr	57.29
Equal tee						
10 mm dia.	3.18	3.99	0.25	8.04	nr	12.03
15 mm dia.	0.32	0.40	0.36	11.58	nr	11.98
22 mm dia.	0.84	1.05	0.39	12.54	nr	13.59
28 mm dia.	6.20	7.77	0.43	13.83	nr	21.60
35 mm dia.	15.81	19.84	0.57	18.32	nr	38.16
42 mm dia.	25.36	31.81	0.60	19.30	nr	51.11
54 mm dia.	51.13	64.14	0.65	20.91	nr	85.05
67 mm dia.	69.56	87.26	0.78	25.09	nr	112.35
Female tee, reducing branch FI						
15 × 15 mm × ¼" dia.	7.65	9.60	0.36	11.58	nr	21.18
22 × 22 mm × ½" dia.	9.36	11.74	0.39	12.56	nr	24.30
28 × 28 mm × ¾" dia.	18.33	22.99	0.43	13.83	nr	36.82
35 × 35 mm × ¾" dia.	26.47	33.21	0.47	15.12	nr	48.33
42 × 42 mm × ½" dia.	31.80	39.89	0.60	19.30	nr	59.19
Backplate tee						
15 × 15 mm × ½" dia.	14.47	18.16	0.62	19.95	nr	38.11
Heater tee						
½ × ½" × 15 mm dia.	13.01	16.32	0.36	11.58	nr	27.90
Union heater tee						
½ × ½" × 15 mm dia.	13.01	16.32	0.36	11.58	nr	27.90
Sweep tee – equal						
15 mm dia.	10.34	12.97	0.36	11.58	nr	24.55
22 mm dia.	13.31	16.70	0.39	12.54	nr	29.24
28 mm dia.	22.41	28.11	0.43	13.83	nr	41.94
35 mm dia.	31.78	39.86	0.57	18.32	nr	58.18
42 mm dia.	47.12	59.10	0.60	19.30	nr	78.40
54 mm dia.	52.18	65.45	0.65	20.91	nr	86.36
67 mm dia.	71.14	89.24	0.78	25.09	nr	114.33
Sweep tee – reducing						
22 × 22 × 15 mm dia.	11.16	14.00	0.39	12.54	nr	26.54
28 × 28 × 22 mm dia.	18.96	23.79	0.43	13.83	nr	37.62
35 × 35 × 22 mm dia.	31.78	39.86	0.57	18.32	nr	58.18
Sweep tee – double						
15 mm dia.	11.71	14.69	0.36	11.58	nr	26.27
22 mm dia.	15.93	19.98	0.39	12.54	nr	32.52
28 mm dia.	24.21	30.37	0.43	13.83	nr	44.20

38 PIPED SUPPLY SYSTEMS

Item	Net Price £	Material £	Labour hours	Labour £	Unit	Total rate £
Cross						
15 mm dia.	15.49	19.43	0.48	15.44	nr	**34.87**
22 mm dia.	17.29	21.68	0.53	17.05	nr	**38.73**
28 mm dia.	24.83	31.15	0.61	19.62	nr	**50.77**
Extra over copper pipes; high duty capillary fittings; BS 864						
Stop end						
15 mm dia.	7.25	9.09	0.16	5.14	nr	**14.23**
Straight coupling; copper to copper						
15 mm dia.	3.31	4.16	0.27	8.68	nr	**12.84**
22 mm dia.	5.32	6.68	0.32	10.28	nr	**16.96**
28 mm dia.	7.11	8.92	0.37	11.91	nr	**20.83**
35 mm dia.	13.27	16.64	0.43	13.83	nr	**30.47**
42 mm dia.	14.51	18.20	0.50	16.08	nr	**34.28**
54 mm dia.	21.35	26.78	0.54	17.37	nr	**44.15**
Reducing coupling						
15 × 12 mm dia.	6.25	7.84	0.27	8.68	nr	**16.52**
22 × 15 mm dia.	7.25	9.09	0.32	10.28	nr	**19.37**
28 × 22 mm dia.	9.98	12.52	0.37	11.91	nr	**24.43**
Straight female connector						
15 mm × ½" dia.	8.15	10.23	0.32	10.28	nr	**20.51**
22 mm × ¾" dia.	9.20	11.54	0.36	11.58	nr	**23.12**
28 mm × 1" dia.	13.57	17.02	0.42	13.51	nr	**30.53**
Straight male connector						
15 mm × ½" dia.	7.93	9.95	0.32	10.28	nr	**20.23**
22 mm × ¾" dia.	9.20	11.54	0.36	11.58	nr	**23.12**
28 mm × 1" dia.	13.57	17.02	0.42	13.51	nr	**30.53**
42 mm × 1 ½" dia.	26.48	33.22	0.53	17.05	nr	**50.27**
54 mm × 2" dia.	43.04	53.98	0.62	19.95	nr	**73.93**
Reducer						
15 × 12 mm dia.	4.11	5.15	0.27	8.68	nr	**13.83**
22 × 15 mm dia.	4.01	5.03	0.32	10.28	nr	**15.31**
28 × 22 mm dia.	7.25	9.09	0.37	11.91	nr	**21.00**
35 × 28 mm dia.	9.20	11.54	0.43	13.83	nr	**25.37**
42 × 35 mm dia.	11.85	14.86	0.50	16.08	nr	**30.94**
54 × 42 mm dia.	19.11	23.97	0.39	12.54	nr	**36.51**
Straight union adaptor						
15 mm × ¾" dia.	6.63	8.32	0.27	8.68	nr	**17.00**
22 mm × 1" dia.	8.98	11.27	0.32	10.28	nr	**21.55**
28 mm × 1 ¼" dia.	11.85	14.86	0.37	11.91	nr	**26.77**
35 mm × 1 ½" dia.	21.50	26.97	0.43	13.83	nr	**40.80**
42 mm × 2" dia.	27.23	34.16	0.50	16.08	nr	**50.24**
Bent union adaptor						
15 mm × ¾" dia.	17.23	21.62	0.27	8.68	nr	**30.30**
22 mm × 1" dia.	23.26	29.18	0.32	10.28	nr	**39.46**
28 mm × 1 ¼" dia.	31.31	39.28	0.37	11.91	nr	**51.19**
Adaptor; male copper to FI						
15 mm × ½" dia.	7.76	9.73	0.27	8.68	nr	**18.41**
22 mm × ¾" dia.	13.31	16.70	0.32	10.28	nr	**26.98**

38 PIPED SUPPLY SYSTEMS

Item	Net Price £	Material £	Labour hours	Labour £	Unit	Total rate £
COLD WATER PIPELINES: COPPER PIPEWORK – cont						
Extra over copper pipes – cont						
Union coupling						
15 mm dia.	14.94	18.74	0.54	17.37	nr	**36.11**
22 mm dia.	19.11	23.97	0.60	19.30	nr	**43.27**
28 mm dia.	26.55	33.31	0.68	21.87	nr	**55.18**
35 mm dia.	46.34	58.13	0.83	26.69	nr	**84.82**
42 mm dia.	54.57	68.45	0.89	28.63	nr	**97.08**
Elbow						
15 mm dia.	9.61	12.05	0.27	8.68	nr	**20.73**
22 mm dia.	10.28	12.89	0.32	10.28	nr	**23.17**
28 mm dia.	15.27	19.15	0.37	11.91	nr	**31.06**
35 mm dia.	23.87	29.94	0.43	13.83	nr	**43.77**
42 mm dia.	29.74	37.31	0.50	16.08	nr	**53.39**
54 mm dia.	51.72	64.88	0.52	16.72	nr	**81.60**
Return bend						
28 mm dia.	23.03	28.88	0.37	11.91	nr	**40.79**
35 mm dia.	28.38	35.60	0.43	13.83	nr	**49.43**
Bent male union connector						
15 mm × ½" dia.	22.30	27.98	0.54	17.37	nr	**45.35**
22 mm × ¾" dia.	30.04	37.68	0.60	19.30	nr	**56.98**
28 mm × 1" dia.	54.57	68.45	0.68	21.87	nr	**90.32**
Composite flange						
35 mm dia.	4.50	5.64	0.38	12.23	nr	**17.87**
42 mm dia.	5.84	7.32	0.41	13.18	nr	**20.50**
54 mm dia.	7.46	9.36	0.43	13.83	nr	**23.19**
Equal tee						
15 mm dia.	11.05	13.87	0.44	14.15	nr	**28.02**
22 mm dia.	13.90	17.44	0.47	15.12	nr	**32.56**
28 mm dia.	18.32	22.98	0.53	17.05	nr	**40.03**
35 mm dia.	31.31	39.28	0.70	22.51	nr	**61.79**
42 mm dia.	39.87	50.01	0.84	27.01	nr	**77.02**
54 mm dia.	62.78	78.75	0.79	25.41	nr	**104.16**
Reducing tee						
15 × 12 mm dia.	15.15	19.01	0.44	14.15	nr	**33.16**
22 × 15 mm dia.	17.89	22.44	0.47	15.12	nr	**37.56**
28 × 22 mm dia.	25.56	32.07	0.53	17.05	nr	**49.12**
35 × 28 mm dia.	40.52	50.83	0.73	23.49	nr	**74.32**
42 × 28 mm dia.	51.87	65.06	0.84	27.01	nr	**92.07**
54 × 28 mm dia.	81.92	102.76	1.01	32.49	nr	**135.25**
Extra over copper pipes; compression fittings; BS 864						
Stop end						
15 mm dia.	1.70	2.13	0.10	3.21	nr	**5.34**
22 mm dia.	2.02	2.53	0.12	3.86	nr	**6.39**
28 mm dia.	2.31	2.90	0.15	4.83	nr	**7.73**

38 PIPED SUPPLY SYSTEMS

Item	Net Price £	Material £	Labour hours	Labour £	Unit	Total rate £
Straight connector; copper to copper						
15 mm dia.	3.51	4.40	0.18	5.79	nr	10.19
22 mm dia.	4.66	5.85	0.21	6.75	nr	12.60
28 mm dia.	6.19	7.76	0.24	7.72	nr	15.48
Straight connector; copper to imperial copper						
22 mm dia.	5.80	7.28	0.21	6.75	nr	14.03
Male coupling; copper to MI (BSP)						
15 mm dia.	1.00	1.25	0.19	6.10	nr	7.35
22 mm dia.	1.57	1.97	0.23	7.40	nr	9.37
28 mm dia.	3.66	4.59	0.26	8.37	nr	12.96
Male coupling with long thread and backnut						
15 mm dia.	6.44	8.08	0.19	6.10	nr	14.18
22 mm dia.	8.17	10.25	0.23	7.40	nr	17.65
Female coupling; copper to FI (BSP)						
15 mm dia.	1.23	1.55	0.19	6.10	nr	7.65
22 mm dia.	1.77	2.22	0.23	7.40	nr	9.62
28 mm dia.	5.11	6.41	0.27	8.68	nr	15.09
Elbow						
15 mm dia.	1.31	1.65	0.18	5.79	nr	7.44
22 mm dia.	2.24	2.81	0.21	6.75	nr	9.56
28 mm dia.	7.56	9.49	0.24	7.72	nr	17.21
Male elbow; copper to FI (BSP)						
15 mm × ½" dia.	2.52	3.16	0.19	6.10	nr	9.26
22 mm × ¾" dia.	3.28	4.11	0.23	7.40	nr	11.51
28 mm × 1" dia.	7.89	9.90	0.27	8.68	nr	18.58
Female elbow; copper to FI (BSP)						
15 mm × ½" dia.	3.86	4.84	0.19	6.10	nr	10.94
22 mm × ¾" dia.	5.59	7.01	0.23	7.40	nr	14.41
28 mm × 1" dia.	9.85	12.35	0.27	8.68	nr	21.03
Backplate elbow						
15 mm × ½" dia.	5.59	7.01	0.50	16.08	nr	23.09
Tank coupling; long thread						
22 mm dia.	8.17	10.25	0.46	14.78	nr	25.03
Tee equal						
15 mm dia.	1.86	2.33	0.28	9.00	nr	11.33
22 mm dia.	3.12	3.91	0.30	9.64	nr	13.55
28 mm dia.	14.23	17.85	0.34	10.94	nr	28.79
Tee reducing						
22 mm dia.	8.05	10.10	0.30	9.64	nr	19.74
Backplate tee						
15 mm dia.	19.66	24.66	0.62	19.95	nr	44.61
Extra over fittings; silver brazed welded joints						
Reducer						
76 × 67 mm dia.	32.45	40.70	1.40	45.04	nr	85.74
108 × 76 mm dia.	68.01	85.31	1.80	57.89	nr	143.20
133 × 108 mm dia.	135.61	170.11	2.20	70.76	nr	240.87
159 × 133 mm dia.	173.44	217.56	2.60	83.62	nr	301.18

38 PIPED SUPPLY SYSTEMS

Item	Net Price £	Material £	Labour hours	Labour £	Unit	Total rate £
COLD WATER PIPELINES: COPPER PIPEWORK – cont						
Extra over fittings – cont						
90° elbow						
76 mm dia.	78.39	98.34	1.60	51.45	nr	**149.79**
108 mm dia.	144.31	181.03	2.00	64.32	nr	**245.35**
133 mm dia.	305.06	382.67	2.40	77.19	nr	**459.86**
159 mm dia.	378.73	475.08	2.80	90.05	nr	**565.13**
45° elbow						
76 mm dia.	73.46	92.15	1.60	51.45	nr	**143.60**
108 mm dia.	115.19	144.49	2.00	64.32	nr	**208.81**
133 mm dia.	294.56	369.50	2.40	77.19	nr	**446.69**
159 mm dia.	423.99	531.85	2.80	90.05	nr	**621.90**
Equal tee						
76 mm dia.	95.47	119.76	2.40	77.19	nr	**196.95**
108 mm dia.	148.30	186.03	3.00	96.49	nr	**282.52**
133 mm dia.	349.24	438.09	3.60	115.79	nr	**553.88**
159 mm dia.	416.52	522.48	4.20	135.08	nr	**657.56**
Extra over copper pipes; dezincification resistant compression fittings; BS 864						
Stop end						
15 mm dia.	2.43	3.05	0.10	3.21	nr	**6.26**
22 mm dia.	3.51	4.40	0.13	4.18	nr	**8.58**
28 mm dia.	7.53	9.44	0.15	4.83	nr	**14.27**
35 mm dia.	11.80	14.81	0.18	5.79	nr	**20.60**
42 mm dia.	19.66	24.66	0.20	6.44	nr	**31.10**
Straight coupling; copper to copper						
15 mm dia.	1.94	2.43	0.18	5.79	nr	**8.22**
22 mm dia.	3.18	3.99	0.21	6.75	nr	**10.74**
28 mm dia.	7.20	9.03	0.24	7.72	nr	**16.75**
35 mm dia.	15.24	19.12	0.29	9.33	nr	**28.45**
42 mm dia.	20.04	25.13	0.33	10.62	nr	**35.75**
54 mm dia.	29.96	37.59	0.38	12.22	nr	**49.81**
Straight swivel connector; copper to imperial copper						
22 mm dia.	7.17	8.99	0.20	6.44	nr	**15.43**
Male coupling; copper to MI (BSP)						
15 mm × ½" dia.	1.74	2.18	0.19	6.10	nr	**8.28**
22 mm × ¾" dia.	2.63	3.30	0.23	7.40	nr	**10.70**
28 mm × 1" dia.	5.10	6.40	0.26	8.37	nr	**14.77**
35 mm × 1 ¼" dia.	11.57	14.52	0.32	10.28	nr	**24.80**
42 mm × 1 ½" dia.	17.38	21.81	0.37	11.91	nr	**33.72**
54 mm × 2" dia.	25.64	32.17	0.57	18.32	nr	**50.49**
Male coupling with long thread and backnuts						
22 mm dia.	9.85	12.35	0.23	7.40	nr	**19.75**
28 mm dia.	10.91	13.69	0.24	7.72	nr	**21.41**

38 PIPED SUPPLY SYSTEMS

Item	Net Price £	Material £	Labour hours	Labour £	Unit	Total rate £
Female coupling; copper to FI (BSP)						
15 mm × ½" dia.	2.08	2.61	0.19	6.10	nr	8.71
22 mm × ¾" dia.	3.04	3.81	0.23	7.40	nr	11.21
28 mm × 1" dia.	6.61	8.29	0.27	8.68	nr	16.97
35 mm × 1 ¼" dia.	13.92	17.46	0.32	10.28	nr	27.74
42 mm × 1 ½" dia.	18.69	23.44	0.37	11.91	nr	35.35
54 mm × 2" dia.	27.43	34.41	0.42	13.51	nr	47.92
Elbow						
15 mm dia.	2.34	2.93	0.18	5.79	nr	8.72
22 mm dia.	3.74	4.69	0.21	6.75	nr	11.44
28 mm dia.	9.29	11.65	0.24	7.72	nr	19.37
35 mm dia.	20.56	25.79	0.29	9.33	nr	35.12
42 mm dia.	27.85	34.93	0.33	10.62	nr	45.55
54 mm dia.	47.94	60.13	0.38	12.22	nr	72.35
Male elbow; copper to MI (BSP)						
15 mm × ½" dia.	4.06	5.10	0.19	6.10	nr	11.20
22 mm × ¾" dia.	4.54	5.69	0.23	7.40	nr	13.09
28 mm × 1" dia.	8.51	10.67	0.27	8.68	nr	19.35
Female elbow; copper to FI (BSP)						
15 mm × ½" dia.	4.34	5.44	0.19	6.10	nr	11.54
22 mm × ¾" dia.	6.26	7.85	0.23	7.40	nr	15.25
28 mm × 1" dia.	10.38	13.03	0.27	8.68	nr	21.71
Backplate elbow						
15 mm × ½" dia.	6.33	7.94	0.50	16.08	nr	24.02
Straight tap connector						
15 mm dia.	3.59	4.50	0.13	4.18	nr	8.68
22 mm dia.	7.79	9.77	0.15	4.83	nr	14.60
Tank coupling						
15 mm dia.	4.95	6.20	0.19	6.10	nr	12.30
22 mm dia.	5.47	6.87	0.23	7.40	nr	14.27
28 mm dia.	11.55	14.49	0.27	8.68	nr	23.17
35 mm dia.	20.44	25.64	0.32	10.28	nr	35.92
42 mm dia.	33.20	41.64	0.37	11.91	nr	53.55
54 mm dia.	42.78	53.66	0.31	9.97	nr	63.63
Tee equal						
15 mm dia.	3.29	4.12	0.28	9.00	nr	13.12
22 mm dia.	5.43	6.81	0.30	9.64	nr	16.45
28 mm dia.	14.82	18.59	0.34	10.94	nr	29.53
35 mm dia.	26.75	33.56	0.43	13.83	nr	47.39
42 mm dia.	42.07	52.77	0.46	14.78	nr	67.55
54 mm dia.	67.57	84.76	0.54	17.37	nr	102.13
Tee reducing						
22 mm dia.	8.68	10.89	0.30	9.64	nr	20.53
28 mm dia.	14.32	17.96	0.34	10.94	nr	28.90
35 mm dia.	26.14	32.79	0.43	13.83	nr	46.62
42 mm dia.	40.41	50.69	0.46	14.78	nr	65.47
54 mm dia.	67.57	84.76	0.54	17.37	nr	102.13

38 PIPED SUPPLY SYSTEMS

Item	Net Price £	Material £	Labour hours	Labour £	Unit	Total rate £
COLD WATER PIPELINES: COPPER PIPEWORK – cont						
Extra over copper pipes; bronze one piece brazing flanges; metric, including jointing ring and bolts						
Bronze flange; PN6						
15 mm dia.	30.71	38.53	0.27	8.68	nr	**47.21**
22 mm dia.	36.59	45.90	0.32	10.28	nr	**56.18**
28 mm dia.	40.20	50.42	0.36	11.58	nr	**62.00**
35 mm dia.	48.03	60.24	0.47	15.12	nr	**75.36**
42 mm dia.	52.40	65.73	0.54	17.37	nr	**83.10**
54 mm dia.	60.78	76.24	0.63	20.26	nr	**96.50**
67 mm dia.	71.65	89.88	0.77	24.76	nr	**114.64**
76 mm dia.	86.35	108.32	0.93	29.92	nr	**138.24**
108 mm dia.	114.09	143.11	1.14	36.67	nr	**179.78**
133 mm dia.	156.33	196.10	1.41	45.35	nr	**241.45**
159 mm dia.	225.52	282.89	1.74	55.96	nr	**338.85**
Bronze flange; PN10						
15 mm dia.	35.16	44.11	0.27	8.68	nr	**52.79**
22 mm dia.	37.42	46.94	0.32	10.28	nr	**57.22**
28 mm dia.	41.16	51.63	0.38	12.22	nr	**63.85**
35 mm dia.	49.76	62.42	0.47	15.12	nr	**77.54**
42 mm dia.	55.78	69.97	0.54	17.37	nr	**87.34**
54 mm dia.	61.03	76.55	0.63	20.26	nr	**96.81**
67 mm dia.	71.71	89.96	0.77	24.76	nr	**114.72**
76 mm dia.	86.35	108.32	0.93	29.92	nr	**138.24**
108 mm dia.	109.14	136.91	1.14	36.67	nr	**173.58**
133 mm dia.	145.72	182.80	1.41	45.35	nr	**228.15**
159 mm dia.	184.70	231.68	1.74	55.96	nr	**287.64**
Bronze flange; PN16						
15 mm dia.	38.68	48.52	0.27	8.68	nr	**57.20**
22 mm dia.	41.18	51.65	0.32	10.28	nr	**61.93**
28 mm dia.	45.27	56.78	0.38	12.22	nr	**69.00**
35 mm dia.	54.74	68.67	0.47	15.12	nr	**83.79**
42 mm dia.	61.36	76.97	0.54	17.37	nr	**94.34**
54 mm dia.	67.14	84.22	0.63	20.26	nr	**104.48**
67 mm dia.	78.87	98.93	0.77	24.76	nr	**123.69**
76 mm dia.	94.98	119.15	0.93	29.92	nr	**149.07**
108 mm dia.	120.05	150.60	1.14	36.67	nr	**187.27**
133 mm dia.	160.29	201.06	1.41	45.35	nr	**246.41**
159 mm dia.	203.16	254.84	1.74	55.96	nr	**310.80**

38 PIPED SUPPLY SYSTEMS

Item	Net Price £	Material £	Labour hours	Labour £	Unit	Total rate £
Extra over copper pipes; bronze blank flanges; metric, including jointing ring and bolts						
Gunmetal blank flange; PN6						
15 mm dia.	25.78	32.33	0.27	8.68	nr	41.01
22 mm dia.	32.88	41.25	0.27	8.68	nr	49.93
28 mm dia.	33.81	42.41	0.27	8.68	nr	51.09
35 mm dia.	55.32	69.40	0.32	10.28	nr	79.68
42 mm dia.	74.63	93.62	0.32	10.28	nr	103.90
54 mm dia.	82.29	103.22	0.34	10.94	nr	114.16
67 mm dia.	101.25	127.01	0.36	11.58	nr	138.59
76 mm dia.	130.44	163.62	0.37	11.91	nr	175.53
108 mm dia.	207.24	259.96	0.41	13.18	nr	273.14
133 mm dia.	244.48	306.68	0.58	18.65	nr	325.33
159 mm dia.	305.57	383.31	0.61	19.62	nr	402.93
Gunmetal blank flange; PN10						
15 mm dia.	31.17	39.10	0.27	8.68	nr	47.78
22 mm dia.	40.34	50.60	0.27	8.68	nr	59.28
28 mm dia.	44.66	56.02	0.27	8.68	nr	64.70
35 mm dia.	55.32	69.40	0.32	10.28	nr	79.68
42 mm dia.	103.69	130.07	0.32	10.28	nr	140.35
54 mm dia.	118.21	148.29	0.34	10.94	nr	159.23
67 mm dia.	126.81	159.07	0.46	14.78	nr	173.85
76 mm dia.	168.51	211.38	0.47	15.12	nr	226.50
108 mm dia.	197.77	248.08	0.51	16.41	nr	264.49
133 mm dia.	262.73	329.57	0.58	18.65	nr	348.22
159 mm dia.	357.81	448.84	0.71	22.83	nr	471.67
Gunmetal blank flange; PN16						
15 mm dia.	31.17	39.10	0.27	8.68	nr	47.78
22 mm dia.	40.95	51.36	0.27	8.68	nr	60.04
28 mm dia.	44.66	56.02	0.27	8.68	nr	64.70
35 mm dia.	55.32	69.40	0.32	10.28	nr	79.68
42 mm dia.	103.69	130.07	0.32	10.28	nr	140.35
54 mm dia.	118.21	148.29	0.34	10.94	nr	159.23
67 mm dia.	147.29	184.76	0.46	14.78	nr	199.54
76 mm dia.	168.51	211.38	0.47	15.12	nr	226.50
108 mm dia.	197.77	248.08	0.51	16.41	nr	264.49
133 mm dia.	454.23	569.79	0.58	18.65	nr	588.44
159 mm dia.	542.56	680.59	0.71	22.83	nr	703.42
Extra over copper pipes; bronze screwed flanges; metric, including jointing ring and bolts						
Gunmetal screwed flange; 6 BSP						
15 mm dia.	25.78	32.33	0.35	11.26	nr	43.59
22 mm dia.	29.80	37.39	0.47	15.12	nr	52.51
28 mm dia.	31.06	38.96	0.52	16.72	nr	55.68
35 mm dia.	41.87	52.52	0.62	19.95	nr	72.47
42 mm dia.	50.26	63.04	0.70	22.51	nr	85.55
54 mm dia.	68.31	85.69	0.84	27.01	nr	112.70
67 mm dia.	85.72	107.53	1.03	33.13	nr	140.66

38 PIPED SUPPLY SYSTEMS

Item	Net Price £	Material £	Labour hours	Labour £	Unit	Total rate £
COLD WATER PIPELINES: COPPER PIPEWORK – cont						
Extra over copper pipes – cont						
Gunmetal screwed flange – cont						
76 mm dia.	103.47	129.80	1.22	39.23	nr	**169.03**
108 mm dia.	163.86	205.54	1.41	45.35	nr	**250.89**
133 mm dia.	194.25	243.67	1.75	56.28	nr	**299.95**
159 mm dia.	247.27	310.17	2.21	71.08	nr	**381.25**
Gunmetal screwed flange; 10 BSP						
15 mm dia.	31.22	39.17	0.35	11.26	nr	**50.43**
22 mm dia.	36.36	45.61	0.47	15.12	nr	**60.73**
28 mm dia.	40.15	50.37	0.52	16.72	nr	**67.09**
35 mm dia.	57.21	71.77	0.62	19.95	nr	**91.72**
42 mm dia.	69.86	87.63	0.70	22.51	nr	**110.14**
54 mm dia.	99.34	124.61	0.84	27.01	nr	**151.62**
67 mm dia.	116.30	145.89	1.03	33.13	nr	**179.02**
76 mm dia.	131.28	164.67	1.22	39.23	nr	**203.90**
108 mm dia.	174.27	218.60	1.41	45.35	nr	**263.95**
133 mm dia.	211.20	264.92	1.75	56.28	nr	**321.20**
159 mm dia.	374.80	470.15	2.21	71.08	nr	**541.23**
Gunmetal screwed flange; 16 BSP						
15 mm dia.	24.52	30.76	0.35	11.26	nr	**42.02**
22 mm dia.	31.01	38.90	0.47	15.12	nr	**54.02**
28 mm dia.	35.89	45.02	0.52	16.72	nr	**61.74**
35 mm dia.	54.85	68.80	0.62	19.95	nr	**88.75**
42 mm dia.	66.03	82.82	0.70	22.51	nr	**105.33**
54 mm dia.	84.92	106.52	0.84	27.01	nr	**133.53**
67 mm dia.	122.75	153.98	1.03	33.13	nr	**187.11**
76 mm dia.	140.76	176.57	1.22	39.23	nr	**215.80**
108 mm dia.	164.70	206.60	1.41	45.35	nr	**251.95**
133 mm dia.	266.87	334.76	1.75	56.28	nr	**391.04**
159 mm dia.	330.11	414.09	2.21	71.08	nr	**485.17**
Extra over copper pipes; labour						
Made bend						
15 mm dia.	–	–	0.26	8.37	nr	**8.37**
22 mm dia.	–	–	0.28	9.00	nr	**9.00**
28 mm dia.	–	–	0.31	9.97	nr	**9.97**
35 mm dia.	–	–	0.42	13.51	nr	**13.51**
42 mm dia.	–	–	0.51	16.41	nr	**16.41**
54 mm dia.	–	–	0.58	18.65	nr	**18.65**
67 mm dia.	–	–	0.69	22.19	nr	**22.19**
76 mm dia.	–	–	0.80	25.73	nr	**25.73**
Bronze butt weld						
15 mm dia.	–	–	0.25	8.04	nr	**8.04**
22 mm dia.	–	–	0.31	9.97	nr	**9.97**
28 mm dia.	–	–	0.37	11.91	nr	**11.91**
35 mm dia.	–	–	0.49	15.76	nr	**15.76**
42 mm dia.	–	–	0.58	18.65	nr	**18.65**
54 mm dia.	–	–	0.72	23.16	nr	**23.16**
67 mm dia.	–	–	0.88	28.30	nr	**28.30**

38 PIPED SUPPLY SYSTEMS

Item	Net Price £	Material £	Labour hours	Labour £	Unit	Total rate £
76 mm dia.	–	–	1.08	34.73	nr	**34.73**
108 mm dia.	–	–	1.37	44.07	nr	**44.07**
133 mm dia.	–	–	1.73	55.64	nr	**55.64**
159 mm dia.	–	–	2.03	65.30	nr	**65.30**
PRESS FIT (copper fittings); Mechanical press fit joints; butyl rubber O ring						
Coupler						
15 mm dia.	1.10	1.38	0.36	11.58	nr	**12.96**
22 mm dia.	1.74	2.18	0.36	11.58	nr	**13.76**
28 mm dia.	3.55	4.46	0.44	14.15	nr	**18.61**
35 mm dia.	4.49	5.63	0.44	14.15	nr	**19.78**
42 mm dia.	8.07	10.12	0.52	16.72	nr	**26.84**
54 mm dia.	10.29	12.90	0.60	19.30	nr	**32.20**
Stop end						
22 mm dia.	2.78	3.48	0.18	5.79	nr	**9.27**
28 mm dia.	4.40	5.52	0.22	7.08	nr	**12.60**
35 mm dia.	7.55	9.48	0.22	7.08	nr	**16.56**
42 mm dia.	11.34	14.22	0.26	8.37	nr	**22.59**
54 mm dia.	13.66	17.14	0.30	9.64	nr	**26.78**
Reducer						
22 × 15 mm dia.	1.28	1.60	0.36	11.58	nr	**13.18**
28 × 15 mm dia.	3.39	4.26	0.40	12.87	nr	**17.13**
28 × 22 mm dia.	3.51	4.40	0.40	12.87	nr	**17.27**
35 × 22 mm dia.	4.21	5.29	0.40	12.87	nr	**18.16**
35 × 28 mm dia.	4.65	5.84	0.44	14.15	nr	**19.99**
42 × 22 mm dia.	7.27	9.12	0.44	14.15	nr	**23.27**
42 × 28 mm dia.	6.93	8.69	0.48	15.44	nr	**24.13**
42 × 35 mm dia.	6.93	8.69	0.48	15.44	nr	**24.13**
54 × 35 mm dia.	9.38	11.77	0.52	16.72	nr	**28.49**
54 × 42 mm dia.	9.38	11.77	0.56	18.01	nr	**29.78**
90° elbow						
15 mm dia.	1.19	1.49	0.36	11.58	nr	**13.07**
22 mm dia.	2.00	2.51	0.36	11.58	nr	**14.09**
28 mm dia.	4.29	5.38	0.44	14.15	nr	**19.53**
35 mm dia.	8.86	11.11	0.44	14.15	nr	**25.26**
42 mm dia.	16.65	20.89	0.52	16.72	nr	**37.61**
54 mm dia.	23.12	29.00	0.60	19.30	nr	**48.30**
45° elbow						
15 mm dia.	1.47	1.85	0.36	11.58	nr	**13.43**
22 mm dia.	2.03	2.54	0.36	11.58	nr	**14.12**
28 mm dia.	6.09	7.64	0.44	14.15	nr	**21.79**
35 mm dia.	8.68	10.89	0.44	14.15	nr	**25.04**
42 mm dia.	14.45	18.12	0.52	16.72	nr	**34.84**
54 mm dia.	20.55	25.78	0.60	19.30	nr	**45.08**
Equal tee						
15 mm dia.	1.89	2.37	0.54	17.37	nr	**19.74**
22 mm dia.	3.45	4.32	0.54	17.37	nr	**21.69**
28 mm dia.	6.20	7.77	0.66	21.22	nr	**28.99**
35 mm dia.	10.73	13.46	0.66	21.22	nr	**34.68**
42 mm dia.	21.18	26.57	0.78	25.09	nr	**51.66**
54 mm dia.	26.42	33.14	0.90	28.95	nr	**62.09**

38 PIPED SUPPLY SYSTEMS

Item	Net Price £	Material £	Labour hours	Labour £	Unit	Total rate £
COLD WATER PIPELINES: COPPER PIPEWORK – cont						
PRESS FIT (copper fittings) – cont						
Reducing tee						
22 × 15 mm dia.	2.80	3.52	0.54	17.37	nr	**20.89**
28 × 15 mm dia.	5.46	6.85	0.62	19.95	nr	**26.80**
28 × 22 mm dia.	7.32	9.18	0.62	19.95	nr	**29.13**
35 × 22 mm dia.	9.55	11.98	0.62	19.95	nr	**31.93**
35 × 28 mm dia.	10.63	13.34	0.62	19.95	nr	**33.29**
42 × 28 mm dia.	19.23	24.12	0.70	22.51	nr	**46.63**
42 × 35 mm dia.	19.23	24.12	0.70	22.51	nr	**46.63**
54 × 35 mm dia.	32.49	40.76	0.82	26.36	nr	**67.12**
54 × 42 mm dia.	32.49	40.76	0.82	26.36	nr	**67.12**
Male iron connector; BSP thread						
15 mm dia.	4.01	5.03	0.18	5.79	nr	**10.82**
22 mm dia.	5.96	7.48	0.18	5.79	nr	**13.27**
28 mm dia.	7.95	9.97	0.22	7.08	nr	**17.05**
35 mm dia.	14.40	18.07	0.22	7.08	nr	**25.15**
42 mm dia.	19.30	24.21	0.26	8.37	nr	**32.58**
54 mm dia.	37.23	46.70	0.30	9.64	nr	**56.34**
90° elbow; male iron BSP thread						
15 mm dia.	6.49	8.14	0.36	11.58	nr	**19.72**
22 mm dia.	10.17	12.76	0.36	11.58	nr	**24.34**
28 mm dia.	15.57	19.53	0.44	14.15	nr	**33.68**
35 mm dia.	20.25	25.40	0.44	14.15	nr	**39.55**
42 mm dia.	26.40	33.12	0.52	16.72	nr	**49.84**
54 mm dia.	38.57	48.38	0.60	19.30	nr	**67.68**
Female iron connector; BSP thread						
15 mm dia.	4.58	5.75	0.18	5.79	nr	**11.54**
22 mm dia.	6.05	7.59	0.18	5.79	nr	**13.38**
28 mm dia.	8.14	10.21	0.22	7.08	nr	**17.29**
35 mm dia.	15.93	19.98	0.22	7.08	nr	**27.06**
42 mm dia.	22.77	28.56	0.26	8.37	nr	**36.93**
54 mm dia.	39.05	48.99	0.30	9.64	nr	**58.63**
90° elbow; female iron BSP thread						
15 mm dia.	5.47	6.87	0.36	11.58	nr	**18.45**
22 mm dia.	8.07	10.12	0.36	11.58	nr	**21.70**
28 mm dia.	13.31	16.70	0.44	14.15	nr	**30.85**
35 mm dia.	17.21	21.59	0.44	14.15	nr	**35.74**
42 mm dia.	23.43	29.39	0.52	16.72	nr	**46.11**
54 mm dia.	34.45	43.21	0.60	19.30	nr	**62.51**

38 PIPED SUPPLY SYSTEMS

Item	Net Price £	Material £	Labour hours	Labour £	Unit	Total rate £
COLD WATER PIPELINES: STAINLESS STEEL PIPEWORK						
Stainless steel pipes; capillary or compression joints; BS 4127, vertical or at low level, with brackets measured separately						
Grade 304; satin finish						
15 mm dia.	4.00	5.02	0.41	13.18	m	**18.20**
22 mm dia.	5.61	7.03	0.51	16.41	m	**23.44**
28 mm dia.	7.63	9.58	0.58	18.65	m	**28.23**
35 mm dia.	11.53	14.46	0.65	20.91	m	**35.37**
42 mm dia.	14.63	18.36	0.71	22.85	m	**41.21**
54 mm dia.	20.40	25.59	0.80	25.73	m	**51.32**
Grade 316 satin finish						
15 mm dia.	5.13	6.44	0.61	19.62	m	**26.06**
22 mm dia.	9.60	12.04	0.76	24.46	m	**36.50**
28 mm dia.	11.39	14.29	0.87	27.99	m	**42.28**
35 mm dia.	20.66	25.92	0.98	31.54	m	**57.46**
42 mm dia.	26.75	33.56	1.06	34.10	m	**67.66**
54 mm dia.	31.17	39.10	1.16	37.32	m	**76.42**
Fixings						
Single pipe ring						
15 mm dia.	12.46	15.64	0.26	8.37	nr	**24.01**
22 mm dia.	14.50	18.19	0.26	8.37	nr	**26.56**
28 mm dia.	15.19	19.05	0.31	9.97	nr	**29.02**
35 mm dia.	17.26	21.65	0.32	10.28	nr	**31.93**
42 mm dia.	19.61	24.60	0.32	10.28	nr	**34.88**
54 mm dia.	22.39	28.09	0.34	10.94	nr	**39.03**
Screw on backplate, female						
All sizes 15 mm to 54 mm dia.	11.23	14.09	0.10	3.21	nr	**17.30**
Screw on backplate, male						
All sizes 15 mm to 54 mm dia.	12.79	16.04	0.10	3.21	nr	**19.25**
Stainless steel threaded rods; metric thread; including nuts, washers etc.						
10 mm dia. × 600 mm long	13.13	16.48	0.18	5.79	nr	**22.27**
Extra over stainless steel pipes; capillary fittings						
Straight coupling						
15 mm dia.	1.47	1.85	0.25	8.04	nr	**9.89**
22 mm dia.	2.37	2.97	0.28	9.00	nr	**11.97**
28 mm dia.	3.13	3.93	0.33	10.62	nr	**14.55**
35 mm dia.	7.23	9.07	0.37	11.91	nr	**20.98**
42 mm dia.	8.34	10.46	0.42	13.51	nr	**23.97**
54 mm dia.	12.54	15.72	0.45	14.47	nr	**30.19**
45° bend						
15 mm dia.	7.89	9.90	0.25	8.04	nr	**17.94**
22 mm dia.	10.36	12.99	0.30	9.53	nr	**22.52**
28 mm dia.	12.72	15.96	0.33	10.62	nr	**26.58**

38 PIPED SUPPLY SYSTEMS

Item	Net Price £	Material £	Labour hours	Labour £	Unit	Total rate £
COLD WATER PIPELINES: STAINLESS STEEL PIPEWORK – cont						
Extra over stainless steel pipes – cont						
45° bend – cont						
35 mm dia.	15.08	18.92	0.37	11.91	nr	30.83
42 mm dia.	19.41	24.35	0.42	13.51	nr	37.86
54 mm dia.	24.00	30.11	0.45	14.47	nr	44.58
90° bend						
15 mm dia.	4.09	5.13	0.28	9.00	nr	14.13
22 mm dia.	5.50	6.90	0.28	9.00	nr	15.90
28 mm dia.	7.76	9.73	0.33	10.62	nr	20.35
35 mm dia.	18.92	23.73	0.37	11.91	nr	35.64
42 mm dia.	26.04	32.66	0.42	13.51	nr	46.17
54 mm dia.	35.32	44.31	0.45	14.47	nr	58.78
Reducer						
22 × 15 mm dia.	9.40	11.79	0.28	9.00	nr	20.79
28 × 22 mm dia.	10.49	13.16	0.33	10.62	nr	23.78
35 × 28 mm dia.	12.84	16.11	0.37	11.91	nr	28.02
42 × 35 mm dia.	13.82	17.34	0.42	13.51	nr	30.85
54 × 42 mm dia.	41.01	51.44	0.48	15.47	nr	66.91
Tap connector						
15 mm dia.	19.84	24.89	0.13	4.18	nr	29.07
22 mm dia.	26.23	32.91	0.14	4.50	nr	37.41
28 mm dia.	36.41	45.67	0.17	5.47	nr	51.14
Tank connector						
15 mm dia.	25.62	32.13	0.13	4.18	nr	36.31
22 mm dia.	38.13	47.84	0.13	4.18	nr	52.02
28 mm dia.	46.95	58.89	0.15	4.83	nr	63.72
35 mm dia.	67.89	85.16	0.18	5.79	nr	90.95
42 mm dia.	89.66	112.47	0.21	6.75	nr	119.22
54 mm dia.	135.74	170.27	0.24	7.72	nr	177.99
Tee equal						
15 mm dia.	7.30	9.16	0.37	11.91	nr	21.07
22 mm dia.	9.09	11.40	0.40	12.87	nr	24.27
28 mm dia.	10.99	13.79	0.45	14.47	nr	28.26
35 mm dia.	26.38	33.10	0.59	19.00	nr	52.10
42 mm dia.	32.57	40.86	0.62	19.95	nr	60.81
54 mm dia.	65.77	82.50	0.67	21.55	nr	104.05
Unequal tee						
22 × 15 mm dia.	14.81	18.58	0.37	11.91	nr	30.49
28 × 15 mm dia.	16.68	20.92	0.45	14.47	nr	35.39
28 × 22 mm dia.	16.68	20.92	0.45	14.47	nr	35.39
35 × 22 mm dia.	29.16	36.58	0.59	19.00	nr	55.58
35 × 28 mm dia.	29.16	36.58	0.59	19.00	nr	55.58
42 × 28 mm dia.	35.84	44.96	0.62	19.95	nr	64.91
42 × 35 mm dia.	35.84	44.96	0.62	19.95	nr	64.91
54 × 35 mm dia.	74.20	93.07	0.67	21.55	nr	114.62
54 × 42 mm dia.	74.20	93.07	0.67	21.55	nr	114.62

38 PIPED SUPPLY SYSTEMS

Item	Net Price £	Material £	Labour hours	Labour £	Unit	Total rate £
Union, conical seat						
15 mm dia.	32.85	41.20	0.25	8.04	nr	49.24
22 mm dia.	51.74	64.90	0.28	9.00	nr	73.90
28 mm dia.	66.89	83.91	0.33	10.62	nr	94.53
35 mm dia.	87.81	110.15	0.37	11.91	nr	122.06
42 mm dia.	110.75	138.92	0.42	13.51	nr	152.43
54 mm dia.	146.52	183.79	0.45	14.47	nr	198.26
Union, flat seat						
15 mm dia.	34.31	43.04	0.25	8.04	nr	51.08
22 mm dia.	53.42	67.01	0.28	9.00	nr	76.01
28 mm dia.	69.06	86.63	0.33	10.62	nr	97.25
35 mm dia.	90.23	113.19	0.37	11.91	nr	125.10
42 mm dia.	113.69	142.61	0.42	13.51	nr	156.12
54 mm dia.	152.43	191.21	0.45	14.47	nr	205.68
Extra over stainless steel pipes; compression fittings						
Straight coupling						
15 mm dia.	28.56	35.83	0.18	5.79	nr	41.62
22 mm dia.	54.41	68.25	0.22	7.08	nr	75.33
28 mm dia.	73.25	91.88	0.25	8.04	nr	99.92
35 mm dia.	113.08	141.85	0.30	9.64	nr	151.49
42 mm dia.	131.98	165.56	0.40	12.87	nr	178.43
90° bend						
15 mm dia.	36.00	45.16	0.18	5.79	nr	50.95
22 mm dia.	71.52	89.71	0.22	7.08	nr	96.79
28 mm dia.	97.53	122.34	0.25	8.04	nr	130.38
35 mm dia.	197.50	247.74	0.33	10.62	nr	258.36
42 mm dia.	288.63	362.06	0.35	11.26	nr	373.32
Reducer						
22 × 15 mm dia.	51.82	65.00	0.28	9.00	nr	74.00
28 × 22 mm dia.	70.98	89.04	0.28	9.00	nr	98.04
35 × 28 mm dia.	103.71	130.10	0.30	9.64	nr	139.74
42 × 35 mm dia.	137.96	173.06	0.37	11.91	nr	184.97
Stud coupling						
15 mm dia.	29.68	37.23	0.42	13.51	nr	50.74
22 mm dia.	50.20	62.97	0.25	8.04	nr	71.01
28 mm dia.	69.71	87.45	0.25	8.04	nr	95.49
35 mm dia.	111.66	140.07	0.37	11.91	nr	151.98
42 mm dia.	131.98	165.56	0.42	13.51	nr	179.07
Equal tee						
15 mm dia.	50.69	63.58	0.37	11.91	nr	75.49
22 mm dia.	104.72	131.36	0.40	12.87	nr	144.23
28 mm dia.	143.24	179.68	0.45	14.47	nr	194.15
35 mm dia.	284.83	357.29	0.59	19.00	nr	376.29
42 mm dia.	394.95	495.42	0.62	19.95	nr	515.37
Running tee						
15 mm dia.	62.41	78.29	0.37	11.91	nr	90.20
22 mm dia.	112.43	141.03	0.40	12.87	nr	153.90
28 mm dia.	190.35	238.77	0.59	19.00	nr	257.77

38 PIPED SUPPLY SYSTEMS

Item	Net Price £	Material £	Labour hours	Labour £	Unit	Total rate £
COLD WATER PIPELINES: STAINLESS STEEL PIPEWORK – cont						
PRESS FIT (stainless steel); Press fit jointing system; butyl rubber O ring mechanical joint						
Pipework						
15 mm dia.	4.91	6.16	0.46	14.78	m	**20.94**
22 mm dia.	7.85	9.84	0.48	15.44	m	**25.28**
28 mm dia.	9.67	12.13	0.52	16.72	m	**28.85**
35 mm dia.	14.25	17.88	0.56	18.01	m	**35.89**
42 mm dia.	17.53	21.99	0.58	18.65	m	**40.64**
54 mm dia.	22.33	28.01	0.66	21.22	m	**49.23**
Fixings						
For stainless steel pipes						
Refer to fixings for stainless steel pipes;						
capillary or compression joints; BS 4127						
Extra over stainless steel pipes; Press fit jointing system;						
Coupling						
15 mm dia.	5.77	7.24	0.36	11.58	nr	**18.82**
22 mm dia.	7.27	9.12	0.36	11.58	nr	**20.70**
28 mm dia.	8.17	10.25	0.44	14.15	nr	**24.40**
35 mm dia.	10.16	12.75	0.44	14.15	nr	**26.90**
42 mm dia.	13.86	17.38	0.52	16.72	nr	**34.10**
54 mm dia.	16.68	20.92	0.60	19.30	nr	**40.22**
Stop end						
22 mm dia.	5.48	6.88	0.18	5.79	nr	**12.67**
28 mm dia.	6.38	8.01	0.22	7.08	nr	**15.09**
35 mm dia.	10.44	13.09	0.22	7.08	nr	**20.17**
42 mm dia.	14.67	18.40	0.26	8.37	nr	**26.77**
54 mm dia.	16.97	21.29	0.30	9.64	nr	**30.93**
Reducer						
22 × 15 mm dia.	6.87	8.61	0.36	11.58	nr	**20.19**
28 × 15 mm dia.	7.79	9.77	0.40	12.87	nr	**22.64**
28 × 22 mm dia.	8.06	10.11	0.40	12.87	nr	**22.98**
35 × 22 mm dia.	9.84	12.34	0.40	12.87	nr	**25.21**
35 × 28 mm dia.	12.18	15.28	0.44	14.15	nr	**29.43**
42 × 35 mm dia.	12.85	16.12	0.48	15.44	nr	**31.56**
54 × 42 mm dia.	14.70	18.44	0.56	18.01	nr	**36.45**
90° bend						
15 mm dia.	8.25	10.35	0.36	11.58	nr	**21.93**
22 mm dia.	11.53	14.46	0.36	11.58	nr	**26.04**
28 mm dia.	14.55	18.26	0.44	14.15	nr	**32.41**
35 mm dia.	22.90	28.73	0.44	14.15	nr	**42.88**
42 mm dia.	38.24	47.97	0.52	16.72	nr	**64.69**
54 mm dia.	52.82	66.26	0.60	19.30	nr	**85.56**

38 PIPED SUPPLY SYSTEMS

Item	Net Price £	Material £	Labour hours	Labour £	Unit	Total rate £
45° bend						
15 mm dia.	11.19	14.03	0.36	11.58	nr	**25.61**
22 mm dia.	13.92	17.46	0.36	11.58	nr	**29.04**
28 mm dia.	16.20	20.32	0.44	14.15	nr	**34.47**
35 mm dia.	19.01	23.84	0.44	14.15	nr	**37.99**
42 mm dia.	30.58	38.36	0.52	16.72	nr	**55.08**
54 mm dia.	39.74	49.85	0.60	19.30	nr	**69.15**
Equal tee						
15 mm dia.	13.53	16.97	0.54	17.37	nr	**34.34**
22 mm dia.	16.61	20.83	0.54	17.37	nr	**38.20**
28 mm dia.	19.39	24.33	0.66	21.22	nr	**45.55**
35 mm dia.	24.57	30.82	0.66	21.22	nr	**52.04**
42 mm dia.	34.87	43.74	0.78	25.09	nr	**68.83**
54 mm dia.	41.73	52.35	0.90	28.95	nr	**81.30**
Reducing tee						
22 × 15 mm dia.	14.19	17.80	0.54	17.37	nr	**35.17**
28 × 15 mm dia.	17.20	21.57	0.62	19.95	nr	**41.52**
28 × 22 mm dia.	18.62	23.35	0.62	19.95	nr	**43.30**
35 × 22 mm dia.	22.08	27.70	0.62	19.95	nr	**47.65**
35 × 28 mm dia.	23.04	28.90	0.62	19.95	nr	**48.85**
42 × 28 mm dia.	32.76	41.09	0.70	22.51	nr	**63.60**
42 × 35 mm dia.	33.75	42.34	0.70	22.51	nr	**64.85**
54 × 35 mm dia.	38.12	47.81	0.82	26.36	nr	**74.17**
54 × 42 mm dia.	39.19	49.16	0.82	26.36	nr	**75.52**

Fixings

For stainless steel pipes
Refer to fixings for stainless steel pipes;
capillary or compression joints; BS 4127

38 PIPED SUPPLY SYSTEMS

Item	Net Price £	Material £	Labour hours	Labour £	Unit	Total rate £
COLD WATER PIPELINES: MEDIUM DENSITY POLYETHYLENE – BLUE PIPEWORK						
Note: MDPE is sized on Outside Dia., i.e OD not ID						
Pipes for water distribution; laid underground; electrofusion joints in the running length; BS 6572						
Coiled service pipe						
20 mm dia.	0.90	1.13	0.37	11.91	m	13.04
25 mm dia.	1.17	1.47	0.41	13.18	m	14.65
32 mm dia.	1.97	2.48	0.47	15.12	m	17.60
50 mm dia.	4.74	5.95	0.53	17.05	m	23.00
63 mm dia.	7.43	9.32	0.60	19.31	m	28.63
Mains service pipe						
90 mm dia.	11.22	14.08	0.90	28.95	m	43.03
110 mm dia.	16.80	21.08	1.10	35.37	m	56.45
125 mm dia.	21.25	26.66	1.20	38.61	m	65.27
160 mm dia.	33.86	42.47	1.48	47.60	m	90.07
180 mm dia.	44.11	55.33	1.50	48.29	m	103.62
225 mm dia.	67.03	84.08	1.77	56.93	m	141.01
250 mm dia.	84.60	106.12	1.75	56.32	m	162.44
315 mm dia.	130.62	163.84	1.90	61.12	m	224.96
Extra over fittings; MDPE blue; electrofusion joints						
Coupler						
20 mm dia.	7.63	9.58	0.36	11.58	nr	21.16
25 mm dia.	7.63	9.58	0.40	12.87	nr	22.45
32 mm dia.	7.63	9.58	0.44	14.15	nr	23.73
40 mm dia.	11.26	14.12	0.48	15.44	nr	29.56
50 mm dia.	10.82	13.57	0.52	16.72	nr	30.29
63 mm dia.	14.14	17.74	0.58	18.65	nr	36.39
90 mm dia.	20.79	26.07	0.67	21.55	nr	47.62
110 mm dia.	33.44	41.94	0.74	23.80	nr	65.74
125 mm dia.	37.72	47.32	0.83	26.69	nr	74.01
160 mm dia.	60.00	75.26	1.00	32.17	nr	107.43
180 mm dia.	70.54	88.48	1.25	40.21	nr	128.69
225 mm dia.	112.74	141.42	1.35	43.41	nr	184.83
250 mm dia.	165.08	207.08	1.50	48.25	nr	255.33
315 mm dia.	272.23	341.49	1.80	57.89	nr	399.38
Extra over fittings; MDPE blue; butt fused joints						
Cap						
25 mm dia.	14.48	18.17	0.20	6.44	nr	24.61
32 mm dia.	14.48	18.17	0.22	7.08	nr	25.25
40 mm dia.	15.34	19.24	0.24	7.72	nr	26.96

38 PIPED SUPPLY SYSTEMS

Item	Net Price £	Material £	Labour hours	Labour £	Unit	Total rate £
50 mm dia.	22.37	28.06	0.26	8.37	nr	36.43
63 mm dia.	25.66	32.19	0.32	10.28	nr	42.47
90 mm dia.	42.33	53.10	0.37	11.91	nr	65.01
110 mm dia.	85.25	106.94	0.40	12.87	nr	119.81
125 mm dia.	68.06	85.38	0.46	14.78	nr	100.16
160 mm dia.	77.81	97.61	0.50	16.08	nr	113.69
180 mm dia.	130.14	163.25	0.60	19.30	nr	182.55
225 mm dia.	153.72	192.83	0.68	21.87	nr	214.70
250 mm dia.	225.63	283.04	0.75	24.12	nr	307.16
315 mm dia.	290.94	364.95	0.90	28.95	nr	393.90
Reducer						
63 × 32 mm dia.	19.56	24.54	0.54	17.37	nr	41.91
63 × 50 mm dia.	22.87	28.68	0.60	19.30	nr	47.98
90 × 63 mm dia.	28.95	36.31	0.67	21.55	nr	57.86
110 × 90 mm dia.	39.89	50.04	0.74	23.80	nr	73.84
125 × 90 mm dia.	58.02	72.78	0.83	26.69	nr	99.47
125 × 110 mm dia.	63.92	80.18	1.00	32.17	nr	112.35
160 × 110 mm dia.	98.20	123.18	1.10	35.37	nr	158.55
180 × 125 mm dia.	106.66	133.80	1.25	40.21	nr	174.01
225 × 160 mm dia.	173.93	218.18	1.40	45.04	nr	263.22
250 × 180 mm dia.	133.63	167.63	1.80	57.89	nr	225.52
315 × 250 mm dia.	153.73	192.84	2.40	77.19	nr	270.03
Bend; 45°						
50 mm dia.	29.85	37.44	0.50	16.08	nr	53.52
63 mm dia.	36.11	45.29	0.58	18.65	nr	63.94
90 mm dia.	55.92	70.15	0.67	21.55	nr	91.70
110 mm dia.	82.00	102.86	0.74	23.80	nr	126.66
125 mm dia.	91.64	114.96	0.83	26.69	nr	141.65
160 mm dia.	169.84	213.05	1.00	32.17	nr	245.22
180 mm dia.	193.33	242.51	1.25	40.21	nr	282.72
225 mm dia.	246.43	309.12	1.40	45.02	nr	354.14
250 mm dia.	266.66	334.50	1.80	57.89	nr	392.39
315 mm dia.	332.22	416.74	2.40	77.19	nr	493.93
Bend; 90°						
50 mm dia.	29.85	37.44	0.50	16.08	nr	53.52
63 mm dia.	36.11	45.29	0.58	18.65	nr	63.94
90 mm dia.	55.92	70.15	0.67	21.55	nr	91.70
110 mm dia.	82.00	102.86	0.74	23.80	nr	126.66
125 mm dia.	91.64	114.96	0.83	26.69	nr	141.65
160 mm dia.	169.84	213.05	1.00	32.17	nr	245.22
180 mm dia.	329.86	413.77	1.25	40.21	nr	453.98
225 mm dia.	417.07	523.17	1.40	45.04	nr	568.21
250 mm dia.	456.20	572.25	1.80	57.89	nr	630.14
315 mm dia.	571.11	716.40	2.40	77.19	nr	793.59
Equal tee						
50 mm dia.	32.64	40.95	0.70	22.51	nr	63.46
63 mm dia.	35.62	44.68	0.75	24.12	nr	68.80
90 mm dia.	65.06	81.61	0.87	27.99	nr	109.60
110 mm dia.	97.02	121.70	1.00	32.17	nr	153.87
125 mm dia.	125.25	157.11	1.08	34.73	nr	191.84
160 mm dia.	205.95	258.34	1.35	43.41	nr	301.75
180 mm dia.	210.75	264.36	1.63	52.42	nr	316.78

38 PIPED SUPPLY SYSTEMS

Item	Net Price £	Material £	Labour hours	Labour £	Unit	Total rate £
COLD WATER PIPELINES: MEDIUM DENSITY POLYETHYLENE – BLUE PIPEWORK – cont						
Extra over fittings – cont						
Equal tee – cont						
225 mm dia.	255.41	320.39	1.90	61.12	nr	381.51
250 mm dia.	352.42	442.08	2.70	86.84	nr	528.92
315 mm dia.	878.56	1102.07	3.60	115.79	nr	1217.86
Extra over plastic fittings, compression joints						
Straight connector						
20 mm dia.	3.38	4.24	0.38	12.23	nr	16.47
25 mm dia.	3.53	4.42	0.45	14.47	nr	18.89
32 mm dia.	8.40	10.54	0.50	16.08	nr	26.62
50 mm dia.	19.34	24.26	0.68	21.87	nr	46.13
63 mm dia.	29.13	36.55	0.85	27.35	nr	63.90
Reducing connector						
25 mm dia.	6.94	8.70	0.38	12.23	nr	20.93
32 mm dia.	11.19	14.03	0.45	14.47	nr	28.50
50 mm dia.	31.02	38.91	0.50	16.08	nr	54.99
63 mm dia.	43.29	54.30	0.62	19.95	nr	74.25
Straight connector; polyethylene to MI						
20 mm dia.	3.06	3.84	0.31	9.97	nr	13.81
25 mm dia.	5.20	6.52	0.35	11.26	nr	17.78
32 mm dia.	5.64	7.08	0.40	12.87	nr	19.95
50 mm dia.	14.39	18.05	0.55	17.68	nr	35.73
63 mm dia.	20.28	25.44	0.65	20.91	nr	46.35
Straight connector; polyethylene to FI						
20 mm dia.	4.11	5.15	0.31	9.97	nr	15.12
25 mm dia.	4.45	5.58	0.35	11.26	nr	16.84
32 mm dia.	5.32	6.68	0.40	12.87	nr	19.55
50 mm dia.	16.90	21.20	0.55	17.68	nr	38.88
63 mm dia.	23.67	29.69	0.75	24.14	nr	53.83
Elbow						
20 mm dia.	4.50	5.64	0.38	12.23	nr	17.87
25 mm dia.	6.64	8.33	0.45	14.47	nr	22.80
32 mm dia.	9.69	12.15	0.50	16.08	nr	28.23
50 mm dia.	22.48	28.20	0.68	21.87	nr	50.07
63 mm dia.	30.55	38.33	0.80	25.73	nr	64.06
Elbow; polyethylene to MI						
25 mm dia.	5.72	7.18	0.35	11.26	nr	18.44
Elbow; polyethylene to FI						
20 mm dia.	4.09	5.13	0.31	9.97	nr	15.10
25 mm dia.	5.56	6.98	0.35	11.26	nr	18.24
32 mm dia.	8.33	10.45	0.42	13.51	nr	23.96
50 mm dia.	19.79	24.82	0.50	16.08	nr	40.90
63 mm dia.	25.93	32.52	0.55	17.68	nr	50.20

38 PIPED SUPPLY SYSTEMS

Item	Net Price £	Material £	Labour hours	Labour £	Unit	Total rate £
Tank coupling						
25 mm dia.	8.58	10.76	0.42	13.51	nr	**24.27**
Equal tee						
20 mm dia.	6.05	7.59	0.53	17.05	nr	**24.64**
25 mm dia.	9.47	11.88	0.55	17.68	nr	**29.56**
32 mm dia.	11.85	14.86	0.64	20.59	nr	**35.45**
50 mm dia.	27.65	34.69	0.75	24.14	nr	**58.83**
63 mm dia.	42.83	53.73	0.87	27.99	nr	**81.72**
Equal tee; FI branch						
20 mm dia.	5.86	7.35	0.45	14.47	nr	**21.82**
25 mm dia.	9.34	11.72	0.50	16.08	nr	**27.80**
32 mm dia.	11.37	14.26	0.60	19.31	nr	**33.57**
50 mm dia.	26.19	32.85	0.68	21.87	nr	**54.72**
63 mm dia.	36.78	46.13	0.81	26.06	nr	**72.19**
Equal tee; MI branch						
25 mm dia.	9.18	11.51	0.50	16.08	nr	**27.59**

38 PIPED SUPPLY SYSTEMS

Item	Net Price £	Material £	Labour hours	Labour £	Unit	Total rate £
COLD WATER PIPELINES: ABS PIPEWORK						
Pipes; solvent welded joints in the running length, brackets measured separately						
Class C (9 bar pressure)						
1" dia.	6.07	7.62	0.30	9.64	m	17.26
1 ¼" dia.	10.20	12.79	0.33	10.62	m	23.41
1 ½" dia.	12.95	16.24	0.36	11.58	m	27.82
2" dia.	17.45	21.88	0.39	12.54	m	34.42
3" dia.	35.95	45.09	0.46	14.78	m	59.87
4" dia.	59.25	74.32	0.53	17.05	m	91.37
6" dia.	117.14	146.94	0.76	24.45	m	171.39
8" dia.	201.17	252.35	0.97	31.19	m	283.54
Class E (15 bar pressure)						
½" dia.	4.63	5.81	0.24	7.72	m	13.53
¾" dia.	7.13	8.95	0.27	8.68	m	17.63
1" dia.	9.39	11.78	0.30	9.64	m	21.42
1 ¼" dia.	14.02	17.58	0.33	10.62	m	28.20
1 ½" dia.	18.48	23.18	0.36	11.58	m	34.76
2" dia.	23.14	29.03	0.39	12.54	m	41.57
3" dia.	46.52	58.35	0.49	15.76	m	74.11
4" dia.	74.84	93.88	0.57	18.32	m	112.20
Fixings						
Refer to steel pipes; galvanized iron. For minimum fixing dimensions, refer to the Tables and Memoranda at the rear of the book						
Extra over fittings; solvent welded joints						
Cap						
½" dia.	1.98	2.49	0.16	5.14	nr	7.63
¾" dia.	2.29	2.87	0.19	6.10	nr	8.97
1" dia.	2.63	3.30	0.22	7.08	nr	10.38
1 ¼" dia.	4.39	5.51	0.25	8.04	nr	13.55
1 ½" dia.	6.77	8.49	0.28	9.00	nr	17.49
2" dia.	8.57	10.75	0.31	9.97	nr	20.72
3" dia.	25.75	32.30	0.36	11.58	nr	43.88
4" dia.	39.38	49.40	0.44	14.15	nr	63.55
Elbow 90°						
½" dia.	2.76	3.46	0.29	9.33	nr	12.79
¾" dia.	3.31	4.16	0.34	10.94	nr	15.10
1" dia.	4.63	5.81	0.40	12.87	nr	18.68
1 ¼" dia.	7.82	9.81	0.45	14.47	nr	24.28
1 ½" dia.	10.16	12.75	0.51	16.41	nr	29.16
2" dia.	15.46	19.40	0.56	18.01	nr	37.41
3" dia.	44.41	55.71	0.65	20.91	nr	76.62
4" dia.	66.32	83.19	0.80	25.73	nr	108.92
6" dia.	266.98	334.90	1.21	38.91	nr	373.81
8" dia.	407.45	511.10	1.45	46.64	nr	557.74

38 PIPED SUPPLY SYSTEMS

Item	Net Price £	Material £	Labour hours	Labour £	Unit	Total rate £
Elbow 45°						
½" dia.	5.35	6.71	0.29	9.33	nr	16.04
¾" dia.	5.42	6.80	0.34	10.94	nr	17.74
1" dia.	6.77	8.49	0.40	12.87	nr	21.36
1 ¼" dia.	9.91	12.43	0.45	14.47	nr	26.90
1 ½" dia.	12.27	15.39	0.51	16.41	nr	31.80
2" dia.	17.05	21.39	0.56	18.01	nr	39.40
3" dia.	40.16	50.38	0.65	20.91	nr	71.29
4" dia.	83.26	104.44	0.80	25.73	nr	130.17
6" dia.	172.58	216.48	1.21	38.91	nr	255.39
8" dia.	371.36	465.83	1.45	46.64	nr	512.47
Reducing bush						
¾" × ½" dia.	2.04	2.55	0.42	13.51	nr	16.06
1" × ½" dia.	2.63	3.30	0.45	14.47	nr	17.77
1" × ¾" dia.	2.63	3.30	0.45	14.47	nr	17.77
1 ¼" × 1" dia.	3.53	4.42	0.48	15.44	nr	19.86
1 ½" × ¾" dia.	4.63	5.81	0.51	16.41	nr	22.22
1 ½" × 1" dia.	4.63	5.81	0.51	16.41	nr	22.22
1 ½" × 1 ¼" dia.	4.63	5.81	0.51	16.41	nr	22.22
2" × 1" dia.	6.07	7.62	0.56	18.01	nr	25.63
2" × 1 ¼" dia.	6.07	7.62	0.56	18.01	nr	25.63
2" × 1 ½" dia.	6.07	7.62	0.56	18.01	nr	25.63
3" × 1 ½" dia.	17.05	21.39	0.65	20.91	nr	42.30
3" × 2" dia.	17.05	21.39	0.65	20.91	nr	42.30
4" × 3" dia.	23.52	29.50	0.80	25.73	nr	55.23
6" × 4" dia.	72.43	90.85	1.21	38.91	nr	129.76
Union						
½" dia.	10.99	13.79	0.34	10.94	nr	24.73
¾" dia.	11.85	14.86	0.39	12.54	nr	27.40
1" dia.	15.97	20.04	0.43	13.83	nr	33.87
1 ¼" dia.	19.57	24.55	0.50	16.08	nr	40.63
1 ½" dia.	26.96	33.82	0.57	18.32	nr	52.14
2" dia.	35.16	44.11	0.62	19.95	nr	64.06
Sockets						
½" dia.	2.04	2.55	0.34	10.94	nr	13.49
¾" dia.	2.29	2.87	0.39	12.54	nr	15.41
1" dia.	2.63	3.30	0.43	13.83	nr	17.13
1 ¼" dia.	4.63	5.81	0.50	16.08	nr	21.89
1 ½" dia.	5.58	7.00	0.57	18.32	nr	25.32
2" dia.	7.82	9.81	0.62	19.95	nr	29.76
3" dia.	31.46	39.47	0.70	22.51	nr	61.98
4" dia.	44.63	55.99	0.70	22.51	nr	78.50
6" dia.	111.53	139.90	1.26	40.53	nr	180.43
8" dia.	222.77	279.44	1.55	49.85	nr	329.29
Barrel nipple						
½" dia.	3.84	4.82	0.34	10.94	nr	15.76
¾" dia.	5.00	6.27	0.39	12.54	nr	18.81
1" dia.	6.48	8.13	0.43	13.83	nr	21.96
1 ¼" dia.	8.98	11.27	0.50	16.08	nr	27.35
1 ½" dia.	10.59	13.28	0.57	18.32	nr	31.60
2" dia.	12.82	16.08	0.62	19.95	nr	36.03
3" dia.	34.09	42.76	0.70	22.51	nr	65.27

38 PIPED SUPPLY SYSTEMS

Item	Net Price £	Material £	Labour hours	Labour £	Unit	Total rate £
COLD WATER PIPELINES: ABS PIPEWORK – cont						
Extra over fittings – cont						
Tee, 90°						
½" dia.	3.15	3.95	0.41	13.18	nr	17.13
¾" dia.	4.39	5.51	0.47	15.12	nr	20.63
1" dia.	6.07	7.62	0.55	17.68	nr	25.30
1 ¼" dia.	8.73	10.95	0.64	20.59	nr	31.54
1 ½" dia.	12.82	16.08	0.71	22.83	nr	38.91
2" dia.	19.57	24.55	0.78	25.09	nr	49.64
3" dia.	57.09	71.61	0.91	29.27	nr	100.88
4" dia.	83.80	105.12	1.12	36.03	nr	141.15
6" dia.	292.87	367.37	1.69	54.35	nr	421.72
8" dia.	456.66	572.84	2.03	65.30	nr	638.14
Full face flange						
½" dia.	34.12	42.80	0.10	3.21	nr	46.01
¾" dia.	34.91	43.79	0.13	4.18	nr	47.97
1" dia.	37.82	47.44	0.15	4.83	nr	52.27
1 ¼" dia.	42.04	52.73	0.18	5.79	nr	58.52
1 ½" dia.	50.55	63.41	0.21	6.75	nr	70.16
2" dia.	68.43	85.84	0.29	9.33	nr	95.17
3" dia.	117.32	147.17	0.37	11.91	nr	159.08
4" dia.	153.77	192.89	0.41	13.18	nr	206.07

38 PIPED SUPPLY SYSTEMS

Item	Net Price £	Material £	Labour hours	Labour £	Unit	Total rate £
COLD WATER PIPELINES: PVC-u PIPEWORK						
Pipes; solvent welded joints in the running length, brackets Measured separately						
Class C (9 bar pressure)						
2" dia.	14.39	18.05	0.41	13.18	m	**31.23**
3" dia.	27.57	34.59	0.47	15.12	m	**49.71**
4" dia.	48.91	61.35	0.50	16.08	m	**77.43**
6" dia.	105.84	132.76	1.76	56.62	m	**189.38**
Class D (12 bar pressure)						
1 ¼" dia.	8.38	10.52	0.41	13.18	m	**23.70**
1 ½" dia.	11.51	14.44	0.42	13.51	m	**27.95**
2" dia.	17.86	22.40	0.45	14.47	m	**36.87**
3" dia.	38.28	48.01	0.48	15.44	m	**63.45**
4" dia.	64.06	80.36	0.53	17.05	m	**97.41**
6" dia.	118.89	149.14	0.58	18.65	m	**167.79**
Class E (15 bar pressure)						
½" dia.	4.10	5.14	0.38	12.22	m	**17.36**
¾" dia.	5.86	7.35	0.40	12.87	m	**20.22**
1" dia.	6.82	8.56	0.41	13.18	m	**21.74**
1 ¼" dia.	10.00	12.54	0.41	13.18	m	**25.72**
1 ½" dia.	13.02	16.33	0.42	13.51	m	**29.84**
2" dia.	20.34	25.51	0.45	14.47	m	**39.98**
3" dia.	44.08	55.29	0.47	15.12	m	**70.41**
4" dia.	72.36	90.76	0.50	16.08	m	**106.84**
6" dia.	156.75	196.63	0.53	17.05	m	**213.68**
Class 7						
½" dia.	7.25	9.09	0.32	10.28	m	**19.37**
¾" dia.	10.18	12.77	0.33	10.62	m	**23.39**
1" dia.	15.51	19.45	0.40	12.87	m	**32.32**
1 ¼" dia.	21.31	26.73	0.40	12.87	m	**39.60**
1 ½" dia.	26.38	33.10	0.41	13.18	m	**46.28**
2" dia.	43.85	55.00	0.43	13.83	m	**68.83**
Fixings						
Refer to steel pipes; galvanized iron. For minimum fixing dimensions, refer to the Tables and Memoranda at the rear of the book						
Extra over fittings; solvent welded joints						
End cap						
½" dia.	1.40	1.76	0.17	5.47	nr	**7.23**
¾" dia.	1.63	2.05	0.19	6.10	nr	**8.15**
1" dia.	1.84	2.31	0.22	7.08	nr	**9.39**
1 ¼" dia.	2.88	3.62	0.25	8.04	nr	**11.66**
1 ½" dia.	4.82	6.05	0.28	9.00	nr	**15.05**
2" dia.	5.91	7.41	0.31	9.97	nr	**17.38**
3" dia.	18.09	22.69	0.36	11.58	nr	**34.27**
4" dia.	27.93	35.03	0.44	14.15	nr	**49.18**
6" dia.	67.50	84.67	0.67	21.55	nr	**106.22**

38 PIPED SUPPLY SYSTEMS

Item	Net Price £	Material £	Labour hours	Labour £	Unit	Total rate £
COLD WATER PIPELINES: PVC-u PIPEWORK – cont						
Extra over fittings – cont						
Socket						
½" dia.	1.48	1.86	0.31	9.97	nr	11.83
¾" dia.	1.63	2.05	0.35	11.26	nr	13.31
1" dia.	1.90	2.39	0.42	13.51	nr	15.90
1 ¼" dia.	3.43	4.30	0.45	14.47	nr	18.77
1 ½" dia.	4.03	5.05	0.51	16.41	nr	21.46
2" dia.	5.72	7.18	0.56	18.01	nr	25.19
3" dia.	21.86	27.42	0.65	20.91	nr	48.33
4" dia.	31.70	39.76	0.80	25.73	nr	65.49
6" dia.	79.51	99.74	1.21	38.91	nr	138.65
Reducing socket						
¾ × ½" dia.	1.73	2.17	0.31	9.97	nr	12.14
1 × ¾" dia.	2.17	2.72	0.35	11.26	nr	13.98
1 ¼ × 1" dia.	4.15	5.21	0.42	13.51	nr	18.72
1 ½ × 1 ¼" dia.	4.63	5.81	0.45	14.47	nr	20.28
2 × 1 ½" dia.	6.98	8.76	0.51	16.41	nr	25.17
3 × 2" dia.	21.27	26.68	0.56	18.01	nr	44.69
4 × 3" dia.	31.48	39.49	0.65	20.91	nr	60.40
6 × 4" dia.	114.75	143.94	0.80	25.73	nr	169.67
8 × 6" dia.	177.72	222.94	1.21	38.91	nr	261.85
Elbow, 90°						
½" dia.	1.95	2.44	0.31	9.97	nr	12.41
¾" dia.	2.35	2.95	0.35	11.26	nr	14.21
1" dia.	3.28	4.11	0.42	13.51	nr	17.62
1 ¼" dia.	5.72	7.18	0.45	14.47	nr	21.65
1 ½" dia.	7.36	9.23	0.45	14.47	nr	23.70
2" dia.	10.92	13.70	0.56	18.01	nr	31.71
3" dia.	31.48	39.49	0.65	20.91	nr	60.40
4" dia.	47.42	59.48	0.80	25.73	nr	85.21
6" dia.	187.76	235.52	1.21	38.91	nr	274.43
Elbow 45°						
½" dia.	3.73	4.68	0.31	9.97	nr	14.65
¾" dia.	3.97	4.98	0.35	11.26	nr	16.24
1" dia.	4.82	6.05	0.45	14.47	nr	20.52
1 ¼" dia.	6.90	8.66	0.45	14.47	nr	23.13
1 ½" dia.	8.66	10.86	0.51	16.41	nr	27.27
2" dia.	12.20	15.30	0.56	18.01	nr	33.31
3" dia.	28.73	36.04	0.65	20.91	nr	56.95
4" dia.	59.04	74.05	0.80	25.73	nr	99.78
6" dia.	121.81	152.80	1.21	38.91	nr	191.71
Bend 90° (long radius)						
3" dia.	87.78	110.11	0.65	20.91	nr	131.02
4" dia.	177.31	222.42	0.80	25.73	nr	248.15
6" dia.	389.65	488.78	1.21	38.91	nr	527.69

38 PIPED SUPPLY SYSTEMS

Item	Net Price £	Material £	Labour hours	Labour £	Unit	Total rate £
Bend 45° (long radius)						
1 ½" dia.	20.84	26.14	0.51	16.41	nr	**42.55**
2" dia.	34.06	42.73	0.56	18.01	nr	**60.74**
3" dia.	72.80	91.32	0.65	20.91	nr	**112.23**
4" dia.	141.68	177.72	0.80	25.73	nr	**203.45**
Socket union						
½" dia.	7.57	9.50	0.34	10.94	nr	**20.44**
¾" dia.	8.66	10.86	0.39	12.54	nr	**23.40**
1" dia.	11.24	14.10	0.45	14.47	nr	**28.57**
1 ¼" dia.	13.99	17.55	0.50	16.08	nr	**33.63**
1 ½" dia.	19.16	24.04	0.57	18.32	nr	**42.36**
2" dia.	24.79	31.09	0.62	19.95	nr	**51.04**
3" dia.	92.30	115.79	0.70	22.51	nr	**138.30**
4" dia.	124.97	156.77	0.89	28.63	nr	**185.40**
Saddle plain						
2" × 1 ¼" dia.	19.48	24.44	0.42	13.51	nr	**37.95**
3" × 1 ½" dia.	27.37	34.33	0.48	15.44	nr	**49.77**
4" × 2" dia.	30.83	38.67	0.68	21.87	nr	**60.54**
6" × 2" dia.	36.20	45.40	0.91	29.27	nr	**74.67**
Straight tank connector						
½" dia.	5.03	6.31	0.13	4.18	nr	**10.49**
¾" dia.	5.69	7.13	0.14	4.50	nr	**11.63**
1" dia.	12.09	15.16	0.14	4.50	nr	**19.66**
1 ¼" dia.	30.71	38.53	0.16	5.14	nr	**43.67**
1 ½" dia.	33.67	42.24	0.18	5.79	nr	**48.03**
2" dia.	40.32	50.58	0.24	7.72	nr	**58.30**
3" dia.	41.35	51.87	0.29	9.33	nr	**61.20**
Equal tee						
½" dia.	2.27	2.84	0.44	14.15	nr	**16.99**
¾" dia.	2.88	3.62	0.48	15.44	nr	**19.06**
1" dia.	4.32	5.42	0.54	17.37	nr	**22.79**
1 ¼" dia.	6.12	7.67	0.70	22.51	nr	**30.18**
1 ½" dia.	8.82	11.07	0.74	23.80	nr	**34.87**
2" dia.	13.99	17.55	0.80	25.73	nr	**43.28**
3" dia.	40.54	50.85	1.04	33.45	nr	**84.30**
4" dia.	59.43	74.55	1.28	41.17	nr	**115.72**
6" dia.	207.03	259.69	1.93	62.08	nr	**321.77**

38 PIPED SUPPLY SYSTEMS

Item	Net Price £	Material £	Labour hours	Labour £	Unit	Total rate £
COLD WATER PIPELINES: PVC-C PIPEWORK						
Pipes; solvent welded in the running length, brackets measured separately						
Pipe; 3 m long; PN25						
16 × 2.0 mm	4.56	5.72	0.20	6.44	m	12.16
20 × 2.3 mm	6.89	8.65	0.20	6.44	m	15.09
25 × 2.8 mm	8.93	11.20	0.20	6.44	m	17.64
32 × 3.6 mm	13.27	16.64	0.20	6.44	m	23.08
Pipe; 5 m long; PN25						
40 × 4.5 mm	16.18	20.29	0.20	6.44	m	26.73
50 × 5.6 mm	24.36	30.55	0.20	6.44	m	36.99
63 × 7.0 mm	37.61	47.17	0.20	6.44	m	53.61
Fixings						
Refer to steel pipes; galvanized iron. For minimum fixing dimensions, refer to the Tables and Memoranda at the rear of the book						
Extra over fittings; solvent welded joints						
Straight coupling; PN25						
16 mm	0.71	0.90	0.20	6.44	nr	7.34
20 mm	1.00	1.25	0.20	6.44	nr	7.69
25 mm	1.24	1.56	0.20	6.44	nr	8.00
32 mm	3.79	4.75	0.20	6.44	nr	11.19
40 mm	4.86	6.09	0.20	6.44	nr	12.53
50 mm	6.51	8.16	0.20	6.44	nr	14.60
63 mm	11.50	14.43	0.20	6.44	nr	20.87
Elbow; 90°; PN25						
16 mm	1.14	1.43	0.20	6.44	nr	7.87
20 mm	1.75	2.20	0.20	6.44	nr	8.64
25 mm	2.17	2.72	0.20	6.44	nr	9.16
32 mm	4.53	5.68	0.20	6.44	nr	12.12
40 mm	6.98	8.76	0.20	6.44	nr	15.20
50 mm	9.66	12.12	0.20	6.44	nr	18.56
63 mm	16.52	20.72	0.20	6.44	nr	27.16
Elbow; 45°; PN25						
20 mm	1.75	2.20	0.20	6.44	nr	8.64
25 mm	2.17	2.72	0.20	6.44	nr	9.16
32 mm	4.53	5.68	0.20	6.44	nr	12.12
40 mm	6.98	8.76	0.20	6.44	nr	15.20
50 mm	9.66	12.12	0.20	6.44	nr	18.56
63 mm	16.52	20.72	0.20	6.44	nr	27.16
Reducer fitting; single stage reduction						
20/16 mm	1.24	1.56	0.20	6.44	nr	8.00
25/20 mm	1.49	1.87	0.20	6.44	nr	8.31
32/25 mm	2.99	3.75	0.20	6.44	nr	10.19
40/32 mm	3.91	4.91	0.20	6.44	nr	11.35
50/40 mm	4.53	5.68	0.20	6.44	nr	12.12
63/50 mm	6.86	8.60	0.20	6.44	nr	15.04

38 PIPED SUPPLY SYSTEMS

Item	Net Price £	Material £	Labour hours	Labour £	Unit	Total rate £
Equal tee; 90°; PN25						
16 mm	1.89	2.37	0.20	6.44	nr	**8.81**
20 mm	2.61	3.27	0.20	6.44	nr	**9.71**
25 mm	3.31	4.16	0.20	6.44	nr	**10.60**
32 mm	5.38	6.75	0.20	6.44	nr	**13.19**
40 mm	9.30	11.67	0.20	6.44	nr	**18.11**
50 mm	13.93	17.47	0.20	6.44	nr	**23.91**
63 mm	23.48	29.46	0.20	6.44	nr	**35.90**
Cap; PN25						
20 mm	1.30	1.64	0.20	6.44	nr	**8.08**
25 mm	1.75	2.20	0.20	6.44	nr	**8.64**
32 mm	2.53	3.17	0.20	6.44	nr	**9.61**
40 mm	3.48	4.37	0.20	6.44	nr	**10.81**
50 mm	4.86	6.09	0.20	6.44	nr	**12.53**
63 mm	7.75	9.72	0.20	6.44	nr	**16.16**

38 PIPED SUPPLY SYSTEMS

Item	Net Price £	Material £	Labour hours	Labour £	Unit	Total rate £
COLD WATER PIPELINES: SCREWED STEEL PIPEWORK						
Galvanized steel pipes; screwed and socketed joints; BS 1387: 1985						
Galvanized; medium, fixed vertically, with brackets measured separately, screwed joints are within the running length, but any flanges are additional						
10 mm dia.	4.86	6.09	0.51	16.41	m	**22.50**
15 mm dia.	4.38	5.50	0.52	16.72	m	**22.22**
20 mm dia.	4.95	6.20	0.55	17.68	m	**23.88**
25 mm dia.	6.92	8.68	0.60	19.30	m	**27.98**
32 mm dia.	8.57	10.75	0.67	21.55	m	**32.30**
40 mm dia.	9.95	12.48	0.75	24.12	m	**36.60**
50 mm dia.	13.96	17.52	0.85	27.33	m	**44.85**
65 mm dia.	18.92	23.73	0.93	29.92	m	**53.65**
80 mm dia.	24.50	30.73	1.07	34.41	m	**65.14**
100 mm dia.	34.67	43.49	1.46	46.95	m	**90.44**
125 mm dia.	55.13	69.16	1.72	55.32	m	**124.48**
150 mm dia.	64.01	80.29	1.96	63.03	m	**143.32**
Galvanized; heavy, fixed vertically, with brackets measured separately, screwed joints are within the running length, but any flanges are additional						
15 mm dia.	5.20	6.52	0.52	16.72	m	**23.24**
20 mm dia.	5.87	7.36	0.55	17.68	m	**25.04**
25 mm dia.	8.39	10.53	0.60	19.30	m	**29.83**
32 mm dia.	10.41	13.06	0.67	21.55	m	**34.61**
40 mm dia.	12.15	15.24	0.75	24.12	m	**39.36**
50 mm dia.	16.84	21.12	0.85	27.33	m	**48.45**
65 mm dia.	22.87	28.68	0.93	29.92	m	**58.60**
80 mm dia.	29.05	36.44	1.07	34.41	m	**70.85**
100 mm dia.	40.50	50.80	1.46	46.95	m	**97.75**
125 mm dia.	58.69	73.62	1.72	55.32	m	**128.94**
150 mm dia.	68.64	86.11	1.96	63.03	m	**149.14**
Galvanized; medium, fixed horizontaly or suspended at high level, with brackets measured separately, screwed joints are within the running length, but any flanges are additional						
10 mm dia.	4.86	6.09	0.51	16.41	m	**22.50**
15 mm dia.	4.38	5.50	0.52	16.72	m	**22.22**
20 mm dia.	4.95	6.20	0.55	17.68	m	**23.88**
25 mm dia.	6.92	8.68	0.60	19.30	m	**27.98**
32 mm dia.	8.57	10.75	0.67	21.55	m	**32.30**
40 mm dia.	9.95	12.48	0.75	24.12	m	**36.60**
50 mm dia.	13.96	17.52	0.85	27.33	m	**44.85**
65 mm dia.	18.92	23.73	0.93	29.92	m	**53.65**
80 mm dia.	24.50	30.73	1.07	34.41	m	**65.14**
100 mm dia.	34.67	43.49	1.46	46.95	m	**90.44**

38 PIPED SUPPLY SYSTEMS

Item	Net Price £	Material £	Labour hours	Labour £	Unit	Total rate £
125 mm dia.	55.13	69.16	1.72	55.32	m	**124.48**
150 mm dia.	64.01	80.29	1.96	63.03	m	**143.32**
Galvanized; heavy, fixed horizontaly or suspended at high level, with brackets measured separately, screwed joints are within the running length, but any flanges are additional						
15 mm dia.	5.20	6.52	0.52	16.72	m	**23.24**
20 mm dia.	5.87	7.36	0.55	17.68	m	**25.04**
25 mm dia.	8.39	10.53	0.60	19.30	m	**29.83**
32 mm dia.	10.41	13.06	0.67	21.55	m	**34.61**
40 mm dia.	12.15	15.24	0.75	24.12	m	**39.36**
50 mm dia.	16.84	21.12	0.85	27.33	m	**48.45**
65 mm dia.	22.87	28.68	0.93	29.92	m	**58.60**
80 mm dia.	29.05	36.44	1.07	34.41	m	**70.85**
100 mm dia.	40.50	50.80	1.46	46.95	m	**97.75**
125 mm dia.	58.69	73.62	1.72	55.32	m	**128.94**
150 mm dia.	68.64	86.11	1.96	63.03	m	**149.14**
Fixings						
For steel pipes; galvanized iron. For minimum fixing dimensions, refer to the Tables and Memoranda at the rear of the book						
Single pipe bracket, screw on, galvanized iron; screwed to wood						
15 mm dia.	1.27	1.59	0.14	4.50	nr	**6.09**
20 mm dia.	1.42	1.78	0.14	4.50	nr	**6.28**
25 mm dia.	1.65	2.07	0.17	5.47	nr	**7.54**
32 mm dia.	2.25	2.82	0.19	6.10	nr	**8.92**
40 mm dia.	3.35	4.20	0.22	7.08	nr	**11.28**
50 mm dia.	4.44	5.57	0.22	7.08	nr	**12.65**
65 mm dia.	5.24	6.57	0.28	9.00	nr	**15.57**
80 mm dia.	8.21	10.30	0.32	10.28	nr	**20.58**
100 mm dia.	11.86	14.87	0.35	11.26	nr	**26.13**
Single pipe bracket, screw on, galvanized iron; plugged and screwed						
15 mm dia.	1.27	1.59	0.25	8.04	nr	**9.63**
20 mm dia.	1.42	1.78	0.25	8.04	nr	**9.82**
25 mm dia.	1.65	2.07	0.30	9.64	nr	**11.71**
32 mm dia.	2.25	2.82	0.32	10.28	nr	**13.10**
40 mm dia.	3.35	4.20	0.32	10.28	nr	**14.48**
50 mm dia.	4.44	5.57	0.32	10.28	nr	**15.85**
65 mm dia.	5.24	6.57	0.35	11.26	nr	**17.83**
80 mm dia.	8.21	10.30	0.42	13.51	nr	**23.81**
100 mm dia.	11.86	14.87	0.42	13.51	nr	**28.38**

38 PIPED SUPPLY SYSTEMS

Item	Net Price £	Material £	Labour hours	Labour £	Unit	Total rate £
COLD WATER PIPELINES: SCREWED STEEL PIPEWORK – cont						
For steel pipes – cont						
Single pipe bracket for building in, galvanized iron						
15 mm dia.	1.35	1.69	0.10	3.21	nr	**4.90**
20 mm dia.	1.51	1.89	0.11	3.54	nr	**5.43**
25 mm dia.	1.65	2.07	0.12	3.86	nr	**5.93**
32 mm dia.	1.75	2.20	0.14	4.50	nr	**6.70**
40 mm dia.	2.25	2.82	0.15	4.83	nr	**7.65**
50 mm dia.	2.72	3.42	0.16	5.14	nr	**8.56**
Pipe ring, single socket, galvanized iron						
15 mm dia.	1.35	1.69	0.10	3.21	nr	**4.90**
20 mm dia.	1.51	1.89	0.11	3.54	nr	**5.43**
25 mm dia.	1.65	2.07	0.12	3.86	nr	**5.93**
32 mm dia.	1.66	2.08	0.15	4.83	nr	**6.91**
40 mm dia.	2.17	2.72	0.15	4.83	nr	**7.55**
50 mm dia.	2.84	3.56	0.16	5.14	nr	**8.70**
65 mm dia.	3.96	4.97	0.30	9.64	nr	**14.61**
80 mm dia.	4.66	5.85	0.35	11.26	nr	**17.11**
100 mm dia.	7.49	9.40	0.40	12.87	nr	**22.27**
125 mm dia.	15.14	19.00	0.60	19.31	nr	**38.31**
150 mm dia.	18.28	22.93	0.77	24.76	nr	**47.69**
Pipe ring, double socket, galvanized iron						
15 mm dia.	11.91	14.94	0.10	3.21	nr	**18.15**
20 mm dia.	13.71	17.20	0.11	3.54	nr	**20.74**
25 mm dia.	15.11	18.95	0.12	3.86	nr	**22.81**
32 mm dia.	16.66	20.90	0.14	4.50	nr	**25.40**
40 mm dia.	21.04	26.39	0.15	4.83	nr	**31.22**
50 mm dia.	23.94	30.03	0.16	5.14	nr	**35.17**
Screw on backplate (Male), galvanized iron; plugged and screwed						
All sizes 15 mm to 50 mm × M12	1.06	1.33	0.10	3.21	nr	**4.54**
Screw on backplate (Female), galvanized iron; plugged and screwed						
All sizes 15 mm to 50 mm × M12	1.06	1.33	0.10	3.21	nr	**4.54**
Extra over channel sections for fabricated hangers and brackets						
Galvanized steel; including inserts, bolts, nuts, washers; fixed to backgrounds						
41 × 21 mm	6.34	7.95	0.29	9.33	m	**17.28**
41 × 41 mm	7.60	9.53	0.29	9.33	m	**18.86**
Threaded rods; metric thread; including nuts, washers etc.						
10 mm dia. × 600 mm long	2.07	2.60	0.18	5.79	nr	**8.39**
12 mm dia. × 600 mm long	3.20	4.01	0.18	5.79	nr	**9.80**

38 PIPED SUPPLY SYSTEMS

Item	Net Price £	Material £	Labour hours	Labour £	Unit	Total rate £
Extra over steel flanges, screwed and drilled; metric; BS 4504						
Screwed flanges; PN6						
15 mm dia.	16.67	20.91	0.35	11.26	nr	32.17
20 mm dia.	16.67	20.91	0.47	15.12	nr	36.03
25 mm dia.	16.67	20.91	0.53	17.05	nr	37.96
32 mm dia.	16.67	20.91	0.62	19.95	nr	40.86
40 mm dia.	16.67	20.91	0.70	22.51	nr	43.42
50 mm dia.	17.78	22.30	0.84	27.01	nr	49.31
65 mm dia.	24.72	31.01	1.03	33.13	nr	64.14
80 mm dia.	34.91	43.79	1.23	39.56	nr	83.35
100 mm dia.	41.28	51.78	1.41	45.35	nr	97.13
125 mm dia.	75.72	94.99	1.77	56.93	nr	151.92
150 mm dia.	75.72	94.99	2.21	71.08	nr	166.07
Screwed flanges; PN16						
15 mm dia.	21.20	26.59	0.35	11.26	nr	37.85
20 mm dia.	21.20	26.59	0.47	15.12	nr	41.71
25 mm dia.	21.20	26.59	0.53	17.05	nr	43.64
32 mm dia.	22.64	28.40	0.62	19.95	nr	48.35
40 mm dia.	22.64	28.40	0.70	22.51	nr	50.91
50 mm dia.	27.62	34.64	0.84	27.01	nr	61.65
65 mm dia.	34.49	43.27	1.03	33.13	nr	76.40
80 mm dia.	42.00	52.68	1.23	39.56	nr	92.24
100 mm dia.	47.11	59.09	1.41	45.35	nr	104.44
125 mm dia.	82.46	103.44	1.77	56.93	nr	160.37
150 mm dia.	81.13	101.77	2.21	71.08	nr	172.85
Extra over steel flanges, screwed and drilled; imperial; BS 10						
Screwed flanges; table E						
½" dia.	28.48	35.73	0.35	11.26	nr	46.99
¾" dia.	28.48	35.73	0.47	15.12	nr	50.85
1" dia.	28.48	35.73	0.53	17.05	nr	52.78
1 ¼" dia.	28.48	35.73	0.62	19.95	nr	55.68
1 ½" dia.	28.48	35.73	0.70	22.51	nr	58.24
2" dia.	28.48	35.73	0.84	27.01	nr	62.74
2 ½" dia.	33.90	42.53	1.03	33.13	nr	75.66
3" dia.	40.84	51.23	1.23	39.56	nr	90.79
4" dia.	51.96	65.18	1.41	45.35	nr	110.53
5" dia.	110.14	138.16	1.77	56.93	nr	195.09
Extra over steel flange connections						
Bolted connection between pair of flanges; including gasket, bolts, nuts and washers						
50 mm dia.	49.14	61.64	0.53	17.05	nr	78.69
65 mm dia.	61.91	77.66	0.53	17.05	nr	94.71
80 mm dia.	71.03	89.10	0.53	17.05	nr	106.15
100 mm dia.	84.76	106.32	0.53	17.05	nr	123.37
125 mm dia.	160.09	200.82	0.61	19.62	nr	220.44
150 mm dia.	164.66	206.55	0.90	28.95	nr	235.50

38 PIPED SUPPLY SYSTEMS

Item	Net Price £	Material £	Labour hours	Labour £	Unit	Total rate £
COLD WATER PIPELINES: SCREWED STEEL PIPEWORK – cont						
Extra over heavy steel tubular fittings; BS 1387						
Long screw connection with socket and backnut						
15 mm dia.	6.57	8.24	0.63	20.26	nr	28.50
20 mm dia.	8.25	10.35	0.84	27.01	nr	37.36
25 mm dia.	10.79	13.53	0.95	30.55	nr	44.08
32 mm dia.	14.20	17.81	1.11	35.71	nr	53.52
40 mm dia.	17.29	21.68	1.28	41.17	nr	62.85
50 mm dia.	25.51	32.00	1.53	49.21	nr	81.21
65 mm dia.	58.61	73.52	1.87	60.14	nr	133.66
80 mm dia.	75.80	95.09	2.21	71.08	nr	166.17
100 mm dia.	85.99	107.87	3.05	98.09	nr	205.96
Running nipple						
15 mm dia.	1.65	2.07	0.50	16.08	nr	18.15
20 mm dia.	2.05	2.58	0.68	21.87	nr	24.45
25 mm dia.	2.20	2.76	0.77	24.76	nr	27.52
32 mm dia.	3.55	4.46	0.90	28.95	nr	33.41
40 mm dia.	4.80	6.03	1.03	33.13	nr	39.16
50 mm dia.	7.29	9.14	1.23	39.56	nr	48.70
65 mm dia.	15.67	19.66	1.50	48.25	nr	67.91
80 mm dia.	24.44	30.65	1.78	57.25	nr	87.90
100 mm dia.	38.30	48.05	2.38	76.54	nr	124.59
Barrel nipple						
15 mm dia.	1.38	1.74	0.50	16.08	nr	17.82
20 mm dia.	2.08	2.61	0.68	21.87	nr	24.48
25 mm dia.	2.33	2.92	0.77	24.76	nr	27.68
32 mm dia.	3.86	4.84	0.90	28.95	nr	33.79
40 mm dia.	4.30	5.40	1.03	33.13	nr	38.53
50 mm dia.	6.15	7.72	1.23	39.56	nr	47.28
65 mm dia.	13.14	16.49	1.50	48.25	nr	64.74
80 mm dia.	18.34	23.00	1.78	57.25	nr	80.25
100 mm dia.	33.17	41.61	2.38	76.54	nr	118.15
125 mm dia.	61.62	77.29	2.87	92.31	nr	169.60
150 mm dia.	97.08	121.78	3.39	109.03	nr	230.81
Close taper nipple						
15 mm dia.	1.95	2.44	0.50	16.08	nr	18.52
20 mm dia.	2.53	3.17	0.68	21.87	nr	25.04
25 mm dia.	3.32	4.17	0.77	24.76	nr	28.93
32 mm dia.	4.95	6.20	0.90	28.95	nr	35.15
40 mm dia.	6.13	7.69	1.03	33.13	nr	40.82
50 mm dia.	9.42	11.82	1.23	39.56	nr	51.38
65 mm dia.	14.87	18.65	1.50	48.25	nr	66.90
80 mm dia.	24.38	30.59	1.78	57.25	nr	87.84
100 mm dia.	46.35	58.14	2.38	76.54	nr	134.68

38 PIPED SUPPLY SYSTEMS

Item	Net Price £	Material £	Labour hours	Labour £	Unit	Total rate £
90° bend with socket						
15 mm dia.	5.38	6.75	0.64	20.59	nr	27.34
20 mm dia.	7.24	9.08	0.85	27.33	nr	36.41
25 mm dia.	11.08	13.90	0.97	31.19	nr	45.09
32 mm dia.	15.89	19.94	1.12	36.03	nr	55.97
40 mm dia.	19.41	24.35	1.29	41.50	nr	65.85
50 mm dia.	30.18	37.86	1.55	49.85	nr	87.71
65 mm dia.	60.85	76.33	1.89	60.79	nr	137.12
80 mm dia.	90.32	113.30	2.24	72.04	nr	185.34
100 mm dia.	160.16	200.91	3.09	99.39	nr	300.30
125 mm dia.	392.29	492.08	3.92	126.08	nr	618.16
150 mm dia.	588.96	738.80	4.74	152.44	nr	891.24
Extra over heavy steel fittings; BS 1740						
Plug						
15 mm dia.	1.49	1.87	0.28	9.00	nr	10.87
20 mm dia.	2.32	2.91	0.38	12.22	nr	15.13
25 mm dia.	4.08	5.12	0.44	14.15	nr	19.27
32 mm dia.	6.33	7.94	0.51	16.41	nr	24.35
40 mm dia.	6.98	8.76	0.59	18.98	nr	27.74
50 mm dia.	9.98	12.52	0.70	22.51	nr	35.03
65 mm dia.	23.86	29.93	0.85	27.33	nr	57.26
80 mm dia.	44.65	56.01	1.00	32.17	nr	88.18
100 mm dia.	85.72	107.53	1.44	46.31	nr	153.84
Socket						
15 mm dia.	1.65	2.07	0.64	20.59	nr	22.66
20 mm dia.	1.87	2.34	0.85	27.33	nr	29.67
25 mm dia.	2.65	3.33	0.97	31.19	nr	34.52
32 mm dia.	3.82	4.79	1.12	36.03	nr	40.82
40 mm dia.	4.63	5.81	1.29	41.50	nr	47.31
50 mm dia.	7.15	8.97	1.55	49.85	nr	58.82
65 mm dia.	14.19	17.80	1.89	60.79	nr	78.59
80 mm dia.	18.36	23.03	2.24	72.04	nr	95.07
100 mm dia.	34.56	43.36	3.09	99.39	nr	142.75
150 mm dia.	82.50	103.49	4.74	152.44	nr	255.93
Elbow, female/female						
15 mm dia.	9.49	11.91	0.64	20.59	nr	32.50
20 mm dia.	12.37	15.51	0.85	27.33	nr	42.84
25 mm dia.	16.79	21.06	0.97	31.19	nr	52.25
32 mm dia.	31.25	39.20	1.12	36.03	nr	75.23
40 mm dia.	37.27	46.75	1.29	41.50	nr	88.25
50 mm dia.	61.10	76.64	1.55	49.85	nr	126.49
65 mm dia.	149.28	187.25	1.89	60.79	nr	248.04
80 mm dia.	178.00	223.28	2.24	72.04	nr	295.32
100 mm dia.	308.26	386.68	3.09	99.39	nr	486.07
Equal tee						
15 mm dia.	11.80	14.81	0.91	29.27	nr	44.08
20 mm dia.	13.70	17.18	1.22	39.23	nr	56.41
25 mm dia.	20.19	25.32	1.40	45.04	nr	70.36

38 PIPED SUPPLY SYSTEMS

Item	Net Price £	Material £	Labour hours	Labour £	Unit	Total rate £
COLD WATER PIPELINES: SCREWED STEEL PIPEWORK – cont						
Extra over heavy steel fittings – cont						
Equal tee – cont						
32 mm dia.	41.71	52.33	1.62	52.10	nr	104.43
40 mm dia.	45.39	56.94	1.86	59.82	nr	116.76
50 mm dia.	73.81	92.59	2.21	71.08	nr	163.67
65 mm dia.	178.50	223.91	2.72	87.48	nr	311.39
80 mm dia.	191.57	240.31	3.21	103.24	nr	343.55
100 mm dia.	308.31	386.75	4.44	142.80	nr	529.55
Extra over malleable iron fittings; BS 143						
Cap						
15 mm dia.	3.70	4.64	0.32	10.28	nr	14.92
20 mm dia.	3.86	4.84	0.43	13.83	nr	18.67
25 mm dia.	6.99	8.77	0.49	15.76	nr	24.53
32 mm dia.	13.27	16.64	0.58	18.65	nr	35.29
40 mm dia.	14.24	17.86	0.66	21.22	nr	39.08
50 mm dia.	22.64	28.40	0.78	25.09	nr	53.49
65 mm dia.	38.91	48.81	0.96	30.87	nr	79.68
80 mm dia.	64.19	80.52	1.13	36.34	nr	116.86
100 mm dia.	109.43	137.27	1.70	54.68	nr	191.95
Plain plug, hollow						
15 mm dia.	1.80	2.26	0.28	9.00	nr	11.26
20 mm dia.	2.79	3.49	0.38	12.22	nr	15.71
25 mm dia.	4.89	6.14	0.44	14.15	nr	20.29
32 mm dia.	7.58	9.51	0.51	16.41	nr	25.92
40 mm dia.	8.37	10.49	0.59	18.98	nr	29.47
50 mm dia.	11.99	15.04	0.70	22.51	nr	37.55
65 mm dia.	28.63	35.92	0.85	27.33	nr	63.25
80 mm dia.	53.58	67.21	1.00	32.17	nr	99.38
100 mm dia.	102.88	129.06	1.44	46.31	nr	175.37
Plain plug, solid						
15 mm dia.	2.23	2.80	0.29	9.33	nr	12.13
20 mm dia.	2.37	2.97	0.38	12.22	nr	15.19
25 mm dia.	3.15	3.95	0.44	14.15	nr	18.10
32 mm dia.	4.85	6.08	0.51	16.41	nr	22.49
40 mm dia.	6.53	8.19	0.59	18.98	nr	27.17
50 mm dia.	8.58	10.76	0.70	22.51	nr	33.27
Elbow, male/female						
15 mm dia.	1.16	1.46	0.64	20.59	nr	22.05
20 mm dia.	1.56	1.96	0.85	27.33	nr	29.29
25 mm dia.	2.59	3.25	0.97	31.19	nr	34.44
32 mm dia.	5.87	7.36	1.12	36.03	nr	43.39
40 mm dia.	8.12	10.18	1.29	41.50	nr	51.68
50 mm dia.	10.47	13.14	1.55	49.85	nr	62.99
65 mm dia.	23.32	29.25	1.89	60.79	nr	90.04
80 mm dia.	31.89	40.01	2.24	72.04	nr	112.05
100 mm dia.	55.76	69.94	3.09	99.39	nr	169.33

38 PIPED SUPPLY SYSTEMS

Item	Net Price £	Material £	Labour hours	Labour £	Unit	Total rate £
Elbow						
15 mm dia.	1.04	1.30	0.64	20.59	nr	**21.89**
20 mm dia.	1.43	1.79	0.85	27.33	nr	**29.12**
25 mm dia.	2.21	2.78	0.97	31.19	nr	**33.97**
32 mm dia.	4.61	5.78	1.12	36.03	nr	**41.81**
40 mm dia.	6.89	8.65	1.29	41.50	nr	**50.15**
50 mm dia.	8.08	10.14	1.55	49.85	nr	**59.99**
65 mm dia.	18.01	22.59	1.89	60.79	nr	**83.38**
80 mm dia.	26.45	33.17	2.24	72.04	nr	**105.21**
100 mm dia.	45.41	56.96	3.09	99.39	nr	**156.35**
125 mm dia.	108.97	136.70	4.44	142.80	nr	**279.50**
150 mm dia.	202.88	254.50	5.79	186.23	nr	**440.73**
45° elbow						
15 mm dia.	2.68	3.36	0.64	20.59	nr	**23.95**
20 mm dia.	3.32	4.17	0.85	27.33	nr	**31.50**
25 mm dia.	4.56	5.72	0.97	31.19	nr	**36.91**
32 mm dia.	10.61	13.31	1.12	36.03	nr	**49.34**
40 mm dia.	12.49	15.67	1.29	41.50	nr	**57.17**
50 mm dia.	17.11	21.46	1.55	49.85	nr	**71.31**
65 mm dia.	24.05	30.17	1.89	60.79	nr	**90.96**
80 mm dia.	36.18	45.38	2.24	72.04	nr	**117.42**
100 mm dia.	69.75	87.49	3.09	99.39	nr	**186.88**
150 mm dia.	212.27	266.27	5.79	186.23	nr	**452.50**
Bend, male/female						
15 mm dia.	2.05	2.58	0.64	20.59	nr	**23.17**
20 mm dia.	3.37	4.22	0.85	27.33	nr	**31.55**
25 mm dia.	4.74	5.95	0.97	31.19	nr	**37.14**
32 mm dia.	8.01	10.05	1.12	36.03	nr	**46.08**
40 mm dia.	11.75	14.74	1.29	41.50	nr	**56.24**
50 mm dia.	22.08	27.70	1.55	49.85	nr	**77.55**
65 mm dia.	33.79	42.38	1.89	60.79	nr	**103.17**
80 mm dia.	52.03	65.26	2.24	72.04	nr	**137.30**
100 mm dia.	128.84	161.62	3.09	99.39	nr	**261.01**
Bend, male						
15 mm dia.	4.70	5.89	0.64	20.59	nr	**26.48**
20 mm dia.	5.28	6.62	0.85	27.33	nr	**33.95**
25 mm dia.	7.75	9.72	0.97	31.19	nr	**40.91**
32 mm dia.	17.14	21.50	1.12	36.03	nr	**57.53**
40 mm dia.	24.04	30.15	1.29	41.50	nr	**71.65**
50 mm dia.	36.47	45.75	1.55	49.85	nr	**95.60**
Bend, female						
15 mm dia.	2.11	2.64	0.64	20.59	nr	**23.23**
20 mm dia.	3.01	3.77	0.85	27.33	nr	**31.10**
25 mm dia.	4.22	5.30	0.97	31.19	nr	**36.49**
32 mm dia.	8.23	10.33	1.12	36.03	nr	**46.36**
40 mm dia.	9.80	12.30	1.29	41.50	nr	**53.80**
50 mm dia.	15.43	19.35	1.55	49.85	nr	**69.20**
65 mm dia.	33.79	42.38	1.89	60.79	nr	**103.17**
80 mm dia.	50.10	62.84	2.24	72.04	nr	**134.88**
100 mm dia.	105.12	131.86	3.09	99.39	nr	**231.25**
125 mm dia.	212.08	266.03	4.44	142.80	nr	**408.83**
150 mm dia.	465.91	584.44	5.79	186.23	nr	**770.67**

38 PIPED SUPPLY SYSTEMS

Item	Net Price £	Material £	Labour hours	Labour £	Unit	Total rate £
COLD WATER PIPELINES: SCREWED STEEL PIPEWORK – cont						
Extra over malleable iron fittings – cont						
Return bend						
15 mm dia.	10.76	13.50	0.64	20.59	nr	34.09
20 mm dia.	17.39	21.82	0.85	27.33	nr	49.15
25 mm dia.	21.70	27.22	0.97	31.19	nr	58.41
32 mm dia.	30.13	37.80	1.12	36.03	nr	73.83
40 mm dia.	40.81	51.20	1.29	41.50	nr	92.70
50 mm dia.	62.30	78.15	1.55	49.85	nr	128.00
Equal socket, parallel thread						
15 mm dia.	1.01	1.27	0.64	20.59	nr	21.86
20 mm dia.	1.32	1.66	0.85	27.33	nr	28.99
25 mm dia.	1.67	2.09	0.97	31.19	nr	33.28
32 mm dia.	3.72	4.67	1.12	36.03	nr	40.70
40 mm dia.	5.04	6.32	1.29	41.50	nr	47.82
50 mm dia.	7.29	9.14	1.55	49.85	nr	58.99
65 mm dia.	11.57	14.52	1.89	60.79	nr	75.31
80 mm dia.	16.39	20.56	2.24	72.04	nr	92.60
100 mm dia.	27.06	33.95	3.09	99.39	nr	133.34
Concentric reducing socket						
20 × 15 mm dia.	1.61	2.02	0.76	24.45	nr	26.47
25 × 15 mm dia.	2.11	2.64	0.86	27.66	nr	30.30
25 × 20 mm dia.	1.99	2.50	0.86	27.66	nr	30.16
32 × 25 mm dia.	3.82	4.79	1.01	32.49	nr	37.28
40 × 25 mm dia.	5.04	6.32	1.16	37.31	nr	43.63
40 × 32 mm dia.	5.59	7.01	1.16	37.31	nr	44.32
50 × 25 mm dia.	9.67	12.13	1.38	44.39	nr	56.52
50 × 40 mm dia.	7.83	9.82	1.38	44.39	nr	54.21
65 × 50 mm dia.	13.66	17.14	1.69	54.35	nr	71.49
80 × 50 mm dia.	17.04	21.37	2.00	64.32	nr	85.69
100 × 50 mm dia.	33.96	42.60	2.75	88.45	nr	131.05
100 × 80 mm dia.	101.60	127.44	2.75	88.45	nr	215.89
150 × 100 mm dia.	94.41	118.43	4.10	131.86	nr	250.29
Eccentric reducing socket						
20 × 15 mm dia.	3.66	4.59	0.76	24.45	nr	29.04
25 × 15 mm dia.	10.37	13.00	0.86	27.66	nr	40.66
25 × 20 mm dia.	11.78	14.77	0.86	27.66	nr	42.43
32 × 25 mm dia.	15.22	19.10	1.01	32.49	nr	51.59
40 × 25 mm dia.	17.44	21.87	1.16	37.31	nr	59.18
40 × 32 mm dia.	18.99	23.82	1.16	37.31	nr	61.13
50 × 25 mm dia.	19.04	23.88	1.18	37.96	nr	61.84
50 × 40 mm dia.	19.09	23.95	1.28	41.17	nr	65.12
65 × 50 mm dia.	19.34	24.26	1.69	54.35	nr	78.61
80 × 50 mm dia.	20.60	25.84	2.00	64.32	nr	90.16
Hexagon bush						
20 × 15 mm dia.	0.90	1.13	0.37	11.91	nr	13.04
25 × 15 mm dia.	1.24	1.56	0.43	13.83	nr	15.39
25 × 20 mm dia.	1.16	1.46	0.43	13.83	nr	15.29

38 PIPED SUPPLY SYSTEMS

Item	Net Price £	Material £	Labour hours	Labour £	Unit	Total rate £
32 × 25 mm dia.	1.57	1.97	0.51	16.41	nr	18.38
40 × 25 mm dia.	2.35	2.95	0.58	18.65	nr	21.60
40 × 32 mm dia.	2.35	2.95	0.58	18.65	nr	21.60
50 × 25 mm dia.	4.97	6.24	0.71	22.83	nr	29.07
50 × 40 mm dia.	4.63	5.81	0.71	22.83	nr	28.64
65 × 50 mm dia.	8.55	10.73	0.84	27.01	nr	37.74
80 × 50 mm dia.	12.90	16.18	1.00	32.17	nr	48.35
100 × 50 mm dia.	29.89	37.50	1.52	48.89	nr	86.39
100 × 80 mm dia.	24.86	31.18	1.52	48.89	nr	80.07
150 × 100 mm dia.	89.34	112.07	2.48	79.77	nr	191.84
Hexagon nipple						
15 mm dia.	0.99	1.24	0.28	9.00	nr	10.24
20 mm dia.	1.11	1.39	0.38	12.22	nr	13.61
25 mm dia.	1.56	1.96	0.44	14.15	nr	16.11
32 mm dia.	3.33	4.18	0.51	16.41	nr	20.59
40 mm dia.	3.82	4.79	0.59	18.98	nr	23.77
50 mm dia.	6.96	8.74	0.70	22.51	nr	31.25
65 mm dia.	11.66	14.63	0.85	27.33	nr	41.96
80 mm dia.	16.12	20.22	1.00	32.17	nr	52.39
100 mm dia.	28.61	35.88	1.44	46.31	nr	82.19
150 mm dia.	79.19	99.33	2.32	74.61	nr	173.94
Union, male/female						
15 mm dia.	4.89	6.14	0.64	20.59	nr	26.73
20 mm dia.	6.00	7.53	0.85	27.33	nr	34.86
25 mm dia.	6.96	8.74	0.97	31.19	nr	39.93
32 mm dia.	12.31	15.44	1.12	36.03	nr	51.47
40 mm dia.	15.75	19.76	1.29	41.50	nr	61.26
50 mm dia.	24.76	31.06	1.55	49.85	nr	80.91
65 mm dia.	55.42	69.52	1.89	60.79	nr	130.31
Union, female						
15 mm dia.	4.63	5.81	0.64	20.59	nr	26.40
20 mm dia.	5.09	6.38	0.85	27.33	nr	33.71
25 mm dia.	5.97	7.49	0.97	31.19	nr	38.68
32 mm dia.	10.29	12.90	1.12	36.03	nr	48.93
40 mm dia.	11.62	14.57	1.29	41.50	nr	56.07
50 mm dia.	17.32	21.73	1.55	49.85	nr	71.58
65 mm dia.	44.37	55.65	1.89	60.79	nr	116.44
80 mm dia.	58.65	73.57	2.24	72.04	nr	145.61
100 mm dia.	111.68	140.09	3.09	99.39	nr	239.48
Union elbow, male/female						
15 mm dia.	7.38	9.26	0.64	20.59	nr	29.85
20 mm dia.	9.25	11.60	0.85	27.33	nr	38.93
25 mm dia.	13.00	16.31	0.97	31.19	nr	47.50
Twin elbow						
15 mm dia.	7.10	8.90	0.91	29.27	nr	38.17
20 mm dia.	7.85	9.84	1.22	39.23	nr	49.07
25 mm dia.	12.72	15.96	1.39	44.71	nr	60.67
32 mm dia.	25.48	31.96	1.62	52.10	nr	84.06
40 mm dia.	32.25	40.45	1.86	59.82	nr	100.27
50 mm dia.	41.46	52.01	2.21	71.08	nr	123.09
65 mm dia.	67.00	84.04	2.72	87.48	nr	171.52
80 mm dia.	76.04	95.38	3.21	103.24	nr	198.62

38 PIPED SUPPLY SYSTEMS

Item	Net Price £	Material £	Labour hours	Labour £	Unit	Total rate £
COLD WATER PIPELINES: SCREWED STEEL PIPEWORK – cont						
Extra over malleable iron fittings – cont						
Equal tee						
15 mm dia.	1.43	1.79	0.91	29.27	nr	31.06
20 mm dia.	2.08	2.61	1.22	39.23	nr	41.84
25 mm dia.	2.99	3.75	1.39	44.71	nr	48.46
32 mm dia.	6.32	7.93	1.62	52.10	nr	60.03
40 mm dia.	8.63	10.83	1.86	59.82	nr	70.65
50 mm dia.	12.43	15.59	2.21	71.08	nr	86.67
65 mm dia.	29.15	36.57	2.72	87.48	nr	124.05
80 mm dia.	33.96	42.60	3.21	103.24	nr	145.84
100 mm dia.	61.58	77.25	4.44	142.80	nr	220.05
125 mm dia.	114.23	143.29	5.38	173.03	nr	316.32
150 mm dia.	273.17	342.66	6.31	202.94	nr	545.60
Tee reducing on branch						
20 × 15 mm dia.	2.13	2.68	1.22	39.23	nr	41.91
25 × 15 mm dia.	2.91	3.65	1.39	44.71	nr	48.36
25 × 20 mm dia.	3.32	4.17	1.39	44.71	nr	48.88
32 × 25 mm dia.	6.40	8.03	1.62	52.10	nr	60.13
40 × 25 mm dia.	8.12	10.18	1.86	59.82	nr	70.00
40 × 32 mm dia.	11.91	14.94	1.86	59.82	nr	74.76
50 × 25 mm dia.	10.79	13.53	2.21	71.08	nr	84.61
50 × 40 mm dia.	16.77	21.03	2.21	71.08	nr	92.11
65 × 50 mm dia.	25.87	32.45	2.72	87.48	nr	119.93
80 × 50 mm dia.	34.98	43.88	3.21	103.24	nr	147.12
100 × 50 mm dia.	57.90	72.63	4.44	142.80	nr	215.43
100 × 80 mm dia.	89.34	112.07	4.44	142.80	nr	254.87
150 × 100 mm dia.	144.91	181.78	6.28	201.98	nr	383.76
Equal pitcher tee						
15 mm dia.	4.95	6.20	0.91	29.27	nr	35.47
20 mm dia.	6.10	7.65	1.22	39.23	nr	46.88
25 mm dia.	9.13	11.46	1.39	44.71	nr	56.17
32 mm dia.	14.15	17.75	1.62	52.10	nr	69.85
40 mm dia.	21.90	27.47	1.86	59.82	nr	87.29
50 mm dia.	30.73	38.55	2.21	71.08	nr	109.63
65 mm dia.	43.74	54.87	2.72	87.48	nr	142.35
80 mm dia.	68.24	85.60	3.21	103.24	nr	188.84
100 mm dia.	153.55	192.62	4.44	142.80	nr	335.42
Cross						
15 mm dia.	4.10	5.14	1.00	32.17	nr	37.31
20 mm dia.	6.43	8.06	1.33	42.77	nr	50.83
25 mm dia.	8.14	10.21	1.51	48.57	nr	58.78
32 mm dia.	12.11	15.19	1.77	56.93	nr	72.12
40 mm dia.	16.30	20.45	2.02	64.96	nr	85.41
50 mm dia.	25.34	31.79	2.42	77.84	nr	109.63
65 mm dia.	36.18	45.38	2.97	95.52	nr	140.90
80 mm dia.	48.10	60.33	3.50	112.57	nr	172.90
100 mm dia.	99.27	124.52	4.84	155.67	nr	280.19

38 PIPED SUPPLY SYSTEMS

Item	Net Price £	Material £	Labour hours	Labour £	Unit	Total rate £
COLD WATER PIPELINES: PIPEWORK ANCILLARIES – VALVES						
Regulators						
Gunmetal; self-acting two port thermostat; single seat; screwed; normally closed; with adjustable or fixed bleed device						
25 mm dia.	726.41	911.21	1.46	46.95	nr	**958.16**
32 mm dia.	747.37	937.50	1.45	46.64	nr	**984.14**
40 mm dia.	798.85	1002.08	1.55	49.86	nr	**1051.94**
50 mm dia.	961.65	1206.30	1.68	54.04	nr	**1260.34**
Self-acting temperature regulator for storage calorifier; integral sensing element and pocket; screwed ends						
15 mm dia.	702.20	880.84	1.32	42.45	nr	**923.29**
25 mm dia.	770.78	966.86	1.52	48.89	nr	**1015.75**
32 mm dia.	995.35	1248.56	1.79	57.58	nr	**1306.14**
40 mm dia.	1217.62	1527.38	1.99	64.00	nr	**1591.38**
50 mm dia.	1423.31	1785.40	2.26	72.69	nr	**1858.09**
Self-acting temperature regulator for storage calorifier; integral sensing element and pocket; flanged ends; bolted connection						
15 mm dia.	1030.83	1293.07	0.61	19.62	nr	**1312.69**
25 mm dia.	1179.78	1479.91	0.72	23.16	nr	**1503.07**
32 mm dia.	1487.15	1865.48	0.94	30.23	nr	**1895.71**
40 mm dia.	1761.41	2209.51	1.03	33.13	nr	**2242.64**
50 mm dia.	2045.13	2565.42	1.18	37.96	nr	**2603.38**
Chrome plated thermostatic mixing valves including non-return valves and inlet swivel connections with strainers; copper compression fittings						
15 mm dia.	113.78	142.72	0.69	22.19	nr	**164.91**
Chrome plated thermostatic mixing valves including non-return valves and inlet swivel connections with angle pattern combined isolating valves and strainers; copper compression fittings						
15 mm dia.	186.71	234.21	0.69	22.19	nr	**256.40**
Gunmetal thermostatic mixing valves including non-return valves and inlet swivel connections with strainers; copper compression fittings						
15 mm dia.	236.13	296.21	0.69	22.19	nr	**318.40**
Gunmetal thermostatic mixing valves including non-return valves and inlet swivel connections with angle pattern combined isolating valves and strainers; copper compression fittings						
15 mm dia.	248.26	311.42	0.69	22.19	nr	**333.61**

38 PIPED SUPPLY SYSTEMS

Item	Net Price £	Material £	Labour hours	Labour £	Unit	Total rate £
COLD WATER PIPELINES: PIPEWORK ANCILLARIES – VALVES – cont						
Ball float valves						
Bronze, equilibrium; copper float; working pressure cold services up to 16 bar; flanged ends; BS 4504 Table 16/21; bolted connections						
25 mm dia.	154.09	193.29	1.04	33.45	nr	**226.74**
32 mm dia.	213.76	268.14	1.22	39.23	nr	**307.37**
40 mm dia.	289.93	363.69	1.38	44.39	nr	**408.08**
50 mm dia.	455.60	571.50	1.66	53.39	nr	**624.89**
65 mm dia.	488.57	612.86	1.93	62.08	nr	**674.94**
80 mm dia.	544.77	683.36	2.16	69.47	nr	**752.83**
Heavy, equilibrium; with long tail and backnut; copper float; screwed for iron						
25 mm dia.	141.93	178.04	1.58	50.81	nr	**228.85**
32 mm dia.	194.86	244.43	1.78	57.25	nr	**301.68**
40 mm dia.	274.71	344.60	1.90	61.12	nr	**405.72**
50 mm dia.	360.75	452.52	2.65	85.23	nr	**537.75**
Brass, ball valve; BS 1212; copper float; screwed						
15 mm dia.	7.67	9.62	0.25	8.04	nr	**17.66**
22 mm dia.	13.94	17.48	0.29	9.33	nr	**26.81**
28 mm dia.	60.75	76.20	0.35	11.26	nr	**87.46**
Gate valves						
DZR copper alloy wedge non-rising stem; capillary joint to copper						
15 mm dia.	16.12	20.22	0.84	27.01	nr	**47.23**
22 mm dia.	19.77	24.80	1.01	32.49	nr	**57.29**
28 mm dia.	26.74	33.54	1.19	38.27	nr	**71.81**
35 mm dia.	47.99	60.20	1.38	44.39	nr	**104.59**
42 mm dia.	81.54	102.28	1.62	52.10	nr	**154.38**
54 mm dia.	114.25	143.32	1.94	62.40	nr	**205.72**
Cocks; capillary joints to copper						
Stopcock; brass head with gun metal body						
15 mm dia.	7.06	8.86	0.45	14.47	nr	**23.33**
22 mm dia.	13.16	16.51	0.46	14.78	nr	**31.29**
28 mm dia.	37.42	46.94	0.54	17.37	nr	**64.31**
Lockshield stop cocks; brass head with gun metal body						
15 mm dia.	8.14	10.21	0.45	14.47	nr	**24.68**
22 mm dia.	8.14	10.21	0.46	14.78	nr	**24.99**
28 mm dia.	8.14	10.21	0.54	17.37	nr	**27.58**
DZR stopcock; brass head with gun metal body						
15 mm dia.	19.32	24.24	0.45	14.47	nr	**38.71**
22 mm dia.	33.45	41.96	0.46	14.78	nr	**56.74**
28 mm dia.	55.75	69.93	0.54	17.37	nr	**87.30**

38 PIPED SUPPLY SYSTEMS

Item	Net Price £	Material £	Labour hours	Labour £	Unit	Total rate £
Gunmetal stopcock						
35 mm dia.	72.87	91.40	0.69	22.19	nr	**113.59**
42 mm dia.	96.77	121.39	0.71	22.83	nr	**144.22**
54 mm dia.	144.54	181.31	0.81	26.05	nr	**207.36**
Double union stopcock						
15 mm dia.	20.76	26.04	0.60	19.30	nr	**45.34**
22 mm dia.	29.19	36.61	0.60	19.30	nr	**55.91**
28 mm dia.	51.93	65.14	0.69	22.19	nr	**87.33**
Double union DZR stopcock						
15 mm dia.	33.88	42.50	0.60	19.30	nr	**61.80**
22 mm dia.	41.66	52.26	0.61	19.62	nr	**71.88**
28 mm dia.	77.00	96.59	0.69	22.19	nr	**118.78**
Double union gun metal stopcock						
35 mm dia.	128.29	160.92	0.63	20.26	nr	**181.18**
42 mm dia.	176.07	220.86	0.67	21.55	nr	**242.41**
54 mm dia.	276.98	347.45	0.85	27.33	nr	**374.78**
Double union stopcock with easy clean cover						
15 mm dia.	35.97	45.12	0.60	19.30	nr	**64.42**
22 mm dia.	44.88	56.30	0.61	19.62	nr	**75.92**
28 mm dia.	83.42	104.64	0.69	22.19	nr	**126.83**
Combined stopcock and drain						
15 mm dia.	35.68	44.76	0.67	21.55	nr	**66.31**
22 mm dia.	43.53	54.60	0.68	21.87	nr	**76.47**
Combined DZR stopcock and drain						
15 mm dia.	47.15	59.15	0.67	21.55	nr	**80.70**
Gate valves						
DZR copper alloy wedge non-rising stem; compression joint to copper						
15 mm dia.	16.12	20.22	0.84	27.01	nr	**47.23**
22 mm dia.	19.77	24.80	1.01	32.49	nr	**57.29**
28 mm dia.	26.74	33.54	1.19	38.27	nr	**71.81**
35 mm dia.	47.99	60.20	1.38	44.39	nr	**104.59**
42 mm dia.	81.54	102.28	1.62	52.10	nr	**154.38**
54 mm dia.	114.25	143.32	1.94	62.40	nr	**205.72**
Cocks; compression joints to copper						
Stopcock; brass head gun metal body						
15 mm dia.	6.41	8.04	0.42	13.51	nr	**21.55**
22 mm dia.	11.96	15.01	0.42	13.51	nr	**28.52**
28 mm dia.	34.02	42.67	0.45	14.47	nr	**57.14**
Lockshield stopcock; brass head gun metal body						
15 mm dia.	6.79	8.51	0.42	13.51	nr	**22.02**
22 mm dia.	9.72	12.20	0.42	13.51	nr	**25.71**
28 mm dia.	17.29	21.68	0.45	14.47	nr	**36.15**
DZR stopcock						
15 mm dia.	16.10	20.19	0.38	12.22	nr	**32.41**
22 mm dia.	27.88	34.98	0.39	12.54	nr	**47.52**
28 mm dia.	46.44	58.25	0.40	12.87	nr	**71.12**

38 PIPED SUPPLY SYSTEMS

Item	Net Price £	Material £	Labour hours	Labour £	Unit	Total rate £
COLD WATER PIPELINES: PIPEWORK ANCILLARIES – VALVES – cont						
Cocks – cont						
DZR stopcock – cont						
35 mm dia.	72.87	91.40	0.52	16.72	nr	**108.12**
42 mm dia.	96.77	121.39	0.54	17.37	nr	**138.76**
54 mm dia.	144.54	181.31	0.63	20.26	nr	**201.57**
DZR lockshield stopcock						
15 mm dia.	6.79	8.51	0.38	12.22	nr	**20.73**
22 mm dia.	10.71	13.44	0.39	12.54	nr	**25.98**
Combined stop/draincock						
15 mm dia.	35.68	44.76	0.22	7.08	nr	**51.84**
22 mm dia.	45.91	57.59	0.45	14.47	nr	**72.06**
DZR combined stop/draincock						
15 mm dia.	47.15	59.15	0.41	13.18	nr	**72.33**
22 mm dia.	61.10	76.64	0.42	13.51	nr	**90.15**
Stopcock to polyethylene						
15 mm dia.	24.65	30.92	0.38	12.22	nr	**43.14**
20 mm dia.	31.05	38.95	0.39	12.54	nr	**51.49**
25 mm dia.	41.30	51.81	0.40	12.87	nr	**64.68**
Draw off coupling						
15 mm dia.	14.97	18.78	0.38	12.22	nr	**31.00**
DZR draw off coupling						
15 mm dia.	14.97	18.78	0.38	12.22	nr	**31.00**
22 mm dia.	17.33	21.74	0.39	12.54	nr	**34.28**
Draw off elbow						
15 mm dia.	16.43	20.61	0.38	12.22	nr	**32.83**
22 mm dia.	20.24	25.39	0.39	12.54	nr	**37.93**
Lockshield drain cock						
15 mm dia.	17.18	21.55	0.41	13.18	nr	**34.73**
Check valves						
DZR copper alloy and bronze, WRC approved cartridge double check valve; BS 6282; working pressure cold services up to 10 bar at 65°C; screwed ends						
32 mm dia.	106.23	133.26	1.38	44.39	nr	**177.65**
40 mm dia.	121.07	151.87	1.62	52.10	nr	**203.97**
50 mm dia.	198.52	249.02	1.94	62.40	nr	**311.42**

38 PIPED SUPPLY SYSTEMS

Item	Net Price £	Material £	Labour hours	Labour £	Unit	Total rate £
COLD WATER PIPELINES: PIPEWORK ANCILLARIES – PUMPS						
Packaged cold water pressure booster set; fully automatic; 3 phase supply; includes fixing in position; electrical work elsewhere						
Pressure booster set						
0.75 l/s @ 30 m head	4487.28	5628.84	9.38	301.68	nr	5930.52
1.5 l/s @ 30 m head	5328.65	6684.26	9.38	301.68	nr	6985.94
3 l/s @ 30 m head	6510.56	8166.85	10.38	333.85	nr	8500.70
6 l/s @ 30 m head	14703.87	18444.53	10.38	333.85	nr	18778.38
12 l/s @ 30 m head	18369.81	23043.09	12.38	398.17	nr	23441.26
0.75 l/s @ 50 m head	5108.30	6407.86	9.38	301.68	nr	6709.54
1.5 l/s @ 50 m head	6510.56	8166.85	9.38	301.68	nr	8468.53
3 l/s @ 50 m head	7191.68	9021.24	10.38	333.85	nr	9355.09
6 l/s @ 50 m head	16422.65	20600.57	10.38	333.85	nr	20934.42
12 l/s @ 50 m head	20172.72	25304.66	12.38	398.17	nr	25702.83
0.75 l/s @ 70 m head	5589.07	7010.93	9.38	301.68	nr	7312.61
1.5 l/s @ 70 m head	7187.68	9016.22	9.38	301.68	nr	9317.90
3 l/s @ 70 m head	7652.42	9599.20	10.38	333.85	nr	9933.05
6 l/s @ 70 m head	17929.10	22490.26	10.38	333.85	nr	22824.11
12 l/s @ 70 m head	21795.34	27340.07	12.38	398.17	nr	27738.24
Automatic sump pump for clear and drainage water; single stage centrifugal pump, presure tight electric motor; single phase supply; includes fixing in position; electrical work elsewhere						
Single pump						
1 l/s @ 2.68 m total head	242.47	304.16	3.50	112.57	nr	416.73
1 l/s @ 4.68 m total head	266.90	334.80	3.50	112.57	nr	447.37
1 l/s @ 6.68 m total head	352.37	442.01	3.50	112.57	nr	554.58
2 l/s @ 4.38 m total head	352.37	442.01	4.00	128.65	nr	570.66
2 l/s @ 6.38 m total head	352.37	442.01	4.00	128.65	nr	570.66
2 l/s @ 8.38 m total head	453.54	568.92	4.00	128.65	nr	697.57
3 l/s @ 3.7 m total head	352.37	442.01	4.50	144.74	nr	586.75
3 l/s @ 5.7 m total head	453.54	568.92	4.50	144.74	nr	713.66
4 l/s @ 2.9 m total head	352.37	442.01	5.00	160.81	nr	602.82
4 l/s @ 4.9 m total head	453.54	568.92	5.00	160.81	nr	729.73
4 l/s @ 6.9 m total head	1257.70	1577.65	5.00	160.81	nr	1738.46
Extra for high level alarm box with single float switch, local alarm and volt free contacts for remote alarm	369.06	462.95	–	–	nr	462.95
Duty/standby pump unit						
1 l/s @ 2.68 m total head	460.51	577.66	5.00	160.81	nr	738.47
1 l/s @ 4.68 m total head	505.88	634.58	5.00	160.81	nr	795.39
1 l/s @ 6.68 m total head	683.80	857.76	5.00	160.81	nr	1018.57
2 l/s @ 4.38 m total head	683.80	857.76	5.50	176.89	nr	1034.65
2 l/s @ 6.38 m total head	683.80	857.76	5.50	176.89	nr	1034.65
2 l/s @ 8.38 m total head	875.69	1098.46	5.50	176.89	nr	1275.35
3 l/s @ 3.7 m total head	683.80	857.76	6.00	192.98	nr	1050.74

38 PIPED SUPPLY SYSTEMS

Item	Net Price £	Material £	Labour hours	Labour £	Unit	Total rate £
COLD WATER PIPELINES: PIPEWORK ANCILLARIES – PUMPS – cont						
Automatic sump pump for clear and drainage water – cont						
Duty/standby pump unit – cont						
3 l/s @ 5.7 m total head	875.69	1098.46	6.00	192.98	nr	**1291.44**
4 l/s @ 2.9 m total head	683.80	857.76	6.50	209.06	nr	**1066.82**
4 l/s @ 4.9 m total head	875.69	1098.46	6.50	209.06	nr	**1307.52**
4 /s @ 6.9 m total head	2327.01	2919.00	7.00	225.14	nr	**3144.14**
Extra for 4nr float switches to give pump on, off and high level alarm	385.60	483.69	–	–	nr	**483.69**
Extra for dual pump control panel, internal wall mounted IP54, including volt free contacts	1836.20	2303.32	4.00	128.65	nr	**2431.97**

38 PIPED SUPPLY SYSTEMS

Item	Net Price £	Material £	Labour hours	Labour £	Unit	Total rate £
COLD WATER PIPELINES: PIPEWORK ANCILLARIES – TANKS						
Cisterns; fibreglass; complete with ball valve, fixing plate and fitted covers						
Rectangular						
60 litre capacity	417.90	524.22	1.33	42.77	nr	**566.99**
100 litre capacity	438.90	550.56	1.40	45.04	nr	**595.60**
150 litre capacity	469.35	588.75	1.61	51.78	nr	**640.53**
250 litre capacity	504.00	632.22	1.61	51.78	nr	**684.00**
420 litre capacity	690.90	866.67	1.99	64.00	nr	**930.67**
730 litre capacity	971.25	1218.34	3.31	106.46	nr	**1324.80**
800 litre capacity	1001.70	1256.53	3.60	115.79	nr	**1372.32**
1700 litre capacity	1349.25	1692.50	13.32	428.40	nr	**2120.90**
2250 litre capacity	1633.80	2049.44	20.18	649.05	nr	**2698.49**
3400 litre capacity	2000.25	2509.11	24.50	787.99	nr	**3297.10**
4500 litre capacity	2638.65	3309.92	29.91	961.99	nr	**4271.91**
Cisterns; polypropylene; complete with ball valve, fixing plate and cover; includes placing in position						
Rectangular						
18 litre capacity	71.99	90.31	1.00	32.17	nr	**122.48**
68 litre capacity	97.01	121.69	1.00	32.17	nr	**153.86**
91 litre capacity	97.64	122.48	1.00	32.17	nr	**154.65**
114 litre capacity	108.64	136.28	1.00	32.17	nr	**168.45**
182 litre capacity	143.46	179.96	1.00	32.17	nr	**212.13**
227 litre capacity	182.73	229.22	1.00	32.17	nr	**261.39**
Circular						
114 litre capacity	101.61	127.46	1.00	32.17	nr	**159.63**
227 litre capacity	103.63	130.00	1.00	32.17	nr	**162.17**
318 litre capacity	172.36	216.20	1.00	32.17	nr	**248.37**
455 litre capacity	198.10	248.49	1.00	32.17	nr	**280.66**

38 PIPED SUPPLY SYSTEMS

Item	Net Price £	Material £	Labour hours	Labour £	Unit	Total rate £
COLD WATER PIPELINES: PIPEWORK ANCILLARIES – TANKS – cont						
Steel sectional water storage tank; hot pressed steel tank to BS 1564 TYPE 1; 5 mm plate; pre-insulated and complete with all connections and fittings to comply with BSEN 13280; 2001 and WRAS water supply (water fittings) regulations 1999; externally flanged base and sides; cost of erection (on prepared base) is included within the net price, labour cost allows for offloading and positioning materials						
Note: Prices are based on the most economical tank size for each volume, and the cost will vary with differing tank dimensions, for the same volume						
Volume, size						
4,900 litres, 3.66 m × 1.22 m × 1,22 m (h)	5876.13	7371.02	6.00	192.98	nr	**7564.00**
20,300 litres, 3.66 m × 2.4 m × 2.4 m (h)	12519.18	15704.06	12.00	385.95	nr	**16090.01**
52,000 litres, 6.1 m × 3.6 m × 2.4 m (h)	22319.79	27997.94	19.00	611.09	nr	**28609.03**
94,000 litres, 7.3 m × 3.6 m × 3.6 m (h)	39297.64	49294.96	28.00	900.56	nr	**50195.52**
140,000 litres, 9.7 m × 6.1 m × 2.44 m (h)	43846.22	55000.70	28.00	900.56	nr	**55901.26**
GRP sectional water storage tank; pre-insulated and complete with all connections and fittings to comply with BSEN 13280; 2001 and WRAS water supply (water fittings) regulations 1999; externally flanged base and sides; cost of erection (on prepared base) is included within the net price, labour cost allows for offloading and positioning materials						
Note: Prices are based on the most economical tank size for each volume, and the cost will vary with differing tank dimensions, for the same volume						
Volume, size						
4,500 litres, 3 m × 1 m × 1.5 m (h)	5231.00	6561.77	5.00	160.81	nr	**6722.58**
10,000 litres, 2.5 m × 2 m × 2 m (h)	7097.00	8902.48	7.00	225.14	nr	**9127.62**
20,000 litres, 4 m × 2.5 m × 2 m (h)	9123.00	11443.89	10.00	321.63	nr	**11765.52**
30,000 litres 5 m × 3 m × 2 m (h)	11122.00	13951.44	12.00	385.95	nr	**14337.39**
40,000 litres, 5 m × 4 m × 2 m (h)	12747.00	15989.84	12.00	385.95	nr	**16375.79**
50,000 litres, 5 m × 4 m × 2.5 m (h)	15935.00	19988.86	14.00	450.28	nr	**20439.14**
60,000 litres, 6 m × 4 m × 2.5 m (h)	17231.00	21614.57	16.00	514.61	nr	**22129.18**
70,000 litres, 7 m × 4 m × 2.5 m (h)	19524.00	24490.91	16.00	514.61	nr	**25005.52**
80,000 litres, 8 m × 4 m × 2.5 m (h)	20249.00	25400.35	16.00	514.61	nr	**25914.96**
90,000 litres, 6 m × 5 m × 3 m (h)	21626.00	27127.65	16.00	514.61	nr	**27642.26**
105,000 litres, 7 m × 5 m × 3 m (h)	24248.00	30416.69	24.00	771.90	nr	**31188.59**

38 PIPED SUPPLY SYSTEMS

Item	Net Price £	Material £	Labour hours	Labour £	Unit	Total rate £
120,000 litres, 8 m × 5 m × 3 m (h)	25144.00	31540.63	24.00	771.90	nr	**32312.53**
135,000 litres, 9 m × 6 m × 2.5 m (h)	26962.00	33821.13	24.00	771.90	nr	**34593.03**
144,000 litres, 8 m × 6 m × 3 m (h)	27853.00	34938.80	24.00	771.90	nr	**35710.70**
CLEANING AND CHEMICAL TREATMENT						
Electromagnetic water conditioner, complete with control box; maximum inlet pressure 16 bar; electrical work elsewhere						
Connection size, nominal flow rate at 50 mbar						
20 mm dia., 0.3 l/s	1689.41	2119.20	1.25	40.21	nr	**2159.41**
25 mm dia., 0.6 l/s	2218.32	2782.66	1.45	46.64	nr	**2829.30**
32 mm dia., 1.2 l/s	3163.16	3967.87	1.55	49.85	nr	**4017.72**
40 mm dia., 1.7 l/s	3779.36	4740.83	1.65	53.08	nr	**4793.91**
50 mm dia., 3.4 l/s	4950.14	6209.46	1.75	56.28	nr	**6265.74**
65 mm dia., 5.2 l/s	5417.43	6795.62	1.90	61.12	nr	**6856.74**
100 mm dia., 30.5 l/s, 595 mbar	10115.95	12689.44	3.00	96.49	nr	**12785.93**
Ultraviolet water sterillizing unit, complete with control unit; UV lamp housed in quartz tube; unit complete with UV intensity sensor, flushing and discharge valve and facilities for remote alarm; electrical work elsewhere						
Maximum flow rate (@ 250 J/m^2 exposure), connection size						
1.28 l/s, 40 mm dia.	3147.76	3948.55	1.98	63.68	nr	**4012.23**
2.00 l/s, 40 mm dia.	3979.63	4992.05	1.98	63.68	nr	**5055.73**
1.28 l/s, 50 mm dia.	4220.97	5294.79	1.98	63.68	nr	**5358.47**
7.4 l/s, 80 mm dia.	10341.89	12972.87	3.60	115.79	nr	**13088.66**
16.8 l/s 100 mm dia.	12847.77	16116.24	3.60	115.79	nr	**16232.03**
32.3 l/s 100 mm dia.	14449.89	18125.95	3.60	115.79	nr	**18241.74**
Base exchange water softener complete with resin tank, brine tank and consumption data monitoring facilities						
Capacities of softeners are based on 300ppm hardness and quoted in m^3 of softened water produced. Design flow rates are recommended for continuous use						
Simplex configuration						
Design flow rate, min–max softened water produced						
0.97 l/s, 7.7 m^3	3055.32	3832.60	10.00	321.63	nr	**4154.23**
1.25 l/s, 11.7 m^3	3132.35	3929.22	12.00	385.95	nr	**4315.17**
1.8 l/s, 19.8 m^3	4667.72	5855.19	12.00	385.95	nr	**6241.14**
2.23 l/s, 23.4 m^3	5237.70	6570.17	12.00	385.95	nr	**6956.12**
2.92 l/s, 31.6 m^3	5417.43	6795.62	15.00	482.44	nr	**7278.06**
3.75 l/s, 39.8 m^3	6470.10	8116.09	15.00	482.44	nr	**8598.53**
4.72 l/s, 60.3 m^3	7348.19	9217.57	18.00	578.93	nr	**9796.50**
6.66 l/s, 96.6 m^3	8678.15	10885.87	20.00	643.26	nr	**11529.13**

38 PIPED SUPPLY SYSTEMS

Item	Net Price £	Material £	Labour hours	Labour £	Unit	Total rate £
COLD WATER PIPELINES: PIPEWORK ANCILLARIES – TANKS – cont						
Capacities of softeners are based on 300ppm hardness – cont						
Duplex configuration						
Design flow rate, min–max softenend water produced						
0.97 l/s, 7.7 m³	4508.53	5655.50	12.00	385.95	nr	**6041.45**
1.25 l/s, 11.7 m³	4693.39	5887.39	12.00	385.95	nr	**6273.34**
1.81 l/s, 19.8 m³	7871.95	9874.57	15.00	482.44	nr	**10357.01**
2.23 l/s, 23.4 m³	9022.19	11317.43	15.00	482.44	nr	**11799.87**
2.29 l/s, 31.6 m³	9299.49	11665.28	18.00	578.93	nr	**12244.21**
3.75 l/s, 39.8 m³	11564.02	14505.90	18.00	578.93	nr	**15084.83**
4.72 l/s, 60.3 m³	13196.95	16554.25	18.00	578.93	nr	**17133.18**
6.66 l/s, 96.6 m³	16319.03	20470.59	23.00	739.75	nr	**21210.34**
THERMAL INSULATION						
Flexible closed cell walled insulation; Class 1/Class O; adhesive joints; including around fittings						
6 mm wall thickness						
15 mm dia.	1.49	1.87	0.15	4.83	m	**6.70**
22 mm dia.	1.76	2.21	0.15	4.83	m	**7.04**
28 mm dia.	2.22	2.79	0.15	4.83	m	**7.62**
9 mm wall thickness						
15 mm dia.	1.71	2.15	0.15	4.83	m	**6.98**
22 mm dia.	2.14	2.69	0.15	4.83	m	**7.52**
28 mm dia.	2.29	2.87	0.15	4.83	m	**7.70**
35 mm dia.	2.59	3.25	0.15	4.83	m	**8.08**
42 mm dia.	2.96	3.72	0.15	4.83	m	**8.55**
54 mm dia.	3.07	3.85	0.15	4.83	m	**8.68**
13 mm wall thickness						
15 mm dia.	2.18	2.73	0.15	4.83	m	**7.56**
22 mm dia.	2.68	3.36	0.15	4.83	m	**8.19**
28 mm dia.	3.18	3.99	0.15	4.83	m	**8.82**
35 mm dia.	3.44	4.31	0.15	4.83	m	**9.14**
42 mm dia.	4.04	5.06	0.15	4.83	m	**9.89**
54 mm dia.	5.63	7.07	0.15	4.83	m	**11.90**
67 mm dia.	8.32	10.44	0.15	4.83	m	**15.27**
76 mm dia.	9.79	12.28	0.15	4.83	m	**17.11**
108 mm dia.	10.26	12.87	0.15	4.83	m	**17.70**
19 mm wall thickness						
15 mm dia.	3.54	4.44	0.15	4.83	m	**9.27**
22 mm dia.	4.32	5.42	0.15	4.83	m	**10.25**
28 mm dia.	5.86	7.35	0.15	4.83	m	**12.18**
35 mm dia.	6.79	8.51	0.15	4.83	m	**13.34**
42 mm dia.	8.01	10.05	0.15	4.83	m	**14.88**
54 mm dia.	10.14	12.72	0.15	4.83	m	**17.55**
67 mm dia.	12.15	15.24	0.15	4.83	m	**20.07**

38 PIPED SUPPLY SYSTEMS

Item	Net Price £	Material £	Labour hours	Labour £	Unit	Total rate £
76 mm dia.	14.04	17.61	0.22	7.08	m	**24.69**
108 mm dia.	19.15	24.02	0.22	7.08	m	**31.10**
25 mm wall thickness						
15 mm dia.	6.96	8.74	0.15	4.83	m	**13.57**
22 mm dia.	7.61	9.54	0.15	4.83	m	**14.37**
28 mm dia.	8.62	10.81	0.15	4.83	m	**15.64**
35 mm dia.	9.55	11.98	0.15	4.83	m	**16.81**
42 mm dia.	10.19	12.78	0.15	4.83	m	**17.61**
54 mm dia.	11.97	15.02	0.15	4.83	m	**19.85**
67 mm dia.	16.14	20.25	0.15	4.83	m	**25.08**
76 mm dia.	18.19	22.81	0.22	7.08	m	**29.89**
32 mm wall thickness						
15 mm dia.	8.71	10.93	0.15	4.83	m	**15.76**
22 mm dia.	9.55	11.98	0.15	4.83	m	**16.81**
28 mm dia.	11.08	13.90	0.15	4.83	m	**18.73**
35 mm dia.	11.44	14.35	0.15	4.83	m	**19.18**
42 mm dia.	13.32	16.71	0.15	4.83	m	**21.54**
54 mm dia.	16.76	21.02	0.15	4.83	m	**25.85**
76 mm dia.	25.06	31.44	0.22	7.08	m	**38.52**

Note: For mineral fibre sectional insulation; bright class O foil faced; bright class O foil taped joints; 19 mm aluminium bands rates, refer to section – Thermal Insulation

Note: For mineral fibre sectional insulation; bright class O foil faced; bright class O foil taped joints; 22 swg plain/embossed aluminium cladding; pop riveted rates, refer to section – Thermal Insulation

Note: For mineral fibre sectional insulation; bright class O foil faced; bright class O foil taped joints; 0.8 mm polyisobutylene sheeting; welded joints rates, refer to section – Thermal Insulation

38 PIPED SUPPLY SYSTEMS

Item	Net Price £	Material £	Labour hours	Labour £	Unit	Total rate £
HOT WATER						
PIPELINES						
Note: For pipework prices refer to section – Cold Water						
PIPELINE ANCILLARIES						
Note: For prices for ancillaries refer to section – Cold Water						
STORAGE CYLINDERS/CALORIFIERS						
Insulated copper storage cylinders; BS 699; includes placing in position						
Grade 3 (maximum 10 m working head)						
BS size 6; 115 litre capacity; 400 mm dia.; 1050 mm high	197.65	247.93	1.50	48.29	nr	**296.22**
BS size 7; 120 litre capacity; 450 mm dia.; 900 mm high	233.69	293.14	2.00	64.32	nr	**357.46**
BS size 8; 144 litre capacity; 450 mm dia.; 1050 mm high	248.42	311.62	2.80	90.09	nr	**401.71**
Grade 4 (maximum 6 m working head)						
BS size 2; 96 litre capacity; 400 mm dia.; 900 mm high	153.04	191.97	1.50	48.29	nr	**240.26**
BS size 7; 120 litre capacity; 450 mm dia.; 900 mm high	160.83	201.75	1.50	48.29	nr	**250.04**
BS size 8; 144 litre capacity; 450 mm dia.; 1050 mm high	191.98	240.82	1.50	48.29	nr	**289.11**
BS size 9; 166 litre capacity; 450 mm dia.; 1200 mm high	280.71	352.13	1.50	48.29	nr	**400.42**
Storage cylinders; brazed copper construction; to BS 699; screwed bosses; includes placing in position						
Tested to 2.2 bar, 15 m maximum head						
144 litres	870.02	1091.35	3.00	96.59	nr	**1187.94**
160 litres	983.53	1233.74	3.00	96.59	nr	**1330.33**
200 litres	1013.76	1271.66	3.76	120.92	nr	**1392.58**
255 litres	1153.71	1447.22	3.76	120.92	nr	**1568.14**
290 litres	1558.46	1954.94	3.76	120.92	nr	**2075.86**
370 litres	1785.46	2239.69	4.50	144.88	nr	**2384.57**
450 litres	2432.15	3050.89	5.00	160.81	nr	**3211.70**
Tested to 2.55 bar, 17 m maximum head						
550 litres	2633.52	3303.48	5.00	160.81	nr	**3464.29**
700 litres	3078.85	3862.11	6.02	193.76	nr	**4055.87**
800 litres	3561.07	4467.01	6.54	210.21	nr	**4677.22**
900 litres	3853.44	4833.75	8.00	257.30	nr	**5091.05**
1000 litres	4066.46	5100.97	8.00	257.30	nr	**5358.27**
1250 litres	4453.71	5586.74	13.16	423.19	nr	**6009.93**
1500 litres	6821.67	8557.10	15.15	487.31	nr	**9044.41**
2000 litres	8186.61	10269.28	17.24	554.53	nr	**10823.81**
3000 litres	11499.37	14424.80	24.39	784.45	nr	**15209.25**

38 PIPED SUPPLY SYSTEMS

Item	Net Price £	Material £	Labour hours	Labour £	Unit	Total rate £
Indirect cylinders; copper; bolted top; up to 5 tappings for connections; BS 1586; includes placing in position						
Grade 3, tested to 1.45 bar, 10 m maximum head						
74 litre capacity	398.90	500.38	1.50	48.29	nr	548.67
96 litre capacity	406.11	509.42	1.50	48.29	nr	557.71
114 litre capacity	417.00	523.08	1.50	48.29	nr	571.37
117 litre capacity	432.81	542.92	2.00	64.32	nr	607.24
140 litre capacity	445.99	559.45	2.50	80.40	nr	639.85
162 litre capacity	623.69	782.35	3.00	96.59	nr	878.94
190 litre capacity	681.74	855.18	3.51	112.86	nr	968.04
245 litre capacity	797.77	1000.72	3.80	122.29	nr	1123.01
280 litre capacity	1414.18	1773.95	4.00	128.65	nr	1902.60
360 litre capacity	1530.24	1919.53	4.50	144.88	nr	2064.41
440 litre capacity	1776.81	2228.83	4.50	144.88	nr	2373.71
Grade 2, tested to 2.2 bar, 15 m maximum head						
117 litre capacity	576.53	723.20	2.00	64.32	nr	787.52
140 litre capacity	627.34	786.93	2.50	80.40	nr	867.33
162 litre capacity	717.97	900.63	2.80	90.09	nr	990.72
190 litre capacity	834.06	1046.25	3.00	96.59	nr	1142.84
245 litre capacity	1008.09	1264.55	4.00	128.65	nr	1393.20
280 litre capacity	1610.00	2019.58	4.00	128.65	nr	2148.23
360 litre capacity	1776.81	2228.83	4.50	144.88	nr	2373.71
440 litre capacity	2103.14	2638.18	4.50	144.88	nr	2783.06
Grade 1, tested 3.65 bar, 25 m maximum head						
190 litre capacity	1240.17	1555.67	3.00	96.59	nr	1652.26
245 litre capacity	1410.52	1769.35	3.00	96.59	nr	1865.94
280 litre capacity	1994.39	2501.77	4.00	128.65	nr	2630.42
360 litre capacity	2523.84	3165.90	4.50	144.88	nr	3310.78
440 litre capacity	3064.07	3843.57	4.50	144.88	nr	3988.45
Indirect cylinders, including manhole; BS 853						
Grade 3, tested to 1.5 bar, 10 m maximum head						
550 litre capacity	2579.21	3235.37	5.21	167.51	nr	3402.88
700 litre capacity	2855.58	3582.04	6.02	193.76	nr	3775.80
800 litre capacity	3316.14	4159.77	6.54	210.21	nr	4369.98
1000 litre capacity	4145.21	5199.76	7.04	226.50	nr	5426.26
1500 litre capacity	4790.00	6008.58	10.00	321.63	nr	6330.21
2000 litre capacity	6632.34	8319.61	16.13	518.76	nr	8838.37
Grade 2, tested to 2.55 bar, 15 m maximum head						
550 litre capacity	2859.54	3587.00	5.21	167.51	nr	3754.51
700 litre capacity	3578.91	4489.39	6.02	193.76	nr	4683.15
800 litre capacity	3776.73	4737.53	6.54	210.21	nr	4947.74
1000 litre capacity	4675.97	5865.54	7.04	226.50	nr	6092.04
1500 litre capacity	5755.03	7219.11	10.00	321.63	nr	7540.74
2000 litre capacity	7193.81	9023.92	16.13	518.76	nr	9542.68

38 PIPED SUPPLY SYSTEMS

Item	Net Price £	Material £	Labour hours	Labour £	Unit	Total rate £
HOT WATER – cont						
Indirect cylinders, including manhole – cont						
Grade 1, tested to 4 bar, 25 m maximum head						
550 litre capacity	3327.12	4173.53	5.21	167.51	nr	4341.04
700 litre capacity	3776.73	4737.53	6.02	193.76	nr	4931.29
800 litre capacity	4046.52	5075.95	6.54	210.21	nr	5286.16
1000 litre capacity	5395.37	6767.95	7.04	226.50	nr	6994.45
1500 litre capacity	6474.38	8121.47	10.00	321.63	nr	8443.10
2000 litre capacity	7913.16	9926.27	16.13	518.76	nr	10445.03
Storage calorifiers; copper; heater battery capable of raising temperature of contents from 10°C to 65°C in one hour; static head not exceeding 1.35 bar; BS 853; includes fixing in position on cradles or legs						
Horizontal; primary LPHW at 82°C/71°C						
400 litre capacity	3654.69	4584.44	7.04	226.50	nr	4810.94
1000 litre capacity	5847.47	7335.07	8.00	257.30	nr	7592.37
2000 litre capacity	11695.01	14670.22	14.08	453.01	nr	15123.23
3000 litre capacity	14436.00	18108.52	25.00	804.07	nr	18912.59
4000 litre capacity	17542.46	22005.27	40.00	1286.51	nr	23291.78
4500 litre capacity	19769.59	24798.97	50.00	1608.14	nr	26407.11
Vertical; primary LPHW at 82°C/71°C						
400 litre capacity	4118.84	5166.67	7.04	226.50	nr	5393.17
1000 litre capacity	6619.59	8303.61	8.00	257.30	nr	8560.91
2000 litre capacity	12608.68	15816.33	14.08	453.01	nr	16269.34
3000 litre capacity	15760.85	19770.41	25.00	804.07	nr	20574.48
4000 litre capacity	19333.24	24251.62	40.00	1286.51	nr	25538.13
4500 litre capacity	21855.01	27414.92	50.00	1608.14	nr	29023.06
Storage calorifiers; galvanized mild steel; heater battery capable of raising temperature of contents from 10°C to 65°C in one hour; static head not exceeding 1.35 bar; BS 853; includes fixing in position on cradles or legs						
Horizontal; primary LPHW at 82°C/71°C						
400 litre capacity	3654.69	4584.44	7.04	226.50	nr	4810.94
1000 litre capacity	5847.47	7335.07	8.00	257.30	nr	7592.37
2000 litre capacity	11695.01	14670.22	14.08	453.01	nr	15123.23
3000 litre capacity	14436.00	18108.52	25.00	804.07	nr	18912.59
4000 litre capacity	17542.46	22005.27	40.00	1286.51	nr	23291.78
4500 litre capacity	19769.59	24798.97	50.00	1608.14	nr	26407.11
Vertical; primary LPHW at 82°C/71°C						
400 litre capacity	4118.84	5166.67	7.04	226.50	nr	5393.17
1000 litre capacity	6619.59	8303.61	8.00	257.30	nr	8560.91
2000 litre capacity	12608.68	15816.33	14.08	453.01	nr	16269.34
3000 litre capacity	15760.85	19770.41	25.00	804.07	nr	20574.48
4000 litre capacity	19333.24	24251.62	40.00	1286.51	nr	25538.13
4500 litre capacity	21855.01	27414.92	50.00	1608.14	nr	29023.06

38 PIPED SUPPLY SYSTEMS

Item	Net Price £	Material £	Labour hours	Labour £	Unit	Total rate £
Indirect cylinders; mild steel, welded throughout, galvanized, with connections. Tested to 4 bar, 95C. Includes sensors, with full insulation and cases, includes delivery						
222 litre nominal content	990.00	1241.86	2.50	80.40	nr	**1322.26**
278 litre nominal content	1101.00	1381.09	2.80	90.09	nr	**1471.18**
474 litre nominal content	1376.00	1726.05	3.00	96.59	nr	**1822.64**
765 litre nominal content	1541.00	1933.03	3.00	96.56	nr	**2029.59**
956 litre nominal content	1817.00	2279.24	3.00	96.59	nr	**2375.83**
1365 litre nominal content	2356.00	2955.37	4.00	128.65	nr	**3084.02**
2039 litre nominal content	3005.00	3769.47	5.00	160.81	nr	**3930.28**
2361 litre nominal content	3885.00	4873.34	6.02	193.76	nr	**5067.10**
Indirect cylinders; mild steel welded throughout, galvanized, with connections. Tested to 4bar, 95C. Includes sensors. Includes delivery						
4021 litre nominal content	5014.00	6289.56	1.50	48.25	nr	**6337.81**
5897 litre nominal content	5944.00	7456.15	1.50	48.25	nr	**7504.40**
8000 litre nominal content	9355.00	11734.91	1.50	48.25	nr	**11783.16**
10170 litre nominal content	10991.00	13787.11	2.00	64.32	nr	**13851.43**
Local electric hot water heaters						
Unvented multipoint water heater; providing hot water for one or more outlets; used with conventional taps or mixers; factory fitted temperature and pressure relief valve; externally adjustable thermostat; elemental 'on' indicator; fitted with 1 m of 3 core cable; electrical supply and connection excluded						
5 litre capacity; 2.2 kW rating	209.27	262.51	1.50	48.25	nr	**310.76**
10 litre capacity; 2.2 kW rating	259.33	325.30	1.50	48.25	nr	**373.55**
15 litre capacity; 2.2 kW rating	462.21	579.80	1.50	48.25	nr	**628.05**
30 litre capacity; 2.2 kW rating	462.21	579.80	2.00	64.32	nr	**644.12**
50 litre capacity; 2.2 kW rating	486.62	610.41	2.00	64.32	nr	**674.73**
80 litre capacity; 2.2 kW rating	957.53	1201.12	2.00	64.32	nr	**1265.44**
100 litre capacity; 2.2 kW rating	1027.96	1289.48	2.00	64.32	nr	**1353.80**
Accessories						
Pressure reducing valve and expansion kit	170.47	213.84	2.00	64.32	nr	**278.16**
Thermostatic blending valve	90.72	113.80	1.00	32.17	nr	**145.97**

38 PIPED SUPPLY SYSTEMS

Item	Net Price £	Material £	Labour hours	Labour £	Unit	Total rate £
HOT WATER – cont						
SOLAR THERMAL PACKAGES						
Packages include solar collectors, mild steel storage vessel, fresh water module (including plate heat exchanger), roof fixing kit, glycol, stratified pump station (including plate heat exchanger), solar expansion vessel, controls, anti-legionella protection; delivery and commissioning						
No. of panels; storage size						
6 solar collectors, 1000 l storage vessel	8258.00	10358.84	11.00	353.79	nr	**10712.63**
8 solar collectors, 1500 l storage vessel	9548.00	11977.01	13.00	418.12	nr	**12395.13**
10 solar collectors, 2000 l storage vessel	10023.00	12572.85	15.00	482.44	nr	**13055.29**
12 solar collectors, 2000 l storage vessel	10990.00	13785.86	17.00	546.77	nr	**14332.63**
14 solar collectors, 2000 l storage vessel	11964.00	15007.64	18.00	578.93	nr	**15586.57**
16 solar collectors, 2500 l storage vessel	12963.00	16260.79	20.00	643.26	nr	**16904.05**

38 PIPED SUPPLY SYSTEMS

Item	Net Price £	Material £	Labour hours	Labour £	Unit	Total rate £
NATURAL GAS: PIPELINES: MEDIUM DENSITY POLYETHYLENE – YELLOW						
Pipe; laid underground; electrofusion joints in the running length; BS 6572; BGT PL2 standards						
Coiled service pipe						
20 mm dia.	1.66	2.08	0.37	11.91	m	**13.99**
25 mm dia.	2.16	2.71	0.41	13.18	m	**15.89**
32 mm dia.	3.57	4.48	0.47	15.12	m	**19.60**
63 mm dia.	13.58	17.04	0.60	19.30	m	**36.34**
90 mm dia.	17.82	22.36	0.90	28.95	m	**51.31**
Mains service pipe						
63 mm dia.	13.19	16.54	0.60	19.30	m	**35.84**
90 mm dia.	17.30	21.71	0.90	28.95	m	**50.66**
125 mm dia.	33.41	41.91	1.20	38.60	m	**80.51**
180 mm dia.	68.99	86.54	1.50	48.25	m	**134.79**
250 mm dia.	126.98	159.29	1.75	56.28	m	**215.57**
Extra over fittings, electrofusion joints						
Straight connector						
32 mm dia.	10.34	12.97	0.47	15.12	nr	**28.09**
63 mm dia.	19.41	24.35	0.58	18.65	nr	**43.00**
90 mm dia.	28.63	35.92	0.67	21.55	nr	**57.47**
125 mm dia.	53.61	67.24	0.83	26.69	nr	**93.93**
180 mm dia.	96.56	121.13	1.25	40.21	nr	**161.34**
Reducing connector						
90 × 63 mm dia.	39.98	50.15	0.67	21.55	nr	**71.70**
125 × 90 mm dia.	80.18	100.58	0.83	26.69	nr	**127.27**
180 × 125 mm dia.	147.11	184.53	1.25	40.21	nr	**224.74**
Bend; 45°						
90 mm dia.	77.27	96.92	0.67	21.55	nr	**118.47**
125 mm dia.	126.41	158.57	0.83	26.69	nr	**185.26**
180 mm dia.	286.75	359.70	1.25	40.21	nr	**399.91**
Bend; 90°						
63 mm dia.	50.06	62.80	0.58	18.65	nr	**81.45**
90 mm dia.	77.27	96.92	0.67	21.55	nr	**118.47**
125 mm dia.	126.41	158.57	0.83	26.69	nr	**185.26**
180 mm dia.	286.75	359.70	1.25	40.21	nr	**399.91**
Extra over malleable iron fittings, compression joints						
Straight connector						
20 mm dia.	14.74	18.49	0.38	12.23	nr	**30.72**
25 mm dia.	16.06	20.15	0.45	14.47	nr	**34.62**
32 mm dia.	18.03	22.61	0.50	16.08	nr	**38.69**
63 mm dia.	36.20	45.40	0.85	27.35	nr	**72.75**
Straight connector; polyethylene to MI						
20 mm dia.	12.52	15.70	0.31	9.97	nr	**25.67**
25 mm dia.	13.67	17.15	0.35	11.26	nr	**28.41**
32 mm dia.	15.25	19.13	0.40	12.87	nr	**32.00**
63 mm dia.	25.56	32.07	0.65	20.91	nr	**52.98**

38 PIPED SUPPLY SYSTEMS

Item	Net Price £	Material £	Labour hours	Labour £	Unit	Total rate £
NATURAL GAS: PIPELINES: MEDIUM DENSITY POLYETHYLENE – YELLOW – cont						
Extra over malleable iron fittings, compression joints – cont						
Straight connector; polyethylene to FI						
20 mm dia.	12.52	15.70	0.31	9.97	nr	25.67
25 mm dia.	15.25	19.13	0.35	11.26	nr	30.39
32 mm dia.	15.25	19.13	0.40	12.87	nr	32.00
63 mm dia.	25.56	32.07	0.75	24.14	nr	56.21
Elbow						
20 mm dia.	19.17	24.05	0.38	12.23	nr	36.28
25 mm dia.	20.91	26.23	0.45	14.47	nr	40.70
32 mm dia.	23.43	29.39	0.50	16.08	nr	45.47
63 mm dia.	47.08	59.06	0.80	25.73	nr	84.79
Equal tee						
20 mm dia.	25.56	32.07	0.53	17.05	nr	49.12
25 mm dia.	29.87	37.46	0.55	17.68	nr	55.14
32 mm dia.	37.64	47.22	0.64	20.59	nr	67.81
GAS BOOSTERS						
Complete skid mounted gas booster set, including AV mounts, flexible connections, low pressure switch, control panel and NRV (for run/standby unit); 3 phase supply; in accordance with IGE/UP/2; includes delivery, offloading and positioning						
Single unit						
Flow, pressure range						
0–200 m³/hour, 0.1–2.6 kPa	3949.06	4953.70	10.00	321.63	nr	5275.33
0–200 m³/hour, 0.1–4.0 kPa	4505.98	5652.30	10.00	321.63	nr	5973.93
0–200 m³/hour, 0.1–7 kPa	5243.68	6577.67	10.00	321.63	nr	6899.30
0–200 m³/hour, 0.1–9.5 kPa	5460.69	6849.89	10.00	321.63	nr	7171.52
0–200 m³/hour 0.1–11.0 kPa	6100.80	7652.85	10.00	321.63	nr	7974.48
0–400 m³/hour, 0.1–4.0 kPa	6635.59	8323.68	10.00	321.63	nr	8645.31
0–1000 m³/hour, 0.1–7.4 kPa	6892.73	8646.24	10.00	321.63	nr	8967.87
50–1000 m³/hour, 0.1–16.0 kPa	13009.97	16319.71	20.00	643.26	nr	16962.97
50–1000 m³/hour, 0.1–24.5 kPa	14559.36	18263.26	20.00	643.26	nr	18906.52
50–1000 m³/hour, 0.1–31.0 kPa	16559.23	20771.90	20.00	643.26	nr	21415.16
50–1000 m³/hour, 0.1–41.0 kPa	18277.02	22926.69	20.00	643.26	nr	23569.95
50–1000 m³/hour, 0.1–51.0 kPa	18964.06	23788.52	20.00	643.26	nr	24431.78
100–1800 m³/hour, 3.5–23.5 kPa	19459.56	24410.08	20.00	643.26	nr	25053.34
100–1800 m³/hour, 4.5–27.0 kPa	20989.26	26328.93	20.00	643.26	nr	26972.19
100–1800 m³/hour, 6.0–32.5 kPa	23245.85	29159.59	20.00	643.26	nr	29802.85
100–1800 m³/hour, 7.2–39.0 kPa	26742.85	33546.23	20.00	643.26	nr	34189.49
100–1800 m³/hour, 9.0–42.0 kPa	28109.84	35260.98	20.00	643.26	nr	35904.24

38 PIPED SUPPLY SYSTEMS

Item	Net Price £	Material £	Labour hours	Labour £	Unit	Total rate £
Run/Standby unit						
Flow, pressure range						
0–200 m³/hour, 0.1–2.6 kPa	19436.89	24381.64	16.00	514.61	nr	24896.25
0–200 m³/hour, 0.1–4.0 kPa	20028.87	25124.21	16.00	514.61	nr	25638.82
0–200 m³/hour, 0.1–7 kPa	20492.76	25706.12	16.00	514.61	nr	26220.73
0–200 m³/hour, 0.1–9.5 kPa	21000.55	26343.09	16.00	514.61	nr	26857.70
0–200 m³/hour 0.1–11.0 kPa	21348.47	26779.52	16.00	514.61	nr	27294.13
0–400 m³/hour, 0.1–4.0 kPa	23542.56	29531.79	16.00	514.61	nr	30046.40
0–1000 m³/hour, 0.1–7.4 kPa	30008.82	37643.07	25.00	804.07	nr	38447.14
50–1000 m³/hour, 0.1–16.0 kPa	41129.70	51593.09	25.00	804.07	nr	52397.16
50–1000 m³/hour, 0.1–24.5 kPa	46498.96	58328.30	25.00	804.07	nr	59132.37
50–1000 m³/hour, 0.1–31.0 kPa	51855.63	65047.71	25.00	804.07	nr	65851.78
50–1000 m³/hour, 0.1–41.0 kPa	57228.03	71786.84	25.00	804.07	nr	72590.91
50–1000 m³/hour, 0.1–51.0 kPa	57485.03	72109.22	25.00	804.07	nr	72913.29
100–1800 m³/hour, 3.5–23.5 kPa	59808.92	75024.31	25.00	804.07	nr	75828.38
100–1800 m³/hour, 4.5–27.0 kPa	61448.67	77081.21	25.00	804.07	nr	77885.28
100–1800 m³/hour, 6.0–32.5 kPa	65283.40	81891.50	25.00	804.07	nr	82695.57
100–1800 m³/hour, 7.2–39.0 kPa	69389.50	87042.19	25.00	804.07	nr	87846.26
100–1800 m³/hour, 9.0–42.0 kPa	72937.67	91493.01	25.00	804.07	nr	92297.08

38 PIPED SUPPLY SYSTEMS

Item	Net Price £	Material £	Labour hours	Labour £	Unit	Total rate £
NATURAL GAS: PIPELINES: SCREWED STEEL						
For prices for steel pipework refer to section – Low Temperature Hot Water Heating						
PIPE IN PIPE						
Note: for pipe in pipe, a sleeve size two pipe sizes bigger than actual pipe size has been allowed. All rates refer to actual pipe size. Black steel pipes – Screwed and socketed joints; BS 1387: 1985 upto 50 mm pipe size. Butt welded joints; BS 1387: 1985 65 mm pipe size and above						
Pipe dia.						
25 mm	15.74	19.75	1.73	55.64	m	**75.39**
32 mm	20.90	26.22	1.95	62.72	m	**88.94**
40 mm	26.89	33.73	2.16	69.47	m	**103.20**
50 mm	35.28	44.25	2.44	78.48	m	**122.73**
65 mm	48.38	60.69	2.95	94.89	m	**155.58**
80 mm	58.48	73.36	3.42	110.00	m	**183.36**
100 mm	73.04	91.62	4.00	128.65	m	**220.27**
Extra over black steel pipes – Screwed pipework; black malleable iron fittings; BS 143. Welded pipework; butt welded steel fittings; BS 1965						
Bend, 90°						
25 mm	9.08	11.39	2.91	93.59	m	**104.98**
32 mm	13.82	17.34	3.45	110.97	m	**128.31**
40 mm	19.37	24.29	5.34	171.75	m	**196.04**
50 mm	21.82	27.37	6.53	210.02	m	**237.39**
65 mm	29.49	36.99	8.84	284.32	m	**321.31**
80 mm	50.27	63.06	10.73	345.11	m	**408.17**
100 mm	66.75	83.73	12.76	410.40	m	**494.13**
Bend, 45°						
25 mm	11.48	14.40	2.91	93.59	m	**107.99**
32 mm	14.87	18.65	3.45	110.97	m	**129.62**
40 mm	16.62	20.84	5.34	171.75	m	**192.59**
50 mm	19.82	24.86	6.53	210.02	m	**234.88**
65 mm	25.89	32.48	8.84	284.32	m	**316.80**
80 mm	44.29	55.55	10.73	345.11	m	**400.66**
100 mm	56.74	71.18	12.76	410.40	m	**481.58**
Equal tee						
25 mm	76.43	95.87	4.18	134.44	m	**230.31**
32 mm	78.46	98.43	4.94	158.88	m	**257.31**
40 mm	96.53	121.08	7.28	234.15	m	**355.23**
50 mm	99.04	124.23	8.48	272.74	m	**396.97**
65 mm	131.86	165.40	11.47	368.91	m	**534.31**
80 mm	226.24	283.80	14.23	457.68	m	**741.48**
100 mm	256.04	321.17	17.92	576.36	m	**897.53**

38 PIPED SUPPLY SYSTEMS

Item	Net Price £	Material £	Labour hours	Labour £	Unit	Total rate £
Copper pipe; capillary or compression joints in the running length; EN1057 R250 (TX) formerly BS 2871 Table X						
Plastic coated gas service pipe for corrosive environments, fixed vertically or at low level with brackets measured separtely						
15 mm dia. (yellow)	8.90	11.17	0.85	27.35	m	**38.52**
22 mm dia. (yellow)	17.55	22.02	0.96	30.90	m	**52.92**
28 mm dia. (yellow)	22.29	27.96	1.06	34.09	m	**62.05**
Copper pipe; capillary or compression joints in the running length; EN1057 R250 (TY) formerly BS 2871 Table Y						
Plastic coated gas and cold water service pipe for corrosive environments, fixed vertically or at low level with brackets measured separately (Refer to Copper Pipe Table X Section)						
15 mm dia. (yellow)	10.29	12.90	0.61	19.62	m	**32.52**
22 mm dia. (yellow)	18.70	23.45	0.69	22.19	m	**45.64**
Fixings						
Refer to Section – Cold Water						
Extra over copper pipes; capillary fittings; BS 864						
Refer to Section – Cold Water						

38 PIPED SUPPLY SYSTEMS

Item	Net Price £	Material £	Labour hours	Labour £	Unit	Total rate £
FUEL OIL STORAGE/DISTRIBUTION						
PIPELINES						
For pipework prices refer to Section – Low Temperature Hot Water Heating						
TANKS						
Fuel storage tanks; mild steel; with all necessary screwed bosses; oil resistant joint rings; includes placing in position. Rectangular shape						
1360 litre (300 gallon) capacity; 2 mm plate	415.61	521.34	12.03	386.92	nr	908.26
2730 litre (600 gallon) capacity; 2.5 mm plate	555.36	696.64	18.60	598.23	nr	1294.87
4550 litre (1000 gallon) capacity; 3 mm plate	1164.36	1460.57	25.00	804.07	nr	2264.64
Fuel storage tanks; 5 mm plate mild steel to BS 799 type J; complete with raised neck manhole with bolted cover, screwed connections, vent and fill connections, drain valve, gauge and overfill alarm; includes placing in position; excludes pumps and control panel. Nominal capacity size						
5,600 litres, 2.5 m × 1.5 m × 1.5 m high	2717.68	3409.06	20.00	643.26	nr	4052.32
Extra for bund unit (internal use)	1562.87	1960.46	30.00	964.88	nr	2925.34
Extra for external use with bund (watertight)	989.81	1241.62	2.00	64.32	nr	1305.94
10,200 litres, 3.05 m × 1.83 m × 1.83 m high	3377.56	4236.81	30.00	964.88	nr	5201.69
Extra for bund unit (internal use)	2274.86	2853.58	40.00	1286.51	nr	4140.09
Extra for external use with bund (watertight)	1215.57	1524.81	2.00	64.32	nr	1589.13
15,000 litres, 3.75 m × 2 m × 2 m high	4271.87	5358.63	40.00	1286.51	nr	6645.14
Extra for bund unit (internal use)	3056.31	3833.84	55.00	1768.95	nr	5602.79
Extra for external use with bund (watertight)	1423.96	1786.22	2.00	64.32	nr	1850.54
20,000 litres, 4 m × 2.5 m × 2 m high	5635.06	7068.62	50.00	1608.14	nr	8676.76
Extra for bund unit (internal use)	3707.53	4650.72	65.00	2090.58	nr	6741.30
Extra for external use with bund (watertight)	1684.43	2112.95	2.00	64.32	nr	2177.27
Extra for BMS output (all tank sizes)	590.42	740.62	–	–	nr	740.62

38 PIPED SUPPLY SYSTEMS

Item	Net Price £	Material £	Labour hours	Labour £	Unit	Total rate £
Fuel storage tanks; plastic; with all necessary screwed bosses; oil resistant joint rings; includes placing in position						
Cylindrical; horizontal						
1350 litre (300 gallon) capacity	399.00	500.51	4.30	138.30	nr	**638.81**
2500 litre (550 gallon) capacity	689.00	864.28	4.88	156.95	nr	**1021.23**
Cylindrical; vertical						
1365 litre (300 gallon) capacity	308.00	386.36	3.73	119.97	nr	**506.33**
2600 litre (570 gallon) capacity	489.00	613.40	4.88	156.95	nr	**770.35**
Bunded tanks						
1135 litre (250 gallon) capacity	930.00	1166.59	4.30	138.30	nr	**1304.89**
1590 litre (350 gallon) capacity	858.00	1076.28	4.88	156.95	nr	**1233.23**
2500 litre (550 gallon) capacity	1200.00	1505.28	5.95	191.37	nr	**1696.65**

38 PIPED SUPPLY SYSTEMS

Item	Net Price £	Material £	Labour hours	Labour £	Unit	Total rate £
FIRE HOSE REELS: PIPEWORK ANCILLARIES						
PIPELINES						
For pipework prices refer to Section – Cold Water						
PIPELINE ANCILLARIES						
For prices for ancillaries refer to Section – Cold Water						
Hose reels; automatic; connection to 25 mm screwed joint; reel with 30.5 m, 19 mm rubber hose; suitable for working pressure up to 7 bar						
Reels						
Non-swing pattern	190.57	239.05	3.75	120.61	nr	**359.66**
Recessed non-swing pattern	342.54	429.68	3.75	120.61	nr	**550.29**
Swinging pattern	221.79	278.21	3.75	120.61	nr	**398.82**
Recessed swinging pattern	221.79	278.21	3.75	120.61	nr	**398.82**
Hose reels; manual; connection to 25 mm screwed joint; reel with 30.5 m, 19 mm rubber hose; suitable for working pressure up to 7 bar						
Reels						
Non-swing pattern	154.00	193.18	3.25	104.53	nr	**297.71**
Recessed non-swing pattern	342.54	429.68	3.25	104.53	nr	**534.21**
Swinging pattern	182.49	228.92	3.25	104.53	nr	**333.45**
Recessed swinging pattern	182.49	228.92	3.25	104.53	nr	**333.45**

38 PIPED SUPPLY SYSTEMS

Item	Net Price £	Material £	Labour hours	Labour £	Unit	Total rate £
DRY RISERS: PIPEWORK ANCILLARIES						
PIPELINES						
For pipework prices refer to Section – Cold Water						
VALVES (BS 5041, parts 2 and 3)						
Bronze/gunmetal inlet breeching for pumping in with 65 mm dia. instantaneous male coupling; with cap, chain and 25 mm drain valve						
Double inlet with back pressure valve, flanged to steel	281.11	352.62	1.75	56.28	nr	**408.90**
Quadruple inlet with back pressure valve, flanged to steel	629.42	789.54	1.75	56.28	nr	**845.82**
Bronze/gunmetal gate type outlet valve with 65 mm dia. instantaneous female coupling; cap and chain; wheel head secured by padlock and leather strap						
Flanged to BS 4504 PN6 (bolted connection to counter flanges measured separately)	225.44	282.79	1.75	56.28	nr	**339.07**
Bronze/gunmetal landing type outlet valve, with 65 mm dia. instantaneous female coupling; cap and chain; wheelhead secured by padlock and leather strap; bolted connections to counter flanges measured separately						
Horizontal, flanged to BS 4504 PN6	245.52	307.98	1.50	48.25	nr	**356.23**
Oblique, flanged to BS 4504 PN6	245.52	307.98	1.50	48.25	nr	**356.23**
Air valve, screwed joints to steel						
25 mm dia.	48.80	61.22	0.55	17.68	nr	**78.90**
INLET BOXES (BS 5041, part 5)						
Steel dry riser inlet box with hinged wire glazed door suitably lettered (fixing by others)						
610 × 460 × 325 mm; double inlet	299.19	375.30	3.00	96.49	nr	**471.79**
610 × 610 × 356 mm; quadruple inlet	558.38	700.44	3.00	96.49	nr	**796.93**
OUTLET BOXES (BS 5041, part 5)						
Steel dry riser outlet box with hinged wire glazed door suitably lettered (fixing by others)						
610 × 460 × 325 mm; single outlet	292.16	366.49	3.00	96.49	nr	**462.98**

38 PIPED SUPPLY SYSTEMS

Item	Net Price £	Material £	Labour hours	Labour £	Unit	Total rate £
SPRINKLERS						
PIPELINES						
Prefabricated black steel pipework; screwed joints, including all coupliings, unions and the like to BS 1387:1985; includes fixing to backgrounds, with brackets measured separately						
Heavy weight						
25 mm dia. – pipe plus allowance for coupling every 6 m	6.43	8.06	0.47	15.12	m	**23.18**
32 mm dia. – pipe plus allowance for coupling every 6 m	7.99	10.02	0.53	17.05	m	**27.07**
40 mm dia. – pipe plus allowance for coupling every 6 m	9.30	11.67	0.58	18.65	m	**30.32**
50 mm dia. – pipe plus allowance for coupling every 6 m	12.91	16.20	0.63	20.26	m	**36.46**
Fixings						
For steel pipes; black malleable iron. For minimum fixing distances, refer to the Tables and Memoranda at the rear of the book						
Pipe ring, single socket, black malleable iron						
25 mm dia.	1.14	1.43	0.12	3.86	nr	**5.29**
32 mm dia.	1.22	1.53	0.14	4.50	nr	**6.03**
40 mm dia.	1.57	1.97	0.15	4.83	nr	**6.80**
50 mm dia.	1.98	2.49	0.16	5.14	nr	**7.63**
Extra over channel sections for fabricated hangers and brackets						
Galvanized steel; including inserts, bolts, nuts, washers; fixed to backgrounds						
41 × 21 mm	6.34	7.95	0.29	9.33	m	**17.28**
41 × 41 mm	7.60	9.53	0.29	9.33	m	**18.86**
Threaded rods; metric thread; including nuts, washers etc.						
12 mm dia. × 600 mm long	3.20	4.01	0.18	5.79	nr	**9.80**
Extra over for black malleable iron fittings; BS 143						
Plain plug, solid						
25 mm dia.	2.29	2.87	0.40	12.87	nr	**15.74**
32 mm dia.	3.30	4.14	0.44	14.15	nr	**18.29**
40 mm dia.	4.64	5.82	0.48	15.44	nr	**21.26**
50 mm dia.	6.10	7.65	0.56	18.01	nr	**25.66**

38 PIPED SUPPLY SYSTEMS

Item	Net Price £	Material £	Labour hours	Labour £	Unit	Total rate £
Concentric reducing socket						
32 mm dia.	2.72	3.42	0.48	15.44	nr	18.86
40 mm dia.	3.54	4.44	0.55	17.68	nr	22.12
50 mm dia.	4.97	6.24	0.60	19.30	nr	25.54
Elbow; 90° female/female						
25 mm dia.	1.59	1.99	0.44	14.15	nr	16.14
32 mm dia.	2.93	3.67	0.53	17.05	nr	20.72
40 mm dia.	4.90	6.15	0.60	19.30	nr	25.45
50 mm dia.	5.75	7.21	0.65	20.91	nr	28.12
Tee						
25 mm dia. equal	2.18	2.73	0.51	16.41	nr	19.14
32 mm dia. reducing to 25 mm dia.	3.97	4.98	0.54	17.37	nr	22.35
40 mm dia.	6.10	7.65	0.65	20.91	nr	28.56
50 mm dia.	8.77	11.00	0.78	25.09	nr	36.09
Cross tee						
25 mm dia. equal	5.93	7.44	1.16	37.31	nr	44.75
32 mm dia.	8.63	10.83	1.40	45.04	nr	55.87
40 mm dia.	11.09	13.91	1.60	51.45	nr	65.36
50 mm dia.	18.02	22.60	1.68	54.04	nr	76.64
Prefabricated black steel pipework; welded joints, including all coupliings, unions and the like to BS 1387:1985; fixing to backgrounds						
Heavy weight						
65 mm dia. – pipe length only, welded joints on runs assumed	16.88	21.18	0.65	20.91	m	42.09
80 mm dia. – pipe length only, welded joints on runs assumed	21.50	26.97	0.70	22.51	m	49.48
100 mm dia. – pipe length only, welded joints on runs assumed	29.99	37.62	0.85	27.33	m	64.95
150 mm dia. – pipe length only, welded joints on runs assumed	40.58	50.90	1.15	36.99	m	87.89
Fixings						
For steel pipes; black malleable iron. For minimum fixing distances, refer to the Tables and Memoranda at the rear of the book						
Pipe ring, single socket, black malleable iron						
65 mm dia.	2.87	3.60	0.30	9.64	nr	13.24
80 mm dia.	3.43	4.30	0.35	11.26	nr	15.56
100 mm dia.	5.20	6.52	0.40	12.87	nr	19.39
150 mm dia.	11.77	14.76	0.77	24.76	nr	39.52
Extra over fittings						
Reducer (one size down)						
65 mm dia.	8.67	10.88	2.70	86.84	nr	97.72
80 mm dia.	11.28	14.15	2.86	91.99	nr	106.14
100 mm dia.	22.49	28.21	3.22	103.57	nr	131.78
150 mm dia.	62.48	78.38	4.20	135.08	nr	213.46

38 PIPED SUPPLY SYSTEMS

Item	Net Price £	Material £	Labour hours	Labour £	Unit	Total rate £
SPRINKLERS – cont						
Extra over fittings – cont						
Elbow; 90°						
65 mm dia.	11.91	14.94	3.06	98.43	nr	**113.37**
80 mm dia.	17.51	21.96	3.40	109.36	nr	**131.32**
100 mm dia.	21.90	27.47	3.70	119.00	nr	**146.47**
150 mm dia.	32.83	41.18	5.20	167.25	nr	**208.43**
Branch bend						
65 mm dia.	12.13	15.22	3.60	115.79	nr	**131.01**
80 mm dia.	14.21	17.83	3.80	122.21	nr	**140.04**
100 mm dia.	21.86	27.42	5.10	164.02	nr	**191.44**
150 mm dia.	56.56	70.95	7.50	241.23	nr	**312.18**
Prefabricated black steel pipe; Victaulic Firelock joints; including all couplings and the like to BS 1387: 1985; fixing to backgrounds						
Heavy weight						
65 mm dia. – pipe plus allowance for Victaulic coupling every 6 m	14.56	18.27	0.70	22.51	m	**40.78**
80 mm dia. – pipe plus allowance for Victaulic coupling every 6 m	18.92	23.73	0.78	25.09	m	**48.82**
100 mm dia. – pipe plus allowance for Victaulic coupling every 6 m	27.59	34.61	0.93	29.92	m	**64.53**
150 mm dia. – pipe plus allowance for Victaulic coupling every 6 m	54.92	68.89	1.25	40.21	m	**109.10**
Fixings						
For fixings refer to For steel pipes; black malleable iron.						
Extra over fittings						
Coupling						
65 mm dia. – Victaulic coupling	10.66	13.37	0.26	8.37	nr	**21.74**
80 mm dia. – Victaulic coupling	11.42	14.32	0.26	8.37	nr	**22.69**
100 mm dia. – Victaulic coupling	14.97	18.78	0.32	10.28	nr	**29.06**
150 mm dia. – Victaulic coupling	26.36	33.06	0.35	11.26	nr	**44.32**
Reducer						
65 mm dia. – Victaulic excl couplings	8.67	10.88	0.48	15.44	nr	**26.32**
80 mm dia. – Victaulic excl couplings	11.28	14.15	0.43	13.83	nr	**27.98**
100 mm dia. – Victaulic excl couplings	22.49	28.21	0.46	14.78	nr	**42.99**
150 mm dia. – Victaulic excl couplings	33.74	42.32	0.45	14.47	nr	**56.79**
Elbow; any degree						
65 mm dia. – Victaulic excl couplings	11.91	14.94	0.56	18.01	nr	**32.95**
80 mm dia. – Victaulic excl couplings	17.51	21.96	0.63	20.26	nr	**42.22**
100 mm dia. – Victaulic excl couplings	33.78	42.37	0.71	22.83	nr	**65.20**
150 mm dia. – Victaulic excl couplings	50.66	63.55	0.80	25.73	nr	**89.28**

38 PIPED SUPPLY SYSTEMS

Item	Net Price £	Material £	Labour hours	Labour £	Unit	Total rate £
Equal tee						
65 mm dia. – Victaulic excl couplings	18.47	23.17	0.74	23.80	nr	**46.97**
80 mm dia. – Victaulic excl couplings	22.24	27.90	0.83	26.69	nr	**54.59**
100 mm dia. – Victaulic excl couplings	40.76	51.13	0.94	30.23	nr	**81.36**
150 mm dia. – Victaulic excl couplings	61.15	76.71	1.05	33.77	nr	**110.48**
PIPELINE ANCILLARIES						
SPRINKLER HEADS						
Sprinkler heads; brass body; frangible glass bulb; manufactured to standard operating temperature of 57–141°C; quick response; RTI<50						
Conventional pattern; 15 mm dia.	4.64	5.82	0.15	4.83	nr	**10.65**
Sidewall pattern; 15 mm dia.	6.81	8.55	0.15	4.83	nr	**13.38**
Conventional pattern; 15 mm dia.; satin chrome plated	5.44	6.82	0.15	4.83	nr	**11.65**
Sidewall pattern; 15 mm dia.; satin chrome plated	7.07	8.87	0.15	4.83	nr	**13.70**
Fully concealed; fusible link; 15 mm dia.	16.99	21.31	0.15	4.83	nr	**26.14**
VALVES						
Wet system alarm valves; including internal non-return valve; working pressure up to 12.5 bar; BS4504 PN16 flanged ends; bolted connections						
100 mm dia.	1495.26	1875.65	25.00	804.07	nr	**2679.72**
150 mm dia.	1826.50	2291.16	25.00	804.07	nr	**3095.23**
Wet system by-pass alarm valves; including internal non-return valve; working pressure up to 12.5 bar; BS4504 PN16 flanged ends; bolted connections						
100 mm dia.	2635.73	3306.26	25.00	804.07	nr	**4110.33**
150 mm dia.	3332.46	4180.24	25.00	804.07	nr	**4984.31**
Alternate system wet/dry alarm station; including butterfly valve, wet alarm valve, dry pipe differential pressure valve and pressure gauges; working pressure up to 10.5 bar; BS4505 PN16 flanged ends; bolted connections						
100 mm dia.	3266.60	4097.62	40.00	1286.51	nr	**5384.13**
150 mm dia.	3806.84	4775.30	40.00	1286.51	nr	**6061.81**
Alternate system wet/dry alarm station; including electrically operated butterfly valve, water supply accelerator set, wet alarm valve, dry pipe differential pressure valve and pressure gauges; working pressure up to 10.5 bar; BS4505 PN16 flanged ends; bolted connections						
100 mm dia.	3723.52	4670.78	45.00	1447.33	nr	**6118.11**
150 mm dia.	4268.90	5354.91	45.00	1447.33	nr	**6802.24**

38 PIPED SUPPLY SYSTEMS

Item	Net Price £	Material £	Labour hours	Labour £	Unit	Total rate £
SPRINKLERS – cont						
ALARM/GONGS						
Water operated motor alarm and gong; stainless steel and aluminum body and gong; screwed connections						
Connection to sprinkler system and drain pipework	455.78	571.73	6.00	192.98	nr	**764.71**
WATER TANKS						
Note: Prices are based on the most economical tank size for each volume, and the cost will vary with differing tank dimensions, for the same volume						
Steel sectional sprinkler tank; ordinary hazard, life safety classification; two compartment tank, complete with all fittings and accessories to comply with LPCB type A requirements; cost of erection (on prepared supports) is included within net price, labour cost allows for offloading and positioning of materials Volume, size						
70 m³, 6.1 m × 4.88 m × 2.44 m (h)	30431.15	38172.84	24.00	771.90	nr	**38944.74**
105 m³, 7.3 m × 6.1 m × 2.44 m (h)	35709.21	44793.64	28.00	900.56	nr	**45694.20**
168 m³, 9.76 m × 4.88 mx 3.66 m (h)	44119.68	55343.72	28.00	900.56	nr	**56244.28**
211 m³, 9.76 m × 6.1 m × 3.66 m (h)	48708.63	61100.11	32.00	1029.21	nr	**62129.32**
Steel sectional sprinkler tank; ordinary hazard, property protection classification; single compartment tank, complete with all fittings and accessories to comply with LPCB type A requirements; cost of erection (on prepared supports) is included within net price, labour cost allows for offloading and positioning of materials Volume, size						
70 m³, 6.1 m × 4.88 m × 2.44 m (h)	21456.84	26915.46	24.00	771.90	nr	**27687.36**
105 m³, 7.3 m × 6.1 m × 2.44 m (h)	26507.84	33251.43	28.00	900.56	nr	**34151.99**
168 m³, 9.76 m × 4.88 m × 3.66 m (h)	34252.66	42966.54	28.00	900.56	nr	**43867.10**
211 m³, 9.76 m × 6.1 m × 3.66 m (h)	38434.40	48212.11	32.00	1029.21	nr	**49241.32**
GRP sectional sprinkler tank; ordinary hazard, life safety classification; two compartment tank, complete with all fittings and accessories to comply with LPCB type A requirements; cost of erection (on prepared supports) is included within net price, labour cost allows for offloading and positioning of materials Volume, size						
55 m³, 6 m × 4 m × 3 m (h)	29142.59	36556.46	14.00	450.28	nr	**37006.74**
70 m³, 6 m × 5 m × 3 m (h)	32583.54	40872.79	16.00	514.61	nr	**41387.40**
80 m³, 8 m × 4 m × 3 m (h)	33768.20	42358.83	16.00	514.61	nr	**42873.44**

38 PIPED SUPPLY SYSTEMS

Item	Net Price £	Material £	Labour hours	Labour £	Unit	Total rate £
105 m³, 10 m × 4 m × 3 m (h)	38211.35	47932.32	24.00	771.90	nr	**48704.22**
125 m³, 10 m × 5 m × 3 m (h)	43093.93	54057.02	24.00	771.90	nr	**54828.92**
140 m³, 8 m × 7 m × 3 m (h)	44878.66	56295.79	24.00	771.90	nr	**57067.69**
135 m³, 9 m × 6 m × 3 m (h)	46235.49	57997.80	24.00	771.90	nr	**58769.70**
160 m³, 13 m × 5 m × 3 m (h)	50762.15	63676.04	24.00	771.90	nr	**64447.94**
185 m³, 12 m × 6 m × 3 m (h)	53696.84	67357.32	24.00	771.90	nr	**68129.22**
GRP sectional sprinkler tank; ordinary hazard, property protection classification; single compartment tank, complete with all fittings and accessories to comply with LPCB type A requirements; cost of erection (on prepared supports) is within net price, labour cost allows for offloading and positioning of materials						
Volume, size						
55 m³, 6 m × 4 m × 3 m (h)	22420.06	28123.73	14.00	450.28	nr	**28574.01**
70 m³, 6 m × 5 m × 3 m (h)	25329.05	31772.76	16.00	514.61	nr	**32287.37**
80 m³, 8 m × 4 m × 3 m (h)	27035.37	33913.16	16.00	514.61	nr	**34427.77**
105 m³, 10 m × 4 m × 3 m (h)	31672.55	39730.05	24.00	771.90	nr	**40501.95**
125 m³, 10 m × 5 m × 3 m (h)	35815.03	44926.37	24.00	771.90	nr	**45698.27**
140 m³, 8 m × 7 m × 3 m (h)	37076.79	46509.12	24.00	771.90	nr	**47281.02**
135 m³, 9 m × 6 m × 3 m (h)	38955.49	48865.77	24.00	771.90	nr	**49637.67**
160 m³, 13 m × 5 m × 3 m (h)	43672.13	54782.32	24.00	771.90	nr	**55554.22**
185 m³, 12 m × 6 m × 3 m (h)	45878.29	57549.72	24.00	771.90	nr	**58321.62**

38 PIPED SUPPLY SYSTEMS

Item	Net Price £	Material £	Labour hours	Labour £	Unit	Total rate £
FIRE EXTINGUISHERS AND HYDRANTS						
EXTINGUISHERS						
Fire extinguishers; hand held; BS 5423; placed in position						
Water type; cartridge operated; for Class A fires						
Water type, 9 litre capacity; 55 gm CO_2 cartridge; Class A fires (fire-rating 13 A)	35.69	44.77	1.00	32.17	nr	76.94
Foam type, 9 litre capacity; 75 gm CO_2 cartridge; Class A & B fires (fire-rating 13 A:183B)	46.49	58.32	1.00	32.17	nr	90.49
Dry powder type; cartridge operated; for Class A, B & C fires and electrical equipment fires						
Dry powder type, 1 kg capacity; 12 gm CO_2 cartridge; Class A, B & C fires (fire-rating 5 A:34B)	36.30	45.54	1.00	32.17	nr	77.71
Dry powder type, 2 kg capacity; 28 gm CO_2 cartridge; Class A, B & C fires (fire-rating 13 A:55B)	48.88	61.32	1.00	32.17	nr	93.49
Dry powder type, 4 kg capacity; 90 gm CO_2 cartridge; Class A, B & C fires (fire-rating 21 A:183B)	87.08	109.23	1.00	32.17	nr	141.40
Dry powder type, 9 kg capacity; 190 gm CO_2 cartridge; Class A, B & C fires (fire-rating 43 A:233B)	116.68	146.36	1.00	32.17	nr	178.53
Dry powder type; stored pressure type; for Class A, B & C fires and electrical equipment fires						
Dry powder type, 1 kg capacity; Class A, B & C fires (fire-rating 5 A:34B)	15.79	19.80	1.00	32.17	nr	51.97
Dry powder type, 2 kg capacity; Class A, B & C fires (fire-rating 13 A:55B)	14.49	18.18	1.00	32.17	nr	50.35
Dry powder type, 4 kg capacity; Class A, B & C fires (fire-rating 21 A:183B)	29.99	37.62	1.00	32.17	nr	69.79
Dry powder type, 9 kg capacity; Class A, B & C fires (fire-rating 43 A:233B)	31.79	39.87	1.00	32.17	nr	72.04
Carbon dioxide type; for Class B fires and electrical equipment fires						
CO_2 type with hose and horn, 2 kg capacity, Class B fires (fire-rating 34B)	31.19	39.12	1.00	32.17	nr	71.29
CO_2 type with hose and horn, 5 kg capacity, Class B fires (fire-rating 55B)	62.49	78.39	1.00	32.17	nr	110.56
Glass fibre blanket, in GRP container						
1100 × 1100 mm	11.99	15.04	0.50	16.08	nr	31.12
1200 × 1200 mm	13.99	17.55	0.50	16.08	nr	33.63
1800 × 1200 mm	17.59	22.06	0.50	16.08	nr	38.14

38 PIPED SUPPLY SYSTEMS

Item	Net Price £	Material £	Labour hours	Labour £	Unit	Total rate £
HYDRANTS						
Fire hydrants; bolted connections						
Underground hydrants, complete with frost plug to BS 750						
sluice valve pattern type 1	239.69	300.66	4.50	144.74	nr	**445.40**
screw down pattern type 2	174.10	218.39	4.50	144.74	nr	**363.13**
Stand pipe for underground hydrant; screwed base; light alloy						
single outlet	137.85	172.92	1.00	32.17	nr	**205.09**
double outlet	201.71	253.03	1.00	32.17	nr	**285.20**
64 mm dia. bronze/gunmetal outlet valves						
oblique flanged landing valve	150.89	189.28	1.00	32.17	nr	**221.45**
oblique screwed landing valve	150.89	189.28	1.00	32.17	nr	**221.45**
Cast iron surface box; fixing by others						
400 × 200 × 100 mm	132.12	165.73	1.00	32.17	nr	**197.90**
500 × 200 × 150 mm	178.01	223.29	1.00	32.17	nr	**255.46**
frost plug	32.06	40.22	0.25	8.04	nr	**48.26**

38 MECHANICAL/COOLING/HEATING SYSTEMS

Item	Net Price £	Material £	Labour hours	Labour £	Unit	Total rate £
GAS/OIL FIRED BOILERS – DOMESTIC						
Domestic water boilers; stove enamelled casing; electric controls; placing in position; assembling and connecting; electrical work elsewhere						
Gas fired; floor standing; connected to conventional flue						
9–12 kW	647.88	812.71	8.59	276.28	nr	**1088.99**
12–15 kW	681.56	854.95	8.59	276.28	nr	**1131.23**
15–18 kW	725.93	910.60	8.88	285.60	nr	**1196.20**
18–21 kW	843.38	1057.94	9.92	319.05	nr	**1376.99**
21–23 kW	934.97	1172.83	10.66	342.85	nr	**1515.68**
23–29 kW	1212.64	1521.14	11.81	379.85	nr	**1900.99**
29–37 kW	1434.26	1799.13	11.81	379.85	nr	**2178.98**
37–41 kW	1491.57	1871.03	12.68	407.83	nr	**2278.86**
Gas fired; wall hung; connected to conventional flue						
9–12 kW	569.43	714.29	8.59	276.28	nr	**990.57**
12–15 kW	735.15	922.17	8.59	276.28	nr	**1198.45**
13–18 kW	932.97	1170.32	8.59	276.28	nr	**1446.60**
Gas fired; floor standing; connected to balanced flue						
9–12 kW	809.30	1015.19	9.16	294.60	nr	**1309.79**
12–15 kW	842.05	1056.27	10.95	352.18	nr	**1408.45**
15–18 kW	906.39	1136.98	11.98	385.31	nr	**1522.29**
18–21 kW	1068.10	1339.82	12.78	411.04	nr	**1750.86**
21–23 kW	1231.90	1545.30	12.78	411.04	nr	**1956.34**
23–29 kW	1570.21	1969.68	15.45	496.92	nr	**2466.60**
29–37 kW	3018.56	3786.48	17.65	567.68	nr	**4354.16**
Gas fired; wall hung; connected to balanced flue						
6–9 kW	611.06	766.52	9.16	294.60	nr	**1061.12**
9–12 kW	699.25	877.14	9.16	294.60	nr	**1171.74**
12–15 kW	791.80	993.24	9.45	303.95	nr	**1297.19**
15–18 kW	949.41	1190.94	9.74	313.26	nr	**1504.20**
18–22 kW	1007.77	1264.14	9.74	313.26	nr	**1577.40**
Gas fired; wall hung; connected to fan flue (including flue kit)						
6–9 kW	698.87	876.66	9.16	294.60	nr	**1171.26**
9–12 kW	778.82	976.95	9.16	294.60	nr	**1271.55**
12–15 kW	845.43	1060.51	10.95	352.18	nr	**1412.69**
15–18 kW	910.67	1142.34	11.98	385.31	nr	**1527.65**
18–23 kW	1186.58	1488.45	12.78	411.04	nr	**1899.49**
23–29 kW	1536.60	1927.51	15.45	496.92	nr	**2424.43**
29–35 kW	1886.50	2366.43	17.65	567.68	nr	**2934.11**
Oil fired; floor standing; connected to conventional flue						
12–15 kW	1213.44	1522.14	10.38	333.85	nr	**1855.99**
15–19 kW	1265.80	1587.82	12.20	392.39	nr	**1980.21**
21–25 kW	1444.64	1812.16	14.30	459.93	nr	**2272.09**

38 MECHANICAL/COOLING/HEATING SYSTEMS

Item	Net Price £	Material £	Labour hours	Labour £	Unit	Total rate £
26–32 kW	1583.86	1986.79	15.80	508.17	nr	**2494.96**
35–50 kW	1788.23	2243.16	20.46	658.04	nr	**2901.20**
Fire place mounted natural gas fire and back boiler; cast iron water boiler; electric control box; fire output 3 kW with wood surround						
10.50 kW	354.01	444.07	8.88	285.60	nr	**729.67**

38 MECHANICAL/COOLING/HEATING SYSTEMS

Item	Net Price £	Material £	Labour hours	Labour £	Unit	Total rate £
GAS/OIL FIRED BOILERS – COMMERCIAL; FORCED DRAFT						
Commercial steel shell floor standing boilers; pressure jet burner; 8 bar max working pressure; including controls, enamelled jacket, insulation, placing in position and commissioning; electrical work elsewhere						
Natural gas; burner and boiler (high/low type), connected to conventional flue						
411–500 kW	11692.00	14666.44	12.00	385.95	nr	15052.39
601–750 kW	13802.00	17313.23	12.00	385.95	nr	17699.18
1151–1250 kW	18904.00	23713.18	14.00	450.28	nr	24163.46
1426–1500 kW	23146.00	29034.34	14.00	450.28	nr	29484.62
1625–1760 kW	27274.00	34212.51	14.00	450.28	nr	34662.79
Natural gas; burner and boiler (modulating type), connected to conventional flue						
411–500 kW	12531.00	15718.89	12.00	385.95	nr	16104.84
601–750 kW	14479.00	18162.46	12.00	385.95	nr	18548.41
1151–1250 kW	21198.00	26590.77	12.00	385.95	nr	26976.72
1426–1500 kW	25304.00	31741.34	14.00	450.28	nr	32191.62
1625–1760 kW	30433.00	38175.16	14.00	450.28	nr	38625.44
1900–2050 kW	31931.00	40054.25	14.00	450.28	nr	40504.53
2450–2650 kW	37478.00	47012.40	14.00	450.28	nr	47462.68
Natural gas; low Nox burner and boiler (modulating type), connected to conventional flue						
411–500 kW	13125.00	16464.00	12.00	385.95	nr	16849.95
601–750 kW	15635.00	19612.54	12.00	385.95	nr	19998.49
1151–1250 kW	22365.00	28054.66	12.00	385.95	nr	28440.61
1426–1500 kW	27544.00	34551.19	14.00	450.28	nr	35001.47
1625–1760 kW	31595.00	39632.77	14.00	450.28	nr	40083.05
1900–2050 kW	33257.00	41717.58	14.00	450.28	nr	42167.86
2450–2650 kW	38577.00	48390.99	14.00	450.28	nr	48841.27
Oil; burner and boiler (high/low type), connected to conventional flue						
411–500 kW	11292.00	14164.68	12.00	385.95	nr	14550.63
601–750 kW	12745.00	15987.33	12.00	385.95	nr	16373.28
1151–1250 kW	17368.00	21786.42	12.00	385.95	nr	22172.37
1426–1500 kW	21149.00	26529.31	14.00	450.28	nr	26979.59
1625–1760 kW	25194.00	31603.35	14.00	450.28	nr	32053.63
Oil; burner and boiler (modulating type), connected to conventional flue						
411–500 kW	12531.00	15718.89	12.00	385.95	nr	16104.84
601–750 kW	14264.00	17892.76	12.00	385.95	nr	18278.71
1151–1250 kW	20368.00	25549.62	12.00	385.95	nr	25935.57
1426–1500 kW	25151.00	31549.41	14.00	450.28	nr	31999.69
1625–1760 kW	29653.00	37196.72	14.00	450.28	nr	37647.00
1900–2050 kW	30519.00	38283.03	14.00	450.28	nr	38733.31
2450–2650 kW	37247.00	46722.64	14.00	450.28	nr	47172.92

38 MECHANICAL/COOLING/HEATING SYSTEMS

Item	Net Price £	Material £	Labour hours	Labour £	Unit	Total rate £
Dual fuel; burner and boiler (high/low oil and modulating gast type), connected to conventional flue						
411–500 kW	14479.00	18162.46	12.00	385.95	nr	**18548.41**
601–750 kW	16263.00	20400.31	12.00	385.95	nr	**20786.26**
1151–1250 kW	24655.00	30927.23	12.00	385.95	nr	**31313.18**
1426–1500 kW	29653.00	37196.72	14.00	450.28	nr	**37647.00**
1625–1760 kW	34325.00	43057.28	14.00	450.28	nr	**43507.56**
2450–2650-kW	44671.00	56035.30	14.00	450.28	nr	**56485.58**
Dual fuel; low NOX burner and boiler (high/low oil and modulating gast type), connected to conventional flue						
411–500 kW	15123.00	18970.29	12.00	385.95	nr	**19356.24**
601–750 kW	17368.00	21786.42	12.00	385.95	nr	**22172.37**
1151–1250 kW	24715.00	31002.50	12.00	385.95	nr	**31388.45**
1426–1500 kW	29652.00	37195.47	14.00	450.28	nr	**37645.75**
1625–1760 kW	37677.00	47262.03	14.00	450.28	nr	**47712.31**
1900–2050 kW	42525.00	53343.36	14.00	450.28	nr	**53793.64**
2450–2650 kW	44671.00	56035.30	14.00	450.28	nr	**56485.58**

38 MECHANICAL/COOLING/HEATING SYSTEMS

Item	Net Price £	Material £	Labour hours	Labour £	Unit	Total rate £
GAS/OIL FIRED BOILERS – COMMERCIAL; CONDENSING						
Gas boiler; Low Nox wall mounted condensing boiler, with high efficiency modulating premix burner; Aluminium heat exchanger; includes integral control panel; delivery included						
Maximum output						
35 kW	2179.00	2733.34	11.00	353.79	nr	**3087.13**
45 kW	2512.00	3151.05	11.00	353.79	nr	**3504.84**
60 kW	3119.00	3912.47	11.00	353.79	nr	**4266.26**
80 kW	3677.00	4612.43	11.00	353.79	nr	**4966.22**
100 kW	4113.00	5159.35	11.00	353.79	nr	**5513.14**
120 kW	4243.00	5322.42	11.00	353.79	nr	**5676.21**
Gas boiler; Low Nox floor standing condensing boiler with high efficiency modulating premix burner; Stainless steel heat exchanger; includes controls; delivery and commissioning						
Maximum output						
50 kW	4237.00	5314.89	8.00	257.30	nr	**5572.19**
70 kW	4680.00	5870.59	8.00	257.30	nr	**6127.89**
100 kW	5066.00	6354.79	8.00	257.30	nr	**6612.09**
125 kW	5231.00	6561.77	10.00	321.63	nr	**6883.40**
150 kW	5835.00	7319.42	10.00	321.63	nr	**7641.05**
200 kW	6717.00	8425.80	10.00	321.63	nr	**8747.43**
250 kW	7699.00	9657.63	10.00	321.63	nr	**9979.26**
300 kW	8586.00	10770.28	10.00	321.63	nr	**11091.91**
350 kW	9687.00	12151.37	10.00	321.63	nr	**12473.00**
400 kW	10900.00	13672.96	10.00	321.63	nr	**13994.59**
450 kW	12103.00	15182.00	10.00	321.63	nr	**15503.63**
500 kW	13209.00	16569.37	10.00	321.63	nr	**16891.00**
575 kW	14859.00	18639.13	10.00	321.63	nr	**18960.76**
650 kW	15853.00	19886.00	10.00	321.63	nr	**20207.63**
720 kW	18493.00	23197.62	10.00	321.63	nr	**23519.25**
850 kW	22842.00	28653.00	12.00	385.95	nr	**29038.95**
1000 kW	25454.00	31929.50	12.00	385.95	nr	**32315.45**
1150 kW	26485.00	33222.78	12.00	385.95	nr	**33608.73**

38 MECHANICAL/COOLING/HEATING SYSTEMS

Item	Net Price £	Material £	Labour hours	Labour £	Unit	Total rate £
Oil boiler; Low Nox floor standing condensing boiler with high efficiency modulating premix burner; Stainless steel heat exchanger; includes controls; delivery and commissioning						
50 kW	5034.00	6314.65	8.00	257.30	nr	**6571.95**
80 kW	6663.00	8358.07	8.00	257.30	nr	**8615.37**
110 kW	9030.00	11327.23	8.00	257.30	nr	**11584.53**
130 kW	9302.00	11668.43	10.00	321.63	nr	**11990.06**
160 kW	9799.00	12291.87	10.00	321.63	nr	**12613.50**
200 kW	10457.00	13117.26	10.00	321.63	nr	**13438.89**
250 kW	13155.00	16501.63	10.00	321.63	nr	**16823.26**
300 kW	14806.00	18572.65	10.00	321.63	nr	**18894.28**

38 MECHANICAL/COOLING/HEATING SYSTEMS

Item	Net Price £	Material £	Labour hours	Labour £	Unit	Total rate £
GAS/OIL FIRED BOILERS – FLUE SYSTEMS						
Flues; suitable for domestic, medium sized industrial and commercial oil and gas appliances; stainless steel, twin wall, insulated; for use internally or externally						
Straight length; 120 mm long; including one locking band						
127 mm dia.	44.89	56.31	0.49	15.76	nr	72.07
152 mm dia.	50.30	63.10	0.51	16.41	nr	79.51
178 mm dia.	58.38	73.24	0.54	17.37	nr	90.61
203 mm dia.	66.42	83.32	0.58	18.65	nr	101.97
254 mm dia.	78.56	98.55	0.70	22.51	nr	121.06
304 mm dia.	98.42	123.46	0.74	23.80	nr	147.26
355 mm dia.	140.94	176.79	0.80	25.73	nr	202.52
Straight length; 300 mm long; including one locking band						
127 mm dia.	69.25	86.87	0.52	16.72	nr	103.59
152 mm dia.	78.28	98.19	0.52	16.72	nr	114.91
178 mm dia.	89.95	112.83	0.55	17.68	nr	130.51
203 mm dia.	101.79	127.68	0.64	20.59	nr	148.27
254 mm dia.	112.88	141.60	0.79	25.41	nr	167.01
304 mm dia.	135.42	169.87	0.86	27.66	nr	197.53
355 mm dia.	148.61	186.41	0.94	30.23	nr	216.64
400 mm dia.	159.02	199.47	1.03	33.13	nr	232.60
450 mm dia.	181.90	228.18	1.03	33.13	nr	261.31
500 mm dia.	195.10	244.73	1.10	35.37	nr	280.10
550 mm dia.	215.30	270.08	1.10	35.37	nr	305.45
600 mm dia.	237.57	298.01	1.10	35.37	nr	333.38
Straight length; 500 mm long; including one locking band						
127 mm dia.	81.38	102.09	0.55	17.68	nr	119.77
152 mm dia.	90.73	113.81	0.55	17.68	nr	131.49
178 mm dia.	102.17	128.16	0.63	20.26	nr	148.42
203 mm dia.	119.43	149.81	0.63	20.26	nr	170.07
254 mm dia.	138.73	174.03	0.86	27.66	nr	201.69
304 mm dia.	166.19	208.47	0.95	30.55	nr	239.02
355 mm dia.	186.64	234.12	1.03	33.13	nr	267.25
400 mm dia.	203.79	255.63	1.12	36.03	nr	291.66
450 mm dia.	235.36	295.23	1.12	36.03	nr	331.26
500 mm dia.	253.60	318.11	1.19	38.27	nr	356.38
550 mm dia.	279.56	350.68	1.19	38.27	nr	388.95
600 mm dia.	289.59	363.26	1.19	38.27	nr	401.53
Straight length; 1000 mm long; including one locking band						
127 mm dia.	145.47	182.48	0.62	19.95	nr	202.43
152 mm dia.	162.14	203.39	0.68	21.87	nr	225.26
178 mm dia.	182.43	228.84	0.74	23.80	nr	252.64

38 MECHANICAL/COOLING/HEATING SYSTEMS

Item	Net Price £	Material £	Labour hours	Labour £	Unit	Total rate £
203 mm dia.	214.88	269.55	0.80	25.73	nr	295.28
254 mm dia.	243.52	305.47	0.87	27.99	nr	333.46
304 mm dia.	281.10	352.61	1.06	34.09	nr	386.70
355 mm dia.	322.56	404.62	1.16	37.31	nr	441.93
400 mm dia.	345.43	433.31	1.26	40.53	nr	473.84
450 mm dia.	364.49	457.22	1.26	40.53	nr	497.75
500 mm dia.	395.55	496.18	1.33	42.77	nr	538.95
550 mm dia.	435.02	545.69	1.33	42.77	nr	588.46
600 mm dia.	457.02	573.28	1.33	42.77	nr	616.05
Adjustable length; boiler removal; internal use only; including one locking band						
127 mm dia.	66.69	83.65	0.52	16.72	nr	100.37
152 mm dia.	75.35	94.52	0.55	17.68	nr	112.20
178 mm dia.	85.40	107.13	0.59	18.98	nr	126.11
203 mm dia.	97.94	122.85	0.64	20.59	nr	143.44
254 mm dia.	145.21	182.16	0.79	25.41	nr	207.57
304 mm dia.	173.93	218.18	0.86	27.66	nr	245.84
355 mm dia.	195.04	244.65	0.99	31.84	nr	276.49
400 mm dia.	341.30	428.13	0.91	29.27	nr	457.40
450 mm dia.	365.33	458.27	0.91	29.27	nr	487.54
500 mm dia.	398.45	499.81	0.99	31.84	nr	531.65
550 mm dia.	434.68	545.26	0.99	31.84	nr	577.10
600 mm dia.	455.05	570.82	0.99	31.84	nr	602.66
Inspection length; 500 mm long; including one locking band						
127 mm dia.	171.40	215.01	0.55	17.68	nr	232.69
152 mm dia.	177.49	222.64	0.55	17.68	nr	240.32
178 mm dia.	186.94	234.49	0.63	20.26	nr	254.75
203 mm dia.	197.96	248.33	0.63	20.26	nr	268.59
254 mm dia.	255.95	321.06	0.86	27.66	nr	348.72
304 mm dia.	277.39	347.96	0.95	30.55	nr	378.51
355 mm dia.	313.21	392.90	1.03	33.13	nr	426.03
400 mm dia.	502.49	630.32	1.12	36.03	nr	666.35
450 mm dia.	516.16	647.47	1.12	36.03	nr	683.50
500 mm dia.	566.39	710.48	1.19	38.27	nr	748.75
550 mm dia.	592.27	742.94	1.19	38.27	nr	781.21
600 mm dia.	602.33	755.56	1.19	38.27	nr	793.83
Adapters						
127 mm dia.	14.34	17.99	0.49	15.76	nr	33.75
152 mm dia.	15.68	19.67	0.51	16.41	nr	36.08
178 mm dia.	16.78	21.04	0.54	17.37	nr	38.41
203 mm dia.	19.10	23.96	0.58	18.65	nr	42.61
254 mm dia.	21.01	26.35	0.70	22.51	nr	48.86
304 mm dia.	26.21	32.88	0.74	23.80	nr	56.68
355 mm dia.	31.83	39.93	0.80	25.73	nr	65.66
400 mm dia.	35.70	44.78	0.89	28.63	nr	73.41
450 mm dia.	38.13	47.84	0.89	28.63	nr	76.47
500 mm dia.	40.48	50.78	0.96	30.87	nr	81.65
550 mm dia.	47.06	59.04	0.96	30.87	nr	89.91
600 mm dia.	56.56	70.95	0.96	30.87	nr	101.82

38 MECHANICAL/COOLING/HEATING SYSTEMS

Item	Net Price £	Material £	Labour hours	Labour £	Unit	Total rate £
GAS/OIL FIRED BOILERS – FLUE SYSTEMS – cont						
Fittings for flue system						
90° insulated tee; including two locking bands						
127 mm dia.	166.64	209.04	1.89	60.79	nr	269.83
152 mm dia.	192.34	241.27	2.04	65.62	nr	306.89
178 mm dia.	210.22	263.70	2.39	76.87	nr	340.57
203 mm dia.	246.14	308.76	2.56	82.34	nr	391.10
254 mm dia.	249.35	312.78	2.95	94.89	nr	407.67
304 mm dia.	310.08	388.96	3.41	109.67	nr	498.63
355 mm dia.	395.31	495.88	3.77	121.25	nr	617.13
400 mm dia.	520.90	653.42	4.25	136.70	nr	790.12
450 mm dia.	542.93	681.05	4.76	153.10	nr	834.15
500 mm dia.	614.37	770.66	5.12	164.67	nr	935.33
550 mm dia.	660.40	828.41	5.61	180.43	nr	1008.84
600 mm dia.	687.15	861.96	5.98	192.34	nr	1054.30
135° insulated tee; including two locking bands						
127 mm dia.	216.13	271.12	1.89	60.79	nr	331.91
152 mm dia.	233.69	293.14	2.04	65.62	nr	358.76
178 mm dia.	255.50	320.50	2.39	76.87	nr	397.37
203 mm dia.	324.26	406.75	2.56	82.34	nr	489.09
254 mm dia.	368.45	462.18	2.95	94.89	nr	557.07
304 mm dia.	277.62	348.24	3.41	109.67	nr	457.91
355 mm dia.	546.25	685.22	3.77	121.25	nr	806.47
400 mm dia.	719.06	901.99	4.25	136.70	nr	1038.69
450 mm dia.	778.43	976.46	4.76	153.10	nr	1129.56
500 mm dia.	906.86	1137.56	5.12	164.67	nr	1302.23
550 mm dia.	930.44	1167.14	5.61	180.43	nr	1347.57
600 mm dia.	981.98	1231.80	5.98	192.34	nr	1424.14
Wall sleeve; for 135° tee through wall						
127 mm dia.	22.12	27.74	1.89	60.79	nr	88.53
152 mm dia.	29.68	37.23	2.04	65.62	nr	102.85
178 mm dia.	31.08	38.99	2.39	76.87	nr	115.86
203 mm dia.	34.97	43.87	2.56	82.34	nr	126.21
254 mm dia.	39.00	48.92	2.95	94.89	nr	143.81
304 mm dia.	45.45	57.01	3.41	109.67	nr	166.68
355 mm dia.	50.49	63.34	3.77	121.25	nr	184.59
15° insulated elbow; including two locking bands						
127 mm dia.	115.64	145.06	1.57	50.49	nr	195.55
152 mm dia.	128.53	161.22	1.79	57.58	nr	218.80
178 mm dia.	137.46	172.44	2.05	65.93	nr	238.37
203 mm dia.	145.73	182.81	2.33	74.94	nr	257.75
254 mm dia.	150.10	188.28	2.45	78.80	nr	267.08
304 mm dia.	189.75	238.02	3.43	110.32	nr	348.34
355 mm dia.	253.85	318.43	4.71	151.48	nr	469.91

38 MECHANICAL/COOLING/HEATING SYSTEMS

Item	Net Price £	Material £	Labour hours	Labour £	Unit	Total rate £
30° insulated elbow; including two locking bands						
127 mm dia.	115.64	145.06	1.44	46.31	nr	**191.37**
152 mm dia.	128.53	161.22	1.62	52.10	nr	**213.32**
178 mm dia.	137.46	172.44	1.89	60.79	nr	**233.23**
203 mm dia.	145.73	182.81	2.17	69.80	nr	**252.61**
254 mm dia.	150.10	188.28	2.16	69.47	nr	**257.75**
304 mm dia.	189.75	238.02	2.74	88.12	nr	**326.14**
355 mm dia.	253.37	317.82	3.17	101.95	nr	**419.77**
400 mm dia.	253.85	318.43	3.53	113.53	nr	**431.96**
450 mm dia.	294.20	369.04	3.88	124.79	nr	**493.83**
500 mm dia.	308.44	386.90	4.24	136.37	nr	**523.27**
550 mm dia.	331.02	415.23	4.61	148.27	nr	**563.50**
600 mm dia.	360.94	452.76	4.96	159.52	nr	**612.28**
45° insulated elbow; including two locking bands						
127 mm dia.	115.64	145.06	1.44	46.31	nr	**191.37**
152 mm dia.	128.53	161.22	1.51	48.57	nr	**209.79**
178 mm dia.	137.46	172.44	1.58	50.81	nr	**223.25**
203 mm dia.	145.73	182.81	1.66	53.39	nr	**236.20**
254 mm dia.	150.10	188.28	1.72	55.32	nr	**243.60**
304 mm dia.	189.75	238.02	1.80	57.89	nr	**295.91**
355 mm dia.	253.85	318.43	1.94	62.40	nr	**380.83**
400 mm dia.	323.72	406.08	2.01	64.65	nr	**470.73**
450 mm dia.	339.08	425.34	2.09	67.22	nr	**492.56**
500 mm dia.	363.72	456.25	2.16	69.47	nr	**525.72**
550 mm dia.	396.69	497.60	2.23	71.72	nr	**569.32**
600 mm dia.	406.74	510.22	2.30	73.98	nr	**584.20**
Flue supports						
Wall support, galvanized; including plate and brackets						
127 mm dia.	70.21	88.08	2.24	72.04	nr	**160.12**
152 mm dia.	77.19	96.82	2.44	78.48	nr	**175.30**
178 mm dia.	85.22	106.90	2.52	81.04	nr	**187.94**
203 mm dia.	88.74	111.32	2.77	89.08	nr	**200.40**
254 mm dia.	105.83	132.75	2.98	95.85	nr	**228.60**
304 mm dia.	119.65	150.09	3.46	111.28	nr	**261.37**
355 mm dia.	160.16	200.91	4.08	131.22	nr	**332.13**
400 mm dia.; with 300 mm support length and collar	411.40	516.06	4.80	154.38	nr	**670.44**
450 mm dia.; with 300 mm support length and collar	438.62	550.20	5.62	180.76	nr	**730.96**
500 mm dia.; with 300 mm support length and collar	479.18	601.08	6.24	200.69	nr	**801.77**
550 mm dia.; with 300 mm support length and collar	524.26	657.63	6.97	224.18	nr	**881.81**
600 mm dia.; with 300 mm support length and collar	553.42	694.21	7.49	240.89	nr	**935.10**

38 MECHANICAL/COOLING/HEATING SYSTEMS

Item	Net Price £	Material £	Labour hours	Labour £	Unit	Total rate £
GAS/OIL FIRED BOILERS – FLUE SYSTEMS – cont						
Flue supports – cont						
Ceiling/floor support						
127 mm dia.	21.74	27.27	1.86	59.82	nr	87.09
152 mm dia.	24.16	30.31	2.14	68.82	nr	99.13
178 mm dia.	28.93	36.29	1.93	62.08	nr	98.37
203 mm dia.	43.56	54.64	2.74	88.12	nr	142.76
254 mm dia.	49.69	62.33	3.21	103.24	nr	165.57
304 mm dia.	56.79	71.23	3.68	118.36	nr	189.59
355 mm dia.	67.68	84.90	4.28	137.66	nr	222.56
400 mm dia.	86.45	108.44	4.86	156.31	nr	264.75
450 mm dia.	93.88	117.77	5.46	175.60	nr	293.37
500 mm dia.	113.98	142.98	6.04	194.26	nr	337.24
550 mm dia.	139.44	174.91	6.65	213.89	nr	388.80
600 mm dia.	200.10	251.00	7.24	232.86	nr	483.86
Ceiling/floor firestop spacer						
127 mm dia.	4.23	5.31	0.66	21.22	nr	26.53
152 mm dia.	4.71	5.91	0.69	22.19	nr	28.10
178 mm dia.	5.32	6.68	0.70	22.51	nr	29.19
203 mm dia.	6.31	7.92	0.87	27.99	nr	35.91
254 mm dia.	6.52	8.18	0.91	29.27	nr	37.45
304 mm dia.	7.83	9.82	0.95	30.55	nr	40.37
355 mm dia.	14.25	17.88	0.99	31.84	nr	49.72
Wall band; internal or external use						
127 mm dia.	25.59	32.10	1.03	33.13	nr	65.23
152 mm dia.	26.72	33.52	1.07	34.41	nr	67.93
178 mm dia.	27.78	34.84	1.11	35.71	nr	70.55
203 mm dia.	29.33	36.79	1.18	37.96	nr	74.75
254 mm dia.	30.52	38.28	1.30	41.81	nr	80.09
304 mm dia.	32.98	41.37	1.45	46.64	nr	88.01
355 mm dia.	35.09	44.02	1.65	53.08	nr	97.10
400 mm dia.	43.52	54.59	1.85	59.49	nr	114.08
450 mm dia.	46.37	58.16	2.39	76.87	nr	135.03
500 mm dia.	55.74	69.92	2.25	72.36	nr	142.28
550 mm dia.	58.38	73.24	2.45	78.80	nr	152.04
600 mm dia.	61.72	77.43	2.66	85.55	nr	162.98
Flashings and terminals						
Insulated top stub; including one locking band						
127 mm dia.	63.05	79.09	1.49	47.91	nr	127.00
152 mm dia.	71.30	89.44	1.90	61.12	nr	150.56
178 mm dia.	76.76	96.29	1.92	61.76	nr	158.05
203 mm dia.	81.68	102.46	2.20	70.76	nr	173.22
254 mm dia.	86.73	108.80	2.49	80.08	nr	188.88
304 mm dia.	118.55	148.71	2.79	89.73	nr	238.44
355 mm dia.	156.51	196.32	3.19	102.59	nr	298.91
400 mm dia.	150.91	189.30	3.59	115.47	nr	304.77
450 mm dia.	163.95	205.65	3.97	127.68	nr	333.33
500 mm dia.	189.66	237.91	4.38	140.87	nr	378.78
550 mm dia.	198.46	248.95	4.78	153.74	nr	402.69

38 MECHANICAL/COOLING/HEATING SYSTEMS

Item	Net Price £	Material £	Labour hours	Labour £	Unit	Total rate £
600 mm dia.	205.38	257.63	5.17	166.29	nr	423.92
Rain cap; including one locking band						
127 mm dia.	33.71	42.29	1.49	47.91	nr	90.20
152 mm dia.	35.23	44.20	1.54	49.54	nr	93.74
178 mm dia.	38.79	48.65	1.72	55.32	nr	103.97
203 mm dia.	46.38	58.18	2.00	64.32	nr	122.50
254 mm dia.	60.99	76.51	2.49	80.08	nr	156.59
304 mm dia.	41.76	52.38	2.80	90.05	nr	142.43
355 mm dia.	110.17	138.20	3.19	102.59	nr	240.79
400 mm dia.	109.91	137.87	3.45	110.97	nr	248.84
450 mm dia.	119.54	149.95	3.97	127.68	nr	277.63
500 mm dia.	129.24	162.12	4.38	140.87	nr	302.99
550 mm dia.	138.67	173.95	4.78	153.74	nr	327.69
600 mm dia.	148.27	185.99	5.17	166.29	nr	352.28
Round top; including one locking band						
127 mm dia.	64.45	80.84	1.49	47.91	nr	128.75
152 mm dia.	70.27	88.14	1.65	53.08	nr	141.22
178 mm dia.	80.10	100.48	1.92	61.76	nr	162.24
203 mm dia.	94.27	118.25	2.20	70.76	nr	189.01
254 mm dia.	111.46	139.82	2.49	80.08	nr	219.90
304 mm dia.	146.39	183.64	2.80	90.05	nr	273.69
355 mm dia.	195.30	244.99	3.19	102.59	nr	347.58
Coping cap; including one locking band						
127 mm dia.	36.10	45.28	1.49	47.91	nr	93.19
152 mm dia.	37.81	47.43	1.65	53.08	nr	100.51
178 mm dia.	41.60	52.18	1.92	61.76	nr	113.94
203 mm dia.	49.92	62.62	2.20	70.76	nr	133.38
254 mm dia.	60.99	76.51	2.49	80.08	nr	156.59
304 mm dia.	82.23	103.15	2.79	89.73	nr	192.88
355 mm dia.	110.17	138.20	3.19	102.59	nr	240.79
Storm collar						
127 mm dia.	6.84	8.58	0.52	16.72	nr	25.30
152 mm dia.	7.32	9.18	0.55	17.68	nr	26.86
178 mm dia.	8.12	10.18	0.57	18.32	nr	28.50
203 mm dia.	8.50	10.66	0.66	21.22	nr	31.88
254 mm dia.	10.65	13.36	0.66	21.22	nr	34.58
304 mm dia.	11.09	13.91	0.72	23.16	nr	37.07
355 mm dia.	11.86	14.87	0.77	24.76	nr	39.63
400 mm dia.	31.44	39.44	0.82	26.36	nr	65.80
450 mm dia.	34.61	43.41	0.87	27.99	nr	71.40
500 mm dia.	37.73	47.33	0.92	29.59	nr	76.92
550 mm dia.	40.89	51.30	0.98	31.53	nr	82.83
600 mm dia.	44.03	55.23	1.03	33.13	nr	88.36
Flat flashing; including storm collar and sealant						
127 mm dia.	38.43	48.20	1.49	47.91	nr	96.11
152 mm dia.	39.71	49.82	1.65	53.08	nr	102.90
178 mm dia.	41.66	52.26	1.92	61.76	nr	114.02
203 mm dia.	45.63	57.24	2.20	70.76	nr	128.00
254 mm dia.	62.35	78.21	2.49	80.08	nr	158.29
304 mm dia.	74.49	93.44	2.80	90.05	nr	183.49
355 mm dia.	117.39	147.26	3.20	102.93	nr	250.19

38 MECHANICAL/COOLING/HEATING SYSTEMS

Item	Net Price £	Material £	Labour hours	Labour £	Unit	Total rate £
GAS/OIL FIRED BOILERS – FLUE SYSTEMS – cont						
Flashings and terminals – cont						
Flat flashing – cont						
400 mm dia.	163.76	205.42	3.59	115.47	nr	320.89
450 mm dia.	188.13	236.00	3.97	127.68	nr	363.68
500 mm dia.	203.80	255.65	4.38	140.87	nr	396.52
550 mm dia.	216.40	271.45	4.78	153.74	nr	425.19
600 mm dia.	224.22	281.27	5.17	166.29	nr	447.56
5°–30° rigid adjustable flashing; including storm collar and sealant						
127 mm dia.	65.45	82.10	1.49	47.91	nr	130.01
152 mm dia.	68.96	86.51	1.65	53.08	nr	139.59
178 mm dia.	73.33	91.99	1.92	61.76	nr	153.75
203 mm dia.	77.11	96.72	2.20	70.76	nr	167.48
254 mm dia.	81.18	101.83	2.49	80.08	nr	181.91
304 mm dia.	100.45	126.00	2.80	90.05	nr	216.05
355 mm dia.	114.25	143.32	3.19	102.59	nr	245.91
400 mm dia.	368.32	462.02	3.59	115.47	nr	577.49
450 mm dia.	429.85	539.20	3.97	127.68	nr	666.88
500 mm dia.	458.89	575.64	4.38	140.87	nr	716.51
550 mm dia.	480.21	602.38	4.77	153.42	nr	755.80
600 mm dia.	522.46	655.38	5.17	166.29	nr	821.67
Domestic and small commercial; twin walled gas vent system suitable for gas fired appliances; domestic gas boilers; small commercial boilers with internal or external flues						
152 mm long						
100 mm dia.	7.00	8.78	0.52	16.72	nr	25.50
125 mm dia.	8.60	10.79	0.52	16.72	nr	27.51
150 mm dia.	9.32	11.69	0.52	16.72	nr	28.41
305 mm long						
100 mm dia.	10.63	13.34	0.52	16.72	nr	30.06
125 mm dia.	12.47	15.65	0.52	16.72	nr	32.37
150 mm dia.	14.79	18.55	0.52	16.72	nr	35.27
457 mm long						
100 mm dia.	11.75	14.74	0.55	17.68	nr	32.42
125 mm dia.	13.21	16.58	0.55	17.68	nr	34.26
150 mm dia.	16.35	20.51	0.55	17.68	nr	38.19
914 mm long						
100 mm dia.	21.00	26.34	0.62	19.95	nr	46.29
125 mm dia.	24.48	30.71	0.62	19.95	nr	50.66
150 mm dia.	28.07	35.21	0.62	19.95	nr	55.16
1524 mm long						
100 mm dia.	30.33	38.05	0.82	26.36	nr	64.41
125 mm dia.	37.33	46.83	0.84	27.01	nr	73.84
150 mm dia.	40.08	50.28	0.84	27.01	nr	77.29

38 MECHANICAL/COOLING/HEATING SYSTEMS

Item	Net Price £	Material £	Labour hours	Labour £	Unit	Total rate £
Adjustable length 305 mm long						
100 mm dia.	13.45	16.87	0.56	18.01	nr	34.88
125 mm dia.	15.10	18.94	0.56	18.01	nr	36.95
150 mm dia.	19.03	23.87	0.56	18.01	nr	41.88
Adjustable length 457 mm long						
100 mm dia.	18.13	22.75	0.56	18.01	nr	40.76
125 mm dia.	21.99	27.59	0.56	18.01	nr	45.60
150 mm dia.	24.46	30.69	0.56	18.01	nr	48.70
Adjustable elbow 0°–90°						
100 mm dia.	15.34	19.24	0.48	15.44	nr	34.68
125 mm dia.	18.13	22.75	0.48	15.44	nr	38.19
150 mm dia.	22.70	28.47	0.48	15.44	nr	43.91
Draughthood connector						
100 mm dia.	4.73	5.94	0.48	15.44	nr	21.38
125 mm dia.	5.33	6.69	0.48	15.44	nr	22.13
150 mm dia.	5.79	7.26	0.48	15.44	nr	22.70
Adaptor						
100 mm dia.	11.48	14.40	0.48	15.44	nr	29.84
125 mm dia.	11.72	14.71	0.48	15.44	nr	30.15
150 mm dia.	11.97	15.02	0.48	15.44	nr	30.46
Support plate						
100 mm dia.	8.28	10.38	0.48	15.44	nr	25.82
125 mm dia.	8.81	11.05	0.48	15.44	nr	26.49
150 mm dia.	9.43	11.83	0.48	15.44	nr	27.27
Wall band						
100 mm dia.	7.49	9.40	0.48	15.44	nr	24.84
125 mm dia.	7.98	10.01	0.48	15.44	nr	25.45
150 mm dia.	10.12	12.69	0.48	15.44	nr	28.13
Firestop						
100 mm dia.	3.24	4.07	0.48	15.44	nr	19.51
125 mm dia.	3.24	4.07	0.48	15.44	nr	19.51
150 mm dia.	3.73	4.68	0.48	15.44	nr	20.12
Flat flashing						
125 mm dia.	22.07	27.69	0.55	17.68	nr	45.37
150 mm dia.	30.72	38.54	0.55	17.68	nr	56.22
Adjustable flashing 5°–30°						
100 mm dia.	57.53	72.16	0.55	17.68	nr	89.84
125 mm dia.	89.81	112.66	0.55	17.68	nr	130.34
Storm collar						
100 mm dia.	4.59	5.76	0.55	17.68	nr	23.44
125 mm dia.	4.69	5.88	0.55	17.68	nr	23.56
150 mm dia.	4.81	6.04	0.55	17.68	nr	23.72
Gas vent terminal						
100 mm dia.	17.51	21.96	0.55	17.68	nr	39.64
125 mm dia.	19.24	24.14	0.55	17.68	nr	41.82
150 mm dia.	24.68	30.96	0.55	17.68	nr	48.64
Twin wall galvanized steel flue box, 125 mm dia.; fitted for gas fire, where no chimney exists						
Free-standing	111.25	139.55	2.15	69.16	nr	208.71
Recess	111.25	139.55	2.15	69.16	nr	208.71
Back boiler	81.96	102.82	2.40	77.19	nr	180.01

38 MECHANICAL/COOLING/HEATING SYSTEMS

Item	Net Price £	Material £	Labour hours	Labour £	Unit	Total rate £
PACKAGED STEAM GENERATORS						
Packaged steam boilers; boiler mountings centrifugal water feed pump; insulation; and sheet steel wrap around casing; plastic coated						
Gas fired						
104 kW	12532.00	15720.14	71.00	2283.56	nr	**18003.70**
276 kW	17042.00	21377.48	86.00	2766.00	nr	**24143.48**
1384 kW	40022.00	50203.60	148.00	4760.10	nr	**54963.70**
2940 KW	57899.00	72628.51	207.00	6657.71	nr	**79286.22**
4843 kW	68308.00	85685.56	295.00	9488.04	nr	**95173.60**
Gas oil fired						
104 kW	9791.00	12281.83	86.45	2780.48	nr	**15062.31**
276 kW	14083.00	17665.72	148.22	4767.17	nr	**22432.89**
1384 kW	35791.00	44896.23	207.50	6673.79	nr	**51570.02**
2940 kW	54280.00	68088.83	207.00	6657.71	nr	**74746.54**
4843 kW	64804.00	81290.14	295.00	9488.04	nr	**90778.18**

38 MECHANICAL/COOLING/HEATING SYSTEMS

Item	Net Price £	Material £	Labour hours	Labour £	Unit	Total rate £
LOW TEMPERATURE HOT WATER HEATING; PIPELINE: SCREWED STEEL						
Black steel pipes; screwed and socketed joints; BS 1387: 1985. Fixed vertically, brackets measured separately. Screwed joints are within the running length, but any flanges are additional						
Medium weight						
10 mm dia.	4.94	6.19	0.37	11.91	m	18.10
15 mm dia.	3.11	3.90	0.37	11.91	m	15.81
20 mm dia.	3.67	4.60	0.37	11.91	m	16.51
25 mm dia.	5.27	6.61	0.41	13.18	m	19.79
32 mm dia.	6.52	8.18	0.48	15.44	m	23.62
40 mm dia.	7.58	9.51	0.52	16.72	m	26.23
50 mm dia.	10.65	13.36	0.62	19.95	m	33.31
65 mm dia.	14.46	18.14	0.65	20.91	m	39.05
80 mm dia.	18.79	23.56	1.10	35.37	m	58.93
100 mm dia.	26.62	33.39	1.31	42.13	m	75.52
125 mm dia.	33.87	42.48	1.66	53.39	m	95.87
150 mm dia.	39.33	49.34	1.88	60.46	m	109.80
Heavy weight						
15 mm dia.	3.71	4.66	0.37	11.91	m	16.57
20 mm dia.	4.39	5.51	0.37	11.91	m	17.42
25 mm dia.	6.43	8.06	0.41	13.18	m	21.24
32 mm dia.	7.99	10.02	0.48	15.44	m	25.46
40 mm dia.	9.31	11.68	0.52	16.72	m	28.40
50 mm dia.	12.91	16.20	0.62	19.95	m	36.15
65 mm dia.	17.57	22.04	0.64	20.59	m	42.63
80 mm dia.	22.37	28.06	1.10	35.37	m	63.43
100 mm dia.	31.22	39.17	1.31	42.13	m	81.30
125 mm dia.	36.13	45.33	1.66	53.39	m	98.72
150 mm dia.	42.23	52.98	1.88	60.46	m	113.44
200 mm dia.	66.56	83.50	2.99	96.16	m	179.66
250 mm dia.	83.45	104.68	3.49	112.25	m	216.93
300 mm dia.	99.36	124.63	3.91	125.75	m	250.38
Black steel pipes; screwed and socketed joints; BS 1387: 1985. Fixed at high level or suspended, brackets measured separately. Screwed joints are within the running length, but any flanges are additional						
Medium weight						
10 mm dia.	4.94	6.19	0.58	18.65	m	24.84
15 mm dia.	3.11	3.90	0.58	18.65	m	22.55
20 mm dia.	3.67	4.60	0.58	18.65	m	23.25
25 mm dia.	5.27	6.61	0.60	19.30	m	25.91
32 mm dia.	6.52	8.18	0.68	21.87	m	30.05
40 mm dia.	7.58	9.51	0.73	23.49	m	33.00
50 mm dia.	10.65	13.36	0.85	27.33	m	40.69

38 MECHANICAL/COOLING/HEATING SYSTEMS

Item	Net Price £	Material £	Labour hours	Labour £	Unit	Total rate £
LOW TEMPERATURE HOT WATER HEATING; PIPELINE: SCREWED STEEL – cont						
Black steel pipes – cont						
Medium weight – cont						
65 mm dia.	14.46	18.14	0.88	28.30	m	**46.44**
80 mm dia.	18.79	23.56	1.45	46.64	m	**70.20**
100 mm dia.	26.62	33.39	1.74	55.96	m	**89.35**
125 mm dia.	33.87	42.48	2.21	71.08	m	**113.56**
150 mm dia.	39.33	49.34	2.50	80.40	m	**129.74**
Heavy weight						
15 mm dia.	3.71	4.66	0.58	18.65	m	**23.31**
20 mm dia.	4.39	5.51	0.58	18.65	m	**24.16**
25 mm dia.	6.43	8.06	0.60	19.30	m	**27.36**
32 mm dia.	7.99	10.02	0.68	21.87	m	**31.89**
40 mm dia.	9.31	11.68	0.73	23.49	m	**35.17**
50 mm dia.	12.91	16.20	0.85	27.33	m	**43.53**
65 mm dia.	17.57	22.04	0.88	28.30	m	**50.34**
80 mm dia.	22.37	28.06	1.45	46.64	m	**74.70**
100 mm dia.	31.22	39.17	1.74	55.96	m	**95.13**
125 mm dia.	36.13	45.33	2.21	71.08	m	**116.41**
150 mm dia.	42.23	52.98	2.50	80.40	m	**133.38**
200 mm dia.	66.56	83.50	2.99	96.16	m	**179.66**
250 mm dia.	83.45	104.68	3.49	112.25	m	**216.93**
300 mm dia.	99.36	124.63	3.91	125.75	m	**250.38**
Fixings						
For steel pipes; black malleable iron. For minimum fixing distances, refer to the Tables and Memoranda to the rear of the book						
Single pipe bracket, screw on, black malleable iron; screwed to wood						
15 mm dia.	0.81	1.02	0.14	4.50	nr	**5.52**
20 mm dia.	0.90	1.13	0.14	4.50	nr	**5.63**
25 mm dia.	1.06	1.33	0.17	5.47	nr	**6.80**
32 mm dia.	1.92	2.41	0.19	6.10	nr	**8.51**
40 mm dia.	1.83	2.30	0.22	7.08	nr	**9.38**
50 mm dia.	2.57	3.23	0.22	7.08	nr	**10.31**
65 mm dia.	3.37	4.22	0.28	9.00	nr	**13.22**
80 mm dia.	4.61	5.78	0.32	10.28	nr	**16.06**
100 mm dia.	6.75	8.47	0.35	11.26	nr	**19.73**
Single pipe bracket, screw on, black malleable iron; plugged and screwed						
15 mm dia.	0.81	1.02	0.25	8.04	nr	**9.06**
20 mm dia.	0.90	1.13	0.25	8.04	nr	**9.17**
25 mm dia.	1.06	1.33	0.30	9.64	nr	**10.97**
32 mm dia.	1.92	2.41	0.32	10.28	nr	**12.69**
40 mm dia.	1.83	2.30	0.32	10.28	nr	**12.58**
50 mm dia.	2.57	3.23	0.32	10.28	nr	**13.51**
65 mm dia.	3.37	4.22	0.35	11.26	nr	**15.48**

38 MECHANICAL/COOLING/HEATING SYSTEMS

Item	Net Price £	Material £	Labour hours	Labour £	Unit	Total rate £
80 mm dia.	4.61	5.78	0.42	13.51	nr	19.29
100 mm dia.	6.75	8.47	0.42	13.51	nr	21.98
Single pipe bracket for building in, black malleable iron						
15 mm dia.	2.37	2.97	0.10	3.21	nr	6.18
20 mm dia.	2.37	2.97	0.11	3.54	nr	6.51
25 mm dia.	2.37	2.97	0.12	3.86	nr	6.83
32 mm dia.	2.70	3.38	0.14	4.50	nr	7.88
40 mm dia.	2.72	3.42	0.15	4.83	nr	8.25
50 mm dia.	2.72	3.42	0.16	5.14	nr	8.56
Pipe ring, single socket, black malleable iron						
15 mm dia.	0.94	1.18	0.10	3.21	nr	4.39
20 mm dia.	1.05	1.32	0.11	3.54	nr	4.86
25 mm dia.	1.14	1.43	0.12	3.86	nr	5.29
32 mm dia.	1.22	1.53	0.14	4.50	nr	6.03
40 mm dia.	1.57	1.97	0.15	4.83	nr	6.80
50 mm dia.	1.98	2.49	0.16	5.14	nr	7.63
65 mm dia.	2.87	3.60	0.30	9.64	nr	13.24
80 mm dia.	3.43	4.30	0.35	11.26	nr	15.56
100 mm dia.	5.20	6.52	0.40	12.87	nr	19.39
125 mm dia.	10.51	13.18	0.60	19.31	nr	32.49
150 mm dia.	11.77	14.76	0.77	24.76	nr	39.52
200 mm dia.	32.55	40.84	0.90	28.95	nr	69.79
250 mm dia.	40.69	51.04	1.10	35.37	nr	86.41
300 mm dia.	48.94	61.39	1.25	40.21	nr	101.60
350 mm dia.	56.98	71.48	1.50	48.25	nr	119.73
400 mm dia.	67.43	84.58	1.75	56.28	nr	140.86
Pipe ring, double socket, black malleable iron						
15 mm dia.	1.09	1.37	0.10	3.21	nr	4.58
20 mm dia.	1.25	1.57	0.11	3.54	nr	5.11
25 mm dia.	1.40	1.76	0.12	3.86	nr	5.62
32 mm dia.	1.66	2.08	0.14	4.50	nr	6.58
40 mm dia.	1.80	2.26	0.15	4.83	nr	7.09
50 mm dia.	2.81	3.53	0.16	5.14	nr	8.67
Screw on backplate (Male), black malleable iron; plugged and screwed						
M12	0.56	0.71	0.10	3.21	nr	3.92
Screw on backplate (Female), black malleable iron; plugged and screwed						
M12	0.56	0.71	0.10	3.21	nr	3.92
Extra over channel sections for fabricated hangers and brackets						
Galvanized steel; including inserts, bolts, nuts, washers; fixed to backgrounds						
41 × 21 mm	6.34	7.95	0.29	9.33	m	17.28
41 × 41 mm	7.60	9.53	0.29	9.33	m	18.86
Threaded rods; metric thread; including nuts, washers etc.						
12 mm dia. × 600 mm long	3.20	4.01	0.18	5.79	nr	9.80

38 MECHANICAL/COOLING/HEATING SYSTEMS

Item	Net Price £	Material £	Labour hours	Labour £	Unit	Total rate £
LOW TEMPERATURE HOT WATER HEATING; PIPELINE: SCREWED STEEL – cont						
Pipe roller and chair						
Roller and chair; black malleable						
Up to 50 mm dia.	17.15	21.52	0.20	6.44	nr	**27.96**
65 mm dia.	17.66	22.15	0.20	6.44	nr	**28.59**
80 mm dia.	19.18	24.06	0.20	6.44	nr	**30.50**
100 mm dia.	19.52	24.48	0.20	6.44	nr	**30.92**
125 mm dia.	27.34	34.29	0.20	6.44	nr	**40.73**
150 mm dia.	27.34	34.29	0.30	9.64	nr	**43.93**
175 mm dia.	46.14	57.88	0.30	9.64	nr	**67.52**
200 mm dia.	70.67	88.65	0.30	9.64	nr	**98.29**
250 mm dia.	92.18	115.63	0.30	9.64	nr	**125.27**
300 mm dia.	116.08	145.61	0.30	9.64	nr	**155.25**
Roller bracket; black malleable						
25 mm dia.	3.51	4.40	0.20	6.44	nr	**10.84**
32 mm dia.	3.68	4.61	0.20	6.44	nr	**11.05**
40 mm dia.	3.94	4.94	0.20	6.44	nr	**11.38**
50 mm dia.	4.15	5.21	0.20	6.44	nr	**11.65**
65 mm dia.	5.47	6.87	0.20	6.44	nr	**13.31**
80 mm dia.	7.87	9.87	0.20	6.44	nr	**16.31**
100 mm dia.	8.76	10.99	0.20	6.44	nr	**17.43**
125 mm dia.	14.46	18.14	0.20	6.44	nr	**24.58**
150 mm dia.	14.46	18.14	0.30	9.64	nr	**27.78**
175 mm dia.	32.22	40.42	0.30	9.64	nr	**50.06**
200 mm dia.	32.22	40.42	0.30	9.64	nr	**50.06**
250 mm dia.	42.84	53.74	0.30	9.64	nr	**63.38**
300 mm dia.	52.61	65.99	0.30	9.64	nr	**75.63**
350 mm dia.	84.60	106.12	0.30	9.64	nr	**115.76**
400 mm dia.	96.70	121.30	0.30	9.64	nr	**130.94**
Extra over black steel screwed pipes; black steel flanges, screwed and drilled; metric; BS 4504						
Screwed flanges; PN6						
15 mm dia.	8.95	11.22	0.35	11.26	nr	**22.48**
20 mm dia.	8.95	11.22	0.47	15.12	nr	**26.34**
25 mm dia.	8.95	11.22	0.53	17.05	nr	**28.27**
32 mm dia.	8.95	11.22	0.62	19.95	nr	**31.17**
40 mm dia.	8.95	11.22	0.70	22.51	nr	**33.73**
50 mm dia.	9.53	11.95	0.84	27.01	nr	**38.96**
65 mm dia.	13.25	16.62	1.03	33.13	nr	**49.75**
80 mm dia.	18.71	23.48	1.23	39.56	nr	**63.04**
100 mm dia.	22.11	27.73	1.41	45.35	nr	**73.08**
125 mm dia.	40.58	50.90	1.77	56.93	nr	**107.83**
150 mm dia.	40.58	50.90	2.21	71.08	nr	**121.98**

38 MECHANICAL/COOLING/HEATING SYSTEMS

Item	Net Price £	Material £	Labour hours	Labour £	Unit	Total rate £
Screwed flanges; PN16						
15 mm dia.	11.35	14.24	0.35	11.26	nr	25.50
20 mm dia.	11.35	14.24	0.47	15.12	nr	29.36
25 mm dia.	11.35	14.24	0.53	17.05	nr	31.29
32 mm dia.	12.14	15.23	0.62	19.95	nr	35.18
40 mm dia.	12.14	15.23	0.70	22.51	nr	37.74
50 mm dia.	14.80	18.57	0.84	27.01	nr	45.58
65 mm dia.	18.49	23.20	1.03	33.13	nr	56.33
80 mm dia.	22.52	28.25	1.23	39.56	nr	67.81
100 mm dia.	25.25	31.67	1.41	45.35	nr	77.02
125 mm dia.	44.18	55.42	1.77	56.93	nr	112.35
150 mm dia.	43.47	54.53	2.21	71.08	nr	125.61
Extra over black steel screwed pipes; black steel flanges, screwed and drilled; imperial; BS 10						
Screwed flanges; Table E						
½" dia.	16.97	21.29	0.35	11.26	nr	32.55
¾" dia.	16.97	21.29	0.47	15.12	nr	36.41
1" dia.	16.95	21.26	0.53	17.05	nr	38.31
1 ¼" dia.	16.97	21.29	0.62	19.95	nr	41.24
1 ½" dia.	16.97	21.29	0.70	22.51	nr	43.80
2" dia.	16.97	21.29	0.84	27.01	nr	48.30
2 ½" dia.	20.19	25.32	1.03	33.13	nr	58.45
3" dia.	22.87	28.68	1.23	39.56	nr	68.24
4" dia.	30.95	38.82	1.41	45.35	nr	84.17
5" dia.	65.57	82.25	1.77	56.93	nr	139.18
6" dia.	65.57	82.25	2.21	71.08	nr	153.33
Extra over black steel screwed pipes; black steel flange connections						
Bolted connection between pair of flanges; including gasket, bolts, nuts and washers						
50 mm dia.	8.53	10.70	0.53	17.05	nr	27.75
65 mm dia.	10.11	12.68	0.53	17.05	nr	29.73
80 mm dia.	15.17	19.03	0.53	17.05	nr	36.08
100 mm dia.	16.47	20.66	0.61	19.62	nr	40.28
125 mm dia.	19.38	24.32	0.61	19.62	nr	43.94
150 mm dia.	27.69	34.73	0.90	28.95	nr	63.68
Extra over black steel screwed pipes; black heavy steel tubular fittings; BS 1387						
Long screw connection with socket and backnut						
15 mm dia.	5.87	7.36	0.63	20.26	nr	27.62
20 mm dia.	7.39	9.27	0.84	27.01	nr	36.28
25 mm dia.	9.71	12.19	0.95	30.55	nr	42.74
32 mm dia.	13.01	16.32	1.11	35.71	nr	52.03
40 mm dia.	15.65	19.63	1.28	41.17	nr	60.80
50 mm dia.	23.05	28.92	1.53	49.21	nr	78.13
65 mm dia.	32.24	40.44	1.87	60.14	nr	100.58

38 MECHANICAL/COOLING/HEATING SYSTEMS

Item	Net Price £	Material £	Labour hours	Labour £	Unit	Total rate £
LOW TEMPERATURE HOT WATER HEATING; PIPELINE: SCREWED STEEL – cont						
Extra over black steel screwed pipes – cont						
Long screw connection with socket and backnut – cont						
80 mm dia.	73.72	92.48	2.21	71.08	nr	163.56
100 mm dia.	119.43	149.81	3.05	98.09	nr	247.90
Running nipple						
15 mm dia.	1.52	1.90	0.50	16.08	nr	17.98
20 mm dia.	1.92	2.41	0.68	21.87	nr	24.28
25 mm dia.	2.35	2.95	0.77	24.76	nr	27.71
32 mm dia.	3.32	4.17	0.90	28.95	nr	33.12
40 mm dia.	4.47	5.61	1.03	33.13	nr	38.74
50 mm dia.	6.80	8.53	1.23	39.56	nr	48.09
65 mm dia.	14.62	18.33	1.50	48.25	nr	66.58
80 mm dia.	22.81	28.62	1.78	57.25	nr	85.87
100 mm dia.	35.73	44.82	2.38	76.54	nr	121.36
Barrel nipple						
15 mm dia.	0.86	1.08	0.50	16.08	nr	17.16
20 mm dia.	1.38	1.74	0.68	21.87	nr	23.61
25 mm dia.	1.82	2.28	0.77	24.76	nr	27.04
32 mm dia.	2.72	3.42	0.90	28.95	nr	32.37
40 mm dia.	3.35	4.20	1.03	33.13	nr	37.33
50 mm dia.	4.78	5.99	1.23	39.56	nr	45.55
65 mm dia.	10.23	12.84	1.50	48.25	nr	61.09
80 mm dia.	14.25	17.88	1.78	57.25	nr	75.13
100 mm dia.	25.82	32.39	2.38	76.54	nr	108.93
125 mm dia.	47.91	60.10	2.87	92.31	nr	152.41
150 mm dia.	75.52	94.73	3.39	109.03	nr	203.76
Close taper nipple						
15 mm dia.	1.82	2.28	0.50	16.08	nr	18.36
20 mm dia.	2.35	2.95	0.68	21.87	nr	24.82
25 mm dia.	3.08	3.86	0.77	24.76	nr	28.62
32 mm dia.	4.61	5.78	0.90	28.95	nr	34.73
40 mm dia.	5.72	7.18	1.03	33.13	nr	40.31
50 mm dia.	8.81	11.05	1.23	39.56	nr	50.61
65 mm dia.	17.09	21.44	1.50	48.25	nr	69.69
80 mm dia.	21.06	26.42	1.78	57.25	nr	83.67
100 mm dia.	43.27	54.28	2.38	76.54	nr	130.82
Extra over black steel screwed pipes; black malleable iron fittings; BS 143						
Cap						
15 mm dia.	0.73	0.92	0.32	10.28	nr	11.20
20 mm dia.	0.83	1.04	0.43	13.83	nr	14.87
25 mm dia.	1.06	1.33	0.49	15.76	nr	17.09

38 MECHANICAL/COOLING/HEATING SYSTEMS

Item	Net Price £	Material £	Labour hours	Labour £	Unit	Total rate £
32 mm dia.	1.84	2.31	0.58	18.65	nr	**20.96**
40 mm dia.	2.19	2.74	0.66	21.22	nr	**23.96**
50 mm dia.	4.55	5.71	0.78	25.09	nr	**30.80**
65 mm dia.	7.60	9.53	0.96	30.87	nr	**40.40**
80 mm dia.	8.60	10.79	1.13	36.34	nr	**47.13**
100 mm dia.	18.84	23.63	1.70	54.68	nr	**78.31**
Plain plug, hollow						
15 mm dia.	0.51	0.64	0.28	9.00	nr	**9.64**
20 mm dia.	0.66	0.83	0.38	12.22	nr	**13.05**
25 mm dia.	0.89	1.12	0.44	14.15	nr	**15.27**
32 mm dia.	1.24	1.56	0.51	16.41	nr	**17.97**
40 mm dia.	2.34	2.93	0.59	18.98	nr	**21.91**
50 mm dia.	3.28	4.11	0.70	22.51	nr	**26.62**
65 mm dia.	5.35	6.71	0.85	27.33	nr	**34.04**
80 mm dia.	8.35	10.47	1.00	32.17	nr	**42.64**
100 mm dia.	15.37	19.28	1.44	46.31	nr	**65.59**
Plain plug, solid						
15 mm dia.	1.59	1.99	0.28	9.00	nr	**10.99**
20 mm dia.	1.54	1.93	0.38	12.22	nr	**14.15**
25 mm dia.	2.29	2.87	0.44	14.15	nr	**17.02**
32 mm dia.	3.30	4.14	0.51	16.41	nr	**20.55**
40 mm dia.	4.64	5.82	0.59	18.98	nr	**24.80**
50 mm dia.	6.10	7.65	0.70	22.51	nr	**30.16**
90° elbow, male/female						
15 mm dia.	0.83	1.04	0.64	20.59	nr	**21.63**
20 mm dia.	1.14	1.43	0.85	27.33	nr	**28.76**
25 mm dia.	1.88	2.36	0.97	31.19	nr	**33.55**
32 mm dia.	3.45	4.32	1.12	36.03	nr	**40.35**
40 mm dia.	5.78	7.25	1.29	41.50	nr	**48.75**
50 mm dia.	7.44	9.33	1.55	49.85	nr	**59.18**
65 mm dia.	16.07	20.16	1.89	60.79	nr	**80.95**
80 mm dia.	21.96	27.55	2.24	72.04	nr	**99.59**
100 mm dia.	38.40	48.17	3.09	99.39	nr	**147.56**
90° elbow						
15 mm dia.	0.76	0.95	0.64	20.59	nr	**21.54**
20 mm dia.	1.04	1.30	0.85	27.33	nr	**28.63**
25 mm dia.	1.59	1.99	0.97	31.19	nr	**33.18**
32 mm dia.	2.93	3.67	1.12	36.03	nr	**39.70**
40 mm dia.	4.90	6.15	1.29	41.50	nr	**47.65**
50 mm dia.	5.75	7.21	1.55	49.85	nr	**57.06**
65 mm dia.	12.41	15.57	1.89	60.79	nr	**76.36**
80 mm dia.	18.22	22.86	2.24	72.04	nr	**94.90**
100 mm dia.	35.15	44.09	3.09	99.39	nr	**143.48**
125 mm dia.	85.50	107.25	4.44	142.80	nr	**250.05**
150 mm dia.	159.15	199.64	5.79	186.23	nr	**385.87**
45° elbow						
15 mm dia.	1.81	2.27	0.64	20.59	nr	**22.86**
20 mm dia.	2.22	2.79	0.85	27.33	nr	**30.12**
25 mm dia.	3.30	4.14	0.97	31.19	nr	**35.33**

38 MECHANICAL/COOLING/HEATING SYSTEMS

Item	Net Price £	Material £	Labour hours	Labour £	Unit	Total rate £
LOW TEMPERATURE HOT WATER HEATING; PIPELINE: SCREWED STEEL – cont						
Extra over black steel screwed pipes – cont						
45° elbow – cont						
32 mm dia.	6.75	8.47	1.12	36.03	nr	**44.50**
40 mm dia.	8.27	10.37	1.29	41.50	nr	**51.87**
50 mm dia.	11.33	14.21	1.55	49.85	nr	**64.06**
65 mm dia.	16.59	20.81	1.89	60.79	nr	**81.60**
80 mm dia.	24.91	31.25	2.24	72.04	nr	**103.29**
100 mm dia.	54.65	68.56	3.09	99.39	nr	**167.95**
150 mm dia.	152.61	191.43	5.79	186.23	nr	**377.66**
90° bend, male/female						
15 mm dia.	1.48	1.86	0.64	20.59	nr	**22.45**
20 mm dia.	2.17	2.72	0.85	27.33	nr	**30.05**
25 mm dia.	3.19	4.00	0.97	31.19	nr	**35.19**
32 mm dia.	5.07	6.36	1.12	36.03	nr	**42.39**
40 mm dia.	8.01	10.05	1.29	41.50	nr	**51.55**
50 mm dia.	14.00	17.56	1.55	49.85	nr	**67.41**
65 mm dia.	23.28	29.20	1.89	60.79	nr	**89.99**
80 mm dia.	35.82	44.93	2.24	72.04	nr	**116.97**
100 mm dia.	88.78	111.36	3.09	99.39	nr	**210.75**
90° bend, male						
15 mm dia.	3.42	4.29	0.64	20.59	nr	**24.88**
20 mm dia.	3.82	4.79	0.85	27.33	nr	**32.12**
25 mm dia.	5.61	7.03	0.97	31.19	nr	**38.22**
32 mm dia.	12.18	15.28	1.12	36.03	nr	**51.31**
40 mm dia.	17.10	21.45	1.29	41.50	nr	**62.95**
50 mm dia.	25.96	32.57	1.55	49.85	nr	**82.42**
90° bend, female						
15 mm dia.	1.36	1.70	0.64	20.59	nr	**22.29**
20 mm dia.	1.94	2.43	0.85	27.33	nr	**29.76**
25 mm dia.	2.73	3.43	0.97	31.19	nr	**34.62**
32 mm dia.	5.18	6.50	1.12	36.03	nr	**42.53**
40 mm dia.	6.89	8.65	1.29	41.50	nr	**50.15**
50 mm dia.	9.72	12.20	1.55	49.85	nr	**62.05**
65 mm dia.	20.74	26.02	1.89	60.79	nr	**86.81**
80 mm dia.	33.04	41.44	2.24	72.04	nr	**113.48**
100 mm dia.	72.43	90.85	3.09	99.39	nr	**190.24**
125 mm dia.	170.86	214.32	4.44	142.80	nr	**357.12**
150 mm dia.	310.74	389.79	5.79	186.23	nr	**576.02**
Return bend						
15 mm dia.	7.79	9.77	0.64	20.59	nr	**30.36**
20 mm dia.	12.61	15.81	0.85	27.33	nr	**43.14**
25 mm dia.	15.73	19.73	0.97	31.19	nr	**50.92**
32 mm dia.	24.38	30.59	1.12	36.03	nr	**66.62**
40 mm dia.	29.03	36.41	1.29	41.50	nr	**77.91**
50 mm dia.	44.30	55.57	1.55	49.85	nr	**105.42**

38 MECHANICAL/COOLING/HEATING SYSTEMS

Item	Net Price £	Material £	Labour hours	Labour £	Unit	Total rate £
Equal socket, parallel thread						
15 mm dia.	0.73	0.92	0.64	20.59	nr	**21.51**
20 mm dia.	0.89	1.12	0.85	27.33	nr	**28.45**
25 mm dia.	1.23	1.55	0.97	31.19	nr	**32.74**
32 mm dia.	2.46	3.09	1.12	36.03	nr	**39.12**
40 mm dia.	3.58	4.49	1.29	41.50	nr	**45.99**
50 mm dia.	5.19	6.51	1.55	49.85	nr	**56.36**
65 mm dia.	7.97	10.00	1.89	60.79	nr	**70.79**
80 mm dia.	11.30	14.18	2.24	72.04	nr	**86.22**
100 mm dia.	18.63	23.37	3.09	99.39	nr	**122.76**
Concentric reducing socket						
20 × 15 mm dia.	1.16	1.46	0.76	24.45	nr	**25.91**
25 × 15 mm dia.	1.36	1.70	0.85	27.33	nr	**29.03**
25 × 20 mm dia.	1.45	1.81	0.86	27.66	nr	**29.47**
32 × 25 mm dia.	2.72	3.42	1.01	32.49	nr	**35.91**
40 × 25 mm dia.	3.42	4.29	1.16	37.31	nr	**41.60**
40 × 32 mm dia.	3.54	4.44	1.16	37.31	nr	**41.75**
50 × 25 mm dia.	5.09	6.38	1.38	44.39	nr	**50.77**
50 × 40 mm dia.	4.96	6.23	1.38	44.39	nr	**50.62**
65 × 50 mm dia.	9.02	11.31	1.69	54.35	nr	**65.66**
80 × 50 mm dia.	11.74	14.73	2.00	64.32	nr	**79.05**
100 × 50 mm dia.	23.42	29.38	2.75	88.45	nr	**117.83**
100 × 80 mm dia.	32.29	40.50	2.75	88.45	nr	**128.95**
150 × 100 mm dia.	65.03	81.57	4.10	131.86	nr	**213.43**
Eccentric reducing socket						
20 × 15 mm dia.	2.06	2.59	0.73	23.49	nr	**26.08**
25 × 15 mm dia.	5.93	7.44	0.85	27.33	nr	**34.77**
25 × 20 mm dia.	6.71	8.42	0.85	27.33	nr	**35.75**
32 × 25 mm dia.	9.15	11.48	1.01	32.49	nr	**43.97**
40 × 25 mm dia.	10.48	13.15	1.16	37.31	nr	**50.46**
40 × 32 mm dia.	5.26	6.60	1.16	37.31	nr	**43.91**
50 × 25 mm dia.	6.79	8.51	1.38	44.39	nr	**52.90**
50 × 40 mm dia.	9.37	11.75	1.38	44.39	nr	**56.14**
65 × 50 mm dia.	10.90	13.68	1.69	54.35	nr	**68.03**
80 × 50 mm dia.	19.08	23.93	2.00	64.32	nr	**88.25**
Hexagon bush						
20 × 15 mm dia.	0.66	0.83	0.37	11.91	nr	**12.74**
25 × 15 mm dia.	0.80	1.01	0.43	13.83	nr	**14.84**
25 × 20 mm dia.	0.83	1.04	0.43	13.83	nr	**14.87**
32 × 25 mm dia.	1.08	1.36	0.51	16.41	nr	**17.77**
40 × 25 mm dia.	1.72	2.16	0.58	18.65	nr	**20.81**
40 × 32 mm dia.	1.57	1.97	0.58	18.65	nr	**20.62**
50 × 25 mm dia.	3.54	4.44	0.71	22.83	nr	**27.27**
50 × 40 mm dia.	3.28	4.11	0.71	22.83	nr	**26.94**
65 × 50 mm dia.	5.48	6.88	0.85	27.33	nr	**34.21**
80 × 50 mm dia.	8.89	11.16	1.00	32.17	nr	**43.33**
100 × 50 mm dia.	20.57	25.80	1.52	48.89	nr	**74.69**
100 × 80 mm dia.	28.38	35.60	1.52	48.89	nr	**84.49**
150 × 100 mm dia.	61.55	77.21	2.57	82.66	nr	**159.87**

38 MECHANICAL/COOLING/HEATING SYSTEMS

Item	Net Price £	Material £	Labour hours	Labour £	Unit	Total rate £
LOW TEMPERATURE HOT WATER HEATING; PIPELINE: SCREWED STEEL – cont						
Extra over black steel screwed pipes – cont						
Hexagon nipple						
15 mm dia.	0.71	0.90	0.28	9.00	nr	9.90
20 mm dia.	0.80	1.01	0.38	12.22	nr	13.23
25 mm dia.	1.14	1.43	0.44	14.15	nr	15.58
32 mm dia.	2.34	2.93	0.51	16.41	nr	19.34
40 mm dia.	2.72	3.42	0.59	18.98	nr	22.40
50 mm dia.	4.93	6.18	0.70	22.51	nr	28.69
65 mm dia.	8.03	10.07	0.85	27.33	nr	37.40
80 mm dia.	11.61	14.56	1.00	32.17	nr	46.73
100 mm dia.	19.71	24.73	1.44	46.31	nr	71.04
150 mm dia.	43.98	55.17	2.32	74.61	nr	129.78
Union, male/female						
15 mm dia.	3.71	4.66	0.64	20.59	nr	25.25
20 mm dia.	4.53	5.68	0.85	27.33	nr	33.01
25 mm dia.	5.67	7.11	0.97	31.19	nr	38.30
32 mm dia.	8.73	10.95	1.12	36.03	nr	46.98
40 mm dia.	11.17	14.01	1.29	41.50	nr	55.51
50 mm dia.	17.60	22.08	1.55	49.85	nr	71.93
65 mm dia.	33.58	42.12	1.89	60.79	nr	102.91
80 mm dia.	46.18	57.93	2.24	72.04	nr	129.97
Union, female						
15 mm dia.	3.05	3.83	0.64	20.59	nr	24.42
20 mm dia.	3.30	4.14	0.85	27.33	nr	31.47
25 mm dia.	3.84	4.82	0.97	31.19	nr	36.01
32 mm dia.	7.25	9.09	1.12	36.03	nr	45.12
40 mm dia.	8.20	10.28	1.29	41.50	nr	51.78
50 mm dia.	13.57	17.02	1.55	49.85	nr	66.87
65 mm dia.	29.26	36.70	1.89	60.79	nr	97.49
80 mm dia.	40.41	50.69	2.24	72.04	nr	122.73
100 mm dia.	76.94	96.51	3.09	99.39	nr	195.90
Union elbow, male/female						
15 mm dia.	5.36	6.72	0.55	17.68	nr	24.40
20 mm dia.	6.72	8.43	0.85	27.33	nr	35.76
25 mm dia.	9.42	11.82	0.97	31.19	nr	43.01
Twin elbow						
15 mm dia.	4.34	5.44	0.91	29.27	nr	34.71
20 mm dia.	4.81	6.04	1.22	39.23	nr	45.27
25 mm dia.	7.77	9.74	1.39	44.71	nr	54.45
32 mm dia.	15.96	20.03	1.62	52.10	nr	72.13
40 mm dia.	20.21	25.36	1.86	59.82	nr	85.18
50 mm dia.	25.98	32.59	2.21	71.08	nr	103.67
65 mm dia.	46.16	57.90	2.72	87.48	nr	145.38
80 mm dia.	78.66	98.67	3.21	103.24	nr	201.91

38 MECHANICAL/COOLING/HEATING SYSTEMS

Item	Net Price £	Material £	Labour hours	Labour £	Unit	Total rate £
Equal tee						
15 mm dia.	1.04	1.30	0.91	29.27	nr	30.57
20 mm dia.	1.51	1.89	1.22	39.23	nr	41.12
25 mm dia.	2.18	2.73	1.39	44.71	nr	47.44
32 mm dia.	3.96	4.97	1.62	52.10	nr	57.07
40 mm dia.	6.10	7.65	1.86	59.82	nr	67.47
50 mm dia.	8.77	11.00	2.21	71.08	nr	82.08
65 mm dia.	19.22	24.11	2.72	87.48	nr	111.59
80 mm dia.	23.42	29.38	3.21	103.24	nr	132.62
100 mm dia.	42.44	53.23	4.44	142.80	nr	196.03
125 mm dia.	123.26	154.62	5.38	173.03	nr	327.65
150 mm dia.	196.39	246.36	6.31	202.94	nr	449.30
Tee reducing on branch						
20 × 15 mm dia.	1.36	1.70	1.22	39.23	nr	40.93
25 × 15 mm dia.	1.88	2.36	1.39	44.71	nr	47.07
25 × 20 mm dia.	2.24	2.81	1.39	44.71	nr	47.52
32 × 25 mm dia.	3.89	4.88	1.62	52.10	nr	56.98
40 × 25 mm dia.	5.15	6.46	1.86	59.82	nr	66.28
40 × 32 mm dia.	6.76	8.48	1.86	59.82	nr	68.30
50 × 25 mm dia.	7.66	9.61	2.21	71.08	nr	80.69
50 × 40 mm dia.	11.42	14.32	2.21	71.08	nr	85.40
65 × 50 mm dia.	17.83	22.37	2.72	87.48	nr	109.85
80 × 50 mm dia.	24.10	30.23	3.21	103.24	nr	133.47
100 × 50 mm dia.	39.91	50.06	4.44	142.80	nr	192.86
100 × 80 mm dia.	61.55	77.21	4.44	142.80	nr	220.01
150 × 100 mm dia.	144.67	181.47	6.31	202.94	nr	384.41
Equal pitcher tee						
15 mm dia.	3.17	3.98	0.91	29.27	nr	33.25
20 mm dia.	3.94	4.94	1.22	39.23	nr	44.17
25 mm dia.	5.93	7.44	1.39	44.71	nr	52.15
32 mm dia.	8.99	11.28	1.62	52.10	nr	63.38
40 mm dia.	13.87	17.39	1.86	59.82	nr	77.21
50 mm dia.	19.48	24.44	2.21	71.08	nr	95.52
65 mm dia.	30.13	37.80	2.72	87.48	nr	125.28
80 mm dia.	47.03	58.99	3.21	103.24	nr	162.23
100 mm dia.	105.78	132.69	4.44	142.80	nr	275.49
Cross						
15 mm dia.	3.10	3.89	1.00	32.17	nr	36.06
20 mm dia.	4.66	5.85	1.33	42.77	nr	48.62
25 mm dia.	5.93	7.44	1.51	48.57	nr	56.01
32 mm dia.	8.62	10.81	1.77	56.93	nr	67.74
40 mm dia.	11.09	13.91	2.02	64.96	nr	78.87
50 mm dia.	18.02	22.60	2.42	77.84	nr	100.44
65 mm dia.	24.91	31.25	2.97	95.52	nr	126.77
80 mm dia.	33.14	41.57	3.50	112.57	nr	154.14
100 mm dia.	68.39	85.79	4.84	155.67	nr	241.46

38 MECHANICAL/COOLING/HEATING SYSTEMS

Item	Net Price £	Material £	Labour hours	Labour £	Unit	Total rate £
LOW TEMPERATURE HOT WATER HEATING; PIPELINE: BLACK WELDED STEEL						
Black steel pipes; butt welded joints; BS 1387: 1985; including protective painting. Fixed Vertical with brackets measured separately (Refer to Screwed Steel Section). Welded butt joints are within the running length, but any flanges are additional						
Medium weight						
10 mm dia.	2.92	3.66	0.37	11.91	m	15.57
15 mm dia.	3.11	3.90	0.37	11.91	m	15.81
20 mm dia.	3.67	4.60	0.37	11.91	m	16.51
25 mm dia.	5.27	6.61	0.41	13.18	m	19.79
32 mm dia.	6.52	8.18	0.48	15.44	m	23.62
40 mm dia.	7.58	9.51	0.52	16.72	m	26.23
50 mm dia.	10.65	13.36	0.62	19.95	m	33.31
65 mm dia.	14.46	18.14	0.64	20.59	m	38.73
80 mm dia.	18.79	23.56	1.10	35.37	m	58.93
100 mm dia.	26.62	33.39	1.31	42.13	m	75.52
125 mm dia.	33.87	42.48	1.66	53.39	m	95.87
150 mm dia.	39.33	49.34	1.88	60.46	m	109.80
Heavy weight						
15 mm dia.	3.71	4.66	0.37	11.91	m	16.57
20 mm dia.	4.39	5.51	0.37	11.91	m	17.42
25 mm dia.	6.43	8.06	0.41	13.18	m	21.24
32 mm dia.	7.99	10.02	0.48	15.44	m	25.46
40 mm dia.	9.31	11.68	0.52	16.72	m	28.40
50 mm dia.	12.91	16.20	0.62	19.95	m	36.15
65 mm dia.	17.57	22.04	0.64	20.59	m	42.63
80 mm dia.	22.37	28.06	1.10	35.37	m	63.43
100 mm dia.	31.22	39.17	1.31	42.13	m	81.30
125 mm dia.	36.13	45.33	1.66	53.39	m	98.72
150 mm dia.	42.23	52.98	1.88	60.46	m	113.44
Black steel pipes; butt welded joints; BS 1387: 1985; including protective painting. Fixed at High Level or Suspended with brackets measured separately (Refer to Screwed Steel Section). Welded butt joints are within the running length, but any flanges are additional						
Medium weight						
10 mm dia.	2.92	3.66	0.58	18.65	m	22.31
15 mm dia.	3.11	3.90	0.58	18.65	m	22.55
20 mm dia.	3.67	4.60	0.58	18.65	m	23.25
25 mm dia.	5.27	6.61	0.60	19.30	m	25.91
32 mm dia.	6.52	8.18	0.68	21.87	m	30.05
40 mm dia.	7.58	9.51	0.73	23.49	m	33.00
50 mm dia.	10.65	13.36	0.85	27.33	m	40.69

38 MECHANICAL/COOLING/HEATING SYSTEMS

Item	Net Price £	Material £	Labour hours	Labour £	Unit	Total rate £
65 mm dia.	14.46	18.14	0.88	28.30	m	46.44
80 mm dia.	18.79	23.56	1.45	46.64	m	70.20
100 mm dia.	26.62	33.39	1.74	55.96	m	89.35
125 mm dia.	33.87	42.48	2.21	71.08	m	113.56
150 mm dia.	39.33	49.34	2.50	80.40	m	129.74
Heavy weight						
15 mm dia.	3.71	4.66	0.58	18.65	m	23.31
20 mm dia.	4.39	5.51	0.58	18.65	m	24.16
25 mm dia.	6.43	8.06	0.60	19.30	m	27.36
32 mm dia.	7.99	10.02	0.68	21.87	m	31.89
40 mm dia.	9.31	11.68	0.73	23.49	m	35.17
50 mm dia.	12.91	16.20	0.85	27.33	m	43.53
65 mm dia.	17.57	22.04	0.88	28.30	m	50.34
80 mm dia.	22.37	28.06	1.45	46.64	m	74.70
100 mm dia.	31.22	39.17	1.74	55.96	m	95.13
125 mm dia.	36.13	45.33	2.21	71.08	m	116.41
150 mm dia.	42.23	52.98	2.50	80.40	m	133.38
Fixings						
Refer to steel pipes; black malleable iron. For minimum fixing distances, refer to the Tables and Memoranda to the rear of the book						
Extra over black steel butt welded pipes; black steel flanges, welded and drilled; metric; BS 4504						
Welded flanges; PN6						
15 mm dia.	3.86	4.84	0.59	18.98	nr	23.82
20 mm dia.	3.86	4.84	0.69	22.19	nr	27.03
25 mm dia.	3.86	4.84	0.84	27.01	nr	31.85
32 mm dia.	6.40	8.03	1.00	32.17	nr	40.20
40 mm dia.	6.84	8.58	1.11	35.71	nr	44.29
50 mm dia.	7.03	8.81	1.37	44.07	nr	52.88
65 mm dia.	10.33	12.96	1.54	49.54	nr	62.50
80 mm dia.	15.53	19.48	1.67	53.72	nr	73.20
100 mm dia.	16.96	21.28	2.22	71.40	nr	92.68
125 mm dia.	22.23	27.89	2.61	83.94	nr	111.83
150 mm dia.	22.46	28.18	2.99	96.16	nr	124.34
Welded flanges; PN16						
15 mm dia.	7.74	9.71	0.59	18.98	nr	28.69
20 mm dia.	7.74	9.71	0.69	22.19	nr	31.90
25 mm dia.	7.74	9.71	0.84	27.01	nr	36.72
32 mm dia.	10.43	13.08	1.00	32.17	nr	45.25
40 mm dia.	10.43	13.08	1.11	35.71	nr	48.79
50 mm dia.	13.71	17.20	1.37	44.07	nr	61.27
65 mm dia.	15.82	19.85	1.54	49.54	nr	69.39
80 mm dia.	20.11	25.22	1.67	53.72	nr	78.94
100 mm dia.	20.53	25.75	2.22	71.40	nr	97.15
125 mm dia.	32.58	40.87	2.61	83.94	nr	124.81
150 mm dia.	38.90	48.80	2.99	96.16	nr	144.96

38 MECHANICAL/COOLING/HEATING SYSTEMS

Item	Net Price £	Material £	Labour hours	Labour £	Unit	Total rate £
LOW TEMPERATURE HOT WATER HEATING; PIPELINE: BLACK WELDED STEEL – cont						
Extra over black steel butt welded pipes – cont						
Blank flanges, slip on for welding; PN6						
15 mm dia.	2.42	3.04	0.48	15.44	nr	**18.48**
20 mm dia.	2.42	3.04	0.55	17.68	nr	**20.72**
25 mm dia.	2.42	3.04	0.64	20.59	nr	**23.63**
32 mm dia.	4.97	6.24	0.76	24.45	nr	**30.69**
40 mm dia.	4.97	6.24	0.84	27.01	nr	**33.25**
50 mm dia.	4.52	5.67	1.01	32.49	nr	**38.16**
65 mm dia.	8.96	11.24	1.30	41.81	nr	**53.05**
80 mm dia.	9.34	11.72	1.41	45.35	nr	**57.07**
100 mm dia.	9.61	12.05	1.78	57.25	nr	**69.30**
125 mm dia.	19.18	24.06	2.06	66.26	nr	**90.32**
150 mm dia.	18.15	22.77	2.35	75.58	nr	**98.35**
Blank flanges, slip on for welding; PN16						
15 mm dia.	2.20	2.76	0.48	15.44	nr	**18.20**
20 mm dia.	3.37	4.22	0.55	17.68	nr	**21.90**
25 mm dia.	4.10	5.14	0.64	20.59	nr	**25.73**
32 mm dia.	4.20	5.26	0.76	24.45	nr	**29.71**
40 mm dia.	5.43	6.81	0.84	27.01	nr	**33.82**
50 mm dia.	8.42	10.56	1.01	32.49	nr	**43.05**
65 mm dia.	10.33	12.96	1.30	41.81	nr	**54.77**
80 mm dia.	13.41	16.82	1.41	45.35	nr	**62.17**
100 mm dia.	16.62	20.84	1.78	57.25	nr	**78.09**
125 mm dia.	24.39	30.60	2.06	66.26	nr	**96.86**
150 mm dia.	31.48	39.49	2.35	75.58	nr	**115.07**
Extra over black steel butt welded pipes; black steel flanges, welding and drilled; imperial; BS 10						
Welded flanges; Table E						
15 mm dia.	7.94	9.96	0.59	18.98	nr	**28.94**
20 mm dia.	7.94	9.96	0.69	22.19	nr	**32.15**
25 mm dia.	7.94	9.96	0.84	27.01	nr	**36.97**
32 mm dia.	7.94	9.96	1.00	32.17	nr	**42.13**
40 mm dia.	8.15	10.23	1.11	35.71	nr	**45.94**
50 mm dia.	10.42	13.07	1.37	44.07	nr	**57.14**
65 mm dia.	12.68	15.90	1.54	49.54	nr	**65.44**
80 mm dia.	16.67	20.91	1.67	53.72	nr	**74.63**
100 mm dia.	24.62	30.88	2.22	71.40	nr	**102.28**
125 mm dia.	47.89	60.08	2.61	83.94	nr	**144.02**
150 mm dia.	47.89	60.08	2.99	96.16	nr	**156.24**
Blank flanges, slip on for welding; Table E						
15 mm dia.	5.46	6.85	0.48	15.44	nr	**22.29**
20 mm dia.	5.46	6.85	0.55	17.68	nr	**24.53**
25 mm dia.	5.46	6.85	0.64	20.59	nr	**27.44**

38 MECHANICAL/COOLING/HEATING SYSTEMS

Item	Net Price £	Material £	Labour hours	Labour £	Unit	Total rate £
32 mm dia.	6.86	8.60	0.76	24.45	nr	33.05
40 mm dia.	7.47	9.37	0.84	27.01	nr	36.38
50 mm dia.	8.25	10.35	1.01	32.49	nr	42.84
65 mm dia.	9.51	11.93	1.30	41.81	nr	53.74
80 mm dia.	20.75	26.03	1.41	45.35	nr	71.38
100 mm dia.	23.56	29.56	1.78	57.25	nr	86.81
125 mm dia.	38.75	48.61	2.06	66.26	nr	114.87
150 mm dia.	49.82	62.50	2.35	75.58	nr	138.08
Extra over black steel butt welded pipes; black steel flange connections						
Bolted connection between pair of flanges; including gasket, bolts, nuts and washers						
50 mm dia.	8.53	10.70	0.50	16.08	nr	26.78
65 mm dia.	10.11	12.68	0.50	16.08	nr	28.76
80 mm dia.	15.17	19.03	0.50	16.08	nr	35.11
100 mm dia.	16.47	20.66	0.50	16.08	nr	36.74
125 mm dia.	19.38	24.32	0.50	16.08	nr	40.40
150 mm dia.	27.69	34.73	0.88	28.30	nr	63.03
Extra over fittings; BS 1965; butt welded						
Cap						
25 mm dia.	18.72	23.49	0.47	15.12	nr	38.61
32 mm dia.	18.72	23.49	0.59	18.98	nr	42.47
40 mm dia.	18.82	23.61	0.70	22.51	nr	46.12
50 mm dia.	22.01	27.61	0.99	31.84	nr	59.45
65 mm dia.	25.86	32.44	1.35	43.41	nr	75.85
80 mm dia.	26.32	33.02	1.66	53.39	nr	86.41
100 mm dia.	34.41	43.16	2.23	71.72	nr	114.88
125 mm dia.	48.39	60.70	3.03	97.45	nr	158.15
150 mm dia.	55.98	70.22	3.79	121.90	nr	192.12
Concentric reducer						
20 × 15 mm dia.	10.42	13.07	0.69	22.19	nr	35.26
25 × 15 mm dia.	13.01	16.32	0.87	27.99	nr	44.31
25 × 20 mm dia.	10.06	12.62	0.87	27.99	nr	40.61
32 × 25 mm dia.	13.86	17.38	1.08	34.73	nr	52.11
40 × 25 mm dia.	18.08	22.68	1.38	44.39	nr	67.07
40 × 32 mm dia.	12.38	15.53	1.38	44.39	nr	59.92
50 × 25 mm dia.	17.19	21.56	1.82	58.53	nr	80.09
50 × 40 mm dia.	12.18	15.28	1.82	58.53	nr	73.81
65 × 50 mm dia.	16.81	21.09	2.52	81.04	nr	102.13
80 × 50 mm dia.	17.10	21.45	3.24	104.20	nr	125.65
100 × 50 mm dia.	28.32	35.53	4.08	131.22	nr	166.75
100 × 80 mm dia.	19.48	24.44	4.08	131.22	nr	155.66
125 × 80 mm dia.	42.56	53.39	4.71	151.48	nr	204.87
150 × 100 mm dia.	46.03	57.74	5.33	171.43	nr	229.17
Eccentric reducer						
20 × 15 mm dia.	15.51	19.45	0.69	22.19	nr	41.64
25 × 15 mm dia.	20.29	25.45	0.87	27.99	nr	53.44
25 × 20 mm dia.	17.04	21.37	0.87	27.99	nr	49.36

38 MECHANICAL/COOLING/HEATING SYSTEMS

Item	Net Price £	Material £	Labour hours	Labour £	Unit	Total rate £
LOW TEMPERATURE HOT WATER HEATING; PIPELINE: BLACK WELDED STEEL – cont						
Extra over fittings – cont						
Eccentric reducer – cont						
32 × 25 mm dia.	18.82	23.61	1.08	34.73	nr	**58.34**
40 × 25 mm dia.	23.25	29.16	1.38	44.39	nr	**73.55**
40 × 32 mm dia.	22.20	27.84	1.38	44.39	nr	**72.23**
50 × 25 mm dia.	26.49	33.23	1.82	58.53	nr	**91.76**
50 × 40 mm dia.	19.09	23.95	1.82	58.53	nr	**82.48**
65 × 50 mm dia.	22.52	28.25	2.52	81.04	nr	**109.29**
80 × 50 mm dia.	27.57	34.59	3.24	104.20	nr	**138.79**
100 × 50 mm dia.	46.71	58.60	4.08	131.22	nr	**189.82**
100 × 80 mm dia.	34.46	43.23	4.08	131.22	nr	**174.45**
125 × 80 mm dia.	92.05	115.47	4.71	151.48	nr	**266.95**
150 × 100 mm dia.	69.71	87.45	5.33	171.43	nr	**258.88**
45° elbow, long radius						
15 mm dia.	3.91	4.91	0.56	18.01	nr	**22.92**
20 mm dia.	3.96	4.97	0.75	24.12	nr	**29.09**
25 mm dia.	5.15	6.46	0.93	29.92	nr	**36.38**
32 mm dia.	6.25	7.84	1.17	37.63	nr	**45.47**
40 mm dia.	6.33	7.94	1.46	46.95	nr	**54.89**
50 mm dia.	8.66	10.86	1.97	63.36	nr	**74.22**
65 mm dia.	11.15	13.99	2.70	86.84	nr	**100.83**
80 mm dia.	10.30	12.92	3.32	106.77	nr	**119.69**
100 mm dia.	15.58	19.54	4.09	131.54	nr	**151.08**
125 mm dia.	33.14	41.57	4.94	158.88	nr	**200.45**
150 mm dia.	41.17	51.64	5.78	185.90	nr	**237.54**
90° elbow, long radius						
15 mm dia.	4.05	5.08	0.56	18.01	nr	**23.09**
20 mm dia.	3.95	4.95	0.75	24.12	nr	**29.07**
25 mm dia.	5.25	6.59	0.93	29.92	nr	**36.51**
32 mm dia.	6.28	7.87	1.17	37.63	nr	**45.50**
40 mm dia.	6.34	7.95	1.46	46.95	nr	**54.90**
50 mm dia.	8.66	10.86	1.97	63.36	nr	**74.22**
65 mm dia.	11.15	13.99	2.70	86.84	nr	**100.83**
80 mm dia.	12.11	15.19	3.32	106.77	nr	**121.96**
100 mm dia.	18.33	22.99	4.09	131.54	nr	**154.53**
125 mm dia.	38.15	47.86	4.94	158.88	nr	**206.74**
150 mm dia.	48.41	60.73	5.78	185.90	nr	**246.63**
Branch bend (based on branch and pipe sizes being the same)						
15 mm dia.	16.12	20.22	0.85	27.33	nr	**47.55**
20 mm dia.	15.97	20.04	0.85	27.33	nr	**47.37**
25 mm dia.	16.14	20.25	1.02	32.80	nr	**53.05**
32 mm dia.	15.25	19.13	1.11	35.71	nr	**54.84**
40 mm dia.	15.05	18.88	1.36	43.74	nr	**62.62**
50 mm dia.	14.62	18.33	1.70	54.68	nr	**73.01**
65 mm dia.	21.62	27.12	1.78	57.25	nr	**84.37**

38 MECHANICAL/COOLING/HEATING SYSTEMS

Item	Net Price £	Material £	Labour hours	Labour £	Unit	Total rate £
80 mm dia.	33.77	42.36	1.82	58.53	nr	**100.89**
100 mm dia.	44.05	55.26	1.87	60.14	nr	**115.40**
125 mm dia.	77.62	97.36	2.21	71.08	nr	**168.44**
150 mm dia.	119.36	149.72	2.65	85.23	nr	**234.95**
Equal tee						
15 mm dia.	38.20	47.91	0.82	26.36	nr	**74.27**
20 mm dia.	38.20	47.91	1.10	35.37	nr	**83.28**
25 mm dia.	38.20	47.91	1.35	43.41	nr	**91.32**
32 mm dia.	38.20	47.91	1.63	52.42	nr	**100.33**
40 mm dia.	38.20	47.91	2.14	68.82	nr	**116.73**
50 mm dia.	40.24	50.48	3.02	97.13	nr	**147.61**
65 mm dia.	58.70	73.63	3.61	116.11	nr	**189.74**
80 mm dia.	58.81	73.77	4.18	134.44	nr	**208.21**
100 mm dia.	73.15	91.76	5.24	168.53	nr	**260.29**
125 mm dia.	167.43	210.02	6.70	215.49	nr	**425.51**
150 mm dia.	182.90	229.43	8.45	271.78	nr	**501.21**
Extra over black steel butt welded pipes; labour						
Made bend						
15 mm dia.	–	–	0.42	13.51	nr	**13.51**
20 mm dia.	–	–	0.42	13.51	nr	**13.51**
25 mm dia.	–	–	0.50	16.08	nr	**16.08**
32 mm dia.	–	–	0.62	19.95	nr	**19.95**
40 mm dia.	–	–	0.74	23.80	nr	**23.80**
50 mm dia.	–	–	0.89	28.63	nr	**28.63**
65 mm dia.	–	–	1.05	33.77	nr	**33.77**
80 mm dia.	–	–	1.13	36.34	nr	**36.34**
100 mm dia.	–	–	2.90	93.27	nr	**93.27**
125 mm dia.	–	–	3.56	114.50	nr	**114.50**
150 mm dia.	–	–	4.18	134.44	nr	**134.44**
Splay cut end						
15 mm dia.	–	–	0.14	4.50	nr	**4.50**
20 mm dia.	–	–	0.16	5.14	nr	**5.14**
25 mm dia.	–	–	0.18	5.79	nr	**5.79**
32 mm dia.	–	–	0.25	8.04	nr	**8.04**
40 mm dia.	–	–	0.27	8.68	nr	**8.68**
50 mm dia.	–	–	0.31	9.97	nr	**9.97**
65 mm dia.	–	–	0.35	11.26	nr	**11.26**
80 mm dia.	–	–	0.40	12.87	nr	**12.87**
100 mm dia.	–	–	0.48	15.44	nr	**15.44**
125 mm dia.	–	–	0.56	18.01	nr	**18.01**
150 mm dia.	–	–	0.64	20.59	nr	**20.59**
Screwed joint to fitting						
15 mm dia.	–	–	0.30	9.64	nr	**9.64**
20 mm dia.	–	–	0.40	12.87	nr	**12.87**
25 mm dia.	–	–	0.46	14.78	nr	**14.78**
32 mm dia.	–	–	0.53	17.05	nr	**17.05**
40 mm dia.	–	–	0.61	19.62	nr	**19.62**
50 mm dia.	–	–	0.73	23.49	nr	**23.49**
65 mm dia.	–	–	0.89	28.63	nr	**28.63**

38 MECHANICAL/COOLING/HEATING SYSTEMS

Item	Net Price £	Material £	Labour hours	Labour £	Unit	Total rate £
LOW TEMPERATURE HOT WATER HEATING; PIPELINE: BLACK WELDED STEEL – cont						
Extra over black steel butt welded pipes – cont						
Screwed joint to fitting – cont						
80 mm dia.	–	–	1.05	33.77	nr	**33.77**
100 mm dia.	–	–	1.46	46.95	nr	**46.95**
125 mm dia.	–	–	2.10	67.54	nr	**67.54**
150 mm dia.	–	–	2.73	87.81	nr	**87.81**
Straight butt weld						
15 mm dia.	–	–	0.31	9.97	nr	**9.97**
20 mm dia.	–	–	0.42	13.51	nr	**13.51**
25 mm dia.	–	–	0.52	16.72	nr	**16.72**
32 mm dia.	–	–	0.69	22.19	nr	**22.19**
40 mm dia.	–	–	0.83	26.69	nr	**26.69**
50 mm dia.	–	–	1.22	39.23	nr	**39.23**
65 mm dia.	–	–	1.57	50.49	nr	**50.49**
80 mm dia.	–	–	1.95	62.72	nr	**62.72**
100 mm dia.	–	–	2.38	76.54	nr	**76.54**
125 mm dia.	–	–	2.83	91.02	nr	**91.02**
150 mm dia.	–	–	3.27	105.17	nr	**105.17**

38 MECHANICAL/COOLING/HEATING SYSTEMS

Item	Net Price £	Material £	Labour hours	Labour £	Unit	Total rate £
LOW TEMPERATURE HOT WATER HEATING; PIPELINE: CARBON WELDED STEEL						
Hot finished seamless carbon steel pipe; BS 806 and BS 3601; wall thickness to BS 3600; butt welded joints; including protective painting, fixed vertically or at low level, brackets measured separately (Refer to Screwed Pipework Section). Welded butt joints are within the running length, but any flanges are additional						
Pipework						
200 mm dia.	96.30	120.80	2.04	65.62	m	186.42
250 mm dia.	86.71	108.77	2.59	83.31	m	192.08
300 mm dia.	203.83	255.68	2.99	96.16	m	351.84
350 mm dia.	215.13	269.86	3.52	113.21	m	383.07
400 mm dia.	375.07	470.49	4.08	131.22	m	601.71
Hot finished seamless carbon steel pipe; BS 806 and BS 3601; wall thickness to BS 3600; butt welded joints; including protective painting, fixed at high level or suspended, brackets measured separately (Refer to Screwed Pipework Section). Welded butt joints are within the running length, but any flanges are additional						
Pipework						
200 mm dia.	96.30	120.80	3.70	119.00	m	239.80
250 mm dia.	86.71	108.77	4.73	152.13	m	260.90
300 mm dia.	203.83	255.68	5.65	181.72	m	437.40
350 mm dia.	215.13	269.86	6.68	214.85	m	484.71
400 mm dia.	375.07	470.49	7.70	247.65	m	718.14
Fixings						
Refer to steel pipes; black malleable iron. For minimum fixing distances, refer to the Tables and Memoranda to the rear of the book						
Extra over fittings; BS 1965 part 1; butt welded						
Cap						
200 mm dia.	78.84	98.90	3.70	119.00	nr	217.90
250 mm dia.	151.30	189.80	4.73	152.13	nr	341.93
300 mm dia.	165.48	207.58	5.65	181.72	nr	389.30
350 mm dia.	254.95	319.80	6.68	214.85	nr	534.65
400 mm dia.	294.90	369.92	7.70	247.65	nr	617.57
Concentric reducer						
200 mm × 150 mm dia.	86.13	108.05	7.27	233.82	nr	341.87
250 mm × 150 mm dia.	124.90	156.68	9.05	291.07	nr	447.75
250 mm × 200 mm dia.	76.44	95.88	9.10	292.68	nr	388.56

38 MECHANICAL/COOLING/HEATING SYSTEMS

Item	Net Price £	Material £	Labour hours	Labour £	Unit	Total rate £
LOW TEMPERATURE HOT WATER HEATING; PIPELINE: CARBON WELDED STEEL – cont						
Extra over fittings – cont						
Concentric reducer – cont						
300 mm × 150 mm dia.	144.27	180.97	10.75	345.76	nr	526.73
300 mm × 200 mm dia.	149.41	187.42	10.75	345.76	nr	533.18
300 mm × 250 mm dia.	132.82	166.61	11.15	358.62	nr	525.23
350 mm × 200 mm dia.	226.58	284.22	12.50	402.04	nr	686.26
350 mm × 250 mm dia.	189.96	238.29	12.70	408.48	nr	646.77
350 mm × 300 mm dia.	189.96	238.29	13.00	418.12	nr	656.41
400 mm × 250 mm dia.	277.76	348.42	14.46	465.07	nr	813.49
400 mm × 300 mm dia.	343.98	431.49	14.51	466.68	nr	898.17
400 mm × 350 mm dia.	299.82	376.10	15.16	487.58	nr	863.68
Eccentric reducer						
200 mm × 150 mm dia.	155.56	195.14	7.27	233.82	nr	428.96
250 mm × 150 mm dia.	207.28	260.01	9.05	291.07	nr	551.08
250 mm × 200 mm dia.	135.63	170.14	9.10	292.68	nr	462.82
300 mm × 150 mm dia.	248.67	311.93	10.75	345.76	nr	657.69
300 mm × 200 mm dia.	286.80	359.77	10.75	345.76	nr	705.53
300 mm × 250 mm dia.	230.87	289.60	11.15	358.62	nr	648.22
350 mm × 200 mm dia.	263.42	330.43	12.50	402.04	nr	732.47
350 mm × 250 mm dia.	219.00	274.71	12.70	408.48	nr	683.19
350 mm × 300 mm dia.	219.00	274.71	13.00	418.12	nr	692.83
400 mm × 250 mm dia.	351.73	441.21	14.46	465.07	nr	906.28
400 mm × 300 mm dia.	329.37	413.16	14.51	466.68	nr	879.84
400 mm × 350 mm dia.	405.83	509.07	15.16	487.58	nr	996.65
45° elbow						
200 mm dia.	82.38	103.34	7.75	249.27	nr	352.61
250 mm dia.	154.72	194.08	10.05	323.23	nr	517.31
300 mm dia.	226.05	283.56	12.20	392.39	nr	675.95
350 mm dia.	182.94	229.48	14.65	471.20	nr	700.68
400 mm dia.	304.05	381.40	17.12	550.64	nr	932.04
90° elbow						
200 mm dia.	96.92	121.58	7.75	249.27	nr	370.85
250 mm dia.	182.06	228.38	10.05	323.23	nr	551.61
300 mm dia.	265.96	333.63	12.20	392.39	nr	726.02
350 mm dia.	365.62	458.63	14.65	471.20	nr	929.83
400 mm dia.	467.01	585.82	17.12	550.64	nr	1136.46
Equal tee						
200 mm dia.	256.04	321.17	11.25	361.83	nr	683.00
250 mm dia.	439.04	550.73	14.53	467.33	nr	1018.06
300 mm dia.	544.86	683.47	17.55	564.46	nr	1247.93
350 mm dia.	731.01	916.98	20.98	674.78	nr	1591.76
400 mm dia.	835.44	1047.97	24.38	784.12	nr	1832.09

38 MECHANICAL/COOLING/HEATING SYSTEMS

Item	Net Price £	Material £	Labour hours	Labour £	Unit	Total rate £
Extra over black steel butt welded pipes; labour						
Straight butt weld						
200 mm dia.	–	–	4.08	131.22	nr	**131.22**
250 mm dia.	–	–	5.20	167.25	nr	**167.25**
300 mm dia.	–	–	6.22	200.05	nr	**200.05**
350 mm dia.	–	–	7.33	235.75	nr	**235.75**
400 mm dia.	–	–	8.41	270.49	nr	**270.49**
Branch weld						
100 mm dia.	–	–	3.46	111.28	nr	**111.28**
125 mm dia.	–	–	4.23	136.06	nr	**136.06**
150 mm dia.	–	–	5.00	160.81	nr	**160.81**
Extra over black steel butt welded pipes; black steel flanges, welding and drilled; metric; BS 4504						
Welded flanges; PN16						
200 mm dia.	52.72	66.14	4.10	131.86	nr	**198.00**
250 mm dia.	75.72	94.99	5.33	171.43	nr	**266.42**
300 mm dia.	106.07	133.06	6.40	205.84	nr	**338.90**
350 mm dia.	205.15	257.34	7.43	238.97	nr	**496.31**
400 mm dia.	270.86	339.76	8.45	271.78	nr	**611.54**
Welded flanges; PN25						
200 mm dia.	175.04	219.56	4.10	131.86	nr	**351.42**
250 mm dia.	209.82	263.20	5.33	171.43	nr	**434.63**
300 mm dia.	283.73	355.91	6.40	205.84	nr	**561.75**
Blank flanges, slip on for welding; PN16						
200 mm dia.	121.89	152.90	2.70	86.84	nr	**239.74**
250 mm dia.	185.25	232.38	3.48	111.93	nr	**344.31**
300 mm dia.	197.93	248.28	4.20	135.08	nr	**383.36**
350 mm dia.	309.76	388.56	4.78	153.74	nr	**542.30**
400 mm dia.	409.59	513.79	5.35	172.07	nr	**685.86**
Blank flanges, slip on for welding; PN25						
200 mm dia.	121.89	152.90	2.70	86.84	nr	**239.74**
250 mm dia.	185.25	232.38	3.48	111.93	nr	**344.31**
300 mm dia.	197.93	248.28	4.20	135.08	nr	**383.36**
Extra over black steel butt welded pipes; black steel flange connections						
Bolted connection between pair of flanges; including gasket, bolts, nuts and washers						
200 mm dia.	102.29	128.31	3.83	123.18	nr	**251.49**
250 mm dia.	158.31	198.59	4.93	158.57	nr	**357.16**
300 mm dia.	215.85	270.76	5.90	189.76	nr	**460.52**

38 MECHANICAL/COOLING/HEATING SYSTEMS

Item	Net Price £	Material £	Labour hours	Labour £	Unit	Total rate £
LOW TEMPERATURE HOT WATER HEATING; PIPELINE: PRESS FIT						
Press fit jointing system; operating temperature −20°C to +120°C; operating pressure 16 bar; butyl rubber 'O' ring mechanical joint. With brackets measured separately (Refer to Screwed Steel Section)						
Carbon steel						
Pipework						
15 mm dia.	1.66	2.08	0.46	14.78	m	**16.86**
20 mm dia.	2.67	3.35	0.48	15.44	m	**18.79**
25 mm dia.	3.76	4.72	0.52	16.72	m	**21.44**
32 mm dia.	4.86	6.09	0.56	18.01	m	**24.10**
40 mm dia.	6.63	8.32	0.58	18.65	m	**26.97**
50 mm dia.	8.61	10.80	0.66	21.22	m	**32.02**
Extra over for carbon steel press fit fittings						
Coupling						
15 mm dia.	1.13	1.42	0.36	11.58	nr	**13.00**
20 mm dia.	1.38	1.74	0.36	11.58	nr	**13.32**
25 mm dia.	1.75	2.20	0.44	14.15	nr	**16.35**
32 mm dia.	2.91	3.65	0.44	14.15	nr	**17.80**
40 mm dia.	3.89	4.88	0.52	16.72	nr	**21.60**
50 mm dia.	4.60	5.77	0.60	19.30	nr	**25.07**
Reducer						
20 × 15 mm dia.	1.05	1.32	0.36	11.58	nr	**12.90**
25 × 15 mm dia.	1.38	1.74	0.40	12.87	nr	**14.61**
25 × 20 mm dia.	1.44	1.80	0.40	12.87	nr	**14.67**
32 × 20 mm dia.	1.61	2.02	0.40	12.87	nr	**14.89**
32 × 25 mm dia.	1.72	2.16	0.44	14.15	nr	**16.31**
40 × 32 mm dia.	3.73	4.68	0.48	15.44	nr	**20.12**
50 × 20 mm dia.	10.76	13.50	0.48	15.44	nr	**28.94**
50 × 25 mm dia.	10.85	13.61	0.52	16.72	nr	**30.33**
50 × 40 mm dia.	11.41	14.31	0.56	18.01	nr	**32.32**
90° elbow						
15 mm dia.	1.63	2.05	0.36	11.58	nr	**13.63**
20 mm dia.	2.15	2.70	0.36	11.58	nr	**14.28**
25 mm dia.	2.95	3.70	0.44	14.15	nr	**17.85**
32 mm dia.	7.36	9.23	0.44	14.15	nr	**23.38**
40 mm dia.	11.77	14.76	0.52	16.72	nr	**31.48**
50 mm dia.	14.06	17.64	0.60	19.30	nr	**36.94**
45° elbow						
15 mm dia.	1.95	2.44	0.36	11.58	nr	**14.02**
20 mm dia.	2.18	2.73	0.36	11.58	nr	**14.31**
25 mm dia.	2.96	3.72	0.44	14.15	nr	**17.87**
32 mm dia.	5.84	7.32	0.44	14.15	nr	**21.47**
40 mm dia.	7.32	9.18	0.52	16.72	nr	**25.90**
50 mm dia.	8.28	10.38	0.60	19.30	nr	**29.68**

38 MECHANICAL/COOLING/HEATING SYSTEMS

Item	Net Price £	Material £	Labour hours	Labour £	Unit	Total rate £
Equal tee						
15 mm dia.	3.13	3.93	0.54	17.37	nr	**21.30**
20 mm dia.	3.63	4.56	0.54	17.37	nr	**21.93**
25 mm dia.	4.86	6.09	0.66	21.22	nr	**27.31**
32 mm dia.	7.57	9.50	0.66	21.22	nr	**30.72**
40 mm dia.	11.17	14.01	0.78	25.09	nr	**39.10**
50 mm dia.	13.39	16.80	0.90	28.95	nr	**45.75**
Reducing tee						
20 × 15 mm dia.	3.56	4.47	0.54	17.37	nr	**21.84**
25 × 15 mm dia.	4.81	6.04	0.62	19.95	nr	**25.99**
25 × 20 mm dia.	5.23	6.56	0.62	19.95	nr	**26.51**
32 × 15 mm dia.	7.05	8.85	0.62	19.95	nr	**28.80**
32 × 20 mm dia.	7.62	9.55	0.62	19.95	nr	**29.50**
32 × 25 mm dia.	7.74	9.71	0.62	19.95	nr	**29.66**
40 × 20 mm dia.	10.20	12.79	0.70	22.51	nr	**35.30**
40 × 25 mm dia.	10.58	13.27	0.70	22.51	nr	**35.78**
40 × 32 mm dia.	10.33	12.96	0.70	22.51	nr	**35.47**
50 × 20 mm dia.	12.16	15.25	0.82	26.36	nr	**41.61**
50 × 25 mm dia.	12.40	15.56	0.82	26.36	nr	**41.92**
50 × 32 mm dia.	12.77	16.02	0.82	26.36	nr	**42.38**
50 × 40 mm dia.	13.38	16.79	0.82	26.36	nr	**43.15**

38 MECHANICAL/COOLING/HEATING SYSTEMS

Item	Net Price £	Material £	Labour hours	Labour £	Unit	Total rate £
LOW TEMPERATURE HOT WATER HEATING; PIPELINE: MECHANICAL GROOVED						
MECHANICAL GROOVED						
Mechanical grooved jointing system; working temperature not exceeding 82° C BS 5750; pipework complete with grooved joints; painted finish. With brackets measured separately (Refer to Screwed Steel Section)						
Grooved joints						
65 mm	15.23	19.11	0.58	18.65	m	**37.76**
80 mm	15.96	20.03	0.68	21.87	m	**41.90**
100 mm	20.71	25.98	0.79	25.41	m	**51.39**
125 mm	32.51	40.78	1.02	32.80	m	**73.58**
150 mm	33.93	42.56	1.15	36.99	m	**79.55**
Extra over mechanical grooved system fittings						
Couplings						
65 mm	15.23	19.11	0.41	13.18	nr	**32.29**
80 mm	15.96	20.03	0.41	13.18	nr	**33.21**
100 mm	20.71	25.98	0.66	21.22	nr	**47.20**
125 mm	32.51	40.78	0.68	21.87	nr	**62.65**
150 mm	33.93	42.56	0.80	25.73	nr	**68.29**
Concentric reducers (one size down)						
80 mm	24.44	30.65	0.59	18.98	nr	**49.63**
100 mm	24.50	30.73	0.71	22.83	nr	**53.56**
125 mm	42.23	52.98	0.85	27.33	nr	**80.31**
150 mm	54.75	68.68	0.98	31.53	nr	**100.21**
Short radius elbow; 90°						
65 mm	27.99	35.11	0.53	17.05	nr	**52.16**
80 mm	28.56	35.83	0.61	19.62	nr	**55.45**
100 mm	38.24	47.97	0.80	25.73	nr	**73.70**
125 mm	63.06	79.11	0.90	28.95	nr	**108.06**
150 mm	81.84	102.66	0.94	30.23	nr	**132.89**
Short radius elbow; 45°						
65 mm	23.96	30.06	0.53	17.05	nr	**47.11**
80 mm	26.92	33.77	0.61	19.62	nr	**53.39**
100 mm	33.46	41.98	0.80	25.73	nr	**67.71**
125 mm	56.61	71.01	0.90	28.95	nr	**99.96**
150 mm	62.32	78.18	0.94	30.23	nr	**108.41**
Equal tee						
65 mm	50.38	63.20	0.83	26.69	nr	**89.89**
80 mm	53.37	66.94	0.93	29.92	nr	**96.86**
100 mm	59.68	74.86	1.18	37.96	nr	**112.82**
125 mm	158.29	198.55	1.37	44.07	nr	**242.62**
150 mm	146.76	184.09	1.43	46.00	nr	**230.09**

38 MECHANICAL/COOLING/HEATING SYSTEMS

Item	Net Price £	Material £	Labour hours	Labour £	Unit	Total rate £
LOW TEMPERATURE HOT WATER HEATING; PIPELINE: PLASTIC PIPEWORK						
Polypropylene PP-R 80 pipe, mechanically stabilized by fibre compound mixture in middle layer; suitable for continuous working temperatures of 0–90°C; thermally fused joints in the running length						
Pipe; 4 m long; PN 20						
20 mm dia.	3.03	3.80	0.35	11.26	m	**15.06**
25 mm dia.	4.56	5.72	0.39	12.54	m	**18.26**
32 mm dia.	5.21	6.54	0.43	13.83	m	**20.37**
40 mm dia.	6.95	8.71	0.47	15.12	m	**23.83**
50 mm dia.	10.11	12.68	0.51	16.41	m	**29.09**
63 mm dia.	16.69	20.93	0.52	16.72	m	**37.65**
75 mm dia.	21.59	27.08	0.60	19.30	m	**46.38**
90 mm dia.	33.30	41.78	0.69	22.19	m	**63.97**
110 mm dia.	50.06	62.80	0.69	22.19	m	**84.99**
125 mm dia.	53.63	67.28	0.85	27.33	m	**94.61**
Fixings						
Refer to steel pipes; black malleable iron. For minimum fixing distances, refer to the Tables and Memoranda to the rear of the book						
Extra over fittings; thermally fused joints						
Overbridge bow						
20 mm dia.	2.01	2.52	0.51	16.41	nr	**18.93**
25 mm dia.	3.70	4.64	0.56	18.01	nr	**22.65**
32 mm dia.	7.39	9.27	0.65	20.91	nr	**30.18**
Elbow 90°						
20 mm dia.	0.68	0.85	0.44	14.15	nr	**15.00**
25 mm dia.	0.89	1.12	0.52	16.72	nr	**17.84**
32 mm dia.	1.28	1.60	0.59	18.98	nr	**20.58**
40 mm dia.	2.03	2.54	0.66	21.22	nr	**23.76**
50 mm dia.	4.38	5.50	0.73	23.49	nr	**28.99**
63 mm dia.	6.70	8.40	0.85	27.33	nr	**35.73**
75 mm dia.	14.82	18.59	0.85	27.33	nr	**45.92**
90 mm dia.	27.38	34.35	1.04	33.45	nr	**67.80**
110 mm dia.	38.92	48.82	1.04	33.45	nr	**82.27**
125 mm dia.	59.98	75.24	1.30	41.81	nr	**117.05**
Long bend 90°						
20 mm dia.	3.60	4.51	0.48	15.44	nr	**19.95**
25 mm dia.	3.77	4.73	0.57	18.32	nr	**23.05**
32 mm dia.	4.32	5.42	0.65	20.91	nr	**26.33**
40 mm dia.	8.08	10.14	0.73	23.49	nr	**33.63**
Elbow 90°, female/male						
20 mm dia.	0.70	0.87	0.44	14.15	nr	**15.02**
25 mm dia.	0.91	1.14	0.52	16.72	nr	**17.86**
32 mm dia.	1.30	1.64	0.59	18.98	nr	**20.62**

38 MECHANICAL/COOLING/HEATING SYSTEMS

Item	Net Price £	Material £	Labour hours	Labour £	Unit	Total rate £
LOW TEMPERATURE HOT WATER HEATING; PIPELINE: PLASTIC PIPEWORK – cont						
Extra over fittings – cont						
Elbow 45°						
20 mm dia.	0.70	0.87	0.44	14.15	nr	**15.02**
25 mm dia.	0.91	1.14	0.52	16.72	nr	**17.86**
32 mm dia.	1.30	1.64	0.59	18.98	nr	**20.62**
40 mm dia.	2.04	2.55	0.66	21.22	nr	**23.77**
50 mm dia.	4.38	5.50	0.73	23.49	nr	**28.99**
63 mm dia.	6.71	8.42	0.85	27.33	nr	**35.75**
75 mm dia.	14.83	18.60	0.85	27.33	nr	**45.93**
90 mm dia.	27.38	34.35	1.04	33.45	nr	**67.80**
110 mm dia.	38.94	48.84	1.04	33.45	nr	**82.29**
125 mm dia.	59.99	75.25	1.30	41.81	nr	**117.06**
Elbow 45°, female/male						
20 mm dia.	0.70	0.87	0.44	14.15	nr	**15.02**
25 mm dia.	0.91	1.14	0.52	16.72	nr	**17.86**
32 mm dia.	1.30	1.64	0.59	18.98	nr	**20.62**
T-piece 90°						
20 mm dia.	0.94	1.18	0.61	19.62	nr	**20.80**
25 mm dia.	1.27	1.59	0.72	23.16	nr	**24.75**
32 mm dia.	1.66	2.08	0.83	26.69	nr	**28.77**
40 mm dia.	2.57	3.23	0.92	29.59	nr	**32.82**
50 mm dia.	7.29	9.14	1.01	32.49	nr	**41.63**
63 mm dia.	10.44	13.09	1.11	35.71	nr	**48.80**
75 mm dia.	17.41	21.84	1.18	37.96	nr	**59.80**
90 mm dia.	32.05	40.21	1.46	46.95	nr	**87.16**
110 mm dia.	50.00	62.72	1.46	46.95	nr	**109.67**
125 mm dia.	66.45	83.35	1.82	58.53	nr	**141.88**
T-piece 90° reducing						
25 × 20 × 20 mm	1.30	1.64	0.72	23.16	nr	**24.80**
25 × 20 × 25 mm	1.30	1.64	0.72	23.16	nr	**24.80**
32 × 20 × 32 mm	1.66	2.08	0.83	26.69	nr	**28.77**
32 × 25 × 32 mm	1.66	2.08	0.83	26.69	nr	**28.77**
40 × 20 × 40 mm	2.57	3.23	0.92	29.59	nr	**32.82**
40 × 25 × 40 mm	2.57	3.23	0.92	29.59	nr	**32.82**
40 × 32 × 40 mm	2.57	3.23	0.92	29.59	nr	**32.82**
50 × 25 × 50 mm	7.29	9.14	1.01	32.49	nr	**41.63**
50 × 32 × 50 mm	7.29	9.14	1.01	32.49	nr	**41.63**
50 × 40 × 50 mm	7.29	9.14	1.01	32.49	nr	**41.63**
63 × 20 × 63 mm	9.83	12.33	1.11	35.71	nr	**48.04**
63 × 25 × 63 mm	9.83	12.33	1.11	35.71	nr	**48.04**
63 × 32 × 63 mm	9.83	12.33	1.11	35.71	nr	**48.04**
63 × 40 × 63 mm	9.83	12.33	1.01	32.49	nr	**44.82**
63 × 50 × 63 mm	9.83	12.33	1.01	32.49	nr	**44.82**
75 × 20 × 75 mm	15.98	20.05	1.18	37.96	nr	**58.01**
75 × 25 × 75 mm	15.98	20.05	1.18	37.96	nr	**58.01**
75 × 32 × 75 mm	15.98	20.05	1.18	37.96	nr	**58.01**
75 × 40 × 75 mm	15.98	20.05	1.18	37.96	nr	**58.01**

38 MECHANICAL/COOLING/HEATING SYSTEMS

Item	Net Price £	Material £	Labour hours	Labour £	Unit	Total rate £
75 × 50 × 75 mm	15.98	20.05	1.18	37.96	nr	**58.01**
75 × 63 × 75 mm	15.98	20.05	1.18	37.96	nr	**58.01**
90 × 63 × 90 mm	32.05	40.21	1.46	46.95	nr	**87.16**
110 × 75 × 110 mm	50.00	62.72	1.46	46.95	nr	**109.67**
110 × 90 × 110 mm	50.00	62.72	1.46	46.95	nr	**109.67**
125 × 90 × 125 mm	59.48	74.61	1.82	58.53	nr	**133.14**
125 × 110 × 125 mm	60.70	76.14	1.82	58.53	nr	**134.67**
Reducer						
25 × 20 mm	0.75	0.94	0.59	18.98	nr	**19.92**
32 × 20 mm	0.97	1.22	0.62	19.95	nr	**21.17**
32 × 25 mm	0.97	1.22	0.62	19.95	nr	**21.17**
40 × 20 mm	1.53	1.92	0.66	21.22	nr	**23.14**
40 × 25 mm	1.53	1.92	0.66	21.22	nr	**23.14**
40 × 32 mm	1.53	1.92	0.66	21.22	nr	**23.14**
50 × 20 mm	2.45	3.07	0.73	23.49	nr	**26.56**
50 × 25 mm	2.45	3.07	0.73	23.49	nr	**26.56**
50 × 32 mm	2.45	3.07	0.73	23.49	nr	**26.56**
50 × 40 mm	2.45	3.07	0.73	23.49	nr	**26.56**
63 × 40 mm	4.96	6.23	0.78	25.09	nr	**31.32**
63 × 25 mm	4.96	6.23	0.78	25.09	nr	**31.32**
63 × 32 mm	4.96	6.23	0.78	25.09	nr	**31.32**
63 × 50 mm	4.96	6.23	0.78	25.09	nr	**31.32**
75 × 50 mm	5.54	6.94	0.85	27.33	nr	**34.27**
75 × 63 mm	5.54	6.94	0.85	27.33	nr	**34.27**
90 × 63 mm	12.37	15.51	1.04	33.45	nr	**48.96**
90 × 75 mm	12.37	15.51	1.04	33.45	nr	**48.96**
110 × 90 mm	19.99	25.08	1.17	37.63	nr	**62.71**
125 × 110 mm	31.23	39.18	1.43	46.00	nr	**85.18**
Socket						
20 mm dia.	0.67	0.84	0.51	16.41	nr	**17.25**
25 mm dia.	0.75	0.94	0.56	18.01	nr	**18.95**
32 mm dia.	0.96	1.21	0.65	20.91	nr	**22.12**
40 mm dia.	1.18	1.48	0.74	23.80	nr	**25.28**
50 mm dia.	2.44	3.06	0.81	26.05	nr	**29.11**
63 mm dia.	4.96	6.23	0.86	27.66	nr	**33.89**
75 mm dia.	5.54	6.94	0.91	29.27	nr	**36.21**
90 mm dia.	14.29	17.92	0.91	29.27	nr	**47.19**
110 mm dia.	24.29	30.46	0.91	29.27	nr	**59.73**
125 mm dia.	33.83	42.44	1.30	41.81	nr	**84.25**
End cap						
20 mm dia.	1.04	1.30	0.25	8.04	nr	**9.34**
25 mm dia.	1.27	1.59	0.29	9.33	nr	**10.92**
32 mm dia.	1.58	1.98	0.33	10.62	nr	**12.60**
40 mm dia.	2.53	3.17	0.36	11.58	nr	**14.75**
50 mm dia.	3.50	4.39	0.40	12.87	nr	**17.26**
63 mm dia.	5.89	7.39	0.44	14.15	nr	**21.54**
75 mm dia.	8.47	10.63	0.47	15.12	nr	**25.75**
90 mm dia.	19.20	24.08	0.57	18.32	nr	**42.40**
110 mm dia.	23.08	28.95	0.57	18.32	nr	**47.27**
125 mm dia.	35.16	44.11	0.85	27.33	nr	**71.44**

38 MECHANICAL/COOLING/HEATING SYSTEMS

Item	Net Price £	Material £	Labour hours	Labour £	Unit	Total rate £
LOW TEMPERATURE HOT WATER HEATING; PIPELINE: PLASTIC PIPEWORK – cont						
Extra over fittings – cont						
Stub flange with gasket						
32 mm dia.	23.32	29.25	0.23	7.40	nr	36.65
40 mm dia.	29.34	36.80	0.27	8.68	nr	45.48
50 mm dia.	35.49	44.52	0.38	12.22	nr	56.74
63 mm dia.	42.61	53.45	0.43	13.83	nr	67.28
75 mm dia.	50.00	62.72	0.48	15.44	nr	78.16
90 mm dia.	67.67	84.88	0.53	17.05	nr	101.93
110 mm dia.	94.76	118.87	0.53	17.05	nr	135.92
125 mm dia.	135.82	170.37	0.75	24.12	nr	194.49
Weld in saddle with female thread						
40–½"	1.68	2.11	0.36	11.58	nr	13.69
50–½"	1.68	2.11	0.36	11.58	nr	13.69
63–½"	1.68	2.11	0.40	12.87	nr	14.98
75–½"	1.68	2.11	0.40	12.87	nr	14.98
90–½"	1.68	2.11	0.46	14.78	nr	16.89
110–½"	1.68	2.11	0.46	14.78	nr	16.89
Weld in saddle with male thread						
50–½"	1.68	2.11	0.36	11.58	nr	13.69
63–½"	1.68	2.11	0.40	12.87	nr	14.98
75–½"	1.68	2.11	0.40	12.87	nr	14.98
90–½"	1.68	2.11	0.46	14.78	nr	16.89
110–½"	1.68	2.11	0.46	14.78	nr	16.89
Transition piece, round with female thread						
20 × ½"	3.92	4.92	0.29	9.33	nr	14.25
20 × ¾"	5.18	6.50	0.29	9.33	nr	15.83
25 × ½"	3.92	4.92	0.33	10.62	nr	15.54
25 × ¾"	5.18	6.50	0.33	10.62	nr	17.12
Transition piece, hexagon with female thread						
32 × 1"	14.59	18.30	0.36	11.58	nr	29.88
40 × 1 ¼"	23.10	28.97	0.36	11.58	nr	40.55
50 × 1 ½"	26.82	33.64	0.36	11.58	nr	45.22
63 × 2"	41.53	52.09	0.40	12.87	nr	64.96
75 × 2"	43.33	54.35	0.40	12.87	nr	67.22
125 × 5"	83.01	104.13	0.51	16.41	nr	120.54
Stop valve for surface assembly						
20 mm dia.	16.12	20.22	0.25	8.04	nr	28.26
25 mm dia.	16.12	20.22	0.29	9.33	nr	29.55
32 mm dia.	30.35	38.07	0.33	10.62	nr	48.69
Ball valve						
20 mm dia.	56.75	71.19	0.25	8.04	nr	79.23
25 mm dia.	60.81	76.28	0.29	9.33	nr	85.61
32 mm dia.	73.07	91.66	0.33	10.62	nr	102.28
40 mm dia.	93.26	116.98	0.36	11.58	nr	128.56
50 mm dia.	128.03	160.60	0.40	12.87	nr	173.47
63 mm dia.	144.36	181.08	0.44	14.15	nr	195.23

38 MECHANICAL/COOLING/HEATING SYSTEMS

Item	Net Price £	Material £	Labour hours	Labour £	Unit	Total rate £
Floor or ceiling cover plates						
Plastic						
15 mm dia.	0.52	0.65	0.16	5.14	nr	5.79
20 mm dia.	0.59	0.74	0.22	7.08	nr	7.82
25 mm dia.	0.63	0.80	0.22	7.08	nr	7.88
32 mm dia.	0.69	0.86	0.24	7.72	nr	8.58
40 mm dia.	1.53	1.92	0.26	8.37	nr	10.29
50 mm dia.	1.68	2.11	0.26	8.37	nr	10.48
Chromium plated						
15 mm dia.	3.55	4.46	0.16	5.14	nr	9.60
20 mm dia.	3.79	4.75	0.17	5.47	nr	10.22
25 mm dia.	3.92	4.92	0.21	6.75	nr	11.67
32 mm dia.	4.02	5.04	0.22	7.08	nr	12.12
40 mm dia.	4.54	5.69	0.26	8.37	nr	14.06
50 mm dia.	5.42	6.80	0.26	8.37	nr	15.17

38 MECHANICAL/COOLING/HEATING SYSTEMS

Item	Net Price £	Material £	Labour hours	Labour £	Unit	Total rate £
LOW TEMPERATURE HOT WATER HEATING; PIPELINE: PIPELINE ANCILLARIES						
EXPANSION JOINTS						
Axial movement bellows expansion joints; stainless steel						
Screwed ends for steel pipework; up to 6 bar G at 100°C						
15 mm dia.	87.09	109.24	0.68	21.87	nr	**131.11**
20 mm dia.	93.02	116.68	0.81	26.05	nr	**142.73**
25 mm dia.	96.63	121.22	0.93	29.92	nr	**151.14**
32 mm dia.	102.31	128.34	1.06	34.09	nr	**162.43**
40 mm dia.	108.36	135.92	1.16	37.31	nr	**173.23**
50 mm dia.	107.96	135.43	1.19	38.27	nr	**173.70**
Screwed ends for steel pipework; aluminium and steel outer sleeves; up to 16 bar G at 120°C						
20 mm dia.	94.87	119.00	1.32	42.45	nr	**161.45**
25 mm dia.	96.63	121.22	1.52	48.89	nr	**170.11**
32 mm dia.	102.31	128.34	1.80	57.89	nr	**186.23**
40 mm dia.	108.36	135.92	2.03	65.30	nr	**201.22**
50 mm dia.	110.11	138.12	2.26	72.69	nr	**210.81**
Flanged ends for steel pipework; aluminium and steel outer sleeves; up to 16 bar G at 120°C						
20 mm dia.	83.94	105.29	0.53	17.05	nr	**122.34**
25 mm dia.	89.93	112.81	0.64	20.59	nr	**133.40**
32 mm dia.	123.92	155.44	0.74	23.80	nr	**179.24**
40 mm dia.	133.90	167.97	0.82	26.36	nr	**194.33**
50 mm dia.	158.88	199.30	0.89	28.63	nr	**227.93**
Flanged ends for steel pipework; up to 16 bar G at 120°C						
65 mm dia.	146.89	184.26	1.10	35.37	nr	**219.63**
80 mm dia.	157.88	198.05	1.31	42.13	nr	**240.18**
100 mm dia.	180.85	226.86	1.78	57.25	nr	**284.11**
150 mm dia.	276.79	347.20	3.08	99.06	nr	**446.26**
Screwed ends for non-ferrous pipework; up to 6 bar G at 100°C						
20 mm dia.	94.87	119.00	0.72	23.16	nr	**142.16**
25 mm dia.	96.63	121.22	0.84	27.01	nr	**148.23**
32 mm dia.	102.31	128.34	1.02	32.80	nr	**161.14**
40 mm dia.	108.36	135.92	1.11	35.71	nr	**171.63**
50 mm dia.	110.11	138.12	1.18	37.96	nr	**176.08**
Flanged ends for steel, copper or non-ferrous pipework; up to 16 bar G at 120°C						
65 mm dia.	188.85	236.89	0.87	27.99	nr	**264.88**
80 mm dia.	194.85	244.42	0.95	30.55	nr	**274.97**
100 mm dia.	224.84	282.04	1.15	36.99	nr	**319.03**
150 mm dia.	368.72	462.53	1.36	43.74	nr	**506.27**

38 MECHANICAL/COOLING/HEATING SYSTEMS

Item	Net Price £	Material £	Labour hours	Labour £	Unit	Total rate £
Angular movement bellows expansion joints; stainless steel						
Flanged ends for steel pipework; up to 16 bar G at 120°C						
50 mm dia.	217.71	273.10	0.71	22.83	nr	**295.93**
65 mm dia.	264.22	331.44	0.83	26.69	nr	**358.13**
80 mm dia.	308.75	387.30	0.91	29.27	nr	**416.57**
100 mm dia.	398.59	499.99	0.97	31.19	nr	**531.18**
125 mm dia.	526.50	660.44	1.16	37.31	nr	**697.75**
150 mm dia.	626.05	785.32	1.18	37.96	nr	**823.28**

38 MECHANICAL/COOLING/HEATING SYSTEMS

Item	Net Price £	Material £	Labour hours	Labour £	Unit	Total rate £
LOW TEMPERATURE HOT WATER HEATING; PIPELINE: VALVES						
Gate valves						
Bronze gate valve; non-rising stem; BS 5154, series B, PN 16; working pressure up to 16 from −10°C to 100°C; 7 bar for saturated steam; screwed ends to steel						
15 mm dia.	52.42	65.76	1.18	37.96	nr	103.72
20 mm dia.	71.60	89.81	1.24	39.87	nr	129.68
25 mm dia.	91.12	114.30	1.31	42.13	nr	156.43
32 mm dia.	124.12	155.69	1.43	46.00	nr	201.69
40 mm dia.	165.71	207.87	1.53	49.21	nr	257.08
50 mm dia.	228.63	286.80	1.63	52.42	nr	339.22
Bronze gate valve; non-rising stem; BS 5154, series B, PN 32; working pressure up to 14 bar for saturated steam, 32 bar from −10°C to 100°C; screwed ends to steel						
15 mm dia.	45.87	57.53	1.11	35.71	nr	93.24
20 mm dia.	59.50	74.64	1.28	41.17	nr	115.81
25 mm dia.	80.60	101.10	1.49	47.91	nr	149.01
32 mm dia.	111.32	139.64	1.88	60.46	nr	200.10
40 mm dia.	149.47	187.50	2.31	74.30	nr	261.80
50 mm dia.	210.02	263.45	2.80	90.05	nr	353.50
Bronze gate valve; non-rising stem; BS 5154, series B, PN 20; working pressure up to 9 bar for saturated steam, 20 bar from −10°C to 100°C; screwed ends to steel						
15 mm dia.	31.28	39.23	0.84	27.01	nr	66.24
20 mm dia.	44.45	55.75	1.01	32.49	nr	88.24
25 mm dia.	57.52	72.15	1.19	38.27	nr	110.42
32 mm dia.	82.08	102.96	1.38	44.39	nr	147.35
40 mm dia.	112.33	140.91	1.62	52.10	nr	193.01
50 mm dia.	160.53	201.36	1.94	62.40	nr	263.76
Bronze gate valve; non-rising stem; BS 5154, series B, PN 16; working pressure up to 7 bar for saturated steam, 16 bar from −10°C to 100°C; BS4504 flanged ends; bolted connections						
15 mm dia.	140.87	176.70	1.18	37.96	nr	214.66
20 mm dia.	155.34	194.86	1.24	39.87	nr	234.73
25 mm dia.	204.57	256.61	1.31	42.13	nr	298.74
32 mm dia.	256.66	321.96	1.43	46.00	nr	367.96
40 mm dia.	310.30	389.24	1.53	49.21	nr	438.45
50 mm dia.	428.99	538.13	1.63	52.42	nr	590.55
65 mm dia.	666.26	835.76	1.71	54.99	nr	890.75
80 mm dia.	942.97	1182.87	1.88	60.46	nr	1243.33
100 mm dia.	1713.66	2149.62	2.03	65.30	nr	2214.92

38 MECHANICAL/COOLING/HEATING SYSTEMS

Item	Net Price £	Material £	Labour hours	Labour £	Unit	Total rate £
Cast iron gate valve; bronze trim; non-rising stem; BS 5150, PN6; working pressure 6 bar from −10°C to 120°C; BS4504 flanged ends; bolted connections						
50 mm dia.	279.01	349.99	1.85	59.49	nr	409.48
65 mm dia.	291.63	365.83	2.00	64.32	nr	430.15
80 mm dia.	337.03	422.77	2.27	73.00	nr	495.77
100 mm dia.	426.33	534.79	2.76	88.77	nr	623.56
125 mm dia.	610.24	765.49	6.05	194.58	nr	960.07
150 mm dia.	702.20	880.84	8.03	258.27	nr	1139.11
200 mm dia.	1337.51	1677.77	9.17	294.93	nr	1972.70
250 mm dia.	2058.03	2581.59	10.72	344.78	nr	2926.37
300 mm dia.	2440.98	3061.97	11.75	377.91	nr	3439.88
Cast iron gate valve; bronze trim; non-rising stem; BS 5150, PN10; working pressure up to 8.4 bar for saturated steam, 10 bar from −10°C to 120°C; BS4504 flanged ends; bolted connections						
50 mm dia.	292.66	367.11	1.85	59.49	nr	426.60
65 mm dia.	363.70	456.22	2.00	64.32	nr	520.54
80 mm dia.	404.93	507.94	2.27	73.00	nr	580.94
100 mm dia.	539.03	676.16	2.76	88.77	nr	764.93
125 mm dia.	753.62	945.34	6.05	194.58	nr	1139.92
150 mm dia.	889.15	1115.35	8.03	258.27	nr	1373.62
200 mm dia.	1633.63	2049.23	9.17	294.93	nr	2344.16
250 mm dia.	2456.84	3081.86	10.72	344.78	nr	3426.64
300 mm dia.	2764.67	3468.00	11.75	377.91	nr	3845.91
350 mm dia.	3215.62	4033.67	12.67	407.50	nr	4441.17
Cast iron gate valve; bronze trim; non-rising stem; BS 5163 series A, PN16; working pressure for cold water services up to 16 bar; BS4504 flanged ends; bolted connections						
50 mm dia.	292.66	367.11	1.85	59.49	nr	426.60
65 mm dia.	363.70	456.22	2.00	64.32	nr	520.54
80 mm dia.	404.93	507.94	2.27	73.00	nr	580.94
100 mm dia.	539.03	676.16	2.76	88.77	nr	764.93
125 mm dia.	753.62	945.34	6.05	194.58	nr	1139.92
150 mm dia.	889.15	1115.35	8.03	258.27	nr	1373.62
Ball valves						
Malleable iron body; lever operated stainless steel ball and stem; working pressure up to 12 bar; flanged ends to BS 4504 16/11; bolted connections						
40 mm dia.	388.98	487.94	1.54	49.54	nr	537.48
50 mm dia.	490.15	614.85	1.64	52.75	nr	667.60
80 mm dia.	792.86	994.56	1.92	61.76	nr	1056.32
100 mm dia.	819.74	1028.28	2.80	90.05	nr	1118.33
150 mm dia.	1515.25	1900.73	12.05	387.56	nr	2288.29

38 MECHANICAL/COOLING/HEATING SYSTEMS

Item	Net Price £	Material £	Labour hours	Labour £	Unit	Total rate £
LOW TEMPERATURE HOT WATER HEATING; PIPELINE: VALVES – cont						
Ball valves – cont						
Malleable iron body; lever operated stainless steel ball and stem; working pressure up to 16 bar; screwed ends to steel						
15 mm dia.	34.91	43.79	1.34	43.11	nr	86.90
20 mm dia.	46.20	57.95	1.34	43.11	nr	101.06
25 mm dia.	47.52	59.61	1.40	45.05	nr	104.66
32 mm dia.	65.46	82.12	1.46	47.03	nr	129.15
40 mm dia.	65.46	82.12	1.54	49.56	nr	131.68
50 mm dia.	78.27	98.18	1.64	52.81	nr	150.99
Carbon steel body; lever operated stainless steel ball and stem; Class 150; working pressure up to 19 bar; screwed ends to steel						
15 mm dia.	85.48	107.23	0.84	27.04	nr	134.27
20 mm dia.	96.09	120.53	1.14	36.68	nr	157.21
25 mm dia.	104.04	130.50	1.30	41.82	nr	172.32
32 mm dia.	134.16	168.29	1.42	45.67	nr	213.96
40 mm dia.	167.70	210.36	1.56	50.18	nr	260.54
50 mm dia.	209.64	262.98	1.68	54.04	nr	317.02
Globe valves						
Bronze; rising stem; renewable disc; BS 5154 series B, PN32; working pressure up to 14 bar for saturated steam, 32 bar from −10°C to 100°C; screwed ends to steel						
15 mm dia.	46.56	58.41	0.77	24.76	nr	83.17
20 mm dia.	63.67	79.87	1.03	33.13	nr	113.00
25 mm dia.	91.34	114.58	1.19	38.27	nr	152.85
32 mm dia.	128.60	161.31	1.38	44.39	nr	205.70
40 mm dia.	179.09	224.65	1.62	52.10	nr	276.75
50 mm dia.	274.89	344.83	1.61	51.78	nr	396.61
Bronze; needle valve; rising stem; BS 5154, series B, PN32; working pressure up to 14 bar for saturated steam, 32 bar from −10°C to 100°C; screwed ends to steel						
15 mm dia.	35.40	44.41	1.07	34.41	nr	78.82
20 mm dia.	59.93	75.17	1.18	37.96	nr	113.13
25 mm dia.	84.49	105.99	1.27	40.85	nr	146.84
32 mm dia.	175.87	220.61	1.35	43.41	nr	264.02
40 mm dia.	277.07	347.56	1.47	47.28	nr	394.84
50 mm dia.	350.94	440.22	1.61	51.78	nr	492.00
Bronze; rising stem; renewable disc; BS 5154, series B, PN16; working pressure upto 7 bar for saturated steam, 16 bar from −10°C to 100°C; BS4504 flanged ends; bolted connections						
15 mm dia.	94.08	118.01	1.16	37.31	nr	155.32
20 mm dia.	108.82	136.51	1.26	40.53	nr	177.04
25 mm dia.	190.82	239.37	1.38	44.39	nr	283.76

38 MECHANICAL/COOLING/HEATING SYSTEMS

Item	Net Price £	Material £	Labour hours	Labour £	Unit	Total rate £
32 mm dia.	240.89	302.18	1.47	47.28	nr	349.46
40 mm dia.	273.81	343.47	1.56	50.18	nr	393.65
50 mm dia.	392.85	492.79	1.71	54.99	nr	547.78
Bronze; rising stem; renewable disc; BS 2060, class 250; working pressure up to 24 bar for saturated steam, 38 bar from −10°C to 100°C; flanged ends (BS 10 table H); bolted connections						
15 mm dia.	239.59	300.54	1.16	37.31	nr	337.85
20 mm dia.	278.45	349.28	1.26	40.53	nr	389.81
25 mm dia.	382.64	479.99	1.38	44.39	nr	524.38
32 mm dia.	499.86	627.02	1.47	47.28	nr	674.30
40 mm dia.	598.58	750.86	1.56	50.18	nr	801.04
50 mm dia.	930.77	1167.56	1.71	54.99	nr	1222.55
65 mm dia.	1113.12	1396.29	1.88	60.46	nr	1456.75
80 mm dia.	1414.49	1774.34	2.03	65.30	nr	1839.64
Check valves						
Bronze; swing pattern; BS 5154 series B, PN 25; working pressure up to 10.5 bar for saturated steam, 25 bar from −10°C to 100°C; screwed ends to steel						
15 mm dia.	21.66	27.17	0.77	24.76	nr	51.93
20 mm dia.	25.74	32.29	1.03	33.13	nr	65.42
25 mm dia.	35.68	44.76	1.19	38.27	nr	83.03
32 mm dia.	60.44	75.81	1.38	44.39	nr	120.20
40 mm dia.	75.19	94.32	1.62	52.10	nr	146.42
50 mm dia.	115.31	144.65	1.94	62.40	nr	207.05
65 mm dia.	216.03	270.98	2.45	78.80	nr	349.78
80 mm dia.	305.46	383.17	2.83	91.02	nr	474.19
Bronze; vertical lift pattern; BS 5154 series B, PN32; working pressure up to 14 bar for saturated steam, 32 bar from −10°C to 100°C; screwed ends to steel						
15 mm dia.	27.49	34.48	0.96	30.87	nr	65.35
20 mm dia.	39.09	49.03	1.07	34.41	nr	83.44
25 mm dia.	57.40	72.00	1.17	37.63	nr	109.63
32 mm dia.	87.44	109.68	1.33	42.77	nr	152.45
40 mm dia.	113.94	142.92	1.41	45.35	nr	188.27
50 mm dia.	170.72	214.16	1.55	49.85	nr	264.01
65 mm dia.	471.67	591.66	1.80	57.89	nr	649.55
80 mm dia.	705.14	884.53	1.99	64.00	nr	948.53
Bronze; oblique swing pattern; BS 5154 series A, PN32; working pressure up to 14 bar for saturated steam, 32 bar from −10°C to 120°C; screwed connections to steel						
15 mm dia.	34.92	43.80	0.96	30.87	nr	74.67
20 mm dia.	41.48	52.04	1.07	34.41	nr	86.45
25 mm dia.	57.51	72.14	1.17	37.63	nr	109.77
32 mm dia.	86.00	107.88	1.33	42.77	nr	150.65
40 mm dia.	110.84	139.04	1.41	45.35	nr	184.39
50 mm dia.	183.63	230.35	1.55	49.85	nr	280.20

38 MECHANICAL/COOLING/HEATING SYSTEMS

Item	Net Price £	Material £	Labour hours	Labour £	Unit	Total rate £
LOW TEMPERATURE HOT WATER HEATING; PIPELINE: VALVES – cont						
Check valves – cont						
Cast iron; swing pattern; BS 5153 PN6; working pressure up to 6 bar from −10°C to 120°C; BS 4504 flanged ends; bolted connections						
50 mm dia.	603.02	756.43	1.86	59.82	nr	816.25
65 mm dia.	660.23	828.20	2.00	64.32	nr	892.52
80 mm dia.	743.15	932.21	2.56	82.34	nr	1014.55
100 mm dia.	952.00	1194.19	2.76	88.77	nr	1282.96
125 mm dia.	1420.25	1781.56	6.05	194.58	nr	1976.14
150 mm dia.	1596.89	2003.14	8.11	260.84	nr	2263.98
200 mm dia.	3531.49	4429.90	9.26	297.83	nr	4727.73
250 mm dia.	4414.25	5537.24	10.72	344.78	nr	5882.02
300 mm dia.	5297.11	6644.69	11.75	377.91	nr	7022.60
Cast iron; horizontal lift pattern; BS 5153 PN16; working pressure up to 13 bar for saturated steam, 16 bar from −10°C to 120°C; BS 4504 flanged ends; bolted connections						
50 mm dia.	544.54	683.07	1.86	59.82	nr	742.89
65 mm dia.	544.54	683.07	2.00	64.32	nr	747.39
80 mm dia.	674.19	845.70	2.56	82.34	nr	928.04
100 mm dia.	785.12	984.85	2.96	95.20	nr	1080.05
125 mm dia.	1171.38	1469.38	7.76	249.59	nr	1718.97
150 mm dia.	1316.97	1652.01	10.50	337.71	nr	1989.72
Cast iron; semi-lugged butterfly valve; BS5155 PN16; working pressure 16 bar from −10°C to 120°C; BS 4504 flanged ends; bolted connections						
50 mm dia.	181.74	227.98	2.20	70.76	nr	298.74
65 mm dia.	187.55	235.27	2.31	74.30	nr	309.57
80 mm dia.	220.92	277.12	2.88	92.62	nr	369.74
100 mm dia.	307.11	385.24	3.11	100.03	nr	485.27
125 mm dia.	448.24	562.27	5.02	161.46	nr	723.73
150 mm dia.	514.51	645.40	6.98	224.50	nr	869.90
200 mm dia.	821.32	1030.27	8.25	265.34	nr	1295.61
250 mm dia.	1212.25	1520.65	10.47	336.74	nr	1857.39
300 mm dia.	1818.41	2281.01	11.48	369.23	nr	2650.24
Cast iron; semi-lugged butterfly valve; BS5155 PN20; working pressure up to 20 bar from −10°C to 120°C; BS 4504 flanged ends; bolted connections						
50 mm dia.	1627.47	2041.50	2.20	70.76	nr	2112.26
65 mm dia.	1683.80	2112.16	2.31	74.30	nr	2186.46
80 mm dia.	1797.65	2254.97	2.88	92.62	nr	2347.59
100 mm dia.	2189.06	2745.96	3.11	100.03	nr	2845.99
125 mm dia.	2809.78	3524.58	5.02	161.46	nr	3686.04
150 mm dia.	3188.78	4000.00	5.02	161.46	nr	4161.46
200 mm dia.	3667.55	4600.58	8.25	265.34	nr	4865.92
250 mm dia.	5783.09	7254.31	10.47	336.74	nr	7591.05
300 mm dia.	7007.34	8790.01	11.48	369.23	nr	9159.24

38 MECHANICAL/COOLING/HEATING SYSTEMS

Item	Net Price £	Material £	Labour hours	Labour £	Unit	Total rate £
Cast iron; semi-lugged butterfly valve; BS5155 PN20; working pressure up to 30 bar from −10°C to 120°C; BS 4504 flanged ends; bolted connections						
50 mm dia.	1627.47	2041.50	2.20	70.76	nr	2112.26
65 mm dia.	1683.80	2112.16	2.31	74.30	nr	2186.46
80 mm dia.	1797.65	2254.97	2.88	92.62	nr	2347.59
100 mm dia.	2189.06	2745.96	3.11	100.03	nr	2845.99
125 mm dia.	2809.78	3524.58	5.02	161.46	nr	3686.04
150 mm dia.	3188.78	4000.00	5.02	161.46	nr	4161.46
200 mm dia.	3667.55	4600.58	8.25	265.34	nr	4865.92
250 mm dia.	5783.09	7254.31	10.47	336.74	nr	7591.05
300 mm dia.	7007.34	8790.01	11.48	369.23	nr	9159.24
Commissioning valves						
Bronze commissioning set; metering station; double regulating valve; BS5154 PN20 Series B; working pressure 20 bar from −10°C to 100°C; screwed ends to steel						
15 mm dia.	91.23	114.44	1.08	34.73	nr	149.17
20 mm dia.	146.58	183.87	1.46	46.95	nr	230.82
25 mm dia.	174.30	218.65	1.68	54.04	nr	272.69
32 mm dia.	233.70	293.15	1.95	62.72	nr	355.87
40 mm dia.	338.92	425.14	2.27	73.00	nr	498.14
50 mm dia.	511.50	641.63	2.73	87.81	nr	729.44
Cast iron commissioning set; metering station; double regulating valve; BS5152 PN16; working pressure 16 bar from −10°C to 90°C; flanged ends (BS 4504, Part 1, Table 16); bolted connections						
65 mm dia.	774.92	972.06	1.80	57.89	nr	1029.95
80 mm dia.	927.86	1163.90	2.56	82.34	nr	1246.24
100 mm dia.	1291.75	1620.37	2.30	73.98	nr	1694.35
125 mm dia.	1892.09	2373.44	2.44	78.48	nr	2451.92
150 mm dia.	2492.47	3126.56	2.90	93.27	nr	3219.83
200 mm dia.	6379.72	8002.72	8.26	265.66	nr	8268.38
250 mm dia.	7980.95	10011.30	10.49	337.39	nr	10348.69
300 mm dia.	8667.48	10872.49	11.49	369.54	nr	11242.03
Cast iron variable orifice double regulating valve; orifice valve; BS5152 PN16; working pressure 16 bar from −10° to 90°C; flanged ends (BS 4504, Part 1, Table 16); bolted connections						
65 mm dia.	567.36	711.69	2.00	64.32	nr	776.01
80 mm dia.	694.44	871.10	2.56	82.34	nr	953.44
100 mm dia.	952.99	1195.43	2.96	95.20	nr	1290.63
125 mm dia.	1427.10	1790.15	7.76	249.59	nr	2039.74
150 mm dia.	1833.07	2299.40	10.50	337.71	nr	2637.11
200 mm dia.	4804.30	6026.52	8.26	265.66	nr	6292.18
250 mm dia.	7304.90	9163.27	10.49	337.39	nr	9500.66
300 mm dia.	13103.53	16437.06	11.49	369.54	nr	16806.60

38 MECHANICAL/COOLING/HEATING SYSTEMS

Item	Net Price £	Material £	Labour hours	Labour £	Unit	Total rate £
LOW TEMPERATURE HOT WATER HEATING; PIPELINE: VALVES – cont						
Commissioning valves – cont						
Cast iron globe valve with double regulating feature; BS5152 PN16; working pressure 16 bar from −10°C to 120°C; flanged ends (BS 4504, Part 1, Table 16); bolted connections						
65 mm dia.	675.86	847.80	2.00	64.32	nr	912.12
80 mm dia.	817.33	1025.26	2.56	82.34	nr	1107.60
100 mm dia.	1076.66	1350.56	2.96	95.20	nr	1445.76
125 mm dia.	1501.15	1883.04	7.76	249.59	nr	2132.63
150 mm dia.	1899.92	2383.26	10.50	337.71	nr	2720.97
200 mm dia.	4918.70	6170.01	8.26	265.66	nr	6435.67
250 mm dia.	7441.88	9335.10	10.49	337.39	nr	9672.49
300 mm dia.	11692.28	14666.79	11.49	369.54	nr	15036.33
Bronze autoflow commissioning valve; PN25 ; working pressure 25 bar up to 100°C; screwed ends to steel						
15 mm dia.	103.44	129.75	0.82	26.36	nr	156.11
20 mm dia.	109.53	137.39	1.08	34.73	nr	172.12
25 mm dia.	136.38	171.08	1.27	40.85	nr	211.93
32 mm dia.	213.00	267.19	1.50	48.25	nr	315.44
40 mm dia.	233.23	292.57	1.76	56.62	nr	349.19
50 mm dia.	344.85	432.58	2.13	68.50	nr	501.08
Ductile iron autoflow commissioning valves; PN16; working pressure 16 bar from −10°C to 120°C; for ANSI 150 flanged ends						
65 mm dia.	908.13	1139.16	2.31	74.30	nr	1213.46
80 mm dia.	1004.05	1259.48	2.88	92.62	nr	1352.10
100 mm dia.	1282.66	1608.97	3.11	100.03	nr	1709.00
150 mm dia.	2642.28	3314.47	6.98	224.50	nr	3538.97
200 mm dia.	3848.71	4827.83	8.26	265.66	nr	5093.49
250 mm dia.	5345.29	6705.13	10.49	337.39	nr	7042.52
300 mm dia.	6826.76	8563.49	11.49	369.54	nr	8933.03
Strainers						
Bronze strainer; Y type; PN32 ; working pressure 32 bar from −10°C to 100°C; screwed ends to steel						
15 mm dia.	41.46	52.01	0.82	26.36	nr	78.37
20 mm dia.	52.66	66.06	1.08	34.73	nr	100.79
25 mm dia.	69.86	87.63	1.27	40.85	nr	128.48
32 mm dia.	112.84	141.55	1.50	48.25	nr	189.80
40 mm dia.	145.94	183.06	1.76	56.62	nr	239.68
50 mm dia.	244.43	306.61	2.13	68.50	nr	375.11
Cast iron strainer; Y type; PN16; working pressure 16 bar from −10°C to 120°C; BS 4504 flanged ends						
65 mm dia.	257.87	323.47	2.31	74.30	nr	397.77
80 mm dia.	305.22	382.87	2.88	92.62	nr	475.49
100 mm dia.	446.60	560.21	3.11	100.03	nr	660.24

38 MECHANICAL/COOLING/HEATING SYSTEMS

Item	Net Price £	Material £	Labour hours	Labour £	Unit	Total rate £
125 mm dia.	823.28	1032.72	5.02	161.46	nr	1194.18
150 mm dia.	1056.28	1324.99	6.98	224.50	nr	1549.49
200 mm dia.	1772.92	2223.95	8.26	265.66	nr	2489.61
250 mm dia.	2563.74	3215.96	10.49	337.39	nr	3553.35
300 mm dia.	4174.99	5237.11	11.49	369.54	nr	5606.65
Regulators						
Gunmetal; self-acting two port thermostatic regulator; single seat; screwed ends; complete with sensing element, 2 m long capillary tube						
15 mm dia.	730.95	916.90	1.37	44.07	nr	960.97
20 mm dia.	754.47	946.41	1.24	39.87	nr	986.28
25 mm dia.	774.27	971.24	1.34	43.10	nr	1014.34
Gunmetal; self-acting two port thermostatic regulator; double seat; flanged ends (BS 4504 PN25); with sensing element, 2 m long capillary tube; steel body						
65 mm dia.	2710.80	3400.43	1.23	44.22	nr	3444.65
80 mm dia.	3199.50	4013.45	1.62	58.24	nr	4071.69
Control valves						
Electrically operated (electrical work elsewhere)						
Pressure independent control valve (PICV)						
Brass; pressure independent control valve; rotary type; PN 25; maximum working pressure 25 bar at from −10°C to 120°C; screwed ends to steel; electrical work elsewhere						
15 mm dia.	83.91	105.26	1.08	34.73	nr	139.99
20 mm dia.	100.47	126.03	1.18	37.96	nr	163.99
25 mm dia.	147.94	185.57	1.35	43.41	nr	228.98
32 mm dia.	215.29	270.05	1.46	46.95	nr	317.00
40 mm dia.	238.47	299.14	1.53	49.21	nr	348.35
50 mm dia.	416.21	522.10	1.61	51.78	nr	573.88
Brass; pressure independent control valve; rotary type with electronic actuator; 24 V motor; PN 25; maximum working pressure 25 bar at from −10°C to 120°C; screwed ends to steel; electrical work elsewhere						
15 mm dia.	175.55	220.21	2.06	66.26	nr	286.47
20 mm dia.	192.10	240.97	2.15	69.16	nr	310.13
25 mm dia.	239.58	300.53	2.27	73.00	nr	373.53
32 mm dia.	337.83	423.77	2.35	75.58	nr	499.35
40 mm dia.	361.01	452.85	2.47	79.44	nr	532.29
50 mm dia.	550.90	691.05	2.55	82.02	nr	773.07

38 MECHANICAL/COOLING/HEATING SYSTEMS

Item	Net Price £	Material £	Labour hours	Labour £	Unit	Total rate £
LOW TEMPERATURE HOT WATER HEATING; PIPELINE: VALVES – cont						
Control valves – cont						
Differential pressure control valves (DPCV) Brass; differential control valve; PN 25; maximum working pressure 25 bar at from −20°C to 120°C; screwed ends to steel						
15 mm dia.	230.75	289.45	1.99	64.00	nr	**353.45**
20 mm dia.	248.49	311.71	2.03	65.30	nr	**377.01**
25 mm dia.	276.12	346.36	2.11	67.86	nr	**414.22**
32 mm dia.	312.56	392.08	2.17	69.80	nr	**461.88**
40 m m dia.	538.74	675.80	2.26	72.69	nr	**748.49**
50 mm dia.	646.88	811.45	2.31	74.30	nr	**885.75**
Cast iron; butterfly type; two position electrically controlled 240 V motor and linkage mechanism; for low pressure hot water; maximum pressure 6 bar at 120°C; flanged ends; electrical work elsewhere						
25 mm dia.	898.92	1127.60	1.47	47.28	nr	**1174.88**
32 mm dia.	928.84	1165.14	1.52	48.89	nr	**1214.03**
40 mm dia.	1297.91	1628.10	1.61	51.78	nr	**1679.88**
50 mm dia.	1334.79	1674.36	1.71	54.99	nr	**1729.35**
65 mm dia.	1368.22	1716.30	2.51	80.73	nr	**1797.03**
80 mm dia.	1421.72	1783.41	2.69	86.52	nr	**1869.93**
100 mm dia.	1491.98	1871.54	2.81	90.38	nr	**1961.92**
125 mm dia.	1652.56	2072.97	2.94	94.56	nr	**2167.53**
150 mm dia.	1790.81	2246.40	3.33	107.11	nr	**2353.51**
200 mm dia.	2221.21	2786.29	3.67	118.04	nr	**2904.33**
Cast iron; three way 240 V motorized; for low pressure hot water; maximum pressure 6 bar 120°C; flanged ends, drilled (BS 10, Table F)						
25 mm dia.	1004.69	1260.28	1.99	64.00	nr	**1324.28**
40 mm dia.	1045.32	1311.25	2.13	68.50	nr	**1379.75**
50 mm dia.	1058.77	1328.12	3.21	103.24	nr	**1431.36**
65 mm dia.	1162.44	1458.16	3.23	103.89	nr	**1562.05**
80 mm dia.	1302.53	1633.89	3.50	112.57	nr	**1746.46**
2 port control valves Cast iron; 2 port; motorized control valve; normally open for heating applications; PN 16; screwed ends to steel						
15 mm dia.	39.74	49.85	1.24	39.87	nr	**89.72**
22 mm dia.	60.68	76.12	1.31	42.13	nr	**118.25**
28 mm dia.	72.57	91.03	1.37	44.07	nr	**135.10**
32 mm dia.	84.79	106.36	1.45	46.64	nr	**153.00**
40 mm dia.	105.99	132.96	1.52	48.89	nr	**181.85**
50 mm dia.	132.48	166.19	1.60	51.45	nr	**217.64**

38 MECHANICAL/COOLING/HEATING SYSTEMS

Item	Net Price £	Material £	Labour hours	Labour £	Unit	Total rate £
Two port normally closed motorized valve; electric actuator; spring return; domestic usage						
22 mm dia.	136.82	171.63	1.18	37.96	nr	**209.59**
28 mm dia.	193.08	242.20	1.35	43.41	nr	**285.61**
Two port on/off motorized valve; electric actuator; spring return; domestic usage						
22 mm dia.	136.82	171.63	1.18	37.96	nr	**209.59**
3 port control valves						
Cast iron; 3 port; motorized control valve with temperatyre control unit; PN 25; screwed ends to steel						
22 mm dia.	61.83	77.56	2.00	64.32	nr	**141.88**
25 mm dia.	70.10	87.93	2.13	68.50	nr	**156.43**
32 mm dia.	89.74	112.57	2.19	70.44	nr	**183.01**
40 mm dia.	112.17	140.71	2.26	72.69	nr	**213.40**
Three port motorized valve; electric actuator; spring return; domestic usage						
22 mm dia.	207.85	260.72	1.18	37.96	nr	**298.68**
4 port control valves						
Cast iron; 4 port; fixed orifice double regulating and control valve; PN 16; screwed ends to steel						
22 mm dia.	54.79	68.72	2.89	92.95	nr	**161.67**
Safety and relief valves						
Bronze relief valve; spring type; side outlet; working pressure up to 20.7 bar at 120°C; screwed ends to steel						
15 mm dia.	158.86	199.27	0.26	8.37	nr	**207.64**
20 mm dia.	174.23	218.56	0.36	11.58	nr	**230.14**
Bronze relief valve; spring type; side outlet; working pressure up to 17.2 bar at 120°C; screwed ends to steel						
25 mm dia.	238.28	298.89	0.38	12.22	nr	**311.11**
32 mm dia.	322.84	404.97	0.48	15.44	nr	**420.41**
Bronze relief valve; spring type; side outlet; working pressure up to 13.8 bar at 120°C; screwed ends to steel						
40 mm dia.	421.48	528.71	0.64	20.59	nr	**549.30**
50 mm dia.	549.58	689.39	0.76	24.45	nr	**713.84**
65 mm dia.	1017.57	1276.44	0.94	30.23	nr	**1306.67**
80 mm dia.	1335.04	1674.67	1.10	35.37	nr	**1710.04**

38 MECHANICAL/COOLING/HEATING SYSTEMS

Item	Net Price £	Material £	Labour hours	Labour £	Unit	Total rate £
LOW TEMPERATURE HOT WATER HEATING; PIPELINE: VALVES – cont						
Cocks; screwed joints to steel						
Bronze gland cock; complete with malleable iron lever; working pressure up to 10 bar at 100°C; screwed ends to steel						
15 mm dia.	55.83	70.03	0.77	24.76	nr	94.79
20 mm dia.	79.42	99.62	1.03	33.13	nr	132.75
25 mm dia.	107.02	134.24	1.19	38.27	nr	172.51
32 mm dia.	161.71	202.85	1.38	44.39	nr	247.24
40 mm dia.	225.80	283.25	1.62	52.10	nr	335.35
50 mm dia.	344.61	432.28	1.94	62.40	nr	494.68
Bronze three-way plug cock; complete with malleable iron lever; working pressure up to 10 bar at 100°C; screwed ends to steel						
15 mm dia.	113.18	141.97	0.77	24.76	nr	166.73
20 mm dia.	131.00	164.33	1.03	33.13	nr	197.46
25 mm dia.	183.12	229.70	1.19	38.27	nr	267.97
32 mm dia.	259.71	325.79	1.38	44.39	nr	370.18
40 mm dia.	313.14	392.81	1.62	52.10	nr	444.91
Air vents; including regulating, adjusting and testing						
Automatic air vent; maximum pressure up to 7 bar at 93°C; screwed ends to steel						
15 mm dia.	14.23	17.85	0.80	25.73	nr	43.58
Automatic air vent; maximum pressure up to 7 bar at 93°C; lockhead isolating valve; screwed ends to steel						
15 mm dia.	14.23	17.85	0.83	26.69	nr	44.54
Automatic air vent; maximum pressure up to 17 bar at 200°C; flanged ends (BS10, Table H); bolted connections to counter flange (measured separately)						
15 mm dia.	512.32	642.66	0.83	26.69	nr	669.35
Radiator valves						
Bronze; wheelhead or lockshield; chromium plated finish; screwed joints to steel						
Straight						
15 mm dia.	18.18	22.80	0.59	18.98	nr	41.78
20 mm dia.	28.28	35.47	0.73	23.49	nr	58.96
25 mm dia.	35.50	44.53	0.85	27.33	nr	71.86
Angled						
15 mm dia.	61.51	77.16	0.59	18.98	nr	96.14
20 mm dia.	81.09	101.72	0.73	23.49	nr	125.21
25 mm dia.	104.26	130.78	0.85	27.33	nr	158.11

38 MECHANICAL/COOLING/HEATING SYSTEMS

Item	Net Price £	Material £	Labour hours	Labour £	Unit	Total rate £
Bronze; wheelhead or lockshield; chromium plated finish; compression joints to copper						
Straight						
15 mm dia.	32.56	40.85	0.59	18.98	nr	**59.83**
20 mm dia.	41.03	51.46	0.73	23.49	nr	**74.95**
25 mm dia.	51.56	64.68	0.85	27.33	nr	**92.01**
Angled						
15 mm dia.	24.57	30.82	0.59	18.98	nr	**49.80**
20 mm dia.	25.78	32.33	0.73	23.49	nr	**55.82**
25 mm dia.	31.96	40.10	0.85	27.33	nr	**67.43**
Twin entry						
8 mm dia.	32.89	41.26	0.23	7.40	nr	**48.66**
10 mm dia.	37.33	46.83	0.23	7.40	nr	**54.23**
Bronze; thermostatic head; chromium plated finish; compression joints to copper						
Straight						
15 mm dia.	13.46	16.89	0.59	18.98	nr	**35.87**
20 mm dia.	19.75	24.77	0.73	23.49	nr	**48.26**
Angled						
15 mm dia.	12.52	15.70	0.59	18.98	nr	**34.68**
20 mm dia.	18.35	23.02	0.73	23.49	nr	**46.51**

38 MECHANICAL/COOLING/HEATING SYSTEMS

Item	Net Price £	Material £	Labour hours	Labour £	Unit	Total rate £
LOW TEMPERATURE HOT WATER HEATING; PIPELINE: EQUIPMENT						
GAUGES						
Thermometers and pressure gauges						
Dial thermometer; coated steel case and dial; glass window; brass pocket; BS 5235; pocket length 100 mm; screwed end Back/bottom entry						
100 mm dia. face	70.29	88.17	0.81	26.06	nr	**114.23**
150 mm dia. face	115.21	144.52	0.81	26.06	nr	**170.58**
Dial thermometer; coated steel case and dial; glass window; brass pocket; BS 5235; pocket length 100 mm; screwed end						
100 mm dia. face	104.20	130.70	0.81	26.06	nr	**156.76**
150 mm dia. face	116.69	146.37	0.81	26.06	nr	**172.43**
PRESSURIZATION UNITS						
LTHW pressurization unit complete with expansion vessel(s), interconnecting pipework and all necessary isolating and drain valves; includes placing in position; electrical work elsewhere. Selection based on a final working pressure of 4 bar, a 3 m static head and system operating temperatures of 82/71°C System volume						
2,400 litres	2064.30	2589.46	15.00	482.44	nr	**3071.90**
6,000–20,000 litres	3745.35	4698.16	22.00	707.58	nr	**5405.74**
25,000 litres	4267.20	5352.77	22.00	707.58	nr	**6060.35**
DIRT SEPARATORS						
Dirt seperator; maximum operating temperature and pressure of 110°C and 10 bar; fitted with drain valve Bore size, flow rate (at 1.0 m/s velocity); threaded connections						
32 mm dia., 3.7 m³/h	100.39	125.93	2.29	73.66	nr	**199.59**
40 mm dia., 5.0 m³/h	119.51	149.91	2.45	78.80	nr	**228.71**
Bore size, flow rate (at 1.5 m/s velocity); flanged connections to PN16						
50 mm dia., 13.0 m³/h	675.64	847.53	3.00	96.49	nr	**944.02**
65 mm dia., 21.0 m³/h	704.34	883.52	3.00	96.49	nr	**980.01**
80 mm dia., 29.0 m³/h	991.16	1243.31	3.84	123.51	nr	**1366.82**

38 MECHANICAL/COOLING/HEATING SYSTEMS

Item	Net Price £	Material £	Labour hours	Labour £	Unit	Total rate £
100 mm dia., 49.0 m³/h	1048.53	1315.27	4.44	142.80	nr	1458.07
125 mm dia., 74.0 m³/h	2009.41	2520.60	11.64	374.37	nr	2894.97
150 mm dia., 109.0 m³/h	2095.45	2628.53	15.75	506.56	nr	3135.09
200 mm dia., 181.0 m³/h	3158.31	3961.79	15.75	506.56	nr	4468.35
250 mm dia., 288.0 m³/h	5597.97	7022.10	15.75	506.56	nr	7528.66
300 mm dia., 407.0 m³/h	8182.61	10264.26	17.24	554.48	nr	10818.74
MICROBUBBLE DEAERATORS						
Microbubble deaerator; maximum operating temperature and pressure of 110°C and 10 bar; fitted with drain valve						
Bore size, flow rate (at 1.0 m/s velocity); threaded connections						
32 mm dia., 3.7 m³/h	92.42	115.93	2.29	73.66	nr	189.59
40 mm dia., 5.0 m³/h	109.96	137.94	2.45	78.80	nr	216.74
Bore size, flow rate (at 1.5 m/s velocity); flanged connections to PN16						
50 mm dia., 13.0 m³/h	675.64	847.53	3.00	96.49	nr	944.02
65 mm dia., 21.0 m³/h	702.99	881.83	3.00	96.49	nr	978.32
80 mm dia., 29.0 m³/h	991.16	1243.31	3.84	123.51	nr	1366.82
100 mm dia., 49.0 m³/h	1048.53	1315.27	4.44	142.80	nr	1458.07
125 mm dia., 74.0 m³/h	2009.41	2520.60	11.64	374.37	nr	2894.97
150 mm dia., 109.0 m³/h	2095.45	2628.53	15.75	506.56	nr	3135.09
200 mm dia., 181.0 m³/h	3158.31	3961.79	15.75	506.56	nr	4468.35
250 mm dia., 288.0 m³/h	5597.97	7022.10	15.75	506.56	nr	7528.66
300 mm dia., 407.0 m³/h	8211.30	10300.26	17.24	554.48	nr	10854.74
COMBINED MICROBUBBLE DEAERATORS AND DIRT SEPARATORS						
Combined deaerator and dirt separators; maximum operating temperature and pressure of 110°C and 10 bar; fitted with drain valve						
Bore size, flow rate (at 1.5 m/s velocity); threaded connections						
25 mm dia., 2.0 m³/h	127.48	159.91	2.75	88.45	nr	248.36
Bore size, flow rate (at 1.5 m/s velocity); flanged connections to PN16						
50 mm dia., 13.0 m³/h	862.08	1081.39	3.60	115.79	nr	1197.18
65 mm dia., 21.0 m³/h	890.77	1117.38	3.60	115.79	nr	1233.17
80 mm dia., 29.0 m³/h	1206.28	1513.15	4.61	148.27	nr	1661.42
100 mm dia., 49.0 m³/h	1263.98	1585.54	5.33	171.43	nr	1756.97
125 mm dia., 74.0 m³/h	2410.97	3024.32	13.97	449.31	nr	3473.63
150 mm dia., 109.0 m³/h	2498.61	3134.25	18.90	607.88	nr	3742.13
200 mm dia., 181.0 m³/h	3875.39	4861.29	18.90	607.88	nr	5469.17
250 mm dia., 288.0 m³/h	6832.93	8571.23	18.90	607.88	nr	9179.11
300 mm dia., 407.0 m³/h	10478.85	13144.67	20.69	665.45	nr	13810.12

38 MECHANICAL/COOLING/HEATING SYSTEMS

Item	Net Price £	Material £	Labour hours	Labour £	Unit	Total rate £
LOW TEMPERATURE HOT WATER HEATING; PIPELINE: PUMPS						
Centrifugal heating and chilled water pump; belt drive; 3 phase, 1450 rpm motor; max. pressure 1000 kN/m²; max. temperature 125°C; bed plate; coupling guard; bolted connections; supply only mating flanges;includes fixing on prepared concrete base; electrical work elsewhere						
40 mm pump size; 4.0 l/s at 70 kPa max head; 0.25 kW max motor rating	1609.04	2018.37	7.59	244.12	nr	**2262.49**
40 mm pump size; 4.0 l/s at 130 kPa max head; 1.5 kW max motor rating	2013.69	2525.97	8.09	260.20	nr	**2786.17**
50 mm pump size; 8.5 l/s at 90 kPa max head; 2.2 kW max motor rating	2076.67	2604.97	8.67	278.86	nr	**2883.83**
50 mm pump size; 8.5 l/s at 190 kPa max head; 3 kW max motor rating	3609.36	4527.58	11.20	360.23	nr	**4887.81**
50 mm pump size; 8.5 l/s at 215 kPa max head; 4 kW max motor rating	2359.17	2959.34	11.70	376.31	nr	**3335.65**
65 mm pump size; 14.0 l/s at 90 kPa max head; 3 kW max motor rating	2069.06	2595.43	11.70	376.31	nr	**2971.74**
65 mm pump size; 14.0 l/s at 160 kPa max head; 4 kW max motor rating	2294.27	2877.93	11.70	376.31	nr	**3254.24**
80 mm pump size; 14.5 l/s at 210 kPa max head; 5.5 kW max motor rating	3519.66	4415.06	11.70	376.31	nr	**4791.37**
80 mm pump size; 22.0 l/s at 130 kPa max head; 5.5 kW max motor rating	3519.66	4415.06	13.64	438.70	nr	**4853.76**
80 mm pump size; 22.0 l/s at 200 kPa max head; 7.5 kW max motor rating	3685.71	4623.36	13.64	438.70	nr	**5062.06**
100 mm pump size; 22.0 l/s at 250 kPa max head; 11 kW max motor rating	5172.59	6488.50	13.64	438.70	nr	**6927.20**
100 mm pump size; 30.0 l/s at 100 kPa max head; 4.0 kW max motor rating	2895.51	3632.13	19.15	615.92	nr	**4248.05**
100 mm pump size; 36.0 l/s at 250 kPa max head; 15.0 kW max motor rating	5478.01	6871.61	19.15	615.92	nr	**7487.53**
100 mm pump size; 36.0 l/s at 550 kPa max head; 30.0 kW max motor rating	7434.41	9325.72	19.15	615.92	nr	**9941.64**
Centrifugal heating and chilled water pump; Twin Head Belt Drive; 3 phase, 1450 rpm motor; max. pressure 1000 kN/m²; max. temperature 125°C; bed plate; coupling guard; bolted connections; supply only mating flanges; includes fixing on prepared concrete base; electrical work elsewhere						
40 mm pump size; 4.0 l/s at 70 kPa max head; 0.75 kW max motor rating	3447.13	4324.08	7.59	244.12	nr	**4568.20**
40 mm pump size; 4.0 l/s at 130 kPa max head; 1.5 kW max motor rating	4149.54	5205.18	8.09	260.20	nr	**5465.38**

38 MECHANICAL/COOLING/HEATING SYSTEMS

Item	Net Price £	Material £	Labour hours	Labour £	Unit	Total rate £
50 mm pump size; 8.5 l/s at 90 kPa max head; 2.2 kW max motor rating	4239.26	5317.73	8.67	278.86	nr	5596.59
50 mm pump size; 8.5 l/s at 190 kPa max head; 4 kW max motor rating	4886.29	6129.36	11.20	360.23	nr	6489.59
65 mm pump size; 8.5 l/s at 215 kPa max head; 4 kW max motor rating	6714.79	8423.03	11.70	376.31	nr	8799.34
65 mm pump size; 14.0 l/s at 90 kPa max head; 3 kW max motor rating	7050.54	8844.19	11.70	376.31	nr	9220.50
65 mm pump size; 14.0 l/s at 160 kPa max head; 4 kW max motor rating	7403.08	9286.42	11.70	376.31	nr	9662.73
80 mm pump size; 14.5 l/s at 210 kPa max head; 7.5 kW max motor rating	7655.83	9603.47	13.64	438.70	nr	10042.17
Centrifugal heating and chilled water pump; Close Coupled; 3 phase, 1450 rpm motor; max. pressure 1000 kN/m²; max. temperature 110°C; bed plate; coupling guard; bolted connections; supply only mating flanges; includes fixing on prepared concrete base; electrical work elsewhere						
40 mm pump size; 4.0 l/s at 23 kPa max head; 0.55 kW max motor rating	1079.00	1353.50	7.31	235.11	nr	1588.61
50 mm pump size; 4.0 l/s at 75 kPa max head; 0.75 kW max motor rating	1196.06	1500.34	7.31	235.11	nr	1735.45
50 mm pump size; 7.0 l/s at 65 kPa max head; 0.75 kW max motor rating	1196.06	1500.34	8.01	257.63	nr	1757.97
65 mm pump size; 10.0 l/s at 33 kPa max head; 0.75 kW max motor rating	1355.69	1700.57	8.01	257.63	nr	1958.20
50 mm pump size; 4.0 l/s at 120 kPa max head; 1.5 kW max motor rating	1632.32	2047.58	8.01	257.63	nr	2305.21
80 mm pump size; 16.0 l/s at 80 kPa max head; 2.2 kW max motor rating	2113.31	2650.94	12.35	397.21	nr	3048.15
80 mm pump size; 16.0 l/s at 120 kPa max head; 4.0 kW max motor rating	2060.11	2584.20	12.35	397.21	nr	2981.41
100 mm pump size; 28.0 l/s at 40 kPa max head; 2.2 kW max motor rating	2087.81	2618.95	17.86	574.43	nr	3193.38
100 mm pump size; 28.0 l/s at 90 kPa max head; 4.0 kW max motor rating	2304.84	2891.19	17.86	574.43	nr	3465.62
125 mm pump size; 40.0 l/s at 50 kPa max head; 3.0 kW max motor rating	2198.44	2757.72	25.85	831.40	nr	3589.12
125 mm pump size; 40.0 l/s at 120 kPa max head; 7.5 kW max motor rating	2713.47	3403.78	25.85	831.40	nr	4235.18
150 mm pump size; 70.0 l/s at 75 kPa max head; 11 kW max motor rating	3988.26	5002.87	30.43	978.72	nr	5981.59
150 mm pump size; 70.0 l/s at 120 kPa max head; 15.0 kW max motor rating	4262.81	5347.27	30.43	978.72	nr	6325.99
150 mm pump size; 70.0 l/s at 150 kPa max head; 15.0 kW max motor rating	4262.81	5347.27	30.43	978.72	nr	6325.99

38 MECHANICAL/COOLING/HEATING SYSTEMS

Item	Net Price £	Material £	Labour hours	Labour £	Unit	Total rate £
LOW TEMPERATURE HOT WATER HEATING; PIPELINE: PUMPS – cont						
Centrifugal heating & chilled water pump; close coupled; 3 phase, Variable Speed motor; max. system pressure 1000 kN/m²; max. temperature 110°C; bed plate; coupling guard; bolted connections; supply only mating flanges; includes fixing on prepared concrete base; electrical work elsewhere						
40 mm pump size; 4.0 l/s at 23 kPa max head; 0.55 kW max motor rating	1947.32	2442.72	7.31	235.11	nr	2677.83
40 mm pump size; 4.0 l/s at 75 kPa max head; 0.75 kW max motor rating	2149.50	2696.33	7.31	235.11	nr	2931.44
50 mm pump size; 7.0 l/s at 65 kPa max head; 1.5 kW max motor rating	2636.85	3307.66	8.01	257.63	nr	3565.29
50 mm pump size; 10.0 l/s at 33 kPa max head; 1.5 kW max motor rating	2636.85	3307.66	8.01	257.63	nr	3565.29
50 mm pump size; 4.0 l/s at 120 kPa max head; 1.5 kW max motor rating	2636.85	3307.66	8.01	257.63	nr	3565.29
80 mm pump size; 16.0 l/s at 80 kPa max head; 2.2 kW max motor rating	3437.07	4311.46	12.35	397.21	nr	4708.67
80 mm pump size; 16.0 l/s at 120 kPa max head; 3.0 kW max motor rating	3707.35	4650.50	12.35	397.21	nr	5047.71
100 mm pump size; 28.0 l/s at 40 kPa max head; 2.2 kW max motor rating	3522.18	4418.22	17.86	574.43	nr	4992.65
100 mm pump size; 28.0 l/s at 90 kPa max head; 4.0.kW max motor rating	4201.09	5269.85	17.86	574.43	nr	5844.28
125 mm pump size; 40.0 l/s at 50 kPa max head; 3.0 kW max motor rating	4032.88	5058.85	25.85	831.40	nr	5890.25
125 mm pump size; 40.0 l/s at 120 kPa max head; 7.5 kW max motor rating	1659.67	2081.89	25.85	831.40	nr	2913.29
150 mm pump size; 70.0 l/s at 75 kPa max head; 7.5 kW max motor rating	6561.27	8230.45	30.43	978.72	nr	9209.17
Glandless domestic heating pump; for low pressure domestic hot water heating systems; 240 volt; 50 Hz electric motor; max working pressure 1000 N/m² and max temperature of 130°C; includes fixing in position; electrical work elsewhere						
1" BSP unions – 2 speed	155.39	194.92	1.58	50.81	nr	245.73
1.25" BSP unions – 3 speed	228.98	287.24	1.58	50.81	nr	338.05
Glandless pumps; for hot water secondary supply; silent running; 3 phase; max pressure 1000 kN/m²; max temperature 130°C; bolted connections; supply only mating flanges; including fixing in position; electrical elsewhere						
1" BSP unions – 3 speed	319.08	400.25	1.58	50.81	nr	451.06

38 MECHANICAL/COOLING/HEATING SYSTEMS

Item	Net Price £	Material £	Labour hours	Labour £	Unit	Total rate £
Pipeline mounted circulator; for heating and chilled water; silent running; 3 phase; 1450 rpm motor; max pressure 1000 kN/m²; max temperature 120°C; bolted connections; supply only mating flanges; includes fixing in position; electrical elsewhere						
32 mm pump size; 2.0 l/s at 17 kPa max head; 0.2 kW max motor rating	638.17	800.52	6.44	207.12	nr	**1007.64**
50 mm pump size; 3.0 l/s at 20 kPa max head; 0.2 kW max motor rating	638.17	800.52	6.86	220.64	nr	**1021.16**
65 mm pump size; 5.0 l/s at 30 kPa max head; 0.37 kW max motor rating	1009.52	1266.34	7.48	240.58	nr	**1506.92**
65 mm pump size; 8.0 l/s at 37 kPa max head; 0.75 kW max motor rating	1130.30	1417.85	7.48	240.58	nr	**1658.43**
80 mm pump size; 12.0 l/s at 42 kPa max head; 1.1 kW max motor rating	1314.19	1648.52	8.01	257.63	nr	**1906.15**
100 mm pump size; 25.0 l/s at 37 kPa max head; 2.2 kW max motor rating	1873.03	2349.52	9.11	293.00	nr	**2642.52**
Dual pipeline mounted circulator; for heating & chilled water; silent running; 3 phase; 1450 rpm motor; max pressure 1000 kN/m²; max temperature 120°C; bolted connections; supply only mating flanges; includes fixing in position; electrical work elsewhere						
40 mm pump size; 2.0 l/s at 17 kPa max head; 0.8 kW max motor rating	1177.19	1476.66	7.88	253.43	nr	**1730.09**
50 mm pump size; 3.0 l/s at 20 kPa max head; 0.2 kW max motor rating	1180.78	1481.17	8.01	257.63	nr	**1738.80**
65 mm pump size; 5.0 l/s at 30 kPa max head; 0.37 kW max motor rating	1916.29	2403.79	9.20	295.90	nr	**2699.69**
65 mm pump size; 8.0 l/s at 37 kPa max head; 0.75 kW max motor rating	2215.55	2779.19	9.20	295.90	nr	**3075.09**
100 mm pump size; 12.0 l/s at 42 kPa max head; 1.1 kW max motor rating	2505.77	3143.24	9.45	303.95	nr	**3447.19**
Glandless accelerator pumps; for low and medium pressure heating services; silent running; 3 phase; 1450 rpm motor; max pressure 1000 kN/m²; max temperature 130°C; bolted connections; supply only mating flanges; includes fixing in position; electrical work elsewhere						
40 mm pump size; 4.0 l/s at 15 kPa max head; 0.35 kW max motor rating	492.83	618.21	6.94	223.20	nr	**841.41**
50 mm pump size; 6.0 l/s at 20 kPa max head; 0.45 kW max motor rating	525.21	658.83	7.35	236.39	nr	**895.22**
80 mm pump size; 13.0 l/s at 28 kPa max head; 0.58 kW max motor rating	1034.93	1298.21	7.76	249.59	nr	**1547.80**

38 MECHANICAL/COOLING/HEATING SYSTEMS

Item	Net Price £	Material £	Labour hours	Labour £	Unit	Total rate £
LOW TEMPERATURE HOT WATER HEATING; PIPELINE: HEAT EXCHANGERS						
Plate heat exchanger; for use in LTHW systems; painted carbon steel frame; stainless steel plates, nitrile rubber gaskets; design pressure of 10 bar and operating temperature of 110/135°C						
Primary side; 80°C in, 69°C out; secondary side; 82°C in, 71°C out						
107 kW, 2.38 l/s	2332.43	2925.80	10.00	321.63	nr	**3247.43**
245 kW, 5.46 l/s	3584.13	4495.94	10.00	321.63	nr	**4817.57**
287 kW, 6.38 l/s	3978.56	4990.71	10.00	321.63	nr	**5312.34**
328 kW, 7.31 l/s	4341.25	5445.66	10.00	321.63	nr	**5767.29**
364 kW, 8.11 l/s	5152.11	6462.80	10.00	321.63	nr	**6784.43**
403 kW, 8.96 l/s	5408.56	6784.50	10.00	321.63	nr	**7106.13**
453 kW, 10.09 l/s	5627.15	7058.70	10.00	321.63	nr	**7380.33**
490 kW, 10.89 l/s	5833.52	7317.56	10.00	321.63	nr	**7639.19**
1000 kW, 21.7 l/s	4340.03	5444.13	12.00	385.95	nr	**5830.08**
1500 kW, 32.6 l/s	5529.45	6936.14	12.00	385.95	nr	**7322.09**
2000 kW, 43.4 l/s	6633.38	8320.92	15.00	482.44	nr	**8803.36**
2500 kW, 54.3 l/s	8153.74	10228.05	15.00	482.44	nr	**10710.49**

Note: For temperature conditions different to those above, the cost of the units can vary significantly, and so manufacturers advice should be sought.

38 MECHANICAL/COOLING/HEATING SYSTEMS

Item	Net Price £	Material £	Labour hours	Labour £	Unit	Total rate £
LOW TEMPERATURE HOT WATER HEATING; PIPELINE: CALORIFIERS						
Non-storage calorifiers; mild steel; heater battery duty 116°C/90°C to BS 853, maximum test on shell 11.55 bar, tubes 26.25 bar						
Horizontal or vertical; primary water at 116°C on, 90°C off						
40 kW capacity	1009.78	1266.66	3.00	96.49	nr	**1363.15**
88 kW capacity	1176.39	1475.67	5.00	160.81	nr	**1636.48**
176 kW capacity	1348.09	1691.04	7.04	226.50	nr	**1917.54**
293 kW capacity	1869.53	2345.13	9.01	289.76	nr	**2634.89**
586 kW capacity	2785.21	3493.77	22.22	714.73	nr	**4208.50**
879 kW capacity	3878.94	4865.74	28.57	918.94	nr	**5784.68**
1465 kW capacity	6231.74	7817.10	50.00	1608.14	nr	**9425.24**
2000 kW capacity	11342.58	14228.13	60.00	1929.77	nr	**16157.90**

38 MECHANICAL/COOLING/HEATING SYSTEMS

Item	Net Price £	Material £	Labour hours	Labour £	Unit	Total rate £
LOW TEMPERATURE HOT WATER HEATING; PIPELINE: HEAT EMITTERS						
Perimeter convector heating; metal casing with standard finish; aluminium extruded grille; including backplates						
Top/sloping/flat front outlet						
60 × 200 mm	35.35	44.34	2.00	64.32	m	108.66
60 × 300 mm	38.45	48.23	2.00	64.32	m	112.55
60 × 450 mm	47.66	59.79	2.00	64.32	m	124.11
60 × 525 mm	50.74	63.65	2.00	64.32	m	127.97
60 × 600 mm	55.35	69.43	2.00	64.32	m	133.75
90 × 260 mm	38.45	48.23	2.00	64.32	m	112.55
90 × 300 mm	39.97	50.14	2.00	64.32	m	114.46
90 × 450 mm	49.22	61.75	2.00	64.32	m	126.07
90 × 525 mm	53.82	67.51	2.00	64.32	m	131.83
90 × 600 mm	56.90	71.38	2.00	64.32	m	135.70
Extra over for dampers						
Damper	16.90	21.20	0.25	8.04	nr	29.24
Extra over for fittings						
60 mm end caps	13.53	16.97	0.25	8.04	nr	25.01
90 mm end caps	21.29	26.70	0.25	8.04	nr	34.74
60 mm corners	28.76	36.08	0.25	8.04	nr	44.12
90 mm corners	42.30	53.07	0.25	8.04	nr	61.11

38 MECHANICAL/COOLING/HEATING SYSTEMS

Item	Net Price £	Material £	Labour hours	Labour £	Unit	Total rate £
LOW TEMPERATURE HOT WATER HEATING; PIPELINE: RADIATORS						
Radiant Strip Heaters						
Suitable for connection to hot water system; aluminium sheet panels with steel pipe clamped to upper surface; including insulation, sliding brackets, cover plates, end closures; weld or screwed BSP ends						
One pipe						
1500 mm long	71.50	89.69	3.11	100.03	nr	189.72
3000 mm long	113.08	141.85	3.11	100.03	nr	241.88
4500 mm long	153.40	192.43	3.11	100.03	nr	292.46
6000 mm long	211.46	265.26	3.11	100.03	nr	365.29
Two pipe						
1500 mm long	133.48	167.44	4.15	133.48	nr	300.92
3000 mm long	210.84	264.48	4.15	133.48	nr	397.96
4500 mm long	287.91	361.16	4.15	133.48	nr	494.64
6000 mm long	387.88	486.56	4.15	133.48	nr	620.04
Pressed steel panel type radiators; fixed with and including brackets; taking down once for decoration; refixing						
300 mm high; single panel						
500 mm length	19.52	24.48	2.03	65.30	nr	89.78
1000 mm length	39.00	48.92	2.03	65.30	nr	114.22
1500 mm length	51.11	64.11	2.03	65.30	nr	129.41
2000 mm length	57.34	71.93	2.47	79.44	nr	151.37
2500 mm length	63.55	79.72	2.97	95.52	nr	175.24
3000 mm length	76.05	95.40	3.22	103.57	nr	198.97
300 mm high; double panel; convector						
500 mm length	37.52	47.06	2.13	68.50	nr	115.56
1000 mm length	75.09	94.19	2.13	68.50	nr	162.69
1500 mm length	112.62	141.27	2.13	68.50	nr	209.77
2000 mm length	150.18	188.38	2.57	82.66	nr	271.04
2500 mm length	187.69	235.44	3.07	98.74	nr	334.18
3000 mm length	225.25	282.55	3.31	106.46	nr	389.01
450 mm high; single panel						
500 mm length	18.20	22.83	2.08	66.90	nr	89.73
1000 mm length	36.44	45.71	2.08	66.90	nr	112.61
1600 mm length	58.30	73.14	2.53	81.37	nr	154.51
2000 mm length	72.89	91.44	2.97	95.52	nr	186.96
2400 mm length	87.44	109.68	3.47	111.61	nr	221.29
3000 mm length	109.30	137.11	3.82	122.85	nr	259.96
450 mm high; double panel; convector						
500 mm length	33.37	41.85	2.18	70.12	nr	111.97
1000 mm length	66.72	83.70	2.18	70.12	nr	153.82
1600 mm length	122.20	153.28	2.63	84.58	nr	237.86
2000 mm length	205.43	257.69	3.06	98.43	nr	356.12
2400 mm length	246.54	309.25	3.37	108.39	nr	417.64
3000 mm length	308.17	386.57	3.92	126.08	nr	512.65

38 MECHANICAL/COOLING/HEATING SYSTEMS

Item	Net Price £	Material £	Labour hours	Labour £	Unit	Total rate £
LOW TEMPERATURE HOT WATER HEATING; PIPELINE: RADIATORS – cont						
Pressed steel panel type radiators – cont						
600 mm high; single panel						
500 mm length	24.40	30.61	2.18	70.12	nr	100.73
1000 mm length	48.80	61.22	2.43	78.16	nr	139.38
1600 mm length	78.07	97.93	3.13	100.67	nr	198.60
2000 mm length	97.59	122.42	3.77	121.25	nr	243.67
2400 mm length	117.12	146.91	4.07	130.89	nr	277.80
3000 mm length	146.40	183.65	5.11	164.35	nr	348.00
600 mm high; double panel; convector						
500 mm length	42.01	52.70	2.28	73.34	nr	126.04
1000 mm length	84.02	105.39	2.28	73.34	nr	178.73
1600 mm length	153.88	193.03	3.23	103.89	nr	296.92
2000 mm length	258.70	324.51	3.87	124.48	nr	448.99
2400 mm length	310.44	389.41	4.17	134.12	nr	523.53
3000 mm length	388.05	486.77	5.24	168.53	nr	655.30
700 mm high; single panel						
500 mm length	28.54	35.80	2.23	71.72	nr	107.52
1000 mm length	57.08	71.60	2.83	91.02	nr	162.62
1600 mm length	91.34	114.58	3.73	119.97	nr	234.55
2000 mm length	114.18	143.23	4.46	143.44	nr	286.67
2400 mm length	137.04	171.90	4.48	144.09	nr	315.99
3000 mm length	171.25	214.82	5.24	168.53	nr	383.35
700 mm high; double panel; convector						
500 mm length	54.60	68.49	2.33	74.94	nr	143.43
1000 mm length	146.90	184.27	3.08	99.06	nr	283.33
1600 mm length	235.02	294.81	3.83	123.18	nr	417.99
2000 mm length	293.78	368.51	4.17	134.12	nr	502.63
2400 mm length	352.55	442.24	4.37	140.56	nr	582.80
3000 mm length	440.68	552.79	4.82	155.02	nr	707.81
Flat panel type steel radiators; fixed with and including brackets; taking down once for decoration; refixing						
300 mm high; single panel (44 mm deep)						
500 mm length	200.93	252.04	2.03	65.30	nr	317.34
1000 mm length	315.59	395.88	2.03	65.30	nr	461.18
1500 mm length	430.25	539.71	2.03	65.30	nr	605.01
2000 mm length	544.91	683.54	2.47	79.44	nr	762.98
2400 mm length	636.64	798.60	2.97	95.52	nr	894.12
3000 mm length	915.97	1149.00	3.22	103.57	nr	1252.57
300 mm high; double panel (100 mm deep)						
500 mm length	398.58	499.98	2.03	65.30	nr	565.28
1000 mm length	624.62	783.52	2.03	65.30	nr	848.82
1500 mm length	850.67	1067.08	2.03	65.30	nr	1132.38
2000 mm length	1076.71	1350.63	2.47	79.44	nr	1430.07
2400 mm length	1257.55	1577.48	2.97	95.52	nr	1673.00
3000 mm length	1670.54	2095.52	3.22	103.57	nr	2199.09

38 MECHANICAL/COOLING/HEATING SYSTEMS

Item	Net Price £	Material £	Labour hours	Labour £	Unit	Total rate £
500 mm high; single panel (44 mm deep)						
500 mm length	236.96	297.25	2.13	68.50	nr	**365.75**
1000 mm length	387.66	486.28	2.13	68.50	nr	**554.78**
1500 mm length	538.36	675.32	2.13	68.50	nr	**743.82**
2000 mm length	689.05	864.35	2.57	82.66	nr	**947.01**
2400 mm length	809.61	1015.57	3.07	98.74	nr	**1114.31**
3000 mm length	1132.19	1420.22	3.31	106.46	nr	**1526.68**
500 mm high; double panel (100 mm deep)						
500 mm length	464.10	582.16	2.08	66.90	nr	**649.06**
1000 mm length	755.66	947.90	2.08	66.90	nr	**1014.80**
1500 mm length	1047.20	1313.60	2.53	81.37	nr	**1394.97**
2000 mm length	1338.79	1679.37	2.97	95.52	nr	**1774.89**
2400 mm length	1572.04	1971.96	3.47	111.61	nr	**2083.57**
3000 mm length	2063.66	2588.66	3.82	122.85	nr	**2711.51**
600 mm high; single panel (44 mm deep)						
500 mm length	267.54	335.60	2.18	70.12	nr	**405.72**
1000 mm length	437.89	549.29	2.18	70.12	nr	**619.41**
1500 mm length	608.24	1312.27	2.63	84.58	nr	**1396.85**
2000 mm length	778.60	976.67	3.06	98.43	nr	**1075.10**
2400 mm length	914.88	1147.63	3.37	108.39	nr	**1256.02**
3000 mm length	1261.04	1581.84	3.92	126.08	nr	**1707.92**
600 mm high; double panel (100 mm deep)						
500 mm length	528.31	662.72	2.18	70.12	nr	**732.84**
1000 mm length	862.46	1081.88	2.43	78.16	nr	**1160.04**
1500 mm length	1196.61	1501.02	3.13	100.67	nr	**1601.69**
2000 mm length	1530.77	1920.20	3.77	121.25	nr	**2041.45**
2400 mm length	1798.09	2255.52	4.07	130.89	nr	**2386.41**
3000 mm length	2340.81	2936.32	5.11	164.35	nr	**3100.67**
700 mm high; single panel (44 mm deep)						
500 mm length	290.47	364.37	2.28	73.34	nr	**437.71**
1000 mm length	483.76	606.83	2.28	73.34	nr	**680.17**
1500 mm length	677.04	849.27	3.23	103.89	nr	**953.16**
2000 mm length	870.32	1091.73	3.87	124.48	nr	**1216.21**
2400 mm length	1025.26	1286.08	4.17	134.12	nr	**1420.20**
3000 mm length	1398.63	1754.45	5.24	168.53	nr	**1922.98**
700 mm high; double panel (100 mm deep)						
500 mm length	570.90	716.14	2.23	71.72	nr	**787.86**
1000 mm length	947.64	1188.72	2.83	91.02	nr	**1279.74**
1500 mm length	1324.38	1661.31	3.73	119.97	nr	**1781.28**
2000 mm length	1701.12	2133.88	4.46	143.44	nr	**2277.32**
2400 mm length	2002.51	2511.95	4.48	144.09	nr	**2656.04**

38 MECHANICAL/COOLING/HEATING SYSTEMS

Item	Net Price £	Material £	Labour hours	Labour £	Unit	Total rate £
LOW TEMPERATURE HOT WATER HEATING; PIPELINE: RADIATORS – cont						
Pressed steel panel type radiators – cont						
Fan convector; sheet metal casing with lockable access panel; centrifugal fan; air filter; LPHW heating coil; extruded aluminium grilles; 3 speed; includes fixing in position; electrical work elsewhere						
Free-standing flat top, 695 mm high, medium speed rating						
Entering air temperature, 18°C						
695 mm long, 1 row 1.94 kW, 75 l/s	679.61	852.50	2.73	87.81	nr	**940.31**
695 mm long, 2 row 2.64 kW, 75 l/s	679.61	852.50	2.73	87.81	nr	**940.31**
895 mm long, 1 row 4.02 kW, 150 l/s	765.81	960.64	2.73	87.81	nr	**1048.45**
895 mm long, 2 row 5.62 kW, 150 l/s	765.81	960.64	2.73	87.81	nr	**1048.45**
1195 mm long, 1 row 6.58 kW, 250 l/s	871.88	1093.69	3.00	96.49	nr	**1190.18**
1195 mm long, 2 row 9.27 kW, 250 l/s	871.88	1093.69	3.00	96.49	nr	**1190.18**
1495 mm long, 1 row 9.04 kW, 340 l/s	973.01	1220.54	3.26	104.85	nr	**1325.39**
1495 mm long, 2 row 12.73 kW, 340 l/s	973.01	1220.54	3.26	104.85	nr	**1325.39**
Free-standing flat top, 695 mm high, medium speed rating, c/w floor plinth						
695 mm long, 1 row 1.94 kW, 75 l/s	713.58	895.12	2.73	87.81	nr	**982.93**
695 mm long, 2 row 2.64 kW, 75 l/s	713.58	895.12	2.73	87.81	nr	**982.93**
895 mm long, 1 row 4.02 kW, 150 l/s	804.10	1008.66	2.73	87.81	nr	**1096.47**
895 mm long, 2 row 5.62 kW, 150 l/s	804.10	1008.66	2.73	87.81	nr	**1096.47**
1195 mm long, 1 row 6.58 kW, 250 l/s	915.50	1148.40	3.00	96.49	nr	**1244.89**
1195 mm long, 2 row 9.27 kW, 250 l/s	915.50	1148.40	3.00	96.49	nr	**1244.89**
1495 mm long, 1 row 9.04 kW, 340 l/s	1021.65	1281.56	3.26	104.85	nr	**1386.41**
1495 mm long, 2 row 12.73 kW, 340 l/s	1021.65	1281.56	3.26	104.85	nr	**1386.41**
Free-standing sloping top, 695 mm high, medium speed rating, c/w floor plinth						
695 mm long, 1 row 1.94 kW, 75 l/s	738.45	926.31	2.73	87.81	nr	**1014.12**
695 mm long, 2 row 2.64 kW, 75 l/s	738.45	926.31	2.73	87.81	nr	**1014.12**
895 mm long, 1 row 4.02 kW, 150 l/s	828.96	1039.85	2.73	87.81	nr	**1127.66**
895 mm long, 2 row 5.62 kW, 150 l/s	828.96	1039.85	2.73	87.81	nr	**1127.66**
1195 mm long, 1 row 6.58 kW, 250 l/s	940.36	1179.58	3.00	96.49	nr	**1276.07**
1195 mm long, 2 row 9.27 kW, 250 l/s	940.36	1179.58	3.00	96.49	nr	**1276.07**
1495 mm long, 1 row 9.04 kW, 340 l/s	1046.51	1312.74	3.26	104.85	nr	**1417.59**
1495 mm long, 2 row 12.73 kW, 340 l/s	1046.51	1312.74	3.26	104.85	nr	**1417.59**
Wall mounted high level sloping discharge						
695 mm long, 1 row 1.94 kW, 75 l/s	745.91	935.67	2.73	87.81	nr	**1023.48**
695 mm long, 2 row 2.64 kW, 75 l/s	745.91	935.67	2.73	87.81	nr	**1023.48**
895 mm long, 1 row 4.02 kW, 150 l/s	767.48	962.73	2.73	87.81	nr	**1050.54**
895 mm long, 2 row 5.62 kW, 150 l/s	767.48	962.73	2.73	87.81	nr	**1050.54**
1195 mm long, 1 row 6.58 kW, 250 l/s	934.03	1171.64	3.00	96.49	nr	**1268.13**
1195 mm long, 2 row 9.27 kW, 250 l/s	934.03	1171.64	3.00	96.49	nr	**1268.13**
1495 mm long, 1 row 9.04 kW, 340 l/s	1012.78	1270.43	3.26	104.85	nr	**1375.28**
1495 mm long, 2 row 12.73 kW, 340 l/s	1012.78	1270.43	3.26	104.85	nr	**1375.28**

38 MECHANICAL/COOLING/HEATING SYSTEMS

Item	Net Price £	Material £	Labour hours	Labour £	Unit	Total rate £
Ceiling mounted sloping inlet/outlet 665 mm wide						
895 mm long, 1 row 4.02 kW, 150 l/s	843.70	1058.33	4.15	133.48	nr	**1191.81**
895 mm long, 2 row 5.62 kW, 150 l/s	843.70	1058.33	4.15	133.48	nr	**1191.81**
1195 mm long, 1 row 6.58 kW, 250 l/s	948.15	1189.36	4.15	133.48	nr	**1322.84**
1195 mm long, 2 row 9.27 kW, 250 l/s	948.15	1189.36	4.15	133.48	nr	**1322.84**
1495 mm long, 1 row 9.04 kW, 340 l/s	1040.96	1305.79	4.15	133.48	nr	**1439.27**
1495 mm long, 2 row 12.73 kW, 340 l/s	1040.96	1305.79	4.15	133.48	nr	**1439.27**
Free-standing unit, extended height 1700/1900/2100 mm						
895 mm long, 1 row 4.02 kW, 150 l/s	976.65	1225.11	3.11	100.03	nr	**1325.14**
895 mm long, 2 row 5.62 kW, 150 l/s	976.32	1224.70	3.11	100.03	nr	**1324.73**
1195 mm long, 1 row 6.58 kW, 250 l/s	1140.42	1430.54	3.11	100.03	nr	**1530.57**
1195 mm long, 2 row 9.27 kW, 250 l/s	1140.42	1430.54	3.11	100.03	nr	**1530.57**
1495 mm long, 1 row 9.04 kW, 340 l/s	1254.81	1574.04	3.11	100.03	nr	**1674.07**
1495 mm long, 2 row 12.73 kW, 340 l/s	1254.81	1574.04	3.11	100.03	nr	**1674.07**
LTHW trench heating; water temperatures 90°C/70°C; room air temperature 20°C; convector with copper tubes and aluminium fins within steel duct; Includes fixing within floor screed; electrical work elsewhere						
Natural Convection Type						
Normal capacity, 182 mm width, complete with linear, natural anodized aluminium grille (grille also costed separately below)						
90 mm deep						
1300 mm long, 287 W output	280.02	351.25	2.00	64.32	nr	**415.57**
2300 mm long, 576 W output	457.56	573.97	4.00	128.65	nr	**702.62**
3300 mm long, 864 W output	626.34	785.68	5.00	160.81	nr	**946.49**
4500 mm long, 1210 W output	857.34	1075.45	7.00	225.14	nr	**1300.59**
4900 mm long, 1325 W output	920.46	1154.63	8.00	257.30	nr	**1411.93**
110 mm deep						
1300 mm long, 331 W output	282.42	354.27	2.00	64.32	nr	**418.59**
2300 mm long, 662 W output	461.58	579.01	4.00	128.65	nr	**707.66**
3300 mm long, 993 W output	631.98	792.76	5.00	160.81	nr	**953.57**
4500 mm long, 1389 W output	865.20	1085.30	7.00	225.14	nr	**1310.44**
4900 mm long, 1522 W output	928.62	1164.86	8.00	257.30	nr	**1422.16**
140 mm deep						
1300 mm long, 551 W output	332.52	417.11	2.00	64.32	nr	**481.43**
2300 mm long, 1100 W output	546.72	685.81	4.00	128.65	nr	**814.46**
3300 mm long, 1651 W output	750.96	942.01	5.00	160.81	nr	**1102.82**
4500 mm long, 2310 W output	1023.54	1283.92	7.00	225.14	nr	**1509.06**
4900 mm long, 2530 W output	1100.46	1380.42	8.00	257.30	nr	**1637.72**
190 mm deep						
1300 mm long, 625 W output	343.74	431.19	2.00	64.32	nr	**495.51**
2300 mm long, 1250 W output	566.94	711.17	4.00	128.65	nr	**839.82**
3300 mm long, 1874 W output	780.18	978.66	5.00	160.81	nr	**1139.47**

38 MECHANICAL/COOLING/HEATING SYSTEMS

Item	Net Price £	Material £	Labour hours	Labour £	Unit	Total rate £
LOW TEMPERATURE HOT WATER HEATING; PIPELINE: RADIATORS – cont						
Normal capacity, 182 mm width – cont						
190 mm deep – cont						
4500 mm long, 2624 W output	1064.28	1335.03	7.00	225.14	nr	**1560.17**
4900 mm long, 2875 W output	1144.62	1435.81	8.00	257.30	nr	**1693.11**
Fan assisted type (outputs assume fan at 50%)						
Normal capacity, 182 mm width, complete with Natural anodized aluminium grille						
140 mm deep						
1300 mm long, 1439 W output	546.18	685.13	2.00	64.32	nr	**749.45**
2300 mm long, 3331 W output	903.24	1133.03	4.00	128.65	nr	**1261.68**
3300 mm long, 5225 W output	1246.80	1563.99	5.00	160.81	nr	**1724.80**
4300 mm long, 5753 W output	1463.40	1835.69	7.00	225.14	nr	**2060.83**
4900 mm long, 6070 W output	1578.30	1979.82	8.00	257.30	nr	**2237.12**
Linear grille anodized aluminium, 170 mm width (if supplied as a separate item)	148.67	186.49	–	–	m	**186.49**
Roll up grille, natural anodized aluminium	148.67	186.49	–	–	m	**186.49**
Thermostatic valve with remote regulator (c/w valve body)	61.38	77.00	4.00	128.65	nr	**205.65**
Fan speed controller	30.00	37.63	2.00	64.32	nr	**101.95**
Note: as an alternative to thermostatic control, the system can be controlled via two port valves. Refer to valve section for valve prices.						
LTHW underfloor heating; water flow and return temperatures of 60°C and 70°C; pipework at 300 mm centres; pipe fixings; flow and return manifolds and zone actuators; wiring block; insulation; includes fixing in position; excludes secondary pump, mixing valve, zone thermostats and floor finishes ; electrical work elsewhere						
Note: All rates are expressed on a m² basis, for the following example areas						
Screeded floor with 15–25 mm stone/marble finish (producing 80–100 W/m²)						
250 m² area (single zone)	26.07	32.70	0.14	4.50	m²	**37.20**
1000 m² area (single zone)	25.39	31.85	0.12	3.86	m²	**35.71**
5000 m² area (multi-zone)	25.33	31.77	0.10	3.21	m²	**34.98**

38 MECHANICAL/COOLING/HEATING SYSTEMS

Item	Net Price £	Material £	Labour hours	Labour £	Unit	Total rate £
Screeded floor with 10 mm carpet tile (producing 80–100 W/m²)						
250 m² area (single zone)	26.07	32.70	0.14	4.50	m²	**37.20**
1000 m² area (single zone)	25.39	31.85	0.12	3.86	m²	**35.71**
5000 m² area (multi-zone)	25.33	31.77	0.10	3.21	m²	**34.98**
Floating timber floor with 20 mm timber finish (producing 70–80 W/m²)						
250 m² are (single zone)	32.85	41.20	0.14	4.50	m²	**45.70**
1000 m² area(single zone)	35.21	44.17	0.12	3.86	m²	**48.03**
5000 m² area (multi-zone)	33.63	42.19	0.10	3.21	m²	**45.40**
Floating timber floor with 10 mm carpet tile (producing 70–80 W/m²)						
250 m² are (single zone)	42.75	53.63	0.16	5.14	m²	**58.77**
1000 m² area(single zone)	39.64	49.73	0.12	3.86	m²	**53.59**
5000 m² area (multi-zone)	37.92	47.57	0.10	3.21	m²	**50.78**

38 MECHANICAL/COOLING/HEATING SYSTEMS

Item	Net Price £	Material £	Labour hours	Labour £	Unit	Total rate £
LOW TEMPERATURE HOT WATER HEATING; PIPELINE: PIPE FREEZING						
Freeze isolation of carbon steel or copper pipelines containing static water, either side of work location, freeze duration not exceeding 4 hours assuming that flow and return circuits are treated concurrently and activities undertaken during normal working hours						
Up to 4 freezes						
50 mm dia.	450.77	565.44	1.00	32.17	nr	**597.61**
65 mm dia.	450.77	565.44	1.00	32.17	nr	**597.61**
80 mm dia.	515.32	646.42	1.00	32.17	nr	**678.59**
100 mm dia.	579.90	727.43	1.00	32.17	nr	**759.60**
150 mm dia.	964.91	1210.38	1.00	32.17	nr	**1242.55**
200 mm dia.	1479.03	1855.29	1.00	32.17	nr	**1887.46**

38 MECHANICAL/COOLING/HEATING SYSTEMS

Item	Net Price £	Material £	Labour hours	Labour £	Unit	Total rate £
LOW TEMPERATURE HOT WATER HEATING; PIPELINE: ENERGY METERS						
Ultrasonic						
Energy meter for measuring energy use in LTHW systems; includes ultrasonic flow meter (with sensor and signal converter), energy calculator, pair of temperature sensors with brass pockets, and 3 m of interconnecting cable; includes fixing in position; electrical work elsewhere						
Pipe size (flanged connections to PN16); maximum flow rate						
50 mm, 36 m³/hr	1322.46	1658.90	1.80	57.89	nr	1716.79
65 mm, 60 m³/hr	1457.74	1828.59	2.32	74.61	nr	1903.20
80 mm, 100 m³/hr	1634.02	2049.71	2.56	82.34	nr	2132.05
125 mm, 250 m³/hr	1893.17	2374.79	3.60	115.79	nr	2490.58
150 mm, 360 m³/hr	2055.15	2577.98	4.80	154.38	nr	2732.36
200 mm, 600 m³/hr	2296.20	2880.35	6.24	200.69	nr	3081.04
250 mm, 1000 m³/hr	2649.67	3323.75	9.60	308.76	nr	3632.51
300 mm, 1500 m³/hr	3114.63	3907.00	10.80	347.36	nr	4254.36
350 mm, 2000 m³/hr	3752.99	4707.75	13.20	424.55	nr	5132.30
400 mm, 2500 m³/hr	4296.08	5389.00	15.60	501.74	nr	5890.74
500 mm, 3000 m³/hr	4875.35	6115.64	24.00	771.90	nr	6887.54
600 mm, 3500 m³/hr	5477.51	6870.99	28.00	900.56	nr	7771.55

38 MECHANICAL/COOLING/HEATING SYSTEMS

Item	Net Price £	Material £	Labour hours	Labour £	Unit	Total rate £
LOW TEMPERATURE HOT WATER HEATING: CONTROL COMPONENTS MECHANICAL						
Room thermostats; light and medium duty; installed and connected						
Range 3°C to 27°C; 240 Volt						
1 amp; on/off type	28.89	36.24	0.30	9.64	nr	**45.88**
Range 0°C to +15°C; 240 Volt						
6 amp; frost thermostat	19.48	24.44	0.30	9.64	nr	**34.08**
Range 3°C to 27°C; 250 Volt						
2 amp; changeover type; dead zone	48.33	60.63	0.30	9.64	nr	**70.27**
2 amp; changeover type	22.43	28.13	0.30	9.64	nr	**37.77**
2 amp; changeover type; concealed setting	28.27	35.46	0.30	9.64	nr	**45.10**
6 amp; on/off type	17.45	21.88	0.30	9.64	nr	**31.52**
6 amp; temperature set-back	36.11	45.29	0.30	9.64	nr	**54.93**
16 amp; on/off type	26.88	33.72	0.30	9.64	nr	**43.36**
16 amp; on/off type; concealed setting	29.30	36.76	0.30	9.64	nr	**46.40**
20 amp; on/off type; concealed setting	24.96	31.32	0.30	9.64	nr	**40.96**
20 amp; indicated 'off' position	31.47	39.48	0.30	9.64	nr	**49.12**
20 amp; manual; double pole on/off and neon indicator	57.29	71.86	0.30	9.64	nr	**81.50**
20 amp; indicated 'off' position	37.49	47.03	0.30	9.64	nr	**56.67**
Range 10°C to 40°C; 240 Volt						
20 amp; changeover contacts	34.16	42.85	0.30	9.64	nr	**52.49**
2 amp; 'heating-cooling' switch	73.88	92.68	0.30	9.64	nr	**102.32**
Surface thermostats						
Cylinder thermostat						
6 amp; changeover type; with cable	20.17	25.30	0.25	8.04	nr	**33.34**
Electrical thermostats; installed and connected						
Range 5°C to 30°C; 230 Volt standard port single time						
10 amp with sensor	28.01	35.13	0.30	9.64	nr	**44.77**
Range 5°C to 30°C; 230 Volt standard port double time						
10 amp with sensor	31.73	39.80	0.30	9.64	nr	**49.44**
10 amp with sensor and on/off switch	48.41	60.73	0.30	9.64	nr	**70.37**
Radiator thermostats						
Angled valve body; thermostatic head; built in sensor						
15 mm; liquid filled	17.46	21.91	0.84	27.04	nr	**48.95**
15 mm; wax filled	17.46	21.91	0.84	27.04	nr	**48.95**

38 MECHANICAL/COOLING/HEATING SYSTEMS

Item	Net Price £	Material £	Labour hours	Labour £	Unit	Total rate £
Immersion thermostats; stem type; domestic water boilers; fitted; electrical work elsewhere						
Temperature range 0°C to 40°C						
Non-standard; 280 mm stem	10.62	13.32	0.25	8.04	nr	**21.36**
Temperature range 18°C to 88°C						
13 amp; 178 mm stem	7.06	8.86	0.25	8.04	nr	**16.90**
20 amp; 178 mm stem	10.97	13.76	0.25	8.04	nr	**21.80**
Non-standard; pocket clip; 280 mm stem	10.17	12.76	0.25	8.04	nr	**20.80**
Temperature range 40°C to 80°C						
13 amp; 178 mm stem	3.88	4.87	0.25	8.04	nr	**12.91**
20 amp; 178 mm stem	7.32	9.18	0.25	8.04	nr	**17.22**
Non-standard; pocket clip; 280 mm stem	11.41	14.31	0.25	8.04	nr	**22.35**
13 amp; 457 mm stem	4.59	5.76	0.25	8.04	nr	**13.80**
20 amp; 457 mm stem	8.02	10.06	0.25	8.04	nr	**18.10**
Temperature range 50°C to 100°C						
Non-standard; 1780 mm stem	9.59	12.03	0.25	8.04	nr	**20.07**
Non-standard; 280 mm stem	10.00	12.54	0.25	8.04	nr	**20.58**
Pockets for thermostats						
For 178 mm stem	12.56	15.76	0.25	8.04	nr	**23.80**
For 280 mm stem	12.77	16.02	0.25	8.04	nr	**24.06**
Immersion thermostats; stem type; industrial installations; fitted; electrical work elsewhere						
Temperature range 5°C to 105°C						
For 305 mm stem	157.72	197.85	0.50	16.08	nr	**213.93**

38 MECHANICAL/COOLING/HEATING SYSTEMS

Item	Net Price £	Material £	Labour hours	Labour £	Unit	Total rate £
THERMAL INSULATION						
For flexible closed cell insulation see						
Section – Cold Water						
Mineral fibre sectional insulation; bright class						
O foil faced; bright class O foil taped joints;						
19 mm aluminium bands						
Concealed pipework						
20 mm thick						
15 mm dia.	4.00	5.02	0.15	4.83	m	9.85
20 mm dia.	4.26	5.34	0.15	4.83	m	10.17
25 mm dia.	4.57	5.73	0.15	4.83	m	10.56
32 mm dia.	5.11	6.41	0.15	4.83	m	11.24
40 mm dia.	5.47	6.87	0.15	4.83	m	11.70
50 mm dia.	6.25	7.84	0.15	4.83	m	12.67
Extra over for fittings concealed insulation						
Flange/union						
15 mm dia.	2.00	2.51	0.13	4.18	nr	6.69
20 mm dia.	2.13	2.68	0.13	4.18	nr	6.86
25 mm dia.	2.30	2.89	0.13	4.18	nr	7.07
32 mm dia.	2.56	3.21	0.13	4.18	nr	7.39
40 mm dia.	2.72	3.42	0.13	4.18	nr	7.60
50 mm dia.	3.12	3.91	0.13	4.18	nr	8.09
Valves						
15 mm dia.	4.00	5.02	0.15	4.83	nr	9.85
20 mm dia.	4.57	5.73	0.15	4.83	nr	10.56
25 mm dia.	4.57	5.73	0.15	4.83	nr	10.56
32 mm dia.	5.11	6.41	0.15	4.83	nr	11.24
40 mm dia.	5.47	6.87	0.15	4.83	nr	11.70
50 mm dia.	6.25	7.84	0.15	4.83	nr	12.67
Expansion bellows						
15 mm dia.	7.99	10.02	0.22	7.08	nr	17.10
20 mm dia.	8.54	10.71	0.22	7.08	nr	17.79
25 mm dia.	9.18	11.51	0.22	7.08	nr	18.59
32 mm dia.	10.21	12.81	0.22	7.08	nr	19.89
40 mm dia.	10.93	13.71	0.22	7.08	nr	20.79
50 mm dia.	12.49	15.67	0.22	7.08	nr	22.75
25 mm thick						
15 mm dia.	4.40	5.52	0.15	4.83	m	10.35
20 mm dia.	4.73	5.94	0.15	4.83	m	10.77
25 mm dia.	5.31	6.66	0.15	4.83	m	11.49
32 mm dia.	5.78	7.25	0.15	4.83	m	12.08
40 mm dia.	6.18	7.75	0.15	4.83	m	12.58
50 mm dia.	7.08	8.88	0.15	4.83	m	13.71
65 mm dia.	8.09	10.15	0.15	4.83	m	14.98
80 mm dia.	8.87	11.12	0.22	7.08	m	18.20
100 mm dia.	11.68	14.65	0.22	7.08	m	21.73
125 mm dia.	13.51	16.95	0.22	7.08	m	24.03
150 mm dia.	16.09	20.18	0.22	7.08	m	27.26

38 MECHANICAL/COOLING/HEATING SYSTEMS

Item	Net Price £	Material £	Labour hours	Labour £	Unit	Total rate £
200 mm dia.	22.75	28.54	0.25	8.04	m	36.58
250 mm dia.	27.23	34.16	0.25	8.04	m	42.20
300 mm dia.	29.12	36.52	0.25	8.04	m	44.56
Extra over for fittings concealed insulation						
Flange/union						
15 mm dia.	2.20	2.76	0.13	4.18	nr	6.94
20 mm dia.	2.36	2.96	0.13	4.18	nr	7.14
25 mm dia.	2.65	3.33	0.13	4.18	nr	7.51
32 mm dia.	2.89	3.63	0.13	4.18	nr	7.81
40 mm dia.	3.08	3.86	0.13	4.18	nr	8.04
50 mm dia.	3.53	4.42	0.13	4.18	nr	8.60
65 mm dia.	4.04	5.06	0.13	4.18	nr	9.24
80 mm dia.	4.44	5.57	0.18	5.79	nr	11.36
100 mm dia.	5.85	7.34	0.18	5.79	nr	13.13
125 mm dia.	6.75	8.47	0.18	5.79	nr	14.26
150 mm dia.	8.06	10.11	0.18	5.79	nr	15.90
200 mm dia.	11.37	14.26	0.22	7.08	nr	21.34
250 mm dia.	13.62	17.08	0.22	7.08	nr	24.16
300 mm dia.	14.55	18.26	0.22	7.08	nr	25.34
Valves						
15 mm dia.	4.40	5.52	0.15	4.83	nr	10.35
20 mm dia.	4.73	5.94	0.15	4.83	nr	10.77
25 mm dia.	5.31	6.66	0.15	4.83	nr	11.49
32 mm dia.	5.78	7.25	0.15	4.83	nr	12.08
40 mm dia.	6.18	7.75	0.15	4.83	nr	12.58
50 mm dia.	7.08	8.88	0.15	4.83	nr	13.71
65 mm dia.	8.09	10.15	0.15	4.83	nr	14.98
80 mm dia.	8.87	11.12	0.20	6.44	nr	17.56
100 mm dia.	11.68	14.65	0.20	6.44	nr	21.09
125 mm dia.	13.51	16.95	0.20	6.44	nr	23.39
150 mm dia.	16.09	20.18	0.20	6.44	nr	26.62
200 mm dia.	22.75	28.54	0.25	8.04	nr	36.58
250 mm dia.	27.23	34.16	0.25	8.04	nr	42.20
300 mm dia.	29.12	36.52	0.25	8.04	nr	44.56
Expansion bellows						
15 mm dia.	8.79	11.02	0.22	7.08	nr	18.10
20 mm dia.	9.50	11.92	0.22	7.08	nr	19.00
25 mm dia.	10.59	13.28	0.22	7.08	nr	20.36
32 mm dia.	11.54	14.47	0.22	7.08	nr	21.55
40 mm dia.	12.37	15.51	0.22	7.08	nr	22.59
50 mm dia.	14.16	17.76	0.22	7.08	nr	24.84
65 mm dia.	16.16	20.27	0.22	7.08	nr	27.35
80 mm dia.	17.72	22.23	0.29	9.33	nr	31.56
100 mm dia.	23.36	29.30	0.29	9.33	nr	38.63
125 mm dia.	27.04	33.91	0.29	9.33	nr	43.24
150 mm dia.	32.22	40.42	0.29	9.33	nr	49.75
200 mm dia.	45.50	57.08	0.36	11.58	nr	68.66
250 mm dia.	54.49	68.35	0.36	11.58	nr	79.93
300 mm dia.	58.25	73.07	0.36	11.58	nr	84.65

38 MECHANICAL/COOLING/HEATING SYSTEMS

Item	Net Price £	Material £	Labour hours	Labour £	Unit	Total rate £
THERMAL INSULATION – cont						
30 mm thick						
15 mm dia.	5.72	7.18	0.15	4.83	m	**12.01**
20 mm dia.	6.12	7.67	0.15	4.83	m	**12.50**
25 mm dia.	6.47	8.12	0.15	4.83	m	**12.95**
32 mm dia.	7.05	8.85	0.15	4.83	m	**13.68**
40 mm dia.	7.48	9.39	0.15	4.83	m	**14.22**
50 mm dia.	8.55	10.73	0.15	4.83	m	**15.56**
65 mm dia.	9.68	12.14	0.15	4.83	m	**16.97**
80 mm dia.	10.58	13.27	0.22	7.08	m	**20.35**
100 mm dia.	13.65	17.12	0.22	7.08	m	**24.20**
125 mm dia.	15.75	19.76	0.22	7.08	m	**26.84**
150 mm dia.	18.48	23.18	0.22	7.08	m	**30.26**
200 mm dia.	25.82	32.39	0.25	8.04	m	**40.43**
250 mm dia.	30.70	38.51	0.25	8.04	m	**46.55**
300 mm dia.	32.56	40.85	0.25	8.04	m	**48.89**
350 mm dia.	35.77	44.87	0.25	8.04	m	**52.91**
Extra over for fittings concealed insulation						
Flange/union						
15 mm dia.	2.85	3.57	0.13	4.18	nr	**7.75**
20 mm dia.	3.07	3.85	0.13	4.18	nr	**8.03**
25 mm dia.	3.22	4.04	0.13	4.18	nr	**8.22**
32 mm dia.	3.53	4.42	0.13	4.18	nr	**8.60**
40 mm dia.	3.73	4.68	0.13	4.18	nr	**8.86**
50 mm dia.	4.28	5.36	0.13	4.18	nr	**9.54**
65 mm dia.	4.85	6.08	0.13	4.18	nr	**10.26**
80 mm dia.	5.26	6.60	0.18	5.79	nr	**12.39**
100 mm dia.	6.85	8.59	0.18	5.79	nr	**14.38**
125 mm dia.	7.85	9.84	0.18	5.79	nr	**15.63**
150 mm dia.	9.25	11.60	0.18	5.79	nr	**17.39**
200 mm dia.	12.93	16.22	0.22	7.08	nr	**23.30**
250 mm dia.	15.37	19.28	0.22	7.08	nr	**26.36**
300 mm dia.	16.29	20.43	0.22	7.08	nr	**27.51**
350 mm dia.	17.89	22.44	0.22	7.08	nr	**29.52**
Valves						
15 mm dia.	6.12	7.67	0.15	4.83	nr	**12.50**
20 mm dia.	6.12	7.67	0.15	4.83	nr	**12.50**
25 mm dia.	6.47	8.12	0.15	4.83	nr	**12.95**
32 mm dia.	7.05	8.85	0.15	4.83	nr	**13.68**
40 mm dia.	7.48	9.39	0.15	4.83	nr	**14.22**
50 mm dia.	8.55	10.73	0.15	4.83	nr	**15.56**
65 mm dia.	9.68	12.14	0.15	4.83	nr	**16.97**
80 mm dia.	10.58	13.27	0.20	6.44	nr	**19.71**
100 mm dia.	13.65	17.12	0.20	6.44	nr	**23.56**
125 mm dia.	15.75	19.76	0.20	6.44	nr	**26.20**
150 mm dia.	18.48	23.18	0.20	6.44	nr	**29.62**
200 mm dia.	25.82	32.39	0.25	8.04	nr	**40.43**
250 mm dia.	30.70	38.51	0.25	8.04	nr	**46.55**
300 mm dia.	32.56	40.85	0.25	8.04	nr	**48.89**
350 mm dia.	35.77	44.87	0.25	8.04	nr	**52.91**

38 MECHANICAL/COOLING/HEATING SYSTEMS

Item	Net Price £	Material £	Labour hours	Labour £	Unit	Total rate £
Expansion bellows						
15 mm dia.	11.44	14.35	0.22	7.08	nr	21.43
20 mm dia.	12.23	15.34	0.22	7.08	nr	22.42
25 mm dia.	12.95	16.24	0.22	7.08	nr	23.32
32 mm dia.	14.15	17.75	0.22	7.08	nr	24.83
40 mm dia.	14.97	18.78	0.22	7.08	nr	25.86
50 mm dia.	17.11	21.46	0.22	7.08	nr	28.54
65 mm dia.	19.36	24.28	0.22	7.08	nr	31.36
80 mm dia.	21.11	26.48	0.29	9.33	nr	35.81
100 mm dia.	27.35	34.31	0.29	9.33	nr	43.64
125 mm dia.	31.45	39.45	0.29	9.33	nr	48.78
150 mm dia.	36.95	46.35	0.29	9.33	nr	55.68
200 mm dia.	51.64	64.78	0.36	11.58	nr	76.36
250 mm dia.	61.45	77.08	0.36	11.58	nr	88.66
300 mm dia.	65.15	81.73	0.36	11.58	nr	93.31
350 mm dia.	71.53	89.72	0.36	11.58	nr	101.30
40 mm thick						
15 mm dia.	7.31	9.17	0.15	4.83	m	14.00
20 mm dia.	7.55	9.48	0.15	4.83	m	14.31
25 mm dia.	8.12	10.18	0.15	4.83	m	15.01
32 mm dia.	8.65	10.85	0.15	4.83	m	15.68
40 mm dia.	9.11	11.42	0.15	4.83	m	16.25
50 mm dia.	10.30	12.92	0.15	4.83	m	17.75
65 mm dia.	11.55	14.49	0.15	4.83	m	19.32
80 mm dia.	12.55	15.75	0.22	7.08	m	22.83
100 mm dia.	16.33	20.48	0.22	7.08	m	27.56
125 mm dia.	18.44	23.13	0.22	7.08	m	30.21
150 mm dia.	21.51	26.98	0.22	7.08	m	34.06
200 mm dia.	29.73	37.30	0.25	8.04	m	45.34
250 mm dia.	34.83	43.69	0.25	8.04	m	51.73
300 mm dia.	37.24	46.72	0.25	8.04	m	54.76
350 mm dia.	41.19	51.67	0.25	8.04	m	59.71
400 mm dia.	45.89	57.57	0.25	8.04	m	65.61
Extra over for fittings concealed insulation						
Flange/union						
15 mm dia.	3.68	4.61	0.13	4.18	nr	8.79
20 mm dia.	3.76	4.72	0.13	4.18	nr	8.90
25 mm dia.	4.05	5.08	0.13	4.18	nr	9.26
32 mm dia.	4.32	5.42	0.13	4.18	nr	9.60
40 mm dia.	4.54	5.69	0.13	4.18	nr	9.87
50 mm dia.	5.15	6.46	0.13	4.18	nr	10.64
65 mm dia.	5.78	7.25	0.13	4.18	nr	11.43
80 mm dia.	6.29	7.88	0.18	5.79	nr	13.67
100 mm dia.	8.15	10.23	0.18	5.79	nr	16.02
125 mm dia.	9.21	11.56	0.18	5.79	nr	17.35
150 mm dia.	10.76	13.50	0.18	5.79	nr	19.29
200 mm dia.	14.87	18.65	0.22	7.08	nr	25.73
250 mm dia.	17.41	21.84	0.22	7.08	nr	28.92
300 mm dia.	18.59	23.32	0.22	7.08	nr	30.40
350 mm dia.	20.59	25.83	0.22	7.08	nr	32.91
400 mm dia.	22.97	28.82	0.22	7.08	nr	35.90

38 MECHANICAL/COOLING/HEATING SYSTEMS

Item	Net Price £	Material £	Labour hours	Labour £	Unit	Total rate £
THERMAL INSULATION – cont						
Extra over for fittings – cont						
Valves						
15 mm dia.	7.31	9.17	0.15	4.83	nr	14.00
20 mm dia.	7.55	9.48	0.15	4.83	nr	14.31
25 mm dia.	8.12	10.18	0.15	4.83	nr	15.01
32 mm dia.	8.65	10.85	0.15	4.83	nr	15.68
40 mm dia.	9.11	11.42	0.15	4.83	nr	16.25
50 mm dia.	10.30	12.92	0.15	4.83	nr	17.75
65 mm dia.	11.55	14.49	0.15	4.83	nr	19.32
80 mm dia.	12.55	15.75	0.20	6.44	nr	22.19
100 mm dia.	16.33	20.48	0.20	6.44	nr	26.92
125 mm dia.	18.44	23.13	0.20	6.44	nr	29.57
150 mm dia.	21.51	26.98	0.20	6.44	nr	33.42
200 mm dia.	29.73	37.30	0.25	8.04	nr	45.34
250 mm dia.	34.83	43.69	0.25	8.04	nr	51.73
300 mm dia.	37.24	46.72	0.25	8.04	nr	54.76
350 mm dia.	41.19	51.67	0.25	8.04	nr	59.71
400 mm dia.	45.89	57.57	0.25	8.04	nr	65.61
Expansion bellows						
15 mm dia.	14.68	18.41	0.22	7.08	nr	25.49
20 mm dia.	15.08	18.92	0.22	7.08	nr	26.00
25 mm dia.	16.25	20.38	0.22	7.08	nr	27.46
32 mm dia.	17.30	21.71	0.22	7.08	nr	28.79
40 mm dia.	18.20	22.83	0.22	7.08	nr	29.91
50 mm dia.	20.59	25.83	0.22	7.08	nr	32.91
65 mm dia.	23.08	28.95	0.22	7.08	nr	36.03
80 mm dia.	25.15	31.55	0.29	9.33	nr	40.88
100 mm dia.	32.60	40.89	0.29	9.33	nr	50.22
125 mm dia.	36.86	46.23	0.29	9.33	nr	55.56
150 mm dia.	43.02	53.96	0.29	9.33	nr	63.29
200 mm dia.	59.45	74.57	0.36	11.58	nr	86.15
250 mm dia.	69.67	87.39	0.36	11.58	nr	98.97
300 mm dia.	74.49	93.44	0.36	11.58	nr	105.02
350 mm dia.	41.19	51.67	0.36	11.58	nr	63.25
400 mm dia.	91.80	115.16	0.36	11.58	nr	126.74
50 mm thick						
15 mm dia.	10.19	12.78	0.15	4.83	m	17.61
20 mm dia.	10.71	13.44	0.15	4.83	m	18.27
25 mm dia.	11.39	14.29	0.15	4.83	m	19.12
32 mm dia.	11.90	14.93	0.15	4.83	m	19.76
40 mm dia.	12.50	15.68	0.15	4.83	m	20.51
50 mm dia.	14.04	17.61	0.15	4.83	m	22.44
65 mm dia.	15.34	19.24	0.15	4.83	m	24.07
80 mm dia.	16.41	20.59	0.22	7.08	m	27.67
100 mm dia.	21.00	26.34	0.22	7.08	m	33.42
125 mm dia.	23.56	29.56	0.22	7.08	m	36.64
150 mm dia.	27.19	34.10	0.22	7.08	m	41.18

38 MECHANICAL/COOLING/HEATING SYSTEMS

Item	Net Price £	Material £	Labour hours	Labour £	Unit	Total rate £
200 mm dia.	37.12	46.56	0.25	8.04	m	54.60
250 mm dia.	42.87	53.77	0.25	8.04	m	61.81
300 mm dia.	45.42	56.97	0.25	8.04	m	65.01
350 mm dia.	50.14	62.90	0.25	8.04	m	70.94
400 mm dia.	55.61	69.75	0.25	8.04	m	77.79
Extra over for fittings concealed insulation						
Flange/union						
15 mm dia.	5.10	6.40	0.13	4.18	nr	10.58
20 mm dia.	5.37	6.73	0.13	4.18	nr	10.91
25 mm dia.	5.71	7.17	0.13	4.18	nr	11.35
32 mm dia.	5.96	7.48	0.13	4.18	nr	11.66
40 mm dia.	6.26	7.85	0.13	4.18	nr	12.03
50 mm dia.	7.00	8.78	0.13	4.18	nr	12.96
65 mm dia.	7.67	9.62	0.13	4.18	nr	13.80
80 mm dia.	8.21	10.30	0.18	5.79	nr	16.09
100 mm dia.	10.51	13.18	0.18	5.79	nr	18.97
125 mm dia.	11.78	14.77	0.18	5.79	nr	20.56
150 mm dia.	13.61	17.07	0.18	5.79	nr	22.86
200 mm dia.	18.55	23.27	0.22	7.08	nr	30.35
250 mm dia.	21.43	26.88	0.22	7.08	nr	33.96
300 mm dia.	22.72	28.50	0.22	7.08	nr	35.58
350 mm dia.	25.09	31.47	0.22	7.08	nr	38.55
400 mm dia.	27.81	34.89	0.22	7.08	nr	41.97
Valves						
15 mm dia.	10.19	12.78	0.15	4.83	nr	17.61
20 mm dia.	10.71	13.44	0.15	4.83	nr	18.27
25 mm dia.	11.39	14.29	0.15	4.83	nr	19.12
32 mm dia.	11.90	14.93	0.15	4.83	nr	19.76
40 mm dia.	12.50	15.68	0.15	4.83	nr	20.51
50 mm dia.	14.04	17.61	0.15	4.83	nr	22.44
65 mm dia.	15.34	19.24	0.15	4.83	nr	24.07
80 mm dia.	16.41	20.59	0.20	6.44	nr	27.03
100 mm dia.	21.00	26.34	0.20	6.44	nr	32.78
125 mm dia.	23.56	29.56	0.20	6.44	nr	36.00
150 mm dia.	27.19	34.10	0.20	6.44	nr	40.54
200 mm dia.	37.12	46.56	0.25	8.04	nr	54.60
250 mm dia.	42.87	53.77	0.25	8.04	nr	61.81
300 mm dia.	45.42	56.97	0.25	8.04	nr	65.01
350 mm dia.	50.14	62.90	0.25	8.04	nr	70.94
400 mm dia.	55.61	69.75	0.25	8.04	nr	77.79
Expansion bellows						
15 mm dia.	20.33	25.50	0.22	7.08	nr	32.58
20 mm dia.	21.41	26.86	0.22	7.08	nr	33.94
25 mm dia.	22.76	28.55	0.22	7.08	nr	35.63
32 mm dia.	23.80	29.86	0.22	7.08	nr	36.94
40 mm dia.	25.04	31.40	0.22	7.08	nr	38.48
50 mm dia.	28.06	35.20	0.22	7.08	nr	42.28
65 mm dia.	30.66	38.46	0.22	7.08	nr	45.54

38 MECHANICAL/COOLING/HEATING SYSTEMS

Item	Net Price £	Material £	Labour hours	Labour £	Unit	Total rate £
THERMAL INSULATION – cont						
Extra over for fittings concealed insulation – cont						
Expansion bellows – cont						
80 mm dia.	32.85	41.20	0.29	9.33	nr	50.53
100 mm dia.	41.95	52.62	0.29	9.33	nr	61.95
125 mm dia.	47.10	59.08	0.29	9.33	nr	68.41
150 mm dia.	54.38	68.22	0.29	9.33	nr	77.55
200 mm dia.	74.26	93.15	0.36	11.58	nr	104.73
250 mm dia.	85.73	107.54	0.36	11.58	nr	119.12
300 mm dia.	90.88	114.00	0.36	11.58	nr	125.58
350 mm dia.	100.29	125.80	0.36	11.58	nr	137.38
400 mm dia.	111.21	139.51	0.36	11.58	nr	151.09
Mineral fibre sectional insulation; bright class O foil faced; bright class O foil taped joints; 22 swg plain/embossed aluminium cladding; pop riveted						
Plantroom pipework						
20 mm thick						
15 mm dia.	6.53	8.19	0.44	14.15	m	22.34
20 mm dia.	6.90	8.66	0.44	14.15	m	22.81
25 mm dia.	7.39	9.27	0.44	14.15	m	23.42
32 mm dia.	8.00	10.04	0.44	14.15	m	24.19
40 mm dia.	8.43	10.57	0.44	14.15	m	24.72
50 mm dia.	9.39	11.78	0.44	14.15	m	25.93
Extra over for fittings plantroom insulation						
Flange/union						
15 mm dia.	7.29	9.14	0.58	18.65	nr	27.79
20 mm dia.	7.75	9.72	0.58	18.65	nr	28.37
25 mm dia.	8.29	10.39	0.58	18.65	nr	29.04
32 mm dia.	9.06	11.37	0.58	18.65	nr	30.02
40 mm dia.	9.57	12.01	0.58	18.65	nr	30.66
50 mm dia.	10.74	13.47	0.58	18.65	nr	32.12
Bends						
15 mm dia.	3.57	4.48	0.44	14.15	nr	18.63
20 mm dia.	3.81	4.78	0.44	14.15	nr	18.93
25 mm dia.	4.07	5.11	0.44	14.15	nr	19.26
32 mm dia.	4.40	5.52	0.44	14.15	nr	19.67
40 mm dia.	4.61	5.78	0.44	14.15	nr	19.93
50 mm dia.	5.17	6.48	0.44	14.15	nr	20.63
Tees						
15 mm dia.	2.14	2.69	0.44	14.15	nr	16.84
20 mm dia.	2.29	2.87	0.44	14.15	nr	17.02
25 mm dia.	2.43	3.05	0.44	14.15	nr	17.20
32 mm dia.	2.64	3.32	0.44	14.15	nr	17.47
40 mm dia.	2.79	3.49	0.44	14.15	nr	17.64
50 mm dia.	3.10	3.89	0.44	14.15	nr	18.04

38 MECHANICAL/COOLING/HEATING SYSTEMS

Item	Net Price £	Material £	Labour hours	Labour £	Unit	Total rate £
Valves						
15 mm dia.	4.00	5.02	0.78	25.09	nr	30.11
20 mm dia.	4.57	5.73	0.78	25.09	nr	30.82
25 mm dia.	4.57	5.73	0.78	25.09	nr	30.82
32 mm dia.	5.11	6.41	0.78	25.09	nr	31.50
40 mm dia.	5.47	6.87	0.78	25.09	nr	31.96
50 mm dia.	6.25	7.84	0.78	25.09	nr	32.93
Pumps						
15 mm dia.	21.43	26.88	2.34	75.26	nr	102.14
20 mm dia.	22.73	28.52	2.34	75.26	nr	103.78
25 mm dia.	24.41	30.62	2.34	75.26	nr	105.88
32 mm dia.	26.63	33.41	2.34	75.26	nr	108.67
40 mm dia.	28.13	35.29	2.34	75.26	nr	110.55
50 mm dia.	31.63	39.68	2.34	75.26	nr	114.94
Expansion bellows						
15 mm dia.	17.14	21.50	1.05	33.77	nr	55.27
20 mm dia.	18.19	22.81	1.05	33.77	nr	56.58
25 mm dia.	19.54	24.51	1.05	33.77	nr	58.28
32 mm dia.	21.30	26.72	1.05	33.77	nr	60.49
40 mm dia.	22.48	28.20	1.05	33.77	nr	61.97
50 mm dia.	25.30	31.74	1.05	33.77	nr	65.51
25 mm thick						
15 mm dia.	7.16	8.98	0.44	14.15	m	23.13
20 mm dia.	7.75	9.72	0.44	14.15	m	23.87
25 mm dia.	8.48	10.64	0.44	14.15	m	24.79
32 mm dia.	9.03	11.32	0.44	14.15	m	25.47
40 mm dia.	9.72	12.20	0.44	14.15	m	26.35
50 mm dia.	10.89	13.66	0.44	14.15	m	27.81
65 mm dia.	12.30	15.43	0.44	14.15	m	29.58
80 mm dia.	13.33	16.72	0.52	16.72	m	33.44
100 mm dia.	16.55	20.76	0.52	16.72	m	37.48
125 mm dia.	19.22	24.11	0.52	16.72	m	40.83
150 mm dia.	22.37	28.06	0.52	16.72	m	44.78
200 mm dia.	30.56	38.34	0.60	19.30	m	57.64
250 mm dia.	36.31	45.55	0.60	19.30	m	64.85
300 mm dia.	40.65	50.99	0.60	19.30	m	70.29
Extra over for fittings plantroom insulation						
Flange/union						
15 mm dia.	8.00	10.04	0.58	18.65	nr	28.69
20 mm dia.	8.65	10.85	0.58	18.65	nr	29.50
25 mm dia.	9.55	11.98	0.58	18.65	nr	30.63
32 mm dia.	10.26	12.87	0.58	18.65	nr	31.52
40 mm dia.	11.00	13.80	0.58	18.65	nr	32.45
50 mm dia.	12.40	15.56	0.58	18.65	nr	34.21
65 mm dia.	14.07	17.65	0.58	18.65	nr	36.30
80 mm dia.	15.29	19.17	0.67	21.55	nr	40.72
100 mm dia.	19.36	24.28	0.67	21.55	nr	45.83
125 mm dia.	22.47	28.19	0.67	21.55	nr	49.74
150 mm dia.	26.34	33.04	0.67	21.55	nr	54.59

38 MECHANICAL/COOLING/HEATING SYSTEMS

Item	Net Price £	Material £	Labour hours	Labour £	Unit	Total rate £
THERMAL INSULATION – cont						
Extra over for fittings – cont						
Flange/union – cont						
200 mm dia.	36.40	45.66	0.87	27.99	nr	**73.65**
250 mm dia.	43.35	54.38	0.87	27.99	nr	**82.37**
300 mm dia.	47.92	60.11	0.87	27.99	nr	**88.10**
Bends						
15 mm dia.	3.93	4.93	0.44	14.15	nr	**19.08**
20 mm dia.	4.26	5.34	0.44	14.15	nr	**19.49**
25 mm dia.	4.66	5.85	0.44	14.15	nr	**20.00**
32 mm dia.	4.97	6.24	0.44	14.15	nr	**20.39**
40 mm dia.	5.37	6.73	0.44	14.15	nr	**20.88**
50 mm dia.	5.99	7.52	0.44	14.15	nr	**21.67**
65 mm dia.	6.75	8.47	0.44	14.15	nr	**22.62**
80 mm dia.	7.32	9.18	0.52	16.72	nr	**25.90**
100 mm dia.	9.11	11.42	0.52	16.72	nr	**28.14**
125 mm dia.	10.59	13.28	0.52	16.72	nr	**30.00**
150 mm dia.	12.30	15.43	0.52	16.72	nr	**32.15**
200 mm dia.	16.80	21.08	0.60	19.30	nr	**40.38**
250 mm dia.	19.97	25.05	0.60	19.30	nr	**44.35**
300 mm dia.	22.36	28.04	0.60	19.30	nr	**47.34**
Tees						
15 mm dia.	2.36	2.96	0.44	14.15	nr	**17.11**
20 mm dia.	2.54	3.18	0.44	14.15	nr	**17.33**
25 mm dia.	2.81	3.53	0.44	14.15	nr	**17.68**
32 mm dia.	2.98	3.74	0.44	14.15	nr	**17.89**
40 mm dia.	3.20	4.01	0.44	14.15	nr	**18.16**
50 mm dia.	3.57	4.48	0.44	14.15	nr	**18.63**
65 mm dia.	4.05	5.08	0.44	14.15	nr	**19.23**
80 mm dia.	4.40	5.52	0.52	16.72	nr	**22.24**
100 mm dia.	5.46	6.85	0.52	16.72	nr	**23.57**
125 mm dia.	6.36	7.97	0.52	16.72	nr	**24.69**
150 mm dia.	7.39	9.27	0.52	16.72	nr	**25.99**
200 mm dia.	10.07	12.63	0.60	19.30	nr	**31.93**
250 mm dia.	12.00	15.05	0.60	19.30	nr	**34.35**
300 mm dia.	13.43	16.84	0.60	19.30	nr	**36.14**
Valves						
15 mm dia.	12.72	15.96	0.78	25.09	nr	**41.05**
20 mm dia.	13.75	17.25	0.78	25.09	nr	**42.34**
25 mm dia.	15.18	19.04	0.78	25.09	nr	**44.13**
32 mm dia.	16.28	20.42	0.78	25.09	nr	**45.51**
40 mm dia.	17.47	21.92	0.78	25.09	nr	**47.01**
50 mm dia.	19.66	24.66	0.78	25.09	nr	**49.75**
65 mm dia.	22.34	28.02	0.78	25.09	nr	**53.11**
80 mm dia.	24.30	30.49	0.92	29.59	nr	**60.08**
100 mm dia.	30.72	38.54	0.92	29.59	nr	**68.13**
125 mm dia.	35.72	44.81	0.92	29.59	nr	**74.40**
150 mm dia.	41.84	52.48	0.92	29.59	nr	**82.07**
200 mm dia.	57.81	72.52	1.12	36.03	nr	**108.55**
250 mm dia.	68.87	86.39	1.12	36.03	nr	**122.42**
300 mm dia.	76.09	95.45	1.12	36.03	nr	**131.48**

38 MECHANICAL/COOLING/HEATING SYSTEMS

Item	Net Price £	Material £	Labour hours	Labour £	Unit	Total rate £
Pumps						
15 mm dia.	23.55	29.55	2.34	75.26	nr	**104.81**
20 mm dia.	25.44	31.91	2.34	75.26	nr	**107.17**
25 mm dia.	28.11	35.26	2.34	75.26	nr	**110.52**
32 mm dia.	30.12	37.78	2.34	75.26	nr	**113.04**
40 mm dia.	32.34	40.57	2.34	75.26	nr	**115.83**
50 mm dia.	36.45	45.72	2.34	75.26	nr	**120.98**
65 mm dia.	41.35	51.87	2.34	75.26	nr	**127.13**
80 mm dia.	44.99	56.44	2.76	88.77	nr	**145.21**
100 mm dia.	56.92	71.40	2.76	88.77	nr	**160.17**
125 mm dia.	66.11	82.92	2.76	88.77	nr	**171.69**
150 mm dia.	77.47	97.18	2.76	88.77	nr	**185.95**
200 mm dia.	107.06	134.30	3.36	108.07	nr	**242.37**
250 mm dia.	127.53	159.97	3.36	108.07	nr	**268.04**
300 mm dia.	140.89	176.74	3.36	108.07	nr	**284.81**
Expansion bellows						
15 mm dia.	18.83	23.62	1.05	33.77	nr	**57.39**
20 mm dia.	20.34	25.51	1.05	33.77	nr	**59.28**
25 mm dia.	22.47	28.19	1.05	33.77	nr	**61.96**
32 mm dia.	24.12	30.25	1.05	33.77	nr	**64.02**
40 mm dia.	25.89	32.48	1.05	33.77	nr	**66.25**
50 mm dia.	29.17	36.59	1.05	33.77	nr	**70.36**
65 mm dia.	33.09	41.51	1.05	33.77	nr	**75.28**
80 mm dia.	36.01	45.17	1.26	40.53	nr	**85.70**
100 mm dia.	45.53	57.11	1.26	40.53	nr	**97.64**
125 mm dia.	52.86	66.30	1.26	40.53	nr	**106.83**
150 mm dia.	61.96	77.73	1.26	40.53	nr	**118.26**
200 mm dia.	85.65	107.44	1.53	49.21	nr	**156.65**
250 mm dia.	102.04	127.99	1.53	49.21	nr	**177.20**
300 mm dia.	112.70	141.37	1.53	49.21	nr	**190.58**
30 mm thick						
15 mm dia.	8.83	11.08	0.44	14.15	m	**25.23**
20 mm dia.	9.33	11.70	0.44	14.15	m	**25.85**
25 mm dia.	9.89	12.41	0.44	14.15	m	**26.56**
32 mm dia.	10.84	13.60	0.44	14.15	m	**27.75**
40 mm dia.	11.27	14.13	0.44	14.15	m	**28.28**
50 mm dia.	12.50	15.68	0.44	14.15	m	**29.83**
65 mm dia.	14.12	17.71	0.44	14.15	m	**31.86**
80 mm dia.	15.43	19.35	0.52	16.72	m	**36.07**
100 mm dia.	18.96	23.79	0.52	16.72	m	**40.51**
125 mm dia.	21.65	27.16	0.52	16.72	m	**43.88**
150 mm dia.	25.23	31.65	0.52	16.72	m	**48.37**
200 mm dia.	33.94	42.57	0.60	19.30	m	**61.87**
250 mm dia.	40.13	50.34	0.60	19.30	m	**69.64**
300 mm dia.	44.18	55.42	0.60	19.30	m	**74.72**
350 mm dia.	49.19	61.70	0.60	19.30	m	**81.00**

38 MECHANICAL/COOLING/HEATING SYSTEMS

Item	Net Price £	Material £	Labour hours	Labour £	Unit	Total rate £
THERMAL INSULATION – cont						
Extra over for fittings plantroom insulation						
Flange/union						
15 mm dia.	10.03	12.58	0.58	18.65	nr	31.23
20 mm dia.	10.65	13.36	0.58	18.65	nr	32.01
25 mm dia.	11.29	14.16	0.58	18.65	nr	32.81
32 mm dia.	12.37	15.51	0.58	18.65	nr	34.16
40 mm dia.	12.94	16.23	0.58	18.65	nr	34.88
50 mm dia.	14.48	18.17	0.58	18.65	nr	36.82
65 mm dia.	16.37	20.53	0.58	18.65	nr	39.18
80 mm dia.	17.89	22.44	0.67	21.55	nr	43.99
100 mm dia.	22.34	28.02	0.67	21.55	nr	49.57
125 mm dia.	25.57	32.08	0.67	21.55	nr	53.63
150 mm dia.	29.91	37.52	0.67	21.55	nr	59.07
200 mm dia.	40.73	51.09	0.87	27.99	nr	79.08
250 mm dia.	48.28	60.56	0.87	27.99	nr	88.55
300 mm dia.	52.53	65.89	0.87	27.99	nr	93.88
350 mm dia.	58.28	73.10	0.87	27.99	nr	101.09
Bends						
15 mm dia.	4.85	6.08	0.44	14.15	nr	20.23
20 mm dia.	5.12	6.42	0.44	14.15	nr	20.57
25 mm dia.	5.44	6.82	0.44	14.15	nr	20.97
32 mm dia.	5.94	7.45	0.44	14.15	nr	21.60
40 mm dia.	6.19	7.76	0.44	14.15	nr	21.91
50 mm dia.	6.90	8.66	0.44	14.15	nr	22.81
65 mm dia.	7.77	9.74	0.44	14.15	nr	23.89
80 mm dia.	8.47	10.63	0.52	16.72	nr	27.35
100 mm dia.	10.42	13.07	0.52	16.72	nr	29.79
125 mm dia.	11.91	14.94	0.52	16.72	nr	31.66
150 mm dia.	13.89	17.43	0.52	16.72	nr	34.15
200 mm dia.	18.66	23.41	0.60	19.30	nr	42.71
250 mm dia.	22.09	27.71	0.60	19.30	nr	47.01
300 mm dia.	24.30	30.49	0.60	19.30	nr	49.79
350 mm dia.	27.08	33.97	0.60	19.30	nr	53.27
Tees						
15 mm dia.	2.90	3.64	0.44	14.15	nr	17.79
20 mm dia.	3.08	3.86	0.44	14.15	nr	18.01
25 mm dia.	3.25	4.08	0.44	14.15	nr	18.23
32 mm dia.	3.56	4.47	0.44	14.15	nr	18.62
40 mm dia.	3.72	4.67	0.44	14.15	nr	18.82
50 mm dia.	4.12	5.16	0.44	14.15	nr	19.31
65 mm dia.	4.65	5.84	0.44	14.15	nr	19.99
80 mm dia.	5.10	6.40	0.52	16.72	nr	23.12
100 mm dia.	6.26	7.85	0.52	16.72	nr	24.57
125 mm dia.	7.16	8.98	0.52	16.72	nr	25.70
150 mm dia.	8.35	10.47	0.52	16.72	nr	27.19
200 mm dia.	11.20	14.04	0.60	19.30	nr	33.34
250 mm dia.	13.26	16.63	0.60	19.30	nr	35.93
300 mm dia.	14.58	18.29	0.60	19.30	nr	37.59
350 mm dia.	16.25	20.38	0.60	19.30	nr	39.68

38 MECHANICAL/COOLING/HEATING SYSTEMS

Item	Net Price £	Material £	Labour hours	Labour £	Unit	Total rate £
Valves						
15 mm dia.	15.94	19.99	0.78	25.09	nr	45.08
20 mm dia.	16.91	21.21	0.78	25.09	nr	46.30
25 mm dia.	17.93	22.49	0.78	25.09	nr	47.58
32 mm dia.	19.65	24.65	0.78	25.09	nr	49.74
40 mm dia.	20.55	25.78	0.78	25.09	nr	50.87
50 mm dia.	23.01	28.86	0.78	25.09	nr	53.95
65 mm dia.	26.01	32.63	0.78	25.09	nr	57.72
80 mm dia.	28.40	35.63	0.92	29.59	nr	65.22
100 mm dia.	35.47	44.50	0.92	29.59	nr	74.09
125 mm dia.	40.62	50.95	0.92	29.59	nr	80.54
150 mm dia.	47.49	59.57	0.92	29.59	nr	89.16
200 mm dia.	64.67	81.12	1.12	36.03	nr	117.15
250 mm dia.	76.68	96.19	1.12	36.03	nr	132.22
300 mm dia.	83.43	104.65	1.12	36.03	nr	140.68
350 mm dia.	92.56	116.11	1.12	36.03	nr	152.14
Pumps						
15 mm dia.	29.54	37.05	2.34	75.26	nr	112.31
20 mm dia.	31.34	39.31	2.34	75.26	nr	114.57
25 mm dia.	33.20	41.64	2.34	75.26	nr	116.90
32 mm dia.	36.38	45.64	2.34	75.26	nr	120.90
40 mm dia.	38.06	47.75	2.34	75.26	nr	123.01
50 mm dia.	42.63	53.48	2.34	75.26	nr	128.74
65 mm dia.	48.14	60.39	2.34	75.26	nr	135.65
80 mm dia.	52.57	65.95	2.76	88.77	nr	154.72
100 mm dia.	65.67	82.38	2.76	88.77	nr	171.15
125 mm dia.	75.22	94.36	2.76	88.77	nr	183.13
150 mm dia.	87.96	110.34	2.76	88.77	nr	199.11
200 mm dia.	119.77	150.24	3.36	108.07	nr	258.31
250 mm dia.	142.03	178.16	3.36	108.07	nr	286.23
300 mm dia.	154.51	193.82	3.36	108.07	nr	301.89
350 mm dia.	171.40	215.01	3.36	108.07	nr	323.08
Expansion bellows						
15 mm dia.	23.63	29.65	1.05	33.77	nr	63.42
20 mm dia.	25.08	31.46	1.05	33.77	nr	65.23
25 mm dia.	26.55	33.31	1.05	33.77	nr	67.08
32 mm dia.	29.10	36.50	1.05	33.77	nr	70.27
40 mm dia.	30.44	38.18	1.05	33.77	nr	71.95
50 mm dia.	34.08	42.75	1.05	33.77	nr	76.52
65 mm dia.	38.52	48.32	1.05	33.77	nr	82.09
80 mm dia.	42.09	52.80	1.26	40.53	nr	93.33
100 mm dia.	52.55	65.92	1.26	40.53	nr	106.45
125 mm dia.	60.20	75.51	1.26	40.53	nr	116.04
150 mm dia.	70.38	88.29	1.26	40.53	nr	128.82
200 mm dia.	95.83	120.21	1.53	49.21	nr	169.42
250 mm dia.	113.62	142.52	1.53	49.21	nr	191.73
300 mm dia.	123.60	155.04	1.53	49.21	nr	204.25
350 mm dia.	137.11	171.99	1.53	49.21	nr	221.20

38 MECHANICAL/COOLING/HEATING SYSTEMS

Item	Net Price £	Material £	Labour hours	Labour £	Unit	Total rate £
THERMAL INSULATION – cont						
Extra over for fittings – cont						
40 mm thick						
15 mm dia.	10.79	13.53	0.44	14.15	m	27.68
20 mm dia.	11.33	14.21	0.44	14.15	m	28.36
25 mm dia.	12.07	15.14	0.44	14.15	m	29.29
32 mm dia.	12.65	15.87	0.44	14.15	m	30.02
40 mm dia.	13.44	16.86	0.44	14.15	m	31.01
50 mm dia.	14.89	18.68	0.44	14.15	m	32.83
65 mm dia.	16.58	20.80	0.44	14.15	m	34.95
80 mm dia.	17.84	22.38	0.52	16.72	m	39.10
100 mm dia.	26.13	32.78	0.52	16.72	m	49.50
125 mm dia.	24.48	30.71	0.52	16.72	m	47.43
150 mm dia.	28.55	35.82	0.52	16.72	m	52.54
200 mm dia.	37.76	47.36	0.60	19.30	m	66.66
250 mm dia.	44.23	55.48	0.60	19.30	m	74.78
300 mm dia.	48.89	61.33	0.60	19.30	m	80.63
350 mm dia.	54.64	68.54	0.60	19.30	m	87.84
400 mm dia.	61.46	77.10	0.60	19.30	m	96.40
Extra over for fittings plantroom insulation						
Flange/union						
15 mm dia.	12.46	15.64	0.58	18.65	nr	34.29
20 mm dia.	13.01	16.32	0.58	18.65	nr	34.97
25 mm dia.	13.90	17.44	0.58	18.65	nr	36.09
32 mm dia.	14.65	18.38	0.58	18.65	nr	37.03
40 mm dia.	15.54	19.49	0.58	18.65	nr	38.14
50 mm dia.	17.30	21.71	0.58	18.65	nr	40.36
65 mm dia.	19.34	24.26	0.58	18.65	nr	42.91
80 mm dia.	20.86	26.16	0.67	21.55	nr	47.71
100 mm dia.	26.13	32.78	0.67	21.55	nr	54.33
125 mm dia.	29.26	36.70	0.67	21.55	nr	58.25
150 mm dia.	34.13	42.82	0.67	21.55	nr	64.37
200 mm dia.	45.80	57.46	0.87	27.99	nr	85.45
250 mm dia.	53.69	67.35	0.87	27.99	nr	95.34
300 mm dia.	58.74	73.68	0.87	27.99	nr	101.67
350 mm dia.	65.49	82.15	0.87	27.99	nr	110.14
400 mm dia.	73.42	92.10	0.87	27.99	nr	120.09
Bends						
15 mm dia.	5.93	7.44	0.44	14.15	nr	21.59
20 mm dia.	6.22	7.81	0.44	14.15	nr	21.96
25 mm dia.	6.64	8.33	0.44	14.15	nr	22.48
32 mm dia.	6.94	8.70	0.44	14.15	nr	22.85
40 mm dia.	7.40	9.28	0.44	14.15	nr	23.43
50 mm dia.	8.18	10.26	0.44	14.15	nr	24.41
65 mm dia.	9.14	11.47	0.44	14.15	nr	25.62
80 mm dia.	9.80	12.30	0.52	16.72	nr	29.02
100 mm dia.	12.10	15.18	0.52	16.72	nr	31.90
125 mm dia.	13.46	16.89	0.52	16.72	nr	33.61
150 mm dia.	15.70	19.69	0.52	16.72	nr	36.41

38 MECHANICAL/COOLING/HEATING SYSTEMS

Item	Net Price £	Material £	Labour hours	Labour £	Unit	Total rate £
200 mm dia.	20.75	26.03	0.60	19.30	nr	45.33
250 mm dia.	24.16	30.31	0.60	19.30	nr	49.61
300 mm dia.	26.90	33.75	0.60	19.30	nr	53.05
350 mm dia.	30.08	37.73	0.60	19.30	nr	57.03
400 mm dia.	33.80	42.40	0.60	19.30	nr	61.70
Tees						
15 mm dia.	3.56	4.47	0.44	14.15	nr	18.62
20 mm dia.	3.72	4.67	0.44	14.15	nr	18.82
25 mm dia.	3.98	5.00	0.44	14.15	nr	19.15
32 mm dia.	4.17	5.23	0.44	14.15	nr	19.38
40 mm dia.	4.44	5.57	0.44	14.15	nr	19.72
50 mm dia.	4.90	6.15	0.44	14.15	nr	20.30
65 mm dia.	5.47	6.87	0.44	14.15	nr	21.02
80 mm dia.	5.89	7.39	0.52	16.72	nr	24.11
100 mm dia.	7.26	9.11	0.52	16.72	nr	25.83
125 mm dia.	8.09	10.15	0.52	16.72	nr	26.87
150 mm dia.	9.43	11.83	0.52	16.72	nr	28.55
200 mm dia.	12.46	15.64	0.60	19.30	nr	34.94
250 mm dia.	14.62	18.33	0.60	19.30	nr	37.63
300 mm dia.	16.12	20.22	0.60	19.30	nr	39.52
350 mm dia.	18.05	22.65	0.60	19.30	nr	41.95
400 mm dia.	20.27	25.42	0.60	19.30	nr	44.72
Valves						
15 mm dia.	19.80	24.84	0.78	25.09	nr	49.93
20 mm dia.	20.66	25.92	0.78	25.09	nr	51.01
25 mm dia.	22.09	27.71	0.78	25.09	nr	52.80
32 mm dia.	23.26	29.18	0.78	25.09	nr	54.27
40 mm dia.	24.65	30.92	0.78	25.09	nr	56.01
50 mm dia.	27.48	34.47	0.78	25.09	nr	59.56
65 mm dia.	30.69	38.49	0.78	25.09	nr	63.58
80 mm dia.	33.14	41.57	0.92	29.59	nr	71.16
100 mm dia.	41.53	52.09	0.92	29.59	nr	81.68
125 mm dia.	46.44	58.25	0.92	29.59	nr	87.84
150 mm dia.	54.23	68.03	0.92	29.59	nr	97.62
200 mm dia.	72.74	91.25	1.12	36.03	nr	127.28
250 mm dia.	85.26	106.95	1.12	36.03	nr	142.98
300 mm dia.	93.31	117.05	1.12	36.03	nr	153.08
350 mm dia.	104.01	130.47	1.12	36.03	nr	166.50
400 mm dia.	116.63	146.31	1.12	36.03	nr	182.34
Pumps						
15 mm dia.	36.64	45.96	2.34	75.26	nr	121.22
20 mm dia.	38.24	47.97	2.34	75.26	nr	123.23
25 mm dia.	40.91	51.32	2.34	75.26	nr	126.58
32 mm dia.	43.03	53.97	2.34	75.26	nr	129.23
40 mm dia.	45.67	57.29	2.34	75.26	nr	132.55
50 mm dia.	50.91	63.86	2.34	75.26	nr	139.12
65 mm dia.	56.86	71.32	2.34	75.26	nr	146.58
80 mm dia.	61.38	77.00	2.76	88.77	nr	165.77
100 mm dia.	76.87	96.42	2.76	88.77	nr	185.19
125 mm dia.	86.00	107.88	2.76	88.77	nr	196.65
150 mm dia.	100.41	125.96	2.76	88.77	nr	214.73

38 MECHANICAL/COOLING/HEATING SYSTEMS

Item	Net Price £	Material £	Labour hours	Labour £	Unit	Total rate £
THERMAL INSULATION – cont						
Extra over for fittings – cont						
Pumps – cont						
200 mm dia.	134.70	168.96	3.36	108.07	nr	**277.03**
250 mm dia.	157.91	198.08	3.36	108.07	nr	**306.15**
300 mm dia.	225.93	283.40	3.36	108.07	nr	**391.47**
350 mm dia.	192.59	241.58	3.36	108.07	nr	**349.65**
400 mm dia.	215.97	270.92	3.36	108.07	nr	**378.99**
Expansion bellows						
15 mm dia.	29.30	36.76	1.05	33.77	nr	**70.53**
20 mm dia.	30.62	38.40	1.05	33.77	nr	**72.17**
25 mm dia.	32.72	41.05	1.05	33.77	nr	**74.82**
32 mm dia.	34.44	43.20	1.05	33.77	nr	**76.97**
40 mm dia.	36.54	45.83	1.05	33.77	nr	**79.60**
50 mm dia.	40.73	51.09	1.05	33.77	nr	**84.86**
65 mm dia.	45.50	57.08	1.05	33.77	nr	**90.85**
80 mm dia.	49.09	61.58	1.26	40.53	nr	**102.11**
100 mm dia.	61.50	77.15	1.26	40.53	nr	**117.68**
125 mm dia.	68.78	86.27	1.26	40.53	nr	**126.80**
150 mm dia.	80.32	100.76	1.26	40.53	nr	**141.29**
200 mm dia.	107.74	135.15	1.53	49.21	nr	**184.36**
250 mm dia.	126.35	158.49	1.53	49.21	nr	**207.70**
300 mm dia.	138.22	173.39	1.53	49.21	nr	**222.60**
350 mm dia.	154.07	193.27	1.53	49.21	nr	**242.48**
400 mm dia.	172.77	216.72	1.53	49.21	nr	**265.93**
50 mm thick						
15 mm dia.	14.08	17.66	0.44	14.15	m	**31.81**
20 mm dia.	14.83	18.60	0.44	14.15	m	**32.75**
25 mm dia.	15.72	19.72	0.44	14.15	m	**33.87**
32 mm dia.	16.53	20.73	0.44	14.15	m	**34.88**
40 mm dia.	17.28	21.67	0.44	14.15	m	**35.82**
50 mm dia.	19.08	23.93	0.44	14.15	m	**38.08**
65 mm dia.	20.49	25.70	0.44	14.15	m	**39.85**
80 mm dia.	22.01	27.61	0.52	16.72	m	**44.33**
100 mm dia.	27.19	34.10	0.52	16.72	m	**50.82**
125 mm dia.	30.10	37.76	0.52	16.72	m	**54.48**
150 mm dia.	34.58	43.38	0.52	16.72	m	**60.10**
200 mm dia.	45.32	56.85	0.60	19.30	m	**76.15**
250 mm dia.	52.39	65.72	0.60	19.30	m	**85.02**
300 mm dia.	57.03	71.53	0.60	19.30	m	**90.83**
350 mm dia.	63.56	79.73	0.60	19.30	m	**99.03**
400 mm dia.	71.08	89.16	0.60	19.30	m	**108.46**
Extra over for fittings plantroom insulation						
Flange/union						
15 mm dia.	16.58	20.80	0.58	18.65	nr	**39.45**
20 mm dia.	17.50	21.95	0.58	18.65	nr	**40.60**
25 mm dia.	18.54	23.25	0.58	18.65	nr	**41.90**

38 MECHANICAL/COOLING/HEATING SYSTEMS

Item	Net Price £	Material £	Labour hours	Labour £	Unit	Total rate £
32 mm dia.	19.45	24.39	0.58	18.65	nr	**43.04**
40 mm dia.	20.39	25.58	0.58	18.65	nr	**44.23**
50 mm dia.	22.63	28.39	0.58	18.65	nr	**47.04**
65 mm dia.	24.45	30.67	0.58	18.65	nr	**49.32**
80 mm dia.	26.20	32.86	0.67	21.55	nr	**54.41**
100 mm dia.	32.77	41.10	0.67	21.55	nr	**62.65**
125 mm dia.	36.44	45.71	0.67	21.55	nr	**67.26**
150 mm dia.	41.94	52.61	0.67	21.55	nr	**74.16**
200 mm dia.	55.74	69.92	0.87	27.99	nr	**97.91**
250 mm dia.	64.43	80.82	0.87	27.99	nr	**108.81**
300 mm dia.	69.50	87.18	0.87	27.99	nr	**115.17**
350 mm dia.	77.29	96.95	0.87	27.99	nr	**124.94**
400 mm dia.	86.22	108.16	0.87	27.99	nr	**136.15**
Bend						
15 mm dia.	7.76	9.73	0.44	14.15	nr	**23.88**
20 mm dia.	8.16	10.24	0.44	14.15	nr	**24.39**
25 mm dia.	8.65	10.85	0.44	14.15	nr	**25.00**
32 mm dia.	9.11	11.42	0.44	14.15	nr	**25.57**
40 mm dia.	9.50	11.92	0.44	14.15	nr	**26.07**
50 mm dia.	10.50	13.17	0.44	14.15	nr	**27.32**
65 mm dia.	11.27	14.13	0.44	14.15	nr	**28.28**
80 mm dia.	12.10	15.18	0.52	16.72	nr	**31.90**
100 mm dia.	14.95	18.75	0.52	16.72	nr	**35.47**
125 mm dia.	16.55	20.76	0.52	16.72	nr	**37.48**
150 mm dia.	19.03	23.87	0.52	16.72	nr	**40.59**
200 mm dia.	24.92	31.26	0.60	19.30	nr	**50.56**
250 mm dia.	28.81	36.14	0.60	19.30	nr	**55.44**
300 mm dia.	31.36	39.33	0.60	19.30	nr	**58.63**
350 mm dia.	34.95	43.84	0.60	19.30	nr	**63.14**
400 mm dia.	39.09	49.03	0.60	19.30	nr	**68.33**
Tees						
15 mm dia.	4.65	5.84	0.44	14.15	nr	**19.99**
20 mm dia.	4.89	6.14	0.44	14.15	nr	**20.29**
25 mm dia.	5.18	6.50	0.44	14.15	nr	**20.65**
32 mm dia.	5.46	6.85	0.44	14.15	nr	**21.00**
40 mm dia.	5.71	7.17	0.44	14.15	nr	**21.32**
50 mm dia.	6.29	7.88	0.44	14.15	nr	**22.03**
65 mm dia.	6.75	8.47	0.44	14.15	nr	**22.62**
80 mm dia.	7.26	9.11	0.52	16.72	nr	**25.83**
100 mm dia.	8.97	11.26	0.52	16.72	nr	**27.98**
125 mm dia.	9.93	12.45	0.52	16.72	nr	**29.17**
150 mm dia.	11.40	14.30	0.52	16.72	nr	**31.02**
200 mm dia.	14.95	18.75	0.60	19.30	nr	**38.05**
250 mm dia.	17.29	21.68	0.60	19.30	nr	**40.98**
300 mm dia.	18.83	23.62	0.60	19.30	nr	**42.92**
350 mm dia.	20.98	26.32	0.60	19.30	nr	**45.62**
400 mm dia.	23.46	29.43	0.60	19.30	nr	**48.73**
Valves						
15 mm dia.	26.36	33.06	0.78	25.09	nr	**58.15**
20 mm dia.	27.77	34.83	0.78	25.09	nr	**59.92**
25 mm dia.	29.44	36.93	0.78	25.09	nr	**62.02**

38 MECHANICAL/COOLING/HEATING SYSTEMS

Item	Net Price £	Material £	Labour hours	Labour £	Unit	Total rate £
THERMAL INSULATION – cont						
Extra over for fittings – cont						
Valves – cont						
32 mm dia.	30.91	38.77	0.78	25.09	nr	63.86
40 mm dia.	32.37	40.60	0.78	25.09	nr	65.69
50 mm dia.	35.92	45.06	0.78	25.09	nr	70.15
65 mm dia.	38.81	48.69	0.78	25.09	nr	73.78
80 mm dia.	41.63	52.23	0.92	29.59	nr	81.82
100 mm dia.	52.06	65.31	0.92	29.59	nr	94.90
125 mm dia.	57.88	72.61	0.92	29.59	nr	102.20
150 mm dia.	66.61	83.55	0.92	29.59	nr	113.14
200 mm dia.	88.57	111.10	1.12	36.03	nr	147.13
250 mm dia.	102.33	128.36	1.12	36.03	nr	164.39
300 mm dia.	110.85	139.05	1.12	36.03	nr	175.08
350 mm dia.	122.75	153.98	1.12	36.03	nr	190.01
400 mm dia.	136.95	171.79	1.12	36.03	nr	207.82
Pumps						
15 mm dia.	48.81	61.23	2.34	75.26	nr	136.49
20 mm dia.	51.45	64.53	2.34	75.26	nr	139.79
25 mm dia.	54.50	68.36	2.34	75.26	nr	143.62
32 mm dia.	57.24	71.80	2.34	75.26	nr	147.06
40 mm dia.	59.97	75.23	2.34	75.26	nr	150.49
50 mm dia.	66.50	83.42	2.34	75.26	nr	158.68
65 mm dia.	95.09	119.28	2.34	75.26	nr	194.54
80 mm dia.	77.12	96.73	2.76	88.77	nr	185.50
100 mm dia.	96.38	120.90	2.76	88.77	nr	209.67
125 mm dia.	107.19	134.46	2.76	88.77	nr	223.23
150 mm dia.	123.34	154.72	2.76	88.77	nr	243.49
200 mm dia.	163.99	205.71	3.36	108.07	nr	313.78
250 mm dia.	189.49	237.70	3.36	108.07	nr	345.77
300 mm dia.	204.47	256.49	3.36	108.07	nr	364.56
350 mm dia.	227.29	285.11	3.36	108.07	nr	393.18
400 mm dia.	253.62	318.14	3.36	108.07	nr	426.21
Expansion bellows						
15 mm dia.	39.03	48.96	1.05	33.77	nr	82.73
20 mm dia.	41.14	51.61	1.05	33.77	nr	85.38
25 mm dia.	43.59	54.68	1.05	33.77	nr	88.45
32 mm dia.	45.77	57.41	1.05	33.77	nr	91.18
40 mm dia.	47.96	60.17	1.05	33.77	nr	93.94
50 mm dia.	53.21	66.75	1.05	33.77	nr	100.52
65 mm dia.	57.49	72.12	1.05	33.77	nr	105.89
80 mm dia.	61.71	77.41	1.26	40.53	nr	117.94
100 mm dia.	77.11	96.72	1.26	40.53	nr	137.25
125 mm dia.	85.75	107.56	1.26	40.53	nr	148.09
150 mm dia.	98.69	123.79	1.26	40.53	nr	164.32
200 mm dia.	131.20	164.57	1.53	49.21	nr	213.78
250 mm dia.	151.59	190.15	1.53	49.21	nr	239.36
300 mm dia.	163.61	205.23	1.53	49.21	nr	254.44
350 mm dia.	181.84	228.10	1.53	49.21	nr	277.31
400 mm dia.	202.88	254.50	1.53	49.21	nr	303.71

38 MECHANICAL/COOLING/HEATING SYSTEMS

Item	Net Price £	Material £	Labour hours	Labour £	Unit	Total rate £
Mineral fibre sectional insulation; bright class O foil faced; bright class O foil taped joints; 0.8 mm polyisobutylene sheeting; welded joints						
External pipework						
20 mm thick						
15 mm dia.	6.32	7.93	0.30	9.64	m	17.57
20 mm dia.	6.73	8.44	0.30	9.64	m	18.08
25 mm dia.	7.26	9.11	0.30	9.64	m	18.75
32 mm dia.	7.97	10.00	0.30	9.64	m	19.64
40 mm dia.	8.51	10.67	0.30	9.64	m	20.31
50 mm dia.	9.61	12.05	0.30	9.64	m	21.69
Extra over for fittings external insulation						
Flange/union						
15 mm dia.	9.68	12.14	0.75	24.12	nr	36.26
20 mm dia.	10.28	12.89	0.75	24.12	nr	37.01
25 mm dia.	11.04	13.84	0.75	24.12	nr	37.96
32 mm dia.	12.04	15.10	0.75	24.12	nr	39.22
40 mm dia.	12.71	15.95	0.75	24.12	nr	40.07
50 mm dia.	14.24	17.86	0.75	24.12	nr	41.98
Bends						
15 mm dia.	1.60	2.00	0.30	9.64	nr	11.64
20 mm dia.	1.69	2.12	0.30	9.64	nr	11.76
25 mm dia.	1.82	2.28	0.30	9.64	nr	11.92
32 mm dia.	2.00	2.51	0.30	9.64	nr	12.15
40 mm dia.	2.13	2.68	0.30	9.64	nr	12.32
50 mm dia.	2.40	3.01	0.30	9.64	nr	12.65
Tees						
15 mm dia.	1.60	2.00	0.30	9.64	nr	11.64
20 mm dia.	1.69	2.12	0.30	9.64	nr	11.76
25 mm dia.	1.82	2.28	0.30	9.64	nr	11.92
32 mm dia.	2.00	2.51	0.30	9.64	nr	12.15
40 mm dia.	2.13	2.68	0.30	9.64	nr	12.32
50 mm dia.	2.40	3.01	0.30	9.64	nr	12.65
Valves						
15 mm dia.	15.39	19.31	1.03	33.13	nr	52.44
20 mm dia.	16.34	20.50	1.03	33.13	nr	53.63
25 mm dia.	17.54	22.00	1.03	33.13	nr	55.13
32 mm dia.	19.09	23.95	1.03	33.13	nr	57.08
40 mm dia.	20.16	25.29	1.03	33.13	nr	58.42
50 mm dia.	22.59	28.34	1.03	33.13	nr	61.47
Expansion bellows						
15 mm dia.	22.77	28.56	1.42	45.67	nr	74.23
20 mm dia.	24.22	30.39	1.42	45.67	nr	76.06
25 mm dia.	26.02	32.64	1.42	45.67	nr	78.31
32 mm dia.	28.30	35.50	1.42	45.67	nr	81.17
40 mm dia.	29.87	37.46	1.42	45.67	nr	83.13
50 mm dia.	33.48	42.00	1.42	45.67	nr	87.67

38 MECHANICAL/COOLING/HEATING SYSTEMS

Item	Net Price £	Material £	Labour hours	Labour £	Unit	Total rate £
THERMAL INSULATION – cont						
Extra over for fittings – cont						
25 mm thick						
15 mm dia.	6.97	8.75	0.30	9.64	m	18.39
20 mm dia.	7.48	9.39	0.30	9.64	m	19.03
25 mm dia.	8.23	10.33	0.30	9.64	m	19.97
32 mm dia.	8.91	11.18	0.30	9.64	m	20.82
40 mm dia.	9.48	11.89	0.30	9.64	m	21.53
50 mm dia.	10.71	13.44	0.30	9.64	m	23.08
65 mm dia.	12.14	15.23	0.30	9.64	m	24.87
80 mm dia.	13.29	16.67	0.40	12.87	m	29.54
100 mm dia.	16.77	21.03	0.40	12.87	m	33.90
125 mm dia.	19.32	24.24	0.40	12.87	m	37.11
150 mm dia.	22.66	28.43	0.40	12.87	m	41.30
200 mm dia.	30.72	38.54	0.50	16.08	m	54.62
250 mm dia.	36.69	46.02	0.50	16.08	m	62.10
300 mm dia.	39.90	50.05	0.50	16.08	m	66.13
Extra over for fittings external insulation						
Flange/union						
15 mm dia.	10.65	13.36	0.75	24.12	nr	37.48
20 mm dia.	11.49	14.41	0.75	24.12	nr	38.53
25 mm dia.	12.61	15.81	0.75	24.12	nr	39.93
32 mm dia.	13.48	16.91	0.75	24.12	nr	41.03
40 mm dia.	14.43	18.10	0.75	24.12	nr	42.22
50 mm dia.	16.18	20.29	0.75	24.12	nr	44.41
65 mm dia.	18.29	22.94	0.75	24.12	nr	47.06
80 mm dia.	19.90	24.96	0.89	28.63	nr	53.59
100 mm dia.	24.73	31.02	0.89	28.63	nr	59.65
125 mm dia.	28.63	35.92	0.89	28.63	nr	64.55
150 mm dia.	33.33	41.81	0.89	28.63	nr	70.44
200 mm dia.	45.09	56.56	1.15	36.99	nr	93.55
250 mm dia.	53.69	67.35	1.15	36.99	nr	104.34
300 mm dia.	59.64	74.82	1.15	36.99	nr	111.81
Bends						
15 mm dia.	1.74	2.18	0.30	9.64	nr	11.82
20 mm dia.	1.88	2.36	0.30	9.64	nr	12.00
25 mm dia.	2.04	2.55	0.30	9.64	nr	12.19
32 mm dia.	2.21	2.78	0.30	9.64	nr	12.42
40 mm dia.	2.37	2.97	0.30	9.64	nr	12.61
50 mm dia.	2.68	3.36	0.30	9.64	nr	13.00
65 mm dia.	3.05	3.83	0.30	9.64	nr	13.47
80 mm dia.	3.32	4.17	0.40	12.87	nr	17.04
100 mm dia.	4.19	5.25	0.40	12.87	nr	18.12
125 mm dia.	4.82	6.05	0.40	12.87	nr	18.92
150 mm dia.	5.68	7.12	0.40	12.87	nr	19.99
200 mm dia.	7.67	9.62	0.50	16.08	nr	25.70
250 mm dia.	9.18	11.51	0.50	16.08	nr	27.59
300 mm dia.	9.97	12.51	0.50	16.08	nr	28.59

38 MECHANICAL/COOLING/HEATING SYSTEMS

Item	Net Price £	Material £	Labour hours	Labour £	Unit	Total rate £
Tees						
15 mm dia.	1.74	2.18	0.30	9.64	nr	11.82
20 mm dia.	1.88	2.36	0.30	9.64	nr	12.00
25 mm dia.	2.04	2.55	0.30	9.64	nr	12.19
32 mm dia.	2.21	2.78	0.30	9.64	nr	12.42
40 mm dia.	2.37	2.97	0.30	9.64	nr	12.61
50 mm dia.	2.68	3.36	0.30	9.64	nr	13.00
65 mm dia.	3.05	3.83	0.30	9.64	nr	13.47
80 mm dia.	3.32	4.17	0.40	12.87	nr	17.04
100 mm dia.	4.19	5.25	0.40	12.87	nr	18.12
125 mm dia.	4.82	6.05	0.40	12.87	nr	18.92
150 mm dia.	5.68	7.12	0.40	12.87	nr	19.99
200 mm dia.	7.67	9.62	0.50	16.08	nr	25.70
250 mm dia.	9.18	11.51	0.50	16.08	nr	27.59
300 mm dia.	9.97	12.51	0.50	16.08	nr	28.59
Valves						
15 mm dia.	16.93	21.24	1.03	33.13	nr	54.37
20 mm dia.	18.25	22.89	1.03	33.13	nr	56.02
25 mm dia.	20.01	25.10	1.03	33.13	nr	58.23
32 mm dia.	21.44	26.89	1.03	33.13	nr	60.02
40 mm dia.	22.91	28.74	1.03	33.13	nr	61.87
50 mm dia.	25.73	32.28	1.03	33.13	nr	65.41
65 mm dia.	29.09	36.49	1.03	33.13	nr	69.62
80 mm dia.	31.61	39.65	1.25	40.21	nr	79.86
100 mm dia.	39.30	49.30	1.25	40.21	nr	89.51
125 mm dia.	45.48	57.05	1.25	40.21	nr	97.26
150 mm dia.	52.95	66.42	1.25	40.21	nr	106.63
200 mm dia.	71.65	89.88	1.55	49.85	nr	139.73
250 mm dia.	85.26	106.95	1.55	49.85	nr	156.80
300 mm dia.	94.72	118.82	1.55	49.85	nr	168.67
Expansion bellows						
15 mm dia.	25.09	31.47	1.42	45.67	nr	77.14
20 mm dia.	27.04	33.91	1.42	45.67	nr	79.58
25 mm dia.	29.65	37.20	1.42	45.67	nr	82.87
32 mm dia.	31.80	39.89	1.42	45.67	nr	85.56
40 mm dia.	33.97	42.62	1.42	45.67	nr	88.29
50 mm dia.	38.08	47.77	1.42	45.67	nr	93.44
65 mm dia.	43.08	54.04	1.42	45.67	nr	99.71
80 mm dia.	46.83	58.74	1.75	56.28	nr	115.02
100 mm dia.	58.22	73.04	1.75	56.28	nr	129.32
125 mm dia.	67.38	84.53	1.75	56.28	nr	140.81
150 mm dia.	78.43	98.38	1.75	56.28	nr	154.66
200 mm dia.	106.13	133.13	2.17	69.80	nr	202.93
250 mm dia.	126.31	158.45	2.17	69.80	nr	228.25
300 mm dia.	140.32	176.02	3.17	101.95	nr	277.97
30 mm thick						
15 mm dia.	10.10	12.67	0.30	9.64	m	22.31
20 mm dia.	9.14	11.47	0.30	9.64	m	21.11
25 mm dia.	9.68	12.14	0.30	9.64	m	21.78

38 MECHANICAL/COOLING/HEATING SYSTEMS

Item	Net Price £	Material £	Labour hours	Labour £	Unit	Total rate £
THERMAL INSULATION – cont						
Extra over for fittings – cont						
30 mm thick – cont						
32 mm dia.	10.51	13.18	0.30	9.64	m	**22.82**
40 mm dia.	11.07	13.89	0.30	9.64	m	**23.53**
50 mm dia.	12.46	15.64	0.30	9.64	m	**25.28**
65 mm dia.	14.01	17.57	0.30	9.64	m	**27.21**
80 mm dia.	15.23	19.11	0.40	12.87	m	**31.98**
100 mm dia.	19.05	23.90	0.40	12.87	m	**36.77**
125 mm dia.	21.81	27.36	0.40	12.87	m	**40.23**
150 mm dia.	25.33	31.77	0.40	12.87	m	**44.64**
200 mm dia.	34.11	42.78	0.50	16.08	m	**58.86**
250 mm dia.	40.48	50.78	0.50	16.08	m	**66.86**
300 mm dia.	43.66	54.77	0.50	16.08	m	**70.85**
350 mm dia.	47.74	59.89	0.50	16.08	m	**75.97**
Extra over for fittings external insulation						
Flange/union						
15 mm dia.	13.04	16.35	0.75	24.12	nr	**40.47**
20 mm dia.	13.80	17.32	0.75	24.12	nr	**41.44**
25 mm dia.	14.65	18.38	0.75	24.12	nr	**42.50**
32 mm dia.	15.98	20.05	0.75	24.12	nr	**44.17**
40 mm dia.	16.69	20.93	0.75	24.12	nr	**45.05**
50 mm dia.	18.62	23.35	0.75	24.12	nr	**47.47**
65 mm dia.	20.97	26.31	0.75	24.12	nr	**50.43**
80 mm dia.	22.85	28.66	0.89	28.63	nr	**57.29**
100 mm dia.	28.13	35.29	0.89	28.63	nr	**63.92**
125 mm dia.	32.15	40.33	0.89	28.63	nr	**68.96**
150 mm dia.	37.34	46.84	0.89	28.63	nr	**75.47**
200 mm dia.	49.88	62.57	1.15	36.99	nr	**99.56**
250 mm dia.	59.07	74.10	1.15	36.99	nr	**111.09**
300 mm dia.	64.74	81.21	1.15	36.99	nr	**118.20**
350 mm dia.	71.50	89.69	1.15	36.99	nr	**126.68**
Bends						
15 mm dia.	2.14	2.69	0.30	9.64	nr	**12.33**
20 mm dia.	2.28	2.86	0.30	9.64	nr	**12.50**
25 mm dia.	2.43	3.05	0.30	9.64	nr	**12.69**
32 mm dia.	2.62	3.28	0.30	9.64	nr	**12.92**
40 mm dia.	2.76	3.46	0.30	9.64	nr	**13.10**
50 mm dia.	3.12	3.91	0.30	9.64	nr	**13.55**
65 mm dia.	3.51	4.40	0.30	9.64	nr	**14.04**
80 mm dia.	3.82	4.79	0.40	12.87	nr	**17.66**
100 mm dia.	4.78	5.99	0.40	12.87	nr	**18.86**
125 mm dia.	5.46	6.85	0.40	12.87	nr	**19.72**
150 mm dia.	6.33	7.94	0.40	12.87	nr	**20.81**
200 mm dia.	8.54	10.71	0.50	16.08	nr	**26.79**
250 mm dia.	10.14	12.72	0.50	16.08	nr	**28.80**
300 mm dia.	10.91	13.69	0.50	16.08	nr	**29.77**
350 mm dia.	11.93	14.96	0.50	16.08	nr	**31.04**

38 MECHANICAL/COOLING/HEATING SYSTEMS

Item	Net Price £	Material £	Labour hours	Labour £	Unit	Total rate £
Tees						
15 mm dia.	2.14	2.69	0.30	9.64	nr	**12.33**
20 mm dia.	2.28	2.86	0.30	9.64	nr	**12.50**
25 mm dia.	2.43	3.05	0.30	9.64	nr	**12.69**
32 mm dia.	2.62	3.28	0.30	9.64	nr	**12.92**
40 mm dia.	2.76	3.46	0.30	9.64	nr	**13.10**
50 mm dia.	3.12	3.91	0.30	9.64	nr	**13.55**
65 mm dia.	3.51	4.40	0.30	9.64	nr	**14.04**
80 mm dia.	3.82	4.79	0.40	12.87	nr	**17.66**
100 mm dia.	4.78	5.99	0.40	12.87	nr	**18.86**
125 mm dia.	5.46	6.85	0.40	12.87	nr	**19.72**
150 mm dia.	6.33	7.94	0.40	12.87	nr	**20.81**
200 mm dia.	8.54	10.71	0.50	16.08	nr	**26.79**
250 mm dia.	10.14	12.72	0.50	16.08	nr	**28.80**
300 mm dia.	10.91	13.69	0.50	16.08	nr	**29.77**
350 mm dia.	11.93	14.96	0.50	16.08	nr	**31.04**
Valves						
15 mm dia.	20.71	25.98	1.03	33.13	nr	**59.11**
20 mm dia.	21.94	27.52	1.03	33.13	nr	**60.65**
25 mm dia.	23.27	29.19	1.03	33.13	nr	**62.32**
32 mm dia.	25.34	31.79	1.03	33.13	nr	**64.92**
40 mm dia.	26.52	33.26	1.03	33.13	nr	**66.39**
50 mm dia.	29.58	37.11	1.03	33.13	nr	**70.24**
65 mm dia.	33.30	41.78	1.03	33.13	nr	**74.91**
80 mm dia.	36.28	45.51	1.25	40.21	nr	**85.72**
100 mm dia.	44.64	56.00	1.25	40.21	nr	**96.21**
125 mm dia.	51.03	64.01	1.25	40.21	nr	**104.22**
150 mm dia.	59.28	74.36	1.25	40.21	nr	**114.57**
200 mm dia.	79.25	99.41	1.55	49.85	nr	**149.26**
250 mm dia.	93.83	117.70	1.55	49.85	nr	**167.55**
300 mm dia.	102.80	128.96	1.55	49.85	nr	**178.81**
350 mm dia.	113.57	142.46	1.55	49.85	nr	**192.31**
Expansion bellows						
15 mm dia.	30.66	38.46	1.42	45.67	nr	**84.13**
20 mm dia.	32.52	40.79	1.42	45.67	nr	**86.46**
25 mm dia.	34.45	43.21	1.42	45.67	nr	**88.88**
32 mm dia.	37.56	47.12	1.42	45.67	nr	**92.79**
40 mm dia.	39.30	49.30	1.42	45.67	nr	**94.97**
50 mm dia.	43.81	54.96	1.42	45.67	nr	**100.63**
65 mm dia.	49.34	61.89	1.42	45.67	nr	**107.56**
80 mm dia.	53.78	67.46	1.75	56.28	nr	**123.74**
100 mm dia.	66.15	82.98	1.75	56.28	nr	**139.26**
125 mm dia.	75.60	94.83	1.75	56.28	nr	**151.11**
150 mm dia.	87.82	110.16	1.75	56.28	nr	**166.44**
200 mm dia.	117.39	147.26	2.17	69.80	nr	**217.06**
250 mm dia.	139.01	174.37	2.17	69.80	nr	**244.17**
300 mm dia.	152.30	191.05	2.17	69.80	nr	**260.85**
350 mm dia.	168.26	211.06	2.17	69.80	nr	**280.86**

38 MECHANICAL/COOLING/HEATING SYSTEMS

Item	Net Price £	Material £	Labour hours	Labour £	Unit	Total rate £
THERMAL INSULATION – cont						
Extra over for fittings – cont						
40 mm thick						
15 mm dia.	10.74	13.47	0.30	9.64	m	**23.11**
20 mm dia.	11.11	13.93	0.30	9.64	m	**23.57**
25 mm dia.	11.87	14.88	0.30	9.64	m	**24.52**
32 mm dia.	12.64	15.86	0.30	9.64	m	**25.50**
40 mm dia.	13.21	16.58	0.30	9.64	m	**26.22**
50 mm dia.	14.73	18.48	0.30	9.64	m	**28.12**
65 mm dia.	16.44	20.62	0.30	9.64	m	**30.26**
80 mm dia.	17.82	22.36	0.40	12.87	m	**35.23**
100 mm dia.	22.24	27.90	0.40	12.87	m	**40.77**
125 mm dia.	25.06	31.44	0.40	12.87	m	**44.31**
150 mm dia.	28.93	36.29	0.40	12.87	m	**49.16**
200 mm dia.	38.59	48.41	0.50	16.08	m	**64.49**
250 mm dia.	45.20	56.69	0.50	16.08	m	**72.77**
300 mm dia.	48.92	61.36	0.50	16.08	m	**77.44**
350 mm dia.	53.78	67.46	0.50	16.08	m	**83.54**
400 mm dia.	59.85	75.07	0.50	16.08	m	**91.15**
Extra over for fittings external insulation						
Flange/union						
15 mm dia.	16.05	20.14	0.75	24.12	nr	**44.26**
20 mm dia.	16.76	21.02	0.75	24.12	nr	**45.14**
25 mm dia.	17.87	22.41	0.75	24.12	nr	**46.53**
32 mm dia.	18.83	23.62	0.75	24.12	nr	**47.74**
40 mm dia.	19.88	24.94	0.75	24.12	nr	**49.06**
50 mm dia.	22.02	27.62	0.75	24.12	nr	**51.74**
65 mm dia.	24.54	30.78	0.75	24.12	nr	**54.90**
80 mm dia.	26.45	33.17	0.89	28.63	nr	**61.80**
100 mm dia.	32.58	40.87	0.89	28.63	nr	**69.50**
125 mm dia.	36.44	45.71	0.89	28.63	nr	**74.34**
150 mm dia.	42.23	52.98	0.89	28.63	nr	**81.61**
200 mm dia.	55.68	69.84	1.15	36.99	nr	**106.83**
250 mm dia.	65.27	81.87	1.15	36.99	nr	**118.86**
300 mm dia.	71.72	89.97	1.15	36.99	nr	**126.96**
350 mm dia.	79.55	99.79	1.15	36.99	nr	**136.78**
400 mm dia.	89.08	111.74	1.15	36.99	nr	**148.73**
Bends						
15 mm dia.	2.68	3.36	0.30	9.64	nr	**13.00**
20 mm dia.	2.78	3.48	0.30	9.64	nr	**13.12**
25 mm dia.	2.97	3.73	0.30	9.64	nr	**13.37**
32 mm dia.	3.15	3.95	0.30	9.64	nr	**13.59**
40 mm dia.	3.32	4.17	0.30	9.64	nr	**13.81**
50 mm dia.	3.71	4.66	0.30	9.64	nr	**14.30**
65 mm dia.	4.12	5.16	0.30	9.64	nr	**14.80**
80 mm dia.	4.47	5.61	0.40	12.87	nr	**18.48**
100 mm dia.	5.56	6.98	0.40	12.87	nr	**19.85**
125 mm dia.	6.26	7.85	0.40	12.87	nr	**20.72**
150 mm dia.	7.25	9.09	0.40	12.87	nr	**21.96**

38 MECHANICAL/COOLING/HEATING SYSTEMS

Item	Net Price £	Material £	Labour hours	Labour £	Unit	Total rate £
200 mm dia.	9.65	12.11	0.50	16.08	nr	28.19
250 mm dia.	11.29	14.16	0.50	16.08	nr	30.24
300 mm dia.	12.23	15.34	0.50	16.08	nr	31.42
350 mm dia.	13.43	16.84	0.50	16.08	nr	32.92
400 mm dia.	14.97	18.78	0.50	16.08	nr	34.86
Tees						
15 mm dia.	2.68	3.36	0.30	9.64	nr	13.00
20 mm dia.	2.78	3.48	0.30	9.64	nr	13.12
25 mm dia.	2.97	3.73	0.30	9.64	nr	13.37
32 mm dia.	3.15	3.95	0.30	9.64	nr	13.59
40 mm dia.	3.32	4.17	0.30	9.64	nr	13.81
50 mm dia.	3.71	4.66	0.30	9.64	nr	14.30
65 mm dia.	4.12	5.16	0.30	9.64	nr	14.80
80 mm dia.	4.47	5.61	0.40	12.87	nr	18.48
100 mm dia.	5.56	6.98	0.40	12.87	nr	19.85
125 mm dia.	6.26	7.85	0.40	12.87	nr	20.72
150 mm dia.	7.25	9.09	0.40	12.87	nr	21.96
200 mm dia.	9.65	12.11	0.50	16.08	nr	28.19
250 mm dia.	11.29	14.16	0.50	16.08	nr	30.24
300 mm dia.	12.23	15.34	0.50	16.08	nr	31.42
350 mm dia.	13.43	16.84	0.50	16.08	nr	32.92
400 mm dia.	14.97	18.78	0.50	16.08	nr	34.86
Valves						
15 mm dia.	25.47	31.95	1.03	33.13	nr	65.08
20 mm dia.	26.62	33.39	1.03	33.13	nr	66.52
25 mm dia.	28.37	35.58	1.03	33.13	nr	68.71
32 mm dia.	29.88	37.49	1.03	33.13	nr	70.62
40 mm dia.	31.59	39.63	1.03	33.13	nr	72.76
50 mm dia.	35.01	43.92	1.03	33.13	nr	77.05
65 mm dia.	38.98	48.90	1.03	33.13	nr	82.03
80 mm dia.	42.01	52.70	1.25	40.21	nr	92.91
100 mm dia.	51.75	64.92	1.25	40.21	nr	105.13
125 mm dia.	57.88	72.61	1.25	40.21	nr	112.82
150 mm dia.	67.09	84.16	1.25	40.21	nr	124.37
200 mm dia.	88.44	110.94	1.55	49.85	nr	160.79
250 mm dia.	103.62	129.98	1.55	49.85	nr	179.83
300 mm dia.	113.92	142.90	1.55	49.85	nr	192.75
350 mm dia.	126.35	158.49	1.55	49.85	nr	208.34
400 mm dia.	141.46	177.45	1.55	49.85	nr	227.30
Expansion bellows						
15 mm dia.	37.73	47.33	1.42	45.67	nr	93.00
20 mm dia.	39.44	49.47	1.42	45.67	nr	95.14
25 mm dia.	42.02	52.71	1.42	45.67	nr	98.38
32 mm dia.	44.27	55.53	1.42	45.67	nr	101.20
40 mm dia.	46.80	58.71	1.42	45.67	nr	104.38
50 mm dia.	51.88	65.08	1.42	45.67	nr	110.75
65 mm dia.	57.74	72.43	1.42	45.67	nr	118.10
80 mm dia.	62.24	78.08	1.75	56.28	nr	134.36
100 mm dia.	76.67	96.17	1.75	56.28	nr	152.45
125 mm dia.	85.75	107.56	1.75	56.28	nr	163.84
150 mm dia.	99.41	124.70	1.75	56.28	nr	180.98

38 MECHANICAL/COOLING/HEATING SYSTEMS

Item	Net Price £	Material £	Labour hours	Labour £	Unit	Total rate £
THERMAL INSULATION – cont						
Extra over for fittings – cont						
Expansion bellows – cont						
200 mm dia.	131.03	164.36	2.17	69.80	nr	**234.16**
250 mm dia.	153.51	192.56	2.17	69.80	nr	**262.36**
300 mm dia.	168.77	211.70	2.17	69.80	nr	**281.50**
350 mm dia.	187.19	234.81	2.17	69.80	nr	**304.61**
400 mm dia.	209.59	262.91	2.17	69.80	nr	**332.71**
50 mm thick						
15 mm dia.	14.15	17.75	0.30	9.64	m	**27.39**
20 mm dia.	14.85	18.63	0.30	9.64	m	**28.27**
25 mm dia.	15.70	19.69	0.30	9.64	m	**29.33**
32 mm dia.	16.94	21.25	0.30	9.64	m	**30.89**
40 mm dia.	17.22	21.60	0.30	9.64	m	**31.24**
50 mm dia.	19.08	23.93	0.30	9.64	m	**33.57**
65 mm dia.	20.79	26.07	0.30	9.64	m	**35.71**
80 mm dia.	22.24	27.90	0.40	12.87	m	**40.77**
100 mm dia.	27.54	34.54	0.40	12.87	m	**47.41**
125 mm dia.	30.81	38.65	0.40	12.87	m	**51.52**
150 mm dia.	35.23	44.20	0.40	12.87	m	**57.07**
200 mm dia.	46.64	58.51	0.50	16.08	m	**74.59**
250 mm dia.	53.88	67.59	0.50	16.08	m	**83.67**
300 mm dia.	57.78	72.48	0.50	16.08	m	**88.56**
350 mm dia.	63.42	79.55	0.50	16.08	m	**95.63**
400 mm dia.	70.25	88.12	0.50	16.08	m	**104.20**
Extra over for fittings external insulation						
Flange/union						
15 mm dia.	20.86	26.16	0.75	24.12	nr	**50.28**
20 mm dia.	21.94	27.52	0.75	24.12	nr	**51.64**
25 mm dia.	23.19	29.09	0.75	24.12	nr	**53.21**
32 mm dia.	24.34	30.53	0.75	24.12	nr	**54.65**
40 mm dia.	25.47	31.95	0.75	24.12	nr	**56.07**
50 mm dia.	28.08	35.22	0.75	24.12	nr	**59.34**
65 mm dia.	30.36	38.08	0.75	24.12	nr	**62.20**
80 mm dia.	32.55	40.84	0.89	28.63	nr	**69.47**
100 mm dia.	40.02	50.20	0.89	28.63	nr	**78.83**
125 mm dia.	44.48	55.80	0.89	28.63	nr	**84.43**
150 mm dia.	50.94	63.90	0.89	28.63	nr	**92.53**
200 mm dia.	66.63	83.59	1.15	36.99	nr	**120.58**
250 mm dia.	76.99	96.58	1.15	36.99	nr	**133.57**
300 mm dia.	83.54	104.79	1.15	36.99	nr	**141.78**
350 mm dia.	92.44	115.95	1.15	36.99	nr	**152.94**
400 mm dia.	103.01	129.21	1.15	36.99	nr	**166.20**
Bends						
15 mm dia.	3.54	4.44	0.30	9.64	nr	**14.08**
20 mm dia.	3.72	4.67	0.30	9.64	nr	**14.31**
25 mm dia.	3.92	4.92	0.30	9.64	nr	**14.56**

38 MECHANICAL/COOLING/HEATING SYSTEMS

Item	Net Price £	Material £	Labour hours	Labour £	Unit	Total rate £
32 mm dia.	4.12	5.16	0.30	9.64	nr	14.80
40 mm dia.	4.32	5.42	0.30	9.64	nr	15.06
50 mm dia.	4.78	5.99	0.30	9.64	nr	15.63
65 mm dia.	5.19	6.51	0.30	9.64	nr	16.15
80 mm dia.	5.54	6.94	0.40	12.87	nr	19.81
100 mm dia.	6.89	8.65	0.40	12.87	nr	21.52
125 mm dia.	7.68	9.63	0.40	12.87	nr	22.50
150 mm dia.	8.82	11.07	0.40	12.87	nr	23.94
200 mm dia.	11.66	14.63	0.50	16.08	nr	30.71
250 mm dia.	13.47	16.90	0.50	16.08	nr	32.98
300 mm dia.	14.46	18.14	0.50	16.08	nr	34.22
350 mm dia.	15.86	19.89	0.50	16.08	nr	35.97
400 mm dia.	17.55	22.02	0.50	16.08	nr	38.10
Tees						
15 mm dia.	3.54	4.44	0.30	9.64	nr	14.08
20 mm dia.	3.72	4.67	0.30	9.64	nr	14.31
25 mm dia.	3.92	4.92	0.30	9.64	nr	14.56
32 mm dia.	4.12	5.16	0.30	9.64	nr	14.80
40 mm dia.	4.32	5.42	0.30	9.64	nr	15.06
50 mm dia.	4.78	5.99	0.30	9.64	nr	15.63
65 mm dia.	5.19	6.51	0.30	9.64	nr	16.15
80 mm dia.	5.54	6.94	0.40	12.87	nr	19.81
100 mm dia.	6.89	8.65	0.40	12.87	nr	21.52
125 mm dia.	7.68	9.63	0.40	12.87	nr	22.50
150 mm dia.	8.82	11.07	0.40	12.87	nr	23.94
200 mm dia.	11.66	14.63	0.50	16.08	nr	30.71
250 mm dia.	13.47	16.90	0.50	16.08	nr	32.98
300 mm dia.	14.46	18.14	0.50	16.08	nr	34.22
350 mm dia.	15.86	19.89	0.50	16.08	nr	35.97
400 mm dia.	17.55	22.02	0.50	16.08	nr	38.10
Valves						
15 mm dia.	33.12	41.54	1.03	33.13	nr	74.67
20 mm dia.	34.84	43.70	1.03	33.13	nr	76.83
25 mm dia.	36.83	46.20	1.03	33.13	nr	79.33
32 mm dia.	38.67	48.51	1.03	33.13	nr	81.64
40 mm dia.	40.46	50.76	1.03	33.13	nr	83.89
50 mm dia.	44.62	55.97	1.03	33.13	nr	89.10
65 mm dia.	48.25	60.52	1.03	33.13	nr	93.65
80 mm dia.	51.70	64.85	1.25	40.21	nr	105.06
100 mm dia.	63.56	79.73	1.25	40.21	nr	119.94
125 mm dia.	70.64	88.61	1.25	40.21	nr	128.82
150 mm dia.	80.87	101.44	1.25	40.21	nr	141.65
200 mm dia.	105.81	132.73	1.55	49.85	nr	182.58
250 mm dia.	122.27	153.37	1.55	49.85	nr	203.22
300 mm dia.	132.67	166.42	1.55	49.85	nr	216.27
350 mm dia.	146.82	184.17	1.55	49.85	nr	234.02
400 mm dia.	163.58	205.20	1.55	49.85	nr	255.05
Expansion bellows						
15 mm dia.	49.07	61.56	1.42	45.67	nr	107.23
20 mm dia.	51.62	64.75	1.42	45.67	nr	110.42
25 mm dia.	54.57	68.45	1.42	45.67	nr	114.12

38 MECHANICAL/COOLING/HEATING SYSTEMS

Item	Net Price £	Material £	Labour hours	Labour £	Unit	Total rate £
THERMAL INSULATION – cont						
Extra over for fittings – cont						
32 mm dia.	57.32	71.90	1.42	45.67	nr	**117.57**
40 mm dia.	59.93	75.17	1.42	45.67	nr	**120.84**
50 mm dia.	66.07	82.88	1.42	45.67	nr	**128.55**
65 mm dia.	71.47	89.66	1.42	45.67	nr	**135.33**
80 mm dia.	76.60	96.08	1.75	56.28	nr	**152.36**
100 mm dia.	94.16	118.12	1.75	56.28	nr	**174.40**
125 mm dia.	104.66	131.29	1.75	56.28	nr	**187.57**
150 mm dia.	119.82	150.30	1.75	56.28	nr	**206.58**
200 mm dia.	156.77	196.65	2.17	69.80	nr	**266.45**
250 mm dia.	181.16	227.25	2.17	69.80	nr	**297.05**
300 mm dia.	196.55	246.56	2.17	69.80	nr	**316.36**
350 mm dia.	217.51	272.84	2.17	69.80	nr	**342.64**
400 mm dia.	242.36	304.01	2.17	69.80	nr	**373.81**

38 MECHANICAL/COOLING/HEATING SYSTEMS

Item	Net Price £	Material £	Labour hours	Labour £	Unit	Total rate £
STEAM HEATING						
PIPELINES						
For pipework prices refer to Section – Low Temperature Hot Water Heating						
PIPELINE ANCILLARIES						
Steam traps and accessories						
Cast iron; inverted bucket type; steam trap pressure range up to 17 bar at 210°C; screwed ends						
½" dia.	106.96	134.18	0.85	27.33	nr	**161.51**
¾" dia.	106.96	134.18	1.13	36.34	nr	**170.52**
1" dia.	180.23	226.08	1.35	43.41	nr	**269.49**
1½" dia.	556.43	697.98	1.80	57.89	nr	**755.87**
2" dia.	622.84	781.29	2.18	70.12	nr	**851.41**
Cast iron; inverted bucket type; steam trap pressure range up to 17 bar at 210°C; flanged ends to BS 4504 PN16; bolted connections						
15 mm dia.	179.15	224.73	1.15	36.99	nr	**261.72**
20 mm dia.	216.50	271.58	1.25	40.21	nr	**311.79**
25 mm dia.	268.76	337.13	1.33	42.77	nr	**379.90**
40 mm dia.	570.15	715.20	1.46	46.95	nr	**762.15**
50 mm dia.	636.56	798.50	1.60	51.45	nr	**849.95**
Steam traps and strainers						
Stainless steel; thermodynamic trap with pressure range up to 42 bar; temperature range to 400°C; screwed ends to steel						
15 mm dia.	71.62	89.84	0.84	27.04	nr	**116.88**
20 mm dia.	81.06	101.68	1.14	36.68	nr	**138.36**
Stainless steel; thermodynamic trap with pressure range up to 24 bar; temperature range to 288°C; flanged ends to DIN 2456 PN64; bolted connections						
15 mm dia.	149.94	188.08	1.24	39.91	nr	**227.99**
20 mm dia.	164.36	206.17	1.34	43.11	nr	**249.28**
25 mm dia.	190.00	238.34	1.40	45.05	nr	**283.39**
Malleable iron pipeline strainer; max steam working pressure 14 bar and temperature range to 230°C; screwed ends to steel						
½" dia.	20.86	26.16	0.84	27.04	nr	**53.20**
¾" dia.	26.07	32.70	1.14	36.68	nr	**69.38**
1" dia.	40.67	51.02	1.30	41.82	nr	**92.84**
1½" dia.	83.53	104.78	1.50	48.29	nr	**153.07**
2" dia.	117.28	147.11	1.74	56.03	nr	**203.14**

38 MECHANICAL/COOLING/HEATING SYSTEMS

Item	Net Price £	Material £	Labour hours	Labour £	Unit	Total rate £
STEAM HEATING – cont						
Steam traps and strainers – cont						
Bronze pipeline strainer; max steam working pressure 25 bar; flanged ends to BS 4504 PN25; bolted connections						
15 mm dia.	179.16	224.74	1.24	39.91	nr	264.65
20 mm dia.	218.54	274.13	1.34	43.11	nr	317.24
25 mm dia.	250.56	314.31	1.40	45.05	nr	359.36
32 mm dia.	388.96	487.92	1.46	47.03	nr	534.95
40 mm dia.	441.41	553.71	1.54	49.56	nr	603.27
50 mm dia.	678.88	851.59	1.64	52.81	nr	904.40
65 mm dia.	751.70	942.93	2.50	80.40	nr	1023.33
80 mm dia.	935.26	1173.19	2.91	93.50	nr	1266.69
100 mm dia.	1619.96	2032.08	3.51	112.86	nr	2144.94
Balanced pressure thermostatic steam trap and strainer; max working pressure up to 13 bar; screwed ends to steel						
½" dia.	148.43	186.19	1.26	40.56	nr	226.75
¾" dia.	148.82	186.68	1.71	55.07	nr	241.75
Bimetallic thermostatic steam trap and strainer; max working pressure up to 21 bar; flanged ends						
15 mm	193.09	242.21	1.24	39.91	nr	282.12
20 mm	212.13	266.10	1.34	43.11	nr	309.21
Sight glasses						
Pressed brass; straight; single window; screwed ends to steel						
15 mm dia.	45.15	56.64	0.84	27.04	nr	83.68
20 mm dia.	50.12	62.87	1.14	36.68	nr	99.55
25 mm dia.	62.63	78.57	1.30	41.82	nr	120.39
Gunmetal; straight; double window; screwed ends to steel						
15 mm dia.	72.82	91.35	0.84	27.04	nr	118.39
20 mm dia.	80.13	100.52	1.14	36.68	nr	137.20
25 mm dia.	99.05	124.25	1.30	41.82	nr	166.07
32 mm dia.	163.17	204.68	1.35	43.47	nr	248.15
40 mm dia.	163.17	204.68	1.74	56.03	nr	260.71
50 mm dia.	198.13	248.54	2.08	67.01	nr	315.55
SG Iron flanged; BS 4504, PN 25						
15 mm dia.	148.59	186.39	1.00	32.17	nr	218.56
20 mm dia.	174.83	219.31	1.25	40.21	nr	259.52
25 mm dia.	222.91	279.62	1.50	48.29	nr	327.91
32 mm dia.	246.18	308.81	1.70	54.70	nr	363.51
40 mm dia.	321.95	403.85	2.00	64.32	nr	468.17
50 mm dia.	388.96	487.92	2.30	74.11	nr	562.03
Check valve and sight glass; gun metal; screwed						
15 mm dia.	73.55	92.27	0.84	27.04	nr	119.31
20 mm dia.	77.93	97.75	1.14	36.68	nr	134.43
25 mm dia.	131.11	164.46	1.30	41.82	nr	206.28

38 MECHANICAL/COOLING/HEATING SYSTEMS

Item	Net Price £	Material £	Labour hours	Labour £	Unit	Total rate £
Pressure reducing valves						
Pressure reducing valve for steam; maximum range of 17 bar and 232°C; screwed ends to steel						
15 mm dia.	550.67	690.76	0.87	27.99	nr	718.75
20 mm dia.	595.82	747.40	0.91	29.27	nr	776.67
25 mm dia.	642.44	805.87	1.35	43.41	nr	849.28
Pressure reducing valve for steam; maximum range of 17 bar and 232°C; flanged ends to BS 4504 PN 25						
25 mm dia.	773.57	970.37	1.70	54.68	nr	1025.05
32 mm dia.	878.45	1101.92	1.87	60.14	nr	1162.06
40 mm dia.	1048.91	1315.75	2.12	68.19	nr	1383.94
50 mm dia.	1210.62	1518.60	2.57	82.66	nr	1601.26
Safety and relief valves						
Bronze safety valve; 'pop' type; side outlet; including easing lever; working pressure saturated steam up to 20.7 bar; screwed ends to steel						
15 mm dia.	187.06	234.65	0.32	10.28	nr	244.93
20 mm dia.	211.45	265.24	0.40	12.87	nr	278.11
Bronze safety valve; 'pop' type; side outlet; including easing lever; working pressure saturated steam up to 17.2 bar; screwed ends to steel						
25 mm dia.	300.91	377.46	0.47	15.12	nr	392.58
32 mm dia.	359.50	450.96	0.56	18.01	nr	468.97
Bronze safety valve; 'pop' type; side outlet; including easing lever; working pressure saturated steam up to 13.8 bar; screwed ends to steel						
40 mm dia.	465.22	583.58	0.64	20.59	nr	604.17
50 mm dia.	647.41	812.11	0.76	24.45	nr	836.56
65 mm dia.	920.72	1154.96	0.94	30.23	nr	1185.19
80 mm dia.	1194.00	1497.75	1.10	35.37	nr	1533.12
EQUIPMENT						
CALORIFIERS						
Non-storage calorifiers; mild steel shell construction with indirect steam heating for secondary LPHW at 82°C flow and 71°C return to BS 853; maximum test on shell 11 bar, tubes 26 bar						
Horizontal/vertical, for steam at 3 bar–5.5 bar						
88 kW capacity	736.75	924.18	8.00	257.30	nr	1181.48
176 kW capacity	1072.52	1345.37	12.05	387.51	nr	1732.88
293 kW capacity	1168.46	1465.72	14.08	453.01	nr	1918.73
586 kW capacity	1705.97	2139.97	37.04	1191.22	nr	3331.19
879 kW capacity	2124.16	2664.55	40.00	1286.51	nr	3951.06
1465 kW capacity	2590.30	3249.28	45.45	1461.94	nr	4711.22

38 MECHANICAL/COOLING/HEATING SYSTEMS

Item	Net Price £	Material £	Labour hours	Labour £	Unit	Total rate £
LOCAL HEATING UNITS						
Warm air unit heater for connection to LTHW or steam supplies; suitable for heights upto 3 m; recirculating type; mild steel casing; heating coil; adjustable discharge louvre; axial fan; horizontal or vertical discharge; normal speed; entering air temperature 15°C; complete with enclosures; includes fixing in position; includes connections to primary heating supply; electrical work elsewhere						
Low pressure hot water						
7.5 kW, 265 l/s	785.12	984.85	6.53	210.02	nr	**1194.87**
15.4 kW, 575 l/s	950.38	1192.16	7.54	242.51	nr	**1434.67**
26.9 kW, 1040 l/s	1287.82	1615.44	8.65	278.21	nr	**1893.65**
48.0 kW, 1620 l/s	1697.58	2129.44	9.35	300.72	nr	**2430.16**
Steam, 2 bar						
9.2 kW, 265 l/s	1117.64	1401.97	6.53	210.02	nr	**1611.99**
18.8 kW, 575 l/s	1210.22	1518.10	6.82	219.34	nr	**1737.44**
34.8 kW, 1040 l/s	1391.96	1746.08	6.82	219.34	nr	**1965.42**
51.6 kW, 1625 l/s	1885.68	2365.40	7.10	228.35	nr	**2593.75**

38 MECHANICAL/COOLING/HEATING SYSTEMS

Item	Net Price £	Material £	Labour hours	Labour £	Unit	Total rate £
CENTRAL REFRIGERATION PLANT						
CHILLERS; Air cooled						
Selection of air cooled chillers based on chilled water flow and return temperatures 6°C and 12°C, and an outdoor temperature of 35°C						
Air cooled liquid chiller; refrigerant 407C; scroll compressors; twin circuit; integral controls; includes placing in position; electrical work elsewhere						
Cooling load						
100 kW	17814.60	22346.63	8.00	257.30	nr	**22603.93**
150 kW	22450.86	28162.36	8.00	257.30	nr	**28419.66**
200 kW	27789.21	34858.79	8.00	257.30	nr	**35116.09**
Air cooled liquid chiller; refrigerant 407C; reciprocating compressors; twin circuit; integral controls; includes placing in position; electrical work elsewhere						
Cooling load						
400 kW	49573.52	62185.02	8.00	257.30	nr	**62442.32**
550 kW	70690.09	88673.65	8.00	257.30	nr	**88930.95**
700 kW	82123.30	103015.47	9.00	289.46	nr	**103304.93**
Air cooled liquid chiller; refrigerant R134a; screw compressors; twin circuit; integral controls; includes placing in position; electrical work elsewhere						
Cooling load						
250 kW	49985.52	62701.83	8.00	257.30	nr	**62959.13**
400 kW	59883.18	75117.46	8.00	257.30	nr	**75374.76**
600 kW	80388.85	100839.77	8.00	257.30	nr	**101097.07**
800 kW	96620.53	121200.79	9.00	289.46	nr	**121490.25**
1000 kW	113290.26	142111.30	9.00	289.46	nr	**142400.76**
1200 kW	143904.32	180513.58	10.00	321.63	nr	**180835.21**
Air cooled liquid chiller; ductable for indoor installation; refrigerant 407C; scroll compressors; integral controls; includes placing in position; electrical work elsewhere						
Cooling load						
40 kW	10594.99	13290.36	6.00	192.98	nr	**13483.34**
80 kW	16591.60	20812.50	6.00	192.98	nr	**21005.48**
CHILLERS; Higher efficiency air cooled						
Selection of air cooled chillers based on chilled water flow and return temperatures of 6°C and 12°C and an outdoor temperature of 25°C						

38 MECHANICAL/COOLING/HEATING SYSTEMS

Item	Net Price £	Material £	Labour hours	Labour £	Unit	Total rate £
CENTRAL REFRIGERATION PLANT – cont						
CHILLERS – cont						
These machines have significantly higher part load operating efficiencies than conventional air cooled machines						
Air cooled liquid chiller, refrigerant R410 A; scroll compressors; complete with free cooling facility; integral controls; including placing in position; electrical work elsewhere						
Cooling load						
250 kW	45239.14	56747.98	8.00	257.30	nr	57005.28
300 kW	49914.49	62612.74	8.00	257.30	nr	62870.04
350 kW	57679.89	72353.66	8.00	257.30	nr	72610.96
400 kW	65097.21	81657.95	8.00	257.30	nr	81915.25
450 kW	67641.47	84849.46	8.00	257.30	nr	85106.76
500 kW	73532.71	92239.44	8.00	257.30	nr	92496.74
600 kW	75460.15	94657.21	8.00	257.30	nr	94914.51
650 kW	88135.30	110556.92	9.00	289.46	nr	110846.38
700 kW	93346.96	117094.43	9.00	289.46	nr	117383.89
CHILLERS; Water cooled						
Selection of water cooled chillers based on chilled water flow and return temperatures of 6°C and 12°C, and condenser entering and leaving temperatures of 27°C and 33°C						
Water cooled liquid chiller; refrigerant 407C; reciprocating compressors; twin circuit; integral controls; includes placing in position; electrical work elsewhere						
Cooling load						
200 kw	22899.58	28725.23	8.00	257.30	nr	28982.53
350 kW	40524.74	50834.24	8.00	257.30	nr	51091.54
500 kW	45981.47	57679.16	8.00	257.30	nr	57936.46
650 kW	53294.61	66852.76	9.00	289.46	nr	67142.22
750 kW	64916.06	81430.71	9.00	289.46	nr	81720.17
Water cooled condenserless liquid chiller; refrigerant 407C; reciprocating compressors; twin circuit; integral controls; includes placing in position; electrical work elsewhere						
Cooling load						
200 kW	19763.90	24791.84	8.00	257.30	nr	25049.14
350 kW	34694.30	43520.53	8.00	257.30	nr	43777.83
500 kW	43849.17	55004.40	8.00	257.30	nr	55261.70
650 kW	53628.57	67271.68	9.00	289.46	nr	67561.14
750 kW	61023.66	76548.08	9.00	289.46	nr	76837.54

38 MECHANICAL/COOLING/HEATING SYSTEMS

Item	Net Price £	Material £	Labour hours	Labour £	Unit	Total rate £
Water cooled liquid chiller; refrigerant R134a; screw compressors; twin circuit; integral controls; includes placing in position; electrical work elsewhere						
Cooling load						
300 kW	38185.29	47899.62	8.00	257.30	nr	48156.92
500 kW	48668.99	61050.38	8.00	257.30	nr	61307.68
700 kW	61371.38	76984.26	9.00	289.46	nr	77273.72
900 kW	86024.35	107908.94	9.00	289.46	nr	108198.40
1100 kW	95026.96	119201.82	10.00	321.63	nr	119523.45
1300 kW	109871.07	137822.27	10.00	321.63	nr	138143.90
Water cooled liquid chiller; refrigerant R134a; centrifugal compressors; twin circuit; integral controls; includes placing in position; electrical work elsewhere						
Cooling load						
1600 kW	155000.26	194432.32	11.00	353.79	nr	194786.11
1900 kW	188775.74	236800.29	11.00	353.79	nr	237154.08
2200 kW	224907.52	282123.99	13.00	418.12	nr	282542.11
2500 kW	232369.58	291484.40	13.00	418.12	nr	291902.52
3000 kW	282770.81	354707.71	15.00	482.44	nr	355190.15
3500 kW	384882.49	482796.60	15.00	482.44	nr	483279.04
4000 kW	408779.22	512772.66	20.00	643.26	nr	513415.92
4500 kW	436124.22	547074.23	20.00	643.26	nr	547717.49
5000 kW	531893.20	667206.83	25.00	804.07	nr	668010.90
CHILLERS; Absorption						
Absorption chiller, for operation using low pressure steam; selection based on chilled water flow and return temperatures of 6°C and 12°C, steam at 1 bar gauge and condenser entering and leaving temperatures of 27°C and 33°C; integral controls; includes placing in position; electrical work elsewhere						
Cooling load						
400 kW	66781.06	83770.16	8.00	257.30	nr	84027.46
700 kW	84795.91	106367.99	9.00	289.46	nr	106657.45
1000 kW	96740.35	121351.09	10.00	321.63	nr	121672.72
1300 kW	115449.06	144819.30	12.00	385.95	nr	145205.25
1600 kW	136633.31	171392.83	14.00	450.28	nr	171843.11
2000 kW	161936.09	203132.63	15.00	482.44	nr	203615.07

38 MECHANICAL/COOLING/HEATING SYSTEMS

Item	Net Price £	Material £	Labour hours	Labour £	Unit	Total rate £
CENTRAL REFRIGERATION PLANT – cont						
CHILLERS – cont						
Absorption chiller, for operation using low pressure hot water; selection based on chilled water flow and return temperatures of 6°C and 12°C, cooling water temperatures of 27°C and 33°C and hot water at 90°C; integral controls; includes placing in position; electrical work elsewhere						
Cooling load						
700 kW	96740.35	121351.09	9.00	289.46	nr	**121640.55**
1000 kW	120929.00	151693.34	10.00	321.63	nr	**152014.97**
1300 kW	140648.34	176429.28	12.00	385.95	nr	**176815.23**
1600 kW	161936.09	203132.63	14.00	450.28	nr	**203582.91**
HEAT REJECTION						
Dry air liquid coolers						
Flat coil configuration						
500 kW	20896.25	26212.26	15.00	482.44	nr	**26694.70**
Extra for inverter panels (factory wired and mounted on units)	7985.23	10016.68	15.00	482.44	nr	**10499.12**
800 kW	26907.50	33752.77	15.00	482.44	nr	**34235.21**
Extra for inverter panels (factory wired and mounted on units)	20896.25	26212.26	15.00	482.44	nr	**26694.70**
1100 kW	32635.94	40938.52	15.00	482.44	nr	**41420.96**
Extra for inverter panels (factory wired and mounted on units)	15971.60	20034.77	15.00	482.44	nr	**20517.21**
1400 kW	53819.58	67511.28	15.00	482.44	nr	**67993.72**
Extra for inverter panels (factory wired and mounted on units)	22014.92	27615.52	15.00	482.44	nr	**28097.96**
1700 kW	53819.58	67511.28	15.00	482.44	nr	**67993.72**
Extra for inverter panels (factory wired and mounted on units)	22383.60	28077.99	15.00	482.44	nr	**28560.43**
2000 kW	64120.00	80432.13	15.00	482.44	nr	**80914.57**
Extra for inverter panels (factory wired and mounted on units)	29460.85	36955.69	15.00	482.44	nr	**37438.13**
Note: heat rejection capacities above 500 kW require multiple units. Prices are therefore for total number of units						
Vee type coil configuration						
500 kW	19179.67	24058.98	15.00	482.44	nr	**24541.42**
Extra for inverter panels (factory wired and mounted on units)	5808.36	7286.00	15.00	482.44	nr	**7768.44**
800 kW	27822.35	34900.35	15.00	482.44	nr	**35382.79**

38 MECHANICAL/COOLING/HEATING SYSTEMS

Item	Net Price £	Material £	Labour hours	Labour £	Unit	Total rate £
Extra for inverter panels (factory wired and mounted on units)	13449.44	16870.97	15.00	482.44	nr	**17353.41**
1100 kW	38491.93	48284.28	15.00	482.44	nr	**48766.72**
Extra for inverter panels (factory wired and mounted on units)	16264.83	20402.60	15.00	482.44	nr	**20885.04**
1400 kW	54815.96	68761.15	15.00	482.44	nr	**69243.59**
Extra for inverter panels (factory wired and mounted on units)	20225.07	25370.33	15.00	482.44	nr	**25852.77**
1700 kW	77062.00	96666.57	15.00	482.44	nr	**97149.01**
Extra for inverter panels (factory wired and mounted on units)	26902.44	33746.42	15.00	482.44	nr	**34228.86**
2000 kW	76955.45	96532.91	15.00	482.44	nr	**97015.35**
Extra for inverter panels (factory wired and mounted on units)	30337.03	38054.77	15.00	482.44	nr	**38537.21**

Note: Heat rejection capacities above 1100 kW require multiple units. Prices are for total number of units.

Air cooled condensers

Air cooled condenser; refrigerant 407C; selection based on condensing temperature of 45°C at 32°C dry bulb ambient; includes placing in position; electrical work elsewhere

Item	Net Price £	Material £	Labour hours	Labour £	Unit	Total rate £
Flat coil configuration						
500 kW	17395.87	21821.37	15.00	482.44	nr	**22303.81**
Extra for inverter panels (factory wired and mounted on units)	7959.86	9984.84	15.00	482.44	nr	**10467.28**
800 kW	28539.67	35800.16	15.00	482.44	nr	**36282.60**
Extra for inverter panels (factory wired and mounted on units)	12185.80	15285.87	15.00	482.44	nr	**15768.31**
1100 kW	38207.29	47927.22	15.00	482.44	nr	**48409.66**
Extra for inverter panels (factory wired and mounted on units)	17323.53	21730.63	15.00	482.44	nr	**22213.07**
1400 kW	47701.22	59836.41	15.00	482.44	nr	**60318.85**
Extra for inverter panels (factory wired and mounted on units)	21303.46	26723.07	15.00	482.44	nr	**27205.51**
1700 kW	62115.78	77918.03	15.00	482.44	nr	**78400.47**
Extra for inverter panels (factory wired and mounted on units)	31955.19	40084.59	15.00	482.44	nr	**40567.03**
2000 kW	71551.83	89754.62	15.00	482.44	nr	**90237.06**
Extra for inverter panels (factory wired and mounted on units)	31955.19	40084.59	15.00	482.44	nr	**40567.03**

Note: Heat rejection capacities above 500 kW require multiple units. Prices are for total number of units

38 MECHANICAL/COOLING/HEATING SYSTEMS

Item	Net Price £	Material £	Labour hours	Labour £	Unit	Total rate £
CENTRAL REFRIGERATION PLANT – cont						
Air cooled condenser – cont						
Vee type coil configuration						
500 kW	17077.49	21422.00	15.00	482.44	nr	21904.44
Extra for inverter panels (factory wired and mounted on units)	6092.92	7642.96	15.00	482.44	nr	8125.40
800 kW	25688.60	32223.78	15.00	482.44	nr	32706.22
Extra for inverter panels (factory wired and mounted on units)	12475.26	15648.96	15.00	482.44	nr	16131.40
1100 kW	37700.76	47291.83	15.00	482.44	nr	47774.27
Extra for inverter panels (factory wired and mounted on units)	21216.63	26614.15	15.00	482.44	nr	27096.59
1400 kW	43894.96	55061.84	15.00	482.44	nr	55544.28
Extra for inverter panels (factory wired and mounted on units)	24950.51	31297.92	15.00	482.44	nr	31780.36
1700 kW	56551.13	70937.74	15.00	482.44	nr	71420.18
Extra for inverter panels (factory wired and mounted on units)	31839.40	39939.35	15.00	482.44	nr	40421.79
2000 kW	65842.46	82592.79	15.00	482.44	nr	83075.23
Extra for inverter panels (factory wired and mounted on units)	37425.77	46946.88	15.00	482.44	nr	47429.32
Note: Heat rejection capacities above 1100 kW require multiple units. Prices are for total number of units.						
Cooling towers						
Cooling towers; forced draught, centrifugal fan, conterflow design; based on water temperatures of 35°C on and 29°C off at 21°C wet bulb ambient temperature; includes placing in position; electrical work elsewhere						
Cooling towers: Open circuit type						
900 kW	11587.54	14535.40	20.00	643.26	nr	15178.66
Extra for stainless steel construction	5195.07	6516.70	–	–	nr	6516.70
Extra for intake and discharge sound attenuation	5591.25	7013.66	–	–	nr	7013.66
Extra for fan dampers for capacity control	1297.30	1627.34	–	–	nr	1627.34
1500 kW	17978.37	22552.06	20.00	643.26	nr	23195.32
Extra for stainless steel construction	8321.68	10438.71	–	–	nr	10438.71
Extra for intake and discharge sound attenuation	8378.65	10510.18	–	–	nr	10510.18
Extra for fan dampers for capacity control	1316.73	1651.71	–	–	nr	1651.71
2100 kW	25505.12	31993.62	20.00	643.26	nr	32636.88
Extra for stainless steel construction	11802.64	14805.24	–	–	nr	14805.24
Extra for intake and discharge sound attenuation	12440.95	15605.92	–	–	nr	15605.92

38 MECHANICAL/COOLING/HEATING SYSTEMS

Item	Net Price £	Material £	Labour hours	Labour £	Unit	Total rate £
Extra for fan dampers for capacity control	1503.56	1886.07	–	–	nr	1886.07
2700 kW	30801.93	38637.94	20.00	643.26	nr	39281.20
Extra for stainless steel construction	14397.78	18060.57	–	–	nr	18060.57
Extra for intake and discharge sound attenuation	16760.29	21024.10	–	–	nr	21024.10
Extra for fan dampers for capacity control	2340.52	2935.95	–	–	nr	2935.95
3300 kW	37957.99	47614.50	20.00	643.26	nr	48257.76
Extra for stainless steel construction	17572.28	22042.66	–	–	nr	22042.66
Extra for intake and discharge sound attenuation	15806.74	19827.98	–	–	nr	19827.98
Extra for fan dampers for capacity control	1914.57	2401.64	–	–	nr	2401.64
3900 kW	42782.51	53666.38	20.00	643.26	nr	54309.64
Extra for stainless steel construction	19806.71	24845.54	–	–	nr	24845.54
Extra for intake and discharge sound attenuation	16110.14	20208.56	–	–	nr	20208.56
Extra for fan dampers for capacity control	1914.57	2401.64	–	–	nr	2401.64
4500 kW	48148.07	60396.94	23.00	739.75	nr	61136.69
Extra for stainless steel construction	22452.91	28164.93	–	–	nr	28164.93
Extra for intake and discharge sound attenuation	21366.60	26802.26	–	–	nr	26802.26
Extra for fan dampers for capacity control	3013.09	3779.62	–	–	nr	3779.62
5100 kW	55618.02	69767.24	23.00	739.75	nr	70506.99
Extra for stainless steel construction	25625.83	32145.04	–	–	nr	32145.04
Extra for intake and discharge sound attenuation	21309.80	26731.02	–	–	nr	26731.02
Extra for fan dampers for capacity control	3013.09	3779.62	–	–	nr	3779.62
5700 kW	58799.98	73758.70	30.00	964.88	nr	74723.58
Extra for stainless steel construction	27456.48	34441.41	–	–	nr	34441.41
Extra for intake and discharge sound attenuation	31556.68	39584.70	–	–	nr	39584.70
Extra for fan dampers for capacity control	3313.50	4156.45	–	–	nr	4156.45
6300 kW	70897.16	88933.40	30.00	964.88	nr	89898.28
Extra for stainless steel construction	32822.33	41172.33	–	–	nr	41172.33
Extra for intake and discharge sound attenuation	32012.52	40156.50	–	–	nr	40156.50
Extra for fan dampers for capacity control	3313.50	4156.45	–	–	nr	4156.45
Cooling towers: Closed circuit type (includes 20% ethylene glycol)						
900 kW	32302.49	40520.24	20.00	643.26	nr	41163.50
Extra for stainless steel construction	26302.53	32993.89	–	–	nr	32993.89
Extra for intake and discharge sound attenuation	9720.23	12193.06	–	–	nr	12193.06
Extra for fan dampers for capacity control	1316.73	1651.71	–	–	nr	1651.71
1500 kW	60055.40	75333.50	20.00	643.26	nr	75976.76
Extra for stainless steel construction	47880.84	60061.72	–	–	nr	60061.72
Extra for intake and discharge sound attenuation	15515.30	19462.40	–	–	nr	19462.40

38 MECHANICAL/COOLING/HEATING SYSTEMS

Item	Net Price £	Material £	Labour hours	Labour £	Unit	Total rate £
CENTRAL REFRIGERATION PLANT – cont						
Cooling towers – cont						
Cooling towers: Closed circuit type (includes 20% ethylene glycol) – cont						
Extra for fan dampers for capacity control	1914.57	2401.64	–	–	nr	**2401.64**
2100 kW	73186.84	91805.57	20.00	643.26	nr	**92448.83**
Extra for stainless steel construction	65049.30	81597.85	–	–	nr	**81597.85**
Extra for intake and discharge sound attenuation	18676.35	23427.61	–	–	nr	**23427.61**
Extra for fan dampers for capacity control	1914.57	2401.64	–	–	nr	**2401.64**
2700 kW	78908.13	98982.36	25.00	804.07	nr	**99786.43**
Extra for stainless steel construction	80599.41	101103.90	–	–	nr	**101103.90**
Extra for intake and discharge sound attenuation	27875.54	34967.07	–	–	nr	**34967.07**
Extra for fan dampers for capacity control	2896.51	3633.38	–	–	nr	**3633.38**
3300 kW	121577.05	152506.26	25.00	804.07	nr	**153310.33**
Extra for stainless steel construction	100549.27	126129.00	–	–	nr	**126129.00**
Extra for intake and discharge sound attenuation	37080.66	46513.98	–	–	nr	**46513.98**
Extra for fan dampers for capacity control	3313.50	4156.45	–	–	nr	**4156.45**
3900 kW	133862.52	167917.14	25.00	804.07	nr	**168721.21**
Extra for stainless steel construction	115145.47	144438.48	–	–	nr	**144438.48**
Extra for intake and discharge sound attenuation	37687.47	47275.17	–	–	nr	**47275.17**
Extra for fan dampers for capacity control	3313.50	4156.45	–	–	nr	**4156.45**
4500 kW	160259.87	201029.98	40.00	1286.51	nr	**202316.49**
Extra for stainless steel construction	119453.62	149842.62	–	–	nr	**149842.62**
Extra for intake and discharge sound attenuation	53866.35	67569.95	–	–	nr	**67569.95**
Extra for fan dampers for capacity control	5793.00	7266.74	–	–	nr	**7266.74**
5100 kW	178740.50	224212.08	40.00	1286.51	nr	**225498.59**
Extra for stainless steel construction	154368.37	193639.68	–	–	nr	**193639.68**
Extra for intake and discharge sound attenuation	53866.35	67569.95	–	–	nr	**67569.95**
Extra for fan dampers for capacity control	5793.00	7266.74	–	–	nr	**7266.74**
5700 kW	220397.58	276466.72	40.00	1286.51	nr	**277753.23**
Extra for stainless steel construction	159451.78	200016.31	–	–	nr	**200016.31**
Extra for intake and discharge sound attenuation	60218.34	75537.88	–	–	nr	**75537.88**
Extra for fan dampers for capacity control	6627.00	8312.91	–	–	nr	**8312.91**
6300 kW	235421.11	295312.24	40.00	1286.51	nr	**296598.75**
Extra for stainless steel construction	182387.02	228786.28	–	–	nr	**228786.28**
Extra for intake and discharge sound attenuation	60218.34	75537.88	–	–	nr	**75537.88**
Extra for fan dampers for capacity control	6627.00	8312.91	–	–	nr	**8312.91**

38 MECHANICAL/COOLING/HEATING SYSTEMS

Item	Net Price £	Material £	Labour hours	Labour £	Unit	Total rate £
CHILLED WATER						
SCREWED STEEL						
PIPELINES						
For pipework prices refer to Section – Low Temperature Hot Water Heating, with the exception of chilled water blocks within brackets as detailed hereafter. For minimum fixing distances, refer to the Tables and Memoranda to the rear of the book						
Fixings						
For steel pipes; black malleable iron						
Oversized pipe clip, to contain 30 mm insulation block for vapour barrier						
15 mm dia.	4.10	5.14	0.10	3.21	nr	8.35
20 mm dia.	4.13	5.19	0.11	3.54	nr	8.73
25 mm dia.	4.34	5.44	0.12	3.86	nr	9.30
32 mm dia.	4.47	5.61	0.14	4.50	nr	10.11
40 mm dia.	4.71	5.91	0.15	4.83	nr	10.74
50 mm dia.	6.74	8.46	0.16	5.14	nr	13.60
65 mm dia.	7.27	9.12	0.30	9.64	nr	18.76
80 mm dia.	7.71	9.68	0.35	11.26	nr	20.94
100 mm dia.	12.23	15.34	0.40	12.87	nr	28.21
125 mm dia.	14.19	17.80	0.60	19.31	nr	37.11
150 mm dia.	22.20	27.84	0.77	24.76	nr	52.60
200 mm dia.	28.33	35.54	0.90	28.95	nr	64.49
250 mm dia.	36.80	46.17	1.10	35.37	nr	81.54
300 mm dia.	43.80	54.95	1.25	40.21	nr	95.16
350 mm dia.	50.82	63.75	1.50	48.25	nr	112.00
400 mm dia.	56.90	71.38	1.75	56.28	nr	127.66
Screw on backplate (Male), black malleable iron; plugged and screwed						
M12	0.56	0.71	0.10	3.21	nr	3.92
Screw on backplate (Female), black malleable iron; plugged and screwed						
M12	0.56	1.42	0.10	3.21	nr	4.63
Extra over channel sections for fabricated hangers and brackets						
Galvanized steel; including inserts, bolts, nuts, washers; fixed to backgrounds						
41 × 21 mm	6.34	7.95	0.29	9.33	m	17.28
41 × 41 mm	7.60	9.53	0.29	9.33	m	18.86
Threaded rods; metric thread; including nuts, washers etc.						
12 mm dia. × 600 mm long	3.20	4.01	0.18	5.79	nr	9.80

38 MECHANICAL/COOLING/HEATING SYSTEMS

Item	Net Price £	Material £	Labour hours	Labour £	Unit	Total rate £
CHILLED WATER – cont						
For plastic pipework suitable for chilled water systems, refer to ABS pipework details in Section – Cold Water, with the exception of chilled water blocks within brackets as detailed for the aforementioned steel pipe. For minimum fixing distances refer to the Tables and Memoranda to the rear of the book						
For copper pipework, refer to Section – Cold Water with the exception of chilled water blocks within brackets as detailed hereafter. For minimum fixing distances refer to the Tables and Memoranda to the rear of the book						
Fixings						
For copper pipework						
Oversized pipe clip, to contain 30 mm insulation block for vapour barrier						
15 mm dia.	4.10	5.14	0.10	3.21	nr	8.35
22 mm dia.	4.13	5.19	0.11	3.54	nr	8.73
28 mm dia.	4.34	5.44	0.12	3.86	nr	9.30
35 mm dia.	4.47	5.61	0.14	4.50	nr	10.11
42 mm dia.	4.71	5.91	0.15	4.83	nr	10.74
54 mm dia.	6.74	8.46	0.16	5.14	nr	13.60
67 mm dia.	7.27	9.12	0.30	9.64	nr	18.76
76 mm dia.	7.71	9.68	0.35	11.26	nr	20.94
108 mm dia.	8.52	26.04	0.40	12.87	nr	38.91
133 mm dia.	14.19	17.80	0.60	19.31	nr	37.11
159 mm dia.	22.20	27.84	0.77	24.76	nr	52.60
Screw on backplate, female						
All sizes 15 mm to 54 mm × 10 mm	1.63	2.05	0.10	3.21	nr	5.26
Screw on backplate, male						
All sizes 15 mm to 54 mm × 10 mm	2.18	2.73	0.10	3.21	nr	5.94
Extra over channel sections for fabricated hangers and brackets						
Galvanized steel; including inserts, bolts, nuts, washers; fixed to backgrounds						
41 × 21 mm	6.34	7.95	0.29	9.33	m	17.28
41 × 41 mm	7.60	9.53	0.29	9.33	m	18.86
Threaded rods; metric thread; including nuts, washers etc.						
10 mm dia. × 600 mm long for ring clips up to 54 mm	2.07	2.60	0.18	5.79	nr	8.39
12 mm dia. × 600 mm long for ring clips from 54 mm	3.20	4.01	0.18	5.79	nr	9.80

38 MECHANICAL/COOLING/HEATING SYSTEMS

Item	Net Price £	Material £	Labour hours	Labour £	Unit	Total rate £
PIPELINE ANCILLARIES						
For prices for ancillaries refer to Section: Low Temperature Hot Water Heating						
HEAT EXCHANGERS						
Plate heat exchanger; for use in CHW systems; painted carbon steel frame, stainless steel plates, nitrile rubber gaskets, design pressure of 10 bar and operating temperature of 110/135°C						
Primary side; 13°C in, 8°C out; secondary side; 6°C in, 11°C out						
264 kW, 12.60 l/s	5824.25	7305.94	10.00	321.63	nr	7627.57
290 kW, 13.85 l/s	6139.30	7701.14	12.00	385.95	nr	8087.09
316 kW, 15.11 l/s	6391.06	8016.95	12.00	385.95	nr	8402.90
350 kW, 16.69 l/s	6873.05	8621.56	16.00	514.61	nr	9136.17
395 kW, 18.88 l/s	7355.04	9226.16	16.00	514.61	nr	9740.77
454 kW, 21.69 l/s	7989.16	10021.60	16.00	514.61	nr	10536.21
475 kW, 22.66 l/s	8144.00	10215.83	16.00	514.61	nr	10730.44
527 kW, 25.17 l/s	8825.24	11070.38	16.00	514.61	nr	11584.99
554 kW, 26.43 l/s	9059.50	11364.24	16.00	514.61	nr	11878.85
580 kW, 27.68 l/s	9369.92	11753.63	16.00	514.61	nr	12268.24
633 kW, 30.19 l/s	9933.28	12460.30	16.00	514.61	nr	12974.91
661 kW, 31.52 l/s	10172.93	12760.92	16.00	514.61	nr	13275.53
713 kW, 34.04 l/s	10859.54	13622.20	18.00	578.93	nr	14201.13
740 kW, 35.28 l/s	11096.51	13919.46	18.00	578.93	nr	14498.39
804 kW, 38.33 l/s	11815.46	14821.32	18.00	578.93	nr	15400.25
1925 kW, 91.82 l/s	19834.26	24880.09	20.00	643.26	nr	25523.35
2710 kW, 129.26 l/s	26345.13	33047.34	20.00	643.26	nr	33690.60
3100 kW, 147.87 l/s	29350.15	36816.83	20.00	643.26	nr	37460.09
Note: For temperature conditions different to those above, the cost of the units can vary significantly, and therefore the manufacturers advice should be sought.						
TRACE HEATING						
Trace heating; for freeze protection or temperature maintenance of pipework; to BS 6351; including fixing to parent structures by plastic pull ties						
Straight laid F-S-C Wintergaurd						
15 mm	22.09	27.71	0.27	8.68	m	36.39
25 mm	22.09	27.71	0.27	8.68	m	36.39
28 mm	22.09	27.71	0.27	8.68	m	36.39
32 mm	22.09	27.71	0.30	9.64	m	37.35
35 mm	22.09	27.71	0.31	9.97	m	37.68
50 mm	22.09	27.71	0.34	10.94	m	38.65
100 mm	22.09	27.71	0.40	12.87	m	40.58
150 mm F-S-B Wintergaurd (to larger sizes)	22.09	27.71	0.40	12.87	m	40.58

38 MECHANICAL/COOLING/HEATING SYSTEMS

Item	Net Price £	Material £	Labour hours	Labour £	Unit	Total rate £
CHILLED WATER – cont						
Trace heating – cont						
Helically wound F-S-C Wintergaurd						
15 mm	28.15	35.31	1.00	32.17	m	**67.48**
25 mm	28.15	35.31	1.00	32.17	m	**67.48**
28 mm	28.15	35.31	1.00	32.17	m	**67.48**
32 mm	28.15	35.31	1.00	32.17	m	**67.48**
35 mm	28.15	35.31	1.00	32.17	m	**67.48**
50 mm	28.15	35.31	1.00	32.17	m	**67.48**
100 mm	28.15	35.31	1.00	32.17	m	**67.48**
150 mm F-S-B Wintergaurd (to larger sizes)	28.15	35.31	1.00	32.17	m	**67.48**
Accessories for trace heating; weatherproof; polycarbonate enclosure to IP standards; fully installed						
Connection junction box						
100 × 100 × 75 mm junction box incl power connecting kit	66.16	82.99	1.40	45.04	nr	**128.03**
Single air thermostat						
150 × 150 × 75 mm AT-TS−13	144.93	181.80	1.42	45.67	nr	**227.47**
Single capillary thermostat						
150 × 150 × 75 mm AT-TS−13	208.52	261.56	1.46	46.95	nr	**308.51**
Twin capillary thermostat						
150 × 150 × 75 mm	374.10	469.27	1.46	46.95	nr	**516.22**
PRESSURIZATION UNITS						
Chilled water packaged pressurization unit complete with expansion vessel(s), interconnecting pipework and necessary isolating and drain valves; includes placing in position; electrical work elsewhere. Selection based on a final working pressure of 4 bar, a 3 m static head and system operating temperatures of 6°/12°C						
System volume						
1800 litres	1159.62	1454.62	8.00	257.30	nr	**1711.92**
4500 litres	1262.50	1583.68	8.00	257.30	nr	**1840.98**
7200 litres	1329.60	1667.85	10.00	321.63	nr	**1989.48**
9900 litres	1782.49	2235.96	10.00	321.63	nr	**2557.59**
15300 litres	1864.13	2338.37	13.00	418.12	nr	**2756.49**
22500 litres	1947.99	2443.56	20.00	643.26	nr	**3086.82**
27000 litres	1947.99	2443.56	20.00	643.26	nr	**3086.82**

38 MECHANICAL/COOLING/HEATING SYSTEMS

Item	Net Price £	Material £	Labour hours	Labour £	Unit	Total rate £
CHILLED BEAMS						
Static (passive) beams; cooling data based on 9 K waterside cooling dt (e.g. water 14°C flow and 16°C return, room temperature 24°C)						
Static passive beam in perforated metal casing for exposed mounting from solid ceiling or roof soffit, 2 pipe cooling only	–	–	–	–	–	–
Length/Cooling Output						
1200 mm / 300 W	395.00	495.49	3.00	96.49	m	591.98
1800 mm / 450 W	515.00	646.02	3.50	112.57	m	758.59
2400 mm / 600 W	615.00	771.46	4.00	128.65	m	900.11
3000 mm / 750 W	730.00	915.71	4.60	147.94	m	1063.65
3600 mm / 900 W	900.00	1128.96	5.00	160.81	m	1289.77
Static passive beam, chassis type for mounting above open grid or perforated metal ceiling or roof soffit, 2 pipe cooling only	–	–	–	–	–	–
Length/Cooling Output						
1200 mm / 300 W	280.00	351.23	2.50	80.40	m	431.63
1800 mm / 450 W	385.00	482.94	2.90	93.27	m	576.21
2400 mm / 600 W	475.00	595.84	3.20	102.93	m	698.77
3000 mm / 750 W	575.00	721.28	3.50	112.57	m	833.85
3600 mm / 900 W	680.00	852.99	3.90	125.44	m	978.43
Multi-service beams; for exposed mounting on solid ceiling slabs, with integrated lighting and lighting control, provision of fascia openings for other services (electrical work and other services elsewhere). Performance criteria equivalent to standard beams as listed above						
2 way discharge exposed linear multi-service climate beam for exposed mounting on solid ceiling; closed type with integrated secondary air circulation, 2 pipe cooling only, with primary air	–	–	–	–	–	–
Length/Cooling Output						
1200 mm / 550 W	800.00	1003.52	3.40	109.36	m	1112.88
1800 mm / 900 W	1100.00	1379.84	3.90	125.44	m	1505.28
2400 mm / 1200 W	1400.00	1756.16	4.40	141.52	m	1897.68
3000 mm / 1550 W	1700.00	2132.48	4.90	157.61	m	2290.09
3600 mm / 1850 W	2050.00	2571.52	5.40	173.69	m	2745.21
2 way discharge exposed linear multi-service climate beam for exposed mounting on solid ceiling; closed type with integrated secondary air circulation, 4 pipe cooling and heating, with primary air	–	–	–	–	–	–

38 MECHANICAL/COOLING/HEATING SYSTEMS

Item	Net Price £	Material £	Labour hours	Labour £	Unit	Total rate £
CHILLED WATER – cont						
Multi-service beams – cont						
Length/Cooling Output/Heating Output						
1200 mm / 550 W / 495 W	830.00	1041.15	3.40	109.36	m	**1150.51**
1800 mm / 900 W / 810 W	1135.00	1423.74	3.90	125.44	m	**1549.18**
2400 mm / 1200 W / 1080 W	1440.00	1806.34	4.40	141.52	m	**1947.86**
3000 mm / 1550 W / 1395 W	1745.00	2188.93	4.90	157.61	m	**2346.54**
3600 mm / 1850 W / 1665 W	2100.00	2634.24	5.40	173.69	m	**2807.93**
Static passive multi-service beam for exposed mounting on solid ceiling, 2 pipe cooling only	–	–	–	–	–	–
Length/Cooling Output						
1200 mm / 300 W	815.00	1022.34	3.40	109.36	m	**1131.70**
1800 mm / 450 W	1115.00	1398.66	3.90	125.44	m	**1524.10**
2400 mm / 600 W	1415.00	1774.98	4.40	141.52	m	**1916.50**
3000 mm / 750 W	1710.00	2145.02	4.90	157.61	m	**2302.63**
3600 mm / 900 W	2060.00	2584.06	5.40	173.69	m	**2757.75**
Ventilated (active) beams; cooling data based on 9 K airside and 9 K waterside cooling dt (e.g. water 14°C flow and 16°C return, primary air 15°C, room temperature 24°C), heating data based on −4 K airside and 20 K waterside heating dt (e.g. water at 45°C flow and 35°C return, primary air 16° C, room temperature 20°C)						
1 way discharge recessed linear active climate beam mounted within a false ceiling; closed type with integrated secondary air circulation; 300 mm wide, 2 pipe cooling only, with 5 l/s/m primary air	–	–	–	–	–	–
Length/Cooling Output						
1200 mm / 340 W	306.00	383.85	2.90	93.27	m	**477.12**
1800 mm / 530 W	378.00	474.16	3.00	96.49	m	**570.65**
2400 mm / 730 W	445.00	558.21	3.20	102.93	m	**661.14**
3000 mm / 900 W	513.00	643.51	3.40	109.36	m	**752.87**
3600 mm / 1100 W	554.00	694.94	3.50	112.57	m	**807.51**
1 way discharge recessed linear active climate beam mounted within a false ceiling; closed type with integrated secondary air circulation; 300 mm wide, 4 pipe cooling and heating, with 5 l/s/m primary air	–	–	–	–	–	–
Length/Cooling Output/Heating Output						
1200 mm / 340 W / 180 W	320.00	401.41	2.90	93.27	m	**494.68**
1800 mm / 530 W / 310 W	396.00	496.74	3.00	96.49	m	**593.23**
2400 mm / 730 W / 450 W	468.00	587.06	3.20	102.93	m	**689.99**

38 MECHANICAL/COOLING/HEATING SYSTEMS

Item	Net Price £	Material £	Labour hours	Labour £	Unit	Total rate £
3000 mm / 900 W / 600 W	531.00	666.09	3.40	109.36	m	775.45
3600 mm / 1100 W / 750 W	573.00	718.77	3.50	112.57	m	831.34
2 way discharge recessed linear active climate beam mounted within a false ceiling; closed type with integrated secondary air circulation; 600 mm wide, 2 pipe cooling only, with 10 l/s/m primary air	–	–	–	–	–	–
Length/Cooling Output						
1200 mm / 550 W	310.00	388.86	3.00	96.49	m	485.35
1800 mm / 900 W	425.00	533.12	3.40	109.36	m	642.48
2400 mm / 1200 W	530.00	664.83	3.70	119.00	m	783.83
3000 mm / 1550 W	620.00	777.73	4.00	128.65	m	906.38
3600 mm / 1850 W	710.00	890.62	4.40	141.52	m	1032.14
2 way discharge recessed linear active climate beam mounted within a false ceiling; closed type with integrated secondary air circulation; 600 mm wide, 4 pipe cooling and heating, with 10 l/s/m primary air	–	–	–	–	–	–
Length/Cooling Output/Heating Output						
1200 mm / 550 W / 495 W	340.00	426.50	3.00	96.49	m	522.99
1800 mm / 900 W / 810 W	460.00	577.02	3.40	109.36	m	686.38
2400 mm / 1200 W / 1080 W	570.00	715.01	3.70	119.00	m	834.01
3000 mm / 1550 W / 1395 W	665.00	834.18	4.00	128.65	m	962.83
3600 mm / 1850 W / 1665 W	760.00	953.34	4.40	141.52	m	1094.86
4 way discharge recessed modular active climate beam mounted within a false ceiling; closed type with integrated secondary air circulation; 600 mm wide, 2 pipe cooling only, with primary air at 10 l/s/m of beam discharge (600 × 600 unit 12 l/s, 1200 × 600 unit 18 l/s)	–	–	–	–	–	–
Length/Cooling Output						
600 × 600 / 500 W	325.00	407.68	2.90	93.27	m	500.95
1200 × 600 / 850 W	375.00	470.40	3.20	102.93	m	573.33
4 way discharge recessed modular active climate beam mounted within a false ceiling; closed type with integrated secondary air circulation; 600 mm wide, 4 pipe cooling and heating, with primary air at 10 l/s/m of beam discharge (600 × 600 unit 12 l/s, 1200 × 600 unit 18 l/s)	–	–	–	–	–	–
Length/Cooling Output/Heating Output						
600 × 600 / 500 W / 450 W	350.00	439.04	2.90	93.27	m	532.31
1200 × 600 / 850 W / 765 W	405.00	508.03	3.20	102.93	m	610.96
1 way sidewall discharge active climate beam mounted within a ceiling bulkhead; closed type with integrated secondary air circulation, 2 pipe cooling only, with 20 l/s/m primary air	–	–	–	–	–	–
Length/Cooling Output						
900 mm / 700 W	360.00	451.58	2.50	80.40	m	531.98
1100 mm / 900 W	390.00	489.22	2.50	80.40	m	569.62
1300 mm / 1100 W	460.00	577.02	2.70	86.84	m	663.86
1500 mm / 1200 W	490.00	614.66	2.70	86.84	m	701.50

38 MECHANICAL/COOLING/HEATING SYSTEMS

Item	Net Price £	Material £	Labour hours	Labour £	Unit	Total rate £
CHILLED WATER – cont						
Ventilated (active) beams – cont						
1 way sidewall discharge active climate beam mounted within a ceiling bulkhead; closed type with integrated secondary air circulation, 4 pipe cooling only, with 20 l/s/m primary air Length/Cooling Output/Heating Output	–	–	–	–	–	–
900 mm / 700 W / 400 W	390.00	489.22	2.50	80.40	m	**569.62**
1100 mm / 900 W / 600 W	420.00	526.85	2.50	80.40	m	**607.25**
1300 mm / 1100 W / 800 W	490.00	614.66	2.70	86.84	m	**701.50**
1500 mm / 1200 W / 900 W	520.00	652.29	2.70	86.84	m	**739.13**
CHILLED CEILINGS						
Traditional (60% Radiant Absorption, 40% Convection) with bonded elements in perforated metal ceiling tiles (typically 30% free area and 2.5 mm perforated hole sizes) / tartan grid arrangement. Maximum waterside cooling effect approx. 65 W/m² at 9dTK						
Radiant elements and metal ceilings combined	140.00	175.62	1.20	38.60	m	**214.22**
Hybrid (40% Radiant Absorption, 60% Convection) with radiant cooling chilled beam elements positioned above a perforated metal ceiling tiles (typically 30% free area and 2.5 mm perforated hole sizes) / tartan grid arrangement. Maximum waterside cooling effect approx. 100 W/m² at 9dTK						
Radiant chilled beams	60.00	75.26	0.35	11.26	m	**86.52**
Metal ceilings	35.00	43.90	0.70	22.51	m	**66.41**
Convective beam (5% Radiant Absorption, 95% Convection) with convective cooling fin coil batteries positioned above a perforated metal ceiling tiles (typically 50% free area and 5.0 mm perforated hole sizes) / tartan grid arrangement. Maximum waterside cooling effect approx. 100 W/m² at 9dTK						
Convective fin coil batteries	65.00	81.54	0.40	12.87	m	**94.41**
Metal ceilings	35.00	43.90	0.70	22.51	m	**66.41**
LEAK DETECTION						
Leak detection system consisting of a central control module connected by a leader cable to water sensing cables						
Control modules						
Alarm only (1 zone)	553.00	693.68	4.00	143.81	nr	**837.49**
Alarm only (8 zone)	1040.00	1304.58	4.00	143.81	nr	**1448.39**
Alarm and location	2690.00	3374.34	8.00	287.60	nr	**3661.94**

38 MECHANICAL/COOLING/HEATING SYSTEMS

Item	Net Price £	Material £	Labour hours	Labour £	Unit	Total rate £
Cables						
Sensing – 3 m length	110.00	137.98	4.00	143.81	nr	**281.79**
Sensing – 5 m length	127.00	159.31	4.00	143.81	nr	**303.12**
Sensing – 7.5 m length	178.00	223.28	4.00	143.81	nr	**367.09**
Sensing – 15 m length	270.00	338.69	8.00	287.60	nr	**626.29**
Leader – 3.5 m length	48.00	60.21	2.00	71.90	nr	**132.11**
End terminal						
End terminal	24.00	30.11	0.05	1.79	nr	**31.90**
ENERGY METERS						
Ultrasonic						
Energy meter for measuring energy use in chilled water systems; includes ultrasonic flow meter (with sensor and signal converter), energy calculator, pair of temperature sensors with brass pockets, and 3 m of interconnecting cable; includes fixing in position; electrical work elsewhere						
Pipe size (flanged connections to PN16); maximum flow rate						
50 mm, 36 m³/hr	1322.46	1658.90	1.80	57.89	nr	**1716.79**
65 mm, 60 m³/hr	1457.74	1828.59	2.32	74.61	nr	**1903.20**
80 mm, 100 m³/hr	1634.02	2049.71	2.56	82.34	nr	**2132.05**
125 mm, 250 m³/hr	1893.17	2374.79	3.60	115.79	nr	**2490.58**
150 mm, 360 m³/hr	2055.15	2577.98	4.80	154.38	nr	**2732.36**
200 mm, 600 m³/hr	2296.20	2880.35	6.24	200.69	nr	**3081.04**
250 mm, 1000 m³/hr	2649.67	3323.75	9.60	308.76	nr	**3632.51**
300 mm, 1500 m³/hr	3114.63	3907.00	10.80	347.36	nr	**4254.36**
350 mm, 2000 m³/hr	3752.99	4707.75	13.20	424.55	nr	**5132.30**
400 mm, 2500 m³/hr	4296.08	5389.00	15.60	501.74	nr	**5890.74**
500 mm, 3000 m³/hr	4875.35	6115.64	24.00	771.90	nr	**6887.54**
600 mm, 3500 m³/hr	5477.51	6870.99	28.00	900.56	nr	**7771.55**

38 MECHANICAL/COOLING/HEATING SYSTEMS

Item	Net Price £	Material £	Labour hours	Labour £	Unit	Total rate £
CHILLED WATER – cont						
Electromagnetic						
Energy meter for measuring energy use in chilled water systems; includes electromagnetic flow meter (with sensor and signal converter), energy calculator, pair of temperature sensors with brass pockets, and 3 m of interconnecting cable; includes fixing in position; electrical work elsewhere						
Pipe size (flanged connections to PN40); maximum flow rate						
25 mm, 17.7 m³/hr	948.11	1189.31	1.48	47.60	nr	**1236.91**
40 mm, 45 m³/hr	957.17	1200.67	1.55	49.85	nr	**1250.52**
Pipe size (flanged connections to PN16); maximum flow rate						
50 mm, 70 m³/hr	967.89	1214.12	1.80	57.89	nr	**1272.01**
65 mm, 120 m³/hr	972.02	1219.30	2.32	74.61	nr	**1293.91**
80 mm, 180 m³/hr	976.97	1225.52	2.56	82.34	nr	**1307.86**
125 mm, 450 m³/hr	1063.54	1334.10	3.60	115.79	nr	**1449.89**
150 mm, 625 m³/hr	1128.67	1415.80	4.80	154.38	nr	**1570.18**
200 mm, 1100 m³/hr	1214.41	1523.36	6.24	200.69	nr	**1724.05**
250 mm, 1750 m³/hr	1362.81	1709.51	9.60	308.76	nr	**2018.27**
300 mm, 2550 m³/hr	1737.11	2179.03	10.80	347.36	nr	**2526.39**
350 mm, 3450 m³/hr	2259.80	2834.70	13.20	424.55	nr	**3259.25**
400 mm, 4500 m³/hr	2578.03	3233.88	15.60	501.74	nr	**3735.62**

38 MECHANICAL/COOLING/HEATING SYSTEMS

Item	Net Price £	Material £	Labour hours	Labour £	Unit	Total rate £
LOCAL COOLING UNITS						
Split system with ceiling void evaporator unit and external condensing unit						
Ceiling mounted 4 way blow cassette heat pump unit with remote fan speed and load control; refrigerant 407C; includes outdoor unit						
Cooling 3.6 kW, heating 4.1 kW	1823.63	2287.57	35.00	1125.70	nr	**3413.27**
Cooling 4.9 kW, heating 5.5 kW	2009.61	2520.85	35.00	1125.70	nr	**3646.55**
Cooling 7.1 kW, heating 8.2 kW	2432.45	3051.26	35.00	1125.70	nr	**4176.96**
Cooling 10 kW, heating 11.2 kW	2877.10	3609.03	35.00	1125.70	nr	**4734.73**
Cooling 12.20 kW, heating 14.60 kW	3147.37	3948.06	35.00	1125.70	nr	**5073.76**
Ceiling mounted 4 way blow cooling only unit with remote fan speed and load control; refrigerant 407C; includes outdoor unit						
Cooling 3.80 kW	1650.70	2070.63	35.00	1125.70	nr	**3196.33**
Cooling 5.20 kW	1843.95	2313.05	35.00	1125.70	nr	**3438.75**
Cooling 7.10 kW	2274.07	2852.60	35.00	1125.70	nr	**3978.30**
Cooling 10 kW	3641.70	4568.14	35.00	1125.70	nr	**5693.84**
Cooling 12.2 kW	2809.09	3523.72	35.00	1125.70	nr	**4649.42**
In ceiling, ducted heat pump unit with remote fan speed and load control; refrigerant 407C; includes outdoor unit						
Cooling 3.60 kW, heating 4.10 kW	1336.84	1676.93	35.00	1125.70	nr	**2802.63**
Cooling 4.90 kW, heating 5.50 kW	1537.36	1928.46	35.00	1125.70	nr	**3054.16**
Cooling 7.10 kW, heating 8.20 kW	1537.36	1928.46	35.00	1125.70	nr	**3054.16**
Cooling 10 kW, heating 11.20 kW	1537.36	1928.46	35.00	1125.70	nr	**3054.16**
Cooling 12.20 kW, heating 14.50 kW	3099.43	3887.92	35.00	1125.70	nr	**5013.62**
In ceiling, ducted cooling only unit with remote fan speed and load control; refrigerant 407C; includes outdoor unit						
Cooling 3.70 kW	1206.06	1512.88	35.00	1125.70	nr	**2638.58**
Cooling 4.90 kW	1434.20	1799.06	35.00	1125.70	nr	**2924.76**
Cooling 7.10 kW	2165.10	2715.90	35.00	1125.70	nr	**3841.60**
Cooling 10 kW	2468.79	3096.84	35.00	1125.70	nr	**4222.54**
Cooling 12.3 kW	2752.13	3452.28	35.00	1125.70	nr	**4577.98**
Room units						
Ceiling mounted 4 way blow cassette heat pump unit with remote fan speed and load control; refrigerant 407C; excludes outdoor unit						
Cooling 3.6 kW, heating 4.1 kW	1150.84	1443.61	17.00	546.77	nr	**1990.38**
Cooling 4.9 kW, heating 5.5 kW	1178.45	1478.24	17.00	546.77	nr	**2025.01**
Cooling 7.1 kW, heating 8.2 kW	1275.81	1600.38	17.00	546.77	nr	**2147.15**
Cooling 10 kW, heating 11.2 kW	1381.89	1733.45	17.00	546.77	nr	**2280.22**
Cooling 12.20 kW, heating 14.60 kW	1498.14	1879.27	17.00	546.77	nr	**2426.04**
Ceiling mounted 4 way blow cooling unit with remote fan speed and load control; refrigerant 407C; excludes outdoor unit						
Cooling 3.80 kW	1086.89	1363.40	17.00	546.77	nr	**1910.17**
Cooling 5.20 kW	1095.62	1374.34	17.00	546.77	nr	**1921.11**
Cooling 7.10 kW	1275.81	1600.38	17.00	546.77	nr	**2147.15**

38 MECHANICAL/COOLING/HEATING SYSTEMS

Item	Net Price £	Material £	Labour hours	Labour £	Unit	Total rate £
LOCAL COOLING UNITS – cont						
Room units – cont						
Cooling 10 kW	1381.89	1733.45	17.00	546.77	nr	**2280.22**
Cooling 12.2 kW	1498.14	1879.27	17.00	546.77	nr	**2426.04**
In ceiling, ducted heat pump unit with remote fan speed and load control; refrigerant 407C; excludes outdoor unit						
Cooling 3.60 kW, heating 4.10 kW	664.06	833.00	17.00	546.77	nr	**1379.77**
Cooling 4.90 kW, heating 5.50 kW	706.20	885.85	17.00	546.77	nr	**1432.62**
Cooling 7.10 kW, heating 8.20 kW	1166.83	1463.67	17.00	546.77	nr	**2010.44**
Cooling 10 kW, heating 11.20 kW	1208.97	1516.54	17.00	546.77	nr	**2063.31**
Cooling 12.20 kW, heating 14.50 kW	1450.14	1819.06	17.00	546.77	nr	**2365.83**
In ceiling, ducted cooling unit only with remote fan speed and load control; refrigerant 407C; excludes outdoor unit						
Cooling 3.70 kW	642.26	805.65	17.00	546.77	nr	**1352.42**
Cooling 4.90 kW	685.86	860.34	17.00	546.77	nr	**1407.11**
Cooling 7.10 kW	1166.83	1463.67	17.00	546.77	nr	**2010.44**
Cooling 10 kW	1208.97	1516.54	17.00	546.77	nr	**2063.31**
Cooling 12.3 kW	1450.19	1819.12	17.00	546.77	nr	**2365.89**
External condensing units suitable for connection to multiple indoor units; inverter driven; refrigerant 407C						
Cooling only						
9 kW	2512.38	3151.53	17.00	546.77	nr	**3698.30**
Heat pump						
Cooling 5.20 kW, heating 6.10 kW	1790.22	2245.66	17.00	546.77	nr	**2792.43**
Cooling 6.80 kW, heating 2.50 kW	2282.79	2863.53	17.00	546.77	nr	**3410.30**
Cooling 8 kW, heating 9.60 kW	2643.17	3315.59	17.00	546.77	nr	**3862.36**
Cooling 14.50 kW, heating 16.50 kW	4386.86	5502.87	21.00	675.42	nr	**6178.29**

38 VENTILATION/AIR CONDITIONING SYSTEMS

Item	Net Price £	Material £	Labour hours	Labour £	Unit	Total rate £
DUCTWORK: CIRCULAR						
AIR DUCTLINES						
Galvanized sheet metal DW144 class B spirally wound circular section ductwork; including all necessary stiffeners, joints, couplers in the running length and duct supports						
Straight duct						
80 mm dia.	5.90	7.40	0.87	26.73	m	34.13
100 mm dia.	6.11	7.66	0.87	26.73	m	34.39
160 mm dia.	8.37	10.49	0.87	26.73	m	37.22
200 mm dia.	10.62	13.32	0.87	26.73	m	40.05
250 mm dia.	12.04	15.10	1.21	37.17	m	52.27
315 mm dia.	15.41	19.33	1.21	37.17	m	56.50
355 mm dia.	22.23	27.89	1.21	37.17	m	65.06
400 mm dia.	22.74	28.53	1.21	37.17	m	65.70
450 mm dia.	27.21	34.14	1.21	37.17	m	71.31
500 mm dia.	29.64	37.18	1.21	37.17	m	74.35
630 mm dia.	57.07	71.59	1.39	42.69	m	114.28
710 mm dia.	63.74	79.96	1.39	42.69	m	122.65
800 mm dia.	67.75	84.99	1.44	44.24	m	129.23
900 mm dia.	80.34	100.78	1.46	44.86	m	145.64
1000 mm dia.	103.24	129.51	1.65	50.69	m	180.20
1120 mm dia.	123.23	154.58	2.43	74.65	m	229.23
1250 mm dia.	134.30	168.47	2.43	74.65	m	243.12
1400 mm dia.	150.86	189.24	2.77	85.10	m	274.34
1600 mm dia.	168.84	211.79	3.06	94.00	m	305.79
Extra over fittings; circular duct class B						
End cap						
80 mm dia.	2.29	2.87	0.15	4.60	nr	7.47
100 mm dia.	2.49	3.12	0.15	4.60	nr	7.72
160 mm dia.	3.75	4.70	0.15	4.60	nr	9.30
200 mm dia.	4.75	5.96	0.20	6.15	nr	12.11
250 mm dia.	7.42	9.31	0.29	8.90	nr	18.21
315 mm dia.	8.66	10.86	0.29	8.90	nr	19.76
355 mm dia.	12.97	16.27	0.44	13.52	nr	29.79
400 mm dia.	14.22	17.84	0.44	13.52	nr	31.36
450 mm dia.	16.02	20.09	0.44	13.52	nr	33.61
500 mm dia.	16.88	21.18	0.44	13.52	nr	34.70
630 mm dia.	61.28	76.87	0.58	17.81	nr	94.68
710 mm dia.	69.57	87.27	0.69	21.20	nr	108.47
800 mm dia.	74.71	93.72	0.81	24.89	nr	118.61
900 mm dia.	95.33	119.58	0.92	28.26	nr	147.84
1000 mm dia.	140.91	176.76	1.04	31.95	nr	208.71
1120 mm dia.	167.88	210.59	1.16	35.64	nr	246.23
1250 mm dia.	189.86	238.16	1.16	35.64	nr	273.80
1400 mm dia.	244.12	306.22	1.16	35.64	nr	341.86
1600 mm dia.	271.44	340.49	1.16	35.64	nr	376.13

38 VENTILATION/AIR CONDITIONING SYSTEMS

Item	Net Price £	Material £	Labour hours	Labour £	Unit	Total rate £
DUCTWORK: CIRCULAR – cont						
Extra over fittings – cont						
Reducer						
80 mm dia.	11.58	14.53	0.29	8.90	nr	**23.43**
100 mm dia.	11.95	14.99	0.29	8.90	nr	**23.89**
160 mm dia.	15.98	20.05	0.29	8.90	nr	**28.95**
200 mm dia.	19.47	24.43	0.44	13.52	nr	**37.95**
250 mm dia.	21.84	27.40	0.58	17.81	nr	**45.21**
315 mm dia.	22.47	28.19	0.58	17.81	nr	**46.00**
355 mm dia.	24.55	30.80	0.87	26.73	nr	**57.53**
400 mm dia.	28.64	35.93	0.87	26.73	nr	**62.66**
450 mm dia.	30.95	38.82	0.87	26.73	nr	**65.55**
500 mm dia.	33.18	41.62	0.87	26.73	nr	**68.35**
630 mm dia.	92.31	115.80	0.87	26.73	nr	**142.53**
710 mm dia.	113.95	142.93	0.96	29.49	nr	**172.42**
800 mm dia.	124.22	155.83	1.06	32.57	nr	**188.40**
900 mm dia.	151.12	189.56	1.16	35.64	nr	**225.20**
1000 mm dia.	219.68	275.56	1.25	38.39	nr	**313.95**
1120 mm dia.	228.21	286.27	3.47	106.60	nr	**392.87**
1250 mm dia.	292.00	366.28	3.47	106.60	nr	**472.88**
1400 mm dia.	313.33	393.04	4.05	124.41	nr	**517.45**
1600 mm dia.	413.91	519.21	4.62	141.93	nr	**661.14**
90° segmented radius bend						
80 mm dia.	4.99	6.26	0.29	8.90	nr	**15.16**
100 mm dia.	5.29	6.63	0.29	8.90	nr	**15.53**
160 mm dia.	8.96	11.24	0.29	8.90	nr	**20.14**
200 mm dia.	11.12	13.94	0.44	13.52	nr	**27.46**
250 mm dia.	15.47	19.41	0.58	17.81	nr	**37.22**
315 mm dia.	14.65	18.38	0.58	17.81	nr	**36.19**
355 mm dia.	14.77	18.52	0.87	26.73	nr	**45.25**
400 mm dia.	18.18	22.80	0.87	26.73	nr	**49.53**
450 mm dia.	20.74	26.02	0.87	26.73	nr	**52.75**
500 mm dia.	26.21	32.88	0.87	26.73	nr	**59.61**
630 mm dia.	77.14	96.77	0.87	26.73	nr	**123.50**
710 mm dia.	101.06	126.77	0.96	29.49	nr	**156.26**
800 mm dia.	106.51	133.60	1.06	32.57	nr	**166.17**
900 mm dia.	149.08	187.01	1.16	35.64	nr	**222.65**
1000 mm dia.	232.34	291.45	1.25	38.39	nr	**329.84**
1120 mm dia.	277.71	348.36	3.47	106.60	nr	**454.96**
1250 mm dia.	299.30	375.45	3.47	106.60	nr	**482.05**
1400 mm dia.	611.90	767.57	4.05	124.41	nr	**891.98**
1600 mm dia.	684.13	858.18	4.62	141.93	nr	**1000.11**
45° radius bend						
80 mm dia.	4.27	5.35	0.29	8.90	nr	**14.25**
100 mm dia.	4.63	5.81	0.29	8.90	nr	**14.71**
160 mm dia.	7.75	9.72	0.29	8.90	nr	**18.62**
200 mm dia.	9.62	12.06	0.40	12.30	nr	**24.36**
250 mm dia.	13.32	16.71	0.58	17.81	nr	**34.52**
315 mm dia.	14.25	17.88	0.58	17.81	nr	**35.69**
355 mm dia.	16.50	20.70	0.87	26.73	nr	**47.43**

38 VENTILATION/AIR CONDITIONING SYSTEMS

Item	Net Price £	Material £	Labour hours	Labour £	Unit	Total rate £
400 mm dia.	17.88	22.43	0.87	26.73	nr	49.16
450 mm dia.	19.19	24.07	0.87	26.73	nr	50.80
500 mm dia.	22.47	28.19	0.87	26.73	nr	54.92
630 mm dia.	76.21	95.60	0.87	26.73	nr	122.33
710 mm dia.	88.69	111.25	0.96	29.49	nr	140.74
800 mm dia.	102.70	128.82	1.06	32.57	nr	161.39
900 mm dia.	125.46	157.38	1.16	35.64	nr	193.02
1000 mm dia.	210.14	263.60	1.25	38.39	nr	301.99
1120 mm dia.	255.70	320.75	3.47	106.60	nr	427.35
1250 mm dia.	278.00	348.72	3.47	106.60	nr	455.32
1400 mm dia.	414.81	520.34	4.05	124.41	nr	644.75
1600 mm dia.	445.99	559.45	4.62	141.93	nr	701.38
90° equal twin bend						
80 mm dia.	26.87	33.70	0.58	17.81	nr	51.51
100 mm dia.	27.33	34.28	0.58	17.81	nr	52.09
160 mm dia.	35.96	45.11	0.58	17.81	nr	62.92
200 mm dia.	46.76	58.65	0.87	26.73	nr	85.38
250 mm dia.	51.98	65.21	1.16	35.64	nr	100.85
315 mm dia.	80.01	100.36	1.16	35.64	nr	136.00
355 mm dia.	94.57	118.63	1.73	53.14	nr	171.77
400 mm dia.	109.79	137.72	1.73	53.14	nr	190.86
450 mm dia.	128.68	161.41	1.73	53.14	nr	214.55
500 mm dia.	136.06	170.68	1.73	53.14	nr	223.82
630 mm dia.	240.64	301.86	1.73	53.14	nr	355.00
710 mm dia.	297.36	373.00	1.82	55.91	nr	428.91
800 mm dia.	408.85	512.86	1.93	59.29	nr	572.15
900 mm dia.	536.22	672.64	2.02	62.06	nr	734.70
1000 mm dia.	825.61	1035.64	2.11	64.81	nr	1100.45
1120 mm dia.	1194.91	1498.90	4.62	141.93	nr	1640.83
1250 mm dia.	1205.53	1512.21	4.62	141.93	nr	1654.14
1400 mm dia.	1632.82	2048.21	4.62	141.93	nr	2190.14
1600 mm dia.	1781.85	2235.15	4.62	141.93	nr	2377.08
Conical branch						
80 mm dia.	20.17	25.30	0.58	17.81	nr	43.11
100 mm dia.	20.87	26.17	0.58	17.81	nr	43.98
160 mm dia.	23.85	29.92	0.58	17.81	nr	47.73
200 mm dia.	27.77	34.83	0.87	26.73	nr	61.56
250 mm dia.	30.76	38.58	1.16	35.64	nr	74.22
315 mm dia.	48.36	60.66	1.16	35.64	nr	96.30
355 mm dia.	57.08	71.60	1.73	53.14	nr	124.74
400 mm dia.	67.36	84.49	1.73	53.14	nr	137.63
450 mm dia.	75.23	94.37	1.73	53.14	nr	147.51
500 mm dia.	83.36	104.56	1.73	53.14	nr	157.70
630 mm dia.	122.16	153.24	1.73	53.14	nr	206.38
710 mm dia.	130.85	164.14	1.82	55.91	nr	220.05
800 mm dia.	140.60	176.37	1.93	59.29	nr	235.66
900 mm dia.	153.16	192.12	2.02	62.06	nr	254.18
1000 mm dia.	213.88	268.30	2.11	64.81	nr	333.11
1120 mm dia.	260.40	326.65	4.62	141.93	nr	468.58
1250 mm dia.	336.44	422.03	5.20	159.75	nr	581.78
1400 mm dia. .	373.23	468.18	5.20	159.75	nr	627.93
1600 mm dia.	427.43	536.17	5.20	159.75	nr	695.92

38 VENTILATION/AIR CONDITIONING SYSTEMS

Item	Net Price £	Material £	Labour hours	Labour £	Unit	Total rate £
DUCTWORK: CIRCULAR – cont						
Extra over fittings – cont						
45° branch						
80 mm dia.	22.44	28.15	0.58	17.81	nr	**45.96**
100 mm dia.	22.63	28.39	0.58	17.81	nr	**46.20**
160 mm dia.	24.05	30.17	0.58	17.81	nr	**47.98**
200 mm dia.	25.96	32.57	0.87	26.73	nr	**59.30**
250 mm dia.	29.06	36.46	1.16	35.64	nr	**72.10**
315 mm dia.	33.59	42.13	1.16	35.64	nr	**77.77**
355 mm dia.	36.62	45.93	1.73	53.14	nr	**99.07**
400 mm dia.	41.24	51.73	1.73	53.14	nr	**104.87**
450 mm dia.	46.40	58.21	1.73	53.14	nr	**111.35**
500 mm dia.	51.55	64.67	1.73	53.14	nr	**117.81**
630 mm dia.	91.23	114.44	1.73	53.14	nr	**167.58**
710 mm dia.	106.70	133.84	1.82	55.91	nr	**189.75**
800 mm dia.	116.89	146.63	2.13	65.43	nr	**212.06**
900 mm dia.	129.59	162.56	2.31	70.96	nr	**233.52**
1000 mm dia.	192.44	241.39	2.31	70.96	nr	**312.35**
1120 mm dia.	237.17	297.51	4.62	141.93	nr	**439.44**
1250 mm dia.	316.75	397.33	4.62	141.93	nr	**539.26**
1400 mm dia.	354.92	445.21	4.62	141.93	nr	**587.14**
1600 mm dia.	406.57	510.00	4.62	141.93	nr	**651.93**
For galvanized sheet metal DW144 class C rates, refer to galvanized sheet metal DW144 class B						

38 VENTILATION/AIR CONDITIONING SYSTEMS

Item	Net Price £	Material £	Labour hours	Labour £	Unit	Total rate £
DUCTWORK: FLAT OVAL						
AIR DUCTLINES						
Galvanized sheet metal DW144 class B spirally wound flat oval section ductwork; including all necessary stiffeners, joints, couplers in the running length and duct supports						
Straight duct						
345 × 102 mm	71.06	89.14	2.71	83.25	m	172.39
427 × 102 mm	75.38	94.56	2.99	91.86	m	186.42
508 × 102 mm	79.73	100.02	3.14	96.47	m	196.49
559 × 152 mm	84.43	105.91	3.43	105.37	m	211.28
531 × 203 mm	84.43	105.91	3.43	105.37	m	211.28
851 × 203 mm	126.33	158.47	5.72	175.72	m	334.19
582 × 254 mm	89.44	112.19	3.62	111.20	m	223.39
823 × 254 mm	126.43	158.59	5.80	178.17	m	336.76
1303 × 254 mm	232.58	291.75	8.13	249.75	m	541.50
632 × 305 mm	93.81	117.68	3.93	120.74	m	238.42
1275 × 305 mm	226.83	284.54	8.13	249.75	m	534.29
765 × 356 mm	107.04	134.27	5.72	175.72	m	309.99
1247 × 356 mm	234.14	293.71	8.13	249.75	m	543.46
1727 × 356 mm	279.32	350.38	10.41	319.79	m	670.17
737 × 406 mm	102.79	128.93	5.72	175.72	m	304.65
818 × 406 mm	124.88	156.65	6.21	190.77	m	347.42
978 × 406 mm	164.11	205.86	6.92	212.59	m	418.45
1379 × 406 mm	245.68	308.18	8.75	268.80	m	576.98
1699 × 406 mm	281.28	352.83	10.41	319.79	m	672.62
709 × 457 mm	111.32	139.64	5.72	175.72	m	315.36
1189 × 457 mm	214.25	268.76	8.80	270.33	m	539.09
1671 × 457 mm	278.48	349.33	10.31	316.72	m	666.05
678 × 508 mm	111.08	139.34	5.72	175.72	m	315.06
919 × 508 mm	139.97	175.58	7.30	224.26	m	399.84
1321 × 508 mm	245.05	307.40	8.75	268.80	m	576.20
Extra over fittings; flat oval duct class B						
End cap						
345 × 102 mm	34.49	43.27	0.20	6.15	nr	49.42
427 × 102 mm	43.25	54.25	0.20	6.15	nr	60.40
508 × 102 mm	46.56	58.41	0.20	6.15	nr	64.56
559 × 152 mm	48.40	60.72	0.29	8.90	nr	69.62
531 × 203 mm	48.42	60.74	0.29	8.90	nr	69.64
851 × 203 mm	59.80	75.02	0.44	13.52	nr	88.54
582 × 254 mm	51.55	64.67	0.44	13.52	nr	78.19
823 × 254 mm	59.82	75.04	0.44	13.52	nr	88.56
1303 × 254 mm	177.19	222.26	0.69	21.20	nr	243.46
632 × 305 mm	53.81	67.50	0.69	21.20	nr	88.70
1275 × 305 mm	173.83	218.05	0.69	21.20	nr	239.25

38 VENTILATION/AIR CONDITIONING SYSTEMS

Item	Net Price £	Material £	Labour hours	Labour £	Unit	Total rate £
DUCTWORK: FLAT OVAL – cont						
Extra over fittings – cont						
End cap – cont						
765 × 356 mm	59.82	75.04	0.69	21.20	nr	**96.24**
1247 × 356 mm	174.15	218.46	0.69	21.20	nr	**239.66**
1727 × 356 mm	275.39	345.45	0.69	21.20	nr	**366.65**
737 × 406 mm	59.88	75.12	1.04	31.95	nr	**107.07**
818 × 406 mm	65.35	81.97	0.69	21.20	nr	**103.17**
978 × 406 mm	88.73	111.31	0.69	21.20	nr	**132.51**
1379 × 406 mm	207.02	259.68	1.04	31.95	nr	**291.63**
1699 × 406 mm	275.68	345.81	1.04	31.95	nr	**377.76**
709 × 457 mm	59.83	75.05	1.04	31.95	nr	**107.00**
1189 × 457 mm	171.02	214.52	1.04	31.95	nr	**246.47**
1671 × 457 mm	270.52	339.34	1.04	31.95	nr	**371.29**
678 × 508 mm	59.82	75.04	1.04	31.95	nr	**106.99**
919 × 508 mm	88.72	111.29	1.04	31.95	nr	**143.24**
1321 × 508 mm	206.54	259.08	1.04	31.95	nr	**291.03**
Reducer						
345 × 102 mm	72.81	91.34	0.95	29.19	nr	**120.53**
427 × 102 mm	78.95	99.03	1.06	32.57	nr	**131.60**
508 × 102 mm	91.08	114.25	1.13	34.71	nr	**148.96**
559 × 152 mm	107.05	134.29	1.26	38.71	nr	**173.00**
531 × 203 mm	108.26	135.80	1.26	38.71	nr	**174.51**
851 × 203 mm	119.00	149.27	1.34	41.17	nr	**190.44**
582 × 254 mm	111.67	140.08	1.34	41.17	nr	**181.25**
823 × 254 mm	120.77	151.49	1.34	41.17	nr	**192.66**
1303 × 254 mm	321.43	403.20	1.34	41.17	nr	**444.37**
632 × 305 mm	114.02	143.02	0.70	21.50	nr	**164.52**
1275 × 305 mm	321.52	403.31	1.16	35.64	nr	**438.95**
765 × 356 mm	132.19	165.82	1.16	35.64	nr	**201.46**
1247 × 356 mm	320.59	402.15	1.16	35.64	nr	**437.79**
1727 × 356 mm	367.93	461.53	1.25	38.39	nr	**499.92**
737 × 406 mm	135.41	169.86	1.16	35.64	nr	**205.50**
818 × 406 mm	147.07	184.49	1.27	39.01	nr	**223.50**
978 × 406 mm	160.85	201.77	1.44	44.24	nr	**246.01**
1379 × 406 mm	321.21	402.93	1.44	44.24	nr	**447.17**
1699 × 406 mm	370.54	464.80	1.44	44.24	nr	**509.04**
709 × 457 mm	133.64	167.64	1.16	35.64	nr	**203.28**
1189 × 457 mm	333.56	418.42	1.34	41.17	nr	**459.59**
1671 × 457 mm	366.33	459.52	1.44	44.24	nr	**503.76**
678 × 508 mm	135.37	169.80	1.16	35.64	nr	**205.44**
919 × 508 mm	177.72	222.94	1.26	38.71	nr	**261.65**
1321 × 508 mm	328.02	411.47	1.44	44.24	nr	**455.71**
90° radius bend						
345 × 102 mm	186.76	234.27	0.29	8.90	nr	**243.17**
427 × 102 mm	207.24	259.96	0.58	17.81	nr	**277.77**
508 × 102 mm	255.57	320.59	0.58	17.81	nr	**338.40**
559 × 152 mm	296.87	372.39	0.58	17.81	nr	**390.20**
531 × 203 mm	300.62	377.09	0.87	26.73	nr	**403.82**
851 × 203 mm	305.72	383.50	0.87	26.73	nr	**410.23**
582 × 254 mm	315.47	395.73	0.87	26.73	nr	**422.46**

38 VENTILATION/AIR CONDITIONING SYSTEMS

Item	Net Price £	Material £	Labour hours	Labour £	Unit	Total rate £
823 × 254 mm	311.26	390.44	0.87	26.73	nr	**417.17**
1303 × 254 mm	327.21	410.46	0.96	29.49	nr	**439.95**
632 × 305 mm	330.54	414.62	0.87	26.73	nr	**441.35**
1275 × 305 mm	342.58	429.73	0.96	29.49	nr	**459.22**
765 × 356 mm	347.17	435.49	0.87	26.73	nr	**462.22**
1247 × 356 mm	338.20	424.23	0.96	29.49	nr	**453.72**
1727 × 356 mm	655.95	822.82	1.25	38.39	nr	**861.21**
737 × 406 mm	357.24	448.12	0.96	29.49	nr	**477.61**
818 × 406 mm	340.98	427.73	0.87	26.73	nr	**454.46**
978 × 406 mm	281.66	353.32	0.96	29.49	nr	**382.81**
1379 × 406 mm	548.20	687.66	1.16	35.64	nr	**723.30**
1699 × 406 mm	662.88	831.52	1.25	38.39	nr	**869.91**
709 × 457 mm	351.63	441.09	0.87	26.73	nr	**467.82**
1189 × 457 mm	393.04	493.02	0.96	29.49	nr	**522.51**
1671 × 457 mm	954.40	1197.20	1.25	38.39	nr	**1235.59**
678 × 508 mm	357.12	447.97	0.87	26.73	nr	**474.70**
919 × 508 mm	334.68	419.82	0.96	29.49	nr	**449.31**
1321 × 508 mm	638.04	800.35	1.16	35.64	nr	**835.99**
45° radius bend						
345 × 102 mm	92.36	115.85	0.79	24.27	nr	**140.12**
427 × 102 mm	121.22	152.06	0.85	26.12	nr	**178.18**
508 × 102 mm	149.36	187.35	0.95	29.19	nr	**216.54**
559 × 152 mm	173.44	217.56	0.79	24.27	nr	**241.83**
531 × 203 mm	175.32	219.92	0.85	26.12	nr	**246.04**
851 × 203 mm	185.90	233.20	0.98	30.11	nr	**263.31**
582 × 254 mm	184.77	231.77	0.76	23.34	nr	**255.11**
823 × 254 mm	188.68	236.68	0.95	29.19	nr	**265.87**
1303 × 254 mm	352.44	442.10	1.16	35.64	nr	**477.74**
632 × 305 mm	194.28	243.70	0.58	17.81	nr	**261.51**
1275 × 305 mm	356.60	447.32	1.16	35.64	nr	**482.96**
765 × 356 mm	206.65	259.22	0.87	26.73	nr	**285.95**
1247 × 356 mm	354.76	445.01	1.16	35.64	nr	**480.65**
1727 × 356 mm	627.27	786.84	1.26	38.71	nr	**825.55**
737 × 406 mm	211.69	265.54	0.69	21.20	nr	**286.74**
818 × 406 mm	205.56	257.86	0.78	23.96	nr	**281.82**
978 × 406 mm	179.00	224.54	0.87	26.73	nr	**251.27**
1379 × 406 mm	508.17	637.45	1.16	35.64	nr	**673.09**
1699 × 406 mm	631.02	791.55	1.27	39.01	nr	**830.56**
709 × 457 mm	208.58	261.64	0.81	24.89	nr	**286.53**
1189 × 457 mm	378.89	475.28	0.95	29.19	nr	**504.47**
1671 × 457 mm	668.40	838.44	1.26	38.71	nr	**877.15**
678 × 508 mm	211.62	265.45	0.92	28.26	nr	**293.71**
919 × 508 mm	205.50	257.78	1.10	33.79	nr	**291.57**
1321 × 508 mm	519.44	651.58	1.25	38.39	nr	**689.97**
90° hard bend with turning vanes						
345 × 102 mm	79.23	99.39	0.55	16.90	nr	**116.29**
427 × 102 mm	136.36	171.05	1.16	35.64	nr	**206.69**
508 × 102 mm	185.08	232.16	1.16	35.64	nr	**267.80**
559 × 152 mm	208.08	261.02	1.16	35.64	nr	**296.66**
531 × 203 mm	208.48	261.52	1.73	53.14	nr	**314.66**
851 × 203 mm	304.59	382.08	1.73	53.14	nr	**435.22**
582 × 254 mm	223.30	280.11	1.73	53.14	nr	**333.25**

38 VENTILATION/AIR CONDITIONING SYSTEMS

Item	Net Price £	Material £	Labour hours	Labour £	Unit	Total rate £
DUCTWORK: FLAT OVAL – cont						
Extra over fittings – cont						
90° hard bend with turning vanes – cont						
823 × 254 mm	304.59	382.08	1.73	53.14	nr	**435.22**
1303 × 254 mm	540.20	677.62	1.82	55.91	nr	**733.53**
632 × 305 mm	238.05	298.61	1.73	53.14	nr	**351.75**
1275 × 305 mm	536.76	673.31	1.82	55.91	nr	**729.22**
765 × 356 mm	290.32	364.18	1.73	53.14	nr	**417.32**
1247 × 356 mm	537.09	673.72	1.82	55.91	nr	**729.63**
1727 × 356 mm	709.33	889.78	1.82	55.91	nr	**945.69**
737 × 406 mm	290.32	364.18	1.73	53.14	nr	**417.32**
818 × 406 mm	329.58	413.43	1.73	53.14	nr	**466.57**
978 × 406 mm	380.47	477.27	1.73	53.14	nr	**530.41**
1379 × 406 mm	591.96	742.56	1.82	55.91	nr	**798.47**
1699 × 406 mm	709.61	890.13	2.11	64.81	nr	**954.94**
709 × 457 mm	290.33	364.19	1.73	53.14	nr	**417.33**
1189 × 457 mm	533.88	669.70	1.82	55.91	nr	**725.61**
1671 × 457 mm	704.33	883.51	2.11	64.81	nr	**948.32**
678 × 508 mm	290.32	364.18	1.82	55.91	nr	**420.09**
919 × 508 mm	380.47	477.27	1.82	55.91	nr	**533.18**
1321 × 508 mm	591.47	741.94	2.11	64.81	nr	**806.75**
90° branch						
345 × 102 mm	149.39	187.40	0.58	17.81	nr	**205.21**
427 × 102 mm	151.50	190.04	0.58	17.81	nr	**207.85**
508 × 102 mm	159.85	200.51	1.16	35.64	nr	**236.15**
559 × 152 mm	186.42	233.84	1.16	35.64	nr	**269.48**
531 × 203 mm	188.81	236.85	1.16	35.64	nr	**272.49**
851 × 203 mm	276.56	346.92	1.73	53.14	nr	**400.06**
582 × 254 mm	198.33	248.79	1.73	53.14	nr	**301.93**
823 × 254 mm	280.08	351.33	1.73	53.14	nr	**404.47**
1303 × 254 mm	384.27	482.03	1.82	55.91	nr	**537.94**
632 × 305 mm	208.07	261.00	1.73	53.14	nr	**314.14**
1275 × 305 mm	392.08	491.83	1.82	55.91	nr	**547.74**
765 × 356 mm	304.96	382.55	1.73	53.14	nr	**435.69**
1247 × 356 mm	388.75	487.65	1.82	55.91	nr	**543.56**
1727 × 356 mm	582.91	731.20	2.11	64.81	nr	**796.01**
737 × 406 mm	311.84	391.17	1.73	53.14	nr	**444.31**
818 × 406 mm	348.29	436.89	1.73	53.14	nr	**490.03**
978 × 406 mm	306.73	384.76	1.82	55.91	nr	**440.67**
1379 × 406 mm	392.46	492.31	1.93	59.29	nr	**551.60**
1699 × 406 mm	587.31	736.72	2.11	64.81	nr	**801.53**
709 × 457 mm	307.36	385.55	1.73	53.14	nr	**438.69**
1189 × 457 mm	423.37	531.07	1.82	55.91	nr	**586.98**
1671 × 457 mm	817.11	1024.98	2.11	64.81	nr	**1089.79**
678 × 508 mm	310.84	389.92	1.73	53.14	nr	**443.06**
919 × 508 mm	343.06	430.34	1.82	55.91	nr	**486.25**
1321 × 508 mm	496.73	623.10	2.11	64.81	nr	**687.91**

38 VENTILATION/AIR CONDITIONING SYSTEMS

Item	Net Price £	Material £	Labour hours	Labour £	Unit	Total rate £
45° branch						
345 × 102 mm	94.53	118.57	0.58	17.81	nr	136.38
427 × 102 mm	162.99	204.46	0.58	17.81	nr	222.27
508 × 102 mm	221.36	277.67	1.16	35.64	nr	313.31
559 × 152 mm	249.33	312.76	1.73	53.14	nr	365.90
531 × 203 mm	249.34	312.77	1.73	53.14	nr	365.91
851 × 203 mm	347.09	435.39	1.73	53.14	nr	488.53
582 × 254 mm	267.00	334.92	1.73	53.14	nr	388.06
823 × 254 mm	347.10	435.40	1.82	55.91	nr	491.31
1303 × 254 mm	605.48	759.52	1.92	58.98	nr	818.50
632 × 305 mm	284.60	357.00	1.73	53.14	nr	410.14
1275 × 305 mm	602.04	755.19	1.82	55.91	nr	811.10
765 × 356 mm	347.10	435.40	1.73	53.14	nr	488.54
1247 × 356 mm	602.37	755.61	1.82	55.91	nr	811.52
1727 × 356 mm	796.45	999.06	1.82	55.91	nr	1054.97
737 × 406 mm	347.10	435.40	1.73	53.14	nr	488.54
818 × 406 mm	375.57	471.12	1.73	53.14	nr	524.26
978 × 406 mm	433.56	543.86	1.73	53.14	nr	597.00
1379 × 406 mm	664.52	833.57	1.93	59.29	nr	892.86
1699 × 406 mm	796.74	999.43	2.19	67.28	nr	1066.71
709 × 457 mm	347.11	435.41	1.73	53.14	nr	488.55
1189 × 457 mm	599.16	751.59	1.82	55.91	nr	807.50
1671 × 457 mm	791.46	992.81	2.11	64.81	nr	1057.62
678 × 508 mm	347.10	435.40	1.73	53.14	nr	488.54
919 × 508 mm	433.55	543.85	1.82	55.91	nr	599.76
1321 × 508 mm	664.02	832.94	1.93	59.29	nr	892.23

**For rates for access doors refer to
ancillaries in Ductwork Ancillaries: Access
Hatches**

38 VENTILATION/AIR CONDITIONING SYSTEMS

Item	Net Price £	Material £	Labour hours	Labour £	Unit	Total rate £
DUCTWORK: FLEXIBLE						
AIR DUCTLINES						
Aluminium foil flexible ductwork, DW 144 class B; multiply aluminium polyester laminate fabric, with high tensile steel wire helix						
Duct						
102 mm dia.	1.51	1.89	0.33	10.14	m	**12.03**
152 mm dia.	2.22	2.79	0.33	10.14	m	**12.93**
203 mm dia.	2.91	3.65	0.33	10.14	m	**13.79**
254 mm dia.	3.66	4.59	0.33	10.14	m	**14.73**
304 mm dia.	4.48	5.62	0.33	10.14	m	**15.76**
355 mm dia.	6.10	7.65	0.33	10.14	m	**17.79**
406 mm dia.	6.79	8.51	0.33	10.14	m	**18.65**
Insulated aluminium foil flexible ductwork, DW144 class B; laminate construction of aluminium and polyester multiply inner core with 25 mm insulation; outer layer of multiply aluminium polyester laminate, with high tensile steel wire helix						
Duct						
102 mm dia.	3.38	4.24	0.50	15.36	m	**19.60**
152 mm dia.	4.42	5.54	0.50	15.36	m	**20.90**
203 mm dia.	5.36	6.72	0.50	15.36	m	**22.08**
254 mm dia.	6.59	8.27	0.50	15.36	m	**23.63**
304 mm dia.	8.48	10.64	0.50	15.36	m	**26.00**
355 mm dia.	10.52	13.19	0.50	15.36	m	**28.55**
406 mm dia.	11.65	14.62	0.50	15.36	m	**29.98**

38 VENTILATION/AIR CONDITIONING SYSTEMS

Item	Net Price £	Material £	Labour hours	Labour £	Unit	Total rate £
DUCTWORK: PLASTIC						
AIR DUCTLINES						
Rigid grey PVC DW 154 circular section ductwork; solvent welded or filler rod welded joints; excludes couplers and supports (these are detailed separately); ductwork to conform to curent HSE regulations						
Straight duct (standard length 6 m)						
110 mm	4.32	5.42	1.74	53.45	m	58.87
160 mm	8.15	10.23	1.71	52.54	m	62.77
200 mm	10.72	13.45	1.79	54.99	m	68.44
225 mm	15.28	19.16	1.96	60.21	m	79.37
250 mm	15.03	18.85	1.98	60.83	m	79.68
315 mm	21.50	26.97	2.23	68.50	m	95.47
355 mm	28.78	36.10	2.25	69.12	m	105.22
400 mm	36.18	45.38	2.39	73.42	m	118.80
450 mm	44.72	56.10	2.95	90.63	m	146.73
500 mm	55.06	69.07	2.98	91.55	m	160.62
600 mm	85.12	106.77	3.12	95.85	m	202.62
Extra for supports (BZP finish); Horizontal – Maximum 2.4 m centres; Vertical – Maximum 4.0 m centres						
Duct size						
110 mm	26.69	33.48	0.59	18.12	m	51.60
160 mm	26.78	33.59	0.59	18.12	m	51.71
200 mm	26.87	33.70	0.59	18.12	m	51.82
225 mm	28.53	35.78	0.63	19.35	m	55.13
250 mm	28.54	35.80	0.63	19.35	m	55.15
315 mm	28.74	36.05	0.63	19.35	m	55.40
355 mm	27.49	34.48	0.81	24.89	m	59.37
400 mm	37.33	46.83	0.78	23.96	m	70.79
450 mm	38.37	48.13	0.80	24.57	m	72.70
500 mm	38.55	48.36	0.80	24.57	m	72.93
600 mm	40.93	51.34	0.85	26.12	m	77.46
Note: These are maximum figures and may be reduced subject to local conditions (i.e. a high number of changes of direction)						
Extra over fittings; Rigid grey PVC						
90° bend						
110 mm	22.08	27.70	1.11	34.09	m	61.79
160 mm	24.37	30.56	1.34	41.17	m	71.73
200 mm	20.39	25.58	2.00	61.44	m	87.02
225 mm	31.30	39.27	1.79	54.99	m	94.26
250 mm	33.32	41.80	2.01	61.75	m	103.55
315 mm	57.53	72.16	2.56	78.64	m	150.80
355 mm	68.50	85.93	2.71	83.25	m	169.18

38 VENTILATION/AIR CONDITIONING SYSTEMS

Item	Net Price £	Material £	Labour hours	Labour £	Unit	Total rate £
DUCTWORK: PLASTIC – cont						
Extra over fittings – cont						
90° bend – cont						
400 mm	87.01	109.14	3.58	109.97	m	**219.11**
450 mm	263.00	329.91	4.55	139.78	m	**469.69**
500 mm	305.25	382.91	5.01	153.90	m	**536.81**
600 mm	522.52	655.45	5.78	177.56	m	**833.01**
45° bend						
110 mm	9.10	11.41	0.71	21.82	m	**33.23**
160 mm	10.56	13.25	0.93	28.57	m	**41.82**
200 mm	11.60	14.55	1.14	35.02	m	**49.57**
225 mm	13.08	16.41	1.41	43.31	m	**59.72**
250 mm	14.01	17.57	1.62	49.76	m	**67.33**
315 mm	22.36	28.04	1.93	59.29	m	**87.33**
355 mm	29.10	36.50	2.22	68.20	m	**104.70**
400 mm	33.45	41.96	2.85	87.56	m	**129.52**
450 mm	128.84	161.62	3.77	115.82	m	**277.44**
500 mm	144.06	180.71	4.16	127.80	m	**308.51**
600 mm	234.70	294.40	4.81	147.77	m	**442.17**
Tee						
110 mm	19.29	24.19	1.04	31.95	m	**56.14**
160 mm	23.54	29.52	1.38	42.40	m	**71.92**
200 mm	31.30	39.27	1.79	54.99	m	**94.26**
225 mm	36.07	45.25	2.25	69.12	m	**114.37**
250 mm	44.14	55.37	2.62	80.48	m	**135.85**
315 mm	73.53	92.23	3.17	97.38	m	**189.61**
355 mm	102.85	129.01	3.73	114.59	m	**243.60**
400 mm	100.26	125.76	4.71	144.69	m	**270.45**
450 mm	197.15	247.31	5.72	175.72	m	**423.03**
500 mm	224.06	281.06	6.33	194.45	m	**475.51**
Coupler						
110 mm	5.15	6.46	0.70	21.50	m	**27.96**
160 mm	6.39	8.02	0.91	27.96	m	**35.98**
200 mm	7.92	9.93	1.12	34.41	m	**44.34**
225 mm	10.37	13.00	1.41	43.31	m	**56.31**
250 mm	11.55	14.49	1.60	49.15	m	**63.64**
315 mm	13.77	17.27	1.86	57.14	m	**74.41**
355 mm	14.47	18.16	2.43	74.65	m	**92.81**
400 mm	15.41	19.33	2.66	81.72	m	**101.05**
450 mm	37.09	46.52	3.20	98.30	m	**144.82**
500 mm	41.85	52.49	3.52	108.12	m	**160.61**
Damper						
110 mm	43.20	54.19	0.96	29.49	m	**83.68**
160 mm	43.86	55.01	1.26	38.71	m	**93.72**
200 mm	44.79	56.18	1.52	46.69	m	**102.87**
225 mm	46.30	58.08	1.88	57.75	m	**115.83**
250 mm	46.13	57.87	2.16	66.36	m	**124.23**
315 mm	52.92	66.38	2.94	90.32	m	**156.70**
355 mm	56.53	70.91	3.42	105.07	m	**175.98**
400 mm	60.69	76.13	3.79	116.44	m	**192.57**

38 VENTILATION/AIR CONDITIONING SYSTEMS

Item	Net Price £	Material £	Labour hours	Labour £	Unit	Total rate £
Reducer						
160 × 110 mm	11.29	14.16	0.85	26.12	m	40.28
200 × 110 mm	16.43	20.61	0.98	30.11	m	50.72
200 × 160 mm	14.24	17.86	1.07	32.86	m	50.72
225 × 200 mm	18.68	23.43	1.38	42.40	m	65.83
250 × 160 mm	19.75	24.77	1.38	42.40	m	67.17
250 × 200 mm	18.75	23.52	1.48	45.47	m	68.99
250 × 225 mm	18.41	23.09	1.58	48.53	m	71.62
315 × 200 mm	20.09	25.20	1.63	50.08	m	75.28
315 × 250 mm	20.66	25.92	1.84	56.53	m	82.45
355 × 200 mm	30.62	38.40	1.99	61.14	m	99.54
355 × 250 mm	22.84	28.65	1.97	60.52	m	89.17
355 × 315 mm	28.46	35.71	2.16	66.36	m	102.07
400 × 225 mm	34.56	43.36	2.47	75.88	m	119.24
400 × 315 mm	34.91	43.79	2.75	84.48	m	128.27
400 × 355 mm	36.76	46.11	2.73	83.87	m	129.98
450 × 315 mm	39.68	49.77	2.91	89.40	m	139.17
Flange						
110 mm	7.45	9.34	0.70	21.50	m	30.84
160 mm	7.54	9.45	0.91	27.96	m	37.41
200 mm	8.13	10.20	1.12	34.41	m	44.61
225 mm	7.90	9.91	1.35	41.47	m	51.38
250 mm	8.43	10.57	1.55	47.62	m	58.19
315 mm	12.02	15.08	1.85	56.84	m	71.92
355 mm	12.72	15.96	2.06	63.28	m	79.24
400 mm	12.75	15.99	2.62	80.48	m	96.47
Polypropylene (PPS) DW154 circular section ductwork; filler rod welded joints; excludes couplers and supports (these are detailed separately); ductwork to conform to current HSE regulations						
Straight duct (standard length 6 m)						
110 mm	20.21	25.36	0.65	19.97	m	45.33
160 mm	22.15	27.79	0.75	23.04	m	50.83
200 mm	23.51	29.49	0.85	26.12	m	55.61
225 mm	30.15	37.82	1.07	32.86	m	70.68
250 mm	30.71	38.53	1.18	36.25	m	74.78
315 mm	56.52	70.90	1.41	43.31	m	114.21
355 mm	60.04	75.31	1.53	47.01	m	122.32
400 mm	73.27	91.91	1.73	53.14	m	145.05
Extra for supports (BZP finish); Horizontal – Maximum 2.4 m centre; Vertical – Maximum 4.0 m centre						
Duct size						
110 mm	36.99	46.40	0.42	12.90	m	59.30
160 mm	36.99	46.40	0.42	12.90	m	59.30
200 mm	36.99	46.40	0.42	12.90	m	59.30

38 VENTILATION/AIR CONDITIONING SYSTEMS

Item	Net Price £	Material £	Labour hours	Labour £	Unit	Total rate £
DUCTWORK: PLASTIC – cont						
Extra for supports (BZP finish) – cont						
Duct size – cont						
225 mm	36.99	46.40	0.42	12.90	m	**59.30**
250 mm	36.99	46.40	0.42	12.90	m	**59.30**
315 mm	36.99	46.40	0.42	12.90	m	**59.30**
355 mm	36.99	46.40	0.55	16.90	m	**63.30**
400 mm	52.36	65.68	0.55	16.90	m	**82.58**
Note: These are maximum figures and may be reduced subject to local conditions (i.e. a high number of changes of direction)						
Extra over fittings; Polypropylene (DW 154)						
90° bend						
110 mm	15.14	19.00	1.01	31.02	m	**50.02**
160 mm	18.87	23.67	1.30	39.94	m	**63.61**
200 mm	20.73	26.01	1.58	48.53	m	**74.54**
225 mm	28.90	36.25	2.10	64.51	m	**100.76**
250 mm	29.30	36.76	2.36	72.51	m	**109.27**
315 mm	87.36	109.58	3.14	96.47	m	**206.05**
355 mm	98.12	123.08	3.52	108.12	m	**231.20**
400 mm	96.15	120.61	4.39	134.86	m	**255.47**
45° bend						
110 mm	9.45	11.85	0.84	25.80	m	**37.65**
160 mm	13.78	17.28	1.13	34.71	m	**51.99**
200 mm	14.66	18.39	1.39	42.69	m	**61.08**
225 mm	17.98	22.56	1.73	53.14	m	**75.70**
250 mm	19.35	24.27	1.99	61.14	m	**85.41**
315 mm	54.85	68.80	2.57	78.95	m	**147.75**
355 mm	61.14	76.70	2.93	90.01	m	**166.71**
400 mm	58.67	73.60	3.60	110.59	m	**184.19**
Tee						
110 mm	42.26	53.01	1.55	47.62	m	**100.63**
160 mm	52.33	65.64	2.17	66.66	m	**132.30**
200 mm	60.51	75.90	2.68	82.33	m	**158.23**
225 mm	72.06	90.40	3.38	103.84	m	**194.24**
250 mm	83.64	104.92	3.95	121.35	m	**226.27**
315 mm	118.29	148.38	4.69	144.08	m	**292.46**
355 mm	147.89	185.52	5.44	167.13	m	**352.65**
400 mm	172.80	216.76	6.09	187.08	m	**403.84**
Coupler						
110 mm	9.31	11.68	0.84	25.80	m	**37.48**
160 mm	10.04	12.59	1.10	33.79	m	**46.38**
200 mm	11.24	14.10	1.36	41.79	m	**55.89**
225 mm	12.12	15.20	1.67	51.31	m	**66.51**
250 mm	12.29	15.41	1.91	58.68	m	**74.09**
315 mm	16.97	21.29	2.24	68.81	m	**90.10**
355 mm	21.82	27.37	2.58	79.25	m	**106.62**
400 mm	22.29	27.96	3.30	101.38	m	**129.34**

38 VENTILATION/AIR CONDITIONING SYSTEMS

Item	Net Price £	Material £	Labour hours	Labour £	Unit	Total rate £
Damper						
110 mm	88.26	110.71	0.79	24.27	m	**134.98**
160 mm	100.27	125.78	1.13	34.71	m	**160.49**
200 mm	108.01	135.49	1.44	44.24	m	**179.73**
225 mm	113.43	142.28	1.81	55.61	m	**197.89**
250 mm	116.02	145.53	2.08	63.90	m	**209.43**
315 mm	132.43	166.12	2.40	73.73	m	**239.85**
355 mm	143.23	179.67	2.75	84.48	m	**264.15**
400 mm	149.57	187.62	3.57	109.67	m	**297.29**
Reducer						
160 × 110 mm	31.81	39.91	0.87	26.73	m	**66.64**
200 × 160 mm	32.93	41.31	1.16	35.64	m	**76.95**
225 × 200 mm	38.39	48.16	1.46	44.86	m	**93.02**
250 × 200 mm	45.23	56.74	1.61	49.46	m	**106.20**
250 × 225 mm	55.43	69.53	1.70	52.23	m	**121.76**
315 × 200 mm	85.27	106.96	1.66	50.99	m	**157.95**
315 × 250 mm	77.11	96.72	2.03	62.36	m	**159.08**
355 × 200 mm	92.06	115.48	2.08	63.90	m	**179.38**
355 × 250 mm	97.98	122.91	2.05	62.97	m	**185.88**
355 × 315 mm	112.06	140.57	2.22	68.20	m	**208.77**
400 × 315 mm	133.19	167.07	2.88	88.47	m	**255.54**
400 × 355 mm	142.54	178.80	2.92	89.70	m	**268.50**
Flange						
110 mm	17.90	22.46	0.76	23.34	m	**45.80**
160 mm	20.11	25.22	1.02	31.34	m	**56.56**
200 mm	22.24	27.90	1.27	39.01	m	**66.91**
225 mm	22.70	28.47	1.55	47.62	m	**76.09**
250 mm	24.52	30.76	1.79	54.99	m	**85.75**
315 mm	27.72	34.78	2.08	63.90	m	**98.68**
355 mm	30.32	38.04	2.36	72.51	m	**110.55**
400 mm	32.21	40.41	3.07	94.30	m	**134.71**

38 VENTILATION/AIR CONDITIONING SYSTEMS

Item	Net Price £	Material £	Labour hours	Labour £	Unit	Total rate £
DUCTWORK: RECTANGULAR – CLASS B						
AIR DUCTLINES						
Galvanized sheet metal DW144 class B rectangular section ductwork; including all necessary stiffeners, joints, couplers in the running length and duct supports						
Ductwork up to 400 mm longest side						
Sum of two sides 200 mm	24.15	30.30	1.16	35.64	m	**65.94**
Sum of two sides 300 mm	25.51	32.00	1.16	35.64	m	**67.64**
Sum of two sides 400 mm	22.01	27.61	1.19	36.56	m	**64.17**
Sum of two sides 500 mm	23.90	29.98	1.19	36.56	m	**66.54**
Sum of two sides 600 mm	25.55	32.05	1.27	39.01	m	**71.06**
Sum of two sides 700 mm	27.19	34.10	1.27	39.01	m	**73.11**
Sum of two sides 800 mm	28.95	36.31	1.27	39.01	m	**75.32**
Extra over fittings; Rectangular ductwork class B; upto 400 mm longest side						
End cap						
Sum of two sides 200 mm	14.72	18.47	0.38	11.68	nr	**30.15**
Sum of two sides 300 mm	16.40	20.57	0.38	11.68	nr	**32.25**
Sum of two sides 400 mm	18.08	22.68	0.38	11.68	nr	**34.36**
Sum of two sides 500 mm	19.76	24.79	0.38	11.68	nr	**36.47**
Sum of two sides 600 mm	21.44	26.89	0.38	11.68	nr	**38.57**
Sum of two sides 700 mm	23.12	29.00	0.38	11.68	nr	**40.68**
Sum of two sides 800 mm	24.81	31.12	0.38	11.68	nr	**42.80**
Reducer						
Sum of two sides 200 mm	24.03	30.14	1.40	43.01	nr	**73.15**
Sum of two sides 300 mm	27.14	34.05	1.40	43.01	nr	**77.06**
Sum of two sides 400 mm	44.44	55.74	1.42	43.62	nr	**99.36**
Sum of two sides 500 mm	48.11	60.35	1.42	43.62	nr	**103.97**
Sum of two sides 600 mm	51.77	64.94	1.69	51.92	nr	**116.86**
Sum of two sides 700 mm	55.43	69.53	1.69	51.92	nr	**121.45**
Sum of two sides 800 mm	59.06	74.09	1.92	58.98	nr	**133.07**
Offset						
Sum of two sides 200 mm	35.47	44.50	1.63	50.08	nr	**94.58**
Sum of two sides 300 mm	40.09	50.29	1.63	50.08	nr	**100.37**
Sum of two sides 400 mm	59.30	74.39	1.65	50.69	nr	**125.08**
Sum of two sides 500 mm	64.47	80.88	1.65	50.69	nr	**131.57**
Sum of two sides 600 mm	68.70	86.17	1.92	58.98	nr	**145.15**
Sum of two sides 700 mm	73.40	92.08	1.92	58.98	nr	**151.06**
Sum of two sides 800 mm	77.50	97.22	1.92	58.98	nr	**156.20**
Square to round						
Sum of two sides 200 mm	31.19	39.12	1.63	50.08	nr	**89.20**
Sum of two sides 300 mm	35.05	43.97	1.63	50.08	nr	**94.05**
Sum of two sides 400 mm	46.67	58.54	1.65	50.69	nr	**109.23**
Sum of two sides 500 mm	50.78	63.69	1.65	50.69	nr	**114.38**
Sum of two sides 600 mm	54.90	68.87	1.92	58.98	nr	**127.85**
Sum of two sides 700 mm	59.02	74.03	1.92	58.98	nr	**133.01**
Sum of two sides 800 mm	63.10	79.15	1.92	58.98	nr	**138.13**

38 VENTILATION/AIR CONDITIONING SYSTEMS

Item	Net Price £	Material £	Labour hours	Labour £	Unit	Total rate £
90° radius bend						
Sum of two sides 200 mm	23.47	29.44	1.22	37.49	nr	66.93
Sum of two sides 300 mm	25.32	31.76	1.22	37.49	nr	69.25
Sum of two sides 400 mm	42.67	53.52	1.25	38.39	nr	91.91
Sum of two sides 500 mm	45.40	56.95	1.25	38.39	nr	95.34
Sum of two sides 600 mm	49.02	61.49	1.33	40.86	nr	102.35
Sum of two sides 700 mm	52.18	65.45	1.33	40.86	nr	106.31
Sum of two sides 800 mm	55.74	69.92	1.40	43.01	nr	112.93
45° radius bend						
Sum of two sides 200 mm	25.34	31.79	0.89	27.35	nr	59.14
Sum of two sides 300 mm	27.82	34.90	1.12	34.41	nr	69.31
Sum of two sides 400 mm	44.72	56.10	1.10	33.79	nr	89.89
Sum of two sides 500 mm	47.92	60.11	1.10	33.79	nr	93.90
Sum of two sides 600 mm	51.58	64.70	1.16	35.64	nr	100.34
Sum of two sides 700 mm	55.00	68.99	1.16	35.64	nr	104.63
Sum of two sides 800 mm	58.61	73.52	1.22	37.49	nr	111.01
90° mitre bend						
Sum of two sides 200 mm	39.23	49.21	1.29	39.63	nr	88.84
Sum of two sides 300 mm	43.10	54.06	1.29	39.63	nr	93.69
Sum of two sides 400 mm	62.67	78.61	1.29	39.63	nr	118.24
Sum of two sides 500 mm	67.68	84.90	1.29	39.63	nr	124.53
Sum of two sides 600 mm	74.11	92.96	1.39	42.69	nr	135.65
Sum of two sides 700 mm	80.08	100.45	1.39	42.69	nr	143.14
Sum of two sides 800 mm	86.60	108.63	1.46	44.86	nr	153.49
Branch						
Sum of two sides 200 mm	38.85	48.73	0.92	28.26	nr	76.99
Sum of two sides 300 mm	43.18	54.16	0.92	28.26	nr	82.42
Sum of two sides 400 mm	54.16	67.94	0.95	29.19	nr	97.13
Sum of two sides 500 mm	58.99	74.00	0.95	29.19	nr	103.19
Sum of two sides 600 mm	63.69	79.89	1.03	31.64	nr	111.53
Sum of two sides 700 mm	68.39	85.79	1.03	31.64	nr	117.43
Sum of two sides 800 mm	73.09	91.68	1.03	31.64	nr	123.32
Grille neck						
Sum of two sides 200 mm	44.56	55.90	1.10	33.79	nr	89.69
Sum of two sides 300 mm	49.89	62.59	1.10	33.79	nr	96.38
Sum of two sides 400 mm	55.23	69.28	1.16	35.64	nr	104.92
Sum of two sides 500 mm	60.56	75.97	1.16	35.64	nr	111.61
Sum of two sides 600 mm	65.90	82.67	1.18	36.25	nr	118.92
Sum of two sides 700 mm	71.23	89.35	1.18	36.25	nr	125.60
Sum of two sides 800 mm	76.56	96.04	1.18	36.25	nr	132.29
Ductwork 401–600 mm longest side						
Sum of two sides 600 mm	29.79	37.36	1.27	39.01	m	76.37
Sum of two sides 700 mm	32.18	40.36	1.27	39.01	m	79.37
Sum of two sides 800 mm	34.51	43.29	1.27	39.01	m	82.30
Sum of two sides 900 mm	36.72	46.07	1.27	39.01	m	85.08
Sum of two sides 1000 mm	38.94	48.84	1.37	42.09	m	90.93
Sum of two sides 1100 mm	41.43	51.97	1.37	42.09	m	94.06
Sum of two sides 1200 mm	43.64	54.75	1.37	42.09	m	96.84

38 VENTILATION/AIR CONDITIONING SYSTEMS

Item	Net Price £	Material £	Labour hours	Labour £	Unit	Total rate £
DUCTWORK: RECTANGULAR – CLASS B – cont						
Extra over fittings; Ductwork 401–600 mm longest side						
End cap						
Sum of two sides 600 mm	22.05	27.66	0.38	11.68	nr	39.34
Sum of two sides 700 mm	23.80	29.86	0.38	11.68	nr	41.54
Sum of two sides 800 mm	25.55	32.05	0.38	11.68	nr	43.73
Sum of two sides 900 mm	27.30	34.25	0.58	17.81	nr	52.06
Sum of two sides 1000 mm	29.06	36.46	0.58	17.81	nr	54.27
Sum of two sides 1100 mm	30.81	38.65	0.58	17.81	nr	56.46
Sum of two sides 1200 mm	32.56	40.85	0.58	17.81	nr	58.66
Reducer						
Sum of two sides 600 mm	50.93	63.88	1.69	51.92	nr	115.80
Sum of two sides 700 mm	54.61	68.50	1.69	51.92	nr	120.42
Sum of two sides 800 mm	58.23	73.05	1.92	58.98	nr	132.03
Sum of two sides 900 mm	61.90	77.65	1.92	58.98	nr	136.63
Sum of two sides 1000 mm	65.58	82.26	2.18	66.98	nr	149.24
Sum of two sides 1100 mm	69.56	87.26	2.18	66.98	nr	154.24
Sum of two sides 1200 mm	73.24	91.87	2.18	66.98	nr	158.85
Offset						
Sum of two sides 600 mm	70.82	88.84	1.92	58.98	nr	147.82
Sum of two sides 700 mm	76.04	95.38	1.92	58.98	nr	154.36
Sum of two sides 800 mm	80.18	100.58	1.92	58.98	nr	159.56
Sum of two sides 900 mm	84.36	105.82	1.92	58.98	nr	164.80
Sum of two sides 1000 mm	89.03	111.68	2.18	66.98	nr	178.66
Sum of two sides 1100 mm	93.36	117.11	2.18	66.98	nr	184.09
Sum of two sides 1200 mm	97.44	122.23	2.18	66.98	nr	189.21
Square to round						
Sum of two sides 600 mm	53.89	67.60	1.33	40.86	nr	108.46
Sum of two sides 700 mm	58.03	72.79	1.33	40.86	nr	113.65
Sum of two sides 800 mm	62.11	77.91	1.40	43.01	nr	120.92
Sum of two sides 900 mm	66.26	83.12	1.40	43.01	nr	126.13
Sum of two sides 1000 mm	70.40	88.31	1.82	55.91	nr	144.22
Sum of two sides 1100 mm	74.62	93.60	1.82	55.91	nr	149.51
Sum of two sides 1200 mm	78.76	98.80	1.82	55.91	nr	154.71
90° radius bend						
Sum of two sides 600 mm	49.62	62.24	1.16	35.64	nr	97.88
Sum of two sides 700 mm	52.36	65.68	1.16	35.64	nr	101.32
Sum of two sides 800 mm	56.43	70.78	1.22	37.49	nr	108.27
Sum of two sides 900 mm	60.60	76.01	1.22	37.49	nr	113.50
Sum of two sides 1000 mm	64.07	80.37	1.40	43.01	nr	123.38
Sum of two sides 1100 mm	68.46	85.88	1.40	43.01	nr	128.89
Sum of two sides 1200 mm	72.65	91.13	1.40	43.01	nr	134.14
45° bend						
Sum of two sides 600 mm	52.54	65.90	1.16	35.64	nr	101.54
Sum of two sides 700 mm	55.78	69.97	1.39	42.69	nr	112.66
Sum of two sides 800 mm	59.72	74.92	1.46	44.86	nr	119.78

38 VENTILATION/AIR CONDITIONING SYSTEMS

Item	Net Price £	Material £	Labour hours	Labour £	Unit	Total rate £
Sum of two sides 900 mm	63.73	79.95	1.46	44.86	nr	124.81
Sum of two sides 1000 mm	67.36	84.49	1.88	57.75	nr	142.24
Sum of two sides 1100 mm	71.67	89.90	1.88	57.75	nr	147.65
Sum of two sides 1200 mm	75.68	94.93	1.88	57.75	nr	152.68
90° mitre bend						
Sum of two sides 600 mm	83.45	104.68	1.39	42.69	nr	147.37
Sum of two sides 700 mm	88.41	110.90	2.16	66.36	nr	177.26
Sum of two sides 800 mm	95.57	119.88	2.26	69.43	nr	189.31
Sum of two sides 900 mm	102.82	128.98	2.26	69.43	nr	198.41
Sum of two sides 1000 mm	109.35	137.17	3.01	92.47	nr	229.64
Sum of two sides 1100 mm	116.96	146.72	3.01	92.47	nr	239.19
Sum of two sides 1200 mm	124.40	156.05	3.01	92.47	nr	248.52
Branch						
Sum of two sides 600 mm	65.77	82.50	1.03	31.64	nr	114.14
Sum of two sides 700 mm	70.69	88.67	1.03	31.64	nr	120.31
Sum of two sides 800 mm	75.61	94.84	1.03	31.64	nr	126.48
Sum of two sides 900 mm	80.54	101.02	1.03	31.64	nr	132.66
Sum of two sides 1000 mm	85.46	107.21	1.29	39.63	nr	146.84
Sum of two sides 1100 mm	90.60	113.65	1.29	39.63	nr	153.28
Sum of two sides 1200 mm	95.52	119.82	1.29	39.63	nr	159.45
Grille neck						
Sum of two sides 600 mm	68.28	85.65	1.18	36.25	nr	121.90
Sum of two sides 700 mm	73.93	92.74	1.18	36.25	nr	128.99
Sum of two sides 800 mm	79.58	99.83	1.18	36.25	nr	136.08
Sum of two sides 900 mm	85.23	106.92	1.18	36.25	nr	143.17
Sum of two sides 1000 mm	90.88	114.00	1.44	44.24	nr	158.24
Sum of two sides 1100 mm	96.53	121.08	1.44	44.24	nr	165.32
Sum of two sides 1200 mm	102.18	128.17	1.44	44.24	nr	172.41
Ductwork 601–800 mm longest side						
Sum of two sides 900 mm	40.99	51.42	1.27	39.01	m	90.43
Sum of two sides 1000 mm	43.20	54.19	1.37	42.09	m	96.28
Sum of two sides 1100 mm	45.40	56.95	1.37	42.09	m	99.04
Sum of two sides 1200 mm	47.91	60.10	1.37	42.09	m	102.19
Sum of two sides 1300 mm	50.12	62.87	1.40	43.01	m	105.88
Sum of two sides 1400 mm	52.33	65.64	1.40	43.01	m	108.65
Sum of two sides 1500 mm	54.55	68.43	1.48	45.47	m	113.90
Sum of two sides 1600 mm	56.76	71.20	1.55	47.62	m	118.82
Extra over fittings: Ductwork 601–800 mm longest side						
End cap						
Sum of two sides 900 mm	27.30	34.25	0.58	17.81	nr	52.06
Sum of two sides 1000 mm	29.06	36.46	0.58	17.81	nr	54.27
Sum of two sides 1100 mm	30.81	38.65	0.58	17.81	nr	56.46
Sum of two sides 1200 mm	32.56	40.85	0.58	17.81	nr	58.66
Sum of two sides 1300 mm	34.32	43.05	0.58	17.81	nr	60.86
Sum of two sides 1400 mm	35.82	44.93	0.58	17.81	nr	62.74
Sum of two sides 1500 mm	41.62	52.20	0.58	17.81	nr	70.01
Sum of two sides 1600 mm	47.42	59.48	0.58	17.81	nr	77.29

38 VENTILATION/AIR CONDITIONING SYSTEMS

Item	Net Price £	Material £	Labour hours	Labour £	Unit	Total rate £
DUCTWORK: RECTANGULAR – CLASS B – cont						
Extra over fittings – cont						
Reducer						
Sum of two sides 900 mm	62.49	78.39	1.92	58.98	nr	**137.37**
Sum of two sides 1000 mm	66.17	83.00	2.18	66.98	nr	**149.98**
Sum of two sides 1100 mm	69.85	87.62	2.18	66.98	nr	**154.60**
Sum of two sides 1200 mm	73.81	92.59	2.18	66.98	nr	**159.57**
Sum of two sides 1300 mm	77.49	97.20	2.30	70.66	nr	**167.86**
Sum of two sides 1400 mm	80.67	101.19	2.30	70.66	nr	**171.85**
Sum of two sides 1500 mm	92.45	115.96	2.47	75.88	nr	**191.84**
Sum of two sides 1600 mm	104.22	130.74	2.47	75.88	nr	**206.62**
Offset						
Sum of two sides 900 mm	86.74	108.81	1.92	58.98	nr	**167.79**
Sum of two sides 1000 mm	90.62	113.67	2.18	66.98	nr	**180.65**
Sum of two sides 1100 mm	94.49	118.53	2.18	66.98	nr	**185.51**
Sum of two sides 1200 mm	98.52	123.58	2.18	66.98	nr	**190.56**
Sum of two sides 1300 mm	102.30	128.33	2.30	70.66	nr	**198.99**
Sum of two sides 1400 mm	106.03	133.00	2.30	70.66	nr	**203.66**
Sum of two sides 1500 mm	121.91	152.92	2.47	75.88	nr	**228.80**
Sum of two sides 1600 mm	137.75	172.79	2.47	75.88	nr	**248.67**
Square to round						
Sum of two sides 900 mm	65.72	82.44	1.40	43.01	nr	**125.45**
Sum of two sides 1000 mm	69.86	87.63	1.82	55.91	nr	**143.54**
Sum of two sides 1100 mm	74.01	92.84	1.82	55.91	nr	**148.75**
Sum of two sides 1200 mm	78.23	98.13	1.82	55.91	nr	**154.04**
Sum of two sides 1300 mm	82.37	103.32	2.15	66.05	nr	**169.37**
Sum of two sides 1400 mm	85.89	107.74	2.15	66.05	nr	**173.79**
Sum of two sides 1500 mm	100.15	125.63	2.38	73.11	nr	**198.74**
Sum of two sides 1600 mm	114.40	143.51	2.38	73.11	nr	**216.62**
90° radius bend						
Sum of two sides 900 mm	56.75	71.19	1.22	37.49	nr	**108.68**
Sum of two sides 1000 mm	60.99	76.51	1.40	43.01	nr	**119.52**
Sum of two sides 1100 mm	65.22	81.82	1.40	43.01	nr	**124.83**
Sum of two sides 1200 mm	69.63	87.35	1.40	43.01	nr	**130.36**
Sum of two sides 1300 mm	73.87	92.66	1.91	58.68	nr	**151.34**
Sum of two sides 1400 mm	76.74	96.26	1.91	58.68	nr	**154.94**
Sum of two sides 1500 mm	89.09	111.75	2.11	64.81	nr	**176.56**
Sum of two sides 1600 mm	101.44	127.24	2.11	64.81	nr	**192.05**
45° bend						
Sum of two sides 900 mm	62.64	78.58	1.22	37.49	nr	**116.07**
Sum of two sides 1000 mm	66.68	83.64	1.40	43.01	nr	**126.65**
Sum of two sides 1100 mm	70.71	88.70	1.88	57.75	nr	**146.45**
Sum of two sides 1200 mm	75.04	94.12	1.88	57.75	nr	**151.87**
Sum of two sides 1300 mm	19.07	23.92	2.26	69.43	nr	**93.35**
Sum of two sides 1400 mm	82.16	103.06	2.26	69.43	nr	**172.49**
Sum of two sides 1500 mm	94.29	118.27	2.49	76.50	nr	**194.77**
Sum of two sides 1600 mm	100.43	125.98	2.49	76.50	nr	**202.48**

38 VENTILATION/AIR CONDITIONING SYSTEMS

Item	Net Price £	Material £	Labour hours	Labour £	Unit	Total rate £
90° mitre bend						
Sum of two sides 900 mm	96.83	121.46	1.22	37.49	nr	158.95
Sum of two sides 1000 mm	104.78	131.43	1.40	43.01	nr	174.44
Sum of two sides 1100 mm	112.74	141.42	3.01	92.47	nr	233.89
Sum of two sides 1200 mm	120.83	151.57	3.01	92.47	nr	244.04
Sum of two sides 1300 mm	128.79	161.55	3.67	112.75	nr	274.30
Sum of two sides 1400 mm	135.03	169.38	3.67	112.75	nr	282.13
Sum of two sides 1500 mm	155.28	194.78	4.07	125.03	nr	319.81
Sum of two sides 1600 mm	175.55	220.21	4.07	125.03	nr	345.24
Branch						
Sum of two sides 900 mm	82.91	104.00	1.22	37.49	nr	141.49
Sum of two sides 1000 mm	87.97	110.35	1.40	43.01	nr	153.36
Sum of two sides 1100 mm	93.02	116.68	1.29	39.63	nr	156.31
Sum of two sides 1200 mm	98.30	123.31	1.29	39.63	nr	162.94
Sum of two sides 1300 mm	103.37	129.66	1.39	42.69	nr	172.35
Sum of two sides 1400 mm	107.76	135.17	1.39	42.69	nr	177.86
Sum of two sides 1500 mm	123.60	155.04	1.64	50.38	nr	205.42
Sum of two sides 1600 mm	139.44	174.91	1.64	50.38	nr	225.29
Grille neck						
Sum of two sides 900 mm	85.23	106.92	1.22	37.49	nr	144.41
Sum of two sides 1000 mm	90.88	114.00	1.40	43.01	nr	157.01
Sum of two sides 1100 mm	96.53	121.08	1.44	44.24	nr	165.32
Sum of two sides 1200 mm	102.18	128.17	1.44	44.24	nr	172.41
Sum of two sides 1300 mm	107.83	135.26	1.69	51.92	nr	187.18
Sum of two sides 1400 mm	113.48	142.35	1.69	51.92	nr	194.27
Sum of two sides 1500 mm	119.13	149.44	1.79	54.99	nr	204.43
Sum of two sides 1600 mm	124.78	156.52	1.79	54.99	nr	211.51
Ductwork 801–1000 mm longest side						
Sum of two sides 1100 mm	62.16	77.97	1.37	42.09	m	120.06
Sum of two sides 1200 mm	65.82	82.57	1.37	42.09	m	124.66
Sum of two sides 1300 mm	69.49	87.17	1.40	43.01	m	130.18
Sum of two sides 1400 mm	73.45	92.13	1.40	43.01	m	135.14
Sum of two sides 1500 mm	77.11	96.72	1.48	45.47	m	142.19
Sum of two sides 1600 mm	80.78	101.33	1.55	47.62	m	148.95
Sum of two sides 1700 mm	84.44	105.92	1.55	47.62	m	153.54
Sum of two sides 1800 mm	88.40	110.89	1.61	49.46	m	160.35
Sum of two sides 1900 mm	92.05	115.47	1.61	49.46	m	164.93
Sum of two sides 2000 mm	95.72	120.08	1.61	49.46	m	169.54
Extra over fittings; Ductwork 801–1000 mm longest side						
End cap						
Sum of two sides 1100 mm	30.81	38.65	1.44	44.24	nr	82.89
Sum of two sides 1200 mm	32.56	40.85	1.44	44.24	nr	85.09
Sum of two sides 1300 mm	34.32	43.05	1.44	44.24	nr	87.29
Sum of two sides 1400 mm	35.82	44.93	1.44	44.24	nr	89.17
Sum of two sides 1500 mm	41.62	52.20	1.44	44.24	nr	96.44
Sum of two sides 1600 mm	47.42	59.48	1.44	44.24	nr	103.72
Sum of two sides 1700 mm	53.22	66.76	1.44	44.24	nr	111.00
Sum of two sides 1800 mm	59.02	74.03	1.44	44.24	nr	118.27
Sum of two sides 1900 mm	64.82	81.31	1.44	44.24	nr	125.55
Sum of two sides 2000 mm	70.62	88.58	1.44	44.24	nr	132.82

38 VENTILATION/AIR CONDITIONING SYSTEMS

Item	Net Price £	Material £	Labour hours	Labour £	Unit	Total rate £
DUCTWORK: RECTANGULAR – CLASS B – cont						
Extra over fittings – cont						
Reducer						
Sum of two sides 1100 mm	58.10	72.88	1.44	44.24	nr	**117.12**
Sum of two sides 1200 mm	60.89	76.38	1.44	44.24	nr	**120.62**
Sum of two sides 1300 mm	63.69	79.89	1.69	51.92	nr	**131.81**
Sum of two sides 1400 mm	66.24	83.09	1.69	51.92	nr	**135.01**
Sum of two sides 1500 mm	77.12	96.73	2.47	75.88	nr	**172.61**
Sum of two sides 1600 mm	88.01	110.40	2.47	75.88	nr	**186.28**
Sum of two sides 1700 mm	98.89	124.05	2.47	75.88	nr	**199.93**
Sum of two sides 1800 mm	110.03	138.02	2.59	79.56	nr	**217.58**
Sum of two sides 1900 mm	120.91	151.67	2.71	83.25	nr	**234.92**
Sum of two sides 2000 mm	131.81	165.35	2.71	83.25	nr	**248.60**
Offset						
Sum of two sides 1100 mm	89.22	111.92	1.44	44.24	nr	**156.16**
Sum of two sides 1200 mm	90.97	114.12	1.44	44.24	nr	**158.36**
Sum of two sides 1300 mm	92.46	115.99	1.69	51.92	nr	**167.91**
Sum of two sides 1400 mm	93.15	116.85	1.69	51.92	nr	**168.77**
Sum of two sides 1500 mm	106.26	133.29	2.47	75.88	nr	**209.17**
Sum of two sides 1600 mm	119.10	149.40	2.47	75.88	nr	**225.28**
Sum of two sides 1700 mm	131.71	165.22	2.59	79.56	nr	**244.78**
Sum of two sides 1800 mm	144.16	180.84	2.61	80.18	nr	**261.02**
Sum of two sides 1900 mm	158.06	198.27	2.71	83.25	nr	**281.52**
Sum of two sides 2000 mm	169.94	213.17	2.71	83.25	nr	**296.42**
Square to round						
Sum of two sides 1100 mm	62.04	77.82	1.44	44.24	nr	**122.06**
Sum of two sides 1200 mm	65.29	81.89	1.44	44.24	nr	**126.13**
Sum of two sides 1300 mm	68.56	86.00	1.69	51.92	nr	**137.92**
Sum of two sides 1400 mm	71.23	89.35	1.69	51.92	nr	**141.27**
Sum of two sides 1500 mm	84.61	106.13	2.38	73.11	nr	**179.24**
Sum of two sides 1600 mm	97.98	122.91	2.38	73.11	nr	**196.02**
Sum of two sides 1700 mm	111.36	139.69	2.55	78.33	nr	**218.02**
Sum of two sides 1800 mm	124.75	156.49	2.55	78.33	nr	**234.82**
Sum of two sides 1900 mm	138.13	173.28	2.83	86.95	nr	**260.23**
Sum of two sides 2000 mm	151.50	190.04	2.83	86.95	nr	**276.99**
90° radius bend						
Sum of two sides 1100 mm	47.06	59.04	1.44	44.24	nr	**103.28**
Sum of two sides 1200 mm	50.79	63.71	1.44	44.24	nr	**107.95**
Sum of two sides 1300 mm	54.52	68.39	1.69	51.92	nr	**120.31**
Sum of two sides 1400 mm	57.88	72.61	1.69	51.92	nr	**124.53**
Sum of two sides 1500 mm	69.70	87.43	2.11	64.81	nr	**152.24**
Sum of two sides 1600 mm	81.52	102.26	2.11	64.81	nr	**167.07**
Sum of two sides 1700 mm	93.33	117.07	2.26	69.43	nr	**186.50**
Sum of two sides 1800 mm	105.28	132.06	2.26	69.43	nr	**201.49**
Sum of two sides 1900 mm	114.99	144.24	2.48	76.19	nr	**220.43**
Sum of two sides 2000 mm	126.81	159.07	2.48	76.19	nr	**235.26**

38 VENTILATION/AIR CONDITIONING SYSTEMS

Item	Net Price £	Material £	Labour hours	Labour £	Unit	Total rate £
45° bend						
Sum of two sides 1100 mm	61.03	76.55	1.44	44.24	nr	**120.79**
Sum of two sides 1200 mm	64.79	81.27	1.44	44.24	nr	**125.51**
Sum of two sides 1300 mm	68.54	85.97	1.69	51.92	nr	**137.89**
Sum of two sides 1400 mm	72.07	90.41	1.69	51.92	nr	**142.33**
Sum of two sides 1500 mm	83.91	105.26	2.49	76.50	nr	**181.76**
Sum of two sides 1600 mm	95.77	120.13	2.49	76.50	nr	**196.63**
Sum of two sides 1700 mm	107.62	134.99	2.67	82.03	nr	**217.02**
Sum of two sides 1800 mm	119.73	150.19	2.67	82.03	nr	**232.22**
Sum of two sides 1900 mm	130.47	163.67	3.06	94.00	nr	**257.67**
Sum of two sides 2000 mm	142.32	178.53	3.06	94.00	nr	**272.53**
90° mitre bend						
Sum of two sides 1100 mm	97.38	122.16	1.44	44.24	nr	**166.40**
Sum of two sides 1200 mm	104.45	131.02	1.44	44.24	nr	**175.26**
Sum of two sides 1300 mm	111.52	139.89	1.69	51.92	nr	**191.81**
Sum of two sides 1400 mm	117.92	147.92	1.69	51.92	nr	**199.84**
Sum of two sides 1500 mm	137.12	172.00	2.07	63.58	nr	**235.58**
Sum of two sides 1600 mm	156.32	196.09	2.07	63.58	nr	**259.67**
Sum of two sides 1700 mm	175.53	220.18	2.80	86.02	nr	**306.20**
Sum of two sides 1800 mm	194.81	244.37	2.67	82.03	nr	**326.40**
Sum of two sides 1900 mm	211.77	265.64	2.95	90.63	nr	**356.27**
Sum of two sides 2000 mm	231.08	289.87	2.95	90.63	nr	**380.50**
Branch						
Sum of two sides 1100 mm	93.24	116.96	1.44	44.24	nr	**161.20**
Sum of two sides 1200 mm	98.30	123.31	1.44	44.24	nr	**167.55**
Sum of two sides 1300 mm	103.37	129.66	1.64	50.38	nr	**180.04**
Sum of two sides 1400 mm	107.98	135.45	1.64	50.38	nr	**185.83**
Sum of two sides 1500 mm	123.82	155.32	1.64	50.38	nr	**205.70**
Sum of two sides 1600 mm	139.06	174.44	1.64	50.38	nr	**224.82**
Sum of two sides 1700 mm	155.50	195.06	1.69	51.92	nr	**246.98**
Sum of two sides 1800 mm	171.56	215.21	1.69	51.92	nr	**267.13**
Sum of two sides 1900 mm	187.41	235.09	1.85	56.84	nr	**291.93**
Sum of two sides 2000 mm	203.25	254.96	1.85	56.84	nr	**311.80**
Grille neck						
Sum of two sides 1100 mm	96.53	121.08	1.44	44.24	nr	**165.32**
Sum of two sides 1200 mm	102.18	128.17	1.44	44.24	nr	**172.41**
Sum of two sides 1300 mm	107.83	135.26	1.69	51.92	nr	**187.18**
Sum of two sides 1400 mm	112.73	141.41	1.69	51.92	nr	**193.33**
Sum of two sides 1500 mm	130.52	163.72	1.79	54.99	nr	**218.71**
Sum of two sides 1600 mm	148.30	186.03	1.79	54.99	nr	**241.02**
Sum of two sides 1700 mm	166.09	208.34	1.86	57.14	nr	**265.48**
Sum of two sides 1800 mm	183.87	230.64	2.02	62.06	nr	**292.70**
Sum of two sides 1900 mm	201.66	252.96	2.02	62.06	nr	**315.02**
Sum of two sides 2000 mm	219.44	275.26	2.02	62.06	nr	**337.32**
Ductwork 1001–1250 mm longest side						
Sum of two sides 1300 mm	84.01	105.38	1.40	43.01	m	**148.39**
Sum of two sides 1400 mm	88.60	111.14	1.40	43.01	m	**154.15**
Sum of two sides 1500 mm	92.86	116.48	1.48	45.47	m	**161.95**

38 VENTILATION/AIR CONDITIONING SYSTEMS

Item	Net Price £	Material £	Labour hours	Labour £	Unit	Total rate £
DUCTWORK: RECTANGULAR – CLASS B – cont						
Ductwork 1001–1250 mm – cont						
Sum of two sides 1600 mm	97.43	122.21	1.55	47.62	m	**169.83**
Sum of two sides 1700 mm	101.70	127.57	1.55	47.62	m	**175.19**
Sum of two sides 1800 mm	105.97	132.93	1.61	49.46	m	**182.39**
Sum of two sides 1900 mm	110.25	138.30	1.61	49.46	m	**187.76**
Sum of two sides 2000 mm	114.81	144.02	1.61	49.46	m	**193.48**
Sum of two sides 2100 mm	119.08	149.37	2.17	66.66	m	**216.03**
Sum of two sides 2200 mm	123.36	154.74	2.19	67.28	m	**222.02**
Sum of two sides 2300 mm	128.24	160.87	2.19	67.28	m	**228.15**
Sum of two sides 2400 mm	132.50	166.21	2.38	73.11	m	**239.32**
Sum of two sides 2500 mm	137.07	171.94	2.38	73.11	m	**245.05**
Extra over fittings; Ductwork 1001– 1250 mm longest side						
End cap						
Sum of two sides 1300 mm	35.02	43.93	1.69	51.92	nr	**95.85**
Sum of two sides 1400 mm	36.58	45.89	1.69	51.92	nr	**97.81**
Sum of two sides 1500 mm	42.43	53.22	1.69	51.92	nr	**105.14**
Sum of two sides 1600 mm	48.29	60.57	1.69	51.92	nr	**112.49**
Sum of two sides 1700 mm	54.14	67.92	1.69	51.92	nr	**119.84**
Sum of two sides 1800 mm	59.99	75.25	1.69	51.92	nr	**127.17**
Sum of two sides 1900 mm	65.85	82.60	1.69	51.92	nr	**134.52**
Sum of two sides 2000 mm	71.70	89.94	1.69	51.92	nr	**141.86**
Sum of two sides 2100 mm	77.55	97.28	1.69	51.92	nr	**149.20**
Sum of two sides 2200 mm	83.41	104.63	1.69	51.92	nr	**156.55**
Sum of two sides 2300 mm	89.26	111.97	1.69	51.92	nr	**163.89**
Sum of two sides 2400 mm	95.11	119.30	1.69	51.92	nr	**171.22**
Sum of two sides 2500 mm	100.97	126.66	1.69	51.92	nr	**178.58**
Reducer						
Sum of two sides 1300 mm	61.74	77.45	1.69	51.92	nr	**129.37**
Sum of two sides 1400 mm	64.03	80.32	1.69	51.92	nr	**132.24**
Sum of two sides 1500 mm	74.89	93.95	2.47	75.88	nr	**169.83**
Sum of two sides 1600 mm	86.00	107.88	2.47	75.88	nr	**183.76**
Sum of two sides 1700 mm	96.86	121.50	2.47	75.88	nr	**197.38**
Sum of two sides 1800 mm	107.72	135.13	2.59	79.56	nr	**214.69**
Sum of two sides 1900 mm	118.58	148.75	2.71	83.25	nr	**232.00**
Sum of two sides 2000 mm	129.70	162.69	2.59	79.56	nr	**242.25**
Sum of two sides 2100 mm	140.56	176.32	2.92	89.70	nr	**266.02**
Sum of two sides 2200 mm	151.43	189.95	2.92	89.70	nr	**279.65**
Sum of two sides 2300 mm	162.34	203.64	2.92	89.70	nr	**293.34**
Sum of two sides 2400 mm	173.40	217.52	3.12	95.85	nr	**313.37**
Sum of two sides 2500 mm	184.51	231.45	3.12	95.85	nr	**327.30**
Offset						
Sum of two sides 1300 mm	120.76	151.48	1.69	51.92	nr	**203.40**
Sum of two sides 1400 mm	124.91	156.69	1.69	51.92	nr	**208.61**
Sum of two sides 1500 mm	139.86	175.44	2.47	75.88	nr	**251.32**
Sum of two sides 1600 mm	154.74	194.11	2.47	75.88	nr	**269.99**
Sum of two sides 1700 mm	169.14	212.17	2.59	79.56	nr	**291.73**
Sum of two sides 1800 mm	183.30	229.94	2.61	80.18	nr	**310.12**
Sum of two sides 1900 mm	197.18	247.34	2.71	83.25	nr	**330.59**

38 VENTILATION/AIR CONDITIONING SYSTEMS

Item	Net Price £	Material £	Labour hours	Labour £	Unit	Total rate £
Sum of two sides 2000 mm	210.93	264.59	2.71	83.25	nr	347.84
Sum of two sides 2100 mm	224.28	281.33	2.92	89.70	nr	371.03
Sum of two sides 2200 mm	239.60	300.55	3.26	100.15	nr	400.70
Sum of two sides 2300 mm	257.16	322.58	3.26	100.15	nr	422.73
Sum of two sides 2400 mm	274.66	344.53	3.48	106.92	nr	451.45
Sum of two sides 2500 mm	292.22	366.56	3.48	106.92	nr	473.48
Square to round						
Sum of two sides 1300 mm	66.05	82.86	1.69	51.92	nr	134.78
Sum of two sides 1400 mm	68.67	86.14	1.69	51.92	nr	138.06
Sum of two sides 1500 mm	82.02	102.88	2.38	73.11	nr	175.99
Sum of two sides 1600 mm	95.41	119.68	2.38	73.11	nr	192.79
Sum of two sides 1700 mm	108.76	136.43	2.55	78.33	nr	214.76
Sum of two sides 1800 mm	122.10	153.16	2.55	78.33	nr	231.49
Sum of two sides 1900 mm	135.45	169.90	2.83	86.95	nr	256.85
Sum of two sides 2000 mm	148.84	186.70	2.83	86.95	nr	273.65
Sum of two sides 2100 mm	162.19	203.45	3.85	118.27	nr	321.72
Sum of two sides 2200 mm	175.54	220.19	4.18	128.41	nr	348.60
Sum of two sides 2300 mm	188.92	236.98	4.22	129.64	nr	366.62
Sum of two sides 2400 mm	202.27	253.72	4.68	143.76	nr	397.48
Sum of two sides 2500 mm	215.65	270.51	4.70	144.38	nr	414.89
90° radius bend						
Sum of two sides 1300 mm	44.57	55.91	1.69	51.92	nr	107.83
Sum of two sides 1400 mm	45.61	57.21	1.69	51.92	nr	109.13
Sum of two sides 1500 mm	57.87	72.59	2.11	64.81	nr	137.40
Sum of two sides 1600 mm	70.19	88.04	2.11	64.81	nr	152.85
Sum of two sides 1700 mm	82.45	103.42	2.19	67.28	nr	170.70
Sum of two sides 1800 mm	94.69	118.78	2.19	67.28	nr	186.06
Sum of two sides 1900 mm	106.95	134.15	2.48	76.19	nr	210.34
Sum of two sides 2000 mm	119.29	149.63	2.26	69.43	nr	219.06
Sum of two sides 2100 mm	131.53	164.99	2.48	76.19	nr	241.18
Sum of two sides 2200 mm	143.79	180.36	2.48	76.19	nr	256.55
Sum of two sides 2300 mm	150.72	189.07	2.48	76.19	nr	265.26
Sum of two sides 2400 mm	163.00	204.47	3.90	119.81	nr	324.28
Sum of two sides 2500 mm	175.31	219.91	3.90	119.81	nr	339.72
45° bend						
Sum of two sides 1300 mm	63.78	80.00	1.69	51.92	nr	131.92
Sum of two sides 1400 mm	65.92	82.69	1.69	51.92	nr	134.61
Sum of two sides 1500 mm	78.00	97.84	2.49	76.50	nr	174.34
Sum of two sides 1600 mm	90.34	113.32	2.49	76.50	nr	189.82
Sum of two sides 1700 mm	102.43	128.49	2.67	82.03	nr	210.52
Sum of two sides 1800 mm	114.51	143.64	2.67	82.03	nr	225.67
Sum of two sides 1900 mm	126.61	158.82	3.06	94.00	nr	252.82
Sum of two sides 2000 mm	138.93	174.27	3.06	94.00	nr	268.27
Sum of two sides 2100 mm	151.03	189.45	4.05	124.41	nr	313.86
Sum of two sides 2200 mm	163.12	204.61	4.05	124.41	nr	329.02
Sum of two sides 2300 mm	172.63	216.55	4.39	134.86	nr	351.41
Sum of two sides 2400 mm	184.73	231.73	4.85	148.99	nr	380.72
Sum of two sides 2500 mm	197.06	247.20	4.85	148.99	nr	396.19

38 VENTILATION/AIR CONDITIONING SYSTEMS

Item	Net Price £	Material £	Labour hours	Labour £	Unit	Total rate £
DUCTWORK: RECTANGULAR – CLASS B – cont						
Extra over fittings – cont						
90° mitre bend						
Sum of two sides 1300 mm	132.02	165.60	1.69	51.92	nr	217.52
Sum of two sides 1400 mm	137.76	172.80	1.69	51.92	nr	224.72
Sum of two sides 1500 mm	160.02	200.73	2.80	86.02	nr	286.75
Sum of two sides 1600 mm	182.30	228.68	2.80	86.02	nr	314.70
Sum of two sides 1700 mm	204.55	256.59	2.95	90.63	nr	347.22
Sum of two sides 1800 mm	226.82	284.52	2.95	90.63	nr	375.15
Sum of two sides 1900 mm	249.08	312.45	4.05	124.41	nr	436.86
Sum of two sides 2000 mm	271.36	340.39	4.05	124.41	nr	464.80
Sum of two sides 2100 mm	293.62	368.31	4.07	125.03	nr	493.34
Sum of two sides 2200 mm	315.87	396.22	4.07	125.03	nr	521.25
Sum of two sides 2300 mm	332.26	416.79	4.39	134.86	nr	551.65
Sum of two sides 2400 mm	354.72	444.96	4.85	148.99	nr	593.95
Sum of two sides 2500 mm	377.16	473.11	4.85	148.99	nr	622.10
Branch						
Sum of two sides 1300 mm	106.03	133.00	1.44	44.24	nr	177.24
Sum of two sides 1400 mm	110.58	138.71	1.44	44.24	nr	182.95
Sum of two sides 1500 mm	126.59	158.79	1.64	50.38	nr	209.17
Sum of two sides 1600 mm	142.81	179.14	1.64	50.38	nr	229.52
Sum of two sides 1700 mm	158.81	199.21	1.64	50.38	nr	249.59
Sum of two sides 1800 mm	174.82	219.30	1.64	50.38	nr	269.68
Sum of two sides 1900 mm	190.82	239.37	1.69	51.92	nr	291.29
Sum of two sides 2000 mm	207.04	259.71	1.69	51.92	nr	311.63
Sum of two sides 2100 mm	223.05	279.80	1.85	56.84	nr	336.64
Sum of two sides 2200 mm	239.05	299.87	1.85	56.84	nr	356.71
Sum of two sides 2300 mm	255.27	320.21	2.61	80.18	nr	400.39
Sum of two sides 2400 mm	271.27	340.28	2.61	80.18	nr	420.46
Sum of two sides 2500 mm	287.50	360.64	2.61	80.18	nr	440.82
Grille neck						
Sum of two sides 1300 mm	111.33	139.65	1.79	54.99	nr	194.64
Sum of two sides 1400 mm	116.50	146.14	1.79	54.99	nr	201.13
Sum of two sides 1500 mm	134.53	168.75	1.79	54.99	nr	223.74
Sum of two sides 1600 mm	152.61	191.43	1.79	54.99	nr	246.42
Sum of two sides 1700 mm	170.66	214.08	1.86	57.14	nr	271.22
Sum of two sides 1800 mm	188.72	236.73	2.02	62.06	nr	298.79
Sum of two sides 1900 mm	206.78	259.38	2.02	62.06	nr	321.44
Sum of two sides 2000 mm	224.82	282.02	2.02	62.06	nr	344.08
Sum of two sides 2100 mm	242.88	304.67	2.61	80.18	nr	384.85
Sum of two sides 2200 mm	260.93	327.31	2.61	80.18	nr	407.49
Sum of two sides 2300 mm	278.99	349.97	2.61	80.18	nr	430.15
Sum of two sides 2400 mm	297.04	372.60	2.88	88.47	nr	461.07
Sum of two sides 2500 mm	315.10	395.26	2.88	88.47	nr	483.73
Ductwork 1251–1600 mm longest side						
Sum of two sides 1700 mm	128.65	161.38	1.55	47.62	m	209.00
Sum of two sides 1800 mm	133.71	167.73	1.61	49.46	m	217.19
Sum of two sides 1900 mm	138.82	174.14	1.61	49.46	m	223.60

38 VENTILATION/AIR CONDITIONING SYSTEMS

Item	Net Price £	Material £	Labour hours	Labour £	Unit	Total rate £
Sum of two sides 2000 mm	143.64	180.19	1.61	49.46	m	229.65
Sum of two sides 2100 mm	148.47	186.24	2.17	66.66	m	252.90
Sum of two sides 2200 mm	153.29	192.28	2.19	67.28	m	259.56
Sum of two sides 2300 mm	158.40	198.70	2.19	67.28	m	265.98
Sum of two sides 2400 mm	163.46	205.05	2.38	73.11	m	278.16
Sum of two sides 2500 mm	168.29	211.10	2.38	73.11	m	284.21
Sum of two sides 2600 mm	173.11	217.15	2.64	81.10	m	298.25
Sum of two sides 2700 mm	177.93	223.19	2.66	81.72	m	304.91
Sum of two sides 2800 mm	182.99	229.54	2.95	90.63	m	320.17
Sum of two sides 2900 mm	203.35	255.08	2.96	90.93	m	346.01
Sum of two sides 3000 mm	208.18	261.14	3.15	96.77	m	357.91
Sum of two sides 3100 mm	213.24	267.49	3.15	96.77	m	364.26
Sum of two sides 3200 mm	218.06	273.54	3.18	97.70	m	371.24
Extra over fittings; Ductwork 1251–1600 mm longest side						
End cap						
Sum of two sides 1700 mm	54.14	67.92	0.58	17.81	nr	85.73
Sum of two sides 1800 mm	59.99	75.25	0.58	17.81	nr	93.06
Sum of two sides 1900 mm	65.85	82.60	0.58	17.81	nr	100.41
Sum of two sides 2000 mm	71.70	89.94	0.58	17.81	nr	107.75
Sum of two sides 2100 mm	77.55	97.28	0.87	26.73	nr	124.01
Sum of two sides 2200 mm	83.41	104.63	0.87	26.73	nr	131.36
Sum of two sides 2300 mm	89.26	111.97	0.87	26.73	nr	138.70
Sum of two sides 2400 mm	95.11	119.30	0.87	26.73	nr	146.03
Sum of two sides 2500 mm	100.97	126.66	0.87	26.73	nr	153.39
Sum of two sides 2600 mm	106.82	134.00	0.87	26.73	nr	160.73
Sum of two sides 2700 mm	112.67	141.33	0.87	26.73	nr	168.06
Sum of two sides 2800 mm	118.53	148.68	1.16	35.64	nr	184.32
Sum of two sides 2900 mm	124.38	156.03	1.16	35.64	nr	191.67
Sum of two sides 3000 mm	130.24	163.37	1.30	39.94	nr	203.31
Sum of two sides 3100 mm	134.89	169.21	1.80	55.29	nr	224.50
Sum of two sides 3200 mm	139.11	174.50	1.80	55.29	nr	229.79
Reducer						
Sum of two sides 1700 mm	66.12	82.94	2.47	75.88	nr	158.82
Sum of two sides 1800 mm	76.27	95.67	2.59	79.56	nr	175.23
Sum of two sides 1900 mm	86.54	108.55	2.71	83.25	nr	191.80
Sum of two sides 2000 mm	96.62	121.20	2.71	83.25	nr	204.45
Sum of two sides 2100 mm	106.71	133.86	2.92	89.70	nr	223.56
Sum of two sides 2200 mm	116.80	146.52	2.92	89.70	nr	236.22
Sum of two sides 2300 mm	127.06	159.39	2.92	89.70	nr	249.09
Sum of two sides 2400 mm	137.21	172.12	3.12	95.85	nr	267.97
Sum of two sides 2500 mm	147.29	184.76	3.12	95.85	nr	280.61
Sum of two sides 2600 mm	157.38	197.42	3.12	95.85	nr	293.27
Sum of two sides 2700 mm	167.46	210.07	3.12	95.85	nr	305.92
Sum of two sides 2800 mm	177.61	222.79	3.95	121.35	nr	344.14
Sum of two sides 2900 mm	175.91	220.66	3.97	121.97	nr	342.63
Sum of two sides 3000 mm	186.00	233.32	4.52	138.85	nr	372.17
Sum of two sides 3100 mm	193.46	242.68	4.52	138.85	nr	381.53
Sum of two sides 3200 mm	200.29	251.24	4.52	138.85	nr	390.09

38 VENTILATION/AIR CONDITIONING SYSTEMS

Item	Net Price £	Material £	Labour hours	Labour £	Unit	Total rate £
DUCTWORK: RECTANGULAR – CLASS B – cont						
Extra over fittings – cont						
Offset						
Sum of two sides 1700 mm	167.28	209.83	2.59	79.56	nr	**289.39**
Sum of two sides 1800 mm	185.86	233.14	2.61	80.18	nr	**313.32**
Sum of two sides 1900 mm	198.37	248.83	2.71	83.25	nr	**332.08**
Sum of two sides 2000 mm	210.33	263.84	2.71	83.25	nr	**347.09**
Sum of two sides 2100 mm	221.94	278.40	2.92	89.70	nr	**368.10**
Sum of two sides 2200 mm	233.20	292.52	3.26	100.15	nr	**392.67**
Sum of two sides 2300 mm	244.00	306.07	3.26	100.15	nr	**406.22**
Sum of two sides 2400 mm	261.96	328.61	3.47	106.60	nr	**435.21**
Sum of two sides 2500 mm	272.35	341.63	3.48	106.92	nr	**448.55**
Sum of two sides 2600 mm	289.84	363.57	3.49	107.22	nr	**470.79**
Sum of two sides 2700 mm	307.33	385.52	3.50	107.53	nr	**493.05**
Sum of two sides 2800 mm	324.92	407.58	4.34	133.32	nr	**540.90**
Sum of two sides 2900 mm	324.65	407.24	4.76	146.23	nr	**553.47**
Sum of two sides 3000 mm	342.14	429.18	5.32	163.43	nr	**592.61**
Sum of two sides 3100 mm	355.71	446.21	5.35	164.35	nr	**610.56**
Sum of two sides 3200 mm	368.30	462.00	5.35	164.35	nr	**626.35**
Square to round						
Sum of two sides 1700 mm	78.02	97.87	2.55	78.33	nr	**176.20**
Sum of two sides 1800 mm	90.66	113.72	2.55	78.33	nr	**192.05**
Sum of two sides 1900 mm	103.19	129.44	2.83	86.95	nr	**216.39**
Sum of two sides 2000 mm	115.76	145.21	2.83	86.95	nr	**232.16**
Sum of two sides 2100 mm	128.34	160.99	3.85	118.27	nr	**279.26**
Sum of two sides 2200 mm	140.91	176.76	4.18	128.41	nr	**305.17**
Sum of two sides 2300 mm	153.44	192.47	4.22	129.64	nr	**322.11**
Sum of two sides 2400 mm	166.08	208.33	4.68	143.76	nr	**352.09**
Sum of two sides 2500 mm	178.64	224.09	4.70	144.38	nr	**368.47**
Sum of two sides 2600 mm	191.22	239.87	4.70	144.38	nr	**384.25**
Sum of two sides 2700 mm	203.79	255.63	4.71	144.69	nr	**400.32**
Sum of two sides 2800 mm	216.43	271.49	8.19	251.60	nr	**523.09**
Sum of two sides 2900 mm	217.00	272.20	8.62	264.80	nr	**537.00**
Sum of two sides 3000 mm	229.57	287.97	8.75	268.80	nr	**556.77**
Sum of two sides 3100 mm	238.92	299.70	8.75	268.80	nr	**568.50**
Sum of two sides 3200 mm	247.41	310.35	8.75	268.80	nr	**579.15**
90° radius bend						
Sum of two sides 1700 mm	166.39	208.72	2.19	67.28	nr	**276.00**
Sum of two sides 1800 mm	181.92	228.20	2.19	67.28	nr	**295.48**
Sum of two sides 1900 mm	205.74	258.08	2.26	69.43	nr	**327.51**
Sum of two sides 2000 mm	229.11	287.39	2.26	69.43	nr	**356.82**
Sum of two sides 2100 mm	252.48	316.71	2.48	76.19	nr	**392.90**
Sum of two sides 2200 mm	275.86	346.04	2.48	76.19	nr	**422.23**
Sum of two sides 2300 mm	299.67	375.91	2.48	76.19	nr	**452.10**
Sum of two sides 2400 mm	314.68	394.73	3.90	119.81	nr	**514.54**
Sum of two sides 2500 mm	338.03	424.02	3.90	119.81	nr	**543.83**
Sum of two sides 2600 mm	361.38	453.32	4.26	130.87	nr	**584.19**
Sum of two sides 2700 mm	384.73	482.61	4.55	139.78	nr	**622.39**

38 VENTILATION/AIR CONDITIONING SYSTEMS

Item	Net Price £	Material £	Labour hours	Labour £	Unit	Total rate £
Sum of two sides 2800 mm	399.33	500.92	4.55	139.78	nr	640.70
Sum of two sides 2900 mm	400.40	502.26	6.87	211.05	nr	713.31
Sum of two sides 3000 mm	423.71	531.51	7.00	215.04	nr	746.55
Sum of two sides 3100 mm	441.81	554.21	7.00	215.04	nr	769.25
Sum of two sides 3200 mm	458.61	575.28	7.00	215.04	nr	790.32
45° bend						
Sum of two sides 1700 mm	79.95	100.28	2.67	82.03	nr	182.31
Sum of two sides 1800 mm	87.62	109.91	2.67	82.03	nr	191.94
Sum of two sides 1900 mm	99.51	124.82	3.06	94.00	nr	218.82
Sum of two sides 2000 mm	111.22	139.52	3.06	94.00	nr	233.52
Sum of two sides 2100 mm	122.92	154.19	4.05	124.41	nr	278.60
Sum of two sides 2200 mm	134.62	168.86	4.05	124.41	nr	293.27
Sum of two sides 2300 mm	146.53	183.80	4.39	134.86	nr	318.66
Sum of two sides 2400 mm	153.92	193.08	4.85	148.99	nr	342.07
Sum of two sides 2500 mm	165.61	207.74	4.85	148.99	nr	356.73
Sum of two sides 2600 mm	177.29	222.39	4.87	149.61	nr	372.00
Sum of two sides 2700 mm	188.99	237.07	4.87	149.61	nr	386.68
Sum of two sides 2800 mm	196.18	246.09	8.81	270.65	nr	516.74
Sum of two sides 2900 mm	195.96	245.82	8.81	270.65	nr	516.47
Sum of two sides 3000 mm	207.64	260.47	9.31	286.00	nr	546.47
Sum of two sides 3100 mm	216.69	271.81	9.31	286.00	nr	557.81
Sum of two sides 3200 mm	225.10	282.36	9.39	288.47	nr	570.83
90° mitre bend						
Sum of two sides 1700 mm	179.40	225.04	2.67	82.03	nr	307.07
Sum of two sides 1800 mm	191.53	240.25	2.80	86.02	nr	326.27
Sum of two sides 1900 mm	214.29	268.80	2.95	90.63	nr	359.43
Sum of two sides 2000 mm	237.11	297.43	2.95	90.63	nr	388.06
Sum of two sides 2100 mm	259.94	326.07	4.05	124.41	nr	450.48
Sum of two sides 2200 mm	282.76	354.69	4.05	124.41	nr	479.10
Sum of two sides 2300 mm	205.52	257.80	4.39	134.86	nr	392.66
Sum of two sides 2400 mm	317.89	398.76	4.85	148.99	nr	547.75
Sum of two sides 2500 mm	340.86	427.57	4.85	148.99	nr	576.56
Sum of two sides 2600 mm	363.82	456.38	4.87	149.61	nr	605.99
Sum of two sides 2700 mm	386.79	485.18	4.87	149.61	nr	634.79
Sum of two sides 2800 mm	399.16	500.71	8.81	270.65	nr	771.36
Sum of two sides 2900 mm	402.75	505.21	14.81	454.97	nr	960.18
Sum of two sides 3000 mm	425.87	534.21	15.20	466.95	nr	1001.16
Sum of two sides 3100 mm	444.89	558.07	15.60	479.23	nr	1037.30
Sum of two sides 3200 mm	463.11	580.92	15.60	479.23	nr	1060.15
Branch						
Sum of two sides 1700 mm	158.81	199.21	1.69	51.92	nr	251.13
Sum of two sides 1800 mm	174.82	219.30	1.69	51.92	nr	271.22
Sum of two sides 1900 mm	191.04	239.64	1.85	56.84	nr	296.48
Sum of two sides 2000 mm	207.04	259.71	1.85	56.84	nr	316.55
Sum of two sides 2100 mm	223.05	279.80	2.61	80.18	nr	359.98
Sum of two sides 2200 mm	239.05	299.87	2.61	80.18	nr	380.05
Sum of two sides 2300 mm	255.27	320.21	2.61	80.18	nr	400.39
Sum of two sides 2400 mm	271.27	340.28	2.88	88.47	nr	428.75
Sum of two sides 2500 mm	287.28	360.36	2.88	88.47	nr	448.83
Sum of two sides 2600 mm	303.28	380.43	2.88	88.47	nr	468.90
Sum of two sides 2700 mm	319.28	400.50	2.88	88.47	nr	488.97

38 VENTILATION/AIR CONDITIONING SYSTEMS

Item	Net Price £	Material £	Labour hours	Labour £	Unit	Total rate £
DUCTWORK: RECTANGULAR – CLASS B – cont						
Extra over fittings – cont						
Branch – cont						
Sum of two sides 2800 mm	335.29	420.58	3.94	121.04	nr	541.62
Sum of two sides 2900 mm	351.51	440.93	3.94	121.04	nr	561.97
Sum of two sides 3000 mm	367.51	461.00	4.83	148.38	nr	609.38
Sum of two sides 3100 mm	380.30	477.05	4.83	148.38	nr	625.43
Sum of two sides 3200 mm	391.96	491.68	4.83	148.38	nr	640.06
Grille neck						
Sum of two sides 1700 mm	170.66	214.08	1.86	57.14	nr	271.22
Sum of two sides 1800 mm	188.72	236.73	2.02	62.06	nr	298.79
Sum of two sides 1900 mm	206.78	259.38	2.02	62.06	nr	321.44
Sum of two sides 2000 mm	224.82	282.02	2.02	62.06	nr	344.08
Sum of two sides 2100 mm	242.88	304.67	2.61	80.18	nr	384.85
Sum of two sides 2200 mm	260.93	327.31	2.61	80.18	nr	407.49
Sum of two sides 2300 mm	278.99	349.97	2.61	80.18	nr	430.15
Sum of two sides 2400 mm	297.04	372.60	2.88	88.47	nr	461.07
Sum of two sides 2500 mm	315.10	395.26	2.88	88.47	nr	483.73
Sum of two sides 2600 mm	333.16	417.92	2.88	88.47	nr	506.39
Sum of two sides 2700 mm	351.21	440.56	2.88	88.47	nr	529.03
Sum of two sides 2800 mm	369.26	463.20	3.94	121.04	nr	584.24
Sum of two sides 2900 mm	387.31	485.84	4.12	126.57	nr	612.41
Sum of two sides 3000 mm	405.37	508.49	5.00	153.60	nr	662.09
Sum of two sides 3100 mm	419.80	526.60	5.00	153.60	nr	680.20
Sum of two sides 3200 mm	432.96	543.11	5.00	153.60	nr	696.71
Ductwork 1601–2000 mm longest side						
Sum of two sides 2100 mm	165.21	207.24	2.17	66.66	m	273.90
Sum of two sides 2200 mm	170.47	213.84	2.17	66.66	m	280.50
Sum of two sides 2300 mm	175.78	220.49	2.19	67.28	m	287.77
Sum of two sides 2400 mm	181.00	227.05	2.38	73.11	m	300.16
Sum of two sides 2500 mm	186.30	233.70	2.38	73.11	m	306.81
Sum of two sides 2600 mm	191.33	240.00	2.64	81.10	m	321.10
Sum of two sides 2700 mm	196.35	246.30	2.66	81.72	m	328.02
Sum of two sides 2800 mm	201.37	252.59	2.95	90.63	m	343.22
Sum of two sides 2900 mm	206.68	259.26	2.96	90.93	m	350.19
Sum of two sides 3000 mm	211.71	265.57	2.96	90.93	m	356.50
Sum of two sides 3100 mm	228.35	286.44	2.96	90.93	m	377.37
Sum of two sides 3200 mm	233.38	292.76	3.15	96.77	m	389.53
Sum of two sides 3300 mm	254.24	318.92	3.15	96.77	m	415.69
Sum of two sides 3400 mm	259.27	325.23	3.15	96.77	m	422.00
Sum of two sides 3500 mm	264.29	331.52	3.15	96.77	m	428.29
Sum of two sides 3600 mm	269.31	337.83	3.18	97.70	m	435.53
Sum of two sides 3700 mm	274.63	344.50	3.18	97.70	m	442.20
Sum of two sides 3800 mm	279.65	350.80	3.18	97.70	m	448.50
Sum of two sides 3900 mm	284.67	357.09	3.18	97.70	m	454.79
Sum of two sides 4000 mm	289.70	363.40	3.18	97.70	m	461.10

38 VENTILATION/AIR CONDITIONING SYSTEMS

Item	Net Price £	Material £	Labour hours	Labour £	Unit	Total rate £
Extra over fittings; Ductwork 1601–2000 mm longest side						
End cap						
Sum of two sides 2100 mm	77.55	97.28	0.87	26.73	nr	**124.01**
Sum of two sides 2200 mm	83.41	104.63	0.87	26.73	nr	**131.36**
Sum of two sides 2300 mm	89.26	111.97	0.87	26.73	nr	**138.70**
Sum of two sides 2400 mm	95.11	119.30	0.87	26.73	nr	**146.03**
Sum of two sides 2500 mm	100.97	126.66	0.87	26.73	nr	**153.39**
Sum of two sides 2600 mm	106.82	134.00	0.87	26.73	nr	**160.73**
Sum of two sides 2700 mm	112.67	141.33	0.87	26.73	nr	**168.06**
Sum of two sides 2800 mm	118.53	148.68	1.16	35.64	nr	**184.32**
Sum of two sides 2900 mm	124.38	156.03	1.16	35.64	nr	**191.67**
Sum of two sides 3000 mm	130.24	163.37	1.73	53.14	nr	**216.51**
Sum of two sides 3100 mm	134.89	169.21	1.80	55.29	nr	**224.50**
Sum of two sides 3200 mm	139.11	174.50	1.80	55.29	nr	**229.79**
Sum of two sides 3300 mm	143.33	179.79	1.80	55.29	nr	**235.08**
Sum of two sides 3400 mm	147.55	185.09	1.80	55.29	nr	**240.38**
Sum of two sides 3500 mm	151.77	190.38	1.80	55.29	nr	**245.67**
Sum of two sides 3600 mm	155.99	195.68	1.80	55.29	nr	**250.97**
Sum of two sides 3700 mm	160.21	200.97	1.80	55.29	nr	**256.26**
Sum of two sides 3800 mm	164.43	206.26	1.80	55.29	nr	**261.55**
Sum of two sides 3900 mm	168.65	211.56	1.80	55.29	nr	**266.85**
Sum of two sides 4000 mm	172.88	216.87	1.80	55.29	nr	**272.16**
Reducer						
Sum of two sides 2100 mm	106.09	133.08	2.61	80.18	nr	**213.26**
Sum of two sides 2200 mm	116.07	145.60	2.61	80.18	nr	**225.78**
Sum of two sides 2300 mm	126.23	158.35	2.61	80.18	nr	**238.53**
Sum of two sides 2400 mm	139.03	174.40	2.88	88.47	nr	**262.87**
Sum of two sides 2500 mm	146.18	183.37	3.12	95.85	nr	**279.22**
Sum of two sides 2600 mm	156.15	195.88	3.12	95.85	nr	**291.73**
Sum of two sides 2700 mm	166.12	208.38	3.12	95.85	nr	**304.23**
Sum of two sides 2800 mm	176.08	220.88	3.95	121.35	nr	**342.23**
Sum of two sides 2900 mm	186.24	233.62	3.97	121.97	nr	**355.59**
Sum of two sides 3000 mm	196.22	246.14	4.52	138.85	nr	**384.99**
Sum of two sides 3100 mm	194.72	244.26	4.52	138.85	nr	**383.11**
Sum of two sides 3200 mm	201.42	252.66	4.52	138.85	nr	**391.51**
Sum of two sides 3300 mm	196.08	245.96	4.52	138.85	nr	**384.81**
Sum of two sides 3400 mm	202.79	254.37	4.52	138.85	nr	**393.22**
Sum of two sides 3500 mm	209.49	262.79	4.52	138.85	nr	**401.64**
Sum of two sides 3600 mm	216.20	271.20	4.52	138.85	nr	**410.05**
Sum of two sides 3700 mm	233.78	293.25	4.52	138.85	nr	**432.10**
Sum of two sides 3800 mm	240.49	301.67	4.52	138.85	nr	**440.52**
Sum of two sides 3900 mm	247.19	310.07	4.52	138.85	nr	**448.92**
Sum of two sides 4000 mm	253.90	318.49	4.52	138.85	nr	**457.34**
Offset						
Sum of two sides 2100 mm	248.75	312.03	2.61	80.18	nr	**392.21**
Sum of two sides 2200 mm	267.48	335.53	2.61	80.18	nr	**415.71**
Sum of two sides 2300 mm	278.22	349.00	2.61	80.18	nr	**429.18**
Sum of two sides 2400 mm	304.98	382.57	2.88	88.47	nr	**471.04**
Sum of two sides 2500 mm	315.34	395.56	3.48	106.92	nr	**502.48**
Sum of two sides 2600 mm	325.12	407.83	3.49	107.22	nr	**515.05**
Sum of two sides 2700 mm	334.53	419.63	3.50	107.53	nr	**527.16**

38 VENTILATION/AIR CONDITIONING SYSTEMS

Item	Net Price £	Material £	Labour hours	Labour £	Unit	Total rate £
DUCTWORK: RECTANGULAR – CLASS B – cont						
Extra over fittings – cont						
Offset – cont						
Sum of two sides 2800 mm	343.54	430.93	4.34	133.32	nr	**564.25**
Sum of two sides 2900 mm	352.27	441.88	4.76	146.23	nr	**588.11**
Sum of two sides 3000 mm	360.50	452.21	5.32	163.43	nr	**615.64**
Sum of two sides 3100 mm	360.99	452.83	5.35	164.35	nr	**617.18**
Sum of two sides 3200 mm	373.69	468.75	5.35	164.35	nr	**633.10**
Sum of two sides 3300 mm	368.23	461.91	5.35	164.35	nr	**626.26**
Sum of two sides 3400 mm	380.93	477.84	5.35	164.35	nr	**642.19**
Sum of two sides 3500 mm	393.64	493.79	5.35	164.35	nr	**658.14**
Sum of two sides 3600 mm	406.35	509.72	5.35	164.35	nr	**674.07**
Sum of two sides 3700 mm	428.82	537.91	5.35	164.35	nr	**702.26**
Sum of two sides 3800 mm	442.53	555.11	5.35	164.35	nr	**719.46**
Sum of two sides 3900 mm	455.23	571.04	5.35	164.35	nr	**735.39**
Sum of two sides 4000 mm	467.94	586.98	5.35	164.35	nr	**751.33**
Square to round						
Sum of two sides 2100 mm	157.87	198.03	2.61	80.18	nr	**278.21**
Sum of two sides 2200 mm	173.13	217.18	2.61	80.18	nr	**297.36**
Sum of two sides 2300 mm	188.34	236.25	2.61	80.18	nr	**316.43**
Sum of two sides 2400 mm	203.67	255.48	2.88	88.47	nr	**343.95**
Sum of two sides 2500 mm	218.87	274.55	4.70	144.38	nr	**418.93**
Sum of two sides 2600 mm	234.10	293.65	4.70	144.38	nr	**438.03**
Sum of two sides 2700 mm	249.33	312.76	4.71	144.69	nr	**457.45**
Sum of two sides 2800 mm	264.56	331.87	8.19	251.60	nr	**583.47**
Sum of two sides 2900 mm	279.76	350.93	8.19	251.60	nr	**602.53**
Sum of two sides 3000 mm	294.99	370.04	8.19	251.60	nr	**621.64**
Sum of two sides 3100 mm	297.55	373.25	8.19	251.60	nr	**624.85**
Sum of two sides 3200 mm	307.88	386.21	8.19	251.60	nr	**637.81**
Sum of two sides 3300 mm	305.96	383.80	8.19	251.60	nr	**635.40**
Sum of two sides 3400 mm	316.29	396.75	8.62	264.80	nr	**661.55**
Sum of two sides 3500 mm	326.62	409.71	8.62	264.80	nr	**674.51**
Sum of two sides 3600 mm	336.95	422.67	8.62	264.80	nr	**687.47**
Sum of two sides 3700 mm	352.61	442.31	8.62	264.80	nr	**707.11**
Sum of two sides 3800 mm	362.94	455.27	8.75	268.80	nr	**724.07**
Sum of two sides 3900 mm	373.27	468.23	8.75	268.80	nr	**737.03**
Sum of two sides 4000 mm	383.61	481.20	8.75	268.80	nr	**750.00**
90° radius bend						
Sum of two sides 2100 mm	229.39	287.75	2.61	80.18	nr	**367.93**
Sum of two sides 2200 mm	439.66	551.51	2.61	80.18	nr	**631.69**
Sum of two sides 2300 mm	476.41	597.61	2.61	80.18	nr	**677.79**
Sum of two sides 2400 mm	487.68	611.74	2.88	88.47	nr	**700.21**
Sum of two sides 2500 mm	524.38	657.79	3.90	119.81	nr	**777.60**
Sum of two sides 2600 mm	560.34	702.89	4.26	130.87	nr	**833.76**
Sum of two sides 2700 mm	596.30	748.00	4.55	139.78	nr	**887.78**
Sum of two sides 2800 mm	632.26	793.11	4.55	139.78	nr	**932.89**
Sum of two sides 2900 mm	668.95	839.13	6.87	211.05	nr	**1050.18**
Sum of two sides 3000 mm	704.91	884.24	6.87	211.05	nr	**1095.29**
Sum of two sides 3100 mm	712.95	894.32	6.87	211.05	nr	**1105.37**

38 VENTILATION/AIR CONDITIONING SYSTEMS

Item	Net Price £	Material £	Labour hours	Labour £	Unit	Total rate £
Sum of two sides 3200 mm	739.91	928.14	6.87	211.05	nr	1139.19
Sum of two sides 3300 mm	738.07	925.84	6.87	211.05	nr	1136.89
Sum of two sides 3400 mm	764.23	958.65	7.00	215.04	nr	1173.69
Sum of two sides 3500 mm	790.41	991.49	7.00	215.04	nr	1206.53
Sum of two sides 3600 mm	816.57	1024.31	7.00	215.04	nr	1239.35
Sum of two sides 3700 mm	875.58	1098.33	7.00	215.04	nr	1313.37
Sum of two sides 3800 mm	901.74	1131.14	7.00	215.04	nr	1346.18
Sum of two sides 3900 mm	927.91	1163.97	7.00	215.04	nr	1379.01
Sum of two sides 4000 mm	954.07	1196.79	7.00	215.04	nr	1411.83
45° bend						
Sum of two sides 2100 mm	110.46	138.57	2.61	80.18	nr	218.75
Sum of two sides 2200 mm	314.26	394.21	2.61	80.18	nr	474.39
Sum of two sides 2300 mm	338.85	425.05	2.61	80.18	nr	505.23
Sum of two sides 2400 mm	349.89	438.91	2.88	88.47	nr	527.38
Sum of two sides 2500 mm	374.44	469.69	4.85	148.99	nr	618.68
Sum of two sides 2600 mm	398.45	499.81	4.87	149.61	nr	649.42
Sum of two sides 2700 mm	422.44	529.91	4.87	149.61	nr	679.52
Sum of two sides 2800 mm	446.44	560.01	8.81	270.65	nr	830.66
Sum of two sides 2900 mm	471.01	590.83	8.81	270.65	nr	861.48
Sum of two sides 3000 mm	495.00	620.93	9.31	286.00	nr	906.93
Sum of two sides 3100 mm	503.26	631.29	9.31	286.00	nr	917.29
Sum of two sides 3200 mm	520.72	653.20	9.31	286.00	nr	939.20
Sum of two sides 3300 mm	524.02	657.33	9.31	286.00	nr	943.33
Sum of two sides 3400 mm	541.49	679.25	9.31	286.00	nr	965.25
Sum of two sides 3500 mm	558.96	701.16	9.39	288.47	nr	989.63
Sum of two sides 3600 mm	576.42	723.06	9.39	288.47	nr	1011.53
Sum of two sides 3700 mm	615.87	772.54	9.39	288.47	nr	1061.01
Sum of two sides 3800 mm	633.33	794.45	9.39	288.47	nr	1082.92
Sum of two sides 3900 mm	650.80	816.37	9.39	288.47	nr	1104.84
Sum of two sides 4000 mm	668.27	838.28	9.39	288.47	nr	1126.75
90° mitre bend						
Sum of two sides 2100 mm	510.92	640.90	2.61	80.18	nr	721.08
Sum of two sides 2200 mm	539.79	677.11	2.61	80.18	nr	757.29
Sum of two sides 2300 mm	583.20	731.56	2.61	80.18	nr	811.74
Sum of two sides 2400 mm	596.13	747.79	2.88	88.47	nr	836.26
Sum of two sides 2500 mm	639.72	802.47	4.85	148.99	nr	951.46
Sum of two sides 2600 mm	683.57	857.47	4.87	149.61	nr	1007.08
Sum of two sides 2700 mm	727.42	912.48	4.87	149.61	nr	1062.09
Sum of two sides 2800 mm	771.27	967.48	8.81	270.65	nr	1238.13
Sum of two sides 2900 mm	814.86	1022.16	14.81	454.97	nr	1477.13
Sum of two sides 3000 mm	858.71	1077.17	15.20	466.95	nr	1544.12
Sum of two sides 3100 mm	870.41	1091.84	15.20	466.95	nr	1558.79
Sum of two sides 3200 mm	904.46	1134.56	15.20	466.95	nr	1601.51
Sum of two sides 3300 mm	914.02	1146.54	15.20	466.95	nr	1613.49
Sum of two sides 3400 mm	948.07	1189.26	15.20	466.95	nr	1656.21
Sum of two sides 3500 mm	982.12	1231.97	15.60	479.23	nr	1711.20
Sum of two sides 3600 mm	1016.17	1274.68	15.60	479.23	nr	1753.91
Sum of two sides 3700 mm	1060.68	1330.52	15.60	479.23	nr	1809.75
Sum of two sides 3800 mm	1094.73	1373.23	15.60	479.23	nr	1852.46
Sum of two sides 3900 mm	1128.78	1415.94	15.60	479.23	nr	1895.17
Sum of two sides 4000 mm	1162.83	1458.65	15.60	479.23	nr	1937.88

38 VENTILATION/AIR CONDITIONING SYSTEMS

Item	Net Price £	Material £	Labour hours	Labour £	Unit	Total rate £
DUCTWORK: RECTANGULAR – CLASS B – cont						
Extra over fittings – cont						
Branch						
Sum of two sides 2100 mm	229.05	287.32	2.61	80.18	nr	367.50
Sum of two sides 2200 mm	245.06	307.41	2.61	80.18	nr	387.59
Sum of two sides 2300 mm	261.30	327.78	2.61	80.18	nr	407.96
Sum of two sides 2400 mm	277.06	347.55	2.88	88.47	nr	436.02
Sum of two sides 2500 mm	293.31	367.93	2.88	88.47	nr	456.40
Sum of two sides 2600 mm	309.35	388.05	2.88	88.47	nr	476.52
Sum of two sides 2700 mm	325.38	408.16	2.88	88.47	nr	496.63
Sum of two sides 2800 mm	341.41	428.27	3.94	121.04	nr	549.31
Sum of two sides 2900 mm	357.67	448.66	3.94	121.04	nr	569.70
Sum of two sides 3000 mm	373.69	468.75	3.94	121.04	nr	589.79
Sum of two sides 3100 mm	386.52	484.85	3.94	121.04	nr	605.89
Sum of two sides 3200 mm	398.20	499.50	3.94	121.04	nr	620.54
Sum of two sides 3300 mm	410.09	514.42	3.94	121.04	nr	635.46
Sum of two sides 3400 mm	421.77	529.07	4.83	148.38	nr	677.45
Sum of two sides 3500 mm	433.45	543.72	4.83	148.38	nr	692.10
Sum of two sides 3600 mm	445.14	558.39	4.83	148.38	nr	706.77
Sum of two sides 3700 mm	462.38	580.01	4.83	148.38	nr	728.39
Sum of two sides 3800 mm	474.07	594.68	4.83	148.38	nr	743.06
Sum of two sides 3900 mm	485.75	609.32	4.83	148.38	nr	757.70
Sum of two sides 4000 mm	497.43	623.97	4.83	148.38	nr	772.35
Grille neck						
Sum of two sides 2100 mm	242.88	304.67	2.61	80.18	nr	384.85
Sum of two sides 2200 mm	260.93	327.31	2.61	80.18	nr	407.49
Sum of two sides 2300 mm	278.99	349.97	2.61	80.18	nr	430.15
Sum of two sides 2400 mm	297.04	372.60	2.88	88.47	nr	461.07
Sum of two sides 2500 mm	315.10	395.26	2.88	88.47	nr	483.73
Sum of two sides 2600 mm	333.16	417.92	2.88	88.47	nr	506.39
Sum of two sides 2700 mm	351.21	440.56	2.88	88.47	nr	529.03
Sum of two sides 2800 mm	369.26	463.20	3.94	121.04	nr	584.24
Sum of two sides 2900 mm	387.31	485.84	4.12	126.57	nr	612.41
Sum of two sides 3000 mm	405.37	508.49	4.12	126.57	nr	635.06
Sum of two sides 3100 mm	419.80	526.60	4.12	126.57	nr	653.17
Sum of two sides 3200 mm	432.96	543.11	4.12	126.57	nr	669.68
Sum of two sides 3300 mm	446.10	559.59	4.12	126.57	nr	686.16
Sum of two sides 3400 mm	459.28	576.12	5.00	153.60	nr	729.72
Sum of two sides 3500 mm	472.44	592.63	5.00	153.60	nr	746.23
Sum of two sides 3600 mm	485.59	609.12	5.00	153.60	nr	762.72
Sum of two sides 3700 mm	498.75	625.63	5.00	153.60	nr	779.23
Sum of two sides 3800 mm	511.90	642.13	5.00	153.60	nr	795.73
Sum of two sides 3900 mm	525.06	658.64	5.00	153.60	nr	812.24
Sum of two sides 4000 mm	538.23	675.16	5.00	153.60	nr	828.76
Ductwork 2001–2500 mm longest side						
Sum of two sides 2500 mm	239.34	300.23	2.38	73.11	m	373.34
Sum of two sides 2600 mm	245.62	308.10	2.64	81.10	m	389.20
Sum of two sides 2700 mm	250.90	314.73	2.66	81.72	m	396.45

38 VENTILATION/AIR CONDITIONING SYSTEMS

Item	Net Price £	Material £	Labour hours	Labour £	Unit	Total rate £
Sum of two sides 2800 mm	257.72	323.29	2.95	90.63	m	**413.92**
Sum of two sides 2900 mm	263.01	329.92	2.96	90.93	m	**420.85**
Sum of two sides 3000 mm	268.02	336.20	3.15	96.77	m	**432.97**
Sum of two sides 3100 mm	273.27	342.79	3.15	96.77	m	**439.56**
Sum of two sides 3200 mm	278.28	349.07	3.15	96.77	m	**445.84**
Sum of two sides 3300 mm	283.57	355.71	3.15	96.77	m	**452.48**
Sum of two sides 3400 mm	288.58	362.00	2.66	81.72	m	**443.72**
Sum of two sides 3500 mm	307.65	385.92	3.15	96.77	m	**482.69**
Sum of two sides 3600 mm	312.67	392.21	3.18	97.70	m	**489.91**
Sum of two sides 3700 mm	333.83	418.76	3.18	97.70	m	**516.46**
Sum of two sides 3800 mm	338.83	425.03	3.18	97.70	m	**522.73**
Sum of two sides 3900 mm	343.84	431.31	3.18	97.70	m	**529.01**
Sum of two sides 4000 mm	350.47	439.63	3.18	97.70	m	**537.33**
Extra over fittings; Ductwork 2001–2500 mm longest side						
End cap						
Sum of two sides 2500 mm	100.97	126.66	0.87	26.73	nr	**153.39**
Sum of two sides 2600 mm	106.82	134.00	0.87	26.73	nr	**160.73**
Sum of two sides 2700 mm	112.67	141.33	0.87	26.73	nr	**168.06**
Sum of two sides 2800 mm	118.53	148.68	1.16	35.64	nr	**184.32**
Sum of two sides 2900 mm	124.38	156.03	1.16	35.64	nr	**191.67**
Sum of two sides 3000 mm	130.24	163.37	1.73	53.14	nr	**216.51**
Sum of two sides 3100 mm	134.89	169.21	1.73	53.14	nr	**222.35**
Sum of two sides 3200 mm	139.11	174.50	1.73	53.14	nr	**227.64**
Sum of two sides 3300 mm	143.33	179.79	1.73	53.14	nr	**232.93**
Sum of two sides 3400 mm	147.55	185.09	1.80	55.29	nr	**240.38**
Sum of two sides 3500 mm	151.77	190.38	1.80	55.29	nr	**245.67**
Sum of two sides 3600 mm	155.99	195.68	1.80	55.29	nr	**250.97**
Sum of two sides 3700 mm	160.21	200.97	1.80	55.29	nr	**256.26**
Sum of two sides 3800 mm	164.43	206.26	1.80	55.29	nr	**261.55**
Sum of two sides 3900 mm	168.65	211.56	1.80	55.29	nr	**266.85**
Sum of two sides 4000 mm	172.88	216.87	1.80	55.29	nr	**272.16**
Reducer						
Sum of two sides 2500 mm	119.13	149.44	3.12	95.85	nr	**245.29**
Sum of two sides 2600 mm	129.05	161.88	3.12	95.85	nr	**257.73**
Sum of two sides 2700 mm	139.20	174.61	3.12	95.85	nr	**270.46**
Sum of two sides 2800 mm	148.90	186.78	3.95	121.35	nr	**308.13**
Sum of two sides 2900 mm	159.04	199.49	3.97	121.97	nr	**321.46**
Sum of two sides 3000 mm	168.99	211.98	3.97	121.97	nr	**333.95**
Sum of two sides 3100 mm	176.33	221.19	3.97	121.97	nr	**343.16**
Sum of two sides 3200 mm	183.00	229.56	3.97	121.97	nr	**351.53**
Sum of two sides 3300 mm	189.90	238.21	3.97	121.97	nr	**360.18**
Sum of two sides 3400 mm	196.57	246.58	4.52	138.85	nr	**385.43**
Sum of two sides 3500 mm	193.46	242.68	4.52	138.85	nr	**381.53**
Sum of two sides 3600 mm	200.15	251.07	4.52	138.85	nr	**389.92**
Sum of two sides 3700 mm	194.53	244.01	4.52	138.85	nr	**382.86**
Sum of two sides 3800 mm	201.22	252.41	4.52	138.85	nr	**391.26**
Sum of two sides 3900 mm	207.91	260.80	4.52	138.85	nr	**399.65**
Sum of two sides 4000 mm	214.70	269.32	4.52	138.85	nr	**408.17**

38 VENTILATION/AIR CONDITIONING SYSTEMS

Item	Net Price £	Material £	Labour hours	Labour £	Unit	Total rate £
DUCTWORK: RECTANGULAR – CLASS B – cont						
Extra over fittings – cont						
Offset						
Sum of two sides 2500 mm	299.64	375.87	3.48	106.92	nr	482.79
Sum of two sides 2600 mm	318.34	399.32	3.49	107.22	nr	506.54
Sum of two sides 2700 mm	325.53	408.34	3.50	107.53	nr	515.87
Sum of two sides 2800 mm	355.72	446.22	4.34	133.32	nr	579.54
Sum of two sides 2900 mm	362.54	454.76	4.76	146.23	nr	600.99
Sum of two sides 3000 mm	368.76	462.57	5.32	163.43	nr	626.00
Sum of two sides 3100 mm	370.73	465.05	5.35	164.35	nr	629.40
Sum of two sides 3200 mm	371.25	465.70	5.32	163.43	nr	629.13
Sum of two sides 3300 mm	371.52	466.03	5.32	163.43	nr	629.46
Sum of two sides 3400 mm	371.25	465.70	5.35	164.35	nr	630.05
Sum of two sides 3500 mm	368.77	462.58	5.32	163.43	nr	626.01
Sum of two sides 3600 mm	381.44	478.48	5.35	164.35	nr	642.83
Sum of two sides 3700 mm	375.57	471.12	5.35	164.35	nr	635.47
Sum of two sides 3800 mm	388.24	487.01	5.35	164.35	nr	651.36
Sum of two sides 3900 mm	400.93	502.92	5.35	164.35	nr	667.27
Sum of two sides 4000 mm	413.58	518.80	5.35	164.35	nr	683.15
Square to round						
Sum of two sides 2500 mm	191.29	239.95	4.70	144.38	nr	384.33
Sum of two sides 2600 mm	206.50	259.03	4.70	144.38	nr	403.41
Sum of two sides 2700 mm	221.70	278.10	4.71	144.69	nr	422.79
Sum of two sides 2800 mm	236.93	297.20	8.19	251.60	nr	548.80
Sum of two sides 2900 mm	252.12	316.25	8.19	251.60	nr	567.85
Sum of two sides 3000 mm	267.33	335.34	8.19	251.60	nr	586.94
Sum of two sides 3100 mm	278.71	349.62	8.19	251.60	nr	601.22
Sum of two sides 3200 mm	289.02	362.54	8.62	264.80	nr	627.34
Sum of two sides 3300 mm	299.33	375.48	8.62	264.80	nr	640.28
Sum of two sides 3400 mm	309.63	388.40	8.62	264.80	nr	653.20
Sum of two sides 3500 mm	309.59	388.35	8.62	264.80	nr	653.15
Sum of two sides 3600 mm	319.90	401.28	8.75	268.80	nr	670.08
Sum of two sides 3700 mm	317.71	398.54	8.75	268.80	nr	667.34
Sum of two sides 3800 mm	328.01	411.45	8.75	268.80	nr	680.25
Sum of two sides 3900 mm	338.33	424.40	8.75	268.80	nr	693.20
Sum of two sides 4000 mm	348.56	437.24	8.75	268.80	nr	706.04
90° radius bend						
Sum of two sides 2500 mm	448.06	562.05	3.90	119.81	nr	681.86
Sum of two sides 2600 mm	468.23	587.35	4.26	130.87	nr	718.22
Sum of two sides 2700 mm	504.85	633.28	4.55	139.78	nr	773.06
Sum of two sides 2800 mm	507.42	636.51	4.55	139.78	nr	776.29
Sum of two sides 2900 mm	543.98	682.37	6.87	211.05	nr	893.42
Sum of two sides 3000 mm	579.87	727.38	6.87	211.05	nr	938.43
Sum of two sides 3100 mm	607.97	762.64	6.87	211.05	nr	973.69
Sum of two sides 3200 mm	634.06	795.37	6.87	211.05	nr	1006.42
Sum of two sides 3300 mm	660.83	828.95	6.87	211.05	nr	1040.00
Sum of two sides 3400 mm	686.94	861.69	6.87	211.05	nr	1072.74
Sum of two sides 3500 mm	687.14	861.95	7.00	215.04	nr	1076.99

38 VENTILATION/AIR CONDITIONING SYSTEMS

Item	Net Price £	Material £	Labour hours	Labour £	Unit	Total rate £
Sum of two sides 3600 mm	713.25	894.70	7.00	215.04	nr	1109.74
Sum of two sides 3700 mm	706.56	886.31	7.00	215.04	nr	1101.35
Sum of two sides 3800 mm	732.65	919.04	7.00	215.04	nr	1134.08
Sum of two sides 3900 mm	758.75	951.78	7.00	215.04	nr	1166.82
Sum of two sides 4000 mm	766.00	960.87	7.00	215.04	nr	1175.91
45° bend						
Sum of two sides 2500 mm	334.87	420.06	4.85	148.99	nr	569.05
Sum of two sides 2600 mm	350.76	439.99	4.87	149.61	nr	589.60
Sum of two sides 2700 mm	375.30	470.78	4.87	149.61	nr	620.39
Sum of two sides 2800 mm	381.96	479.14	8.81	270.65	nr	749.79
Sum of two sides 2900 mm	406.46	509.87	8.81	270.65	nr	780.52
Sum of two sides 3000 mm	430.42	539.92	9.31	286.00	nr	825.92
Sum of two sides 3100 mm	449.27	563.56	9.31	286.00	nr	849.56
Sum of two sides 3200 mm	466.69	585.41	9.31	286.00	nr	871.41
Sum of two sides 3300 mm	484.67	607.97	9.31	286.00	nr	893.97
Sum of two sides 3400 mm	502.10	629.83	9.31	286.00	nr	915.83
Sum of two sides 3500 mm	506.47	635.32	9.31	286.00	nr	921.32
Sum of two sides 3600 mm	523.91	657.19	9.39	288.47	nr	945.66
Sum of two sides 3700 mm	524.37	657.76	9.39	288.47	nr	946.23
Sum of two sides 3800 mm	541.79	679.62	9.39	288.47	nr	968.09
Sum of two sides 3900 mm	559.22	701.49	9.39	288.47	nr	989.96
Sum of two sides 4000 mm	567.21	711.51	9.39	288.47	nr	999.98
90° mitre bend						
Sum of two sides 2500 mm	543.44	681.69	4.85	148.99	nr	830.68
Sum of two sides 2600 mm	576.48	723.14	4.87	149.61	nr	872.75
Sum of two sides 2700 mm	620.31	778.12	4.87	149.61	nr	927.73
Sum of two sides 2800 mm	621.60	779.73	8.81	270.65	nr	1050.38
Sum of two sides 2900 mm	665.61	834.94	14.81	454.97	nr	1289.91
Sum of two sides 3000 mm	709.04	889.41	14.81	454.97	nr	1344.38
Sum of two sides 3100 mm	746.36	936.23	15.20	466.95	nr	1403.18
Sum of two sides 3200 mm	780.89	979.55	15.20	466.95	nr	1446.50
Sum of two sides 3300 mm	815.11	1022.47	15.20	466.95	nr	1489.42
Sum of two sides 3400 mm	849.66	1065.81	15.20	466.95	nr	1532.76
Sum of two sides 3500 mm	849.82	1066.02	15.20	466.95	nr	1532.97
Sum of two sides 3600 mm	884.35	1109.33	15.60	479.23	nr	1588.56
Sum of two sides 3700 mm	889.23	1115.45	15.60	479.23	nr	1594.68
Sum of two sides 3800 mm	923.76	1158.76	15.60	479.23	nr	1637.99
Sum of two sides 3900 mm	958.30	1202.10	15.60	479.23	nr	1681.33
Sum of two sides 4000 mm	968.53	1214.92	15.60	479.23	nr	1694.15
Branch						
Sum of two sides 2500 mm	293.85	368.60	2.88	88.47	nr	457.07
Sum of two sides 2600 mm	309.85	388.67	2.88	88.47	nr	477.14
Sum of two sides 2700 mm	326.10	409.06	2.88	88.47	nr	497.53
Sum of two sides 2800 mm	341.86	428.83	3.94	121.04	nr	549.87
Sum of two sides 2900 mm	358.11	449.21	3.94	121.04	nr	570.25
Sum of two sides 3000 mm	374.14	469.32	3.94	121.04	nr	590.36
Sum of two sides 3100 mm	386.95	485.39	3.94	121.04	nr	606.43
Sum of two sides 3200 mm	398.64	500.06	3.94	121.04	nr	621.10
Sum of two sides 3300 mm	410.54	514.98	3.94	121.04	nr	636.02
Sum of two sides 3400 mm	422.22	529.64	3.94	121.04	nr	650.68
Sum of two sides 3500 mm	434.47	545.00	4.83	148.38	nr	693.38

38 VENTILATION/AIR CONDITIONING SYSTEMS

Item	Net Price £	Material £	Labour hours	Labour £	Unit	Total rate £
DUCTWORK: RECTANGULAR – CLASS B – cont						
Extra over fittings – cont						
Branch – cont						
Sum of two sides 3600 mm	446.15	559.65	4.83	148.38	nr	708.03
Sum of two sides 3700 mm	458.05	574.58	4.83	148.38	nr	722.96
Sum of two sides 3800 mm	469.73	589.23	4.83	148.38	nr	737.61
Sum of two sides 3900 mm	481.41	603.88	4.83	148.38	nr	752.26
Sum of two sides 4000 mm	493.27	618.76	4.83	148.38	nr	767.14
Grille neck						
Sum of two sides 2500 mm	315.10	395.26	2.88	88.47	nr	483.73
Sum of two sides 2600 mm	333.16	417.92	2.88	88.47	nr	506.39
Sum of two sides 2700 mm	351.21	440.56	2.88	88.47	nr	529.03
Sum of two sides 2800 mm	369.26	463.20	3.94	121.04	nr	584.24
Sum of two sides 2900 mm	387.21	485.72	3.94	121.04	nr	606.76
Sum of two sides 3000 mm	405.37	508.49	3.94	121.04	nr	629.53
Sum of two sides 3100 mm	419.80	526.60	4.12	126.57	nr	653.17
Sum of two sides 3200 mm	432.96	543.11	4.12	126.57	nr	669.68
Sum of two sides 3300 mm	446.11	559.60	4.12	126.57	nr	686.17
Sum of two sides 3400 mm	459.28	576.12	4.12	126.57	nr	702.69
Sum of two sides 3500 mm	472.44	592.63	5.00	153.60	nr	746.23
Sum of two sides 3600 mm	485.59	609.12	5.00	153.60	nr	762.72
Sum of two sides 3700 mm	498.75	625.63	5.00	153.60	nr	779.23
Sum of two sides 3800 mm	511.90	642.13	5.00	153.60	nr	795.73
Sum of two sides 3900 mm	525.06	658.64	5.00	153.60	nr	812.24
Sum of two sides 4000 mm	538.23	675.16	5.00	153.60	nr	828.76
Ductwork 2501–4000 mm longest side						
Sum of two sides 3000 mm	379.37	475.88	2.38	73.11	m	548.99
Sum of two sides 3100 mm	388.44	487.26	2.38	73.11	m	560.37
Sum of two sides 3200 mm	397.01	498.01	2.38	73.11	m	571.12
Sum of two sides 3300 mm	405.57	508.75	2.38	73.11	m	581.86
Sum of two sides 3400 mm	416.45	522.39	2.64	81.10	m	603.49
Sum of two sides 3500 mm	425.01	533.13	2.66	81.72	m	614.85
Sum of two sides 3600 mm	433.59	543.89	2.95	90.63	m	634.52
Sum of two sides 3700 mm	444.73	557.87	2.96	90.93	m	648.80
Sum of two sides 3800 mm	453.30	568.62	3.15	96.77	m	665.39
Sum of two sides 3900 mm	477.37	598.81	3.15	96.77	m	695.58
Sum of two sides 4000 mm	485.94	609.56	3.15	96.77	m	706.33
Sum of two sides 4100 mm	494.78	620.65	3.35	102.91	m	723.56
Sum of two sides 4200 mm	503.35	631.40	3.35	102.91	m	734.31
Sum of two sides 4300 mm	511.91	642.14	3.60	110.59	m	752.73
Sum of two sides 4400 mm	524.17	657.52	3.60	110.59	m	768.11
Sum of two sides 4500 mm	532.75	668.28	3.60	110.59	m	778.87
Extra over fittings; Ductwork 2501– 4000 mm longest side						
End cap						
Sum of two sides 3000 mm	131.67	165.17	1.73	53.14	nr	218.31
Sum of two sides 3100 mm	136.36	171.05	1.73	53.14	nr	224.19
Sum of two sides 3200 mm	140.63	176.41	1.73	53.14	nr	229.55

38 VENTILATION/AIR CONDITIONING SYSTEMS

Item	Net Price £	Material £	Labour hours	Labour £	Unit	Total rate £
Sum of two sides 3300 mm	144.81	181.65	1.73	53.14	nr	**234.79**
Sum of two sides 3400 mm	149.17	187.12	1.73	53.14	nr	**240.26**
Sum of two sides 3500 mm	153.44	192.47	1.73	53.14	nr	**245.61**
Sum of two sides 3600 mm	157.71	197.84	1.73	53.14	nr	**250.98**
Sum of two sides 3700 mm	161.98	203.19	1.73	53.14	nr	**256.33**
Sum of two sides 3800 mm	166.25	208.54	1.73	53.14	nr	**261.68**
Sum of two sides 3900 mm	170.52	213.90	1.80	55.29	nr	**269.19**
Sum of two sides 4000 mm	174.78	219.24	1.80	55.29	nr	**274.53**
Sum of two sides 4100 mm	179.06	224.62	1.80	55.29	nr	**279.91**
Sum of two sides 4200 mm	183.32	229.96	1.80	55.29	nr	**285.25**
Sum of two sides 4300 mm	187.59	235.31	1.88	57.75	nr	**293.06**
Sum of two sides 4400 mm	191.86	240.67	1.88	57.75	nr	**298.42**
Sum of two sides 4500 mm	196.13	246.03	1.88	57.75	nr	**303.78**
Reducer						
Sum of two sides 3000 mm	133.94	168.01	3.12	95.85	nr	**263.86**
Sum of two sides 3100 mm	140.25	175.93	3.12	95.85	nr	**271.78**
Sum of two sides 3200 mm	145.70	182.76	3.12	95.85	nr	**278.61**
Sum of two sides 3300 mm	151.17	189.63	3.12	95.85	nr	**285.48**
Sum of two sides 3400 mm	156.56	196.39	3.12	95.85	nr	**292.24**
Sum of two sides 3500 mm	162.02	203.24	3.12	95.85	nr	**299.09**
Sum of two sides 3600 mm	167.48	210.09	3.95	121.35	nr	**331.44**
Sum of two sides 3700 mm	173.09	217.12	3.97	121.97	nr	**339.09**
Sum of two sides 3800 mm	178.54	223.96	4.52	138.85	nr	**362.81**
Sum of two sides 3900 mm	174.15	218.46	4.52	138.85	nr	**357.31**
Sum of two sides 4000 mm	179.60	225.29	4.52	138.85	nr	**364.14**
Sum of two sides 4100 mm	185.25	232.38	4.52	138.85	nr	**371.23**
Sum of two sides 4200 mm	190.72	239.24	4.52	138.85	nr	**378.09**
Sum of two sides 4300 mm	196.17	246.08	4.92	151.14	nr	**397.22**
Sum of two sides 4400 mm	201.75	253.08	4.92	151.14	nr	**404.22**
Sum of two sides 4500 mm	207.20	259.91	5.12	157.29	nr	**417.20**
Offset						
Sum of two sides 3000 mm	372.21	466.91	3.48	106.92	nr	**573.83**
Sum of two sides 3100 mm	366.22	459.39	3.48	106.92	nr	**566.31**
Sum of two sides 3200 mm	358.05	449.14	3.48	106.92	nr	**556.06**
Sum of two sides 3300 mm	349.01	437.80	3.48	106.92	nr	**544.72**
Sum of two sides 3400 mm	382.36	479.63	3.50	107.53	nr	**587.16**
Sum of two sides 3500 mm	372.45	467.20	3.49	107.22	nr	**574.42**
Sum of two sides 3600 mm	361.66	453.67	4.25	130.56	nr	**584.23**
Sum of two sides 3700 mm	396.05	496.81	4.76	146.23	nr	**643.04**
Sum of two sides 3800 mm	384.37	482.15	5.32	163.43	nr	**645.58**
Sum of two sides 3900 mm	407.56	511.25	5.35	164.35	nr	**675.60**
Sum of two sides 4000 mm	393.93	494.14	5.35	164.35	nr	**658.49**
Sum of two sides 4100 mm	379.54	476.09	5.85	179.72	nr	**655.81**
Sum of two sides 4200 mm	364.15	456.79	5.85	179.72	nr	**636.51**
Sum of two sides 4300 mm	374.58	469.87	6.15	188.92	nr	**658.79**
Sum of two sides 4400 mm	412.22	517.09	6.30	193.54	nr	**710.63**
Sum of two sides 4500 mm	395.50	496.12	6.15	188.92	nr	**685.04**
Square to round						
Sum of two sides 3000 mm	272.72	342.10	4.70	144.38	nr	**486.48**
Sum of two sides 3100 mm	284.43	356.79	4.70	144.38	nr	**501.17**
Sum of two sides 3200 mm	294.86	369.87	4.70	144.38	nr	**514.25**

38 VENTILATION/AIR CONDITIONING SYSTEMS

Item	Net Price £	Material £	Labour hours	Labour £	Unit	Total rate £
DUCTWORK: RECTANGULAR – CLASS B – cont						
Extra over fittings – cont						
Square to round – cont						
Sum of two sides 3300 mm	305.30	382.97	4.70	144.38	nr	527.35
Sum of two sides 3400 mm	315.75	396.08	4.71	144.69	nr	540.77
Sum of two sides 3500 mm	326.18	409.16	4.71	144.69	nr	553.85
Sum of two sides 3600 mm	336.61	422.24	8.19	251.60	nr	673.84
Sum of two sides 3700 mm	347.04	435.32	8.62	264.80	nr	700.12
Sum of two sides 3800 mm	357.48	448.43	8.62	264.80	nr	713.23
Sum of two sides 3900 mm	437.04	548.22	8.62	264.80	nr	813.02
Sum of two sides 4000 mm	449.47	563.82	8.75	268.80	nr	832.62
Sum of two sides 4100 mm	407.87	511.63	11.23	344.98	nr	856.61
Sum of two sides 4200 mm	474.30	594.97	11.23	344.98	nr	939.95
Sum of two sides 4300 mm	486.72	610.55	11.25	345.60	nr	956.15
Sum of two sides 4400 mm	499.14	626.12	11.25	345.60	nr	971.72
Sum of two sides 4500 mm	511.57	641.72	11.26	345.91	nr	987.63
90° radius bend						
Sum of two sides 3000 mm	791.25	992.54	3.90	119.81	nr	1112.35
Sum of two sides 3100 mm	828.87	1039.73	3.90	119.81	nr	1159.54
Sum of two sides 3200 mm	862.62	1082.07	4.26	130.87	nr	1212.94
Sum of two sides 3300 mm	896.37	1124.40	4.26	130.87	nr	1255.27
Sum of two sides 3400 mm	875.36	1098.05	4.26	130.87	nr	1228.92
Sum of two sides 3500 mm	908.67	1139.84	4.55	139.78	nr	1279.62
Sum of two sides 3600 mm	941.99	1181.63	4.55	139.78	nr	1321.41
Sum of two sides 3700 mm	918.48	1152.14	6.87	211.05	nr	1363.19
Sum of two sides 3800 mm	951.36	1193.38	6.87	211.05	nr	1404.43
Sum of two sides 3900 mm	890.46	1117.00	6.87	211.05	nr	1328.05
Sum of two sides 4000 mm	922.92	1157.71	7.00	215.04	nr	1372.75
Sum of two sides 4100 mm	956.30	1199.59	7.20	221.19	nr	1420.78
Sum of two sides 4200 mm	988.75	1240.29	7.20	221.19	nr	1461.48
Sum of two sides 4300 mm	1021.20	1280.99	7.41	227.64	nr	1508.63
Sum of two sides 4400 mm	954.04	1196.74	7.41	227.64	nr	1424.38
Sum of two sides 4500 mm	985.85	1236.65	7.55	231.94	nr	1468.59
45° bend						
Sum of two sides 3000 mm	383.86	481.51	4.85	148.99	nr	630.50
Sum of two sides 3100 mm	402.59	505.01	4.85	148.99	nr	654.00
Sum of two sides 3200 mm	419.40	526.10	4.85	148.99	nr	675.09
Sum of two sides 3300 mm	430.22	539.67	4.87	149.61	nr	689.28
Sum of two sides 3400 mm	425.22	533.40	4.87	149.61	nr	683.01
Sum of two sides 3500 mm	441.82	554.22	4.87	149.61	nr	703.83
Sum of two sides 3600 mm	458.42	575.04	8.81	270.65	nr	845.69
Sum of two sides 3700 mm	446.16	559.66	8.81	270.65	nr	830.31
Sum of two sides 3800 mm	462.54	580.20	9.31	286.00	nr	866.20
Sum of two sides 3900 mm	430.91	540.53	9.31	286.00	nr	826.53
Sum of two sides 4000 mm	447.08	560.82	9.31	286.00	nr	846.82
Sum of two sides 4100 mm	463.70	581.66	10.01	307.51	nr	889.17
Sum of two sides 4200 mm	479.87	601.94	9.31	286.00	nr	887.94
Sum of two sides 4300 mm	496.04	622.23	10.01	307.51	nr	929.74
Sum of two sides 4400 mm	461.72	579.19	10.52	323.18	nr	902.37
Sum of two sides 4500 mm	477.58	599.08	10.52	323.18	nr	922.26

38 VENTILATION/AIR CONDITIONING SYSTEMS

Item	Net Price £	Material £	Labour hours	Labour £	Unit	Total rate £
90° mitre bend						
Sum of two sides 3000 mm	930.76	1167.54	4.85	148.99	nr	1316.53
Sum of two sides 3100 mm	976.64	1225.10	4.85	148.99	nr	1374.09
Sum of two sides 3200 mm	413.21	518.34	4.87	149.61	nr	667.95
Sum of two sides 3300 mm	1062.94	1333.35	4.87	149.61	nr	1482.96
Sum of two sides 3400 mm	1036.65	1300.38	4.87	149.61	nr	1449.99
Sum of two sides 3500 mm	1079.41	1354.01	8.81	270.65	nr	1624.66
Sum of two sides 3600 mm	1122.18	1407.66	8.81	270.65	nr	1678.31
Sum of two sides 3700 mm	1100.94	1381.02	14.81	454.97	nr	1835.99
Sum of two sides 3800 mm	1143.38	1434.26	14.81	454.97	nr	1889.23
Sum of two sides 3900 mm	1063.23	1333.72	14.81	454.97	nr	1788.69
Sum of two sides 4000 mm	1105.29	1386.47	15.20	466.95	nr	1853.42
Sum of two sides 4100 mm	1146.79	1438.53	16.30	500.74	nr	1939.27
Sum of two sides 4200 mm	1188.85	1491.29	16.50	506.88	nr	1998.17
Sum of two sides 4300 mm	1230.92	1544.07	17.01	522.55	nr	2066.62
Sum of two sides 4400 mm	1153.98	1447.56	17.01	522.55	nr	1970.11
Sum of two sides 4500 mm	1195.52	1499.66	17.01	522.55	nr	2022.21
Branch						
Sum of two sides 3000 mm	399.39	501.00	2.88	88.47	nr	589.47
Sum of two sides 3100 mm	1019.78	1279.21	2.88	88.47	nr	1367.68
Sum of two sides 3200 mm	425.67	533.96	2.88	88.47	nr	622.43
Sum of two sides 3300 mm	438.12	549.57	2.88	88.47	nr	638.04
Sum of two sides 3400 mm	450.53	565.14	2.88	88.47	nr	653.61
Sum of two sides 3500 mm	462.99	580.78	3.94	121.04	nr	701.82
Sum of two sides 3600 mm	475.46	596.42	3.94	121.04	nr	717.46
Sum of two sides 3700 mm	488.07	612.24	3.94	121.04	nr	733.28
Sum of two sides 3800 mm	500.53	627.86	4.83	148.38	nr	776.24
Sum of two sides 3900 mm	513.49	644.12	4.83	148.38	nr	792.50
Sum of two sides 4000 mm	525.96	659.77	4.83	148.38	nr	808.15
Sum of two sides 4100 mm	538.63	675.66	5.44	167.13	nr	842.79
Sum of two sides 4200 mm	551.09	691.29	5.44	167.13	nr	858.42
Sum of two sides 4300 mm	563.54	706.90	5.85	179.72	nr	886.62
Sum of two sides 4400 mm	567.14	711.42	5.85	179.72	nr	891.14
Sum of two sides 4500 mm	588.60	738.34	5.85	179.72	nr	918.06
Grille neck						
Sum of two sides 3000 mm	412.54	517.48	2.88	88.47	nr	605.95
Sum of two sides 3100 mm	427.22	535.91	2.88	88.47	nr	624.38
Sum of two sides 3200 mm	440.61	552.70	2.88	88.47	nr	641.17
Sum of two sides 3300 mm	454.01	569.51	2.88	88.47	nr	657.98
Sum of two sides 3400 mm	467.41	586.32	3.94	121.04	nr	707.36
Sum of two sides 3500 mm	480.81	603.13	3.94	121.04	nr	724.17
Sum of two sides 3600 mm	414.19	519.56	3.94	121.04	nr	640.60
Sum of two sides 3700 mm	507.59	636.72	4.12	126.57	nr	763.29
Sum of two sides 3800 mm	520.99	653.53	4.12	126.57	nr	780.10
Sum of two sides 3900 mm	534.39	670.34	4.12	126.57	nr	796.91
Sum of two sides 4000 mm	547.78	687.13	5.00	153.60	nr	840.73
Sum of two sides 4100 mm	561.18	703.94	5.00	153.60	nr	857.54
Sum of two sides 4200 mm	574.58	720.75	5.00	153.60	nr	874.35
Sum of two sides 4300 mm	587.98	737.56	5.23	160.66	nr	898.22
Sum of two sides 4400 mm	601.37	754.35	5.23	160.66	nr	915.01
Sum of two sides 4500 mm	614.77	771.16	5.39	165.58	nr	936.74

38 VENTILATION/AIR CONDITIONING SYSTEMS

Item	Net Price £	Material £	Labour hours	Labour £	Unit	Total rate £
DUCTWORK: RECTANGULAR – CLASS C						
AIR DUCTLINES						
Galvanized sheet metal DW144 class C rectangular section ductwork; including all necessary stiffeners, joints, couplers in the running length and duct supports						
Ductwork up to 400 mm longest side						
Sum of two sides 200 mm	24.90	31.24	1.17	35.94	m	67.18
Sum of two sides 300 mm	26.78	33.59	1.17	35.94	m	69.53
Sum of two sides 400 mm	24.50	30.73	1.19	36.56	m	67.29
Sum of two sides 500 mm	26.96	33.82	1.16	35.64	m	69.46
Sum of two sides 600 mm	29.17	36.59	1.19	36.56	m	73.15
Sum of two sides 700 mm	31.38	39.37	1.19	36.56	m	75.93
Sum of two sides 800 mm	33.71	42.29	1.19	36.56	m	78.85
Extra over fittings; Ductwork up to 400 mm longest side						
End cap						
Sum of two sides 200 mm	15.02	18.84	0.38	11.68	nr	30.52
Sum of two sides 300 mm	16.78	21.04	0.38	11.68	nr	32.72
Sum of two sides 400 mm	18.53	23.24	0.38	11.68	nr	34.92
Sum of two sides 500 mm	20.28	25.44	0.38	11.68	nr	37.12
Sum of two sides 600 mm	22.05	27.66	0.38	11.68	nr	39.34
Sum of two sides 700 mm	23.80	29.86	0.38	11.68	nr	41.54
Sum of two sides 800 mm	25.55	32.05	0.38	11.68	nr	43.73
Reducer						
Sum of two sides 200 mm	24.39	30.60	1.40	43.01	nr	73.61
Sum of two sides 300 mm	27.53	34.53	1.40	43.01	nr	77.54
Sum of two sides 400 mm	45.06	56.53	1.42	43.62	nr	100.15
Sum of two sides 500 mm	48.76	61.16	1.42	43.62	nr	104.78
Sum of two sides 600 mm	52.47	65.82	1.69	51.92	nr	117.74
Sum of two sides 700 mm	56.17	70.46	1.69	51.92	nr	122.38
Sum of two sides 800 mm	59.83	75.05	1.92	58.98	nr	134.03
Offset						
Sum of two sides 200 mm	35.98	45.14	1.63	50.08	nr	95.22
Sum of two sides 300 mm	40.68	51.03	1.63	50.08	nr	101.11
Sum of two sides 400 mm	60.09	75.38	1.65	50.69	nr	126.07
Sum of two sides 500 mm	65.32	81.94	1.65	50.69	nr	132.63
Sum of two sides 600 mm	69.58	87.28	1.92	58.98	nr	146.26
Sum of two sides 700 mm	74.31	93.22	1.92	58.98	nr	152.20
Sum of two sides 800 mm	78.44	98.39	1.92	58.98	nr	157.37
Square to round						
Sum of two sides 200 mm	31.65	39.70	1.22	37.49	nr	77.19
Sum of two sides 300 mm	35.56	44.61	1.22	37.49	nr	82.10
Sum of two sides 400 mm	47.31	59.35	1.25	38.39	nr	97.74
Sum of two sides 500 mm	51.48	64.58	1.25	38.39	nr	102.97
Sum of two sides 600 mm	55.64	69.80	1.33	40.86	nr	110.66
Sum of two sides 700 mm	59.82	75.04	1.33	40.86	nr	115.90
Sum of two sides 800 mm	63.94	80.20	1.40	43.01	nr	123.21

38 VENTILATION/AIR CONDITIONING SYSTEMS

Item	Net Price £	Material £	Labour hours	Labour £	Unit	Total rate £
90° radius bend						
Sum of two sides 200 mm	23.81	29.87	1.10	33.79	nr	63.66
Sum of two sides 300 mm	25.69	32.22	1.10	33.79	nr	66.01
Sum of two sides 400 mm	43.24	54.24	1.12	34.41	nr	88.65
Sum of two sides 500 mm	45.97	57.67	1.12	34.41	nr	92.08
Sum of two sides 600 mm	49.64	62.27	1.16	35.64	nr	97.91
Sum of two sides 700 mm	52.81	66.25	1.16	35.64	nr	101.89
Sum of two sides 800 mm	56.31	70.64	1.22	37.49	nr	108.13
45° radius bend						
Sum of two sides 200 mm	25.71	32.26	1.29	39.63	nr	71.89
Sum of two sides 300 mm	28.23	35.41	1.29	39.63	nr	75.04
Sum of two sides 400 mm	45.43	56.99	1.29	39.63	nr	96.62
Sum of two sides 500 mm	48.58	60.94	1.29	39.63	nr	100.57
Sum of two sides 600 mm	52.28	65.58	1.39	42.69	nr	108.27
Sum of two sides 700 mm	55.74	69.92	1.39	42.69	nr	112.61
Sum of two sides 800 mm	59.39	74.50	1.46	44.86	nr	119.36
90° mitre bend						
Sum of two sides 200 mm	39.80	49.93	2.04	62.68	nr	112.61
Sum of two sides 300 mm	43.73	54.86	2.04	62.68	nr	117.54
Sum of two sides 400 mm	63.53	79.69	2.09	64.20	nr	143.89
Sum of two sides 500 mm	68.58	86.03	2.09	64.20	nr	150.23
Sum of two sides 600 mm	75.08	94.18	2.15	66.05	nr	160.23
Sum of two sides 700 mm	81.11	101.74	2.15	66.05	nr	167.79
Sum of two sides 800 mm	87.71	110.03	2.26	69.43	nr	179.46
Branch						
Sum of two sides 200 mm	39.56	49.63	0.92	28.26	nr	77.89
Sum of two sides 300 mm	44.03	55.23	0.92	28.26	nr	83.49
Sum of two sides 400 mm	55.03	69.03	0.95	29.19	nr	98.22
Sum of two sides 500 mm	60.32	75.67	0.95	29.19	nr	104.86
Sum of two sides 600 mm	65.19	81.77	1.03	31.64	nr	113.41
Sum of two sides 700 mm	70.06	87.89	1.03	31.64	nr	119.53
Sum of two sides 800 mm	74.92	93.98	1.03	31.64	nr	125.62
Grille neck						
Sum of two sides 200 mm	45.68	57.30	1.10	33.79	nr	91.09
Sum of two sides 300 mm	51.33	64.39	1.10	33.79	nr	98.18
Sum of two sides 400 mm	56.98	71.48	1.16	35.64	nr	107.12
Sum of two sides 500 mm	62.63	78.57	1.16	35.64	nr	114.21
Sum of two sides 600 mm	68.28	85.65	1.18	36.25	nr	121.90
Sum of two sides 700 mm	73.93	92.74	1.18	36.25	nr	128.99
Sum of two sides 800 mm	79.58	99.83	1.18	36.25	nr	136.08
Ductwork 401–600 mm longest side						
Sum of two sides 600 mm	38.51	48.31	1.17	35.94	m	84.25
Sum of two sides 700 mm	42.35	53.12	1.17	35.94	m	89.06
Sum of two sides 800 mm	46.13	57.87	1.17	35.94	m	93.81
Sum of two sides 900 mm	49.80	62.47	1.27	39.01	m	101.48
Sum of two sides 1000 mm	53.46	67.07	1.48	45.47	m	112.54
Sum of two sides 1100 mm	57.42	72.03	1.49	45.77	m	117.80
Sum of two sides 1200 mm	61.09	76.63	1.49	45.77	m	122.40

38 VENTILATION/AIR CONDITIONING SYSTEMS

Item	Net Price £	Material £	Labour hours	Labour £	Unit	Total rate £
DUCTWORK: RECTANGULAR – CLASS C – cont						
Extra over fittings: Ductwork 401–600 mm longest side						
End cap						
Sum of two sides 600 mm	22.05	27.66	0.38	11.68	nr	39.34
Sum of two sides 700 mm	23.80	29.86	0.38	11.68	nr	41.54
Sum of two sides 800 mm	25.55	32.05	0.38	11.68	nr	43.73
Sum of two sides 900 mm	27.30	34.25	0.38	11.68	nr	45.93
Sum of two sides 1000 mm	29.06	36.46	0.38	11.68	nr	48.14
Sum of two sides 1100 mm	30.81	38.65	0.38	11.68	nr	50.33
Sum of two sides 1200 mm	32.56	40.85	0.38	11.68	nr	52.53
Reducer						
Sum of two sides 600 mm	47.08	59.06	1.69	51.92	nr	110.98
Sum of two sides 700 mm	50.11	62.85	1.69	51.92	nr	114.77
Sum of two sides 800 mm	53.08	66.58	1.92	58.98	nr	125.56
Sum of two sides 900 mm	56.13	70.41	1.92	58.98	nr	129.39
Sum of two sides 1000 mm	59.16	74.21	2.18	66.98	nr	141.19
Sum of two sides 1100 mm	62.49	78.39	2.18	66.98	nr	145.37
Sum of two sides 1200 mm	65.52	82.19	2.18	66.98	nr	149.17
Offset						
Sum of two sides 600 mm	66.32	83.19	1.92	58.98	nr	142.17
Sum of two sides 700 mm	70.80	88.82	1.92	58.98	nr	147.80
Sum of two sides 800 mm	73.30	91.95	1.92	58.98	nr	150.93
Sum of two sides 900 mm	75.60	94.83	1.92	58.98	nr	153.81
Sum of two sides 1000 mm	78.75	98.78	2.18	66.98	nr	165.76
Sum of two sides 1100 mm	80.80	101.36	2.18	66.98	nr	168.34
Sum of two sides 1200 mm	82.39	103.35	2.18	66.98	nr	170.33
Square to round						
Sum of two sides 600 mm	50.04	62.76	1.33	40.86	nr	103.62
Sum of two sides 700 mm	53.53	67.14	1.33	40.86	nr	108.00
Sum of two sides 800 mm	56.98	71.48	1.40	43.01	nr	114.49
Sum of two sides 900 mm	60.47	75.86	1.40	43.01	nr	118.87
Sum of two sides 1000 mm	63.98	80.26	1.82	55.91	nr	136.17
Sum of two sides 1100 mm	67.55	84.74	1.82	55.91	nr	140.65
Sum of two sides 1200 mm	71.06	89.14	1.82	55.91	nr	145.05
90° radius bend						
Sum of two sides 600 mm	44.40	55.70	1.16	35.64	nr	91.34
Sum of two sides 700 mm	45.47	57.04	1.16	35.64	nr	92.68
Sum of two sides 800 mm	48.56	60.92	1.22	37.49	nr	98.41
Sum of two sides 900 mm	51.75	64.92	1.22	37.49	nr	102.41
Sum of two sides 1000 mm	53.68	67.33	1.40	43.01	nr	110.34
Sum of two sides 1100 mm	57.03	71.53	1.40	43.01	nr	114.54
Sum of two sides 1200 mm	60.17	75.48	1.40	43.01	nr	118.49
45° bend						
Sum of two sides 600 mm	49.50	62.09	1.39	42.69	nr	104.78
Sum of two sides 700 mm	51.84	65.03	1.39	42.69	nr	107.72
Sum of two sides 800 mm	55.22	69.27	1.46	44.86	nr	114.13

38 VENTILATION/AIR CONDITIONING SYSTEMS

Item	Net Price £	Material £	Labour hours	Labour £	Unit	Total rate £
Sum of two sides 900 mm	58.65	73.57	1.46	44.86	nr	118.43
Sum of two sides 1000 mm	61.44	77.07	1.88	57.75	nr	134.82
Sum of two sides 1100 mm	65.17	81.75	1.88	57.75	nr	139.50
Sum of two sides 1200 mm	68.59	86.04	1.88	57.75	nr	143.79
90° mitre bend						
Sum of two sides 600 mm	77.88	97.70	2.15	66.05	nr	163.75
Sum of two sides 700 mm	80.92	101.51	2.15	66.05	nr	167.56
Sum of two sides 800 mm	87.00	109.13	2.26	69.43	nr	178.56
Sum of two sides 900 mm	93.20	116.91	2.26	69.43	nr	186.34
Sum of two sides 1000 mm	97.93	122.84	3.03	93.07	nr	215.91
Sum of two sides 1100 mm	104.40	130.96	3.03	93.07	nr	224.03
Sum of two sides 1200 mm	110.70	138.86	3.03	93.07	nr	231.93
Branch						
Sum of two sides 600 mm	65.77	82.50	1.03	31.64	nr	114.14
Sum of two sides 700 mm	70.69	88.67	1.03	31.64	nr	120.31
Sum of two sides 800 mm	75.61	94.84	1.03	31.64	nr	126.48
Sum of two sides 900 mm	80.54	101.02	1.03	31.64	nr	132.66
Sum of two sides 1000 mm	85.46	107.21	1.29	39.63	nr	146.84
Sum of two sides 1100 mm	90.60	113.65	1.29	39.63	nr	153.28
Sum of two sides 1200 mm	95.52	119.82	1.29	39.63	nr	159.45
Grille neck						
Sum of two sides 600 mm	68.28	85.65	1.18	36.25	nr	121.90
Sum of two sides 700 mm	73.93	92.74	1.18	36.25	nr	128.99
Sum of two sides 800 mm	79.58	99.83	1.18	36.25	nr	136.08
Sum of two sides 900 mm	85.23	106.92	1.18	36.25	nr	143.17
Sum of two sides 1000 mm	90.88	114.00	1.44	44.24	nr	158.24
Sum of two sides 1100 mm	96.53	121.08	1.44	44.24	nr	165.32
Sum of two sides 1200 mm	102.18	128.17	1.44	44.24	nr	172.41
Ductwork 601–800 mm longest side						
Sum of two sides 900 mm	54.06	67.82	1.27	39.01	m	106.83
Sum of two sides 1000 mm	57.73	72.42	1.48	45.47	m	117.89
Sum of two sides 1100 mm	61.40	77.02	1.49	45.77	m	122.79
Sum of two sides 1200 mm	65.35	81.97	1.49	45.77	m	127.74
Sum of two sides 1300 mm	69.02	86.58	1.51	46.39	m	132.97
Sum of two sides 1400 mm	72.08	90.42	1.55	47.62	m	138.04
Sum of two sides 1500 mm	76.35	95.77	1.61	49.46	m	145.23
Sum of two sides 1600 mm	80.02	100.37	1.62	49.76	m	150.13
Extra over fittings; Ductwork 601–800 mm longest side						
End cap						
Sum of two sides 900 mm	27.30	34.25	0.38	11.68	nr	45.93
Sum of two sides 1000 mm	29.06	36.46	0.38	11.68	nr	48.14
Sum of two sides 1100 mm	30.81	38.65	0.38	11.68	nr	50.33
Sum of two sides 1200 mm	32.56	40.85	0.38	11.68	nr	52.53
Sum of two sides 1300 mm	34.32	43.05	0.38	11.68	nr	54.73
Sum of two sides 1400 mm	35.82	44.93	0.38	11.68	nr	56.61
Sum of two sides 1500 mm	41.60	52.18	0.38	11.68	nr	63.86
Sum of two sides 1600 mm	47.42	59.48	0.38	11.68	nr	71.16

38 VENTILATION/AIR CONDITIONING SYSTEMS

Item	Net Price £	Material £	Labour hours	Labour £	Unit	Total rate £
DUCTWORK: RECTANGULAR – CLASS C – cont						
Extra over fittings – cont						
Reducer						
Sum of two sides 900 mm	56.72	71.15	1.92	58.98	nr	130.13
Sum of two sides 1000 mm	59.75	74.95	2.18	66.98	nr	141.93
Sum of two sides 1100 mm	62.78	78.75	2.18	66.98	nr	145.73
Sum of two sides 1200 mm	66.11	82.92	2.18	66.98	nr	149.90
Sum of two sides 1300 mm	69.14	86.73	2.30	70.66	nr	157.39
Sum of two sides 1400 mm	71.69	89.92	2.30	70.66	nr	160.58
Sum of two sides 1500 mm	82.82	103.89	2.47	75.88	nr	179.77
Sum of two sides 1600 mm	93.95	117.85	2.47	75.88	nr	193.73
Offset						
Sum of two sides 900 mm	79.99	100.34	1.92	58.98	nr	159.32
Sum of two sides 1000 mm	82.01	102.87	2.18	66.98	nr	169.85
Sum of two sides 1100 mm	83.79	105.10	2.18	66.98	nr	172.08
Sum of two sides 1200 mm	85.49	107.24	2.18	66.98	nr	174.22
Sum of two sides 1300 mm	86.74	108.81	2.47	75.88	nr	184.69
Sum of two sides 1400 mm	88.48	110.99	2.47	75.88	nr	186.87
Sum of two sides 1500 mm	101.42	127.22	2.47	75.88	nr	203.10
Sum of two sides 1600 mm	114.10	143.12	2.47	75.88	nr	219.00
Square to round						
Sum of two sides 900 mm	59.94	75.19	1.40	43.01	nr	118.20
Sum of two sides 1000 mm	63.44	79.58	1.82	55.91	nr	135.49
Sum of two sides 1100 mm	66.94	83.97	1.82	55.91	nr	139.88
Sum of two sides 1200 mm	70.52	88.46	1.82	55.91	nr	144.37
Sum of two sides 1300 mm	74.02	92.85	2.32	71.28	nr	164.13
Sum of two sides 1400 mm	76.90	96.47	2.32	71.28	nr	167.75
Sum of two sides 1500 mm	90.51	113.53	2.56	78.64	nr	192.17
Sum of two sides 1600 mm	104.13	130.63	2.58	79.25	nr	209.88
90° radius bend						
Sum of two sides 900 mm	45.88	57.56	1.22	37.49	nr	95.05
Sum of two sides 1000 mm	48.91	61.35	1.40	43.01	nr	104.36
Sum of two sides 1100 mm	51.94	65.15	1.40	43.01	nr	108.16
Sum of two sides 1200 mm	55.15	69.18	1.40	43.01	nr	112.19
Sum of two sides 1300 mm	58.18	72.98	1.91	58.68	nr	131.66
Sum of two sides 1400 mm	59.06	74.09	1.91	58.68	nr	132.77
Sum of two sides 1500 mm	70.14	87.99	2.11	64.81	nr	152.80
Sum of two sides 1600 mm	81.22	101.89	2.11	64.81	nr	166.70
45° bend						
Sum of two sides 900 mm	56.57	70.96	1.46	44.86	nr	115.82
Sum of two sides 1000 mm	59.93	75.17	1.88	57.75	nr	132.92
Sum of two sides 1100 mm	63.29	79.39	1.88	57.75	nr	137.14
Sum of two sides 1200 mm	66.94	83.97	1.88	57.75	nr	141.72
Sum of two sides 1300 mm	70.30	88.19	2.26	69.43	nr	157.62
Sum of two sides 1400 mm	72.32	90.72	2.44	74.96	nr	165.68
Sum of two sides 1500 mm	83.74	105.04	2.44	74.96	nr	180.00
Sum of two sides 1600 mm	95.18	119.39	2.68	82.33	nr	201.72

38 VENTILATION/AIR CONDITIONING SYSTEMS

Item	Net Price £	Material £	Labour hours	Labour £	Unit	Total rate £
90° mitre bend						
Sum of two sides 900 mm	84.63	106.16	2.26	69.43	nr	175.59
Sum of two sides 1000 mm	91.22	114.43	3.03	93.07	nr	207.50
Sum of two sides 1100 mm	97.83	122.72	3.03	93.07	nr	215.79
Sum of two sides 1200 mm	104.56	131.16	3.03	93.07	nr	224.23
Sum of two sides 1300 mm	111.16	139.44	3.85	118.27	nr	257.71
Sum of two sides 1400 mm	115.04	144.30	3.85	118.27	nr	262.57
Sum of two sides 1500 mm	133.88	167.94	4.25	130.56	nr	298.50
Sum of two sides 1600 mm	152.72	191.58	4.26	130.87	nr	322.45
Branch						
Sum of two sides 900 mm	82.91	104.00	1.03	31.64	nr	135.64
Sum of two sides 1000 mm	87.97	110.35	1.29	39.63	nr	149.98
Sum of two sides 1100 mm	93.02	116.68	1.29	39.63	nr	156.31
Sum of two sides 1200 mm	98.30	123.31	1.29	39.63	nr	162.94
Sum of two sides 1300 mm	103.37	129.66	1.39	42.69	nr	172.35
Sum of two sides 1400 mm	107.76	135.17	1.39	42.69	nr	177.86
Sum of two sides 1500 mm	123.60	155.04	1.64	50.38	nr	205.42
Sum of two sides 1600 mm	139.44	174.91	1.64	50.38	nr	225.29
Grille neck						
Sum of two sides 900 mm	85.23	106.92	1.18	36.25	nr	143.17
Sum of two sides 1000 mm	90.88	114.00	1.44	44.24	nr	158.24
Sum of two sides 1100 mm	96.53	121.08	1.44	44.24	nr	165.32
Sum of two sides 1200 mm	102.18	128.17	1.44	44.24	nr	172.41
Sum of two sides 1300 mm	107.83	135.26	1.69	51.92	nr	187.18
Sum of two sides 1400 mm	112.73	141.41	1.69	51.92	nr	193.33
Sum of two sides 1500 mm	130.52	163.72	1.79	54.99	nr	218.71
Sum of two sides 1600 mm	198.30	248.75	1.79	54.99	nr	303.74
Ductwork 801–1000 mm longest side						
Sum of two sides 1100 mm	62.16	77.97	1.49	45.77	m	123.74
Sum of two sides 1200 mm	65.82	82.57	1.49	45.77	m	128.34
Sum of two sides 1300 mm	69.49	87.17	1.51	46.39	m	133.56
Sum of two sides 1400 mm	73.44	92.12	1.55	47.62	m	139.74
Sum of two sides 1500 mm	77.12	96.73	1.61	49.46	m	146.19
Sum of two sides 1600 mm	80.78	101.33	1.62	49.76	m	151.09
Sum of two sides 1700 mm	84.45	105.93	1.74	53.45	m	159.38
Sum of two sides 1800 mm	88.40	110.89	1.76	54.06	m	164.95
Sum of two sides 1900 mm	92.05	115.47	1.81	55.61	m	171.08
Sum of two sides 2000 mm	95.72	120.08	1.82	55.91	m	175.99
Extra over fittings; Ductwork 801–1000 mm longest side						
End cap						
Sum of two sides 1100 mm	30.80	38.64	0.38	11.68	nr	50.32
Sum of two sides 1200 mm	32.56	40.85	0.38	11.68	nr	52.53
Sum of two sides 1300 mm	34.32	43.05	0.38	11.68	nr	54.73
Sum of two sides 1400 mm	35.83	44.95	0.38	11.68	nr	56.63
Sum of two sides 1500 mm	41.62	52.20	0.38	11.68	nr	63.88
Sum of two sides 1600 mm	47.42	59.48	0.38	11.68	nr	71.16
Sum of two sides 1700 mm	53.22	66.76	0.58	17.81	nr	84.57

38 VENTILATION/AIR CONDITIONING SYSTEMS

Item	Net Price £	Material £	Labour hours	Labour £	Unit	Total rate £
DUCTWORK: RECTANGULAR – CLASS C – cont						
Extra over fittings – cont						
End cap – cont						
Sum of two sides 1800 mm	59.02	74.03	0.58	17.81	nr	91.84
Sum of two sides 1900 mm	64.82	81.31	0.58	17.81	nr	99.12
Sum of two sides 2000 mm	70.62	88.58	0.58	17.81	nr	106.39
Reducer						
Sum of two sides 1100 mm	58.09	72.87	2.18	66.98	nr	139.85
Sum of two sides 1200 mm	60.89	76.38	2.18	66.98	nr	143.36
Sum of two sides 1300 mm	63.69	79.89	2.30	70.66	nr	150.55
Sum of two sides 1400 mm	66.24	83.09	2.30	70.66	nr	153.75
Sum of two sides 1500 mm	77.13	96.76	2.47	75.88	nr	172.64
Sum of two sides 1600 mm	88.00	110.39	2.47	75.88	nr	186.27
Sum of two sides 1700 mm	98.89	124.05	2.59	79.56	nr	203.61
Sum of two sides 1800 mm	110.03	138.02	2.59	79.56	nr	217.58
Sum of two sides 1900 mm	120.91	151.67	2.71	83.25	nr	234.92
Sum of two sides 2000 mm	131.81	165.35	2.71	83.25	nr	248.60
Offset						
Sum of two sides 1100 mm	89.22	111.92	2.18	66.98	nr	178.90
Sum of two sides 1200 mm	90.97	114.12	2.18	66.98	nr	181.10
Sum of two sides 1300 mm	92.46	115.99	2.47	75.88	nr	191.87
Sum of two sides 1400 mm	93.15	116.85	2.47	75.88	nr	192.73
Sum of two sides 1500 mm	106.26	133.29	2.47	75.88	nr	209.17
Sum of two sides 1600 mm	119.10	149.40	2.47	75.88	nr	225.28
Sum of two sides 1700 mm	131.71	165.22	2.61	80.18	nr	245.40
Sum of two sides 1800 mm	144.15	180.82	2.61	80.18	nr	261.00
Sum of two sides 1900 mm	158.06	198.27	2.71	83.25	nr	281.52
Sum of two sides 2000 mm	169.94	213.17	2.71	83.25	nr	296.42
Square to round						
Sum of two sides 1100 mm	62.05	77.84	1.82	55.91	nr	133.75
Sum of two sides 1200 mm	65.29	81.89	1.82	55.91	nr	137.80
Sum of two sides 1300 mm	68.56	86.00	2.32	71.28	nr	157.28
Sum of two sides 1400 mm	71.23	89.35	2.32	71.28	nr	160.63
Sum of two sides 1500 mm	84.60	106.12	2.56	78.64	nr	184.76
Sum of two sides 1600 mm	97.98	122.91	2.58	79.25	nr	202.16
Sum of two sides 1700 mm	111.35	139.68	2.84	87.25	nr	226.93
Sum of two sides 1800 mm	124.75	156.49	2.84	87.25	nr	243.74
Sum of two sides 1900 mm	138.13	173.28	3.13	96.15	nr	269.43
Sum of two sides 2000 mm	151.51	190.05	3.13	96.15	nr	286.20
90° radius bend						
Sum of two sides 1100 mm	47.06	59.04	1.40	43.01	nr	102.05
Sum of two sides 1200 mm	50.79	63.71	1.40	43.01	nr	106.72
Sum of two sides 1300 mm	54.53	68.40	1.91	58.68	nr	127.08
Sum of two sides 1400 mm	57.88	72.61	1.91	58.68	nr	131.29
Sum of two sides 1500 mm	69.70	87.43	2.11	64.81	nr	152.24
Sum of two sides 1600 mm	81.52	102.26	2.11	64.81	nr	167.07
Sum of two sides 1700 mm	93.33	117.07	2.55	78.33	nr	195.40
Sum of two sides 1800 mm	105.28	132.06	2.55	78.33	nr	210.39
Sum of two sides 1900 mm	114.98	144.23	2.80	86.02	nr	230.25
Sum of two sides 2000 mm	126.79	159.04	2.80	86.02	nr	245.06

38 VENTILATION/AIR CONDITIONING SYSTEMS

Item	Net Price £	Material £	Labour hours	Labour £	Unit	Total rate £
45° bend						
Sum of two sides 1100 mm	61.03	76.55	1.88	57.75	nr	134.30
Sum of two sides 1200 mm	64.78	81.26	1.88	57.75	nr	139.01
Sum of two sides 1300 mm	68.55	85.99	2.26	69.43	nr	155.42
Sum of two sides 1400 mm	72.07	90.41	2.44	74.96	nr	165.37
Sum of two sides 1500 mm	83.92	105.27	2.68	82.33	nr	187.60
Sum of two sides 1600 mm	95.76	120.12	2.69	82.64	nr	202.76
Sum of two sides 1700 mm	107.62	134.99	2.96	90.93	nr	225.92
Sum of two sides 1800 mm	119.72	150.18	2.96	90.93	nr	241.11
Sum of two sides 1900 mm	130.47	163.67	3.26	100.15	nr	263.82
Sum of two sides 2000 mm	142.32	178.53	3.26	100.15	nr	278.68
90° mitre bend						
Sum of two sides 1100 mm	97.39	122.17	3.03	93.07	nr	215.24
Sum of two sides 1200 mm	104.45	131.02	3.03	93.07	nr	224.09
Sum of two sides 1300 mm	111.52	139.89	3.85	118.27	nr	258.16
Sum of two sides 1400 mm	117.92	147.92	3.85	118.27	nr	266.19
Sum of two sides 1500 mm	137.12	172.00	4.25	130.56	nr	302.56
Sum of two sides 1600 mm	156.32	196.09	4.26	130.87	nr	326.96
Sum of two sides 1700 mm	175.53	220.18	4.68	143.76	nr	363.94
Sum of two sides 1800 mm	194.81	244.37	4.68	143.76	nr	388.13
Sum of two sides 1900 mm	211.76	265.63	4.87	149.61	nr	415.24
Sum of two sides 2000 mm	231.08	289.87	4.87	149.61	nr	439.48
Branch						
Sum of two sides 1100 mm	93.25	116.97	1.29	39.63	nr	156.60
Sum of two sides 1200 mm	98.30	123.31	1.29	39.63	nr	162.94
Sum of two sides 1300 mm	103.37	129.66	1.39	42.69	nr	172.35
Sum of two sides 1400 mm	107.97	135.44	1.39	42.69	nr	178.13
Sum of two sides 1500 mm	123.82	155.32	1.64	50.38	nr	205.70
Sum of two sides 1600 mm	139.66	175.19	1.64	50.38	nr	225.57
Sum of two sides 1700 mm	155.50	195.06	1.69	51.92	nr	246.98
Sum of two sides 1800 mm	171.57	215.22	1.69	51.92	nr	267.14
Sum of two sides 1900 mm	187.41	235.09	1.85	56.84	nr	291.93
Sum of two sides 2000 mm	203.25	254.96	1.85	56.84	nr	311.80
Grille neck						
Sum of two sides 1100 mm	96.53	121.08	1.44	44.24	nr	165.32
Sum of two sides 1200 mm	102.18	128.17	1.44	44.24	nr	172.41
Sum of two sides 1300 mm	107.83	135.26	1.69	51.92	nr	187.18
Sum of two sides 1400 mm	112.73	141.41	1.69	51.92	nr	193.33
Sum of two sides 1500 mm	130.52	163.72	1.79	54.99	nr	218.71
Sum of two sides 1600 mm	148.29	186.01	1.79	54.99	nr	241.00
Sum of two sides 1700 mm	166.09	208.34	1.86	57.14	nr	265.48
Sum of two sides 1800 mm	183.86	230.63	1.86	57.14	nr	287.77
Sum of two sides 1900 mm	201.66	252.96	2.02	62.06	nr	315.02
Sum of two sides 2000 mm	219.44	275.26	2.02	62.06	nr	337.32
Ductwork 1001–1250 mm longest side						
Sum of two sides 1300 mm	104.32	130.86	1.51	46.39	m	177.25
Sum of two sides 1400 mm	109.43	137.27	1.55	47.62	m	184.89
Sum of two sides 1500 mm	114.25	143.32	1.61	49.46	m	192.78

38 VENTILATION/AIR CONDITIONING SYSTEMS

Item	Net Price £	Material £	Labour hours	Labour £	Unit	Total rate £
DUCTWORK: RECTANGULAR – CLASS C – cont						
Ductwork 1001–1250 mm – cont						
Sum of two sides 1600 mm	119.36	149.72	1.62	49.76	m	199.48
Sum of two sides 1700 mm	124.19	155.78	1.74	53.45	m	209.23
Sum of two sides 1800 mm	129.01	161.83	1.76	54.06	m	215.89
Sum of two sides 1900 mm	133.83	167.88	1.81	55.61	m	223.49
Sum of two sides 2000 mm	138.94	174.28	1.82	55.91	m	230.19
Sum of two sides 2100 mm	143.77	180.34	2.53	77.73	m	258.07
Sum of two sides 2200 mm	163.50	205.09	2.55	78.33	m	283.42
Sum of two sides 2300 mm	168.92	211.89	2.56	78.64	m	290.53
Sum of two sides 2400 mm	173.74	217.94	2.76	84.78	m	302.72
Sum of two sides 2500 mm	178.85	224.35	2.77	85.10	m	309.45
Extra over fittings; Ductwork 1001– 1250 mm longest side						
End cap						
Sum of two sides 1300 mm	35.02	43.93	0.38	11.68	nr	55.61
Sum of two sides 1400 mm	36.59	45.90	0.38	11.68	nr	57.58
Sum of two sides 1500 mm	42.43	53.22	0.38	11.68	nr	64.90
Sum of two sides 1600 mm	48.29	60.57	0.38	11.68	nr	72.25
Sum of two sides 1700 mm	54.14	67.92	0.58	17.81	nr	85.73
Sum of two sides 1800 mm	59.99	75.25	0.58	17.81	nr	93.06
Sum of two sides 1900 mm	65.84	82.59	0.58	17.81	nr	100.40
Sum of two sides 2000 mm	71.71	89.96	0.58	17.81	nr	107.77
Sum of two sides 2100 mm	77.55	97.28	0.87	26.73	nr	124.01
Sum of two sides 2200 mm	83.41	104.63	0.87	26.73	nr	131.36
Sum of two sides 2300 mm	89.26	111.97	0.87	26.73	nr	138.70
Sum of two sides 2400 mm	95.11	119.30	0.87	26.73	nr	146.03
Sum of two sides 2500 mm	100.96	126.65	0.87	26.73	nr	153.38
Reducer						
Sum of two sides 1300 mm	49.69	62.33	2.30	70.66	nr	132.99
Sum of two sides 1400 mm	51.67	64.81	2.30	70.66	nr	135.47
Sum of two sides 1500 mm	62.18	78.00	2.47	75.88	nr	153.88
Sum of two sides 1600 mm	72.95	91.50	2.47	75.88	nr	167.38
Sum of two sides 1700 mm	83.48	104.72	2.59	79.56	nr	184.28
Sum of two sides 1800 mm	94.01	117.92	2.59	79.56	nr	197.48
Sum of two sides 1900 mm	104.53	131.12	2.71	83.25	nr	214.37
Sum of two sides 2000 mm	115.29	144.61	2.71	83.25	nr	227.86
Sum of two sides 2100 mm	125.83	157.84	2.92	89.70	nr	247.54
Sum of two sides 2200 mm	127.60	160.06	2.92	89.70	nr	249.76
Sum of two sides 2300 mm	138.37	173.57	2.92	89.70	nr	263.27
Sum of two sides 2400 mm	148.90	186.78	3.12	95.85	nr	282.63
Sum of two sides 2500 mm	159.66	200.28	3.12	95.85	nr	296.13
Offset						
Sum of two sides 1300 mm	118.36	148.47	2.47	75.88	nr	224.35
Sum of two sides 1400 mm	121.88	152.89	2.47	75.88	nr	228.77
Sum of two sides 1500 mm	135.49	169.96	2.47	75.88	nr	245.84
Sum of two sides 1600 mm	148.96	186.86	2.47	75.88	nr	262.74
Sum of two sides 1700 mm	161.84	203.01	2.61	80.18	nr	283.19
Sum of two sides 1800 mm	174.39	218.76	2.61	80.18	nr	298.94
Sum of two sides 1900 mm	186.60	234.07	2.71	83.25	nr	317.32

38 VENTILATION/AIR CONDITIONING SYSTEMS

Item	Net Price £	Material £	Labour hours	Labour £	Unit	Total rate £
Sum of two sides 2000 mm	198.56	249.08	2.71	83.25	nr	332.33
Sum of two sides 2100 mm	210.04	263.47	2.92	89.70	nr	353.17
Sum of two sides 2200 mm	207.01	259.67	3.26	100.15	nr	359.82
Sum of two sides 2300 mm	224.57	281.70	3.26	100.15	nr	381.85
Sum of two sides 2400 mm	242.06	303.64	3.47	106.60	nr	410.24
Sum of two sides 2500 mm	259.62	325.66	3.48	106.92	nr	432.58
Square to round						
Sum of two sides 1300 mm	54.01	67.75	2.32	71.28	nr	139.03
Sum of two sides 1400 mm	56.32	70.65	2.32	71.28	nr	141.93
Sum of two sides 1500 mm	69.32	86.96	2.56	78.64	nr	165.60
Sum of two sides 1600 mm	82.36	103.31	2.58	79.25	nr	182.56
Sum of two sides 1700 mm	95.38	119.65	2.84	87.25	nr	206.90
Sum of two sides 1800 mm	108.38	135.96	2.84	87.25	nr	223.21
Sum of two sides 1900 mm	121.40	152.29	3.13	96.15	nr	248.44
Sum of two sides 2000 mm	134.44	168.64	3.13	96.15	nr	264.79
Sum of two sides 2100 mm	147.44	184.95	4.26	130.87	nr	315.82
Sum of two sides 2200 mm	151.70	190.29	4.27	131.17	nr	321.46
Sum of two sides 2300 mm	164.75	206.66	4.30	132.10	nr	338.76
Sum of two sides 2400 mm	177.76	222.98	4.77	146.54	nr	369.52
Sum of two sides 2500 mm	190.80	239.34	4.79	147.16	nr	386.50
90° radius bend						
Sum of two sides 1300 mm	22.07	27.69	1.91	58.68	nr	86.37
Sum of two sides 1400 mm	21.69	27.20	1.91	58.68	nr	85.88
Sum of two sides 1500 mm	33.30	41.78	2.11	64.81	nr	106.59
Sum of two sides 1600 mm	44.97	56.41	2.11	64.81	nr	121.22
Sum of two sides 1700 mm	56.57	70.96	2.55	78.33	nr	149.29
Sum of two sides 1800 mm	68.16	85.50	2.55	78.33	nr	163.83
Sum of two sides 1900 mm	79.75	100.04	2.80	86.02	nr	186.06
Sum of two sides 2000 mm	91.44	114.70	2.80	86.02	nr	200.72
Sum of two sides 2100 mm	103.02	129.23	2.80	86.02	nr	215.25
Sum of two sides 2200 mm	97.71	122.57	2.80	86.02	nr	208.59
Sum of two sides 2300 mm	100.81	126.46	4.35	133.63	nr	260.09
Sum of two sides 2400 mm	112.38	140.97	4.35	133.63	nr	274.60
Sum of two sides 2500 mm	123.99	155.53	4.35	133.63	nr	289.16
45° bend						
Sum of two sides 1300 mm	51.53	64.64	2.26	69.43	nr	134.07
Sum of two sides 1400 mm	52.93	66.39	2.44	74.96	nr	141.35
Sum of two sides 1500 mm	64.66	81.11	2.68	82.33	nr	163.44
Sum of two sides 1600 mm	76.64	96.14	2.69	82.64	nr	178.78
Sum of two sides 1700 mm	88.37	110.85	2.96	90.93	nr	201.78
Sum of two sides 1800 mm	100.10	125.56	2.96	90.93	nr	216.49
Sum of two sides 1900 mm	111.84	140.29	3.26	100.15	nr	240.44
Sum of two sides 2000 mm	123.81	155.31	3.26	100.15	nr	255.46
Sum of two sides 2100 mm	135.54	170.02	7.50	230.41	nr	400.43
Sum of two sides 2200 mm	138.09	173.22	7.50	230.41	nr	403.63
Sum of two sides 2300 mm	145.65	182.71	7.55	231.94	nr	414.65
Sum of two sides 2400 mm	157.38	197.42	8.13	249.75	nr	447.17
Sum of two sides 2500 mm	169.32	212.40	8.30	254.98	nr	467.38

38 VENTILATION/AIR CONDITIONING SYSTEMS

Item	Net Price £	Material £	Labour hours	Labour £	Unit	Total rate £
DUCTWORK: RECTANGULAR – CLASS C – cont						
Extra over fittings – cont						
90° mitre bend						
Sum of two sides 1300 mm	117.70	147.64	3.85	118.27	nr	265.91
Sum of two sides 1400 mm	122.63	153.83	3.85	118.27	nr	272.10
Sum of two sides 1500 mm	145.04	181.93	4.25	130.56	nr	312.49
Sum of two sides 1600 mm	167.45	210.04	4.26	130.87	nr	340.91
Sum of two sides 1700 mm	189.85	238.15	4.68	143.76	nr	381.91
Sum of two sides 1800 mm	212.25	266.25	4.68	143.76	nr	410.01
Sum of two sides 1900 mm	234.66	294.36	4.87	149.61	nr	443.97
Sum of two sides 2000 mm	257.08	322.48	4.87	149.61	nr	472.09
Sum of two sides 2100 mm	279.48	350.58	7.50	230.41	nr	580.99
Sum of two sides 2200 mm	282.19	353.98	7.50	230.41	nr	584.39
Sum of two sides 2300 mm	294.69	369.66	7.55	231.94	nr	601.60
Sum of two sides 2400 mm	317.24	397.95	8.13	249.75	nr	647.70
Sum of two sides 2500 mm	339.78	426.22	8.30	254.98	nr	681.20
Branch						
Sum of two sides 1300 mm	106.03	133.00	1.39	42.69	nr	175.69
Sum of two sides 1400 mm	110.58	138.71	1.39	42.69	nr	181.40
Sum of two sides 1500 mm	126.59	158.79	1.64	50.38	nr	209.17
Sum of two sides 1600 mm	142.81	179.14	1.64	50.38	nr	229.52
Sum of two sides 1700 mm	158.81	199.21	1.69	51.92	nr	251.13
Sum of two sides 1800 mm	174.81	219.28	1.69	51.92	nr	271.20
Sum of two sides 1900 mm	190.82	239.37	1.85	56.84	nr	296.21
Sum of two sides 2000 mm	207.04	259.71	1.85	56.84	nr	316.55
Sum of two sides 2100 mm	223.05	279.80	2.61	80.18	nr	359.98
Sum of two sides 2200 mm	239.04	299.85	2.61	80.18	nr	380.03
Sum of two sides 2300 mm	255.27	320.21	2.61	80.18	nr	400.39
Sum of two sides 2400 mm	271.27	340.28	2.88	88.47	nr	428.75
Sum of two sides 2500 mm	287.50	360.64	2.88	88.47	nr	449.11
Grille neck						
Sum of two sides 1300 mm	111.33	139.65	1.69	51.92	nr	191.57
Sum of two sides 1400 mm	116.50	146.14	1.69	51.92	nr	198.06
Sum of two sides 1500 mm	134.56	168.80	1.79	54.99	nr	223.79
Sum of two sides 1600 mm	152.61	191.43	1.79	54.99	nr	246.42
Sum of two sides 1700 mm	170.66	214.08	1.86	57.14	nr	271.22
Sum of two sides 1800 mm	188.72	236.73	1.86	57.14	nr	293.87
Sum of two sides 1900 mm	206.77	259.37	2.02	62.06	nr	321.43
Sum of two sides 2000 mm	224.82	282.02	2.02	62.06	nr	344.08
Sum of two sides 2100 mm	242.88	304.67	2.61	80.18	nr	384.85
Sum of two sides 2200 mm	260.94	327.32	2.80	86.02	nr	413.34
Sum of two sides 2300 mm	278.99	349.97	2.80	86.02	nr	435.99
Sum of two sides 2400 mm	297.08	372.66	3.06	94.00	nr	466.66
Sum of two sides 2500 mm	315.10	395.26	3.06	94.00	nr	489.26
Ductwork 1251–1600 mm longest side						
Sum of two sides 1700 mm	132.04	165.63	1.74	53.45	m	219.08
Sum of two sides 1800 mm	137.30	172.23	1.76	54.06	m	226.29
Sum of two sides 1900 mm	142.62	178.90	1.81	55.61	m	234.51

38 VENTILATION/AIR CONDITIONING SYSTEMS

Item	Net Price £	Material £	Labour hours	Labour £	Unit	Total rate £
Sum of two sides 2000 mm	147.63	185.19	1.82	55.91	m	241.10
Sum of two sides 2100 mm	152.65	191.49	2.53	77.73	m	269.22
Sum of two sides 2200 mm	157.68	197.79	2.55	78.33	m	276.12
Sum of two sides 2300 mm	162.99	204.46	2.56	78.64	m	283.10
Sum of two sides 2400 mm	168.24	211.04	2.76	84.78	m	295.82
Sum of two sides 2500 mm	173.27	217.35	2.77	85.10	m	302.45
Sum of two sides 2600 mm	193.47	242.69	2.97	91.25	m	333.94
Sum of two sides 2700 mm	198.49	248.99	2.99	91.86	m	340.85
Sum of two sides 2800 mm	203.84	255.70	3.30	101.38	m	357.08
Sum of two sides 2900 mm	209.14	262.35	3.31	101.68	m	364.03
Sum of two sides 3000 mm	214.17	268.65	3.53	108.44	m	377.09
Sum of two sides 3100 mm	219.43	275.25	3.55	109.05	m	384.30
Sum of two sides 3200 mm	224.45	281.55	3.56	109.36	m	390.91
Extra over fittings; Ductwork 1251–1600 mm longest side						
End cap						
Sum of two sides 1700 mm	54.14	67.92	0.58	17.81	nr	85.73
Sum of two sides 1800 mm	60.17	75.48	0.58	17.81	nr	93.29
Sum of two sides 1900 mm	65.84	82.59	0.58	17.81	nr	100.40
Sum of two sides 2000 mm	77.55	97.28	0.58	17.81	nr	115.09
Sum of two sides 2100 mm	83.41	104.63	0.87	26.73	nr	131.36
Sum of two sides 2200 mm	89.26	111.97	0.87	26.73	nr	138.70
Sum of two sides 2300 mm	95.11	119.30	0.87	26.73	nr	146.03
Sum of two sides 2400 mm	100.96	126.65	0.87	26.73	nr	153.38
Sum of two sides 2500 mm	106.83	134.01	0.87	26.73	nr	160.74
Sum of two sides 2600 mm	112.67	141.33	0.87	26.73	nr	168.06
Sum of two sides 2700 mm	118.53	148.68	0.87	26.73	nr	175.41
Sum of two sides 2800 mm	124.38	156.03	1.16	35.64	nr	191.67
Sum of two sides 2900 mm	130.24	163.37	1.16	35.64	nr	199.01
Sum of two sides 3000 mm	201.66	252.96	1.73	53.14	nr	306.10
Sum of two sides 3100 mm	134.89	169.21	1.73	53.14	nr	222.35
Sum of two sides 3200 mm	139.11	174.50	1.73	53.14	nr	227.64
Reducer						
Sum of two sides 1700 mm	72.98	91.55	2.59	79.56	nr	171.11
Sum of two sides 1800 mm	82.95	104.05	2.59	79.56	nr	183.61
Sum of two sides 1900 mm	93.10	116.78	2.71	83.25	nr	200.03
Sum of two sides 2000 mm	71.71	89.96	2.71	83.25	nr	173.21
Sum of two sides 2100 mm	103.06	129.28	2.92	89.70	nr	218.98
Sum of two sides 2200 mm	113.04	141.79	2.92	89.70	nr	231.49
Sum of two sides 2300 mm	123.01	154.30	2.92	89.70	nr	244.00
Sum of two sides 2400 mm	133.16	167.04	3.12	95.85	nr	262.89
Sum of two sides 2500 mm	143.15	179.57	3.12	95.85	nr	275.42
Sum of two sides 2600 mm	153.11	192.06	3.16	97.08	nr	289.14
Sum of two sides 2700 mm	151.17	189.63	3.16	97.08	nr	286.71
Sum of two sides 2800 mm	161.15	202.15	4.00	122.89	nr	325.04
Sum of two sides 2900 mm	171.07	214.59	4.01	123.18	nr	337.77
Sum of two sides 3000 mm	191.68	240.44	4.56	140.08	nr	380.52
Sum of two sides 3100 mm	209.00	262.17	4.56	140.08	nr	402.25
Sum of two sides 3200 mm	215.72	270.60	4.56	140.08	nr	410.68

38 VENTILATION/AIR CONDITIONING SYSTEMS

Item	Net Price £	Material £	Labour hours	Labour £	Unit	Total rate £
DUCTWORK: RECTANGULAR – CLASS C – cont						
Extra over fittings – cont						
Offset						
Sum of two sides 1700 mm	179.63	225.33	2.61	80.18	nr	305.51
Sum of two sides 1800 mm	198.37	248.83	2.61	80.18	nr	329.01
Sum of two sides 1900 mm	210.77	264.39	2.71	83.25	nr	347.64
Sum of two sides 2000 mm	222.57	279.19	2.71	83.25	nr	362.44
Sum of two sides 2100 mm	234.00	293.53	2.92	89.70	nr	383.23
Sum of two sides 2200 mm	245.06	307.41	3.26	100.15	nr	407.56
Sum of two sides 2300 mm	255.81	320.89	3.26	100.15	nr	421.04
Sum of two sides 2400 mm	273.63	343.25	3.47	106.60	nr	449.85
Sum of two sides 2500 mm	283.71	355.89	3.48	106.92	nr	462.81
Sum of two sides 2600 mm	283.59	355.73	3.49	107.22	nr	462.95
Sum of two sides 2700 mm	301.19	377.81	3.50	107.53	nr	485.34
Sum of two sides 2800 mm	318.73	399.82	4.33	133.01	nr	532.83
Sum of two sides 2900 mm	346.86	435.10	4.74	145.61	nr	580.71
Sum of two sides 3000 mm	364.47	457.20	5.31	163.12	nr	620.32
Sum of two sides 3100 mm	378.14	474.34	5.34	164.05	nr	638.39
Sum of two sides 3200 mm	390.85	490.28	5.35	164.35	nr	654.63
Square to round						
Sum of two sides 1700 mm	79.99	100.34	2.84	87.25	nr	187.59
Sum of two sides 1800 mm	92.45	115.96	2.84	87.25	nr	203.21
Sum of two sides 1900 mm	104.84	131.51	3.13	96.15	nr	227.66
Sum of two sides 2000 mm	117.27	147.10	3.13	96.15	nr	243.25
Sum of two sides 2100 mm	129.69	162.68	4.26	130.87	nr	293.55
Sum of two sides 2200 mm	142.12	178.27	4.27	131.17	nr	309.44
Sum of two sides 2300 mm	154.51	193.82	4.30	132.10	nr	325.92
Sum of two sides 2400 mm	166.97	209.45	4.77	146.54	nr	355.99
Sum of two sides 2500 mm	179.40	225.04	4.79	147.16	nr	372.20
Sum of two sides 2600 mm	179.93	225.70	4.95	152.07	nr	377.77
Sum of two sides 2700 mm	192.37	241.30	4.95	152.07	nr	393.37
Sum of two sides 2800 mm	204.76	256.85	8.49	260.81	nr	517.66
Sum of two sides 2900 mm	222.38	278.96	8.88	272.80	nr	551.76
Sum of two sides 3000 mm	234.82	294.56	9.02	277.10	nr	571.66
Sum of two sides 3100 mm	244.02	306.10	9.02	277.10	nr	583.20
Sum of two sides 3200 mm	252.37	316.57	9.09	279.24	nr	595.81
90° radius bend						
Sum of two sides 1700 mm	181.25	227.36	2.55	78.33	nr	305.69
Sum of two sides 1800 mm	196.21	246.13	2.55	78.33	nr	324.46
Sum of two sides 1900 mm	219.86	275.79	2.80	86.02	nr	361.81
Sum of two sides 2000 mm	243.05	304.89	2.80	86.02	nr	390.91
Sum of two sides 2100 mm	266.25	333.98	2.61	80.18	nr	414.16
Sum of two sides 2200 mm	289.44	363.07	2.62	80.48	nr	443.55
Sum of two sides 2300 mm	313.08	392.73	2.63	80.80	nr	473.53
Sum of two sides 2400 mm	327.40	410.69	4.34	133.32	nr	544.01
Sum of two sides 2500 mm	350.55	439.73	4.35	133.63	nr	573.36
Sum of two sides 2600 mm	352.36	442.00	4.53	139.16	nr	581.16
Sum of two sides 2700 mm	375.51	471.04	4.53	139.16	nr	610.20

38 VENTILATION/AIR CONDITIONING SYSTEMS

Item	Net Price £	Material £	Labour hours	Labour £	Unit	Total rate £
Sum of two sides 2800 mm	388.03	486.74	7.13	219.03	nr	705.77
Sum of two sides 2900 mm	432.47	542.49	7.17	220.26	nr	762.75
Sum of two sides 3000 mm	455.58	571.48	7.26	223.04	nr	794.52
Sum of two sides 3100 mm	473.47	593.92	7.26	223.04	nr	816.96
Sum of two sides 3200 mm	490.06	614.73	7.31	224.56	nr	839.29
45° bend						
Sum of two sides 1700 mm	87.20	109.38	2.96	90.93	nr	200.31
Sum of two sides 1800 mm	94.56	118.62	2.96	90.93	nr	209.55
Sum of two sides 1900 mm	106.37	133.43	3.26	100.15	nr	233.58
Sum of two sides 2000 mm	117.98	148.00	3.26	100.15	nr	248.15
Sum of two sides 2100 mm	129.57	162.53	7.50	230.41	nr	392.94
Sum of two sides 2200 mm	141.17	177.08	7.50	230.41	nr	407.49
Sum of two sides 2300 mm	152.99	191.91	7.55	231.94	nr	423.85
Sum of two sides 2400 mm	160.02	200.73	8.13	249.75	nr	450.48
Sum of two sides 2500 mm	171.60	215.25	8.30	254.98	nr	470.23
Sum of two sides 2600 mm	140.58	176.34	8.56	262.96	nr	439.30
Sum of two sides 2700 mm	183.35	229.99	8.62	264.80	nr	494.79
Sum of two sides 2800 mm	189.48	237.69	9.09	279.24	nr	516.93
Sum of two sides 2900 mm	211.69	265.54	9.09	279.24	nr	544.78
Sum of two sides 3000 mm	223.26	280.06	9.62	295.52	nr	575.58
Sum of two sides 3100 mm	232.19	291.26	9.62	295.52	nr	586.78
Sum of two sides 3200 mm	240.48	301.66	9.62	295.52	nr	597.18
90° mitre bend						
Sum of two sides 1700 mm	190.74	239.27	4.68	143.76	nr	383.03
Sum of two sides 1800 mm	202.56	254.09	4.68	143.76	nr	397.85
Sum of two sides 1900 mm	225.45	282.80	4.87	149.61	nr	432.41
Sum of two sides 2000 mm	248.40	311.60	4.87	149.61	nr	461.21
Sum of two sides 2100 mm	271.36	340.39	7.50	230.41	nr	570.80
Sum of two sides 2200 mm	294.31	369.19	7.50	230.41	nr	599.60
Sum of two sides 2300 mm	317.20	397.89	7.55	231.94	nr	629.83
Sum of two sides 2400 mm	329.12	412.84	8.13	249.75	nr	662.59
Sum of two sides 2500 mm	352.20	441.80	8.30	254.98	nr	696.78
Sum of two sides 2600 mm	349.99	439.03	8.56	262.96	nr	701.99
Sum of two sides 2700 mm	373.06	467.97	8.62	264.80	nr	732.77
Sum of two sides 2800 mm	383.24	480.74	15.20	466.95	nr	947.69
Sum of two sides 2900 mm	424.30	532.25	15.20	466.95	nr	999.20
Sum of two sides 3000 mm	447.50	561.34	15.60	479.23	nr	1040.57
Sum of two sides 3100 mm	466.61	585.31	16.04	492.76	nr	1078.07
Sum of two sides 3200 mm	484.91	608.27	16.04	492.76	nr	1101.03
Branch						
Sum of two sides 1700 mm	163.69	205.33	1.69	51.92	nr	257.25
Sum of two sides 1800 mm	179.70	225.41	1.69	51.92	nr	277.33
Sum of two sides 1900 mm	195.95	245.80	1.85	56.84	nr	302.64
Sum of two sides 2000 mm	211.97	265.90	1.85	56.84	nr	322.74
Sum of two sides 2100 mm	228.01	286.01	2.61	80.18	nr	366.19
Sum of two sides 2200 mm	244.06	306.15	2.61	80.18	nr	386.33
Sum of two sides 2300 mm	260.30	326.52	2.61	80.18	nr	406.70
Sum of two sides 2400 mm	276.30	346.60	2.88	88.47	nr	435.07
Sum of two sides 2500 mm	292.34	366.71	2.88	88.47	nr	455.18
Sum of two sides 2600 mm	308.37	386.81	2.88	88.47	nr	475.28
Sum of two sides 2700 mm	324.41	406.94	2.88	88.47	nr	495.41

38 VENTILATION/AIR CONDITIONING SYSTEMS

Item	Net Price £	Material £	Labour hours	Labour £	Unit	Total rate £
DUCTWORK: RECTANGULAR – CLASS C – cont						
Extra over fittings – cont						
Branch – cont						
Sum of two sides 2800 mm	340.40	427.00	3.94	121.04	nr	**548.04**
Sum of two sides 2900 mm	361.89	453.96	3.94	121.04	nr	**575.00**
Sum of two sides 3000 mm	377.93	474.07	4.83	148.38	nr	**622.45**
Sum of two sides 3100 mm	390.75	490.16	4.83	148.38	nr	**638.54**
Sum of two sides 3200 mm	402.43	504.81	4.83	148.38	nr	**653.19**
Grille neck						
Sum of two sides 1700 mm	170.66	214.08	1.86	57.14	nr	**271.22**
Sum of two sides 1800 mm	188.72	236.73	1.86	57.14	nr	**293.87**
Sum of two sides 1900 mm	206.77	259.37	2.02	62.06	nr	**321.43**
Sum of two sides 2000 mm	224.82	282.02	2.02	62.06	nr	**344.08**
Sum of two sides 2100 mm	242.88	304.67	2.80	86.02	nr	**390.69**
Sum of two sides 2200 mm	260.94	327.32	2.80	86.02	nr	**413.34**
Sum of two sides 2300 mm	278.99	349.97	2.80	86.02	nr	**435.99**
Sum of two sides 2400 mm	297.04	372.60	3.06	94.00	nr	**466.60**
Sum of two sides 2500 mm	315.10	395.26	3.06	94.00	nr	**489.26**
Sum of two sides 2600 mm	333.15	417.91	3.08	94.62	nr	**512.53**
Sum of two sides 2700 mm	351.21	440.56	3.08	94.62	nr	**535.18**
Sum of two sides 2800 mm	369.25	463.19	4.13	126.87	nr	**590.06**
Sum of two sides 2900 mm	387.31	485.84	4.13	126.87	nr	**612.71**
Sum of two sides 3000 mm	405.37	508.49	5.02	154.21	nr	**662.70**
Sum of two sides 3100 mm	419.81	526.61	5.02	154.21	nr	**680.82**
Sum of two sides 3200 mm	432.96	543.11	5.02	154.21	nr	**697.32**
Ductwork 1601–2000 mm longest side						
Sum of two sides 2100 mm	176.80	221.78	2.53	77.73	m	**299.51**
Sum of two sides 2200 mm	182.61	229.06	2.55	78.33	m	**307.39**
Sum of two sides 2300 mm	188.47	236.42	2.55	78.33	m	**314.75**
Sum of two sides 2400 mm	194.23	243.64	2.56	78.64	m	**322.28**
Sum of two sides 2500 mm	200.09	250.99	2.76	84.78	m	**335.77**
Sum of two sides 2600 mm	205.67	257.99	2.77	85.10	m	**343.09**
Sum of two sides 2700 mm	211.23	264.97	2.97	91.25	m	**356.22**
Sum of two sides 2800 mm	216.82	271.98	2.99	91.86	m	**363.84**
Sum of two sides 2900 mm	222.67	279.32	3.30	101.38	m	**380.70**
Sum of two sides 3000 mm	228.26	286.33	3.31	101.68	m	**388.01**
Sum of two sides 3100 mm	261.00	327.40	3.53	108.44	m	**435.84**
Sum of two sides 3200 mm	266.58	334.40	3.53	108.44	m	**442.84**
Sum of two sides 3300 mm	272.44	341.75	3.53	108.44	m	**450.19**
Sum of two sides 3400 mm	278.02	348.75	3.55	109.05	m	**457.80**
Sum of two sides 3500 mm	283.60	355.75	3.55	109.05	m	**464.80**
Sum of two sides 3600 mm	289.17	362.73	3.55	109.05	m	**471.78**
Sum of two sides 3700 mm	295.03	370.08	3.56	109.36	m	**479.44**
Sum of two sides 3800 mm	300.60	377.07	3.56	109.36	m	**486.43**
Sum of two sides 3900 mm	306.19	384.08	3.56	109.36	m	**493.44**
Sum of two sides 4000 mm	311.76	391.07	3.56	109.36	m	**500.43**

38 VENTILATION/AIR CONDITIONING SYSTEMS

Item	Net Price £	Material £	Labour hours	Labour £	Unit	Total rate £
Extra over fittings; Ductwork 1601–2000 mm longest side						
End cap						
Sum of two sides 2100 mm	78.56	98.55	0.87	26.73	nr	125.28
Sum of two sides 2200 mm	84.46	105.95	0.87	26.73	nr	132.68
Sum of two sides 2300 mm	90.36	113.34	0.87	26.73	nr	140.07
Sum of two sides 2400 mm	96.26	120.75	0.87	26.73	nr	147.48
Sum of two sides 2500 mm	102.17	128.16	0.87	26.73	nr	154.89
Sum of two sides 2600 mm	108.07	135.56	0.87	26.73	nr	162.29
Sum of two sides 2700 mm	113.96	142.96	0.87	26.73	nr	169.69
Sum of two sides 2800 mm	119.86	150.35	1.16	35.64	nr	185.99
Sum of two sides 2900 mm	125.77	157.76	1.16	35.64	nr	193.40
Sum of two sides 3000 mm	131.67	165.17	1.73	53.14	nr	218.31
Sum of two sides 3100 mm	136.36	171.05	1.73	53.14	nr	224.19
Sum of two sides 3200 mm	140.63	176.41	1.73	53.14	nr	229.55
Sum of two sides 3300 mm	144.90	181.76	1.73	53.14	nr	234.90
Sum of two sides 3400 mm	149.18	187.13	1.73	53.14	nr	240.27
Sum of two sides 3500 mm	153.44	192.47	1.73	53.14	nr	245.61
Sum of two sides 3600 mm	157.71	197.84	1.73	53.14	nr	250.98
Sum of two sides 3700 mm	161.98	203.19	1.73	53.14	nr	256.33
Sum of two sides 3800 mm	166.24	208.53	1.73	53.14	nr	261.67
Sum of two sides 3900 mm	170.52	213.90	1.73	53.14	nr	267.04
Sum of two sides 4000 mm	174.78	219.24	1.73	53.14	nr	272.38
Reducer						
Sum of two sides 2100 mm	105.52	132.36	2.92	89.70	nr	222.06
Sum of two sides 2200 mm	115.47	144.85	2.92	89.70	nr	234.55
Sum of two sides 2300 mm	125.59	157.54	2.92	89.70	nr	247.24
Sum of two sides 2400 mm	135.38	169.83	3.12	95.85	nr	265.68
Sum of two sides 2500 mm	145.50	182.52	3.12	95.85	nr	278.37
Sum of two sides 2600 mm	155.44	194.98	3.16	97.08	nr	292.06
Sum of two sides 2700 mm	165.38	207.46	3.16	97.08	nr	304.54
Sum of two sides 2800 mm	175.32	219.92	4.00	122.89	nr	342.81
Sum of two sides 2900 mm	185.44	232.61	4.01	123.18	nr	355.79
Sum of two sides 3000 mm	195.40	245.11	4.01	123.18	nr	368.29
Sum of two sides 3100 mm	181.65	227.86	4.01	123.18	nr	351.04
Sum of two sides 3200 mm	188.31	236.22	4.56	140.08	nr	376.30
Sum of two sides 3300 mm	205.88	258.26	4.56	140.08	nr	398.34
Sum of two sides 3400 mm	212.56	266.64	4.56	140.08	nr	406.72
Sum of two sides 3500 mm	219.23	275.00	4.56	140.08	nr	415.08
Sum of two sides 3600 mm	225.91	283.38	4.56	140.08	nr	423.46
Sum of two sides 3700 mm	232.77	291.98	4.56	140.08	nr	432.06
Sum of two sides 3800 mm	239.45	300.36	4.56	140.08	nr	440.44
Sum of two sides 3900 mm	246.13	308.75	4.56	140.08	nr	448.83
Sum of two sides 4000 mm	252.80	317.12	4.56	140.08	nr	457.20
Offset						
Sum of two sides 2100 mm	248.23	311.38	2.92	89.70	nr	401.08
Sum of two sides 2200 mm	266.94	334.85	3.26	100.15	nr	435.00
Sum of two sides 2300 mm	277.60	348.22	3.26	100.15	nr	448.37
Sum of two sides 2400 mm	304.39	381.83	3.47	106.60	nr	488.43
Sum of two sides 2500 mm	314.66	394.71	3.48	106.92	nr	501.63
Sum of two sides 2600 mm	324.34	406.85	3.49	107.22	nr	514.07
Sum of two sides 2700 mm	333.64	418.52	3.50	107.53	nr	526.05

38 VENTILATION/AIR CONDITIONING SYSTEMS

Item	Net Price £	Material £	Labour hours	Labour £	Unit	Total rate £
DUCTWORK: RECTANGULAR – CLASS C – cont						
Extra over fittings – cont						
Offset – cont						
Sum of two sides 2800 mm	342.56	429.71	4.33	133.01	nr	562.72
Sum of two sides 2900 mm	351.17	440.51	4.74	145.61	nr	586.12
Sum of two sides 3000 mm	359.28	450.68	5.34	164.05	nr	614.73
Sum of two sides 3100 mm	341.49	428.37	5.35	164.35	nr	592.72
Sum of two sides 3200 mm	354.16	444.26	5.31	163.12	nr	607.38
Sum of two sides 3300 mm	377.59	473.65	5.35	164.35	nr	638.00
Sum of two sides 3400 mm	390.25	489.53	5.35	164.35	nr	653.88
Sum of two sides 3500 mm	402.93	505.43	5.35	164.35	nr	669.78
Sum of two sides 3600 mm	415.58	521.30	5.35	164.35	nr	685.65
Sum of two sides 3700 mm	428.31	537.28	5.35	164.35	nr	701.63
Sum of two sides 3800 mm	440.98	553.17	5.35	164.35	nr	717.52
Sum of two sides 3900 mm	453.64	569.05	5.35	164.35	nr	733.40
Sum of two sides 4000 mm	466.30	584.93	5.35	164.35	nr	749.28
Square to round						
Sum of two sides 2100 mm	157.29	197.30	4.26	130.87	nr	328.17
Sum of two sides 2200 mm	172.54	216.43	4.27	131.17	nr	347.60
Sum of two sides 2300 mm	187.71	235.47	4.30	132.10	nr	367.57
Sum of two sides 2400 mm	203.02	254.67	4.77	146.54	nr	401.21
Sum of two sides 2500 mm	218.19	273.69	4.79	147.16	nr	420.85
Sum of two sides 2600 mm	233.39	292.77	4.95	152.07	nr	444.84
Sum of two sides 2700 mm	248.59	311.83	4.95	152.07	nr	463.90
Sum of two sides 2800 mm	263.80	330.92	8.49	260.81	nr	591.73
Sum of two sides 2900 mm	278.97	349.94	8.88	272.80	nr	622.74
Sum of two sides 3000 mm	294.17	369.01	9.02	277.10	nr	646.11
Sum of two sides 3100 mm	284.46	356.83	9.02	277.10	nr	633.93
Sum of two sides 3200 mm	294.77	369.76	9.09	279.24	nr	649.00
Sum of two sides 3300 mm	310.40	389.37	9.09	279.24	nr	668.61
Sum of two sides 3400 mm	320.70	402.28	9.09	279.24	nr	681.52
Sum of two sides 3500 mm	331.01	415.22	9.09	279.24	nr	694.46
Sum of two sides 3600 mm	341.32	428.15	9.09	279.24	nr	707.39
Sum of two sides 3700 mm	351.59	441.03	9.09	279.24	nr	720.27
Sum of two sides 3800 mm	361.90	453.97	9.09	279.24	nr	733.21
Sum of two sides 3900 mm	372.21	466.91	9.09	279.24	nr	746.15
Sum of two sides 4000 mm	382.51	479.82	9.09	279.24	nr	759.06
90° radius bend						
Sum of two sides 2100 mm	231.43	290.30	2.61	80.18	nr	370.48
Sum of two sides 2200 mm	441.73	554.11	2.62	80.48	nr	634.59
Sum of two sides 2300 mm	479.26	601.18	2.63	80.80	nr	681.98
Sum of two sides 2400 mm	489.80	614.41	4.34	133.32	nr	747.73
Sum of two sides 2500 mm	527.34	661.49	4.35	133.63	nr	795.12
Sum of two sides 2600 mm	564.15	707.67	4.53	139.16	nr	846.83
Sum of two sides 2700 mm	600.94	753.82	4.53	139.16	nr	892.98
Sum of two sides 2800 mm	637.75	799.99	7.13	219.03	nr	1019.02
Sum of two sides 2900 mm	675.28	847.07	7.17	220.26	nr	1067.33
Sum of two sides 3000 mm	712.09	893.24	7.26	223.04	nr	1116.28
Sum of two sides 3100 mm	693.01	869.31	7.26	223.04	nr	1092.35

38 VENTILATION/AIR CONDITIONING SYSTEMS

Item	Net Price £	Material £	Labour hours	Labour £	Unit	Total rate £
Sum of two sides 3200 mm	720.02	903.19	7.31	224.56	nr	1127.75
Sum of two sides 3300 mm	779.86	978.25	7.31	224.56	nr	1202.81
Sum of two sides 3400 mm	806.87	1012.13	7.31	224.56	nr	1236.69
Sum of two sides 3500 mm	833.88	1046.02	7.31	224.56	nr	1270.58
Sum of two sides 3600 mm	860.90	1079.92	7.31	224.56	nr	1304.48
Sum of two sides 3700 mm	888.64	1114.71	7.31	224.56	nr	1339.27
Sum of two sides 3800 mm	915.65	1148.59	7.31	224.56	nr	1373.15
Sum of two sides 3900 mm	942.66	1182.47	7.31	224.56	nr	1407.03
Sum of two sides 4000 mm	969.66	1216.34	7.31	224.56	nr	1440.90
45° bend						
Sum of two sides 2100 mm	111.58	139.97	7.50	230.41	nr	370.38
Sum of two sides 2200 mm	315.40	395.64	7.50	230.41	nr	626.05
Sum of two sides 2300 mm	340.42	427.02	7.55	231.94	nr	658.96
Sum of two sides 2400 mm	351.05	440.36	8.13	249.75	nr	690.11
Sum of two sides 2500 mm	376.05	471.72	8.30	254.98	nr	726.70
Sum of two sides 2600 mm	400.51	502.40	8.56	262.96	nr	765.36
Sum of two sides 2700 mm	424.96	533.08	8.62	264.80	nr	797.88
Sum of two sides 2800 mm	449.43	563.76	9.09	279.24	nr	843.00
Sum of two sides 2900 mm	474.44	595.13	9.09	279.24	nr	874.37
Sum of two sides 3000 mm	498.90	625.82	9.62	295.52	nr	921.34
Sum of two sides 3100 mm	492.87	618.25	9.62	295.52	nr	913.77
Sum of two sides 3200 mm	510.80	640.75	9.62	295.52	nr	936.27
Sum of two sides 3300 mm	550.68	690.77	9.62	295.52	nr	986.29
Sum of two sides 3400 mm	568.61	713.26	9.62	295.52	nr	1008.78
Sum of two sides 3500 mm	586.54	735.75	9.62	295.52	nr	1031.27
Sum of two sides 3600 mm	604.45	758.22	9.62	295.52	nr	1053.74
Sum of two sides 3700 mm	622.95	781.42	9.62	295.52	nr	1076.94
Sum of two sides 3800 mm	640.87	803.90	9.62	295.52	nr	1099.42
Sum of two sides 3900 mm	658.79	826.38	9.62	295.52	nr	1121.90
Sum of two sides 4000 mm	676.71	848.87	9.62	295.52	nr	1144.39
90° mitre bend						
Sum of two sides 2100 mm	509.55	639.18	7.50	230.41	nr	869.59
Sum of two sides 2200 mm	538.28	675.21	7.50	230.41	nr	905.62
Sum of two sides 2300 mm	581.63	729.60	7.55	231.94	nr	961.54
Sum of two sides 2400 mm	594.33	745.53	8.13	249.75	nr	995.28
Sum of two sides 2500 mm	637.85	800.12	8.30	254.98	nr	1055.10
Sum of two sides 2600 mm	681.63	855.04	8.56	262.96	nr	1118.00
Sum of two sides 2700 mm	725.39	909.93	8.62	264.80	nr	1174.73
Sum of two sides 2800 mm	769.15	964.82	15.20	466.95	nr	1431.77
Sum of two sides 2900 mm	812.68	1019.42	15.20	466.95	nr	1486.37
Sum of two sides 3000 mm	856.45	1074.33	16.04	492.76	nr	1567.09
Sum of two sides 3100 mm	834.45	1046.73	15.60	479.23	nr	1525.96
Sum of two sides 3200 mm	868.43	1089.36	16.04	492.76	nr	1582.12
Sum of two sides 3300 mm	922.24	1156.86	16.04	492.76	nr	1649.62
Sum of two sides 3400 mm	956.22	1199.49	16.04	492.76	nr	1692.25
Sum of two sides 3500 mm	990.19	1242.09	16.04	492.76	nr	1734.85
Sum of two sides 3600 mm	1024.17	1284.72	16.04	492.76	nr	1777.48
Sum of two sides 3700 mm	1057.90	1327.03	16.04	492.76	nr	1819.79
Sum of two sides 3800 mm	1091.88	1369.66	16.04	492.76	nr	1862.42
Sum of two sides 3900 mm	1125.85	1412.26	16.04	492.76	nr	1905.02
Sum of two sides 4000 mm	1159.82	1454.88	16.04	492.76	nr	1947.64

38 VENTILATION/AIR CONDITIONING SYSTEMS

Item	Net Price £	Material £	Labour hours	Labour £	Unit	Total rate £
DUCTWORK: RECTANGULAR – CLASS C – cont						
Extra over fittings – cont						
Branch						
Sum of two sides 2100 mm	232.07	291.11	2.61	80.18	nr	371.29
Sum of two sides 2200 mm	248.22	311.37	2.61	80.18	nr	391.55
Sum of two sides 2300 mm	264.60	331.91	2.61	80.18	nr	412.09
Sum of two sides 2400 mm	280.51	351.87	2.88	88.47	nr	440.34
Sum of two sides 2500 mm	296.91	372.44	2.88	88.47	nr	460.91
Sum of two sides 2600 mm	313.08	392.73	2.88	88.47	nr	481.20
Sum of two sides 2700 mm	329.25	413.01	2.88	88.47	nr	501.48
Sum of two sides 2800 mm	345.44	433.32	3.94	121.04	nr	554.36
Sum of two sides 2900 mm	361.82	453.87	3.94	121.04	nr	574.91
Sum of two sides 3000 mm	378.01	474.17	4.83	148.38	nr	622.55
Sum of two sides 3100 mm	390.97	490.44	4.83	148.38	nr	638.82
Sum of two sides 3200 mm	402.78	505.24	4.83	148.38	nr	653.62
Sum of two sides 3300 mm	420.18	527.07	4.83	148.38	nr	675.45
Sum of two sides 3400 mm	432.01	541.91	4.83	148.38	nr	690.29
Sum of two sides 3500 mm	443.83	556.74	4.83	148.38	nr	705.12
Sum of two sides 3600 mm	455.66	571.58	4.83	148.38	nr	719.96
Sum of two sides 3700 mm	467.70	586.68	4.83	148.38	nr	735.06
Sum of two sides 3800 mm	479.51	601.50	4.83	148.38	nr	749.88
Sum of two sides 3900 mm	491.35	616.35	4.83	148.38	nr	764.73
Sum of two sides 4000 mm	503.17	631.18	4.83	148.38	nr	779.56
Grille neck						
Sum of two sides 2100 mm	247.90	310.97	2.80	86.02	nr	396.99
Sum of two sides 2200 mm	266.19	333.91	2.80	86.02	nr	419.93
Sum of two sides 2300 mm	284.49	356.87	2.80	86.02	nr	442.89
Sum of two sides 2400 mm	302.79	379.81	3.06	94.00	nr	473.81
Sum of two sides 2500 mm	321.07	402.75	3.06	94.00	nr	496.75
Sum of two sides 2600 mm	339.37	425.70	3.08	94.62	nr	520.32
Sum of two sides 2700 mm	357.66	448.65	3.08	94.62	nr	543.27
Sum of two sides 2800 mm	375.95	471.59	4.13	126.87	nr	598.46
Sum of two sides 2900 mm	394.25	494.55	4.13	126.87	nr	621.42
Sum of two sides 3000 mm	412.54	517.48	5.02	154.21	nr	671.69
Sum of two sides 3100 mm	427.21	535.90	5.02	154.21	nr	690.11
Sum of two sides 3200 mm	440.62	552.71	5.02	154.21	nr	706.92
Sum of two sides 3300 mm	454.01	569.51	5.02	154.21	nr	723.72
Sum of two sides 3400 mm	467.41	586.32	5.02	154.21	nr	740.53
Sum of two sides 3500 mm	480.80	603.12	5.02	154.21	nr	757.33
Sum of two sides 3600 mm	494.20	619.92	5.02	154.21	nr	774.13
Sum of two sides 3700 mm	507.59	636.72	5.02	154.21	nr	790.93
Sum of two sides 3800 mm	520.99	653.53	5.02	154.21	nr	807.74
Sum of two sides 3900 mm	534.38	670.33	5.02	154.21	nr	824.54
Sum of two sides 4000 mm	547.78	687.13	5.02	154.21	nr	841.34
Ductwork 2001–2500 mm longest side						
Sum of two sides 2500 mm	330.61	414.71	2.77	85.10	m	499.81
Sum of two sides 2600 mm	340.44	427.04	2.97	91.25	m	518.29
Sum of two sides 2700 mm	349.27	438.12	2.99	91.86	m	529.98

38 VENTILATION/AIR CONDITIONING SYSTEMS

Item	Net Price £	Material £	Labour hours	Labour £	Unit	Total rate £
Sum of two sides 2800 mm	359.65	451.15	3.30	101.38	m	**552.53**
Sum of two sides 2900 mm	368.49	462.24	3.31	101.68	m	**563.92**
Sum of two sides 3000 mm	377.06	472.99	3.53	108.44	m	**581.43**
Sum of two sides 3100 mm	385.87	484.03	3.55	109.05	m	**593.08**
Sum of two sides 3200 mm	394.44	494.78	3.56	109.36	m	**604.14**
Sum of two sides 3300 mm	403.27	505.86	3.56	109.36	m	**615.22**
Sum of two sides 3400 mm	411.85	516.62	3.56	109.36	m	**625.98**
Sum of two sides 3500 mm	450.32	564.88	3.56	109.36	m	**674.24**
Sum of two sides 3600 mm	458.89	575.64	3.56	109.36	m	**685.00**
Sum of two sides 3700 mm	467.73	586.72	3.56	109.36	m	**696.08**
Sum of two sides 3800 mm	476.30	597.48	3.56	109.36	m	**706.84**
Sum of two sides 3900 mm	484.87	608.22	3.56	109.36	m	**717.58**
Sum of two sides 4000 mm	495.05	621.00	3.56	109.36	m	**730.36**
Extra over fittings; Ductwork 2001– 2500 mm longest side						
End cap						
Sum of two sides 2500 mm	102.17	128.16	0.87	26.73	nr	**154.89**
Sum of two sides 2600 mm	108.07	135.56	0.87	26.73	nr	**162.29**
Sum of two sides 2700 mm	113.96	142.96	0.87	26.73	nr	**169.69**
Sum of two sides 2800 mm	119.86	150.35	1.16	35.64	nr	**185.99**
Sum of two sides 2900 mm	125.77	157.76	1.16	35.64	nr	**193.40**
Sum of two sides 3000 mm	131.67	165.17	1.73	53.14	nr	**218.31**
Sum of two sides 3100 mm	136.36	171.05	1.73	53.14	nr	**224.19**
Sum of two sides 3200 mm	140.63	176.41	1.73	53.14	nr	**229.55**
Sum of two sides 3300 mm	144.90	181.76	1.73	53.14	nr	**234.90**
Sum of two sides 3400 mm	149.18	187.13	1.73	53.14	nr	**240.27**
Sum of two sides 3500 mm	153.44	192.47	1.73	53.14	nr	**245.61**
Sum of two sides 3600 mm	157.71	197.84	1.73	53.14	nr	**250.98**
Sum of two sides 3700 mm	161.98	203.19	1.73	53.14	nr	**256.33**
Sum of two sides 3800 mm	166.24	208.53	1.73	53.14	nr	**261.67**
Sum of two sides 3900 mm	170.52	213.90	1.73	53.14	nr	**267.04**
Sum of two sides 4000 mm	174.78	219.24	1.73	53.14	nr	**272.38**
Reducer						
Sum of two sides 2500 mm	90.28	113.24	3.12	95.85	nr	**209.09**
Sum of two sides 2600 mm	98.98	124.16	3.16	97.08	nr	**221.24**
Sum of two sides 2700 mm	107.90	135.35	3.16	97.08	nr	**232.43**
Sum of two sides 2800 mm	116.36	145.96	4.00	122.89	nr	**268.85**
Sum of two sides 2900 mm	125.28	157.15	4.01	123.18	nr	**280.33**
Sum of two sides 3000 mm	134.00	168.09	4.56	140.08	nr	**308.17**
Sum of two sides 3100 mm	140.10	175.74	4.56	140.08	nr	**315.82**
Sum of two sides 3200 mm	145.57	182.60	4.56	140.08	nr	**322.68**
Sum of two sides 3300 mm	151.22	189.69	4.56	140.08	nr	**329.77**
Sum of two sides 3400 mm	156.68	196.54	4.56	140.08	nr	**336.62**
Sum of two sides 3500 mm	139.84	175.41	4.56	140.08	nr	**315.49**
Sum of two sides 3600 mm	145.29	182.25	4.56	140.08	nr	**322.33**
Sum of two sides 3700 mm	161.88	203.07	4.56	140.08	nr	**343.15**
Sum of two sides 3800 mm	167.35	209.92	4.56	140.08	nr	**350.00**
Sum of two sides 3900 mm	172.80	216.76	4.56	140.08	nr	**356.84**
Sum of two sides 4000 mm	178.42	223.81	4.56	140.08	nr	**363.89**

38 VENTILATION/AIR CONDITIONING SYSTEMS

Item	Net Price £	Material £	Labour hours	Labour £	Unit	Total rate £
DUCTWORK: RECTANGULAR – CLASS C – cont						
Extra over fittings – cont						
Offset						
Sum of two sides 2500 mm	282.38	354.22	3.48	106.92	nr	**461.14**
Sum of two sides 2600 mm	300.30	376.70	3.49	107.22	nr	**483.92**
Sum of two sides 2700 mm	299.90	376.20	3.50	107.53	nr	**483.73**
Sum of two sides 2800 mm	336.14	421.66	4.33	133.01	nr	**554.67**
Sum of two sides 2900 mm	334.87	420.06	4.74	145.61	nr	**565.67**
Sum of two sides 3000 mm	332.51	417.10	5.31	163.12	nr	**580.22**
Sum of two sides 3100 mm	325.41	408.20	5.34	164.05	nr	**572.25**
Sum of two sides 3200 mm	316.38	396.87	5.35	164.35	nr	**561.22**
Sum of two sides 3300 mm	306.58	384.57	5.35	164.35	nr	**548.92**
Sum of two sides 3400 mm	295.78	371.02	5.35	164.35	nr	**535.37**
Sum of two sides 3500 mm	272.43	341.73	5.35	164.35	nr	**506.08**
Sum of two sides 3600 mm	282.86	354.82	5.35	164.35	nr	**519.17**
Sum of two sides 3700 mm	304.30	381.72	5.35	164.35	nr	**546.07**
Sum of two sides 3800 mm	314.75	394.82	5.35	164.35	nr	**559.17**
Sum of two sides 3900 mm	325.18	407.90	5.35	164.35	nr	**572.25**
Sum of two sides 4000 mm	335.64	421.03	5.35	164.35	nr	**585.38**
Square to round						
Sum of two sides 2500 mm	144.59	181.37	4.79	147.16	nr	**328.53**
Sum of two sides 2600 mm	157.94	198.12	4.95	152.07	nr	**350.19**
Sum of two sides 2700 mm	171.29	214.86	4.95	152.07	nr	**366.93**
Sum of two sides 2800 mm	184.65	231.63	8.49	260.81	nr	**492.44**
Sum of two sides 2900 mm	197.97	248.34	8.88	272.80	nr	**521.14**
Sum of two sides 3000 mm	211.33	265.09	9.02	277.10	nr	**542.19**
Sum of two sides 3100 mm	220.84	277.02	9.02	277.10	nr	**554.12**
Sum of two sides 3200 mm	229.30	287.64	9.09	279.24	nr	**566.88**
Sum of two sides 3300 mm	237.73	298.21	9.09	279.24	nr	**577.45**
Sum of two sides 3400 mm	246.19	308.82	9.09	279.24	nr	**588.06**
Sum of two sides 3500 mm	231.77	290.73	9.09	279.24	nr	**569.97**
Sum of two sides 3600 mm	240.23	301.35	9.09	279.24	nr	**580.59**
Sum of two sides 3700 mm	254.13	318.79	9.09	279.24	nr	**598.03**
Sum of two sides 3800 mm	262.58	329.38	9.09	279.24	nr	**608.62**
Sum of two sides 3900 mm	271.03	339.98	9.09	279.24	nr	**619.22**
Sum of two sides 4000 mm	279.43	350.52	9.09	279.24	nr	**629.76**
90° radius bend						
Sum of two sides 2500 mm	405.05	508.10	4.35	133.63	nr	**641.73**
Sum of two sides 2600 mm	416.63	522.63	4.53	139.16	nr	**661.79**
Sum of two sides 2700 mm	451.84	566.79	4.53	139.16	nr	**705.95**
Sum of two sides 2800 mm	437.18	548.40	7.13	219.03	nr	**767.43**
Sum of two sides 2900 mm	471.93	591.99	7.17	220.26	nr	**812.25**
Sum of two sides 3000 mm	505.99	634.72	7.26	223.04	nr	**857.76**
Sum of two sides 3100 mm	532.25	667.65	7.26	223.04	nr	**890.69**
Sum of two sides 3200 mm	556.52	698.10	7.31	224.56	nr	**922.66**
Sum of two sides 3300 mm	581.47	729.40	7.31	224.56	nr	**953.96**
Sum of two sides 3400 mm	605.74	759.84	7.31	224.56	nr	**984.40**
Sum of two sides 3500 mm	570.66	715.84	7.31	224.56	nr	**940.40**

38 VENTILATION/AIR CONDITIONING SYSTEMS

Item	Net Price £	Material £	Labour hours	Labour £	Unit	Total rate £
Sum of two sides 3600 mm	594.93	746.28	7.31	224.56	nr	970.84
Sum of two sides 3700 mm	652.67	818.71	7.31	224.56	nr	1043.27
Sum of two sides 3800 mm	676.94	849.15	7.31	224.56	nr	1073.71
Sum of two sides 3900 mm	701.20	879.58	7.31	224.56	nr	1104.14
Sum of two sides 4000 mm	696.28	873.41	7.31	224.56	nr	1097.97
45° bend						
Sum of two sides 2500 mm	327.32	410.59	8.30	254.98	nr	665.57
Sum of two sides 2600 mm	339.39	425.73	8.56	262.96	nr	688.69
Sum of two sides 2700 mm	363.73	456.27	8.62	264.80	nr	721.07
Sum of two sides 2800 mm	362.23	454.38	9.09	279.24	nr	733.62
Sum of two sides 2900 mm	386.33	484.61	9.09	279.24	nr	763.85
Sum of two sides 3000 mm	409.90	514.18	9.62	295.52	nr	809.70
Sum of two sides 3100 mm	428.33	537.30	9.62	295.52	nr	832.82
Sum of two sides 3200 mm	445.38	558.69	9.62	295.52	nr	854.21
Sum of two sides 3300 mm	462.95	580.72	9.62	295.52	nr	876.24
Sum of two sides 3400 mm	479.98	602.09	9.62	295.52	nr	897.61
Sum of two sides 3500 mm	466.43	585.09	9.62	295.52	nr	880.61
Sum of two sides 3600 mm	483.45	606.44	9.62	295.52	nr	901.96
Sum of two sides 3700 mm	522.90	655.93	9.62	295.52	nr	951.45
Sum of two sides 3800 mm	539.94	677.30	9.62	295.52	nr	972.82
Sum of two sides 3900 mm	556.96	698.66	9.62	295.52	nr	994.18
Sum of two sides 4000 mm	559.37	701.67	9.62	295.52	nr	997.19
90° mitre bend						
Sum of two sides 2500 mm	406.44	509.84	8.30	254.98	nr	764.82
Sum of two sides 2600 mm	425.70	533.99	8.56	262.96	nr	796.95
Sum of two sides 2700 mm	463.83	581.83	8.62	264.80	nr	846.63
Sum of two sides 2800 mm	441.41	553.71	15.20	466.95	nr	1020.66
Sum of two sides 2900 mm	479.09	600.97	15.20	466.95	nr	1067.92
Sum of two sides 3000 mm	517.08	648.63	15.20	466.95	nr	1115.58
Sum of two sides 3100 mm	547.18	686.38	16.04	492.76	nr	1179.14
Sum of two sides 3200 mm	575.37	721.74	16.04	492.76	nr	1214.50
Sum of two sides 3300 mm	603.26	756.73	16.04	492.76	nr	1249.49
Sum of two sides 3400 mm	631.47	792.12	16.04	492.76	nr	1284.88
Sum of two sides 3500 mm	584.69	733.43	16.04	492.76	nr	1226.19
Sum of two sides 3600 mm	612.89	768.81	16.04	492.76	nr	1261.57
Sum of two sides 3700 mm	662.97	831.63	16.04	492.76	nr	1324.39
Sum of two sides 3800 mm	691.18	867.01	16.04	492.76	nr	1359.77
Sum of two sides 3900 mm	719.38	902.40	16.04	492.76	nr	1395.16
Sum of two sides 4000 mm	710.58	891.35	16.04	492.76	nr	1384.11
Branch						
Sum of two sides 2500 mm	315.27	395.47	2.88	88.47	nr	483.94
Sum of two sides 2600 mm	332.05	416.53	2.88	88.47	nr	505.00
Sum of two sides 2700 mm	349.08	437.89	2.88	88.47	nr	526.36
Sum of two sides 2800 mm	365.61	458.62	3.94	121.04	nr	579.66
Sum of two sides 2900 mm	382.64	479.99	3.94	121.04	nr	601.03
Sum of two sides 3000 mm	399.45	501.07	4.83	148.38	nr	649.45
Sum of two sides 3100 mm	413.05	518.13	4.83	148.38	nr	666.51
Sum of two sides 3200 mm	425.51	533.76	4.83	148.38	nr	682.14
Sum of two sides 3300 mm	438.19	549.66	4.83	148.38	nr	698.04
Sum of two sides 3400 mm	450.64	565.29	4.83	148.38	nr	713.67
Sum of two sides 3500 mm	463.67	581.63	4.83	148.38	nr	730.01

38 VENTILATION/AIR CONDITIONING SYSTEMS

Item	Net Price £	Material £	Labour hours	Labour £	Unit	Total rate £
DUCTWORK: RECTANGULAR – CLASS C – cont						
Extra over fittings – cont						
Branch – cont						
Sum of two sides 3600 mm	476.12	597.24	4.83	148.38	nr	**745.62**
Sum of two sides 3700 mm	494.27	620.01	4.83	148.38	nr	**768.39**
Sum of two sides 3800 mm	506.74	635.66	4.83	148.38	nr	**784.04**
Sum of two sides 3900 mm	519.19	651.27	4.83	148.38	nr	**799.65**
Sum of two sides 4000 mm	531.88	667.20	4.83	148.38	nr	**815.58**
Grille neck						
Sum of two sides 2500 mm	321.07	402.75	3.06	94.00	nr	**496.75**
Sum of two sides 2600 mm	339.37	425.70	3.08	94.62	nr	**520.32**
Sum of two sides 2700 mm	357.66	448.65	3.08	94.62	nr	**543.27**
Sum of two sides 2800 mm	375.95	471.59	4.13	126.87	nr	**598.46**
Sum of two sides 2900 mm	394.25	494.55	4.13	126.87	nr	**621.42**
Sum of two sides 3000 mm	412.54	517.48	5.02	154.21	nr	**671.69**
Sum of two sides 3100 mm	427.21	535.90	5.02	154.21	nr	**690.11**
Sum of two sides 3200 mm	440.62	552.71	5.02	154.21	nr	**706.92**
Sum of two sides 3300 mm	454.01	569.51	5.02	154.21	nr	**723.72**
Sum of two sides 3400 mm	467.41	586.32	5.02	154.21	nr	**740.53**
Sum of two sides 3500 mm	480.80	603.12	5.02	154.21	nr	**757.33**
Sum of two sides 3600 mm	494.20	619.92	5.02	154.21	nr	**774.13**
Sum of two sides 3700 mm	507.59	636.72	5.02	154.21	nr	**790.93**
Sum of two sides 3800 mm	520.99	653.53	5.02	154.21	nr	**807.74**
Sum of two sides 3900 mm	534.38	670.33	5.02	154.21	nr	**824.54**
Sum of two sides 4000 mm	547.78	687.13	5.02	154.21	nr	**841.34**
Ductwork 2501–4000 mm longest side						
Sum of two sides 3000 mm	379.37	475.88	2.38	73.11	m	**548.99**
Sum of two sides 3100 mm	388.44	487.26	2.38	73.11	m	**560.37**
Sum of two sides 3200 mm	397.01	498.01	2.38	73.11	m	**571.12**
Sum of two sides 3300 mm	405.57	508.75	2.38	73.11	m	**581.86**
Sum of two sides 3400 mm	416.45	522.39	2.64	81.10	m	**603.49**
Sum of two sides 3500 mm	425.01	533.13	2.66	81.72	m	**614.85**
Sum of two sides 3600 mm	433.59	543.89	2.95	90.63	m	**634.52**
Sum of two sides 3700 mm	444.73	557.87	2.96	90.93	m	**648.80**
Sum of two sides 3800 mm	453.30	568.62	3.15	96.77	m	**665.39**
Sum of two sides 3900 mm	477.37	598.81	3.15	96.77	m	**695.58**
Sum of two sides 4000 mm	485.94	609.56	3.15	96.77	m	**706.33**
Sum of two sides 4100 mm	494.78	620.65	3.35	102.91	m	**723.56**
Sum of two sides 4200 mm	503.35	631.40	3.35	102.91	m	**734.31**
Sum of two sides 4300 mm	511.91	642.14	3.60	110.59	m	**752.73**
Sum of two sides 4400 mm	524.17	657.52	3.60	110.59	m	**768.11**
Sum of two sides 4500 mm	532.75	668.28	3.60	110.59	m	**778.87**
Extra over fittings; Ductwork 2501– 4000 mm longest side						

38 VENTILATION/AIR CONDITIONING SYSTEMS

Item	Net Price £	Material £	Labour hours	Labour £	Unit	Total rate £
End cap						
Sum of two sides 3000 mm	131.67	165.17	1.73	53.14	nr	218.31
Sum of two sides 3100 mm	136.36	171.05	1.73	53.14	nr	224.19
Sum of two sides 3200 mm	140.63	176.41	1.73	53.14	nr	229.55
Sum of two sides 3300 mm	144.81	181.65	1.73	53.14	nr	234.79
Sum of two sides 3400 mm	149.17	187.12	1.73	53.14	nr	240.26
Sum of two sides 3500 mm	153.44	192.47	1.73	53.14	nr	245.61
Sum of two sides 3600 mm	157.71	197.84	1.73	53.14	nr	250.98
Sum of two sides 3700 mm	161.98	203.19	1.73	53.14	nr	256.33
Sum of two sides 3800 mm	166.25	208.54	1.73	53.14	nr	261.68
Sum of two sides 3900 mm	170.52	213.90	1.80	55.29	nr	269.19
Sum of two sides 4000 mm	174.78	219.24	1.80	55.29	nr	274.53
Sum of two sides 4100 mm	179.06	224.62	1.80	55.29	nr	279.91
Sum of two sides 4200 mm	183.32	229.96	1.80	55.29	nr	285.25
Sum of two sides 4300 mm	187.59	235.31	1.88	57.75	nr	293.06
Sum of two sides 4400 mm	191.86	240.67	1.88	57.75	nr	298.42
Sum of two sides 4500 mm	196.13	246.03	1.88	57.75	nr	303.78
Reducer						
Sum of two sides 3000 mm	133.94	168.01	3.12	95.85	nr	263.86
Sum of two sides 3100 mm	140.25	175.93	3.12	95.85	nr	271.78
Sum of two sides 3200 mm	145.70	182.76	3.12	95.85	nr	278.61
Sum of two sides 3300 mm	151.17	189.63	3.12	95.85	nr	285.48
Sum of two sides 3400 mm	156.56	196.39	3.12	95.85	nr	292.24
Sum of two sides 3500 mm	162.02	203.24	3.12	95.85	nr	299.09
Sum of two sides 3600 mm	167.48	210.09	3.95	121.35	nr	331.44
Sum of two sides 3700 mm	173.09	217.12	3.97	121.97	nr	339.09
Sum of two sides 3800 mm	178.54	223.96	4.52	138.85	nr	362.81
Sum of two sides 3900 mm	174.15	218.46	4.52	138.85	nr	357.31
Sum of two sides 4000 mm	179.60	225.29	4.52	138.85	nr	364.14
Sum of two sides 4100 mm	185.25	232.38	4.52	138.85	nr	371.23
Sum of two sides 4200 mm	190.72	239.24	4.52	138.85	nr	378.09
Sum of two sides 4300 mm	196.17	246.08	4.92	151.14	nr	397.22
Sum of two sides 4400 mm	201.75	253.08	4.92	151.14	nr	404.22
Sum of two sides 4500 mm	207.20	259.91	5.12	157.29	nr	417.20
Offset						
Sum of two sides 3000 mm	372.21	466.91	3.48	106.92	nr	573.83
Sum of two sides 3100 mm	366.22	459.39	3.48	106.92	nr	566.31
Sum of two sides 3200 mm	358.05	449.14	3.48	106.92	nr	556.06
Sum of two sides 3300 mm	349.01	437.80	3.48	106.92	nr	544.72
Sum of two sides 3400 mm	382.36	479.63	3.50	107.53	nr	587.16
Sum of two sides 3500 mm	372.45	467.20	3.49	107.22	nr	574.42
Sum of two sides 3600 mm	361.66	453.67	4.25	130.56	nr	584.23
Sum of two sides 3700 mm	396.05	496.81	4.76	146.23	nr	643.04
Sum of two sides 3800 mm	384.37	482.15	5.32	163.43	nr	645.58
Sum of two sides 3900 mm	407.56	511.25	5.35	164.35	nr	675.60
Sum of two sides 4000 mm	393.93	494.14	5.35	164.35	nr	658.49
Sum of two sides 4100 mm	379.54	476.09	5.85	179.72	nr	655.81
Sum of two sides 4200 mm	364.15	456.79	5.85	179.72	nr	636.51
Sum of two sides 4300 mm	374.58	469.87	6.15	188.92	nr	658.79
Sum of two sides 4400 mm	412.22	517.09	6.30	193.54	nr	710.63
Sum of two sides 4500 mm	395.50	496.12	6.15	188.92	nr	685.04

38 VENTILATION/AIR CONDITIONING SYSTEMS

Item	Net Price £	Material £	Labour hours	Labour £	Unit	Total rate £
DUCTWORK: RECTANGULAR – CLASS C – cont						
Extra over fittings – cont						
Square to round						
Sum of two sides 3000 mm	272.72	342.10	4.70	144.38	nr	486.48
Sum of two sides 3100 mm	284.43	356.79	4.70	144.38	nr	501.17
Sum of two sides 3200 mm	294.86	369.87	4.70	144.38	nr	514.25
Sum of two sides 3300 mm	305.30	382.97	4.70	144.38	nr	527.35
Sum of two sides 3400 mm	315.75	396.08	4.71	144.69	nr	540.77
Sum of two sides 3500 mm	326.18	409.16	4.71	144.69	nr	553.85
Sum of two sides 3600 mm	336.61	422.24	8.19	251.60	nr	673.84
Sum of two sides 3700 mm	347.04	435.32	8.62	264.80	nr	700.12
Sum of two sides 3800 mm	357.48	448.43	8.62	264.80	nr	713.23
Sum of two sides 3900 mm	437.04	548.22	8.62	264.80	nr	813.02
Sum of two sides 4000 mm	449.47	563.82	8.75	268.80	nr	832.62
Sum of two sides 4100 mm	407.87	511.63	11.23	344.98	nr	856.61
Sum of two sides 4200 mm	474.30	594.97	11.23	344.98	nr	939.95
Sum of two sides 4300 mm	486.72	610.55	11.25	345.60	nr	956.15
Sum of two sides 4400 mm	499.14	626.12	11.25	345.60	nr	971.72
Sum of two sides 4500 mm	511.57	641.72	11.26	345.91	nr	987.63
90° radius bend						
Sum of two sides 3000 mm	791.25	992.54	3.90	119.81	nr	1112.35
Sum of two sides 3100 mm	828.87	1039.73	3.90	119.81	nr	1159.54
Sum of two sides 3200 mm	862.62	1082.07	4.26	130.87	nr	1212.94
Sum of two sides 3300 mm	896.37	1124.40	4.26	130.87	nr	1255.27
Sum of two sides 3400 mm	875.36	1098.05	4.26	130.87	nr	1228.92
Sum of two sides 3500 mm	908.67	1139.84	4.55	139.78	nr	1279.62
Sum of two sides 3600 mm	941.99	1181.63	4.55	139.78	nr	1321.41
Sum of two sides 3700 mm	918.48	1152.14	6.87	211.05	nr	1363.19
Sum of two sides 3800 mm	951.36	1193.38	6.87	211.05	nr	1404.43
Sum of two sides 3900 mm	890.46	1117.00	6.87	211.05	nr	1328.05
Sum of two sides 4000 mm	922.92	1157.71	7.00	215.04	nr	1372.75
Sum of two sides 4100 mm	956.30	1199.59	7.20	221.19	nr	1420.78
Sum of two sides 4200 mm	988.75	1240.29	7.20	221.19	nr	1461.48
Sum of two sides 4300 mm	1021.20	1280.99	7.41	227.64	nr	1508.63
Sum of two sides 4400 mm	954.04	1196.74	7.41	227.64	nr	1424.38
Sum of two sides 4500 mm	985.85	1236.65	7.55	231.94	nr	1468.59
45° bend						
Sum of two sides 3000 mm	383.86	481.51	4.85	148.99	nr	630.50
Sum of two sides 3100 mm	402.59	505.01	4.85	148.99	nr	654.00
Sum of two sides 3200 mm	419.40	526.10	4.85	148.99	nr	675.09
Sum of two sides 3300 mm	430.22	539.67	4.87	149.61	nr	689.28
Sum of two sides 3400 mm	425.22	533.40	4.87	149.61	nr	683.01
Sum of two sides 3500 mm	441.82	554.22	4.87	149.61	nr	703.83
Sum of two sides 3600 mm	458.42	575.04	8.81	270.65	nr	845.69
Sum of two sides 3700 mm	446.16	559.66	8.81	270.65	nr	830.31
Sum of two sides 3800 mm	462.54	580.20	9.31	286.00	nr	866.20
Sum of two sides 3900 mm	430.91	540.53	9.31	286.00	nr	826.53
Sum of two sides 4000 mm	447.08	560.82	9.31	286.00	nr	846.82

38 VENTILATION/AIR CONDITIONING SYSTEMS

Item	Net Price £	Material £	Labour hours	Labour £	Unit	Total rate £
Sum of two sides 4100 mm	463.70	581.66	10.01	307.51	nr	889.17
Sum of two sides 4200 mm	479.87	601.94	9.31	286.00	nr	887.94
Sum of two sides 4300 mm	496.04	622.23	10.01	307.51	nr	929.74
Sum of two sides 4400 mm	461.72	579.19	10.52	323.18	nr	902.37
Sum of two sides 4500 mm	477.58	599.08	10.52	323.18	nr	922.26
90° mitre bend						
Sum of two sides 3000 mm	930.76	1167.54	4.85	148.99	nr	1316.53
Sum of two sides 3100 mm	976.64	1225.10	4.85	148.99	nr	1374.09
Sum of two sides 3200 mm	413.21	518.34	4.87	149.61	nr	667.95
Sum of two sides 3300 mm	1062.94	1333.35	4.87	149.61	nr	1482.96
Sum of two sides 3400 mm	1036.65	1300.38	4.87	149.61	nr	1449.99
Sum of two sides 3500 mm	1079.41	1354.01	8.81	270.65	nr	1624.66
Sum of two sides 3600 mm	1122.18	1407.66	8.81	270.65	nr	1678.31
Sum of two sides 3700 mm	1100.94	1381.02	14.81	454.97	nr	1835.99
Sum of two sides 3800 mm	1143.38	1434.26	14.81	454.97	nr	1889.23
Sum of two sides 3900 mm	1063.23	1333.72	14.81	454.97	nr	1788.69
Sum of two sides 4000 mm	1105.29	1386.47	15.20	466.95	nr	1853.42
Sum of two sides 4100 mm	1146.79	1438.53	16.30	500.74	nr	1939.27
Sum of two sides 4200 mm	1188.85	1491.29	16.50	506.88	nr	1998.17
Sum of two sides 4300 mm	1230.92	1544.07	17.01	522.55	nr	2066.62
Sum of two sides 4400 mm	1153.98	1447.56	17.01	522.55	nr	1970.11
Sum of two sides 4500 mm	1195.52	1499.66	17.01	522.55	nr	2022.21
Branch						
Sum of two sides 3000 mm	399.39	501.00	2.88	88.47	nr	589.47
Sum of two sides 3100 mm	1019.78	1279.21	2.88	88.47	nr	1367.68
Sum of two sides 3200 mm	425.67	533.96	2.88	88.47	nr	622.43
Sum of two sides 3300 mm	438.12	549.57	2.88	88.47	nr	638.04
Sum of two sides 3400 mm	450.53	565.14	2.88	88.47	nr	653.61
Sum of two sides 3500 mm	462.99	580.78	3.94	121.04	nr	701.82
Sum of two sides 3600 mm	475.46	596.42	3.94	121.04	nr	717.46
Sum of two sides 3700 mm	488.07	612.24	3.94	121.04	nr	733.28
Sum of two sides 3800 mm	500.53	627.86	4.83	148.38	nr	776.24
Sum of two sides 3900 mm	513.49	644.12	4.83	148.38	nr	792.50
Sum of two sides 4000 mm	525.96	659.77	4.83	148.38	nr	808.15
Sum of two sides 4100 mm	538.63	675.66	5.44	167.13	nr	842.79
Sum of two sides 4200 mm	551.09	691.29	5.44	167.13	nr	858.42
Sum of two sides 4300 mm	563.54	706.90	5.85	179.72	nr	886.62
Sum of two sides 4400 mm	567.14	711.42	5.85	179.72	nr	891.14
Sum of two sides 4500 mm	588.60	738.34	5.85	179.72	nr	918.06
Grille neck						
Sum of two sides 3000 mm	412.54	517.48	2.88	88.47	nr	605.95
Sum of two sides 3100 mm	427.22	535.91	2.88	88.47	nr	624.38
Sum of two sides 3200 mm	440.61	552.70	2.88	88.47	nr	641.17
Sum of two sides 3300 mm	454.01	569.51	2.88	88.47	nr	657.98
Sum of two sides 3400 mm	467.41	586.32	3.94	121.04	nr	707.36
Sum of two sides 3500 mm	480.81	603.13	3.94	121.04	nr	724.17
Sum of two sides 3600 mm	414.19	519.56	3.94	121.04	nr	640.60
Sum of two sides 3700 mm	507.59	636.72	4.12	126.57	nr	763.29
Sum of two sides 3800 mm	520.99	653.53	4.12	126.57	nr	780.10
Sum of two sides 3900 mm	534.39	670.34	4.12	126.57	nr	796.91
Sum of two sides 4000 mm	547.78	687.13	5.00	153.60	nr	840.73

38 VENTILATION/AIR CONDITIONING SYSTEMS

Item	Net Price £	Material £	Labour hours	Labour £	Unit	Total rate £
DUCTWORK: RECTANGULAR – CLASS C – cont						
Extra over fittings – cont						
Sum of two sides 4100 mm	561.18	703.94	5.00	153.60	nr	857.54
Sum of two sides 4200 mm	574.58	720.75	5.00	153.60	nr	874.35
Sum of two sides 4300 mm	587.98	737.56	5.23	160.66	nr	898.22
Sum of two sides 4400 mm	601.37	754.35	5.23	160.66	nr	915.01
Sum of two sides 4500 mm	614.77	771.16	5.39	165.58	nr	936.74
DUCTWORK ANCILLARIES: ACCESS DOORS						
Refer to ancillaries in Ductwork Ancillaries: Access Hatches						

38 VENTILATION/AIR CONDITIONING SYSTEMS

Item	Net Price £	Material £	Labour hours	Labour £	Unit	Total rate £
DUCTWORK ANCILLARIES: VOLUME CONTROL AND FIRE DAMPERS						
Volume control damper; opposed blade; galvanized steel casing; aluminium aerofoil blades; manually operated						
Rectangular						
Sum of two sides 200 mm	21.82	27.37	1.60	49.15	nr	76.52
Sum of two sides 300 mm	23.34	29.28	1.60	49.15	nr	78.43
Sum of two sides 400 mm	25.52	32.01	1.60	49.15	nr	81.16
Sum of two sides 500 mm	27.93	35.03	1.60	49.15	nr	84.18
Sum of two sides 600 mm	30.88	38.74	1.70	52.25	nr	90.99
Sum of two sides 700 mm	33.94	42.57	2.10	64.53	nr	107.10
Sum of two sides 800 mm	37.08	46.51	2.15	66.05	nr	112.56
Sum of two sides 900 mm	40.69	51.04	2.30	70.66	nr	121.70
Sum of two sides 1000 mm	44.18	55.42	2.40	73.73	nr	129.15
Sum of two sides 1100 mm	48.13	60.38	2.60	79.87	nr	140.25
Sum of two sides 1200 mm	54.44	68.29	2.80	86.02	nr	154.31
Sum of two sides 1300 mm	58.70	73.63	3.10	95.23	nr	168.86
Sum of two sides 1400 mm	63.29	79.39	3.25	99.84	nr	179.23
Sum of two sides 1500 mm	68.51	85.94	3.40	104.45	nr	190.39
Sum of two sides 1600 mm	73.65	92.39	3.45	105.99	nr	198.38
Sum of two sides 1700 mm	78.56	98.55	3.60	110.59	nr	209.14
Sum of two sides 1800 mm	84.44	105.92	3.90	119.81	nr	225.73
Sum of two sides 1900 mm	89.90	112.77	4.20	129.08	nr	241.85
Sum of two sides 2000 mm	96.44	120.97	4.33	133.01	nr	253.98
Circular						
100 mm dia.	29.13	36.55	0.80	24.57	nr	61.12
150 mm dia.	33.28	41.74	0.80	24.57	nr	66.31
160 mm dia.	34.69	43.51	0.90	27.64	nr	71.15
200 mm dia.	37.63	47.21	1.05	32.27	nr	79.48
250 mm dia.	41.90	52.56	1.20	36.88	nr	89.44
300 mm dia.	46.38	58.18	1.35	41.47	nr	99.65
315 mm dia.	48.44	60.76	1.35	41.52	nr	102.28
350 mm dia.	50.96	63.93	1.65	50.69	nr	114.62
400 mm dia.	55.64	69.80	1.90	58.41	nr	128.21
450 mm dia.	60.11	75.40	2.10	64.53	nr	139.93
500 mm dia.	65.48	82.14	2.95	90.62	nr	172.76
550 mm dia.	70.71	88.70	2.94	90.32	nr	179.02
600 mm dia.	76.59	96.07	2.94	90.32	nr	186.39
650 mm dia.	82.70	103.73	4.55	139.78	nr	243.51
700 mm dia.	89.36	112.09	5.20	159.75	nr	271.84
750 mm dia.	96.44	120.97	5.20	159.75	nr	280.72
800 mm dia.	103.77	130.17	5.80	178.17	nr	308.34
850 mm dia.	111.40	139.74	5.80	178.17	nr	317.91
900 mm dia.	119.36	149.72	6.40	196.62	nr	346.34
950 mm dia.	127.55	160.00	6.40	196.62	nr	356.62
1000 mm dia.	135.84	170.40	7.00	215.04	nr	385.44

38 VENTILATION/AIR CONDITIONING SYSTEMS

Item	Net Price £	Material £	Labour hours	Labour £	Unit	Total rate £
DUCTWORK ANCILLARIES: VOLUME CONTROL AND FIRE DAMPERS – cont						
Volume control damper – cont						
Flat oval						
345 × 102 mm	49.21	61.73	1.20	36.87	nr	**98.60**
508 × 102 mm	55.43	69.53	1.60	49.15	nr	**118.68**
559 × 152 mm	62.63	78.57	1.90	58.41	nr	**136.98**
531 × 203 mm	68.42	85.83	1.90	58.41	nr	**144.24**
851 × 203 mm	72.34	90.74	4.55	139.78	nr	**230.52**
582 × 254 mm	75.95	95.27	2.10	64.53	nr	**159.80**
823 × 254 mm	83.15	104.31	4.10	125.96	nr	**230.27**
632 × 305 mm	83.69	104.98	2.95	90.62	nr	**195.60**
765 × 356 mm	93.62	117.43	4.55	139.78	nr	**257.21**
737 × 406 mm	99.72	125.09	4.55	139.78	nr	**264.87**
818 × 406 mm	101.81	127.71	5.20	159.75	nr	**287.46**
978 × 406 mm	110.10	138.11	5.50	168.95	nr	**307.06**
709 × 457 mm	103.77	130.17	4.50	138.39	nr	**268.56**
678 × 508 mm	109.77	137.69	4.55	139.78	nr	**277.47**
919 × 508 mm	123.19	154.53	6.00	184.32	nr	**338.85**
Fire damper; galvanized steel casing; stainless steel folding shutter; fusible link with manual reset; BS 476 4 hour fire-rated						
Rectangular						
Sum of two sides 200 mm	39.98	50.15	1.60	49.15	nr	**99.30**
Sum of two sides 300 mm	41.25	51.74	1.60	49.15	nr	**100.89**
Sum of two sides 400 mm	43.02	53.96	1.60	49.15	nr	**103.11**
Sum of two sides 500 mm	47.06	59.04	1.60	49.15	nr	**108.19**
Sum of two sides 600 mm	51.56	64.68	1.70	52.25	nr	**116.93**
Sum of two sides 700 mm	56.17	70.46	2.10	64.53	nr	**134.99**
Sum of two sides 800 mm	60.97	76.48	2.15	66.07	nr	**142.55**
Sum of two sides 900 mm	65.79	82.52	2.30	70.78	nr	**153.30**
Sum of two sides 1000 mm	70.71	88.70	2.40	73.84	nr	**162.54**
Sum of two sides 1100 mm	75.10	94.20	2.60	80.00	nr	**174.20**
Sum of two sides 1200 mm	80.33	100.77	2.80	86.05	nr	**186.82**
Sum of two sides 1300 mm	85.68	107.48	3.10	95.23	nr	**202.71**
Sum of two sides 1400 mm	91.45	114.71	3.25	99.84	nr	**214.55**
Sum of two sides 1500 mm	96.47	121.02	3.40	104.50	nr	**225.52**
Sum of two sides 1600 mm	102.01	127.96	3.45	105.99	nr	**233.95**
Sum of two sides 1700 mm	107.58	134.95	3.60	110.59	nr	**245.54**
Sum of two sides 1800 mm	113.44	142.30	3.90	119.81	nr	**262.11**
Sum of two sides 1900 mm	119.00	149.27	4.20	129.08	nr	**278.35**
Sum of two sides 2000 mm	124.23	155.84	4.33	133.01	nr	**288.85**
Sum of two sides 2100 mm	132.20	165.83	4.43	136.09	nr	**301.92**
Sum of two sides 2200 mm	139.87	175.45	4.55	139.78	nr	**315.23**
Circular						
100 mm dia.	48.70	61.08	0.80	24.57	nr	**85.65**
150 mm dia.	51.25	64.29	0.80	24.57	nr	**88.86**
160 mm dia.	51.56	64.68	0.90	27.64	nr	**92.32**

38 VENTILATION/AIR CONDITIONING SYSTEMS

Item	Net Price £	Material £	Labour hours	Labour £	Unit	Total rate £
200 mm dia.	54.11	67.87	1.05	32.26	nr	100.13
250 mm dia.	60.30	75.64	1.20	36.88	nr	112.52
300 mm dia.	67.02	84.07	1.35	41.47	nr	125.54
315 mm dia.	69.93	87.72	1.35	41.52	nr	129.24
350 mm dia.	74.17	93.04	1.65	50.69	nr	143.73
400 mm dia.	81.95	102.79	1.90	58.41	nr	161.20
450 mm dia.	89.62	112.41	2.10	64.53	nr	176.94
500 mm dia.	98.03	122.96	2.95	90.62	nr	213.58
550 mm dia.	106.46	133.55	2.94	90.32	nr	223.87
600 mm dia.	115.39	144.75	4.55	139.78	nr	284.53
650 mm dia.	124.87	156.63	4.55	139.78	nr	296.41
700 mm dia.	134.57	168.81	5.20	159.75	nr	328.56
750 mm dia.	144.77	181.60	5.20	159.75	nr	341.35
800 mm dia.	155.19	194.67	5.80	178.17	nr	372.84
850 mm dia.	166.16	208.43	5.80	178.17	nr	386.60
900 mm dia.	177.44	222.58	6.40	196.62	nr	419.20
950 mm dia.	189.02	237.10	6.40	196.62	nr	433.72
1000 mm dia.	201.16	252.34	7.00	215.04	nr	467.38
Flat oval						
345 × 102 mm	66.98	84.02	1.20	36.88	nr	120.90
427 × 102 mm	72.73	91.24	1.35	41.52	nr	132.76
508 × 102 mm	75.19	94.32	1.60	49.15	nr	143.47
559 × 152 mm	81.97	102.83	1.90	58.41	nr	161.24
531 × 203 mm	86.22	108.16	1.90	58.41	nr	166.57
851 × 203 mm	108.35	135.91	4.55	139.78	nr	275.69
582 × 254 mm	114.33	143.42	2.10	64.53	nr	207.95
632 × 305 mm	119.97	150.49	2.95	90.63	nr	241.12
765 × 356 mm	137.86	172.93	4.55	139.78	nr	312.71
737 × 406 mm	143.81	180.40	4.55	139.78	nr	320.18
818 × 406 mm	152.64	191.48	5.20	159.75	nr	351.23
978 × 406 mm	165.95	208.16	5.50	168.95	nr	377.11
709 × 457 mm	144.89	181.75	4.50	138.39	nr	320.14
678 × 508 mm	150.74	189.09	4.55	139.78	nr	328.87
Smoke/fire damper; galvanized steel casing; stainless steel folding shutter; fusible link and 24 V DC electromagnetic shutter release mechanism; spring operated; BS 476 4 hour fire-rating						
Rectangular						
Sum of two sides 200 mm	306.22	384.13	1.60	49.15	nr	433.28
Sum of two sides 300 mm	307.10	385.22	1.60	49.15	nr	434.37
Sum of two sides 400 mm	307.99	386.34	1.60	49.15	nr	435.49
Sum of two sides 500 mm	315.37	395.60	1.60	49.15	nr	444.75
Sum of two sides 600 mm	323.09	405.28	1.70	52.25	nr	457.53
Sum of two sides 700 mm	331.14	415.39	2.10	64.53	nr	479.92
Sum of two sides 800 mm	339.40	425.75	2.15	66.07	nr	491.82
Sum of two sides 900 mm	348.02	436.55	2.30	70.78	nr	507.33
Sum of two sides 1000 mm	356.93	447.73	2.40	73.84	nr	521.57
Sum of two sides 1100 mm	366.09	459.22	2.60	80.00	nr	539.22
Sum of two sides 1200 mm	375.57	471.12	2.80	86.05	nr	557.17

38 VENTILATION/AIR CONDITIONING SYSTEMS

Item	Net Price £	Material £	Labour hours	Labour £	Unit	Total rate £
DUCTWORK ANCILLARIES: VOLUME CONTROL AND FIRE DAMPERS – cont						
Smoke/fire damper – cont						
Rectangular – cont						
Sum of two sides 1300 mm	385.38	483.43	3.10	95.23	nr	**578.66**
Sum of two sides 1400 mm	395.40	495.99	3.25	99.84	nr	**595.83**
Sum of two sides 1500 mm	405.75	508.97	3.40	104.50	nr	**613.47**
Sum of two sides 1600 mm	416.45	522.39	3.45	105.99	nr	**628.38**
Sum of two sides 1700 mm	427.37	536.09	3.60	110.59	nr	**646.68**
Sum of two sides 1800 mm	438.61	550.19	3.90	119.81	nr	**670.00**
Sum of two sides 1900 mm	450.06	564.56	4.20	129.08	nr	**693.64**
Sum of two sides 2000 mm	461.85	579.34	4.33	133.01	nr	**712.35**
Circular						
100 mm dia.	149.37	187.36	0.80	24.57	nr	**211.93**
150 mm dia.	149.37	187.36	0.90	27.64	nr	**215.00**
200 mm dia.	149.37	187.36	1.05	32.26	nr	**219.62**
250 mm dia.	157.48	197.55	1.20	36.88	nr	**234.43**
300 mm dia.	165.92	208.13	1.35	41.52	nr	**249.65**
350 mm dia.	188.83	236.87	1.65	50.69	nr	**287.56**
400 mm dia.	207.66	260.49	1.90	58.41	nr	**318.90**
450 mm dia.	216.55	271.64	2.10	64.53	nr	**336.17**
500 mm dia.	235.84	295.84	2.95	90.62	nr	**386.46**
550 mm dia.	255.76	320.82	2.95	90.62	nr	**411.44**
600 mm dia.	267.81	335.94	4.55	139.78	nr	**475.72**
650 mm dia.	289.78	363.50	4.55	139.78	nr	**503.28**
700 mm dia.	313.67	393.47	5.20	159.75	nr	**553.22**
750 mm dia.	328.52	412.09	5.20	159.75	nr	**571.84**
800 mm dia.	357.10	447.94	5.80	178.17	nr	**626.11**
850 mm dia.	382.84	480.23	5.80	178.17	nr	**658.40**
900 mm dia.	397.25	498.31	6.40	196.62	nr	**694.93**
950 mm dia.	423.94	531.79	6.40	196.62	nr	**728.41**
1000 mm dia.	443.07	555.79	7.00	215.04	nr	**770.83**
Flat oval						
531 × 203 mm	265.00	332.42	1.90	58.41	nr	**390.83**
851 × 203 mm	295.83	371.09	4.55	139.78	nr	**510.87**
582 × 254 mm	288.47	361.86	2.10	64.53	nr	**426.39**
632 × 305 mm	307.52	385.75	2.95	90.63	nr	**476.38**
765 × 356 mm	331.76	416.16	4.55	139.78	nr	**555.94**
737 × 406 mm	346.73	434.94	4.55	139.78	nr	**574.72**
818 × 406 mm	351.59	441.03	5.20	159.75	nr	**600.78**
978 × 406 mm	371.40	465.89	5.50	168.95	nr	**634.84**
709 × 457 mm	356.68	447.42	4.50	138.39	nr	**585.81**
678 × 508 mm	371.31	465.77	4.55	139.78	nr	**605.55**

38 VENTILATION/AIR CONDITIONING SYSTEMS

Item	Net Price £	Material £	Labour hours	Labour £	Unit	Total rate £
DUCTWORK ANCILLARIES: ACCESS HATCHES						
Access doors, hollow steel construction; 25 mm mineral wool insulation; removeable or hinged; fixed with cams; including subframe and integral sealing gaskets						
Rectangular duct						
150 × 150 mm	17.34	21.75	1.25	38.39	nr	**60.14**
200 × 200 mm	19.08	23.93	1.25	38.39	nr	**62.32**
300 × 150 mm	19.61	24.60	1.25	38.39	nr	**62.99**
300 × 300 mm	22.06	27.68	1.25	38.39	nr	**66.07**
400 × 400 mm	25.16	31.56	1.35	41.52	nr	**73.08**
450 × 300 mm	25.16	31.56	1.50	46.12	nr	**77.68**
450 × 450 mm	27.73	34.79	1.50	46.12	nr	**80.91**
Access doors, hollow steel construction; 25 mm mineral wool insulation; removeable or hinged; fixed with cams; including subframe and integral sealing gaskets						
Flat oval duct						
235 × 90 mm	35.22	44.18	1.25	38.39	nr	**82.57**
235 × 140 mm	37.54	47.08	1.35	41.52	nr	**88.60**
335 × 235 mm	42.89	53.80	1.50	46.12	nr	**99.92**
535 × 235 mm	48.25	60.52	1.50	46.12	nr	**106.64**

38 VENTILATION/AIR CONDITIONING SYSTEMS

Item	Net Price £	Material £	Labour hours	Labour £	Unit	Total rate £
GRILLES/DIFFUSERS/LOUVRES						
Supply grilles; single deflection; extruded aluminium alloy frame and adjustable horizontal vanes; silver grey polyester powder coated; screw fixed						
Rectangular; for duct, ceiling and sidewall applications						
150 × 100 mm	12.14	15.23	0.60	18.42	nr	33.65
150 × 150 mm	14.84	18.61	0.60	18.42	nr	37.03
200 × 150 mm	17.64	22.13	0.65	19.97	nr	42.10
200 × 200 mm	15.26	19.14	0.72	22.12	nr	41.26
300 × 100 mm	13.10	16.43	0.72	22.12	nr	38.55
300 × 150 mm	15.19	19.05	0.80	24.57	nr	43.62
300 × 200 mm	17.27	21.66	0.88	27.04	nr	48.70
300 × 300 mm	21.43	26.88	1.04	31.95	nr	58.83
400 × 100 mm	14.67	18.40	0.88	27.04	nr	45.44
400 × 150 mm	16.96	21.28	0.94	28.87	nr	50.15
400 × 200 mm	19.27	24.17	1.04	31.95	nr	56.12
400 × 300 mm	23.89	29.97	1.12	34.41	nr	64.38
600 × 200 mm	29.40	36.88	1.26	38.71	nr	75.59
600 × 300 mm	47.57	59.67	1.40	43.01	nr	102.68
600 × 400 mm	56.65	71.06	1.61	49.46	nr	120.52
600 × 500 mm	65.77	82.50	1.76	54.06	nr	136.56
600 × 600 mm	74.86	93.90	2.17	66.66	nr	160.56
800 × 300 mm	55.67	69.83	1.76	54.06	nr	123.89
800 × 400 mm	66.28	83.14	2.17	66.66	nr	149.80
800 × 600 mm	87.48	109.74	3.00	92.16	nr	201.90
1000 × 300 mm	63.80	80.04	2.60	79.87	nr	159.91
1000 × 400 mm	75.89	95.20	3.00	92.16	nr	187.36
1000 × 600 mm	100.10	125.56	3.80	116.74	nr	242.30
1000 × 800 mm	124.28	155.89	3.80	116.74	nr	272.63
1200 × 600 mm	112.69	141.36	4.61	141.62	nr	282.98
1200 × 800 mm	139.88	175.47	4.61	141.62	nr	317.09
1200 × 1000 mm	167.07	209.57	4.61	141.62	nr	351.19
Rectangular; for duct, ceiling and sidewall applications; including opposed blade damper volume regulator						
150 × 100 mm	22.63	28.39	0.72	22.13	nr	50.52
150 × 150 mm	26.21	32.88	0.72	22.13	nr	55.01
200 × 150 mm	28.14	35.30	0.83	25.51	nr	60.81
200 × 200 mm	30.42	38.16	0.90	27.64	nr	65.80
300 × 100 mm	32.31	40.53	0.90	27.64	nr	68.17
300 × 150 mm	34.44	43.20	0.98	30.11	nr	73.31
300 × 200 mm	37.11	46.55	1.06	32.58	nr	79.13
300 × 300 mm	42.42	53.21	1.20	36.88	nr	90.09
400 × 100 mm	41.99	52.67	1.06	32.58	nr	85.25
400 × 150 mm	44.34	55.62	1.13	34.71	nr	90.33
400 × 200 mm	44.55	55.89	1.20	36.88	nr	92.77

38 VENTILATION/AIR CONDITIONING SYSTEMS

Item	Net Price £	Material £	Labour hours	Labour £	Unit	Total rate £
400 × 300 mm	50.17	62.93	1.34	41.18	nr	104.11
600 × 200 mm	61.14	76.70	1.50	46.12	nr	122.82
600 × 300 mm	81.22	101.89	1.66	51.03	nr	152.92
600 × 400 mm	92.33	115.82	1.80	55.36	nr	171.18
600 × 500 mm	104.56	131.16	2.00	61.44	nr	192.60
600 × 600 mm	115.67	145.10	2.60	80.00	nr	225.10
800 × 300 mm	111.38	139.72	2.00	61.44	nr	201.16
800 × 400 mm	124.89	156.67	2.60	80.00	nr	236.67
800 × 600 mm	153.43	192.46	3.61	110.90	nr	303.36
1000 × 300 mm	129.60	162.57	3.00	92.24	nr	254.81
1000 × 400 mm	145.21	182.16	3.61	110.90	nr	293.06
1000 × 600 mm	178.31	223.68	4.61	141.57	nr	365.25
1000 × 800 mm	263.70	330.78	4.61	141.57	nr	472.35
1200 × 600 mm	198.08	248.47	5.62	172.58	nr	421.05
1200 × 800 mm	290.65	364.59	6.10	187.40	nr	551.99
1200 × 1000 mm	330.49	414.57	6.50	199.68	nr	614.25
Supply grilles; double deflection; extruded aluminium alloy frame and adjustable horizontal and vertical vanes; white polyester powder coated; screw fixed Rectangular; for duct, ceiling and sidewall applications						
150 × 100 mm	12.14	15.23	0.88	27.04	nr	42.27
150 × 150 mm	15.67	19.66	0.88	27.04	nr	46.70
200 × 150 mm	17.64	22.13	1.08	33.17	nr	55.30
200 × 200 mm	19.66	24.66	1.25	38.39	nr	63.05
300 × 100 mm	19.15	24.02	1.25	38.39	nr	62.41
300 × 150 mm	21.62	27.12	1.50	46.08	nr	73.20
300 × 200 mm	24.10	30.23	1.75	53.76	nr	83.99
300 × 300 mm	29.08	36.48	2.15	66.05	nr	102.53
400 × 100 mm	22.65	28.41	1.75	53.76	nr	82.17
400 × 150 mm	25.60	32.11	1.95	59.91	nr	92.02
400 × 200 mm	28.56	35.83	2.15	66.05	nr	101.88
400 × 300 mm	34.47	43.24	2.55	78.33	nr	121.57
600 × 200 mm	63.67	79.87	3.01	92.47	nr	172.34
600 × 300 mm	76.94	96.51	3.36	103.22	nr	199.73
600 × 400 mm	90.22	113.18	3.80	116.74	nr	229.92
600 × 500 mm	103.48	129.81	4.20	129.02	nr	258.83
600 × 600 mm	116.75	146.45	4.51	138.54	nr	284.99
800 × 300 mm	95.29	119.53	4.20	129.02	nr	248.55
800 × 400 mm	111.78	140.21	4.51	138.54	nr	278.75
800 × 600 mm	144.76	181.59	5.10	156.68	nr	338.27
1000 × 300 mm	113.63	142.54	4.80	147.46	nr	290.00
1000 × 400 mm	133.34	167.26	5.10	156.68	nr	323.94
1000 × 600 mm	172.75	216.70	5.72	175.72	nr	392.42
1000 × 800 mm	212.18	266.16	6.33	194.45	nr	460.61
1200 × 600 mm	200.75	251.82	5.72	175.72	nr	427.54
1200 × 800 mm	246.62	309.36	6.33	194.45	nr	503.81
1200 × 1000 mm	292.48	366.89	6.33	194.45	nr	561.34

38 VENTILATION/AIR CONDITIONING SYSTEMS

Item	Net Price £	Material £	Labour hours	Labour £	Unit	Total rate £
GRILLES/DIFFUSERS/LOUVRES – cont						
Supply grilles – cont						
Rectangular; for duct, ceiling and sidewall applications; including opposed blade damper volume regulator						
150 × 100 mm	24.75	31.05	1.00	30.72	nr	61.77
150 × 150 mm	29.36	36.83	1.00	30.72	nr	67.55
200 × 150 mm	32.38	40.62	1.26	38.71	nr	79.33
200 × 200 mm	34.80	43.66	1.43	43.93	nr	87.59
300 × 100 mm	38.36	48.12	1.43	43.93	nr	92.05
300 × 150 mm	40.88	51.28	1.68	51.61	nr	102.89
300 × 200 mm	43.93	55.10	1.93	59.29	nr	114.39
300 × 300 mm	50.07	62.81	2.31	70.96	nr	133.77
400 × 100 mm	49.98	62.70	1.93	59.29	nr	121.99
400 × 150 mm	52.97	66.45	2.14	65.74	nr	132.19
400 × 200 mm	53.82	67.51	2.31	70.96	nr	138.47
400 × 300 mm	61.21	76.79	2.77	85.10	nr	161.89
600 × 200 mm	95.97	120.39	3.25	99.84	nr	220.23
600 × 300 mm	111.34	139.66	3.62	111.20	nr	250.86
600 × 400 mm	126.73	158.97	3.99	122.58	nr	281.55
600 × 500 mm	143.23	179.67	4.44	136.40	nr	316.07
600 × 600 mm	158.61	198.96	4.94	151.76	nr	350.72
800 × 300 mm	152.02	190.69	4.44	136.40	nr	327.09
800 × 400 mm	171.52	215.15	4.94	151.76	nr	366.91
800 × 600 mm	212.10	266.06	5.71	175.41	nr	441.47
1000 × 300 mm	180.62	226.56	5.20	159.75	nr	386.31
1000 × 400 mm	203.98	255.88	5.71	175.41	nr	431.29
1000 × 600 mm	252.59	316.85	6.53	200.60	nr	517.45
1000 × 800 mm	354.01	444.07	6.53	200.60	nr	644.67
1200 × 600 mm	287.93	361.18	7.34	225.49	nr	586.67
1200 × 800 mm	400.03	501.79	8.80	270.33	nr	772.12
1200 × 1000 mm	458.90	575.65	8.80	270.33	nr	845.98
Floor grille suitable for mounting in raised access floors; heavy duty; extruded alumiinium; standard mill finish; complete with opposed blade volume control damper						
Diffuser						
600 mm × 600 mm	149.84	187.96	0.70	21.50	nr	209.46
Extra for nylon coated black finish	22.93	28.76	–	–	nr	28.76
Exhaust grilles; aluminium						
0° fixed blade core						
150 × 150 mm	15.51	19.45	0.60	18.50	nr	37.95
200 × 200 mm	18.25	22.89	0.72	22.13	nr	45.02
250 × 250 mm	21.24	26.64	0.80	24.57	nr	51.21
300 × 300 mm	24.50	30.73	1.00	30.72	nr	61.45
350 × 350 mm	28.00	35.12	1.20	36.87	nr	71.99

38 VENTILATION/AIR CONDITIONING SYSTEMS

Item	Net Price £	Material £	Labour hours	Labour £	Unit	Total rate £
0° fixed blade core; including opposed blade damper volume regulator						
150 × 150 mm	29.21	36.65	0.62	19.05	nr	55.70
200 × 200 mm	33.39	41.89	0.72	22.13	nr	64.02
250 × 250 mm	38.01	47.68	0.80	24.57	nr	72.25
300 × 300 mm	45.48	57.05	1.00	30.72	nr	87.77
350 × 350 mm	50.91	63.86	1.20	36.87	nr	100.73
45° fixed blade core						
150 × 150 mm	15.51	19.45	0.62	19.05	nr	38.50
200 × 200 mm	18.25	22.89	0.72	22.13	nr	45.02
250 × 250 mm	21.24	26.64	0.80	24.57	nr	51.21
300 × 300 mm	24.50	30.73	1.00	30.72	nr	61.45
350 × 350 mm	28.00	35.12	1.20	36.87	nr	71.99
45° fixed blade core; including opposed blade damper volume regulator						
150 × 150 mm	29.21	36.65	0.62	19.05	nr	55.70
200 × 200 mm	33.39	41.89	0.72	22.13	nr	64.02
250 × 250 mm	38.01	47.68	0.80	24.57	nr	72.25
300 × 300 mm	45.48	57.05	1.00	30.72	nr	87.77
350 × 350 mm	50.91	63.86	1.20	36.87	nr	100.73
Eggcrate core						
150 × 150 mm	10.41	13.06	0.62	19.05	nr	32.11
200 × 200 mm	12.86	16.13	1.00	30.72	nr	46.85
250 × 250 mm	15.67	19.66	0.80	24.57	nr	44.23
300 × 300 mm	19.94	25.01	1.00	30.72	nr	55.73
350 × 350 mm	22.28	27.94	1.20	36.87	nr	64.81
Eggcrate core; including opposed blade damper volume regulator						
150 × 150 mm	24.10	30.23	0.62	19.05	nr	49.28
200 × 200 mm	28.00	35.12	0.72	22.13	nr	57.25
250 × 250 mm	32.41	40.66	0.80	24.57	nr	65.23
300 × 300 mm	40.93	51.34	1.00	30.72	nr	82.06
350 × 350 mm	45.21	56.72	1.20	36.87	nr	93.59
Mesh/perforated plate core						
150 × 150 mm	11.45	14.36	0.62	19.04	nr	33.40
200 × 200 mm	14.15	17.75	0.72	22.12	nr	39.87
250 × 250 mm	17.24	21.63	0.80	24.57	nr	46.20
300 × 300 mm	21.95	27.53	1.00	30.72	nr	58.25
350 × 350 mm	24.53	30.77	1.20	36.87	nr	67.64
Mesh/perforated plate core; including opposed blade damper volume regulator						
150 × 150 mm	25.15	31.55	0.62	19.04	nr	50.59
200 × 200 mm	29.29	36.74	0.72	22.12	nr	58.86
250 × 250 mm	33.99	42.64	0.80	24.57	nr	67.21
300 × 300 mm	42.94	53.86	0.80	24.57	nr	78.43
350 × 350 mm	47.43	59.49	1.20	36.87	nr	96.36

38 VENTILATION/AIR CONDITIONING SYSTEMS

Item	Net Price £	Material £	Labour hours	Labour £	Unit	Total rate £
GRILLES/DIFFUSERS/LOUVRES – cont						
Plastic air diffusion system						
Eggcrate grilles						
150 × 150 mm	4.67	5.86	0.62	19.95	nr	25.81
200 × 200 mm	8.17	10.25	0.72	23.17	nr	33.42
250 × 250 mm	9.33	11.70	0.80	25.73	nr	37.43
300 × 300 mm	9.33	11.70	1.00	32.17	nr	43.87
Single deflection grilles						
150 × 150 mm	7.35	9.22	0.62	19.95	nr	29.17
200 × 200 mm	9.11	11.42	0.72	23.17	nr	34.59
250 × 250 mm	10.02	12.57	0.80	25.73	nr	38.30
300 × 300 mm	12.60	15.80	1.00	32.17	nr	47.97
Double deflection grilles						
150 × 150 mm	9.16	11.49	0.62	19.95	nr	31.44
200 × 200 mm	12.15	15.24	0.72	23.17	nr	38.41
250 × 250 mm	13.88	17.42	0.80	25.73	nr	43.15
300 × 300 mm	19.60	24.58	1.00	32.17	nr	56.75
Door transfer grilles						
150 × 150 mm	13.76	17.26	0.62	19.95	nr	37.21
200 × 200 mm	15.98	20.05	0.72	23.17	nr	43.22
250 × 250 mm	16.55	20.76	0.80	25.73	nr	46.49
300 × 300 mm	17.96	22.53	1.00	32.17	nr	54.70
Opposed blade dampers						
150 × 150 mm	8.33	10.45	0.62	19.95	nr	30.40
200 × 200 mm	9.80	12.30	0.72	23.17	nr	35.47
250 × 250 mm	14.69	18.42	0.80	25.73	nr	44.15
300 × 300 mm	15.98	20.05	1.00	32.17	nr	52.22
Ceiling mounted diffusers; circular aluminium multi-cone diffuser						
Circular; for ceiling mounting						
141 mm dia. neck	53.19	66.72	0.80	24.57	nr	91.29
197 mm dia. neck	77.87	97.68	1.10	33.79	nr	131.47
309 mm dia. neck	105.48	132.32	1.40	43.03	nr	175.35
365 mm dia. neck	90.24	113.20	1.50	46.12	nr	159.32
457 mm dia. neck	142.27	178.46	2.00	61.44	nr	239.90
Circular; for ceiling mounting; including louvre damper volume control						
141 mm dia. neck	79.50	99.72	1.00	30.72	nr	130.44
197 mm dia. neck	104.18	130.68	1.20	36.88	nr	167.56
309 mm dia. neck	137.46	172.44	1.60	49.15	nr	221.59
365 mm dia. neck	124.62	156.32	1.90	58.41	nr	214.73
457 mm dia. neck	182.72	229.21	2.40	73.84	nr	303.05
Ceiling mounted diffusers; rectangular aluminium multi-cone diffuser; four way flow						
Rectangular; for ceiling mounting						
150 × 150 mm neck	21.89	27.46	1.80	55.36	nr	82.82
300 × 150 mm neck	37.34	46.84	2.30	70.66	nr	117.50
300 × 300 mm neck	33.35	41.83	2.80	86.02	nr	127.85

38 VENTILATION/AIR CONDITIONING SYSTEMS

Item	Net Price £	Material £	Labour hours	Labour £	Unit	Total rate £
450 × 150 mm neck	47.58	59.68	2.80	86.02	nr	145.70
450 × 300 mm neck	46.57	58.42	3.20	98.30	nr	156.72
450 × 450 mm neck	46.50	58.33	3.40	104.50	nr	162.83
600 × 150 mm neck	58.28	73.10	3.20	98.30	nr	171.40
600 × 300 mm neck	57.96	72.71	3.50	107.53	nr	180.24
600 × 600 mm neck	64.67	81.12	4.00	122.89	nr	204.01
Rectangular; for ceiling mounting; including opposed blade damper volume regulator						
150 × 150 mm neck	28.10	35.25	1.80	55.36	nr	90.61
300 × 150 mm neck	56.60	71.00	2.30	70.66	nr	141.66
300 × 300 mm neck	41.92	52.58	2.80	86.05	nr	138.63
450 × 150 mm neck	75.88	95.19	2.80	86.02	nr	181.21
450 × 300 mm neck	74.50	93.45	3.30	101.38	nr	194.83
450 × 450 mm neck	58.94	73.93	3.51	107.79	nr	181.72
600 × 150 mm neck	93.05	116.73	3.30	101.38	nr	218.11
600 × 300 mm neck	92.36	115.85	4.00	122.89	nr	238.74
600 × 600 mm neck	95.20	119.41	5.62	172.58	nr	291.99
Slot diffusers; continuous aluminium slot diffuser with flanged frame (1500 mm sections)						
Diffuser						
1 slot	35.98	45.14	3.76	115.49	m	160.63
2 slot	43.80	54.95	3.76	115.45	m	170.40
3 slot	53.90	67.61	3.76	115.45	m	183.06
4 slot	60.67	76.10	4.50	138.39	m	214.49
6 slot	79.93	100.26	4.50	138.39	m	238.65
Diffuser; including equalizing deflector						
1 slot	74.48	93.43	5.26	161.69	m	255.12
2 slot	92.40	115.91	5.26	161.69	m	277.60
3 slot	113.01	141.76	5.26	161.69	m	303.45
4 slot	126.90	159.19	6.33	194.43	m	353.62
6 slot	166.93	209.40	6.33	194.43	m	403.83
Extra over for ends						
1 slot	4.02	5.04	1.00	30.72	nr	35.76
2 slot	4.23	5.31	1.00	30.72	nr	36.03
3 slot	4.49	5.63	1.00	30.72	nr	36.35
4 slot	4.71	5.91	1.30	39.94	nr	45.85
6 slot	5.38	6.75	1.40	43.01	nr	49.76
Plenum boxes; 1.0 m long; circular spigot; including cord operated flap damper						
1 slot	51.27	64.31	2.75	84.64	nr	148.95
2 slot	52.61	65.99	2.75	84.64	nr	150.63
3 slot	53.29	66.84	2.75	84.64	nr	151.48
4 slot	58.14	72.93	3.51	107.79	nr	180.72
6 slot	61.00	76.52	3.51	107.79	nr	184.31
Plenum boxes; 2.0 m long; circular spigot; including cord operated flap damper						
1 slot	56.77	71.21	3.26	100.06	nr	171.27
2 slot	59.52	74.66	3.26	100.06	nr	174.72
3 slot	60.37	75.72	3.26	100.06	nr	175.78

38 VENTILATION/AIR CONDITIONING SYSTEMS

Item	Net Price £	Material £	Labour hours	Labour £	Unit	Total rate £
GRILLES/DIFFUSERS/LOUVRES – cont						
Slot diffusers – cont						
Plenum boxes – cont						
4 slot	67.23	84.34	3.76	115.49	nr	**199.83**
6 slot	71.69	89.92	3.76	115.49	nr	**205.41**
Perforated diffusers; rectangular face aluminium perforated diffuser; quick release face plate; for integration with rectangular ceiling tiles						
Circular spigot; rectangular diffuser						
150 mm dia. spigot; 300 × 300 mm diffuser	52.41	65.74	1.00	30.72	nr	**96.46**
300 mm dia. spigot; 600 × 600 mm diffuser	78.20	98.09	1.40	43.03	nr	**141.12**
Circular spigot; rectangular diffuser; including louvre damper volume regulator						
150 mm dia. spigot; 300 × 300 mm diffuser	78.72	98.75	1.00	30.72	nr	**129.47**
300 mm dia. spigot; 600 × 600 mm diffuser	110.17	138.20	1.60	49.15	nr	**187.35**
Rectangular spigot; rectangular diffuser						
150 × 150 mm dia. spigot; 300 × 300 mm diffuser	52.41	65.74	1.00	32.17	nr	**97.91**
300 × 150 mm dia. spigot; 600 × 300 mm diffuser	70.54	88.48	1.20	38.60	nr	**127.08**
300 × 300 mm dia. spigot; 600 × 600 mm diffuser	78.20	98.09	1.40	45.04	nr	**143.13**
600 × 300 mm dia. spigot; 1200 × 600 mm diffuser	81.68	102.46	1.60	51.45	nr	**153.91**
Rectangular spigot; rectangular diffuser; including opposed blade damper volume regulator						
150 × 150 mm dia. spigot; 300 × 300 mm diffuser	78.72	98.75	1.20	38.60	nr	**137.35**
300 × 150 mm dia. spigot; 600 × 300 mm diffuser	104.94	131.63	1.40	45.05	nr	**176.68**
300 × 300 mm dia. spigot; 600 × 600 mm diffuser	110.17	138.20	1.60	51.45	nr	**189.65**
600 × 300 mm dia. spigot; 1200 × 600 mm diffuser	128.68	161.41	1.80	57.89	nr	**219.30**
Floor swirl diffuser; manual adjustment of air discharge direction; complete with damper and dirt trap						
Plastic diffuser						
150 mm dia.	29.04	36.42	0.50	15.36	nr	**51.78**
200 mm dia.	29.04	36.42	0.50	15.36	nr	**51.78**
Aluminium diffuser						
150 mm dia.	31.09	39.00	0.50	15.36	nr	**54.36**
200 mm dia.	45.44	57.00	0.50	15.36	nr	**72.36**

38 VENTILATION/AIR CONDITIONING SYSTEMS

Item	Net Price £	Material £	Labour hours	Labour £	Unit	Total rate £
Plastic air diffusion system						
Cellular diffusers						
300 × 300 mm	13.35	16.74	2.80	90.09	nr	**106.83**
600 × 600 mm	30.90	38.76	4.00	128.65	nr	**167.41**
Multi-cone diffusers						
300 × 300 mm	10.78	13.52	2.80	90.09	nr	**103.61**
450 × 450 mm	17.73	22.24	3.40	109.40	nr	**131.64**
500 × 500 mm	18.74	23.51	3.80	122.29	nr	**145.80**
600 × 600 mm	25.93	32.52	4.00	128.65	nr	**161.17**
625 × 625 mm	44.61	55.96	4.26	136.86	nr	**192.82**
Opposed blade dampers						
300 × 300 mm	14.37	18.02	1.20	38.61	nr	**56.63**
450 × 450 mm	15.80	19.82	1.50	48.29	nr	**68.11**
600 × 600 mm	32.70	41.01	2.60	83.75	nr	**124.76**
Plenum boxes						
300 mm	9.11	11.42	2.80	90.09	nr	**101.51**
450 mm	14.14	17.74	3.40	109.40	nr	**127.14**
600 mm	22.64	28.40	4.00	128.65	nr	**157.05**
Plenum spigot reducer						
600 mm	17.96	22.53	1.00	32.17	nr	**54.70**
Blanking kits for cellular diffusers						
300 mm	6.36	7.97	0.88	28.31	nr	**36.28**
600 mm	7.55	9.48	1.10	35.39	nr	**44.87**
Blanking kits for multi-cone diffusers						
300 mm	3.36	4.21	0.88	28.31	nr	**32.52**
450 mm	10.06	12.62	0.90	28.95	nr	**41.57**
600 mm	11.74	14.73	1.10	35.39	nr	**50.12**
Acoustic louvres; opening mounted; 300 mm deep steel louvres with blades packed with acoustic infill; 12 mm galvanized mesh birdscreen; screw fixing in opening						
Louvre units; self-finished galvanized steel						
900 mm high × 600 mm wide	131.36	164.77	3.00	96.49	nr	**261.26**
900 mm high × 900 mm wide	162.97	204.43	3.00	96.49	nr	**300.92**
900 mm high × 1200 mm wide	192.95	242.03	3.34	107.43	nr	**349.46**
900 mm high × 1500 mm wide	251.80	315.86	3.34	107.43	nr	**423.29**
900 mm high × 1800 mm wide	282.33	354.16	3.34	107.43	nr	**461.59**
900 mm high × 2100 mm wide	312.86	392.45	3.34	107.43	nr	**499.88**
900 mm high × 2400 mm wide	342.83	430.05	3.68	118.36	nr	**548.41**
900 mm high × 2700 mm wide	387.52	486.10	3.68	118.36	nr	**604.46**
900 mm high × 3000 mm wide	414.77	520.28	3.68	118.36	nr	**638.64**
1200 mm high × 600 mm wide	172.23	216.05	3.00	96.49	nr	**312.54**
1200 mm high × 900 mm wide	212.02	265.96	3.34	107.43	nr	**373.39**
1200 mm high × 1200 mm wide	251.26	315.18	3.34	107.43	nr	**422.61**
1200 mm high × 1500 mm wide	333.02	417.74	3.34	107.43	nr	**525.17**
1200 mm high × 1800 mm wide	372.25	466.95	3.68	118.36	nr	**585.31**
1200 mm high × 2100 mm wide	412.59	517.55	3.68	118.36	nr	**635.91**

38 VENTILATION/AIR CONDITIONING SYSTEMS

Item	Net Price £	Material £	Labour hours	Labour £	Unit	Total rate £
GRILLES/DIFFUSERS/LOUVRES – cont						
Acoustic louvres – cont						
Louvre units – cont						
1200 mm high × 2400 mm wide	451.84	566.79	3.68	118.36	nr	685.15
1500 mm high × 600 mm wide	213.65	268.00	3.00	96.49	nr	364.49
1500 mm high × 900 mm wide	262.18	328.88	3.34	107.43	nr	436.31
1500 mm high × 1200 mm wide	310.67	389.70	3.34	107.43	nr	497.13
1500 mm high × 1500 mm wide	414.24	519.62	3.68	118.36	nr	637.98
1500 mm high × 1800 mm wide	462.75	580.47	3.68	118.36	nr	698.83
1500 mm high × 2100 mm wide	511.81	642.02	4.00	128.65	nr	770.67
1800 mm high × 600 mm wide	253.98	318.60	3.34	107.43	nr	426.03
1800 mm high × 900 mm wide	311.76	391.07	3.34	107.43	nr	498.50
1800 mm high × 1200 mm wide	370.07	464.22	3.68	118.36	nr	582.58
1800 mm high × 1500 mm wide	495.99	622.17	3.68	118.36	nr	740.53
Louvre units; polyester powder coated steel						
900 mm high × 600 mm wide	190.78	239.31	3.00	96.49	nr	335.80
900 mm high × 900 mm wide	251.26	315.18	3.00	96.49	nr	411.67
900 mm high × 1200 mm wide	310.67	389.70	3.34	107.43	nr	497.13
900 mm high × 1500 mm wide	398.97	500.47	3.34	107.43	nr	607.90
900 mm high × 1800 mm wide	458.91	575.66	3.34	107.43	nr	683.09
900 mm high × 2100 mm wide	519.43	651.57	3.34	107.43	nr	759.00
900 mm high × 2400 mm wide	578.82	726.07	3.68	118.36	nr	844.43
900 mm high × 2700 mm wide	652.41	818.38	3.68	118.36	nr	936.74
900 mm high × 3000 mm wide	709.09	889.48	3.68	118.36	nr	1007.84
1200 mm high × 600 mm wide	250.71	314.50	3.00	96.49	nr	410.99
1200 mm high × 900 mm wide	329.74	413.63	3.34	107.43	nr	521.06
1200 mm high × 1200 mm wide	408.78	512.77	3.34	107.43	nr	620.20
1200 mm high × 1500 mm wide	529.22	663.86	3.34	107.43	nr	771.29
1200 mm high × 1800 mm wide	607.71	762.32	3.68	118.36	nr	880.68
1200 mm high × 2100 mm wide	687.30	862.15	3.68	118.36	nr	980.51
1200 mm high × 2400 mm wide	765.77	960.58	3.68	118.36	nr	1078.94
1500 mm high × 600 mm wide	311.76	391.07	3.00	96.49	nr	487.56
1500 mm high × 900 mm wide	409.31	513.44	3.34	107.43	nr	620.87
1500 mm high × 1200 mm wide	506.88	635.84	3.34	107.43	nr	743.27
1500 mm high × 1500 mm wide	659.50	827.28	3.68	118.36	nr	945.64
1500 mm high × 1800 mm wide	757.60	950.33	3.68	118.36	nr	1068.69
1500 mm high × 2100 mm wide	855.17	1072.72	4.00	128.65	nr	1201.37
1800 mm high × 600 mm wide	371.69	466.24	3.34	107.43	nr	573.67
1800 mm high × 900 mm wide	488.35	612.58	3.34	107.43	nr	720.01
1800 mm high × 1200 mm wide	605.54	759.58	3.68	118.36	nr	877.94
1800 mm high × 1500 mm wide	790.31	991.37	3.68	118.36	nr	1109.73
Weather louvres; opening mounted; 300 mm deep galvanized steel louvres; screw fixing in position						
Louvre units; including 12 mm galvanized mesh birdscreen						
900 × 600 mm	120.09	150.64	2.25	72.44	nr	223.08
900 × 900 mm	173.51	217.65	2.25	72.44	nr	290.09
900 × 1200 mm	208.78	261.89	2.50	80.40	nr	342.29

38 VENTILATION/AIR CONDITIONING SYSTEMS

Item	Net Price £	Material £	Labour hours	Labour £	Unit	Total rate £
900 × 1500 mm	254.08	318.72	2.50	80.40	nr	**399.12**
900 × 1800 mm	299.39	375.56	2.50	80.40	nr	**455.96**
900 × 2100 mm	375.28	470.75	2.50	80.40	nr	**551.15**
900 × 2400 mm	403.14	505.70	2.76	88.85	nr	**594.55**
900 × 2700 mm	455.14	570.93	2.76	88.85	nr	**659.78**
900 × 3000 mm	497.05	623.50	2.76	88.85	nr	**712.35**
1200 × 600 mm	162.65	204.03	2.25	72.44	nr	**276.47**
1200 × 900 mm	227.26	285.07	2.50	80.40	nr	**365.47**
1200 × 1200 mm	292.39	366.78	2.50	80.40	nr	**447.18**
1200 × 1500 mm	341.29	428.11	2.50	80.40	nr	**508.51**
1200 × 1800 mm	446.34	559.89	2.76	88.85	nr	**648.74**
1200 × 2100 mm	507.61	636.74	2.76	88.85	nr	**725.59**
1200 × 2400 mm	530.11	664.97	2.76	88.85	nr	**753.82**
1500 × 600 mm	192.23	241.14	2.25	72.44	nr	**313.58**
1500 × 900 mm	264.90	332.29	2.50	80.40	nr	**412.69**
1500 × 1200 mm	348.92	437.68	2.50	80.40	nr	**518.08**
1500 × 1500 mm	409.09	513.16	2.76	88.85	nr	**602.01**
1500 × 1800 mm	482.37	605.08	2.76	88.85	nr	**693.93**
1500 × 2100 mm	603.29	756.76	3.00	96.59	nr	**853.35**
1800 × 600 mm	215.03	269.73	2.50	80.40	nr	**350.13**
1800 × 900 mm	297.26	372.88	2.50	80.40	nr	**453.28**
1800 × 1200 mm	389.79	488.95	2.76	88.85	nr	**577.80**
1800 × 1500 mm	461.08	578.38	3.00	96.59	nr	**674.97**

38 VENTILATION/AIR CONDITIONING SYSTEMS

Item	Net Price £	Material £	Labour hours	Labour £	Unit	Total rate £
PLANT/EQUIPMENT: FANS						
Axial flow fan; including ancillaries, anti-vibration mountings, mounting feet, matching flanges, flexible connectors and clips; 415 V, 3 phase, 50 Hz motor; includes fixing in position; electrical work elsewhere						
Aerofoil blade fan unit; short duct case						
315 mm dia.; 0.47 m³/s duty; 147 Pa	780.69	979.29	4.50	144.74	nr	**1124.03**
500 mm dia.; 1.89 m³/s duty; 500 Pa	1294.77	1624.16	5.00	160.81	nr	**1784.97**
560 mm dia.; 2.36 m³/s duty; 147 Pa	1238.87	1554.03	5.50	176.89	nr	**1730.92**
710 mm dia.; 5.67 m³/s duty; 245 Pa	1663.40	2086.57	6.00	192.98	nr	**2279.55**
Aerofoil blade fan unit; long duct case						
315 mm dia.; 0.47 m³/s duty; 147 Pa	780.69	979.29	4.50	144.74	nr	**1124.03**
500 mm dia.; 1.89 m³/s duty; 500 Pa	1294.77	1624.16	5.00	160.81	nr	**1784.97**
560 mm dia.; 2.36 m³/s duty; 147 Pa	1238.87	1554.03	5.50	176.89	nr	**1730.92**
710 mm dia.; 5.67 m³/s duty; 245 Pa	1663.40	2086.57	6.00	192.98	nr	**2279.55**
Aerofoil blade fan unit; two stage parallel fan arrangement; long duct case						
315 mm; 0.47 m³/s @ 500 Pa	780.69	979.29	4.50	144.74	nr	**1124.03**
355 mm; 0.83 m³/s @ 147 Pa	1294.77	1624.16	4.75	152.77	nr	**1776.93**
710 mm; 3.77 m³/s @ 431 Pa	1238.87	1554.03	6.00	192.98	nr	**1747.01**
710 mm; 6.61 m³/s @ 500 Pa	1663.40	2086.57	6.00	192.98	nr	**2279.55**
Axial flow fan; suitable for operation at 300°C for 90 minutes; including ancillaries, anti-vibration mountings, mounting feet, matching flanges, flexible connectors and clips; 415 V, 3 phase, 50 Hz motor; includes fixing in position; price includes air operated damper; electrical work elsewhere						
450 mm; 2.0 m³/s @ 300 Pa	1667.74	2092.01	5.00	160.81	nr	**2252.82**
630 mm; 4.6 m³/s @ 200 Pa	2345.62	2942.34	5.50	176.89	nr	**3119.23**
900 mm; 9.0 m³/s @ 300 Pa	3733.44	4683.22	6.50	209.06	nr	**4892.28**
1000 mm; 15.0 m³/s @ 400 Pa	5754.27	7218.15	7.50	241.23	nr	**7459.38**
Bifurcated fan; suitable for temperature up to 200°C with motor protection to IP55; including ancillaries, anti-vibration mountings, mounting feet, matching flanges, flexible connectors and clips; 415 V, 3 phase, 50 Hz motor; includes fixing in position; electrical work elsewhere						
300 mm; 0.50 m³/s @ 100 Pa	1521.30	1908.32	4.50	144.74	nr	**2053.06**
400 mm; 1.97 m³/s @ 200 Pa	1892.95	2374.51	5.00	160.81	nr	**2535.32**
800 mm; 3.86 m³/s @ 200 Pa	3032.20	3803.59	5.50	176.89	nr	**3980.48**
1000 mm; 6.10 m³/s @ 400 Pa	4268.54	5354.45	6.50	209.06	nr	**5563.51**

38 VENTILATION/AIR CONDITIONING SYSTEMS

Item	Net Price £	Material £	Labour hours	Labour £	Unit	Total rate £
Duct mounted in line fan with backward curved centrifugal impellor; including ancillaries, matching flanges, flexible connectors and clips; 415 V, 3 phase, 50 Hz motor; includes fixing in position; electrical work elsewhere (NB: inverter included)						
0.5 m³/s @ 200 Pa	3132.05	3928.85	4.50	144.74	nr	4073.59
1.0 m³/s @ 300 Pa	3132.05	3928.85	5.00	160.81	nr	4089.66
3.0 m³/s @ 500 Pa	3019.74	3787.96	5.50	176.89	nr	3964.85
Twin fan extract unit; belt driven; located internally; complete with anti-vibration mounts and non-return shutter; including ancillaries, matching flanges, flexible connectors and clips; 3 phase, 50 Hz motor; includes fixing in position; electrical work elsewhere (NB: inverter and auto change over included)						
0.25 m³/s @ 150 Pa	5257.27	6594.72	4.50	144.74	nr	6739.46
1.00 m³/s @ 200 Pa	5257.27	6594.72	5.00	160.81	nr	6755.53
2.00 m³/s @ 250 Pa	5496.62	6894.96	6.50	209.06	nr	7104.02
Roof mounted extract fan; including ancillaries, fibreglass cowling, fitted shutters and bird guard; 415 V, 3 phase, 50 Hz motor; includes fixing in position; electrical work elsewhere (NB: inverter included)						
Flat roof installation, fixed to curb						
355 mm;	2319.21	2909.22	4.50	144.74	nr	3053.96
400 mm;	2540.96	3187.39	5.50	176.89	nr	3364.28
450 mm;	2540.96	3187.39	7.00	225.14	nr	3412.53
500 mm;	2996.79	3759.17	8.00	257.30	nr	4016.47
Centrifugal fan; single speed for internal domestic kitchens/utility rooms; fitted with standard overload protection; complete with housing; includes placing in position; electrical work elsewhere						
Window mounted						
245 m³/hr	303.55	380.78	0.50	16.08	nr	396.86
500 m³/hr	587.17	736.55	0.50	16.08	nr	752.63
Wall mounted						
245 m³/hr	383.86	481.51	0.83	26.80	nr	508.31
500 m³/hr	743.49	932.64	0.83	26.80	nr	959.44

38 VENTILATION/AIR CONDITIONING SYSTEMS

Item	Net Price £	Material £	Labour hours	Labour £	Unit	Total rate £
PLANT/EQUIPMENT: FANS – cont						
Centrifugal fan; various speeds, simultaneous ventilation from separate areas fitted with standard overload protection; complete with housing; includes placing in position; ducting and electrical work elsewhere						
Fan unit						
147–300 m³/hr	605.64	759.72	1.00	32.17	nr	**791.89**
175–411 m³/hr	605.64	759.72	1.00	32.17	nr	**791.89**
Toilet extract units; centrifugal fan; various speeds for internal domestic bathrooms/ WCs, complete with housing; includes placing in position; electrical work elsewhere						
Fan unit; fixed to wall; including shutter						
Single speed 85 m³/hr	99.84	125.24	0.75	24.12	nr	**149.36**
Two speed 60–85 m³/hr	131.22	164.61	0.83	26.80	nr	**191.41**
Humidity controlled; autospeed; fixed to wall; including shutter						
30–60–85 m³/hr	232.01	291.03	1.00	32.17	nr	**323.20**

38 VENTILATION/AIR CONDITIONING SYSTEMS

Item	Net Price £	Material £	Labour hours	Labour £	Unit	Total rate £
PLANT/EQUIPMENT: AIR FILTRATION						
High efficiency duct mounted filters; 99.997% H13 (EU13); tested to BS 3928 Standard; 1700 m³/hr air volume; continuous rating up to 80°C; sealed wood case, aluminium spacers, neoprene gaskets; water repellant filter media; includes placing in position						
610 × 610 × 292 mm	283.92	356.15	1.00	30.72	nr	386.87
Side withdrawal frame	107.56	134.93	2.50	76.81	nr	211.74
High capacity; 3400 m³/hr air volume; continuous rating up to 80°C; anti-corrosion coated mild steel frame, polyurethane sealant and neoprene gaskets; water repellant filter media; includes placing in position						
610 × 610 × 292 mm	308.71	387.25	1.00	30.72	nr	417.97
Side withdrawal frame	107.56	134.93	2.50	76.81	nr	211.74
Bag filters; 40/60% F5 (EU5); tested to BSEN 779 Duct mounted bag filter; continuous rating up to 60°C; rigid filter assembly; sealed into one piece coated mild steel header with sealed pocket separators; includes placing in position						
6 pocket, 592 × 592 × 25 mm header; pockets 380 mm long; 1690 m³/hr	20.06	25.17	1.00	30.72	nr	55.89
Side withdrawal frame	18.62	23.35	2.00	61.44	nr	84.79
6 pocket, 592 × 592 × 25 mm header; pockets 500 mm long; 2550 m³/hr	20.38	25.57	1.50	46.08	nr	71.65
Side withdrawal frame	18.62	23.35	2.50	76.81	nr	100.16
6 pocket, 592 × 592 × 25 mm header; pockets 600 mm long; 3380 m³/hr	24.05	30.17	1.50	46.08	nr	76.25
Side withdrawal frame	18.62	23.35	3.00	92.16	nr	115.51
Bag filters; 80/90% F7, (EU7); tested to BSEN 779 Duct mounted bag filter; continuous rating up to 60°C; rigid filter assembly; sealed into one piece coated mild steel header with sealed pocket separators; includes placing in position						
6 pocket, 592 × 592 × 25 mm header; pockets 500 mm long; 1688 m³/hr	20.38	25.57	1.00	30.72	nr	56.29
Side withdrawal frame	18.62	23.35	2.00	61.44	nr	84.79
6 pocket, 592 × 592 × 25 mm header; pockets 635 mm long; 2047 m³/hr	24.05	30.17	1.50	46.08	nr	76.25
Side withdrawal frame	18.62	23.35	2.50	76.81	nr	100.16
6 pocket, 592 × 592 × 25 mm header; pockets 762 mm long; 2729 m³/hr	27.73	34.79	1.50	46.08	nr	80.87
Side withdrawal frame	18.62	23.35	3.00	92.16	nr	115.51

38 VENTILATION/AIR CONDITIONING SYSTEMS

Item	Net Price £	Material £	Labour hours	Labour £	Unit	Total rate £
PLANT/EQUIPMENT: AIR FILTRATION – cont						
Grease filters, washable; minimum 65% Double sided extract unit; lightweight stainless steel construction; demountable composite filter media of woven metal mat and expanded metal mesh supports; for mounting on hood and extract systems (hood not included); includes placing in position						
500 × 686 × 565 mm, 4080 m³/hr	418.46	524.92	2.00	61.44	nr	586.36
1000 × 686 × 565 mm, 8160 m³/hr;	638.14	800.49	3.00	92.16	nr	892.65
1500 × 686 × 565 mm, 12240 m³/hr;	875.29	1097.96	3.50	107.53	nr	1205.49
Panel filters; 82% G3 (EU3); tested to BS EN779 Modular duct mounted filter panels; continuous rating up to 100°C; graduated density media; rigid cardboard frame; includes placing in position						
596 × 596 × 47 mm, 2360 m³/hr	5.07	6.36	1.00	30.72	nr	37.08
Side withdrawal frame	36.35	45.60	2.50	76.81	nr	122.41
596 × 287 × 47 mm, 1140 m³/hr	3.50	4.39	1.00	30.72	nr	35.11
Side withdrawal frame	36.35	45.60	2.50	76.81	nr	122.41
Panel filters; 90% G4 (EU4); tested to BS EN779 Modular duct mounted filter panels; continuous rating up to 100°C; pleated media with wire support; rigid cardboard frame; includes placing in position						
596 × 596 × 47 mm, 2560 m³/hr	7.67	9.62	1.00	30.72	nr	40.34
side withdrawal frame	45.88	57.56	3.00	92.16	nr	149.72
596 × 287 × 47 mm, 1230 m³/hr	5.29	6.63	1.00	30.72	nr	37.35
Side withdrawal frame	36.35	45.60	3.00	92.16	nr	137.76
Carbon filters; standard duty disposable carbon filters; steel frame with bonded carbon panels; for fixing to ductwork; including placing in position 12 panels						
597 × 597 × 298 mm, 1460 m³/hr	356.64	447.37	0.33	10.14	nr	457.51
597 × 597 × 451 mm, 2200 m³/hr	402.04	504.31	0.33	10.14	nr	514.45
597 × 597 × 597 mm, 2930 m³/hr	448.35	562.41	0.33	10.14	nr	572.55
8 panels						
451 × 451 × 298 mm, 740 m³/hr	268.86	337.25	0.29	8.90	nr	346.15
451 × 451 × 451 mm, 1105 m³/hr	297.58	373.28	0.29	8.90	nr	382.18
451 × 451 × 597 mm, 1460 m³/hr	324.97	407.65	0.29	8.90	nr	416.55

38 VENTILATION/AIR CONDITIONING SYSTEMS

Item	Net Price £	Material £	Labour hours	Labour £	Unit	Total rate £
6 panels						
298 × 298 × 298 mm, 365 m³/hr	190.52	238.99	0.25	7.67	nr	**246.66**
298 × 298 × 451 mm, 550 m³/hr	203.98	255.88	0.25	7.67	nr	**263.55**
298 × 298 × 597 mm, 780 m³/hr	216.86	272.03	0.25	7.67	nr	**279.70**

38 VENTILATION/AIR CONDITIONING SYSTEMS

Item	Net Price £	Material £	Labour hours	Labour £	Unit	Total rate £
PLANT/EQUIPMENT: AIR CURTAINS						
The selection of air curtains requires consideration of the particular conditions involved; climatic conditions, wind influence, construction and position all influence selection; consultation with a specialist manufacturer is therefore advisable						
Commercial grade air curtains; recessed or exposed units with rigid sheet steel casing; aluminium grilles; high quality motor/centrifugal fan assembly; includes fixing in position; electrical work elsewhere						
Ambient temperature; 240 V single phase supply; mounting height 2.40 m						
1000 × 590 × 270 mm	2695.62	3381.38	12.05	387.51	nr	**3768.89**
1500 × 590 × 270 mm	3450.21	4327.95	12.05	387.51	nr	**4715.46**
2000 × 590 × 270 mm	4173.96	5235.82	12.05	387.56	nr	**5623.38**
2500 × 590 × 270 mm	4639.94	5820.34	13.00	418.12	nr	**6238.46**
Ambient temperature; 240 V single phase supply; mounting height 2.80 m						
1000 × 590 × 270 mm	3114.22	3906.48	16.13	518.76	nr	**4425.24**
1500 × 590 × 270 mm	3975.67	4987.08	16.13	518.76	nr	**5505.84**
2000 × 590 × 270 mm	4956.10	6216.93	16.13	518.76	nr	**6735.69**
2500 × 590 × 270 mm	5730.52	7188.36	17.10	549.98	nr	**7738.34**
Ambient temperature 240 V single phase supply; mounting height 3.30 m						
1000 × 774 × 370 mm	4068.21	5103.17	17.24	554.53	nr	**5657.70**
1500 × 774 × 370 mm	5501.39	6900.95	17.24	554.53	nr	**7455.48**
2000 × 774 × 370 mm	6804.58	8535.67	17.24	554.53	nr	**9090.20**
2500 × 774 × 370 mm	8601.29	10789.45	18.30	588.58	nr	**11378.03**
Ambient temperature; 240 V single phase supply; mounting height 4.00 m						
1000 × 774 × 370 mm	4493.43	5636.56	19.10	614.31	nr	**6250.87**
1500 × 774 × 370 mm	5894.66	7394.26	19.10	614.31	nr	**8008.57**
2000 × 774 × 370 mm	7338.86	9205.86	19.10	614.31	nr	**9820.17**
2500 × 774 × 370 mm	8601.29	10789.45	19.90	640.05	nr	**11429.50**
Water heated; 240 V single phase supply; mounting height 2.40 m						
1000 × 590 × 270 mm; 2.30–9.40 kW output	3171.51	3978.34	12.05	387.51	nr	**4365.85**
1500 × 590 × 270 mm; 3.50–14.20 kW output	4059.40	5092.11	12.05	387.51	nr	**5479.62**
2000 × 590 × 270 mm; 4.70–19.00 kW output	4910.93	6160.27	12.05	387.51	nr	**6547.78**
2500 × 590 × 270 mm; 5.90–23.70 kW output	5459.53	6848.43	13.00	418.12	nr	**7266.55**

38 VENTILATION/AIR CONDITIONING SYSTEMS

Item	Net Price £	Material £	Labour hours	Labour £	Unit	Total rate £
Water heated; 240 V single phase supply; mounting height 2.80 m						
1000 × 590 × 270 mm; 3.30–11.90 kW output	3665.02	4597.40	16.13	518.76	nr	**5116.16**
1500 × 590 × 270 mm; 5.00–17.90 kW output	4678.50	5868.71	16.13	518.76	nr	**6387.47**
2000 × 590 × 270 mm; 6.70–23.90 kW output	5831.87	7315.49	16.13	518.76	nr	**7834.25**
2500 × 590 × 270 mm; 8.30–29.80 kW output	6742.89	8458.28	17.10	549.98	nr	**9008.26**
Water heated; 240 V single phase supply; mounting height 3.30 m						
1000 × 774 × 370 mm; 6.10–21.80 kW output	4786.45	6004.12	17.24	554.53	nr	**6558.65**
1500 × 774 × 370 mm; 9.20–32.80 kW output	6473.00	8119.73	17.24	554.53	nr	**8674.26**
2000 × 774 × 370 mm; 12.30–43.70 kW output	8006.43	10043.26	17.24	554.53	nr	**10597.79**
2500 × 774 × 370 mm; 15.30–54.60 kW output	9477.06	11888.03	18.30	588.58	nr	**12476.61**
Water heated; 240 V single phase supply; mounting height 4.00 m						
1000 × 774 × 370 mm; 7.20–24.20 kW output	5286.58	6631.49	19.10	614.31	nr	**7245.80**
1500 × 774 × 370 mm; 10.90–36.30 kW output	6935.67	8700.10	19.10	614.31	nr	**9314.41**
2000 × 774 × 370 mm; 14.50–48.40 kW output	8634.34	10830.92	19.10	614.31	nr	**11445.23**
2500 × 774 × 370 mm; 18.10–60.60 kW output	10119.30	12693.65	19.90	640.05	nr	**13333.70**
Electrically heated; 415 V three phase supply; mounting height 2.40 m						
1000 × 590 × 270 mm; 2.30–9.40 kW output	3857.80	4839.23	12.05	387.51	nr	**5226.74**
1500 × 590 × 270 mm; 3.50–14.20 kW output	4799.67	6020.71	12.05	387.51	nr	**6408.22**
2000 × 590 × 270 mm; 4.70–19.00 kW output	5720.61	7175.93	12.05	433.15	nr	**7609.08**
2500 × 590 × 270 mm; 5.90–23.70 kW output	6503.85	8158.43	13.00	467.36	nr	**8625.79**
Electrically heated; 415 V three phase supply; mounting height 2.80 m						
1000 × 590 × 270 mm; 3.30–11.90 kW output	4541.90	5697.36	16.13	518.76	nr	**6216.12**
1500 × 590 × 270 mm; 5.00–17.90 kW output	5585.11	7005.96	16.13	518.76	nr	**7524.72**
2000 × 590 × 270 mm; 6.70–23.90 kW output	7052.44	8846.58	16.13	518.76	nr	**9365.34**
2500 × 590 × 270 mm; 8.30–29.80 kW output	7881.95	9887.11	17.10	549.98	nr	**10437.09**

38 VENTILATION/AIR CONDITIONING SYSTEMS

Item	Net Price £	Material £	Labour hours	Labour £	Unit	Total rate £
PLANT/EQUIPMENT: AIR CURTAINS – cont						
Electrically heated; 415 V three phase supply; mounting height 3.30 m						
1000 × 774 × 370 mm; 6.10–21.80 kW output	6901.52	8657.26	17.24	554.48	nr	9211.74
1500 × 774 × 370 mm; 9.20–32.80 kW output	9246.83	11599.22	17.24	554.48	nr	12153.70
2000 × 774 × 370 mm; 12.30–43.70 kW output	11160.31	13999.50	17.24	554.48	nr	14553.98
2500 × 774 × 370 mm; 15.30–54.60 kW output	13211.49	16572.49	18.30	588.58	nr	17161.07
Electrically heated; 415 V three phase supply; mounting height 4.00 m						
1000 × 774 × 370 mm; 7.20–24.20 kW output	7616.46	9554.09	19.10	614.31	nr	10168.40
1500 × 774 × 370 mm; 10.90–36.30 kW output	9881.35	12395.16	19.10	614.31	nr	13009.47
2000 × 774 × 370 mm; 14.50–48.40 kW output	11951.26	14991.66	19.10	614.31	nr	15605.97
2500 × 774 × 370 mm; 18.10–60.60 kW output	14156.66	17758.12	19.90	640.05	nr	18398.17
Industrial grade air curtains; recessed or exposed units with rigid sheet steel casing; aluminium grilles; high quality motor/ centrifugal fan assembly; includes fixing in position; electrical work elsewhere						
Ambient temperature; 230 V single phase supply; including wiring between multiple units; horizontally or vertically mounted; opening maximum 6.00 m						
1500 × 585 × 853 mm; 3.0 A supply	2521.36	3162.79	17.24	554.53	nr	3717.32
2000 × 585 × 853 mm; 4.0 A supply	3055.47	3832.79	17.24	554.53	nr	4387.32
Water heated; 230 V single phase supply; including wiring between multiple units; horizontally or vertically mounted; opening maximum 6.00 m						
1500 × 585 × 956 mm; 3.0 A supply; 34.80 kW output	2890.05	3625.28	17.24	554.53	nr	4179.81
2000 × 585 × 956 mm; 4.0 A supply; 50.70 kW output	3531.20	4429.53	17.24	554.53	nr	4984.06
Water heated; 230 V single phase supply; including wiring between multiple units; vertically mounted in single bank for openings maximum 6.00 m wide or opposing twin banks for openings maximum 10.00 m wide						
1500 × 585 mm; 3.0 A supply; 41.1 kW output	2890.05	3625.28	17.24	554.53	nr	4179.81
2000 × 585 mm; 4.0 A supply; 57.7 kW output	3531.20	4429.53	17.24	554.53	nr	4984.06
Remote mounted electronic controller unit; excluding wiring to units						
0–10 V control	44.33	55.61	5.00	160.81	nr	216.42

38 VENTILATION/AIR CONDITIONING SYSTEMS

Item	Net Price £	Material £	Labour hours	Labour £	Unit	Total rate £
SILENCERS/ACOUSTIC TREATMENT						
Attenuators; DW144 galvanized construction c/w splitters; self-securing; fitted to ductwork						
To suit rectangular ducts; unit length 600 mm						
100 × 100 mm	105.19	131.95	0.75	23.04	nr	**154.99**
150 × 150 mm	109.80	137.74	0.75	23.04	nr	**160.78**
200 × 200 mm	114.40	143.51	0.75	23.04	nr	**166.55**
300 × 300 mm	127.99	160.55	0.75	23.04	nr	**183.59**
400 × 400 mm	143.68	180.23	1.00	30.72	nr	**210.95**
500 × 500 mm	167.69	210.35	1.25	38.39	nr	**248.74**
600 × 300 mm	161.53	202.62	1.25	38.39	nr	**241.01**
600 × 600 mm	226.07	283.58	1.25	38.39	nr	**321.97**
700 × 300 mm	171.27	214.84	1.50	46.08	nr	**260.92**
700 × 700 mm	255.02	319.89	1.50	46.08	nr	**365.97**
800 × 300 mm	177.41	222.54	2.00	61.44	nr	**283.98**
800 × 800 mm	294.22	369.07	2.00	61.44	nr	**430.51**
1000 × 1000 mm	376.62	472.43	3.00	92.16	nr	**564.59**
To suit rectangular ducts; unit length 1200 mm						
200 × 200 mm	126.16	158.26	1.00	30.72	nr	**188.98**
300 × 300 mm	155.97	195.65	1.00	30.72	nr	**226.37**
400 × 400 mm	187.53	235.23	1.33	40.86	nr	**276.09**
500 × 500 mm	226.35	283.93	1.66	50.99	nr	**334.92**
600 × 300 mm	215.82	270.73	1.66	50.99	nr	**321.72**
600 × 600 mm	315.77	396.10	1.66	50.99	nr	**447.09**
700 × 300 mm	231.35	290.20	2.00	61.44	nr	**351.64**
700 × 700 mm	361.35	453.28	2.00	61.44	nr	**514.72**
800 × 300 mm	241.87	303.40	2.66	81.72	nr	**385.12**
800 × 800 mm	416.95	523.02	2.66	81.72	nr	**604.74**
1000 × 1000 mm	564.43	708.02	4.00	122.89	nr	**830.91**
1300 × 1300 mm	968.87	1215.35	8.00	245.76	nr	**1461.11**
1500 × 1500 mm	1108.98	1391.11	8.00	245.76	nr	**1636.87**
1800 × 1800 mm	1517.35	1903.36	10.66	327.48	nr	**2230.84**
2000 × 2000 mm	1702.63	2135.78	13.33	409.49	nr	**2545.27**
To suit rectangular ducts; unit length 1800 mm						
200 × 200 mm	159.22	199.73	1.00	30.72	nr	**230.45**
300 × 300 mm	211.57	265.40	1.00	30.72	nr	**296.12**
400 × 400 mm	252.90	317.24	1.33	40.86	nr	**358.10**
500 × 500 mm	294.73	369.71	1.66	50.99	nr	**420.70**
600 × 300 mm	308.25	386.67	1.66	50.99	nr	**437.66**
600 × 600 mm	430.74	540.32	1.66	50.99	nr	**591.31**
700 × 300 mm	326.79	409.92	2.00	61.44	nr	**471.36**
700 × 700 mm	514.63	645.56	2.00	61.44	nr	**707.00**
800 × 300 mm	345.56	433.47	2.66	81.72	nr	**515.19**
800 × 800 mm	598.79	751.12	2.66	81.72	nr	**832.84**
1000 × 1000 mm	804.88	1009.65	4.00	122.89	nr	**1132.54**
1300 × 1300 mm	1344.58	1686.64	8.00	245.76	nr	**1932.40**
1500 × 1500 mm	1523.77	1911.41	8.00	245.76	nr	**2157.17**
1800 × 1800 mm	2115.47	2653.65	13.33	409.49	nr	**3063.14**
2000 × 2000 mm	2395.47	3004.88	10.66	327.48	nr	**3332.36**

38 VENTILATION/AIR CONDITIONING SYSTEMS

Item	Net Price £	Material £	Labour hours	Labour £	Unit	Total rate £
SILENCERS/ACOUSTIC TREATMENT – cont						
To suit rectangular ducts – cont						
2300 × 2300 mm	2994.29	3756.03	16.00	491.52	nr	**4247.55**
2500 × 2500 mm	4176.67	5239.21	18.66	573.24	nr	**5812.45**
To suit rectangular ducts; unit length 2400 mm						
500 × 500 mm	381.13	478.09	2.08	63.99	nr	**542.08**
600 × 300 mm	363.87	456.43	2.08	63.99	nr	**520.42**
600 × 600 mm	541.94	679.81	2.08	63.99	nr	**743.80**
700 × 300 mm	394.42	494.76	2.50	76.81	nr	**571.57**
700 × 700 mm	640.14	803.00	2.50	76.81	nr	**879.81**
800 × 300 mm	416.95	523.02	3.33	102.30	nr	**625.32**
800 × 800 mm	740.81	929.28	3.33	102.30	nr	**1031.58**
1000 × 1000 mm	999.49	1253.76	5.00	153.60	nr	**1407.36**
1300 × 1300 mm	1618.86	2030.69	10.00	307.20	nr	**2337.89**
1500 × 1500 mm	1847.63	2317.67	10.00	307.20	nr	**2624.87**
1800 × 1800 mm	2636.22	3306.88	13.33	409.44	nr	**3716.32**
2000 × 2000 mm	2958.28	3710.86	16.66	511.80	nr	**4222.66**
2300 × 2300 mm	3660.05	4591.17	20.00	614.41	nr	**5205.58**
2500 × 2500 mm	5139.25	6446.68	23.32	716.51	nr	**7163.19**
To suit circular ducts; unit length 600 mm						
100 mm dia.	91.35	114.59	0.75	23.04	nr	**137.63**
200 mm dia.	107.48	134.83	0.75	23.04	nr	**157.87**
250 mm dia.	121.84	152.84	0.75	23.04	nr	**175.88**
315 mm dia.	141.82	177.90	1.00	30.72	nr	**208.62**
355 mm dia.	164.11	205.86	1.00	30.72	nr	**236.58**
400 mm dia.	176.40	221.28	1.00	30.72	nr	**252.00**
450 mm dia.	213.02	267.21	1.00	30.72	nr	**297.93**
500 mm dia.	234.55	294.22	1.26	38.65	nr	**332.87**
630 mm dia.	276.53	346.88	1.50	46.08	nr	**392.96**
710 mm dia.	301.13	377.74	1.50	46.08	nr	**423.82**
800 mm dia.	334.94	420.15	2.00	61.44	nr	**481.59**
1000 mm dia.	449.61	563.99	3.00	92.16	nr	**656.15**
To suit circular ducts; unit length 1200 mm						
100 mm dia.	134.39	168.58	1.00	30.72	nr	**199.30**
200 mm dia.	147.44	184.95	1.00	30.72	nr	**215.67**
250 mm dia.	165.64	207.78	1.00	30.72	nr	**238.50**
315 mm dia.	189.46	237.66	1.33	40.86	nr	**278.52**
355 mm dia.	227.12	284.89	1.33	40.86	nr	**325.75**
400 mm dia.	240.95	302.24	1.33	40.86	nr	**343.10**
450 mm dia.	279.10	350.10	1.33	40.86	nr	**390.96**
500 mm dia.	286.04	358.80	1.66	50.99	nr	**409.79**
630 mm dia.	378.75	475.10	2.00	61.44	nr	**536.54**
710 mm dia.	418.70	525.21	2.00	61.44	nr	**586.65**
800 mm dia.	417.93	524.25	2.66	81.72	nr	**605.97**
1000 mm dia.	561.05	703.79	4.00	122.89	nr	**826.68**
1250 mm dia.	1056.88	1325.76	8.00	245.76	nr	**1571.52**
1400 mm dia.	1149.08	1441.41	8.00	245.76	nr	**1687.17**
1600 mm dia.	1273.91	1597.99	10.66	327.48	nr	**1925.47**

38 VENTILATION/AIR CONDITIONING SYSTEMS

Item	Net Price £	Material £	Labour hours	Labour £	Unit	Total rate £
To suit circular ducts; unit length 1800 mm						
200 mm dia.	206.62	259.18	1.25	38.39	nr	297.57
250 mm dia.	231.21	290.04	1.25	38.39	nr	328.43
315 mm dia.	274.77	344.67	1.66	50.99	nr	395.66
355 mm dia.	323.94	406.35	1.66	50.99	nr	457.34
400 mm dia.	335.20	420.47	1.66	50.99	nr	471.46
450 mm dia.	342.39	429.50	1.66	50.99	nr	480.49
500 mm dia.	376.40	472.16	2.08	63.90	nr	536.06
630 mm dia.	416.39	522.32	2.50	76.81	nr	599.13
710 mm dia.	487.09	611.00	2.50	76.81	nr	687.81
800 mm dia.	503.99	632.21	3.33	102.30	nr	734.51
1000 mm dia.	1315.84	1650.59	6.66	204.59	nr	1855.18
1250 mm dia.	1337.37	1677.59	6.66	204.59	nr	1882.18
1400 mm dia.	1477.55	1853.44	5.00	153.60	nr	2007.04
1600 mm dia.	1478.44	1854.55	9.90	304.12	nr	2158.67

38 VENTILATION/AIR CONDITIONING SYSTEMS

Item	Net Price £	Material £	Labour hours	Labour £	Unit	Total rate £
THERMAL INSULATION						
Concealed ductwork						
Flexible wrap; 20 kg–45 kg Bright Class O aluminium foil faced; Bright Class O foil taped joints; 62 mm metal pins and washers; aluminium bands						
40 mm thick insulation	15.32	19.22	0.40	12.87	m²	32.09
Semi-rigid slab; 45 kg Bright Class O aluminium foil faced mineral fibre; Bright Class O foil taped joints; 62 mm metal pins and washers; aluminium bands						
40 mm thick insulation	20.04	25.13	0.65	20.91	m²	46.04
Plantroom ductwork						
Semi-rigid slab; 45 kg Bright Class O aluminium foil faced mineral fibre; Bright Class O foil taped joints; 62 mm metal pins and washers; 22 swg plain/embossed aluminium cladding; pop rivited						
50 mm thick insulation	50.56	63.43	1.50	48.25	m²	111.68
External ductwork						
Semi-rigid slab; 45 kg Bright Class O aluminium foil faced mineral fibre; Bright Class O foil taped joints; 62 mm metal pins and washers; 0.8 mm polyisobutylene sheeting; welded joints						
50 mm thick insulation	36.88	46.27	1.25	40.21	m²	86.48

38 VENTILATION/AIR CONDITIONING SYSTEMS

Item	Net Price £	Material £	Labour hours	Labour £	Unit	Total rate £
DUCTWORK: FIRE RATED						
DUCTLINES						
The relevant BS requires that the fire-rating of ductwork meets 3 criteria; stability (hours), integrity (hours) and insulation (hours). The least of the 3 periods defines the fire-rating. The BS does however allow stability and integrity to be considered in isolation. Rates are therefore provided for both types of system.						
Care should to be taken when using the rates within this section to ensure thta the requiremements for stability, integrity and insulation are known and the appropriate rates are used						
High density single layer mineral wool fire-rated ductwork slab, in accordance with BSEN1366, ducts 'Type A' and 'Type B'; 165 kg class O foil faced mineral fibre; 100 mm wide bright class O foil taped joints; welded pins; includes protection to all supports.						
½ hour stability, integrity and insulation						
25 mm thick, vertical and horizontal						
ductwork	89.06	111.72	1.25	40.21	m²	**151.93**
1 hour stability, integrity and insulation						
30 mm thick, vertical ductwork	104.55	131.15	1.50	48.25	m²	**179.40**
40 mm thick, horizontal ductwork	122.96	154.25	1.50	48.25	m²	**202.50**
1½ hour stability, integrity and insulation						
50 mm thick, vertical ductwork	146.89	184.26	1.75	56.28	m²	**240.54**
70 mm thick, horizontal ductwork	183.74	230.48	1.75	56.28	m²	**286.76**
2 hour stability, integrity and insulation						
70 mm, vertical ductwork	189.31	237.47	2.00	64.32	m²	**301.79**
90 mm horizontal ductwork	226.15	283.68	2.00	64.32	m²	**348.00**
Kitchen extract, 1 hour stability, integrity and insulation						
90 mm, vertical and horizontal	226.15	283.68	2.00	64.32	m²	**348.00**
Galvanized sheet metal rectangular section ductwork to BSEN1366, ducts 'Type A' and 'Type B'; provides 2 hours stability and 2 hours integrity at 1100°C (no rating for insulation); including all necessary stiffeners, joints and supports in the running length						
Ductwork up to 600 mm longest side						
Sum of two sides 200 mm	113.53	142.41	2.91	89.40	m	**231.81**
Sum of two sides 300 mm	123.25	154.60	2.99	91.86	m	**246.46**
Sum of two sides 400 mm	132.95	166.77	3.17	97.38	m	**264.15**

38 VENTILATION/AIR CONDITIONING SYSTEMS

Item	Net Price £	Material £	Labour hours	Labour £	Unit	Total rate £
DUCTWORK: FIRE RATED – cont						
Ductwork up to 600 mm – cont						
Sum of two sides 500 mm	142.68	178.98	3.37	103.52	m	**282.50**
Sum of two sides 600 mm	152.38	191.15	3.54	108.74	m	**299.89**
Sum of two sides 700 mm	162.12	203.36	3.72	114.27	m	**317.63**
Sum of two sides 800 mm	171.84	215.56	3.90	119.81	m	**335.37**
Sum of two sides 900 mm	181.54	227.72	5.04	154.83	m	**382.55**
Sum of two sides 1000 mm	191.27	239.93	5.58	171.42	m	**411.35**
Sum of two sides 1100 mm	200.97	252.10	5.84	179.40	m	**431.50**
Sum of two sides 1200 mm	210.67	264.26	6.11	187.70	m	**451.96**
Extra over fittings; Ductwork up to 600 mm longest side						
End cap						
Sum of two sides 200 mm	32.24	40.44	0.81	24.89	nr	**65.33**
Sum of two sides 300 mm	34.20	42.90	0.84	25.80	nr	**68.70**
Sum of two sides 400 mm	36.15	45.35	0.87	26.73	nr	**72.08**
Sum of two sides 500 mm	38.11	47.80	0.90	27.64	nr	**75.44**
Sum of two sides 600 mm	40.06	50.25	0.93	28.57	nr	**78.82**
Sum of two sides 700 mm	42.04	52.73	0.96	29.49	nr	**82.22**
Sum of two sides 800 mm	44.04	55.24	0.98	30.11	nr	**85.35**
Sum of two sides 900 mm	46.00	57.70	1.17	35.94	nr	**93.64**
Sum of two sides 1000 mm	47.92	60.11	1.22	37.49	nr	**97.60**
Sum of two sides 1100 mm	49.91	62.61	1.25	38.39	nr	**101.00**
Sum of two sides 1200 mm	51.86	65.05	1.28	39.32	nr	**104.37**
Reducer						
Sum of two sides 200 mm	127.85	160.37	2.23	68.50	nr	**228.87**
Sum of two sides 300 mm	133.73	167.75	2.37	72.80	nr	**240.55**
Sum of two sides 400 mm	139.58	175.09	2.51	77.11	nr	**252.20**
Sum of two sides 500 mm	145.46	182.47	2.65	81.41	nr	**263.88**
Sum of two sides 600 mm	151.33	189.83	2.79	85.71	nr	**275.54**
Sum of two sides 700 mm	157.20	197.19	2.87	88.17	nr	**285.36**
Sum of two sides 800 mm	163.09	204.58	3.01	92.47	nr	**297.05**
Sum of two sides 900 mm	168.96	211.95	3.06	94.00	nr	**305.95**
Sum of two sides 1000 mm	174.82	219.30	3.17	97.38	nr	**316.68**
Sum of two sides 1100 mm	180.66	226.62	3.22	98.92	nr	**325.54**
Sum of two sides 1200 mm	186.58	234.05	3.28	100.77	nr	**334.82**
Offset						
Sum of two sides 200 mm	308.56	387.06	2.95	90.63	nr	**477.69**
Sum of two sides 300 mm	316.20	396.64	3.11	95.54	nr	**492.18**
Sum of two sides 400 mm	323.86	406.25	3.26	100.15	nr	**506.40**
Sum of two sides 500 mm	331.53	415.87	3.42	105.07	nr	**520.94**
Sum of two sides 600 mm	339.18	425.47	3.57	109.67	nr	**535.14**
Sum of two sides 700 mm	354.50	444.68	3.23	99.22	nr	**543.90**
Sum of two sides 800 mm	346.85	435.09	3.73	114.59	nr	**549.68**
Sum of two sides 900 mm	362.18	454.32	3.45	105.99	nr	**560.31**
Sum of two sides 1000 mm	369.82	463.90	3.67	112.75	nr	**576.65**
Sum of two sides 1100 mm	377.45	473.47	3.78	116.12	nr	**589.59**
Sum of two sides 1200 mm	385.15	483.13	3.89	119.50	nr	**602.63**

38 VENTILATION/AIR CONDITIONING SYSTEMS

Item	Net Price £	Material £	Labour hours	Labour £	Unit	Total rate £
90° radius bend						
Sum of two sides 200 mm	125.11	156.93	2.06	63.28	nr	**220.21**
Sum of two sides 300 mm	134.59	168.83	2.21	67.88	nr	**236.71**
Sum of two sides 400 mm	144.05	180.70	2.36	72.51	nr	**253.21**
Sum of two sides 500 mm	153.56	192.63	2.51	77.11	nr	**269.74**
Sum of two sides 600 mm	163.04	204.51	2.66	81.72	nr	**286.23**
Sum of two sides 700 mm	172.47	216.35	2.81	86.33	nr	**302.68**
Sum of two sides 800 mm	181.98	228.28	2.97	91.25	nr	**319.53**
Sum of two sides 900 mm	191.41	240.11	2.99	91.86	nr	**331.97**
Sum of two sides 1000 mm	200.95	252.07	3.04	93.39	nr	**345.46**
Sum of two sides 1100 mm	210.38	263.91	3.07	94.30	nr	**358.21**
Sum of two sides 1200 mm	219.84	275.77	3.09	94.92	nr	**370.69**
45° radius bend						
Sum of two sides 200 mm	154.26	193.50	1.52	46.69	nr	**240.19**
Sum of two sides 300 mm	158.10	198.32	1.59	48.84	nr	**247.16**
Sum of two sides 400 mm	161.96	203.17	1.66	50.99	nr	**254.16**
Sum of two sides 500 mm	165.78	207.95	1.73	53.14	nr	**261.09**
Sum of two sides 600 mm	169.59	212.73	1.80	55.29	nr	**268.02**
Sum of two sides 700 mm	173.45	217.57	1.87	57.46	nr	**275.03**
Sum of two sides 800 mm	177.24	222.33	1.93	59.29	nr	**281.62**
Sum of two sides 900 mm	181.10	227.17	1.99	61.14	nr	**288.31**
Sum of two sides 1000 mm	184.91	231.95	2.05	62.97	nr	**294.92**
Sum of two sides 1100 mm	188.75	236.77	2.11	64.81	nr	**301.58**
Sum of two sides 1200 mm	192.57	241.56	2.17	66.66	nr	**308.22**
90° mitre bend						
Sum of two sides 200 mm	129.70	162.69	2.47	75.88	nr	**238.57**
Sum of two sides 300 mm	150.33	188.57	2.65	81.41	nr	**269.98**
Sum of two sides 400 mm	170.93	214.41	2.84	87.25	nr	**301.66**
Sum of two sides 500 mm	191.59	240.33	3.02	92.78	nr	**333.11**
Sum of two sides 600 mm	212.19	266.17	3.20	98.30	nr	**364.47**
Sum of two sides 700 mm	232.84	292.07	3.38	103.84	nr	**395.91**
Sum of two sides 800 mm	253.47	317.96	3.56	109.36	nr	**427.32**
Sum of two sides 900 mm	274.09	343.82	3.59	110.29	nr	**454.11**
Sum of two sides 1000 mm	294.75	369.73	3.66	112.44	nr	**482.17**
Sum of two sides 1100 mm	315.33	395.55	3.69	113.36	nr	**508.91**
Sum of two sides 1200 mm	336.02	421.50	3.72	114.27	nr	**535.77**
Branch (Side-on Shoe)						
Sum of two sides 200 mm	48.15	60.40	0.98	30.11	nr	**90.51**
Sum of two sides 300 mm	48.41	60.73	1.06	32.57	nr	**93.30**
Sum of two sides 400 mm	48.68	61.06	1.14	35.02	nr	**96.08**
Sum of two sides 500 mm	48.95	61.40	1.22	37.49	nr	**98.89**
Sum of two sides 600 mm	49.20	61.71	1.30	39.94	nr	**101.65**
Sum of two sides 700 mm	49.46	62.05	1.38	42.40	nr	**104.45**
Sum of two sides 800 mm	50.10	62.84	1.37	42.09	nr	**104.93**
Sum of two sides 900 mm	49.71	62.36	1.46	44.86	nr	**107.22**
Sum of two sides 1000 mm	50.22	63.00	1.46	44.86	nr	**107.86**
Sum of two sides 1100 mm	50.49	63.34	1.50	46.08	nr	**109.42**
Sum of two sides 1200 mm	50.76	63.67	1.55	47.62	nr	**111.29**

38 VENTILATION/AIR CONDITIONING SYSTEMS

Item	Net Price £	Material £	Labour hours	Labour £	Unit	Total rate £
DUCTWORK: FIRE RATED – cont						
Extra over fittings – cont						
Ductwork 601–800 mm longest side						
Sum of two sides 900 mm	181.51	227.68	5.04	154.83	m	**382.51**
Sum of two sides 1000 mm	191.25	239.90	5.58	171.42	m	**411.32**
Sum of two sides 1100 mm	200.97	252.10	5.84	179.40	m	**431.50**
Sum of two sides 1200 mm	210.67	264.26	6.11	187.70	m	**451.96**
Sum of two sides 1300 mm	220.45	276.53	6.38	196.00	m	**472.53**
Sum of two sides 1400 mm	230.15	288.70	6.65	204.29	m	**492.99**
Sum of two sides 1500 mm	239.91	300.94	6.91	212.28	m	**513.22**
Sum of two sides 1600 mm	249.63	313.14	7.18	220.57	m	**533.71**
Extra over fittings; Ductwork 601–800 mm longest side						
End cap						
Sum of two sides 900 mm	42.72	53.59	1.17	35.94	nr	**89.53**
Sum of two sides 1000 mm	45.78	57.42	1.22	37.49	nr	**94.91**
Sum of two sides 1100 mm	48.86	61.29	1.25	38.39	nr	**99.68**
Sum of two sides 1200 mm	51.89	65.09	1.28	39.32	nr	**104.41**
Sum of two sides 1300 mm	54.94	68.91	1.31	40.24	nr	**109.15**
Sum of two sides 1400 mm	58.00	72.76	1.34	41.17	nr	**113.93**
Sum of two sides 1500 mm	61.05	76.59	1.36	41.79	nr	**118.38**
Sum of two sides 1600 mm	64.11	80.42	1.39	42.69	nr	**123.11**
Reducer						
Sum of two sides 900 mm	159.91	200.59	3.06	94.00	nr	**294.59**
Sum of two sides 1000 mm	168.80	211.75	3.17	97.38	nr	**309.13**
Sum of two sides 1100 mm	177.67	222.87	3.22	98.92	nr	**321.79**
Sum of two sides 1200 mm	186.60	234.07	3.28	100.77	nr	**334.84**
Sum of two sides 1300 mm	195.50	245.24	3.33	102.30	nr	**347.54**
Sum of two sides 1400 mm	204.39	256.39	3.38	103.84	nr	**360.23**
Sum of two sides 1500 mm	213.28	267.53	3.44	105.68	nr	**373.21**
Sum of two sides 1600 mm	222.19	278.71	3.49	107.22	nr	**385.93**
Offset						
Sum of two sides 900 mm	340.08	426.60	3.45	105.99	nr	**532.59**
Sum of two sides 1000 mm	356.26	446.89	3.67	112.75	nr	**559.64**
Sum of two sides 1100 mm	372.47	467.23	3.78	116.12	nr	**583.35**
Sum of two sides 1200 mm	388.65	487.52	3.89	119.50	nr	**607.02**
Sum of two sides 1300 mm	404.82	507.81	4.00	122.89	nr	**630.70**
Sum of two sides 1400 mm	421.02	528.12	4.11	126.26	nr	**654.38**
Sum of two sides 1500 mm	437.21	548.44	4.23	129.94	nr	**678.38**
Sum of two sides 1600 mm	453.90	569.37	4.34	133.32	nr	**702.69**
90° radius bend						
Sum of two sides 900 mm	170.03	213.28	2.99	91.86	nr	**305.14**
Sum of two sides 1000 mm	186.74	234.25	3.04	93.39	nr	**327.64**
Sum of two sides 1100 mm	203.39	255.14	3.07	94.30	nr	**349.44**
Sum of two sides 1200 mm	220.11	276.10	3.09	94.92	nr	**371.02**
Sum of two sides 1300 mm	236.78	297.01	3.12	95.85	nr	**392.86**
Sum of two sides 1400 mm	253.47	317.96	3.15	96.77	nr	**414.73**
Sum of two sides 1500 mm	270.16	338.89	3.17	97.38	nr	**436.27**
Sum of two sides 1600 mm	286.87	359.84	3.20	98.30	nr	**458.14**

38 VENTILATION/AIR CONDITIONING SYSTEMS

Item	Net Price £	Material £	Labour hours	Labour £	Unit	Total rate £
45° bend						
Sum of two sides 900 mm	168.83	211.78	1.74	53.45	nr	265.23
Sum of two sides 1000 mm	176.75	221.72	1.85	56.84	nr	278.56
Sum of two sides 1100 mm	184.69	231.67	1.91	58.68	nr	290.35
Sum of two sides 1200 mm	192.59	241.58	1.96	60.21	nr	301.79
Sum of two sides 1300 mm	200.50	251.51	2.02	62.06	nr	313.57
Sum of two sides 1400 mm	208.43	261.45	2.07	63.58	nr	325.03
Sum of two sides 1500 mm	216.34	271.38	2.13	65.43	nr	336.81
Sum of two sides 1600 mm	224.27	281.32	2.18	66.98	nr	348.30
90° mitre bend						
Sum of two sides 900 mm	254.57	319.33	3.59	110.29	nr	429.62
Sum of two sides 1000 mm	281.72	353.39	3.66	112.44	nr	465.83
Sum of two sides 1100 mm	308.85	387.42	3.69	113.36	nr	500.78
Sum of two sides 1200 mm	336.02	421.50	3.72	114.27	nr	535.77
Sum of two sides 1300 mm	363.11	455.48	3.75	115.20	nr	570.68
Sum of two sides 1400 mm	390.26	489.54	3.78	116.12	nr	605.66
Sum of two sides 1500 mm	417.38	523.57	3.81	117.05	nr	640.62
Sum of two sides 1600 mm	444.54	557.63	3.84	117.96	nr	675.59
Branch (Side-on Shoe)						
Sum of two sides 900 mm	47.25	59.27	1.37	42.09	nr	101.36
Sum of two sides 1000 mm	48.41	60.73	1.46	44.86	nr	105.59
Sum of two sides 1100 mm	49.61	62.23	1.50	46.08	nr	108.31
Sum of two sides 1200 mm	50.78	63.69	1.55	47.62	nr	111.31
Sum of two sides 1300 mm	51.96	65.18	1.59	48.84	nr	114.02
Sum of two sides 1400 mm	51.05	64.04	1.68	51.61	nr	115.65
Sum of two sides 1500 mm	53.19	66.72	1.63	50.08	nr	116.80
Sum of two sides 1600 mm	55.53	69.65	1.72	52.83	nr	122.48
Ductwork 801–1000 mm longest side						
Sum of two sides 1100 mm	200.97	252.10	5.84	179.40	m	431.50
Sum of two sides 1200 mm	210.67	264.26	6.11	187.70	m	451.96
Sum of two sides 1300 mm	220.40	276.47	6.38	196.00	m	472.47
Sum of two sides 1400 mm	230.13	288.68	6.65	204.29	m	492.97
Sum of two sides 1500 mm	239.86	300.88	6.91	212.28	m	513.16
Sum of two sides 1600 mm	249.58	313.07	7.18	220.57	m	533.64
Sum of two sides 1700 mm	259.33	325.30	7.45	228.86	m	554.16
Sum of two sides 1800 mm	269.06	337.51	7.71	236.86	m	574.37
Sum of two sides 1900 mm	278.78	349.70	7.98	245.15	m	594.85
Sum of two sides 2000 mm	288.52	361.92	8.25	253.43	m	615.35
Extra over fittings; Ductwork 801–1000 mm longest side						
End cap						
Sum of two sides 1100 mm	48.86	61.29	1.25	38.39	nr	99.68
Sum of two sides 1200 mm	51.89	65.09	1.28	39.32	nr	104.41
Sum of two sides 1300 mm	54.94	68.91	1.31	40.24	nr	109.15
Sum of two sides 1400 mm	58.00	72.76	1.34	41.17	nr	113.93
Sum of two sides 1500 mm	61.05	76.59	1.36	41.79	nr	118.38
Sum of two sides 1600 mm	64.08	80.38	1.39	42.69	nr	123.07
Sum of two sides 1700 mm	67.13	84.21	1.42	43.62	nr	127.83

38 VENTILATION/AIR CONDITIONING SYSTEMS

Item	Net Price £	Material £	Labour hours	Labour £	Unit	Total rate £
DUCTWORK: FIRE RATED – cont						
Extra over fittings – cont						
End cap – cont						
Sum of two sides 1800 mm	70.19	88.04	1.45	44.54	nr	132.58
Sum of two sides 1900 mm	73.23	91.86	1.48	45.47	nr	137.33
Sum of two sides 2000 mm	76.27	95.67	1.51	46.39	nr	142.06
Reducer						
Sum of two sides 1100 mm	177.67	222.87	3.22	98.92	nr	321.79
Sum of two sides 1200 mm	186.63	234.11	3.28	100.77	nr	334.88
Sum of two sides 1300 mm	195.57	245.32	3.33	102.30	nr	347.62
Sum of two sides 1400 mm	204.54	256.57	3.38	103.84	nr	360.41
Sum of two sides 1500 mm	213.44	267.74	3.44	105.68	nr	373.42
Sum of two sides 1600 mm	222.42	279.00	3.49	107.22	nr	386.22
Sum of two sides 1700 mm	231.35	290.20	3.54	108.74	nr	398.94
Sum of two sides 1800 mm	240.29	301.41	3.60	110.59	nr	412.00
Sum of two sides 1900 mm	249.27	312.68	3.65	112.13	nr	424.81
Sum of two sides 2000 mm	262.65	329.47	3.70	113.66	nr	443.13
Offset						
Sum of two sides 1100 mm	369.53	463.53	3.78	116.12	nr	579.65
Sum of two sides 1200 mm	385.15	483.13	3.89	119.50	nr	602.63
Sum of two sides 1300 mm	400.76	502.71	4.00	122.89	nr	625.60
Sum of two sides 1400 mm	416.38	522.31	4.11	126.26	nr	648.57
Sum of two sides 1500 mm	432.01	541.91	4.23	129.94	nr	671.85
Sum of two sides 1600 mm	447.65	561.53	4.34	133.32	nr	694.85
Sum of two sides 1700 mm	463.26	581.11	4.45	136.71	nr	717.82
Sum of two sides 1800 mm	478.88	600.71	4.56	140.08	nr	740.79
Sum of two sides 1900 mm	494.47	620.27	4.67	143.46	nr	763.73
Sum of two sides 2000 mm	517.91	649.67	5.53	169.88	nr	819.55
90° radius bend						
Sum of two sides 1100 mm	203.09	254.76	3.07	94.30	nr	349.06
Sum of two sides 1200 mm	219.81	275.73	3.09	94.92	nr	370.65
Sum of two sides 1300 mm	236.56	296.74	3.12	95.85	nr	392.59
Sum of two sides 1400 mm	253.29	317.72	3.15	96.77	nr	414.49
Sum of two sides 1500 mm	270.02	338.71	3.17	97.38	nr	436.09
Sum of two sides 1600 mm	286.73	359.68	3.20	98.30	nr	457.98
Sum of two sides 1700 mm	303.45	380.64	3.22	98.92	nr	479.56
Sum of two sides 1800 mm	320.18	401.63	3.25	99.84	nr	501.47
Sum of two sides 1900 mm	336.91	422.62	3.28	100.77	nr	523.39
Sum of two sides 2000 mm	353.13	442.97	3.30	101.38	nr	544.35
45° bend						
Sum of two sides 1100 mm	185.89	233.18	1.91	58.68	nr	291.86
Sum of two sides 1200 mm	193.66	242.93	1.96	60.21	nr	303.14
Sum of two sides 1300 mm	201.44	252.68	2.02	62.06	nr	314.74
Sum of two sides 1400 mm	209.23	262.46	2.07	63.58	nr	326.04
Sum of two sides 1500 mm	217.01	272.22	2.13	65.43	nr	337.65
Sum of two sides 1600 mm	224.78	281.96	2.18	66.98	nr	348.94
Sum of two sides 1700 mm	232.54	291.69	2.24	68.81	nr	360.50
Sum of two sides 1800 mm	240.32	301.46	2.30	70.66	nr	372.12
Sum of two sides 1900 mm	248.07	311.18	2.35	72.20	nr	383.38
Sum of two sides 2000 mm	255.89	320.99	2.76	84.78	nr	405.77

38 VENTILATION/AIR CONDITIONING SYSTEMS

Item	Net Price £	Material £	Labour hours	Labour £	Unit	Total rate £
90° mitre bend						
Sum of two sides 1100 mm	308.85	387.42	3.69	113.36	nr	**500.78**
Sum of two sides 1200 mm	336.02	421.50	3.72	114.27	nr	**535.77**
Sum of two sides 1300 mm	363.11	455.48	3.75	115.20	nr	**570.68**
Sum of two sides 1400 mm	390.26	489.54	3.78	116.12	nr	**605.66**
Sum of two sides 1500 mm	417.44	523.63	3.81	117.05	nr	**640.68**
Sum of two sides 1600 mm	444.54	557.63	3.84	117.96	nr	**675.59**
Sum of two sides 1700 mm	471.69	591.68	3.87	118.89	nr	**710.57**
Sum of two sides 1800 mm	498.85	625.76	3.90	119.81	nr	**745.57**
Sum of two sides 1900 mm	525.97	659.78	3.93	120.74	nr	**780.52**
Sum of two sides 2000 mm	553.12	693.83	3.96	121.65	nr	**815.48**
Branch (Side-on Shoe)						
Sum of two sides 1100 mm	49.57	62.18	1.50	46.08	nr	**108.26**
Sum of two sides 1200 mm	50.71	63.62	1.55	47.62	nr	**111.24**
Sum of two sides 1300 mm	51.94	65.15	1.59	48.84	nr	**113.99**
Sum of two sides 1400 mm	53.15	66.67	1.63	50.08	nr	**116.75**
Sum of two sides 1500 mm	54.35	68.17	1.68	51.61	nr	**119.78**
Sum of two sides 1600 mm	55.53	69.65	1.72	52.83	nr	**122.48**
Sum of two sides 1700 mm	56.75	71.19	1.77	54.38	nr	**125.57**
Sum of two sides 1800 mm	57.95	72.69	1.81	55.61	nr	**128.30**
Sum of two sides 1900 mm	59.14	74.19	1.86	57.14	nr	**131.33**
Sum of two sides 2000 mm	60.93	76.43	1.90	58.36	nr	**134.79**
Ductwork 1001–1250 mm longest side						
Sum of two sides 1300 mm	249.83	313.39	6.38	196.00	m	**509.39**
Sum of two sides 1400 mm	260.60	326.89	6.65	204.29	m	**531.18**
Sum of two sides 1500 mm	271.38	340.42	6.91	212.28	m	**552.70**
Sum of two sides 1600 mm	279.45	350.54	6.91	212.28	m	**562.82**
Sum of two sides 1700 mm	287.51	360.65	7.45	228.86	m	**589.51**
Sum of two sides 1800 mm	298.24	374.11	7.71	236.86	m	**610.97**
Sum of two sides 1900 mm	308.99	387.60	7.98	245.15	m	**632.75**
Sum of two sides 2000 mm	319.77	401.12	8.25	253.43	m	**654.55**
Sum of two sides 2100 mm	330.52	414.60	9.62	295.52	m	**710.12**
Sum of two sides 2200 mm	338.59	424.73	9.62	295.52	m	**720.25**
Sum of two sides 2300 mm	346.65	434.84	10.07	309.34	m	**744.18**
Sum of two sides 2400 mm	357.41	448.34	10.47	321.64	m	**769.98**
Sum of two sides 2500 mm	368.16	461.82	10.87	333.94	m	**795.76**
Extra over fittings; Ductwork 1001–1250 mm longest side						
End cap						
Sum of two sides 1300 mm	54.96	68.95	1.31	40.24	nr	**109.19**
Sum of two sides 1400 mm	58.00	72.76	1.34	41.17	nr	**113.93**
Sum of two sides 1500 mm	61.01	76.53	1.36	41.79	nr	**118.32**
Sum of two sides 1600 mm	64.08	80.38	1.39	42.69	nr	**123.07**
Sum of two sides 1700 mm	67.13	84.21	1.42	43.62	nr	**127.83**
Sum of two sides 1800 mm	70.14	87.99	1.45	44.54	nr	**132.53**
Sum of two sides 1900 mm	73.23	91.86	1.48	45.47	nr	**137.33**
Sum of two sides 2000 mm	76.27	95.67	1.51	46.39	nr	**142.06**
Sum of two sides 2100 mm	79.33	99.51	2.66	81.72	nr	**181.23**
Sum of two sides 2200 mm	82.37	103.32	2.80	86.02	nr	**189.34**
Sum of two sides 2300 mm	85.42	107.15	2.95	90.63	nr	**197.78**

38 VENTILATION/AIR CONDITIONING SYSTEMS

Item	Net Price £	Material £	Labour hours	Labour £	Unit	Total rate £
DUCTWORK: FIRE RATED – cont						
Extra over fittings – cont						
End cap – cont						
Sum of two sides 2400 mm	88.45	110.95	3.10	95.23	nr	**206.18**
Sum of two sides 2500 mm	91.50	114.78	3.24	99.53	nr	**214.31**
Reducer						
Sum of two sides 1300 mm	195.50	245.24	3.33	102.30	nr	**347.54**
Sum of two sides 1400 mm	204.30	256.28	3.38	103.84	nr	**360.12**
Sum of two sides 1500 mm	213.10	267.31	3.44	105.68	nr	**372.99**
Sum of two sides 1600 mm	221.94	278.40	3.49	107.22	nr	**385.62**
Sum of two sides 1700 mm	230.81	289.53	3.54	108.74	nr	**398.27**
Sum of two sides 1800 mm	239.66	300.63	3.60	110.59	nr	**411.22**
Sum of two sides 1900 mm	248.48	311.70	3.65	112.13	nr	**423.83**
Sum of two sides 2000 mm	257.33	322.80	3.70	113.66	nr	**436.46**
Sum of two sides 2100 mm	266.16	333.87	3.75	115.20	nr	**449.07**
Sum of two sides 2200 mm	274.98	344.94	3.80	116.74	nr	**461.68**
Sum of two sides 2300 mm	283.83	356.04	3.85	118.27	nr	**474.31**
Sum of two sides 2400 mm	292.67	367.12	3.90	119.81	nr	**486.93**
Sum of two sides 2500 mm	301.54	378.25	3.95	121.35	nr	**499.60**
Offset						
Sum of two sides 1300 mm	399.42	501.03	4.00	122.89	nr	**623.92**
Sum of two sides 1400 mm	415.28	520.92	4.11	126.26	nr	**647.18**
Sum of two sides 1500 mm	431.12	540.79	4.23	129.94	nr	**670.73**
Sum of two sides 1600 mm	446.96	560.67	4.34	133.32	nr	**693.99**
Sum of two sides 1700 mm	462.80	580.54	4.45	136.71	nr	**717.25**
Sum of two sides 1800 mm	478.63	600.40	4.56	140.08	nr	**740.48**
Sum of two sides 1900 mm	494.45	620.23	4.67	143.46	nr	**763.69**
Sum of two sides 2000 mm	510.29	640.10	5.53	169.88	nr	**809.98**
Sum of two sides 2100 mm	526.12	659.96	5.76	176.95	nr	**836.91**
Sum of two sides 2200 mm	541.68	679.48	5.99	184.02	nr	**863.50**
Sum of two sides 2300 mm	557.79	699.69	6.22	191.08	nr	**890.77**
Sum of two sides 2400 mm	573.60	719.52	6.45	198.15	nr	**917.67**
Sum of two sides 2500 mm	589.47	739.44	6.68	205.21	nr	**944.65**
90° radius bend						
Sum of two sides 1300 mm	236.24	296.34	3.12	95.85	nr	**392.19**
Sum of two sides 1400 mm	262.88	329.76	3.15	96.77	nr	**426.53**
Sum of two sides 1500 mm	269.82	338.46	3.17	97.38	nr	**435.84**
Sum of two sides 1600 mm	286.56	359.46	3.20	98.30	nr	**457.76**
Sum of two sides 1700 mm	303.35	380.52	3.22	98.92	nr	**479.44**
Sum of two sides 1800 mm	320.11	401.54	3.25	99.84	nr	**501.38**
Sum of two sides 1900 mm	336.88	422.59	3.28	100.77	nr	**523.36**
Sum of two sides 2000 mm	353.67	443.64	3.30	101.38	nr	**545.02**
Sum of two sides 2100 mm	370.44	464.68	3.32	102.00	nr	**566.68**
Sum of two sides 2200 mm	387.20	485.70	3.34	102.61	nr	**588.31**
Sum of two sides 2300 mm	404.00	506.78	3.36	103.22	nr	**610.00**
Sum of two sides 2400 mm	420.78	527.82	3.38	103.84	nr	**631.66**
Sum of two sides 2500 mm	437.55	548.87	3.40	104.45	nr	**653.32**

38 VENTILATION/AIR CONDITIONING SYSTEMS

Item	Net Price £	Material £	Labour hours	Labour £	Unit	Total rate £
45° bend						
Sum of two sides 1300 mm	191.36	240.04	2.02	62.06	nr	302.10
Sum of two sides 1400 mm	199.97	250.85	2.07	63.58	nr	314.43
Sum of two sides 1500 mm	208.57	261.63	2.13	65.43	nr	327.06
Sum of two sides 1600 mm	217.19	272.44	2.18	66.98	nr	339.42
Sum of two sides 1700 mm	225.80	283.25	2.24	68.81	nr	352.06
Sum of two sides 1800 mm	234.43	294.07	2.30	70.66	nr	364.73
Sum of two sides 1900 mm	243.03	304.85	2.35	72.20	nr	377.05
Sum of two sides 2000 mm	251.66	315.68	2.76	84.78	nr	400.46
Sum of two sides 2100 mm	260.29	326.50	2.89	88.78	nr	415.28
Sum of two sides 2200 mm	268.91	337.32	3.01	92.47	nr	429.79
Sum of two sides 2300 mm	277.51	348.11	3.13	96.15	nr	444.26
Sum of two sides 2400 mm	286.15	358.95	3.26	100.15	nr	459.10
Sum of two sides 2500 mm	294.75	369.73	3.37	103.52	nr	473.25
90° mitre bend						
Sum of two sides 1300 mm	362.29	454.45	3.75	115.20	nr	569.65
Sum of two sides 1400 mm	389.18	488.19	3.78	116.12	nr	604.31
Sum of two sides 1500 mm	417.14	523.26	3.81	117.05	nr	640.31
Sum of two sides 1600 mm	444.52	557.60	3.84	117.96	nr	675.56
Sum of two sides 1700 mm	471.89	591.94	3.87	118.89	nr	710.83
Sum of two sides 1800 mm	499.26	626.27	3.90	119.81	nr	746.08
Sum of two sides 1900 mm	526.66	660.64	3.93	120.74	nr	781.38
Sum of two sides 2000 mm	554.08	695.04	3.96	121.65	nr	816.69
Sum of two sides 2100 mm	581.45	729.37	3.99	122.58	nr	851.95
Sum of two sides 2200 mm	608.85	763.74	4.02	123.49	nr	887.23
Sum of two sides 2300 mm	636.25	798.11	4.05	124.41	nr	922.52
Sum of two sides 2400 mm	663.59	832.41	4.08	125.34	nr	957.75
Sum of two sides 2500 mm	690.97	866.76	4.11	126.26	nr	993.02
Branch (Side-on Shoe)						
Sum of two sides 1300 mm	52.03	65.26	1.59	48.84	nr	114.10
Sum of two sides 1400 mm	53.23	66.77	1.63	50.08	nr	116.85
Sum of two sides 1500 mm	54.40	68.24	1.68	51.61	nr	119.85
Sum of two sides 1600 mm	55.58	69.72	1.72	52.83	nr	122.55
Sum of two sides 1700 mm	56.79	71.23	1.77	54.38	nr	125.61
Sum of two sides 1800 mm	58.00	72.76	1.81	55.61	nr	128.37
Sum of two sides 1900 mm	59.17	74.22	1.86	57.14	nr	131.36
Sum of two sides 2000 mm	60.39	75.76	1.90	58.36	nr	134.12
Sum of two sides 2100 mm	61.57	77.24	2.58	79.25	nr	156.49
Sum of two sides 2200 mm	62.73	78.69	2.61	80.18	nr	158.87
Sum of two sides 2300 mm	63.96	80.24	2.64	81.10	nr	161.34
Sum of two sides 2400 mm	65.16	81.74	2.88	88.47	nr	170.21
Sum of two sides 2500 mm	66.33	83.20	2.91	89.40	nr	172.60
Ductwork 1251–2000 mm longest side						
Sum of two sides 1700 mm	273.95	343.64	7.45	228.86	m	572.50
Sum of two sides 1800 mm	291.52	365.68	7.71	236.86	m	602.54
Sum of two sides 1900 mm	309.07	387.70	7.98	245.15	m	632.85
Sum of two sides 2000 mm	326.61	409.70	8.25	253.43	m	663.13
Sum of two sides 2100 mm	344.19	431.75	9.62	295.52	m	727.27
Sum of two sides 2200 mm	361.73	453.76	9.66	296.76	m	750.52
Sum of two sides 2300 mm	379.26	475.74	10.07	309.34	m	785.08

38 VENTILATION/AIR CONDITIONING SYSTEMS

Item	Net Price £	Material £	Labour hours	Labour £	Unit	Total rate £
DUCTWORK: FIRE RATED – cont						
Ductwork 1251–2000 mm – cont						
Sum of two sides 2400 mm	396.81	497.76	10.47	321.64	m	819.40
Sum of two sides 2500 mm	414.36	519.77	10.87	333.94	m	853.71
Sum of two sides 2600 mm	431.95	541.83	11.27	346.21	m	888.04
Sum of two sides 2700 mm	449.48	563.83	11.67	358.51	m	922.34
Sum of two sides 2800 mm	467.04	585.85	12.08	371.10	m	956.95
Sum of two sides 2900 mm	484.57	607.85	12.48	383.40	m	991.25
Sum of two sides 3000 mm	502.12	629.85	12.88	395.67	m	1025.52
Sum of two sides 3100 mm	519.69	651.90	13.26	407.36	m	1059.26
Sum of two sides 3200 mm	537.24	673.92	13.69	420.56	m	1094.48
Sum of two sides 3300 mm	380.11	476.81	14.09	432.85	m	909.66
Sum of two sides 3400 mm	392.11	491.86	14.49	445.13	m	936.99
Sum of two sides 3500 mm	404.13	506.95	14.89	457.43	m	964.38
Sum of two sides 3600 mm	416.14	522.01	15.29	469.71	m	991.72
Sum of two sides 3700 mm	428.20	537.13	15.69	482.00	m	1019.13
Sum of two sides 3800 mm	440.20	552.18	16.09	494.28	m	1046.46
Sum of two sides 3900 mm	452.23	567.28	16.49	506.58	m	1073.86
Sum of two sides 4000 mm	464.25	582.36	16.89	518.87	m	1101.23
Extra over fittings; Ductwork 1251–2000 mm longest side						
End cap						
Sum of two sides 1700 mm	99.72	125.09	1.42	43.62	nr	168.71
Sum of two sides 1800 mm	106.14	133.15	1.45	44.54	nr	177.69
Sum of two sides 1900 mm	112.55	141.19	1.48	45.47	nr	186.66
Sum of two sides 2000 mm	118.98	149.25	1.51	46.39	nr	195.64
Sum of two sides 2100 mm	125.37	157.26	2.66	81.72	nr	238.98
Sum of two sides 2200 mm	131.81	165.35	2.80	86.02	nr	251.37
Sum of two sides 2300 mm	138.18	173.33	2.95	90.63	nr	263.96
Sum of two sides 2400 mm	144.61	181.40	3.10	95.23	nr	276.63
Sum of two sides 2500 mm	151.02	189.44	3.24	99.53	nr	288.97
Sum of two sides 2600 mm	157.45	197.50	3.39	104.14	nr	301.64
Sum of two sides 2700 mm	163.82	205.50	3.54	108.74	nr	314.24
Sum of two sides 2800 mm	170.25	213.56	3.68	113.04	nr	326.60
Sum of two sides 2900 mm	176.71	221.67	3.83	117.67	nr	339.34
Sum of two sides 3000 mm	183.10	229.68	3.98	122.27	nr	351.95
Sum of two sides 3100 mm	189.51	237.72	4.12	126.57	nr	364.29
Sum of two sides 3200 mm	195.94	245.78	4.27	131.17	nr	376.95
Sum of two sides 3300 mm	147.21	184.67	4.42	135.79	nr	320.46
Sum of two sides 3400 mm	151.89	190.53	4.57	140.39	nr	330.92
Sum of two sides 3500 mm	155.73	195.35	4.72	145.00	nr	340.35
Sum of two sides 3600 mm	158.86	199.27	4.87	149.61	nr	348.88
Sum of two sides 3700 mm	163.54	205.14	5.02	154.21	nr	359.35
Sum of two sides 3800 mm	245.54	308.00	5.17	158.82	nr	466.82
Sum of two sides 3900 mm	175.21	219.79	5.32	163.43	nr	383.22
Sum of two sides 4000 mm	179.89	225.66	5.47	168.04	nr	393.70

38 VENTILATION/AIR CONDITIONING SYSTEMS

Item	Net Price £	Material £	Labour hours	Labour £	Unit	Total rate £
Reducer						
Sum of two sides 1700 mm	215.36	270.14	3.54	108.74	nr	378.88
Sum of two sides 1800 mm	241.59	303.05	3.60	110.59	nr	413.64
Sum of two sides 1900 mm	267.79	335.91	3.65	112.13	nr	448.04
Sum of two sides 2000 mm	320.24	401.71	2.92	89.70	nr	491.41
Sum of two sides 2100 mm	294.04	368.84	3.70	113.66	nr	482.50
Sum of two sides 2200 mm	346.47	434.62	3.10	95.23	nr	529.85
Sum of two sides 2300 mm	372.68	467.49	3.29	101.07	nr	568.56
Sum of two sides 2400 mm	398.91	500.39	3.48	106.92	nr	607.31
Sum of two sides 2500 mm	425.11	533.25	3.66	112.44	nr	645.69
Sum of two sides 2600 mm	451.36	566.18	3.85	118.27	nr	684.45
Sum of two sides 2700 mm	477.58	599.08	4.03	123.79	nr	722.87
Sum of two sides 2800 mm	503.81	631.98	4.22	129.64	nr	761.62
Sum of two sides 2900 mm	530.03	664.87	4.40	135.17	nr	800.04
Sum of two sides 3000 mm	569.31	714.15	4.59	141.01	nr	855.16
Sum of two sides 3100 mm	569.31	714.15	4.78	146.84	nr	860.99
Sum of two sides 3200 mm	595.56	747.07	4.96	152.38	nr	899.45
Sum of two sides 3300 mm	461.98	579.51	5.15	158.20	nr	737.71
Sum of two sides 3400 mm	481.06	603.44	5.34	164.05	nr	767.49
Sum of two sides 3500 mm	500.14	627.38	5.53	169.88	nr	797.26
Sum of two sides 3600 mm	519.20	651.28	5.72	175.72	nr	827.00
Sum of two sides 3700 mm	538.30	675.25	5.91	181.56	nr	856.81
Sum of two sides 3800 mm	557.37	699.16	6.10	187.40	nr	886.56
Sum of two sides 3900 mm	576.47	723.13	6.29	193.22	nr	916.35
Sum of two sides 4000 mm	595.55	747.06	6.48	199.07	nr	946.13
Offset						
Sum of two sides 1700 mm	215.36	270.14	4.45	136.71	nr	406.85
Sum of two sides 1800 mm	233.11	292.41	4.56	140.08	nr	432.49
Sum of two sides 1900 mm	250.88	314.71	4.67	143.46	nr	458.17
Sum of two sides 2000 mm	268.55	336.87	5.53	169.88	nr	506.75
Sum of two sides 2100 mm	286.31	359.15	5.76	176.95	nr	536.10
Sum of two sides 2200 mm	304.05	381.40	5.99	184.02	nr	565.42
Sum of two sides 2300 mm	321.83	403.70	6.22	191.08	nr	594.78
Sum of two sides 2400 mm	339.55	425.94	6.45	198.15	nr	624.09
Sum of two sides 2500 mm	357.29	448.18	6.68	205.21	nr	653.39
Sum of two sides 2600 mm	375.06	470.48	6.91	212.28	nr	682.76
Sum of two sides 2700 mm	392.80	492.73	7.14	219.34	nr	712.07
Sum of two sides 2800 mm	410.57	515.02	7.37	226.41	nr	741.43
Sum of two sides 2900 mm	428.26	537.21	7.60	233.46	nr	770.67
Sum of two sides 3000 mm	446.04	559.51	7.83	240.54	nr	800.05
Sum of two sides 3100 mm	463.80	581.80	8.06	247.61	nr	829.41
Sum of two sides 3200 mm	481.54	604.04	8.29	254.67	nr	858.71
Sum of two sides 3300 mm	363.29	455.71	8.52	261.73	nr	717.44
Sum of two sides 3400 mm	376.19	471.89	8.75	268.80	nr	740.69
Sum of two sides 3500 mm	389.12	488.11	8.98	275.87	nr	763.98
Sum of two sides 3600 mm	402.02	504.29	9.21	282.93	nr	787.22
Sum of two sides 3700 mm	414.94	520.50	9.44	290.00	nr	810.50
Sum of two sides 3800 mm	427.85	536.69	9.67	297.07	nr	833.76
Sum of two sides 3900 mm	440.76	552.89	9.90	304.12	nr	857.01
Sum of two sides 4000 mm	453.69	569.11	10.13	311.19	nr	880.30

38 VENTILATION/AIR CONDITIONING SYSTEMS

Item	Net Price £	Material £	Labour hours	Labour £	Unit	Total rate £
DUCTWORK: FIRE RATED – cont						
Extra over fittings – cont						
90° radius bend						
Sum of two sides 1700 mm	299.71	375.96	3.22	98.92	nr	**474.88**
Sum of two sides 1800 mm	326.79	409.92	3.25	99.84	nr	**509.76**
Sum of two sides 1900 mm	344.90	432.64	3.32	102.00	nr	**534.64**
Sum of two sides 2000 mm	353.89	443.92	3.28	100.77	nr	**544.69**
Sum of two sides 2100 mm	380.97	477.89	3.30	101.38	nr	**579.27**
Sum of two sides 2200 mm	435.88	546.77	3.32	102.00	nr	**648.77**
Sum of two sides 2300 mm	462.24	579.84	3.36	103.22	nr	**683.06**
Sum of two sides 2400 mm	489.27	613.74	3.38	103.84	nr	**717.58**
Sum of two sides 2500 mm	516.39	647.76	3.40	104.45	nr	**752.21**
Sum of two sides 2600 mm	543.49	681.76	3.42	105.07	nr	**786.83**
Sum of two sides 2700 mm	570.60	715.76	3.44	105.68	nr	**821.44**
Sum of two sides 2800 mm	597.68	749.73	3.46	106.30	nr	**856.03**
Sum of two sides 2900 mm	624.71	783.64	3.48	106.92	nr	**890.56**
Sum of two sides 3000 mm	651.82	817.64	3.50	107.53	nr	**925.17**
Sum of two sides 3100 mm	678.89	851.60	3.52	108.12	nr	**959.72**
Sum of two sides 3200 mm	682.66	856.33	3.54	108.74	nr	**965.07**
Sum of two sides 3300 mm	498.18	624.92	3.56	109.36	nr	**734.28**
Sum of two sides 3400 mm	733.86	920.55	3.58	109.97	nr	**1030.52**
Sum of two sides 3500 mm	515.58	646.74	3.60	110.59	nr	**757.33**
Sum of two sides 3600 mm	533.30	668.98	3.62	111.20	nr	**780.18**
Sum of two sides 3700 mm	551.02	691.20	3.64	111.82	nr	**803.02**
Sum of two sides 3800 mm	568.78	713.47	3.66	112.44	nr	**825.91**
Sum of two sides 3900 mm	586.53	735.74	3.68	113.04	nr	**848.78**
Sum of two sides 4000 mm	604.26	757.98	3.70	113.66	nr	**871.64**
45° bend						
Sum of two sides 1700 mm	222.19	278.71	2.24	68.81	nr	**347.52**
Sum of two sides 1800 mm	246.75	309.52	2.30	70.66	nr	**380.18**
Sum of two sides 1900 mm	271.23	340.23	2.35	72.20	nr	**412.43**
Sum of two sides 2000 mm	295.82	371.08	2.76	84.78	nr	**455.86**
Sum of two sides 2100 mm	380.97	477.89	3.01	92.47	nr	**570.36**
Sum of two sides 2200 mm	369.48	463.48	3.13	96.15	nr	**559.63**
Sum of two sides 2300 mm	408.05	511.86	2.89	88.78	nr	**600.64**
Sum of two sides 2400 mm	393.99	494.22	3.26	100.15	nr	**594.37**
Sum of two sides 2500 mm	418.54	525.01	3.37	103.52	nr	**628.53**
Sum of two sides 2600 mm	443.05	555.77	3.49	107.22	nr	**662.99**
Sum of two sides 2700 mm	467.64	586.61	3.62	111.20	nr	**697.81**
Sum of two sides 2800 mm	492.17	617.38	3.74	114.89	nr	**732.27**
Sum of two sides 2900 mm	516.71	648.17	3.86	118.57	nr	**766.74**
Sum of two sides 3000 mm	541.25	678.94	3.98	122.27	nr	**801.21**
Sum of two sides 3100 mm	565.81	709.76	4.10	125.96	nr	**835.72**
Sum of two sides 3200 mm	590.32	740.50	4.22	129.64	nr	**870.14**
Sum of two sides 3300 mm	447.43	561.25	4.34	133.32	nr	**694.57**
Sum of two sides 3400 mm	465.29	583.65	4.46	137.02	nr	**720.67**
Sum of two sides 3500 mm	483.15	606.07	4.58	140.69	nr	**746.76**
Sum of two sides 3600 mm	501.03	628.49	4.70	144.38	nr	**772.87**
Sum of two sides 3700 mm	518.85	650.84	4.82	148.06	nr	**798.90**
Sum of two sides 3800 mm	536.72	673.27	4.94	151.76	nr	**825.03**
Sum of two sides 3900 mm	554.56	695.64	5.06	155.44	nr	**851.08**

38 VENTILATION/AIR CONDITIONING SYSTEMS

Item	Net Price £	Material £	Labour hours	Labour £	Unit	Total rate £
Sum of two sides 4000 mm	572.45	718.08	5.18	159.13	nr	877.21
90° mitre bend						
Sum of two sides 1700 mm	448.88	563.08	3.87	118.89	nr	681.97
Sum of two sides 1800 mm	526.46	660.40	3.90	119.81	nr	780.21
Sum of two sides 1900 mm	604.06	757.74	3.93	120.74	nr	878.48
Sum of two sides 2000 mm	681.58	854.97	3.96	121.65	nr	976.62
Sum of two sides 2100 mm	759.18	952.31	3.82	117.35	nr	1069.66
Sum of two sides 2200 mm	836.75	1049.62	3.83	117.67	nr	1167.29
Sum of two sides 2300 mm	914.35	1146.96	3.83	117.67	nr	1264.63
Sum of two sides 2400 mm	991.92	1244.26	3.84	117.96	nr	1362.22
Sum of two sides 2500 mm	1069.49	1341.57	3.85	118.27	nr	1459.84
Sum of two sides 2600 mm	1147.02	1438.82	3.85	118.27	nr	1557.09
Sum of two sides 2700 mm	1224.61	1536.15	3.86	118.57	nr	1654.72
Sum of two sides 2800 mm	1302.18	1633.45	3.86	118.57	nr	1752.02
Sum of two sides 2900 mm	1379.76	1730.77	3.87	118.89	nr	1849.66
Sum of two sides 3000 mm	1457.33	1828.08	3.87	118.89	nr	1946.97
Sum of two sides 3100 mm	1534.93	1925.41	3.88	119.19	nr	2044.60
Sum of two sides 3200 mm	1612.48	2022.70	3.88	119.19	nr	2141.89
Sum of two sides 3300 mm	1229.80	1542.67	3.89	119.50	nr	1662.17
Sum of two sides 3400 mm	1286.24	1613.46	3.90	119.81	nr	1733.27
Sum of two sides 3500 mm	1342.69	1684.27	3.91	120.12	nr	1804.39
Sum of two sides 3600 mm	1399.14	1755.08	3.92	120.42	nr	1875.50
Sum of two sides 3700 mm	1455.60	1825.90	3.93	120.74	nr	1946.64
Sum of two sides 3800 mm	1512.04	1896.70	3.94	121.04	nr	2017.74
Sum of two sides 3900 mm	1568.50	1967.53	3.95	121.35	nr	2088.88
Sum of two sides 4000 mm	1624.94	2038.32	3.96	121.65	nr	2159.97
Branch (Side-on Shoe)						
Sum of two sides 1700 mm	50.89	63.84	1.77	54.38	nr	118.22
Sum of two sides 1800 mm	55.60	69.74	1.81	55.61	nr	125.35
Sum of two sides 1900 mm	60.37	75.72	1.86	57.14	nr	132.86
Sum of two sides 2000 mm	65.10	81.66	1.90	58.36	nr	140.02
Sum of two sides 2100 mm	69.83	87.60	2.58	79.25	nr	166.85
Sum of two sides 2200 mm	74.54	93.50	2.61	80.18	nr	173.68
Sum of two sides 2300 mm	79.31	99.49	2.65	81.41	nr	180.90
Sum of two sides 2400 mm	84.02	105.39	2.68	82.33	nr	187.72
Sum of two sides 2500 mm	88.76	111.34	2.71	83.25	nr	194.59
Sum of two sides 2600 mm	93.51	117.30	2.75	84.48	nr	201.78
Sum of two sides 2700 mm	98.20	123.18	2.78	85.40	nr	208.58
Sum of two sides 2800 mm	102.97	129.17	2.81	86.33	nr	215.50
Sum of two sides 2900 mm	107.70	135.09	2.84	87.25	nr	222.34
Sum of two sides 3000 mm	112.42	141.02	2.87	88.17	nr	229.19
Sum of two sides 3100 mm	117.15	146.96	2.90	89.08	nr	236.04
Sum of two sides 3200 mm	121.91	152.92	2.93	90.01	nr	242.93
Sum of two sides 3300 mm	92.14	115.58	2.93	90.01	nr	205.59
Sum of two sides 3400 mm	95.59	119.91	3.00	92.16	nr	212.07
Sum of two sides 3500 mm	99.04	124.23	3.03	93.07	nr	217.30
Sum of two sides 3600 mm	102.45	128.51	3.06	94.00	nr	222.51
Sum of two sides 3700 mm	105.92	132.87	3.09	94.92	nr	227.79
Sum of two sides 3800 mm	109.38	137.21	3.12	95.85	nr	233.06
Sum of two sides 3900 mm	112.82	141.52	3.15	96.77	nr	238.29
Sum of two sides 4000 mm	116.24	145.81	3.18	97.70	nr	243.51

38 VENTILATION/AIR CONDITIONING SYSTEMS

Item	Net Price £	Material £	Labour hours	Labour £	Unit	Total rate £
DUCTWORK: FIRE RATED – cont						
Rectangular section ductwork to BSEN1366 (Duraduct LT), ducts 'Type A' and 'Type B'; manufactured from 6 mm thick laminate fire board consisting of steel circular hole punched facings pressed to a fibre cement core; provides upto 4 hours stability, 4 hours integrity and 32 minutes insulation; including all necessary stiffeners, joints and supports in the running length						
Ductwork up to 600 mm longest side						
Sum of two sides 200 mm	485.06	608.46	4.00	122.89	m	731.35
Sum of two sides 400 mm	496.89	623.30	4.00	122.89	m	746.19
Sum of two sides 600 mm	519.25	651.35	4.00	122.89	m	774.24
Sum of two sides 800 mm	685.20	859.51	5.50	168.95	m	1028.46
Sum of two sides 1000 mm	923.10	1157.93	5.50	168.95	m	1326.88
Sum of two sides 1200 mm	954.14	1196.88	6.00	184.32	m	1381.20
Extra over fittings; Ductwork up to 600 mm longest side						
End cap						
Sum of two sides 200 mm	115.12	144.40	0.81	24.89	m	169.29
Sum of two sides 400 mm	115.12	144.40	0.87	26.73	m	171.13
Sum of two sides 600 mm	115.12	144.40	0.93	28.57	m	172.97
Sum of two sides 800 mm	185.94	233.24	0.98	30.11	m	263.35
Sum of two sides 1000 mm	185.94	233.24	1.22	37.49	m	270.73
Sum of two sides 1200 mm	283.31	355.39	1.28	39.32	m	394.71
Reducer						
Sum of two sides 200 mm	101.80	127.70	2.23	68.50	m	196.20
Sum of two sides 400 mm	101.80	127.70	2.51	77.11	m	204.81
Sum of two sides 600 mm	101.80	127.70	2.79	85.71	m	213.41
Sum of two sides 800 mm	199.21	249.89	3.01	92.47	m	342.36
Sum of two sides 1000 mm	199.21	249.89	3.17	97.38	m	347.27
Sum of two sides 1200 mm	252.35	316.55	3.28	100.77	m	417.32
Offset						
Sum of two sides 200 mm	101.80	127.70	2.95	90.63	m	218.33
Sum of two sides 400 mm	101.80	127.70	3.26	100.15	m	227.85
Sum of two sides 600 mm	101.80	127.70	3.57	109.67	m	237.37
Sum of two sides 800 mm	199.21	249.89	3.23	99.22	m	349.11
Sum of two sides 1000 mm	199.21	249.89	3.67	112.75	m	362.64
Sum of two sides 1200 mm	252.35	316.55	3.89	119.50	m	436.05
90° radius bend						
Sum of two sides 200 mm	340.86	427.57	2.06	63.28	m	490.85
Sum of two sides 400 mm	340.86	427.57	2.36	72.51	m	500.08
Sum of two sides 600 mm	340.86	427.57	2.66	81.72	m	509.29
Sum of two sides 800 mm	398.42	499.78	2.97	91.25	m	591.03
Sum of two sides 1000 mm	398.42	499.78	3.04	93.39	m	593.17
Sum of two sides 1200 mm	451.56	566.44	3.09	94.92	m	661.36

38 VENTILATION/AIR CONDITIONING SYSTEMS

Item	Net Price £	Material £	Labour hours	Labour £	Unit	Total rate £
45° radius bend						
Sum of two sides 200 mm	340.86	427.57	1.52	46.69	m	**474.26**
Sum of two sides 400 mm	340.86	427.57	1.66	50.99	m	**478.56**
Sum of two sides 600 mm	340.86	427.57	1.80	55.29	m	**482.86**
Sum of two sides 800 mm	398.42	499.78	1.93	59.29	m	**559.07**
Sum of two sides 1000 mm	398.42	499.78	2.05	62.97	m	**562.75**
Sum of two sides 1200 mm	451.56	566.44	2.17	66.66	m	**633.10**
90° mitre bend						
Sum of two sides 200 mm	340.86	427.57	2.47	75.88	m	**503.45**
Sum of two sides 400 mm	340.86	427.57	2.84	87.25	m	**514.82**
Sum of two sides 600 mm	340.86	427.57	3.20	98.30	m	**525.87**
Sum of two sides 800 mm	398.42	499.78	3.56	109.36	m	**609.14**
Sum of two sides 1000 mm	398.42	499.78	3.66	112.44	m	**612.22**
Sum of two sides 1200 mm	451.56	566.44	3.72	114.27	m	**680.71**
Branch						
Sum of two sides 200 mm	123.96	155.50	0.98	30.11	m	**185.61**
Sum of two sides 400 mm	123.96	155.50	1.14	35.02	m	**190.52**
Sum of two sides 600 mm	123.96	155.50	1.30	39.94	m	**195.44**
Sum of two sides 800 mm	154.93	194.34	1.46	44.86	m	**239.20**
Sum of two sides 1000 mm	168.93	211.90	1.46	44.86	m	**256.76**
Sum of two sides 1200 mm	208.03	260.95	1.55	47.62	m	**308.57**
Ductwork 601–1000 mm longest side						
Sum of two sides 1000 mm	685.20	859.51	6.00	184.32	m	**1043.83**
Sum of two sides 1100 mm	954.14	1196.88	6.00	184.32	m	**1381.20**
Sum of two sides 1300 mm	970.07	1216.86	6.00	184.32	m	**1401.18**
Sum of two sides 1500 mm	992.68	1245.22	6.00	184.32	m	**1429.54**
Sum of two sides 1700 mm	1199.73	1504.94	6.00	184.32	m	**1689.26**
Sum of two sides 1900 mm	1374.87	1724.63	6.00	184.32	m	**1908.95**
Extra over fittings; Ductwork 601–1000 mm longest side						
End cap						
Sum of two sides 1000 mm	185.94	233.24	1.25	38.39	m	**271.63**
Sum of two sides 1100 mm	283.31	355.39	1.25	38.39	m	**393.78**
Sum of two sides 1300 mm	283.31	355.39	1.31	40.24	m	**395.63**
Sum of two sides 1500 mm	283.51	355.63	1.36	41.79	m	**397.42**
Sum of two sides 1700 mm	500.24	627.50	1.42	43.62	m	**671.12**
Sum of two sides 1900 mm	500.24	627.50	1.48	45.47	m	**672.97**
Reducer						
Sum of two sides 1000 mm	199.21	249.89	3.22	98.92	m	**348.81**
Sum of two sides 1100 mm	252.35	316.55	3.22	98.92	m	**415.47**
Sum of two sides 1300 mm	252.35	316.55	3.33	102.30	m	**418.85**
Sum of two sides 1500 mm	252.35	316.55	3.44	105.68	m	**422.23**
Sum of two sides 1700 mm	309.90	388.74	3.54	108.74	m	**497.48**
Sum of two sides 1900 mm	309.90	388.74	3.65	112.13	m	**500.87**
Offset						
Sum of two sides 1000 mm	199.21	249.89	3.78	116.12	m	**366.01**
Sum of two sides 1100 mm	252.35	316.55	3.78	116.12	m	**432.67**
Sum of two sides 1300 mm	252.35	316.55	4.00	122.89	m	**439.44**

38 VENTILATION/AIR CONDITIONING SYSTEMS

Item	Net Price £	Material £	Labour hours	Labour £	Unit	Total rate £
DUCTWORK: FIRE RATED – cont						
Extra over fittings – cont						
Offset – cont						
Sum of two sides 1500 mm	252.35	316.55	4.23	129.94	m	446.49
Sum of two sides 1700 mm	309.90	388.74	4.45	136.71	m	525.45
Sum of two sides 1900 mm	309.90	388.74	4.67	143.46	m	532.20
90° radius bend						
Sum of two sides 1000 mm	398.42	499.78	3.78	116.12	m	615.90
Sum of two sides 1100 mm	451.56	566.44	3.07	94.30	m	660.74
Sum of two sides 1300 mm	451.56	566.44	3.12	95.85	m	662.29
Sum of two sides 1500 mm	451.56	566.44	3.17	97.38	m	663.82
Sum of two sides 1700 mm	593.21	744.13	3.22	98.92	m	843.05
Sum of two sides 1900 mm	593.21	744.13	3.28	100.77	m	844.90
45° bend						
Sum of two sides 1000 mm	451.56	566.44	1.91	58.68	m	625.12
Sum of two sides 1100 mm	451.56	566.44	2.00	61.44	m	627.88
Sum of two sides 1300 mm	451.56	566.44	2.02	62.06	m	628.50
Sum of two sides 1500 mm	451.56	566.44	2.13	65.43	m	631.87
Sum of two sides 1700 mm	593.21	744.13	2.24	68.81	m	812.94
Sum of two sides 1900 mm	593.21	744.13	2.35	72.20	m	816.33
90° mitre bend						
Sum of two sides 1000 mm	398.42	499.78	3.78	116.12	m	615.90
Sum of two sides 1100 mm	451.56	566.44	3.69	113.36	m	679.80
Sum of two sides 1300 mm	451.56	566.44	3.75	115.20	m	681.64
Sum of two sides 1500 mm	451.56	566.44	3.81	117.05	m	683.49
Sum of two sides 1700 mm	593.21	744.13	3.87	118.89	m	863.02
Sum of two sides 1900 mm	593.21	744.13	3.93	120.74	m	864.87
Branch						
Sum of two sides 1000 mm	161.09	202.07	2.38	73.11	m	275.18
Sum of two sides 1100 mm	208.03	260.95	1.50	46.08	m	307.03
Sum of two sides 1300 mm	208.03	260.95	1.59	48.84	m	309.79
Sum of two sides 1500 mm	208.03	260.95	1.68	51.61	m	312.56
Sum of two sides 1700 mm	265.62	333.19	1.77	54.38	m	387.57
Sum of two sides 1900 mm	265.62	333.19	1.86	57.14	m	390.33
Ductwork 1001–1250 mm longest side						
Sum of two sides 1300 mm	954.14	1196.88	6.00	184.32	m	1381.20
Sum of two sides 1500 mm	1014.14	1272.14	6.00	184.32	m	1456.46
Sum of two sides 1700 mm	1199.73	1504.94	6.00	184.32	m	1689.26
Sum of two sides 1900 mm	1219.72	1530.02	6.00	184.32	m	1714.34
Sum of two sides 2100 mm	1598.38	2005.01	6.50	199.68	m	2204.69
Sum of two sides 2300 mm	1685.22	2113.94	6.50	199.68	m	2313.62
Sum of two sides 2500 mm	2006.18	2516.55	6.50	199.68	m	2716.23
Extra over fittings; Ductwork 1001– 1250 mm longest side						
End cap						
Sum of two sides 1300 mm	283.31	355.39	1.31	40.24	m	395.63
Sum of two sides 1500 mm	283.31	355.39	1.36	41.79	m	397.18
Sum of two sides 1700 mm	500.24	627.50	1.42	43.62	m	671.12

38 VENTILATION/AIR CONDITIONING SYSTEMS

Item	Net Price £	Material £	Labour hours	Labour £	Unit	Total rate £
Sum of two sides 1900 mm	500.24	627.50	1.48	45.47	m	672.97
Sum of two sides 2100 mm	681.73	855.16	2.66	81.72	m	936.88
Sum of two sides 2300 mm	681.73	855.16	2.95	90.63	m	945.79
Sum of two sides 2500 mm	681.73	855.16	3.24	99.53	m	954.69
Reducer						
Sum of two sides 1300 mm	252.35	316.55	3.33	102.30	m	418.85
Sum of two sides 1500 mm	252.35	316.55	3.44	105.68	m	422.23
Sum of two sides 1700 mm	309.90	388.74	3.54	108.74	m	497.48
Sum of two sides 1900 mm	309.90	388.74	3.65	112.13	m	500.87
Sum of two sides 2100 mm	367.41	460.88	3.75	115.20	m	576.08
Sum of two sides 2300 mm	367.41	460.88	3.85	118.27	m	579.15
Sum of two sides 2500 mm	367.41	460.88	3.95	121.35	m	582.23
Offset						
Sum of two sides 1300 mm	252.35	316.55	4.00	122.89	m	439.44
Sum of two sides 1500 mm	252.35	316.55	4.23	129.94	m	446.49
Sum of two sides 1700 mm	309.90	388.74	4.45	136.71	m	525.45
Sum of two sides 1900 mm	309.90	388.74	4.67	143.46	m	532.20
Sum of two sides 2100 mm	367.41	460.88	5.76	176.95	m	637.83
Sum of two sides 2300 mm	367.41	460.88	6.22	191.08	m	651.96
Sum of two sides 2500 mm	367.41	460.88	6.68	205.21	m	666.09
90° radius bend						
Sum of two sides 1300 mm	451.56	566.44	3.12	95.85	m	662.29
Sum of two sides 1500 mm	451.56	566.44	3.17	97.38	m	663.82
Sum of two sides 1700 mm	593.21	744.13	3.22	98.92	m	843.05
Sum of two sides 1900 mm	593.21	744.13	3.28	100.77	m	844.90
Sum of two sides 2100 mm	606.48	760.77	3.32	102.00	m	862.77
Sum of two sides 2300 mm	606.48	760.77	3.36	103.22	m	863.99
Sum of two sides 2500 mm	606.48	760.77	3.40	104.45	m	865.22
45° bend						
Sum of two sides 1300 mm	451.56	566.44	2.02	62.06	m	628.50
Sum of two sides 1500 mm	451.56	566.44	2.13	65.43	m	631.87
Sum of two sides 1700 mm	593.21	744.13	2.24	68.81	m	812.94
Sum of two sides 1900 mm	593.21	744.13	2.35	72.20	m	816.33
Sum of two sides 2100 mm	606.48	760.77	2.89	88.78	m	849.55
Sum of two sides 2300 mm	606.48	760.77	3.13	96.15	m	856.92
Sum of two sides 2500 mm	606.48	760.77	3.37	103.52	m	864.29
90° mitre bend						
Sum of two sides 1300 mm	451.56	566.44	3.75	115.20	m	681.64
Sum of two sides 1500 mm	451.56	566.44	3.81	117.05	m	683.49
Sum of two sides 1700 mm	593.21	744.13	3.87	118.89	m	863.02
Sum of two sides 1900 mm	593.21	744.13	3.93	120.74	m	864.87
Sum of two sides 2100 mm	606.48	760.77	3.99	122.58	m	883.35
Sum of two sides 2300 mm	606.48	760.77	4.05	124.41	m	885.18
Sum of two sides 2500 mm	606.48	760.77	4.11	126.26	m	887.03
Branch						
Sum of two sides 1300 mm	208.03	260.95	1.59	48.84	m	309.79
Sum of two sides 1500 mm	208.03	260.95	1.68	51.61	m	312.56
Sum of two sides 1700 mm	265.62	333.19	1.77	54.38	m	387.57
Sum of two sides 1900 mm	265.62	333.19	1.86	57.14	m	390.33
Sum of two sides 2100 mm	278.86	349.80	2.58	79.25	m	429.05
Sum of two sides 2300 mm	278.86	349.80	2.64	81.10	m	430.90
Sum of two sides 2500 mm	278.86	349.80	2.91	89.40	m	439.20

38 VENTILATION/AIR CONDITIONING SYSTEMS

Item	Net Price £	Material £	Labour hours	Labour £	Unit	Total rate £
DUCTWORK: FIRE RATED – cont						
Ductwork 1251–2000 mm longest side						
Sum of two sides 1800 mm	1199.73	1504.94	6.00	184.32	m	**1689.26**
Sum of two sides 2000 mm	1399.65	1755.72	6.00	184.32	m	**1940.04**
Sum of two sides 2200 mm	1598.38	2005.01	6.50	199.68	m	**2204.69**
Sum of two sides 2400 mm	1797.46	2254.74	6.50	199.68	m	**2454.42**
Sum of two sides 2600 mm	1893.66	2375.41	6.66	204.59	m	**2580.00**
Sum of two sides 2800 mm	2077.15	2605.58	6.66	204.59	m	**2810.17**
Sum of two sides 3000 mm	2259.52	2834.34	6.66	204.59	m	**3038.93**
Sum of two sides 3200 mm	2276.41	2855.53	9.00	276.48	m	**3132.01**
Sum of two sides 3400 mm	2504.05	3141.08	9.00	276.48	m	**3417.56**
Sum of two sides 3600 mm	2549.26	3197.79	11.70	359.42	m	**3557.21**
Sum of two sides 3800 mm	2776.09	3482.33	11.70	359.42	m	**3841.75**
Sum of two sides 4000 mm	3211.51	4028.52	11.70	359.42	m	**4387.94**
Extra over fittings; Ductwork 1251– 2000 mm longest sides						
End cap						
Sum of two sides 1800 mm	500.24	627.50	1.45	44.54	m	**672.04**
Sum of two sides 2000 mm	500.24	627.50	1.51	46.39	m	**673.89**
Sum of two sides 2200 mm	681.73	855.16	2.80	86.02	m	**941.18**
Sum of two sides 2400 mm	681.73	855.16	3.10	95.23	m	**950.39**
Sum of two sides 2600 mm	934.02	1171.63	3.39	104.14	m	**1275.77**
Sum of two sides 2800 mm	934.02	1171.63	3.68	113.04	m	**1284.67**
Sum of two sides 3000 mm	934.02	1171.63	3.98	122.27	m	**1293.90**
Sum of two sides 3200 mm	1261.61	1582.56	4.27	131.17	m	**1713.73**
Sum of two sides 3400 mm	1261.61	1582.56	4.57	140.39	m	**1722.95**
Sum of two sides 3600 mm	1553.75	1949.02	4.87	149.61	m	**2098.63**
Sum of two sides 3800 mm	1553.75	1949.02	5.17	158.82	m	**2107.84**
Sum of two sides 4000 mm	1553.75	1949.02	5.47	168.04	m	**2117.06**
Reducer						
Sum of two sides 1800 mm	309.90	388.74	3.60	110.59	m	**499.33**
Sum of two sides 2000 mm	309.90	388.74	3.70	113.66	m	**502.40**
Sum of two sides 2200 mm	367.41	460.88	3.10	95.23	m	**556.11**
Sum of two sides 2400 mm	367.41	460.88	3.48	106.92	m	**567.80**
Sum of two sides 2600 mm	509.06	638.57	3.85	118.27	m	**756.84**
Sum of two sides 2800 mm	509.06	638.57	4.22	129.64	m	**768.21**
Sum of two sides 3000 mm	509.06	638.57	4.59	141.01	m	**779.58**
Sum of two sides 3200 mm	566.61	710.75	4.96	152.38	m	**863.13**
Sum of two sides 3400 mm	566.61	710.75	5.34	164.05	m	**874.80**
Sum of two sides 3600 mm	619.75	777.41	5.72	175.72	m	**953.13**
Sum of two sides 3800 mm	619.75	777.41	6.10	187.40	m	**964.81**
Sum of two sides 4000 mm	619.75	777.41	6.48	199.07	m	**976.48**
Offset						
Sum of two sides 1800 mm	309.90	388.74	4.56	140.08	m	**528.82**
Sum of two sides 2000 mm	309.90	388.74	5.53	169.88	m	**558.62**
Sum of two sides 2200 mm	367.41	460.88	5.99	184.02	m	**644.90**
Sum of two sides 2400 mm	367.41	460.88	6.45	198.15	m	**659.03**
Sum of two sides 2600 mm	509.06	638.57	6.91	212.28	m	**850.85**
Sum of two sides 2800 mm	509.06	638.57	7.37	226.41	m	**864.98**
Sum of two sides 3000 mm	509.06	638.57	7.83	240.54	m	**879.11**

38 VENTILATION/AIR CONDITIONING SYSTEMS

Item	Net Price £	Material £	Labour hours	Labour £	Unit	Total rate £
Sum of two sides 3200 mm	566.61	710.75	8.29	254.67	m	965.42
Sum of two sides 3400 mm	566.61	710.75	8.75	268.80	m	979.55
Sum of two sides 3600 mm	619.75	777.41	9.21	282.93	m	1060.34
Sum of two sides 3800 mm	619.75	777.41	9.67	297.07	m	1074.48
Sum of two sides 4000 mm	619.75	777.41	10.13	311.19	m	1088.60
90° radius bend						
Sum of two sides 1800 mm	593.21	744.13	3.25	99.84	m	843.97
Sum of two sides 2000 mm	593.21	744.13	3.30	101.38	m	845.51
Sum of two sides 2200 mm	606.48	760.77	3.34	102.61	m	863.38
Sum of two sides 2400 mm	606.48	760.77	3.38	103.84	m	864.61
Sum of two sides 2600 mm	929.62	1166.11	3.42	105.07	m	1271.18
Sum of two sides 2800 mm	929.62	1166.11	3.46	106.30	m	1272.41
Sum of two sides 3000 mm	929.62	1166.11	3.50	107.53	m	1273.64
Sum of two sides 3200 mm	1115.54	1399.33	3.54	108.74	m	1508.07
Sum of two sides 3400 mm	1115.54	1399.33	3.58	109.97	m	1509.30
Sum of two sides 3600 mm	1305.86	1638.07	3.62	111.20	m	1749.27
Sum of two sides 3800 mm	1305.86	1638.07	3.66	112.44	m	1750.51
Sum of two sides 4000 mm	1305.86	1638.07	3.70	113.66	m	1751.73
45° bend						
Sum of two sides 1800 mm	593.21	744.13	2.30	70.66	m	814.79
Sum of two sides 2000 mm	593.21	744.13	2.76	84.78	m	828.91
Sum of two sides 2200 mm	606.48	760.77	3.01	92.47	m	853.24
Sum of two sides 2400 mm	606.48	760.77	3.26	100.15	m	860.92
Sum of two sides 2600 mm	929.62	1166.11	3.49	107.22	m	1273.33
Sum of two sides 2800 mm	929.62	1166.11	3.74	114.89	m	1281.00
Sum of two sides 3000 mm	929.62	1166.11	3.98	122.27	m	1288.38
Sum of two sides 3200 mm	1115.54	1399.33	4.22	129.64	m	1528.97
Sum of two sides 3400 mm	1115.54	1399.33	4.46	137.02	m	1536.35
Sum of two sides 3600 mm	1305.86	1638.07	4.70	144.38	m	1782.45
Sum of two sides 3800 mm	1305.86	1638.07	4.94	151.76	m	1789.83
Sum of two sides 4000 mm	1305.86	1638.07	5.18	159.13	m	1797.20
90° mitre band						
Sum of two sides 1800 mm	593.21	744.13	3.90	119.81	m	863.94
Sum of two sides 2000 mm	593.21	744.13	3.96	121.65	m	865.78
Sum of two sides 2200 mm	606.48	760.77	3.83	117.67	m	878.44
Sum of two sides 2400 mm	806.40	1011.55	3.84	117.96	m	1129.51
Sum of two sides 2600 mm	929.62	1166.11	3.85	118.27	m	1284.38
Sum of two sides 2800 mm	929.62	1166.11	3.86	118.57	m	1284.68
Sum of two sides 3000 mm	929.62	1166.11	3.87	118.89	m	1285.00
Sum of two sides 3200 mm	1115.54	1399.33	3.88	119.19	m	1518.52
Sum of two sides 3400 mm	1115.54	1399.33	3.90	119.81	m	1519.14
Sum of two sides 3600 mm	1305.86	1638.07	3.92	120.42	m	1758.49
Sum of two sides 3800 mm	1305.86	1638.07	3.94	121.04	m	1759.11
Sum of two sides 4000 mm	1305.86	1638.07	3.96	121.65	m	1759.72
Branch						
Sum of two sides 1800 mm	265.62	333.19	1.81	55.61	m	388.80
Sum of two sides 2000 mm	265.62	333.19	1.90	58.36	m	391.55
Sum of two sides 2200 mm	278.86	349.80	2.61	80.18	m	429.98
Sum of two sides 2400 mm	278.86	349.80	2.68	82.33	m	432.13
Sum of two sides 2600 mm	331.99	416.45	2.75	84.48	m	500.93
Sum of two sides 2800 mm	331.99	416.45	2.81	86.33	m	502.78
Sum of two sides 3000 mm	331.99	416.45	2.87	88.17	m	504.62

38 VENTILATION/AIR CONDITIONING SYSTEMS

Item	Net Price £	Material £	Labour hours	Labour £	Unit	Total rate £
DUCTWORK: FIRE RATED – cont						
Extra over fittings – cont						
Branch – cont						
Sum of two sides 3200 mm	345.31	433.16	2.93	90.01	m	**523.17**
Sum of two sides 3400 mm	345.31	433.16	3.00	92.16	m	**525.32**
Sum of two sides 3600 mm	442.69	555.31	3.06	94.00	m	**649.31**
Sum of two sides 3800 mm	442.69	555.31	3.12	95.85	m	**651.16**
Sum of two sides 4000 mm	442.69	555.31	3.18	97.70	m	**653.01**

38 VENTILATION/AIR CONDITIONING SYSTEMS

Item	Net Price £	Material £	Labour hours	Labour £	Unit	Total rate £
LOW VELOCITY AIR CONDITIONING: AIR HANDLING UNITS						
Supply air handling unit; inlet with motorized damper, LTHW frost coil (at −5°C to +5°C), panel filter (EU4), bag filter (EU6), cooling coil (at 28°C db/20°C wb to 12°C db/11.5°C wb), LTHW heating coil (at 5°C to 21°C), supply fan, outlet plenum; includes access sections; all units located internally; Includes placing in position and fitting of sections together; electrical work elsewhere.						
Volume, external pressure						
2 m³/s @ 350 Pa	7861.96	9862.05	40.00	1286.51	nr	**11148.56**
2 m³/s @ 700 Pa	8399.38	10536.19	40.00	1286.51	nr	**11822.70**
5 m³/s @ 350 Pa	13256.84	16629.38	65.00	2090.58	nr	**18719.96**
5 m³/s @ 700 Pa	13720.59	17211.11	65.00	2090.58	nr	**19301.69**
8 m³/s @ 350 Pa	19855.93	24907.28	77.00	2476.53	nr	**27383.81**
8 m/³s @ 700 Pa	20165.46	25295.56	77.00	2476.53	nr	**27772.09**
10 m/³ @ 350 Pa	21865.20	27427.70	100.00	3216.28	nr	**30643.98**
10 m³/s @ 700 Pa	22909.11	28737.18	100.00	3216.28	nr	**31953.46**
13 m³/s @ 350 Pa	28052.45	35188.99	108.00	3473.58	nr	**38662.57**
13 m³/s @ 700 Pa	28918.86	36275.81	108.00	3473.58	nr	**39749.39**
15 m³/s @ 350 Pa	32126.24	40299.16	120.00	3859.54	nr	**44158.70**
15 m³/s @ 700 Pa	32710.79	41032.41	120.00	3859.54	nr	**44891.95**
18 m³/s @ 350 Pa	36455.03	45729.19	133.00	4277.65	nr	**50006.84**
18 m³/s @ 700 Pa	37125.19	46569.84	133.00	4277.65	nr	**50847.49**
20 m³/s @ 350 Pa	42245.19	52992.36	142.00	4567.12	nr	**57559.48**
20 m³/s @ 700 Pa	42852.72	53754.46	142.00	4567.12	nr	**58321.58**
Extra for inlet and discharge attenuators at 900 mm long						
2 m³/s @ 350 Pa	2429.05	3047.00	5.00	160.81	nr	**3207.81**
5 m³/s @ 350 Pa	5098.00	6394.93	10.00	321.63	nr	**6716.56**
10 m³/s @ 700 Pa	7253.37	9098.62	13.00	418.12	nr	**9516.74**
15 m³/s @ 700 Pa	10998.88	13797.00	16.00	514.61	nr	**14311.61**
20 m³/s @ 700 Pa	14194.13	17805.12	20.00	643.26	nr	**18448.38**
Extra for locating units externally						
2 m³/s @ 350 Pa	1799.81	2257.68	–	–	nr	**2257.68**
5 m³/s @ 350 Pa	2283.60	2864.55	–	–	nr	**2864.55**
10 m³/s @ 700 Pa	3653.33	4582.74	–	–	nr	**4582.74**
15 m³/s @ 700 Pa	5279.60	6622.73	–	–	nr	**6622.73**
20 m³/s @ 700 Pa	9151.24	11479.32	–	–	nr	**11479.32**

38 VENTILATION/AIR CONDITIONING SYSTEMS

Item	Net Price £	Material £	Labour hours	Labour £	Unit	Total rate £
LOW VELOCITY AIR CONDITIONING: AIR HANDLING UNITS – cont						
Modular air handling unit with supply and extract sections. Supply side; inlet with motorized damper, LTHW frost coil (at −5° C to 5°C), panel filter (EU4), bag filter (EU6), cooling coil at 28°Cdb/20°Cwb to 12° Cdb/11.5°C wb), LTHW heating coil (at 5°C to 21°C), supply fan, outlet plenum. Extract side; inlet with motorized damper, extract fan; includes access sections; placing in position and fitting of sections together; electrical work elsewhere						
Volume, external pressure						
2 m^3/s @ 350 Pa	10982.87	13776.91	50.00	1608.14	nr	**15385.05**
2 m^3/s @ 700 Pa	11514.29	14443.52	50.00	1608.14	nr	**16051.66**
5 m^3/s @ 350 Pa	18928.42	23743.81	86.00	2766.00	nr	**26509.81**
5 m^3/s @ 700 Pa	19283.71	24189.49	86.00	2766.00	nr	**26955.49**
8 m^3/s @ 350 Pa	26284.84	32971.70	105.00	3377.09	nr	**36348.79**
8 m/3 s @ 700 Pa	26915.79	33763.16	105.00	3377.09	nr	**37140.25**
10 m/3 @ 350 Pa	30351.15	38072.48	120.00	3859.54	nr	**41932.02**
10 m/s @ 700 Pa	31005.01	38892.68	120.00	3859.54	nr	**42752.22**
13 m^3/s @ 350 Pa	39057.44	48993.65	130.00	4181.16	nr	**53174.81**
13 m^3/s @ 700 Pa	40282.49	50530.36	130.00	4181.16	nr	**54711.52**
15 m^3/s @ 350 Pa	43643.94	54746.96	145.00	4663.61	nr	**59410.57**
15 m^3/s @ 700 Pa	45708.66	57336.94	145.00	4663.61	nr	**62000.55**
18 m^3/s @ 350 Pa	49890.28	62582.36	160.00	5146.05	nr	**67728.41**
18 m^3/s @ 700 Pa	51058.33	64047.57	160.00	5146.05	nr	**69193.62**
20 m^3/s @ 350 Pa	54528.05	68399.99	175.00	5628.49	nr	**74028.48**
20 m^3/s @ 700 Pa	56423.68	70777.86	175.00	5628.49	nr	**76406.35**
Extra for inlet and discharge attenuators at 900 mm long						
2 m^3/s @ 350 Pa	4428.28	5554.83	8.00	257.30	nr	**5812.13**
5 m^3/s @ 350 Pa	8744.59	10969.21	10.00	321.63	nr	**11290.84**
10 m^3/s @ 700 Pa	12512.78	15696.03	13.00	418.12	nr	**16114.15**
15 m^3/s @ 700 Pa	21403.66	26848.75	16.00	514.61	nr	**27363.36**
20 m^3/s @ 700 Pa	24985.34	31341.61	20.00	643.26	nr	**31984.87**
Extra for locating units externally						
2 m^3/s @ 350 Pa	3544.64	4446.40	–	–	nr	**4446.40**
5 m^3/s @ 350 Pa	5033.37	6313.85	–	–	nr	**6313.85**
10 m^3/s @ 700 Pa	7184.79	9012.60	–	–	nr	**9012.60**
15 m^3/s @ 700 Pa	13642.03	17112.56	–	–	nr	**17112.56**
20 m^3/s @ 700 Pa	20971.98	26307.25	–	–	nr	**26307.25**
Extra for humidifier, self-generating type						
2 m^3/s @ 350 Pa (10 kg/hr)	3088.91	3874.73	5.00	160.81	nr	**4035.54**
5 m^3/s @ 350 Pa (18 kg/hr)	3822.59	4795.06	5.00	160.81	nr	**4955.87**
10 m^3/s @ 700 Pa (30 kg/hr)	4349.91	5456.53	6.00	192.98	nr	**5649.51**
15 m^3/s @ 700 Pa (60 kg/hr)	8055.76	10105.14	8.00	257.30	nr	**10362.44**
20 m^3/s @ 700 Pa (90 kg/hr)	12111.79	15193.02	10.00	321.63	nr	**15514.65**

38 VENTILATION/AIR CONDITIONING SYSTEMS

Item	Net Price £	Material £	Labour hours	Labour £	Unit	Total rate £
Extra for mixing box						
2 m³/s @ 350 Pa	1483.23	1860.57	4.00	128.65	nr	1989.22
5 m³/s @ 350 Pa	2090.69	2622.56	4.00	128.65	nr	2751.21
10 m³/s @ 700 Pa	2929.05	3674.20	5.00	160.81	nr	3835.01
15 m³/s @ 700 Pa	3909.18	4903.67	6.00	192.98	nr	5096.65
20 m³/s @ 700 Pa	5959.44	7475.52	6.00	192.98	nr	7668.50
Extra for runaround coil, including pump and associated pipework; typical outputs in brackets (based on minimal distance between the supply and extract units)						
2 m³/s @350 Pa (26 kW)	4440.32	5569.94	30.00	964.88	nr	6534.82
5 m³/s @ 350 Pa (37 kW)	7872.78	9875.61	30.00	964.88	nr	10840.49
10 m³/s @ 700 Pa (85 kW)	13630.52	17098.12	40.00	1286.51	nr	18384.63
15 m³/s @ 700 Pa (151 kW)	19178.63	24057.68	50.00	1608.14	nr	25665.82
20 m³/s @ 700 Pa (158 kW)	26143.19	32794.01	60.00	1929.77	nr	34723.78
Extra for thermal wheel (typical outputs in brackets)						
2 m³/s @ 350 Pa (37 kW)	9184.21	11520.68	12.00	385.95	nr	11906.63
5 m³/s @ 350 Pa (65 kW)	13190.75	16546.48	12.00	385.95	nr	16932.43
10 m³/s @ 700 Pa (127 kW)	17590.84	22065.95	15.00	482.44	nr	22548.39
15 m³/s @ 700 Pa (160 kW)	32957.76	41342.21	17.00	546.77	nr	41888.98
20 m³/s @ 700 Pa (262 kW)	38498.49	48292.51	19.00	611.09	nr	48903.60
Extra for plate heat exchanger, including additional filtration in extract leg (typical outputs in brackets)						
2 m³/s @ 350 Pa (25 kW)	5820.78	7301.58	12.00	385.95	nr	7687.53
5 m³/s @ 350 Pa (51 kW)	11378.89	14273.68	12.00	385.95	nr	14659.63
10 m³/s @ 700 Pa (98 kW)	16695.99	20943.45	15.00	482.44	nr	21425.89
15 m³/s @ 700 Pa (160 kW)	28832.60	36167.61	17.00	546.77	nr	36714.38
20 m³/s @ 700 Pa (190 kW)	37234.88	46707.44	19.00	611.09	nr	47318.53
Extra for electric heating in lieu of LTHW						
2 m³/s @ 350 Pa	2017.41	2530.64	–	–	nr	2530.64
5 m³/s @ 350 Pa	3779.85	4741.44	–	–	nr	4741.44
10 m³/s @ 700 Pa	5103.38	6401.68	–	–	nr	6401.68
15 m³/s @ 700 Pa	5351.38	6712.78	–	–	nr	6712.78
20 m³/s @ 700 Pa	6315.39	7922.03	–	–	nr	7922.03

38 VENTILATION/AIR CONDITIONING SYSTEMS

Item	Net Price £	Material £	Labour hours	Labour £	Unit	Total rate £
VAV AIR CONDITIONING						
VAV TERMINAL BOXES						
VAV terminal box; integral acoustic silencer; factory installed and prewired control components (excluding electronic controller); selected at 200 Pa at entry to unit; includes fixing in position; electrical work elsewhere						
80 l/s – 110 l/s	499.24	626.25	2.00	61.44	nr	**687.69**
Extra for secondary silencer	121.41	152.30	0.50	15.36	nr	**167.66**
Extra for 2 row LTHW heating coil	299.94	376.24	–	–	nr	**376.24**
150 l/s – 190 l/s	525.34	658.99	2.00	61.44	nr	**720.43**
Extra for secondary silencer	133.32	167.24	0.50	15.36	nr	**182.60**
Extra for 2 row LTHW heating coil	321.36	403.11	–	–	nr	**403.11**
250 l/s – 310 l/s	586.05	735.15	2.00	61.44	nr	**796.59**
Extra for secondary silencer	176.44	221.32	0.50	15.36	nr	**236.68**
Extra for 2 row LTHW heating coil	364.21	456.87	–	–	nr	**456.87**
420 l/s – 520 l/s	634.84	796.34	2.00	61.44	nr	**857.78**
Extra for secondary silencer	200.83	251.92	0.50	15.36	nr	**267.28**
Extra for 2 row LTHW heating coil	443.48	556.30	–	–	nr	**556.30**
650 l/s – 790 l/s	768.72	964.29	2.00	61.44	nr	**1025.73**
Extra for secondary silencer	257.00	322.38	0.50	15.36	nr	**337.74**
Extra for 2 row LTHW heating coil	464.90	583.17	–	–	nr	**583.17**
1130 l/s – 1370 l/s	892.97	1120.15	2.00	61.44	nr	**1181.59**
Extra for secondary silencer	359.12	450.48	0.50	15.36	nr	**465.84**
Extra for 2 row LTHW heating coil	486.32	610.04	–	–	nr	**610.04**
Extra for electric heater & thyristor controls, 3 kW/1ph (per box)	396.34	497.17	–	–	nr	**497.17**
Extra for fitting free issue box controller	–	–	–	–	nr	**–**
Fan assisted VAV terminal box; factory installed and prewired control components (excluding electronic controller); selected at 40 Pa external static pressure; includes fixing in position, electrical work elsewhere						
100 l/s – 175 l/s	1105.15	1386.30	3.00	92.16	nr	**1478.46**
Extra for secondary silencer	131.62	165.10	0.50	15.36	nr	**180.46**
Extra for 1 row LTHW heating coil	353.50	443.43	–	–	nr	**443.43**
170 l/s – 360 l/s	1205.56	1512.26	3.00	92.16	nr	**1604.42**
Extra for secondary silencer	166.80	209.24	0.50	15.36	nr	**224.60**
Extra for 1 row LTHW heating coil	374.92	470.30	–	–	nr	**470.30**
300 l/s – 640 l/s	1382.00	1733.58	3.00	92.16	nr	**1825.74**
Extra for secondary silencer	255.30	320.25	0.50	15.36	nr	**335.61**
Extra for 1 row LTHW heating coil	407.06	510.62	–	–	nr	**510.62**
620 l/s – 850 l/s	1382.00	1733.58	3.00	92.16	nr	**1825.74**
Extra for secondary silencer	255.30	320.25	0.50	15.36	nr	**335.61**
Extra for 1 row LTHW heating coil	439.19	550.92	–	–	nr	**550.92**
Extra for electric heater plus thyristor controls, 3 kW/lph (per box)	407.06	510.62	–	–	nr	**510.62**
Extra for fitting free issue controller	–	–	2.13	65.43	nr	**65.43**

38 VENTILATION/AIR CONDITIONING SYSTEMS

Item	Net Price £	Material £	Labour hours	Labour £	Unit	Total rate £
FAN COIL AIR CONDITIONING						
All selections based on summer return air condition of 23°C @ 50% RH, CHW @ 6°/ 12°C, LTHW @ 82°/71°C (where applicable), medium speed, external resistance of 30 Pa						
All selections are based on heating and cooling units. For waterside control units there is no significant reduction in cost between 4 pipe heating and cooling and 2 pipe cooling only units (excluding controls). For airside control units, there is a marginal reduction (less than 5%) between 4 pipe heating and cooling units and 2 pipe cooling only units (excluding controls)						
Ceiling void mounted horizontal waterside control fan coil unit; cooling coil; LTHW heating coil; multi-tapped speed transformer; fine wire mesh filter; includes fixing in position; electrical work elsewhere						
Total cooling load, heating load						
2800 W, 1000 W	509.36	638.94	4.00	128.65	nr	**767.59**
4000 W, 1700 W	484.79	608.12	4.00	128.65	nr	**736.77**
4500 W, 1900 W	751.90	943.19	4.00	128.65	nr	**1071.84**
6000 W, 2600 W	983.43	1233.61	4.00	128.65	nr	**1362.26**
Ceiling void mounted horizontal waterside control fan coil unit; cooling coil; electric heating coil; multi-tapped speed transformer; fine wire mesh filter; includes fixing in position; electrical work elsewhere + thyristor and 2 No. HTCOs						
Total cooling load, heating load						
2800 W, 1500 W	706.50	886.23	4.00	128.65	nr	**1014.88**
4000 W, 2000 W	730.98	916.94	4.00	128.65	nr	**1045.59**
4500 W, 2000 W	930.75	1167.53	4.00	128.65	nr	**1296.18**
6000 W, 3000 W	1149.88	1442.41	4.00	128.65	nr	**1571.06**
Ceiling void mounted horizontal airside control fan coil unit; cooling coil; LHTW heating coil; multi-tapped speed transformer; fine wire mesh filter, damper actuator & fixing kit; includes fixing in position; electrical work elsewhere						
Total cooling load, heating load						
2600 W, 2200 W	800.41	1004.04	4.00	128.65	nr	**1132.69**
3600 W, 3200 W	823.57	1033.09	4.00	128.65	nr	**1161.74**
4000 W, 3600 W	921.69	1156.16	4.00	128.65	nr	**1284.81**
5400 W, 5000 W	1100.30	1380.22	4.00	128.65	nr	**1508.87**

38 VENTILATION/AIR CONDITIONING SYSTEMS

Item	Net Price £	Material £	Labour hours	Labour £	Unit	Total rate £
FAN COIL AIR CONDITIONING – cont						
Ceiling void mounted horizontal airside control fan coil unit; cooling coil; electric heating coil; multi-tapped speed transformer; fine wire mesh filter, damper actuator & fixing kit; includes fixing in position; electrical work elsewhere + thyristor and 2 No. HTCOs						
Total cooling load, heating load						
2600 W, 1500 W.	852.23	1069.04	4.00	128.65	nr	**1197.69**
3600 W, 2000 W	874.28	1096.69	4.00	128.65	nr	**1225.34**
4000 W, 2000 W	966.89	1212.87	4.00	128.65	nr	**1341.52**
5400 W, 3000 W	1148.81	1441.07	4.00	128.65	nr	**1569.72**
Ceiling void mounted slimline horizontal waterside control fan coil unit, 170 mm deep; cooilng coil; LTHW heating coil; multi-tapped speed transformer; fine wire mesh filter; includes fixing in position; electrical work elsewhere						
Total cooling load, heating load						
1100 W, 1500 W.	628.42	788.29	3.50	112.57	nr	**900.86**
3200 W, 3700 W	1200.62	1506.05	4.00	128.65	nr	**1634.70**
Ceiling void mounted slimline horizontal waterside control fan coil unit, 170 mm deep; cooilng coil; electric heating coil; multi-tapped speed transformer; fine wire mesh filter; includes fixing in position; electrical work elsewhere						
Total cooling load, heating load						
1100 W, 1000 W.	464.15	582.23	3.50	112.57	nr	**694.80**
3400 W, 2000 W	1307.57	1640.22	4.00	128.65	nr	**1768.87**
Ceiling void mounted slimline horizontal airside control fan coil unit, 170 mm deep; cooling coil; multi-tapped speed transformer; fine wire mesh filter,damper actuator & fixing kit; includes fixing in position; electrical work elsewhere						
Total cooling load						
1000 W.	815.85	1023.40	3.50	112.57	nr	**1135.97**
3000 W	1088.17	1365.00	4.00	128.65	nr	**1493.65**
Ceiling void mounted slimline horizontal airside control fan coil unit, 170 mm deep; cooilng coil; electric heating coil; multi-tapped speed transformer; fine wire mesh filter, damper actuator & fixing kit; includes fixing in position; electrical work elsewhere						
Total cooling load, heating load						
1000 W, 1000 W.	604.17	757.87	3.50	112.57	nr	**870.44**
3000 W, 2000 W	1195.11	1499.14	4.00	128.65	nr	**1627.79**

38 VENTILATION/AIR CONDITIONING SYSTEMS

Item	Net Price £	Material £	Labour hours	Labour £	Unit	Total rate £
Low level perimeter waterside control fan coil unit; cooling coil; LTHW heating coil; multi-tapped speed transformer; fine wire mesh filter; includes fixing in position; electrical work elsewhere						
Total cooling load, heating load						
1700 W, 1400 W	553.46	694.27	3.50	112.57	nr	806.84
Extra over for standard cabinet	219.08	274.81	1.00	32.17	nr	306.98
2200 W, 1900 W	715.52	897.55	3.50	112.57	nr	1010.12
Extra over for standard cabinet	266.86	334.75	1.00	32.17	nr	366.92
2600 W, 2200 W	715.52	897.55	3.50	112.57	nr	1010.12
Extra over for standard cabinet	273.44	343.00	1.00	32.17	nr	375.17
3900 W, 3200 W	911.77	1143.72	3.50	112.57	nr	1256.29
Extra over for standard cabinet	298.15	374.00	1.00	32.17	nr	406.17
Low level perimeter waterside control fan coil unit; cooling coil; electric heating coil; multi-tapped speed transformer; fine wire mesh filter; includes fixing in position; electrical work elsewhere						
Total cooling load, heating load						
1700 W, 1500 W	658.19	825.63	3.50	112.57	nr	938.20
Extra over for standard cabinet	219.08	274.81	1.00	32.17	nr	306.98
2200 W, 2000 W	822.47	1031.71	3.50	112.57	nr	1144.28
Extra over for standard cabinet	266.86	334.75	1.00	32.17	nr	366.92
2600 W, 2000 W	822.47	1031.71	3.50	112.57	nr	1144.28
Extra over for standard cabinet	273.44	343.00	1.00	32.17	nr	375.17
3800 W, 3000 W	1020.91	1280.63	3.50	112.57	nr	1393.20
Extra over for standard cabinet	298.15	374.00	1.00	32.17	nr	406.17

PART 4

Material Costs/Measured Work Prices – Electrical Installation

Contractual Procedures in the Construction Industry, 7th edition

Allan Ashworth and Srinath Perera

Contractual Procedures in the Construction Industry 7th edition aims to provide students with a comprehensive understanding of the subject, and reinforces the changes that are taking place within the construction industry. The book looks at contract law within the context of construction contracts, it examines the different procurement routes that have evolved over time and the particular aspects relating to design and construction, lean methods of construction and the advantages and disadvantages of PFI/PPP and its variants. It covers the development of partnering, supply chain management, design and build and the way that the clients and professions have adapted to change in the procurement of buildings and engineering projects.

Key features of the new edition include:

- A revised chapter covering the concept of value for money in line with the greater emphasis on added value throughout the industry today.
- A new chapter covering developments in information technology applications (building information modelling, blockchains, data analytics, smart contracts and others) and construction procurement.
- Deeper coverage of the strategies that need to be considered in respect of contract selection.
- Improved discussion of sustainability and the increasing importance of resilience in the built environment.
- Concise descriptions of some the more important construction case laws.

March 2018: 246 x 174 mm: 458 pp
Pb: 978-1-138-69393-7 : £45.99

To Order: Tel: +44 (0) 1235 400524 Fax: +44 (0) 1235 400525
or Post: Taylor and Francis Customer Services,
Bookpoint Ltd, Unit T1, 200 Milton Park, Abingdon, Oxon, OX14 4TA UK
Email: book.orders@tandf.co.uk

For a complete listing of all our titles visit:
www.tandf.co.uk

Taylor & Francis
Taylor & Francis Group

Material Costs/Measured Work Prices

DIRECTIONS

The following explanations are given for each of the column headings and letter codes.

Unit — Prices for each unit are given as singular (i.e. 1 metre, 1 nr) unless stated otherwise.

Net price — Industry tender prices, plus nominal allowance for fixings, waste and applicable trade discounts.

Material cost — Net price plus percentage allowance for overheads (7%), profit (5%) and preliminaries (12%).

Labour norms — In man-hours for each operation.

Labour cost — Labour constant multiplied by the appropriate all-in man-hour cost based on gang rate (See also relevant Rates of Wages Section) plus percentage allowance for overheads, profit and preliminaries.

Measured work — Material cost plus Labour cost.

Price (total rate)

MATERIAL COSTS

The Material Costs given are based at Second Quarter 2018 but exclude any charges in respect of VAT. The average rate of copper during this quarter is US$6,930/UK£5,131 per tonne. Users of the book are advised to register on the SPON's website www.pricebooks.co.uk/updates to receive the free quarterly updates – alerts will then be provided by e-mail as changes arise.

MEASURED WORK PRICES

These prices are intended to apply to new work in the London area. The prices are for reasonable quantities of work and the user should make suitable adjustments if the quantities are especially small or especially large. Adjustments may also be required for locality (e.g. outside London – refer to cost indices in approximate estimating section for details of adjustment factors) and for the market conditions (e.g. volume of work secured or being tendered) at the time of use.

ELECTRICAL INSTALLATIONS

The labour rate has been based on average gang rates per man hour effective from 1 January 2018 including allowances for all other emoluments and expenses. Future changes will be published in the free SPON's quarterly update by registering on their website. To this rate, has been added 12% and 7% to cover site and head office overheads and preliminary items together with a further 5% for profit, resulting in an inclusive rate of £35.95 per man hour. The rate has been calculated on a working year of 2,025 hours; a detailed build-up of the rate is given at the end of these directions.

DIRECTIONS

In calculating the 'Measured Work Prices' the following assumptions have been made:

(a)   That the work is carried out as a subcontract under the Standard Form of Building Contract.
(b)   That, unless otherwise stated, the work is being carried out in open areas at a height which would not require more than simple scaffolding.
(c)   That the building in which the work is being carried out is no more than six storey's high.

Where these assumptions are not valid, as for example where work is carried out in ducts and similar confined spaces or in multi-storey structures when additional time is needed to get to and from upper floors, then an appropriate adjustment must be made to the prices. Such adjustment will normally be to the labour element only.

DIRECTIONS

LABOUR RATE – ELECTRICAL

The annual cost of a notional eleven man gang

	TECHNICIAN 1 NR	APPROVED ELECTRICIANS 4 NR	ELECTRICIANS 4 NR	LABOURERS 2 NR	SUB-TOTALS
Hourly Rate from 1 January 2018	20.57	18.28	16.45	13.53	
Working hours per annum per man	1,687.50	1,687.50	1,687.50	1,687.50	
x Hourly rate × nr of men = £ per annum	34,711.88	123,390.00	113,805.00	45,663.75	317,570.63
Overtime Rate	30.86	27.42	25.29	20.30	
Overtime hours per annum per man	337.50	337.50	337.50	337.50	
x Hourly rate × nr of men = £ per annum	10,162.13	36,112.52	33,318.00	13,699.13	93,291.78
Total	44,874.01	159,502.52	147,123.00	59,362.88	410,862.40
Incentive schemes (insert percentage) 0.00%	0.00	0.00	0.00	0.00	0.00
Daily Travel Allowance (15–20 miles each way) effective from 1 January 2018	6.72	6.72	6.72	6.72	
Days per annum per man	225.00	225.00	225.00	225.00	
x nr of men = £ per annum	1,512.00	6,048.00	6,048.00	3,024.00	16,632.00
Daily Travel Allowance (15–20 miles each way) effective from 1 January 2018	8.40	8.40	8.40	8.40	
Days per annum per man	225.00	225.00	225.00	225.00	
x nr of men = £ per annum	1,890.00	7,560.00	7,560.00	3,780.00	20,790.00
JIB Pension Scheme @ 3.0%	1,593.46	5,694.41	5,273.98	2,158.26	14,720.11
JIB combined benefits scheme (nr of weeks per man)	52.00	52.00	52.00	52.00	
Benefit Credit effective from 1 January 2018	70.65	64.21	60.08	50.82	
x nr of men = £ per annum	3,673.80	13,355.68	12,496.64	5,285.28	34,811.40
Holiday Top-up Funding	60.21	53.77	49.91	41.06	
x nr of men @ 7.5 hrs per day = £ per annum	3,130.73	11,183.12	10,380.76	4,269.85	28,964.46
National Insurance Contributions:					
Annual gross pay (subject to NI) each	55,080.53	197,649.32	183,608.40	75,722.01	
% of NI Contributions	13.80	13.80	13.80	13.80	
£ Contributions/annum	6,260.92	21,914.85	19,977.20	7,769.26	55,922.23

	SUBTOTAL		582,720.59
TRAINING (INCLUDING ANY TRADE REGISTRATIONS) – SAY		1.00%	5,827.21
SEVERANCE PAY AND SUNDRY COSTS – SAY		1.50%	8,828.22
EMPLOYER'S LIABILITY AND THIRD PARTY INSURANCE – SAY		2.00%	11,947.52
ANNUAL COST OF NOTIONAL GANG			609,323.53
MEN ACTUALLY WORKING = 10.5	THEREFORE ANNUAL COST PER PRODUCTIVE MAN		58,030.81
AVERAGE NR OF HOURS WORKED PER MAN = 2025			28.66
THEREFORE ALL-IN MAN HOUR			28.66
PRELIMINARY ITEMS – SAY		12%	3.44
SITE AND HEAD OFFICE OVERHEADS AND PROFIT (7% & 5% RESPECTIVELY) – SAY		12%	3.85
THEREFORE INCLUSIVE MAN-HOUR			35.95

DIRECTIONS

Notes:

(1)   Hourly wage rates are those effective from 1 January 2018.

(2)   The following assumptions have been made in the above calculations:

          (a)   Hourly rates are based on London rate and reporting at job using own transport.

          (b)   The working week of 37.5 hours is made up of 7.5 hours Monday to Friday.

          (c)   Five days in the year are lost through sickness or similar reason.

          (d)   A working year of 2,025 hours.

(3)   The incentive scheme addition of 0% is intended to reflect bonus schemes typically in use.

(4)   National insurance contributions are those effective from 6 April 2018.

(6)   Weekly JIB Combined Benefit Credit Scheme are those effective from 1 January 2018 – Contact JIB for ECI (0330 221 0240).

(7)   Overtime is paid after 37.5 hours.

ELECTRICAL SUPPLY/POWER/LIGHTING

Item	Net Price £	Material £	Labour hours	Labour £	Unit	Total rate £
ELECTRICAL GENERATION PLANT						
STANDBY GENERATORS						
Standby diesel generating sets; supply and installation; fixing to base; all supports and fixings; all necessary connections to equipment						
Three phase, 400 Volt, four wire 50 Hz packaged standby diesel generating set, complete with radio and television suppressors, daily service fuel tank and associated piping, 4 metres of exhaust pipe and primary exhaust silencer, control panel, mains failure relay, starting battery with charger, all internal wiring, interconnections, earthing and labels. Rated for standby duty; including UK delivery, installation and commissioning						
60 kVA	18336.09	23000.79	100.00	3595.11	nr	**26595.90**
100 kVA	25705.33	32244.77	100.00	3595.11	nr	**35839.88**
150 kVA	29398.20	36877.10	100.00	3595.11	nr	**40472.21**
315 kVA	53003.23	66487.25	120.00	4314.13	nr	**70801.38**
500 kVA	62474.47	78367.98	120.00	4314.13	nr	**82682.11**
750 kVA	106315.63	133362.33	140.00	5033.16	nr	**138395.49**
1000 kVA	141302.58	177249.96	140.00	5033.16	nr	**182283.12**
1500 kVA	166299.48	208606.07	170.00	6111.68	nr	**214717.75**
2000 kVA	256163.39	321331.36	170.00	6111.68	nr	**327443.04**
2500 kVA	350722.28	439946.02	210.00	7549.73	nr	**447495.75**
Extra for residential silencer; peformance 75dBA at 1 m; including connection to exhaust pipe						
60 kVA	984.47	1234.92	10.00	359.51	nr	**1594.43**
100 kVA	1176.35	1475.61	10.00	359.51	nr	**1835.12**
150 kVA	1349.03	1692.22	10.00	359.51	nr	**2051.73**
315 kVA	2624.26	3291.87	15.00	539.27	nr	**3831.14**
500 kVA	3356.35	4210.20	15.00	539.27	nr	**4749.47**
750 kVA	5000.54	6272.67	20.00	719.02	nr	**6991.69**
1000 kVA	6734.82	8448.16	20.00	719.02	nr	**9167.18**
1500 kVA	10728.75	13458.14	20.00	719.02	nr	**14177.16**
2000 kVA	11620.24	14576.43	30.00	1078.54	nr	**15654.97**
2500 kVA	13286.60	16666.71	30.00	1078.54	nr	**17745.25**

ELECTRICAL SUPPLY/POWER/LIGHTING

Item	Net Price £	Material £	Labour hours	Labour £	Unit	Total rate £
ELECTRICAL GENERATION PLANT – cont						
Standby diesel generating sets – cont						
Synchronization panel for paralleling generators – not generators to mains; including interconnecting cables; commissioning and testing; fixing to backgrounds						
2 × 60 kVA	8200.43	10286.62	80.00	2876.09	nr	13162.71
2 × 100 kVA	8815.91	11058.68	80.00	2876.09	nr	13934.77
2 × 150 kVA	11090.36	13911.74	80.00	2876.09	nr	16787.83
2 × 315 kVA	18853.92	23650.36	80.00	2876.09	nr	26526.45
2 × 500 kVA	27219.67	34144.35	80.00	2876.09	nr	37020.44
2 × 750 kVA	30716.21	38530.42	80.00	2876.09	nr	41406.51
2 × 1000 kVA	33104.32	41526.06	120.00	4314.13	nr	45840.19
2 × 1500 kVA	39189.73	49159.60	120.00	4314.13	nr	53473.73
2 × 2000 kVA	42461.92	53264.23	120.00	4314.13	nr	57578.36
2 × 2500 kVA	49503.74	62097.49	120.00	4314.13	nr	66411.62
Prefabricated drop-over acoustic housing; performance 85dBA at 1 m over the range from 60 kVA to 315 kVA, 75 dBA from 500 kVA to 2500 kVA.						
100 kVA	3664.80	4597.13	7.00	251.65	nr	4848.78
150 kVA	5951.08	7465.04	15.00	539.27	nr	8004.31
315 kVA	12697.67	15927.96	25.00	898.78	nr	16826.74
500 kVA	20971.92	26307.18	40.00	1438.05	nr	27745.23
750 kVA	33049.72	41457.57	40.00	1438.05	nr	42895.62
1000 kVA	41812.50	52449.60	40.00	1438.05	nr	53887.65
1500 kVA	41812.50	52449.60	40.00	1438.05	nr	53887.65
2000 kVA	82938.66	104038.26	60.00	2157.06	nr	106195.32
2500 kVA	99690.82	125052.17	70.00	2516.57	nr	127568.74
COMBINED HEAT AND POWER (CHP) UNITS						
Gas fired engine; acoustic enclosure complete with exhaust fan and attenuators; exhaust gas attenuation; includes 6 m long pipe connections; dry air cooler for secondary water circuit to reject excess heat; controls and panel; commissioning						
Electrical output; heat output						
20 kW; 43 kW	43303.00	54319.28	–	–	nr	54319.28
43 kW; 65 kW	84245.00	105676.93	–	–	nr	105676.93
50 kW; 81 kW	86424.00	108410.27	–	–	nr	108410.27
70 kW; 114 kW	95294.00	119536.79	–	–	nr	119536.79
104 kW, 142 kW	125382.00	157279.18	–	–	nr	157279.18
133 kW, 193 kW	126108.00	158189.88	–	–	nr	158189.88
210 kW, 253 kW	173211.00	217275.88	–	–	nr	217275.88
263 kW, 375 kW	198681.00	249225.45	–	–	nr	249225.45
356 kW, 426 kW	236135.00	296207.74	–	–	nr	296207.74
532 kW, 665 kW	277739.00	348395.80	–	–	nr	348395.80

ELECTRICAL SUPPLY/POWER/LIGHTING

Item	Net Price £	Material £	Labour hours	Labour £	Unit	Total rate £
Heat dump; emergency cooling						
20 kW; 43 kW	3490.00	4377.86	–	–	nr	**4377.86**
43 kW; 65 kW	3632.00	4555.98	–	–	nr	**4555.98**
50 kW; 81 kW	3908.00	4902.20	–	–	nr	**4902.20**
70 kW; 114 kW	4937.00	6192.97	–	–	nr	**6192.97**
104 kW, 142 kW	5454.00	6841.50	–	–	nr	**6841.50**
133 kW, 193 kW	6482.00	8131.02	–	–	nr	**8131.02**
210 kW, 253 kW	9290.00	11653.38	–	–	nr	**11653.38**
263 kW, 375 kW	13765.00	17266.82	–	–	nr	**17266.82**
356 kW, 426 kW	15304.00	19197.34	–	–	nr	**19197.34**
532 kW, 665 kW	19323.00	24238.77	–	–	nr	**24238.77**
Note: The costs detailed are based on a specialist subcontract package, as part of the M&E contract works, and include installation						
Upgraded catalytic converter						
43 kW; 65 kW	1121.00	1406.18	–	–	nr	**1406.18**
50 kW; 81 kW	1121.00	1406.18	–	–	nr	**1406.18**
70 kW; 114 kW	2454.00	3078.30	–	–	nr	**3078.30**
133 kW, 193 kW	3328.00	4174.64	–	–	nr	**4174.64**
263 kW, 375 kW	5929.00	7437.34	–	–	nr	**7437.34**
Flexible connections						
20 kW; 43 kW	472.00	592.08	–	–	nr	**592.08**
43 kW; 65 kW	636.00	797.80	–	–	nr	**797.80**
50 kW; 81 kW	636.00	797.80	–	–	nr	**797.80**
70 kW; 114 kW	815.00	1022.34	–	–	nr	**1022.34**
104 kW, 142 kW	1098.00	1377.33	–	–	nr	**1377.33**
133 kW, 193 kW	1154.00	1447.58	–	–	nr	**1447.58**
210 kW, 253 kW	1602.00	2009.55	–	–	nr	**2009.55**
263 kW, 375 kW	1554.00	1949.34	–	–	nr	**1949.34**
356 kW, 426 kW	1753.00	2198.96	–	–	nr	**2198.96**
532 kW, 665 kW	2260.00	2834.94	–	–	nr	**2834.94**

ELECTRICAL SUPPLY/POWER/LIGHTING

Item	Net Price £	Material £	Labour hours	Labour £	Unit	Total rate £
HV SUPPLY						
Cable; 6350/11000 volts, 3 core, XLPE; stranded copper conductors; steel wire armoured; LSOH to BS 7835						
Laid in trench/duct including marker tape (cable tiles measured elsewhere)						
95 mm²	27.02	33.89	0.23	8.27	m	42.16
120 mm²	35.75	44.84	0.23	8.27	m	53.11
150 mm²	40.28	50.52	0.25	8.99	m	59.51
185 mm²	47.52	59.61	0.25	8.99	m	68.60
240 mm²	65.62	82.31	0.27	9.71	m	92.02
300 mm²	78.54	98.52	0.29	10.43	m	108.95
Pile tape; ES 1–12–23; 1 m	35.83	44.95	0.01	0.36	m	45.31
Clipped direct to backgrounds including cleats						
95 mm²	27.02	33.89	0.47	16.90	m	50.79
120 mm²	35.76	44.86	0.50	17.98	m	62.84
150 mm²	40.28	50.52	0.53	19.05	m	69.57
185 mm²	47.52	59.61	0.55	19.77	m	79.38
240 mm²	65.62	82.31	0.60	21.57	m	103.88
300 mm²	78.54	98.52	0.68	24.45	m	122.97
Terminations for above cables, including heat-shrink kit and glanding off						
95 × 3 Core XLPE SWA LSF 11 kV Termination	724.10	908.31	4.75	170.77	nr	1079.08
120 × 3 Core XLPE SWA LSF 11 kV Termination	756.08	948.43	5.00	179.76	nr	1128.19
150 × 3 Core XLPE SWA LSF 11 kV Termination	789.44	990.27	6.00	215.71	nr	1205.98
185 × 3 Core XLPE SWA LSF 11 kV Termination	801.26	1005.10	6.90	248.06	nr	1253.16
240 × 3 Core XLPE SWA LSF 11 kV Termination	835.45	1047.98	7.50	269.63	nr	1317.61
300 × 3 Core XLPE SWA LSF 11 kV Termination	893.33	1120.59	8.75	314.57	nr	1435.16
Cable tiles; single width; laid in trench above cables on prepared sand bed (cost of excavation excluded); reinforced concrete covers; concave/convex ends						
914 × 152 × 63/38 mm	35.83	44.95	0.11	3.95	m	48.90
914 × 229 × 63/38 mm	38.04	47.71	0.11	3.95	m	51.66
914 × 305 × 63/38 mm	40.24	50.48	0.11	3.95	m	54.43

ELECTRICAL SUPPLY/POWER/LIGHTING

Item	Net Price £	Material £	Labour hours	Labour £	Unit	Total rate £
HV SWITCHGEAR AND TRANSFORMERS						
HV circuit breakers; installed on prepared foundations including all supports, fixings and inter panel connections where relevant. Excludes main and multi-core cabling and heat shrink cable termination kits						
Three phase 11 kV, 630 amp, Air or SF6 insulated, with fixed pattern vacuum or SF6 circuit breaker panels; hand charged spring closing operation; prospective fault level up to 25 kA for 3 seconds. Feeders include ammeter with selector switch,VIP relays, overcurrent and earth fault relays with necessary current relays with necessary current transformers; incomers include 3 phase VT, voltmeter and phase selector switch; Includes IDMT overcurrent and earth fault relays/CTs						
Single panel with cable chamber	26820.00	33643.01	31.70	1139.64	nr	**34782.65**
Three panel with one incomer and two feeders; with cable chambers	49555.00	62161.79	67.83	2438.56	nr	**64600.35**
Five panel with two incoming, two feeders and a bus section; with cable chambers	88775.00	111359.36	99.17	3565.27	nr	**114924.63**
Tripping Batteries						
Battery chargers; switchgear tripping and closing; double wound transfomer and earth screen; including fixing to background, commissioning and testing						
Valve regulated lead acid battery NGTS 3.12.2; BS 6290 Part2; TPS 9/3; IEEE485						
30 volt; 19 Ah; 3 A	3021.45	3790.10	6.50	233.68	nr	**4023.78**
40 volt; 29 Ah; 3 A	4840.59	6072.04	8.50	305.58	nr	**6377.62**
110 volt; 19 Ah; 3 A	4937.89	6194.09	6.50	233.68	nr	**6427.77**
110 volt; 29 Ah; 3 A	5387.81	6758.47	8.50	305.58	nr	**7064.05**
110 volt; 38 Ah; 3 A	5757.37	7222.04	10.00	359.51	nr	**7581.55**
Step down transformers; 11/0.415 kV, Dyn 11, 50 Hz. Complete with lifting lugs, mounting skids, provisions for wheels, undrilled gland plates to air-filled cable boxes, off load tapping facility, including UK delivery						
Oil-filled in free breathing ventialted steel tank						
500 kVA	11162.15	14001.80	30.00	1078.54	nr	**15080.34**
800 kVA	12711.93	15945.84	30.00	1078.54	nr	**17024.38**
1000 kVA	14407.32	18072.54	30.00	1078.54	nr	**19151.08**
1250 kVA	17506.83	21960.57	35.00	1258.29	nr	**23218.86**
1500 kVA	20745.72	26023.44	35.00	1258.29	nr	**27281.73**
2000 kVA	26944.73	33799.47	35.00	1258.29	nr	**35057.76**

ELECTRICAL SUPPLY/POWER/LIGHTING

Item	Net Price £	Material £	Labour hours	Labour £	Unit	Total rate £
HV SUPPLY – cont						
Step down transformers – cont						
MIDEL-filled in gasket-sealed steel tank						
500 kVA	14784.43	18545.59	30.00	1078.54	nr	**19624.13**
800 kVA	16850.78	21137.61	30.00	1078.54	nr	**22216.15**
1000 kVA	19185.70	24066.54	30.00	1078.54	nr	**25145.08**
1250 kVA	23189.21	29088.55	35.00	1258.29	nr	**30346.84**
1500 kVA	27590.46	34609.48	35.00	1258.29	nr	**35867.77**
2000 kVA	35726.64	44815.50	40.00	1438.05	nr	**46253.55**
Extra for						
Fluid temperature indicator with 2 N/O contacts	414.74	520.25	2.00	71.90	nr	**592.15**
Winding temperature indicator with 2 N/O contacts	1016.13	1274.64	2.00	71.90	nr	**1346.54**
Dehydrating breather	103.69	130.07	2.00	71.90	nr	**201.97**
Plain rollers	311.06	390.20	2.00	71.90	nr	**462.10**
Pressure relief device with 1 N/O contact	622.10	780.36	2.00	71.90	nr	**852.26**
Step down transformers; 11/0.415 kV, Dyn 11, 50 Hz. Complete with lifting lugs, mounting skids, provisions for wheels, undrilled gland plates to air-filled cable boxes, off load tapping facility, including UK delivery						
Cast resin type in ventilated steel encloure, AN – Air Natural including winding temperture indicator with 2 N/O contacts						
500 kVA	15817.59	19841.58	40.00	1438.05	nr	**21279.63**
800 kVA	18400.50	23081.59	40.00	1438.05	nr	**24519.64**
1000 kVA	22259.37	27922.15	40.00	1438.05	nr	**29360.20**
1250 kVA	24093.23	30222.55	45.00	1617.80	nr	**31840.35**
1600 kVA	27332.18	34285.48	45.00	1617.80	nr	**35903.28**
2000 kVA	30948.29	38821.53	50.00	1797.56	nr	**40619.09**
Cast resin type in ventilated steel enclosure with temperature controlled fans to achieve 40% increase to AN/AF rating. Includes winding temperature indicator with 2 N/O contacts						
500/700 kVA	17367.32	21785.57	42.00	1509.95	nr	**23295.52**
800/1120 kVA	20208.55	25349.61	42.00	1509.95	nr	**26859.56**
1000/1400 kVA	25003.44	31364.31	42.00	1509.95	nr	**32874.26**
12501750 kVA	26036.59	32660.30	47.00	1689.70	nr	**34350.00**
1600/2240 kVA	29414.99	36898.16	47.00	1689.70	nr	**38587.86**
2000/2800 kVA	33547.65	42082.17	52.00	1869.46	nr	**43951.63**

ELECTRICAL SUPPLY/POWER/LIGHTING

Item	Net Price £	Material £	Labour hours	Labour £	Unit	Total rate £
LV DISTRIBUTION: CONDUIT AND CABLE TRUNKING						
Heavy gauge, screwed drawn steel; surface fixed on saddles to backgrounds,with standard pattern boxes and fittings including all fixings and supports. (forming holes, conduit entry, draw wires etc. and components for earth continuity are included)						
Black enamelled						
20 mm dia.	4.07	5.11	0.49	17.61	m	**22.72**
25 mm dia.	5.91	7.41	0.56	20.14	m	**27.55**
32 mm dia.	12.08	15.15	0.64	23.00	m	**38.15**
38 mm dia.	14.37	18.02	0.73	26.24	m	**44.26**
50 mm dia.	22.80	28.60	1.04	37.40	m	**66.00**
Galvanized						
20 mm dia.	3.78	4.74	0.49	17.61	m	**22.35**
25 mm dia.	5.17	6.48	0.56	20.14	m	**26.62**
32 mm dia.	9.90	12.42	0.64	23.00	m	**35.42**
38 mm dia.	9.90	12.42	0.73	26.24	m	**38.66**
50 mm dia.	24.66	30.93	1.04	37.40	m	**68.33**
High impact PVC; surface fixed on saddles to backgrounds; with standard pattern boxes and fittings; including all fixings and supports						
Light gauge						
16 mm dia.	0.66	0.83	0.27	9.71	m	**10.54**
20 mm dia.	0.75	0.94	0.28	10.06	m	**11.00**
25 mm dia.	1.02	1.28	0.33	11.87	m	**13.15**
32 mm dia.	1.63	2.05	0.38	13.66	m	**15.71**
38 mm dia.	2.17	2.72	0.44	15.81	m	**18.53**
50 mm dia.	3.58	4.49	0.48	17.26	m	**21.75**
Heavy gauge						
16 mm dia.	0.93	1.16	0.27	9.71	m	**10.87**
20 mm dia.	1.05	1.32	0.28	10.06	m	**11.38**
25 mm dia.	1.43	1.79	0.33	11.87	m	**13.66**
32 mm dia.	2.27	2.84	0.38	13.66	m	**16.50**
38 mm dia.	3.01	3.77	0.44	15.81	m	**19.58**
50 mm dia.	4.98	6.25	0.48	17.26	m	**23.51**
Flexible conduits; including adaptors and locknuts (for connections to equipment)						
Metallic, PVC covered conduit; not exceeding 1 m long; including zinc plated mild steel adaptors, lock nuts and earth conductor						
16 mm dia.	10.46	13.13	0.42	15.10	nr	**28.23**
20 mm dia.	9.85	12.35	0.46	16.53	nr	**28.88**
25 mm dia.	15.51	19.45	0.43	15.46	nr	**34.91**
32 mm dia.	24.02	30.13	0.51	18.33	nr	**48.46**
38 mm dia.	30.35	38.07	0.56	20.14	nr	**58.21**
50 mm dia.	105.92	132.87	0.82	29.48	nr	**162.35**

ELECTRICAL SUPPLY/POWER/LIGHTING

Item	Net Price £	Material £	Labour hours	Labour £	Unit	Total rate £
LV DISTRIBUTION: CONDUIT AND CABLE TRUNKING – cont						
Flexible conduits – cont						
PVC conduit; not exceeding 1 m long; including nylon adaptors, lock nuts						
16 mm dia.	5.54	6.94	0.46	16.53	nr	23.47
20 mm dia.	5.54	6.94	0.48	17.26	nr	24.20
25 mm dia.	6.95	8.71	0.50	17.98	nr	26.69
32 mm.dia.	10.09	12.66	0.58	20.84	nr	33.50
PVC adaptable boxes; fixed to backgrounds; including all supports and fixings (cutting and connecting conduit to boxes is included)						
Square pattern						
75 × 75 × 53 mm	2.71	3.40	0.69	24.81	nr	28.21
100 × 100 × 75 mm	4.58	5.75	0.71	25.52	nr	31.27
150 × 150 × 75 mm	5.86	7.35	0.80	28.76	nr	36.11
Terminal strips to be fixed in metal or polythene adaptable boxes						
20 amp high density polythene						
2 way	1.11	1.39	0.23	8.27	nr	9.66
3 way	1.40	1.76	0.23	8.27	nr	10.03
4 way	1.86	2.33	0.23	8.27	nr	10.60
5 way	2.33	2.92	0.23	8.27	nr	11.19
6 way	2.79	3.49	0.25	8.99	nr	12.48
7 way	3.25	4.08	0.25	8.99	nr	13.07
8 way	3.72	4.67	0.29	10.43	nr	15.10
9 way	4.19	5.25	0.30	10.79	nr	16.04
10 way	4.65	5.84	0.34	12.22	nr	18.06
11 way	5.12	6.42	0.34	12.22	nr	18.64
13 way	6.04	7.57	0.37	13.29	nr	20.86
14 way	6.51	8.16	0.37	13.29	nr	21.45
15 way	6.98	8.76	0.39	14.02	nr	22.78
16 way	7.44	9.33	0.45	16.18	nr	25.51
18 way	8.37	10.49	0.45	16.18	nr	26.67

ELECTRICAL SUPPLY/POWER/LIGHTING

Item	Net Price £	Material £	Labour hours	Labour £	Unit	Total rate £
LV DISTRIBUTION: TRUNKING						
Galvanized steel trunking; fixed to backgrounds; jointed with standard connectors (including plates for air gap between trunking and background); earth continuity straps included						
Single compartment						
50 × 50 mm	7.81	9.80	0.39	14.02	m	23.82
75 × 50 mm	11.91	14.94	0.44	15.81	m	30.75
75 × 75 mm	12.70	15.93	0.47	16.90	m	32.83
100 × 50 mm	14.81	18.58	0.50	17.98	m	36.56
100 × 75 mm	15.99	20.06	0.57	20.50	m	40.56
100 × 100 mm	16.23	20.36	0.62	22.29	m	42.65
150 × 50 mm	21.56	27.05	0.78	28.03	m	55.08
150 × 100 mm	26.13	32.78	0.78	28.03	m	60.81
150 × 150 mm	27.90	35.00	0.86	30.92	m	65.92
225 × 75 mm	35.37	44.36	0.88	31.64	m	76.00
225 × 150 mm	44.20	55.44	0.84	30.20	m	85.64
225 × 225 mm	52.76	66.18	0.99	35.58	m	101.76
300 × 75 mm	36.36	45.61	0.96	34.51	m	80.12
300 × 100 mm	41.87	52.52	0.99	35.58	m	88.10
300 × 150 mm	32.03	40.17	0.99	35.58	m	75.75
300 × 225 mm	27.26	34.19	1.09	39.19	m	73.38
300 × 300 mm	69.50	87.18	1.16	41.71	m	128.89
Double compartment						
100 × 50 mm	10.43	13.08	0.54	19.42	m	32.50
100 × 75 mm	11.24	14.10	0.62	22.29	m	36.39
100 × 100 mm	23.24	29.15	0.66	23.73	m	52.88
150 × 50 mm	15.83	19.86	0.70	25.17	m	45.03
150 × 100 mm	19.63	24.63	0.83	29.84	m	54.47
150 × 150 mm	35.79	44.89	0.92	33.07	m	77.96
Triple compartment						
150 × 50 mm	18.23	22.87	0.79	28.40	m	51.27
150 × 100 mm	18.72	23.49	0.78	28.03	m	51.52
150 × 150 mm	43.00	53.94	1.01	36.31	m	90.25
Galvanized steel trunking fittings; cutting and jointing trunking to fittings is included						
Stop end						
50 × 50 mm	2.51	3.15	0.19	6.83	nr	9.98
75 × 50 mm	2.65	3.33	0.20	7.19	nr	10.52
75 × 75 mm	2.65	3.33	0.21	7.55	nr	10.88
100 × 50 mm	2.76	3.46	0.31	11.14	nr	14.60
100 × 75 mm	2.98	3.74	0.27	9.71	nr	13.45
100 × 100 mm	2.92	3.66	0.27	9.71	nr	13.37
150 × 50 mm	3.25	4.08	0.28	10.06	nr	14.14
150 × 100 mm	3.54	4.44	0.30	10.79	nr	15.23
150 × 150 mm	3.65	4.58	0.32	11.50	nr	16.08
225 × 75 mm	3.77	4.73	0.35	12.58	nr	17.31
225 × 150 mm	4.19	5.25	0.37	13.29	nr	18.54

ELECTRICAL SUPPLY/POWER/LIGHTING

Item	Net Price £	Material £	Labour hours	Labour £	Unit	Total rate £
LV DISTRIBUTION: TRUNKING – cont						
Galvanized steel trunking fittings – cont						
Stop end – cont						
225 × 225 mm	6.50	8.15	0.38	13.66	nr	21.81
300 × 75 mm	6.54	8.20	0.42	15.10	nr	23.30
300 × 100 mm	5.60	7.02	0.42	15.10	nr	22.12
300 × 150 mm	6.54	8.20	0.43	15.46	nr	23.66
300 × 225 mm	7.85	9.84	0.45	16.18	nr	26.02
300 × 300 mm	8.24	10.34	0.48	17.26	nr	27.60
Flanged connector						
50 × 50 mm	2.65	3.33	0.19	6.83	nr	10.16
75 × 50 mm	4.08	5.12	0.20	7.19	nr	12.31
75 × 75 mm	3.29	4.12	0.21	7.55	nr	11.67
100 × 50 mm	4.23	5.31	0.26	9.34	nr	14.65
100 × 75 mm	4.37	5.48	0.27	9.71	nr	15.19
100 × 100 mm	4.23	5.31	0.27	9.71	nr	15.02
150 × 50 mm	4.28	5.36	0.28	10.06	nr	15.42
150 × 100 mm	4.57	5.73	0.30	10.79	nr	16.52
150 × 150 mm	4.63	5.81	0.32	11.50	nr	17.31
225 × 75 mm	3.66	4.59	0.35	12.58	nr	17.17
225 × 150 mm	4.47	5.61	0.37	13.29	nr	18.90
225 × 225 mm	6.56	8.23	0.38	13.66	nr	21.89
300 × 75 mm	5.50	6.90	0.42	15.10	nr	22.00
300 × 100 mm	6.34	7.95	0.42	15.10	nr	23.05
300 × 150 mm	7.19	9.02	0.43	15.46	nr	24.48
300 × 225 mm	7.19	9.02	0.45	16.18	nr	25.20
300 × 300 mm	8.24	10.34	0.48	17.26	nr	27.60
Bends 90°; single compartment						
50 × 50 mm	11.02	13.82	0.42	15.10	nr	28.92
75 × 50 mm	13.96	17.52	0.45	16.18	nr	33.70
75 × 75 mm	13.32	16.71	0.48	17.26	nr	33.97
100 × 50 mm	14.27	17.90	0.53	19.05	nr	36.95
100 × 75 mm	14.63	18.36	0.56	20.14	nr	38.50
100 × 100 mm	14.63	18.36	0.58	20.84	nr	39.20
150 × 50 mm	18.10	22.70	0.64	23.00	nr	45.70
150 × 100 mm	23.09	28.96	0.91	32.72	nr	61.68
150 × 150 mm	22.03	27.63	0.89	32.00	nr	59.63
225 × 75 mm	34.31	43.04	0.76	27.32	nr	70.36
225 × 150 mm	42.64	53.49	0.82	29.48	nr	82.97
225 × 225 mm	49.22	61.75	0.83	29.84	nr	91.59
300 × 75 mm	45.80	57.46	0.85	30.55	nr	88.01
300 × 100 mm	46.34	58.13	0.90	32.35	nr	90.48
300 × 150 mm	50.98	63.95	0.96	34.51	nr	98.46
300 × 225 mm	56.07	70.34	0.98	35.24	nr	105.58
300 × 300 mm	57.33	71.92	1.06	38.11	nr	110.03
Bends 90°; double compartment						
100 × 50 mm	13.24	16.61	0.53	19.05	nr	35.66
100 × 75 mm	13.94	17.48	0.56	20.14	nr	37.62
100 × 100 mm	22.32	28.00	0.58	20.84	nr	48.84
150 × 50 mm	14.55	18.26	0.65	23.37	nr	41.63
150 × 100 mm	15.32	19.22	0.69	24.81	nr	44.03
150 × 150 mm	38.27	48.00	0.73	26.24	nr	74.24

ELECTRICAL SUPPLY/POWER/LIGHTING

Item	Net Price £	Material £	Labour hours	Labour £	Unit	Total rate £
Bends 90°; triple compartment						
150 × 50 mm	19.13	24.00	0.68	24.45	nr	**48.45**
150 × 100 mm	26.74	33.54	0.73	26.24	nr	**59.78**
150 × 150 mm	45.75	57.39	0.77	27.69	nr	**85.08**
Tees; single compartment						
50 × 50 mm	13.14	16.49	0.56	20.14	nr	**36.63**
75 × 50 mm	15.63	19.61	0.57	20.50	nr	**40.11**
75 × 75 mm	15.12	18.96	0.60	21.57	nr	**40.53**
100 × 50 mm	18.30	22.96	0.65	23.37	nr	**46.33**
100 × 75 mm	18.41	23.09	0.72	25.89	nr	**48.98**
100 × 100 mm	17.66	22.15	0.71	25.52	nr	**47.67**
150 × 50 mm	20.93	26.25	0.82	29.48	nr	**55.73**
150 × 100 mm	26.00	32.61	0.84	30.20	nr	**62.81**
150 × 150 mm	25.22	31.64	0.91	32.72	nr	**64.36**
225 × 75 mm	44.79	56.18	0.94	33.79	nr	**89.97**
225 × 150 mm	60.50	75.89	1.01	36.31	nr	**112.20**
225 × 225 mm	69.41	87.07	1.02	36.67	nr	**123.74**
300 × 75 mm	64.65	81.10	1.07	38.47	nr	**119.57**
300 × 100 mm	66.98	84.02	1.07	38.47	nr	**122.49**
300 × 150 mm	72.16	90.52	1.14	40.98	nr	**131.50**
300 × 225 mm	78.15	98.03	1.19	42.78	nr	**140.81**
300 × 300 mm	82.88	103.97	1.26	45.29	nr	**149.26**
Tees; double compartment						
100 × 50 mm	12.85	16.12	0.65	23.37	nr	**39.49**
100 × 75 mm	30.28	37.98	0.71	25.52	nr	**63.50**
100 × 100 mm	30.25	37.95	0.72	25.89	nr	**63.84**
150 × 50 mm	16.89	21.19	0.82	29.48	nr	**50.67**
150 × 100 mm	17.69	22.19	0.85	30.55	nr	**52.74**
150 × 150 mm	51.28	64.32	0.91	32.72	nr	**97.04**
Tees; triple compartment						
150 × 50 mm	19.32	24.24	0.87	31.27	nr	**55.51**
150 × 100 mm	21.24	26.64	0.89	32.00	nr	**58.64**
150 × 150 mm	60.75	76.20	0.96	34.51	nr	**110.71**
Crossovers; single compartment						
50 × 50 mm	16.99	21.31	0.65	23.37	nr	**44.68**
75 × 50 mm	23.02	28.87	0.66	23.73	nr	**52.60**
75 × 75 mm	23.55	29.55	0.69	24.81	nr	**54.36**
100 × 50 mm	29.09	36.49	0.74	26.61	nr	**63.10**
100 × 75 mm	21.44	26.89	0.80	28.76	nr	**55.65**
100 × 100 mm	29.32	36.78	0.81	29.12	nr	**65.90**
150 × 50 mm	34.73	43.57	0.91	32.72	nr	**76.29**
150 × 100 mm	35.23	44.20	0.94	33.79	nr	**77.99**
150 × 150 mm	33.76	42.35	0.99	35.58	nr	**77.93**
225 × 75 mm	61.83	77.56	1.01	36.31	nr	**113.87**
225 × 150 mm	67.49	84.66	1.08	38.82	nr	**123.48**
225 × 225 mm	91.82	115.18	1.09	39.19	nr	**154.37**
300 × 75 mm	86.45	108.44	1.14	40.98	nr	**149.42**
300 × 100 mm	89.05	111.71	1.16	41.71	nr	**153.42**
300 × 150 mm	96.15	120.61	1.19	42.78	nr	**163.39**
300 × 225 mm	101.73	127.61	1.21	43.50	nr	**171.11**
300 × 300 mm	108.45	136.04	1.29	46.38	nr	**182.42**

ELECTRICAL SUPPLY/POWER/LIGHTING

Item	Net Price £	Material £	Labour hours	Labour £	Unit	Total rate £
LV DISTRIBUTION: TRUNKING – cont						
Galvanized steel trunking fittings – cont						
Stop end – cont						
Crossovers; double compartment						
100 × 50 mm	17.07	21.41	0.74	26.61	nr	48.02
100 × 75 mm	17.91	22.47	0.80	28.76	nr	51.23
100 × 100 mm	34.09	42.76	0.81	29.12	nr	71.88
150 × 50 mm	20.22	25.37	0.86	30.92	nr	56.29
150 × 100 mm	20.29	25.45	0.94	33.79	nr	59.24
150 × 150 mm	30.86	38.71	1.00	35.95	nr	74.66
Crossovers; triple compartment						
150 × 50 mm	22.13	27.76	0.97	34.88	nr	62.64
150 × 100 mm	24.34	30.53	0.99	35.58	nr	66.11
150 × 150 mm	33.91	42.54	1.06	38.11	nr	80.65
Galvanized steel flush floor trunking; fixed to backgrounds; supports and fixings; standard coupling joints; earth continuity straps included						
Triple compartment						
350 × 60 mm	46.42	58.23	1.32	47.45	m	105.68
Four compartment						
350 × 60 mm	48.33	60.63	1.32	47.45	m	108.08
Galvanized steel flush floor trunking; fittings (cutting and jointing trunking to fittings is included)						
Stop end; triple compartment						
350 × 60 mm	4.87	6.10	0.53	19.05	nr	25.15
Stop end; four compartment						
350 × 60 mm	11.29	14.16	0.53	19.05	nr	33.21
Rising bend; standard; triple compartment						
350 × 60 mm	40.15	50.37	1.30	46.74	nr	97.11
Rising bend; standard; four compartment						
350 × 60 mm	42.07	52.77	1.30	46.74	nr	99.51
Rising bend; skirting; triple compartment						
350 × 60 mm	78.77	98.81	1.33	47.81	nr	146.62
Rising bend; skirting; four compartment						
350 × 60 mm	89.40	112.15	1.33	47.81	nr	159.96
Junction box; triple compartment						
350 × 60 mm	52.35	65.67	1.16	41.71	nr	107.38
Junction box; four compartment						
350 × 60 mm	54.45	68.30	1.16	41.71	nr	110.01
Body coupler (pair)						
3 and 4 compartment	2.70	3.38	0.16	5.76	nr	9.14
Service outlet module comprising flat lid with flanged carpet trim; twin 13 A outlet and drilled plate for mounting 2 telephone outlets; one blank plate; triple compartment						
3 compartment	64.94	81.46	0.47	16.90	nr	98.36

ELECTRICAL SUPPLY/POWER/LIGHTING

Item	Net Price £	Material £	Labour hours	Labour £	Unit	Total rate £
Service outlet module comprising flat lid with flanged carpet trim; twin 13 A outlet and drilled plate for mounting 2 telephone outlets; two blank plates; four compartment						
4 compartment	72.33	90.73	0.47	16.90	nr	**107.63**
Single compartment PVC trunking; grey finish; clip on lid; fixed to backgrounds; including supports and fixings (standard coupling joints)						
Dimensions						
50 × 50 mm	11.18	14.02	0.27	9.71	m	**23.73**
75 × 50 mm	12.12	15.20	0.28	10.06	m	**25.26**
75 × 75 mm	13.73	17.23	0.29	10.43	m	**27.66**
100 × 50 mm	15.55	19.51	0.34	12.22	m	**31.73**
100 × 75 mm	17.06	21.40	0.37	13.29	m	**34.69**
100 × 100 mm	18.01	22.59	0.37	13.29	m	**35.88**
150 × 50 mm	15.63	19.61	0.41	14.74	m	**34.35**
150 × 75 mm	27.86	34.94	0.44	15.81	m	**50.75**
150 × 100 mm	33.53	42.06	0.44	15.81	m	**57.87**
150 × 150 mm	34.30	43.03	0.48	17.26	m	**60.29**
Single compartment PVC trunking; fittings (cutting and jointing trunking to fittings is included)						
Crossover						
50 × 50 mm	18.32	22.98	0.29	10.43	nr	**33.41**
75 × 50 mm	20.31	25.48	0.30	10.79	nr	**36.27**
75 × 75 mm	22.04	27.64	0.31	11.14	nr	**38.78**
100 × 50 mm	29.38	36.86	0.35	12.58	nr	**49.44**
100 × 75 mm	37.06	46.49	0.36	12.95	nr	**59.44**
100 × 100 mm	32.78	41.12	0.40	14.38	nr	**55.50**
150 × 75 mm	41.99	52.67	0.45	16.18	nr	**68.85**
150 × 100 mm	50.37	63.18	0.46	16.53	nr	**79.71**
150 × 150 mm	78.19	98.08	0.47	16.90	nr	**114.98**
Stop end						
50 × 50 mm	0.76	0.95	0.12	4.31	nr	**5.26**
75 × 50 mm	1.43	1.79	0.12	4.31	nr	**6.10**
75 × 75 mm	1.09	1.37	0.13	4.68	nr	**6.05**
100 × 50 mm	1.95	2.44	0.16	5.76	nr	**8.20**
100 × 75 mm	3.05	3.83	0.16	5.76	nr	**9.59**
100 × 100 mm	3.06	3.84	0.18	6.47	nr	**10.31**
150 × 75 mm	7.97	10.00	0.20	7.19	nr	**17.19**
150 × 100 mm	9.87	12.38	0.21	7.55	nr	**19.93**
150 × 150 mm	10.07	12.63	0.22	7.92	nr	**20.55**

ELECTRICAL SUPPLY/POWER/LIGHTING

Item	Net Price £	Material £	Labour hours	Labour £	Unit	Total rate £
LV DISTRIBUTION: TRUNKING – cont						
Single compartment PVC trunking – cont						
Flanged coupling						
50 × 50 mm	4.59	5.76	0.32	11.50	nr	17.26
75 × 50 mm	5.25	6.59	0.33	11.87	nr	18.46
75 × 75 mm	6.30	7.91	0.34	12.22	nr	20.13
100 × 50 mm	7.09	8.89	0.44	15.81	nr	24.70
100 × 75 mm	8.03	10.07	0.45	16.18	nr	26.25
100 × 100 mm	8.55	10.73	0.46	16.53	nr	27.26
150 × 75 mm	9.06	11.37	0.57	20.50	nr	31.87
150 × 100 mm	9.55	11.98	0.57	20.50	nr	32.48
150 × 150 mm	10.02	12.57	0.59	21.21	nr	33.78
Internal coupling						
50 × 50 mm	1.62	2.03	0.07	2.52	nr	4.55
75 × 50 mm	1.92	2.41	0.07	2.52	nr	4.93
75 × 75 mm	1.93	2.42	0.07	2.52	nr	4.94
100 × 50 mm	2.58	3.24	0.08	2.87	nr	6.11
100 × 75 mm	2.89	3.63	0.08	2.87	nr	6.50
100 × 100 mm	3.19	4.00	0.08	2.87	nr	6.87
External coupling						
50 × 50 mm	1.77	2.22	0.09	3.24	nr	5.46
75 × 50 mm	2.12	2.65	0.09	3.24	nr	5.89
75 × 75 mm	2.10	2.63	0.09	3.24	nr	5.87
100 × 50 mm	2.82	3.54	0.10	3.60	nr	7.14
100 × 75 mm	3.21	4.03	0.10	3.60	nr	7.63
100 × 100 mm	3.50	4.39	0.10	3.60	nr	7.99
150 × 75 mm	4.01	5.03	0.11	3.95	nr	8.98
150 × 100 mm	4.18	5.24	0.11	3.95	nr	9.19
150 × 150 mm	4.32	5.42	0.11	3.95	nr	9.37
Angle; flat cover						
50 × 50 mm	6.50	8.15	0.18	6.47	nr	14.62
75 × 50 mm	8.66	10.86	0.19	6.83	nr	17.69
75 × 75 mm	9.96	12.50	0.20	7.19	nr	19.69
100 × 50 mm	14.88	18.67	0.23	8.27	nr	26.94
100 × 75 mm	22.66	28.43	0.26	9.34	nr	37.77
100 × 100 mm	20.67	25.93	0.26	9.34	nr	35.27
150 × 75 mm	28.17	35.34	0.30	10.79	nr	46.13
150 × 100 mm	33.51	42.03	0.33	11.87	nr	53.90
150 × 150 mm	50.78	63.69	0.34	12.22	nr	75.91
Angle; internal or external cover						
50 × 50 mm	7.91	9.92	0.18	6.47	nr	16.39
75 × 50 mm	10.90	13.68	0.19	6.83	nr	20.51
75 × 75 mm	14.00	17.56	0.20	7.19	nr	24.75
100 × 50 mm	15.08	18.92	0.23	8.27	nr	27.19
100 × 75 mm	24.25	30.42	0.26	9.34	nr	39.76
100 × 100 mm	24.36	30.55	0.26	9.34	nr	39.89
150 × 75 mm	29.78	37.35	0.30	10.79	nr	48.14
150 × 100 mm	35.10	44.03	0.33	11.87	nr	55.90
150 × 150 mm	49.30	61.85	0.34	12.22	nr	74.07

ELECTRICAL SUPPLY/POWER/LIGHTING

Item	Net Price £	Material £	Labour hours	Labour £	Unit	Total rate £
Tee; flat cover						
50 × 50 mm	5.44	6.82	0.24	8.64	nr	**15.46**
75 × 50 mm	8.30	10.42	0.25	8.99	nr	**19.41**
75 × 75 mm	8.53	10.70	0.26	9.34	nr	**20.04**
100 × 50 mm	18.43	23.12	0.32	11.50	nr	**34.62**
100 × 75 mm	19.58	24.56	0.33	11.87	nr	**36.43**
100 × 100 mm	25.55	32.05	0.34	12.22	nr	**44.27**
150 × 75 mm	33.79	42.38	0.41	14.74	nr	**57.12**
150 × 100 mm	43.34	54.36	0.42	15.10	nr	**69.46**
150 × 150 mm	58.96	73.96	0.44	15.81	nr	**89.77**
Tee; internal or external cover						
50 × 50 mm	15.42	19.34	0.24	8.64	nr	**27.98**
75 × 50 mm	16.94	21.25	0.25	8.99	nr	**30.24**
75 × 75 mm	18.84	23.63	0.26	9.34	nr	**32.97**
100 × 50 mm	24.18	30.33	0.32	11.50	nr	**41.83**
100 × 75 mm	27.62	34.64	0.33	11.87	nr	**46.51**
100 × 100 mm	30.99	38.88	0.34	12.22	nr	**51.10**
150 × 75 mm	40.15	50.37	0.41	14.74	nr	**65.11**
150 × 100 mm	48.39	60.70	0.42	15.10	nr	**75.80**
150 × 150 mm	64.61	81.04	0.44	15.81	nr	**96.85**
Division Strip (1.8 m long)						
50 mm	8.36	10.48	0.07	2.52	nr	**13.00**
75 mm	10.71	13.44	0.07	2.52	nr	**15.96**
100 mm	13.61	17.07	0.08	2.87	nr	**19.94**
PVC miniature trunking; white finish; fixed to backgrounds; including supports and fixing; standard coupling joints						
Single compartment						
16 × 16 mm	1.51	1.89	0.20	7.19	m	**9.08**
25 × 16 mm	1.84	2.31	0.21	7.55	m	**9.86**
38 × 16 mm	2.30	2.89	0.24	8.64	m	**11.53**
38 × 25 mm	2.75	3.45	0.25	8.99	m	**12.44**
Compartmented						
38 × 16 mm	2.69	3.37	0.24	8.64	m	**12.01**
38 × 25 mm	3.21	4.03	0.25	8.99	m	**13.02**
PVC miniature trunking fittings; single compartment; white finish; cutting and jointing trunking to fittings is included						
Coupling						
16 × 16 mm	0.47	0.59	0.10	3.60	nr	**4.19**
25 × 16 mm	0.47	0.59	0.12	4.14	nr	**4.73**
38 × 16 mm	0.47	0.59	0.12	4.31	nr	**4.90**
38 × 25 mm	1.10	1.38	0.14	5.03	nr	**6.41**
Stop end						
16 × 16 mm	0.47	0.59	0.12	4.31	nr	**4.90**
25 × 16 mm	0.47	0.59	0.13	4.57	nr	**5.16**
38 × 16 mm	0.47	0.59	0.15	5.40	nr	**5.99**
38 × 25 mm	1.10	1.38	0.17	6.10	nr	**7.48**

ELECTRICAL SUPPLY/POWER/LIGHTING

Item	Net Price £	Material £	Labour hours	Labour £	Unit	Total rate £
LV DISTRIBUTION: TRUNKING – cont						
PVC miniature trunking fittings – cont						
Bend; flat, internal or external						
16 × 16 mm	0.47	0.59	0.18	6.47	nr	**7.06**
25 × 16 mm	0.47	0.59	0.20	7.11	nr	**7.70**
38 × 16 mm	0.47	0.59	0.21	7.55	nr	**8.14**
38 × 25 mm	1.10	1.38	0.23	8.27	nr	**9.65**
Tee						
16 × 16 mm	0.78	0.97	0.19	6.83	nr	**7.80**
25 × 16 mm	0.78	0.97	0.23	8.27	nr	**9.24**
38 × 16 mm	0.78	0.97	0.26	9.34	nr	**10.31**
38 × 25 mm	1.09	1.37	0.29	10.43	nr	**11.80**
PVC bench trunking; white or grey finish; fixed to backgrounds; including supports and fixings; standard coupling joints						
Trunking						
90 × 90 mm	24.36	30.55	0.33	11.87	m	**42.42**
PVC bench trunking fittings; white or grey finish; cutting and jointing trunking to fittings is included						
Stop end						
90 × 90 mm	5.36	6.72	0.09	3.24	nr	**9.96**
Coupling						
90 × 90 mm	3.32	4.17	0.09	3.24	nr	**7.41**
Internal or external bend						
90 × 90 mm	18.21	22.85	0.28	10.06	nr	**32.91**
Socket plate						
90 × 90 mm – 1 gang	1.03	1.29	0.10	3.60	nr	**4.89**
90 × 90 mm – 2 gang	1.23	1.55	0.10	3.60	nr	**5.15**
PVC underfloor trunking; single compartment; fitted in floor screed; standard coupling joints						
Trunking						
60 × 25 mm	12.42	15.58	0.22	7.92	m	**23.50**
90 × 35 mm	17.89	22.44	0.27	9.71	m	**32.15**
PVC underfloor trunking fittings; single compartment; fitted in floor screed; (cutting and jointing trunking to fittings is included)						
Jointing sleeve						
60 × 25 mm	0.83	1.04	0.08	2.87	nr	**3.91**
90 × 35 mm	1.38	1.74	0.10	3.60	nr	**5.34**
Duct connector						
90 × 35 mm	0.64	0.81	0.17	6.10	nr	**6.91**

ELECTRICAL SUPPLY/POWER/LIGHTING

Item	Net Price £	Material £	Labour hours	Labour £	Unit	Total rate £
Socket reducer						
90 × 35 mm	1.14	1.43	0.12	4.31	nr	**5.74**
Vertical access box; 2 compartment						
Shallow	64.25	80.60	0.37	13.29	nr	**93.89**
Duct bend; vertical						
60 × 25 mm	12.53	15.71	0.27	9.71	nr	**25.42**
90 × 35 mm	14.14	17.74	0.35	12.58	nr	**30.32**
Duct bend; horizontal						
60 × 25 mm	14.78	18.54	0.30	10.79	nr	**29.33**
90 × 35 mm	15.00	18.82	0.37	13.29	nr	**32.11**
Zinc coated steel underfloor ducting; fixed to backgrounds; standard coupling joints; earth continuity straps; (Including supports and fixing, packing shims where required)						
Double compartment						
150 × 25 mm	10.11	12.68	0.57	20.50	m	**33.18**
Triple compartment						
225 × 25 mm	17.91	22.47	0.93	33.43	m	**55.90**
Zinc coated steel underfloor ducting fittings; (cutting and jointing to fittings is included)						
Stop end; double compartment						
150 × 25 mm	3.18	3.99	0.31	11.14	nr	**15.13**
Stop end; triple compartment						
225 × 25 mm	3.64	4.57	0.37	13.29	nr	**17.86**
Rising bend; double compartment; standard trunking						
150 × 25 mm	20.47	25.68	0.71	25.52	nr	**51.20**
Rising bend; triple compartment; standard trunking						
225 × 25 mm	36.78	46.13	0.85	30.55	nr	**76.68**
Rising bend; double compartment; to skirting						
150 × 25 mm	45.56	57.15	0.90	32.37	nr	**89.52**
Rising bend; triple compartment; to skirting						
225 × 25 mm	53.72	67.39	0.95	34.16	nr	**101.55**
Horizontal bend; double compartment						
150 × 25 mm	31.12	39.03	0.64	23.00	nr	**62.03**
Horizontal bend; triple compartment						
225 × 25 mm	35.89	45.02	0.77	27.69	nr	**72.71**
Junction or service outlet boxes; terminal; double compartment						
150 mm	32.20	40.39	0.91	32.72	nr	**73.11**
Junction or service outlet boxes; terminal; triple compartment						
225 mm	36.80	46.17	1.11	39.91	nr	**86.08**

ELECTRICAL SUPPLY/POWER/LIGHTING

Item	Net Price £	Material £	Labour hours	Labour £	Unit	Total rate £
LV DISTRIBUTION: TRUNKING – cont						
Zinc coated steel underfloor ducting fittings – cont						
Junction or service outlet boxes; through or angle; double compartment						
150 mm	42.76	53.64	0.97	34.88	nr	**88.52**
Junction or service outlet boxes; through or angle; triple compartment						
225 mm	47.50	59.58	1.17	42.06	nr	**101.64**
Junction or service outlet boxes; tee; double compartment						
150 mm	42.76	53.64	1.02	36.67	nr	**90.31**
Junction or service outlet boxes; tee; triple compartment						
225 mm	47.50	59.58	1.22	43.87	nr	**103.45**
Junction or service outlet boxes; cross; double compartment						
up to 150 mm	42.76	53.64	1.03	37.03	nr	**90.67**
Junction or service outlet boxes; cross;triple compartment						
225 mm	47.50	59.58	1.23	44.22	nr	**103.80**
Plates for junction/inspection boxes; double and triple compartment						
Blank plate	9.34	11.72	0.92	33.07	nr	**44.79**
Conduit entry plate	11.72	14.71	0.86	30.92	nr	**45.63**
Trunking entry plate	11.72	14.71	0.86	30.92	nr	**45.63**
Service outlet box comprising flat lid with flanged carpet trim; twin 13 A outlet and drilled plate for mounting 2 telephone outlets and terminal blocks; terminal outlet box; double compartment						
150 × 25 mm trunking	66.68	83.64	1.68	60.40	nr	**144.04**
Service outlet box comprising flat lid with flanged carpet trim; twin 13 A outlet and drilled plate for mounting 2 telephone outlets and terminal blocks; terminal outlet box; triple compartment						
225 × 25 mm trunking	74.22	93.11	1.93	69.38	nr	**162.49**
PVC skirting/dado modular trunking; white (cutting and jointing trunking to fittings and backplates for fixing to walls is included)						
Main carrier/backplate						
50 × 170 mm	19.85	24.90	0.22	7.92	m	**32.82**
62 × 190 mm	22.04	27.64	0.22	7.92	m	**35.56**
Extension carrier/backplate						
50 × 42 mm	12.06	15.13	0.58	20.84	m	**35.97**

ELECTRICAL SUPPLY/POWER/LIGHTING

Item	Net Price £	Material £	Labour hours	Labour £	Unit	Total rate £
Carrier/backplate						
Including cover seal	7.04	8.83	0.53	19.05	m	**27.88**
Chamfered covers for fixing to backplates						
50 × 42 mm	4.09	5.13	0.33	11.87	m	**17.00**
Square covers for fixing to backplates						
50 × 42 mm	8.20	10.28	0.33	11.87	m	**22.15**
Plain covers for fixing to backplates						
85 mm	4.09	5.13	0.34	12.22	m	**17.35**
Retainers-clip to backplates to hold cables						
For chamfered covers	1.13	1.42	0.07	2.52	m	**3.94**
For square-recessed covers	0.98	1.23	0.07	2.52	m	**3.75**
For plain covers	3.77	4.73	0.07	2.52	m	**7.25**
Prepackaged corner assemblies						
Internal; for 170 × 50 Assy	7.93	9.95	0.51	18.33	nr	**28.28**
Internal; for 190 × 62 Assy	8.81	11.05	0.51	18.33	nr	**29.38**
Internal; for 215 × 50 Assy	9.86	12.36	0.53	19.05	nr	**31.41**
Internal; for 254 × 50 Assy	11.74	14.73	0.53	19.05	nr	**33.78**
External; for 170 × 50 Assy	7.93	9.95	0.56	20.14	nr	**30.09**
External; for 190 × 62 Assy	8.81	11.05	0.56	20.14	nr	**31.19**
External; for 215 × 50 Assy	9.86	12.36	0.58	20.84	nr	**33.20**
External; for 254 × 50 Assy	11.74	14.73	0.58	20.84	nr	**35.57**
Clip on end caps						
170 × 50 Assy	4.75	5.96	0.11	3.95	nr	**9.91**
215 × 50 Assy	5.62	7.04	0.11	3.95	nr	**10.99**
254 × 50 Assy	6.59	8.27	0.11	3.95	nr	**12.22**
190 × 62 Assy	6.64	8.33	0.11	3.95	nr	**12.28**
Outlet box						
1 Gang; in horizontal trunking; clip in	4.24	5.32	0.34	12.22	nr	**17.54**
2 Gang; in horizontal trunking; clip in	5.29	6.63	0.34	12.22	nr	**18.85**
1 Gang; in vertical trunking; clip in	4.24	5.32	0.34	12.22	nr	**17.54**
Sheet steel adaptable boxes; with plain or knockout sides; fixed to backgrounds; including supports and fixings (cutting and connecting conduit to boxes is included)						
Square pattern – black						
75 × 75 × 37 mm	2.21	2.78	0.69	24.81	nr	**27.59**
75 × 75 × 50 mm	2.67	3.35	0.69	24.81	nr	**28.16**
75 × 75 × 75 mm	2.40	3.01	0.69	24.81	nr	**27.82**
100 × 100 × 50 mm	4.52	5.67	0.71	25.52	nr	**31.19**
150 × 150 × 50 mm	3.35	4.20	0.79	28.40	nr	**32.60**
150 × 150 × 75 mm	5.79	7.26	0.80	28.76	nr	**36.02**
150 × 150 × 100 mm	6.55	8.22	0.80	28.76	nr	**36.98**
200 × 200 × 50 mm	5.25	6.59	0.80	28.76	nr	**35.35**
225 × 225 × 50 mm	6.71	8.42	0.93	33.43	nr	**41.85**
225 × 225 × 100 mm	8.90	11.17	0.94	33.79	nr	**44.96**
300 × 300 × 100 mm	9.55	11.98	0.99	35.58	nr	**47.56**
Square pattern – galvanized						
75 × 75 × 37 mm	2.03	2.54	0.69	24.81	nr	**27.35**
75 × 75 × 50 mm	2.37	2.97	0.70	25.17	nr	**28.14**
75 × 75 × 75 mm	2.76	3.46	0.69	24.81	nr	**28.27**

ELECTRICAL SUPPLY/POWER/LIGHTING

Item	Net Price £	Material £	Labour hours	Labour £	Unit	Total rate £
LV DISTRIBUTION: TRUNKING – cont						
Sheet steel adaptable boxes – cont						
Square pattern – galvanized – cont						
100 × 100 × 50 mm	2.51	3.15	0.71	25.52	nr	**28.67**
150 × 150 × 50 mm	2.93	3.67	0.84	30.20	nr	**33.87**
150 × 150 × 75 mm	3.53	4.42	0.80	28.76	nr	**33.18**
150 × 150 × 100 mm	4.27	5.35	0.80	28.76	nr	**34.11**
225 × 225 × 50 mm	5.52	6.92	0.93	33.43	nr	**40.35**
225 × 225 × 100 mm	6.71	8.42	0.94	33.79	nr	**42.21**
300 × 300 × 100 mm	11.41	–	–	–	nr	–
Rectangular pattern – black						
100 × 75 × 50 mm	3.74	4.69	0.69	24.81	nr	**29.50**
150 × 75 × 50 mm	3.92	4.92	0.70	25.17	nr	**30.09**
150 × 75 × 75 mm	4.25	5.33	0.71	25.52	nr	**30.85**
150 × 100 × 75 mm	9.47	11.88	0.71	25.52	nr	**37.40**
225 × 75 × 50 mm	8.18	10.26	0.78	28.03	nr	**38.29**
225 × 150 × 75 mm	13.07	16.40	0.81	29.12	nr	**45.52**
225 × 150 × 100 mm	24.53	30.77	0.81	29.12	nr	**59.89**
300 × 150 × 50 mm	24.53	30.77	0.93	33.43	nr	**64.20**
300 × 150 × 75 mm	24.53	30.77	0.94	33.79	nr	**64.56**
300 × 150 × 100 mm	24.53	30.77	0.96	34.51	nr	**65.28**
Rectangular pattern – galvanized						
100 × 75 × 50 mm	3.55	4.46	0.69	24.81	nr	**29.27**
150 × 75 × 50 mm	3.72	4.67	0.70	25.17	nr	**29.84**
150 × 75 × 75 mm	4.04	5.06	0.71	25.52	nr	**30.58**
150 × 100 × 75 mm	8.98	11.27	0.71	25.52	nr	**36.79**
225 × 75 × 50 mm	7.78	9.76	0.89	32.00	nr	**41.76**
225 × 150 × 75 mm	12.41	15.57	0.81	29.12	nr	**44.69**
225 × 150 × 100 mm	23.31	29.24	0.81	29.12	nr	**58.36**
300 × 150 × 50 mm	23.31	29.24	0.93	33.43	nr	**62.67**
300 × 150 × 75 mm	23.31	29.24	0.94	33.79	nr	**63.03**
300 × 150 × 100 mm	23.31	29.24	0.96	34.51	nr	**63.75**

ELECTRICAL SUPPLY/POWER/LIGHTING

Item	Net Price £	Material £	Labour hours	Labour £	Unit	Total rate £
LV DISTRIBUTION: CABLES AND WIRING						
ARMOURED CABLE						
Cable; XLPE insulated; PVC sheathed; copper stranded conductors to BS 5467; laid in trench/duct including marker tape; (cable tiles measured elsewhere)						
600/1000 volt grade; single core (aluminium wire armour)						
25 mm²	0.45	0.56	0.15	5.40	m	**5.96**
35 mm²	0.50	0.63	0.15	5.40	m	**6.03**
50 mm²	0.74	0.93	0.17	6.10	m	**7.03**
70 mm²	6.20	7.77	0.18	6.47	m	**14.24**
95 mm²	8.05	10.10	0.20	7.19	m	**17.29**
120 mm²	10.03	12.58	0.22	7.92	m	**20.50**
150 mm²	12.16	15.25	0.24	8.64	m	**23.89**
185 mm²	14.85	18.63	0.26	9.34	m	**27.97**
240 mm²	18.36	23.03	0.30	10.79	m	**33.82**
300 mm²	23.94	30.03	0.31	11.14	m	**41.17**
400 mm²	30.36	38.08	0.38	13.66	m	**51.74**
500 mm²	38.58	48.40	0.44	15.81	m	**64.21**
630 mm²	49.36	61.91	0.52	18.69	m	**80.60**
800 mm²	31.99	40.13	0.62	22.29	m	**62.42**
1000 mm²	37.05	46.48	0.65	23.37	m	**69.85**
600/1000 volt grade; two core (galvanized steel wire armour)						
1.5 mm²	1.07	1.34	0.06	2.16	m	**3.50**
2.5 mm²	1.27	1.59	0.06	2.16	m	**3.75**
4 mm²	1.55	1.95	0.08	2.87	m	**4.82**
6 mm²	1.96	2.46	0.08	2.87	m	**5.33**
10 mm²	2.60	3.26	0.10	3.60	m	**6.86**
16 mm²	3.71	4.66	0.10	3.60	m	**8.26**
25 mm²	4.89	6.14	0.15	5.40	m	**11.54**
35 mm²	6.71	8.42	0.15	5.40	m	**13.82**
50 mm²	8.40	10.54	0.17	6.10	m	**16.64**
70 mm²	11.47	14.39	0.18	6.47	m	**20.86**
95 mm²	15.58	19.54	0.20	7.19	m	**26.73**
120 mm²	19.64	24.64	0.22	7.92	m	**32.56**
150 mm²	23.66	29.68	0.24	8.64	m	**38.32**
185 mm²	29.87	37.46	0.26	9.34	m	**46.80**
240 mm²	35.49	44.52	0.30	10.79	m	**55.31**
300 mm²	46.90	58.83	0.31	11.14	m	**69.97**
400 mm²	49.52	62.12	0.35	12.58	m	**74.70**

ELECTRICAL SUPPLY/POWER/LIGHTING

Item	Net Price £	Material £	Labour hours	Labour £	Unit	Total rate £
LV DISTRIBUTION: CABLES AND WIRING – cont						
Cable – cont						
600/1000 volt grade; three core (galvanized steel wire armour)						
1.5 mm²	1.16	1.46	0.07	2.52	m	3.98
2.5 mm²	1.43	1.79	0.07	2.52	m	4.31
4 mm²	1.79	2.24	0.09	3.24	m	5.48
6 mm²	2.02	2.53	0.10	3.60	m	6.13
10 mm²	3.02	3.79	0.11	3.95	m	7.74
16 mm²	4.31	5.41	0.11	3.95	m	9.36
25 mm²	6.39	8.02	0.16	5.76	m	13.78
35 mm²	8.45	10.60	0.16	5.76	m	16.36
50 mm²	10.89	13.66	0.19	6.83	m	20.49
70 mm²	15.25	19.13	0.21	7.55	m	26.68
95 mm²	21.01	26.35	0.23	8.27	m	34.62
120 mm²	25.93	32.52	0.24	8.64	m	41.16
150 mm²	32.01	40.15	0.27	9.71	m	49.86
185 mm²	39.90	50.05	0.30	10.79	m	60.84
240 mm²	51.38	64.46	0.33	11.87	m	76.33
300 mm²	64.93	81.45	0.35	12.58	m	94.03
400 mm²	64.93	81.45	0.41	14.74	m	96.19
600/1000 volt grade; four core (galvanized steel wire armour)						
1.5 mm²	1.16	1.46	0.08	2.87	m	4.33
2.5 mm²	1.52	1.90	0.09	3.24	m	5.14
4 mm²	2.01	2.52	0.10	3.60	m	6.12
6 mm²	2.84	3.56	0.10	3.60	m	7.16
10 mm²	4.06	5.10	0.12	4.31	m	9.41
16 mm²	5.90	7.40	0.12	4.31	m	11.71
25 mm²	8.60	10.79	0.18	6.47	m	17.26
35 mm²	11.55	14.49	0.19	6.83	m	21.32
50 mm²	15.14	19.00	0.21	7.55	m	26.55
70 mm²	21.49	26.96	0.23	8.27	m	35.23
95 mm²	29.03	36.41	0.26	9.34	m	45.75
120 mm²	36.32	45.56	0.28	10.06	m	55.62
150 mm²	44.86	56.27	0.32	11.50	m	67.77
185 mm²	55.43	69.53	0.35	12.58	m	82.11
240 mm²	71.85	90.13	0.36	12.95	m	103.08
300 mm²	90.43	113.43	0.40	14.38	m	127.81
400 mm²	100.08	125.54	0.45	16.18	m	141.72
600/1000 volt grade; seven core (galvanized steel wire armour)						
1.5 mm²	1.77	2.22	0.10	3.60	m	5.82
2.5 mm²	2.31	2.90	0.10	3.60	m	6.50
4 mm²	2.35	2.95	0.11	3.95	m	6.90
600/1000 volt grade; twelve core (galvanized steel wire armour)						
1.5 mm²	2.72	3.42	0.11	3.95	m	7.37
2.5 mm²	3.83	4.80	0.11	3.95	m	8.75

ELECTRICAL SUPPLY/POWER/LIGHTING

Item	Net Price £	Material £	Labour hours	Labour £	Unit	Total rate £
600/1000 volt grade; nineteen core (galvanized steel wire armour)						
1.5 mm²	3.87	4.85	0.13	4.68	m	9.53
2.5 mm²	5.46	6.85	0.14	5.03	m	11.88
600/1000 volt grade; twenty seven core (galvanized steel wire armour)						
1.5 mm²	5.29	6.63	0.14	5.03	m	11.66
2.5 mm²	7.44	9.33	0.16	5.76	m	15.09
600/1000 volt grade; thirty seven core (galvanized steel wire armour)						
1.5 mm²	7.09	8.89	0.15	5.40	m	14.29
2.5 mm²	9.85	12.35	0.17	6.10	m	18.45
Cable; XLPE insulated; PVC sheathed copper stranded conductors to BS 5467; clipped direct to backgrounds including cleat						
600/1000 volt grade; single core (aluminium wire armour)						
25 mm²	2.05	2.58	0.35	12.58	m	15.16
35 mm²	2.25	2.82	0.36	12.95	m	15.77
50 mm²	2.47	3.10	0.37	13.29	m	16.39
70 mm²	8.51	10.67	0.39	14.02	m	24.69
95 mm²	11.71	14.69	0.42	15.10	m	29.79
120 mm²	13.98	17.54	0.47	16.90	m	34.44
150 mm²	16.10	20.19	0.51	18.33	m	38.52
185 mm²	18.80	23.59	0.59	21.21	m	44.80
240 mm²	22.50	28.22	0.68	24.45	m	52.67
300 mm²	28.08	35.22	0.74	26.61	m	61.83
400 mm²	34.23	42.94	0.88	31.64	m	74.58
500 mm²	44.18	55.42	0.88	31.64	m	87.06
630 mm²	55.56	69.70	1.05	37.74	m	107.44
800 mm²	38.80	48.68	1.33	47.81	m	96.49
1000 mm²	43.39	54.43	1.40	50.32	m	104.75
600/1000 volt grade; two core (galvanized steel wire armour)						
1.5 mm²	1.45	1.81	0.20	7.19	m	9.00
2.5 mm²	1.65	2.07	0.20	7.19	m	9.26
4.0 mm²	1.93	2.42	0.21	7.55	m	9.97
6.0 mm²	2.38	2.99	0.22	7.92	m	10.91
10.0 mm²	3.25	4.08	0.24	8.64	m	12.72
16.0 mm²	4.37	5.48	0.25	8.99	m	14.47
25 mm²	5.93	7.44	0.35	12.58	m	20.02
35 mm²	7.55	9.48	0.36	12.95	m	22.43
50 mm²	9.32	11.69	0.37	13.29	m	24.98
70 mm²	11.92	14.95	0.39	14.02	m	28.97
95 mm²	16.07	20.16	0.42	15.10	m	35.26
120 mm²	20.12	25.23	0.47	16.90	m	42.13
150 mm²	24.44	30.65	0.51	18.33	m	48.98
185 mm²	30.79	38.62	0.59	21.21	m	59.83
240 mm²	39.98	50.15	0.68	24.45	m	74.60
300 mm²	52.14	65.41	0.74	26.61	m	92.02
400 mm²	39.48	49.53	0.88	31.64	m	81.17

ELECTRICAL SUPPLY/POWER/LIGHTING

Item	Net Price £	Material £	Labour hours	Labour £	Unit	Total rate £
LV DISTRIBUTION: CABLES AND WIRING – cont						
Cable – cont						
600/1000 volt grade; three core (galvanized steel wire armour)						
1.5 mm²	1.57	1.97	0.20	7.19	m	**9.16**
2.5 mm²	1.87	2.34	0.21	7.55	m	**9.89**
4.0 mm²	2.25	2.82	0.22	7.92	m	**10.74**
6.0 mm²	2.79	3.49	0.22	7.92	m	**11.41**
10.0 mm²	4.12	5.16	0.25	8.99	m	**14.15**
16.0 mm²	5.51	6.91	0.26	9.34	m	**16.25**
25 mm²	7.92	9.93	0.37	13.29	m	**23.22**
35 mm²	10.24	12.85	0.39	14.02	m	**26.87**
50 mm²	12.85	16.12	0.40	14.38	m	**30.50**
70 mm²	16.92	21.22	0.42	15.10	m	**36.32**
95 mm²	23.44	29.40	0.45	16.18	m	**45.58**
120 mm²	28.77	36.09	0.52	18.69	m	**54.78**
150 mm²	35.85	44.97	0.55	19.77	m	**64.74**
185 mm²	45.00	56.45	0.63	22.66	m	**79.11**
240 mm²	57.94	72.68	0.71	25.52	m	**98.20**
300 mm²	71.96	90.27	0.78	28.03	m	**118.30**
400 mm²	77.59	97.33	0.87	31.27	m	**128.60**
600/1000 volt grade; four core (galvanized steel wire armour)						
1.5 mm²	1.76	2.21	0.21	7.55	m	**9.76**
2.5 mm²	2.12	2.65	0.22	7.92	m	**10.57**
4.0 mm²	2.65	3.33	0.22	7.92	m	**11.25**
6.0 mm²	3.65	4.58	0.23	8.27	m	**12.85**
10.0 mm²	5.12	6.42	0.26	9.34	m	**15.76**
16.0 mm²	6.96	8.74	0.26	9.34	m	**18.08**
25 mm²	9.74	12.22	0.39	14.02	m	**26.24**
35 mm²	12.69	15.92	0.40	14.38	m	**30.30**
50 mm²	16.38	20.55	0.41	14.74	m	**35.29**
70 mm²	22.25	27.91	0.45	16.18	m	**44.09**
95 mm²	29.81	37.40	0.50	17.98	m	**55.38**
120 mm²	37.54	47.08	0.54	19.42	m	**66.50**
150 mm²	46.77	58.67	0.60	21.57	m	**80.24**
185 mm²	57.59	72.24	0.67	24.08	m	**96.32**
240 mm²	74.24	93.13	0.75	26.97	m	**120.10**
300 mm²	93.20	116.91	0.83	29.84	m	**146.75**
400 mm²	92.00	115.40	0.91	32.72	m	**148.12**
600/1000 volt grade; seven core (galvanized steel wire armour)						
1.5 mm²	2.41	3.02	0.20	7.19	m	**10.21**
2.5 mm²	2.95	3.70	0.20	7.19	m	**10.89**
4.0 mm²	3.83	4.80	0.23	8.27	m	**13.07**
600/1000 volt grade; twelve core (galvanized steel wire armour)						
1.5 mm²	4.32	11.60	0.23	8.27	m	**19.87**
2.5 mm²	4.89	6.14	0.24	8.64	m	**14.78**

ELECTRICAL SUPPLY/POWER/LIGHTING

Item	Net Price £	Material £	Labour hours	Labour £	Unit	Total rate £
600/1000 volt grade; nineteen core (galvanized steel wire armour)						
1.5 mm²	4.93	6.18	0.26	9.34	m	**15.52**
2.5 mm²	6.61	8.29	0.28	10.06	m	**18.35**
600/1000 volt grade; twenty seven core (galvanized steel wire armour)						
1.5 mm²	6.43	8.06	0.29	10.43	m	**18.49**
2.5 mm²	8.68	10.89	0.30	10.79	m	**21.68**
600/1000 volt grade; thirty seven core (galvanized steel wire armour)						
1.5 mm²	7.83	9.82	0.32	11.50	m	**21.32**
2.5 mm²	11.09	13.91	0.33	11.87	m	**25.78**
Cable termination; brass weatherproof gland with inner and outer seal, shroud, brass locknut and earth ring (including drilling and cutting mild steel gland plate)						
600/1000 volt grade; single core (aluminium wire armour)						
25 mm²	7.17	8.99	1.70	61.12	nr	**70.11**
35 mm²	7.17	8.99	1.79	64.36	nr	**73.35**
50 mm²	7.85	9.84	2.06	74.05	nr	**83.89**
70 mm²	8.62	10.81	2.12	76.22	nr	**87.03**
95 mm²	8.67	10.88	2.39	85.93	nr	**96.81**
120 mm²	8.70	10.91	2.47	88.79	nr	**99.70**
150 mm²	17.79	22.31	2.73	98.15	nr	**120.46**
185 mm²	18.07	22.67	3.05	109.65	nr	**132.32**
240 mm²	17.98	22.56	3.45	124.04	nr	**146.60**
300 mm²	18.38	23.06	3.84	138.05	nr	**161.11**
400 mm²	26.35	33.05	4.21	151.36	nr	**184.41**
500 mm²	28.13	35.29	5.70	204.92	m	**240.21**
630 mm²	34.35	43.09	6.20	222.89	m	**265.98**
800 mm²	70.34	88.23	7.50	269.63	m	**357.86**
1000 mm²	84.89	106.49	10.00	359.51	m	**466.00**
600/1000 volt grade; two core (galvanized steel wire armour)						
1.5 mm²	6.61	8.29	0.58	20.84	nr	**29.13**
2.5 mm²	6.61	8.29	0.58	20.89	nr	**29.18**
4 mm²	6.61	8.29	0.58	20.84	nr	**29.13**
6 mm²	6.63	8.32	0.67	24.08	nr	**32.40**
10 mm²	7.39	9.27	1.00	35.95	nr	**45.22**
16 mm²	7.57	9.50	1.11	39.91	nr	**49.41**
25 mm²	8.73	10.95	1.70	61.12	nr	**72.07**
35 mm²	8.66	10.86	1.79	64.36	nr	**75.22**
50 mm²	16.85	21.13	2.06	74.05	nr	**95.18**
70 mm²	16.99	21.31	2.12	76.22	nr	**97.53**
95 mm²	16.16	20.27	2.39	85.93	nr	**106.20**
120 mm²	16.24	20.37	2.47	88.79	nr	**109.16**
150 mm²	25.48	31.96	2.73	98.15	nr	**130.11**
185 mm²	26.06	32.69	3.05	109.65	nr	**142.34**
240 mm²	27.72	34.78	3.45	124.04	nr	**158.82**

ELECTRICAL SUPPLY/POWER/LIGHTING

Item	Net Price £	Material £	Labour hours	Labour £	Unit	Total rate £
LV DISTRIBUTION: CABLES AND WIRING – cont						
Cable termination – cont						
600/1000 volt grade – cont						
300 mm²	40.47	50.77	3.84	138.05	nr	188.82
400 mm²	52.50	65.86	4.21	151.36	nr	217.22
600/1000 volt grade; three core (galvanized steel wire armour)						
1.5 mm²	6.74	8.46	0.61	22.10	nr	30.56
2.5 mm²	6.74	8.46	0.62	22.29	nr	30.75
4 mm²	6.74	8.46	0.62	22.29	nr	30.75
6 mm²	7.53	9.44	0.71	25.52	nr	34.96
10 mm²	8.75	10.98	1.06	38.11	nr	49.09
16 mm²	9.02	11.31	1.19	42.78	nr	54.09
25 mm²	17.01	21.34	1.81	65.06	nr	86.40
35 mm²	17.04	21.37	1.99	71.53	nr	92.90
50 mm²	17.07	21.41	2.23	80.17	nr	101.58
70 mm²	16.35	20.51	2.40	86.27	nr	106.78
95 mm²	16.50	20.70	2.63	94.56	nr	115.26
120 mm²	16.59	20.81	2.83	101.74	nr	122.55
150 mm²	26.86	33.69	3.22	115.76	nr	149.45
185 mm²	27.70	34.74	3.44	123.67	nr	158.41
240 mm²	42.17	52.90	3.83	137.69	nr	190.59
300 mm²	43.37	54.40	4.28	153.87	nr	208.27
400 mm²	46.51	58.34	5.00	179.76	nr	238.10
600/1000 volt grade; four core (galvanized steel wire armour)						
1.5 mm²	6.87	8.61	0.67	24.08	nr	32.69
2.5 mm²	6.87	8.61	0.67	24.08	nr	32.69
4 mm²	6.87	8.61	0.71	25.52	nr	34.13
6 mm²	8.89	11.16	0.76	27.32	nr	38.48
10 mm²	8.89	11.16	1.14	40.98	nr	52.14
16 mm²	9.25	11.60	1.29	46.38	nr	57.98
25 mm²	17.21	21.59	1.99	71.53	nr	93.12
35 mm²	17.25	21.64	2.16	77.66	nr	99.30
50 mm²	16.36	20.52	2.49	89.51	nr	110.03
70 mm²	16.64	20.88	2.65	95.27	nr	116.15
95 mm²	24.08	30.21	2.98	107.14	nr	137.35
120 mm²	24.20	30.35	3.15	113.24	nr	143.59
150 mm²	28.24	35.43	3.50	125.83	nr	161.26
185 mm²	41.31	51.82	3.72	133.74	nr	185.56
240 mm²	44.67	56.03	4.33	155.67	nr	211.70
300 mm²	64.64	81.09	4.86	174.72	nr	255.81
400 mm²	49.09	61.58	5.46	196.29	nr	257.87
600/1000 volt grade; seven core (galvanized steel wire armour)						
1.5 mm²	7.26	9.11	0.81	29.12	nr	38.23
2.5 mm²	7.26	9.11	0.85	30.55	nr	39.66
4 mm²	8.27	10.37	0.93	33.43	nr	43.80

ELECTRICAL SUPPLY/POWER/LIGHTING

Item	Net Price £	Material £	Labour hours	Labour £	Unit	Total rate £
600/1000 volt grade; twelve core (galvanized steel wire armour)						
1.5 mm²	8.67	10.88	1.14	40.98	nr	**51.86**
2.5 mm²	9.89	12.41	1.13	40.63	nr	**53.04**
600/1000 volt grade; nineteen core (galvanized steel wire armour)						
1.5 mm²	10.80	13.55	1.54	55.37	nr	**68.92**
2.5 mm²	10.80	13.55	1.54	55.37	nr	**68.92**
600/1000 volt grade; twenty seven core (galvanized steel wire armour)						
1.5 mm²	11.84	14.85	1.94	69.74	nr	**84.59**
2.5 mm²	19.92	24.99	2.31	83.04	nr	**108.03**
600/1000 volt grade; thirty seven core (galvanized steel wire armour)						
1.5 mm²	21.22	26.62	2.53	90.96	nr	**117.58**
2.5 mm²	21.22	26.62	2.87	103.17	nr	**129.79**
Cable; XLPE insulated; LSOH sheathed (LSF); copper stranded conductors to BS 6724; laid in trench/duct including marker tape (cable tiles measured elsewhere)						
600/1000 volt grade; single core (aluminium wire armour)						
50 mm²	5.08	6.37	0.17	6.10	m	**12.47**
70 mm²	7.00	8.78	0.18	6.47	m	**15.25**
95 mm²	9.20	11.54	0.20	7.19	m	**18.73**
120 mm²	11.03	13.83	0.22	7.92	m	**21.75**
150 mm²	13.17	16.52	0.24	8.64	m	**25.16**
185 mm²	15.86	19.89	0.26	9.34	m	**29.23**
240 mm²	21.19	26.58	0.30	10.79	m	**37.37**
300 mm²	25.74	32.29	0.31	11.14	m	**43.43**
400 mm²	33.64	42.20	0.35	12.58	m	**54.78**
500 mm²	42.56	53.39	0.44	15.81	m	**69.20**
630 mm²	56.54	70.92	0.52	18.69	m	**89.61**
800 mm²	71.11	89.20	0.62	22.29	m	**111.49**
1000 mm²	90.76	113.85	0.65	23.37	m	**137.22**
600/1000 volt grade; two core (galvanized steel wire armour)						
1.5 mm²	1.04	1.30	0.06	2.16	m	**3.46**
2.5 mm²	1.24	1.56	0.06	2.16	m	**3.72**
4 mm²	1.63	2.05	0.08	2.87	m	**4.92**
6 mm²	2.24	2.81	0.08	2.87	m	**5.68**
10 mm²	3.32	4.17	0.10	3.60	m	**7.77**
16 mm²	4.54	5.69	0.10	3.60	m	**9.29**
25 mm²	6.09	7.64	0.15	5.40	m	**13.04**
35 mm²	7.69	9.64	0.17	6.10	m	**15.74**
50 mm²	9.65	12.11	0.15	5.40	m	**17.51**
70 mm²	13.16	16.51	0.18	6.47	m	**22.98**
95 mm²	17.25	21.64	0.20	7.19	m	**28.83**

ELECTRICAL SUPPLY/POWER/LIGHTING

Item	Net Price £	Material £	Labour hours	Labour £	Unit	Total rate £
LV DISTRIBUTION: CABLES AND WIRING – cont						
Cable – cont						
600/1000 volt grade – cont						
120 mm²	21.91	27.48	0.22	7.92	m	**35.40**
150 mm²	26.49	33.23	0.24	8.64	m	**41.87**
185 mm²	33.11	41.53	0.26	9.34	m	**50.87**
240 mm²	42.13	52.85	0.30	10.79	m	**63.64**
300 mm²	39.35	49.36	0.31	11.14	m	**60.50**
400 mm²	60.78	76.24	0.35	12.58	m	**88.82**
600/1000 volt grade; three core (galvanized steel wire armour)						
1.5 mm²	1.18	1.48	0.07	2.52	m	**4.00**
2.5 mm²	1.54	1.93	0.07	2.52	m	**4.45**
4 mm²	2.01	2.52	0.09	3.24	m	**5.76**
6 mm²	2.72	3.42	0.10	3.60	m	**7.02**
10 mm²	4.18	5.24	0.11	3.95	m	**9.19**
16 mm²	5.72	7.18	0.11	3.95	m	**11.13**
25 mm²	7.97	10.00	0.16	5.76	m	**15.76**
35 mm²	10.28	12.89	0.16	5.76	m	**18.65**
50 mm²	13.28	16.65	0.19	6.83	m	**23.48**
70 mm²	17.90	22.46	0.21	7.55	m	**30.01**
95 mm²	24.85	31.17	0.23	8.27	m	**39.44**
120 mm²	30.55	38.33	0.24	8.64	m	**46.97**
150 mm²	37.74	47.34	0.27	9.71	m	**57.05**
185 mm²	46.53	58.36	0.30	10.79	m	**69.15**
240 mm²	61.05	76.59	0.33	11.87	m	**88.46**
300 mm²	74.78	93.80	0.35	12.58	m	**106.38**
400 mm²	105.17	131.92	0.41	14.74	m	**146.66**
600/1000 volt grade; four core (galvanized steel wire armour)						
1.5 mm²	1.34	1.68	0.08	2.87	m	**4.55**
2.5 mm²	1.80	2.26	0.09	3.24	m	**5.50**
4 mm²	2.56	3.21	0.10	3.60	m	**6.81**
6 mm²	3.64	4.57	0.10	3.60	m	**8.17**
10 mm²	5.07	6.36	0.12	4.31	m	**10.67**
16 mm²	7.15	8.97	0.12	4.31	m	**13.28**
25 mm²	10.41	13.06	0.18	6.47	m	**19.53**
35 mm²	13.75	17.25	0.19	6.83	m	**24.08**
50 mm²	18.00	22.58	0.21	7.55	m	**30.13**
70 mm²	23.61	29.61	0.23	8.27	m	**37.88**
95 mm²	31.63	39.68	0.26	9.34	m	**49.02**
120 mm²	39.54	49.59	0.28	10.06	m	**59.65**
150 mm²	53.69	67.35	0.32	11.50	m	**78.85**
185 mm²	66.34	83.22	0.35	12.58	m	**95.80**
240 mm²	85.81	107.64	0.36	12.95	m	**120.59**
300 mm²	106.66	133.80	0.40	14.38	m	**148.18**
400 mm²	142.41	178.64	0.45	16.18	m	**194.82**

ELECTRICAL SUPPLY/POWER/LIGHTING

Item	Net Price £	Material £	Labour hours	Labour £	Unit	Total rate £
600/1000 volt grade; seven core (galvanized steel wire armour)						
1.5 mm²	1.94	2.43	0.10	3.60	m	**6.03**
2.5 mm²	2.62	3.28	0.10	3.60	m	**6.88**
4 mm²	4.84	6.07	0.11	3.95	m	**10.02**
600/1000 volt grade; twelve core (galvanized steel wire armour)						
1.5 mm²	3.22	4.04	0.11	3.95	m	**7.99**
2.5 mm²	4.60	5.77	0.11	3.95	m	**9.72**
600/1000 volt grade; nineteen core (galvanized steel wire armour)						
1.5 mm²	4.68	5.87	0.13	4.68	m	**10.55**
2.5 mm²	6.73	8.44	0.14	5.03	m	**13.47**
600/1000 volt grade; twenty seven core (galvanized steel wire armour)						
1.5 mm²	6.43	8.06	0.14	5.03	m	**13.09**
2.5 mm²	9.12	11.44	0.16	5.76	m	**17.20**
600/1000 volt grade; thirty seven core (galvanized steel wire armour)						
1.5 mm²	8.16	10.24	0.15	5.40	m	**15.64**
2.5 mm²	11.60	14.55	0.17	6.10	m	**20.65**
Cable; XLPE insulated; LSOH sheathed (LSF) copper stranded conductors to BS 6724; clipped direct to backgrounds including cleat						
600/1000 volt grade; single core (aluminium wire armour)						
50 mm²	4.03	5.05	0.37	13.29	m	**18.34**
70 mm²	6.12	7.67	0.39	14.02	m	**21.69**
95 mm²	8.23	10.33	0.42	15.10	m	**25.43**
120 mm²	9.81	12.31	0.47	16.90	m	**29.21**
150 mm²	12.12	15.20	0.51	18.33	m	**33.53**
185 mm²	14.56	18.27	0.59	21.21	m	**39.48**
240 mm²	18.77	23.54	0.68	24.45	m	**47.99**
300 mm²	22.87	28.68	0.74	26.61	m	**55.29**
400 mm²	29.72	37.28	0.81	29.12	m	**66.40**
500 mm²	38.13	47.84	0.88	31.64	m	**79.48**
630 mm²	47.82	59.99	1.05	37.74	m	**97.73**
800 mm²	61.62	77.29	1.33	47.81	m	**125.10**
1000 mm²	71.35	89.50	1.40	50.32	m	**139.82**
600/1000 volt grade; two core (galvanized steel wire armour)						
1.5 mm²	0.88	1.11	0.20	7.19	m	**8.30**
2.5 mm²	1.01	1.27	0.20	7.19	m	**8.46**
4.0 mm²	1.31	1.65	0.21	7.55	m	**9.20**
6.0 mm²	1.59	1.99	0.22	7.92	m	**9.91**
10.0 mm²	2.13	2.68	0.24	8.64	m	**11.32**
16.0 mm²	3.21	4.03	0.25	8.99	m	**13.02**
25 mm²	3.21	4.03	0.35	12.58	m	**16.61**
35 mm²	6.48	8.13	0.36	12.95	m	**21.08**
50 mm²	8.03	10.07	0.37	13.29	m	**23.36**
70 mm²	11.93	14.96	0.39	14.02	m	**28.98**
95 mm²	16.03	20.10	0.42	15.10	m	**35.20**

ELECTRICAL SUPPLY/POWER/LIGHTING

Item	Net Price £	Material £	Labour hours	Labour £	Unit	Total rate £
LV DISTRIBUTION: CABLES AND WIRING – cont						
Cable – cont						
600/1000 volt grade – cont						
120 mm²	19.63	24.63	0.47	16.90	m	41.53
150 mm²	24.35	30.54	0.51	18.33	m	48.87
185 mm²	31.02	38.91	0.59	21.21	m	60.12
240 mm²	39.72	49.83	0.68	24.45	m	74.28
300 mm²	53.34	66.91	0.74	26.61	m	93.52
400 mm²	70.57	88.52	0.81	29.12	m	117.64
600/1000 volt grade; three core (galvanized steel wire armour)						
1.5 mm²	0.96	1.21	0.20	7.19	m	8.40
2.5 mm²	1.22	1.53	0.21	7.55	m	9.08
4.0 mm²	1.50	1.88	0.22	7.92	m	9.80
6.0 mm²	1.94	2.43	0.22	7.92	m	10.35
10.0 mm²	2.96	3.72	0.25	8.99	m	12.71
16.0 mm²	4.15	5.21	0.26	9.34	m	14.55
25 mm²	5.60	7.02	0.37	13.29	m	20.31
35 mm²	7.79	9.77	0.39	14.02	m	23.79
50 mm²	11.12	13.94	0.40	14.38	m	28.32
70 mm²	16.51	20.71	0.42	15.10	m	35.81
95 mm²	22.73	28.52	0.45	16.18	m	44.70
120 mm²	28.26	35.45	0.52	18.69	m	54.14
150 mm²	35.11	44.04	0.55	19.77	m	63.81
185 mm²	43.26	54.26	0.63	22.66	m	76.92
240 mm²	55.35	69.43	0.71	25.52	m	94.95
300 mm²	70.15	88.00	0.78	28.03	m	116.03
400 mm²	83.31	104.51	0.87	31.27	m	135.78
600/1000 volt grade; four core (galvanized steel wire armour)						
1.5 mm²	1.13	1.42	0.21	7.55	m	8.97
2.5 mm²	1.41	1.77	0.22	7.92	m	9.69
4.0 mm²	1.75	2.20	0.22	7.92	m	10.12
6.0 mm²	2.57	3.23	0.23	8.27	m	11.50
10.0 mm²	3.61	4.52	0.26	9.34	m	13.86
16.0 mm²	5.21	6.54	0.26	9.34	m	15.88
25 mm²	7.52	9.43	0.39	14.02	m	23.45
35 mm²	9.81	12.31	0.40	14.38	m	26.69
50 mm²	14.20	17.81	0.41	14.74	m	32.55
70 mm²	21.90	27.47	0.45	16.18	m	43.65
95 mm²	29.34	36.80	0.50	17.98	m	54.78
120 mm²	36.85	46.22	0.54	19.42	m	65.64
150 mm²	44.61	55.96	0.60	21.57	m	77.53
185 mm²	54.74	68.67	0.67	24.08	m	92.75
240 mm²	70.76	88.76	0.75	26.97	m	115.73
300 mm²	88.78	111.36	0.83	29.84	m	141.20
400 mm²	92.30	115.79	0.91	32.72	m	148.51

ELECTRICAL SUPPLY/POWER/LIGHTING

Item	Net Price £	Material £	Labour hours	Labour £	Unit	Total rate £
600/1000 volt grade; seven core (galvanized steel wire armour)						
1.5 mm²	1.62	2.03	0.20	7.19	m	9.22
2.5 mm²	2.26	2.83	0.20	7.19	m	10.02
4.0 mm²	4.10	5.14	0.23	8.27	m	13.41
600/1000 volt grade; twelve core (galvanized steel wire armour)						
1.5 mm²	2.57	3.23	0.23	8.27	m	11.50
2.5 mm²	3.86	4.84	0.24	8.64	m	13.48
600/1000 volt grade; nineteen core (galvanized steel wire armour)						
1.5 mm²	3.89	4.88	0.26	9.34	m	14.22
2.5 mm²	5.83	7.31	0.28	10.06	m	17.37
600/1000 volt grade; twenty seven core (galvanized steel wire armour)						
1.5 mm²	5.34	6.70	0.29	10.43	m	17.13
2.5 mm²	7.45	9.34	0.30	10.79	m	20.13
600/1000 volt grade; thirty seven core (galvanized steel wire armour)						
1.5 mm²	6.67	8.37	0.32	11.50	m	19.87
2.5 mm²	9.44	11.84	0.33	11.87	m	23.71
Cable termination; brass weatherproof gland with inner and outer seal, shroud, brass locknut and earth ring (including drilling and cutting mild steel gland plate)						
600/1000 volt grade; single core (aluminium wire armour)						
25 mm²	3.71	4.66	1.70	61.12	nr	65.78
35 mm²	4.04	5.06	1.79	64.36	nr	69.42
50 mm²	4.48	5.62	2.06	74.05	nr	79.67
70 mm²	6.41	8.04	2.12	76.22	nr	84.26
95 mm²	8.56	10.74	2.39	85.93	nr	96.67
120 mm²	10.14	12.72	2.47	88.79	nr	101.51
150 mm²	12.47	15.65	2.73	98.15	nr	113.80
185 mm²	14.85	18.63	3.05	109.65	nr	128.28
240 mm²	19.46	24.42	3.45	124.04	nr	148.46
300 mm²	23.52	29.50	3.84	138.05	nr	167.55
400 mm²	34.34	43.08	4.21	151.36	nr	194.44
500 mm²	43.27	54.28	5.70	204.92	m	259.20
630 mm²	56.28	70.59	6.20	222.89	m	293.48
800 mm²	69.98	87.79	7.50	269.63	m	357.42
1000 mm²	78.26	98.17	10.00	359.51	m	457.68
600/1000 volt grade; two core (galvanized steel wire armour)						
1.5 mm²	1.09	1.37	0.52	18.59	nr	19.96
2.5 mm²	1.26	1.58	0.58	20.84	nr	22.42
4 mm²	1.59	1.99	0.58	20.84	nr	22.83
6 mm²	1.85	2.32	0.67	24.08	nr	26.40
10 mm²	2.40	3.01	1.00	35.95	nr	38.96
16 mm²	3.48	4.37	1.11	39.91	nr	44.28
25 mm²	3.03	3.80	1.70	61.12	nr	64.92

ELECTRICAL SUPPLY/POWER/LIGHTING

Item	Net Price £	Material £	Labour hours	Labour £	Unit	Total rate £
LV DISTRIBUTION: CABLES AND WIRING – cont						
Cable termination – cont						
600/1000 volt grade – cont						
35 mm²	6.76	8.48	1.79	64.36	nr	**72.84**
50 mm²	8.32	10.44	2.06	74.05	nr	**84.49**
70 mm²	12.37	15.51	2.12	76.22	nr	**91.73**
95 mm²	16.11	20.20	2.39	85.93	nr	**106.13**
120 mm²	19.82	24.86	2.47	88.79	nr	**113.65**
150 mm²	24.54	30.78	2.73	98.15	nr	**128.93**
185 mm²	31.24	39.19	3.05	109.65	nr	**148.84**
240 mm²	38.88	48.78	3.45	124.04	nr	**172.82**
300 mm²	54.44	68.29	3.84	138.05	nr	**206.34**
400 mm²	67.78	85.02	4.21	151.36	nr	**236.38**
600/1000 volt grade; three core (galvanized steel wire armour)						
1.5 mm²	1.17	1.47	0.62	22.29	nr	**23.76**
2.5 mm²	1.40	1.76	0.62	22.29	nr	**24.05**
4 mm²	1.78	2.23	0.62	22.29	nr	**24.52**
6 mm²	2.20	2.76	0.71	25.52	nr	**28.28**
10 mm²	3.22	4.04	1.06	38.11	nr	**42.15**
16 mm²	4.43	5.56	1.19	42.78	nr	**48.34**
25 mm²	6.21	7.80	1.81	65.06	nr	**72.86**
35 mm²	8.07	10.12	1.99	71.53	nr	**81.65**
50 mm²	11.49	14.41	2.23	80.17	nr	**94.58**
70 mm²	16.68	20.92	2.40	86.27	nr	**107.19**
95 mm²	22.91	28.74	2.63	94.56	nr	**123.30**
120 mm²	28.44	35.67	2.83	101.74	nr	**137.41**
150 mm²	35.40	44.41	3.22	115.76	nr	**160.17**
185 mm²	43.53	54.60	3.44	123.67	nr	**178.27**
240 mm²	57.79	72.49	3.83	137.69	nr	**210.18**
300 mm²	72.75	91.26	4.28	153.87	nr	**245.13**
400 mm²	86.36	108.33	5.00	179.76	nr	**288.09**
600/1000 volt grade; four core (galvanized steel wire armour)						
1.5 mm²	1.35	1.69	0.67	24.08	nr	**25.77**
2.5 mm²	1.67	2.09	0.67	24.08	nr	**26.17**
4 mm²	2.04	2.55	0.71	25.52	nr	**28.07**
6 mm²	2.85	3.57	0.76	27.32	nr	**30.89**
10 mm²	3.91	4.91	1.14	40.98	nr	**45.89**
16 mm²	5.48	6.88	1.29	46.38	nr	**53.26**
25 mm²	7.81	9.80	1.99	71.53	nr	**81.33**
35 mm²	10.19	12.78	2.16	77.66	nr	**90.44**
50 mm²	14.58	18.29	2.49	89.51	nr	**107.80**
70 mm²	22.08	27.70	2.65	95.27	nr	**122.97**
95 mm²	29.60	37.13	2.98	107.14	nr	**144.27**
120 mm²	37.09	46.52	3.15	113.24	nr	**159.76**
150 mm²	46.80	58.71	3.50	125.83	nr	**184.54**
185 mm²	57.19	71.74	3.72	133.74	nr	**205.48**
240 mm²	74.04	92.87	4.33	155.67	nr	**248.54**
300 mm²	92.92	116.56	4.86	174.72	nr	**291.28**
400 mm²	96.39	120.92	5.46	196.29	nr	**317.21**

ELECTRICAL SUPPLY/POWER/LIGHTING

Item	Net Price £	Material £	Labour hours	Labour £	Unit	Total rate £
600/1000 volt grade; seven core (galvanized steel wire armour)						
1.5 mm²	1.88	2.36	0.81	29.12	nr	**31.48**
2.5 mm²	2.58	3.24	0.85	30.55	nr	**33.79**
4 mm²	4.68	5.87	0.93	33.43	nr	**39.30**
600/1000 volt grade; twelve core (galvanized steel wire armour)						
1.5 mm²	2.75	3.45	1.14	40.98	nr	**44.43**
2.5 mm²	4.13	5.19	1.13	40.63	nr	**45.82**
600/1000 volt grade; nineteen core (galvanized steel wire armour)						
1.5 mm²	4.16	5.22	1.54	55.37	nr	**60.59**
2.5 mm²	6.11	7.66	1.54	55.37	nr	**63.03**
600/1000 volt grade; twenty seven core (galvanized steel wire armour)						
1.5 mm²	5.61	7.03	1.94	69.74	nr	**76.77**
2.5 mm²	7.81	9.80	2.31	83.04	nr	**92.84**
600/1000 volt grade; thirty seven core (galvanized steel wire armour)						
1.5 mm²	7.01	8.79	2.53	90.96	nr	**99.75**
2.5 mm²	9.80	12.30	2.87	103.17	nr	**115.47**
UN-ARMOURED CABLE						
Cable: XLPE insulated; PVC sheathed 90c copper to CMA Code 6181e; for internal wiring; clipped to backgrounds; (Supports and fixings included)						
300/500 volt grade; single core						
6.0 mm²	0.71	0.90	0.09	3.24	m	**4.14**
10 mm²	1.06	1.33	0.10	3.60	m	**4.93**
16 mm²	1.78	2.23	0.12	4.31	m	**6.54**
Cable; LSF insulated to CMA Code 6491B; non-sheathed copper; laid/drawn in trunking/conduit						
450/750 volt grade; single core						
1.5 mm²	0.13	0.17	0.03	1.08	m	**1.25**
2.5 mm²	0.21	0.27	0.03	1.08	m	**1.35**
4.0 mm²	0.32	0.40	0.03	1.08	m	**1.48**
6.0 mm²	0.46	0.58	0.04	1.44	m	**2.02**
10.0 mm²	0.80	1.01	0.04	1.44	m	**2.45**
16.0 mm²	1.25	1.57	0.05	1.79	m	**3.36**
25.0 mm²	1.93	2.42	0.06	2.16	m	**4.58**
35.0 mm²	2.64	3.32	0.06	2.16	m	**5.48**
50.0 mm²	3.68	4.61	0.07	2.52	m	**7.13**
70.0 mm²	5.20	6.52	0.08	2.87	m	**9.39**
95.0 mm²	7.11	8.92	0.08	2.87	m	**11.79**
120.0 mm²	8.97	11.26	0.10	3.60	m	**14.86**
150.0 mm²	11.11	13.93	0.13	4.68	m	**18.61**
185.0 mm²	13.82	17.34	0.13	4.68	m	**22.02**
240.0 mm²	18.07	22.67	0.13	4.68	m	**27.35**
300.0 mm²	22.54	28.27	0.13	4.68	m	**32.95**

ELECTRICAL SUPPLY/POWER/LIGHTING

Item	Net Price £	Material £	Labour hours	Labour £	Unit	Total rate £
LV DISTRIBUTION: CABLES AND WIRING – cont						
Cable; twin & earth to CMA code 6242Y; clipped to backgrounds						
300/500 volt grade; PVC/PVC						
1.5 mm² 2C+E	0.36	0.45	0.01	0.36	m	**0.81**
1.5 mm² 3C+E	0.44	0.55	0.02	0.72	m	**1.27**
2.5 mm² 2C+E	0.67	0.84	0.02	0.72	m	**1.56**
4.0 mm² 2C+E	1.01	1.27	0.02	0.72	m	**1.99**
6.0 mm² 2C+E	1.44	1.80	0.02	0.72	m	**2.52**
10.0 mm² 2C+E	2.42	3.04	0.03	1.08	m	**4.12**
16.0 mm² 2C+E	3.63	4.56	0.03	1.08	m	**5.64**
300/500 volt grade; LSF/LSF						
1.5 mm² 2C+E	0.49	0.62	0.01	0.36	m	**0.98**
1.5 mm² 3C+E	0.70	0.87	0.02	0.72	m	**1.59**
2.5 mm² 2C+E	0.87	1.09	0.02	0.72	m	**1.81**
4.0 mm² 2C+E	1.04	1.30	0.02	0.72	m	**2.02**
6.0 mm² 2C+E	1.51	1.89	0.02	0.72	m	**2.61**
10.0 mm² 2C+E	3.46	4.35	0.03	1.08	m	**5.43**
16.0 mm² 2C+E	5.54	6.94	0.03	1.08	m	**8.02**
EARTH CABLE						
Cable; LSF insulated to CMA Code 6491B; non-sheathed copper; laid/drawn in trunking/conduit						
450/750 volt grade; single core						
1.5 mm²	0.13	0.17	0.03	1.08	m	**1.25**
2.5 mm²	0.21	0.27	0.03	1.08	m	**1.35**
4.0 mm²	0.32	0.40	0.03	1.08	m	**1.48**
6.0 mm²	0.46	0.58	0.04	1.44	m	**2.02**
10.0 mm²	0.80	1.01	0.04	1.44	m	**2.45**
16.0 mm²	1.25	1.57	0.05	1.79	m	**3.36**
25.0 mm²	1.93	2.42	0.06	2.16	m	**4.58**
35.0 mm²	2.64	3.32	0.06	2.16	m	**5.48**
50.0 mm²	3.68	4.61	0.07	2.52	m	**7.13**
70.0 mm²	5.20	6.52	0.08	2.87	m	**9.39**
95.0 mm²	7.11	8.92	0.08	2.87	m	**11.79**
120.0 mm²	8.97	11.26	0.10	3.60	m	**14.86**
150.0 mm²	11.11	13.93	0.13	4.68	m	**18.61**
185.0 mm²	13.82	17.34	0.16	5.76	m	**23.10**
240.0 mm²	18.07	22.67	0.20	7.19	m	**29.86**
300.0 mm²	22.54	28.27	0.13	4.68	m	**32.95**

ELECTRICAL SUPPLY/POWER/LIGHTING

Item	Net Price £	Material £	Labour hours	Labour £	Unit	Total rate £
FLEXIBLE CABLE						
Flexible cord; PVC insulated; PVC sheathed; copper stranded to CMA Code 218Y (laid loose)						
300 volt grade; two core						
0.50 mm²	0.16	0.20	0.07	2.52	m	**2.72**
0.75 mm²	0.21	0.27	0.07	2.52	m	**2.79**
300 volt grade; three core						
0.50 mm²	0.24	0.30	0.07	2.52	m	**2.82**
0.75 mm²	0.33	0.41	0.07	2.52	m	**2.93**
1.0 mm²	0.35	0.44	0.07	2.52	m	**2.96**
1.5 mm²	0.55	0.69	0.07	2.52	m	**3.21**
2.5 mm²	0.74	0.93	0.08	2.87	m	**3.80**
Flexible cord; PVC insulated; PVC sheathed; copper stranded to CMA Code 318Y (laid loose)						
300/500 volt grade; two core						
1.0 mm²	0.65	0.82	0.07	2.52	m	**3.34**
1.5 mm²	0.93	1.16	0.07	2.52	m	**3.68**
2.5 mm²	1.96	2.46	0.07	2.52	m	**4.98**
300/500 volt grade; three core						
0.75 mm²	0.39	0.49	0.07	2.52	m	**3.01**
1.5 mm²	0.65	0.82	0.07	2.52	m	**3.34**
2.5 mm²	1.69	2.12	0.08	2.87	m	**4.99**
300/500 volt grade; four core						
0.75 mm²	1.36	1.70	0.08	2.87	m	**4.57**
1.0 mm²	1.59	1.99	0.08	2.87	m	**4.86**
1.5 mm²	2.26	2.83	0.08	2.87	m	**5.70**
2.5 mm²	3.50	4.39	0.09	3.24	m	**7.63**
Flexible cord; PVC insulated; PVC sheathed for use in high temperature zones; copper stranded to CMA Code 309Y (laid loose)						
300/500 volt grade; two core						
0.50 mm²	0.78	0.97	0.07	2.52	m	**3.49**
0.75 mm²	0.85	1.06	0.07	2.52	m	**3.58**
1.0 mm²	0.35	0.44	0.07	2.52	m	**2.96**
1.5 mm²	0.48	0.60	0.07	2.52	m	**3.12**
2.5 mm²	0.71	0.90	0.07	2.52	m	**3.42**
300/500 volt grade; three core						
0.50 mm²	0.35	0.44	0.07	2.52	m	**2.96**
0.75 mm²	1.03	1.29	0.07	2.52	m	**3.81**
1.0 mm²	1.50	1.88	0.07	2.52	m	**4.40**
1.5 mm²	2.08	2.61	0.07	2.52	m	**5.13**
2.5 mm²	3.12	3.91	0.07	2.52	m	**6.43**

ELECTRICAL SUPPLY/POWER/LIGHTING

Item	Net Price £	Material £	Labour hours	Labour £	Unit	Total rate £
LV DISTRIBUTION: CABLES AND WIRING – cont						
Flexible cord; rubber insulated; rubber sheathed; copper stranded to CMA code 318 (laid loose)						
300/500 volt grade; two core						
0.50 mm²	0.41	0.52	0.07	2.52	m	**3.04**
0.75 mm²	0.51	0.64	0.07	2.52	m	**3.16**
1.0 mm²	0.62	0.77	0.07	2.52	m	**3.29**
1.5 mm²	0.78	0.97	0.07	2.52	m	**3.49**
2.5 mm²	1.16	1.46	0.07	2.52	m	**3.98**
300/500 volt grade; three core						
0.50 mm²	0.51	0.64	0.07	2.52	m	**3.16**
0.75 mm²	1.03	1.29	0.07	2.52	m	**3.81**
1.0 mm²	1.30	1.64	0.07	2.52	m	**4.16**
1.5 mm²	1.64	2.06	0.07	2.52	m	**4.58**
2.5 mm²	2.04	2.55	0.07	2.52	m	**5.07**
300/500 volt grade; four core						
0.50 mm²	0.58	0.73	0.08	2.87	m	**3.60**
0.75 mm²	1.37	1.71	0.08	2.87	m	**4.58**
1.0 mm²	1.59	1.99	0.08	2.87	m	**4.86**
1.5 mm²	2.11	2.64	0.08	2.87	m	**5.51**
2.5 mm²	2.38	2.99	0.08	2.87	m	**5.86**
Flexible cord; rubber insulated; rubber sheathed; for 90C operation; copper stranded to CMA Code 318 (laid loose)						
450/750 volt grade; two core						
0.50 mm²	0.36	0.45	0.07	2.52	m	**2.97**
0.75 mm²	0.56	0.71	0.07	2.52	m	**3.23**
1.0 mm²	0.57	0.72	0.07	2.52	m	**3.24**
1.5 mm²	0.72	0.91	0.07	2.52	m	**3.43**
2.5 mm²	1.05	1.32	0.07	2.52	m	**3.84**
450/750 volt grade; three core						
0.50 mm²	0.58	0.73	0.07	2.52	m	**3.25**
0.75 mm²	0.66	0.83	0.07	2.52	m	**3.35**
1.0 mm²	0.65	0.82	0.07	2.52	m	**3.34**
1.5 mm²	0.89	1.12	0.07	2.52	m	**3.64**
2.5 mm²	1.35	1.69	0.07	2.52	m	**4.21**
450/750 volt grade; four core						
0.75 mm²	0.63	0.80	0.07	2.52	m	**3.32**
1.0 mm²	0.82	1.03	0.08	2.87	m	**3.90**
1.5 mm²	1.25	1.57	0.08	2.87	m	**4.44**
2.5 mm²	1.75	2.20	0.08	2.87	m	**5.07**

ELECTRICAL SUPPLY/POWER/LIGHTING

Item	Net Price £	Material £	Labour hours	Labour £	Unit	Total rate £
Heavy flexible cable; rubber insulated; rubber sheathed; copper stranded to CMA Code 638P (laid loose)						
450/750 volt grade; two core						
1.0 mm²	0.56	0.71	0.08	2.87	m	3.58
1.5 mm²	0.67	0.84	0.08	2.87	m	3.71
2.5 mm²	0.91	1.14	0.08	2.87	m	4.01
450/750 volt grade; three core						
1.0 mm²	0.67	0.84	0.08	2.87	m	3.71
1.5 mm²	0.78	0.97	0.08	2.87	m	3.84
2.5 mm²	0.90	1.13	0.08	2.87	m	4.00
450/750 volt grade; four core						
1.0 mm²	0.91	1.14	0.08	2.87	m	4.01
1.5 mm²	1.00	1.25	0.08	2.87	m	4.12
2.5 mm²	1.37	1.71	0.08	2.87	m	4.58
FIRE RATED CABLE						
Cable, mineral insulated; copper sheathed with copper conductors; fixed with clips to backgrounds. BASEC approval to BS 6207 Part 1 1995; complies with BS 6387 Category CWZ						
Light duty 500 volt grade; bare						
2L 1.0	4.82	6.05	0.23	8.27	m	14.32
2L 1.5	5.50	6.90	0.23	8.27	m	15.17
2L 2.5	6.62	8.30	0.25	8.99	m	17.29
2L 4.0	9.36	11.74	0.25	8.99	m	20.73
3L 1.0	5.69	7.13	0.24	8.64	m	15.77
3L 1.5	6.77	8.49	0.25	8.99	m	17.48
3L 2.5	9.67	12.13	0.25	8.99	m	21.12
4L 1.0	6.50	8.15	0.25	8.99	m	17.14
4L 1.5	7.29	9.14	0.25	8.99	m	18.13
4L 2.5	11.46	14.38	0.26	9.34	m	23.72
7L 1.5	11.55	14.49	0.28	10.06	m	24.55
7L 2.5	14.64	18.37	0.27	9.71	m	28.08
Light duty 500 volt grade; LSF sheathed						
2L 1.0	3.76	4.72	0.23	8.27	m	12.99
2L 1.5	4.30	5.40	0.23	8.27	m	13.67
2L 2.5	4.80	6.03	0.25	8.99	m	15.02
2L 4.0	7.08	8.88	0.25	8.99	m	17.87
3L 1.0	4.45	5.58	0.24	8.64	m	14.22
3L 1.5	5.32	6.68	0.25	8.99	m	15.67
3L 2.5	6.97	8.75	0.25	8.99	m	17.74
4L 1.0	5.01	6.28	0.25	8.99	m	15.27
4L 1.5	6.10	7.65	0.25	8.99	m	16.64
4L 2.5	8.56	10.74	0.26	9.34	m	20.08
7L 1.5	8.84	11.09	0.28	10.06	m	21.15
7L 2.5	11.02	13.82	0.27	9.71	m	23.53
7L 1.0	7.21	9.05	0.27	9.71	m	18.76

ELECTRICAL SUPPLY/POWER/LIGHTING

Item	Net Price £	Material £	Labour hours	Labour £	Unit	Total rate £
LV DISTRIBUTION: CABLES AND WIRING – cont						
Cable, mineral insulated – cont						
Heavy duty 750 volt grade; bare						
1H 10	9.69	12.15	0.25	8.99	m	**21.14**
1H 16	13.11	16.44	0.26	9.34	m	**25.78**
1H 25	18.23	22.87	0.27	9.71	m	**32.58**
1H 35	26.31	33.01	0.32	11.50	m	**44.51**
1H 50	28.80	36.13	0.35	12.58	m	**48.71**
1H 70	38.44	48.22	0.38	13.66	m	**61.88**
1H 95	57.00	71.50	0.41	14.74	m	**86.24**
1H 120	59.96	75.22	0.46	16.53	m	**91.75**
1H 150	73.55	92.27	0.50	17.98	m	**110.25**
1H 185	89.69	112.50	0.56	20.14	m	**132.64**
1H 240	116.69	146.37	0.69	24.81	m	**171.18**
2H 1.5	8.94	11.21	0.25	8.99	m	**20.20**
2H 2.5	10.79	13.53	0.26	9.34	m	**22.87**
2H 4.0	13.21	16.58	0.26	9.34	m	**25.92**
2H 6.0	16.99	21.31	0.29	10.43	m	**31.74**
2H 10.0	21.71	27.24	0.34	12.22	m	**39.46**
2H 16.0	31.06	38.96	0.40	14.38	m	**53.34**
2H 25.0	42.82	53.72	0.44	15.81	m	**69.53**
3H 1.5	9.93	12.45	0.25	8.99	m	**21.44**
3H 2.5	12.30	15.43	0.25	8.99	m	**24.42**
3H 4.0	14.99	18.80	0.27	9.71	m	**28.51**
3H 6.0	18.51	23.22	0.30	10.79	m	**34.01**
3H 10.0	27.45	34.43	0.35	12.58	m	**47.01**
3H 16.0	36.98	46.39	0.41	14.74	m	**61.13**
3H 25.0	54.25	68.05	0.47	16.90	m	**84.95**
4H 1.5	12.06	15.13	0.24	8.64	m	**23.77**
4H 2.5	14.62	18.33	0.26	9.34	m	**27.67**
4H 4.0	18.06	22.66	0.29	10.43	m	**33.09**
4H 6.0	22.84	28.65	0.31	11.14	m	**39.79**
4H 10.0	32.44	40.69	0.37	13.29	m	**53.98**
4H 16.0	46.49	58.32	0.44	15.81	m	**74.13**
4H 25.0	66.38	83.27	0.52	18.69	m	**101.96**
7H 1.5	16.26	20.40	0.30	10.79	m	**31.19**
7H 2.5	21.68	27.19	0.32	11.50	m	**38.69**
12H 2.5	36.98	46.39	0.39	14.02	m	**60.41**
19H 1.5	54.21	68.01	0.42	15.10	m	**83.11**
Heavy duty 750 volt grade; LSF sheathed						
1H 10	7.52	9.43	0.25	8.99	m	**18.42**
1H 16	10.23	12.84	0.26	9.34	m	**22.18**
1H 25	14.06	17.64	0.27	9.71	m	**27.35**
1H 35	19.07	23.92	0.32	11.50	m	**35.42**
1H 50	23.14	29.03	0.35	12.58	m	**41.61**
1H 70	29.42	36.90	0.38	13.66	m	**50.56**
1H 95	41.25	51.74	0.41	14.74	m	**66.48**

ELECTRICAL SUPPLY/POWER/LIGHTING

Item	Net Price £	Material £	Labour hours	Labour £	Unit	Total rate £
1H 120	46.86	58.78	0.46	16.53	m	**75.31**
1H 150	57.00	71.50	0.50	17.98	m	**89.48**
1H 185	69.19	86.79	0.56	20.14	m	**106.93**
1H 240	89.67	112.48	0.68	24.45	m	**136.93**
2H 1.5	7.03	8.81	0.25	8.99	m	**17.80**
2H 2.5	8.42	10.56	0.26	9.34	m	**19.90**
2H 4.0	10.32	12.95	0.26	9.34	m	**22.29**
2H 6.0	13.20	16.55	0.29	10.43	m	**26.98**
2H 10.0	17.44	21.87	0.34	12.22	m	**34.09**
2H 16.0	23.98	30.08	0.40	14.38	m	**44.46**
2H 25.0	33.35	41.83	0.44	15.81	m	**57.64**
3H 1.5	7.65	9.60	0.25	8.99	m	**18.59**
3H 2.5	9.54	11.96	0.25	8.99	m	**20.95**
3H 4.0	11.72	14.71	0.27	9.71	m	**24.42**
3H 6.0	15.09	18.93	0.30	10.79	m	**29.72**
3H 10.0	19.98	25.07	0.35	12.58	m	**37.65**
3H 16.0	27.97	35.09	0.41	14.74	m	**49.83**
3H 25.0	40.20	50.42	0.47	16.90	m	**67.32**
4H 1.5	9.39	11.78	0.24	8.64	m	**20.42**
4H 2.5	11.40	14.30	0.26	9.34	m	**23.64**
4H 4.0	14.52	18.21	0.29	10.43	m	**28.64**
4H 6.0	18.40	23.08	0.31	11.14	m	**34.22**
4H 10.0	24.28	30.45	0.37	13.29	m	**43.74**
4H 16.0	34.65	43.47	0.44	15.81	m	**59.28**
4H 25.0	49.00	61.47	0.52	18.69	m	**80.16**
7H 1.5	12.08	15.15	0.30	10.79	m	**25.94**
7H 2.5	16.54	20.74	0.32	11.50	m	**32.24**
12H 1.5	20.80	26.10	0.39	14.02	m	**40.12**
12H 2.5	27.68	34.72	0.39	14.02	m	**48.74**
19H 1.5	40.14	50.36	0.42	15.10	m	**65.46**
Cable terminations for MI cable; Polymeric one piece moulding; containing grey sealing compound; testing; phase marking and connection Light duty 500 volt grade; brass gland; polymeric one moulding containing grey sealing compound; coloured conductor sleeving; Earth tag; plastic gland shroud						
2L 1.5	12.27	15.39	0.27	9.71	m	**25.10**
2L 2.5	12.27	15.39	0.27	9.71	m	**25.10**
3L 1.5	12.31	15.44	0.27	9.71	m	**25.15**
4L 1.5	12.31	15.44	0.27	9.71	m	**25.15**

ELECTRICAL SUPPLY/POWER/LIGHTING

Item	Net Price £	Material £	Labour hours	Labour £	Unit	Total rate £
LV DISTRIBUTION: CABLES AND WIRING – cont						
Cable terminations; for MI copper sheathed cable. Certified for installation in potentially explosive atmospheres; testing; phase marking and connection; BS 6207 Part 2 1995						
Light duty 500 volt grade; brass gland; brass pot with earth tail; pot closure; sealing compound; conductor sleeving; plastic gland shroud; identification markers						
2L 1.0	12.02	15.08	0.39	14.02	nr	29.10
2L 1.5	12.02	15.08	0.41	14.74	nr	29.82
2L 2.5	12.02	15.08	0.44	15.68	nr	30.76
2L 4.0	12.02	15.08	0.46	16.53	nr	31.61
3L 1.0	12.04	15.10	0.43	15.46	nr	30.56
3L 1.5	12.04	15.10	0.44	15.72	nr	30.82
3L 2.5	12.04	15.10	0.44	15.81	nr	30.91
4L 1.0	12.04	15.10	0.47	16.90	nr	32.00
4L 1.5	12.04	15.10	0.47	17.06	nr	32.16
4L 2.5	12.04	15.10	0.50	17.98	nr	33.08
7L 1.0	25.74	32.29	0.69	24.81	nr	57.10
7L 1.5	25.74	32.29	0.70	25.17	nr	57.46
7L 2.5	25.74	32.29	0.74	26.61	nr	58.90
Heavy duty 750 volt grade; brass gland; brass pot with earth tail; pot closure; sealing compound; conductor sleeving; plastic gland shroud; identification markers						
1H 10	12.08	15.15	0.37	13.29	nr	28.44
1H 16	12.08	15.15	0.39	14.02	nr	29.17
1H 25	12.08	15.15	0.56	20.14	nr	35.29
1H 35	12.08	15.15	0.57	20.50	nr	35.65
1H 50	25.85	32.42	0.60	21.57	nr	53.99
1H 70	25.85	32.42	0.67	24.08	nr	56.50
1H 95	25.85	32.42	0.75	26.97	nr	59.39
1H 120	45.72	57.36	0.94	33.79	nr	91.15
1H 150	45.72	57.36	0.99	35.58	nr	92.94
1H 185	45.72	57.36	1.26	45.29	nr	102.65
1H 240	72.13	90.48	1.37	49.25	nr	139.73
2H 1.5	12.08	15.15	0.42	15.10	nr	30.25
2H 2.5	12.08	15.15	0.44	15.93	nr	31.08
2H 4	12.08	15.15	0.47	16.90	nr	32.05
2H 6	12.08	15.15	0.54	19.42	nr	34.57
2H 10	25.85	32.42	0.58	20.84	nr	53.26
2H 16	25.85	32.42	0.69	24.81	nr	57.23
2H 25	40.17	50.39	0.77	27.69	nr	78.08
3H 1.5	12.09	15.16	0.44	15.81	nr	30.97
3H 2.5	12.09	15.16	0.47	17.01	nr	32.17
3H 4	12.09	15.16	0.57	20.50	nr	35.66
3H 6	25.91	32.50	0.61	21.93	nr	54.43
3H 10	25.91	32.50	0.65	23.37	nr	55.87

ELECTRICAL SUPPLY/POWER/LIGHTING

Item	Net Price £	Material £	Labour hours	Labour £	Unit	Total rate £
3H 16	25.91	32.50	0.78	28.03	nr	60.53
3H 25	61.89	77.64	0.85	30.55	nr	108.19
4H 1.5	12.09	15.16	0.52	18.69	nr	33.85
4H 2.5	12.09	15.16	0.53	19.05	nr	34.21
4H 4	25.91	32.50	0.60	21.57	nr	54.07
4H 6	25.91	32.50	0.65	23.37	nr	55.87
4H 10	25.91	32.50	0.69	24.81	nr	57.31
4H 16	40.17	50.39	0.88	31.64	nr	82.03
4H 25	61.89	77.64	0.93	33.43	nr	111.07
7H 1.5	25.85	32.42	0.71	25.52	nr	57.94
7H 2.5	25.85	32.42	0.74	26.61	nr	59.03
12H 1.5	40.17	50.39	0.85	30.55	nr	80.94
12H 2.5	40.17	50.39	1.00	35.95	nr	86.34
19H 2.5	61.89	77.64	1.11	39.91	nr	117.55
Cable; FP100; LOSH insulated; non-sheathed fire-resistant to LPCB Approved to BS 6387 Catergory CWZ; in conduit or trunking including terminations						
450/750 volt grade; single core						
1.0 mm²	0.74	0.93	0.13	4.68	m	5.61
1.5 mm²	0.33	0.41	0.13	4.68	m	5.09
2.5 mm²	0.46	0.58	0.13	4.68	m	5.26
4.0 mm²	0.67	0.84	0.14	4.93	m	5.77
6.0 mm²	0.83	1.04	0.13	4.68	m	5.72
10 mm²	1.52	1.90	0.16	5.76	m	7.66
16 mm²	2.32	2.91	0.16	5.76	m	8.67
Cable; FP200; Insudite insulated; LSOH sheathed screened fire-resistant BASEC Approved to BS 7629; fixed with clips to backgrounds						
300/500 volt grade; two core						
1.0 mm²	2.65	3.33	0.20	7.19	m	10.52
1.5 mm²	3.26	4.09	0.20	7.19	m	11.28
2.5 mm²	4.38	5.50	0.20	7.19	m	12.69
4.0 mm²	6.86	8.60	0.23	8.27	m	16.87
300/500 volt grade; three core						
1.0 mm²	3.71	4.66	0.23	8.27	m	12.93
1.5 mm²	4.46	5.60	0.23	8.27	m	13.87
2.5 mm²	5.39	6.76	0.23	8.27	m	15.03
4.0 mm²	8.83	11.08	0.25	8.99	m	20.07
300/500 volt grade; four core						
1.0 mm²	4.36	5.47	0.25	8.99	m	14.46
1.5 mm²	5.28	6.62	0.25	8.99	m	15.61
2.5 mm²	7.27	9.12	0.25	8.99	m	18.11
4.0 mm	11.12	13.94	0.28	10.06	m	24.00

ELECTRICAL SUPPLY/POWER/LIGHTING

Item	Net Price £	Material £	Labour hours	Labour £	Unit	Total rate £
LV DISTRIBUTION: CABLES AND WIRING – cont						
Terminations; including glanding-off, connection to equipment						
Two core						
1.0 mm²	0.65	0.82	0.35	12.58	nr	**13.40**
1.5 mm²	0.65	0.82	0.35	12.58	nr	**13.40**
2.5 mm²	0.65	0.82	0.35	12.58	nr	**13.40**
4.0 mm²	0.65	0.82	0.35	12.58	nr	**13.40**
Three core						
1.0 mm²	0.65	0.82	0.35	12.58	nr	**13.40**
1.5 mm²	0.65	0.82	0.35	12.58	nr	**13.40**
2.5 mm²	0.65	0.82	0.35	12.71	nr	**13.53**
4.0 mm²	0.65	0.82	0.35	12.58	nr	**13.40**
Four core						
1.0 mm²	0.65	0.82	0.35	12.58	nr	**13.40**
1.5 mm²	0.65	0.82	0.35	12.58	nr	**13.40**
2.5 mm²	0.65	0.82	0.35	12.58	nr	**13.40**
4.0 mm²	0.65	0.82	0.35	12.58	nr	**13.40**
Cable; FP400; polymeric insulated; LSOH sheathed fire-resistant; armoured; with copper stranded copper conductors; BASEC Approved to BS 7846; fixed with clips to backgrounds						
600/1000 volt grade; two core						
1.5 mm²	3.23	4.05	0.20	7.19	m	**11.24**
2.5 mm²	3.74	4.69	0.20	7.19	m	**11.88**
4.0 mm²	4.15	5.21	0.21	7.55	m	**12.76**
6.0 mm²	5.11	6.41	0.22	7.92	m	**14.33**
10 mm²	5.51	6.91	0.24	8.64	m	**15.55**
16 mm²	8.38	10.52	0.25	8.99	m	**19.51**
25 mm²	11.59	14.54	0.35	12.58	m	**27.12**
600/1000 volt grade; three core						
1.5 mm²	3.58	4.49	0.20	7.19	m	**11.68**
2.5 mm²	4.18	5.24	0.21	7.55	m	**12.79**
4.0 mm²	4.93	6.18	0.22	7.92	m	**14.10**
6.0 mm²	5.22	6.55	0.22	7.92	m	**14.47**
10 mm²	6.34	7.95	0.25	8.99	m	**16.94**
16 mm²	10.38	13.03	0.26	9.34	m	**22.37**
25 mm²	12.96	16.26	0.37	13.29	m	**29.55**
600/1000 volt grade; four core						
1.5 mm²	4.78	5.99	0.21	7.55	m	**13.54**
2.5 mm²	3.49	4.38	0.22	7.92	m	**12.30**
4.0 mm²	6.56	8.23	0.22	7.92	m	**16.15**
6.0 mm²	6.95	8.71	0.23	8.27	m	**16.98**
10 mm²	7.07	8.87	0.26	9.34	m	**18.21**
16 mm²	10.05	12.61	0.26	9.34	m	**21.95**
25 mm²	11.62	14.57	0.39	14.02	m	**28.59**

ELECTRICAL SUPPLY/POWER/LIGHTING

Item	Net Price £	Material £	Labour hours	Labour £	Unit	Total rate £
Terminations; including glanding-off, connection to equipment						
Two core						
1.5 mm²	2.03	2.54	0.58	20.84	nr	**23.38**
2.5 mm²	2.15	2.70	0.58	20.84	nr	**23.54**
4.0 mm²	2.35	2.95	0.61	22.04	nr	**24.99**
6.0 mm²	2.60	3.26	0.67	24.08	nr	**27.34**
10 mm²	2.86	3.58	1.00	35.95	nr	**39.53**
16 mm²	3.14	3.94	1.11	39.91	nr	**43.85**
25 mm²	3.45	4.32	1.70	61.12	nr	**65.44**
Three core						
1.5 mm²	3.06	3.84	0.62	22.29	nr	**26.13**
2.5 mm²	3.21	4.03	0.62	22.29	nr	**26.32**
4.0 mm²	3.54	4.44	0.66	23.87	nr	**28.31**
6.0 mm²	3.89	4.88	0.71	25.52	nr	**30.40**
10 mm²	4.27	5.35	1.06	38.11	nr	**43.46**
16 mm²	4.71	5.91	1.19	42.78	nr	**48.69**
25 mm²	5.17	6.48	1.81	65.06	nr	**71.54**
Four core						
1.5 mm²	4.07	5.11	0.67	24.08	nr	**29.19**
2.5 mm²	5.90	7.40	0.69	24.63	nr	**32.03**
4.0 mm²	4.50	5.64	0.71	25.52	nr	**31.16**
6.0 mm²	4.77	5.98	0.76	27.32	nr	**33.30**
10 mm²	6.76	8.48	1.14	40.98	nr	**49.46**
16 mm²	7.12	8.93	1.29	46.38	nr	**55.31**
25 mm²	12.64	15.86	1.35	48.53	nr	**64.39**
Cable; FP600; polymeric insulated; LSOH sheathed fire-resistant; armoured; with copper stranded copper conductors; BASEC Approved to BS 7846; fixed with clips to backgrounds						
600/1000 volt grade; two core						
4.0 mm²	5.66	7.10	0.21	7.55	m	**14.65**
10 mm²	6.58	8.25	0.24	8.64	m	**16.89**
16 mm²	7.78	9.76	0.25	8.99	m	**18.75**
600/1000 volt grade; three core						
4.0 mm²	6.58	8.25	0.22	7.92	m	**16.17**
6.0 mm²	6.49	8.14	0.22	7.92	m	**16.06**
10 mm²	8.24	10.34	0.25	8.99	m	**19.33**
16 mm²	9.90	12.42	0.26	9.34	m	**21.76**
25 mm²	13.18	16.53	0.37	13.29	m	**29.82**
35 mm²	15.89	19.94	0.38	13.66	m	**33.60**
50 mm²	21.12	26.49	0.40	14.38	m	**40.87**
70 mm²	25.83	32.40	0.42	15.10	m	**47.50**
95 mm²	38.17	47.88	0.45	16.18	m	**64.06**
600/1000 volt grade; four core						
4.0 mm²	6.46	8.11	0.22	7.92	nr	**16.03**
6.0 mm²	7.12	8.93	0.23	8.27	nr	**17.20**
10 mm²	7.95	9.97	0.26	9.34	nr	**19.31**

ELECTRICAL SUPPLY/POWER/LIGHTING

Item	Net Price £	Material £	Labour hours	Labour £	Unit	Total rate £
LV DISTRIBUTION: CABLES AND WIRING – cont						
Cable – cont						
600/1000 volt grade – cont						
16 mm²	9.13	11.46	0.26	9.34	nr	**20.80**
25 mm²	11.14	13.98	0.39	14.02	nr	**28.00**
35 mm²	14.26	17.89	0.40	14.38	m	**32.27**
50 mm²	17.71	22.22	0.42	15.10	m	**37.32**
70 mm²	22.88	28.71	0.45	16.18	m	**44.89**
95 mm²	30.28	37.98	0.47	16.90	m	**54.88**
120 mm²	37.60	47.16	0.48	17.26	m	**64.42**
150 mm²	45.19	56.68	0.51	18.33	m	**75.01**
185 mm²	55.57	69.71	0.52	18.69	m	**88.40**
240 mm²	71.44	89.61	0.55	19.77	m	**109.38**
300 mm²	87.53	109.79	0.59	21.21	m	**131.00**
400 mm²	109.64	137.54	0.63	22.66	m	**160.20**
Terminations; including glanding-off, connection to equipment						
Two core						
4.0 mm²	3.51	4.40	0.35	12.58	nr	**16.98**
10 mm²	4.08	5.12	0.35	12.58	nr	**17.70**
16 mm²	4.82	6.05	0.35	12.58	nr	**18.63**
Three core						
4.0 mm²	4.08	5.12	0.35	12.58	nr	**17.70**
6.0 mm²	4.02	5.04	0.35	12.58	nr	**17.62**
10 mm²	5.11	6.41	0.35	12.58	nr	**18.99**
16 mm²	6.14	7.71	0.35	12.58	nr	**20.29**
25 mm²	8.17	10.25	0.35	12.58	nr	**22.83**
35 mm²	9.85	12.35	0.35	12.58	m	**24.93**
50 mm²	13.10	16.43	0.35	12.58	m	**29.01**
70 mm²	16.01	20.08	0.35	12.58	m	**32.66**
95 mm²	23.66	29.68	0.35	12.58	m	**42.26**
Four core						
4.0 mm²	4.00	5.02	0.35	12.58	nr	**17.60**
6.0 mm²	4.42	5.54	0.35	12.58	nr	**18.12**
10 mm²	4.93	6.18	0.35	12.58	nr	**18.76**
16 mm²	5.66	7.10	0.35	12.58	nr	**19.68**
25 mm²	6.91	8.67	0.35	12.58	nr	**21.25**
35 mm²	8.84	11.09	0.35	12.58	m	**23.67**
50 mm²	10.98	13.78	0.35	12.58	m	**26.36**
70 mm²	14.19	17.80	0.35	12.58	m	**30.38**
95 mm²	18.78	23.55	0.35	12.58	m	**36.13**
120 mm²	23.31	29.24	0.35	12.58	m	**41.82**
150 mm²	28.02	35.15	0.35	12.58	m	**47.73**
185 mm²	34.46	43.23	0.35	12.58	m	**55.81**
240 mm²	44.29	55.55	0.35	12.58	m	**68.13**
300 mm²	54.26	68.06	0.35	12.58	m	**80.64**
400 mm²	67.97	85.27	0.35	12.58	m	**97.85**

ELECTRICAL SUPPLY/POWER/LIGHTING

Item	Net Price £	Material £	Labour hours	Labour £	Unit	Total rate £
Cable; Firetuff fire-resistant to BS 6387; fixed with clips to backgrounds						
Two core						
1.5 mm²	2.44	3.06	0.20	7.19	m	**10.25**
2.5 mm²	2.95	3.70	0.20	7.19	m	**10.89**
4.0 mm²	3.83	4.80	0.21	7.55	m	**12.35**
Three core						
1.5 mm²	2.98	3.74	0.20	7.19	m	**10.93**
2.5 mm²	3.37	4.22	0.21	7.55	m	**11.77**
4.0 mm²	4.87	6.10	0.22	7.92	m	**14.02**
Four core						
1.5 mm²	3.33	4.18	0.21	7.55	m	**11.73**
2.5 mm²	4.08	5.12	0.22	7.92	m	**13.04**
4.0 mm²	5.56	6.98	0.22	7.92	m	**14.90**

ELECTRICAL SUPPLY/POWER/LIGHTING

Item	Net Price £	Material £	Labour hours	Labour £	Unit	Total rate £
LV DISTRUBITION: MODULAR WIRING						
Modular wiring systems; including commissioning						
Master distribution box; steel; fixed to backgrounds; 6 Port						
4.0 mm 18 core armoured home run cable	171.18	214.73	0.90	32.35	nr	**247.08**
4.0 mm 24 core armoured home run cable	171.18	214.73	0.95	34.16	nr	**248.89**
4.0 mm 18 core armoured home run cable & data cable	181.86	228.12	0.95	34.16	nr	**262.28**
6.0 mm 18 core armoured home run cable	171.18	214.73	1.00	35.95	nr	**250.68**
6.0 mm 24 core armoured home run cable	171.18	214.73	1.10	39.55	nr	**254.28**
6.0 mm 18 core armoured home run cable & data cable	181.86	228.12	1.10	39.55	nr	**267.67**
Master distribution box; steel; fixed to backgrounds; 9 Port						
4.0 mm 27 core armoured home run cable	213.96	268.40	1.30	46.74	nr	**315.14**
4.0 mm 27 core armoured home run cable & data cable	213.96	268.40	1.45	52.14	nr	**320.54**
6.0 mm 27 core armoured home run cable	224.66	281.81	1.45	52.14	nr	**333.95**
6.0 mm 27 core armoured home run cable & data cable	224.66	281.81	1.55	55.72	nr	**337.53**
Metal clad cable; BSEN 60439 Part 2 1993; BASEC approved						
4.0 mm 18 core	14.94	18.74	0.30	10.79	m	**29.53**
4.0 mm 24 core	23.00	28.85	0.32	11.50	m	**40.35**
4.0 mm 27 core	23.17	29.06	0.35	12.58	m	**41.64**
6.0 mm 18 core	19.90	24.96	0.32	11.50	m	**36.46**
6.0 mm 27 core	27.72	34.78	0.35	12.58	m	**47.36**
Metal clad data cable						
Single twisted pair	2.91	3.65	0.18	6.47	m	**10.12**
Twin twisted pair	4.69	5.88	0.18	6.47	m	**12.35**
Distribution cables; armoured; BSEN 60439 Part 2 1993; BASEC approved						
3 wire; 6.1 metre long	65.61	82.30	0.92	33.07	nr	**115.37**
4 wire; 6.1 metre long	74.01	92.84	0.96	34.51	nr	**127.35**
Extender cables; armoured; BSEN 60439 Part 2 1993; BASEC approved						
3 wire						
0.9 m long	41.18	51.65	0.13	4.68	nr	**56.33**
1.5 m long	45.35	56.88	0.23	8.27	nr	**65.15**
2.1 m long	49.51	62.10	0.31	11.14	nr	**73.24**
2.7 m long	53.66	67.31	0.40	14.38	nr	**81.69**
3.4 m long	58.52	73.40	0.51	18.33	nr	**91.73**
4.6 m long	66.83	83.83	0.69	24.81	nr	**108.64**
6.1 m long	77.22	96.87	0.92	33.07	nr	**129.94**

ELECTRICAL SUPPLY/POWER/LIGHTING

Item	Net Price £	Material £	Labour hours	Labour £	Unit	Total rate £
7.6 m long	87.62	109.91	1.14	40.98	nr	**150.89**
9.1 m long	98.02	122.95	1.37	49.25	nr	**172.20**
10.7 m long	103.92	130.36	1.61	57.88	nr	**188.24**
4 wire						
0.9 m long	44.08	55.29	0.14	5.03	nr	**60.32**
1.5 m long	49.05	61.53	0.24	8.64	nr	**70.17**
2.1 m long	54.03	67.77	0.32	11.50	nr	**79.27**
2.7 m long	59.00	74.01	0.43	15.46	nr	**89.47**
3.4 m long	64.80	81.29	0.51	18.33	nr	**99.62**
4.6 m long	74.74	93.76	0.67	24.08	nr	**117.84**
6.1 m long	87.44	109.68	0.92	33.07	nr	**142.75**
7.6 m long	99.59	124.92	1.22	43.87	nr	**168.79**
9.1 m long	112.03	140.53	1.46	52.48	nr	**193.01**
10.7 m long	125.30	157.18	1.71	61.48	nr	**218.66**
3 wire; including twisted pair						
0.9 m long	56.45	70.81	0.13	4.68	nr	**75.49**
1.5 m long	63.43	79.56	0.23	8.27	nr	**87.83**
2.1 m long	70.39	88.30	0.31	11.14	nr	**99.44**
2.7 m long	77.36	97.04	0.40	14.38	nr	**111.42**
3.4 m long	85.49	107.24	0.51	18.33	nr	**125.57**
4.6 m long	99.42	124.71	0.69	24.81	nr	**149.52**
6.1 m long	115.80	145.26	0.92	33.07	nr	**178.33**
7.6 m long	134.24	168.39	1.14	40.98	nr	**209.37**
9.1 m long	151.67	190.25	1.37	49.25	nr	**239.50**
10.7 m long	170.24	213.55	1.61	57.88	nr	**271.43**
Extender whip ended cables; armoured;						
BSEN 60439 Part 2 1993; BASEC approved						
3 wire; 3.0 m long	39.43	49.46	0.30	10.79	nr	**60.25**
4 wire; 3.0 m long	43.71	54.84	0.30	10.79	nr	**65.63**
T Connectors						
3 wire						
6 Pole 20a Tee	48.79	61.20	0.10	3.60	nr	**64.80**
Snap fix	24.86	31.18	0.10	3.60	nr	**34.78**
1.5 m flexible cable	28.83	36.16	0.10	3.60	nr	**39.76**
1.5 m armoured cable	32.55	40.84	0.15	5.40	nr	**46.24**
1.5 m armoured cable with twisted pair	41.49	52.05	0.15	5.40	nr	**57.45**
4 wire						
Snap fix	25.57	32.08	0.10	3.60	nr	**35.68**
1.5 m flexible cable	30.02	37.65	0.10	3.60	nr	**41.25**
1.5 m armoured cable	33.75	42.34	0.18	6.47	nr	**48.81**
Splitters						
5 wire	23.74	29.78	0.20	7.19	nr	**36.97**
5 wire converter	30.61	38.39	0.20	7.19	nr	**45.58**
Switch modules						
3 wire; 6.1 m long armoured cable	56.11	70.38	0.75	26.97	nr	**97.35**
4 wire; 6.1 m long armoured cable	63.14	79.21	0.80	28.76	nr	**107.97**

ELECTRICAL SUPPLY/POWER/LIGHTING

Item	Net Price £	Material £	Labour hours	Labour £	Unit	Total rate £
LV DISTRUBITION: MODULAR WIRING – **cont**						
Distribution cables; unarmoured; IEC 998 **DIN/VDE 0628**						
3 wire; 6.1 m long	25.61	32.12	0.70	25.17	nr	**57.29**
4 wire; 6.1 m long	32.04	40.19	0.75	26.97	nr	**67.16**
Extender cables; unarmoured; IEC 998 **DIN/VDE 0628**						
3 wire						
0.9 m long	15.65	19.63	0.07	2.52	nr	**22.15**
1.5 m long	17.12	21.47	0.12	4.31	nr	**25.78**
2.1 m long	18.61	23.34	0.17	6.10	nr	**29.44**
2.7 m long	24.64	30.91	0.27	9.71	nr	**40.62**
3.4 m long	21.83	27.38	0.22	7.92	nr	**35.30**
4.6 m long	24.83	31.15	0.37	13.29	nr	**44.44**
6.1 m long	28.55	35.82	0.49	17.61	nr	**53.43**
7.6 m long	32.25	40.45	0.61	21.93	nr	**62.38**
9.1 m long	35.97	45.12	0.73	26.24	nr	**71.36**
10.7 m long	39.96	50.13	0.86	30.92	nr	**81.05**
4 wire						
0.9 m long	18.83	23.62	0.08	2.87	nr	**26.49**
1.5 m long	20.77	26.05	0.14	5.03	nr	**31.08**
2.1 m long	22.71	28.49	0.19	6.83	nr	**35.32**
2.7 m long	24.64	30.91	0.24	8.64	nr	**39.55**
3.4 m long	26.83	33.66	0.31	11.14	nr	**44.80**
4.6 m long	30.77	38.60	0.41	14.74	nr	**53.34**
6.1 m long	35.61	44.67	0.55	19.77	nr	**64.44**
7.6 m long	40.44	50.72	0.68	24.45	nr	**75.17**
9.1 m long	45.28	56.80	0.82	29.48	nr	**86.28**
10.7 m long	50.44	63.27	0.96	34.51	nr	**97.78**
5 wire						
0.9 m long	28.57	35.84	0.09	3.24	nr	**39.08**
1.5 m long	31.07	38.98	0.15	5.40	nr	**44.38**
2.1 m long	33.56	42.10	0.21	7.55	nr	**49.65**
2.7 m long	36.08	45.26	0.27	9.71	nr	**54.97**
3.4 m long	38.99	48.91	0.34	12.22	nr	**61.13**
4.6 m long	43.99	55.18	0.46	16.53	nr	**71.71**
6.1 m long	50.24	63.02	0.61	21.93	nr	**84.95**
7.6 m long	56.51	70.88	0.76	27.32	nr	**98.20**
9.1 m long	62.75	78.71	0.91	32.72	nr	**111.43**
10.7 m long	69.42	87.08	1.07	38.47	nr	**125.55**
Extender whip ended cables; armoured; **IEC 998 DIN/VDE 0628**						
3 wire; 2.5 mm; 3.0 m long	15.82	19.85	0.30	10.79	nr	**30.64**
4 wire; 2.5 mm; 3.0 m long	18.61	23.34	0.30	10.79	nr	**34.13**

ELECTRICAL SUPPLY/POWER/LIGHTING

Item	Net Price £	Material £	Labour hours	Labour £	Unit	Total rate £
T Connectors						
3 wire						
5 pin; direct fix	14.60	18.31	0.10	3.60	nr	**21.91**
5 pin; 1.5 mm flexible cable; 0.3 m long	20.44	25.64	0.15	5.40	nr	**31.04**
4 wire						
5 pin; direct fix	18.61	23.34	0.20	7.19	nr	**30.53**
5 pin; 1.5 mm flexible cable; 0.3 m long	23.16	29.05	0.20	7.19	nr	**36.24**
5 wire						
5 pin; direct fix	21.18	26.57	0.20	7.19	nr	**33.76**
Splitters						
3 way; 5 pin	12.20	15.30	0.25	8.99	nr	**24.29**
Switch modules						
3 wire	34.78	43.62	0.20	7.19	nr	**50.81**
4 wire	35.74	44.83	0.22	7.92	nr	**52.75**

ELECTRICAL SUPPLY/POWER/LIGHTING

Item	Net Price £	Material £	Labour hours	Labour £	Unit	Total rate £
LV DISTRUBITION: BUSBAR TRUNKING						
MAINS BUSBAR						
Low impedance busbar trunking; fixed to backgrounds including supports, fixings and connections/jointing to equipment						
Straight copper busbar						
1000 amp TP&N	539.84	677.17	3.41	122.60	m	**799.77**
1350 amp TP&N	663.54	832.34	3.58	128.70	m	**961.04**
2000 amp TP&N	832.25	1043.97	5.00	179.76	m	**1223.73**
2500 amp TP&N	1257.39	1577.27	5.90	212.11	m	**1789.38**
Extra for fittings mains bus bar						
IP54 protection						
1000 amp TP&N	36.81	46.18	2.16	77.66	m	**123.84**
1350 amp TP&N	39.56	49.63	2.61	93.83	m	**143.46**
2000 amp TP&N	54.88	68.85	3.51	126.19	m	**195.04**
2500 amp TP&N	65.16	81.74	3.96	142.36	m	**224.10**
End cover						
1000 amp TP&N	42.34	53.11	0.56	20.14	nr	**73.25**
1350 amp TP&N	44.16	55.40	0.56	20.14	nr	**75.54**
2000 amp TP&N	70.86	88.88	0.66	23.73	nr	**112.61**
2500 amp TP&N	71.77	90.03	0.66	23.73	nr	**113.76**
Edge elbow						
1000 amp TP&N	606.45	760.73	2.01	72.26	nr	**832.99**
1350 amp TP&N	682.51	856.14	2.01	72.26	nr	**928.40**
2000 amp TP&N	1058.58	1327.88	2.40	86.27	nr	**1414.15**
2500 amp TP&N	1390.37	1744.08	2.40	86.27	nr	**1830.35**
Flat elbow						
1000 amp TP&N	526.15	660.00	2.01	72.26	nr	**732.26**
1350 amp TP&N	570.51	715.65	2.01	72.26	nr	**787.91**
2000 amp TP&N	809.27	1015.15	2.40	86.27	nr	**1101.42**
2500 amp TP&N	1007.90	1264.31	2.40	86.27	nr	**1350.58**
Offset						
1000 amp TP&N	1056.51	1325.28	3.00	107.86	nr	**1433.14**
1350 amp TP&N	1299.48	1630.07	3.00	107.86	nr	**1737.93**
2000 amp TP&N	2047.51	2568.40	3.50	125.83	nr	**2694.23**
2500 amp TP&N	2330.62	2923.52	3.50	125.83	nr	**3049.35**
Edge Z unit						
1000 amp TP&N	1582.64	1985.27	3.00	107.86	nr	**2093.13**
1350 amp TP&N	2026.37	2541.87	3.00	107.86	nr	**2649.73**
2000 amp TP&N	3036.40	3808.86	3.50	125.83	nr	**3934.69**
2500 amp TP&N	3467.44	4349.55	3.50	125.83	nr	**4475.38**
Flat Z unit						
1000 amp TP&N	1711.15	2146.47	3.00	107.86	nr	**2254.33**
1350 amp TP&N	1711.54	2146.95	3.00	107.86	nr	**2254.81**
2000 amp TP&N	2548.28	3196.56	3.50	125.83	nr	**3322.39**
2500 amp TP&N	3063.85	3843.29	3.50	125.83	nr	**3969.12**

ELECTRICAL SUPPLY/POWER/LIGHTING

Item	Net Price £	Material £	Labour hours	Labour £	Unit	Total rate £
Edge tee						
1000 amp TP&N	1582.64	1985.27	2.20	79.09	nr	**2064.36**
1350 amp TP&N	2026.37	2541.87	2.20	79.09	nr	**2620.96**
2000 amp TP&N	3036.40	3808.86	2.60	93.48	nr	**3902.34**
2500 amp TP&N	3468.47	4350.85	2.60	93.48	nr	**4444.33**
Tap off; TP&N integral contactor/breaker						
18 amp	257.42	322.91	0.82	29.48	nr	**352.39**
Tap off; TP&N fusable with on-load switch;						
excludes fuses						
32 amp	728.35	913.64	0.82	29.48	nr	**943.12**
63 amp	744.21	933.54	0.88	31.64	nr	**965.18**
100 amp	910.12	1141.65	1.18	42.43	nr	**1184.08**
160 amp	1034.57	1297.77	1.41	50.69	nr	**1348.46**
250 amp	1335.90	1675.76	1.76	63.27	nr	**1739.03**
315 amp	1577.46	1978.77	2.06	74.05	nr	**2052.82**
Tap off; TP&N MCCB						
63 amp	921.11	1155.44	0.88	31.64	nr	**1187.08**
125 amp	1104.11	1384.99	1.18	42.43	nr	**1427.42**
160 amp	1205.37	1512.01	1.41	50.69	nr	**1562.70**
250 amp	1550.63	1945.12	1.76	63.27	nr	**2008.39**
400 amp	1976.41	2479.21	2.06	74.05	nr	**2553.26**
RISING MAINS BUSBAR						
Rising mains busbar; insulated supports,						
earth continuity bar; including couplers;						
fixed to backgrounds						
Straight aluminium bar						
200 amp TP&N	188.19	236.06	2.13	76.59	m	**312.65**
315 amp TP&N	210.83	264.47	2.15	77.29	m	**341.76**
400 amp TP&N	244.17	306.29	2.15	77.29	m	**383.58**
630 amp TP&N	301.35	378.01	2.47	88.79	m	**466.80**
800 amp TP&N	452.61	567.75	2.88	103.53	m	**671.28**
Extra for fittings rising busbar						
End feed unit						
200 amp TP&N	376.39	472.15	2.57	92.40	nr	**564.55**
315 amp TP&N	377.57	473.63	2.76	99.22	nr	**572.85**
400 amp TP&N	421.65	528.92	2.76	99.22	nr	**628.14**
630 amp TP&N	422.83	530.40	3.64	130.86	nr	**661.26**
800 amp TP&N	469.29	588.67	4.54	163.22	nr	**751.89**
Top feeder unit						
200 amp TP&N	376.39	472.15	2.57	92.40	nr	**564.55**
315 amp TP&N	377.57	473.63	2.76	99.22	nr	**572.85**
400 amp TP&N	421.65	528.92	2.76	99.22	nr	**628.14**
630 amp TP&N	422.83	530.40	3.64	130.86	nr	**661.26**
800 amp TP&N	469.29	588.67	4.54	163.22	nr	**751.89**
End cap						
200 amp TP&N	32.16	40.34	0.18	6.47	nr	**46.81**
315 amp TP&N	32.16	40.34	0.27	9.71	nr	**50.05**
400 amp TP&N	35.73	44.82	0.27	9.71	nr	**54.53**

ELECTRICAL SUPPLY/POWER/LIGHTING

Item	Net Price £	Material £	Labour hours	Labour £	Unit	Total rate £
LV DISTRUBITION: BUSBAR TRUNKING – cont						
Extra for fittings rising busbar – cont						
End cap – cont						
630 amp TP&N	35.73	44.82	0.41	14.74	nr	**59.56**
800 amp TP&N	103.62	129.98	0.41	14.74	nr	**144.72**
Edge elbow						
200 amp TP&N	44.07	55.28	0.55	19.77	nr	**75.05**
315 amp TP&N	45.26	56.77	0.94	33.79	nr	**90.56**
400 amp TP&N	337.08	422.83	0.94	33.79	nr	**456.62**
630 amp TP&N	338.27	424.32	1.45	52.14	nr	**476.46**
800 amp TP&N	320.40	401.91	1.45	52.14	nr	**454.05**
Flat elbow						
200 amp TP&N	145.31	182.28	0.55	19.77	nr	**202.05**
315 amp TP&N	145.31	182.28	0.94	33.79	nr	**216.07**
400 amp TP&N	196.53	246.52	0.94	33.79	nr	**280.31**
630 amp TP&N	197.72	248.02	1.45	52.14	nr	**300.16**
800 amp TP&N	272.76	342.15	1.45	52.14	nr	**394.29**
Edge tee						
200 amp TP&N	203.67	255.48	0.61	21.93	nr	**277.41**
315 amp TP&N	204.87	256.98	1.02	36.67	nr	**293.65**
400 amp TP&N	287.06	360.09	1.02	36.67	nr	**396.76**
630 amp TP&N	287.06	360.09	1.57	56.45	nr	**416.54**
800 amp TP&N	469.29	588.67	1.57	56.45	nr	**645.12**
Flat tee						
200 amp TP&N	262.04	328.70	0.61	21.93	nr	**350.63**
315 amp TP&N	204.87	256.98	1.02	36.67	nr	**293.65**
400 amp TP&N	412.12	516.96	1.02	36.67	nr	**553.63**
630 amp TP&N	413.31	518.46	1.57	56.45	nr	**574.91**
800 amp TP&N	577.68	724.64	1.57	56.45	nr	**781.09**
Tap off units						
TP&N fusable with on-load switch; excludes fuses						
32 amp	191.77	240.55	0.82	29.48	nr	**270.03**
63 amp	253.70	318.24	0.88	31.64	nr	**349.88**
100 amp	340.65	427.31	1.18	42.43	nr	**469.74**
250 amp	510.98	640.98	1.41	50.69	nr	**691.67**
400 amp	744.43	933.81	2.06	74.05	nr	**1007.86**
TP&N MCCB						
32 amp	195.34	245.03	0.82	29.48	nr	**274.51**
63 amp	270.38	339.17	0.88	31.64	nr	**370.81**
100 amp	433.55	543.85	1.18	42.43	nr	**586.28**
250 amp	734.90	921.86	1.41	50.69	nr	**972.55**
400 amp	1291.15	1619.62	2.06	74.05	nr	**1693.67**

ELECTRICAL SUPPLY/POWER/LIGHTING

Item	Net Price £	Material £	Labour hours	Labour £	Unit	Total rate £
LIGHTING BUSBAR						
Prewired busbar, plug-in trunking for lighting; galvanized sheet steel housing (PE); tin-plated copper conductors with tap-off units at 1 m intervals						
Straight lengths – 25 amp						
2 Pole & PE	31.28	39.23	0.16	5.76	m	**44.99**
4 Pole & PE	34.75	43.59	0.16	5.76	m	**49.35**
Straight lengths – 40 amp						
2 Pole & PE	31.28	39.23	0.16	5.76	m	**44.99**
4 Pole & PE	41.70	52.30	0.16	5.76	m	**58.06**
Components for prewired busbars, plug-in trunking for lighting						
Plug-in tap off units						
10 amp 4 Pole & PE; 3 m of cable	26.64	33.42	0.10	3.60	nr	**37.02**
16 amp 4 Pole & PE; 3 m of cable	28.96	36.33	0.10	3.60	nr	**39.93**
16 amp with phase selection, 2 Pole & PE; no cable	22.01	27.61	0.10	3.60	nr	**31.21**
Trunking components						
End feed unit & cover; 4 Pole & PE	33.59	42.13	0.23	8.27	nr	**50.40**
Centre feed unit	167.98	210.72	0.29	10.43	nr	**221.15**
Right hand, intermediate terminal box feed unit	35.91	45.05	0.23	8.27	nr	**53.32**
End cover (for R/hand feed)	9.27	11.63	0.06	2.16	nr	**13.79**
Flexible elbow unit	79.94	100.27	0.12	4.31	nr	**104.58**
Fixing bracket – universal	5.80	7.28	0.10	3.60	nr	**10.88**
Suspension Bracket – flat	4.63	5.81	0.10	3.60	nr	**9.41**
UNDERFLOOR BUSBAR						
Prewired busbar, plug-in trunking for underfloor power distribution; galvanized sheet steel housing (PE); copper conductors with tap-off units at 300 mm intervals						
Straight lengths – 63 amp						
2 Pole & PE	19.69	24.70	0.28	10.06	m	**34.76**
3 Pole & PE; Clean Earth System	25.49	31.98	0.28	10.06	m	**42.04**
Components for prewired busbars, plug-in trunking for underfloor power distribution						
Plug-in tap off units						
32 amp 2 Pole & PE; 3 m metal flexible prewired conduit	32.44	40.69	0.25	8.99	nr	**49.68**
32 amp 3 Pole & PE; clean earth; 3 m metal flexible prewired conduit	39.39	49.41	0.28	10.06	nr	**59.47**

ELECTRICAL SUPPLY/POWER/LIGHTING

Item	Net Price £	Material £	Labour hours	Labour £	Unit	Total rate £
LV DISTRUBITION: BUSBAR TRUNKING – cont						
Components for prewired busbars, plug-in trunking for underfloor power distribution – cont						
Trunking components						
End feed unit & cover; 2 Pole & PE	34.75	43.59	0.35	12.58	nr	**56.17**
End feed unit & cover; 3 Pole & PE; clean earth	38.23	47.96	0.38	13.66	nr	**61.62**
End cover; 2 Pole & PE	10.43	13.08	0.11	3.95	nr	**17.03**
End cover; 3 Pole & PE	11.58	14.53	0.11	3.95	nr	**18.48**
Flexible interlink/corner; 2 Pole & PE; 1 m long	60.24	75.57	0.34	12.22	nr	**87.79**
Flexible interlink/corner; 3 Pole & PE; 1 m long	68.35	85.74	0.35	12.58	nr	**98.32**
Flexible interlink/corner; 2 Pole & PE; 2 m long	74.14	93.00	0.37	13.29	nr	**106.29**
Flexible interlink/corner; 3 Pole & PE; 2 m long	81.09	101.72	0.37	13.29	nr	**115.01**

ELECTRICAL SUPPLY/POWER/LIGHTING

Item	Net Price £	Material £	Labour hours	Labour £	Unit	Total rate £
LV DISTRUBITION: CABLE SUPPORTS						
LADDER RACK						
Light duty Galvanized Steel Ladder Rack; fixed to backgrounds; including supports, fixings and brackets; earth continuity straps						
Straight lengths						
150 mm wide ladder	29.31	36.77	0.69	24.81	m	**61.58**
300 mm wide ladder	31.75	39.83	0.88	31.64	m	**71.47**
450 mm wide ladder	19.66	24.66	1.26	45.29	m	**69.95**
600 mm wide ladder	36.01	45.17	1.51	54.29	m	**99.46**
750 mm wide ladder	27.14	34.05	1.69	60.76	m	**94.81**
900 mm wide ladder	28.72	36.03	1.75	62.92	m	**98.95**
Extra over (cutting and jointing racking to fittings is included)						
Inside riser bend						
150 mm wide ladder	110.45	138.54	0.33	11.87	nr	**150.41**
300 mm wide ladder	112.67	141.33	0.56	20.14	nr	**161.47**
450 mm wide ladder	94.79	118.90	0.85	30.55	nr	**149.45**
600 mm wide ladder	152.90	191.80	0.99	35.58	nr	**227.38**
750 mm wide ladder	82.22	103.14	1.07	38.47	nr	**141.61**
900 mm wide ladder	85.85	107.69	1.12	40.26	nr	**147.95**
Outside riser bend						
150 mm wide ladder	110.45	138.54	0.43	15.46	nr	**154.00**
300 mm wide ladder	112.67	141.33	0.43	15.46	nr	**156.79**
450 mm wide ladder	74.89	93.95	0.73	26.24	nr	**120.19**
600 mm wide ladder	152.90	191.80	0.86	30.92	nr	**222.72**
750 mm wide ladder	82.22	103.14	0.97	34.88	nr	**138.02**
900 mm wide ladder	85.85	107.69	1.15	41.35	nr	**149.04**
Equal tee						
150 mm wide ladder	91.73	115.07	0.62	22.29	nr	**137.36**
300 mm wide ladder	128.67	161.40	0.62	22.29	nr	**183.69**
450 mm wide ladder	118.27	148.36	1.09	39.19	nr	**187.55**
600 mm wide ladder	105.72	132.62	1.12	40.26	nr	**172.88**
750 mm wide ladder	173.21	217.28	1.16	41.71	nr	**258.99**
900 mm wide ladder	179.16	224.74	1.21	43.50	nr	**268.24**
Unequal tee						
150 mm wide ladder	81.10	101.73	0.57	20.50	nr	**122.23**
300 mm wide ladder	75.31	94.47	0.57	20.50	nr	**114.97**
450 mm wide ladder	95.76	120.12	1.17	42.06	nr	**162.18**
600 mm wide ladder	81.98	102.84	1.17	42.06	nr	**144.90**
750 mm wide ladder	98.79	123.92	1.37	49.25	nr	**173.17**
900 mm wide ladder	99.61	124.95	1.37	49.25	nr	**174.20**
4 way crossovers						
150 mm wide ladder	234.97	294.75	0.72	25.89	nr	**320.64**
300 mm wide ladder	241.79	303.30	0.72	25.89	nr	**329.19**
450 mm wide ladder	172.74	216.69	1.13	40.63	nr	**257.32**

ELECTRICAL SUPPLY/POWER/LIGHTING

Item	Net Price £	Material £	Labour hours	Labour £	Unit	Total rate £
LV DISTRUBITION: CABLE SUPPORTS – cont						
Extra over (cutting and jointing racking to fittings is included) – cont						
4 way crossovers – cont						
600 mm wide ladder	268.71	337.08	1.29	46.38	nr	**383.46**
750 mm wide ladder	166.70	209.10	1.41	50.69	nr	**259.79**
900 mm wide ladder	171.20	214.75	1.64	58.96	nr	**273.71**
Flat bend (light duty)						
150 mm wide ladder	95.76	120.12	0.36	12.95	nr	**133.07**
300 mm wide ladder	99.82	125.22	0.40	14.38	nr	**139.60**
450 mm wide ladder	76.98	96.57	0.42	15.10	nr	**111.67**
600 mm wide ladder	124.98	156.78	0.59	21.21	nr	**177.99**
750 mm wide ladder	102.41	128.46	0.78	28.03	nr	**156.49**
900 mm wide ladder	111.81	140.26	0.86	30.92	nr	**171.18**
Heavy duty galvanized steel ladder rack; fixed to backgrounds; including supports, fixings and brackets; earth continuity straps						
Straight lengths						
150 mm wide ladder	38.94	48.84	0.68	24.45	m	**73.29**
300 mm wide ladder	42.00	52.68	0.79	28.40	m	**81.08**
450 mm wide ladder	44.91	56.34	1.07	38.47	m	**94.81**
600 mm wide ladder	48.59	60.95	1.24	44.58	m	**105.53**
750 mm wide ladder	56.31	70.64	1.49	53.56	m	**124.20**
900 mm wide ladder	48.72	61.12	1.67	60.03	m	**121.15**
Extra over (cutting and jointing racking to fittings is included)						
Flat bend						
150 mm wide ladder	105.20	131.96	0.34	12.22	nr	**144.18**
300 mm wide ladder	111.05	139.31	0.39	14.02	nr	**153.33**
450 mm wide ladder	123.50	154.92	0.43	15.46	nr	**170.38**
600 mm wide ladder	137.60	172.60	0.61	21.93	nr	**194.53**
750 mm wide ladder	154.97	194.40	0.82	29.48	nr	**223.88**
900 mm wide ladder	169.77	212.96	0.97	34.88	nr	**247.84**
Inside riser bend						
150 mm wide ladder	141.63	177.67	0.27	9.71	nr	**187.38**
300 mm wide ladder	143.60	180.13	0.45	16.18	nr	**196.31**
450 mm wide ladder	153.80	192.93	0.65	23.37	nr	**216.30**
600 mm wide ladder	164.14	205.90	0.81	29.12	nr	**235.02**
750 mm wide ladder	170.56	213.95	0.92	33.07	nr	**247.02**
900 mm wide ladder	187.99	235.82	1.06	38.11	nr	**273.93**
Outside riser bend						
150 mm wide ladder	141.63	177.67	0.27	9.71	nr	**187.38**
300 mm wide ladder	143.60	180.13	0.33	11.87	nr	**192.00**
450 mm wide ladder	153.80	192.93	0.61	21.93	nr	**214.86**
600 mm wide ladder	164.14	205.90	0.76	27.32	nr	**233.22**
750 mm wide ladder	170.56	213.95	0.94	33.79	nr	**247.74**
900 mm wide ladder	187.99	235.82	1.05	37.74	nr	**273.56**

ELECTRICAL SUPPLY/POWER/LIGHTING

Item	Net Price £	Material £	Labour hours	Labour £	Unit	Total rate £
Equal tee						
150 mm wide ladder	160.64	201.51	0.37	13.29	nr	**214.80**
300 mm wide ladder	183.89	230.68	0.57	20.50	nr	**251.18**
450 mm wide ladder	198.28	248.72	0.83	29.84	nr	**278.56**
600 mm wide ladder	218.69	274.32	0.92	33.07	nr	**307.39**
750 mm wide ladder	273.58	343.18	1.13	40.63	nr	**383.81**
900 mm wide ladder	287.98	361.24	1.20	43.14	nr	**404.38**
Unequal tee						
300 mm wide ladder	173.36	217.46	0.57	20.50	nr	**237.96**
450 mm wide ladder	185.17	232.28	1.17	42.06	nr	**274.34**
600 mm wide ladder	195.63	245.40	1.17	42.06	nr	**287.46**
750 mm wide ladder	242.88	304.67	1.25	44.95	nr	**349.62**
900 mm wide ladder	246.98	309.81	1.33	47.81	nr	**357.62**
4 way crossovers						
150 mm wide ladder	250.24	313.90	0.50	17.98	nr	**331.88**
300 mm wide ladder	262.66	329.48	0.67	24.08	nr	**353.56**
450 mm wide ladder	324.44	406.97	0.92	33.07	nr	**440.04**
600 mm wide ladder	345.20	433.01	1.07	38.47	nr	**471.48**
750 mm wide ladder	361.41	453.35	1.25	44.95	nr	**498.30**
900 mm wide ladder	394.73	495.15	1.36	48.90	nr	**544.05**
Double set						
150 mm wide ladder	210.42	263.95	0.51	18.33	nr	**282.28**
300 mm wide ladder	222.10	278.60	0.63	22.66	nr	**301.26**
450 mm wide ladder	247.00	309.84	0.86	30.92	nr	**340.76**
600 mm wide ladder	275.20	345.21	0.99	35.58	nr	**380.79**
750 mm wide ladder	309.94	388.79	1.16	41.71	nr	**430.50**
900 mm wide ladder	139.54	175.03	1.31	47.08	nr	**222.11**
Extra heavy duty galvanized steel ladder rack; fixed to backgrounds; including supports, fixings and brackets; earth continuity straps						
Straight lengths						
150 mm wide ladder	50.17	62.93	0.63	22.66	m	**85.59**
300 mm wide ladder	52.65	66.05	0.70	25.17	m	**91.22**
450 mm wide ladder	55.26	69.32	0.83	29.84	m	**99.16**
600 mm wide ladder	56.00	70.25	0.89	32.00	m	**102.25**
750 mm wide ladder	56.15	70.44	1.22	43.87	m	**114.31**
900 mm wide ladder	58.60	73.51	1.44	51.77	m	**125.28**
Extra over (cutting and jointing racking to fittings is included)						
Flat bend						
150 mm wide ladder	121.03	151.82	0.36	12.95	nr	**164.77**
300 mm wide ladder	125.94	157.98	0.39	14.02	nr	**172.00**
450 mm wide ladder	135.99	170.59	0.43	15.46	nr	**186.05**
600 mm wide ladder	153.84	192.98	0.61	21.93	nr	**214.91**
750 mm wide ladder	166.78	209.20	0.82	29.48	nr	**238.68**
900 mm wide ladder	180.63	226.59	0.97	34.88	nr	**261.47**

ELECTRICAL SUPPLY/POWER/LIGHTING

Item	Net Price £	Material £	Labour hours	Labour £	Unit	Total rate £
LV DISTRUBITION: CABLE SUPPORTS – cont						
Extra over (cutting and jointing racking to fittings is included) – cont						
Inside riser bend						
150 mm wide ladder	139.38	174.84	0.36	12.95	nr	**187.79**
300 mm wide ladder	141.64	177.68	0.39	14.02	nr	**191.70**
450 mm wide ladder	149.72	187.81	0.43	15.46	nr	**203.27**
600 mm wide ladder	165.13	207.14	0.61	21.93	nr	**229.07**
750 mm wide ladder	169.41	212.51	0.82	29.48	nr	**241.99**
900 mm wide ladder	182.67	229.14	0.97	34.88	nr	**264.02**
Outside riser bend						
150 mm wide ladder	139.75	175.30	0.36	12.95	nr	**188.25**
300 mm wide ladder	141.64	177.68	0.39	14.02	nr	**191.70**
450 mm wide ladder	149.72	187.81	0.41	14.74	nr	**202.55**
600 mm wide ladder	165.13	207.14	0.57	20.50	nr	**227.64**
750 mm wide ladder	169.41	212.51	0.82	29.48	nr	**241.99**
900 mm wide ladder	182.67	229.14	0.93	33.43	nr	**262.57**
Equal tee						
150 mm wide ladder	180.48	226.40	0.37	13.29	nr	**239.69**
300 mm wide ladder	199.26	249.95	0.57	20.50	nr	**270.45**
450 mm wide ladder	211.05	264.75	0.83	29.84	nr	**294.59**
600 mm wide ladder	237.87	298.38	0.92	33.07	nr	**331.45**
750 mm wide ladder	277.94	348.64	1.13	40.63	nr	**389.27**
900 mm wide ladder	288.00	361.27	1.20	43.14	nr	**404.41**
Unequal tee						
150 mm wide ladder	179.63	225.33	0.37	13.29	nr	**238.62**
300 mm wide ladder	197.82	248.15	0.57	20.50	nr	**268.65**
450 mm wide ladder	209.25	262.48	1.17	42.06	nr	**304.54**
600 mm wide ladder	235.41	295.30	1.17	42.06	nr	**337.36**
750 mm wide ladder	287.52	360.66	1.25	44.95	nr	**405.61**
900 mm wide ladder	289.38	363.00	1.33	47.81	nr	**410.81**
4 way crossovers						
150 mm wide ladder	234.23	293.82	0.50	17.98	nr	**311.80**
300 mm wide ladder	243.42	305.35	0.67	24.08	nr	**329.43**
450 mm wide ladder	284.76	357.20	0.92	33.07	nr	**390.27**
600 mm wide ladder	316.57	397.11	1.07	38.47	nr	**435.58**
750 mm wide ladder	329.08	412.80	1.25	44.95	nr	**457.75**
900 mm wide ladder	343.85	431.32	1.36	48.90	nr	**480.22**

ELECTRICAL SUPPLY/POWER/LIGHTING

Item	Net Price £	Material £	Labour hours	Labour £	Unit	Total rate £
LV DISTRUBITION: CABLE TRAYS						
Galvanized steel cable tray to BS 729; including standard coupling joints, fixings and earth continuity straps (supports and hangers are excluded)						
Light duty tray						
Straight lengths						
50 mm wide	3.75	4.70	0.19	6.83	m	**11.53**
75 mm wide	4.72	5.92	0.23	8.27	m	**14.19**
100 mm wide	5.82	7.30	0.31	11.14	m	**18.44**
150 mm wide	7.62	9.55	0.33	11.87	m	**21.42**
225 mm wide	14.55	18.26	0.39	14.02	m	**32.28**
300 mm wide	20.47	25.68	0.49	17.61	m	**43.29**
450 mm wide	11.53	14.46	0.60	21.57	m	**36.03**
600 mm wide	29.13	36.55	0.79	28.40	m	**64.95**
750 mm wide	37.00	46.41	1.04	37.40	m	**83.81**
900 mm wide	46.01	57.71	1.26	45.29	m	**103.00**
Extra over (cutting and jointing tray to fittings is included)						
Straight reducer						
75 mm wide	15.26	19.14	0.22	7.92	nr	**27.06**
100 mm wide	15.56	19.52	0.25	8.99	nr	**28.51**
150 mm wide	20.34	25.51	0.27	9.71	nr	**35.22**
225 mm wide	26.70	33.49	0.34	12.22	nr	**45.71**
300 mm wide	33.99	42.64	0.39	14.02	nr	**56.66**
450 mm wide	26.62	33.39	0.49	17.61	nr	**51.00**
600 mm wide	42.98	53.92	0.54	19.42	nr	**73.34**
750 mm wide	56.46	70.83	0.61	21.93	nr	**92.76**
900 mm wide	65.60	82.29	0.69	24.81	nr	**107.10**
Flat bend; 90°						
50 mm wide	9.16	11.49	0.19	6.83	nr	**18.32**
75 mm wide	9.53	11.95	0.24	8.64	nr	**20.59**
100 mm wide	10.72	13.45	0.28	10.06	nr	**23.51**
150 mm wide	11.59	14.54	0.30	10.79	nr	**25.33**
225 mm wide	15.05	18.88	0.36	12.95	nr	**31.83**
300 mm wide	22.01	27.61	0.44	15.81	nr	**43.42**
450 mm wide	16.05	20.14	0.57	20.50	nr	**40.64**
600 mm wide	35.48	44.51	0.69	24.81	nr	**69.32**
750 mm wide	50.25	63.03	0.81	29.12	nr	**92.15**
900 mm wide	73.90	92.70	0.94	33.79	nr	**126.49**
Adjustable riser						
50 mm wide	18.40	23.08	0.26	9.34	nr	**32.42**
75 mm wide	19.90	24.96	0.29	10.43	nr	**35.39**
100 mm wide	19.27	24.17	0.32	11.50	nr	**35.67**
150 mm wide	22.30	27.98	0.36	12.95	nr	**40.93**
225 mm wide	29.37	36.84	0.44	15.81	nr	**52.65**
300 mm wide	31.47	39.48	0.52	18.69	nr	**58.17**
450 mm wide	27.19	34.10	0.66	23.73	nr	**57.83**

Material Costs/Measured Work Prices – Electrical Installations

ELECTRICAL SUPPLY/POWER/LIGHTING

Item	Net Price £	Material £	Labour hours	Labour £	Unit	Total rate £
LV DISTRUBITION: CABLE TRAYS – cont						
Extra over (cutting and jointing tray to fittings is included) – cont						
Adjustable riser – cont						
600 mm wide	42.03	52.72	0.79	28.40	nr	81.12
750 mm wide	54.09	67.85	1.03	37.03	nr	104.88
900 mm wide	63.66	79.86	1.10	39.55	nr	119.41
Inside riser; 90°						
50 mm wide	5.78	7.25	0.28	10.06	nr	17.31
75 mm wide	6.04	7.57	0.31	11.14	nr	18.71
100 mm wide	6.64	15.90	0.33	11.87	nr	27.77
150 mm wide	7.89	9.90	0.37	13.29	nr	23.19
225 mm wide	9.70	12.16	0.44	15.81	nr	27.97
300 mm wide	10.26	12.87	0.53	19.05	nr	31.92
450 mm wide	16.42	20.60	0.67	24.08	nr	44.68
600 mm wide	41.93	52.60	0.79	28.40	nr	81.00
750 mm wide	51.72	64.88	0.95	34.16	nr	99.04
900 mm wide	61.45	77.08	1.11	39.91	nr	116.99
Outside riser; 90°						
50 mm wide	5.78	7.25	0.28	10.06	nr	17.31
75 mm wide	6.04	7.57	0.31	11.14	nr	18.71
100 mm wide	6.64	8.33	0.33	11.87	nr	20.20
150 mm wide	7.89	9.90	0.37	13.29	nr	23.19
225 mm wide	9.70	12.16	0.44	15.81	nr	27.97
300 mm wide	10.26	12.87	0.53	19.05	nr	31.92
450 mm wide	16.42	20.60	0.67	24.08	nr	44.68
600 mm wide	23.32	29.25	0.79	28.40	nr	57.65
750 mm wide	31.66	39.72	0.95	34.16	nr	73.88
900 mm wide	31.42	39.41	1.11	39.91	nr	79.32
Equal tee						
50 mm wide	13.83	17.35	0.30	10.79	nr	28.14
75 mm wide	14.38	18.04	0.31	11.14	nr	29.18
100 mm wide	14.81	18.58	0.35	12.58	nr	31.16
150 mm wide	17.97	22.55	0.36	12.95	nr	35.50
225 mm wide	23.79	29.84	0.44	15.81	nr	45.65
300 mm wide	33.05	41.46	0.54	19.42	nr	60.88
450 mm wide	23.52	29.50	0.71	25.52	nr	55.02
600 mm wide	46.81	58.72	0.92	33.07	nr	91.79
750 mm wide	70.17	88.02	1.19	42.78	nr	130.80
900 mm wide	99.80	125.19	1.44	51.77	nr	176.96
Unequal tee						
75 mm wide	8.91	11.18	0.38	13.66	nr	24.84
100 mm wide	8.91	11.18	0.39	14.02	nr	25.20
150 mm wide	9.17	11.50	0.43	15.46	nr	26.96
225 mm wide	10.65	13.36	0.50	17.98	nr	31.34
300 mm wide	13.41	16.82	0.63	22.66	nr	39.48
450 mm wide	17.79	22.31	0.80	28.76	nr	51.07
600 mm wide	27.90	35.00	1.02	36.67	nr	71.67
750 mm wide	79.69	99.96	1.12	40.26	nr	140.22
900 mm wide	105.05	131.78	1.35	48.53	nr	180.31

ELECTRICAL SUPPLY/POWER/LIGHTING

Item	Net Price £	Material £	Labour hours	Labour £	Unit	Total rate £
4 way crossovers						
50 mm wide	20.30	25.47	0.38	13.66	nr	**39.13**
75 mm wide	20.68	25.94	0.40	14.38	nr	**40.32**
100 mm wide	21.76	27.29	0.40	14.38	nr	**41.67**
150 mm wide	25.78	32.33	0.44	15.81	nr	**48.14**
225 mm wide	34.73	43.57	0.53	19.05	nr	**62.62**
300 mm wide	47.80	59.96	0.64	23.00	nr	**82.96**
450 mm wide	41.57	52.15	0.84	30.20	nr	**82.35**
600 mm wide	46.81	58.72	1.03	37.03	nr	**95.75**
750 mm wide	70.17	88.02	1.13	40.63	nr	**128.65**
900 mm wide	99.80	125.19	1.36	48.90	nr	**174.09**
Medium duty tray with return flange						
Straight lengths						
50 mm wide	7.08	8.88	0.33	11.87	m	**20.75**
75 mm wide	8.28	10.38	0.33	11.87	m	**22.25**
100 mm wide	8.81	11.05	0.35	12.58	m	**23.63**
150 mm wide	10.67	13.38	0.39	14.02	m	**27.40**
225 mm wide	13.80	17.32	0.45	16.18	m	**33.50**
300 mm wide	21.30	26.72	0.57	20.50	m	**47.22**
450 mm wide	29.77	37.34	0.69	24.81	m	**62.15**
600 mm wide	39.86	50.00	0.91	32.72	m	**82.72**
Extra over (cutting and jointing tray to fittings is included)						
Straight reducer						
100 mm wide	29.73	37.30	0.25	8.99	nr	**46.29**
150 mm wide	32.48	40.75	0.27	9.71	nr	**50.46**
225 mm wide	37.42	46.94	0.34	12.22	nr	**59.16**
300 mm wide	43.33	54.35	0.39	14.02	nr	**68.37**
450 mm wide	56.00	70.25	0.49	17.61	nr	**87.86**
600 mm wide	69.38	87.04	0.54	19.42	nr	**106.46**
Flat bend; 90°						
75 mm wide	41.32	51.83	0.24	8.64	nr	**60.47**
100 mm wide	43.54	54.61	0.28	10.06	nr	**64.67**
150 mm wide	46.58	58.43	0.30	10.79	nr	**69.22**
225 mm wide	53.60	67.23	0.36	12.95	nr	**80.18**
300 mm wide	64.30	80.66	0.44	15.81	nr	**96.47**
450 mm wide	68.97	86.52	0.57	20.50	nr	**107.02**
600 mm wide	110.20	138.23	0.69	24.81	nr	**163.04**
Adjustable bend						
75 mm wide	45.33	56.86	0.29	10.43	nr	**67.29**
100 mm wide	49.34	61.89	0.32	11.50	nr	**73.39**
150 mm wide	56.42	70.77	0.36	12.95	nr	**83.72**
225 mm wide	63.23	79.32	0.44	15.81	nr	**95.13**
300 mm wide	71.46	89.64	0.52	18.69	nr	**108.33**
Adjustable riser						
75 mm wide	40.43	50.71	0.29	10.43	nr	**61.14**
100 mm wide	41.10	51.55	0.32	11.50	nr	**63.05**
150 mm wide	45.30	56.83	0.36	12.95	nr	**69.78**

ELECTRICAL SUPPLY/POWER/LIGHTING

Item	Net Price £	Material £	Labour hours	Labour £	Unit	Total rate £
LV DISTRUBITION: CABLE TRAYS – cont						
Extra over (cutting and jointing tray to fittings is included) – cont						
Adjustable riser – cont						
225 mm wide	48.05	60.28	0.44	15.81	nr	**76.09**
300 mm wide	51.23	64.27	0.52	18.69	nr	**82.96**
450 mm wide	67.20	84.29	0.66	23.73	nr	**108.02**
600 mm wide	82.43	103.40	0.79	28.40	nr	**131.80**
Inside riser; 90°						
75 mm wide	24.29	30.46	0.31	11.14	nr	**41.60**
100 mm wide	24.53	30.77	0.33	11.87	nr	**42.64**
150 mm wide	27.94	35.04	0.37	13.29	nr	**48.33**
225 mm wide	34.13	42.82	0.44	15.81	nr	**58.63**
300 mm wide	41.46	52.01	0.53	19.05	nr	**71.06**
450 mm wide	59.84	75.06	0.67	24.08	nr	**99.14**
600 mm wide	96.03	120.46	0.79	28.40	nr	**148.86**
Outside riser; 90°						
75 mm wide	24.29	30.46	0.31	11.14	nr	**41.60**
100 mm wide	24.53	30.77	0.33	11.87	nr	**42.64**
150 mm wide	27.94	35.04	0.37	13.29	nr	**48.33**
225 mm wide	34.13	42.82	0.44	15.81	nr	**58.63**
300 mm wide	41.46	52.01	0.53	19.05	nr	**71.06**
450 mm wide	59.84	75.06	0.67	24.08	nr	**99.14**
600 mm wide	96.03	120.46	0.79	28.40	nr	**148.86**
Equal tee						
75 mm wide	58.79	73.74	0.31	11.14	nr	**84.88**
100 mm wide	61.17	76.73	0.35	12.58	nr	**89.31**
150 mm wide	64.77	81.24	0.36	12.95	nr	**94.19**
225 mm wide	69.93	87.72	0.74	26.61	nr	**114.33**
300 mm wide	85.87	107.71	0.54	19.42	nr	**127.13**
450 mm wide	111.09	139.35	0.71	25.52	nr	**164.87**
600 mm wide	158.78	199.17	0.92	33.07	nr	**232.24**
Unequal tee						
100 mm wide	57.57	72.22	0.39	14.02	nr	**86.24**
150 mm wide	57.57	72.22	0.43	15.46	nr	**87.68**
225 mm wide	67.34	84.47	0.50	17.98	nr	**102.45**
300 mm wide	81.74	102.54	0.63	22.66	nr	**125.20**
450 mm wide	107.68	135.07	0.80	28.76	nr	**163.83**
600 mm wide	154.69	194.04	1.02	36.67	nr	**230.71**
4 way crossovers						
75 mm wide	76.56	96.04	0.40	14.38	nr	**110.42**
100 mm wide	82.51	103.50	0.40	14.38	nr	**117.88**
150 mm wide	71.71	89.96	0.44	15.81	nr	**105.77**
225 mm wide	103.68	130.05	0.53	19.05	nr	**149.10**
300 mm wide	120.64	151.33	0.64	23.00	nr	**174.33**
450 mm wide	153.54	192.60	0.84	30.20	nr	**222.80**
600 mm wide	223.82	280.76	1.03	37.03	nr	**317.79**

ELECTRICAL SUPPLY/POWER/LIGHTING

Item	Net Price £	Material £	Labour hours	Labour £	Unit	Total rate £
Heavy duty tray with return flange						
Straight lengths						
75 mm	14.80	18.57	0.34	12.22	m	**30.79**
100 mm	15.01	18.83	0.36	12.95	m	**31.78**
150 mm	16.59	20.81	0.40	14.38	m	**35.19**
225 mm	20.13	25.26	0.46	16.53	m	**41.79**
300 mm	22.84	28.65	0.58	20.84	m	**49.49**
450 mm	40.18	50.40	0.70	25.17	m	**75.57**
600 mm	56.00	70.25	0.92	33.07	m	**103.32**
750 mm	77.45	97.15	1.01	36.31	m	**133.46**
900 mm	81.08	101.71	1.14	40.98	m	**142.69**
Extra over (cutting and jointing tray to fittings is included)						
Straight reducer						
100 mm wide	43.73	54.86	0.25	8.99	nr	**63.85**
150 mm wide	45.39	56.94	0.27	9.71	nr	**66.65**
225 mm wide	51.55	64.67	0.34	12.22	nr	**76.89**
300 mm wide	57.47	72.09	0.39	14.02	nr	**86.11**
450 mm wide	82.19	103.10	0.49	17.61	nr	**120.71**
600 mm wide	91.38	114.63	0.54	19.42	nr	**134.05**
750 mm wide	115.89	145.38	0.60	21.57	nr	**166.95**
900 mm wide	127.13	159.48	0.66	23.73	nr	**183.21**
Flat bend; 90°						
75 mm wide	54.93	68.90	0.24	8.64	nr	**77.54**
100 mm wide	59.51	74.65	0.28	10.06	nr	**84.71**
150 mm wide	64.82	81.31	0.30	10.79	nr	**92.10**
225 mm wide	65.80	82.54	0.36	12.95	nr	**95.49**
300 mm wide	74.71	93.72	0.44	15.81	nr	**109.53**
450 mm wide	119.99	150.52	0.57	20.50	nr	**171.02**
600 mm wide	163.32	204.87	0.69	24.81	nr	**229.68**
750 mm wide	208.42	261.44	0.83	29.84	nr	**291.28**
900 mm wide	217.86	273.28	1.01	36.31	nr	**309.59**
Adjustable bend						
75 mm wide	50.87	63.81	0.29	10.43	nr	**74.24**
100 mm wide	56.23	70.54	0.32	11.50	nr	**82.04**
150 mm wide	59.41	74.52	0.36	12.95	nr	**87.47**
225 mm wide	65.39	82.03	0.44	15.81	nr	**97.84**
300 mm wide	77.58	97.32	0.52	18.69	nr	**116.01**
Adjustable riser						
75 mm wide	44.82	56.22	0.29	10.43	nr	**66.65**
100 mm wide	43.30	54.32	0.32	11.50	nr	**65.82**
150 mm wide	51.60	64.72	0.36	12.95	nr	**77.67**
225 mm wide	55.26	69.32	0.44	15.81	nr	**85.13**
300 mm wide	59.85	75.07	0.52	18.69	nr	**93.76**
450 mm wide	72.84	91.37	0.66	23.73	nr	**115.10**
600 mm wide	86.78	108.85	0.79	28.40	nr	**137.25**
750 mm wide	105.09	131.82	1.03	37.03	nr	**168.85**
900 mm wide	122.04	153.08	1.10	39.55	nr	**192.63**

ELECTRICAL SUPPLY/POWER/LIGHTING

Item	Net Price £	Material £	Labour hours	Labour £	Unit	Total rate £
LV DISTRUBITION: CABLE TRAYS – cont						
Extra over (cutting and jointing tray to fittings is included) – cont						
Inside riser; 90°						
75 mm wide	39.97	50.14	0.31	11.14	nr	**61.28**
100 mm wide	40.50	50.80	0.33	11.87	nr	**62.67**
150 mm wide	43.87	55.03	0.37	13.29	nr	**68.32**
225 mm wide	45.86	57.52	0.44	15.81	nr	**73.33**
300 mm wide	47.31	59.35	0.53	19.05	nr	**78.40**
450 mm wide	81.17	101.82	0.67	24.08	nr	**125.90**
600 mm wide	99.58	124.91	0.79	28.40	nr	**153.31**
750 mm wide	123.25	154.60	0.95	34.16	nr	**188.76**
900 mm wide	145.31	182.28	1.11	39.91	nr	**222.19**
Outside riser; 90°						
75 mm wide	39.97	50.14	0.31	11.14	nr	**61.28**
100 mm wide	40.50	50.80	0.33	11.87	nr	**62.67**
150 mm wide	43.87	55.03	0.37	13.29	nr	**68.32**
225 mm wide	45.86	57.52	0.44	15.81	nr	**73.33**
300 mm wide	47.31	59.35	0.53	19.05	nr	**78.40**
450 mm wide	81.17	101.82	0.67	24.08	nr	**125.90**
600 mm wide	99.58	124.91	0.79	28.40	nr	**153.31**
750 mm wide	123.25	154.60	0.95	34.16	nr	**188.76**
900 mm wide	145.31	182.28	1.11	39.91	nr	**222.19**
Equal tee						
75 mm wide	72.79	91.30	0.31	11.14	nr	**102.44**
100 mm wide	77.60	97.34	0.35	12.58	nr	**109.92**
150 mm wide	84.26	105.69	0.36	12.95	nr	**118.64**
225 mm wide	93.52	117.31	0.44	15.81	nr	**133.12**
300 mm wide	106.08	133.07	0.54	19.42	nr	**152.49**
450 mm wide	154.03	193.21	0.71	25.52	nr	**218.73**
600 mm wide	200.18	251.10	0.92	33.07	nr	**284.17**
750 mm wide	246.94	309.76	1.19	42.78	nr	**352.54**
900 mm wide	301.36	378.02	1.45	52.14	nr	**430.16**
Unequal tee						
75 mm wide	69.63	87.35	0.38	13.66	nr	**101.01**
100 mm wide	77.78	97.56	0.39	14.02	nr	**111.58**
150 mm wide	85.26	106.95	0.43	15.46	nr	**122.41**
225 mm wide	98.11	123.07	0.50	17.98	nr	**141.05**
300 mm wide	106.08	133.07	0.63	22.66	nr	**155.73**
450 mm wide	154.03	193.21	0.80	28.76	nr	**221.97**
600 mm wide	203.36	255.09	1.02	36.67	nr	**291.76**
750 mm wide	268.22	336.46	1.12	40.26	nr	**376.72**
900 mm wide	324.45	406.99	1.35	48.53	nr	**455.52**
4 way crossovers						
75 mm wide	70.67	88.65	0.40	14.38	nr	**103.03**
100 mm wide	101.30	127.08	0.40	14.38	nr	**141.46**
150 mm wide	101.30	127.08	0.44	15.81	nr	**142.89**
225 mm wide	140.85	176.68	0.53	19.05	nr	**195.73**
300 mm wide	155.31	194.82	0.64	23.00	nr	**217.82**
450 mm wide	223.41	280.25	0.84	30.20	nr	**310.45**
600 mm wide	303.65	380.90	1.03	37.03	nr	**417.93**

ELECTRICAL SUPPLY/POWER/LIGHTING

Item	Net Price £	Material £	Labour hours	Labour £	Unit	Total rate £
750 mm wide	370.35	464.56	1.13	40.63	nr	**505.19**
900 mm wide	446.74	560.39	1.36	48.90	nr	**609.29**
GRP cable tray including standard coupling joints and fixings (supports and hangers excluded)						
Tray						
100 mm wide	42.63	53.48	0.34	12.22	m	**65.70**
200 mm wide	54.59	68.48	0.39	14.02	m	**82.50**
400 mm wide	85.61	107.39	0.53	19.05	m	**126.44**
Cover						
100 mm wide	24.02	30.13	0.10	3.60	m	**33.73**
200 mm wide	31.82	39.92	0.11	3.95	m	**43.87**
400 mm wide	54.33	68.15	0.14	5.03	m	**73.18**
Extra for (cutting and jointing to fittings included)						
Reducer						
200 mm wide	113.28	142.09	0.23	8.27	nr	**150.36**
400 mm wide	147.99	185.64	0.30	10.79	nr	**196.43**
Reducer cover						
200 mm wide	70.90	88.94	0.25	8.99	nr	**97.93**
400 mm wide	103.62	129.98	0.28	10.06	nr	**140.04**
Bend						
100 mm wide	94.72	118.82	0.34	12.22	nr	**131.04**
200 mm wide	109.95	137.92	0.40	14.38	nr	**152.30**
400 mm wide	141.31	177.26	0.32	11.50	nr	**188.76**
Bend cover						
100 mm wide	47.07	59.05	0.10	3.60	nr	**62.65**
200 mm wide	62.87	78.86	0.10	3.60	nr	**82.46**
400 mm wide	86.85	108.94	0.13	4.68	nr	**113.62**
Tee						
100 mm wide	120.56	151.23	0.37	13.29	nr	**164.52**
200 mm wide	132.93	166.75	0.43	15.46	nr	**182.21**
400 mm wide	164.89	206.84	0.56	20.14	nr	**226.98**
Tee cover						
100 mm wide	60.74	76.19	0.27	9.71	nr	**85.90**
200 mm wide	73.11	91.71	0.31	11.14	nr	**102.85**
400 mm wide	103.13	129.37	0.37	13.29	nr	**142.66**

ELECTRICAL SUPPLY/POWER/LIGHTING

Item	Net Price £	Material £	Labour hours	Labour £	Unit	Total rate £
LV DISTRUBITION: BASKET TRAY						
Mild steel cable basket; zinc plated including standard coupling joints, fixings and earth continuity straps (supports and hangers are excluded)						
Basket 54 mm deep						
100 mm wide	6.25	7.84	0.22	7.92	m	**15.76**
150 mm wide	6.97	8.75	0.25	8.99	m	**17.74**
200 mm wide	7.62	9.55	0.28	10.06	m	**19.61**
300 mm wide	8.94	11.21	0.34	12.22	m	**23.43**
450 mm wide	10.84	13.60	0.44	15.81	m	**29.41**
600 mm wide	13.26	16.63	0.70	25.17	m	**41.80**
Extra for (cutting and jointing to fittings is included)						
Reducer						
150 mm wide	20.46	25.67	0.25	8.99	nr	**34.66**
200 mm wide	24.01	30.12	0.28	10.06	nr	**40.18**
300 mm wide	24.40	30.61	0.38	13.66	nr	**44.27**
450 mm wide	26.78	33.59	0.48	17.26	nr	**50.85**
600 mm wide	33.36	41.84	0.48	17.26	nr	**59.10**
Bend						
100 mm wide	19.54	24.51	0.23	8.27	nr	**32.78**
150 mm wide	22.89	28.72	0.26	9.34	nr	**38.06**
200 mm wide	23.28	29.20	0.30	10.79	nr	**39.99**
300 mm wide	25.49	31.98	0.35	12.58	nr	**44.56**
450 mm wide	31.76	39.84	0.50	17.98	nr	**57.82**
600 mm wide	37.36	46.86	0.58	20.84	nr	**67.70**
Tee						
100 mm wide	24.56	30.81	0.28	10.06	nr	**40.87**
150 mm wide	26.05	32.68	0.30	10.79	nr	**43.47**
200 mm wide	26.60	33.36	0.33	11.87	nr	**45.23**
300 mm wide	33.45	41.96	0.39	14.02	nr	**55.98**
450 mm wide	43.81	54.96	0.56	20.14	nr	**75.10**
600 mm wide	45.29	56.81	0.65	23.37	nr	**80.18**
Cross over						
100 mm wide	33.51	42.03	0.40	14.38	nr	**56.41**
150 mm wide	34.07	42.74	0.42	15.10	nr	**57.84**
200 mm wide	37.01	46.42	0.46	16.53	nr	**62.95**
300 mm wide	42.04	52.73	0.51	18.33	nr	**71.06**
450 mm wide	49.30	61.85	0.74	26.61	nr	**88.46**
600 mm wide	50.90	63.85	0.82	29.48	nr	**93.33**

ELECTRICAL SUPPLY/POWER/LIGHTING

Item	Net Price £	Material £	Labour hours	Labour £	Unit	Total rate £
Mild steel cable basket; expoxy coated including standard coupling joints, fixings and earth continuity straps (supports and hangers are excluded)						
Basket 54 mm deep						
100 mm wide	14.00	17.56	0.22	7.92	m	**25.48**
150 mm wide	15.83	19.86	0.25	8.99	m	**28.85**
200 mm wide	17.63	22.12	0.28	10.06	m	**32.18**
300 mm wide	20.02	25.11	0.34	12.22	m	**37.33**
450 mm wide	24.71	31.00	0.44	15.81	m	**46.81**
600 mm wide	28.14	35.30	0.70	25.17	m	**60.47**
Extra for (cutting and jointing to fittings is included)						
Reducer						
150 mm wide	29.28	36.72	0.28	10.06	nr	**46.78**
200 mm wide	32.81	41.16	0.28	10.06	nr	**51.22**
300 mm wide	35.00	43.90	0.38	13.66	nr	**57.56**
450 mm wide	40.91	51.32	0.48	17.26	nr	**68.58**
600 mm wide	50.99	63.96	0.48	17.26	nr	**81.22**
Bend						
100 mm wide	28.39	35.62	0.23	8.27	nr	**43.89**
150 mm wide	31.72	39.79	0.26	9.34	nr	**49.13**
200 mm wide	32.12	40.29	0.30	10.79	nr	**51.08**
300 mm wide	36.14	45.34	0.35	12.58	nr	**57.92**
450 mm wide	45.86	57.52	0.50	17.98	nr	**75.50**
600 mm wide	55.07	69.08	0.58	20.84	nr	**89.92**
Tee						
100 mm wide	33.37	41.85	0.28	10.06	nr	**51.91**
150 mm wide	34.91	43.79	0.30	10.79	nr	**54.58**
200 mm wide	35.41	44.42	0.33	11.87	nr	**56.29**
300 mm wide	44.04	55.24	0.39	14.02	nr	**69.26**
450 mm wide	56.12	70.39	0.56	20.14	nr	**90.53**
600 mm wide	62.95	78.96	0.65	23.37	nr	**102.33**
Cross over						
100 mm wide	42.33	53.10	0.40	14.38	nr	**67.48**
150 mm wide	42.81	53.70	0.42	15.10	nr	**68.80**
200 mm wide	45.86	57.52	0.46	16.53	nr	**74.05**
300 mm wide	52.64	66.04	0.51	18.33	nr	**84.37**
450 mm wide	63.40	79.53	0.74	26.61	nr	**106.14**
600 mm wide	68.50	85.93	0.82	29.48	nr	**115.41**

ELECTRICAL SUPPLY/POWER/LIGHTING

Item	Net Price £	Material £	Labour hours	Labour £	Unit	Total rate £
LV DISTRIBUTION: SWITCHGEAR AND DISTRIBUTION BOARDS						
LV switchboard components, factory-assembled modular construction to IP41; form 4, type 5; 2400 mm high, with front and rear access; top cable entry/exit; includes delivery, offloading, positioning and commissioning (hence separate labour costs are not detailed below); excludes cabling and cable terminations						
Air circuit breakers (ACBs) to BSEN 60947–2, withdrawable type, fitted with adjustable instantaneous and overload protection. Includes enclosure and copper links, assembled into LV switchboard						
ACB–100 kA fault rated						
4 pole, 6300 A (1600 mm wide)	40196.12	50422.01	–	–	nr	**50422.01**
4 pole, 5000 A (1600 mm wide)	31901.68	40017.47	–	–	nr	**40017.47**
4 pole, 4000 A (1600 mm wide)	18279.98	22930.41	–	–	nr	**22930.41**
4 pole, 3200 A (1600 mm wide)	14623.98	18344.32	–	–	nr	**18344.32**
4 pole, 2500 A (1600 mm wide)	10626.24	13329.56	–	–	nr	**13329.56**
4 pole, 2000 A (1600 mm wide) 85ka	8500.99	10663.64	–	–	nr	**10663.64**
4 pole, 1600 A (1600 mm wide) 85ka	7725.62	9691.01	–	–	nr	**9691.01**
4 pole, 1250 A (1600 mm wide)	6035.64	7571.11	–	–	nr	**7571.11**
4 pole, 1000 A (1600 mm wide)	7062.54	8859.24	–	–	nr	**8859.24**
4 pole, 800 A (1600 mm wide) 85ka	5650.03	7087.39	–	–	nr	**7087.39**
3 pole, 6300 A (1600 mm wide)	33270.33	41734.30	–	–	nr	**41734.30**
3 pole, 5000 A (1600 mm wide)	26405.02	33122.45	–	–	nr	**33122.45**
3 pole, 4000 A (1600 mm wide)	14049.76	17624.02	–	–	nr	**17624.02**
3 pole, 3200 A (1600 mm wide)	11239.81	14099.22	–	–	nr	**14099.22**
3 pole, 2500 A (1600 mm wide)	8781.09	11015.00	–	–	nr	**11015.00**
3 pole, 2000 A (1600 mm wide)	7781.23	9760.78	–	–	nr	**9760.78**
3 pole, 1600 A (1600 mm wide)	6224.98	7808.62	–	–	nr	**7808.62**
3 pole, 1250 A (1600 mm wide)	4863.28	6100.49	–	–	nr	**6100.49**
3 pole, 1000 A (1600 mm wide)	4791.60	6010.58	–	–	nr	**6010.58**
3 pole, 800 A (1600 mm wide)	3833.28	4808.46	–	–	nr	**4808.46**
ACB–65 kA fault rated						
4 pole, 4000 A (1600 mm wide)	40196.12	50422.01	–	–	nr	**50422.01**
4 pole, 3200 A (1600 mm wide)	13490.76	16922.81	–	–	nr	**16922.81**
4 pole, 2500 A (1600 mm wide)	10104.35	12674.89	–	–	nr	**12674.89**
4 pole, 2000 A (1600 mm wide)	8083.48	10139.92	–	–	nr	**10139.92**
4 pole, 1600 A (1600 mm wide)	7308.12	9167.30	–	–	nr	**9167.30**
4 pole, 1250 A (1600 mm wide)	6892.43	8645.86	–	–	nr	**8645.86**
4 pole, 1000 A (1600 mm wide)	6697.23	8401.01	–	–	nr	**8401.01**
4 pole, 800 A (1600 mm wide)	5357.78	6720.80	–	–	nr	**6720.80**
3 pole, 6300 A (1600 mm wide)	33270.33	41734.30	–	–	nr	**41734.30**
3 pole, 4000 A (1600 mm wide)	21124.01	26497.96	–	–	nr	**26497.96**
3 pole, 3200 A (1600 mm wide)	10404.79	13051.76	–	–	nr	**13051.76**

ELECTRICAL SUPPLY/POWER/LIGHTING

Item	Net Price £	Material £	Labour hours	Labour £	Unit	Total rate £
3 pole, 2500 A (1600 mm wide)	8128.74	10196.69	–	–	nr	**10196.69**
3 pole, 2000 A (1600 mm wide)	7348.84	9218.38	–	–	nr	**9218.38**
3 pole, 1600 A (1600 mm wide)	5879.06	7374.70	–	–	nr	**7374.70**
3 pole, 1250 A (1600 mm wide)	4593.01	5761.47	–	–	nr	**5761.47**
3 pole, 1000 A (1600 mm wide)	4575.40	5739.38	–	–	nr	**5739.38**
3 pole, 800 A (1600 mm wide)	3660.32	4591.51	–	–	nr	**4591.51**
Extra for						
Cable box (one per ACB for form 4, types 6 & 7)	352.72	442.46	–	–	nr	**442.46**
Opening coil (shunt trip)	160.44	201.25	–	–	nr	**201.25**
Closing coil	106.29	133.32	–	–	nr	**133.32**
Undervoltage release	190.26	238.66	–	–	nr	**238.66**
Motor operator	640.19	803.05	–	–	nr	**803.05**
Mechnical interlock (per ACB)	442.56	555.15	–	–	nr	**555.15**
ACB fortress/Castell adaptor kit (one per ACB)	95.43	119.71	–	–	nr	**119.71**
Fortress/Castell ACB lock (one per ACB)	300.77	377.28	–	–	nr	**377.28**
Fortress/Castell key	74.89	93.95	–	–	nr	**93.95**
Moulded case circuit breakers (MCCBs) to BS EN 60947–2; plug-in type, fitted with electronic trip unit. Includes metalwork section and copper links, assembled into LV switchboard						
MCCB–150 kA fault rated						
4 Pole, 630 A (800 mm wide, 600 mm high)	2710.00	3399.42	–	–	nr	**3399.42**
4 Pole, 400 A (800 mm wide, 400 mm high)	1814.65	2276.30	–	–	nr	**2276.30**
4 Pole, 250 A (800 mm wide, 400 mm high)	1675.62	2101.89	–	–	nr	**2101.89**
4 Pole, 160 A (800 mm wide, 300 mm high)	1147.79	1439.78	–	–	nr	**1439.78**
4 Pole, 100 A (800 mm wide, 200 mm high)	1809.31	2269.60	–	–	nr	**2269.60**
3 Pole, 630 A (800 mm wide, 600 mm high)	2141.38	2686.15	–	–	nr	**2686.15**
3 Pole, 400 A (800 mm wide, 400 mm high)	1457.54	1828.33	–	–	nr	**1828.33**
3 Pole, 250 A (800 mm wide, 400 mm high)	953.19	1195.68	–	–	nr	**1195.68**
3 Pole, 160 A (800 mm wide, 300 mm high)	684.88	859.12	–	–	nr	**859.12**
3 Pole, 100 A (800 mm wide, 200 mm high)	557.32	699.10	–	–	nr	**699.10**
MCCB–70kA fault rated						
4 Pole, 630 A (800 mm wide, 600 mm high)	2263.27	2839.04	–	–	nr	**2839.04**
4 Pole, 400 A (800 mm wide, 400 mm high)	1555.76	1951.54	–	–	nr	**1951.54**
4 Pole, 250 A (800 mm wide, 400 mm high)	1132.96	1421.19	–	–	nr	**1421.19**
4 Pole, 160 A (800 mm wide, 300 mm high)	853.91	1071.15	–	–	nr	**1071.15**
4 Pole, 100 A (800 mm wide, 200 mm high)	703.66	882.67	–	–	nr	**882.67**
3 Pole, 630 A (800 mm wide, 600 mm high)	1658.73	2080.71	–	–	nr	**2080.71**
3 Pole, 400 A (800 mm wide, 400 mm high)	1119.53	1404.33	–	–	nr	**1404.33**
3 Pole, 250 A (800 mm wide, 400 mm high)	780.74	979.36	–	–	nr	**979.36**
3 Pole, 160 A (800 mm wide, 300 mm high)	579.56	727.00	–	–	nr	**727.00**
3 Pole, 100 A (800 mm wide, 200 mm high)	459.00	575.77	–	–	nr	**575.77**

ELECTRICAL SUPPLY/POWER/LIGHTING

Item	Net Price £	Material £	Labour hours	Labour £	Unit	Total rate £
LV DISTRIBUTION: SWITCHGEAR AND DISTRIBUTION BOARDS – cont						
Moulded case circuit breakers (MCCBs) to BS EN 60947–2 – cont						
MCCB–45kA fault rated						
4 Pole, 630 A (800 mm wide, 600 mm, LI, SSKA high)	2084.42	2614.70	–	–	nr	2614.70
4 Pole, 400 A (800 mm wide, 400 mm, LI, SSKA high)	1447.02	1815.14	–	–	nr	1815.14
3 Pole, 630 A (800 mm wide, 600 mm, LI, SSKA high)	1553.90	1949.21	–	–	nr	1949.21
3 Pole, 400 A (800 mm wide, 400 mm, LI, SSKA high)	1067.00	1338.44	–	–	nr	1338.44
MCCB–36kA fault rated						
4 Pole, 250 A (800 mm wide, 400 mm high)	1735.30	2176.76	–	–	nr	2176.76
4 Pole, 160 A (800 mm wide, 300 mm high)	1314.53	1648.94	–	–	nr	1648.94
3 Pole, 250 A (800 mm wide, 400 mm high)	985.89	1236.70	–	–	nr	1236.70
3 Pole, 160 A (800 mm wide, 300 mm high)	868.25	1089.13	–	–	nr	1089.13
Extra for						
Cable box (one per MCCB for form 4, types 6 & 7)	149.17	187.12	–	–	nr	187.12
Shunt trip (for ratings 100 A to 630 A)	45.98	57.68	–	–	nr	57.68
Undervoltage release (for ratings 100 A to 630 A)	59.64	74.82	–	–	nr	74.82
Motor operator for 630 A MCCB	109.75	137.67	–	–	nr	137.67
Motor operator for 400 A MCCB	109.75	137.67	–	–	nr	137.67
Motor operator for 250 A MCCB	53.27	66.82	–	–	nr	66.82
Motor operator for 160 A/100 A MCCB	47.36	59.40	–	–	nr	59.40
Door handle for 630/400 A MCCB	93.37	117.12	–	–	nr	117.12
Door handle for 250/160/100 A MCCB	73.98	92.80	–	–	nr	92.80
MCCB earth fault protection	562.67	705.81	–	–	nr	705.81
LV switchboard busbar						
Copper busbar assembled into LV switchboard, ASTA type tested to appropriate fault level. Busbar Length may be estimated by adding the widths of the ACB sections to the width of the MCCB sections. ACBs up to 2000 A rating may be stacked two high; larger ratings are one per section. To determine the number of MCCB sections, add together all the MCCB heights and divide by 1800 mm, rounding up as necessary						
6000 A (6 × 10 mm × 100 mm)	2927.86	3672.70	–	–	nr	3672.70
5000 A (4 × 10 mm × 100 mm)	2406.52	3018.74	–	–	nr	3018.74
4000 A (4 × 10 mm × 100 mm)	2406.52	3018.74	–	–	nr	3018.74
3200 A (3 × 10 mm × 100 mm)	1671.49	2096.72	–	–	nr	2096.72
2500 A (2 × 10 mm × 100 mm)	1410.22	1768.98	–	–	nr	1768.98
2000 A (2 × 10 mm × 80 mm)	1036.59	1300.30	–	–	nr	1300.30
1600 A (2 × 10 mm × 50 mm)	752.11	943.44	–	–	nr	943.44

ELECTRICAL SUPPLY/POWER/LIGHTING

Item	Net Price £	Material £	Labour hours	Labour £	Unit	Total rate £
1250 A (2 × 10 mm × 40 mm)	589.72	739.75	–	–	nr	**739.75**
1000 A (2 × 10 mm × 30 mm)	589.72	739.75	–	–	nr	**739.75**
800 A (2 × 10 mm × 20 mm)	459.07	575.86	–	–	nr	**575.86**
630 A (2 × 10 mm × 20 mm)	459.07	575.86	–	–	nr	**575.86**
400 A (2 × 10 mm × 10 mm)	393.15	493.17	–	–	nr	**493.17**
Automatic power factor correction (PFC); floor standing steel enclosure to IP 42, complete with microprocessor based relay and status indication; includes delivery, offloading, positioning and commissioning; excludes cabling and cable terminations						
Standard PFC (no de-tuning)						
100 kVAr	5740.96	7201.47	–	–	nr	**7201.47**
200 kVAr	8029.03	10071.61	–	–	nr	**10071.61**
400 kVAr	13706.48	17193.41	–	–	nr	**17193.41**
600 kVAr	18774.68	23550.96	–	–	nr	**23550.96**
PFC with de-tuning reactors						
100 kVAr	9340.34	11716.52	–	–	nr	**11716.52**
200 kVAr	13053.27	16374.02	–	–	nr	**16374.02**
400 kVAr	22497.38	28220.72	–	–	nr	**28220.72**
600 kVAr	33085.54	41502.50	–	–	nr	**41502.50**

ELECTRICAL SUPPLY/POWER/LIGHTING

Item	Net Price £	Material £	Labour hours	Labour £	Unit	Total rate £
LV DISTRIBUTION: AUTOMATIC TRANSFER SWITCHES						
Automatic transfer switches; steel enclosure; solenoid operating; programmable controller, keypad and LCD display; fixed to backgrounds; including commissioning and testing						
Panel mounting type 4 pole M6 s; non-BMS connection						
40 amp	1998.34	2506.72	2.40	86.27	nr	2592.99
63 amp	2121.24	2660.88	2.50	89.88	nr	2750.76
80 amp	2169.98	2722.03	2.60	93.48	nr	2815.51
100 amp	2201.78	2761.91	2.60	93.48	nr	2855.39
125 amp	2261.11	2836.33	2.70	97.07	nr	2933.40
160 amp	2378.04	2983.01	2.80	100.67	nr	3083.68
Panel mounting type 4 pole M6e; BMS connection						
40 amp	2758.26	3459.96	2.50	89.88	nr	3549.84
63 amp	2921.49	3664.72	2.60	93.48	nr	3758.20
80 amp	2986.46	3746.22	2.70	97.07	nr	3843.29
100 amp	3028.84	3799.38	2.80	100.67	nr	3900.05
125 amp	3107.97	3898.64	2.90	104.25	nr	4002.89
160 amp	3263.84	4094.16	3.00	107.86	nr	4202.02
Enclosed type 3 pole or 4 pole						
125 amp	4094.18	5135.74	2.60	93.48	nr	5229.22
160 amp	4289.16	5380.32	2.90	104.25	nr	5484.57
250 amp	5337.34	6695.16	3.30	118.64	nr	6813.80
400 amp	8393.08	10528.28	4.30	154.59	nr	10682.87
630 amp	11299.91	14174.61	4.84	174.00	nr	14348.61
800 amp	15596.90	19564.75	5.12	184.07	nr	19748.82
1000 amp	17290.05	21688.64	5.50	197.74	nr	21886.38
1250 amp	21612.56	27110.80	6.00	215.71	nr	27326.51
1600 amp	37195.95	46658.60	6.20	222.89	nr	46881.49
2000 amp	51229.74	64262.59	6.90	248.06	nr	64510.65
2500 amp	57633.46	72295.42	7.70	276.82	nr	72572.24
3200 amp	64037.20	80328.26	8.50	305.58	nr	80633.84
Enclosed type 3 pole or 4 pole; with single by-pass						
40 amp	9509.54	11928.76	2.60	93.48	nr	12022.24
63 amp	9694.15	12160.34	2.60	93.48	nr	12253.82
80 amp	9816.91	12314.33	2.60	93.48	nr	12407.81
100 amp	9951.49	12483.15	2.60	93.48	nr	12576.63
125 amp	10466.23	13128.84	2.90	104.25	nr	13233.09
160 amp	11324.14	14205.00	2.90	104.25	nr	14309.25
250 amp	14738.46	18487.93	3.30	118.64	nr	18606.57
400 amp	15486.37	19426.10	4.30	154.59	nr	19580.69
630 amp	21241.88	26645.82	4.84	174.00	nr	26819.82
800 amp	22820.23	28625.70	5.12	184.07	nr	28809.77
1000 amp	32260.32	40467.35	5.50	197.74	nr	40665.09

ELECTRICAL SUPPLY/POWER/LIGHTING

Item	Net Price £	Material £	Labour hours	Labour £	Unit	Total rate £
1250 amp	39350.26	49360.96	6.00	215.71	nr	49576.67
1600 amp	47840.99	60011.74	6.20	222.89	nr	60234.63
2000 amp	58331.50	73171.03	6.90	248.06	nr	73419.09
2500 amp	67782.42	85026.27	7.70	276.82	nr	85303.09
3200 amp	80390.48	100841.82	8.50	305.58	nr	101147.40
Enclosed type 3 pole or 4 pole; with dual by-pass						
40 amp	12451.27	15618.87	2.60	93.48	nr	15712.35
63 amp	12925.52	16213.77	2.60	93.48	nr	16307.25
80 amp	13085.39	16414.32	2.60	93.48	nr	16507.80
100 amp	13268.68	16644.23	2.60	93.48	nr	16737.71
125 amp	13954.98	17505.13	2.90	104.25	nr	17609.38
160 amp	15095.83	18936.21	2.90	104.25	nr	19040.46
250 amp	19651.28	24650.56	3.30	118.64	nr	24769.20
400 amp	20648.51	25901.49	4.30	154.59	nr	26056.08
630 amp	28322.48	35527.72	4.84	174.00	nr	35701.72
800 amp	30426.97	38167.60	5.12	184.07	nr	38351.67
1000 amp	43013.78	53956.48	5.50	197.74	nr	54154.22
1250 amp	50423.93	63251.78	6.00	215.71	nr	63467.49
1600 amp	59056.12	74079.99	6.20	222.89	nr	74302.88
2000 amp	70421.05	88336.17	6.90	248.06	nr	88584.23
2500 amp	80272.70	100694.07	7.70	276.82	nr	100970.89
3200 amp	94867.73	119002.08	8.50	305.58	nr	119307.66

ELECTRICAL SUPPLY/POWER/LIGHTING

Item	Net Price £	Material £	Labour hours	Labour £	Unit	Total rate £
LV DISTRIBUTION: BREAKERS AND FUSES						
MCCB panelboards; IP4X construction, 50 kA busbars and fully-rated neutral; fitted with doorlock, removable glandplate; form 3b Type2; BSEN 60439–1; including fixing to backgrounds						
Panelboards cubicle with MCCB incomer						
Up to 250 A						
4 way TPN	1083.63	1359.31	1.00	35.95	nr	**1395.26**
Extra over for integral incomer metering	1109.86	1392.20	1.50	53.93	nr	**1446.13**
Up to 630 A						
6 way TPN	1949.30	2445.21	2.00	71.90	nr	**2517.11**
12 way TPN	2201.54	2761.61	2.50	89.88	nr	**2851.49**
18 way TPN	2562.75	3214.71	3.00	107.86	nr	**3322.57**
Extra over for integral incomer metering	1311.64	1645.32	1.50	53.93	nr	**1699.25**
Up to 800 A						
6 way TPN	3335.61	4184.19	2.00	71.90	nr	**4256.09**
12 way TPN	4001.55	5019.55	2.50	89.88	nr	**5109.43**
18 way TPN	4259.83	5343.53	3.00	107.86	nr	**5451.39**
Extra over for integral incomer metering	1311.64	1645.32	1.50	53.93	nr	**1699.25**
Up to 1200 A						
20 way TPN	8925.27	11195.86	3.50	125.83	nr	**11321.69**
Up to 1600 A						
28 way TPN	11558.62	14499.13	3.50	125.83	nr	**14624.96**
Up to 2000 A						
28 way TPN	12581.72	15782.51	4.00	143.81	nr	**15926.32**
Feeder MCCBs						
Single pole						
32 A	91.53	114.81	0.75	26.97	nr	**141.78**
63 A	93.61	117.42	0.75	26.97	nr	**144.39**
100 A	95.68	120.02	0.75	26.97	nr	**146.99**
160 A	101.93	127.86	1.00	35.95	nr	**163.81**
Double pole						
32 A	137.30	172.23	0.75	26.97	nr	**199.20**
63 A	139.37	174.82	0.75	26.97	nr	**201.79**
100 A	203.87	255.73	0.75	26.97	nr	**282.70**
160 A	253.78	318.34	1.00	35.95	nr	**354.29**
Triple pole						
32 A	183.08	229.66	0.75	26.97	nr	**256.63**
63 A	187.21	234.84	0.75	26.97	nr	**261.81**
100 A	243.38	305.30	0.75	26.97	nr	**332.27**
160 A	314.13	394.05	1.00	35.95	nr	**430.00**
250 A	472.23	592.37	1.00	35.95	nr	**628.32**
400 A	638.65	801.12	1.25	44.95	nr	**846.07**
630 A	1048.51	1315.25	1.50	53.93	nr	**1369.18**

ELECTRICAL SUPPLY/POWER/LIGHTING

Item	Net Price £	Material £	Labour hours	Labour £	Unit	Total rate £
MCB distribution boards; IP3X external protection enclosure; removable earth and neutral bars and DIN rail; 125/250 amp incomers; including fixing to backgrounds						
SP&N						
6 way	88.98	111.62	2.00	71.90	nr	**183.52**
8 way	105.53	132.37	2.50	89.88	nr	**222.25**
12 way	121.11	151.92	3.00	107.86	nr	**259.78**
16 way	143.96	180.59	4.00	143.81	nr	**324.40**
24 way	303.26	380.41	5.00	179.76	nr	**560.17**
TP&N						
4 way	616.60	773.46	3.00	107.86	nr	**881.32**
6 way	638.27	800.64	3.50	125.83	nr	**926.47**
8 way	668.28	838.29	4.00	143.81	nr	**982.10**
12 way	712.56	893.84	4.00	143.81	nr	**1037.65**
16 way	815.37	1022.80	5.00	179.76	nr	**1202.56**
24 way	1026.81	1288.03	6.40	230.08	nr	**1518.11**
Miniature circuit breakers for distribution boards; BS EN 60 898; DIN rail mounting; including connecting to circuit						
SP&N; including connecting of wiring						
6 amp	10.74	13.47	0.10	3.60	nr	**17.07**
10–40 amp	11.18	14.02	0.10	3.60	nr	**17.62**
50–63 amp	11.70	14.67	0.14	5.03	nr	**19.70**
TP&N; including connecting of wiring						
6 amp	45.55	57.14	0.30	10.79	nr	**67.93**
10–40 amp	47.34	59.38	0.45	16.18	nr	**75.56**
50–63 amp	49.59	62.20	0.45	16.18	nr	**78.38**
Residual current circuit breakers for distribution boards; DIN rail mounting; including connecting to circuit						
SP&N						
10 mA						
6 amp	70.53	88.47	0.21	7.55	nr	**96.02**
10–32 amp	69.19	86.79	0.26	9.34	nr	**96.13**
45 amp	70.32	88.21	0.26	9.34	nr	**97.55**
30 mA						
6 amp	70.53	88.47	0.21	7.55	nr	**96.02**
10–40 amp	69.19	86.79	0.21	7.55	nr	**94.34**
50 –63 amp	71.45	89.62	0.26	9.34	nr	**98.96**
100 mA						
6 amp	130.41	163.59	0.21	7.55	nr	**171.14**
10–40 amp	130.41	163.59	0.23	8.38	nr	**171.97**
50 –63 amp	130.41	163.59	0.26	9.34	nr	**172.93**

ELECTRICAL SUPPLY/POWER/LIGHTING

Item	Net Price £	Material £	Labour hours	Labour £	Unit	Total rate £
LV DISTRIBUTION: BREAKERS AND FUSES – cont						
HRC fused distribution boards; IP4X external protection enclosure; including earth and neutral bars; fixing to backgrounds						
SP&N						
20 amp incomer						
4 way	168.08	210.84	1.00	35.95	nr	**246.79**
6 way	202.92	254.54	1.20	43.14	nr	**297.68**
8 way	237.88	298.40	1.40	50.32	nr	**348.72**
12 way	307.83	386.14	1.80	64.71	nr	**450.85**
32 amp incomer						
4 way	202.37	253.85	1.00	35.95	nr	**289.80**
6 way	266.13	333.84	1.20	43.14	nr	**376.98**
8 way	313.10	392.75	1.40	50.32	nr	**443.07**
12 way	403.11	505.66	1.80	64.71	nr	**570.37**
TP&N						
20 amp incomer						
4 way	317.13	397.81	1.50	53.93	nr	**451.74**
6 way	401.09	503.13	2.10	75.50	nr	**578.63**
8 way	474.23	594.88	2.70	97.07	nr	**691.95**
12 way	665.33	834.59	3.90	140.20	nr	**974.79**
32 amp incomer						
4 way	379.39	475.91	1.50	53.93	nr	**529.84**
6 way	509.35	638.93	2.10	75.50	nr	**714.43**
8 way	622.20	780.48	2.70	97.07	nr	**877.55**
12 way	862.98	1082.52	3.90	140.20	nr	**1222.72**
63 amp incomer						
4 way	806.46	1011.63	2.17	78.01	nr	**1089.64**
6 way	1034.28	1297.40	2.83	101.74	nr	**1399.14**
8 way	1245.51	1562.37	2.57	92.40	nr	**1654.77**
100 amp incomer						
4 way	1275.41	1599.88	2.40	86.27	nr	**1686.15**
6 way	1667.08	2091.19	2.73	98.15	nr	**2189.34**
8 way	2038.42	2556.99	3.87	139.13	nr	**2696.12**
200 amp incomer						
4 way	3159.07	3962.74	5.36	192.70	nr	**4155.44**
6 way	4175.38	5237.60	6.17	221.82	nr	**5459.42**
HRC fuse; includes fixing to fuse holder						
2–30 amp	3.42	4.29	0.10	3.60	nr	**7.89**
35–63 amp	7.40	9.28	0.12	4.31	nr	**13.59**
80 amp	10.92	13.70	0.15	5.40	nr	**19.10**
100 amp	13.14	16.49	0.15	5.40	nr	**21.89**
125 amp	19.85	24.90	0.15	5.40	nr	**30.30**
160 amp	20.82	26.12	0.15	5.40	nr	**31.52**
200 amp	21.57	27.06	0.15	5.40	nr	**32.46**

ELECTRICAL SUPPLY/POWER/LIGHTING

Item	Net Price £	Material £	Labour hours	Labour £	Unit	Total rate £
Consumer units; fixed to backgrounds; including supports, fixings, connections/ jointing to equipment						
Switched and insulated; moulded plastic case, 63 amp 230 Volt SP&N; earth and neutral bars; 30 mA RCCB protection; fitted MCBs						
2 way	126.83	159.10	1.67	60.03	nr	**219.13**
4 way	142.31	178.52	1.59	57.16	nr	**235.68**
6 way	154.70	194.05	2.50	89.88	nr	**283.93**
8 way	166.93	209.40	3.00	107.86	nr	**317.26**
12 way	193.92	243.25	4.00	143.81	nr	**387.06**
16 way	234.31	293.92	5.50	197.74	nr	**491.66**
Switched and insulated; moulded plastic case, 100 amp 230 Volt SP&N; earth and neutral bars; 30 mA RCCB protection; fitted MCBs						
2 way	126.83	159.10	1.67	60.03	nr	**219.13**
4 way	142.31	178.52	1.59	57.16	nr	**235.68**
6 way	154.70	194.05	2.50	89.88	nr	**283.93**
8 way	166.93	209.40	3.00	107.86	nr	**317.26**
12 way	193.92	243.25	4.00	143.81	nr	**387.06**
16 way	234.31	293.92	5.50	197.74	nr	**491.66**
Extra for						
Residual current device; double pole; 230 volt/ 30 mA tripping current						
16 amp	69.34	86.98	0.22	7.92	nr	**94.90**
30 amp	70.44	88.36	0.22	7.92	nr	**96.28**
40 amp	71.54	89.73	0.22	7.92	nr	**97.65**
63 amp	88.58	111.12	0.22	7.92	nr	**119.04**
80 amp	98.52	123.58	0.22	7.92	nr	**131.50**
100 amp	121.26	152.11	0.25	8.99	nr	**161.10**
Residual current device; double pole; 230 volt/ 100 mA tripping current						
63 amp	80.97	101.57	0.22	7.92	nr	**109.49**
80 amp	93.67	117.50	0.22	7.92	nr	**125.42**
100 amp	121.28	152.13	0.25	8.99	nr	**161.12**
Heavy duty fuse switches; with HRC fuses BS 5419; short circuit rating 65 kA, 500 volt; including retractable operating switches						
SP&N						
63 amp	313.26	392.95	1.30	46.74	nr	**439.69**
100 amp	457.92	574.41	1.95	70.11	nr	**644.52**
TP&N						
63 amp	394.66	495.06	1.83	65.79	nr	**560.85**
100 amp	554.58	695.67	2.48	89.16	nr	**784.83**
200 amp	855.47	1073.11	3.13	112.54	nr	**1185.65**
300 amp	1486.19	1864.27	4.45	159.98	nr	**2024.25**
400 amp	1631.56	2046.63	4.45	159.98	nr	**2206.61**
600 amp	2462.76	3089.28	5.72	205.64	nr	**3294.92**
800 amp	3847.12	4825.82	7.88	283.29	nr	**5109.11**

ELECTRICAL SUPPLY/POWER/LIGHTING

Item	Net Price £	Material £	Labour hours	Labour £	Unit	Total rate £
LV DISTRIBUTION: BREAKERS AND FUSES – cont						
Switch disconnectors to BSEN 60947–3; in sheet steel case; IP41 with door interlock fixed to backgrounds						
Double pole						
20 amp	63.39	79.52	1.02	36.67	nr	**116.19**
32 amp	76.41	95.85	1.02	36.67	nr	**132.52**
63 amp	278.81	349.74	1.21	43.50	nr	**393.24**
100 amp	255.84	320.92	1.86	66.88	nr	**387.80**
TP&N						
20 amp	79.58	99.83	1.29	46.38	nr	**146.21**
32 amp	92.60	116.16	1.83	65.79	nr	**181.95**
63 amp	314.23	394.17	2.48	89.16	nr	**483.33**
100 amp	316.83	397.43	2.48	89.16	nr	**486.59**
125 amp	330.46	414.53	2.48	89.16	nr	**503.69**
160 amp	760.33	953.76	2.48	89.16	nr	**1042.92**
Enclosed switch disconnector to BSEN 60947–3; enclosure minimum IP55 rating; complete with earth connection bar; fixed to backgrounds						
TP						
20 amp	63.39	79.52	1.02	36.67	nr	**116.19**
32 amp	76.41	95.85	1.02	36.67	nr	**132.52**
63 amp	278.81	349.74	1.21	43.50	nr	**393.24**
TP&N						
20 amp	79.58	99.83	1.29	46.38	nr	**146.21**
32 amp	92.60	116.16	1.83	65.79	nr	**181.95**
63 amp	314.23	394.17	2.48	89.16	nr	**483.33**
Busbar chambers; fixed to background including all supports, fixings, connections/jointing to equipment						
Sheet steel case enclosing 4 pole 550 volt copper bars, detachable metal end plates						
600 mm long						
200 amp	577.47	724.38	2.62	94.19	nr	**818.57**
300 amp	742.07	930.85	3.03	108.93	nr	**1039.78**
500 amp	1271.52	1594.99	4.48	161.07	nr	**1756.06**
900 mm long						
200 amp	831.80	1043.41	3.04	109.30	nr	**1152.71**
300 amp	980.31	1229.70	3.59	129.07	nr	**1358.77**
500 amp	1450.06	1818.96	4.42	158.91	nr	**1977.87**
1350 mm long						
200 amp	1136.18	1425.22	3.38	121.51	nr	**1546.73**
300 amp	1337.26	1677.46	3.94	141.65	nr	**1819.11**
500 amp	2140.58	2685.14	4.82	173.29	nr	**2858.43**

ELECTRICAL SUPPLY/POWER/LIGHTING

Item	Net Price £	Material £	Labour hours	Labour £	Unit	Total rate £
Contactor relays; pressed steel enclosure; fixed to backgrounds including supports, fixings, connections/jointing to equipment						
Relays						
6 amp, 415/240 volt, 4 pole N/O	66.88	83.90	0.52	18.69	nr	**102.59**
6 amp, 415/240 volt, 8 pole N/O	81.79	102.59	0.85	30.55	nr	**133.14**
Push button stations; heavy gauge pressed steel enclosure; polycarbonate cover; IP65; fixed to backgrounds including supports, fixings,connections/ joining to equipment						
Standard units						
One button (start or stop)	84.65	106.19	0.39	14.02	nr	**120.21**
Two button (start or stop)	89.84	112.69	0.47	16.90	nr	**129.59**
Three button (forward-reverse-stop)	127.33	159.72	0.57	20.50	nr	**180.22**
Weatherproof junction boxes; enclosures with rail mounted terminal blocks; side hung door to receive padlock; fixed to backgrounds, including all supports and fixings (suitable for cable up to 2.5 mm²; including glandplates and gaskets)						
Sheet steel with zinc spray finish enclosure						
Overall size 229 × 152; suitable to receive						
3 × 20(A) glands per gland plate	85.72	107.53	1.43	51.41	nr	**158.94**
Overall size 306 × 306; suitable to receive						
14 × 20(A) glands per gland plate	115.00	144.26	2.17	78.01	nr	**222.27**
Overall size 458 × 382; suitable to receive						
18 × 20(A) glands per gland plate	167.45	210.04	3.51	126.19	nr	**336.23**
Overall size 762 x508; suitable to receive						
26 × 20(A) glands per gland plate	177.10	222.15	4.85	174.36	nr	**396.51**
Overall size 914 × 610; suitable to receive						
45 × 20(A) glands per gland plate	195.75	245.55	7.01	252.02	nr	**497.57**
Weatherproof junction boxes; enclosures with rail mounted terminal blocks; screw fixed lid; fixed to backgrounds, including all supports and fixings (suitable for cable up to 2.5 mm²; including glandplates and gaskets)						
Glassfibre reinforced polycarbonate enclosure						
Overall size 190 × 190 × 130 mm	117.07	146.85	1.43	51.41	nr	**198.26**
Overall size 190 × 190 × 180 mm	171.41	215.02	1.53	55.00	nr	**270.02**
Overall size 280 × 190 × 130 mm	193.60	242.85	2.17	78.01	nr	**320.86**
Overall size 280 × 190 × 180 mm	216.89	272.07	2.37	85.20	nr	**357.27**
Overall size 380 × 190 × 130 mm	242.00	303.56	3.33	119.72	nr	**423.28**
Overall size 380 × 190 × 180 mm	259.90	326.02	3.30	118.64	nr	**444.66**
Overall size 380 × 280 × 130 mm	277.84	348.52	4.66	167.54	nr	**516.06**
Overall size 380 × 280 × 180 mm	299.34	375.49	5.36	192.70	nr	**568.19**
Overall size 560 × 280 × 130 mm	360.29	451.94	7.01	252.02	nr	**703.96**
Overall size 560 × 380 × 180 mm	371.05	465.45	7.67	275.74	nr	**741.19**

ELECTRICAL SUPPLY/POWER/LIGHTING

Item	Net Price £	Material £	Labour hours	Labour £	Unit	Total rate £
GENERAL LIGHTING						
Fluorescent luminaires; surface fixed to backgrounds						
Batten type; surface mounted						
600 mm Single – 18 W	9.77	12.25	0.58	20.84	nr	33.09
600 mm Twin – 18 W	17.11	21.46	0.59	21.21	nr	42.67
1200 mm Single – 36 W	12.91	16.20	0.76	27.32	nr	43.52
1200 mm Twin – 36 W	14.48	18.17	0.84	30.20	nr	48.37
1500 mm Single – 58 W	24.75	31.05	0.77	27.69	nr	58.74
1500 mm Twin – 58 W	28.89	36.24	0.85	30.55	nr	66.79
1800 mm Single – 70 W	17.63	22.12	1.05	37.74	nr	59.86
1800 mm Twin – 70 W	31.93	40.05	1.06	38.11	nr	78.16
2400 mm Single – 100 W	24.14	30.28	1.25	44.95	nr	75.23
2400 mm Twin – 100 W	41.82	52.46	1.27	45.66	nr	98.12
Surface mounted, opal diffuser						
600 mm Twin – 18 W	26.57	33.33	0.62	22.29	nr	55.62
1200 mm Single – 36 W	22.92	28.75	0.79	28.40	nr	57.15
1200 mm Twin – 36 W	35.32	44.31	0.80	28.76	nr	73.07
1500 mm Single – 58 W	26.34	33.04	0.88	31.64	nr	64.68
1500 mm Twin – 58 W	41.31	51.82	0.90	32.35	nr	84.17
1800 mm Single – 70 W	32.87	41.23	1.09	39.19	nr	80.42
1800 mm Twin – 70 W	44.66	56.02	1.10	39.55	nr	95.57
2400 mm Single – 100 W	43.17	54.15	1.30	46.74	nr	100.89
2400 mm Twin – 100 W	61.71	77.41	1.31	47.08	nr	124.49
Surface mounted linear fluorescent; T8 lamp; high frequency control gear; low brightness; 65° cut-off; including wedge style louvre						
1200 mm, 1 × 36 watt	63.17	79.24	1.09	39.19	nr	118.43
1200 mm 2 × 36 watt	66.87	83.88	1.09	39.19	nr	123.07
Extra for emergency pack	57.56	72.21	0.25	8.99	nr	81.20
1500 mm, 1 × 58 watt	73.63	92.37	0.90	32.35	nr	124.72
1500 mm 2 × 58 watt	77.89	97.71	0.90	32.35	nr	130.06
Extra for emergency pack	57.45	72.06	0.25	8.99	nr	81.05
1800 mm, 1 × 70 watt	110.87	139.07	0.90	32.35	nr	171.42
1800 mm, 2 × 70 watt	125.98	158.03	0.90	32.35	nr	190.38
Extra for emergency pack	95.09	119.28	0.25	8.99	nr	128.27
Modular recessed linear fluorescent; high frequency control gear; low brightness; 65° cut off; including wedge style louvre; fitted to exposed T grid ceiling						
600 × 600 mm, 3 × 18 watt T8	49.68	62.32	0.84	30.20	nr	92.52
600 × 600 mm, 4 × 18 watt T8	50.81	63.74	0.87	31.27	nr	95.01
Extra for emergency pack	58.03	72.79	0.25	8.99	nr	81.78
300 × 1200 mm, 2 × 36 watt T8	69.33	86.97	0.87	31.27	nr	118.24
Extra for emergency pack	73.42	92.10	0.25	8.99	nr	101.09
600 × 1200 mm, 3 × 36 watt T8	75.65	94.90	0.89	32.00	nr	126.90
600 × 1200 mm, 4 × 36 watt T8	86.13	108.05	0.91	32.72	nr	140.77
Extra for emergency pack	67.99	85.29	0.25	8.99	nr	94.28
600 × 600 mm, 3 × 14 watt T5	70.58	88.54	0.84	30.20	nr	118.74
600 × 600 mm, 4 × 14 watt T5	72.69	91.18	0.87	31.27	nr	122.45
Extra for emergency pack	48.65	61.03	0.25	8.99	nr	70.02

ELECTRICAL SUPPLY/POWER/LIGHTING

Item	Net Price £	Material £	Labour hours	Labour £	Unit	Total rate £
Modular recessed; T8 lamp; high frequency control gear; cross-blade louvre; fitted to exposed T grid ceiling						
600 × 600 mm, 3 × 18 watt	54.23	68.03	0.84	30.20	nr	**98.23**
600 × 600 mm, 4 × 18 watt	72.65	91.13	0.87	31.27	nr	**122.40**
Extra for emergency pack	65.98	82.77	0.25	8.99	nr	**91.76**
Modular recessed compact fluorescent; TCL lamp; high frequency control gear; low brightness; 65° cut-off; including wedge style louvre; fitted to exposed T grid ceiling						
300 × 300 mm, 2 × 18 watt	108.82	136.51	0.75	26.97	nr	**163.48**
Extra for emergency pack	101.51	127.33	0.25	8.99	nr	**136.32**
500 × 500 mm, 2 × 36 watt	132.20	165.83	0.82	29.48	nr	**195.31**
600 × 600 mm, 2 × 36 watt	122.79	154.02	0.82	29.48	nr	**183.50**
600 × 600 mm, 2 × 40 watt	119.24	149.58	0.82	29.48	nr	**179.06**
Extra for emergency pack	48.18	60.44	0.25	8.99	nr	**69.43**
Ceiling recessed asymetric compact fluorescent downlighter; high frequency control gear; TCD lamp in 200 mm dia. luminaire; for wall-washing application						
1 × 18 watt	172.34	216.18	0.75	26.97	nr	**243.15**
1 × 26 watt	172.34	216.18	0.75	26.97	nr	**243.15**
2 × 18 watt	190.02	238.36	0.75	26.97	nr	**265.33**
2 × 26 watt	190.02	238.36	0.75	26.97	nr	**265.33**
Ceiling recessed asymetric compact fluorescesnt downlights; high frequency control gear; linear 200 mm × 600 mm luminaire with low glare louvre; for wall washing applications						
1 × 55 watt TCL	97.35	122.11	0.75	26.97	nr	**149.08**
Wall mounted compact fluorescent uplighter; high frequency control gear; TCL lamp in 300 mm × 600 mm luminaire						
2 × 36 watt	345.91	433.91	0.84	30.20	nr	**464.11**
2 × 40 watt	360.55	452.28	0.84	30.20	nr	**482.48**
2 × 55 watt	360.55	452.28	0.84	30.20	nr	**482.48**
Suspended linear fluorescent; T5 lamp; high frequency control gear; low brightness; 65° cut-off; 30% uplight, 70% downlight; including wedge style louvre						
1 × 49 watt	115.24	144.56	0.75	26.97	nr	**171.53**
Extra for emergency pack	49.12	61.61	0.25	8.99	nr	**70.60**
Semi-recessed 'architectural' linear fluorescent; T5 lamp; high frequency control gear; low brightness, delivers direct, ceiling and graduated wall washing illumination						
600 × 600 mm, 2 × 24 watt	150.29	188.52	0.87	31.27	nr	**219.79**
600 × 600 mm, 4 × 14 watt	161.73	202.88	0.87	31.27	nr	**234.15**
500 × 500 mm, 2 × 24 watt	147.77	185.36	0.87	31.27	nr	**216.63**
Extra for emergency pack	47.72	59.86	0.25	8.99	nr	**68.85**

ELECTRICAL SUPPLY/POWER/LIGHTING

Item	Net Price £	Material £	Labour hours	Labour £	Unit	Total rate £
GENERAL LIGHTING – cont						
Downlighter, recessed; low voltage; mirror reflector with white/chrome bezel; dimmable transformer; for dichroic lamps						
85 mm dia. × 20/50 watt	16.89	21.19	0.66	23.73	nr	44.92
118 mm dia. × 50 watt	22.11	27.73	0.66	23.73	nr	51.46
165 mm dia. × 100 watt	100.98	126.67	0.66	23.73	nr	150.40
LUMINAIRES						
High/Low bay						
Compact discharge; aluminium reflector						
150 watt	193.16	242.30	1.50	53.93	nr	296.23
250 watt	144.45	181.19	1.50	53.93	nr	235.12
400 watt	144.45	181.19	1.50	53.93	nr	235.12
Sealed discharge; aluminium reflector						
150 watt	202.45	253.95	1.50	53.93	nr	307.88
250 watt	218.50	274.09	1.50	53.93	nr	328.02
400 watt	285.69	358.37	1.50	53.93	nr	412.30
Corrosion resistant GRP body; gasket sealed; acrylic diffuser						
600 mm Single – 18 W	32.97	41.36	0.49	17.61	nr	58.97
600 mm Twin – 18 W	42.92	53.84	0.49	17.61	nr	71.45
1200 mm Single – 36 W	37.23	46.70	0.64	23.00	nr	69.70
1200 mm Twin – 36 W	47.58	59.68	0.64	23.00	nr	82.68
1500 mm Single – 58 W	41.36	51.88	0.72	25.89	nr	77.77
1500 mm Twin – 58 W	51.42	64.50	0.72	25.89	nr	90.39
1800 mm Single – 70 W	61.21	76.79	0.94	33.79	nr	110.58
1800 mm Twin – 70 W	75.97	95.30	0.94	33.79	nr	129.09
Flameproof to IIA/IIB,I.P. 64; Aluminium Body; BS 229 and 899						
600 mm Single – 18 W	368.64	462.43	1.04	37.40	nr	499.83
600 mm Twin – 18 W	453.51	568.88	1.04	37.40	nr	606.28
1200 mm Single – 36 W	402.85	505.33	1.31	47.08	nr	552.41
1200 mm Twin – 36 W	493.03	618.45	1.18	42.43	nr	660.88
1500 mm Single – 58 W	431.29	541.00	1.64	58.96	nr	599.96
1500 mm Twin – 58 W	515.53	646.68	1.64	58.96	nr	705.64
1800 mm Single – 70 W	472.80	593.08	1.97	70.83	nr	663.91
1800 mm Twin – 70 W	539.94	677.30	1.97	70.83	nr	748.13
LED luminaires						
Modular LED luminaire; recessed into ceiling; including driver, polycarbonate optical lenses/ diffuser, heat sink and 4000 K colour temperature						
300 × 300 mm	97.75	122.62	0.75	26.97	nr	149.59
Extra for emergency pack	88.88	111.50	0.25	8.99	nr	120.49
Extra for DALI lighting control	24.69	30.97	0.25	8.99	nr	39.96
Extra for tuneable capability	25.99	32.60	0.25	8.99	nr	41.59
Extra for wireless control	55.49	69.61	0.25	8.99	nr	78.60
600 × 600 mm	195.51	245.25	0.82	29.48	nr	274.73
Extra for emergency pack	88.88	111.50	0.25	8.99	nr	120.49

ELECTRICAL SUPPLY/POWER/LIGHTING

Item	Net Price £	Material £	Labour hours	Labour £	Unit	Total rate £
Extra for DALI lighting control	29.63	37.17	0.25	8.99	nr	46.16
Extra for tuneable capability	28.59	35.86	0.25	8.99	nr	44.85
Extra for wireless control	60.50	75.89	0.25	8.99	nr	84.88
Modular linear LED luminaire; recessed into ceiling; including driver, polycarbonate optical lenses/diffuser, heat sink and 4000 K colour temperature						
600 × 100 mm	195.51	245.25	0.84	30.20	nr	275.45
Extra for emergency pack	79.00	99.10	0.25	8.99	nr	108.09
Extra for DALI lighting control	24.69	30.97	0.25	8.99	nr	39.96
Extra for tuneable capability	26.29	32.97	0.25	8.99	nr	41.96
Extra for wireless control	58.91	73.90	0.25	8.99	nr	82.89
1200 × 100 mm	332.37	416.92	0.89	32.00	nr	448.92
Extra for emergency pack	88.88	111.50	0.25	8.99	nr	120.49
Extra for DALI lighting control	29.63	37.17	0.25	8.99	nr	46.16
Extra for tuneable capability	29.68	37.23	0.25	8.99	nr	46.22
Extra for wireless control	64.80	81.29	0.25	8.99	nr	90.28
LED downlight/spotlight; surface mounted; including driver, polycarbonate optical lenses/diffuser, heat sink and 4000 K colour temperature						
190 mm dia.	174.16	218.47	0.75	26.97	nr	245.44
Extra for emergency pack (remote)	93.81	117.68	0.25	8.99	nr	126.67
Extra for DALI lighting control (remote)	47.88	60.07	0.25	8.99	nr	69.06
Extra for tuneable capability	25.49	31.98	0.25	8.99	nr	40.97
Extra for wireless control	55.10	69.12	0.25	8.99	nr	78.11
LED downlight/spotlight; track mounted; adjustable fixing; including driver, polycarbonate optical lenses/diffuser, heat sink and 4000 K colour temperature						
120 mm dia.	213.89	268.31	0.75	26.97	nr	295.28
Extra for emergency pack (track mounted remote)	98.75	123.87	0.25	8.99	nr	132.86
Extra for DALI lighting control (with DALI track)	50.40	63.22	0.25	8.99	nr	72.21
LED downlight/spotlight; recessed into ceiling; including driver, polycarbonate optical lenses/diffuser, heat sink and 4000 K colour temperature						
1 × spotlight	83.10	104.24	0.66	23.73	nr	127.97
Extra for emergency pack (remote)	93.81	117.68	0.25	8.99	nr	126.67
Extra for DALI lighting control	47.41	59.47	0.25	8.99	nr	68.46
2 × spotlight	146.63	183.94	0.66	23.73	nr	207.67
Extra for emergency pack (remote)	93.81	117.68	0.25	8.99	nr	126.67
Extra for DALI lighting control	54.31	68.13	0.25	8.99	nr	77.12
3 × spotlight	215.06	269.77	0.66	23.73	nr	293.50
Extra for emergency pack (remote)	93.81	117.68	0.25	8.99	nr	126.67
Extra for DALI lighting control	64.19	80.52	0.25	8.99	nr	89.51

ELECTRICAL SUPPLY/POWER/LIGHTING

Item	Net Price £	Material £	Labour hours	Labour £	Unit	Total rate £
GENERAL LIGHTING – cont						
LED luminaires – cont						
LED downlight/spotlight; track mounted; adjustable fixing; including driver, polycarbonate optical lenses/diffuser, heat sink, 4000 K colour temperature						
120 mm dia.	174.29	218.62	4.00	143.81	nr	**362.43**
Extra for emergency pack (remote)	201.26	252.46	4.00	143.81	nr	**396.27**
Extra for DALI lighting control (with DALI track)	48.91	61.35	0.25	8.99	nr	**70.34**
LED downlight/spotlight; track mounted; adjustable fixing; including driver, polycarbonate optical lenses/diffuser, heat sink, RGB colour						
120 mm dia.	234.61	294.29	0.25	8.99	nr	**303.28**
Extra for emergency pack (remote)	203.19	254.88	0.25	8.99	nr	**263.87**
Extra for DALI lighting control (with DALI track)	48.91	61.35	0.25	8.99	nr	**70.34**
Internal flexible linear LED, IP20, including driver and controller, optical lense/diffuser, white 4000 K colour temperature						
10 mm width (remote)	75.81	95.10	0.25	8.99	nr	**104.09**
Extra for emergency pack (remote)	151.62	190.19	0.25	8.99	nr	**199.18**
Extra for DALI lighting control (remote)	24.69	30.97	0.25	8.99	nr	**39.96**
Internal flexible linear LED, IP20, including driver and controller, optical lense/diffuser, RGB						
10 mm width	113.72	142.65	0.25	8.99	nr	**151.64**
Extra for emergency pack (per unit)	151.62	190.19	0.25	8.99	nr	**199.18**
Extra for DALI lighting control	24.69	30.97	0.25	8.99	nr	**39.96**
LED in-ground performance uplighters, IP68, including driver, polycarbonate optical lenses/ diffuser, heat sink, 4000 K colour temperature						
50 mm dia.	360.10	451.71	0.25	8.99	nr	**460.70**
Extra for emergency pack (per unit)	175.96	220.73	0.25	8.99	nr	**229.72**
Extra for DALI lighting control	14.96	18.77	0.25	8.99	nr	**27.76**
LED in-ground performance uplighters, IP68, including driver, polycarbonate optical lenses/ diffuser, heat sink, RGB						
50 mm dia.	426.44	534.92	0.25	8.99	nr	**543.91**
Extra for emergency pack (per unit)	175.96	220.73	0.25	8.99	nr	**229.72**
Extra for DALI lighting control	14.96	18.77	0.25	8.99	nr	**27.76**
LED in-ground decorative uplighters, IP68, including driver, polycarbonate optical lenses/ diffuser, heat sink, 4000 K colour						
50 mm dia.	170.57	213.96	0.25	8.99	nr	**222.95**
Extra for emergency pack (per unit)	175.96	220.73	0.25	8.99	nr	**229.72**
Extra for DALI lighting control	19.95	25.02	0.25	8.99	nr	**34.01**

ELECTRICAL SUPPLY/POWER/LIGHTING

Item	Net Price £	Material £	Labour hours	Labour £	Unit	Total rate £
LED in-ground decorative uplighters, IP68, including driver, polycarbonate optical lenses/ diffuser, heat sink, RGB						
50 mm dia.	189.53	237.74	0.25	8.99	nr	**246.73**
Extra for emergency pack (per unit)	175.96	220.73	0.25	8.99	nr	**229.72**
Extra for DALI lighting control	19.95	25.02	0.25	8.99	nr	**34.01**
External flexible linear LED, IP66, including driver and controller, optical lenses/diffuser, 4000 K colour temperature						
10 mm width	113.72	142.65	0.25	8.99	nr	**151.64**
Extra for emergency pack (per unit)	156.41	196.20	0.25	8.99	nr	**205.19**
Extra for DALI lighting control	24.94	31.28	0.25	8.99	nr	**40.27**
External flexible linear LED, IP66, including driver and controller, optical lenses/diffuser, RGB						
10 mm width	151.62	190.19	0.25	8.99	nr	**199.18**
Extra for emergency pack (per unit)	156.41	196.20	0.25	8.99	nr	**205.19**
Extra for DALI lighting control	24.94	31.28	0.25	8.99	nr	**40.27**
Handrail lighting, LED, IP66, including driver and controller, optical lenses/diffuser 4000 K colour temperature						
10 mm width	138.25	173.42	0.25	8.99	nr	**182.41**
Extra for emergency pack (per unit)	156.41	196.20	0.25	8.99	nr	**205.19**
Extra for DALI lighting control	24.94	31.28	0.25	8.99	nr	**40.27**
Handrail lighting, LED, IP66, including driver and controller, optical lenses/diffuser, RGB						
10 mm width	158.00	198.20	0.25	8.99	nr	**207.19**
Extra for emergency pack (per unit)	156.41	196.20	0.25	8.99	nr	**205.19**
Extra for DALI lighting control	24.94	31.28	0.25	8.99	nr	**40.27**
External lighting						
Flameproof						
Ground mounted 50 W	413.11	518.20	2.25	80.89	nr	**599.09**
Ceiling mounted 50 W	185.86	233.14	2.25	80.89	nr	**314.03**
Bulkhead; aluminium body and polycarbonate bowl; vandal-resistant; IP65						
60 W	36.41	45.67	0.75	26.97	nr	**72.64**
Extra for						
Emergency version	98.87	124.02	0.25	8.99	nr	**133.01**
2D 2 pin 16 W	30.97	38.85	0.66	23.73	nr	**62.58**
2D 2 pin 28 W	62.05	77.84	0.66	23.73	nr	**101.57**
Photocell	22.04	27.64	0.75	26.97	nr	**54.61**
1500 mm high circular bollard; polycarbonate visor; vandal-resistant; IP54						
50 W	232.65	291.84	1.75	62.92	nr	**354.76**
70 W	236.37	296.50	1.75	62.92	nr	**359.42**
80 W	282.87	354.83	1.75	62.92	nr	**417.75**

ELECTRICAL SUPPLY/POWER/LIGHTING

Item	Net Price £	Material £	Labour hours	Labour £	Unit	Total rate £
GENERAL LIGHTING – cont						
External lighting – cont						
Floodlight; enclosed high performance dischargelight; integeral control gear; reflector; toughened glass; IP65						
70 W	103.06	129.28	1.25	44.95	nr	174.23
100 W	107.86	135.30	1.25	44.95	nr	180.25
150 W	114.35	143.44	1.25	44.95	nr	188.39
250 W	200.18	251.10	1.25	44.95	nr	296.05
400 W	207.36	260.11	1.25	44.95	nr	305.06
Extra for photocell	22.02	27.62	0.75	26.97	nr	54.59
Lighting track						
Single circuit; 25 A 2 P&E steel trunking; low voltage with copper conductors; including couplers and supports; fixed to backgrounds						
Straight track	14.87	18.65	0.50	17.98	m	36.63
Live end feed unit complete with end stop	5.41	6.79	0.33	11.87	nr	18.66
Flexible couplers 0.5 m	5.41	6.79	0.33	11.87	nr	18.66
Tap off complete with 0.8 m of cable	5.41	6.79	0.25	8.99	nr	15.78
Three circuit; 25 A 2 P&E steel trunking; low voltage with copper conductors; including couplers and supports incorporating integral twisted pair comms bus bracket; fixed to backgrounds						
Straight track	23.09	28.96	0.75	26.97	nr	55.93
Live end feed unit complete with end stop	42.00	52.68	0.50	17.98	nr	70.66
Flexible couplers 0.5 m	60.41	75.78	0.45	16.18	nr	91.96
Tap off complete with 0.8 m of cable	10.82	13.57	0.25	8.99	nr	22.56
LIGHTING ACCESSORIES						
Switches						
6 amp metal clad surface mounted switch, gridswitch; one way						
1 gang	10.35	12.98	0.43	15.46	nr	28.44
2 gang	14.12	17.71	0.55	19.77	nr	37.48
3 gang	22.49	28.21	0.77	27.69	nr	55.90
4 gang	26.28	32.96	0.88	31.64	nr	64.60
6 gang	45.28	56.80	1.00	35.95	nr	92.75
8 gang	52.94	66.40	1.28	46.01	nr	112.41
10 gang	80.81	101.37	1.67	60.03	nr	161.40
Extra for						
10 amp – Two way switch	3.59	4.50	0.03	1.08	nr	5.58
20 amp – Two way switch	4.81	6.04	0.04	1.44	nr	7.48
20 amp – Intermediate	9.03	11.32	0.08	2.87	nr	14.19
20 amp – One way SP switch	3.70	4.64	0.08	2.87	nr	7.51
Steel blank plate; 1 gang	2.17	2.72	0.07	2.52	nr	5.24
Steel blank plate; 2 gang	3.65	4.58	0.08	2.87	nr	7.45

ELECTRICAL SUPPLY/POWER/LIGHTING

Item	Net Price £	Material £	Labour hours	Labour £	Unit	Total rate £
6 amp modular type switch; galvanized steel box, bronze or satin chrome coverplate; metalclad switches; flush mounting; one way						
1 gang	25.84	32.41	0.43	15.46	nr	**47.87**
2 gang	29.27	36.71	0.55	19.77	nr	**56.48**
3 gang	42.28	53.03	0.77	27.69	nr	**80.72**
4 gang	60.90	76.40	0.88	31.64	nr	**108.04**
6 gang	102.67	128.79	1.18	42.43	nr	**171.22**
8 gang	122.48	153.64	1.63	58.61	nr	**212.25**
9 gang	153.62	192.70	1.83	65.79	nr	**258.49**
12 gang	182.97	229.52	2.29	82.33	nr	**311.85**
6 amp modular type swtich; galvanized steel box; bronze or satin chrome coverplate; flush mounting; two way						
1 gang	26.87	33.70	0.43	15.46	nr	**49.16**
2 gang	36.82	46.19	0.55	19.77	nr	**65.96**
3 gang	53.19	66.72	0.77	27.69	nr	**94.41**
4 gang	63.33	79.44	0.88	31.64	nr	**111.08**
6 gang	106.77	133.93	1.18	42.43	nr	**176.36**
8 gang	127.37	159.77	1.63	58.61	nr	**218.38**
9 gang	159.78	200.42	1.83	65.79	nr	**266.21**
12 gang	190.29	238.69	2.22	79.82	nr	**318.51**
Plate switches; 10 amp flush mounted, white plastic fronted; 16 mm metal box; fitted brass earth terminal						
1 gang, 1 way, single pole	9.94	12.47	0.28	10.06	nr	**22.53**
1 gang, 2 way, single pole	11.14	13.98	0.33	11.87	nr	**25.85**
2 gang, 2 way, single pole	13.91	17.45	0.44	15.81	nr	**33.26**
3 gang, 2 way, single pole	16.33	20.48	0.56	20.14	nr	**40.62**
1 gang, Intermediate	22.02	27.62	0.43	15.46	nr	**43.08**
1 gang, 1 way, double pole	19.04	23.88	0.33	11.87	nr	**35.75**
1 gang, single pole with bell symbol	15.23	19.11	0.23	8.27	nr	**27.38**
1 gang, single pole marked PRESS	12.18	15.28	0.23	8.27	nr	**23.55**
Time delay switch, suppressed	67.45	84.60	0.49	17.61	nr	**102.21**
Plate switches; 6 amp flush mounted white plastic fronted; 25 mm metal box; fitted brass earth terminal						
4 gang, 2 way, single pole	29.73	37.30	0.42	15.10	nr	**52.40**
6 gang, 2 way, single way	47.21	59.23	0.47	16.90	nr	**76.13**
Architrave plate switches; 6 amp flush mounted, white plastic fronted; 27 mm metal box; brass earth terminal						
1 gang, 2 way, single pole	3.98	5.00	0.30	10.79	nr	**15.79**
2 gang, 2 way, single pole	8.07	10.12	0.36	12.95	nr	**23.07**
Ceiling switches, white moulded plastic, pull cord; standard unit						
6 amp, 1 way, single pole	4.50	5.64	0.32	11.50	nr	**17.14**
6 amp, 2 way, single pole	5.25	6.59	0.34	12.22	nr	**18.81**
16 amp, 1 way, double pole	7.82	9.81	0.37	13.29	nr	**23.10**
45 amp, 1 way, double pole with neon indicator	14.40	18.07	0.47	16.90	nr	**34.97**

ELECTRICAL SUPPLY/POWER/LIGHTING

Item	Net Price £	Material £	Labour hours	Labour £	Unit	Total rate £
GENERAL LIGHTING – cont						
Switches – cont						
10 amp splash proof moulded switch with plain, threaded or PVC entry						
1 gang, 2 way single pole	24.48	30.71	0.34	12.22	nr	42.93
2 gang, 1 way single pole	27.66	34.70	0.36	12.95	nr	47.65
2 gang, 2 way single pole	35.04	43.95	0.40	14.38	nr	58.33
6 amp watertight switch; metalclad; BS 3676; ingress protected to IP65 surface mounted						
1 gang, 2 way; terminal entry	20.86	26.16	0.41	14.74	nr	40.90
1 gang, 2 way; through entry	20.86	26.16	0.42	15.10	nr	41.26
2 gang, 2 way; terminal entry	25.44	31.91	0.54	19.42	nr	51.33
2 gang, 2 way; through entry	25.44	31.91	0.53	19.05	nr	50.96
2 way replacement switch	18.76	23.53	0.10	3.60	nr	27.13
15 amp watertight switch; metalclad; BS 3676; ingress protected to IP65; surface mounted						
1 gang, 2 way, terminal entry	31.63	39.68	0.42	15.10	nr	54.78
1 gang, 2 way, through entry	31.63	39.68	0.43	15.46	nr	55.14
2 gang, 2 way, terminal entry	74.24	93.13	0.55	19.77	nr	112.90
2 gang, 2 way, through entry	74.24	93.13	0.54	19.42	nr	112.55
Intermediate interior only	18.76	23.53	0.11	3.95	nr	27.48
2 way interior only	18.76	23.53	0.11	3.95	nr	27.48
Double pole interior only	18.76	23.53	0.11	3.95	nr	27.48
Electrical accessories; fixed to backgrounds (Including fixings)						
Dimmer switches; rotary action; for individual lights; moulded plastic case; metal backbox; flush mounted						
1 gang, 1 way; 250 watt	19.68	24.68	0.28	10.06	nr	34.74
1 gang, 1 way; 400 watt	25.92	32.51	0.28	10.06	nr	42.57
Dimmer switches; push on/off action; for individual lights; moulded plastic case; metal backbox; flush mounted						
1 gang, 2 way; 250 watt	30.08	37.73	0.34	12.22	nr	49.95
3 gang, 2 way; 250 watt	44.33	55.61	0.48	17.26	nr	72.87
4 gang, 2 way; 250 watt	58.33	73.17	0.57	20.50	nr	93.67
Dimmer switches; rotary action; metal cald; metal backbox; BS 5518 and BS 800; flush mounted						
1 gang, 1 way; 400 watt	54.64	68.54	0.33	11.87	nr	80.41
Ceiling roses						
Ceiling rose: white moulded plastic; flush fixed to conduit box						
Plug in type; ceiling socket with 2 terminals, loop-in and ceiling plug with 3 terminals and cover	10.49	13.16	0.34	12.22	nr	25.38

ELECTRICAL SUPPLY/POWER/LIGHTING

Item	Net Price £	Material £	Labour hours	Labour £	Unit	Total rate £
BC lampholder; white moulded plastic; heat resistent PVC insulated and sheathed cable; flush fixed						
2 Core; 0.75 mm²	3.32	4.17	0.33	11.87	nr	**16.04**
Batten holder: white moulded plastic; 3 terminals; BS 5042; fixed to conduit						
Straight pattern; 2 terminals with loop-in and Earth	4.86	6.09	0.29	10.43	nr	**16.52**
Angled pattern; looped in terminal	–	–	0.29	10.43	nr	**10.43**
LIGHTING CONTROL MODULES						
SPV.92 Lighting control module; plug in; 9 output, 9 channel – switching						
Base and lid assembly	180.35	226.23	2.05	73.70	nr	**299.93**
SPV.92 Lighting control module; plug in; 9 output, 9 channel – dimming (DALI, DSI, 1–10V)						
Base and lid assembly	220.56	276.67	2.05	73.70	nr	**350.37**
SPH.12+ Lighting control module; hard wired; 4 circuit switching						
Base and lid assembly	194.62	244.13	1.85	66.51	nr	**310.64**
SPH.27r Lighting control module; hard wired; 2 circuit 40 ballast drive DALI (with relays)						
Base and lid assembly	278.96	349.93	1.85	66.51	nr	**416.44**
SPH.27 Lighting control module; hard wired; 2 circuit 40 ballast drive DALI (without relays)						
Base and lid assembly	265.99	333.66	1.85	66.51	nr	**400.17**
SPV.56 Compact lighting control module; 3 output 18 ballast drive; dimmable (DALI)						
Base and lid assembly	246.51	309.22	1.85	66.51	nr	**375.73**
SPB.82 Blind control module; 8 outputs						
Base and lid assembly	194.62	244.13	1.85	66.51	nr	**310.64**
Interfaces						
SPO.20 DALI to DMX converter	259.48	325.49	1.85	66.51	nr	**392.00**
Dual bus presence detectors (DALI & E-Bus)						
SPU.7-S Dual-bus presence detector and infra-red sensor	64.87	81.37	0.60	21.57	nr	**102.94**
SPU.6-S Dual-bus universal sensor	71.37	89.52	0.60	21.57	nr	**111.09**
SPU.6-C Compact Dual-bus Universal Sensor	72.52	180.50	0.60	21.57	nr	**202.07**
SPU.6-C1 Compact Dual-bus Universal Sensor conduit/surface mount	76.84	96.39	0.60	21.57	nr	**117.96**
Scene switch plate; anodized aluminium finish						
SPK.9–4+4 Eight button intelligent switch plate	129.74	162.75	2.00	71.90	nr	**234.65**
SPK.9–4-S Four button intelligent switch plate	77.85	97.65	1.60	57.52	nr	**155.17**
SPK.9–2-S Two button intelligent switch plate	64.87	81.37	1.20	43.14	nr	**124.51**

ELECTRICAL SUPPLY/POWER/LIGHTING

Item	Net Price £	Material £	Labour hours	Labour £	Unit	Total rate £
GENERAL LIGHTING – cont						
LIGHTING CONTROL MODULES – cont						
Input Devices/Interfaces						
SPF.4d Compact DALI interface for 4 Gang retractive switch plate	32.44	40.69	1.20	43.14	nr	83.83
SPL.20 External Photocell 20,000 Lux range	155.69	195.29	1.50	53.93	nr	249.22
SPF.6-T AV interface; RS232 interface	168.68	211.59	1.85	66.51	nr	278.10
Sub-addressing for DALI strings						
Sub-addressing charge per SPH.27R (40 Sub-addresses)	126.79	159.04	3.00	107.86	nr	266.90
Sub-addressing charge per SPV.56 (20 Sub-addresses)	63.39	238.56	3.00	107.86	nr	346.42
Commissioning						
Not included – worked out based on complexity of project						

ELECTRICAL SUPPLY/POWER/LIGHTING

Item	Net Price £	Material £	Labour hours	Labour £	Unit	Total rate £
GENERAL LV POWER						
ACCESSORIES						
Outlets						
Socket outlet: unswitched; 13 amp metal clad; BS 1363; galvanized steel box and coverplate with white plastic inserts; fixed surface mounted						
1 gang	10.52	13.19	0.41	14.74	nr	**27.93**
2 gang	17.52	21.97	0.41	14.74	nr	**36.71**
Socket outlet: switched; 13 amp metal clad; BS 1363; galvanized steel box and coverplate with white plastic inserts; fixed surface mounted						
1 gang	12.29	15.41	0.43	15.46	nr	**30.87**
2 gang	22.37	28.06	0.45	16.18	nr	**44.24**
Socket outlet: switched with neon indicator; 13 amp metal clad; BS 1363; galvanized steel box and coverplate withwhite plastic inserts; fixed surface mounted						
1 gang	26.00	32.61	0.43	15.46	nr	**48.07**
2 gang	47.28	59.30	0.45	16.18	nr	**75.48**
Socket outlet: unswitched; 13 amp; BS 1363; white moulded plastic box and coverplate; fixed surface mounted						
1 gang	6.66	8.36	0.41	14.74	nr	**23.10**
2 gang	13.14	16.49	0.41	14.74	nr	**31.23**
Socket outlet; switched; 13 amp; BS 1363; white moulded plastic box and coverplate; fixed surface mounted						
1 gang	4.86	6.09	0.43	15.46	nr	**21.55**
2 gang	7.80	9.79	0.45	16.18	nr	**25.97**
Socket outlet: switched with neon indicator; 13 amp; BS 1363; white moulded plastic box and coverplate; fixed surface mounted						
1 gang	22.19	27.83	0.43	15.46	nr	**43.29**
2 gang	29.92	37.53	0.45	16.18	nr	**53.71**
Socket outlet: switched; 13 amp; BS 1363; galvanized steel box, white moulded coverplate; flush fitted						
1 gang	6.05	7.59	0.43	15.46	nr	**23.05**
2 gang	16.70	20.94	0.45	16.18	nr	**37.12**
Socket outlet: switched with neon indicator; 13 amp; BS 1363; galvanized steel box, white moulded coverplate; flush fixed						
1 gang	22.19	27.83	0.43	15.46	nr	**43.29**
2 gang	38.37	48.13	0.45	16.18	nr	**64.31**

ELECTRICAL SUPPLY/POWER/LIGHTING

Item	Net Price £	Material £	Labour hours	Labour £	Unit	Total rate £
GENERAL LV POWER – cont						
Outlets – cont						
Socket outlet: switched; 13 amp; BS 1363; galvanized steel box, satin chrome coverplate; BS 4662; flush fixed						
1 gang	23.40	29.36	0.43	15.46	nr	**44.82**
2 gang	29.59	37.12	0.45	16.18	nr	**53.30**
Socket outlet: switched with neon indicator; 13 amp; BS 1363; steel backbox, satin chrome coverplate; BS 4662; flush fixed						
1 gang	29.47	36.97	0.43	15.46	nr	**52.43**
2 gang	53.14	66.66	0.45	16.18	nr	**82.84**
RCD protected socket outlets, 13 amp, to BS 1363; galvanized steel box, white moulded cover plate; flush fitted						
2 gang, 10 mA tripping (active control)	132.33	166.00	0.45	16.18	nr	**182.18**
2 gang, 30 mA tripping (active control)	117.88	147.87	0.45	16.18	nr	**164.05**
2 gang, 30 mA tripping (passive control)	117.88	147.87	0.45	16.18	nr	**164.05**
Filtered socket outlets, 13 amp, to BS 1363, with separate 'clean earth' terminal; galvanized steel box, white moulded cover plate; flush fitted						
2 gang (spike protected)	99.64	124.99	0.50	17.98	nr	**142.97**
2 gang (spike and RFI protected)	126.12	158.20	0.55	19.77	nr	**177.97**
Replacement filter cassette	34.95	43.84	0.15	5.40	nr	**49.24**
Non-standard socket outlets, 13 amp, to BS 1363, with separate 'clean earth' terminal; for plugs with T-shaped earth pin; galvanized steel box, white moulded cover plate; flush fitted						
1 gang	18.50	23.21	0.43	15.46	nr	**38.67**
2 gang	33.28	41.74	0.43	15.46	nr	**57.20**
2 gang coloured RED	46.22	57.98	0.43	15.46	nr	**73.44**
Weatherproof socket outlet: 40 amp; switched; single gang; RCD protected; water and dust protected to IP66; surface mounted						
40 A 30 mA tripping current protecting 1 socket	150.56	188.87	0.52	18.69	nr	**207.56**
40 A 30 mA tripping current protecting 2 sockets	66.92	83.94	0.64	23.00	nr	**106.94**
Plug for weatherproof socket outlet: protected to IP66						
13 amp plug	6.59	8.27	0.21	7.55	nr	**15.82**
Floor service outlet box; comprising flat lid with flanged carpet trim; twin 13 A switched socket outlets; punched plate for mounting 2 telephone outlets; one blank plate; triple compartment						
3 compartment	67.77	85.01	0.88	31.64	nr	**116.65**

ELECTRICAL SUPPLY/POWER/LIGHTING

Item	Net Price £	Material £	Labour hours	Labour £	Unit	Total rate £
Floor service outlet box; comprising flat lid with flanged carpet trim; 2 twin 13 A switched socket outlets; punched plate for mounting 1 telephone outlet; one blank plate; triple compartment						
3 compartment	60.48	75.87	0.88	31.64	nr	**107.51**
Floor service outlet box; comprising flat lid with flanged carpet trim; twin 13 A switched socket outlets; punched plate for mounting 2 telephone outlets; two blank plates; four compartment						
4 compartment	79.54	99.77	0.88	31.64	nr	**131.41**
Floor service outlet box; comprising flat lid with flanged carpet trim; single 13 A unswitched socket outlet; single compartment; circular						
1 compartment	84.85	106.43	0.79	28.40	nr	**134.83**
Floor service grommet, comprising flat lid with flanged carpet trim; circular						
Floor grommet	38.38	48.15	0.49	17.61	nr	**65.76**
Power posts/poles/pillars						
Power post						
Power post; aluminium painted body; PVC-u cover; 5 nr outlets	423.70	531.48	4.00	143.81	nr	**675.29**
Power pole						
Power pole; 3.6 m high; aluminium painted body; PVC-u cover; 6 nr outlets	543.31	681.53	4.00	143.81	nr	**825.34**
Extra for						
Power pole extension bar; 900 mm long	59.37	74.47	1.50	53.93	nr	**128.40**
Vertical multi-compartment pillar; PVC-u; BS 4678 Part4 EN60529; excludes accessories						
Single						
630 mm long	189.24	237.38	2.00	71.90	nr	**309.28**
3000 mm long	545.35	684.08	2.00	71.90	nr	**755.98**
Double						
630 mm long	189.24	237.38	3.00	107.86	nr	**345.24**
3000 mm long	578.80	726.05	3.00	107.86	nr	**833.91**
Connection units						
Connection units: moulded pattern; BS 5733; moulded plastic box; white coverplate; knockout for flex outlet; surface mounted – standard fused						
DP switched	11.00	13.80	0.49	17.61	nr	**31.41**
Unswitched	10.11	12.68	0.49	17.61	nr	**30.29**
DP switched with neon indicator	13.96	17.52	0.49	17.61	nr	**35.13**

ELECTRICAL SUPPLY/POWER/LIGHTING

Item	Net Price £	Material £	Labour hours	Labour £	Unit	Total rate £
GENERAL LV POWER – cont						
Connection units – cont						
Connection units: moulded pattern; BS 5733; galvanized steel box; white coverplate; knockout for flex outlet; surface mounted						
DP switched	13.86	31.19	0.49	17.61	nr	**48.80**
DP unswitched	12.97	16.27	0.49	17.61	nr	**33.88**
DP switched with neon indicator	16.83	21.11	0.49	17.61	nr	**38.72**
Connection units: galvanized pressed steel pattern; galvanized steel box; satin chrome or satin brass finish; white moulded plastic inserts; flush mounted – standard fused						
DP switched	20.32	25.49	0.49	17.61	nr	**43.10**
Unswitched	19.10	23.96	0.49	17.61	nr	**41.57**
DP switched with neon indicator	27.42	34.40	0.49	17.61	nr	**52.01**
Connection units: galvanized steel box; satin chrome or satin brass finish; white moulded plastic inserts; flex outlet; flush mounted – standard fused						
Switched	19.49	24.45	0.49	17.61	nr	**42.06**
Unswitched	18.50	23.21	0.49	17.61	nr	**40.82**
Switched with neon indicator	25.08	31.46	0.49	17.61	nr	**49.07**
Shaver sockets						
Shaver unit: self-setting overload device; 200/250 voltage supply; white moulded plastic faceplate; unswitched						
Surface type with moulded plastic box	38.19	47.90	0.55	19.77	nr	**67.67**
Flush type with galvanized steel box	40.09	50.29	0.57	20.50	nr	**70.79**
Shaver unit: dual voltage supply unit; white moulded plastic faceplate; unswitched						
Surface type with moulded plastic box	46.05	57.77	0.62	22.29	nr	**80.06**
Flush type with galvanized steel box	47.87	60.04	0.64	23.00	nr	**83.04**
Cooker control units						
Cooker control unit: BS 4177; 45 amp DP main switch; 13 amp switched socket outlet; metal coverplate; plastic inserts; neon indicators						
Surface mounted with mounting box	54.07	67.83	0.61	21.93	nr	**89.76**
Flush mounted with galvanized steel box	51.68	64.83	0.61	21.93	nr	**86.76**
Cooker control unit: BS 4177; 45 amp DP main switch; 13 amp switched socket outlet; moulded plastic box and coverplate; surface mounted						
Standard	37.52	47.06	0.61	21.93	nr	**68.99**
With neon indicators	44.02	55.22	0.61	21.93	nr	**77.15**

ELECTRICAL SUPPLY/POWER/LIGHTING

Item	Net Price £	Material £	Labour hours	Labour £	Unit	Total rate £
Control components						
Connector unit: moulded white plastic cover and block; galvanized steel back box; to immersion heaters						
3 kW up to 915 mm long; fitted to thermostat	40.20	50.42	0.75	26.97	nr	**77.39**
Water heater switch: 20 amp; switched with neon indicator						
DP switched with neon indicator	19.97	25.05	0.45	16.18	nr	**41.23**
Switch disconnectors						
Switch disconnectors; moulded plastic enclosure; fixed to backgrounds						
3 pole; IP54; grey						
16 amp	35.46	44.49	0.80	28.76	nr	**73.25**
25 amp	41.98	52.66	0.80	28.76	nr	**81.42**
40 amp	68.36	85.75	0.80	28.76	nr	**114.51**
63 amp	106.41	133.48	1.00	35.95	nr	**169.43**
80 amp	184.48	231.41	1.25	44.95	nr	**276.36**
6 pole; IP54; grey						
25 amp	58.98	73.99	1.00	35.95	nr	**109.94**
63 amp	99.60	124.94	1.25	44.95	nr	**169.89**
80 amp	188.80	236.84	1.80	64.71	nr	**301.55**
3 pole; IP54; yellow						
16 amp	38.97	48.89	0.80	28.76	nr	**77.65**
25 amp	46.08	57.80	0.80	28.76	nr	**86.56**
40 amp	74.75	93.77	0.80	28.76	nr	**122.53**
63 amp	116.30	145.89	1.00	35.95	nr	**181.84**
6 pole; IP54; yellow						
25 amp	58.98	73.99	1.00	35.95	nr	**109.94**
Industrial sockets/plugs						
Plugs; Splashproof; 100–130 volts, 50–60 Hz; IP 44 (yellow)						
2 pole and earth						
16 amp	3.36	4.21	0.55	19.77	nr	**23.98**
32 amp	12.06	15.13	0.60	21.57	nr	**36.70**
3 pole and earth						
16 amp	13.40	16.81	0.65	23.37	nr	**40.18**
32 amp	17.14	21.50	0.72	25.89	nr	**47.39**
3 pole; neutral and earth						
16 amp	13.93	17.47	0.72	25.89	nr	**43.36**
32 amp	20.29	25.45	0.78	28.03	nr	**53.48**

ELECTRICAL SUPPLY/POWER/LIGHTING

Item	Net Price £	Material £	Labour hours	Labour £	Unit	Total rate £
GENERAL LV POWER – cont						
Industrial sockets/plugs – cont						
Connectors; Splashproof; 100–130 volts, 50–60 Hz; IP 44 (yellow)						
2 pole and earth						
16 amp	8.64	10.84	0.42	15.10	nr	**25.94**
32 amp	15.01	18.83	0.50	17.98	nr	**36.81**
3 pole and earth						
16 amp	16.26	20.40	0.48	17.26	nr	**37.66**
32 amp	17.54	22.00	0.58	20.84	nr	**42.84**
3 pole; neutral and earth						
16 amp	22.79	28.58	0.52	18.69	nr	**47.27**
32 amp	23.10	28.97	0.73	26.24	nr	**55.21**
Angled sockets; surface mounted; Splashproof; 100–130 volts, 50–60 Hz; IP 44 (yellow)						
2 pole and earth						
16 amp	5.28	6.62	0.55	19.77	nr	**26.39**
32 amp	18.30	22.96	0.60	21.57	nr	**44.53**
3 pole and earth						
16 amp	17.44	21.87	0.65	23.37	nr	**45.24**
32 amp	32.75	41.08	0.72	25.89	nr	**66.97**
3 pole; neutral and earth						
16 amp	24.21	30.37	0.72	25.89	nr	**56.26**
32 amp	31.64	39.69	0.78	28.03	nr	**67.72**
Plugs; Watertight; 100–130 volts, 50–60 Hz; IP67 (yellow)						
2 pole and earth						
16 amp	13.65	17.12	0.55	19.77	nr	**36.89**
32 amp	23.79	29.84	0.60	21.57	nr	**51.41**
63 amp	65.22	81.82	0.75	26.97	nr	**108.79**
Connectors; Watertight; 100–130 volts, 50–60 Hz; IP 67 (yellow)						
2 pole and earth						
16 amp	27.88	34.98	0.42	15.10	nr	**50.08**
32 amp	47.25	59.27	0.50	17.98	nr	**77.25**
63 amp	114.37	143.46	0.67	24.08	nr	**167.54**
Angled sockets; surface mounted; Watertight; 100–130 volts, 50–60 Hz; IP 67 (yellow)						
2 pole and earth						
16 amp	23.04	28.90	0.55	19.77	nr	**48.67**
32 amp	45.06	56.53	0.60	21.57	nr	**78.10**

ELECTRICAL SUPPLY/POWER/LIGHTING

Item	Net Price £	Material £	Labour hours	Labour £	Unit	Total rate £
Plugs; Splashproof; 200–250 volts, 50–60 Hz; IP 44 (blue)						
2 pole and earth						
16 amp	3.29	4.12	0.55	19.77	nr	**23.89**
32 amp	11.81	14.82	0.60	21.57	nr	**36.39**
63 amp	56.88	71.36	0.75	26.97	nr	**98.33**
3 pole and earth						
16 amp	13.07	16.40	0.65	23.37	nr	**39.77**
32 amp	17.55	22.02	0.72	25.89	nr	**47.91**
63 amp	57.09	71.61	0.83	29.84	nr	**101.45**
3 pole; neutral and earth						
16 amp	13.93	17.47	0.72	25.89	nr	**43.36**
32 amp	21.14	26.52	0.78	28.03	nr	**54.55**
Connectors; Splashproof; 200–250 volts, 50–60 Hz; IP 44 (blue)						
2 pole and earth						
16 amp	8.13	10.20	0.42	15.10	nr	**25.30**
32 amp	20.18	25.31	0.50	17.98	nr	**43.29**
63 amp	71.28	89.41	0.67	24.08	nr	**113.49**
3 pole and earth						
16 amp	21.51	26.98	0.48	17.26	nr	**44.24**
32 amp	28.43	35.66	0.58	20.84	nr	**56.50**
63 amp	58.64	73.56	0.75	26.97	nr	**100.53**
3 pole; neutral and earth						
16 amp	25.00	31.36	0.52	18.69	nr	**50.05**
32 amp	85.63	107.42	0.73	26.24	nr	**133.66**
Angled sockets; surface mounted; Splashproof; 200–250 volts, 50–60 Hz; IP 44 (blue)						
2 pole and earth						
16 amp	10.68	13.40	0.55	19.77	nr	**33.17**
32 amp	18.50	23.21	0.60	21.57	nr	**44.78**
63 amp	81.26	101.93	0.75	26.97	nr	**128.90**
3 pole and earth						
16 amp	20.78	26.06	0.65	23.37	nr	**49.43**
32 amp	35.95	45.09	0.72	25.89	nr	**70.98**
63 amp	79.15	99.29	0.83	29.84	nr	**129.13**
3 pole; neutral and earth						
16 amp	18.08	22.68	0.72	25.89	nr	**48.57**
32 amp	31.63	39.68	0.78	28.03	nr	**67.71**
Plugs; Watertight; 200–250 volts, 50–60 Hz; IP67 (blue)						
2 pole and earth						
16 amp	13.93	17.47	0.41	14.74	nr	**32.21**
32 amp	24.37	30.56	0.50	17.98	nr	**48.54**
63 amp	66.86	83.87	0.66	23.73	nr	**107.60**
125 amp	171.32	214.91	0.86	30.92	nr	**245.83**

ELECTRICAL SUPPLY/POWER/LIGHTING

Item	Net Price £	Material £	Labour hours	Labour £	Unit	Total rate £
GENERAL LV POWER – cont						
Industrial sockets/plugs – cont						
Connectors; Watertight; 200–250 volts, 50–60 Hz; IP 67 (blue)						
2 pole and earth						
16 amp	13.93	17.47	0.41	14.74	nr	**32.21**
32 amp	23.75	29.79	0.50	17.98	nr	**47.77**
63 amp	60.97	76.48	0.67	24.08	nr	**100.56**
125 amp	242.38	304.05	0.87	31.27	nr	**335.32**
Angled sockets; surface mounted; Watertight; 200–250 volts, 50–60 Hz; IP 67 (blue)						
2 pole and earth						
16 amp	23.04	28.90	0.55	19.77	nr	**48.67**
32 amp	45.05	56.52	0.60	21.57	nr	**78.09**
125 amp	328.12	411.59	1.00	35.95	nr	**447.54**

ELECTRICAL SUPPLY/POWER/LIGHTING

Item	Net Price £	Material £	Labour hours	Labour £	Unit	Total rate £
UNINTERRUPTIBLE POWER SUPPLY						
Uninterruptible power supply; sheet steel enclosure; self-contained battery pack; including installation, testing and commissioning						
Single phase input and output; 5 year battery life; standard 13 A socket outlet connection						
1.0 kVA (10 min supply)	1097.93	1377.24	0.30	10.79	nr	**1388.03**
1.0 kVA (30 min supply)	1564.72	1962.79	0.50	17.98	nr	**1980.77**
2.0 kVA (10 min supply)	1958.81	2457.13	0.50	17.98	nr	**2475.11**
2.0 kVA (60 min supply)	3287.44	4123.76	0.50	17.98	nr	**4141.74**
3.0 kVA (10 min supply)	2588.38	3246.87	0.50	17.98	nr	**3264.85**
3.0 kVA (40 min supply)	3342.23	4192.50	1.00	35.95	nr	**4228.45**
5.0 kVA (30 min supply)	4236.71	5314.53	1.00	35.95	nr	**5350.48**
8.0 kVA (10 min supply)	5096.28	6392.77	2.00	71.90	nr	**6464.67**
8.0 kVA (30 min supply)	6891.35	8644.51	2.00	71.90	nr	**8716.41**
Uninterruptible power supply; including final connections and testing and commissioning						
Medium size static; single phase input and output; 10 year battery life; in cubicle						
10.0 kVA (10 min supply)	5096.00	6392.42	10.00	359.51	nr	**6751.93**
10.0 kVA (30 min supply)	6891.04	8644.12	15.00	539.27	nr	**9183.39**
15.0 kVA (10 min supply)	8317.92	10434.00	10.00	359.51	nr	**10793.51**
15.0 kVA (30 min supply)	11497.20	14422.08	15.00	539.27	nr	**14961.35**
20.0 kVA (10 min supply)	8317.92	10434.00	10.00	359.51	nr	**10793.51**
20.0 kVA (30 min supply)	14799.20	18564.11	15.00	539.27	nr	**19103.38**
Medium size static; three phase input and output; 10 year battery life; in cubicle						
10.0 kVA (10 min supply)	6878.37	8628.22	10.00	359.51	nr	**8987.73**
10.0 kVA (30 min supply)	8616.64	10808.72	15.00	539.27	nr	**11347.99**
15.0 kVA (10 min supply)	7844.69	9840.38	15.00	539.27	nr	**10379.65**
15.0 kVA (30 min supply)	9694.12	12160.30	20.00	719.02	nr	**12879.32**
20.0 kVA (10 min supply)	7920.09	9934.96	20.00	719.02	nr	**10653.98**
20.0 kVA (30 min supply)	11983.44	15032.02	25.00	898.78	nr	**15930.80**
30.0 kVA (10 min supply)	11852.30	14867.53	25.00	898.78	nr	**15766.31**
30.0 kVA (30 min supply)	14899.59	18690.04	30.00	1078.54	nr	**19768.58**
Large size static; three phase input and output; 10 year battery life; in cubicle						
40 kVA (10 min supply)	11852.30	14867.53	30.00	1078.54	nr	**15946.07**
40 kVA (30 min supply)	17902.62	22457.04	30.00	1078.54	nr	**23535.58**
60 kVA (10 min supply)	19064.72	23914.79	35.00	1258.29	nr	**25173.08**
60 kVA (30 min supply)	22205.24	27854.25	35.00	1258.29	nr	**29112.54**
100 kVA (10 min supply)	28017.97	35145.75	40.00	1438.05	nr	**36583.80**
200 kVA (10 min supply)	46239.10	58002.32	40.00	1438.05	nr	**59440.37**
300 kVA (10 min supply)	75250.56	94394.31	40.00	1438.05	nr	**95832.36**
400 kVA (10 min supply)	88293.44	110755.29	50.00	1797.56	nr	**112552.85**
500 kVA (10 min supply)	103619.04	129979.72	60.00	2157.06	nr	**132136.78**
600 kVA (10 min supply)	139403.91	174868.27	70.00	2516.57	nr	**177384.84**
800 kVA (10 min supply)	174098.53	218389.19	80.00	2876.09	nr	**221265.28**

ELECTRICAL SUPPLY/POWER/LIGHTING

Item	Net Price £	Material £	Labour hours	Labour £	Unit	Total rate £
UNINTERRUPTIBLE POWER SUPPLY – cont						
Uninterruptible power supply – cont						
Integral diesel rotary; 400 V three phase input and output; no break supply; including ventilation and acoustic attenuation oil day tank and interconnecting pipework						
300 kVA	1333839.00	1673167.64	120.00	4314.13	nr	**1677481.77**
500 kVA	524157.76	657503.49	120.00	4314.13	nr	**661817.62**
800 kVA	572311.12	717907.06	140.00	5033.16	nr	**722940.22**
1000 kVA	634942.05	796471.31	140.00	5033.16	nr	**801504.47**
1670 kVA	756315.72	948722.44	170.00	6111.68	nr	**954834.12**
2000 kVA	921375.68	1155773.65	200.00	7190.22	nr	**1162963.87**
2500 kVA	1064805.98	1335692.62	230.00	8268.76	nr	**1343961.38**
Commercial battery storage systems						
Battery storage system; self-contained compact battery pack; flexible layout; rated for outdoor use; includes labour, materials, civils, design and project management						
100 kW nominal / 169 kWh batteries (1.6 hr capacity)	127213.00	159575.99	48.00	1725.65	nr	**161301.64**
500 kW nominal / 845 kWh batteries (1.6 hr capacity)	463594.00	581532.31	58.00	2085.16	nr	**583617.47**
1000 kW nominal / 1690 kWh batteries (1.6 hr capacity)	877315.00	1100503.94	76.00	2732.28	nr	**1103236.22**

ELECTRICAL SUPPLY/POWER/LIGHTING

Item	Net Price £	Material £	Labour hours	Labour £	Unit	Total rate £
EMERGENCY LIGHTING						
24 volt/50 volt/110 volt fluorescent slave luminaires						
For use with DC central battery systems						
Indoor, 8 watt	43.19	54.17	0.80	28.76	nr	**82.93**
Indoor, exit sign box	53.80	67.49	0.80	28.76	nr	**96.25**
Outdoor, 8 watt weatherproof	48.13	60.38	0.80	28.76	nr	**89.14**
Conversion module AC/DC	53.80	67.49	0.25	8.99	nr	**76.48**
Self-contained; polycarbonate base and diffuser; LED charging light to European sign directive; 3 hour standby						
Non-maintained						
Indoor, 8 watt	47.68	59.81	1.00	35.95	nr	**95.76**
Outdoor, 8 watt weatherproof, vandal-resistant IP65	68.47	85.89	1.00	35.95	nr	**121.84**
Maintained						
Indoor, 8 watt	34.13	42.82	1.00	35.95	nr	**78.77**
Outdoor, 8 watt weatherproof, vandal-resistant IP65	100.99	126.68	1.00	35.95	nr	**162.63**
Exit signage						
Exit sign; gold effect, pendular including brackets						
Non-maintained, 8 watt	129.68	162.67	1.00	35.95	nr	**198.62**
Maintained, 8 watt	139.58	175.09	1.00	35.95	nr	**211.04**
Modification kit						
Module & battery for 58 W fluorescent modification from mains fitting to emergency; 3 hour standby	44.70	56.07	0.50	17.98	nr	**74.05**
Extra for remote box (when fitting is too small for modification)	22.19	27.83	0.50	17.98	nr	**45.81**
12 volt low voltage lighting; non-maintained; 3 hour standby						
2 × 20 watt lamp load	154.02	193.20	1.20	43.14	nr	**236.34**
1 × 50 watt lamp load	168.13	210.91	1.00	35.95	nr	**246.86**
Maintained 3 hour standby						
2 × 20 watt lamp load	145.80	182.90	1.20	43.14	nr	**226.04**
1 × 50 watt lamp load	159.90	200.58	1.00	35.95	nr	**236.53**
DC central battery systems BS5266 compliant 24/50/110 volt						
DC supply to luminaires on mains failure; metal cubicle with battery charger, changeover device and battery as integral unit; ICEL 1001 compliant; 10 year design life valve regulated lead acid battery; 24 hour recharge; LCD display & LED indication; ICEL alarm pack; Includes on-site commissioning on 110 volt systems only						

ELECTRICAL SUPPLY/POWER/LIGHTING

Item	Net Price £	Material £	Labour hours	Labour £	Unit	Total rate £
EMERGENCY LIGHTING – cont						
DC central battery systems – cont						
24 volt, wall mounted						
300 W maintained, 1 hour	2455.96	3080.76	4.00	143.81	nr	**3224.57**
635 W maintained, 3 hour	2979.53	3737.52	6.00	215.71	nr	**3953.23**
470 W non-maintained, 1 hour	2189.58	2746.61	4.00	143.81	nr	**2890.42**
780 W non-maintained, 3 hour	2586.85	3244.94	6.00	215.71	nr	**3460.65**
50 volt						
935 W maintained, 3 hour	2352.63	2951.14	8.00	287.60	nr	**3238.74**
1965 W maintained, 3 hour	3274.61	4107.67	8.00	287.60	nr	**4395.27**
1311 W non-maintained, 3 hour	3205.72	4021.26	8.00	287.60	nr	**4308.86**
2510 W non-maintained 3, hour	4895.85	6141.35	8.00	287.60	nr	**6428.95**
110 volt						
1603 W maintained, 3 hour	4614.54	5788.47	8.00	287.60	nr	**6076.07**
4446 W maintained, 3 hour	5270.16	6610.89	10.00	359.51	nr	**6970.40**
2492 W non-maintained, 3 hour	5854.58	7343.99	10.00	359.51	nr	**7703.50**
5429 W non-maintained, 3 hour	7991.35	10024.35	12.00	431.41	nr	**10455.76**
DC central battery systems; BS EN 50171 compliant; 24/50/110 volt						
Central power systems						
DC supply to luminaires on mains failure; metal cubicle with battery charger, changeover device and battery as integral unit; 10 year design life valve regulated lead acid battery; 12 hour recharge to 80% of specified duty; low volts discount; LCD display & LED indication; includes on-site commissioning for CPS systems only; battery sized for 'end of life' @ 20°C test pushbutton						
24 volt, floor standing						
400 W non-maintained, 1 hour	2306.70	2893.52	4.00	143.81	nr	**3037.33**
600 W maintained, 3 hour	2965.76	3720.25	6.00	215.71	nr	**3935.96**
50 volt						
2133 W non-maintained, 3 hour	5185.19	6504.30	8.00	287.60	nr	**6791.90**
1900 W maintained, 3 hour	4955.55	6216.25	8.00	287.60	nr	**6503.85**
110 volt						
2200 W non-maintained, 3 hour	4795.95	6016.04	8.00	287.60	nr	**6303.64**
4000 W maintained, 3 hour	5352.83	6714.59	12.00	431.41	nr	**7146.00**
Low power systems						
DC supply to luminaires on mains failure; metal cubicle with battery charger, changeover device and battery as integral unit; 5 Year design life valve regulated lead acid battery; low volts discount; LED display & LED indication; battery sized for 'end of life' @ 20°C test pushbutton						

ELECTRICAL SUPPLY/POWER/LIGHTING

Item	Net Price £	Material £	Labour hours	Labour £	Unit	Total rate £
24 volt, floor standing						
300 W non-maintained, 1 hour	2529.44	3172.93	4.00	143.81	nr	**3316.74**
600 W maintained, 3 hour	2800.41	3512.84	6.00	215.71	nr	**3728.55**
AC static inverter system; BS5266 compliant; one hour standby						
Central system supplying AC power on mains failure to mains luminaires; ICEL 1001 compliant metal cubicle(s) with changeover device, battery charger, battery & static inverter; 10 year design life valve regulated lead acid battery; 24 hour recharge; LED indication and LCD display; pure sinewave output						
One hour						
750 VA, 600 W single phase I/P & O/P	2988.72	3749.05	6.00	215.71	nr	**3964.76**
3 kVA, 2.55 kW single phase I/P & O/P	4902.74	6150.00	8.00	287.60	nr	**6437.60**
5 kVA, 4.25 kW single phase I/P & O/P	6474.61	8121.75	10.00	359.51	nr	**8481.26**
8 kVA, 6.80 kW single phase I/P & O/P	7627.37	9567.77	12.00	431.41	nr	**9999.18**
10 kVA, 8.5 kW single phase I/P & O/P	9846.81	12351.84	14.00	503.32	nr	**12855.16**
13 kVA, 11.05 kW single phase I/P & O/P	13173.09	16524.32	16.00	575.22	nr	**17099.54**
15 kVA, 12.75 kW single phase I/P & O/P	13653.03	17126.36	30.00	1078.54	nr	**18204.90**
20 kVA, 17.0 kW 3 phase I/P & single phase O/P	18856.60	23653.72	40.00	1438.05	nr	**25091.77**
30 kVA, 25.5 kW 3 phase I/P & O/P	25482.75	31965.56	60.00	2157.06	nr	**34122.62**
40 kVA, 34.0 kW 3 phase I/P & O/P	31769.04	39851.08	80.00	2876.09	nr	**42727.17**
50 kVA, 42.5 kW 3 phase I/P & O/P	41434.45	51975.37	90.00	3235.60	nr	**55210.97**
65 kVA, 55.25 kW 3 phase I/P & O/P	50828.88	63759.75	100.00	3595.11	nr	**67354.86**
90 kVA, 68.85 kW 3 phase I/P & O/P	67168.66	84256.37	120.00	4314.13	nr	**88570.50**
120 kVA, 102 kW 3 phase I/P & O/P	74976.29	94050.25	150.00	5392.67	nr	**99442.92**
Three hour						
750 VA, 600 W single phase I/P & O/P	3497.36	4387.08	6.00	215.71	nr	**4602.79**
3 kVA, 2.55 kW single phase I/P & O/P	5885.58	7382.87	8.00	287.60	nr	**7670.47**
5 kVA, 4.25 kW single phase I/P & O/P	9330.13	11703.72	10.00	359.51	nr	**12063.23**
8 kVA, 6.80 kW single phase I/P & O/P	11608.13	14561.24	12.00	431.41	nr	**14992.65**
10 kVA, 8.5 kW single phase I/P & O/P	15312.16	19207.57	14.00	503.32	nr	**19710.89**
13 kVA, 11.05 kW single phase I/P & O/P	18880.71	23683.97	16.00	575.22	nr	**24259.19**
15 kVA, 12.75 kW single phase I/P & O/P	19190.72	24072.84	30.00	1078.54	nr	**25151.38**
20 kVA, 17.0 kW 3 phase I/P & single phase O/P	27790.60	34860.53	40.00	1438.05	nr	**36298.58**
30 kVA, 25.5 kW 3 phase I/P & O/P	41760.53	52384.40	60.00	2157.06	nr	**54541.46**
40 kVA, 34.0 kW 3 phase I/P & O/P	49927.56	62629.13	80.00	2876.09	nr	**65505.22**
50 kVA, 42.5 kW 3 phase I/P & O/P	67340.88	84472.40	90.00	3235.60	nr	**87708.00**
65 kVA, 55.25 kW 3 phase I/P & O/P	81422.18	102135.98	100.00	3595.11	nr	**105731.09**
90 kVA, 68.85 kW 3 phase I/P & O/P	96363.48	120878.35	120.00	4314.13	nr	**125192.48**
120 kVA, 102 kW 3 phase I/P & O/P	118770.26	148985.41	150.00	5392.67	nr	**154378.08**

ELECTRICAL SUPPLY/POWER/LIGHTING

Item	Net Price £	Material £	Labour hours	Labour £	Unit	Total rate £
EMERGENCY LIGHTING – cont						
AC static inverter system; BS EN 50171 compliant; one hour standby; low power system (typically wall mounted)						
Central system supplying AC power on mains failure to mains luminaires; metal cubicle(s) with changeover device, battery charger, battery & static inverter; 5 year design life valve regulated lead acid battery; LED indication and LCD display; 12 hour recharge to 80% duty; inverter rated for 120% of load for 100% of duty; battery sized for 'end of life' @ 20°C test pushbutton						
One hour						
300 VA, 240 W single phase I/P & O/P	1336.49	1676.49	3.00	107.86	nr	**1784.35**
600 VA, 480 W single phase I/P & O/P	1610.90	2020.72	4.00	143.81	nr	**2164.53**
750 VA, 600 W single phase I/P & O/P	2972.64	3728.88	6.00	215.71	nr	**3944.59**
Three hour						
150 VA, 120 W single phase I/P & O/P	1473.12	1847.88	3.00	107.86	nr	**1955.74**
450 VA, 360 W single phase I/P & O/P	1747.53	2192.10	4.00	143.81	nr	**2335.91**
750 VA, 600 W single phase I/P & O/P	3478.98	4364.04	6.00	215.71	nr	**4579.75**
AC static inverter system central power system; CPS BS EN 50171 compliant; one hour standby						
Central system supplying AC power on mains failure to mains luminaires; metal cubicle(s) with changeover device, battery charger, battery & static inverter; LED indication and LCD display; pure sinewave output; 10 year design life valve regulated lead acid battery; 12 hour recharge to 80% duty specified; unverter rated for 120% of load for 100% of duty; battery sized for 'end of life' @ 20°C test push button; includes on-site commissioning						
One hour						
750 VA, 600 W single phase I/P & O/P	3326.28	4172.48	6.00	215.71	nr	**4388.19**
3 kVA, 2.55 kW single phase I/P & O/P	5300.00	6648.32	8.00	287.60	nr	**6935.92**
5 kVA, 4.25 kW single phase I/P & O/P	6862.68	8608.54	10.00	359.51	nr	**8968.05**
8 kVA, 6.80 kW single phase I/P & O/P	8009.72	10047.40	12.00	431.41	nr	**10478.81**
10 kVA, 8.5 kW single phase I/P & O/P	10285.41	12902.02	14.00	503.32	nr	**13405.34**
13 kVA, 11.05 kW single phase I/P & O/P	13594.47	17052.91	16.00	575.22	nr	**17628.13**
15 kVA, 12.75 kW single phase I/P & O/P	14070.97	17650.63	30.00	1078.54	nr	**18729.17**
20 kVA, 17.0 kW 3 phase I/P & single phase O/P	19383.61	24314.80	40.00	1438.05	nr	**25752.85**
30 kVA, 25.5 kW 3 phase I/P & O/P	26110.81	32753.40	60.00	2157.06	nr	**34910.46**
40 kVA, 34.0 kW 3 phase I/P & O/P	32362.66	40595.72	80.00	2876.09	nr	**43471.81**
50 kVA, 42.5 kW 3 phase I/P & O/P	42147.47	52869.79	90.00	3235.60	nr	**56105.39**
65 kVA, 55.25 kW 3 phase I/P & O/P	52057.42	65300.83	100.00	3595.11	nr	**68895.94**
90 kVA, 68.85 kW 3 phase I/P & O/P	68309.94	85687.99	120.00	4314.13	nr	**90002.12**
120 kVA, 102 kW 3 phase I/P & O/P	76076.25	95430.05	150.00	5392.67	nr	**100822.72**

ELECTRICAL SUPPLY/POWER/LIGHTING

Item	Net Price £	Material £	Labour hours	Labour £	Unit	Total rate £
Three hour						
750 VA, 600 W single phase I/P & O/P	3848.71	4827.83	6.00	215.71	nr	**5043.54**
3 kVA, 2.55 kW single phase I/P & O/P	6277.12	7874.01	8.00	287.60	nr	**8161.61**
5 kVA, 4.25 kW single phase I/P & O/P	9726.25	12200.61	10.00	359.51	nr	**12560.12**
8 kVA, 6.80 kW single phase I/P & O/P	11969.80	15014.92	12.00	431.41	nr	**15446.33**
10 kVA, 8.5 kW single phase I/P & O/P	15723.21	19723.20	14.00	503.32	nr	**20226.52**
13 kVA, 11.05 kW single phase I/P & O/P	19272.23	24175.09	16.00	575.22	nr	**24750.31**
15 kVA, 12.75 kW single phase I/P & O/P	19578.79	24559.63	30.00	1078.54	nr	**25638.17**
20 kVA, 17.0 kW 3 phase I/P & single phase O/P	28531.18	35789.51	40.00	1438.05	nr	**37227.56**
30 kVA, 25.5 kW 3 phase I/P & O/P	42300.17	53061.33	60.00	2157.06	nr	**55218.39**
40 kVA, 34.0 kW 3 phase I/P & O/P	50424.71	63252.76	80.00	2876.09	nr	**66128.85**
50 kVA, 42.5 kW 3 phase I/P & O/P	67914.97	85192.54	90.00	3235.60	nr	**88428.14**
65 kVA, 55.25 kW 3 phase I/P & O/P	82487.70	103472.57	100.00	3595.11	nr	**107067.68**
90 kVA, 68.85 kW 3 phase I/P & O/P	102551.02	128640.00	120.00	4314.13	nr	**132954.13**
120 kVA, 102 kW 3 phase I/P & O/P	119633.68	150068.49	150.00	5392.67	nr	**155461.16**

COMMUNICATIONS/SECURITY/CONTROL

Item	Net Price £	Material £	Labour hours	Labour £	Unit	Total rate £
TELECOMMUNICATIONS						
CABLES						
Multipair internal telephone cable; BS 6746; loose laid on tray/basket						
0.5 mm dia. conductor LSZH insulated and sheathed multipair cables; BT specification CW 1308						
3 pair	0.20	0.25	0.03	1.08	m	1.33
4 pair	0.23	0.29	0.03	1.08	m	1.37
6 pair	0.34	0.43	0.03	1.08	m	1.51
10 pair	0.59	0.74	0.05	1.79	m	2.53
15 pair	0.89	1.12	0.06	2.16	m	3.28
20 pair + 1 wire	1.26	1.58	0.06	2.16	m	3.74
25 pair	1.60	2.00	0.08	2.87	m	4.87
40 pair + earth	2.26	2.83	0.08	2.87	m	5.70
50 pair + earth	2.71	3.40	0.10	3.60	m	7.00
80 pair + earth	3.63	4.56	0.10	3.60	m	8.16
100 pair + earth	5.79	7.26	0.12	4.31	m	11.57
Multipair internal telephone cable; BS 6746; installed in conduit/trunking						
0.5 mm dia. conductor LSZH insulated and sheathed multipair cables; BT specification CW 1308						
3 pair	0.20	0.25	0.05	1.79	m	2.04
4 pair	0.23	0.29	0.06	2.16	m	2.45
6 pair	0.34	0.43	0.06	2.16	m	2.59
10 pair	0.59	0.74	0.07	2.52	m	3.26
15 pair	0.89	1.12	0.07	2.52	m	3.64
20 pair + 1 wire	1.26	1.58	0.09	3.24	m	4.82
25 pair	1.60	2.00	0.10	3.60	m	5.60
40 pair + earth	2.26	2.83	0.12	4.31	m	7.14
50 pair + earth	2.71	3.40	0.14	5.03	m	8.43
80 pair + earth	3.63	4.56	0.14	5.03	m	9.59
100 pair + earth	5.79	7.26	0.15	5.40	m	12.66
Low speed data; unshielded twisted pair; solid copper conductors; LSOH sheath; nominal impedance 100 Ohm; Category 3 to ISO IS 1801/EIA/TIA 568B and EN50173/ 50174 standards to current revisions						
Installed in riser						
25 pair 24 AWG	1.75	2.20	0.03	1.08	m	3.28
50 pair 24 AWG	2.95	3.70	0.06	2.16	m	5.86
100 pair 24 AWG	6.58	8.25	0.10	3.60	m	11.85
Installed below floor						
25 pair 24 AWG	1.75	2.20	0.02	0.72	m	2.92
50 pair 24 AWG	2.95	3.70	0.05	1.79	m	5.49
100 pair 24 AWG	6.58	8.25	0.08	2.87	m	11.12

COMMUNICATIONS/SECURITY/CONTROL

Item	Net Price £	Material £	Labour hours	Labour £	Unit	Total rate £
ACCESSORIES						
Telephone outlet: moulded plastic plate with box; fitted and connected; flush or surface mounted						
Single master outlet	7.48	9.39	0.35	12.58	nr	**21.97**
Single secondary outlet	5.52	6.92	0.35	12.58	nr	**19.50**
Telephone outlet: bronze or satin chromeplate; with box; fitted and connected; flush or surface mounted						
Single master outlet	12.16	15.25	0.35	12.58	nr	**27.83**
Single secondary outlet	13.29	16.67	0.35	12.58	nr	**29.25**
Frames and box connections						
Provision and Installation of a Dual Vertical Krone 108 A Voice Distribution Frame that can accommodate a total of 138 × Krone 237 A Strips	252.83	317.15	1.50	53.93	nr	**371.08**
Label Frame (Traffolyte style)	1.60	2.00	0.27	9.71	nr	**11.71**
Provision and Installation of a Box Connection 301 A Voice Termination Unit that can accommodate a total of 10 × Krone 237 A Strips	21.92	27.50	0.25	8.99	nr	**36.49**
Label Frame (Traffolyte style)	0.53	0.66	0.08	2.99	nr	**3.65**
Provision and Installation of a Box Connection 201 Voice Termination Unit that can accommodate 20 pairs	11.67	14.64	0.17	5.97	nr	**20.61**
Label Frame (Traffolyte style)	0.53	0.66	0.08	2.99	nr	**3.65**
Terminate, Test and Label Voice Multi-core System						
Patch panels						
Voice; 19" wide fully loaded, finished in black including termination and forming of cables (assuming 2 pairs per port)						
25 port – RJ45 UTP – Krone	54.86	68.81	2.60	93.48	nr	**162.29**
50 port – RJ45 UTP – Krone	74.78	93.80	4.65	167.17	nr	**260.97**
900 pair fully loaded Systimax style of frame including forming and termination of 9 × 100 pair cables	617.92	775.12	25.00	898.78	nr	**1673.90**
Installation and termination of Krone Strip (10 pair block – 237 A) including designation label strip	3.69	4.63	0.50	17.98	nr	**22.61**
Patch panel and Outlet labelling per port (Traffolyte style)	0.19	0.24	0.02	0.72	nr	**0.96**
Provision and installation of a voice jumper, for cross termination on Krone Termination Strips	0.06	0.08	0.06	2.16	nr	**2.24**
CW1308 / Cat 3 cable circuit test per pair	–	–	0.04	1.44	nr	**1.44**

COMMUNICATIONS/SECURITY/CONTROL

Item	Net Price £	Material £	Labour hours	Labour £	Unit	Total rate £
RADIO AND TELEVISION						
RADIO						
CABLES						
Radio frequency cable; BS 2316 ; PVC sheathed; laid loose						
7/0.41 mm tinned copper inner conductor; solid polyethylene dielectric insulation; bare copper wire braid; PVC sheath; 75 ohm impedance						
Cable	1.09	1.37	0.05	1.79	m	**3.16**
Twin 1/0.58 mm copper covered steel solid core wire conductor; solid polyethylene dielectric insulation; barecopper wire braid; PVC sheath; 75 ohm impedance						
Cable	0.25	0.31	0.05	1.79	m	**2.10**
TELEVISION						
CABLES						
Television aerial cable; coaxial; PVC sheathed; fixed to backgrounds						
General purpose TV aerial downlead; copper stranded inner conductor; cellular polythene insulation; copper braid outer conductor; 75 ohm impedance						
7/0.25 mm	0.24	0.30	0.06	2.16	m	**2.46**
Low loss TV aerial downlead; solid copper inner conductor; cellular polythene insulation; copper braid outer; conductor; 75 ohm impedance						
1/1.12 mm	0.25	0.31	0.06	2.16	m	**2.47**
Low loss air spaced; solid copper inner conductor; air spaced polythene insulation; copper braid outer conductor; 75 ohm impedance						
1/1.00 mm	0.33	0.41	0.06	2.16	m	**2.57**
Satelite aerial downlead; solid copper inner conductor; air spaced polythene insulation; copper tape and braid outer conductor; 75 ohm impedance						
1/1.00 mm	0.29	0.36	0.06	2.16	m	**2.52**
Satelite TV coaxial; solid copper inner conductor; semi-airspaced polyethylene dielectric insulation; plain annealed copper foil and copper braid screen in outer conductor; PVC sheath; 75 ohm impedance						
1/1.25 mm	0.48	0.60	0.08	2.87	m	**3.47**

COMMUNICATIONS/SECURITY/CONTROL

Item	Net Price £	Material £	Labour hours	Labour £	Unit	Total rate £
Satelite TV coaxial; solid copper inner conductor; air spaced polyethylene dielectric insulation; plain annealed copper foil and copper braid screen in outer conductor; PVC sheath; 75 ohm impedance						
1/1.67 mm	0.78	0.97	0.09	3.24	m	**4.21**
Video cable; PVC flame retardant sheath; laid loose						
7/0.1 mm silver coated copper covered annealed steel wire conductor; polyethylene dielectric insulation with tin coated copper wire braid; 75 ohm impedance						
Cable	0.19	0.24	0.05	1.79	m	**2.03**
ACCESSORIES						
TV co-axial socket outlet: moulded plastic box; flush or surface mounted						
One way Direct Connection	8.36	10.48	0.35	12.58	nr	**23.06**
Two way Direct Connection	11.66	14.63	0.35	12.58	nr	**27.21**
One way Isolated UHF/VHF	14.72	18.47	0.35	12.58	nr	**31.05**
Two way Isolated UHF/VHF	20.05	25.16	0.35	12.58	nr	**37.74**

COMMUNICATIONS/SECURITY/CONTROL

Item	Net Price £	Material £	Labour hours	Labour £	Unit	Total rate £
CLOCKS						
Master clock						
Master clock module 230 V with programming circuits, NTP client and server for synchronization over Ethernet network, sounders Harmonys or Melodys, control of DHF and NTP clock systems						
Rack mounted	1045.00	1310.85	3.00	107.86	nr	**1418.71**
Wall mounted	940.00	1179.14	5.00	179.76	nr	**1358.90**
Master clock module 230 V with programming circuits for control of wired and DHF clock systems, relays or sounders, NTP client and server for synchronization over Ethernet network						
Rack mounted	1411.00	1769.96	3.00	107.86	nr	**1877.82**
Wall mounted	1146.00	1437.54	5.00	179.76	nr	**1617.30**
Clocks and bells, master clock module 230 V with 3 programming circuits for control of wired and DHF clock systems, relays and sounders						
Rack mounted	911.00	1142.76	3.00	107.86	nr	**1250.62**
Wall mounted	720.00	903.17	5.00	179.76	nr	**1082.93**
Clocks only, master clock module 230 V for control of wired and DHF clock systems						
Rack mounted	689.00	864.28	3.00	107.86	nr	**972.14**
Wall mounted	535.00	671.10	5.00	179.76	nr	**850.86**
Extra over for						
GPS antenna with 20 m cable	210.00	263.42	2.50	89.88	nr	**353.30**
DHF transmitter	309.00	387.61	1.50	53.93	nr	**441.54**
DHF repeater, signal booster up to 200 m	354.00	444.06	3.00	107.86	nr	**551.92**
UPS	915.00	1147.78	5.00	179.76	nr	**1327.54**
Rack mounted external AC/DC power supply	526.00	659.81	3.00	107.86	nr	**767.67**
Indoor clock, 30 cm dia.						
Stand-alone						
HMS display, quartz movement, battery operated	50.00	62.72	0.80	28.76	nr	**91.48**
DHF wireless						
Hour and minute display, battery operated	114.00	143.00	0.80	28.76	nr	**171.76**
Hour, minute and second display, battery operated	125.00	156.80	0.80	28.76	nr	**185.56**
HM display, TBT	118.00	148.02	0.80	28.76	nr	**176.78**
HMS display, TBT	127.00	159.31	0.80	28.76	nr	**188.07**
Slave wired						
Hour and minute display, 24V	68.00	85.30	1.00	35.95	nr	**121.25**
Hour, minute and second display, 24V	82.00	102.86	1.00	35.95	nr	**138.81**
Hour, minute display, TBT, AFNOR	113.00	141.75	1.00	35.95	nr	**177.70**

COMMUNICATIONS/SECURITY/CONTROL

Item	Net Price £	Material £	Labour hours	Labour £	Unit	Total rate £
Hour, minute and second display, TBT AFNOR	125.00	156.80	1.00	35.95	nr	**192.75**
Hour, minute display, POE, NTP	158.00	198.20	1.00	35.95	nr	**234.15**
Hour, minute and second display, POE, NTP	168.00	210.74	1.00	35.95	nr	**246.69**
Indoor clock, 40 cm dia.						
Stand-alone						
HMS display, quartz movement, battery operated	95.00	119.17	0.80	28.76	nr	**147.93**
DHF wireless						
Hour and minute display, battery operated	167.00	209.48	0.80	28.76	nr	**238.24**
Hour, minute and second display, battery operated	172.00	215.76	0.80	28.76	nr	**244.52**
HM display, TBT	167.00	209.48	0.80	28.76	nr	**238.24**
HMS display, TBT	179.00	224.54	0.80	28.76	nr	**253.30**
Slave wired						
Hour and minute display, 24V	126.00	158.05	1.00	35.95	nr	**194.00**
Hour, minute and second display, 24V	116.00	145.51	1.00	35.95	nr	**181.46**
Hour, minute display, TBT, AFNOR	177.00	222.03	1.00	35.95	nr	**257.98**
Hour, minute and second display, TBT AFNOR	183.00	229.56	1.00	35.95	nr	**265.51**
Hour, minute display, POE, NTP	209.00	262.17	1.00	35.95	nr	**298.12**
Hour, minute and second display, POE, NTP	215.00	269.70	1.00	35.95	nr	**305.65**
Digital clock, 7 cm digit height, 230 V						
Independent quartz movement						
Red display	210.00	263.42	0.80	28.76	nr	**292.18**
Green display	210.00	263.42	0.80	28.76	nr	**292.18**
Yellow display	210.00	263.42	0.80	28.76	nr	**292.18**
Blue display	285.00	357.50	0.80	28.76	nr	**386.26**
White display	285.00	357.50	0.80	28.76	nr	**386.26**
AFNOR, 230 V						
Red display	215.00	269.70	0.80	28.76	nr	**298.46**
Green display	215.00	269.70	0.80	28.76	nr	**298.46**
Yellow display	215.00	269.70	0.80	28.76	nr	**298.46**
Blue display	290.00	363.78	0.80	28.76	nr	**392.54**
White display	290.00	363.78	0.80	28.76	nr	**392.54**
DHF wireless, 230 V						
Red display	230.00	288.51	0.80	28.76	nr	**317.27**
Green display	230.00	288.51	0.80	28.76	nr	**317.27**
Yellow display	230.00	288.51	0.80	28.76	nr	**317.27**
Blue display	305.00	382.59	0.80	28.76	nr	**411.35**
White display	305.00	382.59	0.80	28.76	nr	**411.35**
NTP POE						
Red display	250.00	313.60	0.80	28.76	nr	**342.36**
Green display	250.00	313.60	0.80	28.76	nr	**342.36**
Yellow display	250.00	313.60	0.80	28.76	nr	**342.36**
Blue display	325.00	407.68	0.80	28.76	nr	**436.44**
White display	325.00	407.68	0.80	28.76	nr	**436.44**

COMMUNICATIONS/SECURITY/CONTROL

Item	Net Price £	Material £	Labour hours	Labour £	Unit	Total rate £
CLOCKS – cont						
Digital clock, 5 cm digit height, 230 V						
Independent quartz movement						
Red display	175.00	219.52	0.80	28.76	nr	**248.28**
Green display	175.00	219.52	0.80	28.76	nr	**248.28**
Yellow display	175.00	219.52	0.80	28.76	nr	**248.28**
Blue display	230.00	288.51	0.80	28.76	nr	**317.27**
White display	230.00	288.51	0.80	28.76	nr	**317.27**
AFNOR, 230 V						
Red display	175.00	219.52	0.80	28.76	nr	**248.28**
Green display	175.00	219.52	0.80	28.76	nr	**248.28**
Yellow display	175.00	219.52	0.80	28.76	nr	**248.28**
Blue display	230.00	288.51	0.80	28.76	nr	**317.27**
White display	230.00	288.51	0.80	28.76	nr	**317.27**
DHF wireless, 230 V						
Red display	205.00	257.15	0.80	28.76	nr	**285.91**
Green display	205.00	257.15	0.80	28.76	nr	**285.91**
Yellow display	205.00	257.15	0.80	28.76	nr	**285.91**
Blue display	260.00	326.14	0.80	28.76	nr	**354.90**
White display	260.00	326.14	0.80	28.76	nr	**354.90**
NTP POE						
Red display	220.00	275.97	0.80	28.76	nr	**304.73**
Green display	220.00	275.97	0.80	28.76	nr	**304.73**
Yellow display	220.00	275.97	0.80	28.76	nr	**304.73**
Blue display	275.00	344.96	0.80	28.76	nr	**373.72**
White display	275.00	344.96	0.80	28.76	nr	**373.72**
Digital clock, 10 cm digit height, 230 V						
Independent quartz movement						
Red display	375.00	470.40	0.80	28.76	nr	**499.16**
Green display	375.00	470.40	0.80	28.76	nr	**499.16**
Yellow display	375.00	470.40	0.80	28.76	nr	**499.16**
Blue display	465.00	583.30	0.80	28.76	nr	**612.06**
White display	465.00	583.30	0.80	28.76	nr	**612.06**
AFNOR, 230 V						
Red display	375.00	470.40	0.80	28.76	nr	**499.16**
Green display	375.00	470.40	0.80	28.76	nr	**499.16**
Yellow display	375.00	470.40	0.80	28.76	nr	**499.16**
Blue display	465.00	583.30	0.80	28.76	nr	**612.06**
White display	465.00	583.30	0.80	28.76	nr	**612.06**
DHF wireless, 230 V						
Red display	390.00	489.22	0.80	28.76	nr	**517.98**
Green display	390.00	489.22	0.80	28.76	nr	**517.98**
Yellow display	390.00	489.22	0.80	28.76	nr	**517.98**
Blue display	480.00	602.11	0.80	28.76	nr	**630.87**
White display	480.00	602.11	0.80	28.76	nr	**630.87**
NTP POE						
Red display	405.00	508.03	0.80	28.76	nr	**536.79**
Green display	405.00	508.03	0.80	28.76	nr	**536.79**
Yellow display	405.00	508.03	0.80	28.76	nr	**536.79**
Blue display	495.00	620.93	0.80	28.76	nr	**649.69**
White display	495.00	620.93	0.80	28.76	nr	**649.69**

COMMUNICATIONS/SECURITY/CONTROL

Item	Net Price £	Material £	Labour hours	Labour £	Unit	Total rate £
DATA TRANSMISSION						
Cabinets						
Floor standing; suitable for 19" patch panels with glass lockable doors, metal rear doors, side panels, vertical cable management, 2 × 4 way PDUs, 4 way fan, earth bonding kit; installed on raised floor						
600 mm wide × 800 mm deep – 18U	749.88	940.65	3.00	107.86	nr	**1048.51**
600 mm wide × 800 mm deep – 24U	781.12	979.83	3.00	107.86	nr	**1087.69**
600 mm wide × 800 mm deep – 33U	838.92	1052.34	4.00	143.81	nr	**1196.15**
600 mm wide × 800 mm deep – 42U	899.86	1128.78	4.00	143.81	nr	**1272.59**
600 mm wide × 800 mm deep – 47U	949.93	1191.59	4.00	143.81	nr	**1335.40**
800 mm wide × 800 mm deep – 42U	1018.61	1277.74	4.00	143.81	nr	**1421.55**
800 mm wide × 800 mm deep – 47U	1062.38	1332.65	4.00	143.81	nr	**1476.46**
Label cabinet	2.40	3.01	0.25	8.99	nr	**12.00**
Wall mounted; suitable for 19" patch panels with glass lockable doors, side panels, vertical cable management, 2 × 4 way PDUs, 4 way fan, earth bonding kit; fixed to wall						
19 mm wide × 500 mm deep – 9U	500.15	627.39	3.00	107.86	nr	**735.25**
19 mm wide × 500 mm deep – 12U	512.45	642.81	3.00	107.86	nr	**750.67**
19 mm wide × 500 mm deep – 15U	531.05	666.15	3.00	107.86	nr	**774.01**
19 mm wide × 500 mm deep – 18U	624.90	783.88	3.00	107.86	nr	**891.74**
19 mm wide × 500 mm deep – 21U	650.08	815.46	3.00	107.86	nr	**923.32**
Label cabinet	2.40	3.01	0.25	8.99	nr	**12.00**
Frames						
Floor standing; suitable for 19" patch panels with supports, vertical cable management, earth bonding kit; installed on raised floor						
19 mm wide × 500 mm deep – 25U	562.53	705.63	2.50	89.88	nr	**795.51**
19 mm wide × 500 mm deep – 39U	687.27	862.11	2.50	89.88	nr	**951.99**
19 mm wide × 500 mm deep – 42U	749.93	940.71	2.50	89.88	nr	**1030.59**
19 mm wide × 500 mm deep – 47U	812.60	1019.32	2.50	89.88	nr	**1109.20**
Label frame	2.40	3.01	0.25	8.99	nr	**12.00**
Copper data cabling						
Unshielded twisted pair; solid copper conductors; PVC insulation; nominal impedance 100 Ohm; Cat 5e to ISO 11801, EIA/TIA 568B and EN 50173/50174 standards to the current revisions						
4 pair 24 AWG; nominal outside dia. 5.6 mm; installed above ceiling	0.21	0.27	0.02	0.72	m	**0.99**
4 pair 24 AWG; nominal outside dia. 5.6 mm; installed in riser	0.21	0.27	0.02	0.72	m	**0.99**
4 pair 24 AWG; nominal outside dia. 5.6 mm; installed below floor	0.21	0.27	0.01	0.36	m	**0.63**
4 pair 24 AWG; nominal outside dia. 5.6 mm; installed in trunking	0.21	0.27	0.02	0.72	m	**0.99**

COMMUNICATIONS/SECURITY/CONTROL

Item	Net Price £	Material £	Labour hours	Labour £	Unit	Total rate £
DATA TRANSMISSION – cont						
Copper data cabling – cont						
Unshielded twisted pair; solid copper conductors; LSOH sheathed; nominal impedance 100 Ohm; Cat 5e to ISO 11801, EIA/TIA 568B and EN 50173/50174 standards to the current revisions						
4 pair 24 AWG; nominal outside dia. 5.6 mm; installed above ceiling	0.24	0.30	0.02	0.72	m	**1.02**
4 pair 24 AWG; nominal outside dia. 5.6 mm; installed in riser	0.24	0.30	0.02	0.72	m	**1.02**
4 pair 24 AWG; nominal outside dia. 5.6 mm; installed below floor	0.24	0.30	0.01	0.36	m	**0.66**
4 pair 24 AWG; nominal outside dia. 5.6 mm; installed in trunking	0.24	0.30	0.02	0.72	m	**1.02**
Unshielded twisted pair; solid copper conductors; PVC insulation; nominal impedance 100 Ohm; Cat 6 to ISO 11801, EIA/TIA 568B and EN 50173/50174 standards to the current revisions						
4 pair 24 AWG; nominal outside dia. 5.6 mm; installed above ceiling	0.33	0.41	0.02	0.54	m	**0.95**
4 pair 24 AWG; nominal outside dia. 5.6 mm; installed in riser	0.33	0.41	0.02	0.54	m	**0.95**
4 pair 24 AWG; nominal outside dia. 5.6 mm; installed below floor	0.33	0.41	0.01	0.29	m	**0.70**
4 pair 24 AWG; nominal outside dia. 5.6 mm; installed in trunking	0.33	0.41	0.02	0.72	m	**1.13**
Unshielded twisted pair; solid copper conductors; LSOH sheathed; nominal impedance 100 Ohm; Cat 6 to ISO 11801, EIA/TIA 568B and EN 50173/50174 standards to the current revisions						
4 pair 24 AWG; nominal outside dia. 5.6 mm; installed above ceiling	0.34	0.43	0.02	0.80	m	**1.23**
4 pair 24 AWG; nominal outside dia. 5.6 mm; installed in riser	0.34	0.43	0.02	0.80	m	**1.23**
4 pair 24 AWG; nominal outside dia. 5.6 mm; installed below floor	0.34	0.43	0.01	0.40	m	**0.83**
4 pair 24 AWG; nominal outside dia. 5.6 mm; installed in trunking	0.34	0.43	0.02	0.80	m	**1.23**
Patch panels						
Category 5e; 19" wide fully loaded, finished in black including termination and forming of cables						
24 port – RJ45 UTP – Krone/110	55.83	70.03	4.75	170.77	nr	**240.80**
48 port – RJ45 UTP – Krone/110	104.78	131.43	9.35	336.15	nr	**467.58**
Patch panel labelling per port	0.25	0.31	0.02	0.72	nr	**1.03**

COMMUNICATIONS/SECURITY/CONTROL

Item	Net Price £	Material £	Labour hours	Labour £	Unit	Total rate £
Category 6; 19" wide fully loaded, finished in black including termination and forming of cables						
24 port – RJ45 UTP – Krone/110	106.69	133.83	5.00	179.76	nr	**313.59**
48 port – RJ45 UTP – Krone/110	194.60	244.10	9.80	352.32	nr	**596.42**
Patch panel labelling per port	0.25	0.31	0.02	0.72	nr	**1.03**
Work station						
Category 5e RJ45 data outlet plate and multiway outlet boxes for wall, ceiling and below floor installations including label to ISO 11801 standards						
Wall mounted; fully loaded						
One gang LSOH PVC plate	5.11	6.41	0.15	5.40	nr	**11.81**
Two gang LSOH PVC plate	7.32	9.18	0.20	7.19	nr	**16.37**
Four gang LSOH PVC plate	12.53	15.71	0.40	14.38	nr	**30.09**
One gang satin brass plate	16.59	20.81	0.25	8.99	nr	**29.80**
Two gang satin brass plate	18.71	23.48	0.33	11.87	nr	**35.35**
Ceiling mounted; fully loaded						
One gang metal clad plate	8.81	11.05	0.33	11.87	nr	**22.92**
Two gang metal clad plate	11.96	15.01	0.45	16.18	nr	**31.19**
Below floor; fully loaded						
Four way outlet box, 5 m length 20 mm flexible conduit with glands and starin relief bracket	29.78	37.35	0.80	28.76	nr	**66.11**
Six way outlet box, 5 m length 25 mm flexible conduit with glands and strain relief bracket	38.50	48.29	1.20	43.14	nr	**91.43**
Eight way outlet box, 5 m length 25 mm flexible conduit with glands and strain relief bracket	48.47	60.80	1.40	50.32	nr	**111.12**
Installation of outlet boxes to desks	–	–	0.40	14.38	nr	**14.38**
Category 6 RJ45 data outlet plate and multiway outlet boxes for wall, ceiling and below floor installations including label to ISO 11801 standards						
Wall mounted; fully loaded						
One gang LSOH PVC plate	6.52	8.18	0.17	5.94	nr	**14.12**
Two gang LSOH PVC plate	9.58	12.02	0.22	7.92	nr	**19.94**
Four gang LSOH PVC plate	17.29	21.68	0.44	15.81	nr	**37.49**
One gang satin brass plate	15.94	19.99	0.28	9.89	nr	**29.88**
Two gang satin brass plate	19.95	25.02	0.36	13.05	nr	**38.07**
Ceiling mounted; fully loaded						
One gang metal clad plate	10.57	13.26	0.36	13.05	nr	**26.31**
Two gang metal clad plate	14.92	18.72	0.50	17.80	nr	**36.52**

COMMUNICATIONS/SECURITY/CONTROL

Item	Net Price £	Material £	Labour hours	Labour £	Unit	Total rate £
DATA TRANSMISSION – cont						
Work station – cont						
Below floor; fully loaded						
Four way outlet box, 5 m length 20 mm flexible conduit with glands and starin relief bracket	36.50	45.79	0.88	31.64	nr	**77.43**
Six way outlet box, 5 m length 25 mm flexible conduit with glands and strain relief bracket	48.55	60.91	1.32	47.45	nr	**108.36**
Eight way outlet box, 5 m length 25 mm flexible conduit with glands and strain relief bracket	61.83	77.56	1.54	55.37	nr	**132.93**
Installation of outlet boxes to desks	–	–	0.40	14.38	nr	**14.38**
Category 5e cable test	–	–	0.10	3.60	nr	**3.60**
Category 6 cable test	–	–	0.11	3.95	nr	**3.95**
Copper patch leads						
Category 5e; straight through booted RJ45 UTP – RJ45 UTP						
Patch lead 1 m length	1.83	2.30	0.09	3.24	nr	**5.54**
Patch lead 3 m length	2.17	2.72	0.09	3.24	nr	**5.96**
Patch lead 5 m length	3.09	3.88	0.10	3.60	nr	**7.48**
Patch lead 7 m length	4.53	5.68	0.10	3.60	nr	**9.28**
Category 6; straight through booted RJ45 UTP – RJ45 UTP						
Patch lead 1 m length	4.86	6.09	0.09	3.24	nr	**9.33**
Patch lead 3 m length	7.33	9.20	0.09	3.24	nr	**12.44**
Patch lead 5 m length	8.59	10.77	0.10	3.60	nr	**14.37**
Patch lead 7 m length	10.88	13.65	0.10	3.60	nr	**17.25**
Note 1: With an intelligent Patching System, add a 25% uplift to the Cat5e and Cat 6 System price. This would be dependent on the System and IMS solution chosen, i.e. iPatch, RIT or iTRACs.						
Note 2: For a Cat 6 augmented solution (10Gbps capable), add 25% to a Cat 6 System price.						
Fibre infrastructure						
Fibre-optic cable, tight buffered, internal/ external application, single mode, LSOH sheathed						
4 core fibre-optic cable	1.45	1.81	0.10	3.60	m	**5.41**
8 core fibre-optic cable	2.42	3.04	0.10	3.60	m	**6.64**
12 core fibre-optic cable	3.00	3.76	0.10	3.60	m	**7.36**
16 core fibre-optic cable	3.64	4.57	0.10	3.60	m	**8.17**
24 core fibre-optic cable	4.40	5.52	0.10	3.60	m	**9.12**

COMMUNICATIONS/SECURITY/CONTROL

Item	Net Price £	Material £	Labour hours	Labour £	Unit	Total rate £
Fibre-optic cable OM1 and OM2, tight buffered, internal/external application, 62.5/125 multimode fibre, LSOH sheathed						
4 core fibre-optic cable	1.55	1.95	0.10	3.60	m	**5.55**
8 core fibre-optic cable	2.57	3.23	0.10	3.60	m	**6.83**
12 core fibre-optic cable	3.28	4.11	0.10	3.60	m	**7.71**
16 core fibre-optic cable	4.07	5.11	0.10	3.60	m	**8.71**
24 core fibre-optic cable	4.71	5.91	0.10	3.60	m	**9.51**
Fibre-optic cable OM3, tight buffered, internal only application, 50/125 multimode fibre, LSOH sheathed						
4 core fibre-optic cable	1.92	2.41	0.10	3.60	nr	**6.01**
8 core fibre-optic cable	3.00	3.76	0.10	3.60	nr	**7.36**
12 core fibre-optic cable	4.35	5.45	0.10	3.60	nr	**9.05**
24 core fibre-optic cable	8.31	10.43	0.10	3.60	nr	**14.03**
Fibre-optic single and multimode connectors and couplers including termination						
ST singlemode booted connector	7.32	9.18	0.25	8.99	nr	**18.17**
ST multimode booted connector	3.54	4.44	0.25	8.99	nr	**13.43**
SC simplex singlemode booted connector	10.37	13.00	0.25	8.99	nr	**21.99**
SC simplex multimode booted connector	3.78	4.74	0.25	8.99	nr	**13.73**
SC duplex multimode booted connector	7.54	9.45	0.25	8.99	nr	**18.44**
ST – SC duplex adaptor	18.33	22.99	0.01	0.36	nr	**23.35**
ST inline bulkhead coupler	3.93	4.93	0.01	0.36	nr	**5.29**
SC duplex coupler	7.34	9.21	0.01	0.36	nr	**9.57**
MTRJ small form factor duplex connector	7.31	9.17	0.25	8.99	nr	**18.16**
LC simplex multimode booted connector	–	–	0.25	8.99	nr	**8.99**
Singlemode core test per core	–	–	0.20	7.19	nr	**7.19**
Multimode core test per core	–	–	0.20	7.19	nr	**7.19**
Fibre; 19" wide fully loaded, labled, alluminium alloy c/w couplers, fibre management and glands (excludes termination of fibre cores)						
8 way ST; fixed drawer	78.72	98.75	0.50	17.98	nr	**116.73**
16 way ST; fixed drawer	118.06	148.10	0.50	17.98	nr	**166.08**
24 way ST; fixed drawer	164.03	205.76	0.50	17.98	nr	**223.74**
8 way ST; sliding drawer	100.15	125.63	0.50	17.98	nr	**143.61**
16 way ST; sliding drawer	137.62	172.63	0.50	17.98	nr	**190.61**
24 way ST; sliding drawer	181.11	227.18	0.50	17.98	nr	**245.16**
8 way (4 duplex) SC; fixed drawer	93.85	117.72	0.50	17.98	nr	**135.70**
16 way (8 duplex) SC; fixed drawer	137.34	172.28	0.50	17.98	nr	**190.26**
24 way (12 duplex) SC; fixed drawer	168.82	211.77	0.50	17.98	nr	**229.75**
8 way (4 duplex) SC; sliding drawer	118.74	148.95	0.50	17.98	nr	**166.93**
16 way (8 duplex) SC; sliding drawer	162.52	203.86	0.50	17.98	nr	**221.84**
24 way (12 duplex) SC; sliding drawer	193.70	242.97	0.50	17.98	nr	**260.95**
8 way (4 duplex) MTRJ; fixed drawer	106.44	133.52	0.50	17.98	nr	**151.50**
16 way (8 duplex) MTRJ; fixed drawer	162.52	203.86	0.50	17.98	nr	**221.84**
24 way (12 duplex) MTRJ; fixed drawer	193.70	242.97	0.50	17.98	nr	**260.95**
8 way (4 duplex) FC/PC; fixed drawer	111.59	139.98	0.50	17.98	nr	**157.96**
16 way (8 duplex) FC/PC; fixed drawer	170.53	213.91	0.50	17.98	nr	**231.89**
24 way (12 duplex) FC/PC; fixed drawer	203.44	255.19	0.50	17.98	nr	**273.17**
Patch panel label per way	0.25	0.31	0.02	0.72	nr	**1.03**

COMMUNICATIONS/SECURITY/CONTROL

Item	Net Price £	Material £	Labour hours	Labour £	Unit	Total rate £
DATA TRANSMISSION – cont						
Fibre patch leads						
Single mode						
Duplex OS1 LC – LC						
Fibre patch lead 1 m length	24.14	30.28	0.08	2.87	nr	**33.15**
Fibre patch lead 3 m length	22.59	28.34	0.08	2.87	nr	**31.21**
Fibre patch lead 5 m length	24.02	30.13	0.10	3.60	nr	**33.73**
Multimode						
Duplex 50/125 OM3 MTRJ – MTRJ						
Fibre patch lead 1 m length	21.19	26.58	0.08	2.87	nr	**29.45**
Fibre patch lead 3 m length	23.03	28.88	0.08	2.87	nr	**31.75**
Fibre patch lead 5 m length	24.19	30.34	0.10	3.60	nr	**33.94**
Duplex 50/125 OM3 ST – ST						
Fibre patch lead 1 m length	26.42	33.14	0.08	2.87	nr	**36.01**
Fibre patch lead 3 m length	28.74	36.05	0.08	2.87	nr	**38.92**
Fibre patch lead 5 m length	30.18	37.86	0.10	3.60	nr	**41.46**
Duplex 50/125 OM3 SC – SC						
Fibre patch lead 1 m length	23.71	29.75	0.08	2.87	nr	**32.62**
Firbe patch lead 3 m length	25.80	32.37	0.08	2.87	nr	**35.24**
Fibre patch lead 5 m length	27.09	33.98	0.10	3.60	nr	**37.58**
Duplex 50/125 OM3 LC – LC						
Fibre patch lead 1 m length	25.40	31.86	0.08	2.87	nr	**34.73**
Fibre patch lead 3 m length	27.64	34.68	0.08	2.87	nr	**37.55**
Fibre patch lead 5 m length	29.03	36.41	0.10	3.60	nr	**40.01**

COMMUNICATIONS/SECURITY/CONTROL

Item	Net Price £	Material £	Labour hours	Labour £	Unit	Total rate £
ACCESS CONTROL EQUIPMENT						
Equipment to control the movement of personnel into defined spaces; includes fixing to backgrounds, termination of power and data cables; excludes cable containment and cable installation						
Access control						
Proximity reader	112.20	140.74	1.50	53.93	nr	**194.67**
Exit button	20.40	25.59	1.50	53.93	nr	**79.52**
Exit PIR	134.64	168.90	2.00	71.90	nr	**240.80**
Emergency break glass double pole	30.60	38.38	1.00	35.95	nr	**74.33**
Alarm contact flush	5.10	6.40	1.00	35.95	nr	**42.35**
Alarm contact surface	5.10	6.40	1.00	35.95	nr	**42.35**
Reader controller 16 door	4375.80	5489.01	4.00	143.81	nr	**5632.82**
Reader controller 8 door	2903.94	3642.70	4.00	143.81	nr	**3786.51**
Reader controller 2 door	1016.94	1275.65	4.00	143.81	nr	**1419.46**
Reader interface	412.44	517.36	1.50	53.93	nr	**571.29**
Lock power supply 12 volt 3 amp	245.82	308.36	2.00	71.90	nr	**380.26**
Lock power supply 24 volt 3 amp	429.42	538.66	2.00	71.90	nr	**610.56**
Rechargeable battery 12 volt 7ah	12.24	15.36	0.25	8.99	nr	**24.35**
Lock equipment						
Single slimline magnetic lock, monitored	106.08	133.07	2.00	71.90	nr	**204.97**
Single slimline magnetic lock, unmonitored	91.80	115.16	1.75	62.92	nr	**178.08**
Double slimline magnetic lock, monitored	210.12	263.57	4.00	143.81	nr	**407.38**
Double slimline magnetic lock, unmonitored	180.54	226.46	4.50	161.78	nr	**388.24**
Standard single magnetic lock, monitored	134.64	168.90	1.25	44.95	nr	**213.85**
Standard single magnetic lock, unmonitored	108.12	135.62	1.00	35.95	nr	**171.57**
Standard single magnetic lock, double monitored	180.54	226.46	1.50	53.93	nr	**280.39**
Standard double magnetic lock, monitored	269.28	337.78	1.50	53.93	nr	**391.71**
Standard double magnetic lock, unmonitored	211.14	264.86	1.25	44.95	nr	**309.81**
12 V electric release fail safe, monitored	140.76	176.57	1.25	44.95	nr	**221.52**
12 V electric release fail secure, monitored	140.76	176.57	1.00	35.95	nr	**212.52**
Solenoid bolt	266.22	333.95	1.50	53.93	nr	**387.88**
Electric mortice lock	99.96	125.40	1.00	35.95	nr	**161.35**

COMMUNICATIONS/SECURITY/CONTROL

Item	Net Price £	Material £	Labour hours	Labour £	Unit	Total rate £
SECURITY DETECTION AND ALARM						
Detection and alarm systems for the protection of property and persons; includes fixing of equipment to backgrounds and termination of power and data cabling; excludes cable containment and cable installation						
Detection, alarm equipment						
Alarm contact flush	5.60	7.02	1.00	35.95	nr	42.97
Alarm contact surface	5.60	7.02	1.00	35.95	nr	42.97
Roller shutter contact	17.92	22.48	1.00	35.95	nr	58.43
Personal attack button	19.04	23.88	1.00	35.95	nr	59.83
Acoustic break glass detectors	25.76	32.31	2.00	71.90	nr	104.21
Vibration detectors	33.60	42.15	2.00	71.90	nr	114.05
12 m PIR detector	39.20	49.17	1.50	53.93	nr	103.10
15 m dual detector	64.96	81.49	1.50	53.93	nr	135.42
8 zone alarm panel	344.96	432.72	3.00	107.86	nr	540.58
8–24 zone end station	338.98	425.22	4.00	143.81	nr	569.03
Remote keypad	114.24	143.30	2.00	71.90	nr	215.20
8 zone expansion	87.36	109.58	1.50	53.93	nr	163.51
Final exit set button	9.31	11.68	1.00	35.95	nr	47.63
Self-contained external sounder	84.00	105.37	2.00	71.90	nr	177.27
Internal loudspeaker	16.80	21.08	2.00	71.90	nr	92.98
Rechargeable battery 12 volt 7 AHr (ampere hours)	13.44	16.86	0.25	8.99	nr	25.85
Surveillance equipment						
Vandal-resistant camera, colour	180.00	225.79	3.00	107.86	nr	333.65
External camera, colour	360.00	451.58	5.00	179.76	nr	631.34
Auto dome external, colour	900.00	1128.96	5.00	179.76	nr	1308.72
Auto dome external, colour/monochrome	900.00	1128.96	5.00	179.76	nr	1308.72
Auto dome internal, colour	360.00	451.58	4.00	143.81	nr	595.39
Auto dome internal colour/monochrome	360.00	451.58	4.00	143.81	nr	595.39
Mini internal domes	597.80	749.88	3.00	107.86	nr	857.74
Camera switcher, 32 inputs	1297.36	1627.40	4.00	143.81	nr	1771.21
Full function keyboard	244.00	306.07	1.00	35.95	nr	342.02
16 CH multiplexors, duplex	719.02	901.94	2.00	71.90	nr	973.84
16 way DVR, 250 GB harddrive	900.00	1128.96	3.00	107.86	nr	1236.82
Additional 250 GB harddrive	240.00	301.06	3.00	107.86	nr	408.92
16 way DVR, 500 GB harddrive	1200.00	2634.24	3.00	107.86	nr	2742.10
16 way DVR, 750 GB harddrive	1200.00	4515.84	1.00	35.95	nr	4551.79
10" colour monitor	156.00	195.69	2.00	71.90	nr	267.59
15" colour monitor	180.00	225.79	2.00	71.90	nr	297.69
17" colour monitor, high resolution	216.00	270.95	2.00	71.90	nr	342.85
21" colour monitor, high resolution	240.00	301.06	2.00	71.90	nr	372.96
IP CCTV						
Internal fixed dome IP camera h.264	400.00	501.76	–	–	nr	501.76
Internal fixed dome IP dome camera 1MP	400.00	501.76	–	–	nr	501.76
External fixed dome IP	450.00	564.48	–	–	nr	564.48
External PTZ IP	1600.00	2007.04	–	–	nr	2007.04
24 port PoE switch	1750.00	2195.20	–	–	nr	2195.20
16 base channel network video recorder with 4TB storage	2000.00	2508.80	–	–	nr	2508.80
IP camera channel licence	80.00	100.35	–	–	nr	100.35

COMMUNICATIONS/SECURITY/CONTROL

Item	Net Price £	Material £	Labour hours	Labour £	Unit	Total rate £
FIRE DETECTION AND ALARM						
STANDARD FIRE DETECTION						
Control panel						
Zone control panel; 16 zone, mild steel case						
Single loop	435.52	546.31	3.00	107.86	nr	**654.17**
Single loop (flush mounted)	458.89	575.64	0.25	8.99	nr	**584.63**
Loop extension card (1 loop)	227.75	285.69	3.51	126.15	nr	**411.84**
Two loop	650.55	816.05	4.00	143.81	nr	**959.86**
Two loop (flush mounted)	673.92	845.36	5.00	179.76	nr	**1025.12**
Repeater panels						
Standard network card	377.86	473.98	4.00	143.81	nr	**617.79**
FR – focus repeater	377.86	473.98	4.00	143.81	nr	**617.79**
Focus repeater with controls	416.82	522.86	0.25	8.99	nr	**531.85**
Equipment						
Manual call point units: plastic covered						
Surface mounted						
Call point	41.59	52.17	0.50	17.98	nr	**70.15**
Call point; Weatherproof	150.64	188.97	0.80	28.76	nr	**217.73**
Flush mounted						
Call point	41.59	52.17	0.56	20.15	nr	**72.32**
Call point; Weatherproof	150.64	188.97	0.86	30.93	nr	**219.90**
Detectors						
Smoke, ionization type with mounting base	16.98	21.30	0.75	26.97	nr	**48.27**
Smoke, optical type with mounting base	16.36	20.52	0.75	26.97	nr	**47.49**
Fixed temperature heat detector with mounting base (60°C)	19.49	24.45	0.75	26.97	nr	**51.42**
Rate of Rise heat detector with mounting base (90°C)	22.18	27.82	0.75	26.97	nr	**54.79**
Duct detector including optical smoke detector and base	234.53	294.19	2.00	71.90	nr	**366.09**
Remote smoke detector LED indicator with base	5.42	6.80	0.50	17.98	nr	**24.78**
Sounders						
Intelligent wall sounder beacon	86.48	108.48	0.75	26.97	nr	**135.45**
Waterproof intelligent sounder beacon	146.60	183.89	0.75	26.97	nr	**210.86**
Siren; 230V	71.84	90.12	1.25	44.95	nr	**135.07**
Magnetic Door Holder; 230 V ; surface fixed	67.75	84.99	1.50	53.97	nr	**138.96**
ADDRESSABLE FIRE DETECTION						
Control panel						
Analogue addressable panel; BS EN54 Part 2 and 4 1998; incorporating 120 addresses per loop (maximum 1–2 km length); sounders wired on loop; sealed lead acid integral battery standby providing 48 hour standby; 24 volt DC; mild steel case; surface fixed						
1 loop; 4 × 12 volt batteries	1429.98	1793.77	6.00	215.71	nr	**2009.48**

COMMUNICATIONS/SECURITY/CONTROL

Item	Net Price £	Material £	Labour hours	Labour £	Unit	Total rate £
FIRE DETECTION AND ALARM – cont						
Control panel – cont						
Extra for 1 loop panel						
Loop expander card	80.85	101.42	1.00	35.95	nr	137.37
Repeater panel	992.75	1245.31	6.00	215.71	nr	1461.02
Network nodes	1698.82	2131.00	6.00	215.71	nr	2346.71
Interface unit; for other systems						
Mains powered	508.74	638.16	1.50	53.93	nr	692.09
Loop powered	223.30	280.11	1.00	35.95	nr	316.06
Single channel I/O	116.05	145.58	1.00	35.95	nr	181.53
Zone module	144.33	181.05	1.50	53.93	nr	234.98
4 loop; 4 × 12 volt batteries; 24 hour standby; 30 minute alarm	2397.33	3007.21	8.00	287.60	nr	3294.81
Extra for 4 loop panel						
Loop card	502.71	630.60	1.00	35.95	nr	666.55
Repeater panel	1423.54	1785.68	6.00	215.71	nr	2001.39
Mimic panel	3494.96	4384.08	5.00	179.76	nr	4563.84
Network nodes	1697.50	2129.34	6.00	215.71	nr	2345.05
Interface unit; for other systems						
Mains powered	508.74	638.16	1.50	53.93	nr	692.09
Loop powered	223.30	280.11	1.00	35.95	nr	316.06
Single channel I/O	116.05	145.58	1.00	35.95	nr	181.53
Zone module	144.33	181.05	1.50	53.93	nr	234.98
Line modules	26.61	33.38	1.00	35.95	nr	69.33
8 loop; 4 × 12 volt batteries; 24 hour standby; 30 minute alarm	4887.08	6130.35	12.00	431.41	nr	6561.76
Extra for 8 loop panel						
Loop card	502.71	630.60	1.00	35.95	nr	666.55
Repeater panel	1423.54	1785.68	6.00	215.71	nr	2001.39
Mimic panel	3494.96	4384.08	5.00	179.76	nr	4563.84
Network nodes	1697.50	2129.34	6.00	215.71	nr	2345.05
Interface unit; for other systems						
Mains powered	508.74	638.16	1.50	53.93	nr	692.09
Loop powered	223.34	280.16	1.00	35.95	nr	316.11
Single channel I/O	48.71	61.11	1.00	35.95	nr	97.06
Zone module	144.33	181.05	1.50	53.93	nr	234.98
Line modules	26.61	33.38	1.00	35.95	nr	69.33
Equipment						
Manual call point						
Surface mounted						
Call point	67.03	84.08	1.00	35.95	nr	120.03
Call point; Weatherproof	245.62	308.10	1.25	44.95	nr	353.05
Flush mounted						
Call point	67.03	84.08	1.00	35.95	nr	120.03
Call point; Weatherproof	245.62	308.10	1.25	44.95	nr	353.05
Detectors						
Smoke, ionization type with mounting base	68.26	85.62	0.75	26.97	nr	112.59
Smoke, optical type with mounting base	67.57	84.76	0.75	26.97	nr	111.73
Fixed temperature heat detector with mounting base (60°C)	67.91	85.19	0.75	26.97	nr	112.16

COMMUNICATIONS/SECURITY/CONTROL

Item	Net Price £	Material £	Labour hours	Labour £	Unit	Total rate £
Rate of rise heat detector with mounting base (90°C)	67.91	85.19	0.75	26.97	nr	112.16
Duct detector including optical smoke detector and addressable base	432.80	542.91	2.00	71.90	nr	614.81
Beam smoke detector with transmitter and receiver unit	667.23	836.98	2.00	71.90	nr	908.88
Zone short circuit isolator	42.91	53.83	0.75	26.97	nr	80.80
Plant interface unit	31.62	39.66	0.50	17.98	nr	57.64
Sounders						
Xenon flasher, 24 volt, conduit box	72.07	90.41	0.50	17.98	nr	108.39
Xenon flasher, 24 volt, conduit box; weatherproof	107.28	134.57	0.50	17.98	nr	152.55
6" bell, conduit box	86.48	290.30	0.75	26.97	nr	317.27
6" bell, conduit box; weatherproof	58.47	73.35	0.75	26.97	nr	100.32
Siren; 24 V polarized	72.99	91.56	1.00	35.95	nr	127.51
Siren; 240 V	71.84	90.12	1.25	44.95	nr	135.07
Magnetic Door Holder; 240 V ; surface fixed	67.75	84.99	1.50	53.97	nr	138.96
Addressable wireless fire detection system, BS 5839 and EN54 Part 25 compliant, 868 Mhz operating frequency						
Control panel						
Wireless fire alarm panel, complete with back-up battery						
8 zone	478.59	600.34	4.00	143.81	nr	744.15
20 zone	645.54	809.76	5.00	179.76	nr	989.52
Equipment						
Smoke detector	128.00	160.56	0.50	17.98	nr	178.54
Combined smoke detector and sounder	244.86	307.15	0.50	17.98	nr	325.13
Combined smoke detector, sounder and beacon	289.38	363.00	0.50	17.98	nr	380.98
Heat detector	122.43	153.57	0.50	17.98	nr	171.55
Combined heat detector and sounder	239.29	300.16	0.50	17.98	nr	318.14
Combined heat detector, sounder and beacon	278.25	349.04	0.50	17.98	nr	367.02
Manual call point	128.00	160.56	0.50	17.98	nr	178.54
Manual call point (weatherproof)	300.51	376.96	0.75	26.97	nr	403.93
Sounder	172.51	216.40	0.50	17.98	nr	234.38
Beacon	211.47	265.27	0.50	17.98	nr	283.25
Combined sounder and beacon	217.03	272.24	0.50	17.98	nr	290.22
Remote silence button	122.43	153.57	0.50	17.98	nr	171.55
Output unit	228.16	286.20	0.50	17.98	nr	304.18
Signal booster panel	422.94	530.53	0.50	17.98	nr	548.51
Extra over for additional wired antenna	222.60	279.23	1.00	35.95	nr	315.18

COMMUNICATIONS/SECURITY/CONTROL

Item	Net Price £	Material £	Labour hours	Labour £	Unit	Total rate £
EARTHING AND BONDING						
Earth bar; polymer insulators and base mounting; including connections						
Non-disconnect link						
6 way	175.97	220.74	0.81	29.12	nr	**249.86**
8 way	194.85	244.42	0.81	29.12	nr	**273.54**
10 way	226.17	283.71	0.81	29.12	nr	**312.83**
Disconnect link						
6 way	198.24	248.67	1.01	36.31	nr	**284.98**
8 way	219.18	274.94	1.01	36.31	nr	**311.25**
10 way	243.06	304.90	1.01	36.31	nr	**341.21**
Soild earth bar; including connections						
150 × 50 × 6 mm	53.63	67.28	1.01	36.31	nr	**103.59**
Extra for earthing						
Disconnecting link						
300 × 50 × 6 mm	60.79	76.25	1.16	41.71	nr	**117.96**
500 × 50 × 6 mm	66.88	83.90	1.16	41.71	nr	**125.61**
Crimp lugs; including screws and connections to cable						
25 mm	0.50	0.63	0.31	11.14	nr	**11.77**
35 mm	0.68	0.85	0.31	11.14	nr	**11.99**
50 mm	0.78	0.97	0.32	11.50	nr	**12.47**
70 mm	1.40	1.76	0.32	11.50	nr	**13.26**
95 mm	2.33	2.92	0.46	16.53	nr	**19.45**
120 mm	1.65	2.07	1.25	44.95	nr	**47.02**
Earth clamps; connection to pipework						
15 mm to 32 mm dia.	1.37	1.71	0.15	5.40	nr	**7.11**
32 mm to 50 mm dia.	1.71	2.15	0.18	6.47	nr	**8.62**
50 mm to 75 mm dia.	2.00	2.51	0.20	7.19	nr	**9.70**

COMMUNICATIONS/SECURITY/CONTROL

Item	Net Price £	Material £	Labour hours	Labour £	Unit	Total rate £
LIGHTNING PROTECTION						
Conductor tape						
PVC sheathed copper tape						
25 × 3 mm	22.24	27.90	0.30	10.79	m	**38.69**
25 × 6 mm	42.04	52.73	0.30	10.79	m	**63.52**
50 × 6 mm	85.92	107.78	0.30	10.79	m	**118.57**
PVC sheathed copper solid circular conductor						
8 mm	12.97	16.27	0.50	17.98	m	**34.25**
Bare copper tape						
20 × 3 mm	18.24	22.88	0.30	10.79	m	**33.67**
25 × 3 mm	20.89	26.21	0.30	10.79	m	**37.00**
25 × 6 mm	38.31	48.06	0.40	14.38	m	**62.44**
50 × 6 mm	74.11	92.96	0.50	17.98	m	**110.94**
Bare copper solid circular conductor						
8 mm	11.79	14.78	0.50	17.98	m	**32.76**
Tape fixings; flat; metalic; PVC sheathed copper						
25 × 3 mm	22.24	27.90	0.33	11.87	nr	**39.77**
25 × 6 mm	42.04	52.73	0.33	11.87	nr	**64.60**
50 × 6 mm	85.92	107.78	0.33	11.87	nr	**119.65**
8 mm	7.44	9.33	0.50	17.98	nr	**27.31**
Bare copper						
20 × 3 mm	19.15	24.02	0.30	10.79	nr	**34.81**
25 × 3 mm	20.89	26.21	0.30	10.79	nr	**37.00**
25 × 6 mm	38.31	48.06	0.40	14.38	nr	**62.44**
50 × 6 mm	74.11	92.96	0.50	17.98	nr	**110.94**
8 mm	12.38	15.53	0.50	17.98	nr	**33.51**
Tape fixings; flat; non-metalic; PVC sheathed copper						
25 × 3 mm	2.59	3.25	0.30	10.79	nr	**14.04**
Tape fixings; flat; non-metalic; bare copper						
20 × 3 mm	1.02	1.28	0.30	10.79	nr	**12.07**
25 × 3 mm	1.02	1.28	0.30	10.79	nr	**12.07**
50 × 6 mm	5.41	6.79	0.30	10.79	nr	**17.58**
Puddle flanges; copper						
600 mm long	118.45	148.58	0.93	33.43	nr	**182.01**
Air rods						
Pointed air rod fixed to structure; copper						
10 mm dia.						
500 mm long	19.72	24.74	1.00	35.95	nr	**60.69**
1000 mm long	33.24	41.70	1.50	53.93	nr	**95.63**
Extra for						
Air terminal base	24.61	30.87	0.35	12.58	nr	**43.45**
Strike pad	26.14	32.79	0.35	12.58	nr	**45.37**
16 mm dia.						
500 mm long	31.25	39.20	0.91	32.72	nr	**71.92**
1000 mm long	57.10	71.62	1.75	62.92	nr	**134.54**
2000 mm long	104.26	130.78	2.50	89.88	nr	**220.66**

COMMUNICATIONS/SECURITY/CONTROL

Item	Net Price £	Material £	Labour hours	Labour £	Unit	Total rate £
LIGHTNING PROTECTION – cont						
Air rods – cont						
Extra for						
Multiple point	52.50	65.86	0.35	12.58	nr	**78.44**
Air terminal base	28.85	36.19	0.35	12.58	nr	**48.77**
Ridge saddle	54.17	67.95	0.35	12.58	nr	**80.53**
Side mounting bracket	53.91	67.63	0.50	17.98	nr	**85.61**
Rod to tape coupling	21.75	27.28	0.50	17.98	nr	**45.26**
Strike Pad	26.14	32.79	0.35	12.58	nr	**45.37**
Air terminals						
16 mm dia.						
500 mm long	31.25	39.20	0.65	23.37	nr	**62.57**
1000 mm long	57.10	71.62	0.78	28.03	nr	**99.65**
2000 mm long	104.26	130.78	1.50	53.93	nr	**184.71**
Extra for						
Multiple point	52.50	65.86	0.35	12.58	nr	**78.44**
Flat saddle	53.91	67.63	0.35	12.58	nr	**80.21**
Side bracket	26.94	33.79	0.50	17.98	nr	**51.77**
Rod to cable coupling	22.37	28.06	0.50	17.98	nr	**46.04**
Bonds and clamps						
Bond to flat surface; copper						
26 mm	4.69	5.88	0.45	16.18	nr	**22.06**
8 mm dia.	19.32	24.24	0.33	11.87	nr	**36.11**
Pipe bond						
26 mm	4.69	5.88	0.45	16.18	nr	**22.06**
8 mm dia.	48.04	60.26	0.33	11.87	nr	**72.13**
Rod to tape clamp						
26 mm	8.67	10.88	0.45	16.18	nr	**27.06**
Square clamp; copper						
25 × 3 mm	9.84	12.34	0.33	11.87	nr	**24.21**
50 × 6 mm	50.39	63.21	0.50	17.98	nr	**81.19**
8 mm dia.	10.52	13.19	0.33	11.87	nr	**25.06**
Test clamp; copper						
26 × 8 mm; oblong	14.67	18.40	0.50	17.98	nr	**36.38**
26 × 8 mm; plate type	44.65	56.01	0.50	17.98	nr	**73.99**
26 × 8 mm; screw down	40.13	50.34	0.50	17.98	nr	**68.32**
Cast in earth points						
2 hole	32.61	40.90	0.75	26.97	nr	**67.87**
4 hole	48.40	60.72	1.00	35.95	nr	**96.67**
Extra for cast in earth points						
Cover plate; 25 × 3 mm	36.60	45.91	0.25	8.99	nr	**54.90**
Cover plate; 8 mm	36.60	45.91	0.25	8.99	nr	**54.90**
Rebar clamp; 8 mm	25.21	31.63	0.25	8.99	nr	**40.62**
Static earth receptacle	171.50	215.13	0.50	17.98	nr	**233.11**

COMMUNICATIONS/SECURITY/CONTROL

Item	Net Price £	Material £	Labour hours	Labour £	Unit	Total rate £
Copper braided bonds						
25 × 3 mm						
200 mm hole centres	16.93	21.24	0.33	11.87	nr	**33.11**
400 mm holes centres	25.95	32.55	0.40	14.38	nr	**46.93**
U bolt clamps						
16 mm	11.18	14.02	0.33	11.87	nr	**25.89**
20 mm	12.68	15.90	0.33	11.87	nr	**27.77**
25 mm	15.54	19.49	0.33	11.87	nr	**31.36**
Earth pits/mats						
Earth inspection pit; hand to others for fixing						
Concrete	50.29	63.08	1.00	35.95	nr	**99.03**
Polypropylene	49.15	61.66	1.00	35.95	nr	**97.61**
Extra for						
5 hole copper earth bar; concrete pit	48.76	122.33	0.35	12.58	nr	**134.91**
5 hole earth bar; polypropylene	40.85	51.24	0.35	12.58	nr	**63.82**
Waterproof electrode seal						
Single flange	306.51	384.48	0.93	33.43	nr	**417.91**
Double flange	515.75	646.96	0.93	33.43	nr	**680.39**
Earth electrode mat; laid in ground and connected						
Copper tape lattice						
600 × 600 × 3 mm	138.34	173.53	0.93	33.43	nr	**206.96**
900 × 900 × 3 mm	248.08	311.19	0.93	33.43	nr	**344.62**
Copper tape plate						
600 × 600 × 1.5 mm	119.85	150.34	0.93	33.43	nr	**183.77**
600 × 600 × 3 mm	237.67	298.13	0.93	33.43	nr	**331.56**
900 × 900 × 1.5 mm	266.48	334.28	0.93	33.43	nr	**367.71**
900 × 900 × 3 mm	516.83	648.31	0.93	33.43	nr	**681.74**
Earth rods						
Solid cored copper earth electrodes driven into ground and connected						
15 mm dia.						
1200 mm long	48.69	61.07	0.93	33.43	nr	**94.50**
Extra for						
Coupling	1.47	1.85	0.06	2.16	nr	**4.01**
Driving stud	1.81	2.27	0.06	2.16	nr	**4.43**
Spike	1.67	2.09	0.06	2.16	nr	**4.25**
Rod clamp; flat tape	8.67	10.88	0.25	8.99	nr	**19.87**
Rod clamp; solid conductor	3.88	4.87	0.25	8.99	nr	**13.86**
20 mm dia.						
1200 mm long	94.44	118.46	0.98	35.24	nr	**153.70**
Extra for						
Coupling	1.47	1.85	0.06	2.16	nr	**4.01**
Driving stud	2.96	3.72	0.06	2.16	nr	**5.88**
Spike	2.73	3.43	0.06	2.16	nr	**5.59**
Rod clamp; flat tape	8.67	10.88	0.25	8.99	nr	**19.87**
Rod clamp; solid conductor	4.29	5.38	0.25	8.99	nr	**14.37**

COMMUNICATIONS/SECURITY/CONTROL

Item	Net Price £	Material £	Labour hours	Labour £	Unit	Total rate £
LIGHTNING PROTECTION – cont						
Earth rods – cont						
Stainless steel earth electrodes driven into ground and connected						
16 mm dia.						
1200 mm long	62.20	78.02	0.93	33.43	nr	**111.45**
Extra for						
Coupling	1.90	2.39	0.06	2.16	nr	**4.55**
Driving head	1.81	2.27	0.06	2.16	nr	**4.43**
Spike	1.67	2.09	0.06	2.16	nr	**4.25**
Rod clamp; flat tape	8.67	10.88	0.25	8.99	nr	**19.87**
Rod clamp; solid conductor	3.88	4.87	0.25	8.99	nr	**13.86**
Surge protection						
Single phase; including connection to equipment						
90–150 V	355.49	445.93	5.00	179.76	nr	**625.69**
200–280 V	337.01	422.74	5.00	179.76	nr	**602.50**
Three phase; including connection to equipment						
156–260 V	717.50	900.03	10.00	359.51	nr	**1259.54**
346–484 V	668.58	838.67	10.00	359.51	nr	**1198.18**
349–484 V; remote display	743.59	932.76	10.00	359.51	nr	**1292.27**
346–484 V; 60 kA	1234.97	1549.15	10.00	359.51	nr	**1908.66**
346–484 V; 120 kA	2362.31	2963.28	10.00	359.51	nr	**3322.79**

COMMUNICATIONS/SECURITY/CONTROL

Item	Net Price £	Material £	Labour hours	Labour £	Unit	Total rate £
CENTRAL CONTROL/BUILDING MANAGEMENT						
EQUIPMENT						
Switches/sensors; includes fixing in position; electrical work elsewhere. **Note:** these are normally free issued to the mechanical contractor for fitting. The labour times applied assume the installation has been prepared for the fitting of the component						
Pressure devices						
Liquid differential pressure sensor	160.05	200.77	0.50	17.98	nr	**218.75**
Liquid differential pressure switch	87.71	110.03	0.50	17.98	nr	**128.01**
Air differential pressure transmitter	115.75	145.20	0.50	17.98	nr	**163.18**
Air differential pressure switch	15.82	19.85	0.50	17.98	nr	**37.83**
Liquid level switch	79.87	100.18	0.50	17.98	nr	**118.16**
Static pressure sensor	90.44	113.44	0.50	17.98	nr	**131.42**
High pressure switch	79.87	100.18	0.50	17.98	nr	**118.16**
Low pressure switch	79.87	100.18	0.50	17.98	nr	**118.16**
Water pressure switch	122.41	153.55	0.50	17.98	nr	**171.53**
Duct averaging temperature sensor	53.25	66.80	1.00	35.95	nr	**102.75**
Temperature devices						
Return air sensor (fan coils)	5.21	6.54	1.00	35.95	nr	**42.49**
Frost thermostat	34.15	42.84	0.50	17.98	nr	**60.82**
Immersion thermostat	55.27	69.33	0.50	17.98	nr	**87.31**
Temperature high limit	55.27	69.33	0.50	17.98	nr	**87.31**
Temperature sensor with averaging element	53.25	66.80	0.50	17.98	nr	**84.78**
Immersion temperature sensor	22.62	28.37	0.50	17.98	nr	**46.35**
Space temperature sensor	6.55	8.22	1.00	35.95	nr	**44.17**
Combined space temperature & humidity sensor	90.44	113.44	1.00	35.95	nr	**149.39**
Outside air temperature sensor	12.66	15.88	2.00	71.90	nr	**87.78**
Outside air temperature & humidity sensor	296.61	372.06	2.00	71.90	nr	**443.96**
Duct humidity sensor	116.66	146.34	0.50	17.98	nr	**164.32**
Space humidity sensor	90.44	113.44	1.00	35.95	nr	**149.39**
Immersion water flow sensor	22.62	28.37	0.50	17.98	nr	**46.35**
Rain sensor	211.62	265.45	2.00	71.90	nr	**337.35**
Wind speed and direction sensor	669.20	839.44	2.00	71.90	nr	**911.34**
Controllers; includes fixing in position; electrical work elsewhere						
Zone						
Fan coil controller	170.91	214.39	2.00	71.90	nr	**286.29**
VAV controller	223.50	280.36	2.00	71.90	nr	**352.26**

COMMUNICATIONS/SECURITY/CONTROL

Item	Net Price £	Material £	Labour hours	Labour £	Unit	Total rate £
CENTRAL CONTROL/BUILDING MANAGEMENT – cont						
Controllers – cont						
Plant						
Controller, 96 I/O points (exact configuration is dependent upon the number of I/O boards added)	3137.05	3935.12	0.50	17.98	nr	**3953.10**
Controller, 48 I/O points (exact configuration is dependent upon the number of I/O boards added)	2618.89	3285.14	0.50	17.98	nr	**3303.12**
Controller, 32 I/O points (exact configuration is dependent upon the number of I/O boards added)	2174.80	2728.07	0.50	17.98	nr	**2746.05**
Additional Digital Input Boards (12 DI)	518.16	649.98	0.20	7.19	nr	**657.17**
Additional Digital Output Boards (6 DO)	443.12	555.84	0.20	7.19	nr	**563.03**
Additional analogue Input Boards (8 AI)	443.12	555.84	0.20	7.19	nr	**563.03**
Additional analogue Output Boards (8 AO)	443.12	555.84	0.20	7.19	nr	**563.03**
Outstation Enclosure (fitted in riser with space allowance for controller and network device)	443.12	555.84	5.00	179.76	nr	**735.60**
Damper actuator; electrical work elsewhere						
Damper actuator 0–10 V	94.05	117.98	1.00	35.95	nr	**153.93**
Damper actuator with auxiliary switches	62.39	78.27	1.00	35.95	nr	**114.22**
Frequency inverters: not mounted within MCC; includes fixing in position; electrical work elsewhere						
2.2 kW	715.66	897.72	2.00	71.90	nr	**969.62**
3 kW	791.34	992.66	2.00	71.90	nr	**1064.56**
7.5 kW	1170.88	1468.76	2.00	71.90	nr	**1540.66**
11 kW	1431.32	1795.45	2.00	71.90	nr	**1867.35**
15 kW	1729.60	2169.61	2.00	71.90	nr	**2241.51**
18.5 kW	2021.21	2535.41	2.50	89.88	nr	**2625.29**
20 kW	2281.65	2862.10	2.50	89.88	nr	**2951.98**
30 kW	2764.69	3468.02	2.50	89.88	nr	**3557.90**
55 kW	4992.92	6263.12	3.00	107.86	nr	**6370.98**
Miscellaneous; includes fixing in position; electrical work elsewhere						
1 kW Thyristor	63.72	79.93	2.00	71.90	nr	**151.83**
10 kW Thyristor	122.70	153.91	2.00	71.90	nr	**225.81**
Front end and networking; electrical work elsewhere						
PC/monitor	946.05	1186.73	2.00	71.90	nr	**1258.63**
Printer	278.25	349.04	2.00	71.90	nr	**420.94**
PC software	–	–	–	–	nr	–
Network server software	2274.97	2853.73	–	–	nr	**2853.73**
Router (allows connection to a network)	114.64	143.81	–	–	nr	**143.81**

COMMUNICATIONS/SECURITY/CONTROL

Item	Net Price £	Material £	Labour hours	Labour £	Unit	Total rate £
WI-FI						
Wi-Fi						
Zone director controller, 1000 Mbps, RJ 45, 2 ports, auto MDX and suto sensing						
Up to 2,000 clients	852.60	1069.50	4.00	143.81	nr	**1213.31**
Up to 5,000 clients, 500 access points	5901.00	7402.21	4.00	143.81	nr	**7546.02**
Up to 20,000 clients, 1,000 access points	7001.05	8782.12	4.00	143.81	nr	**8925.93**
Wi-Fi access points, suitable for large developments (commercial buildings, stadiums, hotels, education)						
Indoor access points, 802.11ac Wi-Fi, dual band, 2.4 GHz and 5 GHz						
Up to 100 connected devices	284.55	356.94	1.20	43.14	nr	**400.08**
Up to 100 connected devices, 867 Mbps	385.35	483.38	1.20	43.14	nr	**526.52**
Up to 300 connected devices, 867 Mbps	501.90	629.59	1.20	43.14	nr	**672.73**
Up to 400 connected devices, 1300 Mbps	618.45	775.78	1.20	43.14	nr	**818.92**
Up to 500 connected devices, 1300 Mbps	773.85	970.72	1.20	43.14	nr	**1013.86**
Up to 500 connected devices, 1733 Mbps	1018.50	1277.61	1.20	43.14	nr	**1320.75**
Outdoor access points, 802.11ac Wi-Fi, dual band, 2.4 GHz and 5 GHz, IP67, plastic enclosure, with flexible wall or pole mounting						
Standard	1006.95	1263.11	1.80	64.71	nr	**1327.82**
Up to 500 Mbps	1858.50	2331.30	1.80	64.71	nr	**2396.01**
Wi-Fi access points, suitable for small/medium sized and residential						
Indoor access points, 802.11ac Wi-Fi, dual band, 2.4 GHz and 5 GHz						
Up to 100 connected devices, 867 Mbps	385.35	483.38	1.20	43.14	nr	**526.52**
Up to 300 connected devices, 567 Mbps	501.90	629.59	1.20	43.14	nr	**672.73**
Up to 400 connected devices, 1300 Mbps	618.45	775.78	1.20	43.14	nr	**818.92**
Outdoor access points, 802.11ac Wi-Fi, dual band, 2.4 GHz and 5 GHz, IP67, plastic enclosure, with flexible wall or pole mounting						
Standard	1006.95	1263.11	1.80	64.71	nr	**1327.82**

Quantity Surveyor's Pocket Book, 3rd Edition

D. Cartlidge

The third edition of the Quantity Surveyor's Pocket Book has been updated in line with NRM1, NRM2 and NRM3, and remains a must-have guide for students and qualified practitioners. Its focused coverage of the data, techniques and skills essential to the quantity surveying role makes it an invaluable companion for everything from initial cost advice to the final account stage.

Key features and updates included in this new edition:
- An up-to-date analysis of NRM1, 2 and 3;
- Measurement and estimating examples in NRM2 format;
- Changes in procurement practice;
- Changes in professional development, guidance notes and schemes of work;
- The increased use of NEC3 form of contract;
- The impact of BIM.

This text includes recommended formats for cost plans, developer's budgets, financial reports, financial statements and final accounts. This is the ideal concise reference for quantity surveyors, project and commercial managers, and students of any of the above.

March 2017; 186 × 123 mm, 466 pp
Pbk: 978-1-138-69836-9; £20.99

Taylor & Francis
Taylor & Francis Group

An Introduction to Electrical Science, 2nd Edition

Adrian Waygood

Heavily updated and expanded, this second edition of Adrian Waygood's textbook provides an indispensable introduction to the science behind electrical engineering.
While fully matched to the electrical science requirements of the 2330 levels 2 and 3 Certificates in Electrotechnical Technology from the City and Guilds (Electrical Installation), the main purpose of this book is to develop an easy understanding of the how and why within each topic. It is aimed for those starting careers in electronics, as well as any hobbyists, with an array of new material to reflect changes in the industry.

New chapters include:
- Electrical Drawings
- Practical Resistors
- Measuring Instruments
- Basic Motor Action
- Practical Inductors
- Basic Transformer Theory
- The Electricity Supply Industry
...and more

The author details the historical context of each main principle and offers a wealth of examples, images and diagrams, all whilst maintaining his signature conversational and accessible style. And there is also a companion site with interactive multiple choice quizzes for each chapter and more, at www.routledge.com/cw/waygood

August 2018: 276 × 219 mm: 384 pp
Pb: 978-0-8153-9181-4: £39.99

To Order: Tel: +44 (0) 1235 400524 Fax: +44 (0) 1235 400525
or Post: Taylor and Francis Customer Services,
Bookpoint Ltd, Unit T1, 200 Milton Park, Abingdon, Oxon, OX14 4TA UK
Email: book.orders@tandf.co.uk

For a complete listing of all our titles visit:
www.tandf.co.uk

Taylor & Francis
Taylor & Francis Group

Electrical Circuit Theory and Technology, 6th edition

John Bird

A fully comprehensive text for courses in electrical principles, circuit theory and electrical technology, providing 800 worked examples and over 1,350 further problems for students to work through at their own pace. This book is ideal for students studying engineering for the first time as part of BTEC National and other pre-degree vocational courses, as well as Higher Nationals, Foundation Degrees and first-year undergraduate modules.

March 2017: 276 x 219 mm: 858 pp
Pb: 978-1-138-67349-6 : £38.99

To Order: Tel: +44 (0) 1235 400524 Fax: +44 (0) 1235 400525
or Post: Taylor and Francis Customer Services,
Bookpoint Ltd, Unit T1, 200 Milton Park, Abingdon, Oxon, OX14 4TA UK
Email: book.orders@tandf.co.uk

For a complete listing of all our titles visit:
www.tandf.co.uk

Electricians' On-Site Companion

Christopher Kitcher

This book contains everything electricians need to know about working on site, covering not only the health and safety aspects of site work, but also the techniques and testing knowledge required from the modern-day electrician. Regulations issues are included alongside step-by-step instructions for each task, after which testing information, checklists and example forms are given so that site workers can ensure they have done everything required of them.

October 2017: 234 x 156 mm: 174 pp
Pb: 978-1-138-68332-7 : £24.99

To Order: Tel: +44 (0) 1235 400524 Fax: +44 (0) 1235 400525
or Post: Taylor and Francis Customer Services,
Bookpoint Ltd, Unit T1, 200 Milton Park, Abingdon, Oxon, OX14 4TA UK
Email: book.orders@tandf.co.uk

For a complete listing of all our titles visit:
www.tandf.co.uk

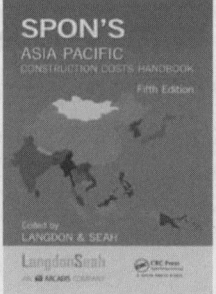

PART 5

Rates of Wages

Estimating and Tendering for Construction Work,
5th edition

Martin Brook

Estimators need to understand the consequences of entering into a contract, often defined by complex conditions and documents, as well as to appreciate the technical requirements of the project. Estimating and Tendering for Construction Work, 5th edition, explains the job of the estimator through every stage, from early cost studies to the creation of budgets for successful tenders.

This new edition reflects recent developments in the field and covers:

- new tendering and procurement methods
- the move from basic estimating to cost-planning and the greater emphasis placed on partnering and collaborative working
- the New Rules of Measurement (NRM1 and 2), and examines ways in which practicing estimators are implementing the guidance
- emerging technologies such as BIM (Building Information Modelling) and estimating systems which can interact with 3D design models

With the majority of projects procured using design-and-build contracts, this edition explains the contractor's role in setting costs, and design statements, to inform and control the development of a project's design.

Clearly-written and illustrated with examples, notes and technical documentation, this book is ideal for students on construction-related courses at HNC/HND and Degree levels. It is also an important source for associated professions and estimators at the outset of their careers.

July 2016; 246 × 189 mm, 334 pp
Pbk: 978-1-138-83806-2; £34.99

To Order: Tel: +44 (0) 1235 400524 Fax: +44 (0) 1235 400525
or Post: Taylor and Francis Customer Services,
Bookpoint Ltd, Unit T1, 200 Milton Park, Abingdon, Oxon, OX14 4TA UK
Email: book.orders@tandf.co.uk

For a complete listing of all our titles visit:
www.tandf.co.uk

Taylor & Francis
Taylor & Francis Group

Mechanical Installations

Rates of Wages

HEATING, VENTILATING, AIR CONDITIONING, PIPING AND DOMESTIC ENGINEERING INDUSTRY

Note – At the time of printing, the October 2018 rates has not been published. We have contacted BESA and approximate uplifts were agreed to calculate the all-in labour rate that is seen in the Mechanical Directions section. This section therefore displays the October 2017 rates for reference.

For full details of the wage agreement and the Heating Ventilating Air Conditioning Piping and Domestic Engineering Industry's National Working Rule Agreement, contact:

Building & Engineering Services Association
Lincoln House,
137–143 Hammersmith Road
London W14 0QL
Tel: 020 7313 4900
www.b-es.org

WAGE RATES, ALLOWANCES AND OTHER PROVISIONS

Hourly rates of wages

All districts of the United Kingdom

Main Grades	From 2 October 2017 p/hr
Foreman	16.81
Senior Craftsman (+2nd welding skill)	14.44
Senior Craftsman	13.89
Craftsman (+2nd welding skill)	13.89
Craftsman	12.79
Operative	11.54
Adult Trainee	9.74
Mate (18 and over)	9.74
Mate (16–17)	4.52
Modern Apprentices	
Junior	6.31
Intermediate	8.96
Senior	11.54

Note: Ductwork Erection Operatives are entitled to the same rates and allowances as the parallel Fitter grades shown.

HEATING, VENTILATING, AIR CONDITIONING, PIPING AND DOMESTIC ENGINEERING INDUSTRY

Trainee *Rates of Pay*

Junior Ductwork Trainees (Probationary)

Age at entry	From 2 October 2017 p/hr
17	5.73
18	5.73
19	5.73
20	5.73

Junior Ductwork Erectors (Year of Training)

	From 2 October 2017		
Age at entry	1 yr p/h	2 yr p/h	3 yr p/h
17	7.13	8.88	10.07
18	7.13	8.88	10.07
19	7.13	8.88	10.07
20	7.13	8.88	10.07

Responsibility Allowance (Craftsmen)	From 2 October 2017 p/hr
Second welding skill or supervisory responsibility (one unit)	0.55
Second welding skill and supervisory responsibility (two units)	0.55

Responsibility Allowance (Senior Craftsmen)	From 2 October 2017 p/hr
Second welding skill	0.55
Supervising responsibility	1.10
Second welding skill and supervisory responsibility	0.55

Daily travelling allowance – Scale 2

C:	Craftsmen including Installers
M&A:	Mates, Apprentices and Adult Trainees

Direct distance from centre to job in miles

		From 2 October 2017	
Over	Not exceeding	C p/hr	M&A p/hr
15	20	2.67	2.29
20	30	6.87	5.94
30	40	9.88	8.55
40	50	13.02	11.13

HEATING, VENTILATING, AIR CONDITIONING, PIPING AND DOMESTIC ENGINEERING INDUSTRY

Daily travelling allowance – Scale 1

C:	Craftsmen including Installers
M&A:	Mates, Apprentices and Adult Trainees

Direct distance from centre to job in miles

		From 2 October 2017	
Over	*Not exceeding*	*C p/hr*	*M&A p/hr*
0	15	7.30	7.30
15	20	9.98	9.60
20	30	14.18	13.25
30	40	17.19	15.86
40	50	20.31	18.46

Weekly Holiday Credit and Welfare Contributions

	From 2 October 2017						
	£ a	£ b	£ c	£ d	£ e	£ f	£ g
Weekly Holiday Credit	78.06	72.21	69.65	67.09	64.51	61.99	59.40
Combined Weekly/Welfare Holiday Credit and Contribution	87.71	81.86	79.30	76.74	74.16	71.64	69.05

	From 2 October 2017			
	£ h	£ i	£ j	£ k
Weekly Holiday Credit	53.58	45.25	41.60	29.31
Combined Weekly/Welfare Holiday Credit and Contribution	63.23	54.90	51.25	38.96

HEATING, VENTILATING, AIR CONDITIONING, PIPING AND DOMESTIC ENGINEERING INDUSTRY

The grades of H&V Operatives entitled to the different rates of Weekly Holiday Credit and Welfare Contribution are as follows:

a	*b*	*c*	*d*
Foreman	Senior Craftsman (RAS & RAW)	Senior Craftsman (RAS)	Senior Craftsman (RAW)
e	*f*	*g*	*h*
Senior Craftsman Craftsman (+2 RA)	Craftsman (+ 1RA)	Craftsman	Installer Senior Modern Apprentice
i	*j*	*k*	*l*
Adult Trainee Mate (over 18)	Intermediate Modern Apprentice	Junior Modern Apprentice	No grade allocated to this Credit Value Category

Daily abnormal conditions money	*From 2 October 2017*
Per day	3.28

Lodging allowance	*From 2 October 2017*
Per night	38.50

Explanatory Notes

1.   Working Hours

    The normal working week (Monday to Friday) shall be 37.5 hours.

2.   Overtime

    Time worked in excess of 37.5 hours during the normal working week shall be paid at time and a half until 12 hours have been worked since the actual starting time. Thereafter double time shall be paid until normal starting time the following morning. Weekend overtime shall be paid at time and a half for the first 5 hours worked on a Saturday and at double time thereafter until normal starting time on Monday morning.

PLUMBING MECHANICAL ENGINEERING SERVICES INDUSTRY

The Joint Industry Board for Plumbing Mechanical Engineering Services has agreed a three year wage agreement for 2018, 2019 and 2020 with effect from 1 January 2018 and subsequently from 7 January 2019, and 6 January 2020.

For full details of this wage agreement and the JIB PMES National Working Rules, contact:

The Joint Industry Board for Plumbing Mechanical Engineering Services in England and Wales

JIB-PMES
PO Box 267
PE19 9DN
Tel: 01480 476925
Email: info@jib-pmes.org.uk

WAGE RATES, ALLOWANCES AND OTHER PROVISIONS
EFFECTIVE FROM **1 January 2018**

Basic Rates of Hourly Pay

Applicable in England and Wales

	Hourly rate £
Operatives	
Technical plumber and gas service technician	16.73
Advanced plumber and gas service engineer	15.07
Trained plumber and gas service fitter	12.93
Apprentices	
4th year of training with NVQ level 3	12.51
4th year of training with NVQ level 2	11.33
4th year of training	9.97
3rd year of training with NVQ level 2	9.85
3rd year of training	8.11
2nd year of training	7.19
1st year of training	6.26
Adult Trainees	
3rd 6 months of employment	11.26
2nd 6 months of employment	10.82
1st 6 months of employment	10.09

PLUMBING MECHANICAL ENGINEERING SERVICES INDUSTRY

Major Projects Agreement

Where a job is designated as being a Major Project then the following Major Project Performance Payment hourly rate supplement shall be payable:

Employee Category

	National Payment £	London* Payment £
Technical Plumber and Gas Service Technician	2.20	3.57
Advanced Plumber and Gas Service Engineer	2.20	3.57
Trained Plumber and Gas Service Fitter	2.20	3.57
All 4th year apprentices	1.76	2.86
All 3rd year apprentices	1.32	2.68
2nd year apprentice	1.21	1.96
1st year apprentice	0.88	1.43
All adult trainees	1.76	2.86

* The London Payment Supplement applies only to designated Major Projects that are within the M25 London orbital motorway and are effective from 1 February 2007.

* National payment hourly rates are unchanged for 2007 and will continue to be at the rates shown in Promulgation 138 A issued on 14 October 2003.

PLUMBING MECHANICAL ENGINEERING SERVICES INDUSTRY

Allowances

Daily travel time allowance plus return fares

All daily travel allowances are to be paid at the daily rate as follows:

Over	Not exceeding	All Operatives	3rd & 4th Year Apprentices	1st & 2nd Year Apprentices
20	30	£4.73	3.05	1.90
30	40	11.04	7.10	4.55
40	50	12.61	7.53	4.73

Responsibility/Incentive Pay Allowance

Employers may, in consultation with the employees concerned, enhance the basic graded rates of pay by the payment of an additional amount, as per the bands shown below, where it is agreed that their work involves extra responsibility, productivity or flexibility.

Band 1   an additional rate of up to £ 0.30 per hour
Band 2   an additional rate of up to £ 0.51 per hour
Band 3   an additional rate of up to £ 0.76 per hour
Band 4   an additional rate of up to £ 0.99 per hour

This allowance forms part of an operative's basic rate of pay and shall be used to calculate premium payments.

Mileage allowance	£0.45 per mile
Lodging allowance	£38.60 per night
Subsistence Allowance (London Only)	£5.43 per night

Plumbers welding supplement

Possession of Gas or Arc Certificate	£0.33 per hour
Possession of Gas and Arc Certificate	£0.53 per hour

PLUMBING MECHANICAL ENGINEERING SERVICES INDUSTRY

Weekly Holiday Credit Contributions (Holiday Credit Period 2017–2018)

	Public & Annual Gross Value, £	Combined Holiday Credit*, £
Technical Plumber and Gas Service Technician	£66.75	£63.20
Advance Plumber and Gas Service Engineer	£60.05	£56.90
Trained Plumber and Gas Service Fitter	£51.40	£48.80
Adult Trainee	£39.35	£38.65
Apprentice in last year of training	£39.35	£38.65
Apprentice 3rd year	£28.25	£27.65
Apprentice 2nd year	£25.00	£24.50
Apprentice 1st year	£21.75	£21.35
Working Principal	£27.80	£26.25
Ancillary Employee	£34.10	£32.55

* Public and Annual gross value is the Combined Holiday credit value less JIB administration value.

Explanatory Notes

1. Working Hours

    The normal working week (Monday to Friday) shall be 37½ hours, with 45 hours to be worked in the same period before overtime rates become applicable.

2. Overtime

    Overtime shall be paid at time and a half up to 8.00pm (Monday to Friday) and up to 1.00pm (Saturday). Overtime worked after these times shall be paid at double time.

3. Major Projects Agreement

    Under the Major Projects Agreement the normal working week shall be 38 hours (Monday to Friday) with overtime rates payable for all hours worked in excess of 38 hours in accordance with 2 above. However, it should be noted that the hourly rate supplement shall be paid for each hour worked but does not attract premium time enhancement.

4. Pension

    In addition to their hourly rates of pay, plumbing employees are entitled to inclusion within the Industry Pension Scheme (or one providing equivalent benefits). The current levels of industry scheme contributions are 6½% (employers) and 3¼% (employees).

5. Additional Holiday Pay (AHP) Contributions

    There has been a fundamental change in the way the JIB-PMES Holiday Pay Schemes shall apply for the Holiday Credit Period 2017–2018, in that 62 credits will be the maximum number payable in a year.

    For full details please refer to the Additional Holiday Pay (AHP) section as published by the JIB for Plumbing Mechanical Engineering Services in England and Wales.

Electrical Installations

Rates of Wages

ELECTRICAL CONTRACTING INDUSTRY

For full details of this wage agreement and the Joint Industry Board for the Electrical Contracting Industry's National Working Rules, contact:

Joint Industry Board
PO Box 127
Swanley
Kent
BR8 9BH
Tel: 0333 321 230
www.jib.org.uk

WAGES (Graded Operatives)

Rates

Since 7 January 2002 two different wage rates have applied to JIB Graded Operatives working on site, depending on whether the Employer transports them to the site or whether they provide their own transport. The two categories are:

Job Employed (Transport Provided)

Payable to an Operative who is transported to and from the job by his Employer. The Operative shall also be entitled to payment for Travel Time, when travelling in his own time, as detailed in the appropriate scale.

Job Employed (Own Transport)

Payable to an Operative who travels by his own means to and from the job. The Operative shall be entitled to payment for Travel Allowance and also Travel Time, when travelling in his own time, as detailed in the appropriate scale.

ELECTRICAL CONTRACTING INDUSTRY

The JIB rates of wages are set out below:

From and including 1 January 2018, the JIB hourly rates of wages shall be as set out below:

(i)   National Standard Rate:

Grade	Transport Provided	Own Transport
Technician (or equivalent specialist grade)	£ 17.49	£ 18.37
Approved Electrician (or equivalent specialist grade)	£ 15.46	£ 16.32
Electrician (or equivalent specialist grade)	£ 14.16	£ 15.05
Senior Graded Electrical Trainee	£ 13.47	£ 14.30
Electrical Improver	£ 12.73	£ 13.56
Labourer	£ 11.24	£ 12.08
Adult Trainee	£ 11.16	£ 12.05

(ii)   London Rate:

Grade	Transport Provided	Own Transport
Technician (or equivalent specialist grade)	£ 19.61	£ 20.57
Approved Electrician (or equivalent specialist grade)	£ 17.31	£ 18.28
Electrician (or equivalent specialist grade)	£ 15.85	£ 16.86
Senior Graded Electrical Trainee	£ 15.06	£ 16.03
Electrical Improver	£ 14.26	£ 15.18
Labourer	£ 12.61	£ 13.53
Adult Trainee	£ 12.61	£ 13.53

From and including 1 January 2018, the JIB hourly rates for Job Employed apprentices shall be:

(i)   National Standard Rates

	Transport Provided	Own Transport
Stage 1	£ 4.95	£ 5.79
Stage 2	£ 7.29	£ 8.16
Stage 3	£ 10.55	£ 11.43
Stage 4	£ 11.16	£ 12.05

(ii)   London Rate

	Transport Provided	Own Transport
Stage 1	£ 5.54	£ 6.49
Stage 2	£ 8.18	£ 9.11
Stage 3	£ 11.83	£ 12.81
Stage 4	£ 12.52	£ 13.50

ELECTRICAL CONTRACTING INDUSTRY

Travelling Time and Travel Allowances

From and including 1 January 2018

With effect from 2 January 2017, the existing Travel Allowance and Travelling Time tables will be replaced by:

(i)   A Mileage Allowance for operatives and apprentices using their own vehicular transport from the shop to their place of work. This allowance will not be taxable because it is within HMRC Approved Mileage Rates.

(ii)   A taxable Mileage Rate for those operatives and apprentices for whom transport has been provided.

These will be introduced on jobs over 15 actual miles travelled from the shop to the job and vice versa using the shortest route. The shortest route will be calculated using the RAC Route Planner and will be paid both ways.

The non-taxable Mileage Allowance payments for operatives and apprentices using their own transport will be:

    21p per mile with effect from Monday 2 January 2017

    22p per mile with effect from 7 January 2019

Operatives and apprentices making their way to the site in transport provided by their employer will receive taxable Mileage Rates of:

    11p per mile with effect from Monday 2 January 2017

    12p per mile with effect from 7 January 2019

For the avoidance of doubt, jobs less than 15 miles from the shop to the job and vice versa will continue to receive no payment.

ELECTRICAL CONTRACTING INDUSTRY

Lodging Allowances

£39.69 from and including 1 January 2018

Lodgings weekend retention fee, maximum reimbursement

£39.69 from and including 1 January 2018

Annual Holiday Lodging Allowance Retention

Maximum £13.06 per night (£91.42 per week) from and including 1 January 2018

Responsibility money

From and including 30 March 1998 the minimum payment increased to 10p per hour and the maximum to £1.00 per hour (no change)

From and including 4 January 1992 responsibility payments are enhanced by overtime and shift premiums where appropriate (no change)

Combined JIB Benefits Stamp Value (from week commencing 1 January 2018)

JIB grade	Weekly JIB combined credit value £	Holiday value £
Technician	£ 60.21	£ 55.50
Approved Electrician	£ 53.77	£ 49.06
Electrician	£ 49.91	£ 44.93
Labourer & Adult Trainee	£ 41.06	£35.05

Explanatory Notes

1.   Working Hours

   The normal working week (Monday to Friday) shall be 37½ hours, with 38 hours to be worked in the same period before overtime rates become applicable.

2.   Overtime

   Overtime shall be paid at time and a half for all weekday overtime. Saturday overtime shall be paid at time and a half for the first 6 hours, or up to 3.00pm (whichever comes first). Thereafter double time shall be paid until normal starting time on Monday.

PART 6

Daywork

When work is carried out in connection with a contract that cannot be valued in any other way, it is usual to assess the value on a cost basis with suitable allowances to cover overheads and profit. The basis of costing is a matter for agreement between the parties concerned but definitions of prime cost for the Heating and Ventilating and Electrical Industries have been published jointly by the Royal Institution of Chartered Surveyors and the appropriate bodies of the industries concerned, for those who wish to use them.

These, together with a schedule of basic plant hire charges are reproduced on the following pages, with the kind permission of the Royal Institution of Chartered Surveyors, who own the copyright.

Building Services Handbook, 9th Edition

Fred Hall and Roger Greeno

The ninth edition of Hall and Greeno's leading textbook has been reviewed and updated in relation to the latest building and water regulations, new technology, and new legislation. For this edition, new updates includes: the reappraisal of CO_2 emissions targets, updates to sections on ventilation, fuel, A/C, refrigeration, water supply, electricity and power supply, sprinkler systems, and much more.

Building Services Handbook summarises the application of all common elements of building services practice, technique and procedure, to provide an essential information resource for students as well as practitioners working in building services, building management and the facilities administration and maintenance sectors of the construction industry. Information is presented in the highly illustrated and accessible style of the best-selling companion title *Building Construction Handbook*.

THE comprehensive reference for all construction and building services students, Building Services Handbook is ideal for a wide range of courses including NVQ and BTEC National through Higher National Certificate and Diploma to Foundation and three-year Degree level. The clear illustrations and complementary references to industry Standards combine essential guidance with a resource base for further reading and development of specific topics.

May 2017; 234 × 156 mm, 786 pp
Pbk: 978-1-138-24435-1; £32.99

To Order: Tel: +44 (0) 1235 400524 Fax: +44 (0) 1235 400525
or Post: Taylor and Francis Customer Services,
Bookpoint Ltd, Unit T1, 200 Milton Park, Abingdon, Oxon, OX14 4TA UK
Email: book.orders@tandf.co.uk

For a complete listing of all our titles visit:
www.tandf.co.uk

Taylor & Francis
Taylor & Francis Group

HEATING AND VENTILATING INDUSTRY

DEFINITION OF PRIME COST OF DAYWORK CARRIED OUT UNDER A HEATING, VENTILATING, AIR CONDITIONING, REFRIGERATION, PIPEWORK AND/OR DOMESTIC ENGINEERING CONTRACT (JULY 1980 EDITION)

This Definition of Prime Cost is published by the Royal Institution of Chartered Surveyors and the Heating and Ventilating Contractors Association for convenience, and for use by people who choose to use it. Members of the Heating and Ventilating Contractors Association are not in any way debarred from defining Prime Cost and rendering accounts for work carried out on that basis in any way they choose. Building owners are advised to reach agreement with contractors on the Definition of Prime Cost to be used prior to entering into a contract or subcontract.

SECTION 1: APPLICATION

1.1   This Definition provides a basis for the valuation of daywork executed under such heating, ventilating, air conditioning, refrigeration, pipework and or domestic engineering contracts as provide for its use.

1.2   It is not applicable in any other circumstances, such as jobbing or other work carried out as a separate or main contract nor in the case of daywork executed after a date of practical completion.

1.3   The terms 'contract' and 'contractor' herein shall be read as 'subcontract' and 'subcontractor' as applicable.

SECTION 2: COMPOSITION OF TOTAL CHARGES

2.1   The Prime Cost of daywork comprises the sum of the following costs:
    (a)   Labour as defined in Section 3.
    (b)   Materials and goods as defined in Section 4.
    (c)   Plant as defined in Section 5.

2.2   Incidental costs, overheads and profit as defined in Section 6, as provided in the contract and expressed therein as percentage adjustments, are applicable to each of 2.1 (a)–(c).

SECTION 3: LABOUR

3.1   The standard wage rates, emoluments and expenses referred to below and the standard working hours referred to in 3.2 are those laid down for the time being in the rules or decisions or agreements of the Joint Conciliation Committee of the Heating, Ventilating and Domestic Engineering Industry applicable to the works (or those of such other body as may be appropriate) and to the grade of operative concerned at the time when and the area where the daywork is executed.

3.2   Hourly base rates for labour are computed by dividing the annual prime cost of labour, based upon the standard working hours and as defined in 3.4, by the number of standard working hours per annum. See example.

3.3   The hourly rates computed in accordance with 3.2 shall be applied in respect of the time spent by operatives directly engaged on daywork, including those operating mechanical plant and transport and erecting and dismantling other plant (unless otherwise expressly provided in the contract) and handling and distributing the materials and goods used in the daywork.

3.4   The annual prime cost of labour comprises the following:
    (a)   Standard weekly earnings (i.e. the standard working week as determined at the appropriate rate for the operative concerned).
    (b)   Any supplemental payments.
    (c)   Any guaranteed minimum payments (unless included in Section 6.1(a)–(p)).
    (d)   Merit money.
    (e)   Differentials or extra payments in respect of skill, responsibility, discomfort, inconvenience or risk (excluding those in respect of supervisory responsibility – see 3.5)
    (f)   Payments in respect of public holidays.
    (g)   Any amounts which may become payable by the contractor to or in respect of operatives arising from the rules etc. referred to in 3.1 which are not provided for in 3.4 (a)–(f) nor in Section 6.1 (a)–(p).
    (h)   Employers contributions to the WELPLAN, the HVACR Welfare and Holiday Scheme or payments in lieu thereof.
    (i)   Employers National Insurance contributions as applicable to 3.4 (a)–(h).
    (j)   Any contribution, levy or tax imposed by Statute, payable by the contractor in his capacity as an employer.

HEATING AND VENTILATING INDUSTRY

3.5 Differentials or extra payments in respect of supervisory responsibility are excluded from the annual prime cost (see Section 6). The time of principals, staff, foremen, chargehands and the like when working manually is admissible under this Section at the rates for the appropriate grades.

SECTION 4: MATERIALS AND GOODS

4.1 The prime cost of materials and goods obtained specifically for the daywork is the invoice cost after deducting all trade discounts and any portion of cash discounts in excess of 5%.

4.2 The prime cost of all other materials and goods used in the daywork is based upon the current market prices plus any appropriate handling charges.

4.3 The prime cost referred to in 4.1 and 4.2 includes the cost of delivery to site.

4.4 Any Value Added Tax which is treated, or is capable of being treated, as input tax (as defined by the Finance Act 1972, or any re-enactment or amendment thereof or substitution therefore) by the contractor is excluded.

SECTION 5: PLANT

5.1 Unless otherwise stated in the contract, the prime cost of plant comprises the cost of the following:

    (a) use or hire of mechanically-operated plant and transport for the time employed on and/or provided or retained for the daywork;

    (b) use of non-mechanical plant (excluding non-mechanical hand tools) for the time employed on and/or provided or retained for the daywork;

    (c) transport to and from the site and erection and dismantling where applicable.

5.2 The use of non-mechanical hand tools and of erected scaffolding, staging, trestles or the like is excluded (see Section 6), unless specifically retained for the daywork.

SECTION 6: INCIDENTAL COSTS, OVERHEADS AND PROFIT

6.1 The percentage adjustments provided in the contract which are applicable to each of the totals of Sections 3, 4 and 5 comprise the following:

    (a) Head office charges.

    (b) Site staff including site supervision.

    (c) The additional cost of overtime (other than that referred to in 6.2).

    (d) Time lost due to inclement weather.

    (e) The additional cost of bonuses and all other incentive payments in excess of any included in 3.4.

    (f) Apprentices' study time.

    (g) Fares and travelling allowances.

    (h) Country, lodging and periodic allowances.

    (i) Sick pay or insurances in respect thereof, other than as included in 3.4.

    (j) Third party and employers' liability insurance.

    (k) Liability in respect of redundancy payments to employees.

    (l) Employer's National Insurance contributions not included in 3.4.

    (m) Use and maintenance of non-mechanical hand tools.

    (n) Use of erected scaffolding, staging, trestles or the like (but see 5.2).

    (o) Use of tarpaulins, protective clothing, artificial lighting, safety and welfare facilities, storage and the like that may be available on site.

    (p) Any variation to basic rates required by the contractor in cases where the contract provides for the use of a specified schedule of basic plant charges (to the extent that no other provision is made for such variation – see 5.1).

    (q) In the case of a subcontract which provides that the subcontractor shall allow a cash discount, such provision as is necessary for the allowance of the prescribed rate of discount.

    (r) All other liabilities and obligations whatsoever not specifically referred to in this Section nor chargeable under any other Section.

    (s) Profit.

6.2 The additional cost of overtime where specifically ordered by the Architect/Supervising Officer shall only be chargeable in the terms of a prior written agreement between the parties.

HEATING AND VENTILATING INDUSTRY

MECHANICAL INSTALLATIONS
Calculation of Hourly Base Rate of Labour for Typical Main Grades applicable from 1 October 2018, refer to notes within Section Three – Rates of Wages.

	Foreman	Senior Craftsman (+ 2nd Welding Skill)	Senior Craftsman	Craftsman	Installer	Mate over 18
Hourly Rate from 1 October 2018	17.15	14.73	14.17	13.05	11.77	9.93
Annual standard earnings excluding all holidays, 45.6 weeks × 38 hours	29,455.13	25,298.78	24,336.98	22,413.38	20,214.98	17,054.78
Employers national insurance contributions for this year	3,559.07	2,928.58	2,757.15	2,455.75	2,111.31	1,616.16
Weekly holiday credit and welfare contributions (52 weeks) for this year	4,652.14	4,341.85	3,933.45	3,662.36	3,353.72	2,911.90
Annual prime cost of labour	37,513.40	32,297.70	30,900.48	28,415.03	25,574.64	21,491.50
Hourly base rate	21.84	18.81	17.99	16.54	14.89	12.51

Notes:

(1) Annual industry holiday (4.6 weeks × 37.5 hours) and public holidays (1.6 weeks × 37.5 hours) are paid through weekly holiday credit and welfare stamp scheme.
(2) Where applicable, Merit money and other variables (e.g. daily abnormal conditions money), which attract Employer's National Insurance contribution, should be included.
(3) Contractors in Northern Ireland should add the appropriate amount of CITB Levy to the annual prime cost of labour prior to calculating the hourly base rate.
(4) Hourly rate based on 1717.5 hours per annum and calculated as follows

52 Weeks @ 37.5 hrs/wk	=		1,950.00
Less			
Sickness = 1 week @ 37.5 hrs/wk	=	37.5	
Public Holiday = 9/5 = 1.6 weeks @ 37.5 hrs/wk	=	60	
Annual holidays = 4.6 weeks @ 3.75 hrs/wk	=	172.5	232.50
Hours	=		1,717.50

(5) For calculation of Holiday Credits and ENI refer to detailed labour rate evaluation.
(6) National Insurance contributions are those effective from 6 April 2018.
(7) Weekly holiday credit/welfare stamp values are those assumed and effective from 1 October 2018.
(8) Hourly rates of wages are those assumed and effective from 1 October 2018.

ELECTRICAL INDUSTRY

DEFINITION OF PRIME COST OF DAYWORK CARRIED OUT UNDER AN ELECTRICAL CONTRACT (MARCH 1981 EDITION)

This Definition of Prime Cost is published by The Royal Institution of Chartered Surveyors and The Electrical Contractors' Associations for convenience and for use by people who choose to use it. Members of The Electrical Contractors' Association are not in any way debarred from defining Prime Cost and rendering accounts for work carried out on that basis in any way they choose. Building owners are advised to reach agreement with contractors on the Definition of Prime Cost to be used prior to entering into a contract or subcontract.

SECTION 1: APPLICATION

1.1   This Definition provides a basis for the valuation of daywork executed under such electrical contracts as provide for its use.
1.2   It is not applicable in any other circumstances, such as jobbing, or other work carried out as a separate or main contract, nor in the case of daywork executed after the date of practical completion.
1.3   The terms 'contract' and 'contractor' herein shall be read as 'subcontract' and 'subcontractor' as the context may require.

SECTION 2: COMPOSITION OF TOTAL CHARGES

2.1   The Prime Cost of daywork comprises the sum of the following costs:
       (a)   Labour as defined in Section 3.
       (b)   Materials and goods as defined in Section 4.
       (c)   Plant as defined in Section 5.
2.2   Incidental costs, overheads and profit as defined in Section 6, as provided in the contract and expressed therein as percentage adjustments, are applicable to each of 2.1 (a)-(c).

SECTION 3: LABOUR

3.1   The standard wage rates, emoluments and expenses referred to below and the standard working hours referred to in 3.2 are those laid down for the time being in the rules and determinations or decisions of the Joint Industry Board or the Scottish Joint Industry Board for the Electrical Contracting Industry (or those of such other body as may be appropriate) applicable to the works and relating to the grade of operative concerned at the time when and in the area where daywork is executed.
3.2   Hourly base rates for labour are computed by dividing the annual prime cost of labour, based upon the standard working hours and as defined in 3.4 by the number of standard working hours per annum. See examples.
3.3   The hourly rates computed in accordance with 3.2 shall be applied in respect of the time spent by operatives directly engaged on daywork, including those operating mechanical plant and transport and erecting and dismantling other plant (unless otherwise expressly provided in the contract) and handling and distributing the materials and goods used in the daywork.
3.4   The annual prime cost of labour comprises the following:
       (a)   Standard weekly earnings (i.e. the standard working week as determined at the appropriate rate for the operative concerned).
       (b)   Payments in respect of public holidays.
       (c)   Any amounts which may become payable by the Contractor to or in respect of operatives arising from operation of the rules etc. referred to in 3.1 which are not provided for in 3.4(a) and (b) nor in Section 6.
       (d)   Employer's National Insurance Contributions as applicable to 3.4 (a)-(c).
       (e)   Employer's contributions to the Joint Industry Board Combined Benefits Scheme or Scottish Joint Industry Board Holiday and Welfare Stamp Scheme, and holiday payments made to apprentices in compliance with the Joint Industry Board National Working Rules and Industrial Determinations as an employer.
       (f)   Any contribution, levy or tax imposed by Statute, payable by the Contractor in his capacity as an employer.

ELECTRICAL INDUSTRY

3.5 Differentials or extra payments in respect of supervisory responsibility are excluded from the annual prime cost (see Section 6). The time of principals and similar categories, when working manually, is admissible under this Section at the rates for the appropriate grades.

SECTION 4: MATERIALS AND GOODS

4.1 The prime cost of materials and goods obtained specifically for the daywork is the invoice cost after deducting all trade discounts and any portion of cash discounts in excess of 5%.
4.2 The prime cost of all other materials and goods used in the daywork is based upon the current market prices plus any appropriate handling charges.
4.3 The prime cost referred to in 4.1 and 4.2 includes the cost of delivery to site.
4.4 Any Value Added Tax which is treated, or is capable of being treated, as input tax (as defined by the Finance Act 1972, or any re-enactment or amendment thereof or substitution therefore) by the Contractor is excluded.

SECTION 5: PLANT

5.1 Unless otherwise stated in the contract, the prime cost of plant comprises the cost of the following:
    (a) Use or hire of mechanically-operated plant and transport for the time employed on and/or provided or retained for the daywork;
    (b) Use of non-mechanical plant (excluding non-mechanical hand tools) for the time employed on and/or provided or retained for the daywork;
    (c) Transport to and from the site and erection and dismantling where applicable.
5.2 The use of non-mechanical hand tools and of erected scaffolding, staging, trestles or the likes is excluded (see Section 6), unless specifically retained for daywork.
5.3 Note: Where hired or other plant is operated by the Electrical Contractor's operatives, such time is to be included under Section 3 unless otherwise provided in the contract.

SECTION 6: INCIDENTAL COSTS, OVERHEADS AND PROFIT

6.1 The percentage adjustments provided in the contract which are applicable to each of the totals of Sections 3, 4 and 5, compromise the following:
    (a) Head Office charges.
    (b) Site staff including site supervision.
    (c) The additional cost of overtime (other than that referred to in 6.2).
    (d) Time lost due to inclement weather.
    (e) The additional cost of bonuses and other incentive payments.
    (f) Apprentices' study time.
    (g) Travelling time and fares.
    (h) Country and lodging allowances.
    (i) Sick pay or insurance in lieu thereof, in respect of apprentices.
    (j) Third party and employers' liability insurance.
    (k) Liability in respect of redundancy payments to employees.
    (l) Employers' National Insurance Contributions not included in 3.4.
    (m) Use and maintenance of non-mechanical hand tools.
    (n) Use of erected scaffolding, staging, trestles or the like (but see 5.2.).
    (o) Use of tarpaulins, protective clothing, artificial lighting, safety and welfare facilities, storage and the like that may be available on site.
    (p) Any variation to basic rates required by the Contractor in cases where the contract provides for the use of a specified schedule of basic plant charges (to the extent that no other provision is made for such variation – see 5.1).
    (q) All other liabilities and obligations whatsoever not specifically referred to in this Section nor chargeable under any other Section.
    (r) Profit.

ELECTRICAL INDUSTRY

(s)   In the case of a subcontract which provides that the subcontractor shall allow a cash discount, such provision as is necessary for the allowance of the prescribed rate of discount.

6.2   The additional cost of overtime where specifically ordered by the Architect/Supervising Officer shall only be chargeable in the terms of a prior written agreement between the parties.

ELECTRICAL INSTALLATIONS

Calculation of Hourly Base Rate of Labour for Typical Main Grades applicable from 1 January 2018.

	Technician	Approved Electrician	Electrician	Labourer
Hourly Rate from 1 January 2018 (London Rates)	20.57	18.28	16.86	13.53
Annual standard earnings excluding all holidays, 45.8 weeks × 37.5 hours	35,483.25	31,533.00	29,083.50	23,339.25
Employers national insurance contributions from 6 April 2018	4,481.52	3,865.28	3,483.16	2,591.50
JIB Combined benefits from 1 January 2018	3,673.80	3,338.92	3,124.16	2,642.64
Holiday top up funding	1,742.25	1,561.88	1,457.14	1,221.65
Annual prime cost of labour	45,380.82	40,299.08	37,147.96	29,795.04
Hourly base rate	26.31	23.36	21.54	17.27

Notes:

(1)   Annual industry holiday (4.4 weeks × 37.5 hours) and public holidays (1.6 weeks × 37.5 hours)
(2)   It should be noted that all labour costs incurred by the Contractor in his capacity as an Employer, other than those contained in the hourly rate above, must be taken into account under Section 6.
(3)   Public Holidays are paid through weekly holiday credit and welfare stamp scheme.
(4)   Contractors in Northern Ireland should add the appropriate amount of CITB Levy to the annual prime cost of labour prior to calculating the hourly base rate.
(5)   Hourly rate based on 1,725 hours per annum and calculated as follows:

52 Weeks @ 37.5 hrs/wk	=		1,950.00
Less			
Public Holiday = 9/5 = 1.6 weeks @ 37.5 hrs/wk	=	60.00	
Annual holidays = 4.4 weeks @ 37.5 hrs/wk	=	165.00	225
Hours	=		1,725.00

(6)   For calculation of holiday credits and ENI refer to detailed labour rate evaluation.
(7)   Hourly wage rates are those effective from 1 January 2018.
(8)   National Insurance contributions are those effective from 6 April 2018.

BUILDING INDUSTRY PLANT HIRE COSTS

SCHEDULE OF BASIC PLANT CHARGES (JULY 2010)

This Schedule is published by the Royal Institution of Chartered Surveyors and is for use in connection with Day-works under a Building Contract.

EXPLANATORY NOTES

1   The rates in the Schedule are intended to apply solely to daywork carried out under and incidental to a Building Contract. They are NOT intended to apply to:
     (i)   jobbing or any other work carried out as a main or separate contract; or
     (ii)   work carried out after the date of commencement of the Defects Liability Period.
2   The rates apply to plant and machinery already on site, whether hired or owned by the Contractor.
3   The rates, unless otherwise stated, include the cost of fuel and power of every description, lubricating oils, grease, maintenance, sharpening of tools, replacement of spare parts, all consumable stores and for licen-ces and insurances applicable to items of plant.
4   The rates, unless otherwise stated, do not include the costs of drivers and attendants (unless otherwise stated).
5   The rates in the Schedule are base costs and may be subject to an overall adjustment for price movement, overheads and profit, quoted by the Contractor prior to the placing of the Contract.
6   The rates should be applied to the time during which the plant is actually engaged in daywork.
7   Whether or not plant is chargeable on daywork depends on the daywork agreement in use and the inclusion of an item of plant in this schedule does not necessarily indicate that item is chargeable.
8   Rates for plant not included in the Schedule or which is not already on site and is specifically provided or hired for daywork shall be settled at prices which are reasonably related to the rates in the Schedule having regard to any overall adjustment quoted by the Contractor in the Conditions of Contract.

BUILDING INDUSTRY PLANT HIRE COSTS

MECHANICAL PLANT AND TOOLS

Item of plant	Size/Rating	Unit	Rate per Hour (£)
PUMPS			
Mobile Pumps			
Including pump hoses, values and strainers, etc.			
Diaphragm	50 mm dia.	Each	1.17
Diaphragm	76 mm dia.	Each	1.89
Diaphragm	102 mm dia.	Each	3.54
Submersible	50 mm dia.	Each	0.76
Submersible	76 mm dia.	Each	0.86
Submersible	102 mm dia.	Each	1.03
Induced Flow	50 mm dia.	Each	0.77
Induced Flow	76 mm dia.	Each	1.67
Centrifugal, self-priming	25 mm dia.	Each	1.30
Centrifugal, self-priming	50 mm dia.	Each	1.92
Centrifugal, self-priming	75 mm dia.	Each	2.74
Centrifugal, self-priming	102 mm dia.	Each	3.35
Centrifugal, self-priming	152 mm dia.	Each	4.27
SCAFFOLDING, SHORING, FENCING			
Complete Scaffolding			
Mobile working towers, single width	2.0 m × 0.72 m base × 7.45 m high	Each	3.36
Mobile working towers, single width	2.0 m × 0.72 m base × 8.84 m high	Each	3.79
Mobile working towers, double width	2.0 m × 1.35 m × 7.45 m high	Each	3.79
Mobile working towers, double width	2.0 m × 1.35 m × 15.8 m high	Each	7.13
Chimney scaffold, single unit		Each	1.92
Chimney scaffold, twin unit		Each	3.59
Push along access platform	1.63–3.1 m	Each	5.00
Push along access platform	1.80 m × 0.70 m	Each	1.79
Trestles			
Trestle, adjustable	Any height	Pair	0.41
Trestle, painters	1.8 m high	Pair	0.31
Trestle, painters	2.4 m high	Pair	0.36
Shoring, Planking and Strutting			
'Acrow' adjustable prop	Sizes up to 4.9 m (open)	Each	0.06
'Strong boy' support attachment		Each	0.22
Adjustable trench strut	Sizes up to 1.67 m (open)	Each	0.16

BUILDING INDUSTRY PLANT HIRE COSTS

Item of plant	Size/Rating	Unit	Rate per Hour (£)
Trench sheet		Metre	0.03
Backhoe trench box	Base unit	Each	1.23
Backhoe trench box	Top unit	Each	0.87
Temporary Fencing			
Including block and coupler			
Site fencing steel grid panel	3.5 m × 2.0 m	Each	0.05
Anti-climb site steel grid fence panel	3.5 m × 2.0 m	Each	0.08
Solid panel Heras	2.0 m × 2.0 m	Each	0.09
Pedestrian gate		Each	0.36
Roadway gate		Each	0.60

LIFTING APPLIANCES AND CONVEYORS

Cranes

Mobile Cranes

Rates are inclusive of drivers

Lorry mounted, telescopic jib

Item of plant	Size/Rating	Unit	Rate per Hour (£)
Two wheel drive	5 tonnes	Each	19.00
Two wheel drive	8 tonnes	Each	42.00
Two wheel drive	10 tonnes	Each	50.00
Two wheel drive	12 tonnes	Each	77.00
Two wheel drive	20 tonnes	Each	89.69
Four wheel drive	18 tonnes	Each	46.51
Four wheel drive	25 tonnes	Each	35.90
Four wheel drive	30 tonnes	Each	38.46
Four wheel drive	45 tonnes	Each	46.15
Four wheel drive	50 tonnes	Each	53.85
Four wheel drive	60 tonnes	Each	61.54
Four wheel drive	70 tonnes	Each	71.79

Static tower crane

Rates inclusive of driver

Note: Capacity equals maximum lift in tonnes times maximum radius at which it can be lifted

	Capacity (Metre/tonnes) Up to	Height under hook above ground (m) Up to	Unit	Rate per Hour (£)
Tower crane	30	22	Each	22.23
Tower crane	40	22	Each	26.62
Tower crane	40	30	Each	33.33

BUILDING INDUSTRY PLANT HIRE COSTS

Item of plant	Size/Rating		Unit	Rate per Hour (£)
Tower crane	50	22	Each	29.16
Tower crane	60	22	Each	35.90
Tower crane	60	36	Each	35.90
Tower crane	70	22	Each	41.03
Tower crane	80	22	Each	39.12
Tower crane	90	42	Each	37.18
Tower crane	110	36	Each	47.62
Tower crane	140	36	Each	55.77
Tower crane	170	36	Each	64.11
Tower crane	200	36	Each	71.95
Tower crane	250	36	Each	84.77
Tower crane with luffing jig	30	25	Each	22.23
Tower crane with luffing jig	40	30	Each	26.62
Tower crane with luffing jig	50	30	Each	29.16
Tower crane with luffing jig	60	36	Each	41.03
Tower crane with luffing jig	65	30	Each	33.13
Tower crane with luffing jig	80	22	Each	48.72
Tower crane with luffing jig	100	45	Each	48.72
Tower crane with luffing jig	125	30	Each	53.85
Tower crane with luffing jig	160	50	Each	53.85
Tower crane with luffing jig	200	50	Each	74.36
Tower crane with luffing jig	300	60	Each	100.00
Crane Equipment				
Muck tipping skip	Up to 200 litres		Each	0.67
Muck tipping skip	500 litres		Each	0.82
Muck tipping skip	750 litres		Each	1.08
Muck tipping skip	1000 litres		Each	1.28
Muck tipping skip	1500 litres		Each	1.41
Muck tipping skip	2000 litres		Each	1.67
Mortar skip	250 litres, plastic		Each	0.41
Mortar skip	350 litres steel		Each	0.77
Boat skip	250 litres		Each	0.92
Boat skip	500 litres		Each	1.08
Boat skip	750 litres		Each	1.23
Boat skip	1000 litres		Each	1.38
Boat skip	1500 litres		Each	1.64
Boat skip	2000 litres		Each	1.90

BUILDING INDUSTRY PLANT HIRE COSTS

Item of plant	Size/Rating	Unit	Rate per Hour (£)
Boat skip	3000 litres	Each	2.82
Boat skip	4000 litres	Each	3.23
Master flow skip	250 litres	Each	0.77
Master flow skip	500 litres	Each	1.03
Master flow skip	750 litres	Each	1.28
Master flow skip	1000 litres	Each	1.44
Master flow skip	1500 litres	Each	1.69
Master flow skip	2000 litres	Each	1.85
Grand master flow skip	500 litres	Each	1.28
Grand master flow skip	750 litres	Each	1.64
Grand master flow skip	1000 litres	Each	1.69
Grand master flow skip	1500 litres	Each	1.95
Grand master flow skip	2000 litres	Each	2.21
Cone flow skip	500 litres	Each	1.33
Cone flow skip	1000 litres	Each	1.69
Geared rollover skip	500 litres	Each	1.28
Geared rollover skip	750 litres	Each	1.64
Geared rollover skip	1000 litres	Each	1.69
Geared rollover skip	1500 litres	Each	1.95
Geared rollover skip	2000 litres	Each	2.21
Multi skip, rope operated	200 mm outlet size, 500 litres	Each	1.49
Multi skip, rope operated	200 mm outlet size, 750 litres	Each	1.64
Multi skip, rope operated	200 mm outlet size, 1000 litres	Each	1.74
Multi skip, rope operated	200 mm outlet size, 1500 litres	Each	2.00
Multi skip, rope operated	200 mm outlet size, 2000 litres	Each	2.26
Multi skip, man riding	200 mm outlet size, 1000 litres	Each	2.00
Multi skip	4 point lifting frame	Each	0.90
Multi skip	Chain brothers	Set	0.87
Crane Accessories			
Multi-purpose crane forks	1.5 and 2 tonnes S.W.L.	Each	1.13
Self-levelling crane forks		Each	1.28
Man cage	1 man, 230 kg S.W.L.	Each	1.90
Man cage	2 man, 500 kg S.W.L.	Each	1.95
Man cage	4 man, 750 kg S.W.L.	Each	2.15
Man cage	8 man, 1000 kg S.W.L.	Each	3.33
Stretcher cage	500 kg, S.W.L.	Each	2.69
Goods carrying cage	1500 kg, S.W.L.	Each	1.33
Goods carrying cage	3000 kg, S.W.L.	Each	1.85

BUILDING INDUSTRY PLANT HIRE COSTS

Item of plant	Size/Rating		Unit	Rate per Hour (£)
Builders' skip lifting cradle	12 tonnes, S.W.L.		Each	2.31
Board/pallet fork	1600 kg, S.W.L.		Each	1.90
Gas bottle carrier	500 kg, S.W.L.		Each	0.92
Hoists				
Scaffold hoist	200 kg		Each	2.46
Rack and pinion (goods only)	500 kg		Each	4.56
Rack and pinion (goods only)	1100 kg		Each	5.90
Rack and pinion (goods and passenger)	8 person, 80 kg		Each	7.44
Rack and pinion (goods and passenger)	14 person, 1400 kg		Each	8.72
Wheelbarrow chain sling			Each	1.67
Conveyors				
Belt conveyors				
Conveyor	8 m long × 450 mm wide		Each	5.90
Miniveyor, control box and loading hopper	3 m unit		Each	4.49
Other Conveying Equipment				
Wheelbarrow			Each	0.62
Hydraulic superlift			Each	4.56
Pavac slab lifter (tile hoist)			Each	4.49
High lift pallet truck			Each	3.08
Lifting Trucks	Payload	Maximum Lift		
Fork lift, two wheel drive	1100 kg	up to 3.0 m	Each	5.64
Fork lift, two wheel drive	2540 kg	up to 3.7 m	Each	5.64
Fork lift, four wheel drive	1524 kg	up to 6.0 m	Each	5.64
Fork lift, four wheel drive	2600 kg	up to 5.4 m	Each	7.44
Fork life, four wheel drive	4000 kg	up to 17 m	Each	10.77
Lifting Platforms				
Hydraulic platform (Cherry picker)	9 m		Each	4.62
Hydraulic platform (Cherry picker)	12 m		Each	7.56
Hydraulic platform (Cherry picker)	15 m		Each	10.13
Hydraulic platform (Cherry picker)	17 m		Each	15.63
Hydraulic platform (Cherry picker)	20 m		Each	18.13
Hydraulic platform (Cherry picker)	25.6 m		Each	32.38
Scissor lift	7.6 m, electric		Each	3.85
Scissor lift	7.8 m, electric		Each	5.13
Scissor lift	9.7 m, electric		Each	4.23
Scissor lift	10 m, diesel		Each	6.41
Telescopic handler	7 m, 2 tonnes		Each	5.13

BUILDING INDUSTRY PLANT HIRE COSTS

Item of plant	Size/Rating	Unit	Rate per Hour (£)
Telescopic handler	13 m, 3 tonnes	Each	7.18
Lifting and Jacking Gear			
Pipe winch including gantry	1 tonne	Set	1.92
Pipe winch including gantry	3 tonnes	Set	3.21
Chain block	1 tonne	Each	0.35
Chain block	2 tonnes	Each	0.58
Chain block	5 tonnes	Each	1.14
Pull lift (Tirfor winch)	1 tonne	Each	0.64
Pull lift (Tirfor winch)	1.6 tonnes	Each	0.90
Pull lift (Tirfor winch)	3.2 tonnes	Each	1.15
Brother or chain slings, two legs	not exceeding 3.1 tonnes	Set	0.21
Brother or chain slings, two legs	not exceeding 4.25 tonnes	Set	0.31
Brother or chain slings, four legs	not exceeding 11.2 tonnes	Set	1.09
CONSTRUCTION VEHICLES			
Lorries			
Plated lorries (Rates are inclusive of driver)			
Platform lorry	7.5 tonnes	Each	16.21
Platform lorry	17 tonnes	Each	22.90
Platform lorry	24 tonnes	Each	30.68
Extra for lorry with crane attachment	up to 2.5 tonnes	Each	3.25
Extra for lorry with crane attachment	up to 5 tonnes	Each	6.00
Extra for lorry with crane attachment	up to 7.5 tonnes	Each	9.10
Tipper Lorries			
(Rates are inclusive of driver)			
Tipper lorry	up to 11 tonnes	Each	15.78
Tipper lorry	up to 17 tonnes	Each	23.95
Tipper lorry	up to 25 tonnes	Each	31.35
Tipper lorry	up to 31 tonnes	Each	37.79
Dumpers			
Site use only (excl. tax, insurance and extra cost of DERV etc. when operating on highway)	*Makers Capacity*		
Two wheel drive	1 tonne	Each	1.71
Four wheel drive	2 tonnes	Each	2.43
Four wheel drive	3 tonnes	Each	2.44

Daywork

BUILDING INDUSTRY PLANT HIRE COSTS

Item of plant	Size/Rating	Unit	Rate per Hour (£)
Four wheel drive	5 tonnes	Each	3.08
Four wheel drive	6 tonnes	Each	3.85
Four wheel drive	9 tonnes	Each	5.65
Tracked	0.5 tonnes	Each	3.33
Tracked	1.5 tonnes	Each	4.23
Tracked	3.0 tonnes	Each	8.33
Tracked	6.0 tonnes	Each	16.03
Dumper Trucks (*Rates are inclusive of drivers*)			
Dumper truck	up to 15 tonnes	Each	28.56
Dumper truck	up to 17 tonnes	Each	32.82
Dumper truck	up to 23 tonnes	Each	54.64
Dumper truck	up to 30 tonnes	Each	63.50
Dumper truck	up to 35 tonnes	Each	73.02
Dumper truck	up to 40 tonnes	Each	87.84
Dumper truck	up to 50 tonnes	Each	133.44
Tractors			
Agricultural Type			
Wheeled, rubber-clad tyred	up to 40 kW	Each	8.63
Wheeled, rubber-clad tyred	up to 90 kW	Each	25.31
Wheeled, rubber-clad tyred	up to 140 kW	Each	36.49
Crawler Tractors			
With bull or angle dozer	up to 70 kW	Each	29.38
With bull or angle dozer	up to 85 kW	Each	38.63
With bull or angle dozer	up to 100 kW	Each	52.59
With bull or angle dozer	up to 115 kW	Each	55.85
With bull or angle dozer	up to 135 kW	Each	60.43
With bull or angle dozer	up to 185 kW	Each	76.44
With bull or angle dozer	up to 200 kW	Each	96.43
With bull or angle dozer	up to 250 kW	Each	117.68
With bull or angle dozer	up to 350 kW	Each	160.03
With bull or angle dozer	up to 450 kW	Each	219.86
With loading shovel	0.8 m³	Each	26.92
With loading shovel	1.0 m³	Each	32.59
With loading shovel	1.2 m³	Each	37.53
With loading shovel	1.4 m³	Each	42.89
With loading shovel	1.8 m³	Each	52.22
With loading shovel	2.0 m³	Each	57.22

BUILDING INDUSTRY PLANT HIRE COSTS

Item of plant	Size/Rating	Unit	Rate per Hour (£)
With loading shovel	2.1 m³	Each	60.12
With loading shovel	3.5 m³	Each	87.26
Light vans			
VW Caddivan or the like		Each	5.26
VW Transport transit or the like	1.0 tonne	Each	6.03
Luton Box Van or the like	1.8 tonnes	Each	9.87
Water/Fuel Storage			
Mobile water container	110 litres	Each	0.62
Water bowser	1100 litres	Each	0.72
Water bowser	3000 litres	Each	0.87
Mobile fuel container	110 litres	Each	0.62
Fuel bowser	1100 litres	Each	1.23
Fuel bowser	3000 litres	Each	1.87
EXCAVATIONS AND LOADERS			
Excavators			
Wheeled, hydraulic	up to 11 tonnes	Each	25.86
Wheeled, hydraulic	up to 14 tonnes	Each	30.82
Wheeled, hydraulic	up to 16 tonnes	Each	34.50
Wheeled, hydraulic	up to 21 tonnes	Each	39.10
Wheeled, hydraulic	up to 25 tonnes	Each	43.81
Wheeled, hydraulic	up to 30 tonnes	Each	55.30
Crawler, hydraulic	up to 11 tonnes	Each	25.86
Crawler, hydraulic	up to 14 tonnes	Each	30.82
Crawler, hydraulic	up to 17 tonnes	Each	34.50
Crawler, hydraulic	up to 23 tonnes	Each	39.10
Crawler, hydraulic	up to 30 tonnes	Each	43.81
Crawler, hydraulic	up to 35 tonnes	Each	55.30
Crawler, hydraulic	up to 38 tonnes	Each	71.73
Crawler, hydraulic	up to 55 tonnes	Each	95.63
Mini excavator	1000/1500 kg	Each	4.87
Mini excavator	2150/2400 kg	Each	6.67
Mini excavator	2700/3500 kg	Each	7.31
Mini excavator	3500/4500 kg	Each	8.21
Mini excavator	4500/6000 kg	Each	9.23
Mini excavator	7000 kg	Each	14.10
Micro excavator	725 mm wide	Each	5.13

BUILDING INDUSTRY PLANT HIRE COSTS

Item of plant	Size/Rating	Unit	Rate per Hour (£)
Loaders			
Shovel loader	0.4 m³	Each	7.69
Shovel loader	1.57 m³	Each	8.97
Shovel loader, four wheel drive	1.7 m³	Each	4.83
Shovel loader, four wheel drive	2.3 m³	Each	4.38
Shovel loader, four wheel drive	3.3 m³	Each	5.06
Skid steer loader wheeled	300/400 kg payload	Each	7.31
Skid steer loader wheeled	625 kg payload	Each	7.67
Tracked skip loader	650 kg	Each	4.42
Excavator Loaders			
Wheeled tractor type with black-hoe excavator			
Four wheel drive			
Four wheel drive, 2 wheel steer	6 tonnes	Each	6.41
Four wheel drive, 2 wheel steer	8 tonnes	Each	8.59
Attachments			
Breakers for excavator		Each	8.72
Breakers for mini excavator		Each	1.75
Breakers for back-hoe excavator/loader		Each	5.13
COMPACTION EQUIPMENT			
Rollers			
Vibrating roller	368–420 kg	Each	1.43
Single roller	533 kg	Each	1.94
Single roller	750 kg	Each	3.43
Twin roller	up to 650 kg	Each	6.03
Twin roller	up to 950 kg	Each	6.62
Twin roller with seat end steering wheel	up to 1400 kg	Each	7.68
Twin roller with seat end steering wheel	up to 2500 kg	Each	10.61
Pavement roller	3–4 tonnes dead weight	Each	6.00
Pavement roller	4–6 tonnes	Each	6.86
Pavement roller	6–10 tonnes	Each	7.17
Pavement roller	10–13 tonnes	Each	19.86
Rammers			
Tamper rammer 2 stroke-petrol	225–275 mm	Each	1.52

BUILDING INDUSTRY PLANT HIRE COSTS

Item of plant	Size/Rating	Unit	Rate per Hour (£)
Soil Compactors			
Plate compactor	75 mm– 400 mm	Each	1.53
Plate compactor rubber pad	375 mm–1400 mm	Each	1.53
Plate compactor reversible plate-petrol	400 mm	Each	2.44
CONCRETE EQUIPEMENT			
Concrete/Mortar Mixers			
Open drum without hopper	0.09/0.06 m³	Each	0.61
Open drum without hopper	0.12/0.09 m³	Each	1.22
Open drum without hopper	0.15/0.10 m³	Each	0.72
Concrete/Mortar Transport Equipment			
Concrete pump incl. hose, valve and couplers			
Lorry mounted concrete pump	24 m max. distance	Each	50.00
Lorry mounted concrete pump	34 m max. distance	Each	66.00
Lorry mounted concrete pump	42 m max. distance	Each	91.50
Concrete Equipment			
Vibrator, poker, petrol type	up to 75 mm dia.	Each	0.69
Air vibrator (*excluding compressor and hose*)	up to 75 mm dia.	Each	0.64
Extra poker heads	25/36/60 mm dia.	Each	0.76
Vibrating screed unit with beam	5.00 m	Each	2.48
Vibrating screed unit with adjustable beam	3.00–5.00 m	Each	3.54
Power float	725–900 mm	Each	2.56
Power float finishing pan		Each	0.62
Floor grinder	660 × 1016 mm, 110 V electric	Each	4.31
Floor plane	450 × 1100 mm	Each	4.31
TESTING EQUIPMENT			
Pipe Testing Equipment			
Pressure testing pump, electric		Set	2.19
Pressure test pump		Set	0.80

SITE ACCOMODATION AND TEMPORARY SERVICES			Rate per Hour (£)
Heating equipment			
Space heater – propane	80,000 Btu/hr	Each	1.03

BUILDING INDUSTRY PLANT HIRE COSTS

Item of plant	Size/Rating	Unit	Rate per Hour (£)
Space heater – propane/electric	125,000 Btu/hr	Each	2.09
Space heater – propane/electric	250,000 Btu/hr	Each	2.33
Space heater – propane	125,000 Btu/hr	Each	1.54
Space heater – propane	260,000 Btu/hr	Each	1.88
Cabinet heater		Each	0.82
Cabinet heater, catalytic		Each	0.57
Electric halogen heater		Each	1.27
Ceramic heater	3 kW	Each	0.99
Fan heater	3 kW	Each	0.66
Cooling fan		Each	1.92
Mobile cooling unit, small		Each	3.60
Mobile cooling unit, large		Each	4.98
Air conditioning unit		Each	2.81
Site Lighting and Equipment			
Tripod floodlight	500 W	Each	0.48
Tripod floodlight	1000 W	Each	0.62
Towable floodlight	4 × 100 W	Each	3.85
Hand held floodlight	500 W	Each	0.51
Rechargeable light		Each	0.41
Inspection light		Each	0.37
Plasterer's light		Each	0.65
Lighting mast		Each	2.87
Festoon light string	25 m	Each	0.55
Site Electrical Equipment			
Extension leads	240 V/14 m	Each	0.26
Extension leads	110 V/14 m	Each	0.36
Cable reel	25 m 110V/240 V	Each	0.46
Cable reel	50 m 110V240 V	Each	0.88
4 way junction box	110 V	Each	0.56
Power Generating Units			
Generator – petrol	2 kVA	Each	1.23
Generator – silenced petrol	2 kVA	Each	2.87
Generator – petrol	3 kVA	Each	1.47
Generator – diesel	5 kVA	Each	2.44
Generator – silenced diesel	10 kVA	Each	1.90
Generator – silenced diesel	15 kVA	Each	2.26
Generator – silenced diesel	30 kVA	Each	3.33
Generator – silenced diesel	50 kVA	Each	4.10

BUILDING INDUSTRY PLANT HIRE COSTS

Item of plant	Size/Rating	Unit	Rate per Hour (£)
Generator – silenced diesel	75 kVA	Each	4.62
Generator – silenced diesel	100 kVA	Each	5.64
Generator – silenced diesel	150 kVA	Each	7.18
Generator – silenced diesel	200 kVA	Each	9.74
Generator – silenced diesel	250 kVA	Each	11.28
Generator – silenced diesel	350 kVA	Each	14.36
Generator – silenced diesel	500 kVA	Each	15.38
Tail adaptor	240 V	Each	0.10
Transformers			
Transformer	3 kVA	Each	0.32
Transformer	5 kVA	Each	1.23
Transformer	7.5 kVA	Each	0.59
Transformer	10 kVA	Each	2.00
Rubbish Collection and Disposal Equipment			
Rubbish Chutes			
Standard plastic module	1 m section	Each	0.15
Steel liner insert		Each	0.30
Steel top hopper		Each	0.22
Plastic side entry hopper		Each	0.22
Plastic side entry hopper liner		Each	0.22
Dust Extraction Plant			
Dust extraction unit, light duty		Each	2.97
Dust extraction unit, heavy duty		Each	2.97
SITE EQUIPMENT – Welding Equipment			
Arc-(Electric) Complete With Leads			
Welder generator – petrol	200 amp	Each	3.53
Welder generator – diesel	300/350 amp	Each	3.78
Welder generator – diesel	4000 amp	Each	7.92
Extra welding lead sets		Each	0.69
Gas-Oxy Welder			
Welding and cutting set (including oxygen and acetylene, excluding underwater equipment and thermic boring)			
Small		Each	2.24
Large		Each	3.75

BUILDING INDUSTRY PLANT HIRE COSTS

Item of plant	Size/Rating	Unit	Rate per Hour (£)
Lead burning gun		Each	0.50
Mig welder		Each	1.38
Fume extractor		Each	2.46
Road Works Equipment			
Traffic lights, mains/generator	2-way	Set	10.94
Traffic lights, mains/generator	3-way	Set	11.56
Traffic lights, mains/generator	4-way	Set	12.19
Flashing light		Each	0.10
Road safety cone	450 mm	Each	0.08
Safety cone	750 mm	Each	0.10
Safety barrier plank	1.25 m	Each	0.13
Safety barrier plank	2 m	Each	0.15
Safety barrier plank post		Each	0.13
Safety barrier plank post base		Each	0.10
Safety four gate barrier	1 m each gate	Set	0.77
Guard barrier	2 m	Each	0.19
Road sign	750 mm	Each	0.23
Road sign	900 mm	Each	0.31
Road sign	1200 mm	Each	0.42
Speed ramp/cable protection	500 mm section	Each	0.14
Hose ramp open top	3 m section	Each	0.07
DPC Equipment			
Damp-proofing injection machine		Each	2.56
Cleaning Equipment			
Vacuum cleaner (industrial wet) single motor		Each	1.08
Vacuum cleaner (industrial wet) twin motor	30 litre capacity	Each	1.79
Vacuum cleaner (industrial wet) twin motor	70 litre capacity	Each	2.21
Steam cleaner	Diesel/electric 1 phase	Each	3.33
Steam cleaner	Diesel/electric 3 phase	Each	3.85
Pressure washer, light duty electric	1450 PSI	Each	0.72
Pressure washer, heavy duty, diesel	2500 PSI	Each	1.33
Pressure washer, heavy duty, diesel	4000 PSI	Each	2.18
Cold pressure washer, electric		Each	2.39
Hot pressure washer, petrol		Each	4.19
Hot pressure washer, electric		Each	5.13

BUILDING INDUSTRY PLANT HIRE COSTS

Item of plant	Size/Rating	Unit	Rate per Hour (£)
Cold pressure washer, petrol		Each	2.92
Sandblast attachment to last washer		Each	1.23
Drain cleaning attachment to last washer		Each	1.03
Surface Preparation Equipment			
Rotavator	5 h.p.	Each	2.46
Rotavator	9 h.p.	Each	5.00
Scabbler, up to three heads		Each	1.53
Scabbler, pole		Each	2.68
Scrabbler, multi-headed floor		Each	3.89
Floor preparation machine		Each	1.05
Compressors and Equipment			
Portable Compressors			
Compressor – electric	4 cfm	Each	1.36
Compressor – electric	8 cfm lightweight	Each	1.31
Compressor – electric	8 cfm	Each	1.36
Compressor – electric	14 cfm	Each	1.56
Compressor – petrol	24 cfm	Each	2.15
Compressor – electric	25 cfm	Each	2.10
Compressor – electric	30 cfm	Each	2.36
Compressor – diesel	100 cfm	Each	2.56
Compressor – diesel	250 cfm	Each	5.54
Compressor – diesel	400 cfm	Each	8.72
Mobile Compressors			
Lorry mounted compressor			
(machine plus lorry only)	up to 3 m³	Each	41.47
(machine plus lorry only)	up to 5 m³	Each	48.94
Tractor mounted compressor			
(machine plus rubber tyred tractor)	Up to 4 m³	Each	21.03
Accessories (Pneumatic Tools) *(with and including up to 15 m of air hose)*			
Demolition pick, medium duty		Each	0.90
Demolition pick, heavy duty		Each	1.03
Breakers (with six steels) light	up to 150 kg	Each	1.19
Breakers (with six steels) medium	295 kg	Each	1.24
Breakers (with six steels) heavy	386 kg	Each	1.44
Rock drill (for use with compressor) hand held		Each	1.18
Additional hoses	15 m	Each	0.09

BUILDING INDUSTRY PLANT HIRE COSTS

Item of plant	Size/Rating	Unit	Rate per Hour (£)
Breakers			
Demolition hammer drill, heavy duty, electric		Each	1.54
Road breaker, electric		Each	2.41
Road breaker, 2 stroke, petrol		Each	4.06
Hydraulic breaker unit, light duty, petrol		Each	3.06
Hydraulic breaker unit, heavy duty, petrol		Each	3.46
Hydraulic breaker unit, heavy duty, diesel		Each	4.62
Quarrying and Tooling Equipment			
Block and stone splitter, hydraulic	600 mm × 600 mm	Each	1.90
Block and stone splitter, manual		Each	1.64
Steel Reinforcement Equipment			
Bar bending machine – manual	up to 13 mm dia. rods	Each	1.03
Bar bending machine – manual	up to 20 mm dia. rods	Each	1.41
Bar shearing machine – electric	up to 38 mm dia. rods	Each	3.08
Bar shearing machine – electric	up to 40 mm dia. rods	Each	4.62
Bar cropper machine – electric	up to 13 mm dia. rods	Each	2.05
Bar cropper machine – electric	up to 20 mm dia. rods	Each	2.56
Bar cropper machine – electric	up to 40 mm dia. rods	Each	4.62
Bar cropper machine – 3 phase	up to 40 mm dia. rods	Each	4.62
Dehumidifiers			
110/240 v Water	68 litre extraction per 24 hours	Each	2.46
110/240 v Water	90 litre extraction per 24 hours	Each	3.38
SMALL TOOLS			
Saws			
Masonry bench saw	350 mm–500 mm dia.	Each	1.13
Floor saw	125 mm max. cut	Each	1.15
Floor saw	150 mm max. cut	Each	3.83
Floor saw, reversible	350 mm max. cut	Each	3.32
Wall saw, electric		Each	2.05
Chop/cut off saw, electric	350 mm dia.	Each	1.79
Circular saw, electric	230 mm dia.	Each	0.72
Tyrannosaw		Each	1.74
Reciprocating saw		Each	0.79
Door trimmer		Each	1.17
Stone saw	300 mm	Each	1.44

BUILDING INDUSTRY PLANT HIRE COSTS

Item of plant	Size/Rating	Unit	Rate per Hour (£)
Chainsaw, petrol	500 mm	Each	3.92
Full chainsaw safety kit		Each	0.41
Worktop jig		Each	1.08
Pipe Work Equipment			
Pipe bender	15 mm–22 mm	Each	0.92
Pipe bender, hydraulic	50 mm	Each	1.76
Pipe bender, electric	50 mm–150 mm dia.	Each	2.19
Pipe cutter, hydraulic		Each	0.46
Tripod pipe vice		Set	0.75
Ratchet threader	12 mm– 32 mm	Each	0.93
Pipe threading machine, electric	12 mm–75 mm	Each	3.07
Pipe threading machine, electric	12 mm–100 mm	Each	4.93
Impact wrench, electric		Each	1.33
Hand-held Drills and equipment			
Impact or hammer drill	up to 25 mm dia.	Each	1.03
Impact or hammer drill	35 mm dia.	Each	1.29
Dry diamond core cutter		Each	0.99
Angle head drill		Each	0.90
Stirrer, mixer drill		Each	1.13
Paint, Insulation Application Equipment			
Airless spray unit		Each	4.13
Portaspray unit		Each	1.16
HPVL turbine spray unit		Each	2.23
Compressor and spray gun		Each	1.91
Other Handtools			
Staple gun		Each	0.96
Air nail gun	110 V	Each	1.01
Cartridge hammer		Each	1.08
Tongue and groove nailer complete with mallet		Each	1.59
Diamond wall chasing machine		Each	2.63
Masonry chain saw	300 mm	Each	5.49
Floor grinder		Each	3.99
Floor plane		Each	1.79
Diamond concrete planer		Each	1.93
Autofeed screwdriver, electric		Each	1.38
Laminate trimmer		Each	0.91
Biscuit jointer		Each	1.49

BUILDING INDUSTRY PLANT HIRE COSTS

Item of plant	Size/Rating	Unit	Rate per Hour (£)
Random orbital sander		Each	0.97
Floor sander		Each	1.54
Palm, delta, flap or belt sander		Each	0.75
Disc cutter, electric	300 mm	Each	1.49
Disc cutter, 2 stroke petrol	300 mm	Each	1.24
Dust suppressor for petrol disc cutter		Each	0.51
Cutter cart for petrol disc cutter		Each	1.21
Grinder, angle or cutter	up to 225 mm	Each	0.50
Grinder, angle or cutter	300 mm	Each	1.41
Mortar raking tool attachment		Each	0.19
Floor/polisher scrubber	325 mm	Each	1.76
Floor tile stripper		Each	2.44
Wallpaper stripper, electric		Each	0.81
Hot air paint stripper		Each	0.50
Electric diamond tile cutter	all sizes	Each	2.42
Hand tile cutter		Each	0.82
Electric needle gun		Each	1.29
Needle chipping gun		Each	1.85
Pedestrian floor sweeper	250 mm dia.	Each	0.82
Pedestrian floor sweeper	Petrol	Each	2.20
Diamond tile saw		Each	1.84
Blow lamp equipment and glass		Set	0.50

SPON'S 2019 PRICEBOOKS from AECOM

Spon's Architects' and Builders' Price Book 2019

Editor: AECOM

Now with Semi and Automatic pedestrian doors -- Revolving, Sliding, and Swing doors; an expanded range of industrial shutter doors; industrial docks and shelters; and an expanded range of aluminium gutters.

Hbk & VitalSource® ebook 840pp approx.
978-1-138-61201-3 £160
VitalSource® ebook
978-0-429-46449-2 £160
(inc. sales tax where appropriate)

Spon's Civil Engineering and Highway Works Price Book 2019

Editor:AECOM

This year gives more emphasis to sampling from major projects of £20-£100m in value run under NEC contracts. Prices are based more strongly on global structural steel grades and rates and extrapolated accordingly. And prices are now given for pipe lining for maintenance drainage works.

Hbk & VitalSource® ebook 712pp approx.
978-1-138-61202-0 £180
VitalSource® ebook
978-0-429-46448-5 £180
(inc. sales tax where appropriate)

Spon's External Works and Landscape Price Book 2019

Editor:AECOM in association with LandPro Ltd
Landscape Surveyors

Now including Corten edgings, glass balustrades; decking pedestals for roof gardens; stone copings; and tier system cladding.

Hbk & VitalSource® ebook 672pp
approx. 978-1-138-61203-7 £150
VitalSource® ebook
978-0-429-46447 £150
(inc. sales tax where appropriate)

Spon's Mechanical and Electrical Services Price Book 2019

Editor:AECOM

This NRM edition includes an updated engineering features section and a section on smart building technology – along with new and significantly developed items: tuneable white luminaires; wireless lighting control; PV cells; and battery storage systems.

Hbk & VitalSource® ebook 872pp approx.
978-1-138-61205-1 £160
VitalSource® ebook
978-0-429-46446-1 £160
(inc. sales tax where appropriate)

> Receive our VitalSource® ebook free when you order any hard copy Spon 2019 Price Book
>
> Visit www.pricebooks.co.uk

To order:

Tel: 01235 400524

Post: Taylor & Francis Customer Services, Bookpoint Ltd, 200 Milton Park, Abingdon, Oxon, OX14 4SB, UK

Email: book.orders@tandf.co.uk

A complete listing of all our books is on www.crcpress.com

CRC Press
Taylor & Francis Group

It is one thing to imagine a better world.

It's another to deliver it.

Creating iconic structures. Planning entire communities. Connecting people with transport systems. Rescuing historic buildings. Our team of circa 900 cost and project managers collaborate to turn imagination into reality, providing innovative solutions to complex challenges across the globe.

Imagine it. Delivered.

New Aspects of Quantity Surveying Practice, 4th edition

Duncan Cartlidge

In this fourth edition of New Aspects of Quantity Surveying Practice, renowned quantity surveying author Duncan Cartlidge reviews the history of the quantity surveyor, examines and reflects on the state of current practice with a concentration on new and innovative practice, and attempts to predict the future direction of quantity surveying practice in the UK and worldwide.

The book champions the adaptability and flexibility of the quantity surveyor, whilst covering the hot topics which have emerged since the previous edition's publication, including:

*   the RICS 'Futures' publication;
*   Building Information Modelling (BIM);
*   mergers and acquisitions;
*   a more informed and critical evaluation of the NRM;
*   greater discussion of ethics to reflect on the renewed industry interest;
*   and a new chapter on Dispute Resolution.

As these issues create waves throughout the industry whilst it continues its global growth in emerging markets, such reflections on QS practice are now more important than ever. The book is essential reading for all Quantity Surveying students, teachers and professionals. It is particularly suited to undergraduate professional skills courses and non-cognate postgraduate students looking for an up to date understanding of the industry and the role.

December 2017: 234 x 156 mm: 282 pp
Pb: 978-1-138-67376-2 : £45.99

To Order: Tel: +44 (0) 1235 400524 Fax: +44 (0) 1235 400525
or Post: Taylor and Francis Customer Services,
Bookpoint Ltd, Unit T1, 200 Milton Park, Abingdon, Oxon, OX14 4TA UK
Email: book.orders@tandf.co.uk

For a complete listing of all our titles visit:
www.tandf.co.uk

Taylor & Francis
Taylor & Francis Group

PART 8

Tables and Memoranda

This part contains the following sections:

BIM and Quantity Surveying

S Pittard *et al.*

The sudden arrival of Building Information Modelling (BIM) as a key part of the building industry is redefining the roles and working practices of its stakeholders. Many clients, designers, contractors, quantity surveyors, and building managers are still finding their feet in an industry where BIM compliance can bring great rewards.

This guide is designed to help quantity surveying practitioners and students understand what BIM means for them, and how they should prepare to work successfully on BIM compliant projects. The case studies show how firms at the forefront of this technology have integrated core quantity surveying responsibilities like cost estimating, tendering, and development appraisal into high profile BIM projects. In addition to this, the implications for project management, facilities management, contract administration and dispute resolution are also explored through case studies, making this a highly valuable guide for those in a range of construction project management roles.

Featuring a chapter describing how the role of the quantity surveyor is likely to permanently shift as a result of this development, as well as descriptions of tools used, this covers both the organisational and practical aspects of a crucial topic.

December 2015: 234 x 156 mm: 258 pp
Pbk: 978-0-415-87043-6; £28.99

To Order: Tel: +44 (0) 1235 400524 Fax: +44 (0) 1235 400525
or Post: Taylor and Francis Customer Services,
Bookpoint Ltd, Unit T1, 200 Milton Park, Abingdon, Oxon, OX14 4TA UK
Email: book.orders@tandf.co.uk

For a complete listing of all our titles visit:
www.tandf.co.uk

Taylor & Francis
Taylor & Francis Group

CONVERSION TABLES

Length	Unit	Conversion factors			
Millimetre	mm	1 in	= 25.4 mm	1 mm	= 0.0394 in
Centimetre	cm	1 in	= 2.54 cm	1 cm	= 0.3937 in
Metre	m	1 ft	= 0.3048 m	1 m	= 3.2808 ft
		1 yd	= 0.9144 m		= 1.0936 yd
Kilometre	km	1 mile	= 1.6093 km	1 km	= 0.6214 mile

Note:

1 cm	= 10 mm	1 ft	= 12 in
1 m	= 1 000 mm	1 yd	= 3 ft
1 km	= 1 000 m	1 mile	= 1 760 yd

Area	Unit	Conversion factors			
Square Millimetre	mm^2	$1\ in^2$	$= 645.2\ mm^2$	$1\ mm^2$	$= 0.0016\ in^2$
Square Centimetre	cm^2	$1\ in^2$	$= 6.4516\ cm^2$	$1\ cm^2$	$= 1.1550\ in^2$
Square Metre	m^2	$1\ ft^2$	$= 0.0929\ m^2$	$1\ m^2$	$= 10.764\ ft^2$
		$1\ yd^2$	$= 0.8361\ m^2$	$1\ m^2$	$= 1.1960\ yd^2$
Square Kilometre	km^2	$1\ mile^2$	$= 2.590\ km^2$	$1\ km^2$	$= 0.3861\ mile^2$

Note:

$1\ cm^2$	$= 100\ mm^2$	$1\ ft^2$	$= 144\ in^2$
$1\ m^2$	$= 10\ 000\ cm^2$	$1\ yd^2$	$= 9\ ft^2$
$1\ km^2$	$= 100$ hectares	1 acre	$= 4\ 840\ yd^2$
		$1\ mile^2$	$= 640$ acres

Volume	Unit	Conversion factors			
Cubic Centimetre	cm^3	$1\ cm^3$	$= 0.0610\ in^3$	$1\ in^3$	$= 16.387\ cm^3$
Cubic Decimetre	dm^3	$1\ dm^3$	$= 0.0353\ ft^3$	$1\ ft^3$	$= 28.329\ dm^3$
Cubic Metre	m^3	$1\ m^3$	$= 35.3147\ ft^3$	$1\ ft^3$	$= 0.0283\ m^3$
		$1\ m^3$	$= 1.3080\ yd^3$	$1\ yd^3$	$= 0.7646\ m^3$
Litre	l	1 l	= 1.76 pint	1 pint	= 0.5683 l
			= 2.113 US pt		= 0.4733 US l

Note:

$1\ dm^3$	$= 1\ 000\ cm^3$	$1\ ft^3$	$= 1\ 728\ in^3$	1 pint	= 20 fl oz
$1\ m^3$	$= 1\ 000\ dm^3$	$1\ yd^3$	$= 27\ ft^3$	1 gal	= 8 pints
1 l	$= 1\ dm^3$				

Neither the Centimetre nor Decimetre are SI units, and as such their use, particularly that of the Decimetre, is not widespread outside educational circles.

Mass	Unit	Conversion factors			
Milligram	mg	1 mg	= 0.0154 grain	1 grain	= 64.935 mg
Gram	g	1 g	= 0.0353 oz	1 oz	= 28.35 g
Kilogram	kg	1 kg	= 2.2046 lb	1 lb	= 0.4536 kg
Tonne	t	1 t	= 0.9842 ton	1 ton	= 1.016 t

Note:

1 g	= 1000 mg	1 oz	= 437.5 grains	1 cwt	= 112 lb
1 kg	= 1000 g	1 lb	= 16 oz	1 ton	= 20 cwt
1 t	= 1000 kg	1 stone	= 14 lb		

Force	Unit	Conversion factors			
Newton	N	1 lbf	= 4.448 N	1 kgf	= 9.807 N
Kilonewton	kN	1 lbf	= 0.004448 kN	1 ton f	= 9.964 kN
Meganewton	MN	100 tonf	= 0.9964 MN		

Tables and Memoranda

CONVERSION TABLES

Pressure and stress	Unit	Conversion factors	
Kilonewton per square metre	kN/m^2	1 lbf/in^2	= 6.895 kN/m^2
		1 bar	= 100 kN/m^2
Meganewton per square metre	MN/m^2	1 tonf/ft^2	= 107.3 kN/m^2 = 0.1073 MN/m^2
		1 kgf/cm^2	= 98.07 kN/m^2
		1 lbf/ft^2	= 0.04788 kN/m^2

Coefficient of consolidation (Cv) or swelling	Unit	Conversion factors	
Square metre per year	m^2/year	1 cm^2/s	= 3 154 m^2/year
		1 ft^2/year	= 0.0929 m^2/year

Coefficient of permeability	Unit	Conversion factors	
Metre per second	m/s	1 cm/s	= 0.01 m/s
Metre per year	m/year	1 ft/year	= 0.3048 m/year
			= 0.9651 × (10)^8 m/s

Temperature	Unit	Conversion factors	
Degree Celsius	°C	°C = 5/9 × (°F − 32)	°F = (9 × °C)/ 5 + 32

SPEED CONVERSION

km/h	m/min	mph	fpm
1	16.7	0.6	54.7
2	33.3	1.2	109.4
3	50.0	1.9	164.0
4	66.7	2.5	218.7
5	83.3	3.1	273.4
6	100.0	3.7	328.1
7	116.7	4.3	382.8
8	133.3	5.0	437.4
9	150.0	5.6	492.1
10	166.7	6.2	546.8
11	183.3	6.8	601.5
12	200.0	7.5	656.2
13	216.7	8.1	710.8
14	233.3	8.7	765.5
15	250.0	9.3	820.2
16	266.7	9.9	874.9
17	283.3	10.6	929.6
18	300.0	11.2	984.3
19	316.7	11.8	1038.9
20	333.3	12.4	1093.6
21	350.0	13.0	1148.3
22	366.7	13.7	1203.0
23	383.3	14.3	1257.7
24	400.0	14.9	1312.3
25	416.7	15.5	1367.0
26	433.3	16.2	1421.7
27	450.0	16.8	1476.4
28	466.7	17.4	1531.1
29	483.3	18.0	1585.7
30	500.0	18.6	1640.4
31	516.7	19.3	1695.1
32	533.3	19.9	1749.8
33	550.0	20.5	1804.5
34	566.7	21.1	1859.1
35	583.3	21.7	1913.8
36	600.0	22.4	1968.5
37	616.7	23.0	2023.2
38	633.3	23.6	2077.9
39	650.0	24.2	2132.5
40	666.7	24.9	2187.2
41	683.3	25.5	2241.9
42	700.0	26.1	2296.6
43	716.7	26.7	2351.3
44	733.3	27.3	2405.9
45	750.0	28.0	2460.6

Tables and Memoranda

CONVERSION TABLES

km/h	m/min	mph	fpm
46	766.7	28.6	2515.3
47	783.3	29.2	2570.0
48	800.0	29.8	2624.7
49	816.7	30.4	2679.4
50	833.3	31.1	2734.0

GEOMETRY

Two dimensional figures

Figure	Diagram of figure	Surface area	Perimeter
Square		a^2	$4a$
Rectangle		ab	$2(a+b)$
Triangle		$\frac{1}{2}ch$	$a+b+c$
Circle		πr^2 $\frac{1}{4}\pi d^2$ where $2r = d$	$2\pi r$ πd
Parallelogram		ah	$2(a+b)$
Trapezium		$\frac{1}{2}h(a+b)$	$a+b+c+d$
Ellipse		Approximately πab	$\pi(a+b)$
Hexagon		$2.6 \times a^2$	

GEOMETRY

Figure	Diagram of figure	Surface area	Perimeter
Octagon		$4.83 \times a^2$	$6a$
Sector of a circle		$\frac{1}{2}rb$ or $\frac{q}{360}\pi r^2$ note $b = $ angle $\frac{q}{360} \times \pi 2r$	
Segment of a circle		$S - T$ where S = area of sector, T = area of triangle	
Bellmouth		$\frac{3}{14} \times r^2$	

GEOMETRY

Three dimensional figures

Figure	Diagram of figure	Surface area	Volume
Cube		$6a^2$	a^3
Cuboid/ rectangular block		$2(ab + ac + bc)$	abc
Prism/ triangular block		$bd + hc + dc + ad$	$\frac{1}{2}hcd$
Cylinder		$2\pi r^2 + 2\pi h$	$\pi r^2 h$
Sphere		$4\pi r^2$	$\frac{4}{3}\pi r^3$
Segment of sphere		$2\pi Rh$	$\frac{1}{6}\pi h(3r^2 + h^2)$ $\frac{1}{3}\pi h^2(3R - H)$
Pyramid		$(a + b)l + ab$	$\frac{1}{3}abh$

GEOMETRY

Figure	Diagram of figure	Surface area	Volume
Frustum of a pyramid		$l(a + b + c + d) + \sqrt{(ab + cd)}$ [rectangular figure only]	$\frac{h}{3}(ab + cd + \sqrt{abcd})$
Cone		πrl (excluding base) $\pi rl + \pi r^2$ (including base)	$\frac{1}{3}\pi r^2 h$ $\frac{1}{12}\pi d^2 h$
Frustum of a cone		$\pi r^2 + \pi R^2 + \pi l(R + r)$	$\frac{1}{3}\pi(R^2 + Rr + r^2)$

FORMULAE

Formulae

Formula	Description
Pythagoras Theorem	$A^2 = B^2 + C^2$ where A is the hypotenuse of a right-angled triangle and B and C are the two adjacent sides
Simpsons Rule	The Area is divided into an even number of strips of equal width, and therefore has an odd number of ordinates at the division points $$\text{area} = \frac{S(A + 2B + 4C)}{3}$$ where S = common interval (strip width) A = sum of first and last ordinates B = sum of remaining odd ordinates C = sum of the even ordinates The Volume can be calculated by the same formula, but by substituting the area of each coordinate rather than its length
Trapezoidal Rule	A given trench is divided into two equal sections, giving three ordinates, the first, the middle and the last $$\text{volume} = \frac{S \times (A + B + 2C)}{2}$$ where S = width of the strips A = area of the first section B = area of the last section C = area of the rest of the sections
Prismoidal Rule	A given trench is divided into two equal sections, giving three ordinates, the first, the middle and the last $$\text{volume} = \frac{L \times (A + 4B + C)}{6}$$ where L = total length of trench A = area of the first section B = area of the middle section C = area of the last section

TYPICAL THERMAL CONDUCTIVITY OF BUILDING MATERIALS

(Always check manufacturer's details – variation will occur depending on product and nature of materials)

	Thermal conductivity (W/mK)		Thermal conductivity (W/mK)
Acoustic plasterboard	0.25	Oriented strand board	0.13
Aerated concrete slab (500 kg/m^3)	0.16	Outer leaf brick	0.77
Aluminium	237	Plasterboard	0.22
Asphalt (1700 kg/m^3)	0.5	Plaster dense (1300 kg/m^3)	0.5
Bitumen-impregnated fibreboard	0.05	Plaster lightweight (600 kg/m^3)	0.16
Blocks (standard grade 600 kg/m^3)	0.15	Plywood (950 kg/m^3)	0.16
Blocks (solar grade 460 kg/m^3)	0.11	Prefabricated timber wall panels (check manufacturer)	0.12
Brickwork (outer leaf 1700 kg/m^3)	0.84	Screed (1200 kg/m^3)	0.41
Brickwork (inner leaf 1700 kg/m^3)	0.62	Stone chippings (1800 kg/m^3)	0.96
Dense aggregate concrete block 1800 kg/m^3 (exposed)	1.21	Tile hanging (1900 kg/m^3)	0.84
Dense aggregate concrete block 1800 kg/m^3 (protected)	1.13	Timber (650 kg/m^3)	0.14
Calcium silicate board (600 kg/m^3)	0.17	Timber flooring (650 kg/m^3)	0.14
Concrete general	1.28	Timber rafters	0.13
Concrete (heavyweight 2300 kg/m^3)	1.63	Timber roof or floor joists	0.13
Concrete (dense 2100 kg/m^3 typical floor)	1.4	Roof tile (1900 kg/m^3)	0.84
Concrete (dense 2000 kg/m^3 typical floor)	1.13	Timber blocks (650 kg/m^3)	0.14
Concrete (medium 1400 kg/m^3)	0.51	Cellular glass	0.045
Concrete (lightweight 1200 kg/m^3)	0.38	Expanded polystyrene	0.034
Concrete (lightweight 600 kg/m^3)	0.19	Expanded polystyrene slab (25 kg/m^3)	0.035
Concrete slab (aerated 500 kg/m^3)	0.16	Extruded polystyrene	0.035
Copper	390	Glass mineral wool	0.04
External render sand/cement finish	1	Mineral quilt (12 kg/m^3)	0.04
External render (1300 kg/m^3)	0.5	Mineral wool slab (25 kg/m^3)	0.035
Felt – Bitumen layers (1700 kg/m^3)	0.5	Phenolic foam	0.022
Fibreboard (300 kg/m^3)	0.06	Polyisocyanurate	0.025
Glass	0.93	Polyurethane	0.025
Marble	3	Rigid polyurethane	0.025
Metal tray used in wriggly tin concrete floors (7800 kg/m^3)	50	Rock mineral wool	0.038
Mortar (1750 kg/m^3)	0.8		

EARTHWORK

Weights of Typical Materials Handled by Excavators

The weight of the material is that of the state in its natural bed and includes moisture.
Adjustments should be made to allow for loose or compacted states

Material	Mass (kg/m³)	Mass (lb/cu yd)
Ashes, dry	610	1028
Ashes, wet	810	1365
Basalt, broken	1954	3293
Basalt, solid	2933	4943
Bauxite, crushed	1281	2159
Borax, fine	849	1431
Caliche	1440	2427
Cement, clinker	1415	2385
Chalk, fine	1221	2058
Chalk, solid	2406	4055
Cinders, coal, ash	641	1080
Cinders, furnace	913	1538
Clay, compacted	1746	2942
Clay, dry	1073	1808
Clay, wet	1602	2700
Coal, anthracite, solid	1506	2538
Coal, bituminous	1351	2277
Coke	610	1028
Dolomite, lumpy	1522	2565
Dolomite, solid	2886	4864
Earth, dense	2002	3374
Earth, dry, loam	1249	2105
Earth, Fullers, raw	673	1134
Earth, moist	1442	2430
Earth, wet	1602	2700
Felsite	2495	4205
Fieldspar, solid	2613	4404
Fluorite	3093	5213
Gabbro	3093	5213
Gneiss	2696	4544
Granite	2690	4534
Gravel, dry ¼ to 2 inch	1682	2835
Gravel, dry, loose	1522	2565
Gravel, wet ¼ to 2 inch	2002	3374
Gypsum, broken	1450	2444
Gypsum, solid	2787	4697
Hardcore (consolidated)	1928	3249
Lignite, dry	801	1350
Limestone, broken	1554	2619
Limestone, solid	2596	4375
Magnesite, magnesium ore	2993	5044
Marble	2679	4515
Marl, wet	2216	3735
Mica, broken	1602	2700
Mica, solid	2883	4859
Peat, dry	400	674
Peat, moist	700	1179

EARTHWORK

Material	Mass (kg/m³)	Mass (lb/cu yd)
Peat, wet	1121	1889
Potash	1281	2159
Pumice, stone	640	1078
Quarry waste	1438	2423
Quartz sand	1201	2024
Quartz, solid	2584	4355
Rhyolite	2400	4045
Sand and gravel, dry	1650	2781
Sand and gravel, wet	2020	3404
Sand, dry	1602	2700
Sand, wet	1831	3086
Sandstone, solid	2412	4065
Shale, solid	2637	4444
Slag, broken	2114	3563
Slag, furnace granulated	961	1619
Slate, broken	1370	2309
Slate, solid	2667	4495
Snow, compacted	481	810
Snow, freshly fallen	160	269
Taconite	2803	4724
Trachyte	2400	4045
Trap rock, solid	2791	4704
Turf	400	674
Water	1000	1685

Transport Capacities

Type of vehicle	Capacity of vehicle	
	Payload	Heaped capacity
Wheelbarrow	150	0.10
1 tonne dumper	1250	1.00
2.5 tonne dumper	4000	2.50
Articulated dump truck (Volvo A20 6 × 4)	18500	11.00
Articulated dump truck (Volvo A35 6 × 6)	32000	19.00
Large capacity rear dumper (Euclid R35)	35000	22.00
Large capacity rear dumper (Euclid R85)	85000	50.00

EARTHWORK

Machine Volumes for Excavating and Filling

Machine type	Cycles per minute	Volume per minute (m³)
1.5 tonne excavator	1	0.04
	2	0.08
	3	0.12
3 tonne excavator	1	0.13
	2	0.26
	3	0.39
5 tonne excavator	1	0.28
	2	0.56
	3	0.84
7 tonne excavator	1	0.28
	2	0.56
	3	0.84
21 tonne excavator	1	1.21
	2	2.42
	3	3.63
Backhoe loader JCB3CX excavator	1	0.28
Rear bucket capacity 0.28 m³	2	0.56
	3	0.84
Backhoe loader JCB3CX loading	1	1.00
Front bucket capacity 1.00 m³	2	2.00

Machine Volumes for Excavating and Filling

Machine type	Loads per hour	Volume per hour (m³)
1 tonne high tip skip loader	5	2.43
Volume 0.485 m³	7	3.40
	10	4.85
3 tonne dumper	4	7.60
Max volume 2.40 m³	5	9.50
Available volume 1.9 m³	7	13.30
	10	19.00
6 tonne dumper	4	15.08
Max volume 3.40 m³	5	18.85
Available volume 3.77 m³	7	26.39
	10	37.70

EARTHWORK

Bulkage of Soils (after excavation)

Type of soil	Approximate bulking of 1 m³ after excavation
Vegetable soil and loam	25–30%
Soft clay	30–40%
Stiff clay	10–20%
Gravel	20–25%
Sand	40–50%
Chalk	40–50%
Rock, weathered	30–40%
Rock, unweathered	50–60%

Shrinkage of Materials (on being deposited)

Type of soil	Approximate bulking of 1 m³ after excavation
Clay	10%
Gravel	8%
Gravel and sand	9%
Loam and light sandy soils	12%
Loose vegetable soils	15%

Voids in Material Used as Subbases or Beddings

Material	m³ of voids/m³
Alluvium	0.37
River grit	0.29
Quarry sand	0.24
Shingle	0.37
Gravel	0.39
Broken stone	0.45
Broken bricks	0.42

Angles of Repose

Type of soil		Degrees
Clay	– dry	30
	– damp, well drained	45
	– wet	15–20
Earth	– dry	30
	– damp	45
Gravel	– moist	48
Sand	– dry or moist	35
	– wet	25
Loam		40

EARTHWORK

Slopes and Angles

Ratio of base to height	Angle in degrees
5:1	11
4:1	14
3:1	18
2:1	27
1½:1	34
1:1	45
1:1½	56
1:2	63
1:3	72
1:4	76
1:5	79

Grades (in Degrees and Percents)

Degrees	Percent	Degrees	Percent
1	1.8	24	44.5
2	3.5	25	46.6
3	5.2	26	48.8
4	7.0	27	51.0
5	8.8	28	53.2
6	10.5	29	55.4
7	12.3	30	57.7
8	14.0	31	60.0
9	15.8	32	62.5
10	17.6	33	64.9
11	19.4	34	67.4
12	21.3	35	70.0
13	23.1	36	72.7
14	24.9	37	75.4
15	26.8	38	78.1
16	28.7	39	81.0
17	30.6	40	83.9
18	32.5	41	86.9
19	34.4	42	90.0
20	36.4	43	93.3
21	38.4	44	96.6
22	40.4	45	100.0
23	42.4		

EARTHWORK

Bearing Powers

Ground conditions		Bearing power		
		kg/m²	lb/in²	Metric t/m²
Rock,	broken	483	70	50
	solid	2415	350	240
Clay,	dry or hard	380	55	40
	medium dry	190	27	20
	soft or wet	100	14	10
Gravel,	cemented	760	110	80
Sand,	compacted	380	55	40
	clean dry	190	27	20
Swamp and alluvial soils		48	7	5

Earthwork Support

Maximum depth of excavation in various soils without the use of earthwork support

Ground conditions	Feet (ft)	Metres (m)
Compact soil	12	3.66
Drained loam	6	1.83
Dry sand	1	0.3
Gravelly earth	2	0.61
Ordinary earth	3	0.91
Stiff clay	10	3.05

It is important to note that the above table should only be used as a guide. Each case must be taken on its merits and, as the limited distances given above are approached, careful watch must be kept for the slightest signs of caving in

CONCRETE WORK

Weights of Concrete and Concrete Elements

Type of material		kg/m³	lb/cu ft
Ordinary concrete (dense aggregates)			
Non-reinforced plain or mass concrete			
Nominal weight		2305	144
Aggregate	– limestone	2162 to 2407	135 to 150
	– gravel	2244 to 2407	140 to 150
	– broken brick	2000 (av)	125 (av)
	– other crushed stone	2326 to 2489	145 to 155
Reinforced concrete			
Nominal weight		2407	150
Reinforcement	– 1%	2305 to 2468	144 to 154
	– 2%	2356 to 2519	147 to 157
	– 4%	2448 to 2703	153 to 163
Special concretes			
Heavy concrete			
Aggregates	– barytes, magnetite	3210 (min)	200 (min)
	– steel shot, punchings	5280	330
Lean mixes			
Dry-lean (gravel aggregate)		2244	140
Soil-cement (normal mix)		1601	100

CONCRETE WORK

Type of material		kg/m² per mm thick	lb/sq ft per inch thick
Ordinary concrete (dense aggregates)			
Solid slabs (floors, walls etc.)			
Thickness:	75 mm or 3 in	184	37.5
	100 mm or 4 in	245	50
	150 mm or 6 in	378	75
	250 mm or 10 in	612	125
	300 mm or 12 in	734	150
Ribbed slabs			
Thickness:	125 mm or 5 in	204	42
	150 mm or 6 in	219	45
	225 mm or 9 in	281	57
	300 mm or 12 in	342	70
Special concretes			
Finishes etc.			
	Rendering, screed etc.		
	Granolithic, terrazzo	1928 to 2401	10 to 12.5
	Glass-block (hollow) concrete	1734 (approx)	9 (approx)
Prestressed concrete		Weights as for reinforced concrete (upper limits)	
Air-entrained concrete		Weights as for plain or reinforced concrete	

CONCRETE WORK

Average Weight of Aggregates

Materials	Voids %	Weight kg/m³
Sand	39	1660
Gravel 10–20 mm	45	1440
Gravel 35–75 mm	42	1555
Crushed stone	50	1330
Crushed granite (over 15 mm)	50	1345
(n.e. 15 mm)	47	1440
'All-in' ballast	32	1800–2000

Material	kg/m³	lb/cu yd
Vermiculite (aggregate)	64–80	108–135
All-in aggregate	1999	125

Applications and Mix Design

Site mixed concrete

Recommended mix	Class of work suitable for	Cement (kg)	Sand (kg)	Coarse aggregate (kg)	Nr 25 kg bags cement per m³ of combined aggregate
1:3:6	Roughest type of mass concrete such as footings, road haunching over 300 mm thick	208	905	1509	8.30
1:2.5:5	Mass concrete of better class than 1:3:6 such as bases for machinery, walls below ground etc.	249	881	1474	10.00
1:2:4	Most ordinary uses of concrete, such as mass walls above ground, road slabs etc. and general reinforced concrete work	304	889	1431	12.20
1:1.5:3	Watertight floors, pavements and walls, tanks, pits, steps, paths, surface of 2 course roads, reinforced concrete where extra strength is required	371	801	1336	14.90
1:1:2	Works of thin section such as fence posts and small precast work	511	720	1206	20.40

CONCRETE WORK

Ready mixed concrete

Application	Designated concrete	Standardized prescribed concrete	Recommended consistence (nominal slump class)
Foundations			
Mass concrete fill or blinding	GEN 1	ST2	S3
Strip footings	GEN 1	ST2	S3
Mass concrete foundations			
Single storey buildings	GEN 1	ST2	S3
Double storey buildings	GEN 3	ST4	S3
Trench fill foundations			
Single storey buildings	GEN 1	ST2	S4
Double storey buildings	GEN 3	ST4	S4
General applications			
Kerb bedding and haunching	GEN 0	ST1	S1
Drainage works – immediate support	GEN 1	ST2	S1
Other drainage works	GEN 1	ST2	S3
Oversite below suspended slabs	GEN 1	ST2	S3
Floors			
Garage and house floors with no embedded steel	GEN 3	ST4	S2
Wearing surface: Light foot and trolley traffic	RC30	ST4	S2
Wearing surface: General industrial	RC40	N/A	S2
Wearing surface: Heavy industrial	RC50	N/A	S2
Paving			
House drives, domestic parking and external parking	PAV 1	N/A	S2
Heavy-duty external paving	PAV 2	N/A	S2

CONCRETE WORK

Prescribed Mixes for Ordinary Structural Concrete

Weights of cement and total dry aggregates in kg to produce approximately one cubic metre of fully compacted concrete together with the percentages by weight of fine aggregate in total dry aggregates

Conc. grade	Nominal max size of aggregate (mm)	40		20		14		10	
	Workability	Med.	High	Med.	High	Med.	High	Med.	High
	Limits to slump that may be expected (mm)	50–100	100–150	25–75	75–125	10–50	50–100	10–25	25–50
7	Cement (kg)	180	200	210	230	–	–	–	–
	Total aggregate (kg)	1950	1850	1900	1800	–	–	–	–
	Fine aggregate (%)	30–45	30–45	35–50	35–50	–	–	–	–
10	Cement (kg)	210	230	240	260	–	–	–	–
	Total aggregate (kg)	1900	1850	1850	1800	–	–	–	–
	Fine aggregate (%)	30–45	30–45	35–50	35–50	–	–	–	–
15	Cement (kg)	250	270	280	310	–	–	–	–
	Total aggregate (kg)	1850	1800	1800	1750	–	–	–	–
	Fine aggregate (%)	30–45	30–45	35–50	35–50	–	–	–	–
20	Cement (kg)	300	320	320	350	340	380	360	410
	Total aggregate (kg)	1850	1750	1800	1750	1750	1700	1750	1650
	Sand								
	Zone 1 (%)	35	40	40	45	45	50	50	55
	Zone 2 (%)	30	35	35	40	40	45	45	50
	Zone 3 (%)	30	30	30	35	35	40	40	45
25	Cement (kg)	340	360	360	390	380	420	400	450
	Total aggregate (kg)	1800	1750	1750	1700	1700	1650	1700	1600
	Sand								
	Zone 1 (%)	35	40	40	45	45	50	50	55
	Zone 2 (%)	30	35	35	40	40	45	45	50
	Zone 3 (%)	30	30	30	35	35	40	40	45
30	Cement (kg)	370	390	400	430	430	470	460	510
	Total aggregate (kg)	1750	1700	1700	1650	1700	1600	1650	1550
	Sand								
	Zone 1 (%)	35	40	40	45	45	50	50	55
	Zone 2 (%)	30	35	35	40	40	45	45	50
	Zone 3 (%)	30	30	30	35	35	40	40	45

REINFORCEMENT

Weights of Bar Reinforcement

Nominal sizes (mm)	Cross-sectional area (mm²)	Mass (kg/m)	Length of bar (m/tonne)
6	28.27	0.222	4505
8	50.27	0.395	2534
10	78.54	0.617	1622
12	113.10	0.888	1126
16	201.06	1.578	634
20	314.16	2.466	405
25	490.87	3.853	260
32	804.25	6.313	158
40	1265.64	9.865	101
50	1963.50	15.413	65

Weights of Bars (at specific spacings)

Weights of metric bars in kilogrammes per square metre

Size (mm)	Spacing of bars in millimetres									
	75	100	125	150	175	200	225	250	275	300
6	2.96	2.220	1.776	1.480	1.27	1.110	0.99	0.89	0.81	0.74
8	5.26	3.95	3.16	2.63	2.26	1.97	1.75	1.58	1.44	1.32
10	8.22	6.17	4.93	4.11	3.52	3.08	2.74	2.47	2.24	2.06
12	11.84	8.88	7.10	5.92	5.07	4.44	3.95	3.55	3.23	2.96
16	21.04	15.78	12.63	10.52	9.02	7.89	7.02	6.31	5.74	5.26
20	32.88	24.66	19.73	16.44	14.09	12.33	10.96	9.87	8.97	8.22
25	51.38	38.53	30.83	25.69	22.02	19.27	17.13	15.41	14.01	12.84
32	84.18	63.13	50.51	42.09	36.08	31.57	28.06	25.25	22.96	21.04
40	131.53	98.65	78.92	65.76	56.37	49.32	43.84	39.46	35.87	32.88
50	205.51	154.13	123.31	102.76	88.08	77.07	68.50	61.65	56.05	51.38

Basic weight of steelwork taken as 7850 kg/m³
Basic weight of bar reinforcement per metre run = 0.00785 kg/mm²
The value of π has been taken as 3.141592654

REINFORCEMENT

Fabric Reinforcement

Preferred range of designated fabric types and stock sheet sizes

Fabric reference	Longitudinal wires			Cross wires			
	Nominal wire size (mm)	Pitch (mm)	Area (mm²/m)	Nominal wire size (mm)	Pitch (mm)	Area (mm²/m)	Mass (kg/m²)
Square mesh							
A393	10	200	393	10	200	393	6.16
A252	8	200	252	8	200	252	3.95
A193	7	200	193	7	200	193	3.02
A142	6	200	142	6	200	142	2.22
A98	5	200	98	5	200	98	1.54
Structural mesh							
B1131	12	100	1131	8	200	252	10.90
B785	10	100	785	8	200	252	8.14
B503	8	100	503	8	200	252	5.93
B385	7	100	385	7	200	193	4.53
B283	6	100	283	7	200	193	3.73
B196	5	100	196	7	200	193	3.05
Long mesh							
C785	10	100	785	6	400	70.8	6.72
C636	9	100	636	6	400	70.8	5.55
C503	8	100	503	5	400	49.0	4.34
C385	7	100	385	5	400	49.0	3.41
C283	6	100	283	5	400	49.0	2.61
Wrapping mesh							
D98	5	200	98	5	200	98	1.54
D49	2.5	100	49	2.5	100	49	0.77

Stock sheet size 4.8 m × 2.4 m, Area 11.52 m²

Average weight kg/m³ of steelwork reinforcement in concrete for various building elements

Substructure	kg/m³ concrete	Substructure	kg/m³ concrete
Pile caps	110–150	Plate slab	150–220
Tie beams	130–170	Cant slab	145–210
Ground beams	230–330	Ribbed floors	130–200
Bases	125–180	Topping to block floor	30–40
Footings	100–150	Columns	210–310
Retaining walls	150–210	Beams	250–350
Raft	60–70	Stairs	130–170
Slabs – one way	120–200	Walls – normal	40–100
Slabs – two way	110–220	Walls – wind	70–125

Note: For exposed elements add the following %:
Walls 50%, Beams 100%, Columns 15%

FORMWORK

Formwork Stripping Times – Normal Curing Periods

Conditions under which concrete is maturing	Minimum periods of protection for different types of cement					
	Number of days (where the average surface temperature of the concrete exceeds 10°C during the whole period)			Equivalent maturity (degree hours) calculated as the age of the concrete in hours multiplied by the number of degrees Celsius by which the average surface temperature of the concrete exceeds 10°C		
	Other	SRPC	OPC or RHPC	Other	SRPC	OPC or RHPC
1. Hot weather or drying winds	7	4	3	3500	2000	1500
2. Conditions not covered by 1	4	3	2	2000	1500	1000

KEY
OPC – Ordinary Portland Cement
RHPC – Rapid-hardening Portland Cement
SRPC – Sulphate-resisting Portland Cement

Minimum Period before Striking Formwork

	Minimum period before striking		
	Surface temperature of concrete		
	16°C	17°C	t°C (0–25)
Vertical formwork to columns, walls and large beams	12 hours	18 hours	300 hours t+10
Soffit formwork to slabs	4 days	6 days	100 days t+10
Props to slabs	10 days	15 days	250 days t+10
Soffit formwork to beams	9 days	14 days	230 days t+10
Props to beams	14 days	21 days	360 days t+10

MASONRY

Number of Bricks Required for Various Types of Work per m² of Walling

Description	Brick size	
	215 × 102.5 × 50 mm	215 × 102.5 × 65 mm
Half brick thick		
Stretcher bond	74	59
English bond	108	86
English garden wall bond	90	72
Flemish bond	96	79
Flemish garden wall bond	83	66
One brick thick and cavity wall of two half brick skins		
Stretcher bond	148	119

Quantities of Bricks and Mortar Required per m² of Walling

	Unit	No of bricks required	Mortar required (cubic metres)		
Standard bricks			**No frogs**	**Single frogs**	**Double frogs**
Brick size 215 × 102.5 × 50 mm					
half brick wall (103 mm)	m²	72	0.022	0.027	0.032
2 × half brick cavity wall (270 mm)	m²	144	0.044	0.054	0.064
one brick wall (215 mm)	m²	144	0.052	0.064	0.076
one and a half brick wall (322 mm)	m²	216	0.073	0.091	0.108
Mass brickwork	m³	576	0.347	0.413	0.480
Brick size 215 × 102.5 × 65 mm					
half brick wall (103 mm)	m²	58	0.019	0.022	0.026
2 × half brick cavity wall (270 mm)	m²	116	0.038	0.045	0.055
one brick wall (215 mm)	m²	116	0.046	0.055	0.064
one and a half brick wall (322 mm)	m²	174	0.063	0.074	0.088
Mass brickwork	m³	464	0.307	0.360	0.413
Metric modular bricks			**Perforated**		
Brick size 200 × 100 × 75 mm					
90 mm thick	m²	67	0.016	0.019	
190 mm thick	m²	133	0.042	0.048	
290 mm thick	m²	200	0.068	0.078	
Brick size 200 × 100 × 100 mm					
90 mm thick	m²	50	0.013	0.016	
190 mm thick	m²	100	0.036	0.041	
290 mm thick	m²	150	0.059	0.067	
Brick size 300 × 100 × 75 mm					
90 mm thick	m²	33	–	0.015	
Brick size 300 × 100 × 100 mm					
90 mm thick	m²	44	0.015	0.018	

Note: Assuming 10 mm thick joints

MASONRY

Mortar Required per m² Blockwork (9.88 blocks/m²)

Wall thickness	75	90	100	125	140	190	215
Mortar m³/m²	0.005	0.006	0.007	0.008	0.009	0.013	0.014

Mortar Group	Cement: lime: sand	Masonry cement: sand	Cement: sand with plasticizer
1	1:0–0.25:3		
2	1:0.5:4–4.5	1:2.5-3.5	1:3–4
3	1:1:5–6	1:4–5	1:5–6
4	1:2:8–9	1:5.5–6.5	1:7–8
5	1:3:10–12	1:6.5–7	1:8

Group 1: strong inflexible mortar
Group 5: weak but flexible

All mixes within a group are of approximately similar strength
Frost resistance increases with the use of plasticizers
Cement: lime: sand mixes give the strongest bond and greatest resistance to rain penetration
Masonry cement equals ordinary Portland cement plus a fine neutral mineral filler and an air entraining agent

Calcium Silicate Bricks

Type	Strength	Location
Class 2 crushing strength	14.0 N/mm²	not suitable for walls
Class 3	20.5 N/mm²	walls above dpc
Class 4	27.5 N/mm²	cappings and copings
Class 5	34.5 N/mm²	retaining walls
Class 6	41.5 N/mm²	walls below ground
Class 7	48.5 N/mm²	walls below ground

The Class 7 calcium silicate bricks are therefore equal in strength to Class B bricks
Calcium silicate bricks are not suitable for DPCs

Durability of Bricks	
FL	Frost resistant with low salt content
FN	Frost resistant with normal salt content
ML	Moderately frost resistant with low salt content
MN	Moderately frost resistant with normal salt content

MASONRY

Brickwork Dimensions

No. of horizontal bricks	Dimensions (mm)	No. of vertical courses	Height of vertical courses (mm)
½	112.5	1	75
1	225.0	2	150
1½	337.5	3	225
2	450.0	4	300
2½	562.5	5	375
3	675.0	6	450
3½	787.5	7	525
4	900.0	8	600
4½	1012.5	9	675
5	1125.0	10	750
5½	1237.5	11	825
6	1350.0	12	900
6½	1462.5	13	975
7	1575.0	14	1050
7½	1687.5	15	1125
8	1800.0	16	1200
8½	1912.5	17	1275
9	2025.0	18	1350
9½	2137.5	19	1425
10	2250.0	20	1500
20	4500.0	24	1575
40	9000.0	28	2100
50	11250.0	32	2400
60	13500.0	36	2700
75	16875.0	40	3000

TIMBER

Weights of Timber

Material	kg/m³	lb/cu ft
General	806 (avg)	50 (avg)
Douglas fir	479	30
Yellow pine, spruce	479	30
Pitch pine	673	42
Larch, elm	561	35
Oak (English)	724 to 959	45 to 60
Teak	643 to 877	40 to 55
Jarrah	959	60
Greenheart	1040 to 1204	65 to 75
Quebracho	1285	80
Material	**kg/m² per mm thickness**	**lb/sq ft per inch thickness**
Wooden boarding and blocks		
Softwood	0.48	2.5
Hardwood	0.76	4
Hardboard	1.06	5.5
Chipboard	0.76	4
Plywood	0.62	3.25
Blockboard	0.48	2.5
Fibreboard	0.29	1.5
Wood-wool	0.58	3
Plasterboard	0.96	5
Weather boarding	0.35	1.8

TIMBER

Conversion Tables (for timber only)

Inches	Millimetres	Feet	Metres
1	25	1	0.300
2	50	2	0.600
3	75	3	0.900
4	100	4	1.200
5	125	5	1.500
6	150	6	1.800
7	175	7	2.100
8	200	8	2.400
9	225	9	2.700
10	250	10	3.000
11	275	11	3.300
12	300	12	3.600
13	325	13	3.900
14	350	14	4.200
15	375	15	4.500
16	400	16	4.800
17	425	17	5.100
18	450	18	5.400
19	475	19	5.700
20	500	20	6.000
21	525	21	6.300
22	550	22	6.600
23	575	23	6.900
24	600	24	7.200

Planed Softwood
The finished end section size of planed timber is usually 3/16" less than the original size from which it is produced. This however varies slightly depending upon availability of material and origin of the species used.

Standards (timber) to cubic metres and cubic metres to standards (timber)

Cubic metres	Cubic metres standards	Standards
4.672	1	0.214
9.344	2	0.428
14.017	3	0.642
18.689	4	0.856
23.361	5	1.070
28.033	6	1.284
32.706	7	1.498
37.378	8	1.712
42.050	9	1.926
46.722	10	2.140
93.445	20	4.281
140.167	30	6.421
186.890	40	8.561
233.612	50	10.702
280.335	60	12.842
327.057	70	14.982
373.779	80	17.122

TIMBER

1 cu metre = 35.3148 cu ft = 0.21403 std

1 cu ft = 0.028317 cu metres

1 std = 4.67227 cu metres

Basic sizes of sawn softwood available (cross-sectional areas)

Thickness (mm)	Width (mm)								
	75	100	125	150	175	200	225	250	300
16	X	X	X	X					
19	X	X	X	X					
22	X	X	X	X					
25	X	X	X	X	X	X	X	X	X
32	X	X	X	X	X	X	X	X	X
36	X	X	X	X					
38	X	X	X	X	X	X	X		
44	X	X	X	X	X	X	X	X	X
47*	X	X	X	X	X	X	X	X	X
50	X	X	X	X	X	X	X	X	X
63	X	X	X	X	X	X	X		
75	X	X	X	X	X	X	X	X	
100		X		X		X		X	X
150				X		X			X
200						X			
250								X	
300									X

* This range of widths for 47 mm thickness will usually be found to be available in construction quality only

Note: The smaller sizes below 100 mm thick and 250 mm width are normally but not exclusively of European origin. Sizes beyond this are usually of North and South American origin

Basic lengths of sawn softwood available (metres)

1.80	2.10	3.00	4.20	5.10	6.00	7.20
	2.40	3.30	4.50	5.40	6.30	
	2.70	3.60	4.80	5.70	6.60	
		3.90			6.90	

Note: Lengths of 6.00 m and over will generally only be available from North American species and may have to be recut from larger sizes

TIMBER

Reductions from basic size to finished size by planning of two opposed faces

Purpose		Reductions from basic sizes for timber			
		15–35 mm	36–100 mm	101–150 mm	over 150 mm
a)	Constructional timber	3 mm	3 mm	5 mm	6 mm
b)	Matching interlocking boards	4 mm	4 mm	6 mm	6 mm
c)	Wood trim not specified in BS 584	5 mm	7 mm	7 mm	9 mm
d)	Joinery and cabinet work	7 mm	9 mm	11 mm	13 mm

Note: The reduction of width or depth is overall the extreme size and is exclusive of any reduction of the face by the machining of a tongue or lap joints

Maximum Spans for Various Roof Trusses

Maximum permissible spans for rafters for Fink trussed rafters

Basic size	Actual size	Pitch (degrees)								
(mm)	(mm)	15 (m)	17.5 (m)	20 (m)	22.5 (m)	25 (m)	27.5 (m)	30 (m)	32.5 (m)	35 (m)
38 × 75	35 × 72	6.03	6.16	6.29	6.41	6.51	6.60	6.70	6.80	6.90
38 × 100	35 × 97	7.48	7.67	7.83	7.97	8.10	8.22	8.34	8.47	8.61
38 × 125	35 × 120	8.80	9.00	9.20	9.37	9.54	9.68	9.82	9.98	10.16
44 × 75	41 × 72	6.45	6.59	6.71	6.83	6.93	7.03	7.14	7.24	7.35
44 × 100	41 × 97	8.05	8.23	8.40	8.55	8.68	8.81	8.93	9.09	9.22
44 × 125	41 × 120	9.38	9.60	9.81	9.99	10.15	10.31	10.45	10.64	10.81
50 × 75	47 × 72	6.87	7.01	7.13	7.25	7.35	7.45	7.53	7.67	7.78
50 × 100	47 × 97	8.62	8.80	8.97	9.12	9.25	9.38	9.50	9.66	9.80
50 × 125	47 × 120	10.01	10.24	10.44	10.62	10.77	10.94	11.00	11.00	11.00

TIMBER

Sizes of Internal and External Doorsets

Description	Internal size (mm)	Permissible deviation	External size (mm)	Permissible deviation
Coordinating dimension: height of door leaf height sets	2100		2100	
Coordinating dimension: height of ceiling height set	2300 2350 2400 2700 3000		2300 2350 2400 2700 3000	
Coordinating dimension: width of all doorsets S = Single leaf set D = Double leaf set	600 S 700 S 800 S&D 900 S&D 1000 S&D 1200 D 1500 D 1800 D 2100 D		900 S 1000 S 1200 D 1800 D 2100 D	
Work size: height of door leaf height set	2090	± 2.0	2095	± 2.0
Work size: height of ceiling height set	2285 2335 2385 2685 2985	± 2.0	2295 2345 2395 2695 2995	± 2.0
Work size: width of all doorsets S = Single leaf set D = Double leaf set	590 S 690 S 790 S&D 890 S&D 990 S&D 1190 D 1490 D 1790 D 2090 D	± 2.0	895 S 995 S 1195 D 1495 D 1795 D 2095 D	± 2.0
Width of door leaf in single leaf sets F = Flush leaf P = Panel leaf	526 F 626 F 726 F&P 826 F&P 926 F&P	± 1.5	806 F&P 906 F&P	± 1.5
Width of door leaf in double leaf sets F = Flush leaf P = Panel leaf	362 F 412 F 426 F 562 F&P 712 F&P 826 F&P 1012 F&P	± 1.5	552 F&P 702 F&P 852 F&P 1002 F&P	± 1.5
Door leaf height for all doorsets	2040	± 1.5	1994	± 1.5

ROOFING

Total Roof Loadings for Various Types of Tiles/Slates

	Roof load (slope) kg/m²		
	Slate/Tile	Roofing underlay and battens²	Total dead load kg/m
Asbestos cement slate (600 × 300)	21.50	3.14	24.64
Clay tile interlocking	67.00	5.50	72.50
plain	43.50	2.87	46.37
Concrete tile interlocking	47.20	2.69	49.89
plain	78.20	5.50	83.70
Natural slate (18" × 10")	35.40	3.40	38.80
	Roof load (plan) kg/m²		
Asbestos cement slate (600 × 300)	28.45	76.50	104.95
Clay tile interlocking	53.54	76.50	130.04
plain	83.71	76.50	60.21
Concrete tile interlocking	57.60	76.50	134.10
plain	96.64	76.50	173.14

ROOFING

Tiling Data

Product		Lap (mm)	Gauge of battens	No. slates per m^2	Battens (m/m^2)	Weight as laid (kg/m^2)
CEMENT SLATES						
Eternit slates	600 × 300 mm	100	250	13.4	4.00	19.50
(Duracem)		90	255	13.1	3.92	19.20
		80	260	12.9	3.85	19.00
		70	265	12.7	3.77	18.60
	600 × 350 mm	100	250	11.5	4.00	19.50
		90	255	11.2	3.92	19.20
	500 × 250 mm	100	200	20.0	5.00	20.00
		90	205	19.5	4.88	19.50
		80	210	19.1	4.76	19.00
		70	215	18.6	4.65	18.60
	400 × 200 mm	90	155	32.3	6.45	20.80
		80	160	31.3	6.25	20.20
		70	165	30.3	6.06	19.60
CONCRETE TILES/SLATES						
Redland Roofing						
Stonewold slate	430 × 380 mm	75	355	8.2	2.82	51.20
Double Roman tile	418 × 330 mm	75	355	8.2	2.91	45.50
Grovebury pantile	418 × 332 mm	75	343	9.7	2.91	47.90
Norfolk pantile	381 × 227 mm	75	306	16.3	3.26	44.01
		100	281	17.8	3.56	48.06
Renown interlocking tile	418 × 330 mm	75	343	9.7	2.91	46.40
'49' tile	381 × 227 mm	75	306	16.3	3.26	44.80
		100	281	17.8	3.56	48.95
Plain, vertical tiling	265 × 165 mm	35	115	52.7	8.70	62.20
Marley Roofing						
Bold roll tile	420 × 330 mm	75	344	9.7	2.90	47.00
		100	–	10.5	3.20	51.00
Modern roof tile	420 × 330 mm	75	338	10.2	3.00	54.00
		100	–	11.0	3.20	58.00
Ludlow major	420 × 330 mm	75	338	10.2	3.00	45.00
		100	–	11.0	3.20	49.00
Ludlow plus	387 × 229 mm	75	305	16.1	3.30	47.00
		100	–	17.5	3.60	51.00
Mendip tile	420 × 330 mm	75	338	10.2	3.00	47.00
		100	–	11.0	3.20	51.00
Wessex	413 × 330 mm	75	338	10.2	3.00	54.00
		100	–	11.0	3.20	58.00
Plain tile	267 × 165 mm	65	100	60.0	10.00	76.00
		75	95	64.0	10.50	81.00
		85	90	68.0	11.30	86.00
Plain vertical tiles (feature)	267 × 165 mm	35	110	53.0	8.70	67.00
		34	115	56.0	9.10	71.00

ROOFING

Slate Nails, Quantity per Kilogram

Length	Type			
	Plain wire	Galvanized wire	Copper nail	Zinc nail
28.5 mm	325	305	325	415
34.4 mm	286	256	254	292
50.8 mm	242	224	194	200

Metal Sheet Coverings

Thicknesses and weights of sheet metal coverings								
Lead to BS 1178								
BS Code No	3	4	5	6	7	8		
Colour code	Green	Blue	Red	Black	White	Orange		
Thickness (mm)	1.25	1.80	2.24	2.50	3.15	3.55		
Density kg/m^2	14.18	20.41	25.40	30.05	35.72	40.26		
Copper to BS 2870								
Thickness (mm)		0.60	0.70					
Bay width								
Roll (mm)		500	650					
Seam (mm)		525	600					
Standard width to form bay	600	750						
Normal length of sheet	1.80	1.80						
Zinc to BS 849								
Zinc Gauge (Nr)	9	10	11	12	13	14	15	16
Thickness (mm)	0.43	0.48	0.56	0.64	0.71	0.79	0.91	1.04
Density (kg/m^2)	3.1	3.2	3.8	4.3	4.8	5.3	6.2	7.0
Aluminium to BS 4868								
Thickness (mm)	0.5	0.6	0.7	0.8	0.9	1.0	1.2	
Density (kg/m^2)	12.8	15.4	17.9	20.5	23.0	25.6	30.7	

ROOFING

Type of felt	Nominal mass per unit area	Nominal mass per unit area of fibre base	Nominal length of roll
	(kg/10 m)	(g/m²)	(m)
Class 1			
1B fine granule	14	220	10 or 20
surfaced bitumen	18	330	10 or 20
	25	470	10
1E mineral surfaced bitumen	38	470	10
1F reinforced bitumen	15	160 (fibre)	15
		110 (hessian)	
1F reinforced bitumen, aluminium faced	13	160 (fibre)	15
		110 (hessian)	
Class 2			
2B fine granule surfaced bitumen asbestos	18	500	10 or 20
2E mineral surfaced bitumen asbestos	38	600	10
Class 3			
3B fine granule surfaced bitumen glass fibre	18	60	20
3E mineral surfaced bitumen glass fibre	28	60	10
3E venting base layer bitumen glass fibre	32	60*	10
3H venting base layer bitumen glass fibre	17	60*	20

* Excluding effect of perforations

GLAZING

Nominal thickness (mm)	Tolerance on thickness (mm)	Approximate weight (kg/m²)	Normal maximum size (mm)
Float and polished plate glass			
3	+ 0.2	7.50	2140 × 1220
4	+ 0.2	10.00	2760 × 1220
5	+ 0.2	12.50	3180 × 2100
6	+ 0.2	15.00	4600 × 3180
10	+ 0.3	25.00)	6000 × 3300
12	+ 0.3	30.00)	
15	+ 0.5	37.50	3050 × 3000
19	+ 1.0	47.50)	3000 × 2900
25	+ 1.0	63.50)	
Clear sheet glass			
2 *	+ 0.2	5.00	1920 × 1220
3	+ 0.3	7.50	2130 × 1320
4	+ 0.3	10.00	2760 × 1220
5 *	+ 0.3	12.50)	2130 × 2400
6 *	+ 0.3	15.00)	
Cast glass			
3	+ 0.4		
	− 0.2	6.00)	2140 × 1280
4	+ 0.5	7.50)	
5	+ 0.5	9.50	2140 × 1320
6	+ 0.5	11.50)	3700 × 1280
10	+ 0.8	21.50)	
Wired glass			
(Cast wired glass)			
6	+ 0.3	−)	
	− 0.7	)	3700 × 1840
7	+ 0.7	−)	
(Polished wire glass)			
6	+ 1.0	−	330 × 1830

* The 5 mm and 6 mm thickness are known as *thick drawn sheet*. Although 2 mm sheet glass is available it is not recommended for general glazing purposes

METAL

Weights of Metals

Material	kg/m³	lb/cu ft
Metals, steel construction, etc.		
Iron		
– cast	7207	450
– wrought	7687	480
– ore – general	2407	150
– (crushed) Swedish	3682	230
Steel	7854	490
Copper		
– cast	8731	545
– wrought	8945	558
Brass	8497	530
Bronze	8945	558
Aluminium	2774	173
Lead	11322	707
Zinc (rolled)	7140	446
	g/mm² per metre	**lb/sq ft per foot**
Steel bars	7.85	3.4
Structural steelwork	Net weight of member @ 7854 kg/m³	
riveted	+ 10% for cleats, rivets, bolts, etc.	
welded	+ 1.25% to 2.5% for welds, etc.	
Rolled sections		
beams	+ 2.5%	
stanchions	+ 5% (extra for caps and bases)	
Plate		
web girders	+ 10% for rivets or welds, stiffeners, etc.	
	kg/m	**lb/ft**
Steel stairs: industrial type		
1 m or 3 ft wide	84	56
Steel tubes		
50 mm or 2 in bore	5 to 6	3 to 4
Gas piping		
20 mm or ¾ in	2	1¼

METAL

Universal Beams BS 4: Part 1: 2005

Designation	Mass (kg/m)	Depth of section (mm)	Width of section (mm)	Thickness Web (mm)	Flange (mm)	Surface area (m²/m)
1016 × 305 × 487	487.0	1036.1	308.5	30.0	54.1	3.20
1016 × 305 × 438	438.0	1025.9	305.4	26.9	49.0	3.17
1016 × 305 × 393	393.0	1016.0	303.0	24.4	43.9	3.15
1016 × 305 × 349	349.0	1008.1	302.0	21.1	40.0	3.13
1016 × 305 × 314	314.0	1000.0	300.0	19.1	35.9	3.11
1016 × 305 × 272	272.0	990.1	300.0	16.5	31.0	3.10
1016 × 305 × 249	249.0	980.2	300.0	16.5	26.0	3.08
1016 × 305 × 222	222.0	970.3	300.0	16.0	21.1	3.06
914 × 419 × 388	388.0	921.0	420.5	21.4	36.6	3.44
914 × 419 × 343	343.3	911.8	418.5	19.4	32.0	3.42
914 × 305 × 289	289.1	926.6	307.7	19.5	32.0	3.01
914 × 305 × 253	253.4	918.4	305.5	17.3	27.9	2.99
914 × 305 × 224	224.2	910.4	304.1	15.9	23.9	2.97
914 × 305 × 201	200.9	903.0	303.3	15.1	20.2	2.96
838 × 292 × 226	226.5	850.9	293.8	16.1	26.8	2.81
838 × 292 × 194	193.8	840.7	292.4	14.7	21.7	2.79
838 × 292 × 176	175.9	834.9	291.7	14.0	18.8	2.78
762 × 267 × 197	196.8	769.8	268.0	15.6	25.4	2.55
762 × 267 × 173	173.0	762.2	266.7	14.3	21.6	2.53
762 × 267 × 147	146.9	754.0	265.2	12.8	17.5	2.51
762 × 267 × 134	133.9	750.0	264.4	12.0	15.5	2.51
686 × 254 × 170	170.2	692.9	255.8	14.5	23.7	2.35
686 × 254 × 152	152.4	687.5	254.5	13.2	21.0	2.34
686 × 254 × 140	140.1	383.5	253.7	12.4	19.0	2.33
686 × 254 × 125	125.2	677.9	253.0	11.7	16.2	2.32
610 × 305 × 238	238.1	635.8	311.4	18.4	31.4	2.45
610 × 305 × 179	179.0	620.2	307.1	14.1	23.6	2.41
610 × 305 × 149	149.1	612.4	304.8	11.8	19.7	2.39
610 × 229 × 140	139.9	617.2	230.2	13.1	22.1	2.11
610 × 229 × 125	125.1	612.2	229.0	11.9	19.6	2.09
610 × 229 × 113	113.0	607.6	228.2	11.1	17.3	2.08
610 × 229 × 101	101.2	602.6	227.6	10.5	14.8	2.07
533 × 210 × 122	122.0	544.5	211.9	12.7	21.3	1.89
533 × 210 × 109	109.0	539.5	210.8	11.6	18.8	1.88
533 × 210 × 101	101.0	536.7	210.0	10.8	17.4	1.87
533 × 210 × 92	92.1	533.1	209.3	10.1	15.6	1.86
533 × 210 × 82	82.2	528.3	208.8	9.6	13.2	1.85
457 × 191 × 98	98.3	467.2	192.8	11.4	19.6	1.67
457 × 191 × 89	89.3	463.4	191.9	10.5	17.7	1.66
457 × 191 × 82	82.0	460.0	191.3	9.9	16.0	1.65
457 × 191 × 74	74.3	457.0	190.4	9.0	14.5	1.64
457 × 191 × 67	67.1	453.4	189.9	8.5	12.7	1.63
457 × 152 × 82	82.1	465.8	155.3	10.5	18.9	1.51
457 × 152 × 74	74.2	462.0	154.4	9.6	17.0	1.50
457 × 152 × 67	67.2	458.0	153.8	9.0	15.0	1.50
457 × 152 × 60	59.8	454.6	152.9	8.1	13.3	1.50
457 × 152 × 52	52.3	449.8	152.4	7.6	10.9	1.48
406 × 178 × 74	74.2	412.8	179.5	9.5	16.0	1.51
406 × 178 × 67	67.1	409.4	178.8	8.8	14.3	1.50
406 × 178 × 60	60.1	406.4	177.9	7.9	12.8	1.49

METAL

Designation	Mass (kg/m)	Depth of section (mm)	Width of section (mm)	Thickness		Surface area (m²/m)
				Web (mm)	Flange (mm)	
406 × 178 × 50	54.1	402.6	177.7	7.7	10.9	1.48
406 × 140 × 46	46.0	403.2	142.2	6.8	11.2	1.34
406 × 140 × 39	39.0	398.0	141.8	6.4	8.6	1.33
356 × 171 × 67	67.1	363.4	173.2	9.1	15.7	1.38
356 × 171 × 57	57.0	358.0	172.2	8.1	13.0	1.37
356 × 171 × 51	51.0	355.0	171.5	7.4	11.5	1.36
356 × 171 × 45	45.0	351.4	171.1	7.0	9.7	1.36
356 × 127 × 39	39.1	353.4	126.0	6.6	10.7	1.18
356 × 127 × 33	33.1	349.0	125.4	6.0	8.5	1.17
305 × 165 × 54	54.0	310.4	166.9	7.9	13.7	1.26
305 × 165 × 46	46.1	306.6	165.7	6.7	11.8	1.25
305 × 165 × 40	40.3	303.4	165.0	6.0	10.2	1.24
305 × 127 × 48	48.1	311.0	125.3	9.0	14.0	1.09
305 × 127 × 42	41.9	307.2	124.3	8.0	12.1	1.08
305 × 127 × 37	37.0	304.4	123.3	7.1	10.7	1.07
305 × 102 × 33	32.8	312.7	102.4	6.6	10.8	1.01
305 × 102 × 28	28.2	308.7	101.8	6.0	8.8	1.00
305 × 102 × 25	24.8	305.1	101.6	5.8	7.0	0.992
254 × 146 × 43	43.0	259.6	147.3	7.2	12.7	1.08
254 × 146 × 37	37.0	256.0	146.4	6.3	10.9	1.07
254 × 146 × 31	31.1	251.4	146.1	6.0	8.6	1.06
254 × 102 × 28	28.3	260.4	102.2	6.3	10.0	0.904
254 × 102 × 25	25.2	257.2	101.9	6.0	8.4	0.897
254 × 102 × 22	22.0	254.0	101.6	5.7	6.8	0.890
203 × 133 × 30	30.0	206.8	133.9	6.4	9.6	0.923
203 × 133 × 25	25.1	203.2	133.2	5.7	7.8	0.915
203 × 102 × 23	23.1	203.2	101.8	5.4	9.3	0.790
178 × 102 × 19	19.0	177.8	101.2	4.8	7.9	0.738
152 × 89 × 16	16.0	152.4	88.7	4.5	7.7	0.638
127 × 76 × 13	13.0	127.0	76.0	4.0	7.6	0.537

METAL

Universal Columns BS 4: Part 1: 2005

Designation	Mass (kg/m)	Depth of section (mm)	Width of section (mm)	Thickness Web (mm)	Thickness Flange (mm)	Surface area (m²/m)
356 × 406 × 634	633.9	474.7	424.0	47.6	77.0	2.52
356 × 406 × 551	551.0	455.6	418.5	42.1	67.5	2.47
356 × 406 × 467	467.0	436.6	412.2	35.8	58.0	2.42
356 × 406 × 393	393.0	419.0	407.0	30.6	49.2	2.38
356 × 406 × 340	339.9	406.4	403.0	26.6	42.9	2.35
356 × 406 × 287	287.1	393.6	399.0	22.6	36.5	2.31
356 × 406 × 235	235.1	381.0	384.8	18.4	30.2	2.28
356 × 368 × 202	201.9	374.6	374.7	16.5	27.0	2.19
356 × 368 × 177	177.0	368.2	372.6	14.4	23.8	2.17
356 × 368 × 153	152.9	362.0	370.5	12.3	20.7	2.16
356 × 368 × 129	129.0	355.6	368.6	10.4	17.5	2.14
305 × 305 × 283	282.9	365.3	322.2	26.8	44.1	1.94
305 × 305 × 240	240.0	352.5	318.4	23.0	37.7	1.91
305 × 305 × 198	198.1	339.9	314.5	19.1	31.4	1.87
305 × 305 × 158	158.1	327.1	311.2	15.8	25.0	1.84
305 × 305 × 137	136.9	320.5	309.2	13.8	21.7	1.82
305 × 305 × 118	117.9	314.5	307.4	12.0	18.7	1.81
305 × 305 × 97	96.9	307.9	305.3	9.9	15.4	1.79
254 × 254 × 167	167.1	289.1	265.2	19.2	31.7	1.58
254 × 254 × 132	132.0	276.3	261.3	15.3	25.3	1.55
254 × 254 × 107	107.1	266.7	258.8	12.8	20.5	1.52
254 × 254 × 89	88.9	260.3	256.3	10.3	17.3	1.50
254 × 254 × 73	73.1	254.1	254.6	8.6	14.2	1.49
203 × 203 × 86	86.1	222.2	209.1	12.7	20.5	1.24
203 × 203 × 71	71.0	215.8	206.4	10.0	17.3	1.22
203 × 203 × 60	60.0	209.6	205.8	9.4	14.2	1.21
203 × 203 × 52	52.0	206.2	204.3	7.9	12.5	1.20
203 × 203 × 46	46.1	203.2	203.6	7.2	11.0	1.19
152 × 152 × 37	37.0	161.8	154.4	8.0	11.5	0.912
152 × 152 × 30	30.0	157.6	152.9	6.5	9.4	0.901
152 × 152 × 23	23.0	152.4	152.2	5.8	6.8	0.889

METAL

Joists BS 4: Part 1: 2005 (retained for reference, Corus have ceased manufacture in UK)

Designation	Mass (kg/m)	Depth of section (mm)	Width of section (mm)	Thickness		Surface area (m²/m)
				Web (mm)	Flange (mm)	
254 × 203 × 82	82.0	254.0	203.2	10.2	19.9	1.210
203 × 152 × 52	52.3	203.2	152.4	8.9	16.5	0.932
152 × 127 × 37	37.3	152.4	127.0	10.4	13.2	0.737
127 × 114 × 29	29.3	127.0	114.3	10.2	11.5	0.646
127 × 114 × 27	26.9	127.0	114.3	7.4	11.4	0.650
102 × 102 × 23	23.0	101.6	101.6	9.5	10.3	0.549
102 × 44 × 7	7.5	101.6	44.5	4.3	6.1	0.350
89 × 89 × 19	19.5	88.9	88.9	9.5	9.9	0.476
76 × 76 × 13	12.8	76.2	76.2	5.1	8.4	0.411

Parallel Flange Channels

Designation	Mass (kg/m)	Depth of section (mm)	Width of section (mm)	Thickness		Surface area (m²/m)
				Web (mm)	Flange (mm)	
430 × 100 × 64	64.4	430	100	11.0	19.0	1.23
380 × 100 × 54	54.0	380	100	9.5	17.5	1.13
300 × 100 × 46	45.5	300	100	9.0	16.5	0.969
300 × 90 × 41	41.4	300	90	9.0	15.5	0.932
260 × 90 × 35	34.8	260	90	8.0	14.0	0.854
260 × 75 × 28	27.6	260	75	7.0	12.0	0.79
230 × 90 × 32	32.2	230	90	7.5	14.0	0.795
230 × 75 × 26	25.7	230	75	6.5	12.5	0.737
200 × 90 × 30	29.7	200	90	7.0	14.0	0.736
200 × 75 × 23	23.4	200	75	6.0	12.5	0.678
180 × 90 × 26	26.1	180	90	6.5	12.5	0.697
180 × 75 × 20	20.3	180	75	6.0	10.5	0.638
150 × 90 × 24	23.9	150	90	6.5	12.0	0.637
150 × 75 × 18	17.9	150	75	5.5	10.0	0.579
125 × 65 × 15	14.8	125	65	5.5	9.5	0.489
100 × 50 × 10	10.2	100	50	5.0	8.5	0.382

METAL

Equal Angles BS EN 10056-1

Designation	Mass (kg/m)	Surface area (m²/m)
200 × 200 × 24	71.1	0.790
200 × 200 × 20	59.9	0.790
200 × 200 × 18	54.2	0.790
200 × 200 × 16	48.5	0.790
150 × 150 × 18	40.1	0.59
150 × 150 × 15	33.8	0.59
150 × 150 × 12	27.3	0.59
150 × 150 × 10	23.0	0.59
120 × 120 × 15	26.6	0.47
120 × 120 × 12	21.6	0.47
120 × 120 × 10	18.2	0.47
120 × 120 × 8	14.7	0.47
100 × 100 × 15	21.9	0.39
100 × 100 × 12	17.8	0.39
100 × 100 × 10	15.0	0.39
100 × 100 × 8	12.2	0.39
90 × 90 × 12	15.9	0.35
90 × 90 × 10	13.4	0.35
90 × 90 × 8	10.9	0.35
90 × 90 × 7	9.61	0.35
90 × 90 × 6	8.30	0.35

Unequal Angles BS EN 10056-1

Designation	Mass (kg/m)	Surface area (m²/m)
200 × 150 × 18	47.1	0.69
200 × 150 × 15	39.6	0.69
200 × 150 × 12	32.0	0.69
200 × 100 × 15	33.7	0.59
200 × 100 × 12	27.3	0.59
200 × 100 × 10	23.0	0.59
150 × 90 × 15	26.6	0.47
150 × 90 × 12	21.6	0.47
150 × 90 × 10	18.2	0.47
150 × 75 × 15	24.8	0.44
150 × 75 × 12	20.2	0.44
150 × 75 × 10	17.0	0.44
125 × 75 × 12	17.8	0.40
125 × 75 × 10	15.0	0.40
125 × 75 × 8	12.2	0.40
100 × 75 × 12	15.4	0.34
100 × 75 × 10	13.0	0.34
100 × 75 × 8	10.6	0.34
100 × 65 × 10	12.3	0.32
100 × 65 × 8	9.94	0.32
100 × 65 × 7	8.77	0.32

METAL

Structural Tees Split from Universal Beams BS 4: Part 1: 2005

Designation	Mass (kg/m)	Surface area (m²/m)
305 × 305 × 90	89.5	1.22
305 × 305 × 75	74.6	1.22
254 × 343 × 63	62.6	1.19
229 × 305 × 70	69.9	1.07
229 × 305 × 63	62.5	1.07
229 × 305 × 57	56.5	1.07
229 × 305 × 51	50.6	1.07
210 × 267 × 61	61.0	0.95
210 × 267 × 55	54.5	0.95
210 × 267 × 51	50.5	0.95
210 × 267 × 46	46.1	0.95
210 × 267 × 41	41.1	0.95
191 × 229 × 49	49.2	0.84
191 × 229 × 45	44.6	0.84
191 × 229 × 41	41.0	0.84
191 × 229 × 37	37.1	0.84
191 × 229 × 34	33.6	0.84
152 × 229 × 41	41.0	0.76
152 × 229 × 37	37.1	0.76
152 × 229 × 34	33.6	0.76
152 × 229 × 30	29.9	0.76
152 × 229 × 26	26.2	0.76

Universal Bearing Piles BS 4: Part 1: 2005

Designation	Mass (kg/m)	Depth of Section (mm)	Width of section (mm)	Thickness	
				Web (mm)	Flange (mm)
356 × 368 × 174	173.9	361.4	378.5	20.3	20.4
356 × 368 × 152	152.0	356.4	376.0	17.8	17.9
356 × 368 × 133	133.0	352.0	373.8	15.6	15.7
356 × 368 × 109	108.9	346.4	371.0	12.8	12.9
305 × 305 × 223	222.9	337.9	325.7	30.3	30.4
305 × 305 × 186	186.0	328.3	320.9	25.5	25.6
305 × 305 × 149	149.1	318.5	316.0	20.6	20.7
305 × 305 × 126	126.1	312.3	312.9	17.5	17.6
305 × 305 × 110	110.0	307.9	310.7	15.3	15.4
305 × 305 × 95	94.9	303.7	308.7	13.3	13.3
305 × 305 × 88	88.0	301.7	307.8	12.4	12.3
305 × 305 × 79	78.9	299.3	306.4	11.0	11.1
254 × 254 × 85	85.1	254.3	260.4	14.4	14.3
254 × 254 × 71	71.0	249.7	258.0	12.0	12.0
254 × 254 × 63	63.0	247.1	256.6	10.6	10.7
203 × 203 × 54	53.9	204.0	207.7	11.3	11.4
203 × 203 × 45	44.9	200.2	205.9	9.5	9.5

METAL

Hot Formed Square Hollow Sections EN 10210 S275J2H & S355J2H

Size (mm)	Wall thickness (mm)	Mass (kg/m)	Superficial area (m²/m)
40 × 40	2.5	2.89	0.154
	3.0	3.41	0.152
	3.2	3.61	0.152
	3.6	4.01	0.151
	4.0	4.39	0.150
	5.0	5.28	0.147
50 × 50	2.5	3.68	0.194
	3.0	4.35	0.192
	3.2	4.62	0.192
	3.6	5.14	0.191
	4.0	5.64	0.190
	5.0	6.85	0.187
	6.0	7.99	0.185
	6.3	8.31	0.184
60 × 60	3.0	5.29	0.232
	3.2	5.62	0.232
	3.6	6.27	0.231
	4.0	6.90	0.230
	5.0	8.42	0.227
	6.0	9.87	0.225
	6.3	10.30	0.224
	8.0	12.50	0.219
70 × 70	3.0	6.24	0.272
	3.2	6.63	0.272
	3.6	7.40	0.271
	4.0	8.15	0.270
	5.0	9.99	0.267
	6.0	11.80	0.265
	6.3	12.30	0.264
	8.0	15.00	0.259
80 × 80	3.2	7.63	0.312
	3.6	8.53	0.311
	4.0	9.41	0.310
	5.0	11.60	0.307
	6.0	13.60	0.305
	6.3	14.20	0.304
	8.0	17.50	0.299
90 × 90	3.6	9.66	0.351
	4.0	10.70	0.350
	5.0	13.10	0.347
	6.0	15.50	0.345
	6.3	16.20	0.344
	8.0	20.10	0.339
100 × 100	3.6	10.80	0.391
	4.0	11.90	0.390
	5.0	14.70	0.387
	6.0	17.40	0.385
	6.3	18.20	0.384
	8.0	22.60	0.379
	10.0	27.40	0.374
120 × 120	4.0	14.40	0.470
	5.0	17.80	0.467
	6.0	21.20	0.465

METAL

Size (mm)	Wall thickness (mm)	Mass (kg/m)	Superficial area (m²/m)
	6.3	22.20	0.464
	8.0	27.60	0.459
	10.0	33.70	0.454
	12.0	39.50	0.449
	12.5	40.90	0.448
140 × 140	5.0	21.00	0.547
	6.0	24.90	0.545
	6.3	26.10	0.544
	8.0	32.60	0.539
	10.0	40.00	0.534
	12.0	47.00	0.529
	12.5	48.70	0.528
150 × 150	5.0	22.60	0.587
	6.0	26.80	0.585
	6.3	28.10	0.584
	8.0	35.10	0.579
	10.0	43.10	0.574
	12.0	50.80	0.569
	12.5	52.70	0.568
Hot formed from seamless hollow	16.0	65.2	0.559
160 × 160	5.0	24.10	0.627
	6.0	28.70	0.625
	6.3	30.10	0.624
	8.0	37.60	0.619
	10.0	46.30	0.614
	12.0	54.60	0.609
	12.5	56.60	0.608
	16.0	70.20	0.599
180 × 180	5.0	27.30	0.707
	6.0	32.50	0.705
	6.3	34.00	0.704
	8.0	42.70	0.699
	10.0	52.50	0.694
	12.0	62.10	0.689
	12.5	64.40	0.688
	16.0	80.20	0.679
200 × 200	5.0	30.40	0.787
	6.0	36.20	0.785
	6.3	38.00	0.784
	8.0	47.70	0.779
	10.0	58.80	0.774
	12.0	69.60	0.769
	12.5	72.30	0.768
	16.0	90.30	0.759
250 × 250	5.0	38.30	0.987
	6.0	45.70	0.985
	6.3	47.90	0.984
	8.0	60.30	0.979
	10.0	74.50	0.974
	12.0	88.50	0.969
	12.5	91.90	0.968
	16.0	115.00	0.959

METAL

Size (mm)	Wall thickness (mm)	Mass (kg/m)	Superficial area (m²/m)
300 × 300	6.0	55.10	1.18
	6.3	57.80	1.18
	8.0	72.80	1.18
	10.0	90.20	1.17
	12.0	107.00	1.17
	12.5	112.00	1.17
	16.0	141.00	1.16
350 × 350	8.0	85.40	1.38
	10.0	106.00	1.37
	12.0	126.00	1.37
	12.5	131.00	1.37
	16.0	166.00	1.36
400 × 400	8.0	97.90	1.58
	10.0	122.00	1.57
	12.0	145.00	1.57
	12.5	151.00	1.57
	16.0	191.00	1.56
(Grade S355J2H only)	20.00*	235.00	1.55

Note: * SAW process

METAL

Hot Formed Square Hollow Sections JUMBO RHS: JIS G3136

Size (mm)	Wall thickness (mm)	Mass (kg/m)	Superficial area (m²/m)
350 × 350	19.0	190.00	1.33
	22.0	217.00	1.32
	25.0	242.00	1.31
400 × 400	22.0	251.00	1.52
	25.0	282.00	1.51
450 × 450	12.0	162.00	1.76
	16.0	213.00	1.75
	19.0	250.00	1.73
	22.0	286.00	1.72
	25.0	321.00	1.71
	28.0 *	355.00	1.70
	32.0 *	399.00	1.69
500 × 500	12.0	181.00	1.96
	16.0	238.00	1.95
	19.0	280.00	1.93
	22.0	320.00	1.92
	25.0	360.00	1.91
	28.0 *	399.00	1.90
	32.0 *	450.00	1.89
	36.0 *	498.00	1.88
550 × 550	16.0	263.00	2.15
	19.0	309.00	2.13
	22.0	355.00	2.12
	25.0	399.00	2.11
	28.0 *	443.00	2.10
	32.0 *	500.00	2.09
	36.0 *	555.00	2.08
	40.0 *	608.00	2.06
600 × 600	25.0 *	439.00	2.31
	28.0 *	487.00	2.30
	32.0 *	550.00	2.29
	36.0 *	611.00	2.28
	40.0 *	671.00	2.26
700 × 700	25.0 *	517.00	2.71
	28.0 *	575.00	2.70
	32.0 *	651.00	2.69
	36.0 *	724.00	2.68
	40.0 *	797.00	2.68

Note: * SAW process

METAL

Hot Formed Rectangular Hollow Sections: EN10210 S275J2h & S355J2H

Size (mm)	Wall thickness (mm)	Mass (kg/m)	Superficial area (m²/m)
50 × 30	2.5	2.89	0.154
	3.0	3.41	0.152
	3.2	3.61	0.152
	3.6	4.01	0.151
	4.0	4.39	0.150
	5.0	5.28	0.147
60 × 40	2.5	3.68	0.194
	3.0	4.35	0.192
	3.2	4.62	0.192
	3.6	5.14	0.191
	4.0	5.64	0.190
	5.0	6.85	0.187
	6.0	7.99	0.185
	6.3	8.31	0.184
80 × 40	3.0	5.29	0.232
	3.2	5.62	0.232
	3.6	6.27	0.231
	4.0	6.90	0.230
	5.0	8.42	0.227
	6.0	9.87	0.225
	6.3	10.30	0.224
	8.0	12.50	0.219
76.2 × 50.8	3.0	5.62	0.246
	3.2	5.97	0.246
	3.6	6.66	0.245
	4.0	7.34	0.244
	5.0	8.97	0.241
	6.0	10.50	0.239
	6.3	11.00	0.238
	8.0	13.40	0.233
90 × 50	3.0	6.24	0.272
	3.2	6.63	0.272
	3.6	7.40	0.271
	4.0	8.15	0.270
	5.0	9.99	0.267
	6.0	11.80	0.265
	6.3	12.30	0.264
	8.0	15.00	0.259
100 × 50	3.0	6.71	0.292
	3.2	7.13	0.292
	3.6	7.96	0.291
	4.0	8.78	0.290
	5.0	10.80	0.287
	6.0	12.70	0.285
	6.3	13.30	0.284
	8.0	16.30	0.279

METAL

Size (mm)	Wall thickness (mm)	Mass (kg/m)	Superficial area (m²/m)
100 × 60	3.0	7.18	0.312
	3.2	7.63	0.312
	3.6	8.53	0.311
	4.0	9.41	0.310
	5.0	11.60	0.307
	6.0	13.60	0.305
	6.3	14.20	0.304
	8.0	17.50	0.299
120 × 60	3.6	9.70	0.351
	4.0	10.70	0.350
	5.0	13.10	0.347
	6.0	15.50	0.345
	6.3	16.20	0.344
	8.0	20.10	0.339
120 × 80	3.6	10.80	0.391
	4.0	11.90	0.390
	5.0	14.70	0.387
	6.0	17.40	0.385
	6.3	18.20	0.384
	8.0	22.60	0.379
	10.0	27.40	0.374
150 × 100	4.0	15.10	0.490
	5.0	18.60	0.487
	6.0	22.10	0.485
	6.3	23.10	0.484
	8.0	28.90	0.479
	10.0	35.30	0.474
	12.0	41.40	0.469
	12.5	42.80	0.468
160 × 80	4.0	14.40	0.470
	5.0	17.80	0.467
	6.0	21.20	0.465
	6.3	22.20	0.464
	8.0	27.60	0.459
	10.0	33.70	0.454
	12.0	39.50	0.449
	12.5	40.90	0.448
200 × 100	5.0	22.60	0.587
	6.0	26.80	0.585
	6.3	28.10	0.584
	8.0	35.10	0.579
	10.0	43.10	0.574
	12.0	50.80	0.569
	12.5	52.70	0.568
	16.0	65.20	0.559
250 × 150	5.0	30.40	0.787
	6.0	36.20	0.785
	6.3	38.00	0.784
	8.0	47.70	0.779
	10.0	58.80	0.774
	12.0	69.60	0.769
	12.5	72.30	0.768
	16.0	90.30	0.759

METAL

Size (mm)	Wall thickness (mm)	Mass (kg/m)	Superficial area (m²/m)
300 × 200	5.0	38.30	0.987
	6.0	45.70	0.985
	6.3	47.90	0.984
	8.0	60.30	0.979
	10.0	74.50	0.974
	12.0	88.50	0.969
	12.5	91.90	0.968
	16.0	115.00	0.959
400 × 200	6.0	55.10	1.18
	6.3	57.80	1.18
	8.0	72.80	1.18
	10.0	90.20	1.17
	12.0	107.00	1.17
	12.5	112.00	1.17
	16.0	141.00	1.16
450 × 250	8.0	85.40	1.38
	10.0	106.00	1.37
	12.0	126.00	1.37
	12.5	131.00	1.37
	16.0	166.00	1.36
500 × 300	8.0	98.00	1.58
	10.0	122.00	1.57
	12.0	145.00	1.57
	12.5	151.00	1.57
	16.0	191.00	1.56
	20.0	235.00	1.55

METAL

Hot Formed Circular Hollow Sections EN 10210 S275J2H & S355J2H

Outside diameter (mm)	Wall thickness (mm)	Mass (kg/m)	Superficial area (m²/m)
21.3	3.2	1.43	0.067
26.9	3.2	1.87	0.085
33.7	3.0	2.27	0.106
	3.2	2.41	0.106
	3.6	2.67	0.106
	4.0	2.93	0.106
42.4	3.0	2.91	0.133
	3.2	3.09	0.133
	3.6	3.44	0.133
	4.0	3.79	0.133
48.3	2.5	2.82	0.152
	3.0	3.35	0.152
	3.2	3.56	0.152
	3.6	3.97	0.152
	4.0	4.37	0.152
	5.0	5.34	0.152
60.3	2.5	3.56	0.189
	3.0	4.24	0.189
	3.2	4.51	0.189
	3.6	5.03	0.189
	4.0	5.55	0.189
	5.0	6.82	0.189
76.1	2.5	4.54	0.239
	3.0	5.41	0.239
	3.2	5.75	0.239
	3.6	6.44	0.239
	4.0	7.11	0.239
	5.0	8.77	0.239
	6.0	10.40	0.239
	6.3	10.80	0.239
88.9	2.5	5.33	0.279
	3.0	6.36	0.279
	3.2	6.76	0.27
	3.6	7.57	0.279
	4.0	8.38	0.279
	5.0	10.30	0.279
	6.0	12.30	0.279
	6.3	12.80	0.279
114.3	3.0	8.23	0.359
	3.2	8.77	0.359
	3.6	9.83	0.359
	4.0	10.09	0.359
	5.0	13.50	0.359
	6.0	16.00	0.359
	6.3	16.80	0.359

METAL

Outside diameter (mm)	Wall thickness (mm)	Mass (kg/m)	Superficial area (m²/m)
139.7	3.2	10.80	0.439
	3.6	12.10	0.439
	4.0	13.40	0.439
	5.0	16.60	0.439
	6.0	19.80	0.439
	6.3	20.70	0.439
	8.0	26.00	0.439
	10.0	32.00	0.439
168.3	3.2	13.00	0.529
	3.6	14.60	0.529
	4.0	16.20	0.529
	5.0	20.10	0.529
	6.0	24.00	0.529
	6.3	25.20	0.529
	8.0	31.60	0.529
	10.0	39.00	0.529
	12.0	46.30	0.529
	12.5	48.00	0.529
193.7	5.0	23.30	0.609
	6.0	27.80	0.609
	6.3	29.10	0.609
	8.0	36.60	0.609
	10.0	45.30	0.609
	12.0	53.80	0.609
	12.5	55.90	0.609
219.1	5.0	26.40	0.688
	6.0	31.50	0.688
	6.3	33.10	0.688
	8.0	41.60	0.688
	10.0	51.60	0.688
	12.0	61.30	0.688
	12.5	63.70	0.688
	16.0	80.10	0.688
244.5	5.0	29.50	0.768
	6.0	35.30	0.768
	6.3	37.00	0.768
	8.0	46.70	0.768
	10.0	57.80	0.768
	12.0	68.80	0.768
	12.5	71.50	0.768
	16.0	90.20	0.768
273.0	5.0	33.00	0.858
	6.0	39.50	0.858
	6.3	41.40	0.858
	8.0	52.30	0.858
	10.0	64.90	0.858
	12.0	77.20	0.858
	12.5	80.30	0.858
	16.0	101.00	0.858
323.9	5.0	39.30	1.02
	6.0	47.00	1.02
	6.3	49.30	1.02
	8.0	62.30	1.02
	10.0	77.40	1.02

METAL

Outside diameter (mm)	Wall thickness (mm)	Mass (kg/m)	Superficial area (m²/m)
	12.0	92.30	1.02
	12.5	96.00	1.02
	16.0	121.00	1.02
355.6	6.3	54.30	1.12
	8.0	68.60	1.12
	10.0	85.30	1.12
	12.0	102.00	1.12
	12.5	106.00	1.12
	16.0	134.00	1.12
406.4	6.3	62.20	1.28
	8.0	79.60	1.28
	10.0	97.80	1.28
	12.0	117.00	1.28
	12.5	121.00	1.28
	16.0	154.00	1.28
457.0	6.3	70.00	1.44
	8.0	88.60	1.44
	10.0	110.00	1.44
	12.0	132.00	1.44
	12.5	137.00	1.44
	16.0	174.00	1.44
508.0	6.3	77.90	1.60
	8.0	98.60	1.60
	10.0	123.00	1.60
	12.0	147.00	1.60
	12.5	153.00	1.60
	16.0	194.00	1.60

METAL

Spacing of Holes in Angles

Nominal leg length (mm)	Spacing of holes						Maximum diameter of bolt or rivet		
	A	B	C	D	E	F	A	B and C	D, E and F
200		75	75	55	55	55		30	20
150		55	55					20	
125		45	60					20	
120									
100	55						24		
90	50						24		
80	45						20		
75	45						20		
70	40						20		
65	35						20		
60	35						16		
50	28						12		
45	25								
40	23								
30	20								
25	15								

KERBS, PAVING, ETC.

KERBS/EDGINGS/CHANNELS

Precast Concrete Kerbs to BS 7263

Straight kerb units: length from 450 to 915 mm

150 mm high × 125 mm thick		
bullnosed	type BN	
half battered	type HB3	
255 mm high × 125 mm thick		
45° splayed	type SP	
half battered	type HB2	
305 mm high × 150 mm thick		
half battered	type HB1	
Quadrant kerb units		
150 mm high × 305 and 455 mm radius to match	type BN	type QBN
150 mm high × 305 and 455 mm radius to match	type HB2, HB3	type QHB
150 mm high × 305 and 455 mm radius to match	type SP	type QSP
255 mm high × 305 and 455 mm radius to match	type BN	type QBN
255 mm high × 305 and 455 mm radius to match	type HB2, HB3	type QHB
225 mm high × 305 and 455 mm radius to match	type SP	type QSP
Angle kerb units		
305 × 305 × 225 mm high × 125 mm thick		
bullnosed external angle	type XA	
splayed external angle to match type SP	type XA	
bullnosed internal angle	type IA	
splayed internal angle to match type SP	type IA	
Channels		
255 mm wide × 125 mm high flat	type CS1	
150 mm wide × 125 mm high flat type	CS2	
255 mm wide × 125 mm high dished	type CD	

KERBS, PAVING, ETC.

Transition kerb units			
from kerb type SP to HB	left handed	type TL	
	right handed	type TR	
from kerb type BN to HB	left handed	type DL1	
	right handed	type DR1	
from kerb type BN to SP	left handed	type DL2	
	right handed	type DR2	

Number of kerbs required per quarter circle (780 mm kerb lengths)

Radius (m)	Number in quarter circle
12	24
10	20
8	16
6	12
5	10
4	8
3	6
2	4
1	2

Precast Concrete Edgings

Round top type ER	Flat top type EF	Bullnosed top type EBN
150 × 50 mm	150 × 50 mm	150 × 50 mm
200 × 50 mm	200 × 50 mm	200 × 50 mm
250 × 50 mm	250 × 50 mm	250 × 50 mm

KERBS, PAVING, ETC.

BASES

Cement Bound Material for Bases and Subbases

CBM1:	very carefully graded aggregate from 37.5–75 mm, with a 7-day strength of 4.5 N/mm^2
CBM2:	same range of aggregate as CBM1 but with more tolerance in each size of aggregate with a 7-day strength of 7.0 N/mm^2
CBM3:	crushed natural aggregate or blast furnace slag, graded from 37.5–150 mm for 40 mm aggregate, and from 20–75 mm for 20 mm aggregate, with a 7-day strength of 10 N/mm^2
CBM4:	crushed natural aggregate or blast furnace slag, graded from 37.5–150 mm for 40 mm aggregate, and from 20–75 mm for 20 mm aggregate, with a 7-day strength of 15 N/mm^2

INTERLOCKING BRICK/BLOCK ROADS/PAVINGS

Sizes of Precast Concrete Paving Blocks

Type R blocks
200 × 100 × 60 mm
200 × 100 × 65 mm
200 × 100 × 80 mm
200 × 100 × 100 mm

Type S
Any shape within a 295 mm space

Sizes of clay brick pavers
200 × 100 × 50 mm
200 × 100 × 65 mm
210 × 105 × 50 mm
210 × 105 × 65 mm
215 × 102.5 × 50 mm
215 × 102.5 × 65 mm

Type PA: 3 kN
Footpaths and pedestrian areas, private driveways, car parks, light vehicle traffic and over-run

Type PB: 7 kN
Residential roads, lorry parks, factory yards, docks, petrol station forecourts, hardstandings, bus stations

KERBS, PAVING, ETC.

PAVING AND SURFACING

Weights and Sizes of Paving and Surfacing

Description of item	Size	Quantity per tonne
Paving 50 mm thick	900 × 600 mm	15
Paving 50 mm thick	750 × 600 mm	18
Paving 50 mm thick	600 × 600 mm	23
Paving 50 mm thick	450 × 600 mm	30
Paving 38 mm thick	600 × 600 mm	30
Path edging	914 × 50 × 150 mm	60
Kerb (including radius and tapers)	125 × 254 × 914 mm	15
Kerb (including radius and tapers)	125 × 150 × 914 mm	25
Square channel	125 × 254 × 914 mm	15
Dished channel	125 × 254 × 914 mm	15
Quadrants	300 × 300 × 254 mm	19
Quadrants	450 × 450 × 254 mm	12
Quadrants	300 × 300 × 150 mm	30
Internal angles	300 × 300 × 254 mm	30
Fluted pavement channel	255 × 75 × 914 mm	25
Corner stones	300 × 300 mm	80
Corner stones	360 × 360 mm	60
Cable covers	914 × 175 mm	55
Gulley kerbs	220 × 220 × 150 mm	60
Gulley kerbs	220 × 200 × 75 mm	120

KERBS, PAVING, ETC.

Weights and Sizes of Paving and Surfacing

Material	kg/m³	lb/cu yd
Tarmacadam	2306	3891
Macadam (waterbound)	2563	4325
Vermiculite (aggregate)	64–80	108–135
Terracotta	2114	3568
Cork – compressed	388	24
	kg/m²	**lb/sq ft**
Clay floor tiles, 12.7 mm	27.3	5.6
Pavement lights	122	25
Damp-proof course	5	1
	kg/m² per mm thickness	**lb/sq ft per inch thickness**
Paving slabs (stone)	2.3	12
Granite setts	2.88	15
Asphalt	2.30	12
Rubber flooring	1.68	9
Polyvinyl chloride	1.94 (avg)	10 (avg)

Coverage (m²) Per Cubic Metre of Materials Used as Subbases or Capping Layers

Consolidated thickness laid in (mm)	Square metre coverage		
	Gravel	Sand	Hardcore
50	15.80	16.50	–
75	10.50	11.00	–
100	7.92	8.20	7.42
125	6.34	6.60	5.90
150	5.28	5.50	4.95
175	–	–	4.23
200	–	–	3.71
225	–	–	3.30
300	–	–	2.47

KERBS, PAVING, ETC.

Approximate Rate of Spreads

Average thickness of course (mm)	Description	Approximate rate of spread			
		Open Textured		Dense, Medium & Fine Textured	
		(kg/m²)	(m²/t)	(kg/m²)	(m²/t)
35	14 mm open textured or dense wearing course	60–75	13–17	70–85	12–14
40	20 mm open textured or dense base course	70–85	12–14	80–100	10–12
45	20 mm open textured or dense base course	80–100	10–12	95–100	9–10
50	20 mm open textured or dense, or 28 mm dense base course	85–110	9–12	110–120	8–9
60	28 mm dense base course, 40 mm open textured of dense base course or 40 mm single course as base course		8–10	130–150	7–8
65	28 mm dense base course, 40 mm open textured or dense base course or 40 mm single course	100–135	7–10	140–160	6–7
75	40 mm single course, 40 mm open textured or dense base course, 40 mm dense roadbase	120–150	7–8	165–185	5–6
100	40 mm dense base course or roadbase	–	–	220–240	4–4.5

KERBS, PAVING, ETC.

Surface Dressing Roads: Coverage (m²) per Tonne of Material

Size in mm	Sand	Granite chips	Gravel	Limestone chips
Sand	168	–	–	–
3	–	148	152	165
6	–	130	133	144
9	–	111	114	123
13	–	85	87	95
19	–	68	71	78

Sizes of Flags

Reference	Nominal size (mm)	Thickness (mm)
A	600 × 450	50 and 63
B	600 × 600	50 and 63
C	600 × 750	50 and 63
D	600 × 900	50 and 63
E	450 × 450	50 and 70 chamfered top surface
F	400 × 400	50 and 65 chamfered top surface
G	300 × 300	50 and 60 chamfered top surface

Sizes of Natural Stone Setts

Width (mm)		Length (mm)		Depth (mm)
100	×	100	×	100
75	×	150 to 250	×	125
75	×	150 to 250	×	150
100	×	150 to 250	×	100
100	×	150 to 250	×	150

SEEDING/TURFING AND PLANTING

Topsoil Quality

Topsoil grade	Properties
Premium	Natural topsoil, high fertility, loamy texture, good soil structure, suitable for intensive cultivation.
General purpose	Natural or manufactured topsoil of lesser quality than Premium, suitable for agriculture or amenity landscape, may need fertilizer or soil structure improvement.
Economy	Selected subsoil, natural mineral deposit such as river silt or greensand. The grade comprises two subgrades; 'Low clay' and 'High clay' which is more liable to compaction in handling. This grade is suitable for low-production agricultural land and amenity woodland or conservation planting areas.

Forms of Trees

Standards:	Shall be clear with substantially straight stems. Grafted and budded trees shall have no more than a slight bend at the union. Standards shall be designated as Half, Extra light, Light, Standard, Selected standard, Heavy, and Extra heavy.
Sizes of Standards	
Heavy standard	12–14 cm girth × 3.50 to 5.00 m high
Extra Heavy standard	14–16 cm girth × 4.25 to 5.00 m high
Extra Heavy standard	16–18 cm girth × 4.25 to 6.00 m high
Extra Heavy standard	18–20 cm girth × 5.00 to 6.00 m high
Semi-mature trees:	Between 6.0 m and 12.0 m tall with a girth of 20 to 75 cm at 1.0 m above ground.
Feathered trees:	Shall have a defined upright central leader, with stem furnished with evenly spread and balanced lateral shoots down to or near the ground.
Whips:	Shall be without significant feather growth as determined by visual inspection.
Multi-stemmed trees:	Shall have two or more main stems at, near, above or below ground.

Seedlings grown from seed and not transplanted shall be specified when ordered for sale as:

1+0	one year old seedling
2+0	two year old seedling
1+1	one year seed bed, one year transplanted = two year old seedling
1+2	one year seed bed, two years transplanted = three year old seedling
2+1	two years seed bed, one year transplanted = three year old seedling
1u1	two years seed bed, undercut after 1 year = two year old seedling
2u2	four years seed bed, undercut after 2 years = four year old seedling

SEEDING/TURFING AND PLANTING

Cuttings

The age of cuttings (plants grown from shoots, stems, or roots of the mother plant) shall be specified when ordered for sale. The height of transplants and undercut seedlings/cuttings (which have been transplanted or undercut at least once) shall be stated in centimetres. The number of growing seasons before and after transplanting or undercutting shall be stated.

0 + 1	one year cutting
0 + 2	two year cutting
0 + 1 + 1	one year cutting bed, one year transplanted = two year old seedling
0 + 1 + 2	one year cutting bed, two years transplanted = three year old seedling

Grass Cutting Capacities in m² per hour

Speed mph	Width of cut in metres												
	0.5	0.7	1.0	1.2	1.5	1.7	2.0	2.0	2.1	2.5	2.8	3.0	3.4
1.0	724	1127	1529	1931	2334	2736	3138	3219	3380	4023	4506	4828	5472
1.5	1086	1690	2293	2897	3500	4104	4707	4828	5069	6035	6759	7242	8208
2.0	1448	2253	3058	3862	4667	5472	6276	6437	6759	8047	9012	9656	10944
2.5	1811	2816	3822	4828	5834	6840	7846	8047	8449	10058	11265	12070	13679
3.0	2173	3380	4587	5794	7001	8208	9415	9656	10139	12070	13518	14484	16415
3.5	2535	3943	5351	6759	8167	9576	10984	11265	11829	14082	15772	16898	19151
4.0	2897	4506	6115	7725	9334	10944	12553	12875	13518	16093	18025	19312	21887
4.5	3259	5069	6880	8690	10501	12311	14122	14484	15208	18105	20278	21726	24623
5.0	3621	5633	7644	9656	11668	13679	15691	16093	16898	20117	22531	24140	27359
5.5	3983	6196	8409	10622	12834	15047	17260	17703	18588	22128	24784	26554	30095
6.0	4345	6759	9173	11587	14001	16415	18829	19312	20278	24140	27037	28968	32831
6.5	4707	7322	9938	12553	15168	17783	20398	20921	21967	26152	29290	31382	35566
7.0	5069	7886	10702	13518	16335	19151	21967	22531	23657	28163	31543	33796	38302

Number of Plants per m² at the following offset spacings

(All plants equidistant horizontally and vertically)

Distance mm	Nr of Plants
100	114.43
200	28.22
250	18.17
300	12.35
400	6.72
500	4.29
600	3.04
700	2.16
750	1.88
900	1.26
1000	1.05
1200	0.68
1500	0.42
2000	0.23

SEEDING/TURFING AND PLANTING

Grass Clippings Wet: Based on 3.5 m³/tonne

Annual kg/100 m²	Average 20 cuts kg/100 m²	m²/tonne	m²/m³
32.0	1.6	61162.1	214067.3

Nr of cuts	22	20	18	16	12	4
kg/cut	1.45	1.60	1.78	2.00	2.67	8.00
Area capacity of 3 tonne vehicle per load						
m²	206250	187500	168750	150000	112500	37500
Load m³	**100 m² units/m³ of vehicle space**					
1	196.4	178.6	160.7	142.9	107.1	35.7
2	392.9	357.1	321.4	285.7	214.3	71.4
3	589.3	535.7	482.1	428.6	321.4	107.1
4	785.7	714.3	642.9	571.4	428.6	142.9
5	982.1	892.9	803.6	714.3	535.7	178.6

Transportation of Trees

To unload large trees a machine with the necessary lifting strength is required. The weight of the trees must therefore be known in advance. The following table gives a rough overview. The additional columns with root ball dimensions and the number of plants per trailer provide additional information, for example about preparing planting holes and calculating unloading times.

Girth in cm	Rootball diameter in cm	Ball height in cm	Weight in kg	Numbers of trees per trailer
16–18	50–60	40	150	100–120
18–20	60–70	40–50	200	80–100
20–25	60–70	40–50	270	50–70
25–30	80	50–60	350	50
30–35	90–100	60–70	500	12–18
35–40	100–110	60–70	650	10–15
40–45	110–120	60–70	850	8–12
45–50	110–120	60–70	1100	5–7
50–60	130–140	60–70	1600	1–3
60–70	150–160	60–70	2500	1
70–80	180–200	70	4000	1
80–90	200–220	70–80	5500	1
90–100	230–250	80–90	7500	1
100–120	250–270	80–90	9500	1

Data supplied by Lorenz von Ehren GmbH
The information in the table is approximate; deviations depend on soil type, genus and weather

FENCING AND GATES

Types of Preservative

Creosote (tar oil) can be 'factory' applied	by pressure to BS 144: pts 1&2 by immersion to BS 144: pt 1 by hot and cold open tank to BS 144: pts 1&2
Copper/chromium/arsenic (CCA)	by full cell process to BS 4072 pts 1&2
Organic solvent (OS)	by double vacuum (vacvac) to BS 5707 pts 1&3 by immersion to BS 5057 pts 1&3
Pentachlorophenol (PCP)	by heavy oil double vacuum to BS 5705 pts 2&3
Boron diffusion process (treated with disodium octaborate to BWPA Manual 1986)	

Note: Boron is used on green timber at source and the timber is supplied dry

Cleft Chestnut Pale Fences

Pales	Pale spacing	Wire lines	
900 mm	75 mm	2	temporary protection
1050 mm	75 or 100 mm	2	light protective fences
1200 mm	75 mm	3	perimeter fences
1350 mm	75 mm	3	perimeter fences
1500 mm	50 mm	3	narrow perimeter fences
1800 mm	50 mm	3	light security fences

Close-Boarded Fences

Close-boarded fences 1.05 to 1.8 m high
Type BCR (recessed) or BCM (morticed) with concrete posts 140 × 115 mm tapered and Type BW with timber posts

Palisade Fences

Wooden palisade fences
Type WPC with concrete posts 140 × 115 mm tapered and Type WPW with timber posts

For both types of fence:
Height of fence 1050 mm: two rails
Height of fence 1200 mm: two rails
Height of fence 1500 mm: three rails
Height of fence 1650 mm: three rails
Height of fence 1800 mm: three rails

FENCING AND GATES

Post and Rail Fences

Wooden post and rail fences Type MPR 11/3 morticed rails and Type SPR 11/3 nailed rails Height to top of rail 1100 mm Rails: three rails 87 mm, 38 mm Type MPR 11/4 morticed rails and Type SPR 11/4 nailed rails Height to top of rail 1100 mm Rails: four rails 87 mm, 38 mm Type MPR 13/4 morticed rails and Type SPR 13/4 nailed rails Height to top of rail 1300 mm Rail spacing 250 mm, 250 mm, and 225 mm from top Rails: four rails 87 mm, 38 mm

Steel Posts

Rolled steel angle iron posts for chain link fencing

Posts	Fence height	Strut	Straining post
1500 × 40 × 40 × 5 mm	900 mm	1500 × 40 × 40 × 5 mm	1500 × 50 × 50 × 6 mm
1800 × 40 × 40 × 5 mm	1200 mm	1800 × 40 × 40 × 5 mm	1800 × 50 × 50 × 6 mm
2000 × 45 × 45 × 5 mm	1400 mm	2000 × 45 × 45 × 5 mm	2000 × 60 × 60 × 6 mm
2600 × 45 × 45 × 5 mm	1800 mm	2600 × 45 × 45 × 5 mm	2600 × 60 × 60 × 6 mm
3000 × 50 × 50 × 6 mm	1800 mm	2600 × 45 × 45 × 5 mm	3000 × 60 × 60 × 6 mm
with arms			

Concrete Posts

Concrete posts for chain link fencing

Posts and straining posts	Fence height	Strut
1570 mm 100 × 100 mm	900 mm	1500 mm × 75 × 75 mm
1870 mm 125 × 125 mm	1200 mm	1830 mm × 100 × 75 mm
2070 mm 125 × 125 mm	1400 mm	1980 mm × 100 × 75 mm
2620 mm 125 × 125 mm	1800 mm	2590 mm × 100 × 85 mm
3040 mm 125 × 125 mm	1800 mm	2590 mm × 100 × 85 mm (with arms)

Rolled Steel Angle Posts

Rolled steel angle posts for rectangular wire mesh (field) fencing

Posts	Fence height	Strut	Straining post
1200 × 40 × 40 × 5 mm	600 mm	1200 × 75 × 75 mm	1350 × 100 × 100 mm
1400 × 40 × 40 × 5 mm	800 mm	1400 × 75 × 75 mm	1550 × 100 × 100 mm
1500 × 40 × 40 × 5 mm	900 mm	1500 × 75 × 75 mm	1650 × 100 × 100 mm
1600 × 40 × 40 × 5 mm	1000 mm	1600 × 75 × 75 mm	1750 × 100 × 100 mm
1750 × 40 × 40 × 5 mm	1150 mm	1750 × 75 × 100 mm	1900 × 125 × 125 mm

Concrete Posts

Concrete posts for rectangular wire mesh (field) fencing

Posts	Fence height	Strut	Straining post
1270 × 100 × 100 mm	600 mm	1200 × 75 × 75 mm	1420 × 100 × 100 mm
1470 × 100 × 100 mm	800 mm	1350 × 75 × 75 mm	1620 × 100 × 100 mm
1570 × 100 × 100 mm	900 mm	1500 × 75 × 75 mm	1720 × 100 × 100 mm
1670 × 100 × 100 mm	600 mm	1650 × 75 × 75 mm	1820 × 100 × 100 mm
1820 × 125 × 125 mm	1150 mm	1830 × 75 × 100 mm	1970 × 125 × 125 mm

Cleft Chestnut Pale Fences

Timber Posts

Timber posts for wire mesh and hexagonal wire netting fences

Round timber for general fences

Posts	Fence height	Strut	Straining post
1300 × 65 mm dia.	600 mm	1200 × 80 mm dia.	1450 × 100 mm dia.
1500 × 65 mm dia.	800 mm	1400 × 80 mm dia.	1650 × 100 mm dia.
1600 × 65 mm dia.	900 mm	1500 × 80 mm dia.	1750 × 100 mm dia.
1700 × 65 mm dia.	1050 mm	1600 × 80 mm dia.	1850 × 100 mm dia.
1800 × 65 mm dia.	1150 mm	1750 × 80 mm dia.	2000 × 120 mm dia.

Squared timber for general fences

Posts	Fence height	Strut	Straining post
1300 × 75 × 75 mm	600 mm	1200 × 75 × 75 mm	1450 × 100 × 100 mm
1500 × 75 × 75 mm	800 mm	1400 × 75 × 75 mm	1650 × 100 × 100 mm
1600 × 75 × 75 mm	900 mm	1500 × 75 × 75 mm	1750 × 100 × 100 mm
1700 × 75 × 75 mm	1050 mm	1600 × 75 × 75 mm	1850 × 100 × 100 mm
1800 × 75 × 75 mm	1150 mm	1750 × 75 × 75 mm	2000 × 125 × 100 mm

FENCING AND GATES

Steel Fences to BS 1722: Part 9: 1992

	Fence height	Top/bottom rails and flat posts	Vertical bars
Light	1000 mm	40 × 10 mm 450 mm in ground	12 mm dia. at 115 mm cs
	1200 mm	40 × 10 mm 550 mm in ground	12 mm dia. at 115 mm cs
	1400 mm	40 × 10 mm 550 mm in ground	12 mm dia. at 115 mm cs
Light	1000 mm	40 × 10 mm 450 mm in ground	16 mm dia. at 120 mm cs
	1200 mm	40 × 10 mm 550 mm in ground	16 mm dia. at 120 mm cs
	1400 mm	40 × 10 mm 550 mm in ground	16 mm dia. at 120 mm cs
Medium	1200 mm	50 × 10 mm 550 mm in ground	20 mm dia. at 125 mm cs
	1400 mm	50 × 10 mm 550 mm in ground	20 mm dia. at 125 mm cs
	1600 mm	50 × 10 mm 600 mm in ground	22 mm dia. at 145 mm cs
	1800 mm	50 × 10 mm 600 mm in ground	22 mm dia. at 145 mm cs
Heavy	1600 mm	50 × 10 mm 600 mm in ground	22 mm dia. at 145 mm cs
	1800 mm	50 × 10 mm 600 mm in ground	22 mm dia. at 145 mm cs
	2000 mm	50 × 10 mm 600 mm in ground	22 mm dia. at 145 mm cs
	2200 mm	50 × 10 mm 600 mm in ground	22 mm dia. at 145 mm cs

Notes: Mild steel fences: round or square verticals; flat standards and horizontals. Tops of vertical bars may be bow-top, blunt, or pointed. Round or square bar railings

Timber Field Gates to BS 3470: 1975

Gates made to this standard are designed to open one way only
All timber gates are 1100 mm high
Width over stiles 2400, 2700, 3000, 3300, 3600, and 4200 mm
Gates over 4200 mm should be made in two leaves

Steel Field Gates to BS 3470: 1975

All steel gates are 1100 mm high
Heavy duty: width over stiles 2400, 3000, 3600 and 4500 mm
Light duty: width over stiles 2400, 3000, and 3600 mm

FENCING AND GATES

Domestic Front Entrance Gates to BS 4092: Part 1: 1966

Metal gates:	Single gates are 900 mm high minimum, 900 mm, 1000 mm and 1100 mm wide

Domestic Front Entrance Gates to BS 4092: Part 2: 1966

Wooden gates:	All rails shall be tenoned into the stiles Single gates are 840 mm high minimum, 801 mm and 1020 mm wide Double gates are 840 mm high minimum, 2130, 2340 and 2640 mm wide

Timber Bridle Gates to BS 5709:1979 (Horse or Hunting Gates)

Gates open one way only Minimum width between posts Minimum height	 1525 mm 1100 mm

Timber Kissing Gates to BS 5709:1979

Minimum width	700 mm
Minimum height	1000 mm
Minimum distance between shutting posts	600 mm
Minimum clearance at mid-point	600 mm

Metal Kissing Gates to BS 5709:1979

Sizes are the same as those for timber kissing gates Maximum gaps between rails 120 mm

Categories of Pedestrian Guard Rail to BS 3049:1976

Class A for normal use Class B where vandalism is expected Class C where crowd pressure is likely

DRAINAGE

Width Required for Trenches for Various Diameters of Pipes

Pipe diameter (mm)	Trench n.e. 1.50 m deep	Trench over 1.50 m deep
n.e. 100 mm	450 mm	600 mm
100–150 mm	500 mm	650 mm
150–225 mm	600 mm	750 mm
225–300 mm	650 mm	800 mm
300–400 mm	750 mm	900 mm
400–450 mm	900 mm	1050 mm
450–600 mm	1100 mm	1300 mm

Weights and Dimensions – Vitrified Clay Pipes

Product	Nominal diameter (mm)	Effective length (mm)	BS 65 limits of tolerance min (mm)	max (mm)	Crushing strength (kN/m)	Weight (kg/pipe)	(kg/m)
Supersleve	100	1600	96	105	35.00	14.71	9.19
	150	1750	146	158	35.00	29.24	16.71
Hepsleve	225	1850	221	236	28.00	84.03	45.42
	300	2500	295	313	34.00	193.05	77.22
	150	1500	146	158	22.00	37.04	24.69
Hepseal	225	1750	221	236	28.00	85.47	48.84
	300	2500	295	313	34.00	204.08	81.63
	400	2500	394	414	44.00	357.14	142.86
	450	2500	444	464	44.00	454.55	181.63
	500	2500	494	514	48.00	555.56	222.22
	600	2500	591	615	57.00	796.23	307.69
	700	3000	689	719	67.00	1111.11	370.45
	800	3000	788	822	72.00	1351.35	450.45
Hepline	100	1600	95	107	22.00	14.71	9.19
	150	1750	145	160	22.00	29.24	16.71
	225	1850	219	239	28.00	84.03	45.42
	300	1850	292	317	34.00	142.86	77.22
Hepduct (conduit)	90	1500	–	–	28.00	12.05	8.03
	100	1600	–	–	28.00	14.71	9.19
	125	1750	–	–	28.00	20.73	11.84
	150	1750	–	–	28.00	29.24	16.71
	225	1850	–	–	28.00	84.03	45.42
	300	1850	–	–	34.00	142.86	77.22

DRAINAGE

Weights and Dimensions – Vitrified Clay Pipes

Nominal internal diameter (mm)	Nominal wall thickness (mm)	Approximate weight (kg/m)
150	25	45
225	29	71
300	32	122
375	35	162
450	38	191
600	48	317
750	54	454
900	60	616
1200	76	912
1500	89	1458
1800	102	1884
2100	127	2619

Wall thickness, weights and pipe lengths vary, depending on type of pipe required

The particulars shown above represent a selection of available diameters and are applicable to strength class 1 pipes with flexible rubber ring joints

Tubes with Ogee joints are also available

DRAINAGE

Weights and Dimensions – PVC-u Pipes

	Nominal size	Mean outside diameter (mm)		Wall thickness	Weight
		min	max	(mm)	(kg/m)
Standard pipes	82.4	82.4	82.7	3.2	1.2
	110.0	110.0	110.4	3.2	1.6
	160.0	160.0	160.6	4.1	3.0
	200.0	200.0	200.6	4.9	4.6
	250.0	250.0	250.7	6.1	7.2
Perforated pipes heavy grade	As above	As above	As above	As above	As above
thin wall	82.4	82.4	82.7	1.7	–
	110.0	110.0	110.4	2.2	–
	160.0	160.0	160.6	3.2	–

Width of Trenches Required for Various Diameters of Pipes

Pipe diameter (mm)	Trench n.e. 1.5 m deep (mm)	Trench over 1.5 m deep (mm)
n.e. 100	450	600
100–150	500	650
150–225	600	750
225–300	650	800
300–400	750	900
400–450	900	1050
450–600	1100	1300

DRAINAGE

DRAINAGE BELOW GROUND AND LAND DRAINAGE

Flow of Water Which Can Be Carried by Various Sizes of Pipe

Clay or concrete pipes

	Gradient of pipeline							
	1:10	1:20	1:30	1:40	1:50	1:60	1:80	1:100
Pipe size	Flow in litres per second							
DN 100 15.0	8.5	6.8	5.8	5.2	4.7	4.0	3.5	
DN 150 28.0	19.0	16.0	14.0	12.0	11.0	9.1	8.0	
DN 225 140.0	95.0	76.0	66.0	58.0	53.0	46.0	40.0	

Plastic pipes

	Gradient of pipeline							
	1:10	1:20	1:30	1:40	1:50	1:60	1:80	1:100
Pipe size	Flow in litres per second							
82.4 mm i/dia.	12.0	8.5	6.8	5.8	5.2	4.7	4.0	3.5
110 mm i/dia.	28.0	19.0	16.0	14.0	12.0	11.0	9.1	8.0
160 mm i/dia.	76.0	53.0	43.0	37.0	33.0	29.0	25.0	22.0
200 mm i/dia.	140.0	95.0	76.0	66.0	58.0	53.0	46.0	40.0

Vitrified (Perforated) Clay Pipes and Fittings to BS En 295-5 1994

Length not specified		
75 mm bore	250 mm bore	600 mm bore
100	300	700
125	350	800
150	400	1000
200	450	1200
225	500	

Precast Concrete Pipes: Prestressed Non-pressure Pipes and Fittings: Flexible Joints to BS 5911: Pt. 103: 1994

Rationalized metric nominal sizes: 450, 500
Length: 500–1000 by 100 increments 1000–2200 by 200 increments 2200–2800 by 300 increments
Angles: length: 450–600 angles 45, 22.5,11.25° 600 or more angles 22.5, 11.25°

DRAINAGE

Precast Concrete Pipes: Unreinforced and Circular Manholes and Soakaways to BS 5911: Pt. 200: 1994

Nominal sizes:	
Shafts:	675, 900 mm
Chambers:	900, 1050, 1200, 1350, 1500, 1800, 2100, 2400, 2700, 3000 mm
Large chambers:	To have either tapered reducing rings or a flat reducing slab in order to accept the standard cover
Ring depths:	1. 300–1200 mm by 300 mm increments except for bottom slab and rings below cover slab, these are by 150 mm increments
	2. 250–1000 mm by 250 mm increments except for bottom slab and rings below cover slab, these are by 125 mm increments
Access hole:	750 × 750 mm for DN 1050 chamber 1200 × 675 mm for DN 1350 chamber

Calculation of Soakaway Depth

The following formula determines the depth of concrete ring soakaway that would be required for draining given amounts of water.

$h = \frac{4ar}{3\pi D^2}$

h = depth of the chamber below the invert pipe
a = the area to be drained
r = the hourly rate of rainfall (50 mm per hour)
π = pi
D = internal diameter of the soakaway

This table shows the depth of chambers in each ring size which would be required to contain the volume of water specified. These allow a recommended storage capacity of ⅓ (one third of the hourly rainfall figure).

Table Showing Required Depth of Concrete Ring Chambers in Metres

Area m²	50	100	150	200	300	400	500
Ring size							
0.9	1.31	2.62	3.93	5.24	7.86	10.48	13.10
1.1	0.96	1.92	2.89	3.85	5.77	7.70	9.62
1.2	0.74	1.47	2.21	2.95	4.42	5.89	7.37
1.4	0.58	1.16	1.75	2.33	3.49	4.66	5.82
1.5	0.47	0.94	1.41	1.89	2.83	3.77	4.72
1.8	0.33	0.65	0.98	1.31	1.96	2.62	3.27
2.1	0.24	0.48	0.72	0.96	1.44	1.92	2.41
2.4	0.18	0.37	0.55	0.74	1.11	1.47	1.84
2.7	0.15	0.29	0.44	0.58	0.87	1.16	1.46
3.0	0.12	0.24	0.35	0.47	0.71	0.94	1.18

DRAINAGE

Precast Concrete Inspection Chambers and Gullies to BS 5911: Part 230: 1994

Nominal sizes:	375 diameter, 750, 900 mm deep
	450 diameter, 750, 900, 1050, 1200 mm deep
Depths:	from the top for trapped or untrapped units:
	centre of outlet 300 mm
	invert (bottom) of the outlet pipe 400 mm
Depth of water seal for trapped gullies:	
	85 mm, rodding eye int. dia. 100 mm
Cover slab:	65 mm min

Bedding Flexible Pipes: PVC-u Or Ductile Iron

Type 1 =	100 mm fill below pipe, 300 mm above pipe: single size material
Type 2 =	100 mm fill below pipe, 300 mm above pipe: single size or graded material
Type 3 =	100 mm fill below pipe, 75 mm above pipe with concrete protective slab over
Type 4 =	100 mm fill below pipe, fill laid level with top of pipe
Type 5 =	200 mm fill below pipe, fill laid level with top of pipe
Concrete =	25 mm sand blinding to bottom of trench, pipe supported on chocks, 100 mm concrete under the pipe, 150 mm concrete over the pipe

DRAINAGE

Bedding Rigid Pipes: Clay or Concrete
(for vitrified clay pipes the manufacturer should be consulted)

Class D:	Pipe laid on natural ground with cut-outs for joints, soil screened to remove stones over 40 mm and returned over pipe to 150 m min depth. Suitable for firm ground with trenches trimmed by hand.
Class N:	Pipe laid on 50 mm granular material of graded aggregate to Table 4 of BS 882, or 10 mm aggregate to Table 6 of BS 882, or as dug light soil (not clay) screened to remove stones over 10 mm. Suitable for machine dug trenches.
Class B:	As Class N, but with granular bedding extending half way up the pipe diameter.
Class F:	Pipe laid on 100 mm granular fill to BS 882 below pipe, minimum 150 mm granular fill above pipe: single size material. Suitable for machine dug trenches.
Class A:	Concrete 100 mm thick under the pipe extending half way up the pipe, backfilled with the appropriate class of fill. Used where there is only a very shallow fall to the drain. Class A bedding allows the pipes to be laid to an exact gradient.
Concrete surround:	25 mm sand blinding to bottom of trench, pipe supported on chocks, 100 mm concrete under the pipe, 150 mm concrete over the pipe. It is preferable to bed pipes under slabs or wall in granular material.

PIPED SUPPLY SYSTEMS

Identification of Service Tubes From Utility to Dwellings

Utility	Colour	Size	Depth
British Telecom	grey	54 mm od	450 mm
Electricity	black	38 mm od	450 mm
Gas	yellow	42 mm od rigid 60 mm od convoluted	450 mm
Water	may be blue	(normally untubed)	750 mm

ELECTRICAL SUPPLY/POWER/LIGHTING SYSTEMS

Electrical Insulation Class En 60.598 BS 4533

Class 1:　　luminaires comply with class 1 (I) earthed electrical requirements
Class 2:　　luminaires comply with class 2 (II) double insulated electrical requirements
Class 3:　　luminaires comply with class 3 (III) electrical requirements

Protection to Light Fittings

BS EN 60529:1992 Classification for degrees of protection provided by enclosures.
(IP Code – International or ingress Protection)

1st characteristic: against ingress of solid foreign objects		
The figure	2	indicates that fingers cannot enter
	3	that a 2.5 mm diameter probe cannot enter
	4	that a 1.0 mm diameter probe cannot enter
	5	the fitting is dust proof (no dust around live parts)
	6	the fitting is dust tight (no dust entry)
2nd characteristic: ingress of water with harmful effects		
The figure	0	indicates unprotected
	1	vertically dripping water cannot enter
	2	water dripping 15° (tilt) cannot enter
	3	spraying water cannot enter
	4	splashing water cannot enter
	5	jetting water cannot enter
	6	powerful jetting water cannot enter
	7	proof against temporary immersion
	8	proof against continuous immersion
Optional additional codes:		A–D protects against access to hazardous parts
	H	high voltage apparatus
	M	fitting was in motion during water test
	S	fitting was static during water test
	W	protects against weather
Marking code arrangement:		(example) IPX5S = IP (International or Ingress Protection)
		X (denotes omission of first characteristic)
		5 = jetting
		S = static during water test

RAIL TRACKS

	kg/m of track	lb/ft of track
Standard gauge Bull-head rails, chairs, transverse timber (softwood) sleepers etc.	245	165
Main lines Flat-bottom rails, transverse prestressed concrete sleepers, etc.	418	280
Add for electric third rail	51	35
Add for crushed stone ballast	2600	1750
	kg/m²	**lb/sq ft**
Overall average weight – rails connections, sleepers, ballast, etc.	733	150
	kg/m of track	**lb/ft of track**
Bridge rails, longitudinal timber sleepers, etc.	112	75

RAIL TRACKS

Heavy Rails

British Standard Section No.	Rail height (mm)	Foot width (mm)	Head width (mm)	Min web thickness (mm)	Section weight (kg/m)
Flat Bottom Rails					
60 A	114.30	109.54	57.15	11.11	30.62
70 A	123.82	111.12	60.32	12.30	34.81
75 A	128.59	114.30	61.91	12.70	37.45
80 A	133.35	117.47	63.50	13.10	39.76
90 A	142.88	127.00	66.67	13.89	45.10
95 A	147.64	130.17	69.85	14.68	47.31
100 A	152.40	133.35	69.85	15.08	50.18
110 A	158.75	139.70	69.85	15.87	54.52
113 A	158.75	139.70	69.85	20.00	56.22
50 'O'	100.01	100.01	52.39	10.32	24.82
80 'O'	127.00	127.00	63.50	13.89	39.74
60R	114.30	109.54	57.15	11.11	29.85
75R	128.59	122.24	61.91	13.10	37.09
80R	133.35	127.00	63.50	13.49	39.72
90R	142.88	136.53	66.67	13.89	44.58
95R	147.64	141.29	68.26	14.29	47.21
100R	152.40	146.05	69.85	14.29	49.60
95N	147.64	139.70	69.85	13.89	47.27
Bull Head Rails					
95R BH	145.26	69.85	69.85	19.05	47.07

Light Rails

British Standard Section No.	Rail height (mm)	Foot width (mm)	Head width (mm)	Min web thickness (mm)	Section weight (kg/m)
Flat Bottom Rails					
20M	65.09	55.56	30.96	6.75	9.88
30M	75.41	69.85	38.10	9.13	14.79
35M	80.96	76.20	42.86	9.13	17.39
35R	85.73	82.55	44.45	8.33	17.40
40	88.11	80.57	45.64	12.3	19.89
Bridge Rails					
13	48.00	92	36.00	18.0	13.31
16	54.00	108	44.50	16.0	16.06
20	55.50	127	50.00	20.5	19.86
28	67.00	152	50.00	31.0	28.62
35	76.00	160	58.00	34.5	35.38
50	76.00	165	58.50	–	50.18
Crane Rails					
A65	75.00	175.00	65.00	38.0	43.10
A75	85.00	200.00	75.00	45.0	56.20
A100	95.00	200.00	100.00	60.0	74.30
A120	105.00	220.00	120.00	72.0	100.00
175CR	152.40	152.40	107.95	38.1	86.92

Fish Plates

British Standard Section No.	Overall plate length		Hole diameter	Finished weight per pair	
	4 Hole (mm)	6 Hole (mm)	(mm)	4 Hole (kg/pair)	6 Hole (kg/pair)
For British Standard Heavy Rails: Flat Bottom Rails					
60 A	406.40	609.60	20.64	9.87	14.76
70 A	406.40	609.60	22.22	11.15	16.65
75 A	406.40	–	23.81	11.82	17.73
80 A	406.40	609.60	23.81	13.15	19.72
90 A	457.20	685.80	25.40	17.49	26.23
100 A	508.00	–	pear	25.02	–
110 A (shallow)	507.00	–	27.00	30.11	54.64
113 A (heavy)	507.00	–	27.00	30.11	54.64
50 'O' (shallow)	406.40	–	–	6.68	10.14
80 'O' (shallow)	495.30	–	23.81	14.72	22.69
60R (shallow)	406.40	609.60	20.64	8.76	13.13
60R (angled)	406.40	609.60	20.64	11.27	16.90
75R (shallow)	406.40	–	23.81	10.94	16.42
75R (angled)	406.40	–	23.81	13.67	–
80R (shallow)	406.40	609.60	23.81	11.93	17.89
80R (angled)	406.40	609.60	23.81	14.90	22.33
For British Standard Heavy Rails: Bull head rails					
95R BH (shallow)	–	457.20	27.00	14.59	14.61
For British Standard Light Rails: Flat Bottom Rails					
30M	355.6	–	–	–	2.72
35M	355.6	–	–	–	2.83
40	355.6	–	–	3.76	–

FRACTIONS, DECIMALS AND MILLIMETRE EQUIVALENTS

Fractions	Decimals	(mm)	Fractions	Decimals	(mm)
1/64	0.015625	0.396875	33/64	0.515625	13.096875
1/32	0.03125	0.79375	17/32	0.53125	13.49375
3/64	0.046875	1.190625	35/64	0.546875	13.890625
1/16	0.0625	1.5875	9/16	0.5625	14.2875
5/64	0.078125	1.984375	37/64	0.578125	14.684375
3/32	0.09375	2.38125	19/32	0.59375	15.08125
7/64	0.109375	2.778125	39/64	0.609375	15.478125
1/8	0.125	3.175	5/8	0.625	15.875
9/64	0.140625	3.571875	41/64	0.640625	16.271875
5/32	0.15625	3.96875	21/32	0.65625	16.66875
11/64	0.171875	4.365625	43/64	0.671875	17.065625
3/16	0.1875	4.7625	11/16	0.6875	17.4625
13/64	0.203125	5.159375	45/64	0.703125	17.859375
7/32	0.21875	5.55625	23/32	0.71875	18.25625
15/64	0.234375	5.953125	47/64	0.734375	18.653125
1/4	0.25	6.35	3/4	0.75	19.05
17/64	0.265625	6.746875	49/64	0.765625	19.446875
9/32	0.28125	7.14375	25/32	0.78125	19.84375
19/64	0.296875	7.540625	51/64	0.796875	20.240625
5/16	0.3125	7.9375	13/16	0.8125	20.6375
21/64	0.328125	8.334375	53/64	0.828125	21.034375
11/32	0.34375	8.73125	27/32	0.84375	21.43125
23/64	0.359375	9.128125	55/64	0.859375	21.828125
3/8	0.375	9.525	7/8	0.875	22.225
25/64	0.390625	9.921875	57/64	0.890625	22.621875
13/32	0.40625	10.31875	29/32	0.90625	23.01875
27/64	0.421875	10.71563	59/64	0.921875	23.415625
7/16	0.4375	11.1125	15/16	0.9375	23.8125
29/64	0.453125	11.50938	61/64	0.953125	24.209375
15/32	0.46875	11.90625	31/32	0.96875	24.60625
31/64	0.484375	12.30313	63/64	0.984375	25.003125
1/2	0.5	12.7	1.0	1	25.4

IMPERIAL STANDARD WIRE GAUGE (SWG)

SWG No.	Diameter (inches)	Diameter (mm)	SWG No.	Diameter (inches)	Diameter (mm)
7/0	0.5	12.7	23	0.024	0.61
6/0	0.464	11.79	24	0.022	0.559
5/0	0.432	10.97	25	0.02	0.508
4/0	0.4	10.16	26	0.018	0.457
3/0	0.372	9.45	27	0.0164	0.417
2/0	0.348	8.84	28	0.0148	0.376
1/0	0.324	8.23	29	0.0136	0.345
1	0.3	7.62	30	0.0124	0.315
2	0.276	7.01	31	0.0116	0.295
3	0.252	6.4	32	0.0108	0.274
4	0.232	5.89	33	0.01	0.254
5	0.212	5.38	34	0.009	0.234
6	0.192	4.88	35	0.008	0.213
7	0.176	4.47	36	0.008	0.193
8	0.16	4.06	37	0.007	0.173
9	0.144	3.66	38	0.006	0.152
10	0.128	3.25	39	0.005	0.132
11	0.116	2.95	40	0.005	0.122
12	0.104	2.64	41	0.004	0.112
13	0.092	2.34	42	0.004	0.102
14	0.08	2.03	43	0.004	0.091
15	0.072	1.83	44	0.003	0.081
16	0.064	1.63	45	0.003	0.071
17	0.056	1.42	46	0.002	0.061
18	0.048	1.22	47	0.002	0.051
19	0.04	1.016	48	0.002	0.041
20	0.036	0.914	49	0.001	0.031
21	0.032	0.813	50	0.001	0.025
22	0.028	0.711			

PIPES, WATER, STORAGE, INSULATION

WATER PRESSURE DUE TO HEIGHT

Imperial

Head (Feet)	Pressure (lb/in^2)	Head (Feet)	Pressure (lb/in^2)
1	0.43	70	30.35
5	2.17	75	32.51
10	4.34	80	34.68
15	6.5	85	36.85
20	8.67	90	39.02
25	10.84	95	41.18
30	13.01	100	43.35
35	15.17	105	45.52
40	17.34	110	47.69
45	19.51	120	52.02
50	21.68	130	56.36
55	23.84	140	60.69
60	26.01	150	65.03
65	28.18		

Metric

Head (m)	Pressure (bar)	Head (m)	Pressure (bar)
0.5	0.049	18.0	1.766
1.0	0.098	19.0	1.864
1.5	0.147	20.0	1.962
2.0	0.196	21.0	2.06
3.0	0.294	22.0	2.158
4.0	0.392	23.0	2.256
5.0	0.491	24.0	2.354
6.0	0.589	25.0	2.453
7.0	0.687	26.0	2.551
8.0	0.785	27.0	2.649
9.0	0.883	28.0	2.747
10.0	0.981	29.0	2.845
11.0	1.079	30.0	2.943
12.0	1.177	32.5	3.188
13.0	1.275	35.0	3.434
14.0	1.373	37.5	3.679
15.0	1.472	40.0	3.924
16.0	1.57	42.5	4.169
17.0	1.668	45.0	4.415

1 bar	=	14.5038 lbf/in^2
1 lbf/in^2	=	0.06895 bar
1 metre	=	3.2808 ft or 39.3701 in
1 foot	=	0.3048 metres
1 in wg	=	2.5 mbar (249.1 N/m^2)

PIPES, WATER, STORAGE, INSULATION

Dimensions and Weights of Copper Pipes to BSEN 1057, BSEN 12499, BSEN 14251

Outside diameter (mm)	Internal diameter (mm)	Weight per metre (kg)	Internal diameter (mm)	Weight per metre (kg)	Internal siameter (mm)	Weight per metre (kg)
	Formerly Table X		Formerly Table Y		Formerly Table Z	
6	4.80	0.0911	4.40	0.1170	5.00	0.0774
8	6.80	0.1246	6.40	0.1617	7.00	0.1054
10	8.80	0.1580	8.40	0.2064	9.00	0.1334
12	10.80	0.1914	10.40	0.2511	11.00	0.1612
15	13.60	0.2796	13.00	0.3923	14.00	0.2031
18	16.40	0.3852	16.00	0.4760	16.80	0.2918
22	20.22	0.5308	19.62	0.6974	20.82	0.3589
28	26.22	0.6814	25.62	0.8985	26.82	0.4594
35	32.63	1.1334	32.03	1.4085	33.63	0.6701
42	39.63	1.3675	39.03	1.6996	40.43	0.9216
54	51.63	1.7691	50.03	2.9052	52.23	1.3343
76.1	73.22	3.1287	72.22	4.1437	73.82	2.5131
108	105.12	4.4666	103.12	7.3745	105.72	3.5834
133	130.38	5.5151	–	–	130.38	5.5151
159	155.38	8.7795	–	–	156.38	6.6056

Dimensions of Stainless Steel Pipes to BS 4127

Outside siameter (mm)	Maximum outside siameter (mm)	Minimum outside diameter (mm)	Wall thickness (mm)	Working pressure (bar)
6	6.045	5.940	0.6	330
8	8.045	7.940	0.6	260
10	10.045	9.940	0.6	210
12	12.045	11.940	0.6	170
15	15.045	14.940	0.6	140
18	18.045	17.940	0.7	135
22	22.055	21.950	0.7	110
28	28.055	27.950	0.8	121
35	35.070	34.965	1.0	100
42	42.070	41.965	1.1	91
54	54.090	53.940	1.2	77

PIPES, WATER, STORAGE, INSULATION

Dimensions of Steel Pipes to BS 1387

Nominal Size (mm)	Approx. Outside Diameter (mm)	Outside diameter				Thickness		
		Light		Medium & Heavy		Light (mm)	Medium (mm)	Heavy (mm)
		Max (mm)	Min (mm)	Max (mm)	Min (mm)			
6	10.20	10.10	9.70	10.40	9.80	1.80	2.00	2.65
8	13.50	13.60	13.20	13.90	13.30	1.80	2.35	2.90
10	17.20	17.10	16.70	17.40	16.80	1.80	2.35	2.90
15	21.30	21.40	21.00	21.70	21.10	2.00	2.65	3.25
20	26.90	26.90	26.40	27.20	26.60	2.35	2.65	3.25
25	33.70	33.80	33.20	34.20	33.40	2.65	3.25	4.05
32	42.40	42.50	41.90	42.90	42.10	2.65	3.25	4.05
40	48.30	48.40	47.80	48.80	48.00	2.90	3.25	4.05
50	60.30	60.20	59.60	60.80	59.80	2.90	3.65	4.50
65	76.10	76.00	75.20	76.60	75.40	3.25	3.65	4.50
80	88.90	88.70	87.90	89.50	88.10	3.25	4.05	4.85
100	114.30	113.90	113.00	114.90	113.30	3.65	4.50	5.40
125	139.70	–	–	140.60	138.70	–	4.85	5.40
150	165.1*	–	–	166.10	164.10	–	4.85	5.40

* 165.1 mm (6.5in) outside diameter is not generally recommended except where screwing to BS 21 is necessary
All dimensions are in accordance with ISO R65 except approximate outside diameters which are in accordance with ISO R64
Light quality is equivalent to ISO R65 Light Series II

Approximate Metres Per Tonne of Tubes to BS 1387

Nom. size (mm)	BLACK						GALVANIZED					
	Plain/screwed ends			Screwed & socketed			Plain/screwed ends			Screwed & socketed		
	L (m)	M (m)	H (m)	L (m)	M (m)	H (m)	L (m)	M (m)	H (m)	L (m)	M (m)	H (m)
6	2765	2461	2030	2743	2443	2018	2604	2333	1948	2584	2317	1937
8	1936	1538	1300	1920	1527	1292	1826	1467	1254	1811	1458	1247
10	1483	1173	979	1471	1165	974	1400	1120	944	1386	1113	939
15	1050	817	688	1040	811	684	996	785	665	987	779	661
20	712	634	529	704	628	525	679	609	512	673	603	508
25	498	410	336	494	407	334	478	396	327	474	394	325
32	388	319	260	384	316	259	373	308	254	369	305	252
40	307	277	226	303	273	223	296	268	220	292	264	217
50	244	196	162	239	194	160	235	191	158	231	188	157
65	172	153	127	169	151	125	167	149	124	163	146	122
80	147	118	99	143	116	98	142	115	97	139	113	96
100	101	82	69	98	81	68	98	81	68	95	79	67
125	–	62	56	–	60	55	–	60	55	–	59	54
150	–	52	47	–	50	46	–	51	46	–	49	45

The figures for 'plain or screwed ends' apply also to tubes to BS 1775 of equivalent size and thickness
Key:
L – Light
M – Medium
H – Heavy

Flange Dimension Chart to BS 4504 & BS 10

Normal Pressure Rating (PN 6) 6 Bar

Nom. size	Flange outside dia.	Table 6/2 Forged Welding Neck	Table 6/3 Plate Slip on	Table 6/4 Forged Bossed Screwed	Table 6/5 Forged Bossed Slip on	Table 6/8 Plate Blank	Raised face Dia.	T'ness	Nr. bolt hole	Size of bolt
15	80	12	12	12	12	12	40	2	4	M10 × 40
20	90	14	14	14	14	14	50	2	4	M10 × 45
25	100	14	14	14	14	14	60	2	4	M10 × 45
32	120	14	16	14	14	14	70	2	4	M12 × 45
40	130	14	16	14	14	14	80	3	4	M12 × 45
50	140	14	16	14	14	14	90	3	4	M12 × 45
65	160	14	16	14	14	14	110	3	4	M12 × 45
80	190	16	18	16	16	16	128	3	4	M16 × 55
100	210	16	18	16	16	16	148	3	4	M16 × 55
125	240	18	20	18	18	18	178	3	8	M16 × 60
150	265	18	20	18	18	18	202	3	8	M16 × 60
200	320	20	22	–	20	20	258	3	8	M16 × 60
250	375	22	24	–	22	22	312	3	12	M16 × 65
300	440	22	24	–	22	22	365	4	12	M20 × 70

Normal Pressure Rating (PN 16) 16 Bar

Nom. size	Flange outside dia.	Table 6/2 Forged Welding Neck	Table 6/3 Plate Slip on	Table 6/4 Forged Bossed Screwed	Table 6/5 Forged Bossed Slip on	Table 6/8 Plate Blank	Raised face Dia.	T'ness	Nr. bolt hole	Size of bolt
15	95	14	14	14	14	14	45	2	4	M12 × 45
20	105	16	16	16	16	16	58	2	4	M12 × 50
25	115	16	16	16	16	16	68	2	4	M12 × 50
32	140	16	16	16	16	16	78	2	4	M16 × 55
40	150	16	16	16	16	16	88	3	4	M16 × 55
50	165	18	18	18	18	18	102	3	4	M16 × 60
65	185	18	18	18	18	18	122	3	4	M16 × 60
80	200	20	20	20	20	20	138	3	8	M16 × 60
100	220	20	20	20	20	20	158	3	8	M16 × 65
125	250	22	22	22	22	22	188	3	8	M16 × 70
150	285	22	22	22	22	22	212	3	8	M20 × 70
200	340	24	24	–	24	24	268	3	12	M20 × 75
250	405	26	26	–	26	26	320	3	12	M24 × 90
300	460	28	28	–	28	28	378	4	12	M24 × 90

PIPES, WATER, STORAGE, INSULATION

Minimum Distances Between Supports/Fixings

Material	BS Nominal pipe size		Pipes – Vertical	Pipes – Horizontal on to low gradients
	(inch)	(mm)	Support distance in metres	Support distance in metres
Copper	0.50	15.00	1.90	1.30
	0.75	22.00	2.50	1.90
	1.00	28.00	2.50	1.90
	1.25	35.00	2.80	2.50
	1.50	42.00	2.80	2.50
	2.00	54.00	3.90	2.50
	2.50	67.00	3.90	2.80
	3.00	76.10	3.90	2.80
	4.00	108.00	3.90	2.80
	5.00	133.00	3.90	2.80
	6.00	159.00	3.90	2.80
muPVC	1.25	32.00	1.20	0.50
	1.50	40.00	1.20	0.50
	2.00	50.00	1.20	0.60
Polypropylene	1.25	32.00	1.20	0.50
	1.50	40.00	1.20	0.50
uPVC	–	82.40	1.20	0.50
	–	110.00	1.80	0.90
	–	160.00	1.80	1.20
Steel	0.50	15.00	2.40	1.80
	0.75	20.00	3.00	2.40
	1.00	25.00	3.00	2.40
	1.25	32.00	3.00	2.40
	1.50	40.00	3.70	2.40
	2.00	50.00	3.70	2.40
	2.50	65.00	4.60	3.00
	3.00	80.40	4.60	3.00
	4.00	100.00	4.60	3.00
	5.00	125.00	5.50	3.70
	6.00	150.00	5.50	4.50
	8.00	200.00	8.50	6.00
	10.00	250.00	9.00	6.50
	12.00	300.00	10.00	7.00
	16.00	400.00	10.00	8.25

Litres of Water Storage Required Per Person Per Building Type

Type of building	Storage (litres)
Houses and flats (up to 4 bedrooms)	120/bedroom
Houses and flats (more than 4 bedrooms)	100/bedroom
Hostels	90/bed
Hotels	200/bed
Nurses homes and medical quarters	120/bed
Offices with canteen	45/person
Offices without canteen	40/person
Restaurants	7/meal
Boarding schools	90/person
Day schools – Primary	15/person
Day schools – Secondary	20/person

Recommended Air Conditioning Design Loads

Building type	Design loading
Computer rooms	500 W/m² of floor area
Restaurants	150 W/m² of floor area
Banks (main area)	100 W/m² of floor area
Supermarkets	25 W/m² of floor area
Large office block (exterior zone)	100 W/m² of floor area
Large office block (interior zone)	80 W/m² of floor area
Small office block (interior zone)	80 W/m² of floor area

PIPES, WATER, STORAGE, INSULATION

Capacity and Dimensions of Galvanized Mild Steel Cisterns – BS 417

Capacity (litres)	BS type (SCM)	Dimensions		
		Length (mm)	Width (mm)	Depth (mm)
18	45	457	305	305
36	70	610	305	371
54	90	610	406	371
68	110	610	432	432
86	135	610	457	482
114	180	686	508	508
159	230	736	559	559
191	270	762	584	610
227	320	914	610	584
264	360	914	660	610
327	450/1	1220	610	610
336	450/2	965	686	686
423	570	965	762	787
491	680	1090	864	736
709	910	1070	889	889

Capacity of Cold Water Polypropylene Storage Cisterns – BS 4213

Capacity (litres)	BS type (PC)	Maximum height (mm)
18	4	310
36	8	380
68	15	430
91	20	510
114	25	530
182	40	610
227	50	660
273	60	660
318	70	660
455	100	760

PIPES, WATER, STORAGE, INSULATION

Minimum Insulation Thickness to Protect Against Freezing for Domestic Cold Water Systems (8 Hour Evaluation Period)

Pipe size (mm)	Insulation thickness (mm)					
	Condition 1			Condition 2		
	λ = 0.020	λ = 0.030	λ = 0.040	λ = 0.020	λ = 0.030	λ = 0.040
Copper pipes						
15	11	20	34	12	23	41
22	6	9	13	6	10	15
28	4	6	9	4	7	10
35	3	5	7	4	5	7
42	3	4	5	8	4	6
54	2	3	4	2	3	4
76	2	2	3	2	2	3
Steel pipes						
15	9	15	24	10	18	29
20	6	9	13	6	10	15
25	4	7	9	5	7	10
32	3	5	6	3	5	7
40	3	4	5	3	4	6
50	2	3	4	2	3	4
65	2	2	3	2	3	3

Condition 1: water temperature 7°C; ambient temperature –6°C; evaluation period 8 h; permitted ice formation 50%; normal installation, i.e. inside the building and inside the envelope of the structural insulation
Condition 2: water temperature 2°C; ambient temperature –6°C; evaluation period 8 h; permitted ice formation 50%; extreme installation, i.e. inside the building but outside the envelope of the structural insulation
λ = thermal conductivity [W/(mK)]

Insulation Thickness for Chilled And Cold Water Supplies to Prevent Condensation

On a Low Emissivity Outer Surface (0.05, i.e. Bright Reinforced Aluminium Foil) with an Ambient Temperature of +25°C and a Relative Humidity of 80%

Steel pipe size (mm)	t = +10			t = +5			t = 0		
	Insulation thickness (mm)			Insulation thickness (mm)			Insulation thickness (mm)		
	λ = 0.030	λ = 0.040	λ = 0.050	λ = 0.030	λ = 0.040	λ = 0.050	λ = 0.030	λ = 0.040	λ = 0.050
15	16	20	25	22	28	34	28	36	43
25	18	24	29	25	32	39	32	41	50
50	22	28	34	30	39	47	38	49	60
100	26	34	41	36	47	57	46	60	73
150	29	38	46	40	52	64	51	67	82
250	33	43	53	46	60	74	59	77	94
Flat surfaces	39	52	65	56	75	93	73	97	122

t = temperature of contents (°C)
λ = thermal conductivity at mean temperature of insulation [W/(mK)]

Insulation Thickness for Non-domestic Heating Installations to Control Heat Loss

Steel pipe size (mm)	t = 75			t = 100			t = 150		
	Insulation thickness (mm)			Insulation thickness (mm)			Insulation thickness (mm)		
	λ = 0.030	λ = 0.040	λ = 0.050	λ = 0.030	λ = 0.040	λ = 0.050	λ = 0.030	λ = 0.040	λ = 0.050
10	18	32	55	20	36	62	23	44	77
15	19	34	56	21	38	64	26	47	80
20	21	36	57	23	40	65	28	50	83
25	23	38	58	26	43	68	31	53	85
32	24	39	59	28	45	69	33	55	87
40	25	40	60	29	47	70	35	57	88
50	27	42	61	31	49	72	37	59	90
65	29	43	62	33	51	74	40	63	92
80	30	44	62	35	52	75	42	65	94
100	31	46	63	37	54	76	45	68	96
150	33	48	64	40	57	77	50	73	100
200	35	49	65	42	59	79	53	76	103
250	36	50	66	43	61	80	55	78	105

t = hot face temperature (°C)

$λ$ = thermal conductivity at mean temperature of insulation [W/(mK)]

Index

Building Regulations Pocket Book

Ray Tricker and Samantha Alford

This handy guide provides you with all the information you need to comply with the UK Building Regulations and Approved Documents. On site, in the van, in the office, wherever you are, this is the book you'll refer to time and time again to double check the regulations on your current job.

The *Building Regulations Pocket Book* is the must have reliable and portable guide to compliance with the Building Regulations.

* Part 1 provides an overview of the Building Act
* Part 2 offers a handy guide to the dos and don'ts of gaining the Local Council's approval for Planning Permission and Building Regulations Approval
* Part 3 presents an overview of the requirements of the Approved Documents associated with the Building Regulations
* Part 4 is an easy to read explanation of the essential requirements of the Building Regulations that any architect, builder or DIYer needs to know to keep their work safe and compliant on both domestic or non-domestic jobs

This book is essential reading for all building contractors and sub-contractors, site engineers, building engineers, building control officers, building surveyors, architects, construction site managers and DIYers. Homeowners will also find it useful to understand what they are responsible for when they have work done on their home (ignorance of the regulations is no defence when it comes to compliance!).

February 2018: 186 x 123 mm: 456 pp
Pb: 978-0-8153-6838-0 : £19.99

To Order: Tel: +44 (0) 1235 400524 Fax: +44 (0) 1235 400525
or Post: Taylor and Francis Customer Services,
Bookpoint Ltd, Unit T1, 200 Milton Park, Abingdon, Oxon, OX14 4TA UK
Email: book.orders@tandf.co.uk

For a complete listing of all our titles visit:
www.tandf.co.uk

Faber & Kell's Heating & Air-conditioning of Buildings

11th Edition

D. Oughton *et al.*

For over 70 years, Faber & Kell's has been the definitive reference text in its field. It provides an understanding of the principles of heating and air-conditioning of buildings in a concise manner, illustrating practical information with simple, easy-to-use diagrams, now in full-colour.

This new-look 11th edition has been re-organised for ease of use and includes fully updated chapters on sustainability and renewable energy sources, as well as information on the new Building Regulations Parts F and L. As well as extensive updates to regulations and codes, it now includes an introduction that explains the role of the building services engineer in the construction process. Its coverage of design calculations, advice on using the latest technologies, building management systems, operation and maintenance makes this an essential reference for all building services professionals.

January 2015: 246 x 189: 968pp
Pbk: 978-0-415-52265-6: £115.00

To Order: Tel: +44 (0) 1235 400524 Fax: +44 (0) 1235 400525
or Post: Taylor and Francis Customer Services,
Bookpoint Ltd, Unit T1, 200 Milton Park, Abingdon, Oxon, OX14 4TA UK
Email: book.orders@tandf.co.uk

For a complete listing of all our titles visit:
www.tandf.co.uk

Building Regulations in Brief, 9th edition

Ray Tricker and Samantha Alford

This ninth edition of the most popular and trusted guide reflects all the latest amendments to the Building Regulations, planning permission and the Approved Documents in England and Wales. This includes coverage of the new Approved Document Q on security, and a second part to Approved Document M which divides the regulations for 'dwellings' and 'buildings other than dwellings'. A new chapter has been added to incorporate these changes and to make the book more user friendly.

Giving practical information throughout on how to work with (and within) the Regulations, this book enables compliance in the simplest and most cost-effective manner possible. The no-nonsense approach of Building Regulations in Brief cuts through any confusion and explains the meaning of the Regulations. Consequently, it has become a favourite for anyone in the building industry or studying, as well as those planning to have work carried out on their home.

December 2017: 234 × 156 mm: 1302 pp
Pb: 978-1-138-28516-3 : £33.99

To Order: Tel: +44 (0) 1235 400524 Fax: +44 (0) 1235 400525
or Post: Taylor and Francis Customer Services,
Bookpoint Ltd, Unit T1, 200 Milton Park, Abingdon, Oxon, OX14 4TA UK
Email: book.orders@tandf.co.uk

For a complete listing of all our titles visit:
www.tandf.co.uk

Ebook Single-User Licence Agreement

We welcome you as a user of this Spon Price Book ebook and hope that you find it a useful and valuable tool. Please read this document carefully. **This is a legal agreement** between you (hereinafter referred to as the "Licensee") and Taylor and Francis Books Ltd. (the "Publisher"), which defines the terms under which you may use the Product. **By accessing and retrieving the access code on the label inside the front cover of this book you agree to these terms and conditions outlined herein. If you do not agree to these terms you must return the Product to your supplier intact, with the seal on the label unbroken and with the access code not accessed**.

1. **Definition of the Product**
   The product which is the subject of this Agreement, (the "Product") consists of online and offline access to the VitalSource ebook edition of *Spon's Mechanical & Electrical Services Price Book 2019*.

2. **Commencement and licence**
   2.1 This Agreement commences upon the breaking open of the document containing the access code by the Licensee (the "Commencement Date").
   2.2 This is a licence agreement (the "Agreement") for the use of the Product by the Licensee, and not an agreement for sale.
   2.3 The Publisher licenses the Licensee on a non-exclusive and non-transferable basis to use the Product on condition that the Licensee complies with this Agreement. The Licensee acknowledges that it is only permitted to use the Product in accordance with this Agreement.

3. **Multiple use**
   Use of the Product is not provided or allowed for more than one user or for a wide area network or consortium.

4. **Installation and Use**
   4.1 The Licensee may provide access to the Product for individual study in the following manner: The Licensee may install the Product on a secure local area network on a single site for use by one user.
   4.2 The Licensee shall be responsible for installing the Product and for the effectiveness of such installation.
   4.3 Text from the Product may be incorporated in a coursepack. Such use is only permissible with the express permission of the Publisher in writing and requires the payment of the appropriate fee as specified by the Publisher and signature of a separate licence agreement.
   4.4 The Product is a free addition to the book and the Publisher is under no obligation to provide any technical support.

5. **Permitted Activities**
   5.1 The Licensee shall be entitled to use the Product for its own internal purposes;
   5.2 The Licensee acknowledges that its rights to use the Product are strictly set out in this Agreement, and all other uses (whether expressly mentioned in Clause 6 below or not) are prohibited.

6. **Prohibited Activities**
   The following are prohibited without the express permission of the Publisher:
   6.1 The commercial exploitation of any part of the Product.
   6.2 The rental, loan, (free or for money or money's worth) or hire purchase of this product, save with the express consent of the Publisher.
   6.3 Any activity which raises the reasonable prospect of impeding the Publisher's ability or opportunities to market the Product.
   6.4 Any networking, physical or electronic distribution or dissemination of the product save as expressly permitted by this Agreement.
   6.5 Any reverse engineering, decompilation, disassembly or other alteration of the Product save in accordance with applicable national laws.
   6.6 The right to create any derivative product or service from the Product save as expressly provided for in this Agreement.
   6.7 Any alteration, amendment, modification or deletion from the Product, whether for the purposes of error correction or otherwise.

7. **General Responsibilities of the License**

7.1   The Licensee will take all reasonable steps to ensure that the Product is used in accordance with the terms and conditions of this Agreement.

7.2   The Licensee acknowledges that damages may not be a sufficient remedy for the Publisher in the event of breach of this Agreement by the Licensee, and that an injunction may be appropriate.

7.3   The Licensee undertakes to keep the Product safe and to use its best endeavours to ensure that the product does not fall into the hands of third parties, whether as a result of theft or otherwise.

7.4   Where information of a confidential nature relating to the product of the business affairs of the Publisher comes into the possession of the Licensee pursuant to this Agreement (or otherwise), the Licensee agrees to use such information solely for the purposes of this Agreement, and under no circumstances to disclose any element of the information to any third party save strictly as permitted under this Agreement. For the avoidance of doubt, the Licensee's obligations under this sub-clause 7.4 shall survive the termination of this Agreement.

8. **Warrant and Liability**

8.1   The Publisher warrants that it has the authority to enter into this agreement and that it has secured all rights and permissions necessary to enable the Licensee to use the Product in accordance with this Agreement.

8.2   The Publisher warrants that the Product as supplied on the Commencement Date shall be free of defects in materials and workmanship, and undertakes to replace any defective Product within 28 days of notice of such defect being received provided such notice is received within 30 days of such supply. As an alternative to replacement, the Publisher agrees fully to refund the Licensee in such circumstances, if the Licensee so requests, provided that the Licensee returns this copy of *Spon's Mechanical & Electrical Services Price Book 2019* to the Publisher. The provisions of this sub-clause 8.2 do not apply where the defect results from an accident or from misuse of the product by the Licensee.

8.3   Sub-clause 8.2 sets out the sole and exclusive remedy of the Licensee in relation to defects in the Product.

8.4   The Publisher and the Licensee acknowledge that the Publisher supplies the Product on an "as is" basis. The Publisher gives no warranties:

8.4.1   that the Product satisfies the individual requirements of the Licensee; or

8.4.2   that the Product is otherwise fit for the Licensee's purpose; or

8.4.3   that the Product is compatible with the Licensee's hardware equipment and software operating environment.

8.5   The Publisher hereby disclaims all warranties and conditions, express or implied, which are not stated above.

8.6   Nothing in this Clause 8 limits the Publisher's liability to the Licensee in the event of death or personal injury resulting from the Publisher's negligence.

8.7   The Publisher hereby excludes liability for loss of revenue, reputation, business, profits, or for indirect or consequential losses, irrespective of whether the Publisher was advised by the Licensee of the potential of such losses.

8.8   The Licensee acknowledges the merit of independently verifying the price book data prior to taking any decisions of material significance (commercial or otherwise) based on such data. It is agreed that the Publisher shall not be liable for any losses which result from the Licensee placing reliance on the data under any circumstances.

8.9   Subject to sub-clause 8.6 above, the Publisher's liability under this Agreement shall be limited to the purchase price.

9. **Intellectual Property Rights**

9.1   Nothing in this Agreement affects the ownership of copyright or other intellectual property rights in the Product.

9.2   The Licensee agrees to display the Publishers' copyright notice in the manner described in the Product.

9.3   The Licensee hereby agrees to abide by copyright and similar notice requirements required by the Publisher, details of which are as follows:

"© 2019 Taylor & Francis. All rights reserved. All materials in *Spon's Mechanical & Electrical Services Price Book 2019* are copyright protected. All rights reserved. No such materials may be used, displayed, modified, adapted, distributed, transmitted, transferred, published or otherwise reproduced in any form or by any means now or hereafter developed other than strictly in accordance with the terms of the licence agreement enclosed with *Spon's Mechanical & Electrical Services Price Book 2019*. However, text and images may be printed and copied for research and private study within the preset program limitations. Please note the copyright notice above, and that any text or images printed or copied must credit the source."

9.4 This Product contains material proprietary to and copyedited by the Publisher and others. Except for the licence granted herein, all rights, title and interest in the Product, in all languages, formats and media throughout the world, including copyrights therein, are and remain the property of the Publisher or other copyright holders identified in the Product.

10. Non-assignment

This Agreement and the licence contained within it may not be assigned to any other person or entity without the written consent of the Publisher.

11. Termination and Consequences of Termination.

11.1 The Publisher shall have the right to terminate this Agreement if:

11.1.1 the Licensee is in material breach of this Agreement and fails to remedy such breach (where capable of remedy) within 14 days of a written notice from the Publisher requiring it to do so; or

11.1.2 the Licensee becomes insolvent, becomes subject to receivership, liquidation or similar external administration; or

11.1.3 the Licensee ceases to operate in business.

11.2 The Licensee shall have the right to terminate this Agreement for any reason upon two month's written notice. The Licensee shall not be entitled to any refund for payments made under this Agreement prior to termination under this sub-clause 11.2.

11.3 Termination by either of the parties is without prejudice to any other rights or remedies under the general law to which they may be entitled, or which survive such termination (including rights of the Publisher under sub-clause 7.4 above).

11.4 Upon termination of this Agreement, or expiry of its terms, the Licensee must destroy all copies and any back up copies of the product or part thereof.

12. General

12.1 *Compliance with export provisions*

The Publisher hereby agrees to comply fully with all relevant export laws and regulations of the United Kingdom to ensure that the Product is not exported, directly or indirectly, in violation of English law.

12.2 *Force majeure*

The parties accept no responsibility for breaches of this Agreement occurring as a result of circumstances beyond their control.

12.3 *No waiver*

Any failure or delay by either party to exercise or enforce any right conferred by this Agreement shall not be deemed to be a waiver of such right.

12.4 *Entire agreement*

This Agreement represents the entire agreement between the Publisher and the Licensee concerning the Product. The terms of this Agreement supersede all prior purchase orders, written terms and conditions, written or verbal representations, advertising or statements relating in any way to the Product.

12.5 *Severability*

If any provision of this Agreement is found to be invalid or unenforceable by a court of law of competent jurisdiction, such a finding shall not affect the other provisions of this Agreement and all provisions of this Agreement unaffected by such a finding shall remain in full force and effect.

12.6 *Variations*

This agreement may only be varied in writing by means of variation signed in writing by both parties.

12.7 *Notices*

All notices to be delivered to: Spon's Price Books, Taylor & Francis Books Ltd., 3 Park Square, Milton Park, Abingdon, Oxfordshire, OX14 4RN, UK.

12.8 *Governing law*

This Agreement is governed by English law and the parties hereby agree that any dispute arising under this Agreement shall be subject to the jurisdiction of the English courts.

If you have any queries about the terms of this licence, please contact:

Spon's Price Books
Taylor & Francis Books Ltd.
3 Park Square, Milton Park, Abingdon, Oxfordshire, OX14 4RN
Tel: +44 (0) 20 7017 6000
www.sponpress.com

Spon Press
an imprint of Taylor & Francis

Ebook options

Your print copy comes with a free ebook on the VitalSource® Bookshelf platform. Further copies of the ebook are available from https://www.crcpress.com/search/results?kw=spon+2019
Pick your price book and select the ebook under the 'Select Format' option.

To buy Spon's ebooks for five or more users of in your organisation please contact:

Spon's Price Books
eBooks & Online Sales
Taylor & Francis Books Ltd.
3 Park Square, Milton Park, Oxfordshire, OX14 4RN
Tel: (020) 337 73480
onlinesales@informa.com